1995 TOYS & PRICES

Edited by

**Roger Case
& Tom Hammel**

© 1994 by
Krause Publications, Inc.

All rights reserved. No portion of this publication may be reproduced or transmitted in any form or by any means, electronic or mechanical, including photocopy, recording, or any information storage and retrieval system, without permission in writing from the author, except by a reviewer who may quote brief passages in a critical article or review to be printed in a magazine or newspaper, or electronically transmitted on radio or television.

Published by

700 E. State Street • Iola, WI 54990-0001

Please call or write for our free catalog of Toys & Comics publications. Our toll-free number to place an order or obtain a free catalog is 800-258-0929 or please use our regular business telephone 715-445-2214 for editorial comment and further information.

Library of Congress Catalog Number: 93-77554
ISBN: 0-87341-313-X

Printed in the United States of America

Contents

Acknowledgements 4
Introduction 5
How To Use the Grading Guides 9
Banks 10
    Mechanical 15
    Still 39
    Die Cast 62
Classic Tin 90
Action Figures 121
    GI Joe 133
Vehicles 168
    Buddy-L 184
    Corgi 215
    Dinky 252
    Hot Wheels 258
    Hubley 268
    Marx 274
    Matchbox 291
    Tootsietoy 321
Games 333
    Pre-War Games 337
    Post-War Games 370
Food Toys 399
    Pez Dispensers 404
    Lunch Kits 414
    Fast Food Toys 431
Space Toys 454
    Star Wars 478
Guns 493
Figural Model Kits 525
Marx 541
    Playsets 544
    Figures 555
Barbie 573
    Dolls 576
    Fashions 580
Erector Sets 596
Character Toys 608

Acknowledgements

This second edition of Toys & Prices is both bigger and better than last year's, mainly because this year we had more help putting it together. Many of our expert contributors will be familiar to you active collectors from their ads in Toy Shop and Toy Collector And Price Guide, and also because you may have seen their names in last year's edition of this book, too.

Also, you "veteran" collectors out there will recognize these people as widely respected authorities in their various fields, from fast food toys and model kits to classic tin and cast iron. They took time out from their busy schedules to sift through every line of this book, over 15,000 toys worth, correcting errors--and there were some doozies!--verifying values, and weeding out duplications.

We then arranged it all in what we hope you will find to be the most attractive, accurate and easy to use reference in the field.

It took the labors and long hours of many people to produce this book, and all of them deserve more thanks than we can possibly express here. But we're going to give it a try anyway, because they deserve it.

First, the editors wish to thank the staff of Krause Publications, particularly Linda Maurer and Brenda Mazemke for data entry, and book editors Mary Sieber and Ron Kowalke for keeping a handle on all the parts. We also wish to thank the production department staff for making the whole huge mess legible.

Dealers and collectors who graciously lent us photos include Wayne Mitchell, Gary Sohmers of Wex Rex, Paul Deion of The Wayback Machine, Bruce Whitehill, Peter Fritz of Toy-A-Day, James Koval, Barry Goodman , George Newcomb of Plymouth Rock Toy Co., Allan Shrem of The Walls of Fame, Al Kasishke, Mike Stannard, Wade Johnson, Vincent Santelmo, The Ertl Company, Apple Patch Toys, R.H. Bruce, The Gene Harris Auction Center, Bob Peirce, Marcie Melillo, Dale Womer of The Hobby Lobby, Mark Huckabone, Anthony Balasco of Figures, Ed and Laura Hock, Sue Sternfeld, Larry Aikins, Jon Shapiro, Dan Casey, Hill Design, Dave Williamson, Mark Arruda, Mike and Debbie Brooks, and Bill Campbell.

We owe a further debt to those expert contributors who assisted in pricing the various sections. They are Wayne Mitchell, Al Kasishke, Paul Fink, Marcie Melillo, Fremont Brown, Bob Peirce, Anthony Balasco, Ed Hock, James Crane, Jon Shapiro, Dale Womer, Sue Sternfeld, George Newcomb, Lyle Anderson, Cotswold Collectibles, Mark Arruda, Bob Chartain, Buzzy Langford of Attic Busters, Peter Fritz of Toy-A-Day, James Koval, Jon Thurmond of Collectorholics, Alexander Edwards of Dream World Collectibles, Gerard Stezelberger of Ancient Idols, Bill Campbell, William Stillman and Matt Crandall.

We wish to express our sincere gratitude to all who helped create this second edition of Toys & Prices. We look forward to bringing you even better editions in the years to come.

We also wish to thank you, the burgeoning thousands of toy collectors all across America and the world, not just because you bought out the first edition in less than five months, but because you care enough about preserving a bit of our shared past and our shared youth to collect it, piece by wonderful piece. May you use this book in good health and good fortune

The Editors

1995 Toys & Prices

New! Improved! And Then Some!

It seems like only yesterday we were working on the first edition of Toys & Prices. That first book, released late last fall, is now long since sold out. Response was so positive and demand so great that we sold the warehouse down to the concrete and were eventually forced to turn away large orders from major chains--and only a handful of people here at Krause Publishing were fortunate enough to get copies for their desks. I was lucky just to be able to squirrel one away for my mom.

Actually, we never stopped working on Toys & Prices, as we have been building new lists throughout the year for inclusion in this 2nd edition and for publication in Toy Collector And Price Guide magazine as well. And the work will continue, so next year's edition will be still bigger, more useful and more fun.

This edition is about 150 pages fatter than last year's, with more listings, more photos, and a few new frills. All chapter introductions have been rewritten and updated, and each intro now also boasts a Trends section, a brief recap of last year's market (between updates of our values, roughly August 1993 through July 1994). The Trends section also notes increases and in some cases percentages of change in our databases from the first edition to this one, and does some educated "crystal ball reading" for the coming year.

The sections that saw the most expansion this year are Vehicles, Character Toys, Games and Action Figures, with other areas seeing increases as well, such as enhanced raygun listings in the Toy Gun section.

Thanks to the miracle of databasing, we are able to bring you another fun bonus in this edition, our Top Toys in each section. The highest value toys in each section are presented in descending order, in a Top 10 or Top 25 listing depending on the size of the section itself. For example, in our bank section, we present the top 10 banks in each of the classes covered, mechanical banks, cast iron still banks and Ertl die cast banks. In future editions, the changing values of these top performers will provide us with a handy way to track the cumulative market appeciation of not only of those individual items, but also of other items in their classes. We will also be able to track relative rates of appreciation between classes, comparing say, action figures to tin toys.

Aren't Computers Wonderful?

We have also revised the presentation of several sections for this edition in order to make them both easier to use and more logically accurate.

Lunch kits, listed last year by maker, are listed in this edition alphabetically by name in just two groupings, metal boxes and plastic boxes, so you can look up your latest box find that much faster.

Games have been broken into two major groupings along a line drawn by World War II, since this is the way games are categorized by game collectors themselves. Games produced before WWII are rarely found in mint condition, and experts in the field understandably balk at arbitrarily assigning a Mint value to a game that may not exist in that condition. So Pre-War Games are graded in no higher than Excellent condition. On the other hand, many Post-War Games do exist in Mint condition, and enough copies of them have traded to allow experts to assign a Mint value--so Post-War Game listings are graded in up to Mint Condition.

Overall, we've worked to make 1995 Toys & Prices more comprehensive, focusing most on the fastest growing and hottest markets while keeping it clean and easy to use. And while it is thicker this year, we have tried to keep the package portable, because when you're ogling a toy at a garage sale, even the best price guide in the world won't do you any good if it's back home on a shelf. We designed it so you can take it with you, and the sooner you get into the habit, the sooner your collecting trips will become both more fun and more profitable.

But we're getting ahead of ourselves. We're assuming you're all veteran collectors. Let's back up and review the field for the benefit of our newest comrades, those beginning collectors who are just now climbing on board.

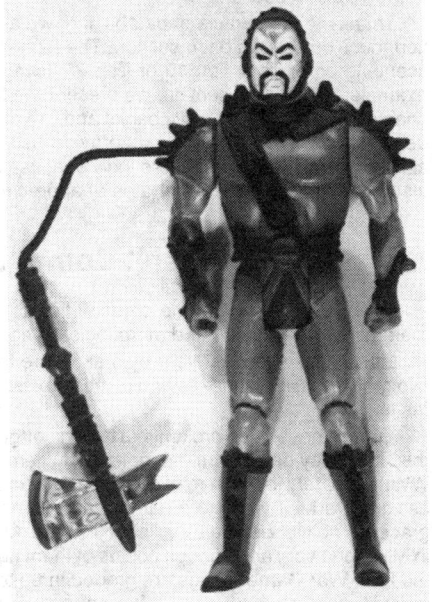

Formed from bronze to tin to plastic and everything in between, collectible toys come in all shapes and sizes. A sample of the toys, and their values, to be found in this book include still banks, action figures, and food-related toys. The banks are represented here by (lower left) The Baseball Player, 1909, A.C. Williams, and (upper right) The Boy Scout "A Scout Is Thrifty," c1910, A.C. Williams; the action figures by (lower right) Steppenwolf from the Super Powers series, 1984-86, Kenner; and the food-related toys by (upper left) Tiny Toon Adventures Flip Cars, 1991, McDonald's.

Toy Collecting--Are You Kidding?

No!

Thousands of people from all walks of life are becoming toy collectors each month, from kids spending their allowances to investment advisors seeking high returns. Toy collectors fit no clear descriptions--we come in all sizes, ages and nationalities and live all over the globe. What we do share is a love of collecting and an appreciation of the artisty, history and just plain fun that toys have always offered, and still do. Sure, many of us are boys who won't grow up, but there are lots of girls who won't stop playing with toys either, whether they be Barbie or GI Joe, Princess Leia or Jack Skellington. We are everywhere, and there are more of us every day.

Toy collecting has been called the hottest collecting field of the 1990s, and the evidence keeps piling up. This hobby has everything going for it, a colorful universe of plentiful and affordable items, the thrill of the hunt, intellectual and emotional gratification, comradery, and the prospect of sometimes dramatic financial gain.

And even though the collecting community grows daily and the values of many toys are now in the hundreds of dollars, our hobby is young enough that you new collectors starting today can still have the feeling of getting in on the ground floor. In a few years, this will no longer be said, but it's still true today.

In addition, the hobby offers almost endless points of entry and foundations on which you can build, regardless of how deep your pockets may be. Model kits, action figures, fast food toys, Hot Wheels and die cast banks--today's "cheap" toys may well be tomorrow's brightest stars.

There is no better example of this than the phenomenal growth in recent years of the Hot Wheels market. According to market observers, Mattel's palm-sized cars have gone from cheap rack toys to big buck collectibles at a rate of 25 percent a year over the last four years.

How big? This book will show you numerous Hot Wheels valued at hundreds of dollars and a few in the thousands. Yes, that's right, thousands of dollars for one Hot Wheels car. And, no, we won't tell you which ones right now--you're buying this book, right?

And there's more. Toy auctions are setting up records and smashing them down like bowling pins. New fields like board games, fast food toys, PEZ dispensers and "garage kits" are blooming all over. Another action figure collector is born every minute.

Toy shows are popping up all over the country--and all over the world--like mechanical rabbits. Toy magazines like Toy Shop and Toy Collector And Price Guide are picking up new readers each issue, and new titles are appearing everywhere you look. It's no longer irrational to consider financing part of your child's college education by shopping at Toys R Us. Just keep those toys in good condition and save those boxes!

It may be that 21st century collectors will look back on the 1990s as the golden age of the hobby, and we wouldn't be a bit surprised. In fact, you could say we're banking on it.

Are These Prices For Real?

Yes.

Our primary sources for prices are print retail ads in magazines like Toy Shop and Toy Collector And Price Guide, dealer price lists, observed prices at toy shows, and prices realized from auction houses across the country.

Information is compiled by category and entered into databases. These databases are then reviewed for accuracy by the editors of Toy Collector And Price Guide.

Before it sees print, each database is then sent out for final review by recognized authorities in the subject field. These contributing experts are some of the leading collectors and dealers in America, and we are proud to list them on our Acknowledgments page.

The final results you see in this book are what we believe to be accurate, current retail prices, listed across a range of grades for each item.

We must also stress here that the values listed in this book are not offers to buy or sell--they are merely guidelines as to what you could expect to pay for an item at a show or by mail.

If you are a student of price guides, you know prices for the same items can vary, sometimes widely, from source to source. There are many reasons for this.

Some things just cost more in some parts of the country than they do in others.

Also, some items are more heavily collected in some areas than others. Farm toys are popular in the Midwest, but go to a toy show in Boston and you might not see three farm toys all day long.

Personal and regional economics can come into play as well. Retail shop owners might put a higher price tag on items because they have rent and utilities to pay. Show dealers might price an item lower because they need to make sales to cover table and travel costs, or simply because they'd rather take home cash than toys.

These and many other factors of location and the market can all impact on the price of any given toy at any time.

Grading

Grading is highly subjective and can often depend on which side of the dealer table you happen to be standing. An Excellent condition toy you are trying to buy might look more like Near Mint if you are trying to sell, and even slight differences in condition can greatly affect prices. Haggling over condition is often as natural as haggling over price, and the argument can be made that they are essentially the same thing.

If you are buying this book to estimate the value of your collection before you sell, please bear this subjectivity of grading in mind. Read the definitions of different grades on the page entitled "How to Use Toys & Prices," and try to grade your toys as conservatively as you can. You also need to remember that grading standards change from class to class and that grading a cast iron bank or a victorian board game has different "rules" than model kits and action figures.

Please also remember that most dealers typically offer between 40 and 60 percent of book value when buying--regardless of condition--as they obviously must be able to turn around and resell your toys at a profit in order to stay in business.

Finally, if you see a price in this or any other book that seems way off base, don't accept it just because it's in print. It may be a typo, or be based on incomplete information. If you do notice what looks like a glaring error, we'd like to know about it, so we can be sure to double check it for the next edition.

Likewise, if you own an item not listed here, we would appreciate the chance to add it to our database, especially if you can give us a clean photo of it along with information on its size, maker, year, and any other descriptions that might apply. Photos are most welcome, but please send only duplicates, as we regret we cannot return them.

In collecting, as in life, knowledge is power. What you don't know might not hurt you too much, but it can sure hurt your pocket book. Our introductory chapters are just that, introductions. If you are seriously interested in building collections in any of the fields covered in this book, there's no better advice we can give you than to read as many books about your subject as you can. Ask your local bookstore what's available, and check Toy Shop and Toy Collector And Price Guide for ads by book dealers who cater to our hobby. We will send you a copy of either magazine for cover price, just for the asking.

Ultimately, the value of a toy is what you are willing to pay for it. Every time someone haggles over and finally settles on the price of a toy, another footnote is added to ongoing story of the hobby.

Toy collecting is a field in which the phrase "Let the buyer beware" is particularly true. So too is "Forewarned is forearmed." Arming you is the purpose of this book. Carry it with you in your toy hunting travels and you will find it more than pays for itelf in ease of use. In time it can prove an invaluable aid in enhancing your enjoyment and appreciation of our most fascinating and profitable of hobbies.

Happy Collecting!

Tom Hammel

Using The Grading Guides In This Book

The prices listed in Toys & Prices are intended only as guidelines as to what you might expect to pay for an item in today's market, either at a show or through the mail. The values listed are not offers to buy or sell toys; the publisher does not engage in buying or selling toys, but in compiling information and prices of issues. The prices in this book represent our best assessment of current market values at press time, and are derived from various printed sources, toy show observations, auction results, and the input of expert contributors.

Prices are generally listed for more than one grade of condition. Toy collectors and dealers alike know grading can be difficult and subjective. It's usually easy to reach agreement on a toy that's "Mint In Box," but, beyond that, grading becomes a somewhat imprecise science.

In order to provide you with general guidelines, the editors of Toys & Prices have adopted the following grade descriptions. Please bear in mind that since various types of toys are looked at in different ways, no single grading system will apply to all of them. Some decriptions may not apply to certain toys, depending on how they are categorized, when they were produced, how they were originally packaged and how they are collected today. For example, mechanical and still cast iron banks rarely appear on the market in any higher than Excellent condition, and when they do they can command astronomical prices. On the other hand, many PEZ collectors view any dispenser in less than Near Mint condition as essentially worthless. Some model kit collectors view any built-up kit as worthless regardless of its age or rarity, but this view is not shared by many toy collectors. Where called for, we will provide further notations on grading specific to each category at the beginning of each grade and price section.

MIB or MIP (Mint In Box, Mint In Package)— Just like new, in the original package, preferably still sealed. The box may have been opened, but any packages inside remain unopened. If on a blister card, the package is intact and unopened. New toys in boxes, where boxes remain factory sealed, can often command higher prices.

MNP (Mint No Package)— This describes a toy typically produced in the 1960s or later that is in mint condition, but is not in its original package.

NM (NEAR MINT)— A toy that appears like new in overall appearance but exhibits very minor wear, and does not have the original box. An exception would be a toy that comes in kit form. A kit toy in near mint condition would be expected to have the original box, but the box would display some wear.

EX (EXCELLENT)— A toy that is complete and has been played with. Signs of minor wear may be evident, but the toy is very clean and well cared for.

VG (Very Good)— A toy that obviously has been played with and shows general wear overall. Paint chipping is readily apparent. In metal toys, some minor rust may be evident. In sets, some minor pieces may be missing.

GD (Good)— A toy with evidence of heavy play, dents, chips and possibly moderate rust. The toy may be missing a major replaceable component, such as a battery compartment door, or may be in need of repair. In sets, several pieces may be missing.

It should also be noted that all pricing sections in this book were generated by computer, and that lower values were calculated on a percentage of the highest known value for items in a given class. For example, Hot Wheels lower values were calculated on a percentage of the Mint In Package value, and mechanical bank values are based on a percentage of the Excellent value. Thus, in most price guides, an item in Very Good condition is generally valued at about one-third of its value in Mint.

When dealing with multiple grade values for tens of thousands of items, such computer aided pricing is essential, but the realities of specific markets can sometimes mean that an item in Mint condition may be worth perhaps 10 times the value of the same item in only Good condition. In the case of extremely low population items, such as rare banks or tin toys, pricing can become quite speculative, even for recognized experts. The same is true for fields experiencing rapid inflation and growth, such as Hot Wheels.

Ultimately, the market is driven by the checkbook, and the bottom line final value of a toy is often the last price at which it was sold. In this sense the toy hobby, like many others, operates in a vacuum, with individual toys setting their own values. Repeated sales of specific items create precedents and establish standards for asking prices across the market. It is by comparison of such multiple transactions that price guides such as this one are created. In the end, the value of a toy is decided by one person, the buyer.

Mechanical Banks, Still Banks and Die Cast Vehicle Banks by Ertl

Mechanical Banks

Mechanical banks are often considered the royalty of American toys. This is in part because they were developed at an eventful time in American history, when industrialization was changing every aspect of life, including the change in toys from wood and tin to the newly discovered cast iron.

The age of mechanical banks began with the end of the Civil War and the dawning of the industrial age, and for collecting purposes ended with the beginning of World War II.

The rise of factories during the Civil War laid the groundwork for the coming industrial revolution and its weapon would be iron. Prior to the Civil War most toys were made of wood, tin, or sheet metal. The new process of "casting" iron, with its durability and lower cost, made it the metal of choice for a wide range of manufacturers, including toy makers. The new process allowed design innovations not previously possible, and toy makers were quick to exploit its potential.

The toys of the day reflected their time, the attitudes, activities, personalities, and morals of the day. Americans believed in frugality both as a morally dictated behavior and as a means toward a secured future, and parents of the day strove to instill that virtue in their children. Many of the nursery rhymes and ditties of the time esposed thrift as their moral, and children were encouraged to save their pennies at every turn. Toy makers saw to it that those parents had clever and colorful allies in the forms of the banks themselves.

In 1869 the first patent was issued for a cast iron mechanical bank, "Hall's Excelsior Bank." It was also the first of many banks designed as buildings, a concept which by imitation and variation grew into the largest category in the related field of nonmechanical or "still" banks. Other main categories of mechanical banks can be roughly broken into animal banks, depicting elephants, birds, and all manner of other creatures; human banks featuring people in various activities such as football players in the "Calamity Bank" or the excessively rare "Girl Skipping Rope"; coin shooting banks such as the "Creedmore Bank" and the "William Tell"; and "darkie" banks including the numerous variations of

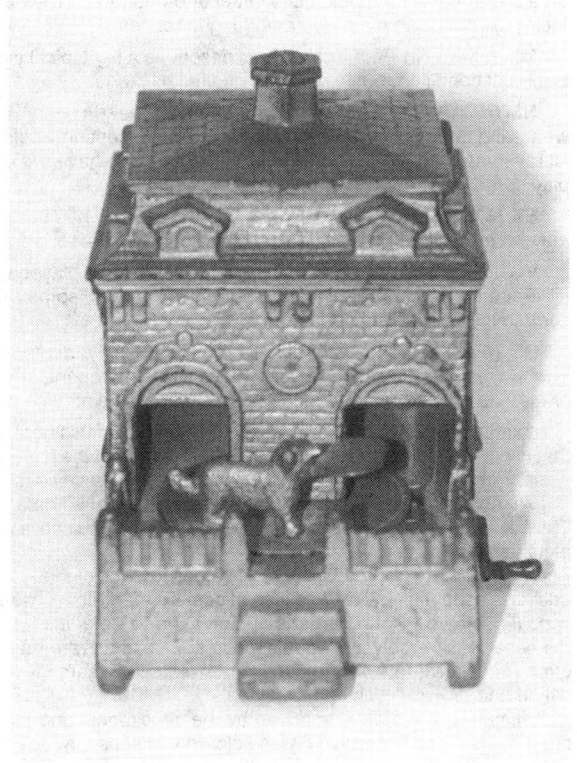

Judd Manufacturing Company's cast iron "Dog on a Turntable," 1870s.

"Jolly Nigger" banks and the "Bad Accident" bank.

Mechanical banks are loosely defined as children's toys in which the coin when deposited sets in motion some mechanism or action. By comparison, nonmechanical or "still" banks are receptacles only; depositing coins in them causes no action. In short, if you drop in the coin and something happens, the bank is most likely mechanical. If nothing happens, the bank is "still."

There is also a gray area. Some banks once considered mechanical are now grouped under the heading of "semi-mechanical." For example, in some banks the coin, while falling into the bank, strikes and rings a bell. Other banks require actions by the depositor to achieve the deposit itself, again, not qualifying the bank for mechanical status.

The degree of action varies from the simple closing of a mouth or nod of a head to complex multiple figure actions involving music, acrobatics, pratfalls, and sporting events. Along with rarity and subject matter, this complexity of design and action is a primary factor in determining the value of a mechanical bank. From a manufacturing standpoint, intricacy of design and action increased production costs, resulting in higher retail prices and lower overall unit sales, which one hundred years later we call rarity.

Some banks never made it into commercial production at all. Bank designers were required to submit working pattern samples with their patent applications, and sometimes these "patterns" are all that remain of a failed patent bid. Other banks were never intended for commercial production, but were instead made for personal reasons. As literal one-of-a-kind examples, surviving banks of this type are obviously rare and command prices in accordance with their rarity. Some banks were produced by manufacturers in an attempt to fool collectors into believing they were legitimate older banks. These are now nearly as old as the originals they cleverly imitated, and have come to be accepted in many collections for what they are, just as ancient counterfeit coins have come to be accepted alongside their contemporary but geniune counterparts. Fakes, yes, but historical fakes.

One other category requires mention here, modern reproductions. However you want to slice it, call it emulation, reproduction, forgery, or just plain faking, from great paintings to toy banks and model kits, if bucks can be made recreating a costly original, someone will do it. And the greater the value of the original, the greater is the likelihood it will be reproduced. Whether intended as outright forgeries or as honest replicas and marked as such, these items can and do find their way from time to time into dealer stock, priced as genuine antique articles. Reproduction labels and imprints can be filed away, metals can be artificially aged and other telltale areas can be altered in hopes of turning an honest $50 profit into $5,000 or $15,000.

Collectors entering this field are well advised to confine their initial dealings to reputable dealers and auction houses with qualified and impartial expert staffs. Reproductions also exist in the area of cast iron still banks, so caution should be the watchword here as well.

Still Banks

The same companies that made mechanical banks also frequently made still banks as less expensive alternatives to their mechanical cousins. Several banks can be found in both still and mechanical versions. The names Arcade, Ives, Kenton, and Stevens among others are familiar to still and mechanical bank collectors alike.

Again, building banks are perhaps the largest single type of still banks, with other main classes being animals, people and busts, and appliances such as safes, clocks, mail boxes, globes, and so on. The range of still banks is much broader than mechanical banks, with over two hundred building banks alone available to collectors. The variety of objects depicted is wonderfully vast, ranging from comic personality caricatures to historical busts to houses, lighthouses, refrigerators, radios, bee hives, purses, fruit, tiny cash registers, loaves of bread, ships, taxicabs and almost every conceivable object known to Americans between the Civil War and World War II.

Building banks span a range from Lincoln's Cabin in pottery to cottages, victorian houses, mansions, and skyscrapers in cast iron. Commemorative banks were particularly popular--banks resembling the Washington Monument, the National Bank of Los Angeles, the Century of Progress building, the Eiffel Tower, and numerous other actual locations. Other building banks offered variations on general themes such as Home Savings Banks, State Banks, and churches.

One notable class of still bank is the registering bank. Often in the shape of a safe or cash register, these banks typically accept only certain coins, such as dimes or nickels. They keep a running tally of deposits and pop open once the bank is filled, typically at $5 or $10. While their delayed reaction mechanism has earned them places in some mechanical collections, they are generally classed as still banks.

Another type of still bank bearing mention is the "conversion." These fall into three main types: commercial factory conversions, individual factory conversions, and home conversions. Many factories engaged in casting iron produced a variety of products, of which toy banks were often a sideline. Another popular side line was in iron door stops of various shapes. At times these door stops would be converted into banks and released for sale to the public. These are now called commercial factory conversions.

Sometimes factory employees would take it upon themselves to convert such objects into banks to take home as gifts for their children or friends. These were converted in the factory of origin, but were not officially authorized and so are today called individual factory conversions.

A.C. Williams' "Three Wise Monkeys," 1900s.

The third type is the home conversion. These were typically also made by fathers as gifts for their children, and generally exhibit care in craftsmanship. Bear in mind that fraudulently intended conversions have also been created, with the sole purpose of ripping off the collecting public.

A note on restoration. As in many other areas of collecting, restoration of banks is strongly discouraged in the marketplace. Unless undertaken by an experienced professional, the restoration of a bank can result in irreparable damage to its collector value.

While most collectible still banks were made of cast iron, other materials were frequently employed as well, including glass, pottery and other ceramics, brass, lead, tin, wood, composition materials, and even plastic. The still bank section of this book concentrates on cast iron banks. Modern ceramic and plastic banks can also be found under their respective character headings in the Character Toys section of this book.

Modern Die-Cast Vehicle Banks by Ertl, 1981-1990.

Still banks of taxicabs and touring cars are the direct ancestors of today's die-cast vehicle banks. The modern market was "pioneered" by the Ertl Company of Dyersville, Iowa. Compared to the history of cast iron banks, die-cast banks were almost literally "born yesterday." Ertl's first bank, a replica 1913 Model T "Parcel Post Mail Service" bank rolled off the assembly line in 1981. Since then thousands of banks have followed, creating in the process a rapidly growing, ravenous market. Prices on early and low production models have risen annually, with a few rarities selling for over $2,000 each today.

Ertl's success has spawned both spinoffs and competitors. Currently the farm town of Dyersville is home to three die-cast vehicle companies, all thriving in a booming national market.

Ertl-types represent one major diversion from classic banks of old in that most are different models only from the skin on out. These banks are collected not by variety of model styles but by variety of corporate sponsors. At present Ertl offers roughly three dozen body styles in scales of 1/25, 1/30, and 1/43, but these can bear thousands of different corporate imprints. Numerous companies such as Amoco, the American Red Cross, Hershey's Chocolate, and Texaco release new models almost annually.

With several hundred models annually released directly into the collector market, collectors are now faced with such variety that they are forced to develop their own collecting specialties. Some collectors confine their efforts to collecting the various banks of Iowa, Iowa State, Nebraska, and other universities. Others collect oil company banks or beverage company banks such as Anheuser-Busch or RC Cola and Clearly Canadian. Harley-Davidson is a particularly popular

Ertl's 1926 Seagrave Pumper Dubuque, Iowa, firetruck bank.

imprint. Here as in other collecting fields, there are as many ways to collect as there are collectors.

Identification of Ertl banks is potentially confusing. All banks are numbered, but the numbers on the banks are not model numbers. Model numbers are printed only on boxes. The numbers on the banks represent the day and year the bank was built. For example, if your bank is numbered 1255 your bank was made on the 125th day of 1985. The first three digits are the days and the last is the year. Most collectors and all dealers identify their banks by the number printed on the box, but banks are also easily identifiable by make of vehicle, color scheme, and corporate sponsor.

Many Ertl banks are also serial numbered by the corporate sponsor after they leave the Ertl factory. This practice is common and re-enforces the limited edition aspect of the banks. Most sponsors will order only one run of banks, again with an eye toward collectibility, but the sizes of individual orders vary by company from as low as 504 banks to over 100,000.

Bank collecting, like other hobbies, is ultimately a matter of discretionary income. The prices of many cast iron mechanical banks have put them in the realm of the wealthy and major investors only. Still banks are also frequently priced beyond the means of casual collectors, but their values are in general a fraction of mechanicals, making them more attractive investment vehicles for collectors of moderate means. Die-cast banks are the last metal banks still affordable by average and beginning collectors, and that fact is part key to their explosive success. All three types offer the collector a piece of history and beauty, and each field has its unique attractions and benefits.

Trends

Roughly fifty percent of our mechanical banks saw increases of an average of twenty-four percent each. The remaining half of the banks listed held their values, with none falling.

In our list of still banks, about twenty-nine percent experienced gains of roughly fourteen percent each. The remaining seventy-one percent held their values, and none fell.

Ertl die-cast banks in our list saw some forty-seven percent of banks experience increases of an average of twenty percent each. The remaining banks generally held their values, with a handful seeing modest losses. However, a few banks saw substantial increases since we last

BANKS

compiled these values about fourteen months ago. For example, the Eastwood Company banks, numbers 1 to 3 rose between two hundred-fifteen and five hundred-sixty percent over our prior valuation. Harley-Davidson has proven to be a premium marque as well, with its first three issues gaining from one hundred-thirty-three and four hundred-eighty percent in the last fourteen months. Other exceptional performers have been the Richard Petty Banks numbers 1 and 2, which have grown two hundred-sixty-five and four hundred-ten percent, respectively.

It should also be noted here that no Ertl banks made after 1991 have been listed in this section.

Top 10 Mechanical Banks
(in Excellent condition)

1. Jonah and the Whale, Jonah Emerges, 1880s, J & E Stevens $35,000
2. Mikado, 1886, Kyser & Rex Co 35,000
3. Calamity, 1905, J & E Stevens 25,000
4. Girl Skipping Rope, 1890, J & E Stevens 25,000
5. Rollerskating Bank, 1880s, Kyser & Rex Co 25,000
6. Circus Bank, 1888, Shepard Hardware Co 22,000
7. Harlequinn, 1907, J & E Stevens..22,000
8. Called Out Bank, 1900, J & E Stevens...20,000
9. Motor Bank, 1889, Kyser & Rex Co..20,000
10. Turtle Bank, 1920s, Kilgore Manufacturing Co...20,000

Top 10 Still Banks
(in Excellent condition)

1. Indiana Paddle Wheeler, 1896, Unknown Maker $8,000
2. San Gabriel Mission, (musical), Unknown Maker...7,500
3. Tug Boat, Unknown Maker..7,500
4. Coin Registering Bank, 1890, Kyser & Rex..6,500
5. Electric Railroad, 1893, Shimer Toy Co 6,000
6. Boston State House, 1880s..6,000
7. Dormer Bank, Unknown Maker 6,000
8. Hippo, Unknown Maker 6,000
9. Carpenter Safe, 1907, J.M. Harper 5,000
10. Mother Hubbard Safe, 1907, J.M.Harper 5,000

Top 10 Die-Cast Vehicle Banks by Ertl 1981-1990
(in Mint condition)

1. Gulf Refining, #9211, 1984..$2,500
2. Texaco #1 Sampler, #2128, 1984 1,550
3. Texaco #2 Sampler, #9238, 1985 1,050
4. Texaco #1, #2128, 1986..825
5. Texaco #1, #2128, 1984..825
6. Durona Productions, #1321, 1982..725
7. Texaco #3 Sampler, #9396, 1986 675
8. Amoco Stanolind #1 Polarine Oil & Grease, #9673, 1989...............................660
9. Eastwood Company #1, #9325, 1989 650
10. Eastwood Company #1, #9325, 1990 650

BANKS

Top to bottom: Santa at the Chimney, 1889, Saalheimer & Strauss; Hall's Liliput, 1877, J&E Stevens Co.; Hen & Chick, 1901, J&E Stevens Co.

Mechanical Banks

NAME	COMPANY	YEAR	DESCRIPTION	GOOD	VERY GOOD	EX
Acrobat	J & E Stevens Co.	1883	deposit coin in opening, press lever which causes the gymnast to kick the clown causing the clown to stand on his head while coin is deposited in bank	2000	4000	9500
Artillery Bank	J & E Stevens Co.	1900s	#24668, 8" long, 6" tall; put coin in barrel, press lever, making soldier drop arm to signal firing, hammer snaps and fires coin through building window	550	900	1800
Artillery Bank	Shepard Hrdwre Co.	1892	nickel plate version, coin is placed in the cannon, the hammer is pushed back; pressing the thumb piece fires the coin into the fort	500	700	1200
Atlas Bank	Unknown		iron base with white metal figure holding wooden globe with paper litho map; put coin in slot, pulling lever makes coin fall into bank, making globe spin	1000	1500	3500
Bad Accident Mule	J & E Stevens Co.	1890s	10" long, painted; deposit coin under the feet of the driver, press lever, boy jumps into the road, frightening the mule; he rears, the cart and driver are thrown backwards and coin falls in cart	850	1500	3500
Bear and Tree Stump	Judd Mfg. Co.	1880s	5" tall; put coin on bear's tongue; pressing lever makes tongue lift coin and drop it into bank	500	700	1200
Bear, Slot in Chest	Kenton Mfg. Co.	1870s	deposit coin in the bear's chest and his mouth opens and closes	750	1000	1500
Bill E. Grin Bank	H.L. Judd Mfg.	1887	4 1/2" tall; dropping coin on top of head makes tongue jut and eyes blink	500	900	2200
Billy Goat Bank	J & E Stevens Co.	1910	deposit coin in slot and pull wire loop forward, goat jumps forward and coin falls in bank	600	1000	1850
Bird on Roof	J & E Stevens Co.	1878	deposit coin in slot on bird's head, pull the wire lever on left side of house, the bird tilts forward, coin rolls into the chimney	650	1000	2000
Bismark Pig	J & E Stevens Co.	1883	lock mechanism, put coin in slot over tail, press the pig's tail, Bismark figure pops up and coin drops	1800	3500	6000
Boy and Bulldog	Judd Mfg. Co.	1870s	deposit coin between the boy and the dog; pulling lever makes boy lean forward and 'push' coin into bank, the dog moves backwards at the same time and coin falls	600	900	1850
Boy on Trapeze	J. Barton & Smith Co.	1891	9 1/2" tall, painted; place a coin in the slot on the boy's head and he revolves	600	1500	4500
Boy Robbing Bird's Nest	J & E Stevens Co.	1906	8" tall; deposit coin in slot and press the lever on the tree; as the boy falls, the coin disappears in the tree	1000	3000	9500
Boy Scout Camp	J & E Stevens Co.	1912	9 1/2" long; drop coin into tent; pressing lever makes Scout raise flag as coin drops	1500	3500	10000
Boys Stealing Water-melon	Kyser & Rex Co.	1880s	6 1/2" long; put coin in slot on top of dog house; pressing lever makes boy move toward watermelon, dog comes out of house, and coin drops	600	1200	3000

BANKS

BANKS

Top to bottom: Boy and Bulldog, 1870s, Judd Mfg. Co.; Brass Jolly Nigger, 1880s, England and Cabin Bank, 1885, J&E Stevens Co.; Chief Big Moon, 1899, J&E Stevens Co.

Mechanical Banks

NAME	COMPANY	YEAR	DESCRIPTION	GOOD	VERY GOOD	EX
Breadwinners	J & E Stevens Co.	1886	put coin in end of club, cock hammer; pressing button makes 'Labor' hit 'Monopoly', "sending the rascals up" and dropping coin into loaf of bread	3500	7000	15000
Bucking Mule, Miniature	Judd Mfg. Co.	1870s	put coin in slot; releasing donkey makes him throw man, knocking coin into bank	600	900	1650
Bull and Bear, Single Pendulum	J & E Stevens Co.	1930s	put coin in pendulum on tree stump; pressing lever releases pendulum to swing, dropping coin into either bull or bear	800	1200	1800
Bull Tosses Boy In Well	Unknown		7" long; put coin in boy's hands; pressing lever makes bull spring and boy jumps back, dropping coin in well	750	1200	1700
Bulldog Bank	J & E Stevens Co.	1880s	put coin on dog's nose, pull his tail, the dog opens his mouth and swallows the coin.	400	800	1400
Bulldog Savings Bank	Ives	1878	8 1/2" long; put coin in man's hand, press lever, making dog jump, bite coin, and fall back, dropping coin into bank	1000	2000	3000
Bulldog Standing	Judd Mfg. Co.	1870s	coin is placed on dog's tongue and tail is lifted; when tail is released the coin is deposited.	350	600	1200
Bureau, Five Knob	James A. Serrill	1869	place coin in open drawer; closing drawer makes coin drop	200	350	500
Bureau, Three Knob	James A. Serrill	1869	place coin in open drawer; closing drawer makes coin drop	300	400	550
Butting Buffalo	Kyser & Rex Co.	1888	place coin into tree trunk; pressing lever makes buffalo 'butt' boy up trunk while raccoon flees into tree	1200	2000	3500
Butting Goat	Judd Mfg. Co.	1870s	deposit coin on holder on tree trunk; lifting the tail causes the goat to spring forward depositing coin in bank. Base length 4 3/4"	600	900	1650
Butting Ram	Ole Storle	1895	put coin on tree limb; pressing lever makes ram butt coin into bank while boy thumbs his nose	2000	6000	12000
Cabin Bank	J & E Stevens Co.	1885	3 1/2" tall, shows darkie in front of cabin; put coin on roof, flip lever to make figure flip and push coin into bank with his feet	300	600	850
Calamity	J & E Stevens Co.	1905	cock tackles into position, put coin in slot in front of fullback; pressing lever activates tackles and coin drops in the collision	3000	8000	25000
Called Out Bank	J & E Stevens Co.	1900	9" tall; push soldier into bank, drop coin into slot, making soldier pop up and drop coin into bank	4000	10000	20000
Calumet, Large	Calumet Baking Powder Co	1924	put coin in slot, making box sway back and forth in 'thanks'	150	250	350
Calumet, Small	Calumet Baking Powder Co	1924	put coin in slot, making box sway back and forth in 'thanks'	175	300	450
Camera	Wrightsville Hrdw. Co.	1888	rotating lever makes picture pop up	1000	3000	5000
Cat and Mouse, Cat Balancing	J & E Stevens Co.	1891	8 1/2" tall; put coin in slot and lock mouse into position; pressing lever makes mouse disappear and kitten appears holding mouse on a ball	1000	1700	2800

BANKS

Mechanical Banks

NAME	COMPANY	YEAR	DESCRIPTION	GOOD	VERY GOOD	EX
Chandlers Bank	National Brass Works	1900s	open the drawer, put in coin, closing drawer drops coin into bank	400	600	1000
Chief Big Moon	J & E Stevens Co.	1899	10" long; put coin in slot in fish tail, push lever, making frog spring up from pond, dropping coin	1200	2200	3750
Chimpanzee	Kyser & Rex Co.	1880	green version; push coin slide toward monkey with log book, making monkey lower head and arm to 'log in' deposit, ringing a bell	1200	2000	3200
Chimpanzee Bank	Kyser & Rex Co.	1880	red variant; push coin slide toward monkey with log book, making monkey lower head and arm to 'log in' deposit, ringing a bell	1400	2200	3200
Cigarette Vending	Unknown (France)	1920s	dispenses real candy cigarettes, illustrated with scenes of children smoking	75	150	250
Circus Bank	Shepard Hrdwre Co.	1888	place coin on money receptacle, turn the crank and pony goes around the ring and the clown deposits coin	5000	8000	22000
Circus Ticket Collector	Judd Mfg. Co.	1830	deposit coin on top of barrel and the man's head nods his thanks.	550	850	1500
Clever Dick	Saalheimer & Strauss	1920s	German bank; put coin on dog's nose; pressing lever makes him toss coin and swallow it	300	600	850
Clown Bank, Arched Top	Unknown (England)	1930s	press lever to make tongue jut out, put coin on tongue; lifting lever makes tongue recede, dropping coin	125	275	450
Clown Bust	Chamberlain & Hill	1880s	English bank; put coin in hand; pressing lever makes arm lift coin and clown swallows it	1700	2500	4000
Clown on Globe	J & E Stevens Co.	1890	clown straddles globe on footed base; pressing lever makes globe and clown move and change positions, leaving clown standing on head; 9" tall	900	1500	3500
Clown, Black Face	Unknown (England)	1930s	press lever to jut out tongue, put coin on tongue and release lever to drop coin into bank	150	300	550
Clown, Chein	Chein	1939	press the lever and the clown sticks out his tongue, place coin on tongue release the lever and he retracts his tongue pulling coin into bank	65	100	150
Coin Registering Bank	Kyser & Rex Co.	1890s	when the last nickel or dime is deposited totaling $5 the door will pop off and the money can be taken out	1200	2500	5500
Columbian Magic	Introduction Co.	1892	swing open the shelf and place a coin on it, close the shelf and the coin is deposited in bank	150	300	550
Confectionery Bank	Kyser & Rex Co.	1881	8 1/2" tall; depicts lady at candy counter; coin dropped into slot, button pushed, making figure slide to receive candy; bell rings as coin drops	2000	4000	8500
Creedmore Bank	J & E Stevens Co.	1877	9 3/4" long, painted; pull back lever on gun, put coin in slot; pressing man's foot shoots coin into tree	600	900	1500

BANKS

BANKS

Top to bottom: (clockwise) Lion & Monkey (unknown), Paddy & the Pig (1882), William Tell (1896), Creedmore Bank (1877), all J&E Stevens Co.; Darktown Battery Bank, J&E Stevens Co.

Mechanical Banks

NAME	COMPANY	YEAR	DESCRIPTION	GOOD	VERY GOOD	EX
Cross Legged-Minstrel	Unknown (Germany)	1909	put coin in slot; pressing lever makes man tip his hat to you	500	800	1200
Crowing Rooster	Keim & Co. (Germany)	1937	push coin through slot into bank; makes a crowing sound	500	800	1200
Cupola Circular Building	J & E Stevens Co.	1874	push the doorbell lever and the top pops up exposing the cashier, who pivots back and returns to his forward position	4000	8000	12000
Dapper Dan	Louis Marx & Co.	1910	deposit coin into the slot and Dapper Dan dances until the spring winds down; to reset wind the key clockwise	300	500	900
Darktown Battery Bank	J & E Stevens Co.		#24670, 10" long, 7 1/4" tall, painted; three lads play ball, red, blue and yellow pitcher's uniform; put coin in pitcher's hand; pressing lever to pitch coin, batter swings and misses, catcher drops	1200	2200	5500
Darky in the Chimney	Unknown		pull the drawer out and the darkey emerges; front door knob when turned counterclockwise allows the trap door in base of bank to be removed	550	1000	1400
Dentist	J & E Stevens Co.	1880s	9 1/2" long; coin drops into dentist's pocket; press lever at figure's feet, making dentist pull patient's tooth, patient falls backward, dropping coin into gas bag.	2000	5500	9500
Dinah	J. Harper & Co.	1911	deposit coin in Dinah's hand and press the lever; she raises her hand, her eyes roll back and her tongue flips in as she swallows the coin	600	900	1450
Dog on Turntable	Judd Mfg. Co.	1870s	5 1/4" long, 5" tall, turn the handle and dog goes in and deposits penny, coming out of the other door for more	400	800	1200
Dog Tray Bank	Kyser & Rex Co.	1880	place coin on plate; the dog faithfully deposits it in the vault	2000	3500	5000
Eagle and Eaglets	J & E Stevens Co.	1883	#24671, 7 3/4" long, 6" tall, painted; put coin in eagle's beak; pressing lever makes eaglets rise, eagle tilts forward and drops coin into nest	600	950	1650
Electric Safe	Louis Mfg. Co.	1904	twist the center knob in a clockwise direction until it stops, coin slot becomes operational, rotate dial counterclockwise and the coin falls into the bank	200	450	650
Elephant and Three Clowns	J & E Stevens Co.	1883	6" tall, painted; place coin between the rings held by acrobat, move the ball on the feet of the other acrobat and the elephant hits coin with his trunk and coin falls into bank	600	1000	1850
Elephant Howdah Bank, Pull Tail	Hubley	1934	put coin in elephant's trunk; pulling tail makes animal lift coin over head, dropping coin in howdah	300	500	1250
Elephant Moves Trunk (Large)	A.C. Williams	1905	the trunk of the elephant moves when coin is inserted, trunk automatically closes the slot as soon as coin is deposited, 6 3/4" long	150	250	400

BANKS

BANKS

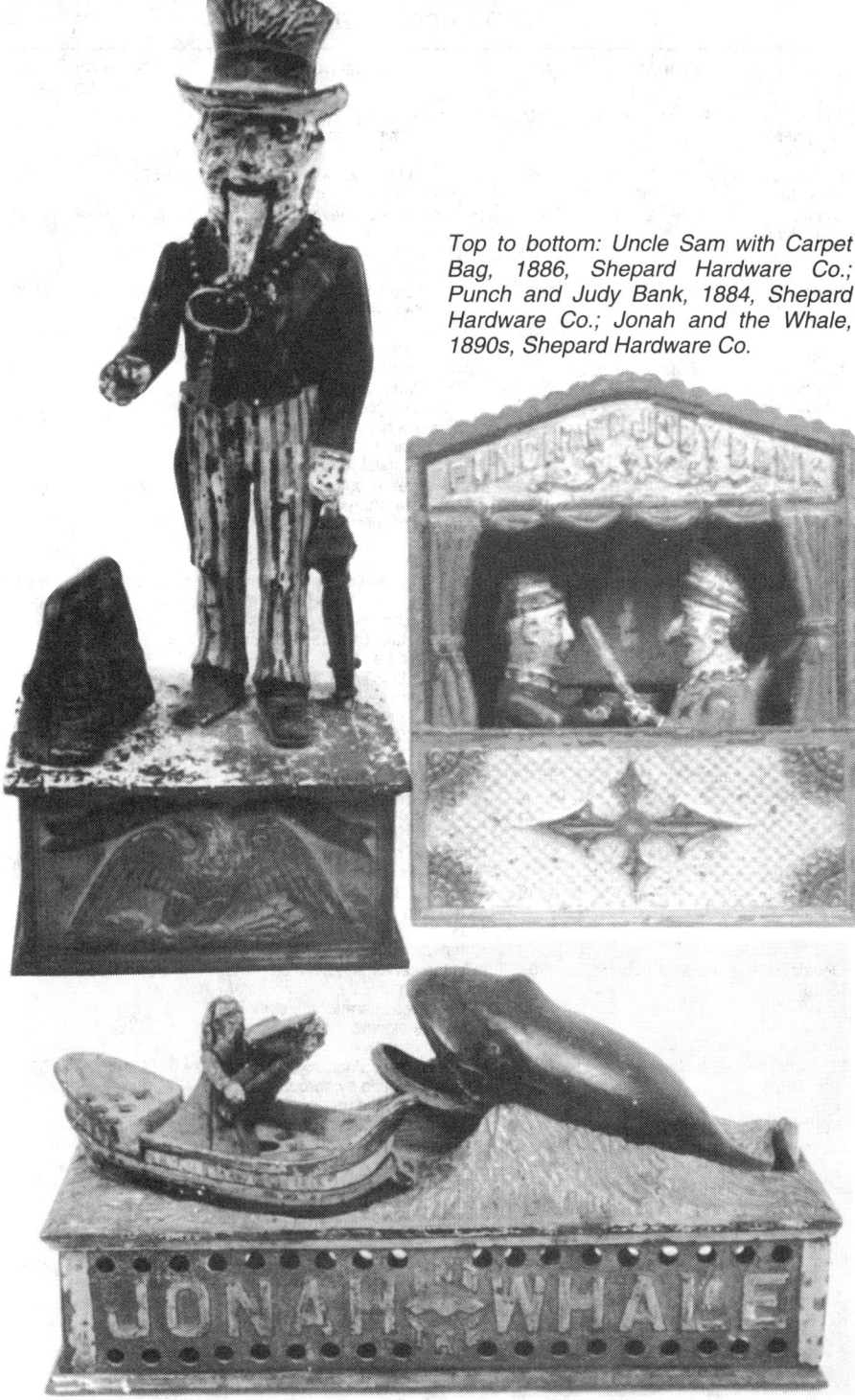

Top to bottom: Uncle Sam with Carpet Bag, 1886, Shepard Hardware Co.; Punch and Judy Bank, 1884, Shepard Hardware Co.; Jonah and the Whale, 1890s, Shepard Hardware Co.

Mechanical Banks

NAME	COMPANY	YEAR	DESCRIPTION	GOOD	VERY GOOD	EX
Elephant Three Stars	Unknown	1884	place coin in the elephant's trunk, touch his tail and the coin will be thrown into his head	300	500	850
Elephant, Man Pops Out	Enterprise Mfg. Co.	1884	push man into howdah or lift trunk to cock mechanism, close howdah lid, put coin in elephant's mouth; pressing lever drops trunk into mouth, dropping coin and man pops up from howdah; 5 3/8" tall	400	800	1250
Feed the Goose	Bankers Thrift Corp.	1927	press the tail feathers lever and the goose opens his mouth; toss in a coin and release, the mouth closes, coin is swallowed and his wings rotate	250	400	600
Football	J. Harper & Co.	1895	place a coin on the platform in front of the player's foot, press the lever and he kicks the coin into the goal net	1200	2200	3500
Fortune Teller Savings Bank	Nickel, Baumgarter & Co.	1901	drop a nickel in the slot of the lever give a sharp jerk backwards, the wheel will spin; when it stops pull lever forward and your fortune will appear in window	500	800	1400
Frog on Rock	Kilgore Mfg. Co.	1920s	press small lever under frog's mouth and he opens to deposit coin	200	400	650
Frog on Round Base	J & E Stevens Co.	1872	press frog's right foot and put coin in his mouth; release lever and he swallows coin and winks	350	550	950
Gem Bank	Judd Mfg. Co.	1878	pull dog back from bank, put coin on tray, lift dog's tail making dog move and drop coin into building	200	550	850
Gem Registering Bank	J & E Stevens Co.	1893	floral embossed rectangular bank with dials at one end; turn thumb piece to top, insert coin, turn thumb piece to bottom, dropping coin into bank	600	1200	2400
General Butler	J & E Stevens Co.	1880s	figure of Butler holds dollar bills in one hand, mouth is coin slot, inscribed "Bullion and yachts for myself and my friends, dry bread and greenbacks for the people."	1000	1800	3500
Germania Exchange	J & E Stevens Co.	1880s	goat on a keg; put coin on goat's tail; turning faucet makes goat drop coin and lifts up a glass of beer	3500	7000	12000
Giant in Tower	J. Harper & Co.	1892	put coin in slot and giant leans forward	6000	10000	15000
Girl in Victorian Chair	W.S. Reed	1880	girl in blue dress sits with dog in her lap in a highback wicker chair; put coin in chair top and press lever, making coin drop and dog lean forward	2000	3000	4500
Girl Skipping Rope	J & E Stevens Co.	1890	blonde girl in light blue dress jumps rope by means of an ornately housed mechanism	10000	15000	25000
Give Me A Penny	F. W. Smith	1870	bureau with drawer; open drawer and picture rises at back, saying 'Give Me A Penny"; putting coin in drawer and closing it makes coin drop and picture fall back into cabinet	1100	2000	3000
Grenadier Bank	J. Harper & Co.	1890s	soldier shoots coin into tree stump	350	650	1100

Mechanical Banks

NAME	COMPANY	YEAR	DESCRIPTION	GOOD	VERY GOOD	EX
Guessing Bank, Man's Figure	McLoughlin Brothers	1877	man sits atop a clock with numbers from 1 to 6 repeated around dial; dropping coin makes dial spin and land on a number; bank reads, "Pays Five For One If You Call the Number"	1200	2000	3000
Haley's Elephant	Unknown		8" long, painted gray with red and gold blanket	350	500	650
Hall's Excelsior Bank	J & E Stevens Co.	1869	value varies according to color of bank; 5" tall, yellow version, monkey sits atop building; put coin on tray in monkey's lap, pull string and monkey disappears inside	300	600	850
Hall's Liliput	J & E Stevens Co.	1877	coin laid on plate is carried around by the cashier and placed in the bank; cashier returns to its place and cycle can begin again	350	500	950
Hall's Liliput, No Tray	J & E Stevens Co.	1877	coin laid on plate is carried around by the cashier and placed in the bank, cashier returns to its place and cycle can begin again	400	550	950
Harlequinn	J & E Stevens Co.	1907	bring figure's hand halfway around to position and place coin in slot and press lever	5000	10000	22000
Hen and Chick	J & E Stevens Co.	1901	place coin in front of hen in slot, raise the lever and as the hen calls, the chicken springs from under her for the coin and disappears	1100	1700	3000
Hindu	Kyser & Rex Co.	1882	deposit coin in mouth and press the lever on the back of the head, his eyes roll down and his tongue swings down, causing the coin to fall into the bank	800	1300	1850
Hold the Fort, Seven Hole	Unknown	1877	pull back the ring until rod is in position, tip the bank, place the coin on target and drop the shot in the cannon; shot follows the coin into the bank and escapes out of the bottom	3000	5000	9500
Home Bank with Dormers	J & E Stevens Co.	1872	pull knob, place penny on its edge in front of the cashier, push the knob to the right, and the coin is deposited in the rear of the bank in vault	800	1200	2200
Home Bank with Tickets, Morrison's	Wm. Morrison	1900s	place 50 tickets in bank, deposit coin in slot and pull desk in front of cashier as far as it will go and coin is deposited in the vault; don't release until you receive your ticket	300	400	550
Home Bank, No Dormers	Wm. Morrison	1872	6" tall, 5" wide, no dormers, bank teller in cage; put coin in side of building and pull tray in front of teller to get a receipt	650	1000	1500
Hoop-La Bank	J. Harper & Co.	1895	place coin in dog's mouth and press the lever, the dog jumps thru the hoop and deposits the coin in the barrel	800	1200	3500
Horse Race, Straight Base	J & E Stevens Co.	1870	pull cord to start spring, place the horses' heads opposite the star, deposit the coin in the opening and the race will begin	3500	5500	8000

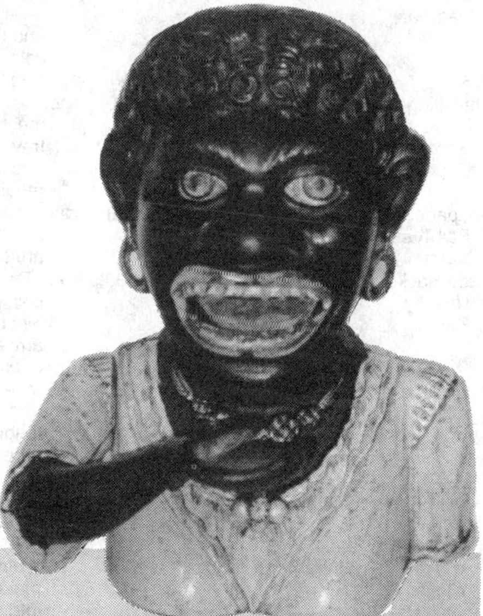

BANKS

Top to bottom: Hold The Fort, five hole, unknown; Dinah, 1911, J. Harper & Co.; United States and Spain, 1898, J&E Stevens Co.

Mechanical Banks

NAME	COMPANY	YEAR	DESCRIPTION	GOOD	VERY GOOD	EX
Humpty Dumpty Bank	Shepard Hdwre Co.	1882	8 1/2" tall; put coin in Humpty's hand, press lever, making him drop coin into bank	800	2200	3800
I Always Did 'Spise a Mule-Jockey	J & E Stevens Co.	1879	#24672, 10 1/2" long, 8" tall; put coin in jockey's mouth; pressing lever makes mule kick, throwing jockey which drops coin into bank	800	1200	2800
I Always Did 'Spise a Mule-Bench	J & E Stevens Co.	1897	put the mule and boy into position; when the knob is touched, the base causes the mule to kick the boy over, throwing the coin from the bench into the receptacle below	800	1200	2800
Independence Hall	Enterprise Mfg. Co.	1875	semi-mechanical bronze finish bank; drop coin in tower and pull lever to make bell ring.	350	550	1050
Indian Shooting Bear	J & E Stevens Co.	1883	10 3/8" long; put coin on rifle barrel; pressing level makes Indian shoot coin into bear	900	2500	4000
Initiating Bank, First Degree	Mechanical Novelty Works	1880	10 1/2" long, Eddy's patent on base; place coin on boy's tray, push lever, the goat butts boy forward and the frog moves upward as the coin slides from the tray into the frog's mouth	5000	7000	9000
Initiating Bank, Second Degree	Mechanical Novelty Works	1880	goat is pressed down to lock mechanism; put coin on man's tray; pressing lever makes goat butt man, dropping coin into frog's mouth	3000	5000	8000
Japanese Ball Tosser Safe	Weeden Mfg. Co.	1888	tin windup; put coin in slot, wind mechanism, making figure move arms, appearing to juggle balls	1000	2200	5000
Joe Socko Novelty Bank	Straits Mfg. Co.	1930s	Joe Palooka is turned clockwise as coin is deposited into slot making him turn quickly, swinging his right arm and knocking down his opponent. Base length 3 1/2"	250	400	650
Jolly Joe Clown, with Verse	Saalheimer & Strauss	1030s	push lever down, making tongue jut and eyes close; put coin on tongue; lifting lever makes tongue recede into mouth, dropping coin, and eyes open	350	600	850
Jolly Nigger in High Hat	Starkie (England)	1920	put coin in hand; pressing lever makes figure swallow coin while eyes roll and ears wiggle	250	400	650
Jolly Nigger in High Hat	J. Harper & Co.	1880s	put coin in hand; pressing lever makes arm raise, dropping coin into mouth, eyes roll and tongue moves back into mouth as coin drops	250	400	650
Jolly Nigger, String Tie	Unknown (England)	1890s	put coin in his hand; pressing level makes him lift coin and swallow it, tongue pulling back and eyes rolling	175	300	500
Jonah and the Whale	Shepard Hdwre Co.	1890s	put coin on Jonah's back; pressing lever makes Jonah turn toward whale's mouth, dropping coin into mouth.	1200	2500	4500
Jonah and the Whale, Jonah Emerges	J & E Stevens Co.	1880s	deposit coin in side of whale; pull Jonah into position by pulling tail backwards; press the lever, the coin is deposited and Jonah will appear	12000	20000	35000

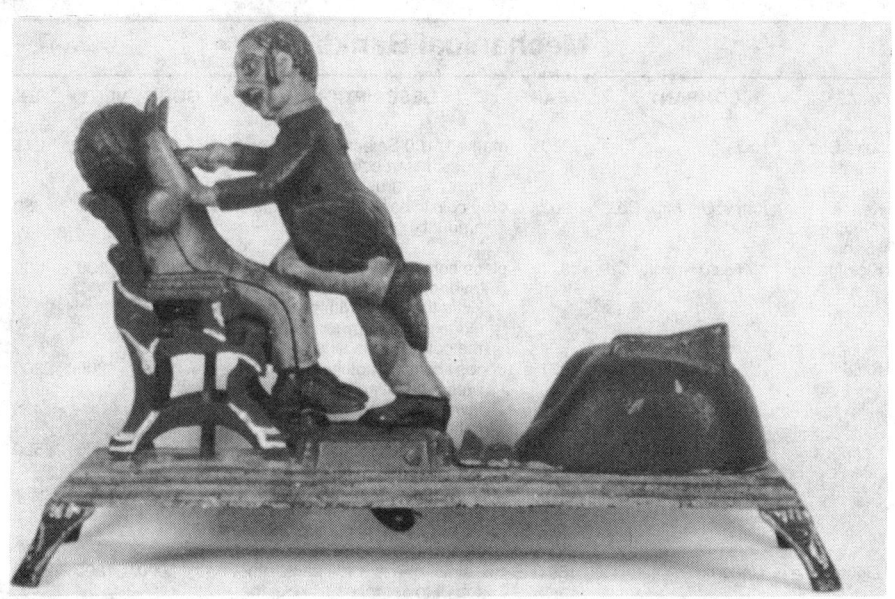

Top to bottom: Dentist, 1880s, J&E Stevens Co.; World's Fair Bank, J&E Stevens Co., 1893; Milking Cow, 1885, J&E Stevens Co.

Mechanical Banks

NAME	COMPANY	YEAR	DESCRIPTION	GOOD	VERY GOOD	EX
Juke Box Bank	Haji	1950s	marked "100 Select-O-Matic" on both sides; insert coin, wind up, turntable turns, music plays	95	200	275
Keene Savings Bank	Kingsbury Mfg. Co.	1902	drop coin in bank, press lever to show amount	350	600	850
Kick Inn	Melvisto Novelty Co.	1921	place coin on the ledge and push the lever on edge of base, the donkey kicks the ledge with his hind feet; ledge moves upward and tosses the coin into the side of the inn	200	500	900
Kiltie	J. Harper & Co.	1931	deposit coin in Scotchman's hand and press the lever; he raises his arm, lowers his eyes and deposits coin in his shirt pocket	750	1200	2200
Leap Frog Bank	Shepard Hdwre Co.	1891	7 1/2" long; put standing boy behind stooping one, put coin in slot; pressing lever makes standing boy leap over other one, who pushes lever on tree, dropping coin into bank	750	1500	4500
Lighthouse	Unknown	1891	two slots, one is a still bank, other on top of tower takes nickels; the button on top of tower will permit the bank to open only after 100 nickels are deposited	800	1200	3000
Lion and Two Monkeys	Kyser & Rex Co.	1883	9" long; put coin in monkey's hand; pressing lever makes monkey lower hand and drop coin into lion's mouth	650	1200	2200
Lion Hunter	J & E Stevens Co.	1911	hunter figure shoots coin into lion's mouth	2000	5000	10000
Little Jocko	Ferdinand Strauss Corp.	1912	drop penny into cup, turn the crank, the monkeys dance and music plays	500	1200	2200
Little Joe (Darkie Bank)	J. Harper & Co.	1910	put coin in Joe's hand; pressing lever makes him lift and swallow coin	150	350	575
Little Miss Muffet	Unknown	1930s	drop coin into bank and spider appears in window	150	400	675
Little Moe	Chamberlin & Hill	1931	place coin in Moe's hand, press the lever on his left shoulder, he raises his right arm, his tongue flips in and he swallows the coin as he lowers his arm, tipping his hat	250	400	600
Lucky Wheel Money Box	Jacob & Co.	1929	deposit coin in top of bank; the wheel spins and stops at one of twelve fortune messages	300	500	700
Magic Bank	J & E Stevens Co.	1873	6" tall; open door to find cashier, put coin on his tray; pressing lever makes cashier disappear and drop coin in building	800	1200	2200
Magician Bank	J & E Stevens Co.	1901	8" tall; put coin on table; pressing lever makes magician lower hat over coin, dropping coin into bank while magicican nods head	2500	3500	5500
Mama Katzenjammer	Kenton Mfg. Co.	1908	deposit coin in Mama's back in slot; her eyes roll up and return to their original position	3000	5000	7500
Mammy and Child	Kyser & Rex Co.	1884	put coin on apron; pressing lever makes Mammy lower spoon to baby, Mammy's head lowers and baby's leg lifts as coin drops	3000	4500	6000

Mechanical Banks

NAME	COMPANY	YEAR	DESCRIPTION	GOOD	VERY GOOD	EX
Mason Bank	Shepard Hdwre Co.	1887	7 1/4" long; drop coin onto hod, press lever, making hod move and drop coin into brick wall	2000	4000	7000
Memorial Money Bank	Enterprise Mfg. Co.	1876	slide the lever forwards to expose the coin slot, deposit coin, release the lever and it snaps back to ring the Liberty Bell	600	1200	1850
Merry-Go-Round	Kyser & Rex Co.	1888	put coin in slot, turn the handle and chimes will ring, the figures will revolve and the attendant turns, raises stick, and coin drops	7500	11000	15000
Mikado	Kyser & Rex Co.	1886	put coin under right hat and turn crank, making coin mysteriously move to left hat, which is lifted to show you this.	15000	25000	35000
Milking Cow	J & E Stevens Co.	1885	deposit coin in cow's back, the lever under the cow's throat is pressed, the cow will kick up its hind leg, upset the boy and the milk pail and deposit coin	3000	5000	7000
Minstrel Bank	Saalheimer & Strauss	1928	7 1/2" tall; press lever making tongue stick out, put coin on tongue, lift lever to make coin retract on tongue into bank	350	600	900
Monkey and Coconut	J & E Stevens Co.	1886	8 1/2" tall; put coin in monkey's hand; pressing lever makes monkey drop coin into coconut	350	900	2500
Monkey and Parrot	Saalheimer & Strauss	1925	put coin into monkey's hands; pressing lever makes monkey toss coin into parrot's mouth	350	450	650
Monkey Bank	Hubley	1920s	9" long, painted dark green base, put coin in monkey's mouth; pressing lever makes monkey spring forward, dropping coin into organ	600	1000	1600
Monkey Tips Hat	Chein	1940s	drop a coin into bank, making monkey tip his hat	100	200	400
Monkey With Tray	German	1900s	unclothed monkey squats on square base with zoo scenes; put coin on tray; pressing tail makes monkey raise tray, head tilt back and swallow coin	300	450	650
Monkey With Tray, Uniformed	German	1900s	uniformed monkey squats on square base with animal scenes; put coin on tray; pressing tail makes monkey raise tray, head tilt back and swallow coin	350	500	750
Mosque Bank	H.L. Judd Mfg.	1880s	9" tall, electroplated; put coin on tray on gorilla's head, turning lever makes gorilla turn, dropping coin into bank	600	1200	2000
Motor Bank	Kyser & Rex Co.	1889	wind the rod with key, drop a coin in slot and the trolley car is set in motion	10000	15000	20000
Mule Entering Barn	Kyser & Rex Co.	1880	8 1/2" long, gray barn version; put coin between mule's hind legs; pushing lever makes mule kick coin into barn and dog appears.	600	1200	1800
Musical Savings Bank, Regina	Regina Music Box Corp.	1894	bank is styled like a mantle clock; wind up mechanism, drop in coin, music plays	2500	4000	6000

BANKS

BANKS

Top to bottom: Monkey Bank, 1920s, Hubley and Trick Dog Bank, 1920s, Hubley; another view of Hubley's Monkey Bank; Speaking Dog Bank, 1885, Shepard Hardware Co., and Clown on Globe, 1890, J&E Stevens Co.

Mechanical Banks

NAME	COMPANY	YEAR	DESCRIPTION	GOOD	VERY GOOD	EX
National Bank	J & E Stevens Co.	1873	place coin on door ledge and push doorbell; the door revolves, slinging the coin into the bank, the man behind the window of the door quickly moves to the right to get out of the way	2000	3500	6000
National, Your Savings	Unknown	1900s	replica cash register, with one slot and key each for pennies, nickels, dimes and quarters; dropping coin and pressing key rings a bell	275	550	850
New Creedmore Bank	J & E Stevens Co.	1891	place coin on barrel of rifle, press right foot and coin is shot into the bull's eye of the target; as coin enters it strikes gong bell	550	950	1850
Novelty Bank	J & E Stevens Co.	1873	open door and put coin on tray, release door which closes by a spring and teller turns and drops coin into vault	550	1000	1850
Octagonal Fort Bank	Unknown	1890	also called Fort Sumter, cock mechanism, put coin in barrel end; pressing lever makes coin fire into tower; 10 3/4" long	1200	2200	3500
Organ Bank	Kyser & Rex Co.	1881	6" tall, brown organ, monkey in blue pants and coat, yellow hat; turn handle and a chime of bells will ring while monkey deposits coins placed in tambourine, tipping his hat in thanks	350	500	850
Organ Bank with Boy and Girl	Kyser & Rex Co.	1882	7 1/2" tall; put coin on tray; turning crank makes monkey lower tray, dropping coin into bank while boy and girl turn.	500	850	1650
Organ Bank with Cat and Dog	Kyser & Rex Co.	1882	#24663, 8 1/2" tall; put coin on tray; turning crank makes monkey lower tray dropping coin into bank while cat and dog rotate	500	850	1650
Organ Bank, Miniature	Kyser & Rex Co.	1890s	put coin in slot; turning crank makes bells ring, monkey turn, and coin drop	500	850	1800
Organ Grinder and Dancing Bear	J & E Stevens Co.	1890s	6 7/8" long, 5 1/2" tall, 4 3/4" wide, painted windup; put coin in slot; pushing button makes grinder deposit coin, play organ, and bear dance	2300	3500	6500
Owl, Slot in Book	Kilgore Mfg. Co.	1926	height 5 3/4"; deposit a coin in the slot and Blinkey's eyes will roll down and back	350	650	950
Owl, Slot in Head	Kilgore Mfg. Co.	1926	deposit coin in slot, eyes roll forward; height 5 5/8"	350	650	950
Owl, Turns Head	J & E Stevens Co.	1880	7 1/2" tall, brown bird with yellow highlights, glass eyes; insert coin in branch; pressing lever makes owl turn head as coin drops	250	500	900
Paddy and the Pig	J & E Stevens Co.	1882	8 1/2" tall; put coin in pig's nose; pressing lever makes pig kick coin into Paddy's mouth	900	2200	3800
Panorama Bank	J & E Stevens Co.	1876	place coin in slot on roof and next picture appears in the window	2500	4500	6500
Patronize the Blind Man	J & E Stevens Co.	1878	place coin in the blind man's hand, the dog takes the coin and deposits it in the bank and returns to his position	2000	3000	4500
Peg-Leg Begger	Judd Mfg. Co.	1875s	insert coin in slot on hat and the begger nods his thanks	750	1200	1700

Mechanical Banks

NAME	COMPANY	YEAR	DESCRIPTION	GOOD	VERY GOOD	EX
Pelican Bank, Baseball Player	J & E Stevens Co.	1878	8" tall; close bird's beak, put coin into top of head which makes beak open revealing a baseball player's head	2000	3500	5000
Picture Gallery Bank	Shepard Hrdwre Co.	1885	place coin in hand of center figure and he deposits coin; all the letters of the alphabet and numbers 1 to 26 are shown in rotation; also 26 animals or objects with short word for each letter	4500	7500	9000
Pig in a High Chair	J & E Stevens Co.	1897	5 1/4" tall; put coin on tray; pressing lever makes tray bring coin to pig's mouth	350	550	1200
Pistol Bank	Richard Elliot Co.	1909	5 1/2" long; pull trigger halfway back, a hook appears from end of the barrel, place a dime on hook and pull trigger; coin is snatched and pistol fires when deposited	450	750	1650
Popeye Knockout Bank with Box	Straits Mfg. Co.	1929	turn Popeye to the right, lift his opponent to his feet and when the coin is deposited Popeye will deliver knockout punch	250	650	950
Presto Bank	Kyser & Rex Co.	1894	pull drawer open and deposit coin, release drawer and coin is deposited in bank	250	525	950
Professor Pug Frog's Great Bicycle Feat	J & E Stevens Co.	1886	place coin on rear wheel; turning crank makes frog spin, dropping coin into bank.	1500	3500	5000
Punch and Judy	Banks & Sons	1929	pressing lever makes figures rise, put coin in slot; releasing lever makes figures fall back into bank as coin drops	2700	5000	7000
Punch and Judy, Large Letters	Shepard Hdwre Co.	1884	7 1/2" tall; put coin on Judy's tray, press lever and Judy deposits coin while Punch tries to hit her with stick	600	1500	2500
Punch and Judy, Small Letters	Shepard Hdwre Co.	1884	7 1/2" tall; put coin on Judy's tray, press lever and Judy deposits coin while Punch tries to hit her with stick	600	1500	2500
Queen Victoria Bust	J. Harper & Co.	1887	drop coin into slot in crown, making eyes roll. Note: This bank was also made in brass.	3000	5000	7500
Rabbit in Cabbage	Kilgore Mfg. Co.	1925	length 4 1/2", white with a green base; press coin into the slot and ears will rise and then flop back down	200	550	1100
Rabbit Standing (small)	Lockwood Mfg. Co.		put coin in rabbit's paws; pressing tail moves ears and drops coin	450	850	1250
Reclining Chinaman	J & E Stevens Co.	1882	8 1/4" long; put coin in slot on log, press lever to make figure raise hand, showing hand of cards and saluting the depositor; as coin falls, rat runs out of the end of the log	1500	2500	6500
Red Riding Hood	W.S. Reed	1880s	put coin in slot in pillow; moving lever makes Grandma's mask shift, revealing the wolf; Red turns her head in 'fear' and coin drops	8500	14000	18000

BANKS

BANKS

Top to bottom: Snapping Bulldog, 1878, Ives; Elephant Howdah Bank, 1934, Hubley; William Tell, 1896, J&E Stevens Co.

Mechanical Banks

NAME	COMPANY	YEAR	DESCRIPTION	GOOD	VERY GOOD	EX
Registering Dime Savings Bank	Ives	1890	deposit dime in chute and pull lever to right, then left and the dime will be deposited and door locked, amount registered on dial; when $10 is deposited door will unlock and pop out	800	1600	2500
Rollerskating Bank	Kyser & Rex Co.	1880s	place coin in roof slot, press lever, the skaters glide to the rear of the rink as the coin is deposited in bank, the man turns to give a little girl a wreath	12000	18000	25000
Rooster Bank	Kyser & Rex Co.	1900s	put coin on rooster's tail; pressing lever causes the rooster to move his head in a crowing position, money is deposited, 6 1/4"	600	900	1500
Safety Locomotive	Edward J. Colby	1887	the weight of the money as it's dropped on the cab will (when it's full), loosen the smokestack, then it can be lifted out and the money poured from the opening	650	1400	2200
Saluting Sailor	Unknown (Germany)		pressing lever makes sailor lift left arm, exposing coin slot while saluting with right arm; dropping coin and releasing lever completes cycle	300	600	900
Santa at the Chimney	Shepard Hdwre Co.	1889	6" tall, painted; put coin in his hand; pressing lever makes him drop coin into chimney	800	1500	2400
Santa at the Desk	S & E		deposit coin in large phone and dial Santa's number, the phone rings and he will pick it up and nod his head, as he's writing Merry Xmas on a lighted piece of paper. Desk 6X8"	250	450	675
Schley Bottling Up Cervera	Unknown	1899	bottle neck shows two portraits, Schley and Cervera; coins can only be dropped when Cervera is visible; shake bank to pull up Cervera picture, dropping coin will make Schley's picture return	4500	8000	18000
Scotchman	Saalheimer & Strauss	1930s	German bank; lifting lever makes eyes blink and tongue jut; put coin on tongue and lower lever to make tongue recede and drop coin into bank	300	500	700
Smyth X-Ray Bank	Henry C. Hart Mfg. Co.	1898	set coin in the path of the scope slot and look into the eyepiece; you will see through the coin and out the other end of the bank; press lever and the coin is deposited in the bank	900	1800	2600
Speaking Dog	Shepard Hrdwre Co.	1885	deposit coin on girl's plate; when thumb piece is pressed the girl's arm moves, depositing coin through trap door on bench and dog wags his tail and moves his mouth.	850	1500	2400
Sportsman (Fowler)	J & E Stevens Co.	1892	place a coin in slot, set the trap, place the bird on the trap and push the lever; bird rises in the air and the sportsman fires, downing bird; bank can use paper caps	7500	12000	18000

Mechanical Banks

NAME	COMPANY	YEAR	DESCRIPTION	GOOD	VERY GOOD	EX
Springing Cat	Charles Bailey	1882	lock cat in place, put coin in slot; pulling lever makes cat move toward coin as a mouse appears, knocks coin into bank, and escapes back into bank, leaving cat with open jaws	3500	6000	8000
Squirrel and Tree Stump	Mechanical Novelty Works	1881	pressing lever makes squirrel move and drop coin into stump	850	1250	2250
Starkie's Aeroplane	Starkie (England)	1919	move plane up pole and lock in place, put coin in slot on plane; pressing lever makes plane coast down pole, dropping coin in bank base	1200	1800	3500
Stollwerck Hand Shadows	Stollwerck (German)	1880s	candy dispenser; deposit coin in slot on top, pull drawer handle to dispense candy; decorated with art of a woman's arm making shadow animal	100	250	400
Stollwerck Postman	Stollwerck (Germany)	1900s	candy vending machine decorated with images on a postal theme; drop coin in slot, pull handle to receive candy	200	350	500
Stollwerck Red Riding Hood	Stollwerck (Germany)	1900s	candy vending machine decorated with images of Red Riding Hood; drop coin in slot, pull handle to receive candy	175	300	450
Stump Speaker Bank	Shepard Hdwre Co.	1886	9 3/4" tall, painted; place coin in hand, press the small knob on top of the box, which lowers the arm and opens the satchel to deposit the coin; release lever and mouth moves up and down	850	2200	4500
Sweet Thrift	Beverly Novelty Corp.	1928	candy dispenser; drop coin in slot and open drawer to receive candy; 5 3/4" tall; red, green or yellow	100	250	450
Tabby Bank	Unknown	1887	cat sits atop large egg, waiting for chick to hatch; drop coin in slot in cat's back and the chick moves its head	250	650	1050
Tammany Bank	J & E Stevens Co.	1873	6" tall; fat politician sits in chair, yellow vest, brown jacket, blue pants; put coin in his hand and he drops it into his pocket	400	800	2200
Tank and Cannon	Starkie (England)	1919	cannon fires coin into tank bank	650	900	1500
Teddy and the Bear	J & E Stevens Co.	1907	10" long, painted; cock gun and put coin in it, push bear into tree and close cover; pressing lever makes Teddy lower head in aim, gun fires coin into tree and bear pops up	1000	2200	4000
Thrifty Animal Bank	Buddy "L" Co.	1940	a registering bank with two coin slots, one takes dimes which become visible as acorns on the tree and the second slot is a still bank; the top of bank pops off when full	250	500	850
Thrifty Tom's Jigger Bank	Ferdinand Strauss Corp.	1910	drop coin into the slot and Thrifty Tom does a dance that lasts until the spring winds down	200	450	850
Toad on Stump	J & E Stevens Co.	1886	press lever to open toad's mouth, put coin on mouth and release lever, dropping coin into bank	550	875	1650

BANKS

BANKS

Top to bottom: Initiating Bank--First Degree, 1880, Mechanical Novelty Works; Dinah, 1911, J. Harper & Co./Hindu, 1882, Kyser & Rex/Uncle Tom, 1882, Kyser & Rex/Jolly Nigger, 1880s; Boy Scout Camp, 1912, J&E Stevens Co.

Mechanical Banks

NAME	COMPANY	YEAR	DESCRIPTION	GOOD	VERY GOOD	EX
Tommy Bank	J. Harper & Co.	1914	cock the rifle, lay a coin in front of the launcher and press the lever on top of soldier's left side; the coin is shot into the tree as his head rises	2500	3500	6500
Treasure Chest Musical Bank	Faith Mfg. Co.	1930	bronze or silver finish domed chest; wind mechanism, dropping coin in slot makes music play	300	500	700
Trick Dog Bank (6 Part Base)	Hubley	1888	deposit coin in dog's mouth; pressing lever makes dog jump through clown's hoop and coin is deposited in his barrel	500	800	1800
Trick Dog Bank, Solid Base	Hubley	1920s	deposit coin in dog's mouth; pressing lever makes dog jump through clown's hoop and coin is deposited in his barrel	200	450	850
Trick Pony	Shepard Hdwre Co.	1885	7" long, 8" tall; put coin in horse's mouth, pulling lever makes pony drop coin in trough	900	1250	3000
Trick Savings Bank	Unknown		deposit coin, coin disappears with drawer closed; height 5 1/2"	55	125	200
Turtle Bank	Kilgore Mfg. Co.	1920s	press a coin in the slot and Pokey's neck extends and then returns	7500	12000	20000
Uncle Remus	Kyser & Rex Co.	1891	5 3/4" long, painted; deposit coin on roof and press the chicken's head, the policeman moves toward Uncle Remus who slams the door to prevent getting caught	1700	2300	3500
Uncle Sam with Carpet Bag	Shepard Hdwre Co.	1886	11 1/2" tall; put coin in Sam's hand; pressing lever lowers coin into his carpet bag.	1000	2800	8000
Uncle Tom, No Star, Lapels	Kyser & Rex Co.	1882	put coin on tongue; pressing lever makes Tom swallow coin and move eyes	250	650	1250
Uncle Tom, With Lapels and Star	Kyser & Rex Co.	1882	put coin on tongue; pressing lever makes Tom swallow coin and move eyes	250	650	1250
United States and Spain	J & E Stevens Co.	1898	U.S. cannon faces Spanish ship; cock cannon and insert paper cap; pressing lever fires cap and shot which strikes ship's mast while coin drops.	2000	4500	7500
United States Safe Bank	J & E Stevens Co.	1880	drop coin in slot, making top flip up showing a small bank book in which to write entry of deposit.	800	2200	4000
Victorian Money Box	Unknown (England)		dropping coin into box makes girl in doorway curtsy	800	1300	1800
Volunteer Bank	J. Harper & Co.	1885	length 10"; cock rifle and put coin in slot; pressing lever makes man fire rifle, shooting coin into tree stump	750	1250	1850
Watch Dog Safe	J & E Stevens Co.	1890s	drop coin in top of bank, lift lever, coin falls into the bank as the dog opens his mouth and barks; release and mouth closes	350	675	1250
Weeden's Plantation Darkie Bank	Weeden Mfg. Co.	1888	5 1/2" tall, windup bank; putting coin in slot at side of house makes banjo player play and other figure dance	750	1250	1850

Mechanical Banks

NAME	COMPANY	YEAR	DESCRIPTION	GOOD	VERY GOOD	EX
William Tell	J & E Stevens Co.	1896	10 1/2" long; lock lever on gun, lower head to aim; lowering boy's arm reveals apple, put coin on gun, press shooter's foot to fire coin into castle, knocking down apple, ringing a gong	600	1250	2500
Wimbledon Bank	J. Harper & Co.	1885	cock rifle in reclining redcoat's hands, put coin on barrel; pressing lever shoots coin into tree as soldier's head rises	3500	5500	8000
Wireless Bank	Hugo Mfg. Co.	1926	battery operated; putting coin on roof of bank and clapping hands makes cover swing over, dropping coin into bank	350	650	850
World's Fair Bank, with Lettering	J & E Stevens Co.	1893	deposit coin on Columbus' feet; pressing lever makes the Indian Chief popup from log, offering peace pipe as Columbus salutes him.	850	1250	1850
Zoo Bank	Kyser & Rex Co.	1890s	building bank; put coin in slot; pressing monkey's face makes coin drop and shutters open on lower windows, and faces of lion and tiger appear through windows	850	1650	2250

Clockwise from upper left: Captain Kidd, unknown; Two Faced Black Boy, 1900s, A.C. Williams; Home Savings Bank, unknown; Multiplying Bank, 1883, J&E Stevens Co.

Still Banks

NAME	COMPANY	YEAR	DESCRIPTION	GOOD	EX
$100,000 Money Bag	Unknown Maker		3 5/8" tall, silver gray finish	300	650
1 Pounder Shell Bank	Grey Iron Casting Co.	1918	8" artillery shell, "1 Pounder Bank"	25	85
1876 Bank, Large	Judd, H.L.	1895	3 3/8" tall, building bank with bronze/copper finish	75	250
1926 Sesquicentenial Bell	Grey Iron Casting Co.	1926	3 3/4" x 3 7/8" diam.	75	200
A.A.O.S.M.S. Shriner's Fez	Allen Mfg. Co.	1920s	2 3/8" red fez with tassle and gold lettering	250	650
Administration Building	Magic Introduction Co.	1893	5", unpainted	250	650
Air Mail Bank on Base	Dent	1920	6 3/8" tall, red	40	95
Alamo	Alamo Iron Works	1930s	1 7/8" tall, 3 3/8" wide, unpainted bronze finish	200	450
Alphabet Bank	Unknown Maker		3 1/2", octagonal.	700	2500
Amherst Buffalo	Unknown Maker	1930s	5 1/4" tall, 8" long	150	350
Amish Boy	Wright, John	1970	5" tall, painted	10	65
Amish Boy in White Shirt	Wright, John	1971	5" tall, blue coveralls, black hat	10	65
Amish Girl	Wright, John	1970	5" tall, painted	10	65
Andy Gump	Arcade	1928	4 3/8" tall, Andy sits reading a paper, painted	500	950
Apollo (plain)	Wright, John	1968	4 1/4", unpainted	20	65
Apollo 8	Wright, John	1968	4 1/4", red, white and blue	20	75
Apple	Kyser & Rex	1882	5 1/4" tall, painted apple on twig with leaves	600	1450
Arabian Safe	Kyser & Rex	1882	4 9/16" x 4 1/4"	100	300
Armoured Car	Williams, A.C.	1900s	3 3/4" tall, 6 3/4" long, red car on gold wheels	650	2200
Art Deco Elephant	Unknown Maker		4 3/8" tall, red	100	225
Aunt Jemima	Williams, A.C.	1900s	5 7/8", also called Mammy with Spoon	125	325
Auto	Williams, A.C.	1910?	5 3/4" long, black, red wheels, 4 passengers	500	1200
Baby in Cradle	Unknown Maker	1890s	3 1/4" tall, rocking cradle	500	1400
Bank of Columbia	Arcade	1800s	4 7/8", unpainted, "Bank of Columbia"	150	375
Bank of England Safe	Kyser & Rex	1882	identical to Egyptian Safe except front is embossed Bank of England	350	650
Barrel	Judd, H.L.	1873	2 3/4" tall	100	225
Baseball on Three Bats	Hubley	1914	5 1/4"	250	950
Baseball Player	Williams, A.C.	1909	5 3/4 inches, gold	100	300
Baseball Player	Williams, A.C.	1910s	5 3/4" tall, in various paint colors	200	450
Basket Puzzle Bank	Nicol	1894	2 3/4" tall, 3 1/2" wide, unpainted	300	550
Basket Registering Bank, Woven	Braun, Chas. A.	1902	2 7/8" x 3 3/4"	50	125
Basset Hound	Unknown Maker		3 1/8", bronze finish	550	1200
Battleship Maine	Grey Iron Casting Co.	1800s	5 1/4" tall, 6 5/8" long, "Maine"	650	3000
Battleship Maine	Stevens, J. & E.	1901	6" tall, 10 1/4" long, white	500	2000
Battleship Oregon	Stevens, J. & E.	1890s	4 7/8" long, silver finish	200	425
Be Wise Owl	Williams, A.C.	1900s	4 7/8" x 2 1/2"	150	375
Bean Pot	Unknown Maker		3", red cooking pot, nickel registering	150	375
Bear Seated on Log	Unknown Maker		7"	400	950
Bear Stealing Pig	Ober Mfg. Co.	1913	Ober #1011, 5 1/2" tall, painted	400	950
Bear with Honey Pot	Hubley		6 1/2" tall, painted	75	150
Bear, Begging	Williams, A.C.	1900s	5 3/8", bronze finish	75	150

BANKS

Top to bottom: Elephant with Howdah, 1900s, A.C. Williams; Everett True The Capitalist, 1913, Ober Mfg. Co.; World's Fair Administration Bldg., 1893, Magic Introduction Co.; Yellow Cab, 1921, Arcade.

Still Banks

NAME	COMPANY	YEAR	DESCRIPTION	GOOD	EX
Beehive Bank	Kyser & Rex	1882	2 3/8"	250	500
Beehive Registering Savings Bank	Unknown Maker	1891	5 3/8" x 6 1/2"	200	425
Beehive with Brass Top	Gobeille, W.M.		5 1/2" tall on base, unpainted	350	750
Bethel College Administration Building	Service Foundry	1935	2 7/8" x 5 1/4"	175	350
Bicentennial Bell	Unknown Maker	1976	4" x 4" diam.	25	45
Billiken	Williams, A.C.	1909	4 1/4" tall, on square base, bronze finish, red cap	55	125
Billiken on Throne	Williams, A.C.	1909	6 1/2" tall	65	175
Billy Bounce	Hubley	1900s	4 11/16" tall, silver painted body	375	850
Billy Possum ("Possum & Taters")	Harper, J.M.	1909	3" x 4 3/4", on base "Billy Possum"	1200	3000
Bird Bank Building	Unknown Maker		5 7/8" unpainted cupola building with bird on top, "Bank New York""	650	1850
Bird Cage Bank, (Crystal Bank)	Arcade	1900s	3 7/8" tall, similar to Crystal Bank #926, but glass is replaced by open mesh	50	125
Bismark Bank,(Pig)	Unknown Maker	1883	3 3/8", "Bismark Bank"	100	300
Bismark Pig with Rider	Unknown Maker	1880s	7 1/4" tall, 6 1/2" long, bronze finish	1000	3500
Boss Tweed	Unknown Maker	1870s	3 7/8" tall	1200	3500
Boston State House	Smith & Egge	1800s	6 3/4" tall, painted	3000	6000
Boxer Bulldog	Hubley	1900s	4 1/2", seated, bronze finish	125	175
Boy Scout	Williams, A.C.	1910s	5 7/8" tall, brown finish	50	125
Boy with Large Football	Hubley	1914	5 1/8" tall, brown	2000	3000
Buckeye (SBCCA)	Filler, Lou	1973	3 1/2", painted "Ohio The Buckeye State", "SBCC 1973"	25	150
Buffalo Bank	Williams, A.C.	1900s	3 1/8" x 4 3/8", gold	50	175
Buffalo Nickel	Knerr, George	1970s	3 7/8"	35	75
Building with Belfry	Kenton		8" tall, in browns	550	2200
Bull on Base	Unknown Maker		4" tall, unpainted	200	450
Bull with Long Horns	Unknown Maker		3 11/16" tall, painted	50	125
Bulldog, Large	Wright, John	1960s	6", painted	25	65
Bulldog, Seated	Hubley	1928	3 7/8"	200	400
Bulldog, Standing	Arcade	1900s	2 1/4", painted	250	450
Bungalow Bank	Grey Iron Casting Co.	1900s	3 3/4" x 3", white cottage with green roof	225	425
Bust of Man	Unknown Maker		5"	100	350
Buster Brown & Tige	Williams, A.C.	1900s	5 1/2"	100	275
Cadet	Hubley	1905	5 3/4" tall, blue uniform with gold trim	300	650
Camel, Kneeling	Kyser & Rex	1889	2 1/2" tall, 4 3/4" long	350	750
Camel, Large	Williams, A.C.	1900s	7 1/4" x 6 1/4"	200	425
Camel, Small	Hubley	1920s	4 3/4" x 3 7/8"	100	225
Camera Bank	Wrightsville Hdw. Co.	1800s	4 5/16" tall, bronze finish bellows camera on tripod	2500	5000
Campbell Kids	Williams, A.C.	1900s	3 5/16" x 4 1/8"	150	350
Cannon	Hubley	1914	3" tall, 6 7/8" long, black cannon on red wheels	2500	5000
Capitalist, The (Everett True)	Ober Mfg. Co.	1913	5" tall, painted	1000	1500
Capitol Bank	Riverside Foundry	1981	5 1/8"	50	100

BANKS

Clockwise from upper left: Novelty Bank, unknown; Pavillion, 1880, Kyser & Rex; Dutch Girl, Grey Iron Casting Co.; Palace, 1885, Ives.

Still Banks

NAME	COMPANY	YEAR	DESCRIPTION	GOOD	EX
Captain Kidd	Unknown Maker	1900s	5 5/8" tall, Kidd stands by tree trunk with shovel, base reads "Captain Kidd"	275	450
Carpenter Safe	Harper, J.M.	1907	4 3/8"	2500	5000
Cash Register Savings Bank	Hubley	1906	4 3/4", unpainted, "Cash Register Savings Bank"	500	750
Cash Register Savings Bank	Unknown Maker	1880s	5 5/8" tall, round face on 3 claw foot feet, "Cash Register Savings Bank"	350	850
Cash Register with Mesh	Arcade	1900s	3 3/4" tall, red finish with gold-bronze mesh	50	125
Castle Bank, Small	Kyser & Rex	1882	3" x 2 13/16"	200	450
Cat on Tub	Williams, A.C.	1920s	4 1/8" tall, bronze finish	100	175
Cat with Ball	Williams, A.C.	1900s	2 1/2" tall	200	375
Cat with Bow	Hubley	1930s	4 1/8"	275	575
Cat with Bow, Seated	Grey Iron Casting Co.	1922	4 3/8" tall, brown finish	225	475
Cat with Bow, Seated	Wright, John		4 3/8" x 2 7/8", painted, white body, red bow	25	50
Cat with Long Tail	Grey Iron Casting Co.	1910s	4 3/8" tall, 6 3/4" long	375	875
Cat with Soft Hair, Seated	Arcade	1900s	4 1/4" x 2 7/8"	85	225
Century of Progress Building	Arcade	1933	4 1/2" x 7", white, "A Century Of Progress" building from Chicago World's Fair	800	1800
Champion Heater	Unknown Maker		4 1/8" green and black, "Champion"	125	375
Chantecleer (Rooster)	Unknown Maker	1911	4 5/8", bronze finish, painted face and comb	850	2800
Chicken Feed Bag	Knerr, George	1973	4 5/8", "Chicken Feed"	35	250
Chipmunk with Nut	Unknown Maker		4 1/16", black	300	950
Church Towers	Unknown Maker		6 3/4"	850	1900
Church Window Safe	Shimer Toy Co.	1890s	3 1/16"	50	125
City Bank with Chimney	Unknown Maker	1870s	6 3/4" tall, painted	650	1400
City Bank with Crown	Unknown Maker	1870s	5 1/2" tall, painted with red "crown" on roof	600	1200
City Bank with Teller	Judd, H.L.		5 1/2", bronze finish	400	700
Clown	Williams, A.C.	1908	6 1/4", gold and red with tall curved hat	125	250
Clown Bust	Knerr, George	1973	4 7/8", painted	100	350
Coca Cola	Unknown Maker		3 3/8" tall, red and green with logo	600	1500
Coin Registering Bank	Kyser & Rex	1890	6 3/4" with red doors and dome, "Coin Registering Bank"	2500	6500
Colonial House with Porch, Large	Williams, A.C.	1900s	4" tall, white	100	225
Colonial House with Porch, Small	Williams, A.C.	1910s	3 " tall, brown finish with red, green or gold roof	75	175
Columbia	Kenton		4 1/2" tall, silver finish building bank	600	900
Columbia Bank	Kenton	1890s	8 3/4" tall, bronze finish	600	1000
Columbia Bank	Kenton	1890s	5 3/4" tall, unpainted silver finish	300	700
Columbia Magic Savings Bank	Magic Introduction Co.	1892	5", unpainted, "Columbia Magic Savings Bank"	300	700
Columbia Tower	Grey Iron Casting Co.	1897	6 7/8", unpainted 3 story tower	450	950
Covered Bridge	Wright, John	1960s	2 1/2" tall, 6 1/8" long, white with red roof	35	75
Covered Wagon	Wilton Products Co.		6 5/8" long, unpainted	10	25
Cow	Williams, A.C.	1920	3 3/8" x 5 1/4", brown or red finish	75	125

Top to bottom: Newfoundland Dog, 1930s, Arcade; Dolphin Boat Bank, 1900s, Grey Iron Casting Co.; Tower Bank, 1891, Kyser & Rex; Minuteman, 1905, Hubley.

Still Banks

NAME	COMPANY	YEAR	DESCRIPTION	GOOD	EX
Crosley Radio, Large	Kenton	1930s	5 1/8" tall, green with gold highlights	650	1400
Crosley Radio, Small	Kenton	1930s	4 5/16" tall, green	150	350
Cross	Unknown Maker		9 1/4" tall, dark finish, "God Is Love" on base	750	1850
Crown Bank on Legs, Small	Unknown Maker		4 5/8", painted	600	850
Cupola Bank	Stevens, J. & E.	1870s	3 1/4" tall, black	75	175
Cupola Bank	Stevens, J. & E.	1872	4 1/4" x 3 3/8" red and gray	100	225
Cupola Bank	Vermong Novelty Works	1869	5 1/2" tall, painted building with center roof cupola	300	550
Cutie Dog	Hubley	1914	3 7/8", painted	65	150
Daisy	Shimer Toy Co.	1899	2 1/8" tall, red safe bank	50	150
Darkey Sharecropper	Williams, A.C.	1900s	5 1/2" tall, toes visible on one foot	75	375
Decker's Iowana, (Pig)	Unknown Maker		2 5/16", unpainted	75	200
Derby, "Pass Around the Hat"	Unknown Maker		1 5/8" tall, 3 1/8" long	100	225
Dime Registering Coin Barrel	Kyser & Rex	1889	4" x 2 1/2", unpainted	125	225
Dime Savings	Shimer Toy Co.	1899	2 1/2" safe, "Dime Savings"	200	425
Dog on Tub	Williams, A.C.	1920s	4 1/16" x 2" diam., bronze finish	125	200
Dog Smoking Cigar	Hubley		4 1/4", painted, white body, red bow tie.	450	850
Dolphin Boat Bank	Grey Iron Casting Co.	1900s	4 1/2" tall, sailor boy in boat holds anchor	500	850
Domed Bank	Williams, A.C.	1899	3" tall	20	55
Domed Mosque Bank	Grey Iron Casting Co.	1900s	3 1/8"tall, bronze finish	65	145
Domed Mosque Bank	Grey Iron Casting Co.	1900s	4 1/4" tall, gold/bronze finish	85	175
Donkey	Unknown Maker		3 1/4" tall, black with red yoke		
Donkey "I Made St. Louis Famous"	Arcade	1903	4 11/16" tall, gray finish	800	1500
Donkey (small)	Arcade	1910s	4 1/2" tall, blue, gold or gray finish	85	175
Donkey on Base	Unknown Maker		6 9/16" tall	250	650
Donkey with Blanket	Kenton	1930s	3 7/8" tall, painted, gray with red blanket	450	850
Donkey, Large	Williams, A.C.	1920s	6 13/16" tall, painted	150	225
Dormer Bank	Unknown Maker		4 3/4" tall, painted building bank with red roof	3500	6000
Double Door	Williams, A.C.	1900s	5 7/16" building with 2 doors, painted white with gold highlights	200	375
Doughboy	Grey Iron Casting Co.	1919	7" tall, painted World War I soldier	350	850
Dry Sink	Wright, John	1970	3" x 2 3/4", dark finish	25	45
Duck	Hubley	1930s	4 3/4", white painted body	150	275
Duck Bank	Williams, A.C.	1900s	4 7/8", unpainted	150	275
Duck on Tub "Save for a Rainy Day"	Hubley	1930s	5 3/8"	95	325
Duck, Round	Kenton	1930s	4"tall, painted, yellow body, red beak and top of head	225	450
Dutch Boy	Grey Iron Casting Co.		6 3/4" tall	600	850
Dutch Boy	Unknown Maker		8 1/4" tall, doorstop conversion	150	275
Dutch Boy on Barrel	Hubley	1930s	5 5/8"	75	150
Dutch Girl	Grey Iron Casting Co.		6 1/2" tall, bronze finish	600	850
Dutch Girl Holding Flowers	Hubley	1930s	5 1/2" tall, painted, iron trap in base	100	175

BANKS

Top to bottom: Boy with Large Football, 1914, Hubley; Indian with Tomahawk, 1900s, Hubley; again, Indian with Tomahawk/Statue of Liberty large, 1900s, Kenton/Statue of Liberty small, 1900s, Kenton/Santa Claus, 1900s, Hubley.

Still Banks

NAME	COMPANY	YEAR	DESCRIPTION	GOOD	EX
Eagle Bank Building	Unknown Maker		9 3/4" tall, painted building with gold eagle on roof	450	950
Eagle with Ball, Building	Unknown Maker		10 3/4" tall, building with eagle and ball on roof	850	1400
Edison Bust	Blevins, Charlotte	1972	5 5/16"	35	65
Eggman (Wm. Howard Taft)	Arcade	1910	4 1/8" tall	850	2000
Egyptian Tomb	Kyser & Rex	1882	6 1/4 " tall, Square safe on base, decorated with Sphinx and obelisk on front, sides show, pyramid, walled ruins and urn with flowers, gold	450	750
Electric Railroad	Shimer Toy Co.	1893	8 1/4"long	2500	6000
Elephant on Bench on Tub	Williams, A.C.	1920s	3 7/8"	125	225
Elephant on Tub	Williams, A.C.	1920s	5 3/8", in bronze finish	100	185
Elephant on Tub, Decorated	Williams, A.C.	1920s	5 3/8", painted version of #483	125	200
Elephant on Wheels	Williams, A.C.	1920s	4" tall, unpainted	150	250
Elephant Trumpeting	Wright, John	1971	7 1/4" tall, black finish	15	35
Elephant with Bent Knee	Kenton	1904	3 1/2", tan finish	200	375
Elephant with Chariot (large)	Hubley	1900s	4 3/4" tall, also produced without chariot	2000	3000
Elephant with Chariot, Small	Hubley	1906	7" long, gray elephant, red chariot, yellow wheels	1400	2200
Elephant with Howdah, Large	Williams, A.C.	1900s	4 7/8" x 6 3/8"	65	125
Elephant with Howdah, Large	Williams, A.C.	1900s	6 3/4", gold	85	150
Elephant with Howdah, Short Trunk	Hubley	1910	3 3/4" tall, painted gray with red belt	125	275
Elephant with Howdah, Small	Williams, A.C.	1900s	3 1/2" x 5"	65	125
Elephant with Raised Slot	Unknown Maker		4 1/2"tall, gray body, gold blanket	150	350
Elephant with Swivel Trunk	Unknown Maker		2 1/2", black finish with gold swivel trunk	125	250
Elephant with Tin Chariot	Wing	1900s	8" long, red chariot	1000	1600
Elephant with Tucked Trunk	Arcade	1900s	2 3/4" x 4 5/8", red or green	65	125
Elephant with Turned Trunk, Seated	Unknown Maker		4 1/4", unpainted	450	850
Elephant, "GOP 1936"	Hubley	1936	3 1/2" tall, "GOP 1936"	750	1200
Elephant, Circus	Hubley	1930s	3 7/8", painted, with lavender pants and red dotted white shirt	150	350
Elf	Unknown Maker		10" tall, painted, converted doorstop	150	450
English Setter	Wright, John	1970	8 1/2" tall, black	125	275
Fidelity Safe, Large	Kyser & Rex	1880	3 5/8" tall, green with gold trim, "Fidelity Safe"	150	300
Fidelity Trust Vault, Lord Fauntleroy	Barton Smith Co.	1890	6 1/2" x 5 7/8"	300	650
Fido	Hubley	1914	5", painted, white body, black eyes and ears, red collar	60	115
Fido on Pillow	Hubley	1920s	7 3/8"long, painted	100	200
Finial Bank	Kyser & Rex	1887	5 3/4" tall, 4 3/8" wide, building bank with single finial on roof	275	650

BANKS

Top to bottom: Camel, Kneeling, 1889, Kyser & Rex; Santa Claus with Tree, 1900s, Hubley; Mean Standing Bear, Hubley.

Still Banks

NAME	COMPANY	YEAR	DESCRIPTION	GOOD	EX
Flags Bank (SBCCA)	Littlestown Hdw & Found.	1976	3 1/4" tall, 6" square white pyramid with color US flags	75	125
Flat Iron Building Bank	Kenton	1900s	5 1/2" tall, silver	85	225
Floral Safe (National Safe)	Stevens, J. & E.	1898	4 5/8" x 4 1/8"	125	275
Football Player	Williams, A.C.	1910s	5 7/8" tall, bronze finish	250	425
Foreman	Grey Iron Casting Co.	1951	4 1/2", painted	175	350
Fort	Unknown Maker	1910s	4 1/8", unpainted bronze finish "Fort"	125	275
Fort Mt. Hope	Unknown Maker		2 7/8" tall	85	385
Four Tower	Ohio Foundry Co.	1949	5 3/8", painted white building with red roof	35	85
Four Tower	Stevens, J. & E.		5 3/4", unpainted with gold highlights	125	375
Foxy Grandpa	Hubley	1920s	5 1/2" tall, painted	150	375
Frog	Iron Art	1973	4 1/8", deep green finish	75	125
Frowning Face	Unknown Maker		5 5/8" tall, hanging bank, chin drops below surface level	850	1600
G. E. Radio Bank	Arcade	1930s	3 3/4" tall, brown cabinet radio on 4 legs	125	325
Gas Pump	Unknown Maker		5 3/4" tall, red	275	650
GE Refrigerator, Small	Hubley	1930s	3 3/4", blue	75	225
Gem Stove	Abendroth Bros.		4 3/4", brown finish	75	175
General Butler	Stevens, J. & E.	1884	6 1/2" tall, painted head on frog body	1200	3500
General Pershing Bust	Grey Iron Casting Co.	1918	7 3/4" tall, bronze finish	75	150
General Sheridan on Base	Arcade	1910s	6" tall, General seated on rearing horse	250	650
George Washington Bust on Safe	Harper, J.M.	1903	5 7/8" tall	1000	2500
Gettysburg Bank	Wilton Products Co.	1960	4 3/4" x 7 1/4" gray monument with reclining soldier	75	200
Give Me A Penny	Hubley	1900s	5 1/2" tall black figure in hat, painted	200	450
Globe Bank With Eagle	Enterprise Mfg. Co.	1875	5 3/4", painted red with eagle atop globe	125	350
Globe on Arc	Grey Iron Casting Co.	1900s	5 1/4" tall, red	100	300
Globe on Claw Feet	Kenton		6"	175	375
Globe on Hand	Unknown Maker	1893	4", bronze finish	375	1275
Globe on Wire Arc	Arcade	1900s	4 5/8"tall, painted spinning globe, red continents	125	350
Globe Safe with Hinged Door	Kenton	1900s	5"	100	250
Globe Savings Fund Bank	Kyser & Rex	1889	7 1/8", painted "Globe Savings Fund 1888"	1800	3000
Gold Eagle	Wright, John	1970	5 3/4"	5	20
Good Luck Horseshoe	Arcade	1908	4 1/4" tall, Buster Brown & Tige with horse inside horseshoe	150	350
Goose Bank	Arcade	1920s	3 3/4", unpainted	85	175
Graf Zeppelin	Williams, A.C.	1920s	6 5/8" long, silver gray finish	85	225
Graf Zeppelin on Wheels	Williams, A.C.	1934	7 3/4" long silver pulltoy bank	150	425
Grandpa's Hat	Unknown Maker		2 1/4" tall, 3 7/8" wide, top hat	225	450
Grenade with Pin	Bartlett Mayward Co.		4 1/4"	85	175
Gunboat	Kenton		8 1/2" long, blue hull, white top, twin masts	650	1400
Hall Clock	Arcade	1923	5 5/8" tall, dark finish with gold highlights	300	650
Hall Clock	Hubley	1900s	5 1/4" tall, brown finish, paper face	275	400
Hall Clock with Cast Face	Hubley	1920s	5 3/26" tall	275	375
Hanging Mailbox	Williams, A.C.	1920s	5 1/8" tall, green, wall mount mailbox replica, gold lettering	45	95

Still Banks

NAME	COMPANY	YEAR	DESCRIPTION	GOOD	EX
Hanging Mailbox on Platform	Unknown Maker	1800s	7 1/4" tall, red box hangs on post in platform base	650	1500
Hard Hat	Knerr, George	1970s	1 15/16" tall, white with red lettering	100	250
Harleysville Bank	Unicast Foundry	1959	2 5/8" tall, 5 1/4" long, white with gray roof	75	225
Hen on Nest	Unknown Maker	1900s	3", bronze finish with red highlights	100	1750
High Rise Building	Kenton		7" tall	200	550
High Rise, Tiered	Kenton		5 3/4"	125	350
Hippo	Unknown Maker		2"tall, 5 3/16" long, bronze with red highlights	3500	6000
Holstein Cow	Arcade	1910s	2 1/2" tall, 4 5/8" long, black finish	125	350
Home Bank	Judd, H.L.	1890s	4" x 3 1/2", dark finish	175	450
Home Bank with Crown	Stevens, J. & E.	1872	5 1/4", painted, "Home Bank"	475	1400
Home Savings	Unknown Maker		10 1/2" painted, "Property of Peoples Savings Bank, Grand Rapids, Mich."	175	550
Home Savings Bank	Shimer Toy Co.	1899	5 7/8", painted	150	525
Home Savings Bank	Unknown Maker		9 5/8" tall, painted	175	550
Home Savings Bank with Dog Finial	Stevens, J. & E.	1891	5 3/4" tall	125	450
Home Savings Bank with Finial	Stevens, J. & E.	1891	3 1/2" tall, mustard finish	125	375
Honey Bear	Unknown Maker		2 1/2", silver finish unpainted bear sits eating honey	675	1200
Hoover/Curtis Elephant "GOP"	Hubley	1928	3 3/8", ivory finish	675	1600
Horse on Tub, Decorated	Williams, A.C.	1920s	5 5/6"	135	300
Horse on Wheels	Williams, A.C.	1920	4 1/4", deep red finish	150	450
Horse, "Beauty"	Arcade	1900s	4 1/8" x 4 3/4", black with raised "Beauty" on side	85	150
Horse, Prancing	Arcade	1910s	4 1/4" tall, black with gray hooves	55	125
Horse, Prancing with Belly Band	Unknown Maker		4 1/2", light bronze finish	175	375
Horse, Prancing, Large	Williams, A.C.	1910s	7 3/16" tall, bronze finish	75	145
Horse, Rearing on Oval Base	Williams, A.C.	1920s	5 1/8" x 4 7/8"	95	250
Horse, Rearing on Pebbled Base	Unknown Maker		7 1/4" x 6 1/2", gold finish	85	165
Horseshoe with Mesh	Williams, A.C.		Horse head inside horseshoe that forms end of mesh coin cage, bronze finish	65	125
Hot Point Electric Stove	Arcade	1925	6", white, on legs	85	150
House with Basement	Ohio Foundry Co.	1893	4 5/8" square, painted	850	1600
House with Bay Window	Unknown Maker	1874	5 5/8" tall, painted	900	1800
House with Chimney Slot	Unknown Maker		2 7/8" x 2 13/16", painted	275	650
House with Knight	Unknown Maker		7 1/4" unpainted "Savings Bank" with knight figure on roof peak	375	950
Hub	Magic Introduction Co.	1892	5" x 5 1/4" x 1 5/8"		
Humphrey-Muskie Donkey	Unknown Maker	1968	4 1/2" tall, pale silver finish, "Humphrey Muskie 68"	10	35
Humpty Dumpty	Unknown Maker	1930s	5 1/2" tall, painted, white egg, red brick wall	375	850

BANKS

BANKS

Top to bottom: Sailor, 1900s, Hubley; Dodge barrel bank, unknown; Rabbit Lying Down, unknown; Columbia bank in all four sizes, 1890s, Kenton.

Still Banks

NAME	COMPANY	YEAR	DESCRIPTION	GOOD	EX
Humpty Dumpty, Seated	Russell, Edward K.	1974	5 3/8" tall, painted	75	125
Husky	Grey Iron Casting Co.	1910s	5"	200	550
I Made Chicago Famous, Large Pig	Harper, J.M.	1902	2 5/8" x 5 5/16"	250	450
I Made Chicago Famous, Small Pig	Harper, J.M.	1902	2 1/8" x 4 1/8"	200	350
Independence Hall	Enterprise Mfg. Co.	1875	10" tall, deep red/brown finish	450	1150
Independence Hall	Unknown Maker	1875	8 1/8" tall, 15 1/2" long, mustard building on base with bell tower	1800	3500
Independence Hall Tower	Enterprise Mfg. Co.	1876	9 1/2"	225	475
Indian Chief Bust	Unknown Maker	1978	4 7/8", unpainted	35	85
Indian Family	Harper, J.M.	1905	3 5/8" X 5 1/8", unpainted	850	1850
Indian Head Penny	Knerr, George	1972	3 1/4" diam.	35	75
Indian Seated on Log	Ouve, A.	1970s	3 5/8" tall, unpainted	85	150
Indian with Tomahawk	Hubley	1900s	5 7/8"	175	425
Indiana Paddle Wheeler	Unknown Maker	1896	7 1/8" long, black with red trim	4000	8000
International Eagle on Globe	Unknown Maker		8" x 8", unpainted	1200	2500
Ironmaster's House	Kyser & Rex	1884	4 1/2", unpainted	600	950
Japanese Safe	Kyser & Rex	1882	5 3/8" tall	100	300
Japanese Safe	Kyser & Rex	1883	5 1/2" tall, painted	125	375
Jarmulowsky Building	Stevens, J. & E.		7 3/4"tall, bronze finish building bank	1200	2000
Jewel Safe	Stevens, J. & E.	1907	5 3/8", unpainted	125	250
John Brown Fort	Unknown Maker		3" tall, red with white cupola	85	135
Junior Cash Register, Small	Stevens, J. & E.	1920s	5 1/4" x 4 5/8" elaborate cast with slot at top	175	375
Kelvinator Bank	Arcade	1930s	#832, 4 1/2" tall, white with grey trim replica refrigerator	150	350
Key	Somerville, W.J.	1905	5 1/2" long, silver finish skeleton key	250	650
Key, St. Louis World's Fair	Unknown Maker	1904	5 3/4" long, dark finish	275	700
King Midas	Hubley	1930s	4 1/2" tall, painted	1250	2500
Kitty Bank	Hubley	1930s	4 3/4" tall, painted, white body with blue bow	65	125
Klondyke	Unknown Maker		3 1/4" cube	650	1400
Kodak Bank	Stevens, J. & E.	1905	4 1/4" tall, 5" wide, "Kodak Bank"	200	450
L'il Tot	Watkins, Bob	1982	5 7/8"	125	175
Labrador Retriever	Unknown Maker		4 1/2" black finish with gold collar	125	375
Lamb	Wright, John	1970	3 1/4" tall, painted white with black highlights	35	75
Lamb, Small	Unknown Maker		3 3/16", painted white	200	375
Laughing Pig	Hubley		2 1/2", painted	125	275
Liberty Bell	Harper, J.M.	1905	3 3/4"	275	550
Liberty Bell with Yoke	Arcade	1920s	3 1/2"	25	65
Liberty Bell, Miniature	Penncraft		3 1/2" x 1 3/4"	20	35
Lighthouse	Lane Art	1950s	9 1/2" tall, "Light of the World"	125	250
Lighthouse	Unknown Maker	1891	10 1/4" tall, red tower rises from unpainted base	1200	2800

Still Banks

NAME	COMPANY	YEAR	DESCRIPTION	GOOD	EX
Limousine	Arcade	1920s	8 1/16" long, black with white rubber tires	Limousine	
Limousine Yellow Cab	Arcade	1921	repaint of version #1478	1400	2800
Lincoln High Hat ("Pass Around the Hat")	Unknown Maker	1880s	2 3/8" tall, black finish	125	200
Lion on Tub, Decorated	Williams, A.C.	1920s	5 1/2" tall	125	225
Lion on Tub, Plain	Williams, A.C.	1920s	7 1/2" tall, bronze finish	100	200
Lion on Tub, small	Williams, A.C.	1920s	4 1/8" tall, brown or green finish	85	175
Lion on Wheels	Williams, A.C.	1920s	4 1/2" x 5 1/2", gold	145	225
Lion, Ears Up	Williams, A.C.	1930s	3 5/8" x 4 1/2"	75	125
Lion, Small	Williams, A.C.	1934	2 1/2" x 3 5/8"	85	150
Lion, Tail Between Legs	Unknown Maker		3" x 5 1/4"	85	145
Lion, Tail Left	Hubley	1910s	3 3/4" tall, bronze finish	100	175
Lion, Tail Right	Arcade	1900s	4" tall	55	100
Lion, Tail Right	Williams, A.C.	1920s	3 1/2" x 4 15/16"	55	100
Lion, Tail Right	Williams, A.C.	1900s	5 1/4" tall, bronze finish	55	150
Little Red Riding Hood Safe	Harper, J.M.	1907	5 1/16" tall, painted	2000	4000
Log Cabin	Kyser & Rex	1882	2 1/2" x 3 1/4", painted	175	400
Lost Dog	Judd H.L.	1890s	5 3/8", unpainted	275	850
Lucky Cabin	Wright, John	1970	4 1/8" tall, painted with horseshoe over door	35	65
Mailbox on Legs, Large	Hubley	1920s	5 1/2" tall, green street corner box replica	85	225
Mailbox on Legs, Small	Hubley	1928	3 3/4" tall, green replica street corner mailbox	35	100
Main Street Trolley with People	Williams, A.C.	1920s	3" x 6 3/4" bronze finish	175	475
Main Street Trolley Without People	Williams, A.C.	1920s	6 3/4" long, no people	175	400
Majestic Radio Bank	Arcade	1930s	4 1/2" tall, mahogany finish replica of a floor standing radio on 4 legs, coin slot in back, with key	125	200
Majestic Refrigerator Bank	Arcade	1930s	4 1/2" tall, red, green or blue with gold trim, replica of single door fridge on 4 legs, coin slot in back, with key lock	375	600
Mammy	Unknown Maker	1970s	8 1/4" tall, doorstop conversion, red dress, white apron	10	25
Mammy with Hands on Hips	Hubley	1900s	5 1/4" tall, red dress, white apron	85	325
Man in Barrel	Stevens, J. & E.	1890s	3 3/4" tall, painted	175	300
Man on Cotton Bale	US Hdwre Co.	1898	4 7/8" tall, painted darkie sits on hay bale, red scarf, yellow pants	1400	2500
Marietta Silo	Unknown Maker		5 1/2" gray finish	275	550
Marshall Stove	Unknown Maker		3 7/8", red	125	225
Mary & Little Lamb	Unknown Maker	1901	4 3/8" tall, painted white with red trim	350	850
Mascot	Hubley	1914	5 3/4" tall, boy stands on baseball	850	1850
McKinley/Teddy Elephant	Unknown Maker	1900	2 1/2" tall, bronze finish	350	650
Mean Standing Bear	Hubley		5 1/2"	100	225
Mellow Furnace	Liberty Toy Co.		3 9/16" x 3 1/8", brown finish	125	225
Mermaid Boat	Grey Iron Casting Co.	1900s	4 1/2" tall, companion piece to Dolphin, girl in boat holds fish	350	850
Merry-Go-Round	Grey Iron Casting Co.	1920s	4 5/8" tall, unpainted	175	550

Still Banks

NAME	COMPANY	YEAR	DESCRIPTION	GOOD	EX
Metropolitan Bank	Stevens, J. & E.	1872	5 7/8", "Metropolitan Bank"	125	275
Mickey Mouse	Wright, John	1970s	5 x 3 3/4" bookend bank, painted	85	125
Mickey Mouse, Hands on Hips	Unknown Maker		9" tall, painted	125	450
Middy with Clapper	Unknown Maker	1887	5 1/4" brown finish	150	350
Minuteman	Hubley	1905	6" tall, painted	175	475
Model T Ford	Arcade	1920s	4" tall, black	650	1250
Moody & Sankey	Smith & Egge	1870	5" painted, 2 oval portraits on front	800	1650
Mosque, Large, (3 Story)	Williams, A.C.	1920s	3 1/2" tall, 3 story building	45	125
Mosque, Small, (2 Story)	Unknown Maker		2 7/8" tall, 2 story building	35	115
Mother Hubbard Safe	Harper, J.M.	1907	4 1/2" tall	1500	5000
Mulligan Policeman (Keystone Cop)	Williams, A.C.	1900s	5 3/4", painted	175	400
Multiplying Bank	Stevens, J. & E.	1883	6 1/2" painted building	700	2200
Mutt & Jeff	Williams, A.C.	1900s	4 1/4" x 3 1/2", gold	75	275
National Safe	Stevens, J. & E.	1800s	3 3/8" tall, unpainted	65	125
Nest Egg, "Horace"	Smith & Egge	1873	3 3/8" tall on base, bronze finish egg on side	450	850
Nesting Doves Safe	Harper, J.M.	1907	5 1/4", bronze finish	1500	3500
New Heatrola Bank	Kenton	1920s	4 1/2" tall, green finish with red trim	85	275
Newfoundland Dog	Arcade	1930s	3 5/8" x 5 3/8", blue or green finish	100	225
Newfoundland Dog with Pack	Unknown Maker		4 11/16" tall	85	175
Nixe	Unknown Maker		4 1/2" tall, silver boy in boat, "Nixe"	350	1450
Nixon Bust	Blevins, Charlotte	1972	5 5/16"	45	85
Nixon/Agnew Elephant	Unknown Maker	1968	2 5/8"	15	35
North Pole Bank	Grey Iron Casting Co.	1920s	4 1/4", unpainted, "Save Your Money And Freeze It"	375	775
Oak Stove	Shimer Toy Co.	1899	2 3/8" tall, unpainted	125	475
Old Abe with Shield, Eagle	Unknown Maker	1880	3 7/8", unpainted	375	1050
Old South Church	Unknown Maker		10" tall, bronze finish	2000	5000
One Car Garage	Williams, A.C.	1920s	2 1/2", painted	125	250
One Story House	Grey Iron Casting Co.	1900s	3" tall	65	175
Oregon Gunboat	Kenton		11" long, blue hull, gray guns, black and red stacks, "Oregon"	850	1800
Organ Grinder	Hubley		6 3/16" x 2 1/8", painted	125	350
Oriental Boy on Pillow	Hubley	1920s	5 1/2" tall, painted	85	200
Oriental Camel	Unknown Maker		3 3/4" tall, on rockers	85	325
Ornate Hall Clock	Hubley	1900s	5 7/8" tall, tan finish, paper face	125	375
Osborn Pig	Unknown Maker		2" x 4", "You can bank on the Osborn..."	100	350
Oscar the Goat	Unknown Maker		7 3/4" tall, black with silver hooves and horns	75	175
Owl	Vindex Toys	1930	4 1/4", painted	75	325
Owl on Stump	Unknown Maker		3 5/8", red	65	125
Ox	Kenton		4 3/8", painted	85	150
Palace	Ives	1885	7 1/2" tall, 8" wide	850	3000
Park Bank Building	Unknown Maker		4 3/8" painted building, "Park Bank"	450	1450
Parlor Stove	Unknown Maker		6 7/8", gray and black	275	425
Parrot on Stump	Unknown Maker		6 1/4", painted	125	450
Pavillion	Kyser & Rex	1880	3 1/8" x 3"	225	500

BANKS

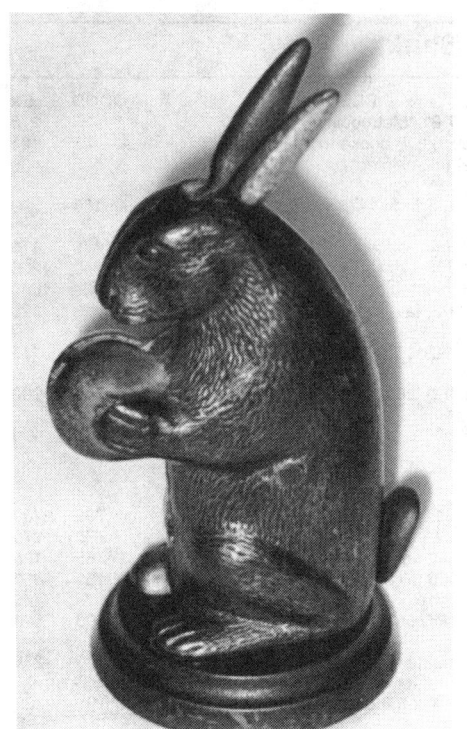
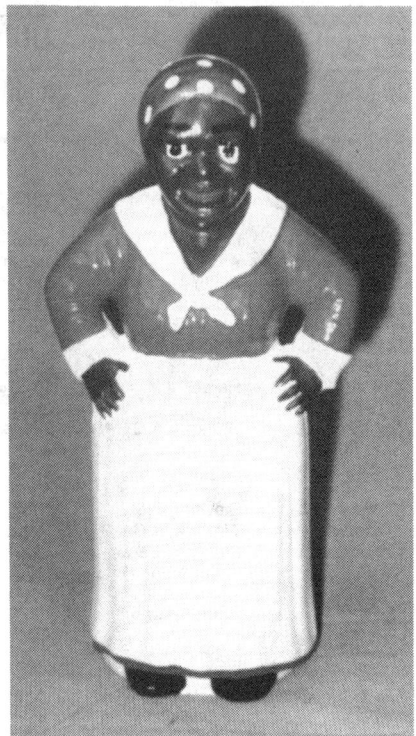

Top to bottom: Rabbit Begging, 1900s, A.C. Williams; Mammy with Hands on Hips, 1900s, Hubley; Elephant with Swivel Trunk, unknown.

Still Banks

NAME	COMPANY	YEAR	DESCRIPTION	GOOD	EX
Pay Phone Bank	Stevens, J. & E.	1926	7 3/16", unpainted	450	1800
Pearl Street Bank	Unknown Maker		4 1/4", unpainted, silver finish, "Pearl Street Bank"	350	850
Peg Legged Pirate	Unknown Maker		5 1/4", unpainted	25	85
Pelican	Hubley	1930s	4 3/4", painted white	350	1000
Penny Register Pail	Kyser & Rex	1889	2 3/4", unpainted	125	250
Penthouse Building	Williams, A.C.		5 7/8" tall, silver finish	350	650
Peters Weatherbird	Arcade		4 1/4" tall	750	2500
Phoenix Dime Register Trunk	Piaget	1890	3 3/4" x 5" steamer trunk	125	250
Pig, A Christmas Roast	Unknown Maker		3 1/4" x 7 1/8"	85	250
Pig, Seated	Williams, A.C.	1900s	3" x 4 9/16"	35	85
Plymouth Rock 1620	Unknown Maker		3 7/8" long, "1620"	650	1850
Polar Bear, Begging	Arcade	1900s	5 1/4", white variant of # 715	275	450
Policeman Bank	Arcade	1930s	5 5/8" tall, blue with aluminum finish on gloves and star, gold buttons, black shoes, flesh face and hands	250	1000
Policeman Safe	Harper, J.M.	1907	5 1/4"	1250	4500
Polish Rooster	Unknown Maker		5 1/2"	850	2200
Polish Rooster	Unknown Maker		5 1/2" tall, painted	850	2200
Pooh Bank	Unknown Maker		5" x 4 7/8"	5	15
Possum	Arcade	1910s	2 3/8" tall, 4 3/8" long, silver finish	125	475
Postal Savings Mailbox	Nicol	1920s	6 3/4"	85	275
Pot Bellied Stove	Knerr, George	1968	5 3/4" tall, flat black finish	5	15
Potato	Martin, Mary A.	1897	5 1/4" long, "Bank"	850	1250
Presto "Bank"	Williams, A.C.	1900s	3 5/8" tall, silver finish with gold dome	85	175
Presto Bank	Unknown Maker		4 1/4" tall building, silver with gold dome	85	175
Presto Bank	Unknown Maker		3 1/4" tall, silver finish, "Bank"	65	150
Presto Trick Bank	Kyser & Rex	1892	4 1/2" tall, red doors and roof	250	850
Professor Pug Frog Bank	Williams, A.C.	1900s	3 1/4"	275	550
Pugdog, seated	Kyser & Rex	1889	3 1/2", painted	250	475
Puppo	Hubley	1920s	4 7/8" tall, painted bee on body	125	225
Puppo on Pillow	Hubley	1920s	5 5/8" x 6", painted brown, cream, black, pink	150	275
Put Money in Thy Purse	Unknown Maker	1886	2 3/4" tall change purse, black, "Put Money in Thy Purse"	625	950
Puzzle Try Me	Unknown Maker	1868	2 11/16" tall, safe, "Puzzle Try Me"	475	975
Quadrafoil House	Various	1900s	3 1/8" tall, made by numerous companies	125	225
Queen Stove	Wright, John	1975	3 3/4" to cook top, "Queen" on oven door	25	65
Quilted Lion	Unknown Maker		3 3/4" tall, 4 3/4" long, bronze finish	185	425
Rabbit Lying Down	Unknown Maker		2 1/8" x 5 1/8", unpainted	175	575
Rabbit Standing, Large	Williams, A.C.	1908	6 1/4" tall, brown metal finish	125	325
Rabbit with Carrot	Knerr, George	1972	3 3/8", painted white, orange and green carrot	85	175
Rabbit, Begging	Williams, A.C.	1900s	5 1/8"	85	275
Rabbit, Large, Seated	Hubley	1900s	4 5/8" tall, painted white with pink highlights	125	375
Rabbit, Small, Seated	Arcade	1910s	3 5/8" tall	125	325
Radio Bank	Hubley	1928	3 5/16" tall, metallic blue	100	375

BANKS

Still Banks

NAME	COMPANY	YEAR	DESCRIPTION	GOOD	EX
Radio Bank with 3 Dials	Kenton	1920s	3" tall, 4 5/8" long, red	100	350
Radio with Combination Door	Kenton	1930s	4 1/2" red, metal sides and back	125	375
Reclining Cow	Unknown Maker		2 1/8" tall, 4" long, black	100	400
Recording Bank	Unknown Maker		6 5/8" x 4 1/4"	150	450
Red Ball Safe	Unknown Maker		3", red ball on base	175	425
Red Goose Shoes on Base	Arcade	1920s	5 1/2", on pedestal w/base	300	750
Red Goose Shoes on Pedestal	Unknown Maker		4 7/16" red goose on bronze base	175	350
Red Goose Shoes, Squatty	Arcade	1920s	4" tall, red body, yellow feet	275	475
Reindeer on Base	Wright, John	1973	10" x 8"	75	125
Reindeer, Large	Williams, A.C.	1900s	9 1/2" tall, bronze finish	125	250
Reindeer, Small	Williams, A.C.		6 1/4" tall, bronze finish	75	135
Reliable Parlor Stove	Schneider & Trenkramp Co.		6 1/4"	425	850
Republic Pig	Wilton Products Co.	1970s	7" tall, painted pig in business suit	35	85
Rhesus Monkey	Unknown Maker		8 1/2" converted doorstop, painted	35	125
Rhino	Arcade	1910s	2 5/8" tall, 5" long, gold	225	525
Rochester Clock	Unknown Maker		5" tall with working clock	225	750
Rocking Chair	Manning, C.J.	1898	6 3/4" tall, brown finish	1500	2750
Rocking Horse (SBCCA)	Knerr, George	1975	5 5/8", white with red saddle, "SBCC"	350	550
Roller Safe	Kyser & Rex	1882	3 11/16" x 2 7/8"4"	125	245
Roof Bank	Grey Iron Casting Co.	1900s	5 1/4"	125	300
Roof Bank	Stevens, J. & E.	1887	5 1/4" x 3 3/4"	125	350
Rooster	Arcade	1910s	4 5/8", black with red comb	125	350
Rooster	Hubley	1910s	4 3/4" by Hubley and A.C. Williams, brown finish with red comb and wattle	125	300
Rooster, Large	Unknown Maker	1913	6 3/4", unpainted except for red comb and wattle	550	1250
Rumplestiltskin	Unknown Maker	1910s	6 x 2 1/4"	200	450
Saddle Horse	Grey Iron Casting Co.	1928	4 3/8" tall	375	650
Safe Deposit	Shimer Toy Co.	1899	3 5/8", "Safe Deposit"	85	150
Safety Locomotive	Unknown Maker	1887	3 1/4" tall, gray	1250	2200
Sailor, Medium	Hubley	1910s	5 1/4" tall	225	475
San Gabriel Mission	Unknown Maker		4 5/8" x 3 3/4", painted, musical building	2000	7500
Santa with (Removable) Wire Tree	Ives	1890s	7 1/4" tall, with removable ornate tree	875	1500
Santa Claus	Hubley	1900s	5 3/4", painted with arms folded in front	450	950
Santa Claus With Tree	Hubley	1910s	5 3/4" tall with arms folded in front, tree at back, painted	450	950
Save for Ice, Ice Box	Arcade		4 1/4" tall, white, "Save For Ice"	175	450
Scottie, seated	Hubley	1930s	4 7/8" x 6, black finish, red collar	125	300
Scrollwork Safe	Unknown Maker	1900s	2 3/4" tall	85	225
Seal on Rock	Arcade	1900s	3 1/2", black	175	475
Security Safe	Unknown Maker	1894	4 1/2" tall, red door	125	275
Security Safe Deposit	Unknown Maker	1881	3 7/8" tall	95	150
Shell Out	Stevens, J. & E.	1882	4 3/4" long, conch shell on base, off white	225	700
Show Horse	Lane Chair Co.	1973	5 7/8" tall	75	150
Six Sided Building, Two Story	Unknown Maker		3 3/8" tall	100	275

BANKS

58

Still Banks

NAME	COMPANY	YEAR	DESCRIPTION	GOOD	EX
Six-Sided Building	Unknown Maker		2 3/8" tall, unpainted	225	575
Skyscraper Bank	Williams, A.C.	1900s	5 1/2" tall, silver building bank with 4 gold posts	85	150
Skyscraper Bank	Williams, A.C.	1900s	4 3/8" tall, silver building bank with 4 gold posts	85	125
Skyscraper with Six Posts	Williams, A.C.	1900s	6 1/2" tall, silver building bank with gold posts	125	450
Songbird on Stump	Williams, A.C.	1900s	4 3/4", bronze finish	300	800
Space Heater with Bird	Chamberlain & Hill	1890s	English, 6 1/2" tall	175	375
Space Heater with Flowers	Unknown Maker	1890s	English, 6 1/2" tall, oriental motif, red finish	175	375
Spaniel, Large	Wright, John	1960s	10 1/2" long, painted	65	125
Spitz	Grey Iron Casting Co.	1928	4 1/4", bronze finish	225	575
Squatty "Red Goose Shoes"	Arcade	1920s	3 7/8"	225	575
Squirrel with Nut	Unknown Maker		4 1/8"	425	1250
St. Bernard with Pack, Large	Williams, A.C.	1900s	5 1/2" x 7 3/4"	125	225
St. Bernard with Pack, Small	Williams, A.C.	1900s	3 3/4" x 5 1/2"	85	175
Star Safe	Kyser & Rex	1882	2 5/8" tall	150	450
State Bank	Arcade	1910s	4 1/8" tall, bronze finish	85	175
State Bank	Kenton	1900	8" x 7"	550	1000
State Bank	Kenton	1890s	3" tall, unpainted building bank	95	200
State Bank	Kyser & Rex	1890s	5 1/2" tall, bronze building bank	125	250
Statue of Liberty	Kenton	1900s	6 1/16" tall, also made by A.C. Williams, #1165, 6 3/8" tall	85	125
Statue of Liberty	Kenton	1900s	6 3/8" tall, silver finish with gold highlights	100	175
Statue of Liberty, Large	Kenton	1900s	9 1/2" tall, silver gray finish, gold highlights	350	950
Steamboat	Williams, A.C.	190s	7 5/8" long, brown finish	125	375
Steamboat with Small Wheels	Kenton		7 7/16" long, silver finish	175	425
Stop Sign	Dent	1920	5 5/8" tall, green with red and gold highlights	325	650
Stork Safe	Harper, J.M.	1907	5 1/2"	850	1750
Street Car	Grey Iron Casting Co.	1891	4 1/2" long, painted	250	650
Sun Dial	Arcade	1900s	4 5/16" tall	650	2000
Sunbonnet Sue	Unknown Maker	1970	7 1/2", painted	65	165
Tabernacle Savings	Keyless Lock Co.		2 1/4" x 5", unpainted	850	1250
Taft-Sherman Bust	Harper, J.M.	1908	4" tall, one side Smiling Jim, other side Peaceful Bill	1000	1850
Tank Bank 1918, Small	Williams, A.C.	1920s	2 3/8"long, gold finish	65	150
Tank Bank 1919	Unknown Maker		3" x 5 1/2", silver finish, "1919"	125	225
Tank Bank USA 1918, Large	Williams, A.C.	1920s	3" tall x 3 11/16" long, gold finish	100	200
Tank Savings Bank	Ferrosteel	1919	9 1/2" long, "Tank Savings Bank"	150	375
Teddy Bear	Arcade	1900	2 1/2" x 3 7/8"	125	275
Teddy Roosevelt Bust	Williams, A.C.	1919	5" tall	175	450
Templetone Radio	Arcade	1930s	4 1/2", red	275	575
Thoroughbred	Hubley	1946	5 1/4", bronze finish	75	150
Three Wise Monkeys	Williams, A.C.	1900s	3 1/4" tall, 3 1/2" wide	225	475
Time Is Money Clock Bank	Williams, A.C.	1910s	3 1/2" tall, alarm clock shaped, gold finish, "Time Is Money"	125	200
Time Safe	Roche, E.M. Co.		7" tall, 3 3/4" wide, unpainted	375	750

BANKS

Still Banks

NAME	COMPANY	YEAR	DESCRIPTION	GOOD	EX
Tower	Kenton	1915	4 1/8", unpainted, "Tower"	175	375
Tower Bank	Harper, J.M.	1900s	9 1/4" tall, unpainted, brown finish	175	375
Tower Bank	Kyser & Rex	1890	6 7/8" building with tower rising from roof, "Tower Bank 1890"	1200	2200
Town Hall Bank	Kyser & Rex	1882	4 5/8", red, "Town Hall Bank"	375	750
Toy Soldier	Worley, Laverne A.	1982	7 1/2" tall, painted, "SBCCA"	15	65
Treasure Chest	Wright, John	1970	2 3/4" x 4", smaller version is #928	60	35
Triangular Building	Hubley	1914	6" tall, "Bank"	325	675
Trick Buffalo	Unknown Maker		5 1/2" tall, black	750	1500
Trolley Car	Kenton	1900s	5 1/4" long, painted silver	225	650
Trunk on Dolly	Piaget	1890	2 5/8" x 3 9/16"	175	350
Trust Bank	Stevens, J. & E.	1800s	7 1/4"	875	2500
Tug Boat	Unknown Maker		5 1/2" long, red, pulltoy	4500	7500
Turkey, Large	Williams, A.C.	1900s	4 1/4" x 4", painted wattle	250	475
Turkey, Small	Williams, A.C.	1900s	3 3/8" tall, red head and wattle	150	275
Turtle Bank	Unknown Maker		1" tall, 3 7/16" long	2000	3500
Two Car Garage	Williams, A.C.	1920s	2 1/2", painted	125	300
Two Faced Black Boy, Large	Williams, A.C.	1900s	4 1/8" tall	125	350
Two Faced Black Boy, Small	Williams, A.C.	1900s	3 1/8" x 2 3/4", negro toy bank, painted	85	300
Two Faced Devil	Williams, A.C.	1004	4 1/4" tall, deep red	550	1250
Two Goats Butting	Harper, J.M.		4 1/2", two goats on tree stump, "Two Kids" on base	950	2000
Two Story House	Williams, A.C.	1930s	3 1/16" tall, brown finish with red roof	75	150
Two-Faced Indian	Williams, A.C.	1900s	4 5/16" tall, bronze finish with painted highlights	1500	2750
U.S. Bank, Eagle Finial	Unknown Maker	1890s	9 1/4" tall, green with gold trim	850	1500
U.S. Mail	Kenton	1900s	4 3/4" tall, silver gray with red lettering	100	275
U.S. Mail Bank with Combination lock	Fish, O.B.	1903	6 7/8" tall, silver gray with red lettering	225	475
U.S. Mail with Eagle	Hubley	1906	4" x 4"	175	325
U.S. Mail with Eagle	Kenton	1930s	4 1/8" x 3 1/2"	85	175
U.S. Mail, Small	Kenton	1900s	3 5/8" x 2 3/4", silver or green mail box with red lettering	75	150
U.S. Navy Akron Zeppelin	Williams, A.C.	1930	6 5/8" long, silver finish, "US Navy Akron"	175	450
U.S. Treasury Bank	Grey Iron Casting Co.	1920s	3 1/4", painted	250	475
Ulysses S. Grant Bust	Unknown Maker	1976	5 1/2" tall	125	250
Ulysses S. Grant Bust on Safe	Harper, J.M.	1903	5 5/8" tall	1750	3000
Uncle Sam Hat	Knerr, George		2" x 3", red, white and blue	125	250
United Banking and Trust, Building Bank	Williams, A.C.		3" tall, bronze finish	225	450
Victorian House	Stevens, J. & E.	1892	4 1/2", unpainted deep gray finish	175	375
Victorian House	Unknown Maker		3 1/4" tall, gray metallic finish	150	275
Villa	Kyser & Rex	1894	5 9/16" unpainted except for red finial	375	700
Villa Bank	Kyser & Rex	1882	3 7/8" x 3 3/8", "1882"	375	700
Vindex Bulldog	Vindex Toys	1931	5 1/4" tall, painted, "Vindex Toys"	125	275
Washington Bell with Yoke	Grey Iron Casting Co.	1932	2 3/4", red	125	250
Washington Monument	Williams, A.C.	1900s	6" tall	150	325

Still Banks

NAME	COMPANY	YEAR	DESCRIPTION	GOOD	EX
Washington, George, Bust	Grey Iron Casting Co.	1920s	8" tall, bronze finish	850	1450
Watch Dog Safe	Unknown Maker		5 1/8", with brass handle, dog stands guard on front	1850	4000
Water Spaniel with Pack (I Hear A Call)	Unknown Maker	1900	5 3/8" x 7 7/8"	225	450
Weaver Hen	Unknown Maker	1970s	6", white with red comb and wattle, "Weaver"	20	50
Westside Presbyterian Church	Unknown Maker	1916	3 3/4" x 3 5/8", silver finish	350	950
Whale of a Bank	Knerr, George	1975	2 3/4" x 5 3/16", "A Whale of a Bank"	85	200
Whippet on Base	Unknown Maker		3 1/2" tall, gold finish	75	125
White City Barrel #1 on Cart	Nicol	1894	5" long, unpainted, "White City Puzzle Savings Bank, A Barrel of Money"	275	475
White City Barrel, Large	Nicol	1893	5 1/8" tall, silver finish barrel	175	275
White City Pail	Nicol	1893	2 5/8" tall, silver finish pail with handle	125	200
White City Puzzle Safe #10	Nicol	1893	4 5/8", unpainted	125	225
White City Puzzle Safe #12	Nicol	1893	4 7/8", unpainted	150	325
White Horse on Base	Knerr, George	1973	9 1/2" tall	125	225
Wirehaired Terrier	Hubley	1920s	4 5/8", painted	125	275
Wisconsin Beggar Boy	Unknown Maker		6 7/8" tall, "Help the Crippled Children of Wisconsin"	525	900
Wisconsin War Eagle	Unknown Maker	1880	2 7/8"	675	1500
Wise Pig, The	Hubley	1930s	6 5/8" tall, painted off white pig holding plaque	85	225
Woolworth Building	Kenton	1915	7 7/8" tall, bronze finish	100	225
Woolworth Building	Kenton	1915	5 3/4" x 1 1/4"	85	150
Work Horse on Base	Unknown Maker		9" tall, painted white	75	125
Work Horse with Flynet	Arcade	1910s	4" tall	300	550
World's Fair Adminstration Building	Unknown Maker	1893	6" x 6", painted	1400	2250
Yellow Cab	Arcade	1921	7 7/8" long, orange and black, rubber tires	1500	2400
York Stove	Abendroth Bros.		4" tall, unpainted, "York Stove"	225	525
Young America	Kyser & Rex	1882	4 3/8" x 3 1/8" safe	125	275

BANKS

BANKS

Top to bottom: Toys R Us 1918 Ford Runabout; Fram Filters 1950 Chevrolet; Coca-Cola 1923 Chevrolet.

Ertl Banks

NAME	NUMBER	YEAR	DESCRIPTION	MINT
A.J. Seibert Co.	1323	1987	1913 Model T Van, White Body, Red Trim	130
AC Rochester #1 - United Auto Workers	9746	1989	1950 Chevy Panel, Red Body, Black Trim	115
ACE Hardware	9019	1989	1918 Runabout, Red Body, Black Trim	75
ACE Hardware #1	9038	1989	1918 Runabout, Red/White Body, Black Trim	25
ACE Hardware #2	7697EO	1990	1926 Mack W/Crates, Red Body, Black/Brown Trim	30
ACE Hardware #3-Marked "3rd Edition"	9459	1989	1932 Ford Panel, Red Body, White Trim	25
Achenbach's Pastry Shop	9643	1989	Step Van, White Body, White Trim	35
Agway #1	9444	1986	1913 Model T Van, White Body, Red Trim	295
Agway #2	9195	1987	1918 Runabout, Black Body, Black Trim	35
Agway #3	9743	1988	1905 Ford Delivery Van, Black Body, Black Trim	35
Agway #4 - LTD Ed	9687	1989	1932 Ford Panel, Black Body, Silver Trim	30
Agway #5 LTD Ed W/Spare	7514	1990	1917 Model T Van, Blue Body, Blue Trim	30
Alberta	9218	1985	1913 Model T Van 1 Of 10 Canadian Provinces, White Body, Brown	30
Alex Cooper Auctioneers	9201	1984	1913 Model T Van, White Body, Red Trim	60
Alka-Seltzer #1	9155	1987	1918 Runabout, Blue Body, Beige/Blue Trim	140
Alka-Seltzer #2	9791	1988	1917 Model T Van, White Body, Blue Trim	45
Allerton, Illinois Centennial	9460	1986	1913 Model T Van, White Body, Red Trim	35
Allied Can Lines #1	1369	1983	1913 Model T Van, Orange Body, Black Trim	80
Allied Van Lines #2	2136	1984	Horse & Carriage, Orange Body, Black Trim	65
Allied Van Lines #3	2119	1985	1917 Model T Van, Orange Body, Black Trim	65
Allied Van Lines #4	9776	1988	1937 Ford Tractor/Trailor, Orange Body, Black Trim	65
Allied Van Lines #5	7517UO	1990	1947 IH Tractor/Trailor, Orange Body, Trim	40
Allis-Chalmers "A-C"	1201	1984	1926 Mack Truck, Tan Body, Black Trim	40
Allis-Chalmers "A-C"	2226EO	1989	1918 Runabout, Orange Body, Black Trim	30
Alzheimer's Association #2-LTD Ed	9594UO	1990	1905 Ford Delivery Van, White Body, Purple Trim	30
Alzheimer's Association #1 LTD Ed	9680	1989	1913 Model T Van, White Body, Purple Trim	95
American Red Cross #1	9294	1987	1913 Model T Van, Black Body, White Trim	75
American Red Cross #2	9294	1988	1913 Model T Van, Black Body, White Trim	35
American Red Cross #3 W/Spare Tire	9294	1989	1913 Model T Van, White Body, Red Trim	50
American Red Cross #4-Gold Spokes L.E.	9294	1989	1913 Model T Van, White Body, Red Trim	45
American Red Cross #5 LTD Ed.	9685	1989	1905 Ford Delivery Van, Black Body, Trim	45
American Red Cross #6 LTD Ed.	2984UO	1990	1950 Chevy Panel, Red Body, Black Trim	43
American Red Cross #7-LTD Ed.	7616	1990	1926 Mack W/Crates, Red Body, Brown Trim	35
Amoco	9150	1987	1913 Model T Van, White Body, Red Trim	155
Amoco	1333	1988	1905 Ford Delivery Van, White Body, Red Trim	90
Amoco #1	9373	1986	1926 Mack Tanker, Silver Body, Red Trim	240
Amoco #2	9173	1987	1926 Mack Tanker, White Body, Red Trim	225
Amoco #3	9447	1987	1926 Mack Tanker, Silver Body, Black Trim	325
Amoco (Certicare)	9151	1987	1913 Model T Van, White Body, Red Trim	50
Amoco (Certicare)	7668UA	1990	1932 Ford Panel, White Body, Black Trim	45
Amoco - Atlas Auto Products	9496	1988	1905 Ford Delivery Van, White Body, Red Trim	45
Amoco 100th Anniversary - LTD Ed.	9745	1989	1917 Model T Van, White Body, Blue Trim	35
Amoco 100th Anniversary- LTD Ed.	9660	1989	1918 Runabout, White Body, Blakc/Blue Trim	55
Amoco Stanolind #1 Polarine Oil & Grease	9673	1989	1917 Model T Van, Orange Body, Orange Trim	660

Ertl Banks

NAME	NUMBER	YEAR	DESCRIPTION	MINT
Amoco Stanolind #2 Polarine LTD. Ed.	9060	1989	1926 Mack Tanker, Orange Body, Black Trim	165
Amoco Stanolind #3-Polarine Lubricants	9383	1989	1926 Mack Tanker, Orange Body, Black Trim	150
Amoco Stanolind #4 LTD. Ed.	7657UA	1990	1932 Ford Panel, Dark Green Body, Black Trim	155
Amoco-Red Crown Gas-Stand. Oil LTD. Ed.	9563UA	1990	Horse Team & Tanker, Black Body, Red Trim	25
Amsouth	9454	1986	1913 Model T Van, White Body, Blue Trim	60
Andrews Toy Shop Ltd. Ed.	1322UA	1990	1913 Model T Van, White Body, Blue Trim	25
Anheuer-Busch #2	9047	1989	1926 Mack W/Crates, Red Body, Black Trim	60
Anheuser-Busch #3	7574EO	1990	1931 Hawkeye Truck, Red Body, Black Trim	55
Anheuser-Busch (Chrome) 1st Issue	9766	1988	1918 Barrel Runabout, Red Body, Black Trim	135
Anheuser-Busch(Chrome) 2nd Re-Issue	9766	1990	1918 Barrel Runabout, Red Body, Black Trim	75
Anthracite Battery	9264	1987	1905 Ford Delivery Van, White Body, Red Trim	90
Arkansas 150th Anniversary	9367	1986	1913 Model T Van, White Body, Red Trim	40
Arkansas Razorbacks	9353	1985	1913 Model T Van, White Body, Red Trim	45
Arm & Hammer	9938UO	1989	1932 Ford Panel, Yellow Body, Red Trim	40
Arm & Hammer W/Spare	7553UO	1990	1932 Ford Panel, Yellow Body, Red Trim	75
Arm And Hammer	9486	1987	1913 Model T Van, Yellow Body, Red Trim	100
Arm And Hammer	9828	1988	1905 Ford Delivery Van, Yellow Body, Red Trim	60
Arm And Hammer (Decal)	9828	1989	1905 Ford Delivery Van, Yellow Body, Red Trim	30
Armour Food	9891	1988	1913 Model T Van, White Body, Red Trim	60
Arrow Distributing #1-LTD Ed.	9270	1987	1932 Ford Panel, White Body, Blue Trim	85
Arrow Distributing #2	9725	1988	1950 Chevy Panel, Silver Body, Trim	125
Arrow Distributing #3	9328	1989	1918 Runabout, Blue Body, White Trim	30
Arrow Distributing #4	7542UO	1990	1905 Ford Delivery Van, White Body, Red Trim	30
Artworks (Donneckers)	7550UO	1990	1905 Ford Delivery Van, White Body, Red Trim	35
Associated Grocers Of Colorado	9212	1984	1913 Model T Van, White Body, Trim	60
Atlanta Falcons	1248	1984	1913 Model T Van, Silver Body, Red Trim	60
Atlas Van Lines #1	9514	1987	1926 Mack Truck, White Body, Blue Trim	100
Atlas Van Lines #2-LTD. Ed.	9771	1988	1932 Ford Panel, White Body, Red Trim	35
Atlas Van Lines #3 W/Spare-LTD. Ed.	9577	1989	1913 Model T Van, White Body, Blue Trim	35
Atlas Van Lines #4-LTD. Ed.	7612UA	1990	1937 Ford Tractor/Trailer, White Body, Blue Trim	35
Baker Oil Tools	9210UP	1990	1926 Mack Truck, Yellow Body, Blue Trim	35
Baltimore Gas And Electric	9153	1987	1932 Ford Panel, Black Body, Light Brown Trim	175
Baltimore Gas And Electric #2	9870	1988	1918 Runabout, Black Body, Trim	45
Baltimore Gas And Electric #3	9752	1989	1950 Chevy Panel, Gold Body, Black Trim	40
Baltimore Gas And Electric #4	2102UO	1990	Step Van, Beige Body, Blue Trim	40
Barq's Rootbeer #1	9826	1988	1913 Model T Van, Met. Silver Body, Trim	65
Barq's Rootbeer #2	9072	1989	1932 Ford Panel W/Spare Tire, Met. Silver Body, Trim	50
Barq's Rootbeer #3	9054UO	1990	1918 Barrel Runabout, Silver Body, Black Trim	50
Barrett Jackson Car Auction	9361UP	1990	1950 Chevy Panel, Black Body, White Trim	40
Barrick's Farm Sales	9271	1987	1913 Model T Van, White Body, Red Trim	25
Basehor, Kansas	9007	1989	1905 Ford Delivery Van, White Body, Red Trim	25
Beckman High School #1	9311	1989	1905 Ford Delivery Van, Green Body, Gold Trim	30
Beckman High School #2	1656UO	1990	1913 Model T Van, Green Body, Yellow Trim	25

Ertl Banks

NAME	NUMBER	YEAR	DESCRIPTION	MINT
Bell System	9801	1988	Horse & Carriage, Dark Brown Body, Light Brown Trim	30
Bell System	9803	1988	1932 Ford Panel, Black Body, Trim	35
Bell Telephone #1	2141	1984	Horse & Carriage, Dark Blue Body, Black Trim	80
Bell Telephone #1	9203	1984	1950 Chevy Panel, Olive Body, Trim	60
Bell Telephone #1	9298	1987	1918 Runabout W/O Ladder, Olive Body, Black Trim	40
Bell Telephone #2	9203	1985	1950 Chevy Panel, Olive Body, Trim	40
Bell Telephone #2	9800	1988	1918 Runabout W/Ladder, Dark Green Body, Black Trim	35
Bell Telephone 70th Anniversary	9695	1981	1913 Model T Van, Grey Body, Black Trim	100
Bell Telephone AT&T	7610IU	1990	1905 Ford Delivery Van, Black Body, Trim	25
Bell Telephone Canada	7609UO	1990	1905 Ford Delivery Van, Black Body, Trim	30
Bell Telephone Of America	9802	1988	1937 Ford Tractor/Trailer, White Body, Blue Trim	40
Bell Telephone Of Canada	1327	1982	1913 Model T Van, Black Body, Trim	65
Bell Telephone Pioneers Of America	2117	1985	1913 Model T Van, Black Body, Trim	25
Bell Telephone System	9646	1981	1913 Model T Van, Black Body, Trim	65
Bell Telephone Yellow Pages	2142	1984	1926 Mack Truck, Yellow Body, Black Trim	55
Ben Franklin	1319	1989	1918 Runabout, Grey Body, Red Trim	25
Ben Franklin	9688	1989	1905 Ford Delivery Van, Grey Body, Red Trim	35
Big "A" -Wagner Brake	1366UA	1990	1918 Runabout, Black Body, Red Trim	115
Big "A" Auto Parts	9482	1987	1905 Ford Delivery Van, White Body, Red Trim	55
Big "A" Auto Parts	9094	1989	1926 Mack Truck, Black Body, Red Trim	25
Big "A" Auto Parts-LTD Ed.	9772	1988	1917 Model T Van, Black Body, Trim	35
Big Bear Family Center	9981	1988	1918 Runabout, White/Black Body, Brown Trim	25
Big Bear Family Center	9006	1989	1905 Ford Delivery Van, White Body, Red Trim	30
Biglerville Hose Co.	9760	1988	1913 Model T Van, White Body, Red Trim	35
Binkley-Hurst Bros. 50th Anniversary	9626	1989	1910 Model T Van, White Body, Red Trim	30
BJR Auto Radiator Service	9500	1986	1913 Model T Van, Silver Body, Blue Trim	75
BJR Auto Radiator Service	9059	1989	1932 Ford Panel, Black Body, Silver Trim	80
BJR Auto Radiator Service	7614UO	1990	1950 Chevy Panel, Red Body, Black Trim	100
Bookmobile (Coos Bay)	9257	1985	1913 Model T Van, Whtie Body, Red Trim	40
Boone Co. Fair	9716	1988	1905 Ford Delivery Van, White Body, Blue Trim	30
Borg Warner #1 - LTD Ed.	9346	1985	1913 Model T Van, White Body, Red Trim	75
Borg Warner #2 - LTD. Ed.	9390	1986	1913 Model T Van, White Body, Red Trim	55
Bost Bakery	9029	1988	1917 Model T Van, White Body, Trim	35
Bost Bakery #1 (Gold Spokes)	9235	1985	1913 Model T Van, White Body, Red Trim	210
Bost Bakery #1 (Red Spokes) Ltd. Ed.	9235	1985	1913 Model T Van, White Body, Red Trim	35
Bost Bakery #2	9437	1986	1913 Model T Van, White Body, Red Trim	40
Bost Bakery Ltd. Ed.	9170	1987	1926 Mack Tanker, White Body, Red Trim	45
Brendle's (Gold Spokes)	9823	1988	1917 Model T Van, White Body, Blue Trim	35
Brendle's (Red Spokes)	9823	1988	1917 Model T Van, White Body, Blue Trim	125
Breyer's Ice Cream	9028	1988	1905 Ford Delivery Van, Black Body, Trim	75
Briggs & Stratton	9986	1988	1918 Runabout, White Body, Black Trim	145
Briggs & Stratton	9509	1989	1937 Ford Tractor/Trailer, White Body, Trim	50
British Columbia (1 Of 10 Canadian Prov)	9221	1985	1913 Model T Van, White Body, Pink Trim	30
Broadlands Centennial	9880	1988	1905 Ford Delivery Van, White Body, Blue Trim	25
Brownberry Bakeries	9441	1986	1913 Model T Van, White Body, Trim	60
Buckeye, Arizona	9287	1987	1905 Ford Delivery Van, White Body, Blue Trim	55

BANKS

Top to bottom: Pennzoil 1932 Ford Panel; Bardahl 1931 Hawkeye Tanker; Check The Oil 1930 Diamond T Tanker.

Ertl Banks

NAME	NUMBER	YEAR	DESCRIPTION	MINT
Budweiser	1315	1983	1913 Model T Van, White Body, Red Trim	210
Bush's Pork & Beans	1357	1990	1905 Ford Delivery Van, White Body, Red Trim	30
Bussmann Fuses 75th Anniversary	9333	1989	1918 Runabout, White Body, Black Trim	65
C.R.'S Friendly Market	9699UO	1989	1917 Model T Van, White Body, Orange Trim	30
Campbell's Pork & Beans	9394	1986	1905 Ford Delivery Van, Red Body, Black Trim	75
Campbell's Pork & Beans	9184	1987	1918 Runabout, White Body, Red Trim	75
Canada	9226	1985	1913 Model T Van, White Body, Red Trim	25
Canada Dry Ginger Ale	2133	1985	1913 Model T Van, Green Body, Trim	135
Canada Dry Ginger Ale	7680UO	1990	1918 Barrel Runabout, Green/White Body, Black Trim	40
Cardinal Foods	2139	1984	1913 Model T Van, White Body, Red Trim	95
Carl Biddig Meats	2106	1984	1913 Model T Van, Red Body, Trim	80
Carl's Chicken Barbeque	9089	1989	Step Van, White Body, Trim	25
Carlisle H.S. Thundering Herd	9682	1989	1913 Model T Van, White Body, Green Trim	25
Carlisle H.S. Thundering Herd	9937UA	1989	1918 Runabout, Green Body	25
Carlisle H.S. Thundering Herd	2140UP	1990	1937 Ford Tractor/Trailer, White Body, Trim	35
Carlisle Productions - Fall Carlisle	7570UO	1990	1950 Chevy Panel, Red Body	70
Carnation	9178	1987	1913 Model T Van, White Body, Red Trim	45
Carnation	9179	1987	1926 Mack Tanker, White Body, Red Trim	75
Castrol #1 - Blk Tire/Wht Spokes	9464	1986	1926 Mack Tanker, White Body, Green Trim	140
Castrol #2 - Wht Tire/Grn SPokess L.E.	9464UP	1987	1926 Mack Tanker. White Body, Green Trim	75
Castrol Motor Oil #1 -Blk Tire/ Wht Spoke	9463	1986	1913 Model T Van, White Body, Green Trim	75
Castrol Oil #2 - Wht Tire/Gld Spoke L.E.	9463	1987	1913 Model T Van, White Body, Green Trim	55
Celotex	9317	1987	1926 Mack Truck, Red Body, Black Trim	565
Celotex	9475	1987	1913 Model T Van, White Body, Red Trim	95
Central Hawkeye Gas Engine	9196	1987	1905 Ford Delivery Van, White Body, Red Trim	30
Champion Sparkplug	9067OU	1990	1918 Runabout, White Body, Black Trim	60
Charter Oak Centennial	9031	1989	1918 Runabout, White Body, Black Trim	20
Chemical Bank	1662UP	1990	1905 Ford Delivery Van, White Body, Blue Trim	30
Chevrolet #1 - Heartbeat Of America	9873	1989	1950 Chevy Panel, White Body, Black Trim	60
Chevrolet #1 - Today's Truck	9561UO	1989	1950 Chevy Panel, Black Body, Trim	35
Chevrolet #2 - Today's Truck	9561UP	1990	1950 Chevy Panel, White Body, Black Trim	25
Chevrolet #2 - Today's Truck	9873UP	1990	1950 Chevy Panel, White Body, Black Trim	25
Chevrolet Barrel 1/43 Dime Bank	9931	1990	1930 Chevy Stake Truck, Blue Body, Graphic Trim	10
Chevrolet Heartbeat Of America	9048	1989	1950 Chevy Tractor Trailer, White On White	35
Chicago Cubs	7545	1990	1926 Mack Truck, White Body, Blue Trim	30
Chicago Tribune	9386	1987	Step Van, White On White	110
Chicago Tribune	9102	1988	1917 Model T Van, Black On Black	70
Chicago Tribune	9017	1989	Horse & Carriage, Black On Black	55
Chicago Tribune	2150	1989	1917 Model T Van, Spare Tire, Black On Black	45
Chipco	9882	1988	1905 Ford Delivery Van, Black Body, Green Trim	35
Chiquita Bananas	9662	1989	1913 Model T Van, Yellow Body, Blue Trim	70
Christmas-Happy Holidays 1989	9584	1989	1913 Model T Van, White Body, Red Trim	35
Christmas-Happy Holidays 1990	7575DO	1990	1905 Ford Delivery Van, Red Body, Green Trim	20
Chrome King-American Bumper	9825	1988	1913 Model T Van, Silver On Silver	30

Ertl Banks

NAME	NUMBER	YEAR	DESCRIPTION	MINT
Cincinnati Bengals	1249	1984	1913 Model T Van, White Body, Orange Trim	55
Cintas	7666UP	1990	Step Van, White	80
Citgo #1 Lubricants	9307	1988	1926 Mack Tanker, White Body, Black Trim	425
Citgo #2 Lubricants	9456EA	1990	1918 Barrel Runabout, Black Body, White Trim	145
Citgo #3	7537	1989	1913 Model T Van, Spare Tire, White Body, Red Trim	75
Classic Motorbooks 25th Anniversary	7567UO	1990	1950 CHevy Panel, Blue Body, Silver Trim	60
Clemson University	9523	1989	1918 Runabout, White Body, Orange, Black Trim	55
CLemson University Limited Edition	9775	1988	1913 Model T Van, White Body, Orange Trim	55
CO-OP, The Farm Store	9245	1985	1913 Model T Van, Tan Body, Green Trim	45
Coast To Coast	9188	1987	1913 Model T Van, White Body, Black Trim	45
Coast To Coast	9742	1988	1905 Ford Delivery Van, White Body, Black Trim	25
Coast To Coast	9049	1989	1926 Mack Truck With Crates, White Body, Black, Brown Trim	25
Coast To Coast Hardware Store	2105EO	1990	1918 Runabout, White Body, Black Trim	25
Cohen & Sons, William	9339	1989	1926 Mack Truck, Red Body, Tan Trim	65
Comet Cleanser	7507UO	1990	1905 Ford Delivery Van, Metallic Green Body, Gold Trim	30
Conoco #1	9750	1989	1926 Mack Truck, Silver Body, Green Trim	250
Conoco #2	7523UA	1990	Horse Team, Tanker, Black Body, White Trim	45
Coos Bay, House Of Books	9256	1985	1013 Model T Van, White Body, Red Trim	30
Country General	1345UO	1990	1918 Runabout, White Body, Red Trim	25
Country Store-Reiman First Edition	7564UO	1990	1926 Mack Truck, Yellow Body, Black Trim	30
Crescent Electric Supply	9008	1989	1913 Model T Van, White Body, Blue Trim	80
Cumberland Valley Tractor Pullers 1988	1324	1988	1926 Mack Tanker, Silver Body, Red Trim	35
Cumberland Valley Tractor Pullers 1989	9657	1989	1932 Ford Panel, Silver Body, Black Trim	40
Cumberland Valley Tractor Pullers 1990	9761UO	1990	1937 Ford Tractor Trailer, Silver Body, Blue Trim	35
Currie's	7529UO	1990	1905 Ford Delivery Van, White Body, Red Trim	30
Cycle-AM Motocross	9204	1984	1913 Model T Van, White Body, Red Trim	95
Daily Press	9056	1989	1905 Ford Delivery Van, White Body, Red Trim	60
Daily Press, Newport News, Virginia	9521	1987	1913 Model T Van, White Body, Red Trim	160
Dairy Farm	9525	1987	1913 Model T Van, Black On Black	250
Dairy Queen	9144	1987	1913 Model T Van, White Body, Red Trim	125
Dairy Queen	9284	1988	1937 Ford Tractor Trailer, White On White	95
Dairy Queen	9033	1989	1918 Runabout, Red Body, Black Trim	85
Dairy Queen	9034	1989	1932 Ford Panel, White Body, Blue Trim	95
Dairy Queen Limited Edition	9285	1988	1917 Model T Van, White On White	85
Dallas Cowboys	1247	1984	1913 Model T Van, Silver Body, Blue Trim	125
Decorah, Iowa	9424	1986	1905 Ford Delivery Van, White Body, Red Trim	30
Decorah, Iowa	9143	1987	1918 Runabout, Blue Body, Beige Trim	30
Decorah, Iowa	9762	1988	1932 Ford Panel, White Body, Blue Trim	30
Decorah, Iowa, Chamber Of Cintas	9255	1985	1913 Model T Van, White Body, Red Trim	45
Decorah, Iowa, Chamber Of Commerce	9677	1989	1926 Mack Truck, White Body, Red Trim	30
Delaval	9681	1989	Step Van, White On White	110
Delaware Valley Old Time Power & Equip.	9522	1986	1905 Ford Delivery Van, White Body, Red Trim	30

Ertl Banks

NAME	NUMBER	YEAR	DESCRIPTION	MINT
Delaware Valley Old Time Power & Equip.	9492	1987	1918 Runabout, White Body, Black Trim	45
Democratic Party, Election '88	9806	1988	1905 Ford Delivery Van, White Body, Blue Trim	105
Detroit News	1667	1983	1913 Model T Van, Red Body, Blue Trim	75
Deutz-Allis	2209	1987	1913 Model T Van, White Body, Blue Trim	25
Deutz-Allis	2217	1989	1905 Ford Delivery Van, White Body, Black Trim	20
Diamond Crystal Salt	9438	1986	1926 Mack Truck, Red Body, Black Trim	260
Diamond Crystal Salt	9414	1987	1913 Model T Van, White Body, Red Trim	75
Dixie Brewing	9728	1988	1937 Ford Tractor Trailer, Both Doors Labelled, White On White	200
Dixie Brewing	9073	1989	1918 Barrel Runabout, Green Body, Black, White Trim	50
Dixie Brewing Limited Edition	9728	1988	1937 Ford Tractor Trailer, White On White	75
Dobyns-Bennett High School, Kingsport, Tennessee	7516UO	1990	1950 Chevy Panel, Maroon Body, White Trim	90
Dolly Madison	9206UP	1990	Step Van, White	65
Domtar Gypsum	9824	1989	1913 Model T Van, White Body, Blue Trim	25
Double "J" Limited Edition	9215	1985	1913 Model T Van, White Body, Red Trim	25
Dr. Pepper	7572	1990	1926 Mack Truck, Red Body, White Trim	60
Dr. Pepper	7573UO	1990	1918 Runabout, White Body, Red Trim	60
Dr. Pepper Special Edition	9739	1988	1905 Ford Delivery Van, White Body, Red Trim	50
Drake Hotel	2113	1984	1913 Model T Van, White Body, Red Trim	125
Drake, The (Hilton Hotels)	7672UO	1990	1932 Ford Panel, White Body, Blue Trim	85
Dreyer's Ice Cream	9617	1989	1905 Ford Delivery Van, Cream Body, Black Trim	30
Dubuque G&CC Invitational Golf #1	1657	1990	1917 Model T Van, White Body, Blue Trim	165
Dubuque, Iowa	9503	1986	1905 Ford Delivery Van, White Body, Red Trim	30
Durona Productions	1321	1982	1913 Model T Van, Cream On Cream	725
Durona Productions	9313	1986	1932 Ford Panel, Whire Body, Blue Trim	345
Dyersville Historical Society	9529	1986	1905 Ford Delivery Van, White Body, Red Trim	35
Dyersville Historical Society	9490	1987	1913 Model T Van, White Body, Red Trim	35
Dyersville Historical Society	9883	1988	1918 Runabout, Red Body, Black Trim	35
Dyersville Historical Society	9037	1989	1932 Ford Panel, White Body, Red Trim	85
Dyersville Historical Society	7571UO	1990	1918 Barrel Runabout, Red, Black Body, White Trim	25
East Buchanan, Iowa	9360	1985	1913 Model T Van, White Body, Red Trim	30
East Tennessee University	7627	1990	1913 Model T Van, White Body, Blue Trim	40
Eastview Pharmacy	1317UP	1990	1950 Chevy Panel, Blue Body, Silver Trim	95
Eastview Pharmacy, Limited Edition	9671	1989	1913 Model T Van, White Body, Blue Trim	125
Eastwood Company #1, 1989	9325	1989	1950 Chevy Panel, Blue On Blue	650
Eastwood Company #1, 1990	9325	1989	1950 CHevy Panel, Blue On Blue	650
Eastwood Company #2	9562UO	1990	1932 Ford Panel With Spare Tire, Tan Body, Maroon Trim	425
Eastwood Company #3	2985UO	1990	1931 Hawkeye Truck, Green Body, Black Trim	100
Eastwood Company #4	7664UO	1990	1937 Ford Tractor Trailer, Red Body, Green Trim	180
Eastwood Company #5	2141UP	1990	1930 Diamond T Tanker, Blue On Blue	100
Edy's Ice Cream	9644	1989	1905 Ford Delivery Van, Cream Body, Black Trim	30
Elma, Iowa	9399	1986	1913 Model T Van, White Body, Blue Trim	225
Elmira Maple Festival Ltd. Ed.	9759UA	1990	1905 Ford Delivery Van, White Body, Red Trim	30
Elmira Syrup Festival	9656	1989	1913 Model T Van, White Body, Blue Trim	25
Entenmann's	9455	1986	1913 Model T Van, White Body, Blue Trim	80

BANKS

Ertl Banks

NAME	NUMBER	YEAR	DESCRIPTION	MINT
Entenmann's	1317	1987	Step Van, White Body, Trim	90
Entenmann's	9780	1988	Step Van, White Body, Trim	100
Ephrata Fair 1989	9141	1989	1913 Model T Van, Black Body, Blue Trim	30
Ephrata Fair 1990	7541UO	1990	1950 Chevy Panel, Red Body, Black Trim	30
Ertl Collector's Club Ltd. Ed.	1660PA	1990	Step Van, Black Body	45
Ertl Collectors Club	1668	1983	1913 Model T Van, White Body, Red Trim	150
Ertl Collectors Club	9064	1989	1950 Chevy Panel, Gold Body, White Trim	95
Ertl N.Y. Premium Incentive Show Ltd. Ed.	9737	1988	1905 Ford Delivery Van, Silver Body, Black Trim	50
Ertl Safety Award	7554UA	1990	1913 Model T Van, White Body, Blue Trim	100
Evers Toy Store	9566UO	1990	Horse Team & Tanker, White Body, Black Trim	25
F-D-R Associates Ltd. Ed.	9378	1987	1905 Ford Delivery Van, Silver Body, Black Trim	45
Fanny Farmer	2104	1983	1913 Model T Van, White Body, Brown Trim	35
Farm Bureau Co-Op	7622	1990	1913 Model T Van, White Body, Red Trim	25
Farm Toy Capital Of The World #3 L.E.	9189	1987	1926 Mack Tanker, Silver Body, Red Trim	70
Farm Toy Capital Of The World #4 L.E.	9779	1988	1932 Ford Panel, Black Body, Silver Trim	50
Farm Toy Capital Of The World #5 L.E.	9107	1989	1905 Ford Delivery Van, Green Body, Black Trim	30
Farm Toy Capital Of The World #6 L.E.	1664UP	1990	1931 Hawkeye Truck, Red Body, Black Trim	35
Farm Toy Capitol Of The World #1 L.E.	9233	1986	1913 Model T Van, White Body, Green Trim	95
Farm Toy Capitol Of The World #2 L.E.	9510	1986	1918 Runabout, Blue Body, Blue/Beige Trim	85
Federal Express	9334	1989	Step Van, White Body, Trim	45
Felix Grundy Days Ltd. Ed.	6125	1989	1913 Model T Van, White Body, Red Trim	25
Field Of Dreams-Universal Studios Ltd. Ed.	7617UA	1990	1905 Ford Delivery Van, White Body, Blue Trim	50
Fina	9186	1987	1926 Mack Tanker, White Body, Blue Trim	90
Fina	9043	1989	1905 Ford Delivery Van, White Body, Blue Trim	35
Fina - Ltd. Ed.	9407	1987	1913 Model T Van, White Body, Blue Trim	45
Fina - Ltd. Ed.	9456	1989	1917 Model T Van, White Body, Trim	150
Firehouse Films (Durona)	9369	1990	1950 Chevy Panel, Red Body, Black Trim	165
First National Bank (Oklahoma) Ltd. Ed.	2988UO	1990	1918 Runabout, White Body, Gold Trim	50
First Tennessee Bank #1 Ltd. Ed.	1318	1988	1917 Model T Van, White Body, Blue Trim	165
First Tennessee Bank #2 Ltd. Ed.	9331	1989	1918 Runabout, White Body, Black/Blue Trim	30
Flav-O-Rich Ltd. Ed.	9044	1989	1913 Model T Van, White Body, Red Trim	30
Flint Piston Service - U.A.W. #2	7551UO	1990	Step Van, White Body	65
Food City - Ltd. Ed.	9857	1988	1905 Ford Delivery Van, Silver Body, Black Trim	30
Food Lion - Ltd. Ed.	9279	1987	1913 Model T Van, Gold Body, Blue Trim	30
Ford	0865	1986	1905 Ford Delivery Van, White Body, Blue Trim	25
Ford	0837EO	1987	1918 Runabout, White Body, Blue Trim	25
Ford #1	1334	1981	1913 Model T Van, White Body, Blue Trim	35
Ford #2 (Nat'l Truck Dlrs) Ltd. Ed.	1322	1983	1913 Model T Van, White Body, Blue Trim	110
Ford Motorsports #1	9871	1988	1905 Ford Delivery Van, White Body, Blue Trim	55
Ford Motorsports #2	2151	1989	1918 Runabout, White Body, Blue Trim	30
Ford Motorsports #3	1658	1990	1913 Model T Van, White Body, Blue Trim	35
Ford New Holland #5	0374	1990	1917 Model T Van, White Body, Blue Trim	20
Four-H Clubs Of America	9379	1987	1913 Model T Van, White Body, Blue Trim	35
Four-H Clubs Of America	9701	1988	1917 Model T Van, White Body, Trim	30

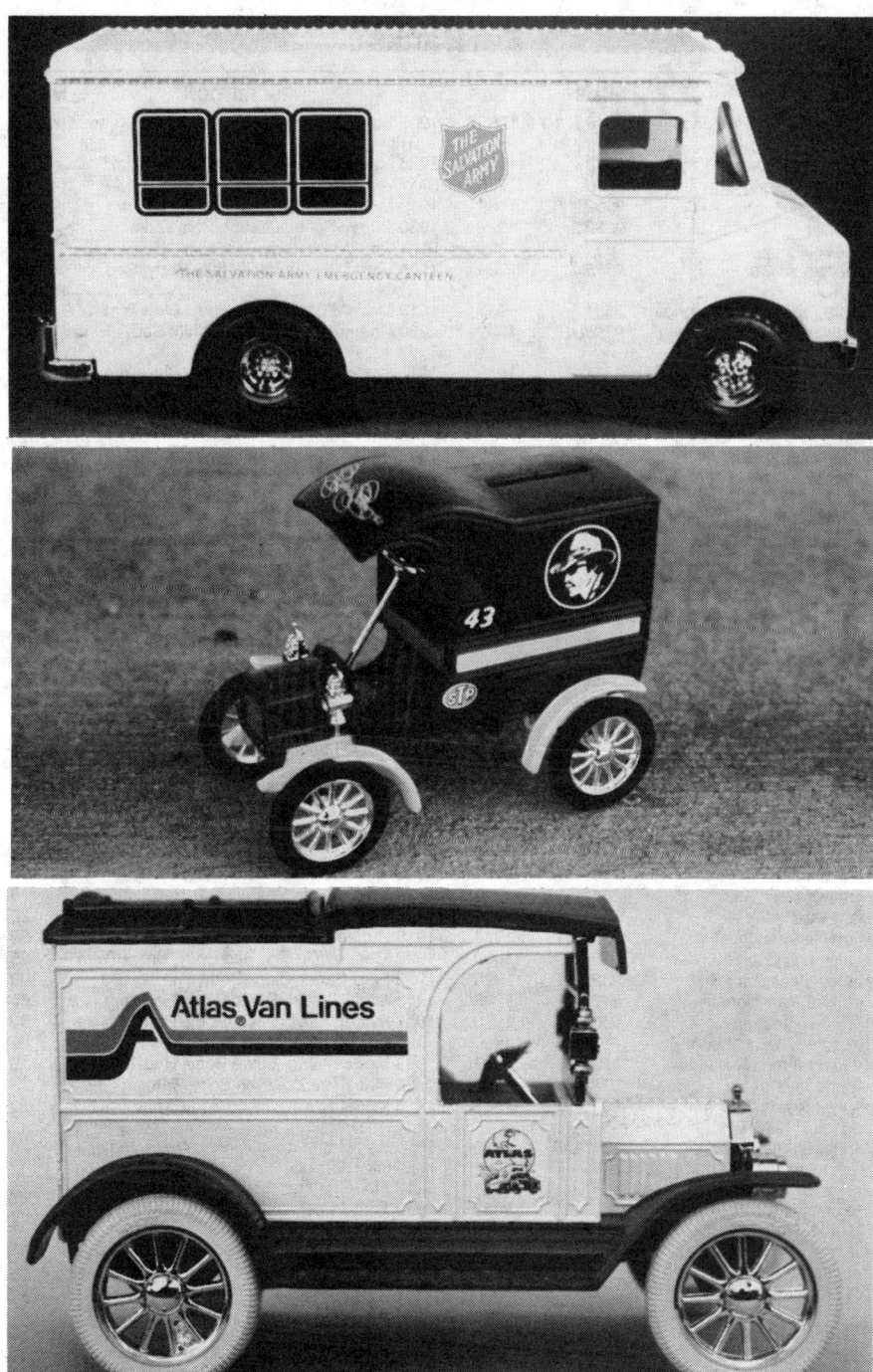

Top to bottom: Salvation Army Step Van; Richard Petty Enterprises 1905 Ford Delivery Van; Atlas Van Lines 1913 Ford Model T Van.

Ertl Banks

NAME	NUMBER	YEAR	DESCRIPTION	MINT
Four-H Clubs Of America	9848	1988	1905 Ford Delivery Van, White Body, Green Trim	35
Franco-American	9302	1986	1926 Mack Truck, Red Body, Green Trim	60
Freihofer Baking Co.	9710	1988	Step Van, Red Body, Trim	50
Frito-Lay	9632	1989	1913 Model T Van, White Body, Red Trim	30
Frito-Lay	9633	1989	1950 Chevy Panel, Orange Body, Trim	40
Frito-Lay	9634	1989	Step Van, White Body, Tan Trim	45
Fuller Brush Co.	9085	1989	1905 Ford Delivery Van, White Body, Red Trim	30
Future Farmers Of America	9531	1987	1913 Model T Van, White Body, Blue Trim	35
Gateway Toy Show #1 -9th Anniversary	9598UO	1989	1950 Chevy Tractor/Trailer, White Body, Trim	55
Genstar (Gypsum Products Co.)	9358	1985	1913 Model T Van, White Body, Blue Trim	25
Georgia Tech - Not A Bank	9251	1985	1932 Ford Roadster, Met. Gold Body, White Trim	125
Gilbertville, Iowa	9368	1986	1932 Ford Panel, White Body, Red Trim	65
Gilbertville, Iowa 3rd Ann	9246	1985	1913 Model T Van, White Body, Red Trim	50
Glaxo	1353UO	1990	1932 Ford Panel, White Body, Blue Trim	30
Glendale Medical Center	9266	1987	1913 Model T Van, White Body, Blue Trim	60
Global Van Lines	1655	1983	1913 Model T Van, Light Blue Body, Black Trim	45
Global Van Lines	1655UO	1990	1913 Model T Van, Light Blue Body, Black Trim	25
Golden Flake	9118	1987	Step Van, White Body, Trim	70
Good (J.F. Good Co.)	9524	1986	1913 Model T Van, White Body, Trim	125
Good (J.F. Good Co.)	9603	1988	1918 Runabout, White Body, Brown Trim	40
Good (J.F. Good Co.)	9332	1989	1926 Mack Truck, White Body, Red Trim	20
Goshen, H. & W. Dairy	9146	1987	Horse & Carriage, White Body, Orange Trim	70
Grauer's Paint	2139UO	1990	1932 Ford Panel, White Body, Red Trimn	40
Gulf - That Good Gulf Gasoline (Reissue)	7652UO	1990	1926 Mack Tanker, Orange Body, Blue Trim	45
Gulf Ohio Gas Marketing	9443	1984	1932 Ford Panel, White Body, Red Trim	410
Gulf Ohio Pipeline	9443	1984	1932 Ford Panel, White Body, Red Trim	465
Gulf Refining	9211	1984	1950 Chevy Panel, Orange Body, Black Trim	2500
H.E. Butts	1365	1983	1913 Model T Van, White Body, Red Trim	25
Hamm's Beer	2145	1984	1913 Model T Van, White Body, Blue Trim	85
Hamm's Beer	7619UO	1990	1926 Mack Truck, White Body, Red Trim	70
Hardware Hank	7635EO	1990	1917 Model T Van, Red Body, Black Trim	20
Harley-Davidson #1	9784	1988	1918 Runabout, Olive Body, Black Trim	650
Harley-Davidson #2	9135UO	1989	1926 Mack W/Crates, Red Body, Brown Trim	525
Harley-Davidson #3	7525UA	1990	1932 Ford Panel, Black Body, Trim	300
Hartford Provisions	2108	1983	1913 Model T Van, White Body, Red Trim	265
Hawkeye Tech	9533	1986	1913 Model T Van, White Body, Red Trim	30
Heatcraft - Lennox Ltd. Ed.	7562UA	1990	1926 Mack Truck, White Body, Red Trim	30
Heating Alternatives Ltd.	1312	1987	1913 Model T Van, White Body, Red Trim	30
Heilig Meyers	9749	1989	1926 Mack Truck, Green Body, Trim	25
Heilig Meyers 75th Ann. (1913-1988)	9700	1988	1913 Model T Van, Green Body, Trim	30
Heineken Beer #1	9570UO	1989	1918 Barrel Runabout, Green/White Body, Black Trim	175
Heinz "57"	1345	1981	1913 Model T Van, White Body, Trim	75
Hemmings Motor News #1 (Irish Green)	9669	1989	1932 Ford Panel, Light Green Body, Black Trim	200
Hemmings Motor News #2 (British Green)	9669	1989	1932 Ford Panel, Green Body, Black Trim	45
Hemmings Motor News #2 (British Green)	9669	1990	1932 Ford Panel, Green Body, Black Trim	25
Henderson Centennial	9370	1986	1913 Model T Van, White Body, Blue Trim	30
Henny Penny	9889	1988	1913 Model T Van, White Body, Red Trim	30
Henny Penny	9890	1988	1905 Ford Delivery Van, White Body, Red Trim	40

Ertl Banks

NAME	NUMBER	YEAR	DESCRIPTION	MINT
Henny Penny	9945UO	1989	1918 Runabout, White Body, Black Trim	25
Henny Penny	9946UO	1989	1932 Ford Panel, White Body, Red Trim	40
Hershey Auto Club	9799	1988	1917 Model T Van, White Body, Trim	90
Hershey Auto Club	9084	1989	1918 Barrel Runabout, Tan Body, Brown Trim	75
Hershey Auto Club	7640UO	1990	1926 Mack Truck, White Body, Black Trim	35
Hershey Auto Club (Regional)	7639UO	1990	1905 Ford Delivery Van, White Body, Maroon Trim	40
Hershey's Chocolate Milk	1349UO	1990	1926 Mack Tanker, Brown Body, Trim	45
Hershey's Cocoa	9665	1989	1905 Ford Delivery Van, White Body, Brown Trim	70
Hershey's Golden Almond	2129	1990	1913 Model T Van 1000+, Gold Plt Body, MetalizedTrim	100
Hershey's Kisses	2126UO	1990	1950 Chevy Panel 1000+, Chrome Body, Metalized Trim	135
Hershey's Milk Chocolate	1350UO	1990	1913 Model T Van, Brown Body, Trim	25
Hershey's Milk Chocolate With Almonds	1351UO	1990	1913 Model T Van, Brown Body, Trim	25
Hills Department Stores	9768	1988	1913 Model T Van, White Body, Red Trim	25
Hinckley & Schmitt	9427	1986	1913 Model T Van, White Body, Blue Trim	25
Hoffman Laroche	9601	1988	1913 Model T Van, White Body, Red Trim	155
Hoffman Laroche	9974	1988	1905 Ford Delivery Van, White Body, Blue Trim	145
Holiday Wholesale	9470	1987	Step Van, White Body, Blue Trim	30
Holly Cliff Farms #1	9477	1987	1926 Mack Tanker, Silver Body, Red Trim	50
Holly Cliff Farms #2	9972	1989	1926 Mack Truck, White Body, Black Trim	40
Holt Mfg. #1 (Caterpillar)	7709DO	1989	1905 Ford Delivery Van, Black On Black	20
Home Federal Savings	2149	1984	1926 Mack Truck, White Body, Black Trim	35
Home Hardware #1	1356	1982	1913 Model T Van, Yellow Body, Black Trim	185
Home Hardware #2	2109	1984	1926 Mack Truck, Yellow Body, Black Trim	125
Home Hardware #3	9250	1985	1932 Ford Panel, Yellow Body, Black Trim	60
Home Hardware #4	9401	1986	1905 Ford Delivery Van, Yellow Body, Black Trim	65
Home Hardware #5	9145	1987	1918 Ford Runabout, Yellow Body, Black Trim	45
Home Hardware #6	9819	1988	1950 Chevy Panel, Yellow Body, Black Trim	35
Home Hardware #7	9012	1989	1926 Mack Truck With Crates, Yellow Body, Black Trim	35
Home Hardware #8	9011	1989	1917 Model T Van, Yellow Body, Black, Tan Trim	30
Home Savings & Loan	9844	1988	1905 Ford Delivery Van, White Body, Blue Trim	20
Home Savings & Loan	9845	1988	1913 Model T Van, White Body, Red Trim	30
Home Savings & Loan	9846	1988	1918 Ford Runabout, White Body, Black Trim	25
Home Savings & Loan	9292	1989	1926 Mack Truck, White Body, Red Trim	40
Home Savings & Loan	9309	1989	1921 Ford Panel, White Body, Blue Trim	40
Homestead Collectibles	9651	1989	1905 Ford Delivery Van, White Body, Red Trim	35
Hostess Cakes #1	1661	1982	1913 Model T Van, White Body, Black Trim	50
Hostess Cakes #2	9422	1986	1913 Model T Van, White Body, Blue Trim	30
Howard Brand Discounts	1366	1983	1913 Model T Van, White Body, Red Trim	30
Hudson Bay Company	9781	1988	1913 Model T Van, Black On Black	45
Husker Harvest Days	1346UO	1990	1918 Ford Runabout, White Body, Red Trim	25
I.B.C.	9610	1988	Step Van, White On White	65
Idaho Centennial 1890-1990	9139	1989	1913 Model T Van, White Body, Red Trim	30
Ideal Trucking	2963UO	1990	1937 Ford Tractor Trailer, White	35
IDED	9851	1988	1918 Ford Runabout, White Body, Black Trim	85
IDED (Ertl Logo)	9849	1988	1918 Ford Runabout, White Body, Black Trim	115
IGA #1	1651	1983	1913 Modle T Van, White Body, Red Trim	55
IGA #2	2138	1984	1926 Mack Truck, White Body, Red Trim	45
IGA #3	2126	1984	1913 Model T Van, White Body, Red Trim	45
IGA #4 (60th Anniversary)	9350	1985	1932 Ford Panel, White Body, Red Trim	50
IGA #5	9120	1987	1905 Ford Delivery Van, White Body, Red Trim	25

Ertl Banks

NAME	NUMBER	YEAR	DESCRIPTION	MINT
IGA #6	9023	1988	1918 Ford Runabout, Red Body, Black Trim	25
IGA #7	9015	1989	1950 Chevy Panel, Red Body, White Trim	25
IGA #8	7696EO	1990	1931 Hawkeye Truck, White Body, Red Trim	25
IGA Credit Union	9794	1988	1905 Ford Delivery Van, White Body, Blue Trim	25
Imperial Palace #1	2107	1983	1913 Model T Van, Grey Body, Black Trim	40
Imperial Palace #2	9943UO	1989	1926 Mack Truck, White Body, Blue Trim	40
Independence, Iowa (4th Of July, 1985)	9253	1985	1913 Model T Van, White Body, Red Trim	30
Independence, Iowa (4th Of July, 1986)	9248	1986	1913 Model T Van, White Body, Blue Trim	30
Independence, Iowa (4th Of July, 1987)	9194	1987	1918 Ford Runabout, White Body, Black Trim	25
Independence, Iowa (Chamber Of Commerce)	9236	1984	1913 Model T Van, White Body, Red Trim	35
Independence, Iowa (Christmas)	9359	1985	1913 Model T Van, White Body, Red Trim	35
Independence, Iowa (Christmas)	9359	1985	1913 Model T Van, White Body, Red Trim	650
Independence, Iowa (Christmas)	9303	1986	1905 Ford Delivery Van, White Body, Red Trim	30
Independence, Iowa (Christmas)	9499	1987	1905 Ford Delivery Van, White Body, Blue Trim	35
Independence, Iowa (Christmas)	9888	1988	1918 Ford Runabout, Red On Red	30
Independence, Iowa (Lion's Club)	9507	1986	1913 Model T Van, Grey Body, Blue Trim	70
Independence, Iowa (Lion's Club)	9288	1987	1917 Model T Van, White Body, Blue Trim	35
Indianapolis 500	9813	1988	1918 Ford Runabout, White Body, Black Trim	45
Iowa Fireman's Assn. 105th	1346	1983	1913 Model T Van, White Body, Red Trim	195
Iowa Fireman's Assn. 106th	2137	1984	1913 Model T Van, White Body, Red Trim	40
Iowa Fireman's Assn. 107th	9237	1985	1913 Model T Van, White Body, Red Trim	60
Iowa Fireman's Assn. 109th	9165	1987	1005 Ford Delivery Van, White Body, Red Trim	55
Iowa Hawkeyes #1	1311	1983	1913 Model T Van, Yellow Body, Black Trim	55
Iowa Hawkeyes #10 (10th Anniversary Edition)	1665UO	1990	1931 Hawyee With Crates, Yellow Body, Black Trim	30
Iowa Hawkeyes #2	1351	1982	1913 Model T Van, Yellow Body, Black Trim	30
Iowa Hawkeyes #3	1355	1983	1913 Model T Van, Black Body, Yellow Trim	55
Iowa Hawkeyes #4	1663	1983	1926 Mack Truck, Yellow Body, Black Trim	60
Iowa Hawkeyes #6	2135	1984	1926 Mack Truck, Yellow Body, Black Trim	35
Iowa Hawkeyes #7	9180	1987	1918 Ford Runabout, Yellow Body, Black Trim	35
Iowa Hawkeyes #8	9810	1988	1905 Ford Delivery Van, Black Body, Yellow Trim	35
Iowa Hawkeyes #9	9748	1989	1932 Ford Panel, Yellow Body, Black Trim	40
Iowa Jaycee Express	9701	1990	1913 Model T Van, Grey Body, Black Trim	90
Iowa Jaycee Express Limited Edition	9457UO	1990	1918 Ford Runbaout, Beige Body, Blue Trim	25
Iowa State Cyclones #1	9259	1985	1913 Model T Van, Yellow Body, Red Trim	30
Iowa State Cyclones #2	9834	1988	1905 Ford Delivery Van, Yellow Body, Red Trim	35
Iowa State Cyclones #3	9127	1989	1950 Chevy Panel. Red Body, Yellow Trim	45
Iowa State Cyclones #4	1312UP	1990	1918 Ford Runabout, Red Body, Yellow Trim	30
J.C. Penney	9232	1985	Horse, Carriage, Tan Body, Black Trim	65
J.C. Penney	9234	1985	1932 Ford Panel, Beige Body, Black Trim	75
J.C. Penney	1328	1988	1918 Ford Runabout, White Body, Blue Trim	35
J.C. Penney	9640	1989	1050 Chevy Panel, Yellow On Yellow	30
J.C. Penney	9641	1989	1926 Mack Truck, White Body, Blue Trim	30
J.C. Penney	2976UO	1990	1918 Ford Barrel Runabout, Red Body, Tan Trim	25
J.C. Penney	2977UO	1990	1926 Mack Truck With Crates, Yellow Body, Black Trim	20

Ertl Banks

NAME	NUMBER	YEAR	DESCRIPTION	MINT
J.C. Penney	2975UO	1990	1932 Ford Panel With Spare Tire, Beige Body, Red Trim	35
J.C. Penney #1	1354	1983	1913 Model T Van, White Body, Red Trim	155
J.C. Penney #2	1354	1983	1913 Model T Van, Grey Body, Black Trim	95
J.C. Penney #3	1354	1985	1913 Model T Van, Yellow Body, Black Trim	40
J.C. Penney (Golden Rule)	1326	1988	1905 Ford Delivery Van, Orange Body, Black Trim	35
J.C. Penney (Golden Rule)	9639	1989	1917 Model T Van, Green Body, Black Trim	25
J.I. Case	0216	1984	1926 Mack Truck, Tan Body, Black Trim	35
J.I. Case	0699	1987	1905 Ford Delivery Van, Tan Body, Grey Trim	20
J.I. Case	0668	1988	1913 Model T Van, Red On Red.	20
J.I. Case	0401	1989	1905 Ford Delivery Van, Red Body, Black Trim	25
J.I. Case	0286	1990	1913 Model T Van,	25
J.L. Kraft	2147	1985	1913 Model T Van, Yellow Body, Black Trim	90
J.T. General Store	1321	1982	1913 Model T Van, Beige On Beige.	35
J.T. General Store (90th Anniversary)	9674	1989	1905 Ford Delivery Van, Beige Body, Brown Trim	25
Jack Daniels	9852	1988	1905 Ford Delivery Van, Black On Black.	75
Jack Daniels	9077	1989	1918 Ford Runabout With Barrels, Black On Black.	50
Jackson Brewery	9050	1989	1905 Ford Delivery Van, White Body, Red Trim	35
Janesville, Iowa	9344	1985	1913 Model T Van, White Body, Red Trim	25
Jesup, Iowa, Chamber Of Commerce	9258	1985	1913 Model T Van, White Body, Red Trim	30
Jim Beam #116 (Northern Ohio)	7661UO	1990	1932 Ford Panel, White Body, Blue Trim	65
Jim Beam (District #1)	1313UO	1990	1905 Ford Delivery Van, White Body, Blue Trim	45
Jim Beam (District #10)	9989	1988	1926 Mack Truck, White Body, Black Trim	35
Jim Beam (District #5) #1	9412	1986	1913 Model T Van, White Body, Red Trim	245
Jim Beam (District #5) #2	9387	1987	1932 Ford Panel, White Body, Blue Trim	95
Jim Beam (District #5) #3	9729	1988	1918 Ford Runabout, White Body, Black Trim	85
Jim Beam (District #5) #4	9676	1989	1905 Ford Delivery Van, Yellow Body, Red Trim	85
Jim Beam (District #5) #5	2964UO	1990	1918 Ford Runabout With Barrels, Red Body, Beige Trim	75
Jim Beam (District #6)	7442	1990	1932 Ford Panel.	30
Jim Beam (District #8)	9683	1989	1913 Model T Van, White Body, Red Trim	35
Jim Beam (District #9)	9647	1989	1917 Model T Van, White Body, Blue Trim	35
Jim Beam (District #9) Susquehanna	1316	1990	1950 Chevy Panel, Blue Body, Silver Trim	75
Jim Beam, Sugar River Beamers	2125UO	1990	1905 Ford Delivery Van, White Body, Red Trim	30
Jim's Auto Sales	9818	1988	1913 Model T Van, White Body, Red Trim	25
John Deere #1	0531	1984	1926 Mack Truck, Green Body, Yellow Trim	115
John Deere #2	5534	1986	1926 Mack Truck, Green Body, Yellow Trim	70
John Deere #3	5564EO	1989	1926 Mack Truck, Yellow Body, Green Trim	40
John Deere I	5621	1989	1950 Chevy Panel, Green Body, Yellow Trim	35
John Deere II	5621	1989	1950 Chevy Panel, Wide Tire Version, Green Body, Yellow Trim	25
Johnson Wax	9459	1987	1913 Model T Van, White Body, Red Trim	185
Kansas State Fair	9272	1987	1913 Model T Van, White Body, Blue Trim	225
Kansas State Fair	9272	1987	1913 Model T Van, White Body, Red Trim	75
Kauffman's Orchard Apple Farm	7560UO	1990	1918 Ford Runabout, White Body, Black Trim	30
Kerr McGee Oil Co.	9773	1988	1926 Mack Tanker, Grey Body, Black TrimT	50
Kerr McGee Oil Co.	7662UA	1990	1932 Ford Panel With Spare Tire, Grey Body, Black Trim	35
Kerr McGee Oil Co. Limited Edition	9130	1989	1913 Model T Van, Grey Body, Black Trim	45
Key Federal Bank	9702	1988	1918 Ford Runabout, White Body, Black Trim	35

BANKS

Top to bottom: Texaco 1939 Dodge Airflow Tank Truck; Sinclair 1930 Diamond T Tanker; Lion Oil 1931 Hawkeye Tanker; Royalite Oil Co. 1931 Hawkeye Tanker.

Ertl Banks

NAME	NUMBER	YEAR	DESCRIPTION	MINT
Key Federal Bank	9703	1988	1932 FOrd Panel, White Body, Blue Trim	35
Key-Aid	9175	1987	1913 Model T Van, White Body, Red Trim	70
Key-Aid	9485	1988	1905 Ford Delivey Van, White Body, Red Trim	30
Key-Aid	9944	1989	1932 Ford Panel, Black Body, Silver Trim	30
Key-Aid Limited Edition	1332UA	1990	1931 Hawkeye With Crates, Green Body, Black Trim	30
Kidde	9351	1985	1913 Model T Van, White Body, Red Trim	55
King Edward Cigars	9854	1988	1913 Model T Van, Red Body, Black Trim	40
Kingsport, Tennessee #1	9174	1987	1918 Ford Runabout, White Body, Black Trim	110
Kingsport, Tennessee (Citivan)	7682UA	1990	1932 Ford Panel, Black Body, Silver Trim	40
Kodak #1	9985	1987	1905 Ford Delivery Van, Gold Spokes, Yellow Body, Red Trim	200
Kodak #2	9985	1987	1905 Ford Delivery Van, Red Spokes, Yellow Body, Red Trim	100
Kraft Dairy Group	9675	1989	1917 Model T Van, White On White.	35
Kroger Foods	9511	1986	1013 Model T Van, White Body, Blue Trim	65
Kuiken Brothers	9362	1985	1913 Model T Van, White Body, Red Trim	30
Lake Odessa, Michigan, Centennial	9519	1986	1913 Model T Van, White Body, Blue Trim	30
Leidy's	9578UO	1989	1937 Ford Tractor Trailer, White.	30
Leinenkugel's	7569UO	1990	1918 Ford Runabout With Barrels, Red, Black Body, White Trim	80
Lennox #1	9461	1986	1913 Model T Van, White Body, Red Trim	135
Lennox #2	9192	1987	1932 Ford Panel, White Body, Red Trim	210
Lennox #3	9793	1988	1918 Ford Runabout, Red Body, White Trim	45
Lennox #4 Limited Edition	9323	1989	1905 Ford Delivery Van, White Body, Red Trim	25
Lennox #5 Limited Edition	7561UA	1990	1926 Mack Truck, White Body, Red Trim	30
LePage Glue	2120	1984	1913 Model T Van, Yellow Body, Red Trim	35
Light Commercial Vehicle Assn.	2123UO	1990	1913 Model T Van, White Body, Green Trim	35
Link-Belt	2107UP	1990	1918 Ford Runabout, Red Body, Black Trim	40
Lion Coffee #1	9306	1988	1913 Model T Van, Red On Red.	65
Lion Coffee #2 (125th Anniversary)	9434	1989	1905 Ford Delivery Van, White Body, Green Trim	35
Lipton Tea #1	7505	1989	1913 Model T Van, White Body, Red Trim	165
Little Debbie #1	9377	1987	Step Van, White On White.	100
Little Debbie #2	9377UO	1990	Step Van, White On White.	70
Lolli Pups	2146	1984	1913 Model T Van, Yellow Body, Brown Trim	40
Longview, Illinois	9704	1988	1917 Model T Van, White Body, Blue Trim	20
Loras College, Limited Edition	9679	1989	1913 Model T Van, White Body, Purple Trim	35
Los Angeles Times	7667UA	1990	1917 Model T Van, Blue On Blue.	45
Louisiana State University	9516	1986	1913 Model T Van, White Body, Purple Trim	50
MAC Tools	9608	1988	1950 Chevy Panel, Grey On Grey.	315
Mace Brothers	9391	1986	1913 Model T Van, White Body, Red Trim	55
Madison Electric, Limited Edition	9589UA	1990	1913 Model T Van, Blue Body, White Trim	105
Manitoba, Canada	9220	1985	1913 Model T Van, White Body, Green Trim	30
Marathon Oil	9783	1988	1926 Mack Truck, Silver Body, Red Trim	375
Marshall Fields	1650	1983	1913 Model T Van, white body, green Trim	65
Marshall Fields	1650	1983	1913 Model T Van, White Body, Green Trim, Boxed With C, Y.	125
Marshfield, Wisconsin, Chamber Of Commerce	9658	1989	1926 Mack Truck, White Body, Green Trim	25
Martin's Potato Chips	9856	1988	Step Van, White On White.	45
Massey Ferguson	1122	1984	1926 Mack Truck, Tan Body, Black Trim	35
Massey Ferguson	1089	1989	1918 Ford Runabout, Red Body, Black, Yellow Trim	25
Massey Ferguson	1348	1990	1913 Model T Van	15
Massey-Harris	1092	1987	1913 Model T Van, Gold Body, Red Trim	25
Matco	9659	1989	Step Van	65

BANKS

Ertl Banks

NAME	NUMBER	YEAR	DESCRIPTION	MINT
Maurice's	9476	1987	1932 Ford Panel, Grey Body, Maroon Trim	35
Mayer's Well Drilling	9267	1987	1913 Model T Van, White Body, Blue Trim	35
McGlynn Bakery	9495	1987	Step Van, White On White.	85
McGlynn Bakery #1	2112	1984	1913 Model T Van, Red Spokes, White Body, Black Trim	45
McGlynn Bakery #2	2112	1986	1913 Model T Van, Black Spokes, White Body, Black Trim	35
McLain Boiler - 1 Of 3 Piece Set	21122R	1990	1905 Ford Delivery Van, White Body, Red Trim	30
McLain Boiler - 1 Of 3 Piece Set	2122RO	1990	1917 Model T Van, White Body, Blue Trim	30
Meijer Foods	9352	1985	1913 Model T Van, Red Body, White Trim	35
Mellon Bank	9318	1987	1913 Model T Van, White Body, Green Trim	25
Mellon Bank	1358UO	1990	1950 Ford Delivery Van, White Body, Green Trim	35
Merit Oil	9980	1988	1926 Mack Tanker, White Body, Red Trim	135
Merita Bread (American Bakeries)	9316	1987	1913 Model T Van, White Body, Red Trim	35
Merita Bread (American Bakeries)	1316	1987	1917 Model T Van, White On White.	35
Meyer's Funeral Home	9568UA	1990	1932 Ford Panel, Black Body, Silver Trim	40
Michigan Milk Producers Assn.	1325	1983	1913 Model T Van, White Body, Red Trim	95
Michigan Sesquicentennial	9172	1987	1913 Model T Van, White Body, Blue Trim	35
Michigan State University, Rose Bowl	9026	1988	1917 Model T Van, White On White.	35
Mike's Trainland #1	9850	1988	1937 Ford Tractor Trailer, White On White.	70
Mike's Trainland #2	9411	1989	1913 Model T Van, White Body, Red Trim	30
Mike's Trainland #3	9221UO	1990	1905 Ford Delivery Van, Blue On Blue.	30
Minnesota Industrial Tools	9308UO	1990	Step Van, silver.	35
Minnesota Vikings	1246	1984	1913 Model T Van, White Body, Purple Trim	55
Missouri Tourism, "Wake Up To Missouri"	2143Uo	1990	1905 Ford Delivery Van, Blue.	25
Mongomery Ward #1	9542	1981	1913 Model T Van, Green Body, Black Trim	200
Mongomery Ward #2	9052	1982	1917 Model T Van, Brown On Brown.	100
Mongomery Ward #3	1363	1983	1926 Mack Truck, Yellow Body, Black Trim	60
Mongomery Ward #4	1367	1983	Horse, Carriage, Blue Body, Black Trim	60
Mongomery Ward #5	2110	1984	1932 Ford Panel, Yellow Body, Green Trim	65
Mongomery Ward #6	9230	1985	1905 Ford Delivery Van, Red Body, Black Trim	60
Monroe Shocks, "Monroe"	7511UO	1990	1913 Model T Van, Yellow Body, Blue Trim	35
Monroe Shocks, "The World Rides Monroe"	7511UP	1990	1913 Model T Van, Yellow Body, Blue Trim	35
Monroe Shocks," America Rides Monroe"	7511UR	1990	1913 Model T Van, Yellow Body, Blue Trim	35
Montana Cantennial	9747	1989	1913 Model T Van, White Body, Red Trim	25
Monticello, Iowa	9364	1986	1913 Model T Van, White Body, Red Trim	40
Moorman Mfg. Co.	9585UO	1990	1905 Ford Delivery Van, White Body, Red Trim	30
Morton Salt	9787	1988	1905 Ford Delivery Van, Blue On Blue.	55
Mountain Dew	7670UO	1990	1950 Chevy Panel, Green Body, White Trim	40
Mrs. Baird's	9474	1987	Horse, Carriage, White Body, Blue Trim	120
Mutual Savings & Loan	9187	1987	1905 Ford Delivery Van, White Body, Blue Trim	25
N.E.W. Hobby	9479	1987	Step Van, White On White.	25
Nabisco Almost Home Cookies	1653	1984	1913 Model T Van, Tan Body, Red Trim	65
Nabisco Premium Saltine Crackers	9699	1981	1913 Model T Van, Yellow Body, Maroon Trim	315
NASA	9467	1989	Step Van, White On White.	40
Nash Finch	9347	1985	1913 Model T Van, White Body, Red Trim	120
National Street Rod Association	7504	1990	1937 Ford Tractor Trailer, White On White.	80

Ertl Banks

NAME	NUMBER	YEAR	DESCRIPTION	MINT
National Van Lines #1	9505	1986	1913 Model T Van, White Body, Blue Trim	65
National Van Lines #2	9119	1987	1932 Ford Panel, White Body, Blue Trim	95
National Van Lines #3	1342	1988	1905 Ford Delivery Van, White Body, Blue Trim	60
Neilson's Ice Cream	2131	1984	1913 Model T Van, White On White.	30
New Brunswick, Canada	9223	1985	1913 Model T Van, White Body, Purple Trim	25
New Holland	9415	1986	1913 Model T Van, Yellow Body, Red Trim	175
New Holland	0379	1987	1905 Ford Delivery Van, Yellow Body, Red Trim	15
New Holland Tractors	9397	1986	1913 Model T Van, Yellow Body, Red Trim	30
New Holland, Limited Edition	9397	1986	1913 Model T Van, Yellow Body, Red Trim	125
Newfoundland, Canada	9224	1985	1913 Model T Van, White Body, Purple Trim	25
Nittany Machinery Association	9642	1989	1913 Model T Van, White Body, Red Trim	30
Norand Data Systems	9207UO	1990	1926 Mack Truck With Crates, White Body, Black Trim	25
North American Van Lines	9030	1988	1937 Ford Tractor Trailer, Red, Tan Body, Red Trim	65
North Dakota Centennial	9045	1989	Horse, Carriage, Tan Body, Black Trim	30
North Dakota Centennial	9690	1989	1913 Model T Van, White Body, Brown Trim	25
Northern Electric	9517	1986	1905 Ford Delivery Van, Black On Black.	40
Nova Scotia, Canada	9222	1985	1913 Model T Van, White Body, Red Trim	25
Oakwood Mobile Homes	9822	1988	1917 Modle T Van, White Body, Blue Trim	30
Oelwein, Iowa, Chamber Of Commerce	9361	1985	1913 Model T Van, White Body, Red Trim	25
Old Country	9478	1987	Step Van, White On White.	95
Old El Paso	7636UO	1990	1905 Ford Delivery Van, Yellow Body, Red Trim	45
Olivet Union, Nazarine University	2115UP	1990	1905 Ford Delivery Van, Yellow Body, Purple Trim	30
Ontario, Canada	9217	1985	1913 Model T Van, White Body, Green Trim	25
Orkin #1	9389	1987	1913 Model T Van, White Body, Red Trim	35
Orkin #2	9842	1988	1905 Ford Delivery Van, White Body, Red Trim	30
Oroweat	9286	1987	Step Van, White On White.	60
Otasco #1	1359	1982	1913 Model T Van, Yellow Body, Black Trim	105
Otasco #2 (65th Anniversary)	1368	1983	1913 Model T Van, Yellow Body, Black Trim	65
Otasco #3	2134	1984	1926 Mack Truck, Yellow Body, Black Trim	100
Otasco #4	9342	1985	1950 Chevy Panel, Yellow On Yellow.	110
Otasco #5	9371	1986	1932 Ford Panel, Yellow Body, Black Trim	110
Otasco #6	9168	1987	1918 Ford Runabout, Yellow Body, Black Trim	75
Otasco #7 (70th Anniversary)	9777	1988	1918 Ford Runabout, White Body, Blue, Black Trim	70
Our Own Hardware	9767	1988	1913 Model T Van, White Body, Red Trim	25
Overnite Trucking	9142	1987	1913 Model T Van, Silver Body, Blue Trim	265
Overnite Trucking	9068	1989	1937 Ford Tractor Trailer, Blue Cab, Silver Trailer.	80
P.J. Valves	7519Uo	1990	1905 Ford Delivery Van, White Body, Blue Trim	30
Parcel Post Mail Service	9467	1981	1913 Modle T Van, Black On Black.	140
Peavey	1357	1982	1913 Model T Van, White Body, Red Trim	25
Penn State	9263	1987	1917 Model T Van, White Body, Blue Trim	160
Penn State	9512	1989	1918 Ford Runabout, White Body, Blue Trim	75
People's National Bank	9877	1988	1913 Model T Van, White Body, Green Trim	145
People's National Bank	9581UO	1990	1918 Ford Runabout, Blue Body, White Trim	45
Pepsi Cola	1314	1987	1917 Model T Van, White Body, Blue Trim	125
Pepsi Cola	9736	1988	1905 Ford Delivery Van, White Body, Blue Trim	80
Pepsi Cola	6936	1989	1918 Ford Runabout, Blue Body, Black, White Trim	55
Pepsi Cola	9635	1989	1950 Chevy Panel, Narrow Tires, Blue Body, White Trim	75

BANKS

BANKS

Top to bottom: Dyersville, Iowa, Chevrolet school bus; New Holland 1905 Ford Delivery Van; Pocono Antique Bazaar 1913 Ford Model T Van; Sunray 1930 Diamond T Tanker.

Ertl Banks

NAME	NUMBER	YEAR	DESCRIPTION	MINT
Pepsi Cola I	9637	1989	1932 Ford Panel, White Body, Blue Trim	55
Pet Milk	1652	1983	1913 Model T Van, Orange Spokes, White Body, Black Trim	85
Pet Milk	1652	1984	1913 Model T Van, Red Spokes, White Body, Black Trim	65
Petty Enterprises #1	9573UO	1989	1913 Model T Van, Blue Body, Red Trim	200
Petty Enterprises #2	9574UO	1989	1913 Model T Van, Blue On Blue.	350
Philadelphia Cream Cheese	9835	1988	1913 Model T Van, Silver On Silver.	45
Phillips 66 #1 (Old Logo)	9407UP	1990	1931 Hawkeye Tanker, Orange Body, Black Trim	185
Pittsburgh Steelers	1298	1984	1913 Model T Van, White Body, Black Trim	55
Pizza Today Magazine	9933UO	1990	1932 Ford Panel, Yellow Body, Black Trim	35
Pocono Antique Bazaar	2917	1990	1913 Model T Van, White Body, Brown Trim	25
Poynors Home & Auto	9273	1987	1932 Ford Panel, White Body, Blue Trim	30
Preston Trucking Co.	9579UO	1990	1937 Ford Tractor Trailer, Orange.	45
Prince Edward Island, Canada	9219	1985	1913 Model T Van, White Body, Orange Trim	25
Protivin, Iowa, Holy Trinity	9784	1988	1918 Ford Runabout, White Body, Black Trim	40
Publix	9431	1986	1950 Chevy Panel, White Body, Green Trim	95
Publix	9436	1986	1905 Ford Delivery Van, White Body, Green Trim	35
Publix	9718	1988	1937 Ford Tractor Trailer, White Body, Green Trim	40
Publix	9693	1989	1926 Mack Truck With Crates, Green Body, Brown Trim	30
Publix	9698	1989	1950 Chevy Tractor Trailer With Reefer, White Body, Green Trim	35
Publix #1	1337	1981	1913 Model T Van, White Body, Green Trim	45
Publix #1	2115	1984	1926 Mack Truck, Black Tires, Red Spokes, White Body, Green Trim	125
Publix #1	9248	1985	1932 Ford Panel, White Body, Green Trim	45
Publix #1	9149	1987	1918 Ford Runabout, White Body, Green Trim	55
Publix #2	1337	1984	1913 Model T Van, Red Wheels, White Body, Green Trim	35
Publix #2	2115	1985	1926 Mack Truck, White Tires, Green Spokes, White Body.	100
Publix #2	9719	1988	1926 Mack Tanker, White Cab, Body, Green Trim	35
Publix #2 (Pleasure)	9723	1988	1918 Ford Runabout, White Body, Green Trim	25
Publix #2 Food & Pharmacy	9147	1987	1932 Ford Panel, White Body, Green Trim	35
Publix #3 (The Deli)	9694	1989	1918 Ford Runabout, White Body, Brown Trim	25
Publix #3 Limited Edition	1337	1985	1913 Modle T Van, White Body, Green Trim	95
Publix Danish Bakery	9249	1985	1913 Model T Van, White Body, Brown Trim	35
Publix Danish Bakery	9435	1986	1932 Ford Panel, White Body, Brown Trim	30
Publix Danish Bakery	9152	1987	1905 Ford Delivery Van, White Body, Brown Trim	30
Publix Danish Bakery	9722	1988	1950 Chevy Panel, Brown Body, White Trim	35
Publix Danish Bakery	9697	1989	1917 Model T Van, White Body, Brown Trim	25
Publix Danish Bakery #2	7690DO	1990	1932 Ford Panel, With Spar Tire, White Body, Orange Trim	35
Publix Dari-Fresh	9430	1986	1926 Mack Tanker, White Body, Green Cab, Trim	45
Publix Dari-Fresh	9001	1989	1905 Ford Delivery Van, White Body, Green Trim	25
Publix Dari-Fresh	9695	1989	1937 Ford Tanker Trailer, White Body, Green Trim	35
Publix Dari-Fresh	7686DO	1990	1931 Hawkeye Tanker, White Body, Green Trim	25
Publix Deli	7689DO	1990	1905 Ford Delivery Van, Orange Body, Brown Trim	20

Ertl Banks

BANKS

NAME	NUMBER	YEAR	DESCRIPTION	MINT
Publix Floral	7687DO	1990	1918 Ford Runabout, White Body, Green Trim	20
Publix Food & Pharmacy	9721	1988	1905 Ford Delivery Van, White Body, Black Trim	25
Publix Food & Pharmacy	9692	1989	1950 Chevy Panel, Green Body, White Trim	35
Publix Food & Pharmacy	7688DO	1990	1913 Model T Van, Green Body, White Trim	20
Publix Produce	9720	1988	1926 Mack Truck, White Body, Green Trim	30
Publix Produce	9696	1989	1932 Ford Panel, Green Body, White Trim	45
Publix Produce	7685DO	1990	1931 Hawkeye Truck, White Body, Green Trim	20
Publix Produce Limited Edition	9148	1987	1917 Model T Van, White Body, Green Trim	35
Quakertown National Bank #1	9515	1986	1913 Model T Van, White Body, Blue Trim	75
Quakertown National Bank #2	9291	1987	1932 Ford Panel, White Body, Blue Trim	55
Quakertown National Bank #3	9979	1988	1905 Ford Delivery Van, White Body, Blue Trim	70
Quakertown National Bank #4	9417	1989	1918 Ford Runabout, Blue Body, White Trim	45
Quakertown National Bank #5	1370UP	1990	1917 Model T Van, White Body, Blue Trim	35
Quality Farm & Fleet #1	9491	1987	1913 Model T Van, White Body, Red Trim	30
Quality Farm & Fleet #2	9609	1988	1905 Ford Delivery Van	30
Quality Farm & Fleet #3	9609	1989	1918 Ford Runabout, White Body, Red Trim	25
Quebec, Canada	9216	1985	1913 Model T Van, White Body, Blue Trim	25
R.C. Cola	9827	1988	1917 Model T Van, White Body, Blue Trim	40
Ragrai XVII	9108	1989	1905 Ford Delivery Van, White Body, Blue Trim	25
RCA #1	9314	1987	1913 Model T Van, White Body, Black Trim	60
RCA #2	9315	1987	1905 Ford Delivery Van, White Body, Red Trim	55
RCA #3	9275	1987	1926 Mack Truck, White Body, Red Trim	40
RCA #4	1344	1988	1918 Ford Runabout, Black Body, White Trim	35
RCA #5	1343	1988	1917 Model T Van, White Body, Red Trim	35
RCA #6	9621	1989	1932 Ford Panel, White Body, Black Trim	35
Red Crown Gasoline #1 (1st Run)	7654UO	1990	1931 Hawkeye Tanker, Red On Red	75
Red Crown Gasoline #1 (2nd Run)	7654UO	1990	1931 Hawkeye Tanker, Red On Red	60
Red Rose Tea	2130	1984	1913 Model T Van, Red On Red	40
Renninger's Antique Market, Adamstown, PA	9712	1988	1905 Ford Delivery Van, White Body, Red Trim	35
Renninger's Antique Market, Adamstown, PA	9894	1989	1918 Ford Runabout With Barrels, White Body, Red Trim	35
Renninger's Antique Market, Adamstown, PA	7556UO	1990	1932 Ford Panel, White Body, Red Trim	30
Renninger's Antique Market, Kutztown, PA	9714	1988	1905 Ford Delivery Van, Black Body, Red Trim	35
Renninger's Antique Market, Kutztown, PA	9895	1989	1918 Ford Runabout With Barrels, Black Body, Red Trim	35
Renninger's Antique Market, Kutztown, PA	7555	1990	1932 Ford Panel, Black Body, Red Trim	30
Renninger's Antique Market, Mt. Dora, FL	9713	1988	1905 Ford Delivery Van, Brown On Brown	35
Renninger's Antique Market, Mt. Dora, FL	9896	1989	1918 Ford Runabout With Barrels, Tan Body, Brown Trim	35
Renninger's Antique Market, Mt. Dora, FL	7557UO	1990	1932 Ford Panel, Beige Body, Brown Trim	35
Republican Central Comm., Douglas County	9369	1986	1913 Model T Van, White Body, Red Trim	75
Republican Party, Election 198	9805	1988	1905 Ford Delivery Van, White Body, Red Trim	150
Reynolds Aluminum	9731	1988	1917 Model T Van, White On White	55
Reynolds Wrap	9162	1987	1913 Model T Van, Silver Body, Blue Trim	285
Richlandtown	9290	1987	1926 Mack Tanker, White Body, Red Trim	65

Ertl Banks

NAME	NUMBER	YEAR	DESCRIPTION	MINT
Richlandtown	2969UO	1990	1905 Ford Delivery Van, White Body, Red Trim	30
Ringling Brothers Circus	9027	1988	1913 Model T Van, Red On Red	175
Ringling Brothers Circus	9726	1988	1937 Ford Tractor Trailer, White Body, Yellow Trim	145
Riverview Nursery	9121	1987	1926 Mack Tanker, White Body, Red Trim	65
Riverview Nursery (Purina)	9804	1988	1937 Ford Tractor Trailer, White On White	45
Rogersville, Tennessee	9082	1989	1932 Ford Panel With Spare Tire, Black Body, Silver Trim	80
Rogersville, Tennessee, Heritage Days	2108UP	1990	1918 Ford Runabout, Red Body, Black Trim	35
Round Top Arms	9418	1989	1918 Ford Runabout, White Body, Black Trim	25
Rubschlager Bakery	9446	1986	1913 Model T Van, White Body, Red Trim	40
S&F Toys #1	9205	1984	1913 Model T Van, Grey Body, Maroon Trim	20
S&F Toys #2	9205	1989	1913 Model T Van, Grey Body, Maroon Trim	20
Sacred Heart Church	9340	1985	1913 Model T Van, White Body, Red Trim	35
Safety Kleen	9289	1987	Step Van, Yellow On Yellow	55
Safety Kleen (Drycleaner Service)	9506	1986	1913 Model T Van, Yellow Body, Black Trim	45
Safety Kleen (Parts Cleaner Service)	9506	1986	1913 Model T Van, Yellow Body, Black Trim	45
Saia Trucking	9462	1986	1913 Model T Van, White Body, Red Trim	415
Salvation Army	2101	1987	Step Van, White On White	110
Salvation Army	9109	1987	1913 Model T Van, White Body, Red Trim	100
San Francisco 49ers	1297	1984	1913 Model T Van, Gold Body, Red Trim	105
Sara Lee	9941	1989	1950 Chevy Panel, Maroon Body, Black Trim	70
Sasco Aloe Vera	9354	1985	1917 Model T Van, Grey Body, Maroon Trim	25
Saskatchewan, Canada	9225	1985	1913 Model T Van, White Body, Orange Trim	20
Schneider Meats	1332	1984	1913 Model T Van, Orange Body, Blue Trim	50
Schneider Meats	9228	1985	1926 Mack Truck, Orange Body, Blue Trim	125
Schneider Meats	9229	1985	1932 Ford Panel, Orange Body, Blue Trim	115
Schneider Meats	9445	1986	1950 Chevy Panel, Orange Body, Black Trim	150
Schwan's Ice Cream	9210	1984	1950 Chevy Panel, Brown On Brown	165
Scott Tissue #1	9652	1989	1917 Model T Van, Gold Spokes, White Body, Blue Trim	85
Scott Tissue #2	9652UP	1990	1917 Model T Van, Silver Spokes, White Body, Blue Trim	50
Sealed Power Piston Rings	2121UO	1990	1905 Ford Delivery Van, White Body, Red Trim	25
Sealed Power Speed-Pro	1663UP	1990	1905 Ford Delivery Van, Black Body, Red Trim	25
Sears	2129	1984	1913 Model T Van, Black Body, White Trim	30
Servistar	9817	1988	1905 Ford Delivery Van, Grey Body, Red Trim	45
Servistar	9036	1989	1913 Model T Van, Grey Body, Blue Trim	25
Servistar	7510UA	1990	1918 Ford Runabout, Grey Body, Black, Red Trim	20
Seven-Up	1662	1988	1913 Model T Van, White Body, Green Trim	180
Shelby Life Insurance	1347	1988	1905 Ford Delivery Van, White Body, Blue Trim	85
Shopko	9039	1989	1913 Model T Van, White Body, Blue Trim	30
Shoprite	9426	1986	1913 Model T Van, Yellow Body, Red Trim	35
Shoprite	9163	1987	1918 Ford Runabout, Yellow Body, Red Trim	30
Shoprite	9711	1988	1905 Ford Delivery Van, Yellow Body, Red Trim	15
Shoprite	9666	1989	1926 Mack Truck With Crates, Yellw Body, Red Trim	25
Sidney Fire	1335	1988	1913 Model T Van, White Body, Red Trim	20
Silver Springs Flea Market	9619	1988	1918 Ford Runabout, White Body, Black Trim	30
Silver Springs Speedway	2972UO	1990	1932 Ford Panel, White Body, Red Trim	55
Sinclair	2119UP	1990	1926 Mack Tanker, White Body, Green Trim	80
Sinclair	2120UO	1990	1926 Mack Tanker, White Body, Green Trim	55
Smoke Craft	9493	1987	1918 Ford Runabout, White Body, Black Trim	30

BANKS

Ertl Banks

NAME	NUMBER	YEAR	DESCRIPTION	MINT
Smoke Craft	9494	1987	1905 Ford Delivery Van, White Body, Red Trim	30
Smoke Craft #1	9164	1987	1913 Model T Van, White Body, Blue Trim	55
Smoke Craft Meats	9653	1989	Step Van, White On Whie	40
Smoke Craft Meats	9654	1989	1932 Ford Panel, White Body, Blue Trim	35
Smokey The Bear #1	9124	1989	1913 Model T Van, White Body, Green Trim	40
Smokey The Bear #2	9123	1989	1918 Ford Runabout, Green, White Body, Black Trim	45
Sohio Gas	9269	1987	1926 Mack Tanker, White Body, Red Trim	365
South Dakota Centennial	9879	1988	Horse, Carriage, Blue Body, Black Trim	35
Southern States Oil	9199	1987	1926 Mack Tanker, Silver Body, Red Trim	275
Southern States Oil	9797	1988	1926 Mack Truck, Grey Body, Red Trim	50
Southern States Oil	9322	1989	1918 Ford Runabout, Red Body, Black Trim	60
Southern States Oil	7628UO	1990	1937 Ford Tractor Trailer, Red Body, Black Trim	45
Sparklettes Water	9741	1988	1926 Mack Truck, Grenn On Green	30
Spartan Food Stores	9247	1985	1913 Model T Van, Green On Green	30
St. Columbkille	9304	1986	1913 Model T Van, White Body, Blue Trim	45
Steamtown, USA	9167	1987	1926 Mack Tanker, White Body, Red Trim	100
Steelcase	9604	1988	1905 Ford Delivery Van, Blue Body, Silver Trim	30
Steelcase	9041	1989	1926 Mack Truck With Crates, Black Body, Brown Trim	40
Steelcase	7502UO	1990	1937 Ford Tractor Trailer, White Body, Black Trim	95
Steelcase	7503UO	1990	1932 Ford Panel, Blue Body, Silver Trim	95
Steelcase #1	1657	1982	1913 Model T Van, Red Spokes, Blue Body, Silver Trim	95
Steelcase #2	9265	1987	1918 Ford Runabout, Blue Body, Black Trim	45
Steelcase #2	1657	1989	1913 Model T Van, Chrome Spokes, Blue Body, Silver Trim	45
Steelcase #3	1657	1989	1913 Model T Van, Red Spokes With Chrome Trim, Blue Body, Silver	100
Stevens Brothers Cartage (Bekins)	9841	1988	1926mack Truck, White Body, Black Trim	35
Still Transfer Company	9224	1990	1937 Ford Tractor Trailer, White Body, Red Trim	25
Strawberry Point, Iowa	9353	1990	1913 Model T Van, White Body, Red Trim	45
Stroh's Beer	7679UO	1990	1918 Ford Runabout With Barrels, Red, Black Body, Red Trim	55
Sunbeam Bread	9518	1986	1913 Model T Van, White Body, Red Trim	100
Sunbeam Bread	9631	1989	1913 Model T Van, White Body, Red Trim	45
Sunbeam Bread	1330	1990	1932 Ford Panel With Spare Tire, Blue Body, Yellow Trim	35
Sunbeam Bread Step Van #1	9638	1989	Step Van, Yellow Body, Blue Trim	60
Sunholidays Travel	9618	1988	1918 Ford Runabout, White Body, Black Trim	35
Sunmaid Raisins	9575	1989	1905 Ford Delivery Van, Red On Red	30
Sunmaid Raisins	9576	1990	Step Van, Red On Red	40
Super Valu #1	9663EO	1990	1932 Ford Panel With Spare Tire, White Body, Red Trim	25
Support Your Local Fire Department	7613UO	1990	1950 Chevy Panel, Red On Red	30
Support Your Local Police	7506UO	1990	1950 Chevy Panel, Black On Black	25
Support Your Local Sheriff	7540UO	1990	1950 Chevy Panel, Black On Black	30
Sussex County Farm & Horse Show (Coors)	7632UO	1990	1905 Ford Delivery Van, White Body, Red Trim	60
Swiss Valley #1	2140	1984	1913 Model T Van, White Body, Red Trim	85
Swiss Valley #3	9847	1988	1095 Ford Delivery Van, White Body, Orange Trim	35
Tabasco (McIlhenny)	9878	1988	1905 Ford Delivery Van, White Body, Orange Trim	25
Tabasco (McIlhenny)	9078	1989	1918 Ford Runabout With Barrels, Orange Body, Black/White Trim	30

BANKS

Top to bottom: Anheuser Busch Inc. 1918 Ford Barrel Runabout; John Deere 1926 Mack truck; Riverview Nursery 1937 Ford Tractor-Trailer; Sasco 1917 Ford Model T Van.

Ertl Banks

NAME	NUMBER	YEAR	DESCRIPTION	MINT
Tennessee Homecoming 1986	9420	1986	1913 Model T Van, White Body, Red Trim	175
Terminix International #1	9465	1986	1913 Model T Van, White Body, Orange Trim	45
Terminix International #2	9840	1988	1917 Model T Van, White On White	25
Terminix International #3	9086	1989	1905 Ford Delivery Van, White Body, Orange Trim	30
Texaco #1	2128	1984	1913 Model T Van, Silk Screened, White Body, Red Trim	825
Texaco #1	2128	1986	1913 Model T Van, Silk Screened, White Body, Red Trim	825
Texaco #1 Sampler	2128	1984	1913 Model T Van, Applied Label, White Body, Red Trim	1550
Texaco #2	9238	1985	1926 Mack Tanker, White Body, Red Trim	525
Texaco #2 Sampler	9238	1985	1926 Mack Tanker, White Body, Red Trim	1050
Texaco #3	9396	1986	1932 Ford Panel, White Body, Red Trim	310
Texaco #3 Sampler	9396	1986	1932 Ford Panel, White Body, Red Trim	675
Texaco #4	9321	1987	1905 Ford Delivery Van, Black Body, Red Trim	165
Texaco #4 Sampler	9376	1987	1905 Ford Delivery Van, White Body, Red Trim	575
Texaco #5	9740	1988	1918 Ford Runabout, Black Body, Red Trim	100
Texaco #5 Sampler	9740	1988	1918 Ford Runabout, Gold Spokes, Black Body, Red Trim	250
Texaco #6	9040VO	1989	1926 Mack Truck, Marker 1925 Mack, Red Body, Black Trim	85
Texaco #7	9330VO	1990	1930 Diamond T Tanker, Embossed, Red Body, Black Trim	75
Texaco #7 Sampler	9330	1990	1930 Diamond T Tanker, No #7 Mark On Base, Red Body, Black Trim	185
Thompson Trucking	9613	1989	1937 Ford Tractor Trailer, White Body, Yellow Trim	30
Thunderhills #1	9268	1987	1913 Model T Van, White Body, Blue Trim	350
Thunderhills #2	9792	1988	1905 Ford Delivery Van, White Body, Red Trim	285
Thunderhills #3	9326	1989	1926 Mack Truck, White Body, Black Trim	265
Thunderhills #4	7566UO	1990	1918 Ford Runabout, White Body, Black Trim	225
Tide Soap	7509	1990	1913 Model T Van, Orange	55
Tioga County, NY, Bicentennial	1321	1990	1913 Model T Van, White Body, Blue Trim	30
Tisco	9948UO	1989	1926 Mack Truck, White Body, Green Trim	100
Tisco	9949UO	1989	1918 Ford Runabout, White Body, Black Trim	30
Tisco	9649UO	1990	Step Van, White Body, Red Trim	35
Tisco	9983UA	1990	1905 Ford Delivery Van, White Body, Red Trim	115
Tisco	9983UO	1990	1917 Model T Van, White Body, Black Trim	50
Titleist Golf Balls	9489	1987	1913 Model T Van, White Body, Red Trim	65
Tom's Snack Foods	1338UP	1990	Step Van, New Logo, Tan Body	30
Tom's Snack Foods	1337UO	1990	Step Van, Old Logo, Tan Body, Red Trim	30
Toy Farmer #1	1664	1983	1913 Model T Van, White Body, Blue Trim	30
Toy Farmer #2 (10th Anniversary)	9483	1987	1913 Model T Van, Red Body, Blue Trim	25
Toy Shop #1	9442	1989	1926 Mack Truck, White Body, Red Trim	30
Toy Shop #2	2118	1990	1913 Model T Van, White Body, Red Trim	25
Toy Tractor Times #1	9480	1988	1937 Ford Tractor Trailer, White On White.	30
Toy Tractor Times #2	9480	1988	1937 Ford Tractor/Trailer, White Body, Trim	30
Toymaster	2105	1984	1913 Model T Van, Yellow Body, Trim	35
Tractor Supply Company #1	1349	1982	1913 Model T Van, White Body, Red Trim	40
Tractor Supply Company #1	9355	1986	1905 Ford Delivery Van, Red Body, White Trim	25
Tractor Supply Company #10	9133	1989	1926 Mack W/Crates, Red Body, Black Trim	35
Tractor Supply Company #2	1349	1983	1913 Model T Van, Red Body, White Trim	40
Tractor Supply Company #3	2121	1984	1926 Mack Truck, Red Body, White Trim	55
Tractor Supply Company #4	9208	1984	1932 Ford Panel, White Body, Red Trim	60

Ertl Banks

NAME	NUMBER	YEAR	DESCRIPTION	MINT
Tractor Supply Company #5	9207	1985	1950 Chevy Panel, White Body, Trim	65
Tractor Supply Company #6	9355	1986	1917 Model T Van, Red Body, White Trim	30
Tractor Supply Company #6	9357	1986	1905 Ford Delivery Van, White Body, Red Trim	25
Tractor Supply Company #7	9356	1986	1917 Model T Van, White Body, Red Trim	25
Tractor Supply Company #8	9530	1986	1918 Runabout, White Body, Red Trim	30
Tractor Supply Company #9	2100	1987	1926 Mack Truck, White Body, Red Trim	40
Traer, Iowa Lions Club	9416	1986	1913 Model T Van, White Body, Purple Trim	45
Trappey "Bull Brand" Hot Sauce	1311	1990	1905 Ford Delivery Van, Cream Body, Red Trim	35
Tremont Area Ambulance Association	9754	1989	1913 Model T Van, White Body, Blue Trim	30
Tremont Area Ambulance Association	7665UO	1990	1932 Ford Panel, White Body, Blue Trim	40
Tropicana Orange Juice	7637UO	1990	1905 Ford Delivery Van, White Body, Orange Trim	75
Trucklite	9898	1988	1937 Ford Tractor/Trailer, White Body, Trim	30
True Value #1	1348	1982	1913 Model T Van, White Body, Blue Trim	225
True Value #2	1362	1983	1926 Mack Truck, Red Body, Black Trim	105
True Value #3	1296	1984	1950 Chevy Panel (Marked 1948 Chevy Panel), Light Brown Body	80
True Value #4	9232	1985	1932 Ford Panel, White Body, Blue Trim	65
True Value #5	9366	1986	1918 Runabout, White Body, Blue Trim	40
True Value #6	9301	1987	1905 Ford Delivery Van, White Body, Red Trim	35
True Value #7	9105	1988	1926 Mack W/Crates, Brown Body, Red Trim	25
True Value #8	9623	1989	1918 Barrel Runabout, Red Body, Black Trim	25
True Value #9	7625EO	1990	Horse Team & Tanker, Red Body, Black Trim	25
Trusthworty Hardware #3 - Ltd. Ed.	9375	1987	1926 Mack Truck, White Body, Brown Trim	40
Trustworthy Hardware #1 - Ltd. Ed.	9260	1985	1917 Model T Van, White Body, Brown Trim	195
Trustworthy Hardware #2	9395	1986	1905 Ford Delivery Van, White Body, Brown Trim	55
Trustworthy Hardware #4 - Ltd. Ed.	9774	1988	1918 Runabout, Brown Body, Black Trim	30
Trustworthy Hardware #5 W/Spare	9744	1989	1913 Model T Van, White Body, Brown Trim	25
Trustworthy Hardware #6 Ltd. Ed. SAMPLE	9100YA	1990	1932 Ford Panel, Brown Body, Silver Trim	75
Trustworthy Hardware #6 W/Spare Ltd. Ed.	9100UA	1990	1932 Ford Panel, White Body, Brown Trim	35
Turner Hydraulics	2106UP	1990	1931 Hawkeye Truck, White Body, Black Trim	30
U.S. Mail	1659	1988	1905 Ford Delivery Van, White Body, Blue Trim	95
U.S. Mail	9727	1988	1937 Ford Tractor/Trailer, White Body, Trim	40
U.S. Mail	9730	1988	1917 Model T Van, White Body, Trim	40
U.S. Mail	9532	1989	1913 Model T Van, White Body, Trim	25
U.S. Mail	9051	1989	1932 Ford Panel, Whtie Body, Dark Blue Trim	90
U.S. Mail	9209UO	1990	Step Van, White Body, Red/White/Black Trim	110
U.S. Mail #1 (Limited Edition)	9532	1987	1913 Model T Van, White Body, Trim	35
U.S. Mail #2 Ltd. Ed. W/Spare	9843	1988	1918 Runabout, Whtie Body, Trim	45
U.S. Mail #3 Ltd. Ed. W/Spare	9052	1989	1932 Ford Panel, White Body, Blue Trim	25
U.S. Mail #4 Ltd. Ed.	7641UA	1990	1905 Ford Delivery Van, Blue Body, Red Trim	30
U.S. Mail (Express Mail)	9169	1987	1926 Mack Truck, White Body, Black Trim	115
U.S. Mail (Express Mail)	9893	1988	1937 Ford Tractor/Trailer, White Body, Trim	85
U.S. Mail (With Reversed Eagle)	9532	1986	1913 Model T Van, White Body, Trim	285
U.S. Mail - 1/43 Dime Bank	2136	1990	1932 Ford Panel, White Body, Red/White/Black Trim	10
U.S. Mail - W/O Spare	9296	1987	1918 Runabout, White Body, Black Trim	85
U.S.A. Baseball Team	9795	1988	1905 Ford Delivery Van, White Body, Navy Blue Trim	60

Ertl Banks

NAME	NUMBER	YEAR	DESCRIPTION	MINT
Unique Gardens	9202	1984	1913 Model T Van, White Body, Red Trim	45
Unique Gardens	9122	1987	1926 Mack Tanker, White Body, Red Trim	65
United Hardware	9299	1987	1950 Chevy Panel, Red Body, White Trim	50
United Van Lines #1	2100	1984	1913 Model T Van, White Body, Black Trim	55
United Van Lines #2	9227	1985	Horse & Carriage, White Body, Black Trim	40
United Van Lines #3	9393	1986	1905 Ford Delivery Van, White Body, Black Trim	25
United Van Lines #4 - Ltd. Ed.	9715	1988	1917 Model T Van, White Body, Black Trim	30
United Van Lines #5 - Ltd. Ed.	9096	1989	1918 Runabout, White Body, Black Trim	25
University Of Florida "Gators"	9821	1988	1905 Ford Delivery Van, White Body, Orange Trim	55
University Of Indiana	9497	1987	1913 Model T Van, White Body, Red Trim	30
University Of Kansas (Jayhawks)	9605	1988	1913 Model T Van, White Body, Red Trim	40
University Of Kansas - '88 Nat'l Champs	9816	1988	1905 Ford Delivery Van, White Body, Blue Trim	40
University Of Michigan "Wolverines"	9513	1989	1913 Model T Van, Yellow Body, Blue Trim	40
University Of Nebraska (Go Big Red)	1330	1982	1913 Model T Van, White Body, Red Trim	70
University Of Northern Iowa	9300	1986	1913 Model T Van, White Body, Trim	400
University Of Wisconsin	9655	1989	1913 Model T Van, White Body, Red Trim	30
V & S Variety Stores #1	9622	1989	1905 Ford Delivery Van, White Body, Orange Trim	35
V & S Variety Stores #2	7625EO	1990	1918 Runabout, Black Body, Orange Trim	25
Valley Forge	9616	1989	1926 Mack Truck, White Body, Red Trim	100
W.R. Meadows Inc.	2968	1990	1916 Model T Van, Green Body, White Trim	90
W.W. Irwin Gasoline Maintenance Company	2142UP	1990	1905 Ford Delivery Van, Light Blue Body	30
Washington State Centennial	9137	1989	1913 Model T Van, White Body, Red Trim	25
Washington Suburban Sanitary Commission	7522UO	1990	1926 Mack Tanker, Black Body, Red Trim	30
Watkin's Inc.	9786	1988	1905 Ford Delivery Van, Black Body, Red Trim	55
Weber's Supermarket	9276	1987	1913 Model T Van, White Body, Red Trim	30
Weil-McLain Boiler - 1 Of 3 Piece Set	2122RO	1990	1926 Mack Truck, Red Body, Black Trim	30
Western Auto	1328	1981	1913 Model T Van, Red Body, Black Trim	110
Wheatbelt Stores - Five Point	9481	1987	1913 Model T Van, Tan Body, Brown Trim	25
Wheelers	1358	1982	1913 Model T Van, White Body, Red Trim	30
White House (National Fruit Products)	9004	1989	1913 Model T Van, Yellow Body, Green Trim	235
White House (National Fruit Products)	9005	1989	1913 Model T Van, Yellow Body, Green Trim	210
Wilson Foods	9897	1988	1918 Runabout, Red Body, Black Trim	35
Winn-Dixie #1 (Red Spokes)	1364	1983	1913 Model T Van, White Body, Green Trim	55
Winn-Dixie #1 (White Spokes)	1364	1984	1913 Model T Van, White Body, Green Trim	45
Winn-Dixie #10 - Ltd. Ed.	9014	1989	1926 Mack W/ Crates, White Body, Green/Brown Trim	30
Winn-Dixie #11 - Ltd. Ed.	9013	1989	1950 Chevy Panel, White Body, Green Trim	35
Winn-Dixie #12 W/Spare Ltd. Ed.	7694	1990	1913 Model T Van, White Body, Green Trim	25
Winn-Dixie #13 Ltd. Ed.	7693	1990	1931 Hawkeye Truck, White Body, Green Trim	25
Winn-Dixie #2	2125	1984	1926 Mack Truck, Green Body, White Trim	60
Winn-Dixie #3	9341	1985	1932 Ford Panel, White Body, Green Trim	55
Winn-Dixie #4	9423	1986	1932 Ford Panel, Green Body, White Trim	45
Winn-Dixie #5	9392	1986	1905 Ford Delivery Van, White Body, Green Trim	55
Winn-Dixie #6	9116	1987	1918 Runabout, Green Body, White Trim	30
Winn-Dixie #7	9117	1987	1905 Ford Delivery Van, Green Body, White Trim	30

Ertl Banks

NAME	NUMBER	YEAR	DESCRIPTION	MINT
Winn-Dixie #8 - Ltd. Ed.	9706	1988	1937 Ford Tractor/Trailer, White Body, Green Trim	30
Winn-Dixie #9 (W/Spare Tire)	9707	1988	1932 Ford Panel, White Body, Green Trim	35
Wonder Bread #1	1660	1982	1913 Model T Van, White Body, Black Trim	75
Wonder Bread #2	9421	1986	1913 Model T Van, White Body, Black Trim	40
Wood Heat	9498	1987	1905 Ford Delivery Van, White Body, Blue Trim	35
Wood Heat	9978	1989	1937 Ford Tractor/Trailer, White Body, Trim	55
Worldwide Products	9114	1987	Step Van, Blue/White Body	75
Worthington Firehouse	7539UO	1990	1905 Ford Delivery Truck, White Body, Red Trim	40
Wwonder Bread #3	9161	1987	1913 Model T Van, White Body, Black Trim	50
Yelton Trucking	9402	1986	1913 Model T Van, White Body, Red Trim	30
Yelton Trucking - Limited Edition	9171	1987	1926 Mack Tanker, White Body, Red Trim	40
Yoder Popcorn	9627	1989	Step Van, White Body, Trim	55
York Peppermint Patties #1 W/ Spare	9069	1990	1932 Ford Panel, Silver Body, Trim	105
York Peppermint Patties #3	9071UO	1990	1937 Ford Tractor/Trailer, Blue Body, Silver Trailer.	105
Yuengling Beer	9429	1986	1913 Model T Van, White Body, Red Trim	160
Yuengling Beer	9176	1987	1905 Ford Delivery Van, White Body, Red Trim	135
Yuengling Beer	9770	1988	1937 Ford Tractor/Trailer	155

Mechanical Tin Toys

Metal toys produced before World War I could be considered works of art as such. If a toy had a multicolored scheme, it was painted by hand.

But the advent of chromolithography changed the way most toys were produced. Chromolithography was actually developed late in the nineteenth century. The technique allowed multicolor illustrations to be printed on tinplate. The flat tin sheets were subsequently molded into toys.

Starting in the 1920s, lithographed tin toys began to dramatically change toy production. American manufacturers began to mass produce these colorful toys and offer them to the buying public at far better prices than those demanded for the classic European toys that had dominated until this time.

With mass production came mass appeal. New tin mechanical toys were based on the characters and celebrities that were popular at the time. The newspaper comic strips and Walt Disney movies provided already popular subject matter for toy marketers.

Among the most well-known makers of mechanical tin toys were Marx, Chein, Lehmann, and Strauss. Others included Courtland, Girard, Ohio Art, Schuco, Unique Art, and Wolverine.

Many of these manufacturers had business relationships with each other. Over the years, some would be found working together, producing toys for others, distributing others' toys or being absorbed by other companies. There appeared to be even some flat-out copying of others' ideas.

One of the advantages of lithography was that it allowed old toys to be recycled in many ways. When a character's public appeal

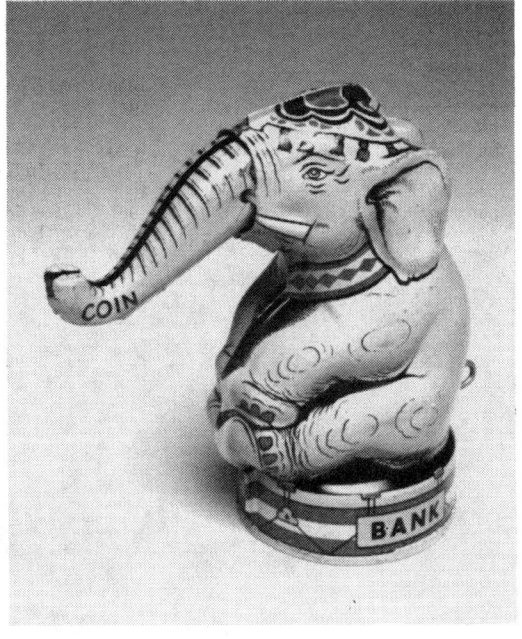

Mechanical Elephant Bank from Chein, 1950s.

began to wane, a new image could be printed on the same old toy and, presto, a new toy. Or when a toy company was absorbed by another, its tired old models could be dusted off and dressed up with new lithography to produce new toys.

Many of the mechanical tin wind-up toys show up in surprisingly similar versions with another manufacturer's name on them.

Of the companies listed here, Marx was no doubt the most prolific. The company's founder, Louis Marx, at one time was employed by another leading toy maker, Ferdinand Strauss. He left Strauss in 1918 to start his own company. Some of his first successes were new versions of old Strauss toys, the Climbing Monkey and Alabama Coon Jigger.

Many of the popular Marx tin wind-ups were based on popular characters. One of the most sought-after is the Merrymakers Band, which was a group of Mickey Mouse-type musicians. Some of the other highly valued character toys are the Amos 'n Andy Fresh Air Taxi, the Donald Duck Duet, Popeye the Champ, and Superman Rollover Airplane.

While Marx went on to produce many, many kinds of toys, other companies, such as Chein, specialized only in inexpensive lithographed tin. And like Marx, Chein also capitalized on popular cartoon characters, producing several Popeye toys, among others. J. Chein and Company, which was founded in 1903, was best known for its carnival-themed mechanical toys. Its Ferris wheel is fairly well known among toy collectors, and was made in several lithographed versions, including one with a Disneyland theme. Chein also produced a number of affordable tin banks.

Girard was founded shortly after Chein, but did not start producing toys until 1918. It subcontracted toys for Marx and Strauss in the 1920s. In fact, several Girard and Marx toys are identical, having been produced in the same plant with different names on them. Marx later took over the company in the 1930s.

New Jersey-based Unique Art is not known for an extensive line of toys, but it produced some that are favorites among tin toy collectors. It, too, reportedly was acquired by Marx at some point.

"Doughboy Tank" tin windup from the 1930s.

There are many other companies that produced lithographed tin toys not included in this section, particularly German and Japanese companies. Lehmann and Schuco, both German firms, are the only non-American toy makers listed here. More lithographed tin toys can be found in the vehicles section of this book.

Prices shown here are for toys in Mint, Excellent, and Good conditions. Toys will usually command a premium over the listed price if they are in their original boxes.

Trends

We talked with Wayne Mitchell of Wayne Mitchell's Toy Shop in Grapevine, Texas, about the state of the tin toy market in 1994. Here is what he had to say.

"The older antique toys are soaring in value," he said. "Three years ago you could buy a Mint-in-the-box Dog Patch Band by Unique Art for $350 to $450. That same toy is now going for $850 to $1,000. Early Marx, Chein, Lindstrom, and Unique Art tin toys, as in the past, continue to bring premium prices. German tin toys by Lehmann, Strauss, and others are by far the most desirable and are heavily sought after by collectors who are able and willing to invest and, in many cases, pay far above current market values.

Penny and Nickel toys are also at a premium, both in price and collectibility.

Marx wind-ups are always in demand, as are Chein. In particular, early Chein toys with the original circle logo have become premium items. Chein Ducks, Easter Bunnies, and so on are very popular.

The market for Japanese toys has softened recently in the U.S. However, Japanese collectors have picked up the slack in both the tin and battery operated toys made in Japan prior to the mid 1960s.

Robots, as always, are hottest in Japan.

Older Japanese celluloid toys, such as the Mickey and Minnie (Mouse) Acrobats, Express Boy, and so on are still nonetheless strongest with American collectors.

Disney is still the king of toys all over the globe in both tin and battery ops. Also highly desirable are all cartoon and character toys — 1960s Flintstone toys are super-strong and will likely soar in value as a result of the new movie.

Tin cars and trucks are a little soft at present. However, the older ones from the 1800s through 1930s are on the increase."

CLASSIC TIN

Roller skating Popeye tin figure from Linemar; Popeye Transit Company tin moving van, also from Linemar.

The Top 25 Tin Toys (in Mint Condition)

1. Popeye the Heavy Hitter, Chein $6,500
2. Popeye the Acrobat, Marx 5,500
3. Popeye the Champ Big Fight Boxing Toy, Marx 4,600
4. Mikado Family, Lehmann 3,900
5. Red the Iceman, Marx 3,500
6. Snappy the Dragon, Marx 3,000
7. Lehmann Autobus, Lehmann 2,500
8. Masuyama, Lehmann 2,500
9. Popeye with Punching Bag, Chein 2,500
10. Hey Hey the Chicken Snatcher, Marx 2,400
11. Mortimer Snerd Hometown Band, Marx 2,400
12. Popeye Express with Airplane, Marx 2,400
13. Hott and Trott, Unique Art 2,300
14. Superman Holding Airplane, Marx 2,250
15. Ajax Acrobat, Lehmann 2,200
16. Chicken Snatcher, Marx 2,200
17. Mortimer Snerd Bass Drummer, Marx 2,200
18. Tut-Tut Car, Lehmann 2,200
19. Popeye Handcar, Marx 2,000
20. Zig-Zag, Lehmann 2,000
21. Ham and Sam, Strauss 1,950
22. Howdy Doody and Buffalo Bob at the Piano, Unique Art 1,950
23. Santee Claus, Strauss 1,895
24. Li-La Car, Lehmann 1,850
25. Butter and Egg Man, Marx 1,850

CLASSIC TIN

Top to bottom: Climbing Monkey, Lehmann; windup Scottie Dog, Lehmann, 1930s, Popeye the Pilot, Marx; Ride 'Em Cowboy, Wyandotte.

CLASSIC TIN

Chein

NAME	DESCRIPTION	GOOD	EX	MINT
Banks				
Cash Box	1930's, 2" high	35	55	85
Cash Box	1930, 2" high, round trap	30	50	75
Child's Safe Bank	1900's, 5 1/2" high	40	65	100
Child's Safe Bank	1910, 4" high, sailboat on front of door	35	60	90
Child's Safe Bank	1910, 3" high, dog on front of door	35	60	95
Church	1930's, 4" high	35	60	90
Drum	1930's, 2 1/2" high	35	60	95
God Bless America	1930's, 2 1/2" high, drum shaped	30	50	75
Happy Days Cash Register	1930's, 4" high	45	80	120
Humpty Dumpty	1934, 5 1/4" high	60	100	150
Log Cabin	1930's, 3" high	80	130	200
Mascot Safe	1914, 5" high, large	35	60	95
Mascot Safe	1914, 4" high, small	35	55	85
New Deal	1930's, 3 1/4" high	50	80	125
Prosperity Bank, with band	1930's, 2 1/4" high, pail shaped	35	60	95
Prosperity Bank, without band	1930's, 2 1/4" high, pail shaped	30	50	75
Roly Poly	1940's, 6" high	135	230	350
Scout	1931, 3 1/4" high, cylinder	100	165	250
Three Little Pigs	1930's, 3" high	60	100	150
Treasure Chest	1930's, 2" high	35	60	90
Uncle Sam	1934, 4" high, hat shaped	35	60	95
Mechanical Banks				
Church	1954, 3 1/2" high	70	115	175
Clown	1931, 5" high	70	115	175
Clown	1949, 5" high, says bank on front	35	60	95
Elephant	1950's, 5" high	55	90	135
Monkey	1950's, 5 1/4" high, tin litho	55	90	135
National Duck	1954, 3 1/2" high	90	145	225
National Duck, Disney characters	6 1/2" high, Donald's tongue receives money	90	145	225
Register	dime bank	35	60	95
Uncle Wiggly	1950's, 5" high	40	65	100
Miscellaneous Toys				
Army Drummer	1930's, 7" high, plunger-activated	70	115	175
Drum		30	50	75
Easter Basket, nursery rhyme figures		35	55	85
Easter Egg with chicken on top	1938, 5 1/2", tin, opens to hold candy	35	55	85
Helicopter, Toy Town Airways	1950's, 13" long, friction drive	55	90	135
Indian in Headdress	1930's, 5 1/2" high	70	115	175
Marine	hand on belt	60	100	150
Melody Organ Player	1 roll	60	100	150
Musical Top Clown	1950's, 7" high, clown head handle	75	125	195
Player Piano	8 rolls	195	325	500
Sand Toy Set	duck mold, sifter, frog on card	30	45	70
Sand Toy, monkey bends and twists	7" high	30	50	75
Scuba Diver	10" long	70	120	185
See-Saw Sand Toy, bright colors	1930's, boy and girl on see-saw move	55	90	135
See-Saw Sand Toy, pastel colors	1930's, boy and girl on see-saw move	70	120	185
Snoopy Bus	1962	135	230	350

CLASSIC TIN

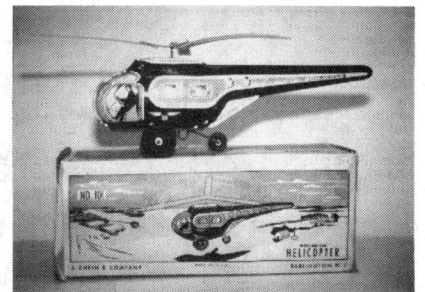

Clockwise from upper left: See-Saw Sand Toy, Chein, 1930s; Toy Town Airways Helicopter, 1950s, Chein; Penny Register Bank, Chein; windup Roller Coaster, Chein; Disneyland Ferris Wheel, 1940s, Chein.

Chein

NAME	DESCRIPTION	GOOD	EX	MINT
Space Ride, lever action with music	tin litho, boxed	135	230	350
Sparkler Toy	5", on original card	30	50	75
Teeter-Totter	11", to work pour water or sand onto board	35	55	85

Tin Windups

NAME	DESCRIPTION	GOOD	EX	MINT
"Hercules" Ferris Wheel		205	340	525
Airplane, square-winged	early tin, 7" wingspan	90	145	225
Alligator with Native on Its Back		165	275	425
Army Cargo Truck	1920's, 8" long	235	390	600
Army Plane	11" wingspan	135	230	350
Army Sergeant		70	120	185
Army Truck, cannon on back	8 1/2" long	50	80	125
Army Truck, open bed	8 1/2" long	40	65	100
Barnacle Bill	1930's, looks like Popeye, waddles	330	550	850
Barnacle Bill in barrel	1930's, 7" high	310	520	800
Bear with hat, pants, shirt, bow tie	1938	55	90	135
Bunny, bright colors	1940's, tin litho	35	60	95
Cabin Cruiser	1940's, 9" long	35	55	85
Cat	with wood wheels	40	65	100
Chick, brightly colored clothes, polka dot bow-tie	4" high	35	55	85
Chicken pushing wheelbarrow	1930's	35	60	95
China Clipper	10" long	135	225	350
Clown Balancing	1930's, 5" tall	50	80	125
Clown Boxing	8" tall, tin	235	390	600
Clown in Barrel	1930's, 8" high, waddles	175	295	450
Clown with Parasol	1920's, 8" tall, springs parasol on nose	105	180	275
Dan-Dee Dump Truck		115	195	300
Disneyland Ferris Wheel	1940's	390	650	1000
Disneyland Roller Coaster		370	620	950
Doughboy, WWI soldier with rifle	1920's, 6" high, tin litho	145	245	375
Drummer Boy	1930's, 9" high, with shako	90	145	225
Duck	1930, 4" high, waddles	35	60	95
Duck, long-beaked in orange sailor suit	1930, 6" high, waddles	50	80	125
Ferris Wheel, 6 compartments	1930's, 16 1/2" high, ringing bell	250	425	650
Ferris Wheel, The Giant Rides	16" high	60	100	150
Greyhound Bus	6" long, wood tires	90	155	235
Handstand Clown		75	125	195
Happy Hooligan	1932, 6" high, tin litho	195	325	500
Hercules Ferris Wheel		145	245	375
Indian	4" high, red with headdress	50	80	125
Jumping Rabbit	1925	100	165	250
Junior Bus	9" long, yellow	70	115	175
Mack "Hercules" Motor Express	19 1/2" long, tin litho	235	385	595
Mack "Hercules" Truck	7 1/2" long	165	275	425
Mark 1 Cabin Cruiser	1957, 9" long	35	60	95
Mechanical Aquaplane, No. 39	1932, 8 1/2" long, boat-like pontoons	100	165	250
Mechanical Fish	1940's, 11" long	30	50	75
Merry-Go-Round with swan chairs	11"	475	775	1200
Motorboat	1950's, 9" long	40	65	100
Motorboat, crank action	1950's, 7" long	35	60	90
Musical Aero Swing	1940's, 10" high	295	490	750
Musical Merry-Go-Round	small version	155	260	400
Musical Toy Church	1937, crank music box	90	145	225
Peggy Jane Speedboat		35	55	85
Pelican		100	165	250

CLASSIC TIN

CLASSIC TIN

Top to bottom: The Giant Ride Ferris Wheel, Chein; Clown in Barrel, 1930s, Chein; SS Candy Land boat, Chein; windup Roller Coaster, 1950s, Chein.

Chein

NAME	DESCRIPTION	GOOD	EX	MINT
Penguin in tuxedo	1940, tin litho	50	80	125
Pig		35	60	95
Playland Merry-Go-Round	1930's, 9 1/2" high	475	775	1200
Playland Whip, No. 340	4 bump cars, driver's head wobbles	490	810	1250
Popeye in barrel		380	625	975
Popeye the heavy hitter	bell and mallet	2550	4225	6500
Popeye with punching bag		975	1625	2500
Rabbit in shirt and pants	1938	35	60	95
Ride-A-Rocket, carnival ride	1950's, 19" high, 4 rockets	510	845	1300
Roadster	1925, 8 1/2" long, tin litho	50	80	125
Roller Coaster	1938, includes 2 cars	255	425	650
Roller Coaster	1950's, includes 2 cars	225	375	575
Royal Blue Line Coast to Coast Service		410	685	1050
Sandmill	beach scene on side	90	145	225
Santa's Elf	1925, 6" high, boxed	235	390	600
Sea Plane	1930's, silver, red, and blue	115	195	300
Seal	balancing barbells	90	145	225
Ski-Boy	1930's, 8" long, tin	165	275	425
Ski-Boy	1930's, 6" long, tin	165	275	425
Speedboat	14" long	90	145	225
Touring Car	7" long, tin litho	60	100	150
Turtle with Native on Its Back	1940's, tin litho	165	275	425
Walk on Hands Clown	striped pants	70	115	175
Walk on Hands Clown	polka dot pants	75	125	195
Woody Car	1940's, 5" long, red	100	165	250
Yellow Cab	7" long	100	165	250
Yellow Taxi	early tin, 6" long, orange and black	135	225	350

Courtland

Bakery Delivery Truck	"Pies, Cakes, Rolls, Fresh Bread," 7" long, 1950s	90	145	225
Bakery Panel Truck #4000	"Hot Buns & Hot Donuts," 7" long	75	130	200
Black Diamond Coal Truck	13" long	75	130	200
Caterpillar Tractor	5-1/2" long	70	115	175
Checker Cab #4000		100	165	250
Circus Elephant and Lions Cart	11-1/2" long, 1940s	75	125	195
City Meat Panel Truck #4000	7" long	90	145	225
Express & Hauling Stake Truck	9" long, 1940s	85	140	215
Farm Tractor	9" long	75	125	195
Fire Patrol No. 2 Truck	9" long, 1940s	80	130	200
Ice Cream Scooter	with ringing bell, 6-1/2" long, 1940s	140	235	360
State Police Car	7" long	90	145	225

Girard

Airplane with Twirling Propellor		370	625	950
Airways Express Air Mail Tri Motor Airplane	13" long	490	825	1250
Cabriolet Coupe	14" long	325	550	850
Coupe	windup with electric lights	370	625	950
Fire Chief Car	10" long	370	625	950
Fire Fighter Steam Boiler	early 1900s	255	425	650
Flasho The Mechanical Grinder	1920s	70	115	175
Gobbling Goose		145	245	375
Railroad Handcar		135	225	350
Whiz Sky Fighter	9" long	215	360	550
"Aha" Truck		470	780	1200

CLASSIC TIN

Top to bottom: Monkey Mechanical Bank, 1950s, Chein; Clown Balancing, 1930s, Chein; windup Roller Coaster, Chein.

Lehmann

NAME	DESCRIPTION	GOOD	EX	MINT
"Auton" Boy & Cart		195	325	495
"Galop" Race Car	1920s	235	390	600
"Ito" Sedan and Driver		525	875	1350
"Wild West" Bucking Bronco		525	875	1350
Ajax Acrobat	does somersaults, 10" tall	850	1425	2200
Alabama Jigger	windup tap dancer on square base, 1920s	490	825	1250
Captain of Kopenick	early 1900s	625	1050	1600
Crocodile	walks, mouth opens	285	475	725
Dancing Sailor		470	780	1200
Dancing Sailor		585	975	1500
Delivery Van	"Huntley & Palmers Biscuits"	650	1075	1650
Express Man & Cart		335	550	850
Flying Bird	flapping tin litho	295	475	750
Gustav The Climbing Miller		390	650	1000
KADI	Chinese men carrying box	585	975	1500
Lehmann's Autobus		975	1625	2500
Li-La Car	driver in rear, women passengers	725	1200	1850
Masuyama		975	1625	2500
Mikado Family		1525	2550	3900
Minstrel Man	early 1900s	335	550	850
New Century Cycle	driver and black man with umbrella	700	1175	1800
Ostrich Cart		380	650	975
Paddy Riding Pig		700	1175	1800
Quack Quack	duck pulling babies	295	495	750
Rooster and Rabbit	rooster pulls rabbit on cart	380	625	975
Sea Lion		145	250	375
Sedan and Garage		335	550	850
Shenandoah Zeppelin		155	255	395
Skier	windup skier, 1920s	510	850	1300
Taxi	10" long, 1920s	450	750	1150
Tut-Tut Car	driver has horn	850	1430	2200
Zebra Cart "Dare Devil"	1920s	255	425	650
Zig-Zag	handcar-type vehicle on oversized wheels	780	1300	2000

Linemar

Buildings and Rooms

Atomic Reactor	battery operated	80	130	200

Marx

Battery Operated Toys

Benjali Prowling Tiger	growls, 12" long	70	115	175
Brewster the Rooster	stop and go action, 10" tall	80	130	200
Disneyland Haunted House Bank	battery operated, 1950's	90	145	225
Drummer Boy	moving eyes, 1930's	105	180	275
Fishing Kitty	tin and cloth, battery operated, 9" tall	100	165	250
Frankenstein	1950's	625	1050	1600
Fred Flintstone on Dino		225	375	575
Hootin' Hollow Haunted House	1960's	585	975	1500
Mighty King Kong		135	225	350
Mighty Long Gorilla	remote controlled, tin, plush	115	195	300
Mister Mercury Robot	remote controlled, 13" tall	335	550	850
Mounted Leopard Head	head moves, roars, eyes light, tin and plastic, 1960's	90	145	225
NASA Moon Helicopter	remote controlled, 7" long	60	100	150
Nutty Mad Indian	12" tall, 1960's	105	180	275

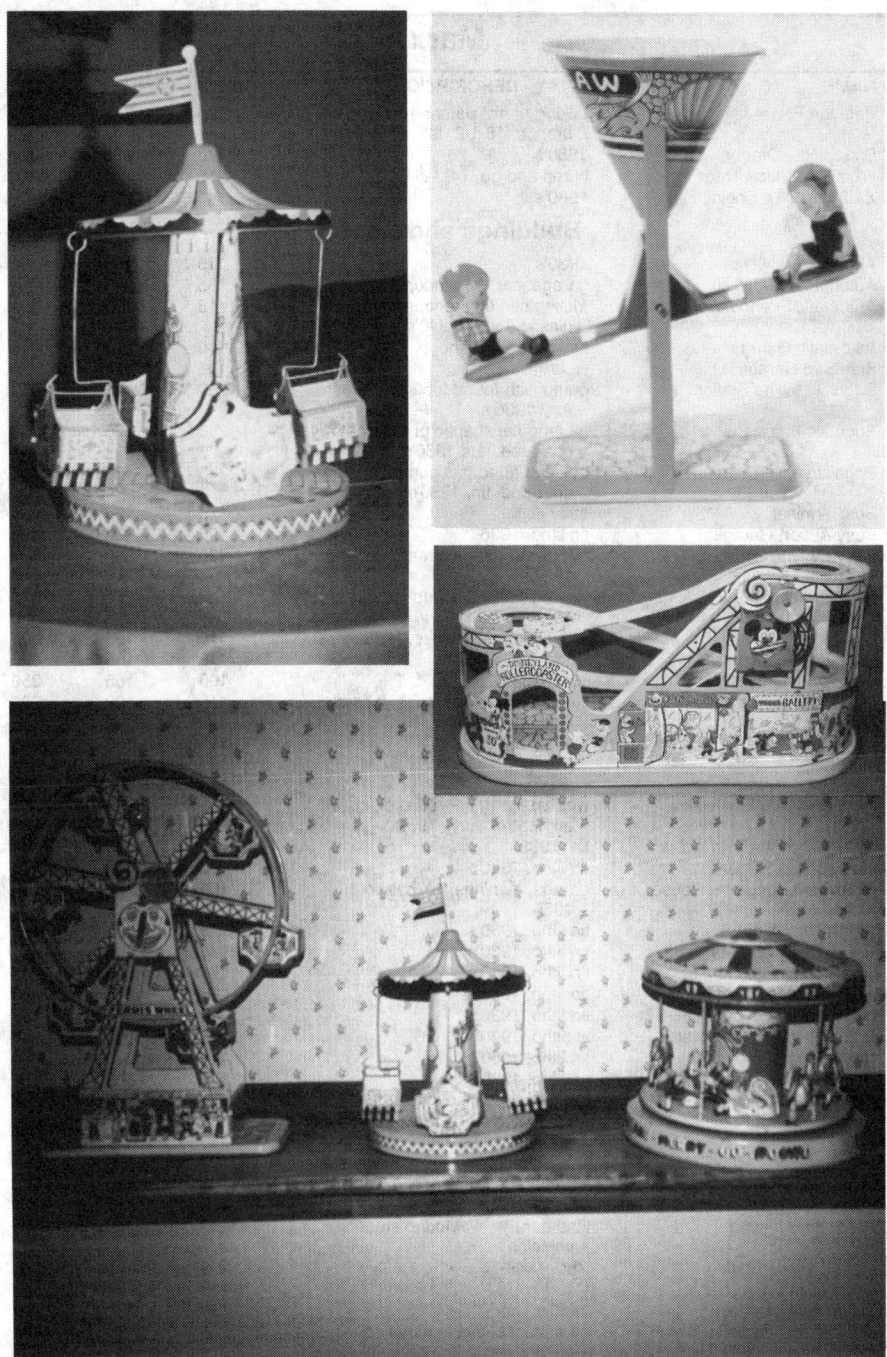

Top to bottom: Musical Aero Swing, 1940s, Chein; See-Saw Sand Toy, 1930s, Chein; Disneyland Roller Coaster, Chein; Hercules Ferris Wheel with Disney cars/Musical Aero Swing/Playland Merry-Go-Round, 1930s, all Chein.

Marx

NAME	DESCRIPTION	GOOD	EX	MINT
Pete the Parrot	talking parrot perched on litho tin branch, 16 1/2" tall	195	325	500
Snappy the Dragon	1960's	1170	1950	3000
Whistling Spook Tree	bump and go, 14" tall	475	775	1200
Za-Zoom Bike Engine	1960's	40	65	100

Buildings and Rooms

NAME	DESCRIPTION	GOOD	EX	MINT
Airport	1930's	115	195	300
Automatic Car Wash	garage, car, tin windup	135	225	350
Automatic Firehouse with Fire Chief Car	friction car, tin firehouse with plastic doors, 1940's	115	195	300
Automatic Garage	family car, tin windup	145	245	375
Blue Bird Garage	tin, 1937	135	225	350
Brightlite Filling Station	pump with round top says "Fresh Air," 1930's	250	425	650
Brightlite Filling Station	rectangular shaped pumps, battery operated, late 1930's	255	425	650
Brightlite Filling Station	bottle shaped gas pumps, battery operated, tin, 1930's	255	425	650
Bus Terminal	1937	175	295	450
Busy Airport Garage	tin litho, 1936	195	325	500
Busy Parking Lot	five heavy gauge streamline autos, 1937	235	390	600
Busy Street	six vehicles waiting to get gas, 1935	145	310	475
City Airport	with two metal planes, 1938	115	195	300
Crossing Gate House		115	195	300
Crossover Speedway	150", 1941	100	165	250
Crossover Speedway	litho buildings on bridge, 2 cars litho drivers, 144", 1938	115	190	295
Dick Tracy Automatic Police Station	station and car	375	625	950
Gas Pump Island		115	195	300
General Alarm Fire House	windup alarm bell, steel chief car and patrol truck, 1938	175	295	450
Greyhound Bus Terminal	tin, 1938	115	195	300
Gull Service Station	tin litho, 1940's	185	310	475
Hollywood Bungalow House	garage, awnings, tin, celluloid, 1935	165	275	425
Home Town Drug Store	tin litho, 1930's	175	295	450
Home Town Favorite Store	tin litho, 1930's	175	295	450
Home Town Fire House	tin litho, 1930's	175	295	450
Home Town Grocery Store	tin litho, 1930's	165	275	425
Home Town Meat Market	tin litho, 1930's	180	300	465
Home Town Movie Theatre	tin litho, 1930's	145	245	375
Home Town Police Station	tin litho, 1930's	145	245	375
Home Town Savings Bank	tin litho, 1930's	145	245	375
Honeymoon Garage	heavy gauge litho steel, 1935	165	275	425
Lincoln Highway Set	pumps, oil-grease rack, traffic light and car, 1933	350	585	900
Loop-the-Loop Auto Racer	1 3/4" long car, 1931	155	260	400
Magic Garage	litho garage, friction town car, 1934	145	250	375
Magic Garage	litho garage, windup car, 1934	145	250	375
Main Street Station	litho garage, 4" windup steel vehicles	165	275	425
Metal Service Station	litho, 1949-50	165	275	425
Military Airport		105	180	275
Model School House	1960's	50	80	125
Mot-O-Run 4 Lane Hi-Way	cars, trucks, buses move on electric track, 27" track, 1949	105	180	275
New York World's Fair Speedway	litho track, two red cars, 1939	295	495	750
Newlywed's Bathroom	tin litho, 1920's	100	165	250
Newlywed's Bedroom	tin litho, 1920's	100	165	250

CLASSIC TIN

Top to bottom: Marx Merry Makers, 1931, Marx; Donald Duck Dipsy Car, 1953, Marx; Dick Tracy Squad Car #1, 1940s, Marx; Cat Pushing Ball, 1938, Marx.

Marx

NAME	DESCRIPTION	GOOD	EX	MINT
Newlywed's Dining Room	tin litho, 1920's	100	165	250
Newlywed's Kitchen	tin litho, 1920's	100	165	250
Newlywed's Library	tin litho, 1920's	100	165	250
Roadside Rest Service Station	Laurel and Hardy at counter with stools in front, 1935	625	1050	1600
Roadside Rest Service Station	Laurel and Hardy at counter no stool in front, 1938	550	950	1450
Service Station	2 pumps, 2 friction vehicles, 1929	295	495	750
Service Station Gas Pumps	tin litho, windup, 9" tall	125	210	325
Sky Hawk Flyer	tin-plated, windup, two planes, tower 7 1/2" tall	125	210	325
Stunt Auto Racer	two blue racers, 1931	135	225	350
Sunnyside Garage	litho cardboard garage, ten vehicles, 1935	255	425	650
Sunnyside Service Station	oil cart, two pumps, litho garage, tin windup, 1934	370	625	950
TV and Radio Station		135	225	350
Universal Motor Repair Shop	tin, 1938	255	425	650
Used Car Market	base, several vehicles and signs, 1939	255	425	650
Whee-Whiz Auto Racer	four 2" multicolored racers with litho driver, 1925	295	495	750

Miscellaneous Toys

NAME	DESCRIPTION	GOOD	EX	MINT
Army Code Sender	pressed steel	20	35	50
Baby Grand Piano	tin, with piano-shaped music books	40	65	100
Big Shot		50	85	130
Cat Pushing Ball	lever action, wood ball, tin, 1938	50	80	125
Cat with Ball	cable-operated, tin litho	50	80	125
Champion Skater	ballet dancer	90	145	225
Flashy Flickers Magic Picture Gun	1960's	50	80	125
Hopalong Cassidy on his Horse	1950's wind-up	225	375	575
Hopping Rabbit	metal and plastic, 4" tall, 1950's	40	65	100
Jack in the Music Box	1950's	40	65	100
Jumping Frog	tin	40	65	100
Jungle Man Spear	tin	80	130	200
King Kong	on wheels, with spring-loaded arms, 6 1/2" tall	35	60	95
Mysterious Woodpecker	tin	50	80	125
Pathe Movie Camera	tin litho, 6" tall, 1930's	55	90	135
Rooster	large	60	100	150
Roy Rogers and Trigger	1950's	165	275	425
Searchlight	tin litho, 3 1/2" tall	40	65	100
Toto the Acrobat		100	165	250
Warriors of the World Historic Trader Cards	1963, boxed complete set	50	80	125

Trains

NAME	DESCRIPTION	GOOD	EX	MINT
Commodore Vanderbilt Train	track, windup	75	125	190
Crazy Express Train	plastic and litho, windup, 12" long, 1960's	115	195	300
Disneyland Express	locomotive and 3 tin cars, tin windup, 21 1/2" long, 1950's	255	425	650
Disneyland Express, Casey Jr. Circus Train	tin windup, 12" long	100	165	250
Disneyland Train	Goofy drives locomotive with 3 tin cars, windup, 1950	135	225	350
Engine Train	ten cars, no track, HO, 1960's	65	110	170
Flintstones Choo Choo Train "Bedrock Express"	tin windup, 13" long, 1950's	235	390	600

CLASSIC TIN

Top to bottom: Dottie the Driver, 1950s, Marx; Porky Pig Rotating Umbrella w/o top hat, 1939, Marx; U.S. Army airplane, Marx.

Marx

NAME	DESCRIPTION	GOOD	EX	MINT
Glendale Depot Railroad Station Train	accessories, tin, 1930's	235	390	600
Mickey Mouse Express Train Set	tin litho, plastic, 1952	235	390	600
Mickey Mouse Meteor Train	four cars/engine, windup, 1950's	255	425	650
Musical Choo-Choo	1966	35	60	90
Mystery Tunnel	tin windup	90	145	225
New York Central Engine Train	four cars, tin litho	195	325	500
New York Circular with Train, with airplane	tin windup, 1928	490	825	1250
New York Circular with Train, without airplane	tin windup, 1928	410	695	1050
Popeye Express, with airplane	1936	925	1550	2400
Railroad Watch Tower	electric light, 9" tall	35	60	90
Roy Rogers Stagecoach Train	hard plastic and tin litho, windup, 14" long, 1950's	135	225	350
Scenic Express Train Set	tin windup, 1950's	60	105	160
Subway Express	with plastic tunnel, 1954	185	310	475
Train Set	plastic locomotive, tin cars, windup, 6" long, 1950's	60	100	150
Trolley No. 200	headlight, bell, tin windup, 9" long, 1920's	195	325	500
Tunnel	depicts farm scene, rolling hills, houses, tin litho	100	165	250
Walt Disney Train Set	ranger-sized train set, tin litho, windup, 21 1/2" long, 1950	175	295	450

Wagons and Carts

NAME	DESCRIPTION	GOOD	EX	MINT
Bluto, Brutus, Horse and Cart	celluloid figure, metal, 1938	350	585	900
Busy Delivery	open three-wheel cart, windup, 9" long, 1939	295	475	750
Farm Wagon	horse pulling wagon, 10" long, 1940's	60	100	150
Horse and Cart	with driver, 9 1/2" long, 1950's	40	65	100
Horse and Cart	windup, 7" long, 1934	80	130	200
Horse and Cart with Clown Driver	windup, 7 5/8" long, 1923	135	225	350
Pinocchio Busy Delivery	on unicycle facing 2-wheel cart, windup, 7 3/4" long, 1939	275	475	750
Popeye Horse and Cart	tin windup	335	550	850
Rooster Pulling Wagon	tin litho, 1930's	135	225	350
Toylands Farm Products Milk Wagon	tin windup, 10 1/2" long, 1930's	145	245	375
Toylands Milk and Cream Wagon	balloon tires, tin litho, windup 10" long, 1931	165	275	425
Toytown Dairy Horsedrawn Cart	tin windup, 10 1/2" long, 1930's	185	310	475
Two Donkeys Pulling Cart	with driver, tin litho, windup, 10 1/4" long, 1940's	105	180	275
Wagon with Two-Horse Team	late 1940's, tin windup	90	145	225
Wagon with Two-Horse Team	1950, tin windup	60	100	150

Windup Toys

NAME	DESCRIPTION	GOOD	EX	MINT
Acrobatic Marvel Monkey	balances on two chairs, tin, 1930's	100	165	250
Acrobatic Marvel Rocking Monkey	tin litho, 13 1/4" tall, 1930's	165	275	425
Acrobatic Pinocchio	tin	265	450	675
Amos and Andy Walkers	walker, tin litho, 11" tall, 1930 sold separately	650	1075	1650
Andy "Andrew Brown" of Amos 'n Andy	walker, tin, 12" tall, 1930's	650	1075	1650
B.O. Plenty Holding Sparkle Plenty	tin litho, 8 1/2" tall, 1940's	185	310	475
Balky Mule	tin litho, 8 3/4" long, 1948	90	155	235
Ballerina	6" tall	90	145	225
Barney Rubble Riding Dino	tin, 8" long, 1960's	185	310	475

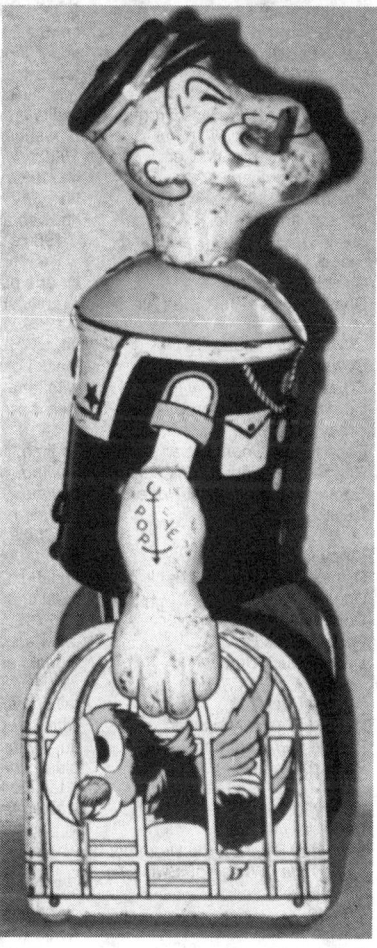

Clockwise from upper left: Milton Berle Crazy Car, 1950s, Marx; Mechanical Honeymoon Express, Marx; Popeye Express, 1932, Marx; Dapper Dan Coon Jigger, 1922, Marx.

Marx

NAME	DESCRIPTION	GOOD	EX	MINT
Bear Cyclist	lever action, metal, litho, 5 3/4" tall, 1934	70	115	175
Bear Waddler	tin, 4" tall, 1960's	50	80	125
Beat!! The Komikal Kop	tin, 1930's	125	210	325
Big Parade	tin litho, moving vehicles, soldiers, etc., 1928	410	685	1050
Big Three Aerial Acrobats	tin, 1920	235	390	600
Black Man with Bananas	tin, 1920's	375	625	950
Boy on Trapeze	tin	90	145	225
Busy Bridge	vehicles on bridge, tin litho, 24" long, 1937	350	575	895
Busy Miners	tin litho miners car, 16 1/2" long, 1930's	175	295	450
Butter and Egg Man	walker, tin litho, windup	700	1175	1800
Cake Nodders, Donald Duck, Mickey, Goofy, Pluto		50	80	125
Captain America	5" tall, 1968	55	90	140
Cat and Ball	tin, 5 1/2"	30	50	75
Cat with Ball	tin, 4" long	25	40	65
Charleston Trio	one adult, two child dancers, 9" tall, 1921	525	875	1350
Charlie McCarthy Bass Drummer	walks fast beating drum that he pushes along, tin litho, 1939	450	750	1150
Charlie McCarthy Walker	1930's	335	550	850
Chicken Snatcher	tin, 1927	850	1450	2200
Chipmunk	tin	50	80	125
Chompy the Beetle	tin, with action and sound, 6" tall, 1960's	60	100	150
Clancy	walker, tin windup, 11" tall, 1931	195	325	500
Climbing Fireman	tin and plastic, 1950's	135	225	350
Coast Defense Revolving Airplane	circular with 3 cannons, tin windup, 1929	390	650	1000
Cowboy on Horse	tin, 6" tall, 1925	70	115	175
Cowboy Rider	black horse version, tin, 7", 1930's	165	275	425
Cowboy Rider	with lariat on black horse, tin, 1941	145	245	375
Cowboy Riding a Horse	with lasso, 1940's	115	195	300
Crazy Dora		165	275	425
Dapper Dan Coon Jigger	tin litho, 10" tall, 1922	375	625	950
Dapper Dan Jigger Bank	tin litho, windup, 10" tall, 1923	375	625	950
Dapper Dan the Jigger Porter	tin litho, 9 1/2" tall, 1924	375	625	950
Daschshund	walker, hard plastic, 3" long, 1950's	25	40	60
Dippy Dumper	tin	135	225	350
Disney Cash Register Bank	tin litho, 1950's	40	65	100
Donald Duck	plastic, 6 1/2" tall, 1960's	90	145	225
Donald Duck and Scooter	1960's	90	145	225
Donald Duck Bank	lever action, tin litho, 1940	70	120	185
Donald Duck Duet	Donald and Goofy, tin, 10 1/2" tall, 1946	470	780	1200
Donald Duck Toy	tail spins, plastic, 7" tall, 1950's	125	210	325
Donald Duck Toy Walker	with 3 nephews	100	165	250
Donald the Drummer	tin, 10" tall, 1940's	135	225	350
Donald the Skier	plastic, wears metal skis, 10 1/2" tall, 1940's	165	275	425
Dopey	walker, tin, 8" tall, 1938	310	525	800
Doughboy Walker		215	360	550
Drummer Boy "Let the Drummer Boy Play While You Swing & Sway	walker, tin litho, windup, 1939	275	475	725
Dumbo	rollover action, tin, 4" tall, 1941	175	295	450
Easter Rabbit	holds litho Easter basket, tin, 5" tall	70	115	175
Ferdinand the Bull	tail spins, tin litho, 4" tall, 1938	205	340	525

Top to bottom: Fred Flintstone on Dino, 1962, Marx; Superman Holding Airplane, 1940, Marx; Rocket Fighter, 1950, Marx; Toytown Dairy Horse Cart, 1930s, Marx.

Marx

NAME	DESCRIPTION	GOOD	EX	MINT
Ferdinand the Bull and the Matador	tin litho, 5 1/2" tall, 1938	295	475	750
Figaro, from Pinocchio	rollover action, tin, 5" long, 1940	135	225	350
Fireman On Ladder	tin, 24" tall	165	275	425
Flipping Monkey		80	130	200
Flippo the Jumping Dog, See Me Jump	tin litho, 3" tall, 1940	165	275	425
Flutterfly	tin litho, 3" long, 1929	90	145	225
Fred Flintstone on Dino	8" long, 1962	215	360	550
George the Drummer Boy	moving eyes, tin, 9" tall, 1930's	195	325	500
George the Drummer Boy	stationary eyes, tin, 9" tall, 1930's	175	295	450
Gobbling Goose	lays golden eggs, tin, 1940's	125	210	325
Golden Pecking Goose	tin litho, 9 1/2" long, 1924	135	225	350
Goofy	tail spins, plastic, 9" tall, 1950's	125	210	325
Goofy the Walking Gardener	holds a wheelbarrow, tin, 9" tall, 1960	250	425	650
Hap/Hop Ramp Walker	hard plastic, 2 1/2" tall, 1950's	40	65	100
Harold Lloyd Funny Face	walker, tin windup, 11" tall, 1928	275	475	725
Hey Hey the Chicken Snatcher	black man with dog hanging behind, tin, 8 1/2" tall, 1926	925	1550	2400
Honeymoon Cottage, Honeymoon Express 7	tin, square base	115	195	300
Honeymoon Express	1940's, circular train and plane	215	360	550
Honeymoon Express	1927, old-fashioned train on circular track, tin	225	375	575
Honeymoon Express	1947, streamlined train on circular track, tin	185	310	475
Honeymoon Express	1920's, tin litho lettering on base	195	325	495
Honeymoon Express	1935, M10000 streamline train, tin windup	175	295	450
Honeymoon Express, with airplane	1940, M10000 train w/ black windows, yellow sides, red bridge	175	295	450
Honeymoon Express, with airplane	1946, red, yellow, black # 6 great northern streamline train	165	275	425
Honeymoon Express, with airplane	1948-52, freight train with No. 4127 Lumar Line caboose	170	280	435
Honeymoon Express, with airplane	1936, M10000 train w/red roof, Wimpy & Sappo lithos on tunnel	175	295	450
Honeymoon Express, with airplane	1937, M10000 train w/red roof, two green tinted tunnels	165	275	425
Honeymoon Express, with airplane	1938, M10000 train w/red roof, green tunnels, copper bridge	175	295	450
Honeymoon Express, with airplane	1939, M10000 train w/red roof, green tunnels, silver bridge	185	310	475
Honeymoon Express, with flagman	1930, British steamer passenger express train	165	275	425
Honeymoon Express, with flagman	1926	165	275	425
Honeymoon Express, without airplane	1940, M10000 train w/ black windows, yellow sides, red bridge	145	250	375
Honeymoon Express, without airplane	1946, red, yellow, black # 6 great northern streamline train	145	250	375
Honeymoon Express, without airplane	1948-52, freight train with No. 4127 Lumar Line caboose	145	250	375
Honeymoon Express, without airplane	1936, M10000 train w/red roof, Wimpy & Sappo lithos on tunnel	195	325	500
Honeymoon Express, without airplane	1937, M10000 train w/red roof, two green tinted tunnels	155	260	400
Honeymoon Express, without airplane	1938, M10000 train w/red roof, green tunnels, copper bridge	155	260	400

CLASSIC TIN

Clockwise from upper left: Sandy the Cat With Ball, Marx; Old Jalopy, Marx; Donald the Drummer, 1940s, Marx; Andy "Andrew Brown" of Amos 'n Andy, 1930s, Marx.

Marx

NAME	DESCRIPTION	GOOD	EX	MINT
Honeymoon Express, without airplane	1939, M10000 train w/red roof, green tunnels, silver bridge	145	250	375
Honeymoon Express, without flagman	1930, British steamer passenger express train	145	250	375
Honeymoon Express, without flagman	1933	145	250	375
Honeymoon Express, without flagman	1926	155	260	400
Hopalong Cassidy Rocking Horse Cowboy	tin, 11 1/4" tall, 1946	295	475	750
Hoppo the Monkey	plays cymbals, tin, 8" tall, 1925	100	165	250
Howdy Doody	plays banjo and moves head, tin 5" tall, 1950	250	425	650
Howdy Doody	does jig and Clarabell sits at piano, tin, 5 1/2" tall, 1950	525	875	1350
Jazzbo Jim Roof Dancer	9" tall, 1920's	375	625	950
Jetsons Figure	4" tall, 1960's	75	130	200
Jiminy Cricket Pushing Bass Fiddle	walker	85	145	220
Jiving Jigger	tin, 1950	135	225	350
Jocko Climbing Monkey	1930's	70	115	175
Jocko Monkey	on string, tin litho, 9 1/2" long, 1950's	70	115	175
Joe Penner and His Duck Goo-Goo	tin, 7 1/2" tall, 1934	375	625	950
Jumbo The Climbing Monkey	litho, 9 3/4" tall, 1923	135	225	350
Jungle Book Dancing Bear	plastic	40	65	100
Knockout Champs Boxing Toy	tin litho, 1930's	175	295	450
Leopard	growls and walks, 1950	65	105	165
Let the Drummer Boy Play	tin, 1930's	145	250	375
Little King Walking Toy	walkers, plastic, 3" tall, 1963	50	80	125
Little Orphan Annie and Sandy	tin, 1930's	275	475	725
Little Orphan Annie Skipping Rope	tin	165	275	425
Little Orphan Annie's Dog Sandy	tin litho	105	180	275
Lone Ranger and Silver	tin, 8" tall, 1938	195	325	500
Mad Russian Drummer	7" tall	225	375	575
Main Street	street scene with moving cars, traffic cop, tin litho, 1927	340	575	875
Mammy's Boy	walker, eyes move, tin litho, windup, 11"tall, 1929	310	525	795
Merry Makers Minstrel Bank	four mice and piano, tin windup, 1930's	585	975	1500
Merrymakers Band	with marquee, mouse band, tin-plated litho, 1931	700	1175	1800
Merrymakers Band	without marquee, mouse band, tin-plated litho, 1931	585	975	1500
Mickey and Donald Handcar	plastic, 1948	125	210	325
Mickey Mouse	tail spins, 7" tall	125	210	325
Mickey Mouse Express	train and plane action, tin	325	550	850
Minnie	tin, 7" tall	295	475	750
Minnie Mouse In Rocker	tin, 1950's	275	450	695
Minstrel Figure	tin, 11" tall	175	295	450
Monkey Cyclist	litho, 9 3/4" tall, 1923	105	180	275
Moon Creature	5-1/2" tall	135	225	350
Moon Mullins and Kayo on Handcar	tin, 6" long, 1930's	335	550	850
Mortimer Snerd Band "Hometown Band"	1935	925	1550	2400
Mortimer Snerd Bass Drummer	walks fast beating drum that he pushes along, tin litho, 1939	850	1450	2200
Mortimer Snerd Walker Toy	walker, tin litho, windup, 1939	185	310	475
Mother Goose	7 1/2" tall, 1920's	115	195	300

CLASSIC TIN

CLASSIC TIN

Top to bottom: Wee Scottie, 1952, Marx; Donald Duck Duet, 1946, Marx; Uncle Wiggily, Marx; Climbing Fireman (tin and plastic), 1950s, Marx; Gobbling Goose, 1940s, Marx.

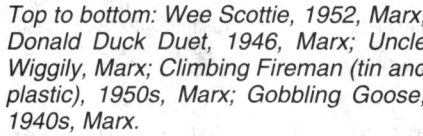

CLASSIC TIN

Marx

NAME	DESCRIPTION	GOOD	EX	MINT
Mother Penguin with Baby Penguin on Sled	walker, hard plastic, 3" long, 1950's	35	55	85
Musical Circus Horse	pull toy, metal drum rolls with chimes, 10 1/2" long, 1939	80	130	200
Mystery Cat	tin litho, 8 1/2" long, 1931	100	165	250
Mystery Pluto	sniffs ground, 8" tall, 1948	100	165	250
Mystery Sandy Dog	Little Orphan Annie's Dog Sandy, tin, 8 1/2" long, 1938	145	250	375
Nodding Goose		30	50	75
Nutty Indian Drummer		60	100	150
Pecos Bill	twirls rope, plastic, 10" tall, 1950's	80	130	200
Pikes Peak Mountain Climber	vehicle on track	325	525	800
Pinched	square based, open circular track, 1927	275	450	695
Pinocchio	5" tall, 1950's	250	425	650
Pinocchio	standing erect, eyes skyward, tin, 9" tall, 1938	275	450	700
Pinocchio the Acrobat	tin, rocking, 16" tall, 1939	335	550	850
Pinocchio Walker	stationary eyes, tin, 1930's	250	425	650
Pinocchio Walker	animated eyes, tin, 8 1/2" tall, 1939	275	450	700
Pluto Drum Major	tin	235	390	595
Pluto Toy	mechanical, plastic, 1950's	105	180	275
Pluto Watch Me Roll-Over	8" long, 1939	165	275	425
Poor Fish	tin litho, 8 1/2" long, 1936	70	115	175
Popeye Acrobat	tin	2150	3600	5500
Popeye and Olive Oyl Jiggers	on cabin roof, tin litho, 10" tall, 1936	700	1175	1800
Popeye Express	carrying parrots in cages, tin ltho, 8 1/4" tall, 1932	375	625	950
Popeye Handcar	Popeye and Olive Oyl rubber, metal handcar, 1935	775	1300	2000
Popeye Jigger	on cabin roof, tin, 10" tall, 1936	475	825	1250
Popeye the Champ Big Fight Boxing Toy	tin and celluloid, 7" long, 1936	1800	3000	4600
Porky Pig Cowboy with Lariat	tin, 8" tall, 1949	280	475	725
Porky Pig Rotating Umbrella, with top hat	tin, 8" tall, 1939	310	525	800
Porky Pig Rotating Umbrella, without top hat	tin, 8" tall, 1939	310	525	800
Red Cap Porter	tin	375	625	950
Red the Iceman	tin with wood ice cube	1350	2275	3500
Ride 'Em Cowboy	tin	90	145	225
Ring-A-Ling Circus	tin litho, 7 1/2" diameter base, 1925	475	775	1200
Rodeo Joe	tin, 1933	165	275	425
Roll Over Cat	black cat pushing ball, tin litho, 8 1/2" long, 1931	100	165	250
Roll Over Pluto		145	250	375
Running Scottie	metal, 12 1/2" long, 1938	70	115	175
Smitty Riding A Scooter	8" tall, 1932	500	850	1300
Smokey Joe The Climbing Fireman	21" tall	215	350	550
Smokey Joe the Climbing Fireman	1930s, 7-1/2" tall	225	375	575
Smokey Sam the World Fireman	7" tall, 1950's	125	210	325
Snappy the Miracle Dog	came with dog house, tin litho, 3 1/2" long, 1931	80	130	200
Snoopy Bus		475	825	1250
Somstepa Jigger	8" tall	155	260	400
Spic and Span Drummer and Dancer	tin litho, 10" tall, 1924	500	850	1300
Spic Coon Drummer	tin, 8 1/2" tall, 1924	425	715	1100
Stop, Look and Listen	1927	295	490	750

CLASSIC TIN

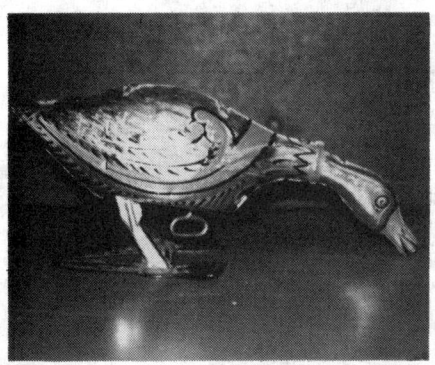

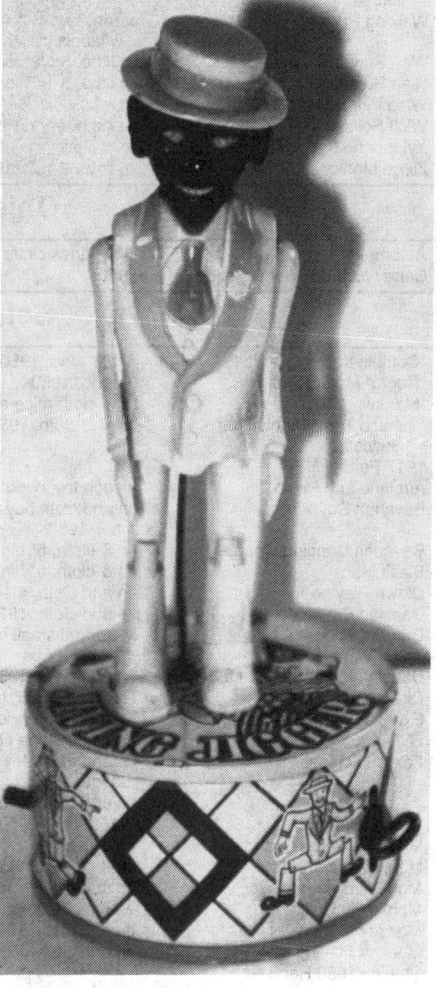
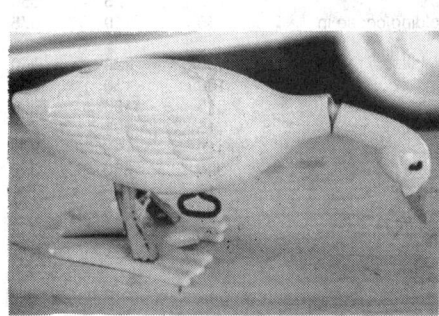

Clockwise from upper left: Golden Pecking Goose, 1924, Marx; College Jalopy, Marx; Jiving Jigger, 1950, Marx; Gobbling Goose, 1940s, Marx; Highway Patrolman on Motorcycle, Marx.

CLASSIC TIN

Marx

NAME	DESCRIPTION	GOOD	EX	MINT
Streamline Speedway	two racers on track	80	130	200
Subway Express	tin, 1950's	65	110	170
Superman Holding Airplane	tin, 6" wingspan on airplane, 1940	875	1450	2250
The Carter Climbing Monkey	8 1/2" tall, 1921	115	190	290
Tidy Tim Streetcleaner	pushes wagon, tin litho, 8" tall, 1933	175	295	450
Tiger	plush with vinyl head, walks, growls	40	65	100
Tom Tom Jungle Boy	tin, 7" tall	40	65	100
Tony the Tiger	plastic, whirling tail action, 7" tall, 1960's	60	100	150
Tumbling Monkey	4 1/2" tall, 1942	75	125	190
Tumbling Monkey and Trapeze	5 3/4" tall, 1932	135	225	350
Walking Popeye	1930's	335	550	850
Walking Popeye Carrying Parrot Cages	tin litho, windup, 8 1/4" tall, 1932	295	475	750
Walking Porter	carries two suitcases covered w/labels, tin, 8" tall, 1930's	205	340	525
Wee Running Scottie	tin litho, 5 1/2" long, 1930's	65	110	165
Wee Running Scottie	tin litho, 5 1/2" long, 1952	30	50	75
Wise Pluto	tin	100	165	250
WWI Soldier	prone position with rifle	50	80	125
Xylophonist	tin, 5" long	20	35	50
Zippo Monkey	tin litho, 9 1/2", 1938	70	115	175

Ohio Art

Automatic Airport	two planes circle tower	195	325	500
Coney Island Roller Coaster	1950s	100	165	250

Schuco

"Combinato" Convertible	7-1/2" long, 1950s	100	165	250
"Curvo" Motorcycle	5" long, 1950s	135	225	350
"Mauswagen"	tin & cloth mice and wagon	195	325	500
"Schuco Turn" Monkey on Suitcase	tin and cloth, 1950s	115	195	300
1917 Ford		50	80	125
Airplane and Pilot	friction toy, oversized pilot, 1930s	135	225	350
Bavarian Boy	tin and cloth boy with beer mug, 5" tall, 1950s	115	195	295
Bavarian Dancing Couple	tin & cloth, 5" high	105	180	275
Black Man	tin & cloth, 5" high	250	425	650
Clown Playing Violin	tin and cloth, 4-1/2" tall, 1950s	135	225	350
Dancing Boy and Girl	tin and cloth, 1930s	115	195	300
Dancing Mice	large and small mouse, tin and cloth, 1950s	135	225	350
Dancing Monkey With Mouse	tin and cloth, 1950s	125	210	325
Drummer	tin and cloth, 5" tall, 1930s	125	210	325
Examico 4001 Convertible	maroon tin windup, 5-1/2"	145	250	375
Flic 4520	traffic cop type figure	135	225	350
Fox And Goose	tin and cloth, fox holding goose in cage, 1950s	145	250	375
Juggling Clown	tin and cloth, 4-1/2" tall	135	225	350
Mercer Car #1225	7 1/2" long, 1950s	100	150	225
Mickey & Minnie Dancing	tin & cloth	675	1150	1750
Monk Drinking Beer	tin & cloth, 5" high	115	195	300
Monkey Drummer	tin and cloth, 1950s	115	195	300
Monkey in Car	1930s	335	550	850
Monkey on Scooter	tin and cloth, 1930s	115	195	300
Monkey Playing Violin	tin and cloth, 1950s	125	210	325
Studio #1050 Race Car		90	145	225

CLASSIC TIN

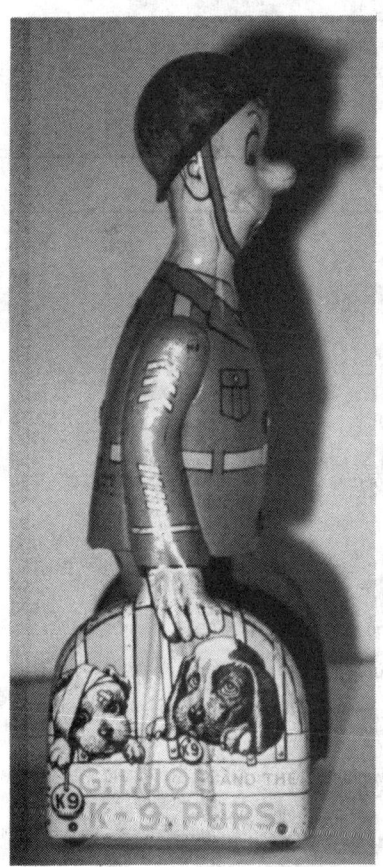

Top to bottom: GI Joe and the K-9 Pups, 1941, Unique Art; Artie the Clown in His Crazy Car, Unique Art; Howdy Doody & Buffalo Bob at the Piano, Unique Art; Lincoln Tunnel, 1935, Unique Art.

CLASSIC TIN

Schuco

NAME	DESCRIPTION	GOOD	EX	MINT
Tumbling Boy	tin and cloth, 1950s	100	165	250
Yes-No Monkey		165	275	425

Strauss

NAME	DESCRIPTION	GOOD	EX	MINT
Boston Confectionery Co. Truck	yellow, 9" long	375	650	975
Bus De Lux	blue, 1920s	475	825	1250
Circus Wagon	8-1/2" long	575	975	1500
Clown Crazy Car	1920s	375	625	950
Dandy Jim Clown Dancer	10" tall	335	550	850
Dizzie Lizzie		375	625	950
Flying Air Ship Los Angeles	aluminum, 10" long, 1920s	375	625	950
Ham and Sam	6"x6-1/2" multicolored	750	1275	1950
Interstate Double Decker Bus	green and yellow, 10-1/2" long	550	900	1400
Interstate Double Decker Bus	brown and yellow, 11" long	375	625	975
Jazzbo Jim	dancer on rooftop	425	725	1100
Jenny The Balking Mule	farmer and reluctant mule, 9" long	175	295	450
Jitney Bus	green and yellow, 9-1/2" long	250	425	650
Knockout Prize Fighters	windup boxers, 1920s	250	400	625
Leaping Lena	9" long	275	450	700
Miami Sea Sled	Yellow and red, 15" long	275	450	700
Play Golf	12" long	295	475	750
Pool Player		235	390	595
Red Cap Porter	with wagon and trunk	325	525	825
Rollo Chair	Windup cart pushed by figure, 7" long	525	875	1350
Santee Clause	Santa in sleigh, two reindeer, clockwork, 11" long	750	1250	1895
Timber King Tractor Trailer		90	145	225
Tip Top Wheelbarrow		250	400	625
Tombo Alabama Coon Jigger	11" high	375	625	950
Travel Chicks	chickens on railraod car, windup	205	340	525
Trik Auto	7" long, says "Trik Auto"	245	400	625
Trik Auto	red and yellow windup, 6-1/2" long	245	400	625
What's It? Car	multicolored litho, 9-1/2" long	625	1025	1575
Yell-O-Taxi	7-1/2" long	380	625	975

Unique Art

NAME	DESCRIPTION	GOOD	EX	MINT
Artie the Clown in his Crazy Car		235	390	595
Bombo the Monk		115	195	295
Butter and Egg Man		475	825	1250
Capitol Hill Race		125	210	325
Casey The Cop		350	575	895
Dandy Jim		295	490	750
Daredevil Motor Cop		205	350	525
Finnegan the Porter		135	225	350
Flying Circus		490	825	1250
G.I. Joe and His Jouncing Jeep		125	210	325
G.I. Joe and His K-9 Pups		105	180	275
Gertie the Galloping Goose		115	190	295
Hee Haw	donkey pulling milk cart	115	190	295
Hillbilly Express		135	225	350
Hobo Train		175	295	450
Hott and Trott		900	1500	2300
Howdy Doody & Buffalo Bob at Piano		750	1275	1950
Jazzbo Jim		295	475	750
Kid-Go-Round		155	260	400
Kiddy Go-Round		155	260	400
Krazy Kar		200	335	515
Li'l Abner and His Dogpatch Band		335	550	850

CLASSIC TIN

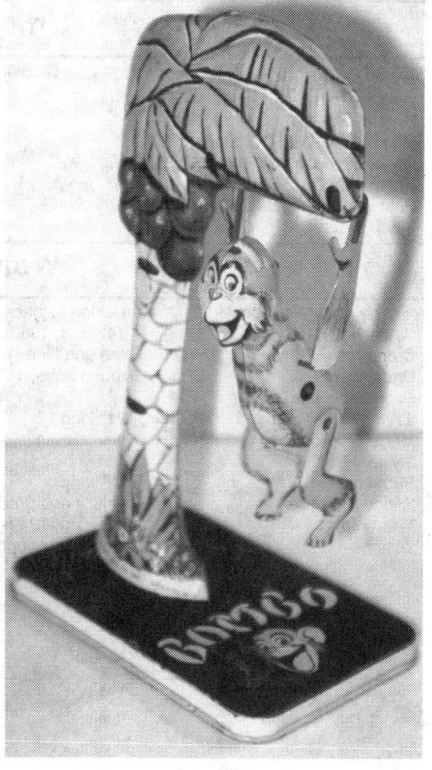

Top to bottom: Jazzbo Jim, Strauss; Bombo the Monk, 1930s, Unique Art; Red Cap Porter, Strauss; Capitol Hill Racer, 1930s, Unique Art.

Unique Art

NAME	DESCRIPTION	GOOD	EX	MINT
Lincoln Tunnel		235	375	595
Motorcycle Cop		75	125	195
Musical Sail-Way Carousel		175	295	450
Pecking Goose, Witch and Cat		175	295	450
Rodeo Joe Crazy Car		145	250	375
Rollover Motorcycle Cop		135	225	350
Sky Rangers		250	400	625

Wolverine

NAME	DESCRIPTION	GOOD	EX	MINT
"Sandy Andy" Fullback	kicking fullback, 8" tall	175	295	450
Battleship	14" long, 1930s	70	120	185
Crane	red and blue, 18" high	40	65	100
Drum Major	round base, 13"	100	165	250
Express Bus		125	210	325
Jet Roller Coaster	21" long	105	180	275
Merry-Go-Round		235	390	595
Mystery Car		100	165	250
Submarine	13" long	100	165	250
Sunny and Tank	yellow and green, 14-1/2" long	75	130	200
Yellow Taxi	13" long, 1940s	135	225	350
Zilotone	clown on xylophone, with musical discs, 1920s	450	750	1150

Action Figures

Specialists in different collecting fields often feel their particular area is the greatest thing since sliced bread. Action figure collectors are no exception to this, but they do have one argument in their favor — they may be right this time. Action figure collecting is one of the fastest growing and potentially largest collectible areas since the baseball card boom of the 1980s. A stroll through the toy section of any department store is proof enough. Plus, it is a given that a percentage of today's teen and preteenaged action figure buyers will in a few years become collectors, and their potential numbers are huge. Action figures could bring more collectors into the hobby than G.I. Joe, Hot Wheels, and model kits combined.

Toy Shop Magazine, the biweekly bible of the toy collecting hobby, presents a universe of action figures for sale in each issue. *Toy Collector And Price Guide* magazine, America's highest circulation toy enthusiast's magazine, is also filled with ads offering action figures to an eager market. In fact, so many figures are for sale that you might conclude that they are common items, and in a way, you would be right. In some places, namely toy stores, action figures are literally climbing the walls.

So why collect them if you can buy them right off the shelves at the store? The answer to this is that for many collectors, that is exactly how it is done.

The Beginning of Time

Some action figure collectors act as if time began in the 1960s. The twentieth century is broken into two great periods, BAF (Before Action Figures) and AAF (After Action Figures.) This has a lot to do with the definition of action figure entailing an articulated, posable body. Of course, boys played with toy soldiers even before the turn of the century. But these were solid, immobile figures that had no realistically movable parts. The same held true for the hard plastic Marx figures of the 1950s. If you are a stickler for semantics, then action figures were created in the 1960s and the story does begin there.

The reality is not quite as epochal as all that, but the 1960s did see the convergence of several cultural influences and technologies that changed American life in many ways, particularly how toys would be sold.

By the late 1950s television had replaced the dinner table and parlor radio as the family hearth. The sturdy cabinet in the corner of the living room glowed with a captivating power only hinted at by radio, and which has never been challenged since. It was a working window not only into a wide world of people and places, but also — increasingly — of neat things to buy. While Mom sat on the couch and Dad in his chair, the youngsters clustered on the floor, bathing in the new light, soaking up the names and lore of their new friends and heroes, Barbie, Superman, Batman, G.I. Joe.

From 1961 to 1963 toy makers watched with envy and despair as Mattel's Barbie, aided by TV, took the world of girl's toys by storm. Of course, no one would dream of selling dolls to boys, so this barrier seemed insurmountable. But wheels of industry would not be easily stopped, and the simple solution to this dilemma ranks as one

Hasbro's The Original GI Joe Action Soldier in two versions.

of the greatest marketing "spins" of all time. If boys won't play with dolls, then call them something else. How about "action figures?"

Hasbro's first test of G.I. Joe, the male answer to Barbie, debuted at the New York International Toy Fair in early 1964. Toy Fair is the annual new products bash for the entire toy industry, where toy makers meet toy retailers and determine what children will ask for on their wish lists for the coming year. Countless toys have died without ever seeing production, all due to lack of buyer interest at Toy Fair.

The Toy Fair buyers met the twelve-inch G.I. Joe with hopes and reservations. They wanted to believe that a sussessful "Barbie for boys" had been created, but as much as Hasbro people touted Joe as "America's Movable Fighting Man," the buyers still heard "doll." They wished Hasbro all the best and retreated in droves.

Virtually no orders were generated at Toy Fair, so in June, with no fanfare or ad support, Hasbro released the new toy directly into a New York test market. Every test store sold out within a week and the invasion of America was on. By year's end Joe had earned Hasbro some $17 million, in spite of sales lost to product shortages.

G.I. Joe was the first true fully articulated action figure for boys, but he would not be the only one for long. A.C. Gilbert launched their James Bond figures in 1965, but for the first time in his career, Ian Fleming's superspy failed in his mission. Marx also entered the ring with their "Best of the West" series, but G.I. Joe had a seemingly limitless arsenal of battle-geared appeal.

The first reasonably successful challange to G.I. Joe came from Ideal's Captain Action. While Joe's identity was well established, Captain Action was a man of many faces. Ideal designed Captain Action to establish not only his own identity, but also to capitalize on those of many popular super heroes. Joe was just Joe but Captain Action was Spider-Man, Batman, Superman, the Lone Ranger and a host of others. Today, Captain Action figures and sets command the second highest prices in the action figure market, second only to the classic G.I. Joes.

Ideal's brief foray into the world of super hero action figures paved the way for many to come. While G.I. Joe was forced to temper his image and soften it from the quintessentially military Green Beret Joe of 1967 into the Adventure Team Joe of 1970, super heroes were largely immune to the Vietnam protests that forced Joe's change of "mission." By 1969 Ideal tired of Captain Action's complex licensing agreements and discontinued the series, but another company was waiting in the wings.

It was Mego. In 1972 Mego released its first super hero series, the six figure set of "World's Greatest Super Heroes." These eight-inch tall

MEGO's eight-inch action figure of Catwoman.

cloth and plastic figures were joined by twenty-eight others by the time the series ended ten years later. Mego supplemented this super hero line with licensed film and TV characters, notably "Planet of the Apes," "Star Trek," and the "Dukes of Hazzard," as well as with historic figures representing the old west and the "World's Greatest Super Knights" of the apparently Super Round Table.

Nineteen seventy-seven saw another milestone in action figure history, one that everybody on the block ironically missed at the time. A fun little outer space movie called Star Wars had from nowhere become a worldwide smash, and nobody had made any toys to support it at all. Only Kenner had bothered to secure rights to merchandise toys, and it had done nothing with them. When Kenner realized the magnitude of Star Wars' potential, it rushed toys through production, but it didn't have time to get action figures on the shelves by Christmas. Kenner instead essentially pre-sold the figures, using a promotion they called the Early Bird Certificate Package, which entitled the owner to mail delivery of the first four figures as soon as they were available.

By Christmas of 1978 the line had grown to seventeen figures and the first wave of a deluge of accessories and related toys. The Star Wars figures also established a third standard size for action figures. G.I. Joe and Captain Action were twelve-inch figures, Megos measured eight and Kenner's Star Wars figures came in at just 3-3/4 inches tall. Their tremendous popularity cemented that size as a new standard that holds to this day.

The next size to be established was the six-inch figure, set by Mattel's highly successful and lucrative 1981 Masters of the Universe series. This series was the first to be reverse licensed, in other words, Mattel made the toys first, and then Mattel sold the licensing to television and film, not the other way around. Mattel also upped the manufacturing ante by endowing the figures with action features such as punching and grabbing movements, thus enhancing their play value and setting another standard in the process.

Hasbro then scored again with the 1985 introduction of the next level in the evolution of action figures, the transforming figure. The aptly named Transformers did just that, changing from innocuous looking vehicles into menacing robots with a few deft twists, and then back again. Hasbro's little mutating robots also transformed the toy industry, spawning numerous competitors and introducing the element of interchangeability of elements into toy design. It should be noted here that Hasbro did not invent the transforming robot. That credit, as far as we are aware, goes to a Japanese line called GoDaiKins. But Hasbro perfected the mass merchandising of the concept like no company before it.

Today's present generation of micro chip powered talking and sound effect laden toys are now the standard of the industry, but this standard too will undoubtedly be made obsolete by future evolutions of controllability and interaction. We are most likely just a few holiday wish lists away from fully remote controlled robotic warriors that will meet in tabletop combat while we control their every move from easy chairs.

Action figures are big business, and hot series like Star Trek and Teenage Mutant Ninja Turtles are now regularly ranked in the top 20 best selling lines by industry trade papers. Back in the department store, hardware and housewares may be slow, but there are always shoppers in the toy aisles. New lines quickly replace slow sellers, many of which soon find new life in the collector market. If it was on the store shelves last week but is gone this week, it will probably show up in a collector magazine next week — at a suitably higher price.

Again, an enduring character identity is a key to continued demand and future appreciation. "Star Trek: The Next Generation" and those Turtles have proven themselves worthy long term franchises, and are joining the ranks of Star Wars and Masters of the Universe as blue chip stocks of the action figure market.

The action figure aisles are now attracting more adults, and they are not always buying for their kids. More adults today are buying action figures as collectibles and investments. And those investments will in years hence feed the needs of tomorrow's collectors, the ones who are, right now, sitting on the floor playing with Captain Picard, Jack Skellington, Donatello and Fester Addams.

Trends

By all reports, 1994 was the year of the action figure, both at Toy Fair and on the retail level. Nineteen ninety-three's phenomenal success of the Mighty Morphin Power Rangers will only provide further inspiration for marketers. The net result will likely be still more new figure lines courting success at next year's Toy Fair, quite a few of which will later be put to the test on store shelves. The action figure market remains strong and as was discussed earlier, the future for this segment of the hobby is exceptionally bright.

Regarding line by line performance, the primary value increases were seen in "blue chip" lines such as Kenner Super Powers, Mego's Official World's Greatest Super Heroes, and the Teenage Mutant Ninja Turtles. Values for most lines held their own this year, with none losing ground. The G.I. Joe market has leveled off this year, with few prices rising but none falling.

The Top 10 Action Figures (in Mint Condition) [No G.I. Joe or Captain Action]

1. Major Matt Mason, Scorpio Figure, Mattel, 1966-69 ....$1,500
2. Official World's Greatest Super Heroes, Batman's Wayne Foundation Penthouse, Mego, 1977 1,200
3. Mad Monster Series, Mad Monster Castle, Mego, 1974 600
4. Star Trek, Romulan Figure, Mego, 1974-80 600
5. Major Matt Mason, Mission Team Four-Pack, Mattel, 1966-69 575
6. Official World's Greatest Super Heroes, Aquaman Vs. The Great White Shark Playset, Mego, 1978 500
7. Adventures of Indiana Jones, Belloq in Ceremonial Robe, on card, Kenner, 1982-83 500
8. Official World's Greatest Super Heroes, Bruce Wayne, Mego 1974 500
9. Official World's Greatest Super Heroes, Clark Kent, Mego, 1974 500
10. Official World's Greatest Super Heroes, Dick Grayson, Mego, 1974 500

The Top 20 G.I. Joe Figures and Sets (in Mint-in-Box condition)

1. Crash Crew Fire Truck Set, #8040, Action Pilot Series, 1967.............................$3,500
2. Foreign Soldiers of the World, #8111-83, Action Soldiers of the World Series, 1968 3,000
3. G.I. Nurse, #8060, Action Girl Series, 1967 3,000
4. Desert Patrol Attack Jeep Set, #8030, Action Soldier Series, 1967 2,500
5. Green Beret, #7536, Action Soldier Series, 1966..2,500
6. Talking Landing Signal Officer, #90621, Action Sailor Series, 1968 2,500
7. Talking Shore Patrol Equipment Set, #90612, Action Sailor Series, 1968 2,500
8. Military Police Uniform Set, #7539, Action Soldier Series, 1968 2,300
9. Aquanaut, #7910, Adventures of G.I.Joe series, 1969..2,000
10. Talking Adventure Pack, Bivouac Equipment, #90513, Action Soldier Series, 1968 2,000
11. Military Police Uniform Set, #7539, Action Soldier Series, 1967 1,800
12. Australian Jungle Fighter, #8105, Action Soldiers of the World, 1966 1,750
13. British Commando, #8104, Action Soldiers of the World, 1966..........................1,750
14. French Resistance Fighter, #8103, Action Soldiers of the World, 1966..............1,750
15. German Storm Trooper, #8100, Action Soldiers of the World, 1966...................1,750
16. Japanese Imperial Soldier, #8101, Action Soldiers of the World, 1966...............1,750
17. Russian Infantry Man, #8102, Action Soldiers of the World, 1966 1,750
18. Uniforms of Six Nations, #5038, Action Soldiers of the World, 1967..................1,700
19. Jungle Fighter Set, #7732, Action Marine Series, 1968 1,650
20. Black Action Soldier, #7900, Action Soldier Series, 1965 1,600

Top to bottom: Princess Wildflower from Best of the West series, 1960s, Marx; General Custer from Fort Apache Fighters series, 1960s, Marx; Forbidden Zone Trap from Planet of the Apes playsets, 1975, Mego.

ACTION FIGURES

A-Team Galoob, 1984

3 3/4" Figures and Accessories

NAME	MNP	MIP
A-Team Four Figure Set	12	30
Armored Attack Adventure with B.A. Figure	8	20
Bad Guys Four-Figure Set: Viper, Rattle, Cobra, Python	10	25
Combat Attack Gyrocopter	10	25
Command Center Playset	14	35
Corvette with Face Figure	8	20
Interceptor Jet Bomber with Murdock	10	25
Tactical Van Playset	6	15

6 1/2" Figures and Accessories

Amy Allen Figure, 6-1/2"	10	25
B.A. Baracus Figure, 6-1/2"	8	20
Cobra Figure, 6-1/2"	6	15
Face Figure, 6-1/2"	6	15
Hannibal Figure, 6-1/2"	6	15
Murdock Figure, 6-1/2"	8	20
Off Road Attack Cycle	8	20
Python Figure, 6-1/2"	6	15
Rattler Figure, 6-1/2"	6	15
Viper Figure, 6-1/2"	6	15

Action Jackson Mego, 1974

8" Figures

Action Jackson, blonde hair, 1974	15	30
Action Jackson, black hair, 1974	15	30
Action Jackson, brown hair, 1974	15	30
Action Jackson, blonde bearded, 1974	15	30
Action Jackson, black bearded, 1974	15	30
Action Jackson, brown bearded, 1974	15	30
Action Jackson, black version, 1974	25	60

Accessories

Parachute Plunge, 1974	5	15
Strap-On Helicopter, 1974	5	15
Water Scooter, 1974	5	15

Outfits

Air Force Pilot, 1974	7	15
Army Outfit, 1974	7	15
Aussie Marine, 1974	7	15
Baseball, 1974	7	15
Fisherman, 1974	7	15
Football, 1974	7	15
Frog Man, 1974	7	15
Hockey, 1974	7	15
Jungle Safari, 1974	7	15
Karate, 1974	7	15
Navy Sailor, 1974	7	15

Action Jackson Mego, 1974

NAME	MNP	MIP
Rescue Squad, 1974	7	15
Scramble Cyclist, 1974	7	15
Secret Agent, 1974	7	15
Ski Patrol, 1974	7	15
Snowmobile Outfit, 1974	7	15
Surf and Scuba Outfit, 1974	7	15
Western Cowboy, 1974	7	15

Playsets

Jungle House, 1974	40	85
Lost Continent Playset, 1974	40	85

Vehicles

Adventure Set, 1974	40	85
Campmobile, 1974	40	85
Dune Buggy, 1974	30	60
Formula Racer, 1974	30	60
Mustang, 1974	30	60
Rescue Helicopter, 1974	40	85
Safari Jeep, 1974	40	85
Scramble Cycle, 1974	20	40
Snow Mobile, 1974	15	30

American West Series Mego, 1973

8" Figures

Buffalo Bill Cody, 1973, boxed	40	75
Buffalo Bill Cody, 1973, carded	40	100
Cochise, 1973, boxed	40	75
Cochise, 1973, carded	40	100
Davy Crockett, 1973, boxed	70	110
Davy Crockett, 1973, carded	70	140
Shadow (Horse), 1973, carded	70	140
Sitting Bull, 1973, boxed	45	90
Sitting Bull, 1973, carded	45	125
Wild Bill Hickok, 1973, boxed	40	75
Wild Bill Hickok, 1973, carded	40	125
Wyatt Earp, 1973, boxed	40	75
Wyatt Earp, 1973, carded	40	125

Playsets

Dodge City Playset, 1973, vinyl	100	200

Archies Marx, 1975

Archie	25	50
Betty	25	50
Jughead	25	50
Veronica	25	50

Astronauts Marx, 1969

Jane Apollo Astronaut	35	75
Johnny Apollo Astronaut	35	75
Kennedy Space Center Astronaut	35	75

ACTION FIGURES

Clockwise from upper left: General Ursus from Planet of the Apes, 1975, Mego; Indiana Jones, Kenner; Buck Rogers from Captain Action series, 1966-67, Ideal; Ivanhoe from World's Greatest Super Knights, 1975, Mego.

ACTION FIGURES

Banana Splits Sutton, 1970

NAME	MNP	MIP
Bingo the Bear	45	125
Drooper the Lion	45	125
Fleagle Beagle	45	125
Snorky the Elephant	45	125

Batman Returns Kenner 1992-93

All Batman's	5	15
Bruce Wayne	9	18
Catwoman	10	20
Penguin	15	30
Robin	10	22

Beetlejuice Kenner, 1989-90

Adam Maitland	8	20
Creepy Cruiser	5	15
Exploding Beetlejuice	5	10
Harry the Haunted Hunter	8	20
Old Buzzard	8	20
Otho the Obnoxious	8	20
Shipwreck Beetlejuice	5	10
Shish Kabab Beetlejuice	5	15
Showtime Beetlejuice	5	15
Spinhead Beetlejuice	5	15
Street Rat	8	20
Talking Beetlejuice, 12" tall	20	40
Teacher Creature	10	20
Vanishing Vault	10	20

Best of the West Marx, 1960s

Best of the West Series, 1960s

Bill Buck, 1967	100	200
Brave Eagle, 1967	45	90
Buckboard with Horse and Harness	35	75
Chief Cherokee, 1965	45	90
Daniel Boone, 1965	100	200
Davy Crockett	100	200
Fighting Eagle, 1967	45	90
General Custer, 1965	40	80
Geronimo, 1967	45	90
Geronimo and Pinto	40	80
Jamie West, 1967	32	65
Jane West, 1966	40	80
Janice West, 1967	32	65
Jay West, 1967	32	65
Johnny West, 1965	40	80
Johnny West Covered Wagon, with horse and harness	35	75
Johnny West with Comanche	0	0
Josie West, 1967	32	65
Pancho Horse, for 9" figures, 1968	20	40
Princess Wild Flower, 1974	50	100
Sam Cobra, 1972	45	90

Best of the West Marx, 1960s

NAME	MNP	MIP
Sheriff Garrett, 1973	40	80
Thunderbolt Horse	35	75
Zeb Zachary, 1967	40	80

Black Hole Mego, 1979-80

12" Figures

Captain Holland, 1979	25	50
Dr. Alan Durant, 1979	25	50
Dr. Hans Reinhardt, 1979	25	50
Harry Booth, 1979	30	60
Kate McCrae, 1979	35	80
Pizer, 1979	25	50

3 3/4" Figures

Captain Holland, 1979	10	20
Dr. Alan Durant, 1979	10	20
Dr. Hans Reinhardt, 1979	10	25
Harry Booth, 1979	10	25
Humanoid, 1980	70	135
Kate McCrae, 1979	10	25
Maximillian, 1979	17	40
Old B.O.B., 1980	60	120
Pizer, 1979	10	20
S.T.A.R., 1980	60	120
Sentry Robot, 1980	25	60
V.I.N.cent., 1979	30	60

Buck Rogers Mego, 1979

12" Figures

Buck Rogers, 1979	25	50
Doctor Huer, 1979	25	50
Draco, 1979	25	50
Droconian Guard, 1979	25	50
Killer Kane, 1979	25	50
Tiger Man, 1979	25	50
Walking Twiki, 1979	30	60

3 3/4" Figures

Ardella, 1979	6	15
Buck Rogers, 1979	15	30
Doctor Huer, 1979	6	15
Draco, 1979	6	15
Draconian Guard, 1979	10	20
Killer Kane, 1979	6	15
Tiger Man, 1979	6	15
Twiki, 1979	15	25
Wilma Deering, 1979	12	25

3 3/4" Playsets

Star Fighter Command Center, 1979	35	75

3 3/4" Vehicles

Draconian Marauder, 1979	25	50
Land Rover, 1979	20	40
Laserscope Fighter, 1979	20	40
Star Fighter, 1979	25	50
Starseeker, 1979	30	60

ACTION FIGURES

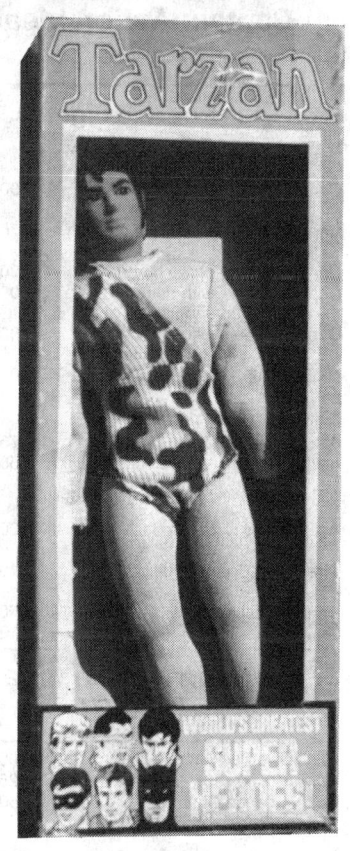

Clockwise from upper left: Ponch figure from CHiPs, 1979, Mego; Captain Action, 1966-67, Ideal; Tarzan, 1972, Mego; Johnny Apollo Astronaut, 1969, Marx.

C.H.I.P.s Mego, 1979

3 3/4" Figures

NAME	MNP	MIP
Jimmy Squeaks, 1979, carded	5	15
Jon, 1979, carded	5	15
Ponch, 1979, carded	5	15
Sarge, 1979, carded	10	20
Wheels Willie, 1979, carded	5	15

8" Figures

Jon, 1979, carded	20	40
Ponch, 1979, carded	15	30
Sarge, 1979, carded	25	50

Accessories, 3 3/4" Figures

Launcher with Motorcycle, 1979, carded	25	50
Motorcycle, 1979, carded	5	15

Accessories, 8" Figures

Motorcycle, 1979, carded	30	60

Captain Action Ideal, 1966-67

12" Posable Figures

*Figures and accessories made by Ideal, 1966-67

Captain Action, box photo Captain Action, 1966	200	375
Captain Action, parachute offer on box, 1967	275	600
Captain Action, photo box, 1967	200	675
Captain Action, with Lone Ranger on box, red shirt, 1966	200	300
Captain Action, with Lone Ranger on box, blue shirt, 1966	200	325
Dr. Evil, 1967	300	600

9" Figures

Action Boy, 1967	275	650
Action Boy, with space suit, 1968	350	825

Accessories

Action Cave Carrying Case, vinyl, 1967	400	0
Directional Communicator Set, 1966	110	300
Dr. Evil Sanctuary, 1967	600	0
Jet Mortar, 1966	110	225

Captain Action Ideal, 1966-67

Parachute Pack, 1966	100	225
Power Pack, 1966	125	250
Quick Change Chamber, Cardboard, Sears Exclusive, 1967	750	0
Silver Streak Amphibian, 1967	500	950

Captain Action Ideal, 1966-67

NAME	MNP	MIP
Silver Streak Garage, cardboard, w/ Silver Streak Vehicle, Sears Exclusive	400	500
Survival Kit, 20 pieces, 1967	125	275
Vinyl Headquarters Carrying Case, Sears Exclusive, 1967	200	0
Weapons Arsenal, ten pieces, 1966	110	225

Action Boy Costumes

Aqualad, 1967	300	525
Robin, 1967	300	625
Superboy, 1967	300	625

Captain Action Costumes

Aquaman, 1966	160	350
Aquaman, with videomatic ring, 1967	180	400
Batman, 1966	225	450
Batman, with videomatic ring, 1967	250	500
Buck Rogers, with videomatic ring, 1967	450	895
Captain America, 1966	220	425
Captain America, with videomatic ring, 1967	225	465
Flash Gordon, 1966	200	425
Flash Gordon, with videomatic ring, 1967	225	475
Green Hornet, with videomatic ring, 1967	1000	3200
Lone Ranger, blue shirt, with videomatic ring, 1967	300	800
Lone Ranger, red shirt, 1966	170	465
Phantom, 1966	150	400
Phantom, with videomatic ring, 1967	175	475
Sergeant Fury, 1966	200	475
Spider-Man, with videomatic ring, 1967	550	1500
Steve Canyon, 1966	150	350
Steve Canyon, with videomatic ring, 1967	175	475
Superman, 1966	200	425
Superman, with videomatic ring, 1967	225	525
Tonto, with videomatic ring, 1967	375	925

Comic Action Heroes Mego, 1975-78

3 3/4" Figures

Aquaman, 1975	30	60
Batman, 1975	20	50
Captain America, 1975	20	50
Green Goblin, 1975	22	55
Hulk, 1975	20	50
Joker, 1975	20	50
Penguin, 1975	20	50
Robin, 1975	20	50
Shazam, 1975	20	50
Spider-Man, 1975	20	50
Superman, 1975	20	45

Comic Action Heroes Mego, 1975-78

NAME	MNP	MIP
Wonder Woman, 1975	20	40

Accessories

Collapsing Tower, 1975, w/Invisble Plane & Wonder Woman	50	125
Exploding Bridge with Batmobile, 1975	75	150
Fortress of Solitude with Superman, 1975	100	200
The Mangler, 1975	55	110

Commander Power Mego, 1975

Figure with Vehicle

Commander Power with Lightning Cycle, 1975, packaged	20	40

DC Comics Super Heroes Toy Biz, 1989

Aquaman	7	15
Aquaman	10	30
Batman	5	10
Batman	5	15
Bob The Goon	7	15
Bob The Goon	10	20
Flash	7	15
Flash	7	15
Flash II	7	15
Flash II	10	25
Green Lantern	10	20
Green Lantern	15	30
Hawkman	7	15
Hawkman	15	30
Joker, no Facial Hair	5	10
Joker, no Facial Hair	5	15
Joker, w/Facial Hair	10	25
Joker, w/Facial Hair	10	25
Lex Luthor	10	
Lex Luthor	5	15
Penguin, short missile	15	30
Penguin, long missile	15	25
Penguin, umbrella-firing	5	10
Penguin, short missile	10	30
Penguin, long missile	10	25
Penguin, umbrella-firing	5	15
Riddler	7	15
Riddler	7	15
Superman	15	35
Superman	15	35
Two Face	7	15
Two Face	15	30
Wonder Woman	5	14
Wonder Woman	5	15

Die-Cast Super Heroes Mego, 1979

6" Figures

NAME	MNP	MIP
Batman, 1979	30	75
Hulk, 1979	25	65
Spider-Man, 1979	30	75
Superman, 1979	30	75

Dukes of Hazzard Mego, 1981-82

3 3/4" Figures

Bo Duke, 1981, carded	8	15
Boss Hogg, 1981, carded	8	15
Cletus, 1981, carded	15	30
Cooter, 1981, carded	15	30
Coy Duke, 1981, carded	15	30
Daisy Duke, 1981, carded	12	25
Luke Duke, 1981, carded	8	15
Rosco Coltrane, 1981, carded	15	30
Uncle Jesse, 1981, carded	15	30
Vance Duke, 1981, carded	15	30

3 3/4" Figures Vehicles

Dasiy Jeep with Daisy, 1981, boxed	25	50
General Lee Car with Bo and Luke, 1981, boxed	25	50

8" Figures

Bo Duke, 1981, carded	15	30
Boss Hogg, 1981, carded	20	35
Coy Duke (card says Bo), 1982, carded	25	50
Daisy Duke, 1981, carded	25	50
Luke Duke, 1981, carded	15	30
Vance Duke (card says Luke), 1982, carded	25	50

Flash Gordon Mego, 1976

9" Figures

Dale Arden, 1976	35	70
Dr. Zarkow, 1976	55	110
Flash, 1976	55	110
Ming, 1976	30	60
Flash Gordon Playset, 1976	55	125

Fort Apache Fighters Marx, 1960s

Captain Maddox, 1967	35	70
Fighting Eagle, 1967	35	70
Fighting Eagle and Comanche	50	100
General Custer, 1967	35	70
Geronimo, 1967	35	70

ACTION FIGURES

Ghostbusters Kenner, 1986-91

NAME	MNP	MIP
*5-1/4" figures & accessories (unless noted) by Kenner	0	0

1986

NAME	MNP	MIP
Bad to the Bone Ghost	5	15
Banshee Bomber Gooper Ghost with Ecto-Plazm	5	15
Bug-Eye Ghost	5	15
Ecto-1	20	40
Egon Spengler & Gulper Ghost	6	15
Firehouse Headquarters	25	50
Ghost Pooper	5	10
Ghost Zapper	5	15
Gooper Ghost Sludge Bucket	5	15
Gooper Ghost Squisher with Ecto-Plazm	5	15
H2 Ghost	5	15
Peter Venkman & Grabber Ghost	6	15
Proton Pack	20	40
Ray Stantz & Wrapper Ghost	6	15
Slimer Plush Figure, 13"	20	35
Slimer with Pizza	20	40
Stay-Puft Marshmallow Man Plush, 13"	15	30
Winston Zeddmore & Chomper Ghost	7	18

1988

NAME	MNP	MIP
Brain Blaster Ghost Haunted Human	5	15
Ecto-2 Helicopter	5	15
Fright Feature Egon	5	15
Fright Feature Janine Melnitz	5	15
Fright Feature Peter	5	15
Fright Feature Ray	5	15
Fright Feature Winston	5	15
Gooper Ghost Slimer	12	25
Granny Gross Haunted Human	5	15
Hard Hat Horror Haunted Human	5	15
Highway Haunter	10	20
Mail Fraud Haunted Human	5	15
Mini Ghost Mini-Gooper	5	10
Mini Ghost Mini-Shooter	5	10
Mini Ghost Mini-Trap	5	10
Pull Speed Ahead Ghost	5	15
Terror Trash Haunted Human	5	15
Tombstone Tackle Haunted Human	5	15
X-Cop Haunted Human	5	15

1989

NAME	MNP	MIP
Dracula	5	15
Ecto-3	5	15
Fearsome Flush	5	10
Frankenstein	3	15
Hunchback	3	15
Mummy	3	15
Screaming Hero Egon	5	15
Screaming Hero Janine Melnitz	5	15
Screaming Hero Peter	5	15
Screaming Hero Ray	5	15

Ghostbusters Kenner, 1986-91

NAME	MNP	MIP
Screaming Hero Winston	5	15
Slimer with Proton Pack, red or blue	15	35
Super Fright Feature Egon Spengler with Slimy Spider	5	15
Super Fright Feature Janine with Boo Fish Ghost	5	15
Super Fright Feature Peter Venkman & Snake Head	5	15
Super Fright Feature Ray	5	15
Super Fright Feature Winston Zeddmore & Meanie Wienie	5	15
Wolfman	5	15
Zombie	5	15

1990

NAME	MNP	MIP
Ecto Bomber with Bomber Ghost	5	15
Ecto-1A with Ambulance Ghost	20	40
Ghost Sweeper, vehicle in box	5	15
Gobblin' Goblin Nasty Neck	6	15
Gobblin' Goblin Terrible Teeth	6	15
Gobblin' Goblin Terror Tongue	6	15
Slimed Hero Egon	5	15
Slimed Hero Louis Tully & Four Eyed Ghost	5	15
Slimed Hero Peter Venkman & Tooth Ghost	5	15
Slimed Hero Ray Stantz & Vapor Ghost	5	15
Slimed Hero Winston	5	15

1991

NAME	MNP	MIP
Ecto-Glow Egon	10	20
Ecto-Glow Louis Tully	10	20
Ecto-Glow Peter	10	20
Ecto-Glow Ray	10	20
Ecto-Glow Winston Zeddmore	10	20

Ghostbusters, Filmation Schaper, 1986

NAME	MNP	MIP
*Produced in 1986 by Schaper	10	15
Belfry and Brat-A-Rat Figures	10	15
Bone Troller	10	15
Eddie	10	15
Fangster	10	15
Fib Face	10	15
Futura	10	15
Ghost Popper Ghost Buggy	20	40
Haunter	10	15
Jake	10	15
Jessica	10	15
Mysteria	10	15
Prime Evil	10	15
Scare Scooter Vehicle	10	20
Scared Stiff	6	15
Time Hopper Vehicle	10	15
Tracy	10	15

ACTION FIGURES

Top to bottom: Demolition set, 1971; Japanese Imperial Soldier, 1966; Forward Observer Set, 1967; Field Communications Set, 1967, all Hasbro.

GI JOE

Accessories

Action Marine Series

NO.	NAME	DESCRIPTION	YEAR	EX	MNP	MIP
7713	Beachhead Assault Field Pack Set	M-1 rifle, bayonet, entrenching shovel and cover, canteen w/cover, belt, mess kit w/cover, field pack, flamethrower, first aid pouch, tent, pegs and poles (complete), tent camo and camo.	1964	94	225	375
7711	Beachhead Assault Tent Set	Tent, flamethrower, pistol belt, first-aid pouch, mess kit with utensils and manual	1964	95	225	375
7715	Beachhead Fatigue Pants		1964	25	55	95
7714	Beachhead Fatigue Shirt	Adventure Pack	1964	25	55	95
7712	Beachhead Field Pack	Cartridge belt, rifle, grenades, field pack, entrenching tool, canteen and manual	1964	20	50	80
7718	Beachhead Flamethrower Set	Reissued Adventure Pack	1967	25	55	90
7718	Beachhead Flamethrower Set	Adventure Pack	1964	25	55	90
7716	Beachhead Mess Kit Set	Adventure Pack	1964	90	210	350
7717	Beachhead Rifle Set	Bayonet, cartridge belt, hand grenades and M-1 rifle.	1964	20	45	75
7717	Beachhead Rifle Set	Reissued Adventure Pack	1967	20	45	75
7703	Communications Field Radio/Telephone Set	Reissued Adventure Pack	1967	20	50	85
7703	Communications Field Set	Adventure Pack	1964	20	50	85
7704	Communications Flag Set	Flags for Army, Navy, Air Corps, Marines and United States.	1964	115	270	450
7702	Communications Poncho	Poncho	1964	40	100	165
7701	Communications Post and Poncho Set	Field radio and telephone, wire roll, carbine, binoculars, map, case, manual, poncho	1964	75	180	300
7710	Dress Parade Set	Reissued Adventure Pack	1968	250	600	1000
7732	Jungle Fighter Set	Reissued Adventure Pack	1968	500	1200	2000
7732	Jungle Fighter Set	Bush hat, jacket w/emblems, pants, flamethrower, field telephone, knife and sheath, pistol belt, pistol, holster, canteen w/cover and knuckle knife.	1967	500	1200	2000
7726	Marine Automatic Machine Gun Set	Adventure Pack	1967	80	195	325
7722	Marine Basics Set	Adventure Pack	1966	25	50	80
7723	Marine Bunk Bed Set	Adventure Pack	1966	90	210	350
7723	Marine Bunk Bed Set	Reissued Adventure Pack	1967	90	210	350
7710	Marine Deluxe Dress Parade Set	Marine jacket, trousers, pistol belt, shoes, hat, M-1 rifle and manual.	1964	235	570	950
7730	Marine Demolition Set	includes figure with camo shirt and pants, cap, boots, mine detector and harness, land mine	1966	65	150	250
7730	Marine Demolition Set	Reissued Adventure Pack	1968	65	150	250
7721	Marine First Aid Set	First-aid pouch, arm band and helmet	1964	25	65	110
7721	Marine First Aid Set	Adventure Pack	1965	25	60	100
7720	Marine Medic Set	Reissued Adventure Pack	1967	20	50	85
7710	Marine Medic Set	Adventure Pack	1965	20	50	85
7720	Marine Medic Set	stretcher, medic bag, Red Cross flag and arm band, helmet, first-aid pouch, stethoscope, plasma bottle, crutches, bandages, splints	1964	25	55	95

Accessories

NO.	NAME	DESCRIPTION	YEAR	EX	MNP	MIP
	Marine Medic Set w/stretcher	First-aid shoulder pouch, stretcher, bandages, arm bands, plasma bottle, stethoscope, Red Cross Flag and manual.	1964	235	570	950
7719	Marine Medic Set w/stretcher	Adventure Pack	1965	235	570	950
7725	Marine Mortar Set	Adventure Pack	1967	55	135	225
7727	Marine Weapons Rack Set	Adventure Pack	1967	90	210	350
7708	Paratrooper Camouflage Set	netting and foliage	1964	15	35	60
7707	Paratrooper Helmet Set	Adventure Pack	1964	15	35	55
7709	Paratrooper Parachute Pack	Adventure Pack	1964	20	50	80
7706	Paratrooper Small Arms Set	Reissued Adventure Pack	1967	375	900	1500
7731	Tank Commander Set	includes figure, "leather" jacket, helmet and visor, insignia, radio with tripod, machine gun, ammo box	1967	425	1020	1700
7731	Tank Commander Set	Reissued Adventure Pack	1968	215	510	850

Action Pilot Series

NO.	NAME	DESCRIPTION	YEAR	EX	MNP	MIP
7822	Air Academy Cadet Set	deluxe set with figure, dress jacket, shoes, and pants, garrison cap, saber and scabbard, white M-1 rifle, chest sash and belt sash	1967	275	660	1100
7822	Air Academy Cadet Set	Reissued Adventure Pack	1968	275	660	1100
7814	Air Force Basics Set	Adventure Pack	1966	30	75	125
7814	Air Force Basics Set	Reissued Adventure Pack	1967	30	75	125
7816	Air Force Mae West Air Vest & Equipment Set	Adventure Pack	1967	45	105	175
7813	Air Force Police Set	Adventure Pack	1965	65	150	250
7813	Air Force Police Set	Reissued Adventure Pack	1967	65	150	250
7815	Air Force Security Set	Air Security radio and helmet, cartridge belt, pistol and holster	1967	215	510	850
7825	Air/Sea Rescue Set	includes figure with black air tanks, rescue ring, buoy, depth gauge, face mask, fins, orange scuba outfit	1967	500	1200	2000
7825	Air/Sea Rescue Set	Reissued Adventure Pack	1968	500	1200	2000
7824	Astronaut Set	Helmet w/visor, foil space suit, booties, gloves, space camera, propellant gun, tether cord, oxygen chest pack, silver boots, white jumpsuit and cloth cap.	1967	250	600	1000
7824	Astronaut Set	Reissued Adventure Pack	1968	250	600	1000
7812	Communications Set		1964	40	90	150
7820	Crash Crew Set	fire proof jacket, hood, pants and gloves, silver boots, belt, flash light, axe, pliers, fire extinguisher, stretcher, strap cutter	1966	75	180	300
7804	Dress Uniform Jacket Set	Adventure Pack	1964	50	120	200
7805	Dress Uniform Pants	Adventure Pack	1964	40	90	150
7803	Dress Uniform Set	Air Force jacket, trousers, shirt, tie, cap and manual	1964	475	1140	1900
7806	Dress Uniform Shirt & Equipment Set	Adventure Pack	1964	50	120	200
7823	Fighter Pilot Set	working parachute and pack, gold helmet, Mae West vest, green pants, flash light, orange jump suit, black boots	1967	550	1320	2200
7823	Fighter Pilot Set	Reissued Adventure Pack	1968	550	1320	2200
7812	Scramble Communications Set	Reissued Adventure Pack	1967	40	90	150

Accessories

NO.	NAME	DESCRIPTION	YEAR	EX	MNP	MIP
7812	Scramble Communications Set	Poncho, field telephone and radio, map w/case, binoculars and wire roll.	1965	40	90	150
7810	Scramble Crash Helmet	helmet, face mask, hose, tinted visor	1964	50	120	200
7810	Scramble Crash Helmet	Reissued Adventure Pack	1967	50	120	200
7808	Scramble Flight Suit	gray flight suit	1964	35	75	130
7808	Scramble Flight Suit		1967	35	75	130
7811	Scramble Parachute Pack	Adventure Pack	1964	20	40	70
7807	Scramble Parachute Set	Deluxe set, gray flight suit, orange air vest, white crash helmet, pistol belt w/.45 pistol, holster, clipboard, flare gun and parachute w/insert.	1964	125	300	500
7809	Scramble Parachute Set	Reissued Adventure Pack	1967	115	270	450
7802	Survival Life Raft Set	Raft with oars and sea anchor	1964	55	135	225
7801	Survival Life Raft Set	Raft with oars, flare gun, knife, air vest, first-aid kit, sea anchor and manual	1964	215	510	850

Action Sailor Series

NO.	NAME	DESCRIPTION	YEAR	EX	MNP	MIP
7624	Annapolis Cadet	Reissued Adventure Pack	1968	300	720	1200
7624	Annapolis Cadet	Garrison cap, dress jacket, pants, shoes, sword, scabbard, belt and white M-1 rifle, (add $200 for photo box)	1967	225	550	900
7625	Breeches Buoy	yellow jacket and pants, chair and pulley, flare gun, blinker light	1967	375	900	1500
7625	Breeches Buoy	Reissued Adventure Pack	1968	375	900	1500
7623	Deep Freeze	Reissued Adventure Pack	1968	325	775	1300
7623	Deep Freeze	White boots, fur parka, pants, snow shoes, ice axe, snow sled w/rope and flare gun.	1967	325	775	1300
7620	Deep Sea Diver Equipment Set	Diving suit, helmet with breast plates, weighted belts and shoes, air pump, hose, tools, signal floats	1964	325	775	1300
7620	Deep Sea Diver Set	Underwater uniform, helmet, upper and lower plate, sledge hammer, buoy w/rope, gloves, compass, hoses, lead boots and weight belt	1965	325	775	1300
7620	Deep Sea Diver Set	Reissued Adventure Pack	1968	325	775	1300
7604	Frogman Scuba Bottom Set	Adventure Pack	1964	15	35	60
7605	Frogman Scuba Equipment Set	Adventure Pack	1964	15	35	60
7606	Frogman Scuba Tank Set	Adventure Pack	1964	15	35	60
7603	Frogman Scuba Top Set	Adventure Pack	1964	30	75	125
7602	Frogman Underwater Demolition Set	Headpiece, face mask, swim fins, rubber suit, scuba tank, depth gauge, knife, dynamite and manual	1964	200	475	800
7621	Landing Signal Officer	jumpsuit, signal paddles, goggles, cloth head gear, headphones, clipbaord (complete), binoculars and flare gun.	1966	135	325	550
7610	Navy Attack Helmet Set	shirt and pants, boots, yellow life vest, blue helmet, flare gun binoculars, signal flags	1964	45	105	175
7611	Navy Attack Life Jacket	Adventure Pack	1964	15	30	50
7607	Navy Attack Set	life jacket, field glasses, blinker light, signal flags, manual	1964	50	120	200
7609	Navy Attack Work Pants Set	Adventure Pack	1964	20	45	75
7608	Navy Attack Work Shirt Set	Adventure Pack	1964	20	45	75

Accessories

NO.	NAME	DESCRIPTION	YEAR	EX	MNP	MIP
7628	Navy Basics Set	Adventure Pack	1966	25	60	100
7619	Navy Dress Parade Set	Billy club, cartridge belt, bayonet and white dress rifle	1964	20	50	85
7619	Navy Dress Parade Rifle Set	Adventure Pack	1965	20	50	85
7602	Navy Frogman	Reissued Adventure Pack	1968	200	475	800
7626	Navy L.S.O. Equipment Set	helmet, headphones, signal paddles, flare gun	1966	30	75	125
7627	Navy Life Ring Set	U.S.N. life ring, helmet sticker	1966	20	45	75
7618	Navy Machine Gun Set	Adventure Pack	1965	25	55	95
7601	Sea Rescue	life raft, oar, anchor, flare gun, first-aid kit, knife, scabbard, manual	1964	175	450	750
7622	Sea Rescue Set	Reissued with life preserver	1966	175	450	750
7612	Shore Patrol	dress shirt, tie and pants, helmet, white belt, .45 and holster, billy club, boots, arm band, sea bag	1964	250	600	1000
7612	Shore Patrol	Adventure Pack	1967	250	600	1000
7613	Shore Patrol Dress Jumper Set	Adventure Pack	1964	90	210	350
7614	Shore Patrol Dress Pant Set	Adventure Pack	1964	20	50	85
7616	Shore Patrol Helmet and Small Arms Set	white belt, billy stick, white helmet, .45 pistol	1964	25	60	100
7615	Shore Patrol Sea Bag Set	Adventure Pack	1964	11	25	45

Action Soldier Series

NO.	NAME	DESCRIPTION	YEAR	EX	MNP	MIP
8007.83	Adventure Pack with 16 items	Adventure Pack	1968	120	285	475
8006.83	Adventure Pack with 12 items	Adventure Pack	1968	120	285	475
8008.83	Adventure Pack with 14 pieces	Adventure Pack	1968	120	285	475
8005.83	Adventure Pack with 12 items	Adventure Pack	1968	120	285	475
7549-83	Adventure Pack, Army Bivouac Series		1968	300	720	1200
7813	Air Police Equipment	gray field phone, carbine, white helmet and bayonet	1964	45	105	175
8000	Basic Footlocker	wood tray	1964	40	90	150
7513	Bivouac Deluxe Pup Tent Set	M-1 rifle and bayonet, shovel and cover, canteen and cover, mess kit, cartridge belt, machine gun, tent, pegs, poles, camoflage, sleeping bag, netting, ammo box	1964	65	150	250
7514	Bivouac Machine Gun Set	Reissue	1967	15	40	65
7514	Bivouac Machine Gun Set	machine gun set and ammo box	1964	15	40	65
7512	Bivouac Sleeping Bag Set	mess kit, canteen, bayonet, cartridge belt, M-1 rifle, manual	1964	50	120	200
7515	Bivouac Sleeping Bag	zippered bag	1964	15	30	50
7511	Combat Camoflaged Netting Set	foliage and posts	1964	6	15	25
7572	Combat Construction Set	orange safety helmet, work gloves, jack hammer	1967	135	325	550
7573	Combat Demolition Set		1967	90	210	350
7571	Combat Engineer Set	helmet, machine gun, tripod and transit	1967	65	150	250
7504	Combat Fatigue Pants Set		1964	25	55	95
7503	Combat Fatigue Shirt Set		1964	25	55	95
7505	Combat Field Jacket		1964	45	105	175
7501	Combat Field Jacket Set	Jacket, bayonet, cartridge belt, hand grenades, M-1 rifle and manual	1964	100	240	400
7506	Combat Field Pack & Entrenching Tool		1964	20	45	75

ACTION FIGURES

Accessories

NO.	NAME	DESCRIPTION	YEAR	EX	MNP	MIP
7502	Combat Field Pack Deluxe Set	field jacket, pack, entrenching shovel w/cover, mess kit, first-aid pouch, canteen w/cover	1964	55	135	225
7507	Combat Helmet Set	with netting and foliage leaves	1964	15	35	55
7509	Combat Mess Kit	plate, fork, knife, spoon, canteen	1964	15	35	55
7705	Combat Paratrooper Parachute Pack Set	first-aid pouch, pistol, belt, carbine, 6 grenades, knife, scabbard, canteen, manual	1964	90	210	350
7510	Combat Rifle and Helmet Set	bayonet, M-1 rifle, belt and grenades	1967	25	55	90
7508	Combat Sandbags Set		1964	8	20	30
7520	Command Post Field Radio and Telephone Set	Field radio, telephone with wire roll and map	1964	20	45	75
7520	Command Post Field Radio and Telephone Set	Reissue	1967	20	45	75
7517	Command Post Poncho Set	Poncho, field radio and telephone, wire roll, pistol, belt and holster, map and case and manual.	1964	45	105	175
7519	Command Post Poncho		1964	15	35	55
7518	Command Post Small Arms Set	Holster and .45 pistol, belt, grenades	1964	20	50	80
8009.83	Dress Parade Adventure Pack with 37 pieces	Adventure Pack	1968	375	900	1500
7533	Green Beret and Small Arms Set	Reissue	1967	50	120	200
7533	Green Beret and Small Arms Set		1966	50	120	200
5978	Green Beret Machine Gun Outpost Set	Sear exclusive	1966	300	725	1200
7538	Heavy Weapons Set	mortar launcher and shells, M-60 machine gun, grenades, flak jacket, shirt and pants	1967	250	600	1000
7538	Heavy Weapons Set	Reissue	1968	250	600	1000
7523	Military Police Duffle Bag Set		1964	10	25	45
7526	Military Police Helmet and Small Arms Set		1964	20	50	85
7526	Military Police Helmet and Small Arms Set	Reissue	1967	20	50	85
7524	Military Police Ike Jacket	Jacket with red scarf and arm band	1964	30	70	115
7525	Military Police Ike Pants	Matches Ike Jacket	1964	20	45	75
7521	Military Police Uniform Set	includes figure, with Ike jacket and pants, scarf, boots, helmet, belt with ammo pouches, .45 pistol and holster, billy club, armband, duffle bag	1964	250	600	1000
7539	Military Police Uniform Set	includes figure in green or tan uniform, black and gold MP Helmet, billy club, belt, pistol and holster, MP armband and red tunic	1967	500	1200	2000
7539	Military Police Uniform Set	Reissue	1968	500	1200	2000
7530	Mountain Troops Set	snow shoes, ice axe, ropes, grenades, camoflage pack, web belt, manual	1964	45	110	185
7516	Sabotage Set	dingy and oar, blinker light, detonator w/strap, TNT, wool stocking cap, gas mask, binoculars, green radio and .45 pistol and holster	1967	235	575	950
7516	Sabotage Set	Reissue in photo box	1968	235	575	950
7531	Ski Patrol Deluxe Set	White parka, boots, goggles, mittens, skis, poles and manual	1964	165	390	650
7527	Ski Patrol Helmet and Small Arms Set		1965	20	50	85

ACTION FIGURES

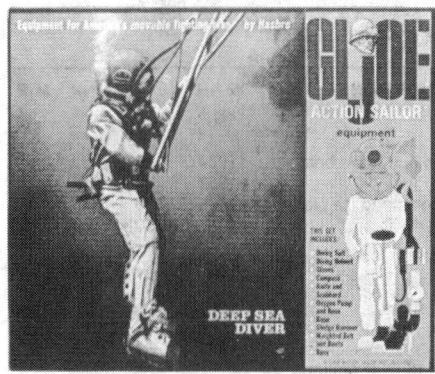

Top to bottom: Deep Sea Diver, 1968; Spacewalk Mystery, 1970; Talking GI Joe series with Kung Fu Grip, 1974; Official Sea Sled and Frogman, 1964, all Hasbro.

Accessories

NO.	NAME	DESCRIPTION	YEAR	EX	MNP	MIP
7527	Ski Patrol Helmet and Small Arms Set	Reissue	1967	20	50	85
7527	Ski Patrol Set	Helmet, grenades and white belt	1964	20	50	85
7529	Snow Troops Set	Reissue	1967	20	45	75
7529	Snow Troops Set	snow shoes, goggles and ice pick	1966	20	45	75
7528	Special Forces Bazooka Set		1966	50	120	200
7528	Special Forces Bazooka Set	Reissue	1967	50	120	200
7532	Special Forces Uniform Set		1966	250	600	1000
7537	West Point Cadet Uniform Set	Dress jacket, pants, shoes, chest and belt sash, parade hat w/plume, saber, scabbard and white M-1 rifle.	1967	250	600	1000
7537	West Point Cadet Uniform Set	Reissue	1968	250	600	1000

Action Soldiers of the World

NO.	NAME	DESCRIPTION	YEAR	EX	MNP	MIP
8305	Australian Jungle Fighter Set		1966	80	195	325
8304	British Commando Set	Sten submachine gun, gas mask and carrier, canteen and cover, cartridge belt, rifle, "Victoria Cross" medal, manual	1966	90	210	350
8303	French Resistance Fighter Set	shoulder holster, Lebel pistol, knife, grenades, radio, 7.65 submachine gun, "Croix de Guerra" medal, counter-intelligence manual	1966	45	105	175
8300	German Storm Trooper		1966	75	175	295
8301	Japanese Imperial Soldier Set	field pack, Nambu pistol and holster, Arisaka rifle with bayonet, cartridge belt, "Order of the Kite" medal, counter-intelligence manual	1966	90	210	350
8302	Russian Infantry Man Set	DP light machine gun, bipod, field glasses and case, anti-tank grenades, ammo box, "Order of Lenin" medal, counter-intelligence medal	1966	65	150	250

Adventure Team

NO.	NAME	DESCRIPTION	YEAR	EX	MNP	MIP
7490	Adventure Team Headquarters Set	Adventure Team playset	1972	30	75	125
7495	Adventure Team Training Center Set	rifle rack, logs, barrel, barber wire, rope ladder, 3 tires, 2 targets, escape slide, tent and poles, first aid kit, respirator and mask, snake, instructions	1973	35	85	140
7345	Aerial Reconnaissance Set	jumpsuit, helmet, aerial recon vehicle with built-in camera	1971	20	45	75
7420	Attack at Vulture Falls	Super Deluxe Set	1975	20	50	85
7414	Black Widow Rendezvous	Super Deluxe Set	1975	25	60	100
7328-5	Buried Bounty	Deluxe Set	1975	15	40	65
7437	Capture of the Pygmy Gorilla Set	Adventure Pack	1970	55	135	225
8032	Challenge of Savage River	Deluxe Set	1975	15	40	65
7313	Chest Winch Set	Reissued Adventure Pack	1974	6	15	25
7313	Chest Winch Set	Adventure Pack	1972	6	15	25
8033	Command Para Drop	Deluxe Set	1975	13	30	50
7308-3	Copter Rescue Set	blue jumpsuit, red binoculars	1973	8	20	30
7412	Danger of the Depths Set	Adventure Pack	1970	45	105	175

Accessories

NO.	NAME	DESCRIPTION	YEAR	EX	MNP	MIP
7338-1	Danger Ray Detection	magnetic ray detector, solar communicator with headphones, 2-piece uniform, instructions and comic	1975	20	50	80
7309-2	Dangerous Climb Set	Adventure Pack	1973	8	20	30
7608-5	Dangerous Mission Set	green shirt, pants, hunting rifle	1973	8	20	30
7371	Demolition Set	Adventure Pack with land mines, mine detector and carrying case.	1971	15	35	60
7370	Demolition Set	armored suit, face shield, bomb, bomb disposal box, extension grips	1971	10	25	45
7309-5	Desert Explorer Set	Adventure Pack	1973	8	20	30
7308-6	Desert Survival Set	Adventure Pack	1973	8	20	30
8031	Dive to Danger	Mike Powers set, orange scuba suit, fins, mask, spear gun, shark, buoy, knife and scabbard, mini sled, air tanks, comic	1975	45	105	175
7328-6	Diver's Distress	Deluxe Set	1975	10	20	35
7364	Drag Bike Set	3-wheel motorcycle brakes down to backpack size	1971	20	45	75
7422	Eight Ropes of Danger Set	Adventure Pack	1970	50	120	200
7374	Emergency Rescue Set	shirt, pants, rope ladder and hook, walkie talkie, safety belt, flashlight, oxygen tank, axe, first aid kit	1971	15	35	60
7319-5	Equipment Tester Set		1972	8	20	30
7360	Escape Car Set		1971	20	45	75
7319-1	Escape Slide Set	Adventure Pack	1972	8	20	30
8028-2	Fangs of the Cobra	Deluxe Set	1975	8	20	30
7423	Fantastic Freefall Set	Adventure Pack	1970	45	105	175
7982	Fight for Survival Set w/Polar Explorer	Adventure Pack	1969	145	345	575
7308-2	Fight for Survival Set	brown shirt and pants, machete	1973	8	20	30
7351	Fire Fighter Set	Adventure Pack	1971	10	25	45
7431	Flight For Survival Set	Adventure Pack	1970	95	225	375
7361	Flying Rescue Set	Adventure Pack	1971	10	25	45
7425	Flying Space Adventure Set	Adventure Pack	1970	100	240	400
8000	Footlocker	Adventure Pack	1974	25	60	100
7328-4	Green Danger		1975	10	20	35
7415	Hidden Missile Discovery Set	Adventure Pack	1970	35	80	135
7308-1	Hidden Treasure Set	shirt, pants, pick axe, shovel	1973	8	20	30
7342	High Voltage Escape Set	net, jumpsuit, hat, wrist meter, wire cutters, wire, warning sign	1971	20	45	75
7343	Hurricane Spotter Set	slicker suit, rain measure, portable radar, map and case, binoculars	1971	15	40	65
7421	Jaws of Death	Super Deluxe Set	1975	35	90	150
7339-2	Jettison to Safety	infrared terrain scanner, mobile rocket pack, 2-piece flight suit, instructions and comic	1975	20	50	80
7309-3	Jungle Ordeal Set	Adventure Pack	1973	8	20	30
7373	Jungle Survival Set	Adventure Pack	1971	25	60	100
7372	Karate Set	Adventure Pack	1971	15	40	65
7311	Laser Rescue Set	Reissued Adventure Pack	1974	10	20	35
7311	Laser Rescue Set	hand-held laser with backpack generator	1972	10	20	30
7353	Life-Line Catapult Set	Adventure Pack	1971	10	25	40
7328-3	Long Range Recon	Deluxe Set	1975	9	20	35
7319-2	Magnetic Flaw Detector Set	Adventure Pack	1972	8	20	30
7339-3	Mine Shaft Breakout	sonic rock blaster, chest winch, 2-piece uniform, netting, instructions, comic	1975	10	25	40
7340	Missile Recovery Set	Adventure Pack	1971	15	35	55

Accessories

NO.	NAME	DESCRIPTION	YEAR	EX	MNP	MIP
	Mystery of the Boiling Lagoon	Sears, pontoon boat, diver's suit, diver's helmet, weighted belt and boots, depth gauge, air hose, buoy, nose cone, pincer arm, instructions	1973	90	210	350
7338-2	Night Surveillance	Deluxe Set	1975	10	25	45
7416	Peril of the Raging Inferno	fire proof suit, hood and boots, breathing apparatus, camera, fire extinguisher, detection meter, gaskets	1975	55	135	225
7309-4	Photo Reconnaissance Set	Adventure Pack	1973	8	20	30
8028-1	Race for Recovery		1975	8	20	30
7341	Radiation Detection Set	jumpsuit with belt, "uranium ore", goggles, container, pincer arm.	1971	20	45	75
7339-1	Raging River Dam Up		1975	20	45	75
7350	Rescue Raft Set	Adventure Pack	1971	20	45	75
7413	Revenge of the Spy Shark	Super Deluxe Set	1975	65	150	250
7312	Rock Blaster	sonic blaster with tripod, backpack generator, face shield	1972	6	15	25
7315	Rocket Pack Set	Adventure Pack	1972	8	20	30
7315	Rocket Pack Set	Reissued Adventure Pack	1974	9	20	35
7319-3	Sample Analyzer Set	Adventure Pack	1972	8	20	30
7439.16	Search for the Abominable Snowman Set	Sears, white suit, belt, goggles, gloves, rifle, skis and poles, show shoes, sled, rope, net, supply chest, binoculars, Abominable Snowman, comic book	1973	80	195	325
7375	Secret Agent Set	Adventure Pack	1971	25	60	100
7328-1	Secret Courier	Deluxe Set	1975	10	20	35
7309-1	Secret Mission Set	Adventure Pack	1973	8	20	30
8030	Secret Mission Set	Deluxe Set	1975	15	30	50
7411	Secret Mission to Spy Island Set	comic, inflatable raft with oar, binoculars, signal light, flare gun, TNT and detonator, wire roll, boots, pants, sweater, black cap, camera, radio with earphones, .45 submachine gun	1970	50	120	200
7308-4	Secret Rendezvous Set	parka, pants, flare gun	1973	8	20	30
7319-6	Seismograph Set	Adventure Pack	1972	8	20	30
7338-3	Shocking Escape	escape slide, chest pack climber, jumpsuit with gloves and belt, high voltage sign, instructions and comic	1975	10	25	45
7362	Signal Flasher Set	large back pack type signal flash unit	1971	15	30	50
7440	Sky Dive to Danger	Super Deluxe Set	1975	45	105	175
7314	Solar Communicator Set	Reissue	1974	10	20	35
7314	Solar Communicator Set	Adventure Pack	1972	10	20	35
7312	Sonic Rock Blaster Set	Adventure Pack	1972	8	20	30
7312	Sonic Rock Blaster Set	Reissued Adventure Pack	1974	10	20	35
8028-3	Special Assignment	Deluxe Set	1975	6	15	25
7319-4	Thermal Terrain Scanner Set	Adventure Pack	1972	8	20	30
7480	Three in One Super Adventure Set	Cold of the Arctic, Heat of the Desert and Danger of the Jungle Adventure Pack	1971	150	360	600
7480	Three in One Super Adventure Set	Danger of the Depths, Secret Mission to Spy Island and Flying Space Adventure Packs	1971	175	425	700
7328-2	Thrust into Danger	Deluxe Set	1975	10	20	35
59289	Trouble at Vulture Pass	Sears exclusive, Super Deluxe Set	1975	40	90	150
7363	Turbo Copter Set	strap-on one man helicopter	1971	10	25	45
7309-6	Undercover Agent Set	trenchcoat and belt, walkie-talkie	1973	8	20	30

Accessories

NO.	NAME	DESCRIPTION	YEAR	EX	MNP	MIP
7310	Underwater Demolition Set	Reissued Adventure Pack	1974	10	20	35
7310	Underwater Demolition Set	hand-held propulsion device, breathing apparatus, dynamite	1972	8	20	30
7354	Underwater Explorer Set	self propelled underwater device	1971	15	30	50
7344	Volcano Jumper Set	jumpsuit with hood, belt, nylon rope, chest pack, TNT pack	1971	15	35	55
7436	White Tiger Hunt Set	hunter's jacket and pants, hat, rifle, tent, cage, chain, campfire, white tiger, comic	1970	65	150	250
7353	Windboat Set	back pack, sled with wheels, sail	1971	10	25	40
7309-4	Winter Rescue Set	Replaced Photo Reconnaissance Set - Adventure Pack	1973	10	25	40

Adventures of GI Joe

NO.	NAME	DESCRIPTION	YEAR	EX	MNP	MIP
7940	Adventure Locker	Footlocker	1969	70	165	275
7941	Aqua Locker	Footlocker	1969	75	180	300
7942	Astro Locker	Footlocker	1969	75	180	300
7920	Danger of the Depths Underwater Diver Set	Basic Adventure Pack	1969	90	210	350
7950	Eight Ropes of Danger Set	Deluxe Adventure Pack, diving suit, treasure chest, octopus	1969	95	225	375
7951	Fantastic Freefall Set	includes figure with parachute and pack, blinker light, air vest, flash light, crash helmet with visor and oxygen mask, dog tags, orange jump suit, black boots	1969	100	235	390
7982.83	Flight for Survival Set w/o Polar Explorer	Reissued Adventure Pack	1969	130	315	525
7952	Hidden Missile Discovery Set	Deluxe Adventure Pack	1969	95	230	385
7953	Mouth of Doom Set	Deluxe Adventure Pack	1969	80	195	325
7921	Mysterious Explosion Set	Basic Adventure Pack	1969	75	180	300
7923	Perilous Rescue Set	Basic Adventure Pack	1969	100	240	400
7922	Secret Mission to Spy Island Set	Basic Adventure Pack	1969	95	225	375

G.I. Joe Action Series, Army, Navy, Marine and Air Force

NO.	NAME	DESCRIPTION	YEAR	EX	MNP	MIP
8000	Basic Footlocker	Adventure Pack	1965	40	90	150
8002.83	Footlocker Adventure Pack with 22 items	Adventure Pack	1968	60	145	240
8001.83	Footlocker Adventure Pack with 15 pieces	Adventure Pack	1968	55	135	225
8002.83	Footlocker Adventure Pack with 15 pieces	Adventure Pack	1968	55	135	225
8000.83	Footlocker Adventure Pack with 16 pieces	Adventure Pack	1968	55	135	225

Figure Sets

Action Girl Series

NO.	NAME	DESCRIPTION	YEAR	EX	MNP	MIP
8060	G.I. Nurse	Red Cross hat and arm band, white dress, stockings, shoes, crutches, medic bag, stethescope, plasma bottle, bandages and splints.	1967	750	1800	3000

Action Marine Series

NO.	NAME	DESCRIPTION	YEAR	EX	MNP	MIP
90711	Marine Medic Series	camo shirt and pants, boots, Red Cross helmet, flag and arm bands, crutch, bandages, splints, first aid pouch, stethoscope, plasma bottle, stretcher, medic bag, belt with ammo pouches	1967	325	810	1350

ACTION FIGURES

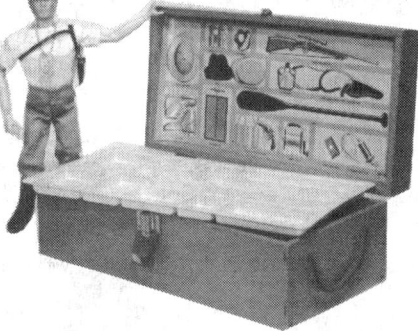

Top to bottom: Black Action Soldier, 1965; USMC Jungle Fighter, 1967; Foot-locker Adventure Pack, 1968; Motorized Battle Tank, 1982, all Hasbro.

Accessories

NO.	NAME	DESCRIPTION	YEAR	EX	MNP	MIP
7790	Talking Action Marine		1967	175	420	700
7790	Talking Action Marine	figure with camo shirt and pants, cap, boots, dog tags	1968	165	395	650
90712	Talking Adventure Pack and Field Pack Equip.	Talking Adventure Pack	1968	300	725	1200
90711	Talking Adventure Pack and Tent Set	Talking Adventure Pack	1968	300	725	1200

Action Pilot Series

NO.	NAME	DESCRIPTION	YEAR	EX	MNP	MIP
7800	Action Pilot	Orange jumpsuit, blue cap, black boots, dog tags, insignias, manual, catalog and club application	1964	115	275	450
7890	Talking Action Pilot	Talking Adventure Pack	1967	235	575	950

Action Sailor Series

NO.	NAME	DESCRIPTION	YEAR	EX	MNP	MIP
7600	Action Sailor	White cap, denim shirt and pants, boots, dog tags, navy manual and insignias.	1964	90	210	350
7643-83	Navy Scuba Set	Adventure Pack	1968	350	850	1400
7690	Talking Action Sailor		1967	165	395	650
90621	Talking Landing Signal Officer	Talking Adventure Pack	1968	625	1500	2500
90612	Talking Shore Patrol Equipment Set	Talking Adventure Pack	1968	625	1500	2500

Action Soldier Series

NO.	NAME	DESCRIPTION	YEAR	EX	MNP	MIP
7700	Action Marine	Fatiques, green cap, boots, dog tags, insignias and manual.	1964	95	225	375
7500	Action Soldier	Fatique cap, shirt, pants, boots, dog tags, army manual and insignias, helmet, belt with pouches, M-1 rifle	1964	75	180	300
7900	Black Action Soldier		1965	325	775	1300
5004	Canadian Mountie Set	Sears exclusive	1967	375	900	1500
8030	Desert Patrol Attack Jeep Set	G.I. Joe Desert Fighter figure, jeep with steering wheel, spare tire, tan tripod, gun and gun mount and ring, black antenna, tan jacket and shorts, socks, goggles	1967	375	900	1500
5969	Forward Observer Set	Sears exclusive.	1966	125	300	500
7536	Green Beret	Field radio, bazooka rocket, bazooka, green beret, jacket, pants, M-16 rifle, grenades, camo scarf, belt pistol and holster.	1966	625	1500	2500
7522	Jungle Fighter		1966	35	90	145
7522	Jungle Fighter	Reissue	1967	35	90	150
7531	Machine Gun Emplacement Set	Sears exclusive.	1965	200	475	800
7590	Talking Action Soldier		1967	100	240	400
7590	Talking Action Soldier		1968	100	240	400
90517	Talking Adventure Pack, Command Post Equip.		1968	300	725	1200
7557-83	Talking Adventure Pack, Mountain Troop Series		1968	300	725	1200
90532	Talking Adventure Pack, Special Forces Equip.		1968	300	725	1200
90513	Talking Aventure Pack, Bivouac Equipment		1968	500	1200	2000

Action Soldiers of the World

NO.	NAME	DESCRIPTION	YEAR	EX	MNP	MIP
8205	Australian Jungle Fighter	Standard set with action figure uniform, no equipment	1966	350	850	1400

ACTION FIGURES

Figure Sets

NO.	NAME	DESCRIPTION	YEAR	EX	MNP	MIP
8105	Australian Jungle Fighter	action figure with jacket, shorts, socks, boots, bush hat, belt, "Victoria Cross" medal, knuckle knife, flamethrower, entrenching tool, bush knife and sheath	1966	1000	2400	4000
8204	British Commando	Standard set with equipment	1966	300	725	1200
8104	British Commando	Deluxe set with action figure, helmet, night raidgreen jacket, pants, boots, canteen and cover, gas mask and cover, belt, Sten sub machine gun, gun clip and "Victoria Cross" medal.	1966	1000	2400	4000
8111-83	Foreign Soldiers of the World	Talking Adventure Pack, Sears exclusive	1968	750	1800	3000
8203	French Resistance Fighter	Standard set with action figure and equipment	1966	315	750	1250
8103	French Resistance Fighter	Deluxe set with figure, beret, short black boots, black sweater, denim pants, "Croix de Guerre" medal, knife, shoulder holster, pistol, radio, sub machine gun and grenades.	1966	1000	2400	4000
8100	German Storm Trooper	Deluxe set with figure, helmet, jacket, pants, boots, Luger pistol, holster, cartridge belt, cartridges, "Iron Cross" medal, stick grenades, 9MM Schmeisser, field pack	1966	1000	2400	4000
8200	German Storm Trooper	Standard set with equipment	1966	315	750	1250
8201	Japanese Imperial Soldier	Standard set with equipment	1966	350	850	1400
8101	Japanese Imperial Soldier	Deluxe set with figure, Arisaka rifle, belt, cartridges, field pack, Nambu pistol, holster, bayonet, "Order of the Kite" medal, helmet, jacket, pants, short brown boots	1966	1000	2400	4000
8202	Russian Infantry Man	Standard set with equipment	1966	350	850	1400
8102	Russian Infantry Man	Deluxe set with action figure, fur cap, tunic, pants, boots, ammo box, ammo rounds, anti-tank grenades, belt, bipod, DP light machine gun, "Order of Lenin" medal, field glasses and case.	1966	1000	2400	4000
5038	Uniforms of Six Nations		1967	425	1025	1700

Adventure Team

NO.	NAME	DESCRIPTION	YEAR	EX	MNP	MIP
7272	Air Adventurer	Adventure Pack with "New" Life-Like action figure, uniform and equipment	1976	60	145	245
7282	Air Adventurer	Adventure Pack action figure with Life-Like Body and Kung Fu Grip	1974	55	135	225
7282	Air Adventurer	Adventure Pack with "New" action figure, uniform and equipment	1976	75	180	300
7403	Air Adventurer	includes figure with kung fu grip, orange flight suit, boots, insignia, dog tags, rifle, boots, warranty, club insert	1970	55	135	225
7273	Black Adventurer	Adventure Pack	1976	50	120	200
7283	Black Adventurer	Adventure Pack	1976	45	105	175
7283	Black Adventurer	Adventure Pack figure with Life-Like Body and Kung Fu Grip	1974	65	150	250
7404	Black Adventurer	includes figure, shirt with insignia, pants, boots, dog tags, shoulder holster with pistol	1970	75	180	300
8026	Bulletman	Adventure Pack	1976	20	50	85

ACTION FIGURES

Accessories

NO.	NAME	DESCRIPTION	YEAR	EX	MNP	MIP
7278	Eagle Eye Black Commando	Adventure Pack	1976	30	75	125
7276	Eagle Eye Land Commander	Adventure Pack	1976	30	70	115
7277	Eagle Eye Man of Action	Adventure Pack	1976	30	75	125
8050	Intruder Commander	"New" Life-Like action figure and equipment	1976	20	50	80
8051	Intruder Warrior	"New" Life-Like action figure with equipment	1976	25	60	100
7270	Land Adventurer	Adventure Pack	1976	45	105	175
7280	Land Adventurer	Action figure with Life-Like Body and Kung Fu Grip	1974	50	120	200
7280	Land Adventurer	Adventure Pack	1976	50	120	200
7401	Land Adventurer	includes figure, camo shirt and pants, boots, insignia, shoulder holster and pistol, dog tags and team inserts	1970	45	105	175
7284	Man of Action	Adventure Pack figure with Life-Like Body and Kung Fu Grip	1974	60	135	230
7284	Man of Action	Adventure Pack	1976	45	105	175
7274	Man of Action	Adventure Pack	1976	45	105	175
7500	Man of Action	includes figure, shirt and pants, boots, insignia, dog tags, team inserts	1970	50	120	200
8025	Mike Powers/Atomic Man	figure with "atomic" flashing eye, arm that spins hand-held helicopter	1975	20	45	75
7271	Sea Adventurer	Adventure Pack	1976	40	90	150
7281	Sea Adventurer	Adventure Pack figure with Life-Like Body and Kung Fu Grip	1974	55	135	225
7281	Sea Adventurer	Adventure Pack	1976	55	125	210
7402	Sea Adventurer	includes figure, shirt, dungarees, insignia, boots, shoulder holster and pistol	1970	60	145	245
8040	Secret Mountain Outpost		1975	20	50	85
7290	Talking Adventure Team Commander	Talking Adventure Pack figure with Life-Like Body and Kung Fu Grip	1974	75	175	300
7291	Talking Adventure Team Black Commander	Talking Adventure Pack figure with Life-Like Body and Kung Fu Grip	1974	125	300	500
7400	Talking Adventure Team Commander	includes figure, 2-pocket green shirt, pants, boots, insignia, instructions, dog tag, shoulder holster and pistol	1970	65	150	250
7406	Talking Adventure Team Black Commander	Talking Adventure Pack	1973	150	360	600
7590	Talking Astronaut	Talking Adventure Pack	1970	90	210	350
7291	Talking Black Commander	Adventure Pack	1976	125	300	500
7290	Talking Commander	Adventure Pack	1976	75	180	300
7292	Talking Man of Action	Adventure Pack	1976	75	180	300
7590	Talking Man of Action	talking figure, shirt, pants, boots, dog tags, rifle, insignia, instructions	1970	75	180	300
7292	Talking Man of Action	Talking Adventure Pack figure with Life-Like Body and Kung Fu Grip	1974	75	180	300

Adventures of GI Joe

NO.	NAME	DESCRIPTION	YEAR	EX	MNP	MIP
7910	Aquanaut	Adventure Pack	1969	500	1200	2000
7905	Negro Adventurer	Sears exclusive, includes painted hair figure, blue jeans, pullover sweater, shoulder holster and pistol, plus product letter from Sears	1969	400	950	1600
7980	Sharks Surprise Set w/Frogman	with figure, orange scuba suit, blue sea sled, air tanks, harpoon, face mask, treasure chest, shark, instructions and comic	1969	125	300	500

ACTION FIGURES

Clockwise from upper left: Windboat, 1971; Flying Rescue, 1971; Combat Field Jacket set, 1964; Official Jeep from Combat Set, 1965; Man of Action, 1964; Combat Field Pack set, 1964, all Hasbro.

Figure Sets

NO.	NAME	DESCRIPTION	YEAR	EX	MNP	MIP
7615	Talking Astronaut	hard-hand figure with white coveralls with insignias, white boots, dog tags	1969	165	395	650

Vehicle Sets

Action Pilot Series

8040	Crash Crew Fire Truck Set	Adventure Pack	1967	875	2100	3500
8020	Official Space Capsule Set	space capsule, record, space suit, cloth space boots, space gloves, helmet with visor	1966	145	345	575
5979	Official Space Capsule Set w/flotation	Sears exclusive with collar, life raft and oars	1966	175	425	700

Action Sailor Series

8050	Official Sea Sled and Frogman Set	Adventure Pack, without cave	1966	90	210	350
5979	Official Sea Sled and Frogman Set	Sears, with figure and underwater cave, orange scuba suit, fins, mask, tanks, sea sled in orange and black	1966	115	275	450

Action Soldier Series

5693	Amphibious Duck	Irwin, 26 inches long	1967	175	425	700
5397	Armored Car	Irwin, friction powered, 20 inches long	1967	125	300	500
5395	Helicopter	Irwin, friction powered, 28 inches long	1967	125	300	500
5396	Jet Fighter Plane	Irwin, friction powered, 30 inches long	1967	200	475	800
5652	Military Staff Car	Irwin, friction powered, 24 inches long	1967	165	395	650
5651	Motorcycle and Sidecar	Irwin, 14 inches long, khaki, with decals	1967	90	210	350
7000	Official Combat Jeep Set	Trailer, steering wheel, spare tire, windshield, cannon, search light, shell, flag, guard rails, tripod, tailgate and hood, without Moto-Rev Sound.	1965	125	300	500
7000	Official Jeep Combat Set	With Moto-Rev Sound	1965	125	300	500
5694	Personnel Carrier/Mine Sweeper	Irwin, 26 inches long	1967	195	450	750

Adventure Team

	Action Sea Sled	J.C. Penney, 13", Adventure Pack	1973	10	25	45
7005	Adventure Team Vehicle Set	Adventure Pack	1970	30	65	1102
3528	All Terrain Vehicle	14", Adventure Pack	1973	15	40	655
9158	Amphicat	by Irwin, scaled to fit 2 figures	1973	25	55	90
	Avenger Pursuit Craft	Sears exclusive	1976	95	225	375
7498	Big Trapper	Adventure set without action figure.	1976	45	105	175
7494	Big Trapper Adventure with Intruder	Adventure set with action figure.	1976	55	135	225
7480	Capture Copter	Vehicle set, no action figure included	1976	55	135	225
7481	Capture Copter Adventure with Intruder	vehicle set with action figure included	1976	70	165	275
59114	Chopper Cycle	15", Adventure Pack	1973	10	25	45
59751	Combat Action Jeep	18", Adventure Pack	1973	25	60	100
7000	Combat Jeep and Trailer		1976	55	135	225
7439	Devil of the Deep	Adventure Pack	1974	45	105	175
7460	Fantastic Sea Wolf Submarine		1975	30	75	125
7450	Fate of the Troubleshooter	Adventure Pack	1974	30	65	110
59189	Giant Air-Sea Helicopter	28", Adventure Pack	1973	20	45	75

ACTION FIGURES

ACTION FIGURES

Vehicle Sets

NO.	NAME	DESCRIPTION	YEAR	EX	MNP	MIP
7380	Helicopter	J.C. Penney, 14" helicopter in yellow with working winch	1973	30	75	125
7380	Helicopter		1976	40	90	150
7499	Mobile Support Vehicle Set	Adventure Pack	1972	90	210	350
	Recovery of the Lost Mummy Adventurer Set	Sears exclusive Adventure Pack	1971	95	225	375
7493	Sandstorm Survival Adventure		1974	50	120	200
7418	Search for the Stolen Idol Set	Adventure Pack	1971	70	165	275
7441	Secret of the Mummy's Tomb Set	with Land Adventurer figure, shirt, pants, boots, insignia, pith helmet, pick, shovel, Mummy's tomb, net, gems, vehicle with winch, comic	1970	95	225	375
7442	Sharks Surprise Set w/Sea Adventurer	Adventure Pack	1970	100	240	400
	Signal All Terrain Vehicle	J.C. Penney, 12", Adventure Pack	1973	10	25	45
7470	Sky Hawk	5 and 3/4 foot wingspan	1975	25	60	100
7445	Spacewalk Mystery Set w/Astronaut	Adventure Pack	1970	120	285	475
79-59301	Trapped in the Coils of Doom	Adventure Pack	1974	35	90	145

Adventures of GI Joe

NO.	NAME	DESCRIPTION	YEAR	EX	MNP	MIP
7980.83	Sharks Surprise Set without Frogman	Adventure Pack	1969	115	270	450
7981	Sharks Surprise Set with Frogman	Adventure Pack	1969	150	360	600
7981.83	Spacewalk Mystery Set without Spaceman	Reissued Adventure Pack	1969	135	325	550

ACTION FIGURES

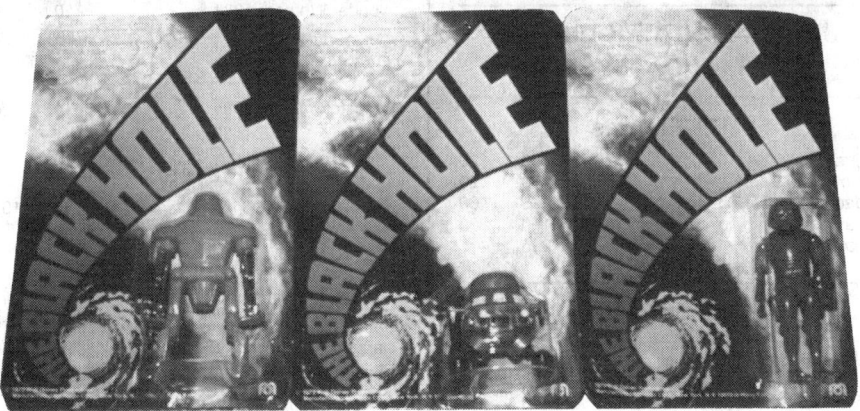

Clockwise from upper left: the Amazing Spider-Man, 1979, Mego; Fonzie's Motorcycle from Happy Days Gang, 1978, Mego; Clubber Lane and Rocky, PAC; The Black Hole figures, 1979-80, Mego; Chuck Norris Karate Kommandos series figures, Kenner.

ACTION FIGURES

Happy Days Mego, 1978

NAME	MNP	MIP
Fonzie, 1978, boxed	30	60
Fonzie, 1978, carded	30	60
Potsy, 1978, carded	30	60
Ralph, 1978, carded	30	60
Richie, 1978, carded	30	60

Playsets

Fonzie's Garage Playsets, 1978	60	125

Vehicles

Fonzie's Jalopy, 1978	40	80
Fonzie's Motorcycle, 1978	40	80

Indiana Jones, Adventures of Kenner, 1982-83

*3-3/4" Figures & accessories made by Kenner, 1982-83	0	0
Belloq	10	25
Belloq in Ceremonial Robe, in mailer box	6	15
Belloq in Ceremonial Robe, on card	200	500
Cairo Swordsman	5	15
Convoy Truck	15	35
German Mechanic	15	35
Indiana Jones	50	100
Indiana Jones, 12" tall	125	250
Indiana Jones in German Uniform	20	45
Map Room Set	20	50
Marion Ravenwood	70	175
Sallah	20	45
Streets of Cairo Set	18	45
Toht	5	15
Well of Souls	30	75

James Bond:Moonraker Mego, 1979

12" Figures

Drax, 1979	60	120
Holly, 1979	60	120
James Bond, 1979	25	60
James Bond, deluxe version, 1979	150	300
Jaws, 1979	215	415

Johnny West Marx, 1975

Jeb Gibson	125	275
Johnny West with Quick Draw	35	70
Sam Cobra with Quick Draw	40	80
Sheriff Garrett	35	70
Thunderbolt, Western ranch horse	25	50

Kenner Super Powers Kenner, 1984-86

5" Figures

NAME	MNP	MIP
*Figures and accessories made by Kenner, 1984-86		
Aquaman, 1984	15	35
Batman, 1984	25	55
Braniac, 1984	15	30
Clark Kent, mail-in figure, 1986	65	0
Cyborg, 1986	75	200
Cyclotron, 1986	35	75
Darkseid, 1985	5	15
Desaad, 1985	10	30
Doctor Fate, 1985	25	50
Firestorm, 1985	15	35
Flash, 1984	10	20
Golden Pharoah, 1986	30	65
Green Arrow, 1985	25	55
Green Lantern, 1984	30	60
Hawkman, 1984	25	50
Joker, 1984	15	30
Kalibak, 1985	5	15
Lex Luthor, 1984	5	15
Mantis, 1985	10	30
Martian Manhunter, 1985	10	30
Mr. Freeze, 1986	15	35
Mr. Miracle, 1986	75	200
Orion, 1986	20	40
Parademon, 1985	15	35
Penguin, 1984	20	40
Plastic Man, 1986	40	80
Red Tornado, 1985	25	55
Robin, 1984	25	50
Samurai, 1986	25	50
Shazam (Captain Marvel), 1986	20	40
Steppenwolf, on card, 1985	15	75
Steppenwolf, in mail-in bag, 1985	15	0
Superman, 1984	20	35
Tyr, 1986	25	50
Wonder Woman, 1984	10	20

Accessories

Collector's Case, 1984	10	20

Playsets

Hall of Justice, 1984	30	100

Vehicles

Batcopter, 1986	40	75
Batmobile, 1984	40	75
Darkseid Destroyer, 1985	25	50
Delta Probe One, 1985	15	30
Justice Jogger Wind-Up, 1986	10	20
Kalibak Boulder Bomber, 1985	10	25
Lex-Soar 7, 1984	10	20
Supermobile, 1984	15	30

Top to bottom: Monstroid from Masters of the Universe The Evil Horde, Mattel; Wizard of Oz characters, 1974, Mego; Samurai/Tyr/Shazam from Super Powers Collection, 1984-86, Kenner.

ACTION FIGURES

Laverne and Shirley Mego, 1978

12" Figures

NAME	MNP	MIP
Laverne and Shirley, 1978, boxed	60	125
Lenny and Squiggy, 1978, boxed	90	175

Love Boat Mego, 1981

4" Figures

Captain Stubing, 1981, carded	5	15
Doc, 1981, carded	5	15
Gopher, 1981, carded	5	15
Isaac, 1981, carded	5	15
Julie, 1981, carded	5	15
Vicki, 1981, carded	5	15

M.A.S.H. Tristar, 1982

3 3/4" Figures and Vehicles

*3-3/4" figures & accessories, unless noted.	0	0
B.J.	5	10
Col. Potter	5	10
Father Mulcahy	5	10
Hawkeye with Ambulance	15	35
Hawkeye	5	10
Hawkeye with Jeep	10	25
Helicopter with Hawkeye	8	20
Hot Lips	10	20
Klinger	5	10
Klinger in Drag	15	35
Mash Figures Collectors Set	26	65
Winchester	5	10

8" Figures

B.J., Large Figure	10	25
Hawkeye, Large Figure	10	25
Hot Lips, Large Figure	12	30

Mad Monster Series Mego, 1974

8" Figures

Mad Monster Castle, 1974, vinyl	300	600
The Dreadful Dracula, 1974	80	160
The Horrible Mummy, 1974	45	90
The Human Wolfman, 1974	75	150
The Monster Frankenstein, 1974	45	90

Major Matt Mason Mattel, 1966-69

Action Figures

Calisto	65	225
Captain Laser	75	275

Major Matt Mason Mattel, 1966-69

NAME	MNP	MIP
Doug Davis	50	150
Jeff Long	75	250
Major Matt Mason	40	125
Mission Team 4-Pack	175	575
Scorpio	350	1500
Sergeant Storm	50	140

Vehicles and Accessories

Astro-Trak	35	65
Firebolt Space Cannon	35	75
Gamma Ray Guard	30	100
Moon Suit Pak	25	55
Reconojet Pak	25	55
Rocket Launch	25	60
Satellite Launch Pak	25	60
Satellite Locker	30	75
Space Power Suit	30	100
Space Probe Pak	25	60
Space Shelter Pak	25	60
Star Seeker	85	175
Supernaut Power Limbs	30	100
Uni-Tred & Space Bubble	50	100
XRG-1 Reentry Glider	75	175

Marvel Secret Wars Mattel, 1984-85

4" Action Figures

*Figures and accessories made by Mattel, 1984-85

Baron Zemo, 1984	15	35
Captain America, 1984	10	25
Constrictor (foreign release), 1984	30	60
Daredevil, 1984	15	35
Doctor Doom, 1984	10	20
Doctor Octopus, 1984	10	20
Electro (foreign release), 1984	30	60
Falcon, 1984	20	40
Hobgoblin, 1984	30	60
Ice Man (foreign release), 1984	30	60
Iron Man, 1984	20	35
Kang, 1984	10	20
Magneto, 1984	10	20
Spider-Man, black outfit, 1984	25	50
Spider-Man, red and blue outfit, 1984	20	40
Three Figure Set, 1985	0	90
Two Figure Set, 1984	0	50
Wolverine, black claws, 1984	25	60
Wolverine, silver claws, 1984	25	50

Accessories

Secret Messages Pack	1	5

Playsets

Tower of Doom, 1984	10	25

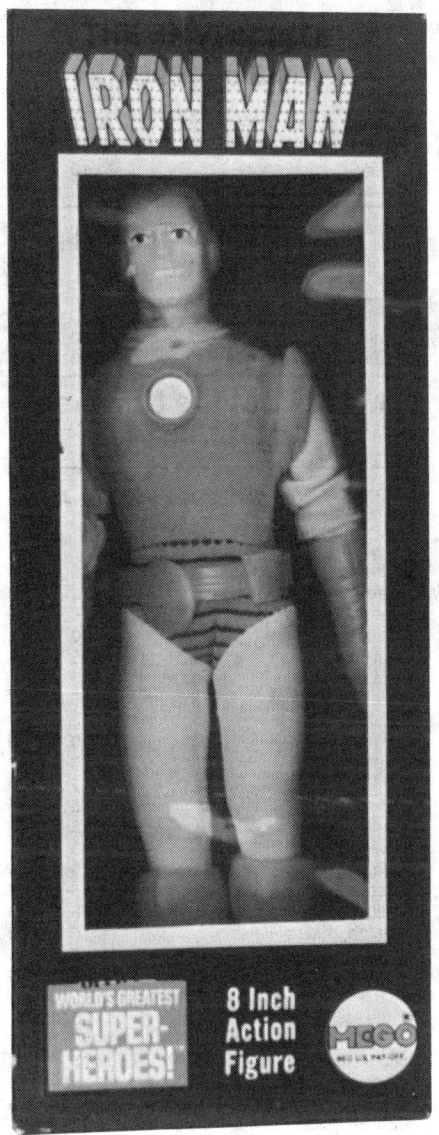

ACTION FIGURES

Top to bottom: The Invincible Iron Man, 1974, Mego; Stony "Stonewall" Smith The Battling Soldier, Marx; Joker from Batman series, 1973, Mego.

Marvel Secret Wars Mattel, 1984-85

Vehicles

NAME	MNP	MIP
Doom Copter, 1984	10	35
Doom Copter with Doctor Doom, 1984	15	55
Doom Cycle, 1984	6	20
Doom Cycle with Doctor Doom, 1985	10	40
Doom Roller, 1984	10	20
Doom Star Glider with Kang, 1984	15	30
Freedom Fighter, 1984	10	30
Star Dart with Spider-Man(black outfit), 1985	25	50
Turbo Copter, 1984	10	40
Turbo Cycle, 1984	5	20

Marvel Super Heroes Toy Biz, 1990-92

Series 1, 1990

NAME	MNP	MIP
Captain America	10	25
Daredevil	15	50
Doctor Doom	10	25
Doctor Octopus	10	25
Hulk	5	15
Punisher, Cap Firing	5	15
Silver Surfer	10	30
Spider-Man, Suction Cups	5	15

Series 2, 1991

NAME	MNP	MIP
Green Goblin, Back Lever	15	40
Green Goblin, no Lever	10	25
Iron Man	10	25
Punisher, Machine Gun Sound	5	15
Spider-Man, Web Shooting	10	30
Spider-Man, Web Climbing	15	35
Thor, Back Lever	15	40
Thor, no Lever	10	25
Venom	10	20

Series 3, 1992

NAME	MNP	MIP
Annihilus	5	15
Deathlok	5	15
Human Torch	5	15
Invisible Woman	75	150
Mister Fantastic	5	15
Silver Surfer, Chrome	5	15
Spider-Man, Ball Joints	5	15
Spider-Man, Web Tracer	5	15
Thing	5	15
Venom, Tongue Flicking	15	20

Talking Heroes

NAME	MNP	MIP
Cyclops	10	20
Hulk	10	20
Magneto	10	20
Punisher	10	20

Marvel Super Heroes Toy Biz, 1990-92

NAME	MNP	MIP
Spider-Man	10	20
Venom	10	20
Wolverine	10	20

Marx Action Figures, Miscellaneous

NAME	MNP	MIP
Dwight D. Eisenhower, 1960s	125	250
Johnny Kolt Cowboy	150	300
Mike Hazard	150	275

Masters of the Universe Mattel, 1981-90

5 3/4" Action Figures

*Figures and accessories made by Mattel, early 1980s

NAME	MNP	MIP
Battle Armor He-Man	10	20
Battle Armor Skeletor	5	20
Beast Man	5	15
Blade	10	25
Blast-Attak	5	15
Buzz-Off	5	15
Buzz-Saw Hordak	5	15
Clamp Champ	5	15
Clawful	10	20
Dragstor	5	15
Evil-Lyn	5	20
Extendar	5	15
Faker	15	40
Faker (re-issue)	5	15
Fisto	5	15
Grizzlor	5	15
Gwildor	5	15
He-Man, original version	15	30
Hordak	5	15
Horde Trooper	5	15
Jitsu	10	20
King Hiss	5	15
King Randor	10	25
Kobra Khan	5	20
Leech	5	15
Man-At-Arms	10	20
Man-E-Faces	10	20
Mantenna	5	15
Mekaneck	5	15
Mer-Man	5	20
Modulok	5	20
Mosquitor	5	15
Moss Man	5	15
Multi-Bot	5	20
Ninjor	5	15
Orko	5	20
Prince Adam	10	25
Ram Man	15	35
Rattlor	5	15
Rio Blast	5	15

Masters of the Universe Mattel, 1981-90

NAME	MNP	MIP
Roboto	5	15
Rokkon	5	15
Rotar	5	15
Saurod	5	20
Scare Glow	10	25
Skeletor, original version	10	25
Snake Face	5	20
Snout Spout	5	15
Sorceress	10	25
Spikor	10	20
SSSqueeze	5	15
Stinkor	5	15
Stonedar	5	15
Stratos, blue wings	10	20
Stratos, red wings	10	20
Sy-Klone	5	15
Teela	5	15
Trap Jaw	10	20
Tri-Klops	10	20
Tung Lashor	5	15
Twistoid	5	15
Two-Bad	5	15
Webstor	5	15
Whiplash	5	15
Zodac	10	20

Accessories

Battle Bones Carrying Case	5	10
Battle Cat	10	25
Battle Cat with He-Man (original version)	20	40
Deam-Blaster and Artilloray	15	30
Jet Sled	5	15
Mantisaur	8	15
Megalaser	5	15
Monstroid Creature	15	30
Night Stalker	5	15
Night Stalker with Jitsu	10	25
Panthor (evil cat)	10	25
Panthor with Skeletor (original version)	15	40
Screech	5	15
Screech with Skeletor (original version)	10	25
Stilt Stalkers	5	15
Stridor Armored Horse	5	15
Stridor with Fisto	10	25
Weapons Pak	2	5
Zoar	5	15
Zoar with Teela	15	30

Deluxe Action Figures

Dragon Blaster Skeletor, 5th Anniversary	10	25
Flying Fists He-Man, 5th Anniversary	10	25
Hurricane Hordak	10	25
Terror Claws Skeletor, 5th Anniversary	10	25

Masters of the Universe Mattel, 1981-90

NAME	MNP	MIP
Thunder Punch He-Man, 5th Anniversary	10	25

GraySkull Dinosaur Series

Bionatops	10	25
Turbodaltyl	10	25
Tyrantisaurus Rex	10	25

Meteorbs

Astro Lion	5	10
Comet Cat	5	10
Cometroid	5	10
Crocobite	5	10
Dinosorb	5	10
Gore-illa	5	10
Orbear	5	10
Rhinorb	5	10
Tuskor	5	10
Ty-Gyr	5	10

Playsets

Castle GraySkull	25	75
Eternia	100	200
Fright Zone	25	50
Slime Pit	10	20
Snake Mountain	25	50

Vehicles

Attak Trak	10	20
Bashasaurus	10	20
Battle Ram	10	30
Blasterhawk	15	30
Dragon Walker	10	20
Fright Fighter	10	25
Land Shark	10	20
Laser Bolt	10	20
Point Dread	10	40
Road Ripper	10	20
Roton	10	20
Spydor	15	35
Wind Raider	10	20

Micronauts Mego, 1976-80

Alien Invaders Accessories

Karrio, 1979	10	20

Alien Invaders Carded

Antron, 1979	15	30
Centaurus, 1980	35	70
Kronos, 1980	35	70
Lobros, 1980	35	70
Membros, 1979	15	30
Repto, 1979	13	25

ACTION FIGURES

Micronauts Mego, 1976-80

Alien Invaders Playsets

NAME	MNP	MIP
Deluxe Rocket Tubes	20	40
Rocket Tubes, 1978	23	50

Alien Invaders Vehicles

Alphatron	5	10
Aquatron, 1977	10	20
Betatron	5	10
Gammatron	5	10
Hornetroid, 1979	20	40
Hydra, 1976	7	15
Mobile Exploration Lab, 1976	17	35
Solarion, 1978	15	30
Star Searcher, 1978	15	40
Taurion, 1978	11	22
Terraphant, 1979	20	40

Boxed Figures

Andromeda, 1977	10	25
Baron Karza, 1977	15	30
Biotron, 1976	10	25
Force Commander, 1977	10	25
Giant Acroyear, 1977	10	25
Megas, 1981	10	25
Microtron, 1976	5	20
Nemesis Robot, 1978	7	15
Oberon, 1977	10	25
Phobos Robot, 1978	12	25

Carded Figures

Acroyear II, 1977; red, blue, orange	7	15
Acroyear, 1976; red, blue, orange	10	20
Galactic Defender, 1978; white, yellow,	7	15
Galactic Warriors, 1976; red, blue, orange	4	10
Pharoid, with Time Chamber, 1977; blue, red, grey	10	20
Space Glider, 1976; blue, green, orange	5	10
Time Traveler, clear plastic, 1976; yellow, orange	3	10
Time Traveler, solid plastic, 1976; yellow, orange	5	15

Micropolis Playsets

Galactic Command Center, 1978	20	40
Interplanetary Headquarters, 1978	20	40
Mega City, 1978	20	30
Microrail City, 1978	20	40

Playsets

Astro Station, 1976	10	20
Stratstation, 1976	15	30

Micronauts Mego, 1976-80

Vehicle

NAME	MNP	MIP
Battle Cruiser, 1977	30	60
Crater Cruncher with figure, 1976	5	15
Galactic Cruiser, 1976	7	17
Hydro Copter, 1976	10	25
Neon Orbiter, 1977	6	20
Photon Sled with figure, 1976	5	15
Rhodium Orbiter, 1977	6	20
Thorium Orbiter, 1977	6	20
Ultronic Scooter with figure, 1976	5	15
Warp Racer with figure, 1976	5	15

Mighty Crusaders Remco, 1980

low demand series of 8 figures, average value	3	15

Noble Knights Marx, 1968

Bravo Armor Horse	100	130
The Black Knight	75	190
The Gold Knight	60	120
The Silver Knight	60	120
Valiant Armor Horse	100	130
Valor Armor Horse	100	130
Victor Armor Horse	100	130

Official World's Greatest Super Heroes Mego, 1972-78

12 1/2" Figures

Amazing Spider-Man, 1978, boxed	35	75
Batman, 1978, boxed	60	125
Captain America, 1978, boxed	75	150
Hulk, 1978, boxed	30	60

8" Figures

Aquaman, 1972, boxed	50	150
Aquaman, 1972, carded	50	150
Batgirl, 1973, boxed	125	300
Batgirl, 1973, carded	125	250
Batman, fist fighting, 1975, boxed	150	350
Batman, painted mask, 1972, carded	60	100
Batman, painted mask, 1972, boxed	60	150
Batman, removable mask, 1972, Kresge card only	200	450
Batman, removable mask, 1972, boxed	200	350
Bruce Wayne, 1974, boxed, Montgomery Ward exclusive	400	500

ACTION FIGURES

Clockwise from upper left: Buck Rogers in the 25th Century, 1979, Mego; Jane West from Best of the West series, 1960s, Marx; Chopper from Starsky & Hutch, 1976, Mego; Captain America, 1970s, Mego.

Official World's Greatest Super Heroes Mego, 1972-78

NAME	MNP	MIP
Captain America, 1972, boxed	60	200
Captain America, 1972, carded	60	150
Catwoman, 1973, boxed	100	200
Catwoman, 1973, carded	100	200
Clark Kent, 1974, boxed, Montgomery Ward exclusive	400	500
Conan, 1975, boxed	120	300
Conan, 1975, carded	120	300
Dick Grayson, 1974, boxed, Montgomery Ward exclusive	400	500
Falcon, 1974, boxed	60	150
Falcon, 1974, carded	60	200
Green Arrow, 1973, boxed	100	250
Green Arrow, 1973, carded	100	400
Green Goblin, 1974, boxed	90	225
Green Goblin, 1974, carded	90	300
Human Torch, Fantastic Four, 1975, boxed	25	90
Human Torch, Fantastic Four, 1975, card	25	50
Incredible Hulk, 1974, boxed	20	100
Incredible Hulk, 1974, carded	20	50
Invisible Girl, Fantastic Four, 1975, boxed	30	150
Invisible Girl, Fantastic Four, 1975, card	30	60
Iron Man, 1974, boxed	75	125
Iron Man, 1974, carded	75	250
Isis, 1076, boxed	75	250
Isis, 1976, carded	75	125
Joker, 1973, boxed	60	150
Joker, 1973, carded	60	150
Joker, fist fighting, 1975, boxed	150	400
Lizard, 1974, boxed	75	200
Lizard, 1974, carded	75	250
Mr. Fantastic, Fantastic Four, 1975, boxed	30	140
Mr. Fantastic, Fantastic Four, 1975, card	30	60
Mr. Mxyzptlk, open mouth, 1973, boxed	50	75
Mr. Mxyzptlk, open mouth, 1973, carded	50	150
Mr. Mxyzptlk, smirk, 1973, boxed	60	150
Penguin, 1973, carded	60	125
Penguin, 1973, boxed	60	150
Peter Parker, 1974, boxed, Montgomery Ward exclusive	400	500
Riddler, 1973, boxed	100	250
Riddler, 1973, carded	100	400
Riddler, fist fighting, 1975, boxed	150	400
Robin, fist fighting, 1975, boxed	125	350
Robin, painted mask, 1972, boxed	60	150
Robin, painted mask, 1972, carded	60	90
Robin, removable mask, 1972, boxed	250	400
Shazam, 1972, boxed	75	200
Shazam, 1972, carded	75	150
Spider-Man, 1972, boxed	20	100

Official World's Greatest Super Heroes Mego, 1972-78

NAME	MNP	MIP
Spider-Man, 1972, carded	20	40
Supergirl, 1973, boxed	150	400
Supergirl, 1973, carded	150	400
Superman, 1972, boxed	50	125
Superman, 1972, carded	50	100
Tarzan, 1972, boxed	50	150
Tarzan, 1976, Kresge card only	60	225
Thing, Fantastic Four, 1975, boxed	40	150
Thing, Fantastic Four, 1975, carded	40	60
Thor, 1975, boxed	150	300
Thor, 1975, carded	150	300
Wonder Woman, boxed	100	250
Wonder Woman, Kresge card only	100	350

Accessories

	MNP	MIP
Super Hero Carry Case, 1973	40	0
Supervator, 1974	60	120

Playsets

	MNP	MIP
Aquaman vs. the Great White Shark, 1978	200	500
Batcave Playset, 1974, vinyl	125	250
Batman's Wayne Foundation Penthouse, 1977, fiberboard	600	1200
Hall of Justice, 1976, vinyl	125	250

Superman Series

	MNP	MIP
General Zod, 1978	50	100
Jor-El, 1978	50	100
Lex Luthor, 1978	50	100
Superman Playset, 1978	75	150
Superman, 1978	50	125

Vehicles

	MNP	MIP
Batcopter, 1974, boxed	75	150
Batcopter, 1974, carded	55	110
Batcycle, black, 1975, carded	60	150
Batcycle, black, 1975, boxed	75	185
Batcycle, blue, 1974, carded	75	135
Batcycle, blue, 1974, boxed	75	170
Batmobile and Batman	40	100
Batmobile, 1974, boxed	50	125
Batmobile, 1974, carded	50	120
Captain Americar, 1976	100	200
Green Arrowcar, 1976	175	350
Jokermobile, 1976	150	300
Mobile Bat Lab, 1975	125	250
Spidercar, 1976	50	125

Wonder Woman Series

	MNP	MIP
Major Steve Trevor, 1978	26	65
Queen Hippolyte, 1978	40	100
Queen Nubia, 1978	40	100
Wonder Woman with Diana Prince outfit, 1978	55	80
Wonder Woman Playset, 1978	50	100

One Million Years, B.C. Mego, 1976

NAME	MNP	MIP
Dimetrodon, 1976, boxed	75	150
Grok, 1976, carded	25	50
Hairy Rino, 1976, boxed	75	150
Mada, 1976, carded	25	50
Orm, 1976, carded	25	50
Trag, 1976, carded	25	50
Tribal Lair Gift Set, 1976, with 5 figures	0	180
Tribal Lair, 1976	60	120
Tyrannosaur, 1976, boxed	75	150
Zon, 1976, carded	25	50

Planet of the Apes Mego, 1973-75

8" Figures

Astronaut Burke, 1975, carded	50	100
Astronaut Burke, 1975, boxed	50	130
Astronaut Verdon, 1975, carded	50	90
Astronaut Verdon, 1975, boxed	50	140
Astronaut, 1975, carded	50	100
Astronaut, 1973, boxed	50	150
Cornelius, 1975, carded	40	75
Cornelius, 1973, boxed	40	140
Dr. Zaius, 1975, carded	40	60
Dr. Zaius, 1973, boxed	40	150
Galen, 1975, carded	40	90
Galon, 1975, boxed	40	140
General Urko, 1975, carded	50	100
General Urko, 1975, boxed	50	130
General Ursus, 1975, carded	50	100
General Ursus, 1975, boxed	50	120
Soldier Ape, 1975, carded	30	60
Soldier Ape, 1973, boxed	30	140
Zira, 1973, boxed	30	150
Zira, 1975, carded	30	60

Accessories

Action Stallion, brown mototrized, 1975, boxed	50	100
Battering Ram, 1975, boxed	20	40
Dr. Zaius' Throne, 1975, boxed	20	40
Jail, 1975, boxed	20	40

Playsets

Forbidden Zone Trap, 1975	65	150
Fortress, 1975	60	120
Treehouse, 1975	50	100
Village, 1975	60	130

Vehicles

Catapult and Wagon, 1975, boxed	25	50

Pocket Super Heroes Mego, 1979

3 3/4" Figures

NAME	MNP	MIP
Aquaman, 1976, white card	50	100
Batman, 1976, red card	20	40
Batman, 1976, white card	20	40
Captain America, 1976, white card	50	100
Captain Marvel, 1979, red card	20	40
General Zod, 1979, red card	5	15
Green Goblin, 1976, white card	50	100
Hulk, 1976, white card	15	40
Hulk, 1979, red card	15	30
Joker, 1979, red card	20	40
Jor-El (Superman), 1979, red card	10	20
Lex Luthor (Superman), 1979, red card	10	20
Penguin, 1979, red card	20	40
Robin, 1979, red card	20	40
Robin, 1976, white card	20	40
Spider-Man, 1976, white card	15	40
Spider-Man, 1979, red card	15	30
Superman, 1979, red card	15	30
Superman, 1976, white card	15	30
Wonder Woman, 1979, white card	20	40

Accessories

Batcave, 1981	120	300

Vehicles

Batmachine, 1979	40	100
Batmobile, 1979, with Batman	80	200
Invisible Jet, 1979	50	125
Spider-Car, 1979, with Spider-Man	30	75
Spider-Machine, 1979	40	100

Robin Hood and His Merry Men Mego, 1974

8" Figures

Friar Tuck, 1974	25	50
Little John, 1974	45	80
Robin Hood, 1974	75	150
Will Scarlet, 1974	75	150

Robin Hood Prince of Thieves Kenner, 1991

Accessories

Battle Wagon	15	30
Bola Bomber	5	10
Net Launcher	5	10
Sherwood Forest Playset	30	60

Action Figures

Azeem	7	15
Friar Tuck	15	30

ACTION FIGURES

ACTION FIGURES

Robin Hood Prince of Thieves Kenner, 1991

NAME	MNP	MIP
Little John	7	15
Robin Hood, Crossbow	5	18
Robin Hood, Crossbow, Costner Head	5	18
Robin Hood, Long Bow	8	17
Robin Hood, Long Bow, Costner Head	5	15
Sheriff of Nottingham	5	15
The Dark Warrior	8	20
Will Scarlett	8	20

RoboCop and the Ultra Police Kenner, 1989-90

Accessories

Robo-Glove	0	0
Robo-Helmet	15	40

Action Figure

Ed-260	10	25
RoboCop, Gatlin' Gun	15	30
Scorcher	6	15
Sgt. Reed	6	15
Toxic Waster	10	20
Wheels Wilson	6	15

Action Figures

Ace Jackson	5	15
Anne Lewis	5	15
Birdman Barnes	8	15
Chainsaw	5	15
Claw Callahan	7	15
Dr. McNamara	5	15
Headhunter	5	15
Nitro	5	15
RoboCop	9	20
RoboCop Night Fighter	6	20

Vehicles

Robo-1	10	25
Robo-Command	10	30
Robo-Copter	15	35
Robo-Cycle	5	10
Robo-Hawk	0	0
Robo-Jailer	15	40
Robo-Tank	0	0
Skull-Hog	5	10
Vandal-1	5	20

Simpsons Mattel, 1990

Bart	5	15
Bartman	5	15
Homer	5	15
Lisa	10	25
Maggie	10	25
Marge	5	15

Simpsons Mattel, 1990

NAME	MNP	MIP
Nelson	5	15
Sofa Set	10	30

Six Million Dollar Man Kenner, 1975-78

Accessories

Back Pack Radio	10	20
Bionic Bigfoot	30	75
Bionic Cycle	10	20
Bionic Mission Vehicle	25	55
Bionic Transport	10	30
Bionic Video Center	25	65
Critical Assignment Arms	15	30
Critical Assignment Legs	15	30
Dual Launch Drag Set	45	80
Flight Suit	15	30
Mission Control Center	25	50
Mission to Mars Space Suit	15	30
OSI Headquarters	30	70
OSI Undercover Blue Denims	15	30
Tower & Cycle Set	25	50
Venus Space Probe	50	80

Action Figures

Maskatron	20	70
Oscar Goldman	20	50
Steve Austin, 1975	20	50

Star Trek Mego, 1974-80

12" Figures

Arcturian, 1979, boxed	30	60
Captain Kirk, 1979, boxed	25	55
Decker, 1979, boxed	45	115
Ilia, 1979, boxed	25	50
Klingon, 1979, boxed	40	85
Mr. Spock, 1979, boxed	30	60

3 3/4" Alien Figures

Acturian, 1980, carded	75	150
Betelgeusian, 1980, carded	75	150
Klingon, 1980, carded	75	150
Megarite, 1980, carded	75	150
Rigellian, 1980, carded	75	150
Zatanite, 1980, carded	75	150

3 3/4" Figures

Captain Kirk, 1979, carded	10	25
Decker, 1979, carded	10	25
Dr. McCoy, 1979, carded	10	25
Ilia, 1979, carded	10	20
Mr. Spock, 1979, carded	10	25
Scotty, 1979, carded	10	25

Clockwise from upper left: Isis, 1976, Mego; Exploding Beetlejuice, Kenner; Klingon from Star Trek series, 1980, Mego; Boss Hogg from The Dukes of Hazzard series, 1981, Mego.

ACTION FIGURES

Star Trek Mego, 1974-80

8" Alien Figures

NAME	MNP	MIP
Andorian, 1976, carded	200	400
Cheron, 1975, carded	75	150
Gorn, 1975, carded	80	180
Mugato, 1976, carded	150	300
Neptunian, 1975, carded	100	225
Romulan, 1976, carded	300	600
Talos, 1976, carded	165	300
The Keeper, 1975, carded	75	175

8" Figures

Captain Kirk, 1974, carded	25	50
Dr. McCoy, 1974, carded	35	75
Klingon, 1974, carded	25	50
Lt. Uhura, 1974, carded	50	100
Mr. Spock, 1974, carded	25	50
Scotty, 1974, carded	35	80

Playsets

Command Bridge (for 3 3/4" figures), 1980	45	105
Enterprise Bridge (for 8" figures), 1976	60	150
Gift set: Enterprise bridge with figures, 1976	95	250
Mission to Gamma VI (for 8" figures), 1976	200	500

Star Trek V Galoob, 1989

Large Action Figure Statues

Captain James T. Kirk	15	30
Dr. McCoy	15	30
Klaa	15	30
Mr. Spock	15	30
Sybok	15	30

Star Trek: The Next Generation Galoob, 1988-89

3 3/4" Figures, Series 1

Data, Blue Face	70	160
Data, Dark Face	25	60
Data, Flesh Face	15	30
Data, Spotted Face	15	30
Gordy La Forge	5	15
Jean-Luc Picard	5	15
Lt. Worf	5	15
William Riker	5	15
Yar	10	25

Star Trek: The Next Generation Galoob, 1988-89

3 3/4" Figures, Series 2

NAME	MNP	MIP
Antican	35	75
Enterprise	10	35
Ferengi	35	75
Ferengi Fighter	15	50
Galileo Shuttle	15	50
Phaser	20	40
Q	35	75
Selay	35	75

Starsky and Hutch Mego, 1976

8" Figures

Captain Dobey, 1976	25	50
Car, 1976	65	125
Chopper, 1976	25	45
Huggy Bear, 1976	25	45
Hutch, 1976	20	45
Starsky, 1976	20	45

Stony Smith Marx, 1960s

Stony Smith, battling soldier	125	250
Stony Smith, paratrooper	115	225
Stony Smith, trooper	125	250

Super Hero Bendables Mego, 1972

5" Figures

Aquaman, 1972	48	120
Batgirl, 1972	48	120
Batman, 1972	36	90
Captain America, 1972	36	90
Catwoman, 1972	70	175
Joker, 1972	60	150
Mr. Mxyzptlk, 1972	50	125
Penguin, 1972	60	150
Riddler, 1972	60	150
Robin, 1972	30	75
Shazam, 1972	50	125
Supergirl, 1972	70	175
Superman, 1972	30	75
Tarzan, 1972	24	60
Wonder Woman, 1972	40	100

Teen Titans Mego, 1976

Aqualad, 1976, carded	175	350
Kid Flash, 1976, carded	175	300

Teen Titans Mego, 1976

NAME	MNP	MIP
Speedy, 1976, carded	300	500
Wondergirl, 1976, carded	200	450

Teenage Mutant Ninja Turtles Playmates, 1988-92

*4-1/2" figures & accessories (unless noted), Playmates

1988, Series 1

April O'Neil, No Stripe	60	150
Bebop Action Figure	3	8
Donatello Figure w/fan club form	8	40
Donatello Figure	8	20
Foot Soldier Action Figure	8	20
Leonardo Figure w/fan club form	8	40
Leonardo Figure	8	20
Michelangelo Figure w/fan club form	8	40
Michelangelo Figure	8	20
Raphael Figure w/fan club form	8	40
Raphael Figure	8	20
Rocksteady Action Figure	8	20
Shredder Action Figure	8	20
Splinter Action Figure	8	20

1989, Series 2

Ace Duck Figure, Hat On	5	15
Ace Duck Figure, Hat Off	5	40
April O'Neil, Blue Stripe	12	30
Baxter Stockman Action Figure	10	25
Genghis Frog Figure, Yellow Belt	30	75
Genghis Frog Figure, Black Belt, bagged weapons	5	30
Genghis Frog Figure, Black Belt	5	15
Krang Action Figure	5	15

1989, Series 3

Casey Jones Action Figure	5	15
General Traag Action Figure	5	15
Leatherhead Action Figure	25	50
Metalhead Action Figure	5	15
Rat King Action Figure	5	15
Usagi Yojimbo Action Figure	5	15

1990, Series 4

Mondo Gecko	5	15
Muckman and Joe Eyeball Action Figure	5	15
Scumbag Action Figure	5	15
Wingnut & Screwloose	5	15

1990, Series 5

Fugitoid	5	15
Slash w/purple belt, red "S"	25	75
Slash w/black belt	3	25

Teenage Mutant Ninja Turtles Playmates, 1988-92

NAME	MNP	MIP
Triceraton Action Figure	5	15

1990, Series 6

Mutagen Man	5	15
Napoleon Bonafrog	5	15
Panda Khan	5	15

1991, Giant Turtles, 13"

Donatello	20	40
Leonardo	20	40
Michelangelo	20	40
Raphael	20	40

1991, Series 10

Grand Slam Raph	5	10
Hose'em Down Don	5	10
Lieutenant Leo	5	10
Make My Day Leo	5	10
Midshipman Mike	5	10
Pro Pilot Don	5	10
Raph the Green Teen Beret	5	10
Slam Dunkin' Don	5	10
Slapshot Leo	5	10
T.D. Tossin' Leonardo	5	10

1991, Series 7

April O'Neil, No "Press"	25	125
April O'Neil, w/"Press"	10	20
Pizza Face	5	15
Ray Fillet, purple body, red V	10	25
Ray Fillet, red body, maroon V	10	30
Ray Fillet, yellow body, blue V	10	15

1991, Series 8

Don The Undercover Turtle	5	10
Leo the Sewer Samurai	5	10
Mike the Sewer Surfer	5	10
Raph The Space Cadet	5	10

1991, Series 9

Chrome Dome	5	10
Dirt Bag	5	10
Ground Chuck	5	10
Storage Shell Don	5	10
Storage Shell Leo	5	10
Storage Shell Michaelangelo	5	10
Storage Shell Raphael	5	10

1991, Wacky Action

Breakfightin' Raphael	5	15
Creepy Crawlin' Splinter	5	15
Headspinnin' Bebop	5	15
Machine Gunnin' Rocksteady	5	15
Rock & Roll Michaelangelo (Wacky Action)	5	15
Sewer Swimmin' Don	5	15
Slice 'n Dice Shredder	10	25
Sword Slicin' Leonardo	8	15

ACTION FIGURES

ACTION FIGURES

Teenage Mutant Ninja Turtles Playmates, 1988-92

NAME	MNP	MIP
Wacky Walkin' Mouser	10	20

1992, Giant Turtles, 13"

	MNP	MIP
Bebop	20	40
Movie Don	20	40
Movie Leo	20	40
Movie Mike	20	40
Movie Raph	20	40
Rocksteady	20	40

1992, Series 11

	MNP	MIP
Rahzer, red nose	7	25
Rahzer, black nose	5	15
Skateboard'N Mike	5	15
Super Shredder	5	15
Tokka, brown trim	9	25
Tokka, grey trim	5	15

1992, Series 12

	MNP	MIP
Movie Don	5	15
Movie Leo	5	15
Movie Mike	5	15
Movie Raph	5	15
Movie Slpinter, no tooth	5	15
Movie Splinter, w/Tooth	25	75

Vehicles/Accessories

	MNP	MIP
Flushomatic	4	10
Foot Cruiser	14	35
Foot Ski	4	10
Mega Mutant Killer Bee	3	8
Mega Mutant Needlenose	8	20
Mike's Pizza Chopper Backpack	4	10
Mutant Sewer Cycle with Sidecar	4	10
Ninja Newscycle	5	12
Oozey	4	10
Pizza Powered Sewer Dragster	5	15
Pizza Thrower	14	35
Psycho Cycle	10	25
Raph's Sewer Dragster	6	16
Raph's Sewer Speedboat	5	12
Retrocatapult	4	10
Retromutagen Ooze	2	4
Sewer Seltzer Cannon	4	10
Sludgemobile	7	18
Technodrome, 22"	24	60
Toilet Taxi	5	12
Turtle Blimp, 30" green vinyl	12	30
Turtle Party Wagon	16	40
Turtle Trooper Parachute, 22"	4	10
Turtlecopter	16	40

Vikings Marx, 1960s

	MNP	MIP
Eric the Viking	35	65
Mighty Viking Horse	30	60
Odin the Viking Chieftan	35	65

Waltons Mego, 1975

8" Figures

NAME	MNP	MIP
Grandma and Grandpa, 1975	25	50
John Boy and Ellen, 1975	25	50
Mom and Pop, 1975	25	50

Accessories

	MNP	MIP
Barn, 1975	50	100
Country Store, 1975	50	100
Truck, 1975	40	80

Playsets

	MNP	MIP
Farm House, 1975	50	100
Farm House with the 6 figures, 1975	0	200

Wizard of Oz Mego, 1974

4" Munchkin Figures

	MNP	MIP
Dancer, 1974, boxed	75	150
Flower Girl, 1974, boxed	75	150
General, 1974, boxed	75	150
Lolly Pop Kid, 1974, boxed	75	150
Mayor, 1974, boxed	75	150

8" Figures

	MNP	MIP
Cowardly Lion, 1974, boxed	25	50
Dorothy with Toto, 1974, boxed	25	50
Glinda the Good Witch, 1974, boxed	25	50
Scarecrow, 1974, boxed	25	50
Tin Woodsman, 1974, boxed	25	25
Wicked Witch, 1974, boxed	50	100
Witch's Monkey, 1974, boxed	75	150
Wizard of Oz, 1974, loose	35	0

Playsets

	MNP	MIP
Emerald City with eight 8" figures, 1974	125	350
Emerald City with Wizard of Oz, 1974	43	100
Munchkin Land, 1974	150	300
Witch's Castle, 1974, Sears Exclusive	250	450

World's Greatest Super Knights Mego, 1975

8" Figures

	MNP	MIP
Castle Playset, 1975	80	160
Ivanhoe, 1975, boxed	60	120
Jousting Horse, 1975, battery operated	40	70
King Arthur, 1975, boxed	60	120
Sir Galahad, 1975, boxed	75	150
Sir Lancelot, 1975, boxed	75	150

World's Greatest Super Knights Mego, 1975

NAME	MNP	MIP
The Black Knight, 1975, boxed	80	160

World's Greatest Super Pirates Mego, 1974

8" Figures

Black Beard, 1974, boxed	70	150
Captain Patch, 1974, boxed	70	150
Jean LaFitte, 1974, boxed	80	160
Long John Silver, 1974, boxed	80	160

X-Force Toy Biz, 1992

Bridge	5	15

X-Force Toy Biz, 1992

NAME	MNP	MIP
Cable	5	15
Deadpool	15	35
Forearm	10	20
Gideon	5	15
Kane	5	15
Shatterstar	5	15
Stryfe	10	25
Warpath	10	25

Zorro Gabriel, 1982

Amigo	10	20
Captain Ramon	10	20
Picaro	15	35
Sergeant Gonzales	10	20
Tempest	10	25
Zorro	10	25

ACTION FIGURES

Toy Vehicles

Modern man has always had a love affair with his machines that move. Partial evidence of this is the amazing number of toy vehicles that have been produced in the twentieth century. In fact, it could be reasonably argued that toy vehicles are collected more than any other type of toy.

With the dawning of the modern industrial age, the mass production of real-life automobiles and their toy counterparts seemed to go hand-in-hand. As cars were rolling off assembly lines, their miniature replicas were not far behind.

It was not just cars and trucks that were among the favorite subjects of toy makers. Any sort of vehicle that moved people or things from one place to another was a natural for miniaturization. Boats, airplanes, and wagons all fall into this category.

And the types and manufacturers of toy vehicles were as varied as the real things. Toy makers built them out of everything from cast iron and tin to paper and plastic.

The earliest toy automobiles came along soon after their big daddy originals in the late nineteenth century and were produced in cast iron. But it was not until after World War I that toy automobile production really began to hit its stride.

Firms such as Arcade and Hubley are among the most well-known and soughtafter manufacturers of early cast iron vehicles.

Buddy L Navy "LST Landing Ship," 1976.

Cars, trucks, and buses produced by Arcade Manufacturing of Freeport, Illinois, are highly valued among toy vehicle collectors. Arcade actually began producing toys in the late 1800s, but it was not until around 1920 when the company reportedly issued its first toy vehicle, a replica of a Chicago Yellow Cab.

After that came more realistic models of actual cars, trucks, and buses. "They Look Real" was the company's slogan.

Hubley is another name associated with quality toy vehicles. This Pennsylvania company began manufacturing cast iron toys in the 1890s, mostly horse-drawn wagons, trains, and guns. By the 1930s, Hubley was producing the cast iron cars that became their most well-known products. Many were patterned from actual automobiles of the day, while others were apparently looser interpretations of reality. Some of the Hubley vehicles also included company names, and some of the most interesting pieces had separate nickel-plated grilles.

As the toy manufacturing world changed, the toy makers either kept pace or became dinosaurs. Hubley began phasing out cast metal in the 1940s, and after a toy-making hiatus during

World War II, came back with authentic die-cast white metal replicas of real cars. Hubley began producing plastic products as well, in the 1950s, and many collectors find the firm's plastic vehicles to be a cut above the typical offerings of the period.

One of the best makers of smaller scale cast iron vehicles was the A.C. Williams Company. The Ohio company began producing toys in the late 1800s, but its toy vehicles sought by today's collectors were generally produced in the 1920s-30s. The smaller cars and airplanes produced by A.C. Williams were intended for the five-and-dime market of the time. Williams' toys are difficult for the novice collector to identify, as there are no company markings on the toys.

Hot Wheels "1932 Vicky," 1969.

While heavy cast iron toys had been the rule at the turn of the century, lithographed tinplate toys began stealing a large part of the market in the 1920s. One of the world's leading producers of these toys was Louis Marx. Over the years, Marx produced an extensive line of toy cars, trucks, airplanes, and farm equipment, not only in tin, but also in steel and later in plastic.

Marx capitalized on the popularity of certain celebrities and comic strip characters, incorporating them into its toy vehicles. With lithography, it was easy to put a new character into a car and thus have a brand new toy ready for market. Characters such as Mickey Mouse, Donald Duck, Dick Tracy, Blondie and Dagwood, Charlie McCarthy, Amos 'n Andy, and Milton Berle show up in Marx cars.

One of the most famous manufacturers of toy cars and (especially) trucks was Buddy L. These large-scale pressed steel toys were not the kind of toys bought for display or quiet play on the living room floor. These were those BIG trucks approaching two feet in length that all boys loved to get down on their knees in the dirt with.

Buddy L toys grew out of the Moline Pressed Steel Company of Moline, Illinois. The name, Buddy L, was for the son of the company's owner, reportedly for whom the first toys were produced. The Buddy L toys most sought by collectors were produced in the 1920s and 1930s and were of heavy-duty construction. Starting in the early 1930s, the company began to use lighter weight materials.

The Buddy L name has remained, but its post-World War II toys are not considered at all in the same league as its early issues, which command high collector prices today.

Buddy L is best remembered for its heavy-duty trucks, but another name that was synonymous with trucks was Smith-Miller. Founded by Bob Smith and Matt Miller, the company specialized in "Famous Trucks In Miniature." Smith-Miller was also known as Miller-Ironson Corporation later in its life, but it is more commonly referred to as Smitty Toys. It produced large cast metal and aluminum trucks.

Because of their outstanding quality, some of the Mack trucks made by Smith-Miller are highly regarded among toy collectors. The Smith-Miller name continues today, with new limited edition trucks produced for collectors.

Wyandotte is another company associated with pressed steel vehicles. Known as both Wyandotte Toys or All Metal Products, this Michigan company produced several large steel vehicles with baked enamel finishes in the 1930s. Not all Wyandotte toys are marked, which tends to cause some confusion among collectors, but the vehicles can often be identified by their art deco-type styling and wooden wheels. Marx is reported to have bought some of the Wyandotte products before the company went out of business in the 1950s.

And yet another company that produced large steel toys was Structo. The company originally produced metal construction sets, but developed a line of vehicles in the 1920s.

While major toy companies were producing vehicles in cast iron, tin, and steel, others began making toys in rubber. Probably the best-known manufacturer of rubber toys is the Auburn Rubber Company of Auburn, Indiana. From the mid-1930s into the 1950s, Auburn produced rubber cars, trucks, tractors, motorcycles, airplanes, trains, and boats. The company is known for producing replicas of actual vehicles as well as race cars of its own design. And there probably is not anyone who has spent much time looking at toys who has not seen an Auburn rubber tractor. You could find replicas of most of the major tractor models and even get the accompanying implements.

In the 1950s, Auburn began to abandon rubber as its material of choice, and soon it was producing vehicles made only of vinyl. It was not many years later before Auburn was out of business.

Another popularly collected type of toy vehicle is the smaller die-cast models, generally in the range of three to six inches in length. Probably the leading producer of this type of vehicle was Tootsietoys.

The company dates back to before the turn of the century to Samuel Dowst of Chicago. The trade name for the toy products, which would eventually become the firm's mainstay, came from Dowst's daughter, Tootsie. Although a few toy cars were produced by the firm before 1920, it was during the Roaring Twenties when the name Tootsietoys began to regularly appear. By the 1930s, the company was producing a wide line of toys, many of which are highly prized by collectors today. Tootsietoys' Federal vans from the 1920s are among the most soughtafter toys, particularly those with company logos.

Trends

Steel and cast iron toys: 1994 started out strong but flattened through summer. However, rare and unusual toys, especially top condition examples, have remained in demand at good prices. Common toys have held their ground. The greatest advance this year and in 1995 is projected to be in mid-1950s to mid-1960s Buddy-L toys such as the GMC 550 Series.

Tin toys: Values on the rise through 1994. Marx has shifted into second gear, with many items soaring in value, particularly older trucks, such as the 1920s and 1930s "Mack." Quality material is getting harder to find, as top condition pieces are being squirreled away in collections and for investment.

Die-cast toys: Hot Wheels leads the pack with continued explosive growth. Other lines, Corgi, Dinky, and Matchbox, remain strong.

The Top 25 Vehicle Toys
(listed alphabetically, in Mint condition)

1. Checker Cab, Arcade, 1932 ....$15,000
2. Packard Straight 8, Hubley, 1927 15,000
3. White Dump Truck, Arcade 15,000
4. White Moving Van, Arcade, 1928 13,500
5. Baggage Truck, Buddy-L, 1930-32 12,000
6. Elgin Street Sweeper, Hubley, 1930 11,500
7. Ingersoll-Rand Compressor, Hubley, 1933 10,000
8. Motorized Sidecar Motorcycle, Hubley, 1932 10,000
9. Pile Driver on Treads, Buddy-L, 1929 10,000
10. Tug Boat, Buddy-L, 1928 10,000
11. Yellow Cab, Arcade, 1936 10,000
12. Coach, Buddy-L, 1927 9,000
13. Dredge on Tread, Buddy-L, 1929 9,000
14. Improved Steam Shovel on Treads, Buddy-L, 1929-30 9,000
15. Ahrens-Fox Fire Engine, Hubley, 1932 8,000
16. Borden's Milk Bottle Truck, Arcade, 1936 8,000
17. Buick Coupe, Arcade, 1927 7,500
18. Buick Sedan, Arcade, 1927 7,500
19. Fire Engine, Hubley, 1920s 7,500
20. Seven Man Fire Patrol, Hubley, 1912 7,500
21. Trench Digger, Buddy-L, 1928-31 7,500
22. "America" Plane, Hubley 7,000
23. Auto Fire Engine, Hubley, 1912 7,000
24. Junior Line City Dray, Buddy-L, 1930 7,000
25. Indian Armored Car (Motorcycle), Hubley, 1928 6,000

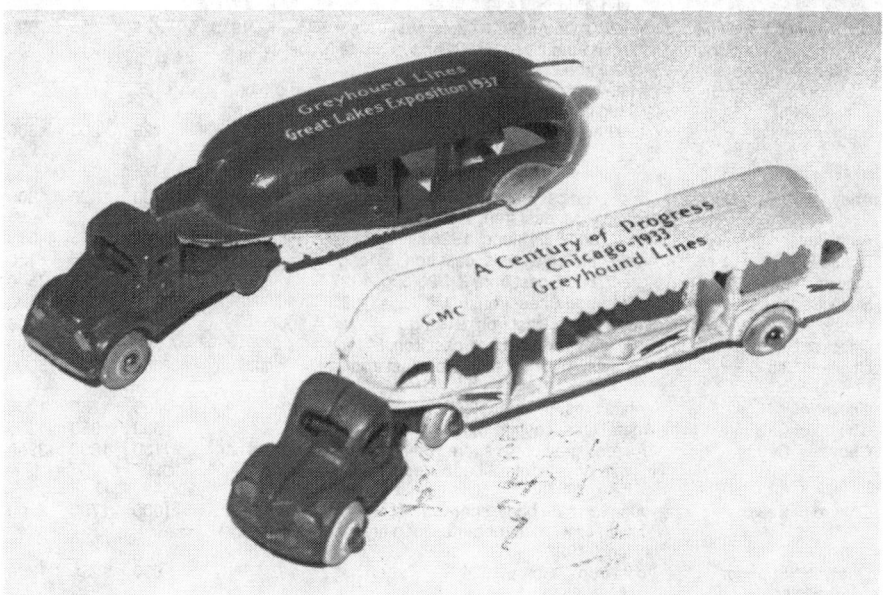

Top to bottom: Austin Transport Set, 1930, A.C. Williams; Batmobile, Japan; Zero Fighter, Japan; Greyhound Lines set, 1930s, Arcade.

VEHICLES

Arcade

Airplanes

NAME	DESCRIPTION	GOOD	EX	MINT
Airplane	cast iron, 6" long	90	135	180
Boeing United Airplane	white rubber wheels, two propellors, cast iron, 5" wingspan, 1936	95	145	200

Boats and Ships

Battleship New York	cast iron, 20" long, 1912	800	1200	1600
Showboat	cast iron, 10 3/4" long, 1934	475	715	950

Buses

ACF Coach	dual rear wheels, front door opens and closes, cast iron, 11 1/2" long, 1925	1200	1600	2200
Bus	cast iron, 8" long	400	600	800
Century of Progress Bus	cast iron, 7 5/8" long, 1933	125	165	265
Century of Progress Bus	cast iron, 10 1/2" long, 1933	145	215	300
Century of Progress Bus	cast iron, 14 1/2" long, 1933	175	250	500
Century of Progress Bus	cast iron, 12" long, 1933	165	235	350
Century of Progress Bus	cast iron, 6" long, 1933	100	150	200
Double Deck Bus	cast iron, 8" long, 1938	350	525	700
Double Deck Bus	green, white rubber wheels with red centers, 'Chicago Motor Coach' on side, cast iron, 8 1/4" long, 1936	425	650	850
Double Deck Bus	rubber wheels, cast iron, 8" long, 1936	100	150	200
Double Deck Yellow Coach Bus	nickel-plated driver, rubber balloon tires, cast iron, 13 1/2" long, 1926	800	1200	1500
Fageol Bus	cast iron, 13" long	335	500	675
Fageol Safety Bus	bright enamel colors, cast iron, 12" long, 1925	350	525	700
Fageol Safety Bus	nickel-plated wheels, with or without driver, cast iron, 7 3/4" long, 1926	200	300	400
Great Lakes Expo Bus	large, white rubber wheels with blue centers, cast iron, 11 1/4" long, 1936	300	575	1000
Great Lakes Expo Bus	small, white rubber wheels, cast iron, 7 1/4" long, 1936	225	385	695
Greyhound Cross Country Bus	"Greyhound Lines GMC" on top, white rubber wheels, cast iron, 7 3/4" long, 1936	125	215	300
Greyhound Trailer Bus	white with blue cab, 'GMC Greyhound Lines' on top, blue centered rubber wheels, cast iron, 10 1/2" long 1933	125	175	265
Sightseeing Bus		125	200	300

Cars

Andy Gump and Old 348	bright red car, green trim, green disc wheels with red hubcaps, cast iron, 7 1/4" long, 1923	800	1350	1800
Auto Racer	cast iron, 7 3/4" long, 1926	200	295	375
Boattail Racer	nickel plated wheels, cast iron, 5" long	75	115	150
Buick Coupe	green body, cast iron, 8 1/2" long, 1927	3000	5000	7500
Buick Sedan	green body, cast iron, 8 1/2" long, 1927	3000	5000	7500
Checker Cab	deep green, cast iron, 9" long	500	750	1200
Checker Cab	plain two row checker, cast iron, 8" long	750	1000	1500
Checker Cab	yellow body with black roof, cast iron, 9 1/4" long, 1932	5000	10000	15000
Chevrolet	white tires, 8" long, 1928	400	750	1000
Chevrolet Cab	metal tires, cast iron, 8" long, 1920's	500	850	1200
Chevrolet Coupe	Arizona gray body, spare wheel and tire on rear of car, with or without rubber balloon tires, cast iron, 8 1/4" long, 1927	1100	1800	3200
Chevrolet Sedan	Algerian blue body, spare wheel and tire on rear of car, with or without rubber balloon tires, cast iron, 8 1/4" long, 1927	1000	1700	2900
Chevrolet Superior Roadster	cast iron, 7" long, 1924	800	1250	1700
Chevrolet Superior Sedan	cast iron, 7" long, 1924			

Arcade

NAME	DESCRIPTION	GOOD	EX	MINT
Chevrolet Superior Touring Car	cast iron, 7" long, 1923	850	1300	2000
Chevrolet Utility Coupe	cast iron, 7" long, 1924	650	1000	1300
Chevy Coupe	cast iron, 8" long, 1929	400	750	1000
Coupe	red painted cast iron, 5" long, 1928	90	135	180
Coupe	two-toned, metal wheels, 5" long, 1920's	80	125	165
Coupe	solid wheels, 6 1/2" long, 1920's	425	650	850
Coupe	solid wheels, no spokes, cast iron, 6 3/4" long, 1920's	125	185	250
Coupe	with or without rubber tires, removable driver, cast iron, 9" long, 1922	375	575	750
DeSoto Sedan	white rubber wheels, cast iron, 4" long, 1936	75	150	250
Ford Coupe	nickel-plated wheels, no driver, cast iron, 5" long, 1926	150	225	325
Ford Coupe	with or without rubber tires, cast iron, 6 1/2" long, 1927	225	350	450
Ford Fordor Sedan	with or without rubber tires, removable driver, 6 1/2" long, 1920's	850	1275	1700
Ford Fordor Sedan	nickel-plated wheels, no driver, cast iron, 5" long, 1926	650	950	1200
Ford Sedan and Covered Wagon Trailer	cast iron, sedan 5 1/2" long, trailer 6 1/2" long, 1937	850	1200	1800
Ford Touring Car	with or without rubber tires, removable driver, cast iron, 6 1/2" long, 1926	450	650	900
Ford Tudor Sedan	with or without rubber tires, removable driver, cast iron, 6 1/2" long, 1920's	400	600	850
Ford Tudor Sedan	visor over front windshield, with or without rubber tires, driver, cast iron, 6 1/2" long, 1926	450	600	865
Ford With Rumble Seat In Back		40	60	80
Ford Yellow Cab	special edition for Chicago World's Fair, cast iron, 6 7/8" long, 1933	700	1000	1500
Limousine Yellow Cab	yellow with black body stripe, nickel-plated driver, spare tire at rear, cast iron, 8 1/2" long, 1930	650	850	1300
Model A	cast iron, 8 1/2" long	225	325	450
Model A	cast iron, 6 3/4" long, 1928	175	225	300
Model A Coupe		425	650	850
Model A Ford	white rubber tires, cast iron, 1929	225	350	450
Model A Sedan	orange, cast iron, 6 3/4" long, 1928	700	1050	1400
Model T	rubber tires, 6" long	125	185	250
Model T Sedan, center door	cast iron, 6 1/2" long, 1923	100	150	200
Pierce Arrow Coupe	'Silver Arrow' on sides, cast iron, 7 1/4" long, 1936	150	225	425
Plymouth Sedan	cast iron, 4 3/4" long, 1933	100	185	275
Pontiac Sedan	white rubber wheels, cast iron, 4 1/4" long, 1936	85	145	185
Racer	driver's head and number highlighted with gold bronze, cast iron, 8" long, 1936			
Racer	white rubber tires, cast iron, 5 3/4" long, 1936	50	75	100
Red Top Cab	cast iron, 8" long, 1924	700	1200	2500
Reo Coupe	cast iron, 9 3/8" long, 1931	100	2000	5000
Runabout Auto	wood, pressed steel, steel, cast iron, 8 1/2" long, 1908			
Sedan	cast iron, 5" long, 1920's	100	150	200
Sedan	cast iron, 8" long, late 1930's	125	185	250
Sedan with Red Cap Trailer	cast iron, sedan 5 7/8" long, trailer 2 1/2" long, 1939	250	400	650
Yellow Cab	rubber tires, cast iron, 8" long, 1924	500	750	1000
Yellow Cab	rubber tires, cast iron, 9" long, 1925	600	825	1200
Yellow Cab	rubber tires, cast iron, 5 1/4" long, 1925	475	650	850
Yellow Cab	bright yellow body, black top, white rubber wheels with black centers, cast iron, 8 1/4" long, 1936	1250	2000	3000
Yellow Cab	yellow with 'Yellow Cab' in black on top, cast iron, 4 1/4" long, 1936, "Darmalee" Cab	5000	7500	10000

Emergency Vehicles

NAME	DESCRIPTION	GOOD	EX	MINT
Ambulance	cast iron, 6" long, 1932	500	750	1000
Ambulance	cast iron, 8" long, 1932			

Arcade

NAME	DESCRIPTION	GOOD	EX	MINT
Fire Engine	red enamel with gold striping, cast iron, 7 1/2" long, 1926	750	1000	1500
Fire Engine	red, white rubber tires with green centers, cast iron, 4 1/2" long, 1936	125	175	250
Fire Engine	red trimmed in gold bronze, white rubber wheels with blue centers, cast iron, 6 1/4" long, 1936	150	225	300
Fire Engine	red trimmed in gold bronze, white rubber wheels with blue centers, cast iron, 9" long, 1936	300	450	750
Fire Ladder Truck	red trimmed in gold bronze, white rubber wheels with blue centers, cast iron, 7" long, 1936	400	600	800
Fire Ladder Truck	red trimmed in gold bronze, ladders yellow, white rubber wheels blue centers, cast iron, 12 1/2" long with ladders, 1936	475	750	950
Fire Pumper	with six firemen, cast iron, 13 1/4" long, 1938	500	800	1000
Fire Trailer Truck	red trimmed in gold bronze, two-piece fire engine, cast iron, truck 16 1/4" long, with ladders 20" long, 1936	500	775	1000
Fire Truck	cast iron, 15" long	175	265	350
Hook and Ladder Fire Truck	cast iron, 16" long	275	415	550
Mack Fire Apparatus Truck	bright red truck, hose reel, removable extention ladders, bell that rings, cast iron, 21" long, 1925	1000	1500	3000
Mack Fire Apparatus Truck	red trimmed in gold, yellow extension ladders, imitation hose, nickeled driver, nickeled bell that rings, 21" long, 1936	1000	1500	3000
Pontiac Boiler Fire Truck	cast iron	200	315	400
Pontiac Ladder Fire Truck	cast iron, 7" long	210	325	425

Farm and Construction Equipment

NAME	DESCRIPTION	GOOD	EX	MINT
'Ten' Caterpillar Tractor	cast iron, 7 1/2" long, 1929	550	850	1200
Allis-Chalmers Tractor	cast iron, 3" long, 1934	65	95	130
Allis-Chalmers Tractor	cast iron, 6" long, 1940	175	250	400
Allis-Chalmers Tractor	cast iron, cast in driver, 7" long, 1940	200	300	500
Allis-Chalmers Tractor	cast iron, separate plated driver, 7" long, 1940	225	350	600
Allis-Chalmers Tractor Trailer	red tractor, trimmed gold bronze, green trailer, cast iron, 13" long, 1936	250	400	750
Allis-Chalmers Tractor with Earth Mover	cast iron, 5" long, 1934	95	145	190
Avery Tractor	gray frame with gold striping, red wheels, flat radiator, cast iron, 4 1/2" long, 1929	100	150	200
Caterpillar Tractor	yellow trimmed in black, black wheels, nickeled steel track and driver, cast iron, 7 3/4" long, 1936	500	850	1000
Caterpillar, No. 270	steel, 8 1/2" long, 1920's	750	1000	1500
Crawler	driver, chain tracks, cast iron, 3" long, 1930			
Crawler	driver, chain tracks, cast iron, 3 7/8" long, 1930	160	245	325
Crawler	driver, chain tracks, cast iron, 5 5/8" long, 1930	325	450	650
Crawler	nickel-plated driver and tracks, cast iron, 6 5/8" long, 1930	425	625	850
Crawler	nickel-plated driver and tracks, cast iron, 7 1/2" long, 1930	625	950	1250
Fairbanks-Moorse	small portable Z-engine, no wheels, cast iron, 3 1/2" long, 1930	200	350	500
Ford Tractor	cast iron, 1/12 scale, 1941	100	150	200
Ford Tractor	cast iron, 3 1/4" long, 1926	100	150	200
Ford Tractor	cast iron, cast in driver, 1/25 scale, 1940	100	150	200
Ford Tractor	cast iron, cast in driver, with plow, 1/25 scale, 1940	200	300	400
Fordson F/Loader Tractor	rear crank operated loader, cast iron, 1/16 scale	200	275	350
Fordson Tractor	disk rubber wheels, cast iron, 3 1/2" long, 1936	75	100	150
Fordson Tractor	cast iron, 4 3/4" long, 1926	75	100	150
Fordson Tractor	cast iron, 5 3/4" long, 1932	100	150	200
Fordson Tractor	with or without lugs on rear wheels, cast iron, 6" long, 1926	275	400	575

Arcade

NAME	DESCRIPTION	GOOD	EX	MINT
International-Harvester A Tractor	cast iron, 1/12 scale, 1941	500	750	900
International-Harvester Crawler	cast iron, 1/16 scale	375	565	750
International-Harvester Crawler	cast iron, plated driver, 1/16 scale, 1941	500	1000	2000
International-Harvester Crawler	cast iron, plated driver, 1/16 scale, 1936	800	1600	3000
International-Harvester M Tractor	rubber wheels, cast iron, 5 1/4" long, 1941	300	400	550
International-Harvester M Tractor	rubber wheels, cast iron, 7" long, 1940	450	700	800
John Deere Open Flywheel Tractor	cast iron, 1/16 scale, 1941	175	265	350
John Deere Wagon	wooden box, iron running gear, cast iron, 1/16 scale, 1940	150	225	300
McCormick-Deering Farmall Tractor	gray body trimmed in gold, red wheels, driver, cast iron, 6 1/4" long, 1936	300	450	600
McCormick-Deering M Tractor	wood wheels, cast iron, 4 1/4" long, 1942	175	250	400
McCormick-Deering Plow	red frame, yellow wheels, aluminum bronze plow, shares and disks, cast iron, 7 3/4" long, 1926	250	400	600
McCormick-Deering Plow	red frame, cream wheels, aluminum plow, shares and disks, cast iron, 7 3/4" long, 1936	175	250	325
McCormick-Deering Thresher	assorted colors with gold striping, red grain pipe, 9 1/2" long, 1936	150	250	400
McCormick-Deering Thresher	gray, red trim, cream colored wheels, cast iron, 12" long, 1929	300	600	1200
McCormick-Deering Tractor	belt pulley, cast iron, 1/16 scale, 1925	450	650	850
McCormick-Deering Tractor	assorted colors trimmed in gold, nickeled driver, cast iron, 7 1/2" long, 1936	300	400	500
Oliver Planter	red or green, 1/25 scale, cast iron, 1950	35	50	75
Oliver Plow	red with nickel-plated wheels, aluminum bronze plow shares, 6 3/4" long, 1926	225	350	475
Oliver Plow	red enamel finish, blades striped with aluminum bronze, 6 1/2" long, 1926	225	350	475
Oliver Plow	red cast iron, 1/16 scale, 1940	75	115	150
Oliver Plow	red or green cast iron, 1/25 scale, 1940	25	35	50
Oliver Spreader	yellow, cast iron, 1/16 scale, 1940	350	650	1200
Oliver Tractor	red or green with rubber tires, cast iron, 5 1/4" long, 1946	75	100	150
Oliver Tractor	red or green, cast iron, 7 1/2" long, 1940	300	600	1200
Threshing Machine	gray and white, red trim, cast iron, 10" long	160	245	325

Tanks

NAME	DESCRIPTION	GOOD	EX	MINT
Army Tank with Gun	shoots steel balls, cast iron, 8" long, 1940	200	350	600

Trucks

NAME	DESCRIPTION	GOOD	EX	MINT
'Yellow Baby' Dump Truck	cast iron, 10 3/4" long, 1935	2000	3500	5500
Auto Express 548 Truck	flatbed, cast iron, 9" long	225	350	450
Borden's Milk Bottle Truck	'Borden's' cast on side, cast iron, 6 1/4" long, 1936	1500	2750	8000
Carry-Car Truck and Trailer Set	red truck, green trailer, with three 3 3/4" Austin vehicles, cast iron, 14 1/4" long, 1936	650	1150	1700
Carry-Car Truck and Trailer Set	red truck, green trailer, with four vehicles, cast iron, 28" long, 1936	1500	3000	4500
Carry-Car Truck Trailer	cast iron, 24 1/2" long, 1928	750	1300	1900
Chevrolet Panel Delivery Van	white rubber tires with colored centers, cast iron, 4" long, 1936	125	200	300
Chevrolet Stake Truck		90	165	225
Chevrolet Utility Express Truck	cast iron, 9 1/4" long, 1923	450	675	1200

Arcade

NAME	DESCRIPTION	GOOD	EX	MINT
Chevrolet Wrecker		90	165	225
Chrome Wheeled Truck	cast iron, 7" long	225	350	450
Delivery Truck	yellow, cast iron, 8 1/4" long, 1926	800	1200	1500
Dump Truck	spoked wheels, low bed, cast iron, 6" long	175	350	500
Dump Truck	cast iron, 6 1/2" long, 1941	175	350	500
Dump Truck	red chassis, green dump body, white rubber tires with green centers, cast iron, 4 1/2" long, 1936	75	100	150
Ford Anthony Dump Truck	nickel-plated spoked wheels, black enamel finish with gray dump body, cast iron, 8 1/2" long, 1926	1250	2000	2500
Ford Weaver Wrecker	cast iron, 8 1/4" long, 1928	700	950	1400
Ford Weaver Wrecker, Model A	cast iron, 11" long, 1926	550	800	1200
Ford Weaver Wrecker, Model T	cast iron, 11" long, 1926	600	875	1300
Ford Wrecker	white rubber wheels, cast iron, 7" long, 1936	200	400	600
Gasoline Truck	cast iron, 13" long, 1920's	600	900	1200
Ice Truck	red, cast iron, 7" long, 1930's	175	225	350
International Delivery Truck	cast iron, 9 1/2" long, 1936	1250	1875	2500
International Dump Truck	cast iron, 10 3/4" long, 1930	1200	1500	2000
International Dump Truck	white rubber wheels with red centers, cast iron, 10 1/2" long, 1935	1200	1500	2000
International Dump Truck	red cab, green dump, cast iron, 9 1/2" long, 1940	600	950	1250
International Harvester Dump Truck	cast iron, 11" long, 1941	300	465	675
International Harvester Pickup	cast iron, 9 1/2" long, 1941	300	465	675
International Stake Truck	white rubber wheels with red centers, cast iron, 12" long, 1935	1200	1500	2000
International Stake Truck	yellow, cast iron, 9 1/2" long, 1940	600	750	1250
International Wrecker Truck	cast iron, 11" long	1250	1875	2500
Mack Dump Truck	assorted colors trimmed in gold bronze, 'Mack' decals on doors, dual rear wheels, cast iron, 12 1/4" long, 1936	750	1125	1500
Mack Dump Truck	cast iron, 13" long	375	565	750
Mack Dump Truck	light grey or blue, gold trim, white tires, cast iron, 12" long, 1925	700	1050	1400
Mack Gasoline Truck	cast iron, 13 1/4" long, 1925	750	1500	2200
Mack High Dump	nickel-plated levers mechanically raise the dump bed, cast iron, 8 1/2" long, 1930	750	1500	2200
Mack High Dump	cast iron, 12 3/8" long, 1931	1000	2000	3000
Mack Ice Truck	ice blocks and tongs, cast iron, 8 1/2" long, 1931	350	525	700
Mack Ice Truck	ice blocks and tongs, cast iron, 10 3/4" long, 1931	550	750	950
Mack Lubrite Tank Truck	cast iron, 13 1/4" long, 1925	1200	2000	3000
Mack Oil Truck	cast iron, 10" long, 1920's	350	525	700
Mack Tank Truck	cast iron with tin tank, 12 3/4" long, 1929	1200	1800	2900
Mack Tank Truck	assorted colors with gold trim, nickeled driver, dual rear wheels, tank holds water, cast iron, 13" long, 1936	1100	1800	2500
Mack Tank Truck	nickel-plated driver, cast iron, 13 1/4" long, 1926	1000	1500	2000
Mack Wrecker Truck	white rubber tires, cast iron, 11" long	500	750	1000
Model-A Stakebody Truck	iron wheels, cast iron, 7 1/2" long, 1920's	150	225	300
Plymouth Wrecker		100	185	275
Pontiac Stake Truck	white rubber tires, cast iron, 6 1/4" long, 1936	225	350	500
Pontiac Stake Truck	white rubber tires, cast iron, 4 1/4" long, 1936	65	95	165
Pontiac Stake Truck		85	145	185
Pontiac Wrecker	white rubber tires, cast iron, 4 1/4" long, 1936	70	100	175
Pontiac Wrecker		85	145	185
Red Baby Dump Truck	bright red truck, white enameled tires, crank dump, cast iron, 10 3/4" long, 1924	1200	1800	2400

Arcade

NAME	DESCRIPTION	GOOD	EX	MINT
Red Baby Truck	bright red truck, white enameled tires, cast iron, 10 1/4" long, 1924	1000	1500	2000
Semi Truck	cast iron, 1920's	500	750	1000
Stake Truck	cast iron, 7" long	110	175	225
Tow Truck	cast iron, 4" long	55	85	110
Transport	double-deck semi-trailer with four sedans, cast iron and pressed steel, 18 1/2" long, sedans 4 3/4" long, 1938	750	1000	2000
White Delivery Truck	cast iron, 8 1/2" long, 1931	1250	2100	3300
White Dump Truck	cast iron, 11 1/2" long	7500	10000	15000
White Moving Van	cast iron, 13 1/2" long, 1928	6000	8500	13500

Vehicle Banks

NAME	DESCRIPTION	GOOD	EX	MINT
Ford Touring Car Bank	removable driver, cast iron, 6 1/2" long, 1925	650	975	1300
Mack Dump Truck Bank	cast iron, 13" long	375	565	750
Yellow Cab Bank	cast iron, 8" long, 1924	475	775	1000
Yellow Cab Bank	with or without rubber tires, cast iron, 9" long, 1926	600	900	1200
Yellow Cab Bank	cast iron, 8 1/2" long, 1930	650	1400	2000

Wagons, Carts and Trailers

NAME	DESCRIPTION	GOOD	EX	MINT
Auto Dump Wagon	red and gold, cast iron, 7" long, 1920	200	400	600
Circus Wagon	circus wagon with driver and two horses, 'Big Six Circus & Wild West' on side of wagon, 14 1/2" long, 1936	300	450	600
Ice Wagon	with black horses trimmed in gold, cast iron, 11 3/4" long, 1926	575	875	1300
McCormick-Deering Weber Wagon	removable wagon seat and box, with two horses, cast iron, overall length 12 1/8" long, 1925	200	400	600
Panama Dump Wagon	gray, nickeled pick and shovel, cast iron, 12 3/4" long, 1926	200	400	600
Panama Dump Wagon	gray, with two horses, cast iron, nickeled pick and shovel, 14 1/4" long, 1926	300	450	600
Whitehead & Kales Truck Trailer	sides and end gates removable by sections, cast iron, 8 1/2" long, 1926	150	250	400

Auburn

Airplanes

NAME	DESCRIPTION	GOOD	EX	MINT
Clipper Plane	rubber, 7" wingspan, 1941	10	20	25
Dive Bomber	rubber, 4" wingspan, 1937	10	20	25
Pursuit Ship Plane	4" wingspan, 1941	10	20	25
Two-Engine Transport Plane	1937	10	20	27

Boats and Ships

NAME	DESCRIPTION	GOOD	EX	MINT
Battleship	rubber, 8 1/4" long, 1941	15	25	30
Cruiser	rubber	15	25	30
Freighter	rubber, 8" long, 1941	15	25	30
Submarine	rubber, 6 1/2" long, 1941	10	20	27

Cars

NAME	DESCRIPTION	GOOD	EX	MINT
1947 Buick Coupe	#100 on license plate, 7" long	75	115	150
Airport Limousine	rubber, 8" long	25	40	50
Fire Chief's Car	red, yellow wheels	7	10	15
Ford	rubber, 1930's	10	20	25
Race Car	rubber, 6" long, 1930's	15	25	30
Race Car	red, rubber, 6" long	40	65	85
Race Car With Goggled Driver	rubber, 10" long	50	75	100
Racer	rubber	20	30	40

Top to bottom: Wrecker and coupe, 1930s, Barclay; 1956 Pontiac Safari station wagon, Tootsietoys; Kenton sulky with driver; Steam Shovel, Structo; Wagon, Arcade; 1932 Ford hot rod, Schuco; T-9 Bulldozer, Tonka.

Auburn

NAME	DESCRIPTION	GOOD	EX	MINT
Racer	red vinyl with white plastic tires	25	40	50
Sedan	green, rubber, license #500R	15	25	30
Sedan	cast iron driver	25	40	50

Emergency Vehicles

Fire Engine	red, rubber, 8" long	15	25	30
Fire Truck	black rubber wheels	7	10	15
Rescue Truck	dark army green	15	25	30

Farm and Construction Equipment

Allis-Chalmers Tractor	red and silver plastic, 1/16 scale, 1950	25	40	50
Earthmover	red front, yellow back, plastic wheels	15	25	30
Giant Tractor	red tractor, silver motor, black tires, 7" long, 1950's	30	45	60
John Deere Tractor	plastic, 1/20 scale	25	40	50
Minneapolis Moline Tractor	red, large rubber tires, rubber 1/16 scale, 1950	25	40	50
Tractor and Wagon	orange tractor, silver motor, black tires, red spreader wagon, yellow spoke tires	75	115	150

Motorcycles

Motorcycle		20	30	40
Police Cycle	red rubber, drive chain, 6" long, 1950's	50	75	100

Trucks

2 1/2 Ton Truck		20	30	40
Army Jeep	olive drab	5	7	10
Army Recon Half Truck	bright green	10	15	20
Stake Truck	rubber	15	25	30
Telephone Truck	6 1/2" long	35	55	75

Bandai

Buses

Volkswagen Bus	red and white, battery operated, 9 1/2" long, 1960's	65	150	275

Cars

1915 Ford Touring Car	7" long	90	135	180
1960 Rolls Royce Silver Cloud	blue body, white top, electric lights, 12" long	350	525	700
Cadillac	gold fins, black top, tin, 11 1/2" long, 1959	150	225	300
Cadillac	8" long	100	175	275
Cadillac	white, hardtop, friction powered, 11" long, 1959	100	175	275
Cadillac	copper, hardtop, friction powered, 11" long, 1960	75	125	175
Cadillac	gold, hardtop, friction powered, 17" long, 1960's	195	350	550
Cadillac Convertible	red, green interior, friction powered, 11" long, 1959	175	300	400
Cadillac Convertible	black, friction powered, 11" long, 1960	180	275	380
Cadillac Convertible	white, red interior, friction powered, 17" long, 1963	200	400	600
Chevrolet Corvette	white and black, battery operated, 8" long, 1962	75	125	175
Chevrolet Corvette	red, friction powered, 8" long, 1963	75	125	175
Chevrolet Impala Convertible	white, friction powered, 11" long, 1961	250	400	550
Chevrolet Impala Sedan	cream, friction powered, 11" long, 1961	225	350	500
Citroen	blue and white, friction powered, 12" long, 1958	300	650	1000
Corvair Bertone	white, battery operated, tin, 12" long, 1963	70	120	175
Cougar	white, battery operated, tin	100	150	200
Excalibur Roadster	white body, red fenders, black top, battery operated, rubber wheels, motor sparks, 11" long	135	200	275
Ferrari	silver with red interior, battery operated, tin, gearshift n floor, working lights, horn, and engine noise, 11" long, 1958	300	550	850

VEHICLES

Bandai

NAME	DESCRIPTION	GOOD	EX	MINT
Ford Convertible	green, friction powered, 12" long, 1955	350	650	900
Ford Convertible	red and black or two tone green, friction powered, 12" long, 1957	150	275	375
Ford Country Sedan	blue and white, friction powered, 10 1/2" long, 1961	65	100	140
Ford F.B.I. Mustang	black and white, friction powered, 11" long, 1965	75	100	140
Ford Flower Delivery Wagon	blue, friction powered, 12" long, 1955	275	400	650
Ford G T	red battery operated, 10" long, 1960's	100	150	200
Ford Mustang	red, battery operated, 11" long, 1965	125	200	275
Ford Mustang	red, battery operated, 13" long, 1967	80	125	175
Ford Mustang	silver and black, battery operated, 11" long, 1965	65	100	145
Ford Ranchero	two tone blue, friction powered, 12"long, 1955	175	250	350
Ford Ranchero	black and red, friction powered, 12" long, 1957	175	250	350
Ford Standard Fresh Coffee Wagon	black and orange, friction powered, 12" long, 1955	350	625	975
Ford Station Wagon	cream and black, two tone green, or red and black, friction powered, 12" long, 1955	75	125	200
Ford Thunderbird	red or red and black, friction powered, 8" long, 1962	65	100	150
Ford Thunderbird	red and black, friction powered, 10 3/4" long, 1965	100	175	225
Ford Wagon	green body, black top, 12" long	300	450	600
Ford Wagon	blue and white, friction powered, 12" long, 1957	90	150	200
GT-40	blue, hood and trunk open, rubber tires, tin, battery operated, 11" long	100	150	200
Isetta	white and two tone green, 3 wheels, friction powered, 6 1/2" long, 1950's	200	300	400
Jaguar XK 140 Convertible	various colors, 9 1/2" long, 1950s	125	200	275
Jaguar XK-E	red, battery operated, 10" long, 1960's	100	150	250
Lincoln Continental	turquoise and white, matching interior, 1958	150	250	375
Lincoln Continental Convertible	white, red interior, 1958	200	300	425
Lincoln Mark III	turquoise and whote, friction powered, 11" long, 1958	150	275	375
Lotus	blue, battery operated, 9 1/2" long	110	175	225
Lotus Elite	red and black, friction powered, 8 1/2" long, 1950's	75	100	150
Mazda 360 Coupe	blue, friction powered, 7" long, 1960	65	100	125
Mercedes Benz 300 SL Coupe	silver and black, friction powered, 8" long, 1950's	160	245	325
Mercedes-Benz Taxi	black, battery operated, 10" long, 1960's	125	250	350
MG 1600 Mark II	red, friction powered, 8 1/2" long, 1950's	75	125	175
MG TF	green, friction powered, 8" long, 1955	75	125	175
Old Timer Police Car	battery operated, 8" long	60	95	125
Olds Toronado	gold, battery operated, 11" long, 1966	30	50	65
Oldsmobile	surrey top, friction, 1900's	75	115	150
Plymouth Valiant	blue, windup, 8" long	50	75	100
Pontiac Firebird	red, battery operated, 9 1/2" long, 1967	75	125	175
Porsche 911	white, battery operated, 10" long, 1960's	90	150	200
Racer with Hand Control		85	130	175
Rolls-Royce Convertible	several colors available, 12" long, friction, 1950's	200	300	450
Rolls-Royce Hardtop Sedan	blue, black, white, rare version with working headlights, battery operated, 12" long, 1950's	275	400	600
Subaru 360	red, friction powered, 8" long, 1959	75	125	175
Taxi	friction, 1950's	75	115	150
Volkswagen	red, battery operated, 8" long, 1960's	35	75	100
Volkswagen	blue, battery operated, 10 1/2" long, 1960's	75	125	175
Volkswagen	red, battery operated, 15" long, 1960's	95	150	225
Volkswagen	red, with sun roof, battery operated, 15" long, 1960's	90	150	225
Volkswagen Convertible	white, battery operated, 7 1/2" long, 1960's	50	95	125

Emergency Vehicles

NAME	DESCRIPTION	GOOD	EX	MINT
Plymouth Ambulance	white, red cross on doors, friction powered, 12" long, 1961	35	50	75
Rambler Ambulance	white, friction powered, 11" long, 1962	40	75	100

Bandai

NAME	DESCRIPTION	GOOD	EX	MINT
### Motorcycles					
Police Auto Cycle	battery operated, hard plastic, 10" long, 1970's	100	150	200
### Sets					
Ferrari and Speed Boat	white car, red and white speed boat with white trailer, tin, overall 23" long, 1958	350	650	900
Lincoln Continental and Cabin Cruiser	turquoise and white car and cruiser, red car interior and cruiser trailer, overall 23" long, 1958	300	550	800
Rambler Wagon and Cabin Cruiser	green and white Rambler wagon, red trailer, friction powered rambler, electric boat motor, overall 23" long, 1959	200	400	600
Rambler Wagon and Shasta Trailer	green and white Rambler wagon, yellow and white trailer, 11" long wagon, 12" long trailer, 1959	200	400	600
Rambler, Trailer and Cabin Cruiser	turquoise and white Rambler, red trailer, cruiser color varies, overall 35" long, 1959	350	650	1000
### Trucks					
1958 Ford Ranchero Pickup Truck	8" long	110	175	225
Land Rover	maroon, 8" long	125	185	250
Land Rover	red, friction powered, 7 1/2" long, 1960	75	100	150
Volkswagen Truck	blue, open flatbed cargo section, battery operated, 8" long, 1960's	90	150	200

Barclay

NAME	DESCRIPTION	GOOD	EX	MINT
### Airplanes					
Dirigible Plane	4 3/8" long, 1930's	7	10	15
Lindy-Type Plane	4" wingspan, 1930's	7	10	15
Monoplane	single engine plane	15	28	35
Monoplane	single engine plane, red propeller, red metal wheels	15	30	40
U.S. Army Single Engine Transport Plane	white rubber wheels, 1940	10	15	20
U.S. Army Small Pursuit Plane	lead, rubber wheels, 1941	10	15	20
### Buses					
Double-Decker Bus	4" long	20	30	40
### Cannons					
Cannon	barrel elevated, 2 1/2" long	15	25	35
Cannon	1931	20	30	40
Cannon	very large wheels, 4" long	15	20	25
Cannon	silver with black rubber wheels, 7 3/4" long	20	30	40
Cannon	spoked wheels, 3" long	10	15	20
Coast Defense Rifle Cannon	4 1/2" long	25	40	50
Howitzer Cannon	horizontal loop hitch, 4 wheels, 3" long	15	25	35
Howitzer Cannon	vertical loop hitch, 4 wheels, 3" long	15	25	35
Mortar Cannon	swivels on base, 3" long	20	30	40
Spring-Firing Cannon	spoked wheels, 4" long	20	25	35
### Cars					
Armoured Car	1937	15	25	35
Car Carrier	with two cars	25	40	55
Coupe	1930's	20	30	40
Coupe	3" long, 1930's	15	25	35
Race Car	white tires, 4" long	25	40	55

VEHICLES

Barclay

NAME	DESCRIPTION	GOOD	EX	MINT
Trucks				
Beer Truck	slush metal, 4" long	15	25	35
Mack Pickup Truck	3 1/2" long	25	40	55
Milk Truck #377	with milk cans	25	40	55
Open Truck	3 1/2" long	10	20	25
Stake Truck	slush metal, 5" long	15	25	35

Brooklin

NAME	DESCRIPTION	GOOD	EX	MINT
Cars				
1932 Packard Light 8 Coupe		35	50	70
1933 Pierce Arrow	silver	35	50	70
1934 Chrysler Airflow Sedan	four door	35	50	70
1935 Dodge "City Ice Delivery" Van		40	60	80
1935 Dodge "Dr.Pepper" Van		45	70	90
1935 Dodge "Sears Roebuck" Van		40	60	80
1940 Ford Sedan Delivery "Ford Service"		40	60	80
1941 Packard Clipper" Van		40	60	80
1948 Tucker Torpedo		35	50	70
1949 Buick Roadmaster		35	55	75
1949 Mercury Coupe	two door	30	45	65
1952 Hudson Hornet Convertible	1/43 scale	30	45	65
1952 Studebaker Champion Starlight Coupe		40	60	80
1953 Buick Skylark		35	50	70
1953 Pontiac Sedan Delivery Gulf Oil Truck	1/43 scale	30	45	60
1953 Pontiac Sedan Delivery Mobil Oil Truck	1/43 scale	30	45	60
1953 Studebaker Commander	1/43 scale	30	45	60
1953 Studebaker Indiana State Police	1/43 scale	50	75	100
1954 Dodge 500 Indy Pace Car	1/43 scale	50	75	100
1955 Chrysler 300		30	45	60
1956 Ford Fairline Victoria	two door	25	40	55
1956 Ford Thunderbird 500	hardtop	30	45	65
1956 Lincoln Continental		30	45	60
1956 Lincoln Continental Mark II Coupe		30	45	65
1956 Lincoln Continental MKII		30	45	65
1957 Ford Fairlane Skyliner Police	1/43 scale	30	45	60
1958 Edsel Citation	two door hardtop	30	45	30
1958 Pontiac Bonneville		30	45	65

Brooklin

NAME	DESCRIPTION	GOOD	EX	MINT
1960 Ford Sunliner Convertible	1/43 scale	30	45	60
1963 Chevrolet Corvette Stingray Coupe		35	55	75
1968 Shelby Mustang GT 500		35	55	75
Lincoln Mark	1/43 scale	30	45	60
Mini Marquee Packard Convertible	1/43 scale	50	75	100
Tucker	1/43 scale	30	45	60

VEHICLES

VEHICLES

Top to bottom: GMC Mobilgas Tanker, 1949, Smith-Miller; Ford Pickup Truck, 1959, Tonka; Auburn Super Racer (rubber), Auburn; Ford Pepsi Cola delivery truck, Ny-lint.

VEHICLES

Buddy L

Airplanes

NAME	DESCRIPTION	GOOD	EX	MINT
5000 Monocoupe "The Lone Eagle"	orange wing, black fuselage and tail with tailskid, all steel high wind cabin monoplane, 9 7/8" wingspan, 1929	250	450	550
Army Tank Transport Plane	low-wing monoplane, two small four-wheel tanks that clip beneath wings, 27" wingspan, 1941	250	375	500
Catapult Airplane and Hangar	5000 Monocoupe with tailwheel, 9 7/8" wingspan, olive-gray hangar, black twin-spring catapult, 1930	1000	1500	2000
Four Motor Air Cruiser	white, red engine cowlings, yellow fuselage and twin tails, four engine monoplane, 27" wingspan, 1952	200	300	400
Four-Engine Transport	green wings, white engine cowlings, yellow fuselage and twin tails, four engine monoplane, 27" wingspan, 1949	200	300	400
Hangar and Three 5000 Monocoupes	olive-gray hangar, windows outlined in red or orange, planes 9 7/8" wingspan, all steel high wing cabin monoplanes, 1930	1500	2250	3000
Transport Airplane	white wings and engine cowlings, red fuselage and twin tails, four engine monoplane, 27" wingspan, 1946	200	300	400

Boats and Ships

LST Landing Ship	navy gray flat-bottomed steel hull, ship 12 3/4" long, with 4 1/4" long tank and 5" long troop transport, 1976	125	185	250
Tug Boat	medium bluish-gray-green hull, keel and rudder, gray pilot house, cabin and deck, 28" long, 1928	5000	7500	10000

Buses

Coach	bluish gray-green, opening front doors, interior has 22 chairs plus 2 benches over back wheels, 29 1/4" long, 1927	3000	6000	9000
Greyhound Bus	white roof, blue and white sides, blue front has yellow headlights, 16 1/2" long, 1938	275	415	550
Greyhound Bus	white roof and back, blue and white sides and front, 16 3/4" long, 1949	200	300	400

Cars

Army Staff Car	olive drab body, 15 3/4" long, 1964	100	150	200
Bloomin' Bus	chartreuse body, white roof and supports, similar to VW minibus, 10 3/4" long, 1969	90	135	180
Buddywagen	red body with white roof, 10 3/4" long, 1966	100	150	200
Buddywagen	red body with white roof, no chrome v on front, 10 3/4" long, 1967	95	145	190
Chrysler Six Passenger Airflow Sedan	pressed steel, 17 1/4" long, 1934	0	0	0
Colt Sportsliner	red open body, white hardtop, off-white seats and interior, 10 1/4" long, 1967	35	50	70
Colt Sportsliner	light blue-green open body, white hardtop, pale tan seats and interior, 10 1/4" long, 1968	30	45	65
Colt Utility Car	red open body, white plastic seats, floor and luggage space, 10 1/4" long, 1967	35	50	70
Colt Utility Car	light orange body, beige-tan interior, 10 1/4" long, 1968	30	45	65
Country Squire Wagon	off-white hood fenders, endgate and roof, brown wood-grain side panels, 15 1/2" long, 1963	85	130	175
Country Squire Wagon	red hood fenders, endgate and roof, brown wood-grain side panels, 15" long, 1965	75	115	150
Deluxe Convertible Coupe	metallic blue enamel front, sides and deck, cream top retracts into rumble seat, 19" long, 1949	300	450	600

Buddy L

NAME	DESCRIPTION	GOOD	EX	MINT
Desert Rats Command Car	light tan open body, light beige interior, black .50-caliber machine gun swivels, tits, on post between seats, 10 1/4" long, 1967	50	75	100
Desert Rats Command Car	light tan open body, light beige interior, black .50-caliber machine gun swivels, tits, on post between seats, 10 1/4" long, blackwall tires, 1968	45	65	90
DeSoto Airflow Sedan	pressed steel, 17" long, 1934	0	0	0
Flivver Coupe	black except for red 8-spoke wheels, black hubs, aluminum tires, flat, hard-top roof on enclosed glass-window-style body, 11" long, 1924	775	1100	1550
Flivver Roadster	black except for red 8-spoke wheels, black hubs, aluminum tires, simulated soft, folding top, 11" long, 1924	1000	1500	2000
Jr. Camaro	metallic blue body, white racing stripes across hood nose, 9" long, 1968	50	75	100
Jr. Flower Power Sportster	purple hood, fenders and body, white roof and supports, white plastic seats, lavender and orange five-petal blossom decals on hood top, roof, and sides, 6" long, 1969	35	55	75
Jr. Sportster	blue hood and open body, white hardtop and upper sides, 6" long, 1968	35	55	75
Mechanical Scarab Automobile	red radically streamlined body, bright metal front and rear bumpers, 10 1/2" long, 1936	200	300	500
Police Colt	deep blue open body, white hardtop, "POLICE" across top of hood, "POLICE 1" on sides, 10 1/4" long, 1968	50	75	100
Ski Bus	white body and roof, similar to VW minibus, 10 3/4" long, 1967	75	115	150
Station Wagon	light blue-green body and roof, 15 1/2" long, 1963	75	115	150
Streamline Scarab	red, radically streamlined body, non-mechanical, 10 1/2" long, 1941	145	225	290
Suburban Wagon	powder blue or white body and roof, 15 1/2" long, 1963	75	115	150
Suburban Wagon	gray-green body and roof, 15 3/4" long, 1964	70	100	140
Town and Country Convertible	maroon front, hood, rear deck and fenders, gray top retracts into rumble seat, 19" long, 1947	300	450	600
Travel Trailer and Station Wagon	red station wagon, 2-wheel trailer with red lower body and white steel camper-style upper body, 27 1/4" long, 1965	150	225	300
Yellow Taxi with Skyview	yellow hood, roof and body, red radiator front and fenders, 18 1/2" long, 1948	325	500	675

Emergency Vehicles

NAME	DESCRIPTION	GOOD	EX	MINT
Aerial Ladder and Emergency Truck	red except for white ladders, bumper and steel disc wheels, three 8-rung steel ladders, 22 1/4" long, 1952	200	300	400
Aerial Ladder and Emergency Truck	red except for white ladders, bumper and steel disc wheels, three 8-rung steel ladders, no rear step, no siren or SIREN decal, 22 1/4" long, 1953	225	345	450
Aerial Ladder Fire Engine	red tractor, wraparound bumper and semi-trailer, 2 aluminum 13-rung extension ladders on sides, swivel-base aluminum central ladder, 26 1/2" long, 1960	125	185	250
Aerial Ladder Fire Engine	red tractor and semi-trailer, white plastic bumper with integral grill guard, 2 aluminum 13-rung extension ladders on sides, swivel-base aluminum central ladder, 26 1/2" long, 1961	125	185	250
Aerial Ladder Fire Engine	red tractor and semi-trailer, chrome one-piece wraparound bumper, slotted grille, 2 aluminum 13-rung extension ladders on sides, swivel-base aluminum central ladder, 26 1/2" long, 1966	125	185	250
Aerial Ladder Fire Engine	red cab-over-engine tractor and semi-trailer units, two 13-rung white sectional ladders and swivel-mounted aerial ladder with side rails, 25 1/2" long, 1968	100	150	200

Buddy L

NAME	DESCRIPTION	GOOD	EX	MINT
Aerial Ladder Fire Engine	snub-nose red tractor and semi-trailer, white swivel-mounted aerial ladder with side rails, 2 white 13-rung sectional ladders, 27 1/2" long, 1970	100	150	200
Aerial Truck	red ecept for nickel ladders, black hand wheel, brass bell, and black hubs, 39" long with ladder down, 1925	850	1300	1700
American LaFrance Aero-Chief Pumper	red cab-over-engine and body, white underbody, rear step and simulated hose reels, black extension ladders on right side, 25 1/2" long, 1972	125	185	250
Brute Fire Pumper	red cab-over-engine body and frame, 2 yellow 5-rung sectional ladders on sides of open body, 5 1/4" long, 1969	50	75	100
Brute Hook-n-Ladder	red cab-over-engine tractor and detachable semi-trailer, white elevating, swveling aerial ladder with side rails, 10" long, 1969	30	40	55
Extension Ladder Fire Truck	all red except for silver -painted ladders and yellow removable rider seat, enclosed cab, 35" long, 1945	200	300	400
Extension Ladder Rider Fire Truck	duo-tone slant design, tractor has white front, lower hood sides and lower doors, red hood top, cab and frame, red semi-trailer, white 10-rung and 8-rung ladders, 32 1/2" long, 1949	150	225	300
Extension Ladder Trailer Fire Truck	red tractor with enclosed cab, boxy fenders, red semi-trailer with fenders, two white 8-rung side ladders, 10-rung central extension ladder, 29 1/2" long, 1955	200	300	400
Extension Ladder Trailer Fire Truck	red tractor unit and semi-trailer, enclosed cab, two white 13-rung side extension ladders on sides, white central ladder on swivel base, 29 1/2" long, 1956	125	185	250
Fire and Chemical Truck	duo-tone slant design, white front, lower hood sides and lower doors, rest is red, bright-metal or white 8-rung ladder on sides, 25" long, 1949	125	185	250
Fire Department Emergency Truck	red streamlined body, enclosed cab, chrome one-piece grille, bumper, and headlights, 12 3/4" long, 1953	100	150	200
Fire Engine	red except for nickel-plated upright broiler, nickel rims and flywheels on dummy water pump, brass bell, 23 1/4" long, 1925-29	3000	4000	5000
Fire Engine	red except for nickel rim flywheels on dummy pump, brass bell, dim-or-bright electric headlights, 25 1/2" long, 1933	1500	2000	2500
Fire Hose and Water Pumper	red except for 2 white 5-rung ladders and 2 removable fire extinguishers, enclosed cab, 12 1/2" long, 1950	100	150	200
Fire Hose and Water Pumper	red except for 2 white 5-rung ladders and 1 red and white removable fire extinguisher, enclosed cab, 12 1/2" long, 1952	100	150	200
Fire Pumper	red cab-over-engine and open body, 11-rung white 10" ladder on each side, 16 1/4" long, 1968	100	150	200
Fire Pumper with Action Hydrant	red wraparound bumper, hood cab and cargo section, aluminum 9-rung ladders, white hose reel, 15" long, 1960	75	115	150
Fire Truck	all red except for white ladders, black rubber wheels, enclosed cab, 12" long, 1945	75	115	150
Fire Truck	duo-tone slant design, tractor has white front, lower hood sides and lower doors, red hood top, cab and frame, red semi-trailer, rubber wheels with black tires, 32 1/2" long, 1953	200	300	400
Fire Truck	bright red, except for black inverted L-shaped crane mounted in socket on seat back, open driver's seat, 26" long, 1924	1500	2500	3500
Fire Truck	bright "fire engine" red, except for black inverted L-shaped crane mounted in socket on seat back, red floor, open driver's seat, 26" long, 1925	1800	3000	4500
Fire Truck	bright "fire engine" red, red floor, open driver's seat, 26" long, 1928	750	1200	1750

VEHICLES

Buddy L

NAME	DESCRIPTION	GOOD	EX	MINT
Fire Truck	red except for black solid-rubber firestone tires on red 7-spoke embossed metal wheels, two 18 1/2" red steel sectional ladders, 26" long, 1930	950	1500	1900
Fire Truck	red except for nickel or white ladders, bright-metal radiator grille and black removable rider saddle, 25 1/2" long, 1935	250	375	500
Fire Truck	duo-tone slant design, yellow front, single-bar bumper, hood sides and removable rider seat, rest is red, 25 1/2" long, 1936	500	750	1000
Fire Truck	duo-tone slant design, yellow front, bumper, hood sides and skirted fenders, rest is red, nickel ladders, 28 1/2" long, 1939	550	850	1100
Fire Truck	red except for 2 white ladders, enclosed cab, bright metal grille and headlights, 25" long, 1948	125	200	250
GMC Deluxe Aerial Ladder Fire Engine	white tractor and semi-trailer units, golden 13-rung extension ladder on sides, golden central aerial ladder, black and white DANGER battery case with 2 flashing lights, 28" long, 1959	225	345	450
GMC Extension Ladder Trailer Fire Engine	red tractor with chrome GMC bar grille, red semi-trailer, white 13-rung extension ladders on sides, white swiveling central ladder with side rails, 27 1/4" long, 1957	100	250	400
GMC Fire Pumper with Horn	red except for aluminum-finish 11-rung side ladders and white reel of black plastic hose in open cargo section, chrome GMC bar grille, 15" long, 1958	150	200	300
GMC Hydraulic Aerial Ladder Fire Engine	red tractor unit with chrome GMC bar grille, red semi-trailer, white 13-rung extension ladders on sides, white swiveling central ladder, 26 1/2" long, 1958	125	185	250
GMC Red Cross Ambulance	all white, removable fabric canopy with a red cross and "Ambulance" in red capitals, 14 1/2" long, 1960	150	250	400
Hook & Ladder Fire Truck	medium-dark red, except for black inverted L-shaped crane mounted in socket on seat back, open driver's seat, 26" long, 1923	1200	1800	2400
Hose Truck	all red except for 2 white hose pipes, white cord hose on reeland brass nozzle, electric headlights with red bulbs, 21 3/4" long, 1933	500	750	1000
Hydraulic Aerial Truck	duo-tone slant design, yellow bumper, radiator front, fenders, lower hood sides and removable rider saddle, rest is red, nickel extension ladders, 41" long with ladders, 1939	1500	2500	3500
Hydraulic Aerial Truck	red except for black removable rider saddle and twisted-wire removble pull-n-ride handle, nickel extension ladders, 40" long, 1933	1000	2000	3000
Hydraulic Aerial Truck	duo-tone slant design, yellow front, single-bar bumper, chassis, radiator, front fender, lower sides and removable rider saddle, rest is red, 40" long with ladders down, 1936	550	825	1100
Hydraulic Aerial Truck	red except for brass bell on cowl, nickel ladders mounted on 5 1/2" turntable rotated by black hand wheel, 39" long, 1927	850	1300	1700
Hydraulic Aerial Truck	red except for brass bell on cowl, nickel ladders mounted on 5 1/2" turntable rotated by black hand wheel, 2-bar nickel front bumper, 39" long, 1930	1000	1500	2500
Hydraulic Aerial Truck	red except for brass bell on cowl, nickel ladders mounted on 5 1/2" turntable rotated by black hand wheel, 2-bar nickel front bumper, nickel-rim headlights in red shells, 39" long, 1931	1200	1750	2500
Hydraulic Aerial Truck	duo-tone slant design, red hood and body, yellow rider seat, nickel extension ladders, 41" long with ladders, 1941	650	975	1300
Hydraulic Snorkel Fire Pumper	red cab-over-engine and open rear body, white 11-rung 10" ladder on each side, snorkel pod with solid sides, 21" long, 1969	100	150	200

VEHICLES

Buddy L

NAME	DESCRIPTION	GOOD	EX	MINT
Hydraulic Water Tower Truck	red except nickel water tower, dim-or -bright electric headlights, brass bell, 44 7/8" long with tower down, 1933	1200	2000	3000
Hydraulic Water Tower Truck	red except nickel water tower, dim-or -bright electric headlights, brass bell, added-on bright-metal grille, 44 7/8" long with tower down, 1935	2000	4000	6000
Hydraulic Water Tower Truck	duo-tone slant design, yellow bumper, hood sides, front fenders, rest is red, electric headlights, added-on bright-metal grille, 44 7/8" long with tower down, 1936	1200	2000	3000
Hydraulic Water Tower Truck	duo-tone slant design, yellow front, single-bar bumper and hood sides, red hood top, enclosed cab and water tank, brass bell, nickel water tower, 46" long with tower down, 1939	1200	2000	3000
Jr. Fire Emergency Truck	red cab-over-engine and body, one-piece chrome wraparound narrow bumper and 24-hole grille with plastic vertical-pair headlights, 6 3/4" long, 1968	50	75	100
Jr. Fire Emergency Truck	red cab-over-engine and body, wider one-piece chrome wraparound narrow bumper and 4-slot grille with 2 square plastic headlights, 6 3/4" long, 1969	50	75	100
Jr. Fire Snorkel Truck	red cab-over-engine and body, chrome one-piece narrow wraparound bumper and 24-hole grille with plastic vertical-pair headlights, 11 1/2" long, 1968	100	150	200
Jr. Fire Snorkel Truck	red cab-over-engine and body, full-width chrome one-piece bumper and 4-slot grille with 2 square plastic headlights, 11" long, 1969	60	150	200
Jr. Hook-n-Ladder Aerial Truck	red cab-over-engine tractor and semi-trailer, white high-sides ladder, chrome one-piece wraparound bumper and 24-hole grille, plastic veritcal-pair headlights, 17" long, 1967	100	150	200
Jr. Hook-n-Ladder Aerial Truck	red cab-over-engine tractor with one-piece 4-slot grille and 2 square plastic headlights, red semi-trailer, white high-sides ladder, plastic veritcal-pair headlights, 17" long, 1969	75	115	150
Ladder Fire Truck	red except for bright-metal V noco radiator, headlights and ladder, black wooden wheels, 12" long, 1941	125	200	250
Ladder Truck	red except for 2 yellow sectional ladders, enclosed square cab, 22 3/4" long, 1933	300	450	600
Ladder Truck	red except for 2 yellow ladders, enclosed square cab with sharply protruding visor, 22 3/4" long, 1934	200	300	400
Ladder Truck	red except for 2 yellow ladders, enclosed square cab with sharply protruding visor, bright-metal radiator front, 22 3/4" long, 1935	250	375	500
Ladder Truck	duo-tone slant design, white front, hood sides, fenders and 2 ladders, rest is red, square enclosed cab with sharply protruding visor, 22 3/4" long, 1936	150	225	300
Ladder Truck	duo-tone slant design, white front, hood sides, fenders and 2 ladders, rest is red, square enclosed cab with sharply protruding visor, no headlights, 22 3/4" long, 1937	135	200	275
Ladder Truck	duo-tone slant design, white front, fenders, hood sides and 2 ladders, rest is red, enclosed cab, 24" long, 1939	200	300	400
Ladder Truck	all red except for bright-metal grille and headlights, 2 white ladders, 24" long, 1939	200	300	400
Ladder Truck	all red except for yellow severely streamlined, skirted fenders and lower doors, white ladders, bright-metal grille, no bumper, 17 1/2" long, 1940	200	300	400
Ladder Truck	modified duo-tone slant design, white front, front fenders and lower doors, white ladders, bright-metal grille, no bumper, 17 1/2" long, 1941	200	300	400
Ladder Truck	red except for 2 white ladders, bright-metal radiator grille and headlights, 24" long, 1941	125	200	250

VEHICLES

Buddy L

NAME	DESCRIPTION	GOOD	EX	MINT
Police Squad Truck	yellow front and front fenders, dark blue-green body, yellow fire extinguisher, 21 1/2" long over ladders, 1947	275	500	750
Pumping Fire Engine	red except for nickel stack on boiler, nickel rims on pump flywheels, nickel-rim headlights and searchlight, 23 1/2" long, 1929	3000	3500	4000
Rear Steer Trailer Fire Truck	red except for 2 white 10-rung ladders, chrome one-piece grille, headlights and bumper, 20" long, 1952	125	200	250
Red Cross Ambulance	all white, removable fabric canopy with a red cross and "Ambulance" in red capitals, 14 1/2" long, 1958	60	95	125
Suburban Pumper	red station-wagon body, white plastic wraparound bumpers, one-piece grille and double headlights, 15" long, 1964	100	150	200
Texaco Fire Chief American LaFrance Pumper	promotional piece, red rounded-front enclosed cab and body, white one-pice underbody, running boards and rear step, 25" long, 1962	200	300	400
Trailer Ladder Truck	duo-tone slant design, tractor unit has yellow front, lower hood sides and lower doors, red hood top, enclosed cab and semi-trailer, nickel 10-rung ladders, 30" long with ladders, 1940	200	300	400
Trailer Ladder Truck	all red except for cream removable rider saddle, 3 bright-metal 10-rung ladders, 20" long over ladders, 1941	150	225	300
Water Tower Truck	red except nickel 2-bar front bumper, red nickel-rim headlights plus searchllight on cowl, nickel latticework water tower, 45 1/2" long with tower down, 1929	3000	4500	6000

Farm and Construction Equipment

NAME	DESCRIPTION	GOOD	EX	MINT
Aerial Tower Tramway	two tapering dark green 33 1/2" tall towers and 12" square bases, black hand crank, 1928	3000	4000	5000
Big Derrick	red mast and 20" boom, black base, 24" tall, 1921	600	900	1200
Brute Articulated Scooper	yellow front-loading scoop, cab, articulated frame and rear power unit, black radiator, exhaust, steering wheel and driver's seat, 5 1/2" long, 1970	50	75	100
Brute Double Dump Train	yellow hood, fenders and back on tractor unit, yellow coupled bottom-dumping earth carriers, 9 1/2" long, 1969	50	75	100
Brute Dumping Scraper	yellow hood, fenders and back on 2-wheel tractor unit, yellow scraper-dump unit, 7" long, 1970	50	75	100
Brute Farm Tractor-n-Cart	bright blue tractor body and rear fenders, green plastic radiator, engine, exhaust and driver's seat, bright blue detachable, square, 2-wheel open cart, 6 1/4" long, 1969	30	40	55
Brute Road Grader	yellow hood, cab, frame and adjustable blade, black radiator, driver's seat and steering wheel, 6 1/2" long, 1970	50	75	100
Cement Mixer on Treads	medium gray except for black treads and water tank, 16" tall, 1929-31	2500	3500	4500
Cement Mixer on Wheels	medium gray except for black cast steel wheels and water tank, 14 1/2" tall, 1926-29	700	900	1500
Concrete Mixer	medium gray except for black cast-steel wheels, black water tank, with wood-handle, steel-blade scoop shovel, 17 3/4" long with tow bar up, 1926	700	900	1500
Concrete Mixer	green except for black cast-steel wheels, crank, gears, and band mixing drum, 10 1/2" long, 1930	175	265	350
Concrete Mixer	yellow-orange frame and base, red hopper and drum, black crank handle, 10 1/2" long, 1936	110	175	225
Concrete Mixer	red frame and base, cream-yellow hopper, drum and crank handle, 10 1/2" long, 1941	125	185	250
Concrete Mixer	green frame, base, crank, crank handle and bottom of mixing drum, gray hopper and top of drum, 9 5/8" long, 1949	100	150	200

Buddy L

NAME	DESCRIPTION	GOOD	EX	MINT
Concrete Mixer on Tread	gray except for black water tank, with wood-handle, steel-blade scoop shovel, 15 3/4" long, 1929	2500	3500	4500
Concrete Mixer with Motor Sound	green frame, base, crank, crank handle and bottom of mixing drum, gray hopper and top of drum, with rat-rat-tat motor-noise sound when crank rotates drum, 9 5/8" long, 1950	75	115	150
Dandy Digger	yellow seat lower control lever and main boom, black underframe, skids, shovel and arm, 38 1/2" long with shovel arm extended, 1953	75	115	150
Dandy Digger	red main frame, operators, seat and boom, black shovel, arm, under frame and twin skids, 27" long, 1931	100	160	215
Dandy Digger	yellow main frame, operators, seat and boom, green shovel, arm, under frame and twin skids, 27" long, 1936	85	130	175
Dandy Digger	yellow main frame, operators, seat and boom, brown shovel, arm, under frame and twin skids, 27" long, 1941	95	145	195
Dandy Digger	yellow seat, lower control lever and main boom, black underframe, skids, shovel, arm and control lever, 38 1/2" long with shovel arm extended, 1953	75	115	150
Digger	red main frame, operators, seat and boom, black shovel, arm, lower frame and twin skids, curved connecting rod, boom tilts down for digging, 11 1/2" long with shovel arm extended, 1935	100	150	200
Dredge	red corrugated roof and base with 4 wide black wheels, red hubs, black boiler, floor, frame boom and clamshell bucket, 19" long, 1924	750	1000	1500
Dredge on Tread	red corrugated roof and base with crawler treads with red side frames, red hubs, black boiler, floor, frame boom and clamshell bucket, 21" long, 1929	5000	7000	9000
Giant Digger	red main frame, operators, seat and boom, black shovel, arm, lower frame and twin skids, boom tilts down for digging, 42" long with shovel arm extended, 1931	275	415	550
Giant Digger	red main frame, operators, seat and boom, black shovel, arm, lower frame and twin skids, curved connecting rod, boom tilts down for digging, 31" long with shovel arm extended, 1933	265	395	525
Giant Digger	yellow main frame, operators, seat and boom, green shovel, arm, lower frame and twin skids, boom tilts down for digging, 11 1/2" long with shovel arm extended, 1936	85	130	175
Giant Digger	yellow main frame, operators, seat and boom, brown shovel, arm, lower frame and twin skids, boom tilts down for digging, 11 1/2" long with shovel arm extended, 1941	75	115	155
Gradall	bright yellow truck and superstructure, black plastic bumper and radiator, 32" long with digging arm extended, 1965	350	750	1000
Hauling Rig with Construction Derrick	duo-tone slant design tractor, yellow bumper, lower hood and cab sides, white upper hood and cab, white trailer except for yellow loading ramp, overall 38 1/2" long, 1953	175	250	400
Hauling Rig with Construction Derrick	yellow tractor unit, green semi-trailer, winch on front of trailer make rat-tat motor sound, 36 3/4" long, 1954	110	175	225
Hoisting Tower	dark green, hoist tower and three distribution chutes, 29" tall, 1928-31	1500	2000	2500
Husky Tractor	bright yellow body and large rear fenders, black engine block, exhaust, steering wheel and driver's seat, 13" long, 1966	50	75	100
Husky Tractor	bright blue body and large rear fenders, black engine block, exhaust, steering wheel and driver's seat, 13" long, 1969	40	60	80

VEHICLES

Buddy L

NAME	DESCRIPTION	GOOD	EX	MINT
Husky Tractor	bright yellow body, red large rear fenders and wheels, black engine block, exhaust, steering wheel and driver's seat, 13" long, 1970	30	45	65
Improved Steam Shovel	black except for red roof and base, 14" tall, 1927-29	1000	1500	2000
Improved Steam Shovel on Treads	black except for red roof and tread frames, 17" tall, 1929-30	5000	7000	9000
Junior Excavator	red shovel, arm, underframe, control lever and twin skids, yellow boom, rear lever, frame and seat, 28" long, 1945	75	115	150
Junior Line Steam Shovel on Treads	black except for red roof and tread frames, 14" tall, 1930-32	1250	1750	2500
Mechanical Crane	orange removable roof, boom and wheels in black cleated rubber crawler treads, olive-green enclosed cab and base, hand crank with rat-tat motor noise, 20" tall, 1950	175	265	350
Mechanical Crane	orange removable roof, boom, yellow wheels in white rubber crawler treads, olive-green enclosed cab and base, hand crank with rat-tat motor noise, 20" tall, 1952	150	225	300
Mobile Construction Derrick	orange laticework main mast, swiveling base, yellow latticework boom, green clamshell bucket and main platform base, 25 1/2" long with boom lowered, 1953	150	250	350
Mobile Construction Derrick	orange laticework main mast, swiveling base, yellow latticework boom, gray clamshell bucket, green main platform base, 25 1/2" long with boom lowered, 1955	150	250	350
Mobile Construction Derrick	orange laticework main mast, swiveling base, yellow latticework boom, gray clamshell bucket, orange main platform base, 25 1/2" long with boom lowered, 1956	150	250	350
Mobile Power Digger Unit	clamshell dredge mounted on 10-wheel truck, orange truck, yellow dredge cab on swivel base, 31 3/4" long with boom lowered, 1955	125	185	250
Mobile Power Digger Unit	clamshell dredge mounted on 6-wheel truck, orange truck, yellow dredge cab on swivel base, 31 3/4" long with boom lowered, 1956	115	175	230
Overhead Crane	black folding end frames and legs, braces, red crossbeams and platform, 45" long, 1924	2000	2500	3000
Pile Driver on Treads	black except for red roof and tread frames, 22 1/2" tall, 1929	5000	7500	10000
Pile Driver on Wheels	black except for red roof and base, 22 1/2" tall, 1924-27	800	1500	3000
Polysteel Farm Tractor	orange molded plastic 4-wheel tractor, silver radiator front, headlights, and motor parts, 12" long, 1961	75	115	150
Pull-n-Ride Horse-Drawn Farm Wagon	red 4-wheel steel hopper-body wagon, detailed litho horse, 22 3/4" long, 1952	150	225	300
Road Roller	dark green except for red roof and rollers, nickel plated steam cylinders, 20" long, 1929-31	3000	4000	5000
Ruff-n-Tuff Tractor	yellow grille, hood and frame, black plastic engine block and driver's seat, 10 1/2" long, 1971	50	75	100
Sand Loader	warm gray except for 12 black buckets, 21" long, 18" high, 1924	350	550	750
Sand Loader	warm gray except for 12 black buckets, chain-tension adjusting device at bottom of elevator side frames, 21" long, 1929	350	500	700
Sand Loader	yellow except for 12 black buckets, chain-tension adjusting device at bottom of elevator side frames, 21" long, 1931	200	250	350
Scoop-n-Load Conveyor	cream body frame, green loading scoop, black circular crank operates black rubber cleated conveyor belt, 18" long, 1953	75	115	150
Scoop-n-Load Conveyor	cream body frame, green loading scoop, black circular crank operates black rubber cleated conveyor belt, "PORTABLE" decal in red, 18" long, 1954	65	100	135

VEHICLES

Buddy L

NAME	DESCRIPTION	GOOD	EX	MINT
Scoop-n-Load Conveyor	cream body frame, red loading scoop and chute, bright-plated circular crank operates black rubber cleated conveyor belt, "PORTABLE" decal in white, 18" long, 1955	60	95	125
Scoop-n-Load Conveyor	cream body frame, red loading scoop and chute, bright-plated circular crank operates black rubber cleated conveyor belt, "PORTABLE" decal in yellow, 18" long, 1956	55	85	115
Side Conveyor Load-n-Dump	yellow plastic front end including cab, yellow steel bumper, green frame and dump body, red conveyor frame with chute, 20 1/2" long, 1953	70	125	145
Side Conveyor Load-n-Dump	all steel yellow cab, hood, bumper and frame, white dump body and tailgate, red conveyor frame with chute, 21 1/4" long, 1954	65	100	135
Side Conveyor Load-n-Dump	all steel yellow cab, hood, bumper and frame, deep blue dump body, white tailgate, red conveyor frame with chute, 21 1/4" long, 1955	60	95	125
Sit-n-Ride Dandy Digger	yellow seat, lower control lever and main boom, green underframe, skids, shovel, arm and control lever, 38 1/2" long with shovel arm extended, 1955	75	115	150
Small Derrick	red 20" movable boom and 3 angle-iron braces, black base and vertical mast, 21 1/2" tall, 1921	500	750	1000
Steam Shovel	black except for red roof and base, 25 1/2" tall, 1921-22	150	200	250
Traveling Crane	red crane, carriage, and long cross beams, hand wheel rotates crane boom, 46" long, 1928	1275	1900	2550
Trench Digger	yellow main frame, base, and motor housing, red elevator and conveyor frame and track frames, 20" tall, 1928-31	3500	5000	7500

Sets

NAME	DESCRIPTION	GOOD	EX	MINT
Army Combination Set	searchlight repair-it truck, transport truck and howitzer, ammunition conveyor, stake delivery truck, ammo, soldiers, 1956	300	400	500
Army Commando Set	14 1/2" truck, searchlight unit, two-wheel howitzer, soldiers, 1957	125	185	250
Big Brute 3-Piece Hi-Way Set	bulldozer, dump truck, yellow 4-wheel trailer, 1971	100	150	200
Big Brute 3-Piece Road Set	cement mixer truck, scooper, dump truck, 1971	125	185	250
Big Brute 4-Piece Freeway Set	scraper, grader, scooper and dump truck, 1971	125	185	250
Brute 5-Piece Hi-Way Set	bulldozer, grader, scraper, dumping scraper and double dump train, 1970	50	75	100
Brute Fire Department Set	semi-trailer aerial ladder truck, fire pumper, fire wrecker, brute tow truck, 1970	75	115	150
Brute Fleet Set	car carrier with 2 plastic coupes, dump truck, pickup truck, cement mixer truck, tow truck, 1969	85	130	175
Delivery Set Combination	16 1/2" long wrigley express truck, 15" long sand and stone dump truck, 14 1/4" long freight conveyor and 14 1/4" long stake delivery truck, 1955	175	265	350
Family Camping Set	camper-n-cruiser truck, 15 1/2" long maroon suburban wagon, and brown/light gray-beige folding teepee camping trailer, 1963	60	95	125
Family Camping Set	blue camping trailer and suburban wagon, blue camper-n-cruiser, 1964	50	75	100
Farm Combination Set	cattle transport stake truck with 6 plastic steers, hydraulic farm supplies trailer dump truck, trailer and 3 farm machines and farm machinery trailer hauler truck, 1956	100	150	200
Fire Department Set	aerial ladder fire engine, fire pumper with action hydrant that squirts water, 2 plastic hoses, 2 plastic firemen, fire chief's badge, 1960	250	375	500

Buddy L

NAME	DESCRIPTION	GOOD	EX	MINT
Freight Conveyor and Stake Delivery Truck	blue frame 14 1/4" long conveyor, red, white and yellow body 14 3/4" long truck, 1955	125	185	250
GMC Air Defense Set	15" long, GMC army searchlight truck, 15" long, GMC signal corps truck, two 4-wheel trailers, plastic soliders, 1957	300	500	700
GMC Brinks Bank Set	silvery gray, barred windows on sides and in double doors, coin slot and hole in roof, brass padlock with 2 keys, pouch, paper play money, 2 gray plastic guard figures, 16" long, 1959	300	350	450
GMC Fire Department Set	red GMC extension ladder trailer and GMC pumper with ladders and hose reel, 4-wheel red electric searchlight trailer, warning barrier, red plastic helmet, 4 firemen, policeman, 1958	300	500	750
GMC Hi-Way Maintenance Fleet	all orange, maintenance truck with trailer, sand & stone dump truck, scoop-n-load conveyor, sand hopper, steel scoop shovel, 4 white steel road barriers, 1957	300	500	700
GMC Livestock Set	red fenders, hood, cab and frame, white flatbed cargo section, 6 section of brown plastic rail fencing, 5 black plastic steers, 14 1/2" long, 1958	300	400	500
GMC Western Roundup Set	blue fenders, hood, cab and frame, white flatbed cargo section, plastic 6 sections of rail fencing with swinging gate, rearing and standing horse, 2 cowboys, calf, steer, 1959	300	400	500
Hi-Way Maintenance Mechanical Truck & Concrete Mixer	20" truck plus movable ramp, with duo-tone slant design, blue lower hood sides, yellow hood top and cab, 10 3/4" blue and yellow mixer, overall 36" long, 1949	160	245	325
Highway Construction Set	orange and black bulldozer and driver, truck with orange pickup body, orange dump truck, 1962	200	300	400
Interstate Highway Set	all orange, parks department dumper, landscape truck, telephone truck, 3 lichen trees, 2 sawhorse barriers, steel scoop shovel, 2 metal drums, 4 orange traffic cones, 4 workmen, 1959	250	400	500
Interstate Highway Set	all orange, husky dumper, contractor's truck and ladder, utility truck, plastic pickaxe, spade, shovel, nail keg, 1960	250	350	500
Jr. Animal Farm Set	6 1/2" long Jr. Giraffe Truck, 6 1/4" long Jr. Kitty Kennel, 11 1/4" long Jr. Pony Trailer with Sportster, 1968	125	185	250
Jr. Fire Department	17" long Jr. hook-n-ladder aerial truck, 11 1/2" long Jr. fire snorkel, 6 3/4" long Jr. fire emergency truck, all have 24-hole chrome grilles, 1968	125	185	250
Jr. Fire Department	17" long Jr. hook-n-ladder aerial truck, 11 1/2" long Jr. fire snorkel, 6 3/4" long Jr. fire emergency truck, all have 4-slot grilles and 2 square plastic headlights, 1969	125	185	250
Jr. Hi-Way Set	yellow and black Jr. scooper tractor, yellow and white Jr. cement mixer truck, yellow Jr. dump truck, 1969	200	300	400
Jr. Sportsman Set	Jr. camper pickup with red cab and body and yellow camper, towing 6" plastic runabout on yellow 2-wheel boat trailer, 1971	50	75	100
Loader, Dump Truck, and Shovel Set	conveyor, green body sand and gravel dump truck, 8 3/4" long green-enameled steel scoop shovel, 1954	100	150	200
Loader, Dump Truck, and Shovel Set	conveyor, blue body sand and gravel dump truck, 8 3/4" long blue-enameled steel scoop shovel, 1955	85	130	175
Mechanical Hauling Truck and Concrete Mixer	truck with duo-tone slant design, red-orange lower hood sides, dark green upper hood, cab, trailer and ramp, 9 5/8" green mixer, gray hopper, 38" long with ramps, 1950	160	245	325

VEHICLES

Buddy L

NAME	DESCRIPTION	GOOD	EX	MINT
Mechanical Hauling Truck and Concrete Mixer	truck with duo-tone slant design, red-orange lower hood sides, dark green upper hood, cab, ramp, yellow trailer, 9 5/8" green mixer, gray hopper, 38" long with ramps, 1951	150	225	300
Polysteel Farm Set	blue milkman truck with rack and 9 milk bottles, red and gray milk tanker, orange farm tractor, 1961	75	115	150
Road Builder Set	green and white cement mixer truck, yellow and black bulldozer, red dump truck, husky dumper, 1963	200	300	400
Truck with Concrete Mixer Trailer	22" truck with duo-tone slant design, green fenders and lower sides, yellow squarish cab and body, 10" mixer with yellow frame and red hopper, overall 34 1/2" long, 1937	175	265	350
Truck with Concrete Mixer Trailer	22" truck with duo-tone slant design, green front and lower hood sides, yellow upper hood, cab and body, 10" mixer with yellow frame and red hopper, overall 32 1/2" long, 1938	165	250	330
Warehouse Set	coca-cola truck, 2 hand trucks, 8 cases coke bottles, store-door delivery truck, lumber, sign, 2 barrels, forklift, 1958	175	265	350
Warehouse Set	coca-cola truck, 2 hand trucks, 8 cases coke bottles, store-door delivery truck, sign, 2 barrels, forklift, 1959	150	225	300
Western Roundup Set	turquoise fenders, hood, cab and frame, white flatbed cargo section, plastic 6 sections of rail fencing with swinging gate, rearing and standing horse, 2 cowboys, calf, steer, 1960	175	250	400

Trucks

NAME	DESCRIPTION	GOOD	EX	MINT
"Big Fella" Hydraulic Rider Dumper	duo-tone slant design, yellow front and lower hood, red upper cab, dump body and upper hood, rider seat has large yellow sunburst-style decal, 26 1/2" long, 1950	110	175	225
"Standard Oil" Tank Truck	duo-tone slant design, white upper cab and hood, red lower cab, grill, fenders and tank, rubber wheels, electric headlights, 26" long, 1936-37	2000	3500	5000
Air Force Supply Transport	blue, blue removable fabric canopy, rubber wheels, decals on cab doors, 14 1/2" long, 1957	125	250	350
Air Mail Truck	black front, hood fenders, enclosed cab and opening doors, red enclosed body and chassis, 24" long, 1930	675	1000	1400
Allied Moving Van	tractor and semi-trailer van, duo-tone slant design, black front and lower sides, orange hood top, cab and van body, 29 1/2" long, 1941	600	900	1200
Army Electric Searchlight Unit	shiny olive drab flatbed truck, battery operated searchlight, 14 3/4" long, 1957	125	225	325
Army Half-Track and Howitzer	olive drab with olive drab carriage, 12 1/2" truck, 9 3/4" gun, overall 22 1/2" long, 1953	100	150	200
Army Half-Track with Howitzer	olive drab with olive drab, red firing knob on gun, 17" truck, 9 3/4" gun, overall 27" long, 1955	100	150	200
Army Medical Corps Truck	white, black rubber tires on white steel disc wheels, 29 1/2" long, 1941	125	185	250
Army Searchlight Repair-It Truck	shiny olive drab truck and flatbed cargo section, 15" long, 1956	125	175	225
Army Suppy Truck	shiny olive drab truck and removable fabric cover, 14 1/2" long, 1956	100	150	175
Army Transport Truck and Trailer	olive drab truck, 20 1/2" long, trailer 34 1/2" long, 1940	250	350	450
Army Transport with Howitzer	olive drab, 12" truck, 9 3/4" gun, overall 28" long, 1953	100	150	200
Army Transport with Howitzer	olive drab steel, 17" truck, 9 3/4" gun, overall 27" long, 1955	150	250	350
Army Transport with Howitzer	olive drab steel, re-firing knob on gun, 17" truck, 9 3/4" gun, overall 27" long, 1954	115	175	230

VEHICLES

Buddy L

NAME	DESCRIPTION	GOOD	EX	MINT
Army Transport with Tank	olive drab, 15 1/2" long truck, 11 1/2" long detachable 2-wheel trailer, overall 26 1/2" long, 7 1/2" long tank, 1959	100	150	200
Army Troop Transport with Howitzer	dark forest green truck and gun, canopy mixture of greens, 14" long truck, 12" long, gun, overall 25 3/4" long, 1965	100	150	200
Army Truck	olive drab, 20 1/2" long, 1939	110	175	225
Army Truck	olive drab, 17" long, 1940	150	200	250
Atlas Van Lines	green tractor unit, chrome one-piece toothed grille and headlights, green lower half of semi-trailer van body, cream upper half, silvery roof, 29" long, 1956	200	300	400
Auto Hauler	yellow cab-over-engine tractor unit and double-deck semi-trailer, three 8" long vehicles, 25 1/2" long, 1968	75	115	150
Auto Hauler	snub-nose medium blue tractor unit and double-deck semi-truck trailer, 3 plstic coupes, overall 27 1/2" long, 1970	65	95	130
Baggage Rider	duo-tone horizontal design, green bumper, fenders and lower half of truck, white upper half, 28" long, 1950	250	175	500
Baggage Truck	black front, hood, and fenders, doorless cab, yellow four-post stake sides, 2 chains across back, 26 1/2" long, 1927	2000	3000	4000
Baggage Truck	black front, hood, and fenders, open door cab, yellow four-post stake sides, 2 chains across back, 26 1/2" long, 1930	1000	2000	3000
Baggage Truck	green front, hood, and fenders, non-open doors, yellow cargo section slat sides, 26 1/2" long, 1933-34	1000	2000	3000
Baggage Truck	green front, hood, and fenders, non-open doors, yellow cargo section solid sides, 26 1/2" long, 1933	1000	2000	3000
Baggage Truck	green front, hood, and fenders, non-open doors, yellow cargo section slat or solid sides, metal grille, 26 1/2" long, 1935	1000	2000	3000
Baggage Truck	duo-tone slant design, yellow fenders, green hood top, cab, and removable rider seat, 26 1/2" long, 1936	1200	2400	3600
Baggage Truck	duo-tone slant design, yellow skirted fenders and cargo section, green hood top, enclosed cab, 27 3/4" long, 1938	1000	2000	3000
Baggage Truck	green hood, fenders, and cab, yellow cargo section, no bumper, 17 1/2" long, 1945	175	265	350
Baggage Truck	black front, hood, and fenders, enclosed cab with opening doors, nickel-rim, red-shell headlights, yellow stake body, 26 1/2" long, 1930-32	6000	9000	12000
Big Brute Dumper	yellow cab-over-engine, frame and tiltback dump section with cab shield, striped black and yellow bumper, black grille, 8" long, 1971	50	75	100
Big Brute Mixer Truck	yellow cab-over-engine, body and frame, white plastic mixing drum, white plastic seats, 7" long, 1971	35	50	70
Big Mack Dumper	off-white front, hood cab and chassis, blue-green tiltback dump section, white plastic bumper, 20 1/2" long, 1964	75	115	150
Big Mack Dumper	yellow front, hood cab, chassis and tiltback dump section, black plastic bumper, 20 1/2" long, 1967	70	100	140
Big Mack Dumper	yellow front, hood cab, chassis and tiltback dump section, black plastic bumper, single rear wheels, 20 1/2" long, 1968	65	95	130
Big Mack Dumper	yellow front, hood cab, chassis and tiltback dump section, black plastic bumper, heavy-duty black ballon tires on yellow plastic 5-spoke wheels, 20 1/2" long, 1971	60	90	120
Big Mack Hydraulic Dumper	red hood, cab and tiltback dump section with cab shield, white plastic bumper, short step ladder on each side, 20 1/2" long, 1968	50	75	100

VEHICLES

Buddy L

NAME	DESCRIPTION	GOOD	EX	MINT
Big Mack Hydraulic Dumper	white hood, cab and tiltback dump section with cab shield, white plastic bumper, short step ladder on each side, 20 1/2" long, 1969	45	65	90
Big Mack Hydraulic Dumper	red hood, cab and tiltback dump section with cab shield, dump body sides have a large circular back, white plastic bumper, short step ladder on each side, 20 1/2" long, 1970	40	60	80
Boat Transport	blue flatbed truck carrying 8" litho metal boat, boat deck white, hull red, truck 15" long, 1959	300	550	750
Borden's Milk Delivery Van	white upper cab-over-engine van body and sliding side doors, yellow lower body, metal-handle yellow plastic tray and 6 white milk bottle with yellow caps, 11 1/2" long, 1965	125	200	275
Brute Car Carrier	bright blue cab-over-engine tractor unit and detachable double-deck semi-trailer, 2 plastic cars, 10" long, 1969	60	95	125
Brute Cement Mixer Truck	sand-beige cab-over-engine body and frame, white plastic mixing drum, white plastic seats, 5 1/4" long, 1968	35	55	75
Brute Cement Mixer Truck	blue cab-over-engine body and frame, white plastic mixing drum, white plastic seats, white-handled crank rotates drum, 5 1/4" long, 1969	30	45	65
Brute Dumper	red cab-over-engine body and cab shield on tiltback dump section, wide chrome wraparound bumper, 5" long, 1968	35	55	75
Brute Monkey House	yellow cab-over-engine body, striped orange and white awning roof, cage on back, 2 plastic monkeys, 5" long, 1968	50	75	100
Brute Monkey House	yellow cab-over-engine body, red and white awning roof, cage on back, 2 plastic monkeys, 5" long, 1969	40	60	80
Brute Sanitation Truck	lime green cab-over-engine and frame, white open-top body, wide chrome wraparound bumper, 5 1/4" long, 1969	50	75	100
Buddy "L" Milk Farms Truck	white body, black roof, short hood with black wooden headlights, 13 1/2" long, 1945	175	350	525
Buddy "L" Milk Farms Truck	light cream body, red roof, nickel glide headlights, sliding doors, 13" long, 1949	200	400	600
Camper	bright medium blue steel truck and camper body, 14 1/2" long, 1964	60	95	125
Camper	medium blue truck and back door, white camper body, 14 1/2" long, 1965	50	75	100
Camper-n-Cruiser	powder blue pickup truck and trailer, pale blue camper body, 24 1/2" long, 1963	60	95	125
Camper-n-Cruiser	bright medium blue camper with matching boat trailer and 8 1/2" long plastic sport cruiser, ovreall 27" long, 1964	50	75	100
Campers Truck	turquoise pickup truck, pale turquoise plastic camper, 14 1/2" long, 1961	55	85	110
Campers Truck with Boat	green-turquoise pickup truck, lime green camper body, red plastic runabout boat on camper roof, 14 1/2" long, 1962	50	100	150
Campers Truck with Boat	green-tuquoise pickup, no side mirror, lime green camper body with red plastic runabout boat on top, 14 1/2" long, 1963	50	100	150
Camping Trailer and Wagon	bright medium blue suburban wagon, matching teepee trailer, overall 24 1/2" long, 1964	60	95	125
Cattle Transport Truck	red with yellow stake sides, 15" long, 1956	75	115	150
Cattle Transport Truck	green and white with white stake sides, 15" long, 1957	75	115	150
Cement Mixer Truck	turquoise body, tank ends, and chute, white side ladder, water tank, mixing drum and loading hopper, 16 1/2" long, 1964	60	95	125
Cement Mixer Truck	red body, tank ends, and chute, white side ladder, water tank, mixing drum and loading hopper, 15 1/2" long, 1965	75	115	150

VEHICLES

Buddy L

NAME	DESCRIPTION	GOOD	EX	MINT
Cement Mixer Truck	red body, tank ends, and chute, white water tank, mixing drum and loading hopper, black wall tires, 15 1/2" long, 1967	60	95	125
Cement Mixer Truck	red body, tank ends, and chute, white water tank, mixing drum and loading hopper, whitewall tires, 15 1/2" long, 1968	50	75	100
Cement Mixer Truck	snub-nosed yellow body, cab, frame and chute, white plastic mixing drum, loading hopper and water tank with yellow ends, 16" long, 1970	35	50	70
Charles Chip Delivery Truck Van	tan-beige body, decal on sides has brown irregular center resembling a large potato chip, 1966	125	200	275
City Baggage Dray	green front, hood, and fenders, non-open doors, yellow stake-side cargo section, 19" long, 1934	200	400	600
City Baggage Dray	green front, hood, and fenders, non-open doors, yellow stake-side cargo section, bright metal grille, 19" long, 1935	200	400	600
City Baggage Dray	duo-tone slant design, green front and fenders, yellow hood top and cargo section, 19" long, 1936	200	400	600
City Baggage Dray	duo-tone slant design, green front and fenders, yellow hood top and cargo section, dummy headlights, 19" long, 1937	200	400	600
City Baggage Dray	duo-tone slant design, green front and skirted fenders, yellow hood top, enclosed cab and cargo section, 20 3/4" long, 1938	175	350	500
City Baggage Dray	light green except for aluminum-finish grille, no bumper, black rubber wheels, 20 3/4" long, 1939	175	350	500
City Baggage Dray	cream except for aluminum-finish grille, no bumper, black rubber wheels, 20 3/4" long, 1940	175	350	500
Coal Truck	black hopper body and fully enclosed cab with opening doors, red wheels, 25" long, 1930	3000	4500	6000
Coal Truck	black front, hood, fenders, doorless cab, red chassis and disc wheels, 25" long, 1926	1500	2500	3500
Coal Truck	black front, hood, fenders, sliding discharge door on each side of hopper body, red chassis and disc wheels, 25" long, 1927	1200	2400	3200
Coca-Cola Bottling Route Truck	bright yellow, with small metal hand truck, 8 or 6 yellow cases of miniature green Coke bottles, 14 3/4" long, 1955	125	175	250
Coca-Cola Delivery Truck	orange-yellow cab and double-deck, open-side cargo, 2 small hand trucks, 4 red and 4 green cases of bottles, 15" long, 1960	100	150	200
Coca-Cola Delivery Truck	orange-yellow cab and double-deck, open-side cargo, 2 small hand trucks, 4 red and 4 green cases of bottles, 15" long, 1963	75	100	150
Coca-Cola Delivery Truck	orange-yellow cab and double-deck, open-side cargo, 2 small hand trucks, 4 red and 4 green cases of bottles, 15" long, 1964	60	95	125
Coca-Cola Delivery Truck	red lowercab-over-engine and van body, white upper cab, left side of van lifts to reveal 10 miniature bottle cases, 9 1/2" long, 1971	25	40	55
Coca-Cola Route Truck	bright yellow, with 2 small metal hand trucks and 8 yellow cases of miniature green Coke bottles, 14 3/4" long, 1957	110	175	225
Coke Coffee Co. Delivery Truck Van	black lower half of body, orange upper half, roof and sliding side doors, 1966	85	130	175
Colt Vacationer	blue/white colt sportsliner with trailer carrying 8 1/2" long red/white plastic sport cruiser, overall 22 1/2" long, 1967	60	95	125
Curtiss Candy Trailer Van	blue tractor and bumper, white semi-trailer van, blue roof, chrome one-piece toothed grille and headlights, white drop-down rear door, 32 3/4" long with tailgate/ramp lowered, 1955	250	400	500

VEHICLES

Clockwise from upper left: Hydraulic Dump Truck, 1960s; Coca Cola Route Truck, 1957; GMC Texaco Tank Truck, 1959; U.S. Army truck with rear canopy cover, 1940s; Station Wagon and Tee Pee Camping Trailer, 1963; Hydraulic Highway Dump Truck with Scraper Blade, 1958; Wrecker, all Buddy L.

VEHICLES

Buddy L

NAME	DESCRIPTION	GOOD	EX	MINT
Dairy Transport Truck	duo-tone slant design, red front and lower hood sides, white hood top, cab and semi-trailer tank body, tank opens in back, 26" long, 1939	150	225	300
Deluxe Auto Carrier	turquoise tractor unit, aluminum loading ramps, 3 plastic cars, overall 34" long including, 1962	100	175	250
Deluxe Camping Outfit	turquoise pickup truck and camper, and 8 1/2" long plastic boat on pale turquoise boat trailer, overall 24" long, 1961	60	95	125
Deluxe Hydraulic Rider Dump Truck	duo-tone slant design, red front and lower hood sides, white upper cab, dump body and chassis, red or black removable rider saddle, 26" long, 1948	175	265	350
Deluxe Motor Market	duo-tone slant design, red front, curved bumper, lower hood and cab sides, white hood top, body and cab, 22 1/4" long, 1950	250	350	500
Deluxe Rider Delivery Truck	duo-tone horizontal design, deep blue lower half, gray upper half, red rubber disc wheels, black barrel skid, 22 3/4" long, 1945	135	200	270
Deluxe Rider Delivery Truck	duo-tone horizontal design, gray lower half, blue upper half, red rubber disc wheels, black barrel skid, 22 3/4" long, 1945	135	200	270
Deluxe Rider Dump Truck	various colors, dual rear wheels, no bumper, 25 1/2" long, 1945	75	115	150
Double Hydraulic Self-Loader-n-Dump	green front loading scoop with yellow arms attached to cab sides, yellow hood and enclosed cab, orange fram and wide dump body, 29" long with scoop lowered, 1956	85	130	175
Double Tandem Hydraulic Dump and Trailer	truck has red bumper, hood, cb and frame, 4-wheel trailer with red tow and frame, both with white tiltback dump bodies, 38" long, 1957	85	130	175
Double-Deck Boat Transport	light blue steel flatbed truck carrying three 8" white plastic boats with red decks, truck 15" long, 1960	150	250	400
Dr. Pepper Delivery Truck Van	white, red, and blue, 1966	85	130	175
Dump Body Truck	black front, hood, open driver's seat and dump section, red chassis, crank windlass with ratchet raises dump bed, 25" long, 1921	800	1400	2000
Dump Body Truck	black front, hood, open driver's seat and dump section, red chassis, chain drive dump mechanism, 25" long, 1923	1200	1800	2500
Dump Truck	black enclosed cab and opening doors, front and hood, red dump body and chassis, crank handle lifts dump bed, 24" long, 1931	750	1125	1500
Dump Truck	yellow upper hood and enclosed cab, red wide-skirt fenders and open-frame chassis, blue dump body, no bumper, 17 1/4" long, 1940	85	130	175
Dump Truck	black enclosed cab and opening doors, front and hood, red dump body and chassis, simple lever arrangement lifts dump bed, 24" long, 1930	650	975	1300
Dump Truck	yellow enclosed cab, front and hood, red dump section, no bumper, 20" long, 1934	275	415	550
Dump Truck	yellow enclosed cab, front and hood, red dump section, no bumper, bright-metal radiator, 20" long, 1935	325	485	650
Dump Truck	duo-tone slant design, yellow enclosed cab and hood, red front and dump body, no bumper, bright-metal headlights, 20" long, 1936	250	375	500
Dump Truck	duo-tone slant design, yellow enclosed cab and hood, red front and dump body, no bumper, dummy headlights, 20" long, 1937	250	375	500
Dump Truck	duo-tone slant design, red lower cab, lower hood, front and dump body, yellow upper hood, upper cab and chassis, no bumper, 22 1/4" long, 1939	250	375	500

VEHICLES

Buddy L

NAME	DESCRIPTION	GOOD	EX	MINT
Dump Truck	duo-tone slant design, red front, fenders and lower doors, white upper, bright radiator grille and headlights, no bumper, 22 1/4" long, 1939	250	375	500
Dump Truck	green except for cream hood top and upper enclosed cab, no bumper, bright-metal headlights and grille, 22 1/4" long, 1941	85	130	175
Dump Truck	white upper hood, enclosed cab, wide-skirt fenders and open-frame chassis, orange dump body, bright-metal grille, no bumper, 17 3/8" long, 1941	85	130	175
Dump Truck	red hood top and cab, white or cream dump body and frame, no bumper, 17 1/2" long, 1945	75	115	150
Dump Truck	various colors, black rubber wheels, 12" long, 1945	50	75	100
Dump Truck	duo-tone slant design, red front, fenders and dump body, yellow hood top, upper sides, uppe cab and chassis, no bumper, 22 1/2" long, 1948	125	185	250
Dump Truck-Economy Line	dark blue dump body, remainder is yellow, bright-metal grille and headlights, no bumper or running boards, 12" long, 1941	75	115	150
Dump-n-Dozer	orange husky dumper truck and orange flatbed 4-wheel trailer carrying orange bulldozer, 23" long including trailer, 1962	75	115	150
Dumper with Shovel	turquoise body, frame and dump section, white one-piece bumper and grille guard, large white steel scoop shovel, 15" long, 1962	75	115	150
Dumper with Shovel	medium green body, frame and dump section, white one-piece bumper and grille guard, no side mirror, large white steel scoop shovel, 15" long, 1963	75	115	150
Dumper with Shovel	medium green body, frame and dump section, white one-piece bumper and grille guard, no side mirror, large white steel scoop shovel, spring suspension on front axle only, 15" long, 1964	75	115	150
Dumper with Shovel	orange body, frame and dump section, chrome one-piece grille, no bumper guard, no side mirror, large white steel scoop shovel, no spring suspension, 15 3/4" long, 1965	75	115	150
Express Trailer Truck	red tractor unit, hood, fenders and enclosed cab, green semi-trailer van with removable roof and drop-down rear door, 23 3/4" long, 1933	350	525	700
Express Trailer Truck	red tractor unit, hood, fenders and enclosed cab, green semi-trailer van with removable roof and drop-down rear door, bright-metal dummy headlights, 23 3/4" long, 1934	350	525	700
Express Truck	all black except red frame, enclosed cab with opening doors, nickel-rim, red-shell headlights, 6 rubber tires, double bar front bumper, 24 1/2" long, 1930-32	3000	4500	6000
Farm Machinery Hauler Trailer Truck	blue tractor unit, yellow flatbed semi-trailer, 31 1/2" long, 1956	125	185	250
Farm Suppies Automatic Dump	duo-tone slant design, blue curved bumper, front, lower hood sides and cab, yellow upper hood, cab and rest of body, 22 1/2" long, 1950	125	185	250
Farm Supplies Dump Truck	duo-tone slant design, red front, fenders and lower hood sides, yellow upper hood, cab and body, 22 3/4" long, 1949	125	185	250
Farm Supplies Hydraulic Dump Trailer	green tractor unit, long cream bocy on semi-trailer, 14 rubber wheels, 26 1/2" long, 1956	100	150	200
Fast Delivery Pickup	yellow hood and cab, red open cargo body, removable chain across open back, 13 1/2" long, 1949	100	150	200
Finger-Tip Steering Hydraulic Dumper	powder blue bumper, fenders, hood, cab and frame, white tiltback dump body, 22" long, 1959	75	115	150
Fisherman	light tan pickup truck with tan steel trailer carrying plastic 8 1/2" long sport crusier, overall 24 1/4" long, 1962	80	120	160

Buddy L

NAME	DESCRIPTION	GOOD	EX	MINT
Fisherman	pale blue-green station wagon with 4-wheel boat trailer carrying plastic 8 1/2" long boat, overall 27 1/2" long, 1963	75	115	150
Fisherman	metallic sage green pickup truck with boat trailer carrying plastic 8 1/2" long sport cruiser, overall 25" long, 1964	70	100	140
Fisherman	sage gray-green and white pickup truck with steel trailer carrying plastic 8 1/2" long sport cruiser, overall 25" long, 1965	65	95	130
Flivver Dump Truck	black except for red 8-spoke wheels, black hubs, aluminum tires, flat, open dump section with squared-off back with latching, drop-down endgate, 11" long, 1926	1500	2500	3500
Flivver Huckster Truck	black except for red 8-spoke wheels, black hubs, aluminum tires, flat, continous hard top canopy extending from enclosed cab over cargo section, 14" long, 1927	2500	4000	5500
Flivver One-Ton Express Truck	black except for red 8-spoke wheels, black hubs, aluminum tires, flat, enclosed cab, operating steering wheel, open cargo section, 14 1/4" long, 1927	3500	5000	6500
Flivver Scoop Dump Truck	black except for red 8-spoke wheels, 12 1/2" long, 1926-27, 1929-30	1500	2500	3500
Flivver Truck	black except for red 8-spoke wheels with aluminum tires, black hubs, 12" long, 1924	1000	1500	2000
Ford Flivver Dump Cart	black except for red 8-spoke wheels, black hubs, aluminum tires, flat, short open dump section tapers to point on each side, 12 1/2" long, 1926	1500	2500	3500
Frederick & Nelson Delivery Truck Van	medium green body, roof and sliding side doors, 1966	125	200	275
Freight Delivery Stake Truck	red hood, bumper, cab and frame, white cargo section, yellow 3-post, 3-slat removable stake sides, 14 3/4" long, 1955	75	125	150
Front Loader Hi-Lift Dump Truck	red scoop and arms attached to white truck at rear fenders, green dump body, 17 3/4" long with scoop down and dump body raised, 1955	85	130	175
Giant Hydraulic Dumper	red bumper, frame, hood and cab, light tan tiltback dump body and cab shield 23 3/4" long, 1960	125	185	250
Giant Hydraulic Dumper	overall color turquoise, dump lever has a red plastic tip, 22 3/4" long, 1961	135	200	275
Giraffe Truck	powder blue hood, white cab roof, high-sided open-top cargo section, 2 orange and yellow plastic giraffes, 13 1/4" long, 1968	60	95	125
GMC Air Force Electric Searchlight Unit	all blue flatbed, off white battery operated searchlight swivel mount, decals on cab doors, 14 3/4" long, 1958	200	300	400
GMC Airway Express Van	green hood, cab and van body, latching double rear doors, shiny metal drum coin bank and metal hand truck, 17 1/2" long with rear doors open, 1957	250	350	450
GMC Anti-Aircraft Unit with Searchlight	15" truck with four-wheel trailer, battery operated, over 25 1/4" long, 1957	250	350	450
GMC Army Hauler with Jeep	shiny olive drab tractor unit and flatbed trailer, 10" long jeep, overall 31 1/2" long, 1958	200	300	400
GMC Army Transport with Howitzer	shiny olive drab, 14 1/2" long, truck, overall with gun 22 1/2" long, 1957	200	300	400
GMC Brinks Armored Truck Van	silvery gray, barred windows on sides and in double doors, coin slot and hole in roof, brass padlock with 2 keys, pouch, paper play money, 3 gray plastic guard figures, 16" long, 1958	300	350	450
GMC Coca-Cola Route Truck	lime-yellow, with small metal hand truck and 8 cases of miniature green Coke bottles, 14 1/8" long, 1957	200	300	400
GMC Coca-Cola Route Truck	orange-yellow, with 2 small metal hand truck and 8 cases of miniature green Coke bottles, 14 1/8" long, 1958	200	300	400

Buddy L

NAME	DESCRIPTION	GOOD	EX	MINT
GMC Construction Company Dumper	pastel blue including control lever onleft and dump section with cab shield, hinged tailgate, chrome GMC bar grille, six wheels, 16" long, 1958	200	300	400
GMC Construction Company Dumper	pastel blue including control lever onleft and dump section with cab shield, hinged tailgate, chrome GMC bar grille, four wheels, 16" long, 1959	150	250	350
GMC Deluxe Hydraulic Dumper	pastel blue, chrome GMC bar grille, attached headlights, yellow steel scoop shovel on left side, 19" long over scraper blade, 1959	200	300	400
GMC Hi-Way Giant Trailer	blue tractor, blue and white van, chrome GMC bar grille and headlights, blue roof on semi-trailer, white tailgate doubles as loading ramp, 18-wheeler, 31 1/4" long, 1957	250	350	450
GMC Hi-Way Giant Trailer Truck	blue tractor, blue and white van, chrome GMC bar grille and headlights, blue roof on semi-trailer, white tailgate doubles as loading ramp, 14-wheeler, 30 3/4" long, 1958	200	300	400
GMC Husky Dumper	red hood, bumper, cab and chassis, chrome GMC bar grille and nose emblem, white oversize dump body, red control lever on right side, 17 1/2" long, 1957	150	250	350
GMC Self-Loading Auto Carrier	yellow tractor and double-deck semi trailer, 3 plastic cars, overall 33 1/4" long, 1959	200	300	400
GMC Signal Corps Unit	both olive drab, 14 1/4" long truck with removable fabric canopy, 8" long four-wheel trailer, 1957	150	200	250
Grocery Motor Market Truck	duo-tone slant design, yellow front, lower hood sides, fenders and lower doors, white hood top, enclosed cab and body, no bumper, 20 1/2" long, 1937	275	415	550
Grocery Motor Market Truck	duo-tone slant design, yellow front, lower hood sides, skirted fenders and lower doors, white hood top, cab and body, no bumper, 21 1/2" long, 1938	275	415	550
Heavy Hauling Dumper	red hood, bumper, cab and frame, cream tiltback dump body, 20 1/2" long, 1955	75	125	150
Heavy Hauling Dumper	red hood, bumper, cab and frame, cream oversize dump body, hinged tailgate, 21 1/2" long, 1956	70	125	140
Heavy Hauling Hydraulic Dumper	green hood, cab and frame, cream tiltback dump body, and cab shield, raising dump body almost to vertical, 23" long, 1956	70	105	140
Hertz Auto Hauler	bright yellow tractor and double-deck semi-trailer, 3 plastic vehicles, 27" long, 1965	100	150	200
Hi-Lift Farm Supplies Dump	red plastic front end including hood and enclosed cab, yellow dump body, cab shield and hinged tailgate, 21 1/2" long, 1953	100	175	225
Hi-Lift Farm Supplies Dump	all steel, red front end including hood and enclosed cab, yellow dump body, cab shield and hinged tailgate, 23 1/2" long, 1954	100	175	225
Hi-Lift Scoop-n-Dump Truck	orange truck with deeply fluted sides, dark green scoop on front rises to empty load into hi-lift cream-yellow dump body, 16" long, 1952	85	130	175
Hi-Lift Scoop-n-Dump Truck	orange truck with deeply fluted sides, dark green scoop on front rises to empty load into hi-lift light cream dump body, 16" long, 1953	80	125	165
Hi-Lift Scoop-n-Dump Truck	orange truck with deeply fluted sides, dark green scoop on front rises to empty load into deep hi-lift slightly orange dump body, 16" long, 1955	75	115	155
Hi-Lift Scoop-n-Dump Truck	orange hood, fenders and cab, yellow front loading scoop and arms attached to fenders, white frame, dump body and cab shield, 17 3/4" long, 1956	70	125	145
Hi-Lift Scoop-n-Dump Truck	blue hood, fenders and cab, yellow front loading scoop and arms attached to fenders, white frame, dump body, cab shield, and running boards, 17 3/4" long, 1957	65	100	135
Hi-Tip Hydraulic Dumper	orange hood, cab and frame, cream tiltback dump body, and cab shield, raising dump body almost to vertical, 23" long, 1957	75	115	150

VEHICLES

Buddy L

NAME	DESCRIPTION	GOOD	EX	MINT
Hi-Way Maintenance Truck with Trailer	all orange except for black rack of 4 simulated floodlights behind cab, 19 1/2" long including small 2-wheel trailer, 1957	100	150	200
Highway Hawk Trailer Van	bronze cab tractor, chrome metallized plastic bumper, grille, air cleaner and exhaust, 19 3/4" long, 1985	50	75	100
Husky Dumper	orange wraparound bumper, body, frame and dump section, hinged tailgate, plated dump lever on left side, 15 1/4" long, 1960	75	115	150
Husky Dumper	white plastic wraparound bumper, tan body, frame and dump section, hinged tailgate, plated dump lever on left side, 15 1/4" long, 1961	70	125	140
Husky Dumper	bright yellow, chrome one piece bumper, slotted rectangular grill and double headlights, 14 1/2" long, 1966,	75	115	150
Husky Dumper	red hood, cab, chassis and dump section, chrome one-piece bumper and slotted grille with double headlights, 14 1/2" long, 1968	60	95	125
Husky Dumper	yellow hood, cab, fram and tiltback dump section with cab shield, crome one-piece wraparound bumper, 14 1/2" long, 1969	50	75	100
Husky Dumper	snub-nose red body, tiltback dump section with cab shield, full-width chrome bumperless grille, deep-tread whitewall tires, 14 1/2" long, 1970	45	70	90
Husky Dumper	snub-nose red body, tiltback dump section with cab shield, full-width chrome bumperless grille, white-tipped dump-control lever on left, deep-tread whitewall tires, 14 1/2" long, 1971	40	60	80
Hydraulic Auto Hauler with Four GMC Cars	powder blue GMC tractor, 7" long plastic cars, overall 33 1/2" long including loading ramp, 1958	250	350	450
Hydraulic Construction Dumper	red front, cab and chassis, large green dump section with cab shield, 15 1/4" long, 1962	65	100	135
Hydraulic Construction Dumper	tan-beige front, cab and chassis, large green dump section with cab shield, 15 1/4" long, 1963	60	95	125
Hydraulic Construction Dumper	bright blue front, cab and chassis, large green dump section with cab shield, 15 1/2" long, 1964	50	75	100
Hydraulic Construction Dumper	bright green front, cab and chassis, large green dump section with cab shield, 14" long, 1965	50	75	100
Hydraulic Construction Dumper	medium blue front, cab and chassis, large green dump section with cab shield, 15 1/4" long, 1967	50	75	100
Hydraulic Dump Truck	black front, hood, fenders, open seat, and dump body, red chassis and disc wheels with aluminum tires, 25" long, 1926	1000	1500	2000
Hydraulic Dump Truck	black front, hood, fenders, dark reddish maroon dump body, red chassis and disc wheels with 7 embossed, simulated spokes, black hubs, 25" long, 1931	1500	2500	3500
Hydraulic Dump Truck	black front, hood, fenders and enclosed cab, red dump body, chassis and wheels with 6 embossed, simulated spokes, bright hubs, 24 3/4" long, 1933	325	485	650
Hydraulic Dump Truck	duo-tone slant design, red hood sides, dump body and chassis, white upper hood, cab and removable rider seat, electric headl-37ghts, 24 3/4" long, 1936	1000	2000	3000
Hydraulic Dump Truck	duo-tone slant design, red front, lower hood sides, dump body and chassis, white upper hood and cab, 26 1/2" long, 1939	1500	2500	3500
Hydraulic Dumper	green, plated dump lever on left side, large hooks on left side hold yellow or off-white steel scoop shovel, white plastic side mirro and grille guard, 17" long, 1961	125	185	250
Hydraulic Dumper with Shovel	green, plated dump lever on left side, large hooks on left side hold yellow or off-white steel scoop shovel, 17" long, 1960	125	185	250
Hydraulic Hi-Lift Dumper	duo-tone slant design, green hood nose and lower cab sides, remainder white except for chrome grille, enclosed cab, 24" long, 1953	75	115	150

VEHICLES

Buddy L

NAME	DESCRIPTION	GOOD	EX	MINT
Hydraulic Hi-Lift Dumper	green hood, fenders, cab, and dump-body supports, white dump body with cab shield, 22 1/2" long, 1954	85	130	175
Hydraulic Hi-Lift Dumper	blue hood, fenders, cab, and dump-body supports, white dump body with cab shield, 22 1/2" long, 1955	75	115	150
Hydraulic Hi-Way Dumper with Scraper Blade	orange except for row of black square across scraper edges, one-piece chrome 8-hole grille and double headlights, 17 3/4" long over blade and raised dump body, 1958	75	115	150
Hydraulic Highway Dumper	orange except for row of black square across scraper edges, one-piece chrome 8-hole grille and double headlights, no scraper blade, 17 3/4" long over blade and raised dump body, 1959	50	75	100
Hydraulic Husky Dumper	red body, frame, dump section and cab shield, 15 1/4" long, 1962	65	100	135
Hydraulic Husky Dumper	red body, white one-piece bumper and grille guard, heavy side braces on dump section, 14" long, 1963	50	75	100
Hydraulic Rider Dumper	duo-tone slant design, yellow front and lower hood, red upper cab, dump body and upper hood, 26 1/2" long, 1949	175	265	350
Hydraulic Sturdy Dumper	lime green hood, cab, fram and tiltback dump section, green lever on left side controls hydraulic dumping, 14 1/2" long, 1969	50	75	100
Hydraulic Sturdy Dumper	yellow hood, cab, fram and tiltback dump section, green lever on left side controls hydraulic dumping, 14 1/2" long, 1969	50	75	100
Hydraulic Sturdy Dumper	snub-nose greenish-yellow body, cab and tiltback dump section, white plastic seats, 14 1/2" long, 1970	45	70	90
Ice Truck	black front, hood, fenders and doorless cab, yellow open cargo section, canvas sliding cover, 26 1/2" long, 1926	2000	3000	4000
Ice Truck	black front, hood, fenders and enclosed cab, yellow open cargo section, canvas, ice cakes, miniature tongs, 26 1/2" long, 1930	3000	4500	6000
Ice Truck	black front, hood, fenders and enclosed cab, yellow ice compartment, canvas, ice cakes, miniature tongs, 26 1/2" long, 1933-34	2000	3000	4000
Ice Truck	black front, hood, fenders and enclosed cab, yellow ice compartment, 26 1/2" long, 1933	2000	3000	4000
IHC "Red Baby" Express Truck	red doorless roofed cab, open pickup body, chassis and fenders, 24 1/4" long, 1928	1500	2000	3000
IHC "Red Baby" Express Truck	red except for black hubs and aluminum tires, 24 1/4" long, 1929	1500	2000	3000
Insurance Patrol	red including open driver's seat and body, brass bell on cowl and full-length handrails, 27" long, 1925	650	1000	1300
Insurance Patrol	red including open driver's seat and body, brass bell on cowl and full-length handrails, no CFD decal, 27" long, 1928	625	950	1250
International Delivery Truck	red except for removable black rider saddle, black-edged yellow horizontal strip on cargo body, 24 1/2" long, 1935	225	350	450
International Delivery Truck	duo-tone slant design, red front, bumper and lower hood sides, yellow hood top, upper sides, cab and open cargo body, 24 1/2" long, 1936	200	300	400
International Delivery Truck	duo-tone slant design, red front, bumper and lower hood sides, yellow hood top, upper sides, cab and open cargo body, bright metal dummy headlights, 24 1/2" long, 1938	150	225	300
International Dump Truck	red except for bright-metal radiator grille, and black removable rider saddle, 25 3/4" long, 1935	325	485	650
International Dump Truck	duo-tone slant design, yellow radiator, fenders, lower hood and detachable rider seat, rest of truck is red, 25 3/4" long, 1936	315	475	630
International Dump Truck	red, with red headlights on radiator, black removable rider saddle, 25 3/4" long, 1938	125	185	250

VEHICLES

Buddy L

NAME	DESCRIPTION	GOOD	EX	MINT
International Railway Express Truck	duo-tone slant design, yellow front, lower hood sides and removable vab top, green hood top, enclosed cab and van body, electric headlights, 25" long, 1937	350	525	700
International Railway Express Truck	duo-tone slant design, yellow front, lower hood sides and removable vab top, green hood top, enclosed cab and van body, dummy headlights, 25" long, 1938	345	525	690
International Wrecker Truck	duo-tone slant design, yellow upper cab, hood, and boom, red lower cab, fenders, grill and body, rubber tires, removable rider seat, 32" long, 1938	1500	2500	3500
Jewel Home Service Truck Van	dark brown body and sliding side doors, 1967	125	200	275
Jewel Home Shopping Truck Van	pale mint green upper body and roof, darker mint green lower half, no sliding doors, 1968	125	200	275
Jolly Joe Ice Cream Truck	white except for black roof, black tires and wooden wheels, 17 1/2" long, 1947	225	350	450
Jolly Joe's Popsicle Truck	white except for black roof, black tires and wooden wheels, 17 1/2" long, 1948	275	425	550
Jr. Animal Ark	fushcia lapstrake hull, four black tires, 10 pairs of plastic animals, 5" long, 1970	40	60	80
Jr. Auto Carrier	yellow cab-over-engine tractor unit and double-deck semi-trailer, 2 red plastic cars, 15 1/2" long, 1967	50	75	100
Jr. Auto Carrier	bright blue cab-over-engine tractor unit and double-deck semi-trailer, 2 plastic cars, 17 1/4" long, 1969	60	95	125
Jr. Beach Buggy	yellow hood, fenders and topless jeep body, red seats, white plastic surfboard that clips to roll bar and windshield, truck 6" long, 1969	45	65	90
Jr. Beach Buggy	lime green hood, fenders and topless jeep body, red plastic seats, lime green plastic surfboard that clips to roll bar and windshield, truck 6" long, 1971	35	50	70
Jr. Buggy Hauler	fuschia jeep body with orange seats, orange 2-wheel trailer tilts to unload sandpiper beach buggy, 12" long including jeep and trailer, 1970	35	55	75
Jr. Camper	red cab and pickup body wih yellow camper body, 7" long, 1971	50	75	100
Jr. Canada Dry Delivery Truck	two-tone green and pale lime green cab-over-engine body, hand truck, 10 cases of green bottles, 9 1/2" long, 1968	100	150	200
Jr. Canada Dry Delivery Truck	two-tone green and pale lime green cab-over-engine body, hand truck, 10 cases of green bottles, 9 1/2" long, 1969	85	130	170
Jr. Cement Mixer Truck	blue cab-over-engine body, frame and hopper, white plastic mixing drum, white plastic seats, 7 1/2" long, 1968	50	75	100
Jr. Cement Mixer Truck	blue cab-over-engine body, frame and hopper, white plastic mixing drum, white plastic seats, wide one-piece chrome bumper, 7 1/2" long, 1969	35	50	70
Jr. Dump Truck	red cab-over-engine, frame and tiltback dump section, plastic vertical headlights, 7 1/2" long, 1967	50	75	100
Jr. Dumper	avocado cab-over-engine, frame and tiltback dump section with cab shield, one-piece chrome bumper and 4-slot grille, 7 1/2" long, 1969	335	55	75
Jr. Giraffe Truck	turquoise cab-over-engine body, white cab roof, plastic giraffe, 6 1/2" long, 1968	50	75	100
Jr. Giraffe Truck	turquoise cab-over-engine body, white cab roof, plastic giraffe, 6 1/4" long, 1969	40	60	80
Jr. Kitty Kennel	pink cab-over-engine body, white cab roof, 4 white plastic cats, 6 1/4" long, 1969	55	85	115
Jr. Kitty Kennel	pink cab-over-engine body, white cab roof, 4 colored plastic cats, 6 1/4" long, 1968	60	95	125
Jr. Sanitation Truck	blue cab-over-engine, white frame, refuse body and loading hopper, 10" long, 1968	75	115	150

VEHICLES

Buddy L

NAME	DESCRIPTION	GOOD	EX	MINT
Jr. Sanitation Truck	yellow cab-over-engine and underframe, refuse body and loading hopper, full width bumper and grille, 10" long, 1969	75	115	150
Junior Line Air Mail Truck	black enclosed cab, red chassis and body, headlights and double bar bumper, 6 rubber tires, 24" long, 1930-32	3000	4000	5000
Junior Line City Dray	black, front, hood and fenders, 5-digit decal license plate, 24" long, 1930	400	5500	7000
Junior Line Dairy Truck	stake-bed style truck, black front, hood, and enclosed cab with opening doors, blue-green stake body, red chassis, 6 miniature milk cans with removable lids, 24" long, 1930	1500	2500	3500
Junior Line Dump Truck	black enclosed cab, front, hood and chassis, red dump body front and back are higher than sides, 21" long, 1933	1500	2500	3500
Kennel Trucks	medium blue pickup body and cab, clear plastic 12 section kennel with 12 plastic dogs fits in cargo box, 13 1/2" long, 1964	60	90	120
Kennel Trucks	turquoise pickup body and cab, clear plastic 12 section kennel with 12 plastic dogs fits in cargo box, 13 1/2" long, 1965	60	95	125
Kennel Trucks	bright blue pickup body and cab, clear plastic 12 section kennel with 12 plastic dogs fits in cargo box, 13 1/4" long, 1966	95	145	190
Kennel Trucks	bright blue pickup body and cab, clear plastic 12 section kennel with 12 plastic dogs fits in cargo box, 13 1/4" long, 1967	85	130	175
Kennel Trucks	cream yellow pickup body and cab, clear plastic 12 section kennel with 12 plastic dogs fits in cargo box, 13 1/4" long, 1968	80	120	160
Kennel Trucks	red-orange pickup body and cab, yellow roof 6 section kennel with 6 plastic dogs fits in cargo box, 13 1/4" long, 1969	65	100	135
Kennel Trucks	snub-nosed red orange body and cab, plastic kennel section in back, 6 kennels with 6 plastic dogs, 13 1/4" long, 1970	60	95	125
Lumber Truck	black front, hood, fenders, cabless open seat and low-sides cargo bed, red bumper, chassis and a pair of removable solid stake sides, load of lumber pieces, 24" long, 1924	1500	2500	4000
Lumber Truck	black front, hood, fenders, doorless cab and low-sides cargo bed, red bumper, chassis and a pair of removable solid stake sides, load of 12 to 16 lumber pieces, 25 1/2" long, 1926	2000	3000	4000
Mack Hydraulic Dumper	red front, hood, cab, chassis and tiltback dump section with cab shield, white plastic bumper, 20 1/2" long, 1965	60	95	125
Mack Hydraulic Dumper	red front, hood, cab, chassis and tiltback dump section with cab shield, white plastic bumper, short step ladder on each side, 20 1/2" long, 1967	50	75	100
Mack Quarry Dumper	orange front, hood cab and chassis, blue-green tiltback dump section, white plastic bumper, 20 1/2" long, 1965	75	115	150
Mammoth Hydraulic Quarry Dumper	deep green hood, cab and chassis, red heavily braced tiltback dump section, black plastic bumper, 23" long, 1962	65	100	135
Mammoth Hydraulic Quarry Dumper	deep green hood, red cab, chassis and heavily braced tiltback dump section, black plastic bumper, 22 1/2" long, 1963	60	90	125
Marshall Field Delivery Truck Van	hunter's green body, sliding doors and roof, 1966	125	200	275
Milkman Truck	medium blue hood, cab and flatbed body, white side rails, eight 3" white plastic milk bottles with red or green caps, 14 1/4" long, 1961	110	175	225

VEHICLES

Buddy L

NAME	DESCRIPTION	GOOD	EX	MINT
Milkman Truck	deep cream hood, cab and flatbed body, white side rails, fourteen 3" white plastic milk bottles with red caps, 14 1/4" long, 1962	100	150	200
Milkman Truck	light blue hood, cab and flatbed body, white side rails, fourteen 3" white plastic milk bottles with red and green caps, 14 1/4" long, 1963	85	130	175
Milkman Truck	light yellow hood, cab and flatbed body, white side rails, fourteen 3" white plastic milk bottles with red and green caps, 14 1/4" long, 1964	75	115	150
Milkman Truck	medium blue hood, cab and flatbed body, white side rails, eight 3" white plastic milk bottles with red or green caps, 14 1/4" long, 1961	110	175	225
Milkman Truck	deep cream hood, cab and flatbed body, white side rails, fourteen 3" white plastic milk bottles with red caps, 14 1/4" long, 1962	100	150	200
Milkman Truck	light blue hood, cab and flatbed body, white side rails, fourteen 3" white plastic milk bottles with red and green caps, 14 1/4" long, 1963	85	130	175
Milkman Truck	lime yellow hood, cab and flatbed body, white side rails, fourteen 3" white plastic milk bottles with red and green caps, 14 1/4" long, 1964	75	115	150
Mister Buddy Ice Cream Truck	white cab-over-engine van body, pale blue or off-white plastic underbody and floor, 11 1/2" long, 1964	75	115	150
Mister Buddy Ice Cream Truck	white cab-over-engine van body, red plastic underbody and floor, 11 1/2" long, 1966	65	100	135
Mister Buddy Ice Cream Truck	white cab-over-engine van body, red plastic underbody and floor, red bell knob, 11 1/2" long, 1967	55	85	115
Model T Flivver Truck	black except for red 8-spoke wheels with aluminum tires, black hubs, 12" long, 1924	1000	1500	2000
Motor Market Truck	duo-tone horizontal design, white hood top, upper cab and high partition in cargo section, yellow-orange grille, fenders, lower hood and cab sides, 21 1/2" long, 1941	200	350	550
Moving Van	black front, hood and seat, red chassis and disc wheels with black hubs, green van body, roof extends forward above open driver's seat, 25" long, 1924	1200	2000	3000
Overland Trailer Truck	yellow tractor unit with encllosed cab, red semi-trailer and 4-wheel full trailer with removable roofs, length of 3 units 39 3/4" long, 1935	350	525	700
Overland Trailer Truck	duo-tone slant design, green and yellow tractor unit with yellow cab, red semi-trailer and 4-wheel full trailer with yellow removable roofs, length of 3 units 39 3/4" long, 1936	325	485	650
Overland Trailer Truck	duo-tone slant design, green and yellow semi-streamlined tractor and green hood sides, yellow hood, chassis, enclosed cab, 40" long over three units, 1939	350	550	700
Overland Trailer Truck	duo-tone horizontal design, red and white tractor has red front, lower half chassis, chassis, enclosed cab, 40" long over three units, 1939	350	550	700
Pepsi Delivery Truck	powder blue hood and lower cab, white upper cab and double-deck cargo section, 2 hand trucks, 4 blue cases of red bottles, 4 red cases of blue bottles, 15" long, 1970	60	95	125
Polysteel Boat Transport	medium blue soft plastic body, steel flatbed carrying 8" white plastic runabout boat with red deck, truck 12 1/2" long, 1960	75	115	150
Polysteel Coca-Cola Delivery Truck	yellow plastic truck, slanted bottle racks, 8 red coke cases with green bottles, small metal hand truck, 12 1/2" long, 1961	50	75	100

VEHICLES

Buddy L

NAME	DESCRIPTION	GOOD	EX	MINT
Polysteel Coca-Cola Delivery Truck	yellow plastic truck, slanted bottle racks, 8 green coke cases with red bottles, small metal hand truck, 12 1/4" long, 1962	60	90	120
Polysteel Dumper	green soft molded plastic front, cab and frame, yellow steel dump body with sides rounded at back, hinged tailgate, 13" long, 1959	100	150	200
Polysteel Dumper	medium blue soft molded plastic front, cab and frame, off-white steel dump body with sides rounded at back, hinged tailgate, 13" long, 1960	87	130	175
Polysteel Dumper	orange plastic body and tiltback dump section with cab shield, "Come-Back Motor" if truck is pulled backward it rolls ahead when released, 13" long, 1961	75	115	150
Polysteel Dumper	orange plastic body and tiltback dump section with cab shield, no "Come-Back Motor", no door decals, 13 1/2" long, 1962	60	95	125
Polysteel Highway Transport	red soft plastic tractor, cab roof lights, double horn, radio antenna and side fuel tanks, white steel semi-trailer van, 20 1/2" long, 1960	100	150	200
Polysteel Hydraulic Dumper	beige soft molded-plastic front, cab and frame, off-white steel dump section with sides rounded at rear, 13" long, 1959	60	95	125
Polysteel Hydraulic Dumper	red soft molded-plastic front, cab and frame, light green steel dump section with sides rounded at rear, 13" long, 1960	80	120	160
Polysteel Hydraulic Dumper	yellow soft plastic body, frame and tiltback ribbed dump section with cab shield, 13" long, 1961	75	115	150
Polysteel Hydraulic Dumper	red soft plastic body, frame and tiltback ribbed dump section with cab shield, 13" long, 1962	65	100	130
Polysteel Milk Tanker	red soft plastic tractor unit, light bluish-gray semi-trailer tank with red ladders and five dooms, 22" long, 1961	60	95	125
Polysteel Milk Tanker	turquoise soft plastic tractor unit, light bluish-gray semi-trailer tank with red ladders and five dooms, 22" long, 1961	60	95	125
Polysteel Milkman Truck	light blue soft plastic front, cab and frame, light yellow steel open cargo section with 9 oversized white plastic milk bottles, 11 3/4" long, 1960	65	100	130
Polysteel Milkman Truck	light blue soft plastic front, cab and frame, light blue steel open cargo section with 9 oversized white plastic milk bottles, 11 3/4" long, 1961	60	95	125
Polysteel Milkman Truck	turquoise soft plastic front, cab and frame, light blue steel open cargo section with 9 oversized white plastic milk bottles with red caps, 11 3/4" long, 1962	35	50	70
Polysteel Supermarket Delivery	medium blue soft molded-plastic front, hood, cab and frame, steel off-white open cargo section, 13" long, 1959	75	115	150
Pull-N-Ride Baggage Truck	duo-tone horizontal design, light cream upper half, off-white lower half and bumper, 24 1/4" long, 1953	150	225	300
R E A Express Truck	dark green cab-over-engine van body, sliding side doors, double rear doors, white plastic one-piece bumper, 11 1/2" long, 1964	200	300	400
R E A Express Truck	dark green cab-over-engine van body, sliding side doors, double rear doors, white plastic one-piece bumper, no spring suspension, 11 1/2" long, 1965	130	195	260
R E A Express Truck	dark green cab-over-engine van body, sliding side doors, double rear doors, white plastic one-piece bumper, no spring suspension, side doors are embossed "BUDDY L", 11 1/2" long, 1966	125	185	250
Railroad Transfer Rider Delivery Truck	duo-tone horizontal design, yellow upper half, hood top, cab and slatted caro sides, green lower half, small hand truck, 2 milk cans with removable lids, 23 1/4" long, 1949	70	100	140

VEHICLES

Buddy L

NAME	DESCRIPTION	GOOD	EX	MINT
Railroad Transfer Store Door Delivery	duo-tone horizontal design, yellow hood top, cab and upper body, red lower half of hood and body, small hand truck, 2 metal drums with coin slots, 23 1/4" long, 1950	90	135	180
Railway Express Truck	red tractor unit, enclosed square cab, green 12 1/4" long 2-wheel semi-trailer van with removable roof, "Wrigley's Spearmint Gum" poster on trailer sides, 23" long, 1935	375	565	750
Railway Express Truck	duo-tone slant design, tractor unit has white skirted fenders and hood sides, green hood top, enclosed cab and chassis, green semi-trailer with white removable roof, 25" long, 1939	350	475	700
Railway Express Truck	duo-tone slant design, tractor has silvery and hood sides, green hood top, enclosed cab, green semi-trailer, "Wrigley's Spearmint Gum" poster on trailer sides, 23" long, 1935	400	600	800
Railway Express Truck	black front hood, fenders, seat and low body sides, dark green van body, red chassis, 25" long, 1926	2200	3500	4500
Railway Express Truck	dark green or light green screen body, double-bar nickel front bumper, brass radiator knob, red wheels, 25" long, 1930	1500	2500	3500
Railway Express Truck	yellow and green tractor unit has white skirted fenders and hood sides, green hood top, enclosed cab and chassis, green semi-trailer with yellow removable roof, 25" long, 1940	330	495	660
Railway Express Truck	duo-tone horizontal design, tractor unit has yellow front, lower door and chassis, green hood top and enclosed upper cab, semi-trailer has yellow lower sides, 25" long, 1941	325	485	650
Railway Express Truck	deep green plastic "Diamond T" hood and cab, deep green steel frame and van body with removable silvery roof, small 2-wheel hand truck, steel 4-rung barrel skid, 21" long, 1952	200	300	400
Railway Express Truck	green plastic hood and cab, green steel high-sides open body, frame and bumper, small 2-wheel hand truck, steel 4-rung barrel skid, 20 3/4" long, 1953	125	185	250
Railway Express Truck	green all-steel hood, cab, frame and high-sides open body, sides have 3 horizontal slots in upper back corners, 22" long, 1954	75	115	150
Ranchero Stake Truck	medium green, white plastic one-piece bumper and grille guard, 4-post, 4-slat fixed stake sides and cargo section, 14" long, 1963	50	75	100
Rider City Special Delivery Truck Van	duo-tone horizontal design, yellow upper half including hood top and cab, brown removable van roof, warm brown front and lower half of van body, 24 1/2" long, 1949	150	225	300
Rider Dump Truck	duo-tone horizontal design, yellow hood top, upper cab and upper dump body, red front, hood sides, lower doors and lower dump body, no bumper, 21 1/2" long, 1945	160	245	325
Rider Dump Truck	duo-tone horizontal design, yellow hood top, upper cab and upper dump body, red front, hood sides, lower doors and lower dump body, no bumper, 23" long, 1947	75	115	150
Rider Van Lines Trailer	duo-tone slant design, black front and lower hood sides and doors, deep red hood top, enclosed cab and chassis, 35 1/2" long, 1949	350	525	700
Rival Dog Food Delivery Van	cream front, cab and boxy van body, metal drum coin bank with "RIVAL DOG FOOD" label in blue, red, white and yellow, 16 1/2" long, 1956	160	245	325
Robotoy	black fenders and chassis, red hood and enclosed cab with small visor, green dump body's front and back are higher than sides, 21 5/8" long, 1932	750	1000	1500

Buddy L

NAME	DESCRIPTION	GOOD	EX	MINT
Rockin' Giraffe Truck	powder blue hood, cab, and high-sided open-top cargo section, 2 orange and yellow plastic giraffes, 13 1/4" long, 1967	75	115	150
Ruff-n-Tuff Cement Mixer Truck	yellow snub-nosed cab-over-engine body, frame and water-tank ends, white plastic water tank and mixing drum, white seats, 16" long, 1971	35	55	75
Ruff-n-Tuff Log Truck	yellow snub-nose cab-over-engine, frame and shallow truck bed, black full-width grille, 16" long, 1971	50	75	100
Saddle Dump Truck	duo-tone slant design, yellow front, fenders and removable rider seat, red enclosed square cab and dump body, no bumper, 19 1/2" long, 1937	200	300	400
Saddle Dump Truck	duo-tone slant design, yellow front, fenders, lower hood and cab, and removable rider seat, rest of body red, no bumper, 21 1/2" long, 1939	125	185	250
Saddle Dump Truck	duo-tone horizontal design, deep blue hood top, upper cab and upper dump body, orange fenders radiator front lower two-thirds of cab and lower half of dump body, 21 1/2" long, 1941	85	130	175
Sand and Gravel Rider Dump	duo-tone horizontal design, blue lower half, yellow upper half including hoop top and enclosed cab, 24" long, 1950	350	525	700
Sand and Gravel Truck	black body, doorless roofed cab and steering wheel, red chassis and disc wheels with black hubs, 25 1/2" long, 1926	1500	2500	3500
Sand and Gravel Truck	dark or medium green hood, cab, roof lights and skirted body, white or cream dump section, 13 1/2" long, 1949	100	150	200
Sand and Gravel Truck	duo-tone horizontal design, red front, bumper, lower hood, cab sides, chassis and lower dump body sides, white hood top, enclosed cab and upper dump body, 23 3/4" long, 1949	350	525	700
Sand and Gravel Truck	black except for red chassis and wheels, nickel-rim, red-shell headlights, enclosed cab with opening doors, 25 1/2" long, 1930-32	2000	3000	5000
Sand Loader and Dump Truck	duo-tone horizontal design, yellow hood top and upper dump blue cab sides, frame and lower dump body, red loader on dump with black rubber conveyor belt, 24 1/2" long, 1950	175	265	350
Sand Loader and Dump Truck	duo-tone horizontal design, yellow hood top and upper dump blue cab sides, frame and lower dump body, red loader on dump with black rubber conveyor belt, 24 1/2" long, 1952	60	95	125
Sanitation Service Truck	blue front fenders, hood, cab and chassis, white encllosed dump section and hinged loading hopper, one-piece chrome bumper, plastic windows in garbage section, 16 1/2" long, 1967	100	150	200
Sanitation Service Truck	blue front fenders, hood, cab and chassis, white encllosed dump section and hinged loading hopper, one-piece chrome bumper, no plastic windows in garbage section, 16 1/2" long, 1968	75	115	150
Sanitation Service Truck	blue snub-nose hood, cab and frame, whote cargo dump body and rear loading unit, 2 round plastic headlights, 17" long, 1972	75	115	150
Sears, Roebuck Delivery Truck Van	gray-green and off-white, no side doors, 1967	125	200	275
Self-Loading Auto Carrier	medium tan tractor unit, 3 plastic cars, overall 34" long including loading ramp, 1960	85	130	175
Self-Loading Boat Hauler	pastel blue tractor and semi-trailer with three 8 1/2" long boats, overall 26 1/2" long, 1962	100	200	300
Self-Loading Boat Hauler	pastel blue tractor and semi-trailer with three 8 1/2" long boats, no side mirror on truck, overall 26 1/2" long, 1963	125	225	350
Self-Loading Car Carrier	lime green tractor unit, 3 plastic cars, overall 33 1/2" long including, 1963	75	115	150

Buddy L

NAME	DESCRIPTION	GOOD	EX	MINT
Self-Loading Car Carrier	beige-yellow tractor unit, 3 plastic cars, overall 33 1/2" long including, 1964	60	95	125
Shell Pickup and Delivery	reddish-orange hood and body, open cargo section with solid sides, chain across back, red coin-slot oil drum with Shell emblem and lettering, 13 1/4" long, 1950	135	200	275
Shell Pickup and Delivery	yellow-orange hood and body, open cargo section, 3-curved slots toward rear in sides, chains across back, red coin-slot oil drum with Shell emblem and lettering, 13 1/4" long, 1952	125	185	250
Shell Pickup and Delivery	yellow-orange hood and body, open cargo section with 3-curved slots toward rear in sides, red coin-slot oil drum with Shell emblem and lettering, 13 1/4" long, 1953	110	175	225
Smoke Patrol	lemon-yellow body, six wheels, garden hose attaches and water squirts through large chrome swivel-mount water cannon on rear deck, 7" long, 1970	50	75	100
Sprinkler Truck	black front, hood, fenders and cabless open driver's seat, red bumper and chassis, bluish-gray-green water tank, 25" long, 1929	2500	3500	4500
Stake Body Truck	black cabless open driver's seat, hood, front fenders and flatbed body, red chassis and 5 removable stake sections, 25" long, 1921	1000	1500	2000
Stake Body Truck	black cabless open driver's seat, hood, front fenders and flatbed body, red chassis and 5 removable stake sections, cargo bed with low sideboards, drop-down tailgate, 25" long, 1924	1000	1500	2000
Standard Coffee Co. Delivery Truck Van	1966	125	200	275
Stor-Dor Delivery	red hood and body, open cargo body with 4 long horizontal slots in sides, plated chains across open back, 14 1/2" long, 1955	125	185	250
Street Sprinkler Truck	black front, hood, front fenders and cabless open driver's seat, red bumper and chassis, bluish-gray-green water tank, 25" long, 1929	1000	1800	2600
Street Sprinkler Truck	black front, hood, and fenders, open cab, nickel-rim, red-shell headlights, double bar front bumper, bluish-gray-green water tank, 6 rubber tires, 25" long, 1930-32	3000	3500	4000
Sunshine Delivery Truck Van	bright, yellow cb-over-engine van body and opening double rear doors, off-white plastic bumper and under body, 11 1/2" long, 1967	125	200	275
Super Motor Market	duo-tone horizontal design, white hood top, upper cab and high partition in cargo section, yellow-orange lower hood and cab sides, semi-trailer carrying supplies, 21 1/2" long, 1942	300	500	700
Supermarket Delivery	all white except for rubber wheels, enclosed cab, pointed nose, bright-metal one-piece grille, 13 3/4" long, 1950	125	185	250
Supermarket Delivery	blue bumper, front, hood, cab and frame, one-piece chrome 4-hole grille and headlights, 14 1/2" long, 1956	75	115	150
Tank and Sprinkler Truck	black front, hood, fenders, doorless cab and seat, dark green tank and side racks, black or dark green sprinkler attachment, 26 1/4" long with sprinkler attachment, 1924	4000	5500	7000
Teepee Camping Trailer and Wagon	maroon suburban wagon, 2-wheel teepee trailer and its beige plastic folding tent, overall 24 1/2" long, 1963	150	225	300
Texaco Tank Truck	red steel GMC 550-series blunt-nose tractor and semi-trailer tank, 25" long, 1959	175	250	400
Tom's Toasted Peanuts Delivery Truck Van	light tan-beige body, no seat or sliding doors, blue bumpers, floor and underbody, 11 1/2" long, 1973	125	200	275

VEHICLES

Buddy L

NAME	DESCRIPTION	GOOD	EX	MINT
Trail Boss	red, square-corner body with sloping sides, open cockpit, white plastic seat, 7" long, 1970	40	60	80
Trail Boss	lime green, square-corner body with sloping sides, open cockpit, yellow plastic seat, 7" long, 1971	35	55	75
Trailer Dump Truck	cream tractor unit with enclosed cab, dark blue semi-trailer dump body with high sides and top-hinged opening endgate, no bumper, 20 3/4" long, 1941	75	115	155
Trailer Van Truck	red tractor and van roof, blue bumper, white semi-trailer van, chrome one-piece toothed grille and headlights, white drop-down rear door, 29" long with tailgate/ramp lowered, 1956	150	225	300
Trailer Van with Tailgate Loader	green high-impacted styrene plastic tractor on steel frame, cream steel detachable semi-trailer van with green roof and crank operated tailgate, 33" long with tailgate lowered, 1953	125	185	250
Trailer Van with Tailgate Loader	green steel tractor, bumper, chrome one-piece toothed grille and headlights, cream van with green roof and tailgate loader, 31 3/4" long, with tailgate down, 1954lgate lowered, 1953	125	185	250
Traveling Zoo	red high side pickup with yellow plastic triple-cage unit, 6 campartments with plastic animals, 13 1/4" long, 1965	85	130	175
Traveling Zoo	red high side pickup with yellow plastic triple-cage unit, 6 campartments with plastic animals, 13 1/4" long, 1967	75	115	150
Traveling Zoo	yellow high side pickup with red plastic triple-cage unit, 6 campartments with plastic animals, 13 1/4" long, 1969	65	95	130
Traveling Zoo	snub-nosed yellow body and cab, 6 red plastic cages with 6 plastic zoo animals, 13 1/4" long, 1970	60	95	125
U.S. Army Half-Track and Howitzer	olive drab, 12 1/2" truck, 9 3/4" gun, overall 22 1/2" long, 1952	125	200	275
U.S. Mail Delivery Truck	blue cab, hood, bumper, frame and removable roof on white van body, 23 1/4" long, 1956	225	400	575
U.S. Mail Delivery Truck	white upper cab-over-engine, sliding side doors and double rear doors, red belt-line stripe on sides and front, blue lower body, 11 1/2" long, 1964	125	200	275
U.S. Mail Truck	shiny olive green body and bumper, yellow-cream removable van roof, enclosed cab, 22 1/2" long, 1953	225	400	575
United Parcel Delivery Van	duo-tone horizontal design, deep cream upper half except for brown removable roof, chocolate brown front and lower half, 25" long, 1941	250	450	650
Utility Delivery Truck	duo-tone slant design, blue front and lower hood sides, gray hood top, cab and open body with red and yellow horizontal stripe, 22 3/4" long, 1940	250	450	650
Utility Delivery Truck	duo-tone horizontal design, green upper half including hood top, dark cream lower half, green wheels, red and yellow horizontal stripe, 22 3/4" long, 1941	125	185	250
Utility Dump Truck	duo-tone slant design, red front, lower doord and fenders, gray chassis and enclosed upper cab, royal blue dump body, yellow removable rider seat, 25 1/2" long, 1940	125	185	250
Utility Dump Truck	duo-tone slant design, red front, lower door and fenders, gray chassis, red upper hood, upper enclosed cab and removable rider seat, yellow body, 25 1/2" long, 1941	85	130	175
Van Freight Carriers Trailer	bright blue streamlined tractor and enclosed cab, cream-yellow semi-trailer van, removable silvery roof, 22" long, 1949	65	100	135
Van Freight Carriers Trailer	red streamlined tractor, bright blue enclosed cab, cream-yellow semi-trailer van, white removable van roof, 22" long, 1952	55	85	115

VEHICLES

Buddy L

NAME	DESCRIPTION	GOOD	EX	MINT
Van Freight Carriers Trailer	red streamlined tractor, bright blue enclosed cab, light cream-white semi-trailer van with removable white roof, 22" long, 1953	125	185	250
Wild Animal Circus	red tractor unit and semi-trailer, three cages with plastic elephant, lion, tiger, 26" long, 1966	150	225	300
Wild Animal Circus	red tractor unit and semi-trailer, three cages with 6 plastic animals (adult and baby), 26" long, 1967	110	175	225
Wild Animal Circus	red tractor unit and semi-trailer, trailer cage doors lighter red than body, 26" long, 1970	100	150	200
Wrecker Truck	black front, hood, and fenders, open cab, 4 ruber tires, red wrecker body, 26 1/2" long, 1930	2500	3500	4500
Wrecker Truck	duo-tone slant design, red upper cab, hood, and boom, white lower cab, grill, fenders, body, rubber wheels, electric headlights, removable rider seat, 31" long, 1936	1000	2000	3000
Wrecker Truck	black open cab, red chassis and bed, disc wheels, 26 1/2" long, 1928-29	2000	4000	6000
Wrigley Express Truck	forest green except for chrome one-piece, three-bar grille and headlights, "Wrigley's Spearmint" poster on sides, 16 1/2" long, 1955	135	200	275
Zoo-A-Rama	lime green Colt Sportsliner with four-wheel trailer cage, cage contains plastic tree, monkeys and bears, 20 3/4" long, 1967	100	150	200
Zoo-A-Rama	sand yellow four-wheel trailer cage, matching Colt Sportsliner with white top, 3 plastic animals, 20 3/4" long, 1968	100	150	200
Zoo-A-Rama	greenish-yellow four-wheel trailer cage, matching Colt Sportsliner with white top, 3 plastic animals, 20 3/4" long, 1969	85	130	175

VEHICLES

Clockwise from upper left: Forward Control Jeep FC-150; E.R.F. Model 64G Earth Dumper; Chitty Chitty Bang Bang with figures; U.S. Army Field Kitchen; rocket firing Batman's Batbike, all Corgi.

VEHICLES

Corgi

NAME	DESCRIPTION	GOOD	EX	MINT
Adams Drag-Star	4 3/8" long, orange body, red nose, gold engines, chrome pipes and hood panels, amber windshield, driver, black catwalk	18	27	45
Adams Probe 16	3 5/8" long, one-piece body, blue sliding canopy, in 3 colors; metallic burgundy, or metallic lime/gold with and without racing stripes	16	24	40
Agricultural Set	#69 Massey Ferguson tractor, #62 trailer, #438 Land Rover, #484 Farm Truck, #71 harrow, #1490 skip, and accessories	120	180	300
Agricultural Set	vehicle and accessory set in 2 versions, with #55 tractor and yellow trailer, 1962-64; with #60 tractor and red trailer, 1965-66	280	420	700
Agricultural Set	with mustard yellow conveyor	60	90	150
Alfa Romeo P33 Pininfarina	3 5/8" long, in two versions, with either gold or black spoiler	16	24	40
All Winners Set	5 vehicle set in 3 versions: white Mustang and Marcos, red Ferrari, silver Corvette and gold Jaguar, 1966 only; prior vehicles in other colors; or Tornado, Ferrari, MGB, Corvette and Jaguar	160	240	400
Allis-Chalmers AFC 60 Fork Lift	4 3/8" long, yellow body with white engine hood, with driver	14	21	35
AMC Pacer	4 3/4" metallic dark red body, white Pacer X decals, working hatch, clear windows, light yellow interior, chrome bumpers, grill & headlights, black plastic grill & tow hook, suspension, chrome wheels	14	21	35
AMC Pacer Rescue Car	4 7/8" long, with chrome roll bars and red roof lights, in white with black engine hood, in 2 versions: with or without Secours decal	16	24	40
American La France Ladder Truck	11 1/8 " long, red working cab, trailer and ladder rack, in either red/chrome body with red wheels or red/white body with unpainted wheels	60	90	150
AMX 30D Recovery Tank	6 7/8" long, olive drab body with black plastic turret and gun, with accessories and 3 figures	32	48	80
Army Equipment Transporter	9 1/2" long, olive drab cab and trailer with white U.S. Army decals	70	105	175
Army Troop Transporter	5 1/2" long, olive drab, with white U.S. Army decals	70	105	175
Aston Martin DB4	3 3/4" long, red or yellow body with working hood, detailed engine, clear windows, plastic interior, silver lights, grill, license plate and bumpers, red taillights, rubber tires, working scoop on early models	44	66	110
Aston Martin DB4	3 3/4" white top & aqua green sides, yellow plastic interior, racing #1, 3 or 7	50	75	125
Austin A40	3 1/8" long, one-piece light blue or red body with clear windows, silver lights, grill and bumpers, smooth wheels, rubber tires	34	51	85
Austin A40-Mechanical	3 1/8" long, same as 216-A but with friction motor and red body, black roof	52	78	130
Austin A60 Driving School (LDH)	3 3/4" medium blue body with silver trim, left hand drive steering wheel, one piece body, clear windows, shaped wheels, rubber tires	44	66	110
Austin A60 Motor School (RHD)	3 3/4" long, light blue body with silver trim, red interior, single body casting, right hand drive steering wheel, two figures, silver bumpers, grill, headlight & trim, red taillights, L plate decals, suspension, shaped wheels, rubber tires	44	66	110
Austin Cambridge	3 1/2" long, one-piece body in several colors, clear windows, silver lights, grill and bumpers, smooth wheels, rubber tires; colors include gray, green/green, silver/green, aqua	40	60	100

Corgi

NAME	DESCRIPTION	GOOD	EX	MINT
Austin Cambridge-Mechanical	3 1/2" long, same as model 201-A but with fly wheel motor, available in orange, cream, light or dark gray body colors	50	75	125
Austin London Taxi	3 7/8" long, one-piece body, clear windows, black body with yellow plastic interior, with or without driver, smooth or rubber wheels	36	54	90
Austin London Taxi	4 5/8" long, black body with 2 working doors, light brown interior	14	21	35
Austin London Taxi/Reissue	3 7/8" long, updated version with Whizz Wheels, black or maroon body	14	21	35
Austin Mini Countryman	3 1/8" long, turquoise body with 2 working doors, in 3 versions, one with shaped wheels, the others with cast wheels and with or without aluminum parts	52	78	130
Austin Mini Van	3 1/8" long, with 2 working doors, clear windows, metallic deep green body	40	60	100
Austin Mini-Metro	3 1/2" three version with plastic interior, working rear hatch and doors, clear windows, folding seats, chrome headlights, orange taillights, black plastic base, grill, bumpers, Whizz Wheels	18	27	45
Austin Mini-Metro Datapost	3 1/2" white body, blue roof, hood & trim, red plastic interior, hepolite & #77 decals, working hatch & doors, clear windows, folding seats, chrome headlights, orange taillights, Whizz Wheels	12	18	30
Austin Police Mini Van	3 1/8" long, dark blue body with policeman and dog figures, white police decals	50	75	125
Austin Seven Mini	2 3/4" red or yellow body, yellow interior, silver bumpers, grill & headlights, orange taillights, suspension, shaped wheels, rubber tires	50	75	125
Austin-Healey	3 1/4" cream body, red seats or red body cream seats, clear windshield, one piece body, sheet metal base, silver grill, bumpers, headlights, smooth wheels, rubber tires	50	75	125
Avengers Set	set of 2 vehicles and 2 figures with umbrellas, in 2 versions: with either green or red Bentley	260	390	650
Basil Brush's Car	3 5/8" long, red body, dark yellow chassis, gold lamps and dash, Basil Brush figure, red plastic wheels, plastic tires	70	105	175
Batbike	4 1/4" black body, one piece body, black & red plastic parts, gold engine & exhaust pipes, clear windshield, chrome stand, black plastic tive spoked wheels, Batman and decals	40	60	100
Batboat	5 1/8" long, black plastic boat, red seats, fin and jet, blue windshield, Batman and Robin figures, gold cast trailer, large black/yellow decals on fin, cast wheels, plastic tires	60	90	150
Batboat	5 1/8" long, black plastic boat with Batman and Robin figures, small decals on fin, Whizz Wheels on trailer	30	45	75
Batcopter	5 1/2" long, black body with yellow/red/black decals, red rotors, Batman figure	26	39	65
Batman Set	3 vehicle set; Batmobile, Batboat with trailer and Batcopter, Whizz Wheels on trailer	90	135	225
Batmobile	5" long, gold hubs, bat logos on door, maroon interior, black body, plastic rockets, gold headlights & rocket control, tinted canopy, working front chain cutter, no tow hook	200	300	500
Batmobile	5" chrome hubs with red bat logos on door, maroon interior, red plastic tires, gold tow hook, plastic rockets, gold headlight & rocket control, tinted canopy with chrome support, chain cutter	140	210	350
Batmobile	5" chrome hubs with red bat logos on door, light red interior, regular wheels, gold tow hook, plastic rockets, gold headlights and rocket control, tinted canopy with chrome support	80	120	200

VEHICLES

Corgi

NAME	DESCRIPTION	GOOD	EX	MINT
Batmobile	5" long, gold hubs, bat logos on door, maroon interior, black body, plastic rockets, gold headlights & rocket control, tinted canopy, working front chain cutter, with tow hook	180	270	450
Batmobile, Batboat & Trailer	4 versions: red bat hubs on wheels, 1967-72; red tires and chrome wheels, 1973; black tires, big decals on boat, 1974-76; chrome wheels, small boat decals, Whizz Wheels on trailer, 1977-81, each set	240	360	600
Beach Buggy & Sailboat	purple buggy, yellow trailer and red/white boat	20	30	50
Beast Carrier Trailer	4 1/2" long, red chassis, yellow body and tailgate, 4 plastic calfs, red plastic wheels, black rubber tires	24	36	60
Beatles' Yellow Submarine	5" long, yellow and white body, working hatches with 2 Beatles in each	180	270	450
Bedford AA Road Service Van	3 5/8" long, dark yellow body in 2 versions, divided windshield, 1957-59, single windshield, 1960-62	50	75	125
Bedford Army Tanker	7 3/8" long, olive drab cab and tanker, with white U.S. Army decals	140	210	350
Bedford Articulated Horse Box	10" long, cast cab, lower body and 3 working ramps, yellow interior, plastic upper body, with horse and Newmarket Racing Stables decals, in 3 colors; dark metallic green or light green body with either orange or yellow upper	32	48	80
Bedford Car Transporter	10 1/4" black diecast cab base with blue cab, yellow semi trailer, blue lettering decals and/or red cab, pale green upper & blue lower semi-trailer, white decals, lower tailgate, clear windshield, silver bumper, grill, headlights & wheel	70	105	175
Bedford Car Transporter	10 5/8" long, red cab with blue lower and light green upper trailer, working ramp, yellow interior, clear windows, white wording and Corgi dog decals	60	90	150
Bedford Carrimore Low Loader	8 1/2" red or yellow cab, metallic blue semi trailer & tailgate, available with smooth and/or shaped wheels	60	90	150
Bedford Carrimore Low Loader	9 1/2" long, yellow cab and working tailgate, red trailer, clear windows, red interior, suspension, shaped wheels, rubber tires	56	84	140
Bedford Corgi Toys Van	3 1/4" long, both with Corgi Toys decals, with either yellow body/blue roof or yellow upper/blue lower body	60	90	150
Bedford Daily Express Van	3 1/4" long, dark blue body with white Daily Express decals, divided windshield, smooth wheels, rubber tires	60	90	150
Bedford Dormobile	3 1/4" long, in 2 versions and several colors: divided windshield with either cream, green or metallic maroon body; or single windshield with yellow body/ blue roof in either shaped or smooth wheels	50	75	125
Bedford Dormobile-Mechanical	3 1/4" long, with friction motor, in either dark metallic red or turquoise body	60	90	150
Bedford Evening Standard Van	3 1/4' long, clear windows, smooth wheels, rubber tires, in 2 colors: black body/silver roof or black lower body/silver upper body and roof, both with same Evening Standard decals	52	78	130
Bedford Fire Tender	3 1/4" long, divided windshield, in either red or green body, each with different decals	60	90	150
Bedford Fire Tender	3 1/4" long, single windshield version, red body with either black ladders and smooth wheels or unpainted ladders and shaped wheels	60	90	150
Bedford Fire Tender-Mechanical	3 1/4" long, with friction motor, red body with Fire Dept. decals	70	105	175
Bedford Giraffe Transporter	red Bedford truck with blue giraffe box with Chipperfield decal, 3 giraffes	60	90	150
Bedford KLG Van-Mechanical	3 1/4" long, with friction motor, in either red body with K.L.G. Spark Plugs decals, or dark blue body with Daily Express decals	70	105	175

VEHICLES

Corgi

NAME	DESCRIPTION	GOOD	EX	MINT
Bedford Machinery Carrier	9 1/4" long, either red or blue cab, both with silver trailer with working ramps, removable fenders, working winch with line, smooth wheels, rubber tires	50	80	130
Bedford Military Ambulance	3 1/4" long, with clear front and white rear windows, olive drab body with Red Cross decals, with or without suspension	56	84	140
Bedford Milk Tanker	7 3/4" long, light blue cab and lower semi, white upper tank with blue/white Milk decals	110	165	275
Bedford Milk Tanker	7 1/2" long, light blue cab and lower semi, white upper tank, with blue/white Milk decals, shaped wheels, rubber tires	100	150	250
Bedford Mobilgas Tanker	7 3/4" long, red cab and tanker with red/white/blue Mobilgas decals, shaped wheels, rubber tires	100	150	250
Bedford Mobilgas Tanker	7 5/8" long, either red or blue cab with Mobilgas decals, shaped wheels, rubber tires	100	150	250
Bedford Tanker	7 1/2" long, red cab with black chassis, plastic tank with chrome catwalk, Corgi Chemco decals	14	21	35
Bedford TK Tipper Truck	4 1/8" long, in 4 colors; red cab with either yellow or gray tipper, or yellow or blue cab	26	39	65
Bedford Utilecon Ambulance	3 1/4" long, divided windshield, cream body with red/white/blue decals, smooth wheels	50	75	125
Beep Beep London Bus	4 3/4" long, battery operated working horn, red body, black windows, BTA decals	26	39	65
Belgian Police Range Rover	4" long, white body, working doors, red interior, with Belgian Police decals	22	33	55
Bell Army Helicopter	5 1/4" long, two-piece olive/tan camo body, clear canopy, olive green rotors, U.S. Army decals	24	36	60
Bell Rescue Helicopter	5 3/4" long, two-piece blue body with working doors, red interior, yellow plastic floats, black rotors, white N428 decals	20	30	50
Bentley Continental	4 1/4" four different versions, red interior, clear windows, chrome grill & bumpers, jewel headlights, red jeweled taillights luggage & spare wheel in trunk, suspension, shaped wheels, gray rubber tires	44	66	110
Bentley T Series	4 1/2" red rose body, cream interior, working hood, trunk & doors, clear windows, folding seats, chrome bumper/grill, jewel headlights, orange taillights, detailed engine, suspension	30	54	00
Berliet Articulated Horse Box	10 7/8" long, bronze cab and lower semi body, cream chassis, white upper body, black interior, 3 working ramps, National Racing Stables decals, horse figures	30	45	75
Berliet Container Truck	US Cines	30	45	75
Berliet Dolphinarium Truck		56	84	140
Berliet Fruehauf Dumper	11 1/4" yellow cab, fenders & dumper, black cab & semi chassis, plastic orange dumper body or dark orange, black interior, stack, dump knob & semi hitch, 2 black plastic trailer rest wheels, chrome headlights with amber lenses, black grill	30	45	75
Berliet Holmes Wrecker	5" red cab & bed, blue rear body, white chassis, black interior, 2 gold booms & hooks, yellow dome light, driver, amber lenses, red-white-blue stripes	30	45	75
Bertone Barchetta Runabout	3 1/4" long, black interior, amber windows, diecast air foil, suspension, red/yellow Runabout decals, Whizz Wheels	12	18	30
Bertone Shake Buggy	3 3/8" long, clear windows, green interior, gold engine, in 4 versions: yellow upper/white lower body with either spoked or solid chromed wheels; or metallic mauve upper/white lower body with either spoked or solid chrome wheels	12	18	30
BL Roadtrain & Trailers		16	24	40
Bloodhound & Launching Platform		110	165	275

VEHICLES

Corgi

VEHICLES

NAME	DESCRIPTION	GOOD	EX	MINT
Bloodhound Launching Ramp		34	51	85
Bloodhound Loading Trolley		40	60	100
Bloodhound Missile		70	105	175
Bloodhound Missile on Trolley		120	180	300
BMC Mini-Cooper	3" white body, black working hood, trunk & 2 drs., red interior, clear windows, folding seats, chrome bumpers, grill, jewel headlights, red taillights, orange/black stripes & #177 decals, suspension, detailed engine, Whizz Wheels	30	45	75
BMC Mini-Cooper Magnifique	2 7/8" long, metallic blue or olive green body versions with working doors, hood and trunk, clear windows and sunroof, cream interior with folding seats, jewel headlights, cast wheels, plastic tires	34	51	85
BMC Mini-Cooper S	3" bright yellow body, red plastic interior, chrome plastic roof rack with two spare wheels, clear windshield, one piece body silver grill, bumpers, headlights, red taillights, suspension, Whizz Wheels	44	66	110
BMC Mini-Cooper S Rally	2 7/8" long, red body, white roof, chrome roof rack with 2 spare tires, Monte Carlo Rally and #177 decals, in 2 versions: shaped wheels/rubber tires or cast wheels/plastic tires	40	60	100
BMC Mini-Cooper S Rally Car	2 7/8" long, red body, white roof, 5 jewel headlights, Monte Carlo Rally decals with either number 52 (1965) or 2 (1966) with drivers' autographs on roof	60	90	150
BMC Mini-Cooper S Rally Car	2 7/8" long, red body, white roof with 6 jewel headlights, RAC Rally and #21 decals	60	90	150
BMW M1	5" yellow body, black plastic base, rear panel & interior, white seats, clear windshield, multicolored stripes, lettering & #25 decal, black grill & headlights, red taillights, four spoked wheels	14	21	35
BMW M1 BASF	4 7/8" long, red body, white trim with black/white BASF and #80 decals	12	18	30
Breakdown Truck	3 7/8" long, red body, black plastic boom with gold hook, yellow interior, amber windows, black/yellow decals, Whizz Wheels	12	18	30
British Leyland Mini 1000	3 3/8" long, metallic blue body, working doors, black base, clear windows, white interior, silver lights, grill and bumper, Union Jack decal on roof, Whizz Wheels	18	27	45
British Leyland Mini 1000	3 1/4" long, red interior, chrome lights, grill and bumper, #8 decal, in 3 colors: silver body with decals, 1978-82; silver body, no decals; orange body with extra hood stripes	16	24	40
British Racing Cars	set of 3 cars, 3 versions: blue Lotus, green BRM, green Vanwall, all with smooth wheels, 1959; same cars with shaped wheels, 1960-61; red Vanwall, green BRM and blue Lotus, 1963, each set	140	210	350
BRM Racing Car	3 1/2' long, silver seat, dash and pipes, smooth wheels, rubber tires, in 3 versions: dark green body, 1958-60; light green body with driver and various number decals 1961-65; light green body, no driver	50	75	125
Buck Rogers Starfighter	6 1/2" long, white body with yellow plastic wings, amber windows, blue jets, color decal, Buck and Wilma figures	32	48	80
Buick & Cabin Cruiser	3 versions: light blue, dark metallic blue or gold metallic Buick	80	120	200
Buick Police Car	4 1/8" long, metallic blue body with white stripes and Police decals, chrome light bar with red lights, orange taillights, chrome spoke wheels	18	27	45

Corgi

NAME	DESCRIPTION	GOOD	EX	MINT
Buick Riviera	4 1/4" long, metallic gold or dark blue, pale blue or bronze body, red interior, gray steering wheel, & tow hook, clear windshield, chrome grill & bumpers, suspension, Tan-o-lite tail & headlights, spoked wheels & rubber tires	30	45	75
Cadillac Superior Ambulance	4 1/2" long, battery operated, in 2 versions; red lower/cream upper body, or white lower/blue upper body	60	90	150
Cafe Racer Motorcycle		12	18	30
Campbell Bluebird	5 1/8" long, blue body, red exhaust, clear windshield, driver, in 2 versions: with black plastic wheels, 1960; with metal wheels and rubber tires	56	84	140
Canadian Mounted Police Set	blue Land Rover with Police sign on roof and RCMP decals, plus mounted Policeman	30	45	75
Captain America Jetmobile	6" white body, metallic blue chassis, black nose cone, red shield & jet, red-white-blue Captain America decals, light blue seats & driver, chrome wheels, red tires	24	36	60
Captain Marvel Porsche	4 3/4" white body, gold parts, red seat, driver, red-yellow-blue Captain Marvel decals, black plastic base, gold wheels	20	30	50
Car Transporter & Cars	Scammell transporter with 5 cars; Ford Capri, the Saint's Volvo, Pontiac Firebird, Lancia Fulvia, MGC GT, Marcos 3 Litre, set	200	300	500
Car Transporter & Four Cars	2 versions: with Fiat 1800, Renault Floride, Mercedes 230SE and Ford Consul, 1963-65; with Chevy Corvair, VW Ghia, Volvo P-1800 and Rover 2000, 1966 only, each set	200	300	500
Carrimore & Six Cars	sold by mail order only	240	360	600
Carrimore Car Transporter	3 versions: transporter with Riley, Jaguar, Austin Healey and Triumph, 1957-60; with 4 American cars, 1959; with Triumph, Mini, Citroen and Plymouth, 1961-62, each set	300	450	750
Carrimore Car Transporter		70	105	175
Carrimore Car Trasporter		60	90	150
Caterpillar Tractor	4 1/4", Tc-12 lime green body with black or white rubber treads, gray plastic seat, driver figure, controls, stacks	70	105	175
Centurion Mark III Tank		30	45	75
Centurion Tank & Transporter	olive colored tank and transport	52	78	130
Chevrolet Astro I	4 1/8" long, dark metallic green/blue body with working rear door, cream interior with 2 passengers, in 2 versions: with either gold gold wheels with red plastic hubs or Whizz Wheels	18	27	45
Chevrolet Camaro SS	4" long, blue or turquoise body with white stripe, cream interior, working doors, white plastic top, clear windshield, folding seats, silver air intakes, red taillights, black grill & headlights, suspension, Whizz Wheels	30	45	75
Chevrolet Camaro SS	4" long, metallic lime-gold body with 2 working doors, black roof and stripes, red interior, cast wheels, plastic tires	30	45	75
Chevrolet Caprice Classic	5 7/8" long, working doors and trunk, whitewall tires, in 2 versions: light metallic green body with green interior or silver on blue body with brown interior	24	36	60
Chevrolet Caprice Classic	6" long, white upper body, red sides with red/white/blue stripes and #43 decals, tan interior	24	36	60
Chevrolet Caprice Fire Chief Car	5 3/4" red body, red-white-orange decals, chrome roof bar, opaque black windows, red dome light, chrome bumpers, grill & headlights, orange taillights, Fire Dept. & Fire Chief decals, chrome wheels	28	42	70
Chevrolet Caprice Police Car	5 7/8" long, black body with white roof, doors and trunk, red interior, silver light bar, Police decals	20	30	50

Corgi

NAME	DESCRIPTION	GOOD	EX	MINT
Chevrolet Caprice Taxi	5 7/8" long, orange body with red interior, white roof sign, Taxi and TWA decals	20	30	50
Chevrolet Charlie's Angeles Van	4 5/8" long, light rose-mauve body with Charlie's Angels decals, in 2 versions: either solid or spoked chrome wheels	10	15	25
Chevrolet Coca-Cola Van	4 5/8" long, red body, white trim, with Coca Cola logos	14	21	35
Chevrolet Corvair	3 3/4" three body versions with yellow interior & working hood, detailed engine, clear windows, silver bumpers, headlights & trim, red taillights, rear window blind, shaped wheels, rubber tires	36	54	90
Chevrolet Impala	4 1/4" long, pink body, yellow plastic interior, clear windows, silver headlights, bumpers, grill and trim, suspension, die cast base with rubber tires	50	75	125
Chevrolet Impala	4 1/4" tan body, cream interior, gray steering wheel, clear windshields, chrome bumpers, grill, headlights, suspension, red taillights, shaped wheels & rubber tires	50	75	125
Chevrolet Impala Fire Chief	4 1/8" long, red body, yellow interior, in 3 versions: with 4 white doors, with round shield decals on 2 doors, or with white decals on doors	52	78	130
Chevrolet Impala Fire Chief	4" long, with Fire Chief decal on hood, yellow interior with driver, in 2 versions: either all red body or red on white body	52	78	130
Chevrolet Impala Police Car	4" long, black lower body and roof, white upper body, yellow interior with driver, Police and Police Patrol decals on doors and hood	52	78	130
Chevrolet Impala Taxi	4 1/4" light orange body, base with hexagonal panel under rear axle & smooth wheels, or two raised lines & shaped wheels, one piece body, clear windows, plastic interior, silver grill, headlights & bumpers, rubber tires	50	75	125
Chevrolet Impala Yellow Cab	4" long, red lower body, yellow upper, red interior with driver, white roof sign, red decals	80	120	200
Chevrolet Kennel Club Van	4" long, white upper, red lower body, working tailgate and rear windows, green interior, dog figures, kennel club decals, cast wheels, rubber tires	56	84	140
Chevrolet Performing Poodles Van	4" blue upper body & tailgate, red lower body & base, clear windshield, pale blue interior with poodles in back & ring of poodles & trainer, plastic tires	160	240	400
Chevrolet Rough Rider Van	4 5/8" long, yellow body with working rear doors, cream interior, amber windows, Rough Rider decals	12	18	30
Chevrolet Spider-Van	4 5/8" long, dark blue body with Spider-Man decals, in 2 versions: with either spoke or solid wheels	26	39	65
Chevrolet State Patrol Car	4" black body, State Patrol decals, smooth wheels with hexagonal panel or raised lines & shaped wheels, yellow plastic interior, gray antenna, clear windows, silver bumpers, grill, headlights & trim, rubber tires	50	75	125
Chevrolet Superior Ambulance	4 3/4" long, white body, orange roof and stripes, 2 working doors, clear windows, red interior with patient on stretcher and attendant, Red Cross decals	30	45	75
Chevrolet Vanatic Van	4 5/8" long, off white body with Vanatic decals	10	15	25
Chevrolet Vantastic Van	4 5/8" long, black body with Vantastic decals	10	15	25
Chieftain Medium Tank		30	45	75
Chitty Chitty Bang Bang	6 1/4" metallic copper body, dark red interior & spoked wheels, figures, black chassis with silver running boards, silver hood, horn, brake, dash, tail- & headlights, gold radiator, red & orange wings, (handbrake operates/sides retract), blk	180	270	450
Chopper Squad Helicopter		20	30	50
Chopper Squad Rescue Set	blue Jeep with Chopper Squad decal and red/white boat with Surf Rescue decal	40	60	100

Corgi

NAME	DESCRIPTION	GOOD	EX	MINT
Chrysler Imperial Convertible	4 1/4" red or blue-green body with gray base, working hood, trunk & doors, golf bag in trunk, detailed engine, clear windshield, aqua interior, driver, chrome bumpers	44	66	110
Chubb Pathfinder Crash Tender		44	66	110
Chubb Pathfinder Crash Truck	9 1/2" red body with either "Airport Fire Brigade" or "New York Airport" decals, upper & lower body, gold water cannon unpainted & sirens, clear windshield, yellow interior, black steering wheel, chrome plastic deck, silver lights, plastic	60	90	150
Circus Cage Wagon		56	84	140
Circus Crane & Cage	red and blue trailer	400	600	1000
Circus Crane & Cage Wagon	crane truck, cage wagon and accessories	150	225	375
Circus Crane Truck		80	120	200
Circus Horse Transporter		80	120	200
Circus Human Cannonball Truck		30	45	75
Circus Land Rover & Elephant Cage	red Range Rover with blue canopy, Chipperfields Circus decal on canopy, burnt orange elephant cage on red bed trailer	90	135	225
Circus Land Rover & Trailer	yellow/red Land Rover with Pinder-Jean Richard decals	30	45	75
Circus Menagerie Transporter		120	180	300
Circus Set	vehicle and accessory set in 2 versions: with #426 Booking Office, 1963-65; with #503 Giraffe Truck, 1966, each set	340	510	850
Citroen 2CV Charleston	4 1/8" long, yellow/black or maroon/black body versions	12	18	30
Citroen Alpine Rescue Safari	4" white body, light blue interior, red roof & rear hatch, yellow roof rack & skis, clear windshield, man & dog, gold die cast bobsled, Alpine Rescue decals	80	120	200
Citroen DS 19 Rally	4" long, light blue body, white roof, yellow interior, 4 jewel headlights, Monte Carlo Rally and #75 decals	70	105	175
Citroen DS19	4" long, one-piece body in several colors, clear windows, silver lights, grill and bumpers, smooth wheels, rubber tires: colors: cream, yellow/black, red, metallic green, yellow	56	84	140
Citroen Dyane	4 1/2" metallic yellow or green body, black roof & interior, working rear hatch, clear windows, black base & tow bar, silver bumpers, grill & headlights, red taillights, marching duck and French flag decals, suspension, chrome wheels	12	18	30
Citroen ID-19 Safari	4"long, orange body with red and brown or red and green luggage on roof rack, green/brown interior, working hatch, 2 passengers, Wildlife Preservation decals	40	60	100
Citroen Le Dandy Coupe	4" metallic maroon body & base, yellow interior, working trunk & 2 doors, clear windows, plastic interior, folding seats, chrome grill & bumpers, jewel headlights, red taillights, suspension, spoked wheels, rubber tires	50	75	125
Citroen Le Dandy Coupe	4" metallic dark blue hood, sides & base, plastic aqua interior, white roof & trunk lid, clear windows, folding seats, chrome grill & bumpers, jewel headlights, red tailllights, suspension, spoked wheels, rubber tires	70	105	175
Citroen SM	4 3/16" metallic lime gold with chrome wheels or mauve body with spoked wheels, pale blue interior & lifting hatch cover, working rear hatch & 2 doors, chrome inner drs., window frames, bumpers, grill, amber headlights, red taillights, blk	16	24	40

VEHICLES

Top to bottom: Bentley Continental Sports Saloon by H.J. Mulliner; Mack Bulldog Buffalo Fire Dept. Search and Rescue Co. 3; Monkeemobile with Monkees figures; London to Sydney Marathon Winner; Rocket Firing Batmobile with Batman and Robin figures, all Corgi.

Corgi

NAME	DESCRIPTION	GOOD	EX	MINT
Citroen Tour de France Car	4 1/4" red body, yellow interior & rear bed, clear windshield & headlights, driver, black plastic rack with four bicycle wheels, swiveling team manager figure with megaphone in back of car, Paramount & Tour de France decals, Whizz Wheels	40	60	100
Citroen Winter Olympics Car	4 1/8" long, white body, blue roof and hatch, blue interior, red roof rack with yellow skis, gold sled with rider, skier, gold Grenoble Olympiade decals on car roof	70	105	175
Citroen Winter Sports Safari	4" long, white body in 3 versions: 2 with Corgi Ski Club decals and either with or without roof ski rack, or 1 with 1964 Winter Olympics decals	56	84	140
Coast Guard Jaguar XJ12C	3 1/4" long, olive drab body either with or without suspension, Red Cross decals, clear front windows, white rear windows	18	27	45
Combine, Tractor & Trailer	set of 3: #1111 combine, #50 Massey Ferguson tractor, and #51 trailer	110	165	275
Commer 3/4 Ton Ambulance	3 1/2" long, in either white or cream body, red interior, blue dome light, red Ambulance decals	36	54	90
Commer 3/4 Ton Milk Float	3 1/2" long, white cab with either light or dark blue body, one version with CO-OP decals	32	48	80
Commer 3/4 ton Pickup	3 1/2" long, either red cab with orange canopy or yellow cab with red canopy, yellow interior in both	30	45	75
Commer 3/4 Ton Police Bus	3 1/2" long, battery operated working dome light, in several color combinations of dark or light metallic blue or green bodies	44	66	110
Commer 3/4 Ton Van	3 1/2" long, either dark blue body with Hammonds decals (1971) or white body with CO-OP decals (1970)	44	66	110
Commer 5 Ton Dropside Truck	4 5/8" long, either blue or red cab, both with cream rear body, sheet metal tow hook, smooth wheels, rubber tires	40	60	100
Commer 5 Ton Platform Truck	4 5/8" long, either yellow or metallic blue cab with silver body	40	60	100
Commer Holiday Mini Bus	3 1/2" white interior, clear windshield, silver bumpers, grill & headlights, Holiday Camp Special decal, roof rack, two working rear doors	30	45	75
Commer Military Ambulance	3 5/8" long, olive drab body, blue rear windows and dome light, driver, Red Cross decals	50	75	125
Commer Military Police Van	3 5/8" long, olive drab body, barred rear windows, white MP decals, driver	52	78	130
Commer Mobile Camera Van	3 1/2" long, metallic blue lower body and roof rack, white upper body, 2 working rear doors, black camera on gold tripod, cameraman	60	90	150
Commer Refrigerator Van	4 5/8" long, either light or dark blue cab, both with cream bodies and red/white/blue Wall's Ice Cream decals	80	120	200
Commuter Dragster	4 7/8" long, maroon body with Ford Commuter, Union Jack and #2 decals, cast silver engine, chrome plastic suspension and pipes, clear windshield, driver, spoke wheels	30	45	75
Concorde-First Issues	BOAC	20	30	50
Concorde-First Issues	Air France	20	30	50
Concorde-First Issues	Air Canada	80	120	200
Concorde-First Issues	Japan Airlines	280	420	700
Concorde-Second Issues		14	21	35
Construction Set	orange tractor and Mazda	32	48	80
Constructor Set	one each red and white cab bodies, with 4 different interchangable rear units; van, pickup, milk truck, and ambulance	48	72	120
Cooper-Maserati Racing Car	3 3/8" long, blue body with red/white/blue Maserati and #7 decals, unpainted engine and suspension, chrome plastic steering wheel, roll bar, mirrors and pipes, driver, cast 8-spoke wheels, plastic tires	26	39	65

VEHICLES

Corgi

NAME	DESCRIPTION	GOOD	EX	MINT
Cooper-Maserati Racing Car	3 3/8" long, yellow/white body with yellow/black stripe and #3 decals, driver tilts to steer car	18	27	45
Corgi Flying Club Set	blue/orange Land Rover with red dome light, blue trailer with either orange/yellow or orange/white plastic airplane	24	36	60
Corporal Missile & Erector		240	360	600
Corporal Missile Launcher		36	54	90
Corporal Missile on Launcher		80	120	200
Corporal Missile Set	missile and ramp, erector vehicle and army truck	340	510	850
Corvette Sting Ray	4" metallic green or red body, yellow interior, black working hood, working headlights, clear windshield, amber roof panel, gold dash, chrome grill & bumpers, flag decals, gray die cast base, Golden jacks, cast wheels, plastic tires	40	60	100
Corvette Sting Ray	3 3/4" metallic silver/red body, 2 working headlights, clear windshield, yellow interior, silver hood panels, 4 jewel headlights, suspension, chrome bumpers, with spoked or shaped wheels, rubber tires	60	90	150
Corvette Sting Ray	3 3/4" long, yellow body, red interior, suspension, #13 decals	56	84	140
Corvette Sting Ray	3 5/8" long, metallic gray body with black hood, Whizz Wheels	50	75	125
Corvette Sting Ray	3 7/8" long, either dark metallic blue or metallic mauve-rose body, chrome dash, Whizz Wheels	50	75	125
Country Farm Set	#50 Massey Ferguson tractor, red hay trailer with load, fences, figures	30	45	75
Country Farm Set	same as 4-B but without hay load on trailer	30	45	75
Daily Planet Helicopter		24	36	60
Daimier 38 1910		20	30	50
Daktari Set	2 versions: cast wheels, 1968-73; Whizz Wheels, 1974-75, each set	50	75	125
Datsun 240Z	3 5/8" long, red body with #11 and other decals, 2 working doors, white interior, orange roll bar and tire rack; one version also has East Africa Rally decals	14	21	35
Datsun 240Z	3 5/8" long, white body with red hood and roof, #46 and other decals	14	21	35
David Brown Combine		30	45	75
David Brown Tractor	4 1/8" long, white body with black/white David Brown #1412 decals, red chassis and plastic engine	12	18	30
David Brown Tractor & Trailer	2 piece set; #55 tractor and #56 trailer	30	45	75
De Tomaso Mangusta	3 7/8" long, metallic dark green body with gold stripes and logo on hood, silver lower body, clear front windows, cream interior, amber rear windows and headlights, gray antenna, spare wheel, Whizz Wheels	26	39	65
De Tomaso Mangusta	5" white upper/light blue lower body/base, black interior, clear windows, silver engine, black grill, amber headlights, red taillights, gray antenna, spare wheel, gold stripes & black logo decal on hood, suspension, removeable metallic gray chassis	32	48	80
Decca Airfield Radar Van		120	180	300
Decca Radar Scanner	3 1/4" long, with either orange or custard colored scanner frame, silver scanner face, with gear on base for turning scanner	34	51	85
Dick Dastardly's Racing Car	5" long, dark blue body, yellow chassis, chrome engine, red wings, Dick and Muttley figures	40	60	100
Dodge Kew Fargo Tipper	5 1/4" long, white cab and working hood, blue tipper, red interior, clear windows, black hydraulic cylinders, cast wheels, plastic tires	34	51	85
Dodge Livestock Truck	5 3/8" long, tan cab and hood, green body, working tailgate and ramps, 5 pigs	34	51	85

Corgi

NAME	DESCRIPTION	GOOD	EX	MINT
Dolphin Cabin Cruiser	5 1/4" long, white hull, blue deck plastic boat with red/white stripe decals, driver, blue motor with white cover, gray prop, cast trailer with smooth wheels, rubber tires	24	36	60
Dougal's Magic Roundabout Car	4 1/2" long, yellow body, red interior, clear windows, dog and snail, red wheels with gold trim, Magic Roundabout decals	70	105	175
Drax Jet Helicopter	5 7/8" long, white body, yellow rotors and fins, yellow/black Drax decals	24	36	60
Dropside Trailer	4 3/8" long, cream body, red chassis in 5 versions: smooth wheels 1957-61; shaped wheels, 1962-1965; white body, cream or blue chassis; or silver gray body, blue chassis, each	10	15	25
Ecurie Ecosse Racing Set	transporter with 3 cars in 2 versions: RRM, Vanwall and Lotus XI, 1961-64; BRM, Vanwall and Ferrari, 1964-66, each set	140	210	350
Ecurie Ecosse Transporter	7 3/4" long, in dark blue body with either blue or yellow lettering, or light blue body with red or yellow lettering, working tailgate and sliding door, yellow interior, shaped wheels, rubber tires	70	105	175
Emergency Set	3 vehicle set with figures and accessories, Ford Cortina Police car, Police Helicopter, Range Rover Ambulance	40	60	100
Emergency Set	Land Rover Police Car and Police Helicopter with figures and accessories	40	60	100
ERF 44G Dropside Truck		36	54	90
ERF 44G Moorhouse Van	4 5/8" long, yellow cab, red body, Moorhouse Lemon Cheese decals	100	150	250
ERF 44G Platform Truck	4 5/8" long, light blue cab with either dark blue or white flatbed body	36	54	90
ERF Dropside Truck & Trailer	#456 truck and #101 trailer with #1488 cement sack load and #1485 plank load	60	90	150
ERF Neville Cement Tipper	3 3/4" long, yellow cab, gray tipper, cement decal, with either smooth or shaped wheels	32	48	80
ERG 64G Earth Dumper	4" long, red cab, yellow tipper, clar windows, unpainted hydraulic cylinder, spare tire, smooth wheels, rubber tires	30	45	75
Euclid Caterpillar Tractor	4 1/4" TC-12 lime green body with black or shite rubber treads, gray plastic seat, driver figure, controls, stacks, silver grill, painted blue engine sides & Euclid decals	70	105	175
Euclid TC-12 Bulldozer	5", yellow body with black treads or lime green body with white treads, silver blade surface, gray plastic seat controls & stacks, silver grill & lights, painted blue engine sides, sheet metal base, rubber treads & Euclid decals	80	120	200
Euclid TC-12 Bulldozer	6 1/8" long, red or green body, metal control rod, driver, black rubber treads	80	120	200
Ferrari 206 Dino	4 1/8" long, black interior and fins, in either red body with #30 and gold or Whizz Wheels, or yellow body with #23 and gold or Whizz Wheels	24	36	60
Ferrari 308GTS	4 5/8" long, red or black body with working rear hood, black interior with tan seats, movable chrome headlights, detailed engine	14	21	35
Ferrari 308GTS Magnum	4 5/8" long, red body with solid chrome wheels	24	36	60
Ferrari 312 B2 Racing Car	4" long, red body, white fin, gold engine, chrome suspension, mirrors and wheels, Ferrari and #5 decals	16	24	40
Ferrari Berlinetta 250LM	3 3/4" red body with yellow stripe, blue windshields, chrome interior, grill & exhaust pipes, detailed engine, #4 Ferrari logo & yellow stripe decals, spoked wheels & spare, rubber tires	30	45	75

VEHICLES

Corgi

NAME	DESCRIPTION	GOOD	EX	MINT
Ferrari Daytona	5" apple green body, black tow hook, red-yellow-silver black Daytona #5 & other racing decals, amber windows, headlights, black plastic interior, base, four spoke chrome wheels	14	21	35
Ferrari Daytona	4 3/4" long, white body with red roof and trunk, black interior, 2 working doors, amber windows and headlights, #81 and other decals	14	21	35
Ferrari Daytona & Racing Car	blue/yellow Ferrari and Surtees on yellow trailer	12	18	30
Ferrari Daytona JCB	4 3/4" long, orange body with #33, Corgi and other decals, chrome spoked wheels	16	24	40
Ferrari Racing Car	3 5/8" long, red body, chrome plastic engine, roll bar and dash, driver, silver cast base and exhaust, Ferrari and #36 decals	24	36	60
Fiat 1800	3 3/4" long, one-piece body in several colors, clear windows, plastic interior, silver lights, grill and bumpers, red taillights, smooth wheels, rubber tires; colors: blue body with light or bright yellow interior, mustard or cream body	24	36	60
Fiat 2100	3 3/4" long, light mauve body, yellow interior, purple roof, clear windows with rear blind, silver grill, licenese plates & bumpers, red taillights, shaped wheels, rubber tires	22	33	55
Fiat X 1/9 & Powerboat	green/white Fiat, with white/gold boat	30	45	75
Fiat X1/9	4 3/4" metallic blue body & base, white Fiat #3, multicolored lettering & stripe decals, black roof, trim, interior, rear panel, grill, bumpers & tow hook, chrome wheels & detailed enginee	14	21	35
Fiat X1/9	4 1/2" metallic light green or silver body with black roof, trim & interior, 2 working doors, rear panel, grill, tow hook & bumpers, detailed engine, suspension, chrome wheels	14	21	35
Fire Bug		20	30	50
Fire Engine		16	24	40
Flying Club Set	green/white Jeep with Corgi Flying Club decals, green trailer, blue/white airplane	36	54	90
Ford 5000 Super Major Tractor	3 3/4" long, blue body/chassis with Ford Super Major 5000 decals, gray cast fenders and rear wheels, gray plastic front wheels, black plastic tires, driver	30	45	75
Ford 5000 Tractor with Scoop	3 1/8" long, blue body/chassis, gray fenders, yellow scoop arm and controls, chrome scoop, black control lines	52	78	130
Ford Aral Tank Truck		20	30	50
Ford Capri	4" orange-red or dark red body, gold wheels with red hubs, 2 working doors, clear windshield & headlights, black interior, folding seats, black grill, silver bumpers	14	21	35
Ford Capri 3 Litre GT		14	21	35
Ford Capri 30 S		14	21	35
Ford Capri S	4 3/4" white body, red lower body & base, red interior, clear windshield, black bumpers, grill & tow hook, chrome headlights & wheels, red taillights, #6 & other racing decals	14	21	35
Ford Capri Santa Pod Gloworm	4 3/8" long, white/blue body with red/white/blue lettering and flag decals, red chassis, amber windows, gold-based black engine, gold scoop, pipes and front suspension	18	27	45
Ford Car Transporter		20	30	50
Ford Car Transporter		20	30	50
Ford Cobra Mustang		12	18	30
Ford Consul	3 5/8" long, one-piece body in several colors, clear windows, silver grill, lights and bumpers, smooth wheels, rubber tires	45	65	110

Corgi

NAME	DESCRIPTION	GOOD	EX	MINT
Ford Consul Classic	3 3/4" long, cream or gold body & base, yellow interior, pink roof, clear windows, gray steering wheel, silver bumpers, grill	35	55	90
Ford Consul-Mechanical	same as model 200-A but with friction motor and blue or green body	55	85	140
Ford Cortina Estate Car	3 1/2" metallic dark blue body & base, brown & cream simulated wood panels, cream interior, chrome bumpers & grill, jewel headlights	35	55	90
Ford Cortina Estate Car	3 3/4" red body & base or metallic charcoal gray body & base, cream interior, chrome bumpers & grill, jewel headlights	35	55	90
Ford Cortina GXL	4" tan or metallic silver blue body, black roof & stripes, red plastic interior, working doors, clear windshield, folding seats	30	45	75
Ford Cortina Police Car	4" white body, red or pink & black stripe labels, red interior, folding seats, blue dome light, clear windows, chrome bumpers	12	18	30
Ford Covered Semi-Trailer		15	25	40
Ford Escort 13 GL		8	12	20
Ford Escort Police Car	4 3/16" blue body & base, tan interior, white doors, blue dome lights, red Police labels, black grill & bumpers	8	12	20
Ford Esso Tank Truck		15	25	40
Ford Express Semi-Trailer		60	90	150
Ford Exxon Tank Truck		15	25	40
Ford GT 70		10	15	25
Ford Guinness Tanker		20	30	50
Ford Gulf Tank Truck		15	25	40
Ford Holmes Wrecker		60	90	150
Ford Michelin Container Truck		15	25	40
Ford Mustang Fastback		30	45	75
Ford Mustang Fastback		25	35	60
Ford Mustang Mach 1	4 1/4" green upper body, white lower body & base, cream interior, folding seat backs, chrome headlights & rear bumper	25	35	60
Ford Sierra	5" many versions with plastic interior, working hatch & 2 doors, clear windows, folding seat back, lifting hatch cover	8	12	20
Ford Sierra and Caravan Trailer	blue #299 Sierra, two-tone blue/white #490 Caravan	15	20	35
Ford Sierra Taxi		8	12	20
Ford Thames Airborne Caravan	3 3/4" different versions of body & plastic interior with table, white blinds, silver bumpers, grill & headlights, two doors	35	55	95
Ford Thames Wall's Ice Cream Van	4" light blue body, cream pillar, chimes, chrome bumpers & grill, crank at rear to operate chimes, no figures	35	50	85
Ford Thunderbird 1957	5 3/16" cream body, dark brown, black or orange plastic hardtop, black interior, open hood & trunk, chrome bumpers, headlights	10	15	25
Ford Thunderbird 1957	5 1/4" white body, black interior & plastic top, amber windows, white seats, chrome bumpers, headlights & spare wheel cover	10	15	25
Ford Thunderbird Hardtop	4 1/8" long, clear windows, silver lights, grill and bumpers, red taillights, rubber tires; light green body 1959-61	50	80	130
Ford Thunderbird Hardtop-Mechanical	4 1/8" long, same as 214-A but with friction motor and pink or light green body	70	105	175
Ford Thunderbird Roadster	4 1/8" long, clear windshield, silver seats, lights, grill and bumpers, red taillights, rubber tires, white body	50	75	125

VEHICLES

Corgi

NAME	DESCRIPTION	GOOD	EX	MINT
Ford Torino Road Hog	5 3/4" orange-red body, yellow and gray chassis, gold lamps, chrome radiator shell, windows & bumpers, one piece body	15	20	35
Ford Tractor & Conveyor	tractor, conveyor with trailer, figures and accessories	60	90	150
Ford Tractor and Beast Carrier		60	90	150
Ford Tractor with Trencher	5 5/8" long, blue body/chassis, gray fenders, cast yellow trencher arm and controls, chrome trencher, black control lines	50	75	125
Ford Transit Milk Float	5 1/2" white one piece body, blue hood & roof, tan interior, chrome & red roof lights, open compartment door & milk cases	15	25	40
Ford Transit Tipper		10	15	25
Ford Transit Wrecker		25	35	60
Ford Wall's Ice Cream Van	3 1/4" light blue body, dark cream pillars, plastic striped rear canopy, white interior, silver bumpers, grill & headlights	50	75	125
Ford Zephyr Estate Car	3 7/8" light blue one piece body, dark blue hood & stripes, red interior, silver bumpers, grill & headlights, red taillights	30	45	75
Ford Zephyr Patrol Car	3 3/4" white or cream body, blue & white Police/Politie/ Rijkspolitie decals, red interior, blue dome light, silver bumpers	35	50	85
Fordson Power Major Halftrack Tractor	3 1/2" long, blue body/chassis, silver steering wheel, seat and grill, 3 versions: orange cast wheels, gray treads	90	135	225
Fordson Power Major Tractor	3 1/4" long, blue body/chassis with Fordson Power Major decals, silver steering wheel, seat, exhaust, grill and lights	45	65	110
Fordson Power Major Tractor	3 3/8" long, blue body with Fordson Power Major decals, driver, blue chassis and steering wheel, silver seat, hitch, exhaust	50	75	125
Fordson Tractor & Plough	tractor and 4-furrow plow	55	85	140
Fordson Tractor & Plough	#55 Fordson Tractor & 56 Four Furrow Plough	55	85	140
Four Furrow Plough	3 5/8" long, red frame, yellow plastic parts	15	20	35
Four Furrow Plough	3 3/4" long, blue frame with chrome plastic parts	15	20	35
French Construction Set		25	35	60
Futuristic Space Vehicle		12	18	30
Futuristic Space Vehicle		12	18	30
Futuristic Space Vehicle		12	18	30
German Life Saving Set	red/white Land Rover and lifeboat, white trailer, German decals	30	45	75
Ghia L64 Chrysler V8	4 1/4", different color versions, plastic interior, hood, trunk & two doors working, detailed engine, clear windshield	25	40	65
Ghia-Fiat 600 Jolly	3 1/4", light or dark blue body, red & silver canopy, red seats, two figures, windshield, chrome dash, floor, steering wheels	45	65	110
Ghia-Fiat 600 Jolly	3 1/4" long, dark yellow body, red seats, two figures and a dog, clear windshield, silver bumpers and headlights, red taillights	60	90	150
Giant Daktari Set	black/green Land Rover, tan Giraffe truck, blue/brown Dodge Livestock truck, tan figures, set	225	350	600
Giant Tower Crane		35	50	85
Glider Set	2 versions: white Honda, 1981-82; yellow Honda, 1983 on, each set	30	45	75
Golden Eagle Jeep	3 3/4" three different versions, tan plastic top, chrome plastic base, bumpers & steps, chrome wheels	8	12	20
Golden Guinea Set	3 vehicle set, gold plated Bentley Continental, Chevy Corvair and Ford Consul	90	135	225
GP Beach Buggy		15	20	35

VEHICLES

Corgi

NAME	DESCRIPTION	GOOD	EX	MINT
Grand Prix Racing Set	4 vehicle set with accessories in 2 versions: with #330 Porsche, 1969; with Porsche #371, 1970-72, each set	135	210	350
Grand Prix Set	sold by mail order only	50	75	125
Green Hornet's Black Beauty	5" black body, green window/interior, 2 figures, working chrome grill & panels with weapons, green headlights, red taillights	175	275	450
Green Line Bus	4 7/8" green body, white interior & stripe, TDK lables, six spoked wheels	10	15	25
Half Track Rocket Launcher & Trailer	6 1/2" two rocket launchers & single trailer castings, gray plastic roll cage, man with machine gun, front wheels & hubs	20	35	55
Hardy Boys' Rolls-Royce	4 5/8" long, red body with yellow hood, roof and window frames, band figures on roof on removable green base	70	105	175
HDL Hovercraft SR-N1	blue superstructure, gray base and deck, clear canopy, red seats, yellow SR-N1 decals	60	90	150
Hesketh-Ford Racing Car	5 5/8" long, white body with red/white/blue Hesketh, stripe and #24 decals, chrome suspension, roll bar, mirrors and pipes	12	18	30
HGB-Angus Firestreak	6 1/4" long, chrome plastic spotlight and ladders, black hose reel, red dome light, white water cannon, in 2 interior versions	35	50	85
Hillman Hunter	4 1/4" blue body, gray interior, black hood, white roof, unpainted spotlights, clear windshield, red radiator screen, black equipment	45	65	110
Hillman Husky	3 1/2" long, one-piece tan or metallic blue/silver body, clear windows, silver lights, grill and bumpers, smooth wheels	40	60	100
Hillman Husky-Mechanical	3 1/2" long, same as 206-A but with friction motor, black base and dark blue, gray or cream body	50	75	125
Hillman Imp	3 1/4" metallic copper, blue, dark blue, gold & maroon one-piece bodies, with white/yellow interior, silver bumpers, headlights	30	45	75
Hillman Imp Rally	3 1/4" long, in various metallic body colors, with cream interior, Monte Carlo Rally and #107 decals	30	45	75
Honda Ballade Driving School	4 3/4" red body/base, tan interior, clear windows, tow hook, mirrors, bumpers	10	15	25
Honda Prelude	4 3/4" long, dark metallic blue body, tan interior, clear windows, folding seats, sunroof, chrome wheels	8	12	20
Hughes Police Helicopter	5 1/2" long, red interior, dark blue rotors, in several international imprints, Netherlands, German, Swiss, in white or yellow	20	30	50
Hyster 800 Stacatruck	8 1/2" long, clear windows, black interior with driver	35	50	85
Incredible Hulk Mazda Pickup	5" metallic light brown body, gray or red plastic cage, black interior, green & red figure, Hulk decal on hood, chrome wheels	20	30	50
Inter-City Mini Bus	4 3/16" long, orange body with brown interior, clear windows, green/yellow/black decals, Whizz Wheels	8	12	20
International 6x6 Army Truck	5 1/2" long, olive drab body with clear windows, red/blue decals, 6 cast olive wheels with rubber tires	70	105	175
Iso Grifo 7 Litre	4" metallic blue body, light blue interior, black hood & stripe, clear windshield, black dash, folding seats, chrome bumpers	12	18	30
Jaguar 1952 XK120 Rally	4 3/4" long, cream body with black top and trim, red interior, Rally des Alps and #414 decals	8	12	20
Jaguar 2.4 Litre	3 7/8" long, one-piece white body with no interior 1957-59, or yellow body with red interior 1960-63, clear windows	50	80	130
Jaguar 2.4 Litre Fire Chief's Car	3 3/4" long, red body with unpainted roof signal/siren, red/white Fire and shield decals on doors, in 2 versions	60	90	150
Jaguar 2.4 Litre-Mechanical	3 7/8" long, same as 208-A but with friction motor and metallic blue body	60	90	150

VEHICLES

Corgi

NAME	DESCRIPTION	GOOD	EX	MINT
Jaguar E Type	3 3/4" maroon or metallic dark gray body, tan interior, red & clear plastic removeable hardtop, clear windshield, folded top	45	65	110
Jaguar E Type 2+2	4 3/16" long, working hood, doors and hatch, black interior with folding seats, copper engine, pipes and suspension, spoked wheels	40	60	100
Jaguar E Type 2+2	4 1/8" long, in 5 versions	35	55	90
Jaguar E Type Competition	3 3/4" gold or chrome plated body, black interior, blue & white stripes & black #2 decals, no top, clear windshield & headlights	45	65	110
Jaguar Mark X Saloon	4 1/4", seven different versions with working front & rear hood castings, clear windshields, plastic interior, gray steering wheel	35	55	90
Jaguar XJ12C	5 1/4" five different metallic versions, working hood & 2 doors, clear windows, tow hook, chrome bumpers, grill & headlights	10	15	25
Jaguar XJ12C Police Car	5 1/8" long, white body with blue and pink stripes, light bar with blue dome light, tan interior, Police decals	12	18	30
Jaguar XJS	5 3/4" long, metallic burgundy body, tan interior, clear windows, working doors, spoked chrome wheels	10	15	25
Jaguar XJS Motul	5 3/4" long, black body with red/white Motul and #4, chrome wheels	8	12	20
Jaguar XJS-HE Supercat	5 1/4" black body with silver stripes & trim, red interior, dark red taillights, light gray antenna, no tow hook, clear windshield	8	12	20
Jaguar XK120 Hardtop	4 3/4" long, red body, black hardtop, working hood and trunk, detailed engine, cream interior, clear windows, chrome wheels	8	12	20
James Bond Aston Martin	3 3/4" metallic gold body, red interior, working roof hatch, clear windows, 2 figures, left seat ejects	70	105	175
James Bond Aston Martin	4" metallic silver body, red interior, 2 figures, working roof hatch, ejector seat, bullet shield and guns, chrome bumpers	100	150	250
James Bond Aston Martin	5" metallic silver body & die cast base, red interior, 2 figures, clear windows, passenger set raises to eject	30	45	75
James Bond Bobsled	2 7/8" long, yellow body, silver base, Bond figure, 007 decals, Whizz Wheels	60	90	150
James Bond Citroen 2CV6	4 1/4" dark yellow body & hood, red interior, clear windows, chrome headlights, red taillights, black plastic grill	15	25	40
James Bond Lotus Esprit	4 3/4" white body & base, black windshield, grill & hood panel, white plastic roof device that triggers fins & tail, rockets	30	45	75
James Bond Moon Buggy	4 3/8" long, white body with blue chassis, amber canopy, yellow tanks, red radar dish, arms and jaws, yellow wheels	175	275	450
James Bond Mustang Mach 1	4 3/8" long, red and white body with black hood	100	150	250
James Bond Set	set of 3; Lotus Esprit, Space Shuttle and Aston Martin	80	120	200
James Bond Space Shuttle	5 7/8" long, white body with yellow/black Moonraker decals	30	45	75
James Bond SPECTRE Bobsled	2 7/8" long, orange body with wild boar decals	60	90	150
James Bond Toyota 2000GT	4" long, white body, black interior with Bond and passenger, working trunk and gun rack, spoked wheels, plastic tires	115	180	300
JCB 110B Crawler Loader	6 1/2" long, white cab, yellow body, working red shovel, red interior with driver, clear windows, black treads, JCB decals	20	30	50
Jean Richard Circus Set	yellow/red Land Rover and cage trailer with Pinder-Jean Richard decals, office van and trailer, Human Cannonball truck, ring	90	135	225
Jeep & Horse Box	metallic painted Jeep and trailer	15	25	40
Jeep & Motorcycle Trailer	red working Jeep with 2 blue/yellow bikes on trailer	15	20	35

Corgi

NAME	DESCRIPTION	GOOD	EX	MINT
Jeep CJ-5	4 " long, dark metallic green body, removable white top, white plastic wheels, spare tire	8	12	20
Jeep FC-150 Covered Truck	issued in 4 versions: blue body, rubber tires (1965-67), yellow/brown body, rubber tires (1965-67), blue body, plastic tires	30	45	75
Jeep FC-150 Pickup	3 1/2" long, blue body, clear windows, sheet metal tow hook, in 2 wheel versions: smooth or shaped wheels	35	55	90
Jeep FC-150 Pickup with Conveyor Belt	7 1/2" long, red body, yellow interior, orange grill, 2 rubber belts, shaped wheels, black rubber tires	45	65	110
Jeep FC-150 Tower Wagon	4 5/8" long, metallic green body, yellow interior and basket with workman figure, clear windows, with either rubber or plastic wheels	40	60	100
Jet Ranger Police Helicopter	5 7/8" long, white body with chrome interior, red floats and rotors, amber windows, Police decals	25	40	65
JPS Lotus Racing Car	10 1/2" long, black body, scoop and wings with gold John Player Special, Texaco and #1 decals, gold suspension, pipes and wheels	30	45	75
Karrier Bantam Two Ton Van	4" long, blue body, red chassis and bed, clear windows, smooth wheels, rubber tires	35	55	95
Karrier Butcher Shop	3 5/8" long, white body, blue roof, butcher shop interior, Home Service decals, in 2 versions: with or without suspension	65	100	165
Karrier Circus Booking Office	3 5/8" long, red body, light blue roof, clear windows, circus decals, shaped wheels, rubber tires	105	165	275
Karrier Dairy Van	4 1/8" long, light blue body with Drive Safely on Milk decals, white roof, with either smooth or shaped wheels	50	75	125
Karrier Field Kitchen	3 5/8" long, olive body, white decals	60	90	150
Karrier Ice Cream Truck	3 5/8" long, cream upper, blue lower body and interior, clear windows, sliding side windows, Mister Softee decals	90	135	225
Karrier Lucozade Van	4 1/8" long, yellow body with gray rear door, Lucozade decals, rubber tires, with either smooth or shaped wheels	70	105	175
Karrier Mobile Canteen	3 5/8" long, blue body, white interior, amber windows, roof knob rotates figure, working side panel counter	60	90	150
Karrier Mobile Grocery	3 5/8" long, light green body, grocery store interior, red/white Home Service decals, friction motor, rubber tires	70	110	185
King Tiger Heavy Tank	6 1/8" long, tan and rust body, working turret and barrel, tan rollers and treads, German decals	30	45	75
Kojak's Buick Regal	5 3/4" metallic bronze brown body, off-white interior, 2 doors, clear windows, chrome bumpers, grill & headlights, red taillights	25	40	65
Lamborghini Miura	3 3/4" long, silver body, black interior, yellow/purple stripes and #7 decal	30	45	75
Lamborghini Miura P400	3 3/4" long, with red or yellow body, working hood, detailed engine, clear windows, jewel headlights, bull, Whizz Wheels	40	60	100
Lancia Fulvia Zagato	3 5/8" long, metallic blue body, light blue interior, working hood and doors, folding seats, amber lights, cast wheels	25	35	60
Lancia Fulvia Zagato	3 5/8' long, orange body, black working hood and interior, Whizz Wheels	15	25	40
Land Rover & Ferrari Racer	red/tan Land Rover, yellow trailer	60	90	150
Land Rover & Horse Box	blue/white Land Rover with horse trailer in 2 versions: cast wheels, 1968-74; Whizz Wheels, 1975-77	50	75	125
Land Rover 109WB	5 1/4" long, working rear doors, tan interior, spare on hood, plastic tow hook	12	18	30
Land Rover and Pony Trailer	2 versions: tan/cream Rover, 1958-62; tan/cream #438 Land Rover, 1963-68, each set	50	75	125

VEHICLES

Top to bottom: James Bond's Aston Martin D.B. 5; The Green Hornet's Black Beauty Crime Fighting Car with figures; The Saint's Volvo P.1800, all Corgi.

Corgi

NAME	DESCRIPTION	GOOD	EX	MINT
Land Rover Breakdown Truck	4 3/8" long, red body with silver boom and yellow canopy, revolving spotlight, Breakdown Service decals	35	55	90
Land Rover Breakdown Truck	4 3/8" long, red body, yellow canopy, chrome revolving spotlight, Breakdown Service decals	25	35	60
Land Rover Circus Vehicle	3 1/2" long, red body, yellow interior, blue rear and speakers, revolving clown, chimp figures, Chipperfield decals	60	90	150
Land Rover Pickup	3 3/4" long, yellow or metallic blue body, spare on hood, clear windows, sheet metal tow hook, rubber tires	45	70	120
Land Rover with Canopy	3 3/4" long, one-piece body with clear windows, plastic interior, spare on hood, issued in numerous colors	35	55	90
Lincoln Continental	5 3/4" metallic gold or blue body, black roof, maroon plastic interior, working hood, trunk & doors, clear windows, TV	60	90	150
Lions of Longleat	black/white Land Rover pickup with lion cages and accessories, 2 versions: cast wheels, 1969-73; Whizz Wheels, 1974, each	60	90	150
London Set	orange Mini, Policeman, London Taxi and Routemaster bus	50	75	125
London Set	London Taxi and Routemaster bus in 2 versions: with mounted Policeman, 1980-81; without Policeman, 1982 on, each set	25	35	60
London Set	taxi and bus with policeman, in 2 versions: "Corgi Toys" on bus, 1964-66; "Outspan Oranges" on bus, 1967-68, each set	55	85	140
London Transport Routemaster Bus	4 1/2" long, clear windows with driver and conductor, released with numerous advertiser logos: Outspan, Corgi Toys	35	50	85
London Transport Routemaster Bus	4 7/8" long, clear windows, interior, some models have driver and conductor, others don't; released in numerous advertiser logos	25	35	60
Lotus Elan S2 Hardtop	2 1/4" long, cream interior with folding seats and tan dash, working hood, separate chrome chassis, issued in blue body	30	45	75
Lotus Elan S2 Roadster	3 3/8" long, working hood, plastic interior with folding seats, shaped wheels and rubber tires, issued in metallic blue or white	30	45	75
Lotus Eleven	2 1/4" long, clear windshield and plastic headlights, smooth wheels, rubber tires, racing decals, in several color variations	60	95	160
Lotus Elite	5 1/8" red body, white interior, 2 working doors, clear windshield, black dash, hood panel, grill, bumpers, base & tow hook	12	18	30
Lotus Elite 22	4 3/4" long, dark blue body with silver trim	12	18	30
Lotus Racing Car	5 5/8" long, black body and base, gold cast engine, roll bar, pipes, dash and mirrors, driver, gold cast wheels, in 2 versions	25	35	60
Lotus Racing Set	3 versions: #3 on Elite and JPS on racer; #7 on Elite and JPS on racer; #7 on ELite and Texaco on racer, each set	30	45	75
Lotus Racing Team Set	4 vehicle set in 2 versions, with accessories: #319 Lotus has either red or black interior, each set	30	45	75
Lotus-Climax Racing Car	3 5/8" long, green body and base with black/white #1 and yellow racing stripe decals, unpainted engine and suspension	25	35	60
Lotus-Climax Racing Car	3 5/8" long, orange/white body with black/white stripe and #8 decals, unpainted cast rear wing, cast 8-spoke wheels	15	25	40
Lunar Bug	5" long, white body with red roof, blue interior and wings, clear and amber windows, red working ramp, Lunar Bug decals	25	40	65

VEHICLES

Corgi

NAME	DESCRIPTION	GOOD	EX	MINT
M60 A1 Medium Tank	4 3/4" long, green/tan camo body, working turret and barrel, green rollers, white decals	30	45	75
Mack Container Truck	11 3/8" long, yellow cab, red interior, white engine, red suspension, white ACL decals	30	50	80
Mack Esso Tank Truck	10 3/4" long, white cab and tank with Esso decals, red tank chassis and fenders	20	30	50
Mack Exxon Tank Truck	10 3/4" long, white cab and tank, red tank chassis and fenders, red interior, chrome catwalk, Exxon decals	15	25	40
Mack Trans Continental Semi	10", orange cab body & semi chassis & fenders, metallic light blue semi body, unpainted trailer rests	35	55	90
Mack-Priestman Crane Truck	9" long, red truck, yellow crane cab, red interior, black engine, Hi Lift and Long Vehicle decals	50	75	125
Magic Roundabout Train		70	105	175
Man From U.N.C.L.E. Olds	4 1/8" long, plastic interior, blue windows, two figures, 2 spotlights, issued in cream body or dark metallic blue	80	120	200
Marcos 3 Litre	3 3/8" long, working hood, detailed engine, black interior, Marcos decal, Whizz Wheels, issued in orange or metallic blue/green	20	30	50
Marcos Matis	4 1/4" metallic red body & doors, cream interior & headlights, silver gray lower body base, bumpers, hood panel	20	35	55
Marcos Volvo 1800 GT	3 5/8" long, issued with either white or blue body, plastic interior with driver, spoked wheels, rubber tires	25	35	60
Massey Ferguson 165 Tractor	3" long, gray engine and chassis, red hood and fenders with black/white Massey Ferguson 165 decals, white grill, red cast wheels	35	55	90
Massey Ferguson 165 Tractor with Saw	3 1/2" long, red hood and fenders, gray engine and seat, cast yellow arm and control, chrome circular saw	55	85	140
Massey Ferguson 165 Tractor with Shovel	5 1/8" long, gray chassis, red hood, fenders and shovel arms, unpainted shovel and cylinder, red cast wheels, black plastic tires	45	65	110
Massey Ferguson 50B Tractor	4" long, yellow body, black interior and roof, red plastic wheels with black plastic tires	12	18	30
Massey Ferguson 65 Tractor	3" long, silver steering wheel, seat and grill, red engine hood, red wheels with black rubber tires	40	60	100
Massey Ferguson 65 Tractor And Shovel	4 3/4" long, 2 versions: either cream or gray chassis, each	55	85	140
Massey Ferguson Combine	6 1/2" long, red body with yellow metal blades, metal tines, black/white decals, orange wheels	70	105	175
Massey Ferguson Combine	6 1/2" long, red body, plastic blades, red wheels	60	90	150
Massey Ferguson Tipping Trailer	3 5/8" long, 2 versions: either yellow or gray tipper and tailgate, each	10	15	25
Massey Ferguson Tractor & Tipping Trailer		50	75	125
Massey Ferguson Tractor & Tipping Trailer	#50 MF tractor with driver, #51 trailer	50	75	125
Massey Ferguson Tractor with Fork	4 7/8" long, red cast body and shovel, arms, cream chassis, red plastic wheels, black rubber tires, Massey Ferguson 65 decals	60	90	150
Massey Ferguson Tractor with Shovel	6" long, 2 versions: either yellow and red or red and white body colors	20	30	50
Massey Ferguson Tractor with Shovel & Trailer	#35 MF tractor with driver and shovel, #62 trailer	30	45	75
Matra & Motorcycle Trailer	red Rancho with 2 yellow/blue bikes on trailer	15	20	35
Matra & Racing Car	black/yellow Rancho and yellow car with Team Corgi decals	15	25	40
Mazda 4X4 Open Truck	4 7/8" long, blue body, white roof, black windows, no interior, white plastic wheels	15	20	35

Corgi

NAME	DESCRIPTION	GOOD	EX	MINT
Mazda B-1600 Pickup Truck	4 7/8" long, issued in either blue and white or blue and silver bodies with working tailgate, black interior, chrome wheels	15	20	35
Mazda Camper Pickup	5 3/8" long, red truck and white camper with red interior and folding supports	15	25	40
Mazda Custom Pickup	4 7/8" long, orange body with red roof	12	18	30
Mazda Motorway Maintenance Truck	6 1/8" long, deep yellow body with red base, black interior and hydraulic cylinder, yellow basket with workman figure	18	25	45
Mazda Pickup & Dinghy	2 versions: red Mazda with "Ford' decals; or with "Sea Spray" decals	25	35	60
McLaren M19A Racing Car	4 5/8" long, white body, orange stripes, chrome engine, exhaust and suspension, black mirrors, driver, Yardley McLaren #55 decals	15	25	40
McLaren M23 Racing Car	10 1/4" long, red/white body and wings with red/white/black Texaco-Marlboro #5 decals, chrome pipes, suspension and mirrors,	30	45	75
Mercedes-Benx 240D	5 1/4" three different versions, working trunk, 2 doors, clear windows, plastic interior, two hook, chrome bumpers, grill & headlights	10	15	25
Mercedes-Benz & Caracan	truck and trailer in 2 versions: with blue Mercedes truck, 1975-79; with brown Mercedes, 1980-81	15	25	40
Mercedes-Benz 220SE Coupe	3 3/4" long, cream, black or dark red body, red plastic interior, clear windows, working trunk, silver bumpers, grill & plate	40	60	100
Mercedes-Benz 220SE Coupe	4" metallic maroon body, yellow plastic interior, light gray base, clear windows, silver bumpers, headlights, grill & license	40	60	100
Mercedes-Benz 220SE Coupe	4" metallic dark blue body, cream plastic interior, medium gray base, clear windows, silver bumpers, headlights, grill & license	40	60	100
Mercedes-Benz 240D Rally	5 1/8" cream or tan body, black, red & blue lettering & dirt, red plastic interior, clear windows, black radiator guard & roof	10	15	25
Mercedes-Benz 240D Taxi	5" orange body, orange interior, black roof sign with red and white Taxi labels, black on door	12	18	30
Mercedes-Benz 300SC Convertible	5" black body, black folded top, white interior, folding seat backs, detailed engine, chrome grill & wheels, lights, bumpers	8	12	20
Mercedes-Benz 300SC Hardtop	5" maroon body, tan top & interior, open hood & trunk, clear windows, folding seat backs, top with chrome side irons	8	12	20
Mercedes-Benz 300SL	5" redy body & base, tan interior, open hood & two gullwing doors, black dash, detailed engine, clear windows, chrome bumpers	8	12	20
Mercedes-Benz 300SL	4 3/4" silver body, tan interior, black dash, clear windows, open hood & two gullwing doors, detailed engine, chrome bumpers,	8	12	20
Mercedes-Benz 300SL Coupe	3 3/4" different body & interior versions, hardtop, clear windows, '59-60 smooth wheels no suspension, '61-65 racing stripes	45	65	110
Mercedes-Benz 300SL Roadster	3 3/4" different body & interior versions, plastic interior, smooth, shaped or cast wheels, racing stripes & number, driver	45	65	110
Mercedes-Benz 350SL	3 3/4" white body, spoke wheels or metallic dark blue body solid wheels, pale blue interior, folding seats, detailed engine	15	25	40
Mercedes-Benz 600 Pullman	4 1/4" metallic maroon body, cream interior & steering wheel, clear windshields, chrome grill, trim & bumpers	40	60	100
Mercedes-Benz Ambulance	5 3/4" three different versions, white interior, open rear & two doors, blue windows & dome lights, chrome bumpers, grill & headlights	15	20	35

VEHICLES

237

Corgi

NAME	DESCRIPTION	GOOD	EX	MINT
Mercedes-Benz Ambulance	5 3/4" white body & base, red stripes & taillights, Red Cross & black & white Ambulance labels, open rear door, white interior	15	20	35
Mercedes-Benz C-111	4" orange main body with black lower & base, black interior, vents, front & rear grilles, silver headlights, red taillights	12	18	30
Mercedes-Benz Fire Chief	5" light red body, black base, tan plastic interior, blue dome light, white Notruf 112 decals, red taillights, no tow hook	15	25	40
Mercedes-Benz Police Car	5" white body with two different hood versions, brown interior, Polizei or Police lettering, blue dome light	12	18	30
Mercedes-Benz Refrigerator		12	18	30
Mercedes-Benz Refrigerator	8" yellow cab & tailgate, red semi-trailer, two piece lowering tailgate & yellow spare wheel base, red interior, clear windows	12	18	30
Mercedes-Benz Semi-Trailer		12	18	30
Mercedes-Benz Semi-Trailer		12	18	30
Mercedes-Benz Semi-Trailer Van	8 1/4" black cab & plastic semi trailer, white chassis & airscreen, red doors, red-blue & yellow stripes, white Corgi lettering	12	18	30
Mercedes-Benz Tanker	7 1/4" tan cab, plastic tank body, black chassis, black & red Guinness labels, with chrome or black plastic catwalk, clear windows	12	18	30
Mercedes-Benz Tanker	7 1/4" two different versions, cab & tank, chassis, chrome or black plastic catwalk, red-white-green 7-up labels	12	18	30
Mercedes-Benz Unimog & Dumper	6 3/4" yellow cab & tipper, red fenders & tipper chassis, charcoal gray cab chassis, black. plastic mirrors or without	25	35	60
Mercedes-Benz Unimog 406	3 3/4" yellow body, red front fenders & bumpers, metallic charcoal gray chassis with olive or tan rear plastic covers, red interior	18	25	45
Mercedes-Faun Street Sweeper	5" orange body with light orange or brown figure, red interior, black chassis & unpainted brushing housing & arm castings	15	25	40
Metropolis Police Car	6" metallic blue body, off white interior, white roof/stripes, 2 working doors, clear windows, chrome bumpers, grill & headlights	20	30	50
MG Maestro	4 1/2" yellow body, black trim, opaque black windows, black plastic grill, bumpers, spoiler, trim & battery hatch, clear headlights	15	20	35
MGA	3 3/4" metallic light brown body, all white interior, black dash, clear windshield, silver bumpers, grill & headlight decals	60	90	150
MGB GT	3 1/2" dark red body, pale blue interior, open hatch & two doors, jewel headlights, chrome grill & bumpers, orange taillights	50	75	125
MGC GT	3 1/2" bright yellow body & base, black interior, hood & hatch, folding seats, luggage, jewel headlights, red taillights	50	75	125
MGC GT	3 1/2" red body, black hood & base, black interior, open hatch & two doors, folding seat backs, luggage, orange taillights	50	75	125
Midland Red Express Coach	5 1/2" red one piece body, black roof with shape or smooth wheels, yellow interior, clear windows, silver grill & headlights	70	105	175
Military Set	set of 3, Tiger tank, Bell Helicopter, Saladin Armored Car	60	90	150
Milk Truck & Trailer	blue/white milk truck with trailer	60	90	150
Mini Camping Set	cream Mini, with red/blue tent, grill and figures	25	40	65

Corgi

NAME	DESCRIPTION	GOOD	EX	MINT
Mini-Marcos GT850	2 1/8" white body, red-white-blue racing stripe & #7 labels, clear headlights, Whizz Wheels	20	30	50
Mini-Marcos GT850	3 1/4" metallic maroon body, white name & trim decals, cream interior, open hood & doors, clear windows & headlights	30	45	75
Minissima	2 1/4" cream upper body & dr., metallic lime green lower body with black stripe centered, black interior, clear windows, headlights	15	20	35
Monkeemobile	4 3/4" red body/base, white roof, yellow interior, clear windows, 4 figures, chrome grill, headlights, engine, orange taillights	145	225	375
Monte Carlo Rally Set	3 vehicle set, Citroen, Mini and Rover rally cars	295	450	750
Morris Cowley	3 1/8" long, one-piece body in several colors, clear windows, silver lights, grill and bumper, smooth wheels, rubber tires	45	65	110
Morris Cowley-Mechanical	3 1/8" long, same as 202-A but with friction motor, available in off-white or green body	55	85	140
Morris Marina	3 3/4" metallic dark red or lime green body, cream interior, working hood & two doors, clear windshield, chrome grill & bumper	15	25	40
Morris Mini-Cooper	2 3/4" yellow or blue body either body & base and/or hood, white roof and/or hood, 2 versions, red plastic interior, jewel headlights	25	40	70
Morris Mini-Cooper	2 7/8" red body & base, white roof, yellow interior, chrome spotlight, No. 37 & Monte Carlo Rally decals	55	85	140
Morris Mini-Cooper Deluxe	2 3/4" black body/base, red roof, yellow & black wicker work decals on sides & rear, yellow interior, gray steering wheel, jewel headlights	45	65	110
Morris Mini-Minor	2 7/8" long, one-piece body in several colors, plastic interior, silver lights, grill and bumpers, red taillights, Whizz Wheels	30	45	75
Morris Mini-Minor	2 3/4" three to four different versions with shaped and/or smooth wheels, plastic interior, silver bumpers, grill & headlights	40	60	100
Motorway Ambulance	4" white body, dark blue interior, red-white-black labels, dark blue windows, clear headlights, red die cast base & bumpers	10	15	25
Mr. McHenry's Trike		70	105	175
Muppet Vehicles		12	18	30
Musical Carousel		275	425	700
Mustang Organ Grinder Dragster	4" long, yellow body with green/yellow name, #39 and racing stripe decals, black base, green windshield, red interior, roll bar	20	30	50
NASA Space Shuttle	6" white body, two open hatches, black plastic interior, jets & base, unpainted retracting gear castings, black plastic wheels	30	45	75
National Express Bus		8	12	20
Noddy's Car	3 3/4" yellow body, red fenders & base, Chubby, Golliwog & Noddy figures, chrome bumpers & grill castings, black grill	107	165	275
Noddy's Car	3 3/4" yellow body, red chassis, Chubby inside 1970, 3 1/2" 1975-1977 Noddy alone, closed trunk with spare tire	60	90	150
NSU Sport Prinz	3 1/4" metallic burgundy or maroon body, yellow interior, one piece body, silver bumpers, headlights & trim, shaped wheels	30	45	75
Off Road Set	#5 decal on Jeep, blue boat	15	20	35
Olds Toronado & Speedboat	blue Tornado, blue/yellow boat with swordfish decals	60	90	150
Oldsmobile 88 Staff Car	4 1/4" drab olive body, four figures, white decals	50	75	125
Oldsmobile Sheriff's Car	4 1/4" long, black upper body with white sides, red interior with red dome light & County Sheriff decals on doors, single body casting	50	75	125

VEHICLES

Corgi

NAME	DESCRIPTION	GOOD	EX	MINT
Oldsmobile Super 88	4 1/4" long, 3 versions: light blue, light or dark metallic blue body with white stripes, red interior, single body casting	40	60	100
Oldsmobile Toronado	4 1/8" metallic peacock blue body, cream interior, one piece body, clear windshield, chrome bumpers, grill, headlight covers	35	55	90
Oldsmobile Toronado	4 3/16" metallic copper red one piece body, cream interior, Golden jacks, gray tow hook, clear windows, bumpers, grill, headlights	35	55	90
Opel Senator Doctor's Car		10	15	25
Open Top Disneyland Bus	4 3/4" yellow body, red interior & stripe, Disneyland labels, eight spoked wheels or orange body, white interior & stripe	30	50	80
OSI DAF City Car	2 3/4" orange/red body, light cream interior, textured black roof, sliding left dr., working hood, hatch & 2 right drs.	18	25	45
Penguinmobile	3 3/4" black & white lettering on orange-yellow-blue decals, gold body panels, seats, air scoop, chrome engine	20	30	50
Pennyburn Workmen's Trailer	3 1/8" long, blue body w/working lids, red plastic interior, chrome tools, red cast wheels, plastic tires	15	20	35
Peugeot 505 STI	4 7/8" cream body & base, red interior, blue-red-white Taxi labels, black grill, bumpers, tow hook, chrome headlights & wheels	8	12	20
Peugeot 505 Taxi	4 7/8" long, cream body, red interior, red/white/blue taxi decals	8	12	20
Platform Trailer	4 3/8" long, in 5 versions	10	15	25
Playground		295	450	750
Plymouth Sports Suburban	4 1/4" long, dark cream body, tan roof, red interior, die cast base, red axle, silver bumpers, trim and grill and rubber tires	40	60	100
Plymouth Sports Suburban	4 1/4" pale blue body with silver trim, red roof, yellow interior, gray diecast base without rear axle bulge, shaped wheels	40	60	100
Plymouth Suburban Mail Car	4 1/4" white upper, blue lower body with red stripes, gray die cast base without rear axle bulge, silver bumpers & grill	55	85	140
Police Land Rover	5" white body, red & blue Police stripes, black lettering, open rear door, opaque black windows, blue dome light, roof light	15	25	40
Police Land Rover & Horse Box	white Land Rover with Police decals and mounted Policeman	30	45	75
Police Vigilant Range Rover	4" white body, red interior, black shutters, blue dome light, 2 chrome & amber spotlights, black grill, silver headlights	25	35	60
Pontiac Firebird	4" metallic silver body & base, red interior, black hood, stripes & convertible top, doors open, clear windows, folding seats	50	75	125
Pony Club Set	brown/white Land Rover with Corgi Pony Club decals, horse box, horse and rider	30	45	75
Pop Art Mini-Motest	2 3/4" light red body & base, yellow interior, jewel headlights, orange taillights, yellow-blue-purple pop art & "Motest" decals	100	150	250
Popeye's Paddle Wagon	4 7/8" yellow body, red chassis, blue rear fenders, bronze & yellow stacks, white plastic deck, blue lifeboat with Swee' Pea	195	300	500
Popeye's Paddle Wagon		70	105	175
Porsche 917	4 1/4" red or blue body, black or gray base, blue or amber tinted windows & headlights, open rear hood, headlights	15	20	35
Porsche 92 Turbo	4 1/2" black body with gold trim, yellow interior, 4 chrome headlights, clear windshield, taillight-license plate decal, blac	15	20	35

VEHICLES

Corgi

NAME	DESCRIPTION	GOOD	EX	MINT
Porsche 924	4 1/2" bright orange body, dark red interior, black plastic grill, multicolored stripes, swivel roof spotlight	10	15	25
Porsche 924	4 7/8" red or metallic light brown body, dark red interior, two doors open & rear window, chrome headlights, black plastic grill	10	15	25
Porsche 924 Police Car	4 1/4" white body with different hood & doors versions, blue & chrome light, Polizei white on green panels or Police labels	15	25	40
Porsche Carrera 6	3 7/8" white body, red or blue trim, blue or amber tinted engine covers, black interior, clear windshield & canopy, red jewel taillights	30	45	75
Porsche Carrera 6	3 3/4" white upper body, red front hood, doors, upper fins & base, black interior, purple rear window, tinted engine cover	25	35	60
Porsche Targa 9111S	3 1/2" three different versions, black roof with or without stripe, orange interior, open hood & two doors, chrome engine & bumpers	25	35	60
Porsche Targa Police Car	3 1/2" white body & base, red doors & hood, black roof & plastic interior also comes with an orange interior, unpainted siren	25	35	60
Porsche-Audi 917	4 3/4" white body, red & black no. 6, L & M, Porsche Audi & stripe labels or orange body, orange two tone green white no. 6,	15	20	35
Powerboat Team	white/red Jaguar with red/white boat on silver trailer, Team Corgi Carlsberg, Union Jack and #1 decals on boat	25	35	60
Priestman Cub Crane	9" orange body, red chassis & two piece bucket, unpainted bucket arms, lower boom, knobs, gears & drum castings, clear window	50	75	125
Priestman Cub Power Shovel	6" orange upper body & panel, yellow lower body, lock rod & chassis, rubber or plastic treads, pulley panel, gray boom	40	60	100
Priestman Shovel & Carrier	cub shovel & low loader machinery carrier	90	135	225
Psychedelic Ford Mustang	3 3/4" light blue body & base, aqua interior, red-orange-yellow No. 20 & flower decals, cast eight spoke wheels, plastic tires	30	45	75
Public Address Land Rover	4" green body, yellow plastic rear body & loudspeakers, red interior, clear windows, silver bumper, grill & headlights	50	75	125
Quartermaster Dragster	5 3/4" long, dark metallic green upper body with green/yellow/black #5 and Quartermaster decals, light green lower body	30	45	75
RAC Land Rover	3 3/4" three versions of body, plastic interior & rear cover, RAC & Radio Rescue decals	60	90	150
Radio Luxembourg Dragster	5 3/4" long, blue body with yellow/white/blue John Wolfe Racing, Radio Luxembourg and #5 decals, silver engine	30	45	75
Radio Roadshow Van	4 3/4" white body, red plastic roof & rear interior, opaque black windows, red-white-black Radio Tele Luxemburg labels, gray	25	35	60
RAF Land Rover	3 3/4" blue body & cover, one piece body, sheet metal rear cover, RAF rondel decal, with or without suspension, silver bumper	60	90	150
RAF Land Rover & Bloodhound	set of three standard colored, Massey Ferguson Tractor, Bloodhound Missile, Ramp & Trolley	150	240	400
RAF Land Rover & Thunderbird	Standard colors, #350 Thunderbird Missile on Trolley & 351 R.A.F. Land Rover	100	150	250
Rambler Marlin Fastback	4 1/8" red or blue body, black roof & trim, cream interior, clear windshield, folding seats, chrome bumpers, grill & headlights	35	55	90
Rambler Marlin with Kayak & Trailer	blue Marlin with roof rack, blue/white trailer	100	150	250

VEHICLES

Corgi

NAME	DESCRIPTION	GOOD	EX	MINT
Range Rover Ambulance	4" two different versions of body sides, red interior, raised roof, open upper & lower doors, black shutters, blue dome light	20	30	50
Raygo Rascal Roller	4 7/8" dark yellow body, base & mounting, green interior & engine, orange & silver roller mounting & castings, clear windshield	15	25	40
Red Wheelie Motorcycle	4" long, red plastic body and fender with black/white/yellow decals, black handlebars, kickstand and seat, chrome engine, pipes	10	15	25
Reliant Bond Rug 700 E.S.	2 1/2" bright orange or lime green body, off white seats, black trim, silver headlights, red taillights	15	25	40
Renault 11 GTL	4 1/4" light tan body & base, red interior, open doors & rear hatch, lifting hatch cover, folding seats, gril	15	25	40
Renault 16	3 3/4" metallic maroon body, dark yellow interior, chrome base, grill & bumpers, clear windows, hatch cover/Renault decal	25	35	60
Renault 16TS	3 7/8" long, metallic blue body with Renault decal on working hatch, clear windows, detailed engine, yellow interior	25	35	60
Renault 5 Police Car	3 7/8" white body, red interior, blue dome light, black hood, hatch & doors with white Police labels, orange taillights	15	20	35
Renault 5 Turbo	3 3/4" bright yellow body, red plastic interior, black roof & hood, working hatch & two doors, black dash, chrome rear engine	12	18	30
Renault 5 Turbo	4" white body, red roof, red & blue trim painted on, No. 5 lettering, blue & white label on windshield	12	18	30
Renault 5TS	3 3/4" metallic golden orange body, black trim, tan plasic interior, working hatch & 2 doors, clear windows & headlights	12	18	30
Renault 5TS	3 3/4" light blue body, red plasic interior, dark blue roof, dome light, S.O.S. Medicins lettering, working hatch & 2 doors	12	18	30
Renault 5TS Fire Chief	3 3/4" red body, tan interior, amber headlights, gray antenna, black/white Sapeurs Pompiers decals, blue dome light	15	25	40
Renault Alpine 5TS	3 3/4" dark blue body, off white interior, red & chrome trim, clear windows & headlights, gray base & bumpers, black grill	15	25	40
Renault Floride	3 5/8" long, one-piece body, clear windows, silver bumper, grill, lights and plates, red taillights, rubber tires	35	55	95
Renegade Jeep	4" dark blue body with no top, white interior, base & bumper, white plastic wheels & rear mounted spare, Renegade	8	12	20
Renegade Jeep with Hood	4" yellow body with removeable hood, red interior, base, bumper, white plastic wheels, side mounted spare, Renegade, number 8	8	12	20
Rice Beaufort Double Horse Box	3 3/8" long, blue body and working gates, white roof, brown plastic interior, 2 horses, cast wheels, plastic tires	15	25	40
Rice Pony Trailer	3 3/8" long, cast body and chassis w/working tailgate, horse, in 4 versions	20	30	50
Riley Pathfinder	4" long, red or dark blue one-piece body, clear windows, silver lights, grill and bumpers, smooth wheels, rubber tires	45	65	110
Riley Pathfinder Police Car	4" long, black body with blue/white Police lettering, unpainted roof sign, gray antenna	50	75	125
Riley Pathfinder-Mechanical	4" long with friction motor and either red or blue body	50	75	125
Riot Police Quad Tractor	3 3/4" white body & chassis, brown interior, red roof with white panel, gold water cannons, gold spotlight with amber lense	15	20	35

Corgi

NAME	DESCRIPTION	GOOD	EX	MINT
Road Repair Unit	10" dark yellow Land Rover with battery hatch & trailer with red plastic interior with sign & open panels, stripe & Roadwork	15	25	40
Rocket Age Set	set of eight standard including colored Thunderbird Missile on Trolley, R.A.F. Land Rover, R.A.F. Staff Car, Radar Scanner, Decca Radar	295	450	750
Rocket Launcher & Trailer		25	35	60
Roger Clark's Capri	4" white body, black hood, grill & interior, open doors, folding seats, chrome bumpers, clear headlights, red taillights	12	18	30
Rolls-Royce Corniche	5 1/2" different versions with light brown interior, working hood, trunk & 2 doors, clear windows, folding seats, chrome bumpers	10	15	25
Rolls-Royce Silver Ghost	4 1/2" silver body/hood, charcoal & silver chassis, bronze interior, gold lights, box & tank, clear windows, dash lights, radiator	15	25	40
Rolls-Royce Silver Shadow	4 3/4" metallic white upper/dusty blue lower body, working hood, trunk & 2 doors, clear windows, folding seats, chrome bumpers	30	45	75
Rolls-Royce Silver Shadow	4 3/4" metallic silver upper/metallic blue lower body, light brown interior, hole in trunk for spare tire mounting	25	40	65
Rolls-Royce Silver Shadow	4 3/4"metallic silver upper/metallic blue lower body, light brown interior, no hole in trunk for spare tire	25	40	65
Rolls-Royce Silver Shadow	4 3/4" metallic blue body, bright blue interior, working hood, trunk & 2 doors, clear windows, folding seats, spare wheel	25	40	65
Routemaster Bus-Promotionals	4 7/8" different body & interior versions promotional	15	25	40
Routemaster Bus-Promotionals		15	20	35
Rover 2000	3 3/4" metallic blue with red interior or maroon body with yellow interior, gray steering wheel, clear windshields	30	45	75
Rover 2000 Rally	3 3/4" two different versions, metallic dark red body, white roof, shaped wheels, No. 136 & Monte Carlo Rally decal	50	75	125
Rover 2000TC	3 3/4" metallic olive green or maroon one piece body, light brown interior, chrome bumpers/grill, jewel headlights, red taillights	30	45	75
Rover 2000TC	3 3/4" metallic purple body, light orange interior, black grill, one piece body, amber windows, chrome bumpers & headlights	25	35	60
Rover 3500	5 1/4" three different body & interior versions, plastic interior, open hood, hatch & two doors, lifting hatch cover	8	12	20
Rover 3500 Police CAr	5 1/4" white body, light red interior, red stripes, white plastic roof sign, blue dome light, red & blue Police & badge label	8	12	20
Rover 3500 Triplex	5 1/4" white sides & hatch, blue roof & hood, red plastic interior & trim, detailed engine, red-white-black no. 1	8	12	20
Rover 90	3 7/8" long, one-piece body in several colors, silver headlights, grill and bumpers, smooth wheels, rubber tires; colors available	50	75	125
Rover 90-Mechanical	3 7/8" long with friction motor and red, green, gray or metallic green body	60	90	150
Safari Land Rover & Trailer	black/white Land Rover in 2 versions: with chrome wheels, 1976; with red wheels, 1977-80	20	30	50
Saladin Armored Car	3 1/4" drab olive body, swiveling turret & raising barrel castings, black plastic barrel end & tires, olive cast wheels	30	45	75

VEHICLES

VEHICLES

Top to bottom: Superior Ambulance on a Cadillac Chassis; Chevrolet Astro 1 Experimental Car; Corgi Rockets autobatic speedset, all Corgi.

Corgi

NAME	DESCRIPTION	GOOD	EX	MINT
Scammell Carrimore Tri-deck Car Transporter	11" orange cab chassis & lower deck, white cab & middle deck, blue top deck (three decks), red interior, black hydraulic cylinders	35	55	95
Scammell Circus Crane Truck	8" red upper cab & silver rear body, light blue crane base & winch crank housing, red interior & tow hook, jewel headlights	175	275	450
Scammell Coop Semi-Trailer Truck	9" white cab & fenders, light blue semi-trailer, red interior, gray bumper base, jewel headlights, black hitch lever, spare wheel	135	210	350
Scammell Ferrymasters Semi-Trailer Truck	9 1/4" long, white cab, red interior, yellow chassis, black fenders, clear windows, jewel headlights, cast wheels, plastic tires	60	90	150
Scania Bulk Carrier	5 5/8" long, white cab, blue and white silos, ladders and catwalk, amber windows, blue British Sugar decals, Whizz Wheels	6	9	15
Scania Bulk Carrier	5 5/8" long, white cab, orange and white silos, clear windows, orange screen, black/orange Spillers Flour decals	6	9	15
Scania Container Truck	5 1/2" long, yellow truck and box with red Ryder Truck rental decals, clear windows, black exhaust stack, red rear doors	6	9	15
Scania Container Truck	5 1/2" long, blue cab with blue and white box and rear doors, white deck, Securicor Parcels Decals, in 2 rear door colors	6	9	15
Scania Container Truck	5 1/2" long, white cab and box with BRS Truck Rental decals, blue windows, red screen, roof and rear doors	6	9	15
Scania Dump Truck	5 3/4" long, white cab with green tipper, black/green Barratt decals, black exhaust and hydraulic cylinders, spoked Whizz Wheels	6	9	15
Scania Dump Truck	5 3/4" long, yellow truck and tipper with black Wimpey decals, in 2 versions: either clear or green windows	6	9	15
Security Van	4" long, black body, blue windows and dome light, yellow/black Security decals, Whizz Wheels	6	9	15
Service Ramp	accessory	30	45	75
Shadow-Ford Racing Car	5 5/8" long, black body and base with white/black #17, UOP and American flag decals, cast chrome suspension and pipes	10	15	25
Shadow-Ford Racing Car	5 5/8" long, white body, red stripes, driver, chrome plastic pipes, mirrors and steering wheel, in 2 versions	10	15	25
Shell or BP Garage	gas station/garage with pumps and other accessories in 2 versions: Shell or B.P., each set	295	450	750
Shelvoke & Drewry Garbage Truck	5 7/8" long, orange cab, silver body with City Sanitation decals, black interior, grill and bumpers, clear windows	15	25	40
Sikorsky Skycrane Army Helicopter	5 1/2" long, olive drab and yellow body with Red Cross and army decals	15	20	35
Sikorsky Skycrane Casualty Helicopter	6 1/8" long, red and white body, black rotors and wheels, orange pipes, working rear hatch, Red Cross decals	15	20	35
Silo & Conveyor Belt	with yellow conveyor and Corgi Harvesting Co. decal on silo	35	50	85
Silver Jubilee Landau	Landau with 4 horses, 2 footmen, 2 riders, Queen and Prince figures, and Corgi dog, in 2 versions	15	25	40
Silver Jubilee London Transport Bus	4 7/8" long, silver body with red interior, no passengers, decal reads "Woolworth Welcomes the World"	12	18	30
Silver Streak Jet Dragster	6 1/4" long, metallic blue body with sponsor and flag decals on tank, silver engine, orange plastic jet and nose cone	12	18	30
Silverstone Racing Layout	7 vehicle set with accessories; Vanwall, Lotus IX, Aston Martin, Mercedes 300SL, BRM, Ford Thunderbird, Land Rover Truck	400	600	1000

VEHICLES

Corgi

NAME	DESCRIPTION	GOOD	EX	MINT
Simca 1000	3 1/2" chrome plated body, #8 & red-white-blue stripe decals, one piece body, clear windshield, red interior	30	45	75
Simon Snorkel Fire Engine	10 1/2" long, red body with yellow interior, blue windows and dome lights, chrome deck, black hose reels and hydraulic cylinders	30	45	75
Simon Snorkel Fire Engine	9 7/8" long, red body with yellow interior, 2 snorkle arms, rotating base, 5 firemen in cab and 1 more in basket	35	55	90
Skyscraper Tower Crane	9 1/8" tall, red body with yellow chassis and booms, gold hook, gray loads of block, black/white Skyscraper decals	30	45	75
Spider-Bike	4 1/2" medium blue body, one piece body, dark blue plastic front body & seat, blue & red Spider-Man figure, amber windshield	40	60	100
Spider-Buggy	5 1/8" red body, blue hood, clear windows, dark blue dash, seat & crane, chrome base with bumper & steps. silver headlights	50	75	125
Spider-Copter	5 5/8" long, blue body with Spider-Man decals, red plastic legs, tongue and tail rotor, black windows and main rotor	30	45	75
Spider-Man Set	set of 3; Spider-Bike, Spider-Copter and Spider-Buggy	80	120	200
Standard Vanguard	3 5/8" long, one-piece red and white body, clear windows, silver lights, grill and bumpers, smooth wheels, rubber tires	50	75	125
Standard Vanguard RAF Staff Car	3 3/4" long, blue body with friction motor, RAF decals	55	85	140
Standard Vanguard-Mechanical	3 5/8" long with friction motor and red/off-white body with black or gray base, or red/gray body	55	85	140
Starsky & Hutch Ford Torino	5 3/4" red one piece body, white trim, light yellow interior, clear windows, chrome bumpers, grill & headlights, orange taillights	35	55	95
STP Patrick Eagle Racing Car	5 5/8" long, red body with red/white/black STP and #20 decals, chrome lower engine and suspension, black plastic upper engine	20	30	50
Stromberg Jet Ranger Helicopter	5 5/8" long, black body with yellow trim and interior, clear windows, black plastic rotors, white/blue decals	30	45	75
Studebaker Golden Hawk	4 1/8" long, one-piece body in several colors, clear windows, silver lights, grill and bumpers, smooth wheels, rubber tires	55	85	140
Studebaker Golden Hawk-Mechanical	4 1/8" long with friction motor and white body with gold trim	70	105	175
Stunt Motorcycle	3" long, made for Corgi Rockets race track, gold cycle, blue rider with yellow helmet, clear windshield, plastic tires	70	105	175
SU-100 Medium Tank	5 5/8" long, olive and cream camo upper body, gray lower, working hatch and barrel, black treads, red star and #103 decals	30	50	80
Sunbeam Imp Police Car	3 1/4" white or light blue body, tan interior, driver, black or white hood & lower doors, dome light, Police decals, cast wheels	25	40	65
Sunbeam Imp Rally	3 3/8" long, metallic blue body with white stripes, Monte Carlo Rally and #77 decals, cast wheels	20	35	55
Super Karts	2 karts, orange and blue, Whizz Wheels in front, slicks on rear, silver and gold drivers	12	18	30
Superman Set	set of 3; Supermobile, Daily Planet Helicopter and Metropolis Police Car	70	120	200
Supermobile	5 1/2" blue body, red, chrome or gray fists, red interior, clear canopy, driver, chrome arms with removeable "striking fists"	30	45	75
Supervan	4 5/8" long, silver van with Superman decals, working rear doors, chrome spoked wheels	15	25	40

Corgi

NAME	DESCRIPTION	GOOD	EX	MINT
Surtees TS9 Racing Car	4 5/8" long, black upper engine, chrome lower engine, pipes and exhaust, driver, Brook Bond Oxo-Rob Walker decals, in 2 versions	12	18	30
Surtees TS9B Racing Car	4 3/8" long, red body with white stripes and wing, black plastic lower engine, driver, chrome upper engine, pipes, suspension	12	18	30
Talbot-Matra Rancho	4 3/4" long, working tailgate and hatch, clear windows, plastic interior, black bumpers, grill and tow hook, in several colors	10	15	25
Tandem Disc Harrow	3 5/8" long, yellow main frame, red upper frame, working wheels linkage, unpainted linkage and cast discs, black plastic tires	15	20	35
Tarzan Set	metallic green Land Rover with trailer, cage and other accessories	100	150	250
The Professionals Ford Capri	5" metallic silver body & base, red interior, black spoiler, grill, bumpers, tow hook & trim, blue windows, chrome wheels	30	45	75
The Saint's Jaguar XJS	5 1/4" white body, red interior, black trim, Saint figure hood label, open doors, black grill, bumpers & tow hook, chrome headlights	30	45	75
The Saint's Volvo P-1800	3 5/8" long, one-piece white body with red Saint decals on hood, gray base, clear windows, black interior with driver	55	85	140
The Saint's Volvo P-1800	3 3/4" three versions of white body with silver trim & different colored Saint decals on hood, driver, one piece body	55	85	140
Thunderbird Bermuda Taxi	4" white body with blue, yellow, green plastic canopy with red fringe, yellow interior, driver, yellow & black labels	50	75	125
Thunderbird Missile & Trolley	5 1/2" ice blue or silver missile, RAF blue trolley, red rubber nose cone, plastic tow bar, steering front & rear axles	55	85	140
Thwaites Tusker Skip Dumper	3 1/8" yellow body, chassis & tipper, driver & seat, hydraulic cylinder, red wheels, black tires two sizes, name labels	10	15	25
Tiger Mark I Tank	6" tan & green camouflage finish, German emblem, swiveling turret & raising barrel castings, black plastic barrel end, antenna	30	45	75
Tipping Farm Trailer	5 1/8" long, cast chassis and tailgate, red plastic tipper and wheels, black tires, in 2 versions	10	15	25
Tipping Farm Trailer	4 1/4" long, red working tipper and tailgates, yellow chassis, red plastic wheels, black tires	10	15	25
Tour de France Set	Renault with Paramount Film roof sign, rear platform with cameraman and black camera on tripod, plus bicycle and rider	60	90	150
Tour de France Set	with white Peugeot	25	40	65
Touring Caravan	4 3/4" white body with blue trim, white plastic open roof & door, pale blue interior, flowerspot, red plastic hitch & awning	15	25	40
Tower Wagon & Lamp Standard	red Jeep Tower wagon with yellow basket, workman figure	40	60	100
Toyota 2000 GT	4" metallic dark blue or purple body, cream interior, one piece body, red gear shift & antenna, 2-red & 2-amber taillights	15	25	40
Tractor & Beast Carrier	Fordson tractor, figures & Beast carrier	65	100	165
Tractor with Shovel & Trailer	standard colors, No. 69 Massey Ferguson Tractor & 62 Tipping Trailer	65	100	165
Tractor, Trailer & Field Gun	10 3/4" tractor body & chassis, trailer body, base & opening doors, gun chassis & raising barrel castings, brown plastic interior	30	50	80
Transporter & Six Cars	Ford transporter with 6 cars, Mini DeLuxe, Mini Rally, Mini, Rover, Sunbeam Imp, Ford Cortina Estate Car	225	365	600

VEHICLES

Corgi

NAME	DESCRIPTION	GOOD	EX	MINT
Transporter & Six Cars	Scammell transporter with 6 cars, Mini DeLuxe, Mini, Mini Rally, The Saint's Volvo, Sunbeam Imp, MGC GT	250	395	650
Triumph Acclaim Driving School	4 3/4" dark yellow body with black trim, black roof mounted steering wheel steers front wheels, clear windows, mirrors, bumpers	15	25	40
Triumph Acclaim Driving School	4 3/4" yellow or red body/base, Corgi Motor School decals, black roof mounted steering wheel steers front wheels, clear windows	15	25	40
Triumph Acclaim HLS	4 3/4" metallic peacock blue body/base, black trim, light brown interior, clear windows, mirrors, bumpers, vents, tow hook	12	18	30
Triumph Herald Coupe	3 1/2" long, blue or gold top and lower body, white upper body, red interior, clear windows, silver bumpers, grill, headlights	35	50	85
Triumph TR2	3 1/4" cream body with red seats, light green body with white or cream seats, one piece body, clear windshield, silver grill	70	105	175
Triumph TR3	2 1/4" metallic olive or cream body, red seats, one piece body, clear windshield, silver grill, bumpers & headlights	60	90	150
Trojan Heinkel	2 1/2" long, issued in mauve, red or orange body, plastic interior, silver bumpers & headlights, red taillights, suspension	35	55	95
Trojan Heinkel	2 1/2" long, red body, yellow plastic interior, clear windows, silver bumpers & headlights, red taillights, suspension	35	55	95
Trojan Heinkel	2 1/2" long, orange body, yellow plastic interior, clear windows, silver bumpers & headlights, red taillights, suspension	35	55	95
Twin Packs and 2601		6	9	15
Tyrrell P34 Racing Car	4 3/8" long, dark blue body and wings with yellow stripes, #4 and white Elf and Union Jack decals, chrome plastic engine	20	30	55
Tyrrell P34 Racing Car	without yellow decals	20	30	55
Tyrrell-Ford Racing Car	4 5/8" long, dark blue body with blue/black/white Elf and #1 decals, chrome suspension, pipes, mirrors, driver	18	25	45
Unimog Dump Truck	3 3/4" blue cab, yellow tipper, fenders & bumpers, metallic charcoal gray chassis, red interior, black mirrors, gray tow hook	20	30	50
Unimog Dump Truck	4" yellow cab, chassis, rear frame & blue tipper, fenders & bumpers, red interior, no mirrors, gray tow hook, hydraulic cylinders	20	30	50
Unimog Dumper & Priestman Cub Shovel	Standard colors, #1145 Mercedes-Benz unimor with Dumper & 1128 Priestman Cub Shovel	70	105	175
Unimog with Snowplow (Mercedes-Benz)	6" four different body versions, red interior, cab, rear body, fender-plow mounting, lower & charcoal upper chassis, rear fenders,	30	45	75
US Racing Buggy	3 3/4" long, white body with red/white/blue stars, stripes and USA #7 decals, red base, gold engine, red plastic panels	18	25	45
Vanwall Racing Car	3 3/4" long, clear windshield, unpainted dash, silver pipes and decals, smooth wheels, rubber tires, in 3 versions: green body	35	55	90
Vauxhall Velox	3 3/4" long, one-piece body in several colors, clear windows, silver lights, grill and bumpers, smooth wheels, rubber tires	50	75	125
Vauxhall Velox-Mechanical	3 3/4" long with friction motor; came in either orange or red body	60	90	150
Vegas Ford Thunderbird	5 1/4" orange/red body & base, black interior & grill, open hood & trunk, amber windshield, white seats, driver, chrome bumper	25	40	65
VM Polo Mail Car		25	35	60

Corgi

NAME	DESCRIPTION	GOOD	EX	MINT
Volkswagen 1200	3 1/2" seven different versions, plastic interior, one piece body, silver headlights, red taillights, die cast base & bumpers	20	30	50
Volkswagen 1200	3 1/2" dark yellow body, white roof, red interior & dome light, unpainted base & bumpers, black & white ADAC Strassenwacht	60	90	150
Volkswagen 1200 Driving School	3 1/2" metallic red or blue body, yellow interior, gold roof mounted steering wheel that steers, silver headlights, red taillights	25	35	60
Volkswagen 1200 Police Car	3 1/2" two different body versions made for Germany, Netherlands & Switzerland, blue dome light in chrome collar	40	60	100
Volkswagen 1200 Rally	3 1/2" light blue body, off-white plastic interior, silver headlights, red taillights, suspension, Whizz Wheels	20	30	50
Volkswagen Breakdown Van	4" tan or white body, red interior & equipment boxes, clear windshield, chrome tools, spare wheels, red VW emblem, no lettering	50	75	125
Volkswagen Delivery Van	3 1/4" white upper & red lower body, plastic red or yellow interior, silver bumpers & headlights, red VW emblem, shaped wheels	55	85	140
Volkswagen Driving School	3 1/2" metallic blue body, yellow interior, gold roff mounted steering wheel that steers, silver headlights, red taillights	25	40	70
Volkswagen East African Safari	3 1/2" light red body, brn interior, working front & rear hood, clear windows, spare wheel on roof steers front wheels, jewel headlights	50	80	130
Volkswagen Kombi Bus	3 3/4" off green upper & olive green lower body, red interior, silver bumpers & headlights, red VW emblem, shaped wheels	50	75	125
Volkswagen Military Personnel Carrier	3 1/2" drab olive body, white decals, driver	55	85	140
Volkswagen Pickup	3 1/2" dark yellow body, red interior & rear plastic cover, silver bumpers & headlights, red VW emblem, shaped wheels	45	65	110
Volkswagen Police Car/ Foreign Issues	3 1/2" five different versions, one piece body, red interior, dome light, silver headlights, red taillights, clear windows	60	90	150
Volkswagen Tobler Van	3 1/2" light blue body, plastic interior, silver bumpers, Trans-o-lite headlights & roof panel, shaped wheels, rubber tires	55	85	140
Volvo Concrete Mixer	8 1/4" yellow or orange cab, red or white mixer with yellow & black stripes, rear chassis, chrome chute & unpainted hitch casings	30	45	75
Volvo P-1800	3 1/2" one piece body with six versions, clear windows, plastic interior, shaped wheels, rubber tires	40	60	100
VW 1500 Karmann-Ghia	3 1/2" three color versions, plastic interior and taillights, front & rear working hoods, clear windshields, silver bumpers	35	55	90
VW Polo	3 3/4" apple green or bright yellow body, black DBP & posthorn (German Post Ofc.) decals, off white interior, black das	25	40	65
VW Polo	3 3/4" metallic light brown body, off-white interior, black dash, clear windows, silver bumpers, grill & headlights	12	18	30
VW Polo Auto Club Car		15	25	40
VW Polo German Auto Club Car	3 1/2" yellow body, off-white interior, black dash, silver bumpers, grill & headlights, white roof, yellow dome light	25	35	60
VW Polo Police Car	3 1/2" white body, green hood & doors, black dash, silver bumpers, grill & headlights, white roof, blue dome light	15	25	40

Corgi

NAME	DESCRIPTION	GOOD	EX	MINT
VW Polo Turbo	3 3/4" cream body, red interior with red & orange trim, working hatch & two door castings, clear windshield, black plastic dash	12	18	30
VW Racing Tender & Cooper	white VW with racing decals, blue Cooper	50	75	125
VW Racing Tender & Cooper Maserati	vehicle set in 2 versions, with either tan or white VW truck, each set	50	75	125
Warner & Swasey Crane	8 1/2" yellow cab & body, blue chassis, blue/yellow stripe decals, red interior, black steering wheel, silver knob, gold hook	30	45	75
White Wheelie Motorcycle	4" long white body with black/white Police decals	15	20	35
Wild Honey Dragster	3" long, yellow body with red/yellow Wild Honey and Jaguar Powered decals, green windows and roof, black grill, driver, Whizz Wheels	25	40	65

Airplanes

NAME	DESCRIPTION	GOOD	EX	MINT
Daily Planet Jet-Copter, No. 929		35	55	75
Spidercopter, No. 928		50	75	100

Boats and Ships

NAME	DESCRIPTION	GOOD	EX	MINT
Beatle's Yellow Submarine, No. 803		75	200	500

Cars

NAME	DESCRIPTION	GOOD	EX	MINT
Arnold Sundquist's Jet Car, No. 169		25	40	55
Batmobile, No. 267		85	130	175
Black Beauty Green Hornet, No. 268		250	375	500
Camaro Convertible	1968	35	55	75
Can-Am Porsche 917-10, No. 397		25	40	50
Dick Dastardly Car, No. 809		50	75	100
Elf Tyrrell Ford F-1, No. 158		20	35	45
Ferrari 312 B2 Formula 1, No. 152		20	30	40
Firebird, No. 343		50	75	100
Hesketh 308 Formula 1 Car, No. 160		20	35	45
James Bond Aston Martin, No. 261	gold	125	185	250
James Bond Lotus Esprit, No. 269		75	115	150
Kojak's Buick, No. 290		40	60	80
Lotus, No. 154		45	65	90
Metropolis Buick, No. 260		37	55	75
Monkeemobile, No. 277		200	300	400
Patrick Eagle Indy Car, No. 159		25	40	50
Penguinmobile, No. 259		35	55	75
Saint's Volvo, No. 257		75	115	150
Starsky & Hutch Ford Torino, No. 292		50	75	100
State Landau-Queen's Jubilee, No. 41		20	30	40
The Professional's Ford Capri, No. 342		50	75	100
Thunderbird	1958	35	55	75
Thunderbird, No. 215		110	165	225

Corgi

NAME	DESCRIPTION	GOOD	EX	MINT
Vegas Thunderbird, No. 348		75	115	150

Motorcycles

NAME	DESCRIPTION	GOOD	EX	MINT
Spiderbike, No. 266	white tires	110	175	225

Sets

NAME	DESCRIPTION	GOOD	EX	MINT
Batman Triple Pack Gift Set, No. GS-40		225	335	450
Batmobile and Batboat, No. GS-3		200	300	400
Giant Daktari, No. 14		200	300	400
Land Rover-Pony Trailer, No. 2		75	115	150
Police Cortina, No. 18	helicopter, ambulance	65	95	130
Tarzan Gift Set, No. GS-36		225	335	450

Space Vehicles

NAME	DESCRIPTION	GOOD	EX	MINT
Buck Rogers Starfighter, No. 647		45	70	90
James Bond Moon Buggy, No. 811		100	150	200
OO7 Space Shuttle, No. 649		45	70	90

Trucks

NAME	DESCRIPTION	GOOD	EX	MINT
Berliet Wrecker Truck, No. 1144		25	40	50
Canteen Truck, No. 471		25	40	50
Decca Radar Van, No. 1106		125	185	250
Holmes Wrecker, No. 1142		85	130	175
Human Cannonball Truck, No. 1163		35	55	75
Ice Cream Truck, No. 428		25	40	50
Jeep Tower Van, No. 65-14		125	185	250
Land Rover and Elephant, No. GS-19		50	75	100
Spider-Man Van, No. 436		25	40	50
Super Van, No. 435		25	40	50
Toy Van, No. 422		175	265	350
Vanwall, No. 150		75	115	150

VEHICLES

Top to bottom: Cadillac Eldorado; Austin Van Raleigh Cycles; Leyland Cement Wagon; UFO Interceptor space craft, all Dinky.

VEHICLES

Dinky

Accessories

NO.	NAME	GOOD	EX	MINT
13	"Halls Distemper" sign	150	300	600
47A	4 face traffic light	10	15	25
766	British Road Signs	70	115	175
F49D/592	Esso Gas Pumps	40	60	85
12A	G.P.O. Pillar Box	20	35	65
45	Garage	75	150	300
752	Goods Yard Crane	50	75	100
994	Loading ramp (for 582/982)	20	35	65
1003	Passengers	55	80	125
42D	Point Duty Policeman	20	35	65
778	Road Repair Boards	10	20	30
786	Tyre rack with tyres "Dunlop"	10	20	30

Aircraft

NO.	NAME	GOOD	EX	MINT
749/992	Avro Vulcan Delta Wing Bomber	100	2000	5000
70A/704	Avro York	60	85	130
710	Beechcraft Bonanza 535	50	75	100
62B	Bristol Blenheim	60	85	130
998	Bristol Britannia	135	200	350
F60Z	Cierva Autogiro	75	100	150
702/999	DH Comet Airline	75	100	150
70E	Gloster Meteor	25	40	70
722	Hawker Harrier	75	100	150
66A	Heavy Bomber	145	225	450
60A	Imperial Airways Liner	135	200	350
F804	Nord 2501 Noratlas	120	175	275
60H	Singapore Flying Boat	115	170	250
F60C/892	Super G Constellation	120	175	275
734	Supermarine Swift	45	65	85

Buses & Taxis

NO.	NAME	GOOD	EX	MINT
40H/254	Austin Taxi	60	95	135
F29D	Autobus Parisien	80	135	200
F29F/571	Autocar Chausson	70	120	175
283	B.O.A.C. Coach	55	80	115
953	Continental Touring Coach	135	200	350
F24XT	Ford Vedette Taxi	60	95	135
284	London Taxi	25	35	50
29F/280	Observation Coach	50	75	100
F1400	Peugeot 404 Taxi	50	75	100
266	Plymouth Canadian Taxi	60	95	135
289	Routemaster Bus "Tern Shirts"	75	100	150
297	Silver Jubilee Bus	25	35	50

Cars

NO.	NAME	GOOD	EX	MINT
36A	Armstrong Siddeley, blue or brown	85	130	200
106/140A	Austin Atlantic Convertible, blue	60	95	150
342	Austin Mini-Moke	20	30	45
131	Cadillac Eldorado	60	95	135
32/30A	Chrysler Airflow	130	250	450
F550	Chrysler Saratoga	70	120	190
F535/24T	Citroen 2 cv	50	70	90
F522/24C	Citroen DS-19	60	80	115
F545	DeSoto Diplomat, green	80	135	215
F545	DeSoto Diplomat, orange	60	85	125
191	Dodge Royal	75	115	150
27D/344	Estate Car	45	70	115
212	Ford Cortina Rally Car	35	55	75

Dinky

NO.	NAME	GOOD	EX	MINT
148	Ford Fairlane	150	300	700
148	Ford Fairlane, pale green	30	55	80
57/005	Ford Thunderbird (Hong Kong)	50	70	100
F565	Ford Thunderbird, S. African Issue, blue	80	135	215
238	Jaguar D-Type	60	86	125
157	Jaguar XK 120, green, yellow, red	50	75	100
157	Jaguar XK 120, turquoise, cerise	80	125	200
157	Jaguar XK 120, white	120	200	400
157	Jaguar XK 120, yellow/grey	80	125	200
241	Lotus Racing Car	20	30	50
231	Maserati Race Car	45	75	110
161	Mustang Fastback	35	55	75
F545	Panhard PL17	45	70	90
F521/24B	Peugeot 403	50	75	100
115	Plymouth Fury Sports	35	55	80
F524/24E	Renault Dauphine	50	75	100
30B	Rolls Royce (1940's Version)	65	100	125
198	Rolls Royce Phantom V	50	75	100
145	Singer Vogue	40	55	80
153	Standard Vanguard	60	85	120
F24Y/540	Studebaker Commander	65	90	130
24C	Town Sedan	85	130	200
105	Triumph TR-2, grey	60	85	120
105	Triumph TR-2, yellow	75	110	175
129	Volkswagen 1300 Sedan	20	35	60
187	VW Karman Ghia	50	75	100

Emergency Vehicles

NO.	NAME	GOOD	EX	MINT
30F	Ambulance	100	160	275
F501	Citroen DS19 Police	75	95	160
F25D/562	Citroen Fire Van	60	85	135
555/955	Commer Fire Engine	60	85	135
F32D/899	Delahaye Fire Truck	100	160	275
195	Fire Chief Land Rover	35	50	85
F551	Ford Taunus Police	50	75	100
255	Mersey Tunnel Police	60	85	135
244	Plymouth Police Car	25	35	50
268	Range Rover Ambulance	25	35	50
25H/25	Streamlined Fire Engine (Post-War)	75	100	175
263	Superior Criterion Ambulance	50	75	100
956	Turntable Fire Escape (Bedford)	75	110	175
251	USA Police Car (Pontiac)	35	50	85
278	Vauxhall Victor Ambulance	55	85	115

Farm & Construction

NO.	NAME	GOOD	EX	MINT
305	"David Brown" Tractor	35	50	75
984	Atlas Digger	30	45	70
561	Blaw Knox Bulldozer	45	75	115
965	Euclid Dump Truck	45	75	115
37N/301	Field Marshall Tractor	60	85	135
105A	Garden Roller	15	25	35
324	Hayrake	30	40	60
27A/300	Massey-Harris Tractor	50	75	100
27G/342	Moto-cart	35	50	75
437	Muir Hill 2wl Loader	30	40	60
F830	Richier Road Roller	75	100	150
963	Road Grader	30	45	70
F595	Salev Crane	65	100	135

Military

NO.	NAME	GOOD	EX	MINT
622	10 Ton Army Wagon	50	85	125

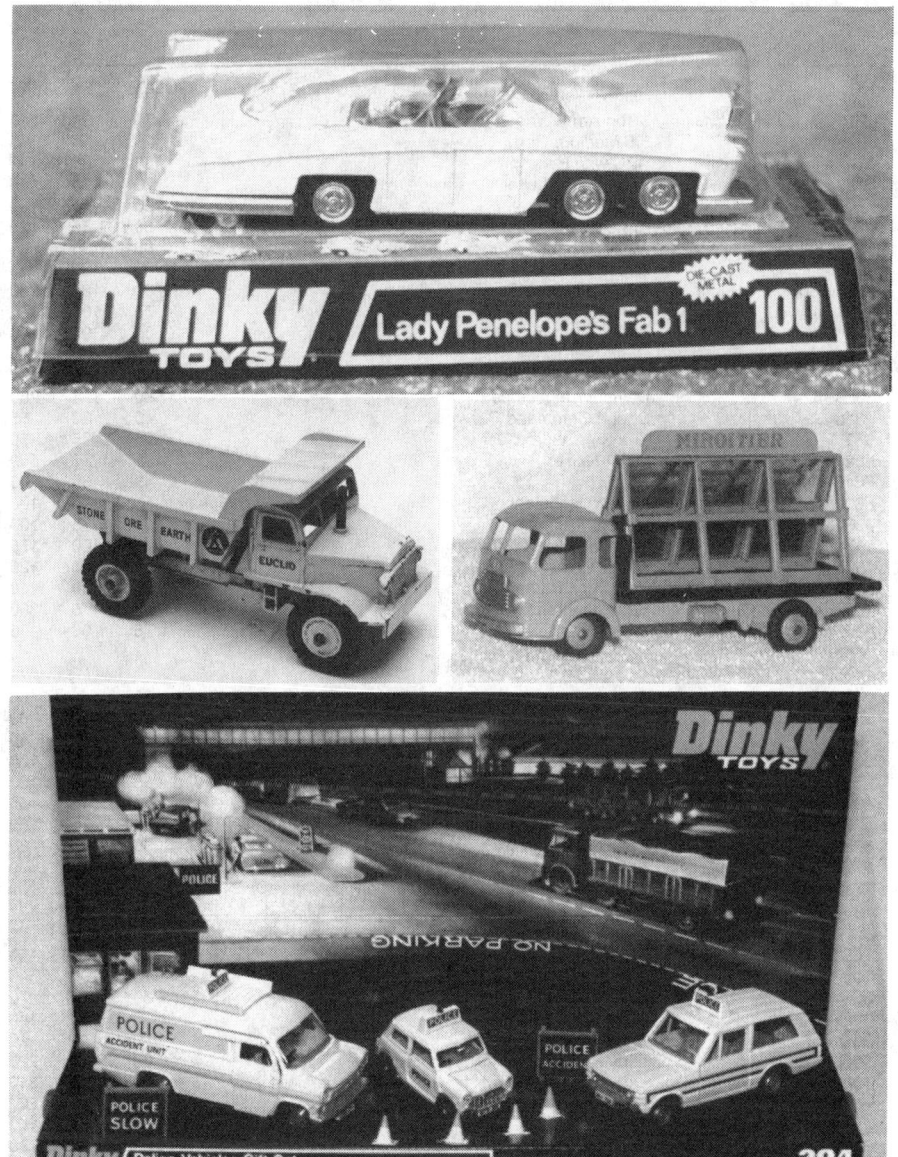

Top to bottom: Lady Penelope's Fab 1 with figures; Euclid Dumper; Simca Glass Truck; Police Vehicles Gift Set, all Dinky.

Dinky

NO.	NAME	GOOD	EX	MINT
621	3 Ton Army Wagon	50	85	125
692	5.5 Medium Gun	25	40	65
618	AEC with Helicopter	50	85	125
F883	AMX Bridge Layer	75	100	175
F80C/817	AMX Tank	50	75	100
677	Armoured Command Vehicle	50	85	125
30SM/625	Austin Covered Truck	85	135	275
601	Austin Paramoke	25	35	50
25WM/60	Bedford Military Truck	80	125	250
620	Berliet Missile Loader	75	100	175
F806	Berliet Wrecker	60	90	140
651	Centurion Tank	40	60	85
612	Commando Jeep	25	35	50
30HM/624	Daimlet Ambulance	80	125	250
F810	Dodge Command Car	40	60	85
630	Ferret Armoured Car	25	35	50
F823	GMC Tanker	125	250	500
F816	Jeep	40	60	85
F80F/820	Military Ambulance	50	75	100
626	Military Ambulance	50	75	100
667	Missile Servicing Platform	75	100	175
152B	Reconnaisance Car	50	85	125
661	Recovery Tractor	60	90	140
161A	Searchlight (Pre-War)	100	175	350
660	Tank Transporter	75	100	175

Motorcycles & Caravans

NO.	NAME	GOOD	EX	MINT
240/44B	A.A. motorcycle patrol (Post-War)	30	45	60
190	Caravan	30	45	60
30G	Caravan (Post-War)	40	60	85
30G	Caravan (Pre-War)	50	75	100
F564	Caravane Caravelair	65	100	150
42B	Police motorcycle patrol (Post-War)	30	45	60
42B	Police motorcycle patrol (Pre-War)	50	75	100
37B	Police motorcyclist (Post-War)	30	45	60
37B	Police motorcyclist (Pre-War)	50	75	100
271	TS motorcycle patrol (Swiss Version)	65	100	150

Space

NO.	NAME	GOOD	EX	MINT
F281	"Pathe News" Camera Car	65	100	150
102	"The Prisoner" Mini-Moke	115	200	340
361	Galactic War Chariot	30	45	70
102	Joe's Car	60	90	140
357	Klingon Battle Cruiser	30	45	70
100	Lady Penelope's Fab 1, pink version	80	135	220
100	Lady Penelope's Fab 1, shocking pink version	115	200	340
F1406	Renault Sinpar	80	135	220
485	Santa Special Model T Ford	65	100	150
350	Tiny's Mini-Moke	60	85	120
371/803	U.S.S. Enterprise	30	45	70

Trucks

NO.	NAME	GOOD	EX	MINT
974	A.E.C. Hoyner Transporter	60	90	130
471	Austin Van "Nestles"	60	100	150
472	Austin Van "Raleigh"	60	100	150
470	Austin Van "Shell/BP"	60	100	150
14A/400	B.E.V. Truck	30	45	90
482	Bedford Van "Dinky Toys"	80	135	200
F898	Berliet Transformer Carrier	135	300	550
923	Big Bedford "Heinz" (Baked Beans)	100	165	275
408/922	Big Bedford (blue/yellow)	90	135	210

Dinky

NO.	NAME	GOOD	EX	MINT
408/922	Big Bedford (maroon/fawn)	80	120	185
449	Chevrolet El Camino	35	65	100
F561	Citroen "Cibie" Delivery Van	65	95	140
F586	Citroen Milk Truck	145	275	600
F35A/582	Citroen Wrecker	75	120	225
571/971	Coles Mobile Crane	40	70	100
25B	Covered Wagon ("Carter Paterson")	150	300	500
25B	Covered Wagon (green, grey)	65	115	160
28N	Delivery Van ("Atco", type 2)	200	375	600
28N	Delivery Van ("Atco", type 3)	135	200	350
28E	Delivery Van ("Ensign", type 1)	300	500	1000
28B	Delivery Van ("Pickfords", type 1)	300	500	1000
28B	Delivery Van ("Pickfords", type 2)	200	375	600
30W/421	Electric Articulated Vehicle	60	85	120
941	Foden "Mobilgas" Tanker	145	300	600
942	Foden "Regent" Tanker	135	250	500
503/903	Foden Flat Truck w/ Tailboard 1, grey/blue	140	210	450
503/903	Foden Flat Truck w/ Tailboard 1, red/black	140	210	450
503/903	Foden Flat Truck w/ Tailboard 2, blue/yellow	90	125	210
503/903	Foden Flat Truck w/ Tailboard 2, green/green	100	150	230
417	Ford Transit Van	15	20	30
25R	Forward Control Wagon	45	65	90
431	Guy 4 Ton Truck	70	110	185
514	Guy Van "Lyons"	275	550	1600
514	Guy Van "Slumberland"	135	300	575
431	Guy Warrior 4 Ton	150	270	450
449/451	Johnston Road Sweeper	25	50	75
419/533	Leland Comet Cement Truck	85	140	210
944	Leland Tanker "Shell/BP"	125	215	400
	Leyland Tanker "Corn Products"	700	1200	3000
25F	Market Gardeners Wagon (yellow)	65	115	160
280	Midland Bank	60	85	120
986	Mighty Antar with Propeller	125	215	400
273	Mini Mino Van ("R.A.C.")	65	115	150
274	Mini Minor Van (Joseph Mason Paints)	150	300	500
260	Morris Royal Mail	65	115	150
22C	Motor Truck (red, green, blue)	80	120	200
22C	Motor Truck (red/blue)	150	350	650
F32C/576	Panhard ("Esso")	75	120	170
F32AJ	Panhard ("Kodak")	140	250	450
F32AB/55	Panhard ("SNCF")	100	165	280
25D	Petrol Wagon ("Power")	150	300	500
982/582	Pullmore Car Transporter	75	125	175
F561	Renault Estafette	50	75	110
F571	Saviem Race Horse Van	125	225	400
F33C/579	Simca Glass Truck (grey/green)	75	120	170
F33C/579	Simca Glass Truck (yellow/green)	100	150	250
30P/440	Studebaker Tanker ("Mobilgas")	70	100	140
422/30R	Thames Flat Truck	45	75	110
31B/451	Trojan ("Dunlop")	70	110	160
F38A/895	Unic Bucket Truck	75	120	170
F36A/897	Willeme Log Truck	75	120	200
F36B/896	Willeme Semi	85	130	225

VEHICLES

VEHICLES

Top to bottom: The Demon, 1970; Olds 442, 1971; Maserati Mistral, 1969; Clip-on Hot Wheels badges, all Mattel.

VEHICLES

Hot Wheels

YEAR	MODEL	NO.	DESCRIPTION	EX	MINT
1984	3 Window '34	4352	black, Real Rider	85	125
1992	#43-STP		petty blue, grey rollbars, blackwalls	20	30
1977	'31 Doozie	9649	orange, redline	25	30
	'31 Doozie	9649	orange, blackwall	8	10
1993	'32 Ford Delivery		white/pink, Early Times logo, blackwalls	18	25
1983	'40 Ford 2-Door	4367	black with white hubs, Real Rider	25	35
1992	'55 Chevy		blue #92, Real Rider	15	20
	'55 Chevy		black #92, Real Rider	15	20
	'55 Nomad		purple, blackwalls	12	20
1993	'55 Nomad		purple, Real Rider	12	22
1977	'56 Hi Tail Hauler	9647	orange, redline	30	40
	'56 Hi Tail Hauler	9647	orange, blackwall	20	30
	'57 Chevy	9638	red, redline	35	45
	'57 Chevy	9638	red, blackwall	12	15
1992	'57 Chevy		white #22, Real Rider	17	22
1986	'57 T-Bird	9522	black with white hubs, Real Rider	80	150
1990	'59 Caddy		pink, Canadian, blackwalls	20	25
1993	'59 Caddy		gold, blackwalls	12	18
1981	A-OK		red	100	250
1988	Alien		blue	10	15
1973	Alive '55	6968	assorted	125	400
1974	Alive '55	6968	green	55	110
	Alive '55	6968	blue	95	350
1977	Alive '55	9210	chrome, redline	25	30
	Alive '55	9210	chrome, blackwall	15	25
1970	Ambulance	6451	assorted	35	45
1976	American Hauler	9118	blue	18	25
	American Tipper	9089	red	15	22
1975	American Victory	7662	light blue	20	35
1971	AMX/2	6460	assorted	35	85
1976	Aw Shoot	9243	olive	18	25
1975	Backwoods Bomb	7670	light blue	40	60
1977	Backwoods Bomb	7670	green, redline or blackwall	30	40
1974	Baja Bruiser	8258	orange	40	55
	Baja Bruiser	8258	yellow, mgnta in tampo	250	500
	Baja Bruiser	8258	yellow, blue in tampo	250	500
1976	Baja Bruiser	8258	light green	300	400
1977	Baja Bruiser	8258	blue, redline or blackwall	35	50
1968	Beatnik Bandit	6217	assorted	12	55
1990	Black Passion		black	15	25
1970	Boss Hoss	6499	chrome, Club Kit	45	120
1971	Boss Hoss	6406	assorted	55	120
1969	Bragham-Repco F1	6264	assorted	8	25
1981	Bronco 4-Wheeler	1690	Toys R Us	45	60
1971	Bugeye	6178	assorted	35	75
1973	Buzz Off	6976	assorted	125	300
1974	Buzz Off	6976	blue	40	90
1977	Buzz Off	6976	gold plated, redline or blackwall	22	30
1971	Bye Focal	6187	assorted	75	225
1979	Bywayman	2509	Toys R Us	65	85
1989	Bywayman	2196	blue, red interior	40	50
1983	Cadillac Seville		grey, French	60	80
1987	Cadillac Seville		gold, Mexican, Real Rider	75	100
1979	Captain America	2879	white, Scene Machine	40	50
1970	Carabo	6420	assorted	30	50
1974	Carabo	7617	light green	35	50
	Carabo	7617	yellow	350	600
1970	Cement Mixer	6452	assorted	25	35
1969	Chapparal 2G	6256	assorted	15	35

Hot Wheels

YEAR	MODEL	NO.	DESCRIPTION	EX	MINT
1975	Chevy Monza 2+2	7671	orange	45	80
	Chevy Monza 2+2	9202	light green	225	300
	Chief's Special Cruiser	7665	red	35	60
1977	Chief's Special Cruiser	7665	red, redline	15	25
	Chief's Special Cruiser	7665	red, blackwall	8	15
1981	Circus Cats	3303	white 60	0	50
1969	Classic '36 Ford Coupe	6253	blue	10	15
	Classic '36 Ford Coupe	6253	assorted	15	50
	Classic 31 Ford Woody	6251	assorted	12	65
	Classic 32 Ford Vicky	6250	assorted	24	60
	Classic 57 T-Bird	6252	assorted	30	70
1992	Classic Caddy	2529	red/white/blue, Museum Exhibit car	20	25
1985	Classic Cobra		blue with white hubs, Real Rider	34	50
1970	Classic Nomad	6404	assorted	40	85
1971	Cockney Cab	6466	assorted	35	75
1969	Continental Mark III	6266	assorted	18	60
1977	Cool One	9120	plum, blackwall	35	40
1976	Corvette Stingray	9241	red	35	40
	Corvette Stingray	9506	chrome	40	50
1977	Corvette Stingray	9506	chrome, blackwall set only	40	0
1969	Custom AMX	6267	assorted	40	80
1968	Custom Barracuda	6211	assorted	55	300
	Custom Camaro	6208	assorted	55	200
	Custom Camaro	6208	white enamel	250	350
1969	Custom Charger	6268	assorted	60	120
1968	Custom Corvette	6215	assorted	55	175
	Custom Cougar	6205	assorted	60	275
	Custom El Dorado	6218	assorted	35	100
	Custom Firebird	6212	assorted	35	220
	Custom Fleetside	6213	assorted	55	160
	Custom Mustang	6206	assorted	65	320
	Custom Mustang	6206	assorted with open hood scoops or ribbed windows	300	800
1969	Custom Police Cruiser	6269	assorted	45	120
1968	Custom T-Bird	6207	assorted	50	165
	Custom VW Bug	6220	assorted	12	55
1982	Datsun 200SX	3255	maroon, Canada	100	125
1968	Deora	6210	assorted	50	350
1973	Double Header	5880	assorted	110	300
	Double Vision	6975	assorted	100	275
1983	Dumpin A			0	0
1973	Dune Daddy	6967	assorted	100	275
1975	Dune Daddy	6967	light green	25	55
	Dune Daddy	6967	orange	150	300
1974	El Rey Special	8273	light green	60	90
	El Rey Special	8273	green	40	70
	El Rey Special	8273	dark blue	150	350
	El Rey Special	8273	light blue	200	550
1975	Emergency Squad	7650	red	15	50
1971	Evil Weevil	6471	assorted	35	65
1970	Ferrari 312P	6417	assorted	20	30
1973	Ferrari 312P	6973	assorted	250	700
1974	Ferrari 312P	6973	red	40	55
1972	Ferrari 512-S	6021	var.	75	250
1970	Fire Chief Cruiser	6469	red	12	22
	Fire Engine	6454	red	25	60
1984	Flat out 442		green, Canada	75	120
1968	Ford J-Car	6214	assorted	10	55
1969	Ford MK IV	6257	assorted	8	30
1976	Formula 5000	9119	white	15	25

VEHICLES

Top to bottom: Classic Cord, 1971; Custom VW, 1968; Fire Chief Cruiser, 1970; Custom Charger, 1969; T42, 1971, all Mattel.

Hot Wheels

YEAR	MODEL	NO.	DESCRIPTION	EX	MINT
	Formula 5000	9511	chrome	25	30
1971	Fuel Tanker	6018	assorted	65	150
1972	Funny Money	6005	gray	60	170
1974	Funny Money	7621	magenta	40	55
1977	Funny Money	7621	grey, redline	35	45
	Funny Money	7621	grey, blackwall	18	35
	GMC Motorhome	9645	orange, blackwall	7	14
	GMC Motorhome	9645	orange, redline	400	500
1992	Gold Passion		gold, Toy Fair promo	20	25
1986	Good Humor Truck		white, Popsicle blacked out	55	125
1992	Goodyear Blimp		chrome, Mattel promo	65	0
1971	Grass Hopper	6461	assorted	30	55
1974	Grass Hopper	7621	light green	45	60
1975	Grass Hopper	7622	light green with no engine	30	45
1987	Greased Gremlin		red, Mexican, Real Rider	125	200
1975	Gremlin Grinder	7652	green	30	45
1977	Gremlin Grinder	9201	chrome, blackwall	20	30
1989	GT Racer	1789	blue	50	60
1976	Gun Bucket	9090	olive	15	25
1977	Gun Bucket	9090	olive, blackwall	15	25
1975	Gun Slinger	7664	olive	25	45
1976	Gun Slinger	7664	olive, blackwall	15	25
1971	Hairy Hauler	6458	assorted	30	45
1980	Hammer Down		red, set	100	0
1970	Heavy Chevy	6408	assorted	30	55
	Heavy Chevy	6189	chrome, Club Kit	45	120
1974	Heavy Chevy	7619	yellow	55	90
	Heavy Chevy	7619	light green	250	350
1977	Heavy Chevy	9212	chrome, redline or blackwall	60	120
1973	Hiway Robber	6979	assorted	100	250
1980	Hot Bird		brown	100	120
	Hot Bird		blue	80	100
1968	Hot Heap	6219	assorted	10	55
1979	Human Torch	2881	black	15	25
1971	Ice "T"	6184	yellow	45	175
1974	Ice "T"	6980	light green	45	60
	Ice "T"	6980	yellow with hood tampo	145	300
1977	Ice "T"	6980	light green, blackwall	25	35
1973	Ice-T	6980	assorted	140	400
1979	Incredbile Hulk Van	2850	white, Scene Machine	40	50
1969	Indy Eagle	6263	gold	75	200
	Indy Eagle	6263	assorted	8	25
1976	Inferno	9186	yellow	35	40
1970	Jack n'Box Promotion	6421	white	150	250
	Jack Rabbit Special	6421	white	10	40
1971	Jet Threat	6179	assorted	60	160
1976	Jet Threat II	8235	magenta	22	30
	Khaki Kooler	9183	olive	20	30
1970	King Kuda	6411	assorted	35	75
	King Kuda	6411	chrome, Club Kit	45	120
1975	Large Charge	8272	green	35	60
1977	Letter Getter	9643	white, redline	450	550
	Letter Getter	9643	white, blackwall	8	10
1970	Light My Firebird	6412	assorted	20	55
1969	Lola GT 70	6254	assorted	8	25
	Lotus Turbine	6262	assorted	8	25
1976	Lowdown	9185	light blue	40	45
1977	Lowdown	9185	gold plated, redline or blackwall	20	30
1970	Mantis	6423	assorted	12	40
1969	Masterati Mistral	6277	assorted	45	100
1976	Maxi Taxi	9184	yellow	30	35

VEHICLES

Hot Wheels

YEAR	MODEL	NO.	DESCRIPTION	EX	MINT
1977	Maxi Taxi	9184	yellow, blackwall	20	30
1969	McClaren M6A	6255	assorted	8	40
	Mercedes 280SL	6275	assorted	15	40
1973	Mercedes 280SL	6962	assorted	100	350
1972	Mercedes C-111	6169	var.	80	250
1973	Mercedes C-111	6978	assorted	225	975
1974	Mercedes C-111	6978	red	40	70
1970	Mighty Maverick	6414	assorted	40	75
1975					
	Mighty Maverick	7653	blue	45	65
	Mighty Maverick	9209	light green	110	140
1977	Mighty Maverick	9209	chrome, blackwall	30	40
1970	Mod-Quad	6456	assorted	18	35
1973	Mongoose	6970	red/blue	750	1400
1970	Mongoose Funny Car	6410	red	55	175
1971	Mongoose II	5954	metallic blue	90	350
	Mongoose Rail Dragster	5952	blue, two pack	75	600
1975	Monte Carlo Stocker	7660	yellow	45	75
1977	Monte Carlo Stocker	7660	yellow, blackwall	35	45
1975	Motocross I	7668	red	80	160
1979	Motocross Team Van	2853	red, Scene Machine	40	50
1980	Movin' On		white, set	100	0
1970	Moving Van	6455	assorted	40	60
1975	Mustang Stocker	7664	yellow with magenta tampo	75	120
	Mustang Stocker	9203	yellow with red in tampo	250	500
	Mustang Stocker	7664	white	300	600
1976	Mustang Stocker	9203	chrome	40	60
1977	Mustang Stocker	9203	chrome, redline or blackwall	40	60
1971	Mutt Mobile	5185	assorted	75	150
1983	NASCAR Stocker	3927	white, NASCAR/Mountain Dew base	125	165
1976	Neet Streeter	9244	blue	20	30
	Neet Streeter	9510	chrome	30	35
1977	Neet Streeter	0244	blue, blackwall	20	30
	Neet Streeter	9510	chrome, blackwall set only	35	0
1970	Nitty Gritty Kitty	6405	assorted	35	65
1971	Noodle Head	6000	assorted	45	90
1973	Odd Job	6981	assorted	110	350
1977	Odd Rod	9642	yellow, redline	40	50
	Odd Rod	9642	yellow, blackwall	20	25
	Odd Rod	9642	plum, blackwall or redline	300	400
1982	Old Number 5	1695	red, no louvers	15	20
1971	Olds 442	6467	assorted	250	600
1972	Open Fire	5881	var.	85	220
1975	P-911	7648	yellow	45	75
	P-911	6972	orange	25	45
1977	P-911	9206	chrome, redline or blackwall	25	40
	P-911	7648	black, six pack blackwall	225	350
	P-917	6972	orange, blackwall	14	22
1970	Paddy Wagon	6402	blue	12	25
1973	Paddy Wagon	6966	blue	30	120
1977	Paddy Wagon	6966	blue, blackwall	8	15
1975	Paramedic	7661	white	35	55
1976	Paramedic	7661	yellow	22	30
1977	Paramedic	7661	yellow, blackwall or redline	22	30
1970	Peepin' Bomb	6419	assorted	8	25
1982	Pepsi Challenger	2023	yellow Funnycar	15	20
1971	Pit Crew Car	6183	white	75	450
1976	Poision Pinto	9508	chrome	30	35
	Poison Pinto	9240	light green	18	25
1977	Poison Pinto	9240	green, blackwall	10	12
	Poison Pinto	9508	chrome, blackwall set only	35	0

Hot Wheels

YEAR	MODEL	NO.	DESCRIPTION	EX	MINT
1973	Police Cruiser	6963	white	125	300
1974	Police Cruiser	6963	white	35	65
1977	Police Cruiser	6963	white, blackwall	35	40
	Police Cruiser	6963	white with blue light	30	35
1970	Porsche 917	6416	assorted	12	35
1973	Porsche 917	6972	assorted	250	750
1974	Porsche 917	6972	orange	40	75
	Porsche 917	6972	red	250	500
1970	Power Pad	6459	assorted	35	65
1973	Prowler	6965	assorted	225	850
1974	Prowler	6965	orange	35	60
	Prowler	6965	light green	350	575
1977	Prowler	9207	chrome, blackwall	35	40
1968	Python	6216	assorted	10	55
1986	Race Ace	2620	white	20	30
1971	Racer Rig	6194	red/white	85	350
1981	Racing Team Van		yellow, Scene Machine	50	60
1975	Ramblin Wrecker	7659	white	25	45
1977	Ramblin' Cruiser	7659	white without phone number	15	20
	Ramblin' Wrecker	7659	white, blackwall	8	12
1975	Ranger Rig	7666	green	30	50
1974	Rash I	7616	green	45	65
	Rash I	7616	blue	250	500
1972	Rear Engine Moongoose	5699	red	165	370
	Rear Engine Snake	5856	yellow	165	370
1970	Red Baron	6400	red	15	30
1973	Red Baron	6964	red	30	175
1977	Red Baron	6964	red, blackwall	9	16
1994	Red Passion		red	10	12
1982	Rescue Squad	3304	red, Scene Machine	75	90
1974	Road King Truck	7615	yellow, set only	500	800
1976	Rock Buster	9088	yellow	18	25
	Rock Buster	9507	chrome	25	30
1977	Rock Buster	9088	yellow, blackwall	10	14
	Rock Buster	9507	chrome, blackwall set only	35	0
1971	Rocket Bye Baby	6186	assorted	45	175
1974	Rodger Dodger	8259	magenta	50	70
	Rodger Dodger	8259	blue	350	550
1977	Rodger Dodger	8259	gold plated, blackwall or redline	22	35
1969	Rolls Royce Silver Shadow	6276	assorted	25	35
1992	Ruby Red Passion		red	15	25
1971	S'Cool Bus	6468	yellow	150	750
1979	S.W.A.T. Van	2854	blue, Scene Machine	40	50
1970	Sand Crab	6403	assorted	8	35
1975	Sand Drifter	7651	yellow	30	50
	Sand Drifter	7651	green	175	350
1973	Sand Witch	6974	assorted	110	300
1971	Scooper	6193	assorted	100	250
1970	Seasider	6413	assorted	55	120
1977	Second Wind	9644	white, blackwall or redline	30	50
1969	Shelby Turbine	6265	assorted	8	25
1971	Short Order	6176	assorted	35	100
1977	Show Hoss II	9646	yellow, redline	300	375
	Show Hoss II	9646	yellow, blackwall	45	60
1973	Show-Off	6982	assorted	110	300
1972	Sidekick	6022	var.	80	195
1968	Silhouette	6209	assorted	10	80
1990	Simpsons Camper		blue, Scene Machine	5	10
	Simpsons Van		yellow, Scene Machine	5	10
1977	Sir Rodney Roadster	8261	yellow, blackwall	24	30

VEHICLES

VEHICLES

Top to bottom: Chrome King Kuda, 1970; Custom Mustang (with open hood scoop), 1968; Snorkle, 1971; Cockney Cab, 1971, all Mattel.

Hot Wheels

YEAR	MODEL	NO.	DESCRIPTION	EX	MINT
1974	Sir Sidney Roadster	8261	yellow	30	65
	Sir Sidney Roadster	8261	light green	200	400
	Sir Sidney Roadster	8261	orange/brown	225	500
1971	Six Shooter	6003	assorted	60	175
1970	Sky Show Fleetside (Aero Launcher)	6436	assorted	400	600
1973	Snake	6969	white/yellow	750	1500
1971	Snake Dragster	5951	white in a two pack	65	0
1970	Snake Funny Car	6409	assorted	60	300
1971	Snake II	5953	white	60	250
	Snorkel	6020	assorted	60	150
1979	Space Van	2855	grey, Scene Machine	50	60
1971	Special Delivery	6006	blue	40	150
1979	Spiderman	2852	black	15	25
	Spiderman Van	2852	white, Scene Machine	40	50
1969	Splittin' Image	6261	assorted	8	35
1977	Spoiler Sport	9641	light green, redline	15	22
	Spoiler Sport	9641	light green, blackwall	650	750
	Staff Car	9521	olive, blackwall	650	750
	Staff Car	9521	olive, six pack only	500	600
1974	Steam Roller	8260	white	25	60
	Steam Roller	8260	white with seven stars	120	200
1977	Steam Roller	9208	chrome, redline or blackwall	25	35
	Steam Roller	9208	chrome with seven stars	125	200
1975	Street Eater	7669	black	30	50
1976	Street Rodder	9242	black	35	40
1977	Street Rodder	9242	black, blackwall	20	25
1973	Street Snorter	6971	assorted	110	350
1971	Strip Teaser	6188	assorted	65	200
	Sugar Caddy	6418	assorted	20	65
1977	Super Chromes	9505	chrome, blackwall six pack	350	0
1975	Super Van	7649	blue	450	0
	Super Van	7649	plum	85	175
	Super Van	7649	Toys-R-Us	175	250
1976	Super Van	9205	chrome	25	30
1977	Super Van	7649	black, blackwall	12	18
1973	Superfine Turbine	6004	assorted	250	750
	Sweet "16"	6007	assorted	110	300
1970					
	Swingin' Wing	6422	assorted	15	35
1971	T-4-2	6177	assorted	35	150
1977	T-Totaller	9648	black, six pack only	300	375
	T-Totaller	9648	black, blackwall	8	12
	T-Totaller	9648	brown, blackwall	8	12
1971	Team Trailer	6019	white/red	75	200
1970	The Demon	6401	assorted	15	35
1971	The Hood	6175	assorted	15	90
1979	Thing, The	2882	dark blue	20	25
	Thor	2880	yellow	10	25
1977	Thrill Driver Torino	9793	red/white, blackwall set of two	250	0
1970	TNT-Bird	6407	assorted	28	65
1974	Top Eliminator	7630	blue	40	65
1977	Top Eliminator	7630	gold plated, redline or blackwall	30	45
1969	Torero	6260	assorted	10	55
1975	Torino Stocker	7647	red	40	60
1977	Torino Stocker	7647	gold plated, redline or blackwall	30	45
1975	Tough Customer	7655	olive	15	40
1970	Tow Truck	6450	assorted	25	45
	Tri-Baby	6424	assorted	12	35
1984	Turbo Mustang		blue	50	60
1969	Turbofire	6259	assorted	10	40

Hot Wheels

YEAR	MODEL	NO.	DESCRIPTION	EX	MINT
	Twinmill	6258	assorted	10	30
1976	Twinmill II	9509	chrome	25	30
	Twinmill II	8240	orange	20	25
1977	Twinmill II	8240	orange, blackwall	9	12
	Twinmill II	9502	chrome, blackwall set only	35	0
1975	Vega Bomb	7658	orange	45	85
	Vega Bomb	7658	green	350	550
1977	Vega Bomb	7654	orange, blackwall	40	55
1974	Volkswagen	7620	orange with bug on roof	35	55
	Volkswagen	7620	orange with stripes on roof	175	400
1969	Volkswagen Beach Bomb	6274	assorted	50	100
	Volkswagen Beach Bomb	6274	boards in rear	2500	0
1993	VW Bug		pink, Real Rider	20	25
1975	Warpath	7654	white	40	65
1971	Waste Wagon	6192	assorted	85	250
	What-4	6001	assorted	55	150
1970					
	Whip Creamer	6457	assorted	15	35
1990	White Passion		white, in box	15	20
1974	Winnipeg	7618	yellow	60	90
1973	Xploder	6977	assorted	125	350
1977	Z Whiz	9639	grey, redline	40	50
	Z Whiz	9639	grey, blackwall	8	10
	Z Whiz	9639	white, redline	900	0
1982	Z Whiz	9639	blue	35	50

VEHICLES

VEHICLES

Top to bottom: Road Tug service truck, 1966, Structo; Kiddie Toys set, Hubley.

VEHICLES

Hubley

Airplanes

NAME	DESCRIPTION	GOOD	EX	MINT
"America" Plane	trimotor, open cockpit, co-pilot, pilot, cast iron, 17" wingspan	2500	5000	7000
American Eagle Airplane	WWII Fighter, 11" wingspan	150	225	300
American Eagle Carrier Plane	cast metal, 11" wingspan, 1971	60	95	125
B-17 Bomber	15" wingspan	125	185	250
Bremen Junkers Monoplane	10" wingspan	1200	1750	2000
Corsair-Type Fighter Plane		30	45	65
Delta Wing Jet		60	95	125
DO-X Plane	six engines, 5 7/8" wingspan, 1935	165	200	285
Flying Circus	12" wingspan	45	70	90
Lindy Plane	cast iron, 13 1/4" long	1200	1750	2000
Lindy Plane	cast iron, 10" wingspan	500	750	1000
Navy WWII Fighter	folding wings and wheels	30	45	65
P-38 Fighter	camouflage paint, 12 1/2" wingspan	75	100	185
P-38 Plane	black rubber tires	70	95	165
Piper Cub Plane	pot metal	50	75	115
Sea Plane	orange and blue, 2 engines	35	55	75
Single Engine Fighter	3 1/2"	60	90	120
U.S. Air Force	12" wingspan	25	45	75
U.S. Army Monoplane	7 5/8" wingspan, 1941	50	75	100
U.S. Army Single Engine Fighter Plane	black rubber tires	20	35	50

Boats and Ships

Penn Yan Motorboat	15" long	1700	2600	3500

Buses

School Bus	metal, wooden wheels	60	85	125
Service Coach	cast iron, 5" long	675	900	1500

Cars

Auto and Trailer	cast iron, 6 3/4" long, 1939	175	235	295
Auto and Trailer	cast iron, sedan 7 1/4" long, trailer 7 1/8" long, 1936	185	245	315
Buick Convertible	opening top, 6 1/2" long	50	65	90
Buick Convertible	top down, 6 1/2" long	45	60	80
Cadillac	black rubber tires, die cast with tin bottom plate	25	40	70
Car Carrier With Four Cars	cast iron, 10" long	300	475	675
Chrysler Airflow	6" long	145	200	350
Chrysler Airflow	battery operated lights, cast iron, 1934	1250	1900	2750
Chrysler Airflow Car	cast iron, 4 1/2" long	110	195	325
Coupe	cast iron, 9 1/2" long, 1928	900	1400	1800
Coupe	cast iron, 8 1/2" long, 1928	600	1100	1500
Coupe	cast iron, 7" long, 1928	400	650	800
Ford Convertible	cast iron, V/8, 1930's	65	100	185
Ford Coupe	cast iron, V/8, 1930's	65	100	185
Ford Model-T	movable parts	100	150	200
Ford Sedan	cast iron, V/8, 1930s	65	100	185
Ford Town Car	cast iron, V/8, 1930s	65	100	185
Limousine	cast iron, 7" long, 1918	250	325	400
Lincoln Zephyr	1937	250	350	450
Mr. Magoo Car	old timer car, battery operated, 9" long, 1961	75	115	150
Open Touring Car	cast iron, 7 1/2" long, 1911	675	900	1250
Packard Roadster	9 1/2" long, 1930	90	145	250

Hubley

NAME	DESCRIPTION	GOOD	EX	MINT
Packard Straight 8	hood raises, detailed cast motor, cast iron, 11" long, 1927	7500	10000	15000
Race Car #22	cast iron, 7 1/2" long	40	55	85
Race Car #2241	7" long, 1930's	50	75	100
Racer	white rubber wheels, cast iron, 7" long	200	300	400
Racer	cast iron, 10 3/4" long, 1931	75	150	250
Racer	nickel-plated driver, cast iron, 4 3/4" long, 1960's	75	115	150
Racer	red with black wheels, silver grille and driver, 7 1/2" long	35	55	75
Racer #12	die cast, prewar	60	95	125
Racer #629	7" long, 1939	50	75	135
Roadster	cast iron, 7 1/2" long, 1920	75	150	225
Sedan	cast iron, 7" long, 1920's	175	265	350
Service Car	5" long, 1930's	75	115	150
Speedster	cast iron, 7" long, 1911	125	225	350
Station Wagon	die cast	50	75	100
Streamlined Racer	cast iron, 5" long	70	125	140
Studebaker	take-apart, 5" long	400	600	800
Tinytown Station Wagon and Boat Trailer		45	70	90
Touring Car	lady and her dog seated in back, driver in front, 10" long, 1920	650	1000	1450
Yellow Cab	with luggage rack, cast iron, 8" long, 1940	325	500	700

Cycles

NAME	DESCRIPTION	GOOD	EX	MINT
Marathon Rider	bicycle, cast iron	200	300	400

Emergency Vehicles

NAME	DESCRIPTION	GOOD	EX	MINT
Ahrens-Fox Fire Engine	cast iron, 11 1/2" long, 1932	5000	6500	8000
Auto Fire Engine	cast iron, 15" long, 1912	4200	5500	7000
Fire Engine	blue and green, large rear wheels with smaller front wheels, cast iron, 10 3/4" long, 1920"s	4500	5750	7500
Fire Engine	cast iron, 14 1/2" long, 1932	1575	2250	3000
Fire Truck	5" long, 1930's	75	100	150
Fire Truck No. 468		60	95	125
Hook and Ladder Fire Truck	rubber wheels, die cast, 18" long	150	250	500
Hook and Ladder Truck	cast iron, 8" long	100	125	200
Hook and Ladder Truck	cast iron, 23" long, 1912	1850	3000	4500
Hook and Ladder Truck	cast iron, 16 1/2" long, 1926	850	1200	1700
Ladder Fire Truck	cast iron, 5 1/2" long	40	60	80
Ladder Truck	14" long, 1940's	150	275	500
Police Patrol	with three policemen, cast iron, 11" long, 1919	900	1450	2500
Pumper Fire Truck	plastic, 1950's	25	45	65
Seven Man Fire Patrol	cast iron, 15" long, 1912	3575	5700	7500
Special Ladder Truck	cast iron, 13" long, 1938	465	575	975

Farm and Construction Equipment

NAME	DESCRIPTION	GOOD	EX	MINT
Avery	round radiator, cast iron, 4 1/2" long, 1920	175	300	450
Diesel Road Roller	plastic, 1950's	15	25	40
Elgin Street Sweeper	brush sweeps dirt into a bin in the body, uniformed driver, cast iron, 8 1/2" long, 1930	4000	8000	11500
Farm Trailer	with gate, 8" long	30	45	60
Ford 4000	blue and gray, die cast, 1/12 scale	100	200	300
Ford 6000	blue and gray, die cast, 1/12 scale, 1963	125	250	400
Ford 961 Powermaster	red and gray, die cast, 1/12 scale, 1961	125	250	400
Ford 961 Powermaster	red and gray, row crop, die cast, 1/12 scale, 1961	100	200	300
Ford 961 Select-O-Speed	red and gray, die cast, 1/12 scale, 1962	125	250	400
Ford Commander 6000	blue and gray, die cast, 1/12 scale, 1963	125	225	300
Fordson	with loader, cast iron, 8 1/2" long, 1938	750	1250	1700
Fordson	cast iron, 5 1/2" long	150	225	300

VEHICLES

Top to bottom: 1955 Corvette, Hubley; Ford Minute Maid delivery truck, 1955, Tonka; Harley-Davidson, 1930s, Hubley; Tractor with Scoop Shovel and Wagon, 1946-52, Tootsietoys; Mercedes-Benz roadster, 1953, Distler.

Hubley

NAME	DESCRIPTION	GOOD	EX	MINT
Fordson F	with crank and driver, cast iron, 5 1/2" long	150	225	300
Huber Road Roller	large, with standing driver, cast iron, 15" long, 1927	1675	2250	3750
Huber Steam Roller	cast iron, 1/25 scale, 1929	365	450	600
Huber Steam Roller	cast iron, 3 1/4" long, 1929	100	150	215
Junior Tractor		60	95	125
Oliver 70 Orchard	fenders over rear wheels, cast iron, 5" long, 1938	175	300	575
Road Scraper	plastic, 1950's	30	45	60
Steam Shovel	red, nickel-plated boom, cast iron, 4 3/4" long	75	115	150
Tractor	yellow, 5 1/4" long	35	55	75
Tractor and Farmer	plastic	25	35	55
Tractor Shovel	cast iron, 8 1/2" long, 1933	1250	1650	2000

Motorcycles

NAME	DESCRIPTION	GOOD	EX	MINT
Crash Car	motorcycle with cart on back, cast iron, 9" long, 1930's	1000	1650	2000
Harley-Davidson Parcel Post	cast iron, 10" long, 1928	1500	2500	4000
Harley-Davidson Sidecar Motorcycle	cast iron, 9" long, 1930	900	1625	1900
Hill Climber	cast iron, 6 3/4" long, 1935	375	500	900
Indian 4-cylinder Motocycle	cast iron, 9" long, 1929	1700	2425	3000
Indian Air Mail	cast iron, 9 1/4" long, 1929	1575	2650	3500
Indian Armored Car	motorcyle police, cast iron, 8 1/2" long, 1928	1750	3500	6000
Indian Motorcycle	cast iron, 9" long	600	850	1500
Motorcycle	Harley-Davidson, cast iron, 7 1/2" long, 1932	300	450	800
Motorcycle	three wheels, cast iron	400	600	800
Motorcycle Cop With Sidecar	Harley-Davidson, cast iron	700	1000	1500
Motorcycle Crash Car	with cart on back, cast iron, 5" long, 1930"s	100	135	195
Motorized Sidecar Motorcycle	clockwork motor, cast iron, 8 1/2" long, 1932	4000	6250	10000
P.D. Motorcycle Cop	red cycle, black cop, plastic	35	65	95
Patrol Motorcycle	green, 6 1/2" long	275	350	475
Popeye Patrol	cast iron, 9" long, 1938	425	600	950
Popeye Spinach Delivery	red motorcycle, cast iron, 6" long, 1938	375	500	750
Traffic Car	three-wheel transport vehicle, cast iron, 12" long, 1930	600	950	1500
Auto Dump Coal Wagon	cast iron, 16 1/4" long, 1920	800	1200	1500
Auto Express with Roof	cast iron, 9 1/2" long, 1910	500	875	1200
Auto Truck	spoke wheels, cast iron, 10" long, 1918	600	1200	1650
Auto Truck	five-ton truck, cast iron, 17 1/2" long, 1920	1000	1650	2250
Bell Telephone Truck	12" long, 1940	50	75	100
Bell Telephone Truck	spoke wheels, cast iron, 5 1/2" long, 1930	225	335	450
Bell Telephone Truck	white tires, winch works, cast iron, 10" long, 1930	500	1000	1500
Bell Telephone Truck	no driver, solid white tires, cast iron, 3 3/4" long, 1930	150	225	300
Bell Telephone Truck	solid white tires, cast iron, 7" long, 1930	300	400	500
Borden's Milk Truck	cast iron, 7 1/2" long, 1930	1650	2850	4250
Compressor Truck	1953 Ford	60	90	120
Delivery Van	cast iron, 4 1/2" long, 1932	365	475	675
Dump Truck	cast iron, 7 1/2" long	225	335	450
Dump Truck	white rubber tires, 4 1/2" long, 1930's	100	150	200
Dump Truck	plastic, 1950's	25	35	55
Gas Tanker	5 1/2" long	115	225	295
General Shovel Truck	dual rear wheels, cast iron, 10" long, 1931	500	750	1000
Ingersoll-Rand Compressor	cast iron, 8 1/4" long, 1933	3250	6500	10000
Lifesaver Truck	cast iron, 4 1/4" long	350	475	700
Long Bed Dump Truck	series 510, Ford, cast iron	110	175	300
Mack Dump Truck	cast iron, 11" long, 1928	600	800	1450
Merchants Delivery Truck	cast iron, 6 1/4" long, 1925	400	600	850
Milk Truck	cast iron, 3 3/4" long, 1930	115	200	295
Model A Cab with Side Dump Trailer	cast iron, 12 7/8" long, 1931	0	0	0

Hubley

NAME	DESCRIPTION	GOOD	EX	MINT
Nucar Transport	with four vehicles, cast iron, 16" long, 1932	675	1200	1500
Open Bed Auto Express	cast iron, 9 1/2" long, 1910	725	1200	1700
Panama Shovel Truck	Mack truck, cast iron, 13" long, 1934	750	1500	2000
Railway Express Truck	cast iron	135	225	275
Shovel Truck	metal, 10" long	250	375	500
Stake Bed Truck	cast iron, 3" long	75	115	150
Stake Truck	die cast, 7" long, 1950's	75	115	150
Stake Truck	cast metal, 7" long	60	95	125
Stake Truck	white rubber tires, 7" long, 1930's	85	130	175
Stake Truck	white cab, blue stake bed, 12" long	100	150	200
Stockyard Truck #851	with three pigs	60	95	165
Tanker	cast iron, 7" long, 1940's	85	130	175
Tow Truck	cast iron	50	75	100
Tow Truck	Ford, cast metal, 7" long, 1950's	50	65	165
Tow Truck	9" long	135	225	275
Truckmixer	Ford, mixer cylinder rotates when truck moves, cast iron, 8" long, 1932	50	85	185
Wrecker	whitewall tires, green and white, 11 1/2" long	25	50	75
Wrecker	cast iron, 5" long	60	95	125
Wrecker	6" long	70	125	145
Wrecker	Ford, die cast	30	55	125
Wrecker	red, die cast, 9 1/2" long, 1940"s	60	95	165

Wagons and Carts

Alphonse in Mule Pulled Wagon	6 1/2" long	225	335	450
Alphonse in Wagon with Two Goats Pulling	13 3/4" long, 1900's	100	150	210

VEHICLES

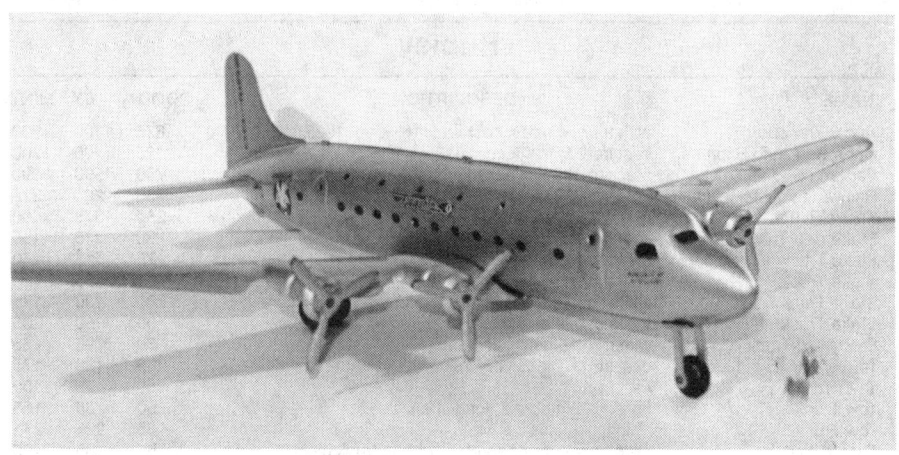

Top to bottom: American Airlines Flagship, 1940; Speed Cop, 1930s; Nutty Mads Car, 1960s; G-Man Pursuit Car, 1935, all Marx.

VEHICLES

Marx

Airplanes

NAME	DESCRIPTION	GOOD	EX	MINT
727 Riding Jet	jet engine sound	150	225	300
Air-Sea Power Bombing Set	12" wingspan, 1940's	325	450	650
Airmail Biplane	4 engines, tin windup, 18" wingspan, 1936	225	325	450
Airmail Monoplane	2 engines, tin windup, 1930	100	150	225
Airplane	light fuselage, tin windup	125	175	250
Airplane	medium fuselage, tin windup	125	150	250
Airplane	monoplane, adjustable rudder, tin windup, 9 1/4" wingspan	150	225	300
Airplane	adjustable rudder, tin windup, 10" wingspan, 1926	150	225	300
Airplane	tin windup, 9 1/4" wingspan, 1926	150	225	300
Airplane	mail biplane, tin windup, 9 3/4" wingspan, 1926	150	225	300
Airplane	two propellers, tin windup, 9 7/8" wingspan, 1927	200	300	400
Airplane	twin engine, tin windup, 9 1/2" wingspan	100	150	200
Airplane	no engines, tin windup, 9 1/2" wingspan	135	200	300
Airplane	monoplane, pressed steel, 9" wingspan, 1942	110	165	300
Airplane #90	tin windup, 5" wingspan, 1930	240	360	480
Airplane with Parachute	monoplane, tin windup, 13" wingspan, 1929	115	170	325
Airways Express Plane	tin windup, 13" wingspan, 1929	200	300	400
American Airlines Airplane	passenger plane, tin windup, 27" wingspan, 1940	130	190	400
American Airlines Flagship	pressed steel, wood wheels, 27" wingspan, 1940	200	300	500
Army Airplane	tin, mechanical fighter, 7" wingspan	125	170	230
Army Airplane	2 engines, tin windup, 18" wingspan, 1938	125	190	250
Army Airplane	biplane, tin windup, 25 3/4" wingspan, 1930	225	340	450
Army Airplane	18" wingspan, 1951	150	225	300
Army Bomber	tri-motor, 25 1/2" wingspan, 1935	250	375	500
Army Bomber	2 engines, tin windup, 18" wingspan, 1940's	250	375	500
Army Bomber	monoplane, litho machine gun & pilot, 25 1/2" wingspan, 1935	300	450	600
Army Bomber with Bombs	camouflage pattern, metal, windup, 12" wingspan, 1930's	100	150	250
Army Fighter Plane	tin windup, 5" wingspan, 1940's	100	150	250
Autogyro	tin windup, 27" wingspan, 1940's	150	225	375
Blue and Silver Bomber	2 engines, tin windup, 18" wingspan, 1940	190	280	375
Bomber	four propellers, metal, windup, 14 1/2" wingspan	100	150	300
Bomber with Tricycle Landing Gear	four engine, tin windup, 18" wingspan, 1940	225	325	425
Camouflage Airplane	four engines, 18" wingspan, 1942	125	200	295
China Clipper	4-engines, tin windup, 18 1/4" wingspan, 1938	100	150	250
City Airport	extra tower and planes, 1930's	125	190	250
Crash-Proof Airplane	monoplane, tin windup, 11 3/4" wingspan, 1933	100	150	200
Cross Country Flyer	19" tall, 1929	375	550	725
Dagwood's Solo Flight Airplane	windup, 9" wingspan, 1935	200	300	950
Dare Devil Flyer	tin windup, 1929	115	170	325
Dare Devil Flyer	Zeppelin shaped, 1928	225	350	475
DC-3 Airplane	aluminum, windup, 9 1/2" wingspan, 1930's	125	190	300
Eagle Air Scout	monoplane, tin windup, 26 1/2" wingspan, 1929	200	300	400
Fighter Jet, USAF	battery operated, 7" wingspan	90	135	250
Fighter Plane	battery operated, remote controlled, 1950's	90	135	250
Fix All Helicopter		275	400	600
Flip-Over Airplane	tin windup	200	300	450
Floor Zeppelin	9 1/2" long, 1931	225	340	500
Floor Zeppelin	16 1/2" long, 1931	350	525	750
Flying Fortress 2095	sparkling, 4 engines, 1940	150	245	400
Flying Zeppelin	windup, 9" long, 1930	225	340	475

Marx

NAME	DESCRIPTION	GOOD	EX	MINT
Flying Zeppelin	windup, 17" long, 1930	350	525	750
Flying Zeppelin	windup, 10" long	275	400	000
Four-Motor Transport Plane	friction, tin litho	120	180	325
Golden Tricky Airplane		75	115	225
Hangar with One Plane	1940's	150	225	500
International Airline Express	monoplane, tin windup, 17 1/2" wingspan, 1931	200	300	425
Jet Plane	friction, 6" wingspan, 1950's	65	90	195
Little Lindy Airplane	friction, 2 1/4" wingspan, 1930	200	300	500
Looping Plane	silver version, tin windup, 7" wingspan, 1941	225	325	525
Lucky Stunt Flyer	tin windup, 6" long, 1928	150	225	350
Mammoth Zeppelin, 1st mammoth	pull toy, 28" long, 1930	400	600	900
Mammoth Zeppelin, 2nd mammoth	pull toy, 28" long, 1930	375	575	775
Municipal Airport Hangar	1929	100	150	425
Overseas Biplane	three propellers, tin windup, 9 7/8" wingspan, 1928	150	275	395
PAA Clipper Plane	pressed steel, 27" wingspan, 1952	125	175	525
PAA Passenger Plane	tin litho, 14" wingspan, 1950's	120	175	275
Pan American	pressed steel, 4 motor, 27" wingspan, 1940	90	150	525
Piggy Back Plane	tin windup, 9" wingspan, 1939	100	150	350
Pioneer Air Express Monoplane	tin litho, pull toy, 25 1/2" wingspan	125	190	300
Popeye Flyer	Popeye and Olive Oyl in plane, tin litho tower, windup, 1936	475	700	1250
Popeye Flyer	Wimpy and Swee'Pea litho on tower, 1936	600	900	1600
Popeye the Pilot	number 47 on side of plane, 8 1/2" wingspan, 1936	300	450	1200
Pursuit Planes	one propeller, 8" wingspan, 1930's	125	200	300
Rollover Airplane	tin windup, forward and reverse, 6" wingspan, 1947	200	300	575
Rollover Airplane	tin windup, 1920's	200	300	575
Rookie Pilot	tin litho, windup, 7" long, 1930's	225	340	550
Seversky P-35	single-engine plane, 16" wingspan, 1940's	125	200	350
Sky Bird Flyer	two planes, 9 1/2" tower, 1947	275	400	550
Sky Cruiser Two-Motored Transport Plane	18" wingspan, 1940's	125	175	325
Sky Flyer	biplane and Zeppelin, 8 1/2" tall tower, 1927	225	340	450
Sky Flyer	9" tall tower, 1937	150	225	375
Spirit of America	monoplane, tin windup, 17 1/2" wingspan, 1930	325	500	650
Spirit of St. Louis	tin windup, 9 1/4" wingspan, 1929	150	225	595
Stunt Pilot	tin windup	175	250	425
Superman Rollover Airplane	tin windup, 1940's	1400	2100	2900
Tower Flyers	1926	175	250	350
Trans-Atlantic Zeppelin	windup, 10" long, 1930	225	350	450
TWA Biplane	4 engine, 18" wingspan	225	350	450
U.S. Marines Plane	monoplane, tin windup, 17 7/8" wingspan, 1930	200	300	400
Zeppelin	friction pull toy, steel, 6" long	100	200	250
Zeppelin	flies in circles, windup, 17" long, 1930	350	525	700
Zeppelin	friction, pull toy, 28" long	375	550	750
Zeppelin	all metal, pull toy, 28" long, 1929	400	600	800
Zeppelin	pull toy, 28" long	375	550	750

Boats and Ships

NAME	DESCRIPTION	GOOD	EX	MINT
Battleship USS Washington	friction, 14" long, 1950's	50	75	225
Caribbean Luxury Liner	sparkling, friction, 15" long	50	75	225
Luxury Liner Boat	tin, friction	100	150	250
Mosquito Fleet Putt Putt Boat		40	55	95
River Queen Paddle Wheel Station	plastic	50	75	150

Marx

NAME	DESCRIPTION	GOOD	EX	MINT
Sparkling Warship	tin windup, 14" long	50	75	195
Tugboat	plastic, battery operated, 6" long, 1966	50	75	195

Buses

NAME	DESCRIPTION	GOOD	EX	MINT
American Van Lines Bus	cream and red, tin windup, 13 1/2" long	65	100	130
Blue Line Tours Bus	tin litho, windup, 9 1/2" long, 1930's	150	225	425
Bus	red bus, 4" long, 1940	35	50	95
Coast to Coast Bus	tin litho, windup, 10" long, 1930's	125	200	325
Greyhound Bus	tin litho, windup, 6" long, 1930's	100	150	200
Liberty Bus	tin litho, windup, 5" long, 1931	75	125	200
Mystery Speedway Bus	tin litho, windup, 14" long, 1938	200	300	550
Royal Bus Lines Bus	tin litho, windup, 10 1/4" long, 1930's	135	200	450
Royal Van Co. Truck "We Haul Anywhere"	tin windup, 9" long, 1920's-30's	140	225	500
School Bus	steel body, wooden wheels, pull toy, 11 1/2" long	125	200	325

Cars

NAME	DESCRIPTION	GOOD	EX	MINT
Amos 'n' Andy Fresh Air Taxicab	tin litho, windup, 8" long, 1930	650	875	1600
Amos 'n' Andy Taxi	tin windup	475	725	950
Anti-Aircraft Gun on Car	5 1/4" long	50	75	225
Army Car	battery operated	65	100	200
Army Staff Car	litho steel, tin windup, 1930's	125	200	425
Army Staff Car	with flasher and siren, tin windup, 11" long, 1940's	75	125	400
Big Lizzie Car	tin windup, 7 1/4" long, 1930's	75	125	235
Blondie's Jalopy	tin litho, 16" long, 1941	325	500	850
Boat Tail Racer #3	tin windup, 5" long, 1930's	40	55	200
Bouncing Benny Car	pull toy, 7" long, 1939	325	500	750
Bumper Auto	large bumpers front and rear, tin windup, 1939	60	100	225
Cadillac Coupe	8 1/2" long, 1931	175	275	400
Cadillac Coupe	trunk w/ tools on luggage carrier, tin windup, 11" long, 1931	200	300	525
Camera Car	heavy guage steel car, 9 1/2" long, 1900	850	1300	1900
Careful Johnnie	plastic driver, 6 1/2" long, 1950's	100	150	350
Charlie McCarthy "Benzine Buggy" Car	with white wheels, tin windup, 7" long, 1938	450	625	950
Charlie McCarthy "Benzine Buggy" Car	with red wheels, tin windup, 7" long, 1938	600	900	1450
Charlie McCarthy and Mortimer Snerd Private Car	tin windup, 16" long, 1939	600	900	1500
Charlie McCarthy Private Car	windup, 1935	1450	2200	3500
College Boy Car	blue car with yellow trim, tin windup, 8" long, 1930's	300	450	525
Convertible Roadster	nickel-plated tin, 11" long, 1930's	175	275	395
Coo Coo Car	8" long, tin windup, 1931	375	575	825
Crazy Dan Car	tin windup, 6" long, 1930's	140	225	395
Dagwood the Driver	tin litho, windup	475	650	900
Dagwood the Driver	8" long, tin windup, 1941	200	300	900
Dan Dipsy Car	nodder, tin windup, 5 3/4" long, 1950's	250	375	450
Dick Tracy Police Car	9" long	150	225	350
Dick Tracy Police Station Riot Car	friction, sparkling, 7 1/2" long, 1946	130	200	375
Dick Tracy Squad Car	yellow flashing light, tin litho, windup, 11" long, 1940's	250	375	525
Dick Tracy Squad Car	battery operated, tin litho, 11 1/4" long, 1949	170	275	475
Dick Tracy Squad Car	friction, 20" long, 1948	125	170	300
Dippy Dumper	Brutus or Popeye, celluloid figure, tin windup, 9", 1930's	350	525	900
Disney Parade Roadster	tin litho, windup, 1950's	100	150	900
Donald Duck Disney Dipsy Car	plastic Donald, tin windup, 5 3/4" long, 1953	425	650	895
Donald Duck Go-Kart	plastic and metal, friction, rubber tires, 1960's	75	130	325

VEHICLES

Marx

VEHICLES

NAME	DESCRIPTION	GOOD	EX	MINT
Donald the Driver	plastic Donald, tin car, windup, 6 1/2" long, 1950's	200	300	595
Dora Dipsy Car	nodder, tin windup, 5 3/4" long, 1953	400	600	725
Dottie the Driver	nodder, tin windup, 6 1/2" long, 1950's	150	225	450
Drive-Up Self Car	turns left, right or straight, 1940	100	150	225
Driver Training Car	tin windup, 1930's	80	120	225
Electric Convertible	tin and plastic, 20" long	65	100	295
Falcon	plastic bubble top, black rubber tires	50	75	175
Funny Fire Fighters	7" long, tin windup, 1941	800	1200	1600
Funny Flivver Car	tin litho, windup, 7" long, 1926	275	425	675
G-Man Pursuit Car	sparks, 14 1/2" long, 1935	190	285	750
Gang Buster Car	tin windup, 14 1/2" long, 1938	200	300	800
Giant King Racer	dark blue, tin windup, 12 1/4" long, 1928	250	375	725
Hot Rod #23	friction motor, tin, 8" long, 1967	45	75	90
Huckleberry Hound Car	friction	125	200	250
International Agent Car	tin windup	30	55	125
International Agent Car	friction, tin litho, 1966	60	100	195
Jaguar	battery operated, 13" long	225	325	450
Jalopy	tin driver, friction, 1950's	125	200	250
Jalopy Car	tin driver, motor sparks, crank, windup	140	225	280
Jolly Joe Jeep	tin litho, 5 3/4" long, 1950's	150	225	375
Joy Riders Crazy Car	tin litho, windup, 8" long, 1928	340	500	675
Jumping Jeep	tin litho, 5 3/4" long, 1947	210	325	425
King Racer	yellow body, red trim, tin windup, 8 1/2" long, 1925	375	575	750
King Racer	yellow with black outlines, 8 1/2" long, 1925	250	430	575
Komical Kop	black car, tin litho, windup, 7 1/2" long, 1930's	450	675	900
Leaping Lizzie Car	tin windup, 7" long, 1927	250	375	500
Learn To Drive Car	windup	110	165	225
Lonesome Pine Trailer and Convertible Sedan	22" long, 1936	375	600	795
Machine Gun on Car	hand crank activation on gun, 3" long	75	130	225
Magic George and Car	litho, 1940's	170	250	350
Mechanical Speed Racer	tin windup, 12" long, 1948	125	200	275
Mickey Mouse Disney Dipsy Car	plastic Mickey, tin windup, 5 3/4" long, 1953	425	655	875
Mickey the Driver	plastic Mickey, tin car, windup, 6 1/2" long, 1950's	170	250	450
Midget Racer "Midget Special" #2	miniature car, clockwork-powered, 5" long, 1930's	125	200	250
Midget Racer "Midget Special" #7	miniature car, tin windup, 5" long, 1930's	125	200	250
Milton Berle Crazy Car	tin litho, windup, 6" long, 1950's	250	375	595
Mortimer Snerd's Tricky Auto	tin litho, windup, 7 1/2" long, 1939	400	600	750
Mystery Car	press down activation, 9" long, 1936	125	200	275
Mystery Taxi	press down activation, steel, 9" long, 1938	160	250	375
Nutty Mad Car	blue car with goggled driver, friction, hard plastic, 1960's	75	130	200
Nutty Mad Car	red tin car, vinyl driver, friction, 4" long, 1960's	100	150	225
Nutty Mad Car	with driver, battery operated, 1960's	100	150	225
Old Jalopy	tin windup, driver, "Old Jalopy" on hood, 7" long, 1950	225	325	425
Parade Roadster	with Disney characters, tin litho, windup, 11" long, 1950	225	325	900
Peter Rabbit Eccentric Car	tin windup, 5 1/2" long, 1950's	250	375	500
Queen of the Campus	with four college students heads, 1950	250	400	525
Race 'N Road Speedway	HO scale racing set, 1950's	60	100	125
Racer #12	tin litho, windup, 16" long, 1942	225	325	625
Racer #3	miniature car, tin windup, 5" long	75	125	195
Racer #4	miniature car, tin windup, 5" long	75	125	195
Racer #5	miniature car, tin windup, 5" long, 1948	75	125	195
Racer #61	miniature car, tin windup, 4 3/4" long, 1930	75	125	195
Racer #7	miniature car, tin windup, 5" long, 1948	75	125	195
Racing Car	two man team, tin litho, windup, 12" long, 1940	125	200	425

Marx

NAME	DESCRIPTION	GOOD	EX	MINT
Racing Car	plastic driver, tin windup, 27" long, 1950	100	175	450
Roadster	11 1/2" long, 1949	100	150	225
Roadster and Cannon Ball Keeper	windup, 9" long	175	250	350
Roadster Convertible with Trailer and Racer	mechanical, 1950	125	200	275
Rocket Racer	tin litho, 1935	275	425	550
Rolls Royce	black plastic, friction, 6" long, 1955	40	60	80
Royal Coupe	tin litho, windup, 9" long, 1930	175	275	375
Secret Sam Agent 012 Car	tin litho, friction, 5" long, 1960's	40	65	165
Sedan	battery operated, plastic, 9 1/2" long	175	275	350
Sheriff Sam and His Whoppee Car	plastic, tin windup, 5 3/4" long, 1949	200	300	475
Siren Police Car	15" long, 1930's	75	125	395
Smokey Sam the Wild Fireman Car	6 1/2" long, 1950	125	200	395
Smokey Stover Whoopee Car	1940's	175	275	525
Snoopy Gus Wild Fireman	7" long, 1926	500	750	1150
Speed Cop	two 4" all tin windup cars, track, 1930's	175	275	795
Speed King Racer	tin litho, windup, 16" long, 1929	325	500	650
Speed Racer	13" long, 1937	250	375	500
Speedway Coupe	tin windup, battery operated headlights, 8" long, 1938	200	300	400
Speedway Set	two windup sedans, figure eight track, 1937	250	375	500
Sports Coupe	tin, 15" long, 1930's	125	200	250
Station Wagon	green with wood-grained pattern, windup, 7" long, 1950	50	75	250
Station Wagon	litho family of four with dogs on back windows, 6 3/4" long	60	100	275
Station Wagon	light purple with wood-grained pattern, windup, 7 1/2" long	60	100	275
Station Wagon	friction, 11" long, 1950	125	200	325
Streamline Speedway	two tin windup racing cars, 1936	175	275	395
Stutz Roadster	driver, 15" long, windup, 1928	325	500	750
Super Hot Rod	"777" on rear door, 11" long, 1940's	200	300	400
Super Streamlined Racer	tin windup, 17" long, 1950's	125	200	400
The Marvel Car, Reversible Coupe	tin windup, 1938	125	200	400
Tricky Safety Car	6 1/2" long, 1950	100	150	200
Tricky Taxi	black and white version, tin windup, 4 1/2" long, 1935	160	250	375
Tricky Taxi	red, black and white, tin windup, 4 1/2" long, 1940's	175	275	375
Uncle Wiggly, He Goes A Ridin' Car	rabbit driving, tin windup, 7 1/2" long, 1935	425	650	850
Walt Disney Television Car	friction, 7 1/2" long, 1950's	125	200	425
Western Auto Track	steel, 24" long	75	125	225
Whoopee Car	witty slogans, tin litho, windup, 7 1/2" long, 1930's	375	580	775
Whoopee Cowboy Car	bucking car, cowboy driver, tin windup, 7 1/2" long, 1930's	400	600	800
Woody Sedan	tin friction, 7 1/2" long	60	100	225
Yellow Taxi	windup, 7" long, 1927	275	425	575
Yogi Bear Car	friction, 1962	50	85	195
<subhead>Emergency Vehicles				
Ambulance	tin litho, 11" long	125	175	375
Ambulance	13 1/2" long, 1937	225	350	650
Ambulance with Siren	tin windup	100	150	400
Army Ambulance	13 1/2" long, 1930's	250	375	750
Boat Tail Racer # 2	litho, 13" long, 1948	125	200	425

VEHICLES

Marx

NAME	DESCRIPTION	GOOD	EX	MINT
Chief-Fire Department No. 1 Truck	friction, 1948	60	90	195
Chrome Racer	miniature racer, 5" long, 1937	85	125	250
City Hospital Mack Ambulance	tin litho, windup, 10" long, 1927	190	280	500
Electric Car	runs on electric power, license # A7132, 1933	225	325	500
Electric Car	windup, 1933	175	250	475
Fire Chief Car	working lights, 16" long	125	200	295
Fire Chief Car	windup, 6 1/2" long, 1949	75	125	275
Fire Chief Car	battery operated headlights, windup, 11" long, 1950	100	150	325
Fire Chief Car	friction, loud fire siren, 8" long, 1936	150	250	425
Fire Chief Car with Bell	10 1/2" long, 1940	175	250	450
Fire Engine	sheet iron, 9" long, 1920's	100	175	335
Fire Truck	battery operated, two celluloid firemen, 12" long	50	75	300
Fire Truck	friction, all metal, 14" long, 1945	90	135	295
Giant King Racer	pale yellow, tin windup, 12 1/2" long, 1928	225	325	575
Giant King Racer	red, 13" long, 1941	200	300	550
Giant Mechanical Racer	tin litho, 12 3/4" long, 1948	100	175	350
H.Q-Staff Car	14 1/2" long, 1930's	325	500	750
Hook and Ladder Fire Truck	3 tin litho firemen, 13 1/2" long	90	135	225
Hook and Ladder Fire Truck	plastic ladder on top, 24" long, 1950	90	135	225
Plastic Racer	6" long, 1948	50	75	150
Racer with Plastic Driver	tin litho car, 16" long, 1950	150	225	375
Rocket-Shaped Racer #12	1930's	275	425	600
Siren Fire Chief Car	red car with siren, 1934	225	325	495
Siren Fire Chief Truck	battery operated, 15" long, 1930's	100	175	325
Tricky Fire Chief Car	4 1/2" long, 1930's	250	375	625
V.F.D. Emergency Squad	with ladder, metal, electrically powered, 14" long, 1940's	70	125	200
V.F.D. Fire Engine	with hoses and siren, 14" long, 1940's	120	180	295
V.F.D. Hook and Ladder Fire Truck	33" long, 1950	140	225	325
War Department Ambulance	1930's	100	150	695

Farm and Construction Equipment

NAME	DESCRIPTION	GOOD	EX	MINT
Aluminum Bulldog Tractor Set	tin windup, 9 1/2" long tractor, 1940	250	375	500
American Tractor	with accessories, tin windup, 8" long, 1926	150	225	300
Army Design Climbing Tractor	tin windup, 7 1/2" long, 1932	80	120	300
Automatic Steel Barn and Mechanical Plastic Tractor	tin windup, 7" long red tractor, 1950	85	125	475
Bulldozer Climbing Tractor	caterpillar type, tin windup, 10 1/2" long, 1950's	45	75	295
Bulldozer Climbing Tractor	bumper auto, large bumpers, tin windup, 1939	50	75	325
Caterpillar Climbing Tractor	yellow tractor, tin windup, 9 1/2" long, 1942	75	125	300
Caterpillar Climbing Tractor	orange tractor, tin windup, 9 1/2" long, 1942	100	175	325
Caterpillar Tractor and Hydraulic Lift	tin windup, 1948	50	75	295
Climbing Tractor	with driver, tin windup, 1920's	50	100	275
Climbing Tractor	tin windup, 8 1/4" long, 1930	100	150	325
Climbing Tractor with Chain Pull	tin windup, 7 1/2" long, 1929	150	225	350
Co-Op Combine	tin friction, 6"	30	45	95
Construction Tractor	reversing, tin windup, 14" long, 1950's	100	150	465

VEHICLES

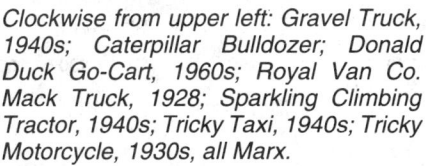

Clockwise from upper left: Gravel Truck, 1940s; Caterpillar Bulldozer; Donald Duck Go-Cart, 1960s; Royal Van Co. Mack Truck, 1928; Sparkling Climbing Tractor, 1940s; Tricky Taxi, 1940s; Tricky Motorcycle, 1930s, all Marx.

Marx

NAME	DESCRIPTION	GOOD	EX	MINT
Copper-Colored Tractor with Scraper	tin windup, 8 1/2" long, 1942	90	135	300
Covered Wagon	friction, tin litho, 9" long	20	30	195
Crawler	with or without blades and drivers, litho, 1/25 scale, 1950	75	125	150
Crawler with Stake Bed	litho, 1/25 scale, 1950	75	130	175
Farm Tractor	tin driver	50	75	100
Farm Tractor and Implement Set	tin windup, tractor mower, hayrake, three-gang plow, 1948	200	300	400
Farm Tractor Set	40 pieces, tin windup, 1939	325	500	650
Farm Tractor Set	40 pieces, 8 1/2" long copper colored tractor, windup, 1940	300	450	600
Farm Tractor Set and Power Plant	32 pieces, tin windup, 1938	200	300	400
Hill Climbing Dump Truck	tin windup, 13 1/2" long, 1932	140	210	375
Industrial Tractor Set	orange and red heavy guage plate tractor, 7 1/2" long, 1930	145	225	395
International Harvester Tractor	diesel, driver and set of tools, 1/12 scale, 1954	75	125	150
Magic Barn and Tractor	plastic tractor, tin litho barn, 1950's	70	100	375
Mechanical Tractor	tin windup, 5 1/2" long, 1942	125	200	275
Midget Climbing Tractor	tin windup, 5 1/4" long, 1935	125	200	250
Midget Road Building Set	tin windup, 5 1/2" long tractor, 1939	200	300	400
Midget Tractor	copper-colored all metal, tin windup, 5 1/4" long, 1940	40	60	125
Midget Tractor	red all metal, tin windup, 5 1/4" long, 1940	30	45	125
Midget Tractor and Plow	tin windup, 1937	70	100	225
Midget Tractor with Driver	red all metal, tin windup, 5 1/4" long, 1940	40	60	220
No. 2 Tractor	red with black wheels, tin windup, 8 1/2" long, 1940	95	150	225
Plastic Sparkling Tractor Set	tin windup, 6 1/2" long tractor with 10 1/2" long wagon, 1950	40	60	195
Plastic Tractor with Scraper	tin windup, 8" long with road scraper, 1949	50	75	195
Power Grader	black or white wheels, 17 1/2" long	50	100	125
Power Shovel		50	75	100
Reversible Six-Wheel Farm Tractor-Truck	tin windup, 13 3/4" steel tractor, 7 1/2" stake truck, 1950	80	125	425
Reversible Six-Wheel Tractor	red steel tractor, tin windup, 11 3/4" long, 1940	200	300	475
Self-Reversing Tractor	tin windup, 10" long, 1936	125	200	350
Sparkling Climbing Tractor	tin windup, 8 1/2" long, 1950's	100	150	300
Sparkling Climbing Tractor	tin windup, 10" long, 1940's	175	275	350
Sparkling Heavy Duty Bulldog Tractor	with road scraper, tin windup, 11" long, 1950's	40	60	225
Sparkling Hi-Boy Climbing Tractor	10 1/2" long, 1950's	25	40	195
Sparkling Tractor	with driver and trailer, tin windup, 16" long, 1950's	60	90	295
Sparkling Tractor	with plow blade, tin windup, 1939	50	75	235
Sparkling Tractor and Trailer Set "Marborook Farms"	tin windup, 21" long, 1950's	55	75	255
Steel Farm Tractor and Implements	tin windup, 15" long steel bulldozer tractor, 1947	160	240	325
Super Power Reversing Tractor	tin windup, 12" long, 1931	100	150	325
Super Power Tractor and Trailer Set	tin windup, 8 1/2" tractor, 1937	125	200	375
Super-Power Bulldog Tractor with V-Shaped Plow	aluminum finish, tin windup, 1938	75	130	200

VEHICLES

Marx

NAME	DESCRIPTION	GOOD	EX	MINT
Super-Power Climbing Tractor and Nine Piece Set	tin windup, 9 1/2" long tractor, 1942	160	240	395
Super-Power Giant Climbing Tractor	tin windup, 13" long, 1939	180	270	450
Tractor	tin windup, 8 1/2" long, 1941	100	175	315
Tractor	red tractor, tin windup, 8 1/2" long, 1941	100	175	300
Tractor and Equipment Set	five pieces, tin windup, 16" long tractor, 1949	160	240	320
Tractor and Mower	tin windup, 5" long litho steel tractor, 1948	50	75	215
Tractor and Six Implement Set	tin windup, 8 1/2" long aluminum tractor, 1948	160	240	425
Tractor and Trailer	tin windup, 16 1/2" long, 1950's	40	60	115
Tractor Road Construction Set	36 pieces, tin windup, 8 1/2" long tractor, 1938	200	325	595
Tractor Set	seven pieces, tin windup, 8 1/2" long tractor, 1932	125	200	400
Tractor Set	five pieces, tin windup, 8 1/2" long tractor, 1935	70	100	220
Tractor Set	four pieces, tin windup, 8 1/2" long, 1936	100	150	325
Tractor Set	32 pieces, tin windup, 1937	180	270	450
Tractor Set	five pieces, tin windup, 8 1/2" tractor, 1938	90	135	295
Tractor Set	40 pieces, tin windup, 8 1/2" long, 1942	300	450	800
Tractor Set	two pieces, tin windup, 19" long steel tractor, 1950	65	100	215
Tractor Trailer and Scraper	tin windup, 8 1/2" long tractor, 1946	80	120	200
Tractor Train with Tractor Shed	tin windup, 8 1/2" long, 1936	100	150	220
Tractor with Airplane	windup, 5 1/2" long tractor, 27" wingspan on airplane, 1941	250	375	675
Tractor with Driver	windup, 1940's	100	150	300
Tractor with Earth Grader	tin windup, mechanical, 21 1/2" long, 1950's	40	60	190
Tractor with Plow and Scraper	aluminum tractor, tin windup, 1938	70	125	295
Tractor with Plow and Wagon	tin windup, 1934	70	125	295
Tractor with Road Scraper	tin windup, 8 1/2" long climbing tractor, 1937	90	125	325
Tractor with Scraper	tin windup, 8 1/2" long, 1933	70	125	295
Tractor with Trailer and Plow	tin windup, 8 1/2" long, 1940	80	120	295
Tractor, Trailer, and V-Shaped Plow	tin windup, 8 1/2" steel tractor, 1939	85	125	325
Tractor-Trailer Set	tin windup, 8 1/2" long copper-colored tractor, 1939	70	125	295
Yellow and Green Tractor	tin windup, 8 1/2" long, 1930	90	135	335
<subhead>Motorcycles				
Motorcycle Cop	tin litho, mechanical, siren, 8 1/4" long	75	125	235
Motorcycle Police	red uniform on cop, tin windup, 8" long, 1930's	125	200	275
Motorcycle Police #3	tin windup, 8 1/2" long	100	150	250
Motorcycle Policeman	orange and blue, tin windup, 8" long, 1920's	100	150	325
Motorcycle Trooper	tin litho, windup, 1935	80	120	275
Motorcyle Delivery Toy	"Speedy Boy Delivery Toy" on rear of cart, tin windup, 1932	175	275	500
Motorcyle Delivery Toy	"Speedy Boy Delivery Toy" on side of cart, tin windup, 1930's	175	280	525
Mystery Police Cycle	yellow, tin windup, 4 1/2" long, 1930's	75	125	225
Mystic Motorcycle	tin litho, windup, 4 1/4" long, 1936	75	130	235
Mystic Motorcycle	tin litho, windup, 4 1/4" long, 1936	225	350	450
P.D. Motorcyclist	tin windup, 4" long	40	60	195
Pinched Roadster Motorcycle Cop	in circular track, tin windup, 1927	125	200	365
Pluto Motorcycle with Siren	1930's	150	225	550
Police Motorcycle with Sidecar	tin windup, 8" long, 1930's	425	650	950

Marx

NAME	DESCRIPTION	GOOD	EX	MINT
Police Motorcycle with Sidecar	tin windup, 3 1/2" long, 1930's	300	450	750
Police Motorcycle with Sidecar	tin litho, windup, 8" long, 1950	175	280	425
Police Patrol Motorcycle with Sidecar	tin windup, 1935	110	165	400
Police Siren Motorcycle	tin litho, windup, 8" long, 1938	150	225	350
Police Squad Motorcycle Sidecar	tin litho, windup, 1950	175	275	395
Police Squad Motorcycle Sidecar	tin litho, windup, 8" long, 1950	100	150	375
Police Tipover Motorcycle	tin litho, windup, 8" long, 1933	200	300	525
Rookie Cop	yellow with driver, tin litho, windup, 8" long, 1940	175	275	425
Sparkling Soldier Motorcycle	tin litho, windup, 8" long, 1940	175	275	425
Speeding Car and Motorcycle Policeman	tin litho, windup, 1939	90	150	350
Tricky Motorcycle	tin windup, 4 1/2" long, 1930's	100	150	250

Tanks

NAME	DESCRIPTION	GOOD	EX	MINT
Anti-aircraft Tank Outfit	4 flat metal soldiers, tank, anti-aircraft gun, etc, 1941	90	135	300
Anti-aircraft Tank Outfit	3 tanks made of cardboard	150	225	350
Army Tank	sparking climbing tank, tin windup, 1940's	150	225	375
Climbing Fighting Tank	tin windup	90	135	250
Climbing Tank	tin windup, 9 1/2" long, 1930	125	200	325
Doughboy Tank	doughboy pops out, tin litho, windup, 9 1/2" long, 1930	150	225	425
Doughboy Tank	sparking tank, tin windup, 10" long, 1937	125	200	400
Doughboy Tank	tin windup, 10" long, 1942	150	225	450
E12 Tank	makes rat-a-tat-tat or rumbling noise, tin windup, 1942	100	150	375
E12 Tank	green tank, 9 1/2" long, tin windup, 1942	150	225	375
M48T Tank	battery operated, 1960's	50	100	225
Midget Climbing Fighting Tank	tin litho, windup, 5 1/4" long, 1931	90	135	235
Midget Climbing Fighting Tank	wide plastic wheels, supergrid tread, 5 1/4" long, 1951	100	150	300
Midget Climbing Fighting Tank	5 1/2" long, 1937	100	175	375
Refrew Tank	tin windup	75	125	200
Rex Mars Planet Patrol Tank	tin windup, 10" long, 1950's	150	225	375
Sparkling Army Tank	tan or khaki hull, tin litho, windup, 1938	95	150	235
Sparkling Army Tank	yellow hull, E12 Tank, tin litho, windup, 1942,	100	150	300
Sparkling Army Tank	camouflage hull, 2 olive guns, tin litho, windup, 5 1/2" long	100	150	275
Sparkling Army Tank	camouflage hull, 2 khaki guns, tin litho, windup, 5 1/2" long	125	200	325
Sparkling Climbing Tank	tin windup, 10" long, 1939	110	165	310
Sparkling Space Tank	tin windup, 1950's	250	375	600
Sparkling Super Power Tank	tin windup, 9 1/2" long, 1950's	75	125	250
Sparkling Tank	tin windup, 4" long, 1948	75	130	195
Superman Turnover Tank	Superman lifts tank, 4" long, tin windup, 1940	250	375	600
Tank	pop up army man shooting	150	225	400
Turnover Army Tank	camouflage, tan or khaki hull, tin windup, 1938	175	275	450
Turnover Army Tank	tin windup, 9" long, 1930	150	225	375
Turnover Tank	tin litho, windup, 4" long, 1942	100	150	275

Trucks

NAME	DESCRIPTION	GOOD	EX	MINT
"Run Right To Read's" Truck	14" long, 1940	125	200	250
A&P Super Market Truck	pressed steel, rubber tires, litho, 19" long	50	100	275
Aero Oil Co. Mack Truck	tin litho, friction, 5 1/2" long, 1930	125	200	350
Air Force Truck	32" long	75	125	300

Marx

NAME	DESCRIPTION	GOOD	EX	MINT
American Railroad Express Agency Inc. Truck	open cab, 7" long, 1930's	100	150	325
American Truck Co. Mack Truck	friction, 5" long	100	150	225
Armored Trucking Co. Mack Truck	black cab, yellow printing, windup, 9 3/4" long	200	300	400
Armored Trucking Co. Truck	tin litho, windup, 10" long, 1927	100	150	325
Army Truck	tin, 12" long, 1950	60	90	150
Army Truck	20" long	50	75	125
Army Truck	canvas top, 20" long, 1940's	150	225	325
Army Truck	olive drab truck, 4 1/2" long, 1930's	125	200	350
Army Truck with Rear Benches and Canopy	olive drab paint, 10" long	75	125	225
Artillery Set	3 piece set, 1930	100	150	225
Auto Carrier	2 yellow plastic cars, 2 ramp tracks. 14" long, 1950	125	200	295
Auto Mac Truck	yellow and red, 12" long, 1950	50	75	175
Auto Transport Mack Truck and Trailer	dark blue cab, dark green trailer, windup, 11 1/2" long, 1932	150	225	375
Auto Transport Mack Truck and Trailer	medium blue cab, friction, 11 1/2" long, 1932	150	225	425
Auto Transport Mack Truck and Trailer	dark blue cab, windup, 11 1/2" long, 1932	150	225	400
Auto Transport Truck	pressed steel, with two plastic cars, 14" long, 1940	75	125	225
Auto Transport Truck	with 2 tin litho cars, 34" long, 1950's	60	90	200
Auto Transport Truck	with 3 windup cars, 22 3/4" long, 1931	250	375	525
Auto Transport Truck	with 3 racing coupes, 22" long, 1933	250	375	525
Auto Transport Truck	double decker transport truck, 24 1/2" long, 1935	275	425	575
Auto Transport Truck	with dump truck, roadster and coupe, 30 1/2" long, 1938	275	425	575
Auto Transport Truck	with 3 cars, 21" long, 1940	250	375	575
Auto Transport Truck	21" long, 1947	175	250	400
Auto Transport Truck	with 2 plastic sedans, wooden wheels, 13 3/4" long, 1950	75	130	225
Auto Transwalk Truck	with 3 cars, 1930's	175	250	395
Bamberger Mack Truck	dark green, windup, 5" long, 1920's	200	300	525
Big Load Van Co. Hauler and Trailer	with little cartons of products, 12 3/4" long, 1927	200	300	525
Big Load Van Co. Mack Truck	windup, 13" long, 1928	225	350	600
Big Shot Cannon Truck	battery operated, 23" long, 1960's	150	250	325
Cannon Army Mack Truck	9" long, 1930's	175	250	325
Carpenter's Truck	stake bed truck, pressed steel, 14" long, 1940's	175	250	350
Carrier with Three Racers	tin litho, windup, 22 3/4" long, 1930	150	225	425
Cement Mixer Truck	red cab, tin finish mixing barrel, 6" long, 1930's	100	150	300
City Coal Co. Mack Dump Truck	14" long, 1934	225	350	525
City Delivery Van	yellow steel truck, 11" long	140	210	280
City Sanitation Dept. "Help Keep Your City Clean" Truck	12 3/4" long, 1940	60	90	250
Coal Truck	battery operated, automatic dump, forward and reverse, tin	75	125	175
Coal Truck	1st version, red cab, litho blue and yellow dumper, 12" long	140	225	280
Coal Truck	2nd version, light blue truck, 12" long	140	225	280
Coal Truck	3rd version, Lumar Co. truck, 10" long, 1939	150	250	310
Coca-Cola Truck	tin, 17" long, 1940's	125	200	250
Coca-Cola Truck	yellow, 20" long, 1950	150	225	300
Coca-Cola Truck	stamped steel, 20" long, 1940's	175	250	350
Coke Truck	red steel, 11 1/2" long, 1940's	100	175	250

VEHICLES

Marx

NAME	DESCRIPTION	GOOD	EX	MINT
Contractors and Builders Truck	10" long	125	200	250
Curtiss Candy Truck	red plastic truck, 10" long, 1950	125	200	250
Dairy Farm Pickup Truck	22" long	60	100	150
Delivery Truck	blue truck, 4" long, 1940	100	175	225
Deluxe Delivery Truck	with six delivery boxes, stamped steel, 13 1/4" long, 1948	100	150	200
Deluxe Trailer Truck	tin and plastic, 14" long, 1950's	70	100	145
Dodge Salerno Engineering Department Truck		600	900	1200
Dump Truck	4 1/2" long, 1930's	100	175	325
Dump Truck	motor, tin friction, 12" long, 1950's	50	75	100
Dump Truck	red cab, gray bumper, yellow bed, 18" long, 1950	100	150	200
Dump Truck	6" long, 1930's	100	150	250
Dump Truck	red cab, green body, 6 1/4" long	50	75	100
Dump Truck	yellow cab, blue bumper, red bed, 18" long, 1950	100	150	200
Emergency Service Truck	friction, tin	125	200	275
Emergency Service Truck	friction, tin, searchlight behind car and siren	150	225	300
Firestone Truck	metal, 14" long, 1950's	50	75	150
Ford Heavy Duty Express Truck	cab with canopy, 1950's	60	100	175
Gas Truck	green truck, 4" long, 1940	60	90	165
Giant Reversing Tractor Truck	with tools, tin windup, 14" long, 1950's	75	130	350
Gravel Truck	1st version, pressed steel cab, red tin dumper, 10" long, 1930	100	150	250
Gravel Truck	2nd version, metal, 8 1/2" long, 1940's	75	125	200
Gravel Truck	3rd version, metal with "Gravel Mixer" drum, 10" long, 1930's	100	150	250
Grocery Truck	cardboard boxes, tinplate and plastic, 14 1/2" long	90	150	225
Guided Missile Truck	blue, red and yellow body, friction, 16" long, 1958	75	125	225
Hi-Way Express Truck	pressed steel	95	150	185
Hi-Way Express Truck	tin, tin tires, 16" long, 1940's	50	75	185
Hi-Way Express Truck	"Nationwide Delivery," 1950's	100	150	225
Hi-Way Express Van Truck	metal, 15 1/2" long, 1940's	100	150	225
Jalopy Pickup Truck	tin windup, 7" long	60	90	175
Jeep	11" long, 1946-50	75	130	200
Jeepster	mechanical, plastic	110	165	250
Lazy Day Dairy Farm Pickup Truck and Trailer	22" long	45	75	275
Lincoln Transfer and Storage Co. Mack Truck	wheels have cut out spokes, tin litho, windup, 13" long, 1928	350	525	750
Lone Eagle Oil Co. Mack Truck	bright blue cab, green tank, windup, 12" long, 1930	225	350	550
Lumar Contractors Scoop/Dump Truck	17 1/2" long, 1940's	75	125	200
Lumar Lines and Gasoline Set	1948	250	375	725
Lumar Lines Truck	red cab, aluminum finished trailer, 14" long	125	200	300
Lumar Motor Transport Truck	litho, 13" long, 1942	75	130	250
Machinery Moving Truck		60	90	200
Mack Army Truck	pressed steel, windup, 7 1/2" long	70	100	250
Mack Army Truck	friction, 5" long, 1930	125	200	350
Mack Army Truck	khaki brown body, windup, 10 1/2" long	125	225	400
Mack Army Truck	13 1/2" long, 1929	225	350	600
Mack Dump Truck	dark red cab, medium blue bed, windup, 10" long, 1928	150	225	475

VEHICLES

Marx

NAME	DESCRIPTION	GOOD	EX	MINT
Mack Dump Truck	medium blue truck, windup, 13" long, 1934	200	300	525
Mack Dump Truck	silver cab, medium blue dump, windup, 12 3/4" long, 1936	225	355	525
Mack Dump Truck	no driver, 19" long, 1930	300	450	650
Mack Dump Truck	tin litho, windup, 13 1/2" long, 1926	225	350	575
Mack Railroad Express Truck #7	tin, 1930's	75	125	450
Mack Towing Truck	dark green cab, windup, 8" long, 1926	175	275	450
Mack U.S. Mail Truck	black body, windup, 9 1/2" long	250	375	500
Magnetic Crane and Truck	1950	175	275	350
Mammouth Truck Train	truck with 5 trailers, 1930's	150	250	450
Meadowbrook Dairy Truck	14" long, 1940	150	275	375
Mechanical Sand Dump Truck	steel, 1940's	100	150	275
Medical Corps Ambulance Truck	olive drab paint, 1940's	100	150	275
Merchants Transfer Mack Truck	red open-stake truck, 10" long	225	350	450
Merchants Transfer Mack Truck	13 1/3" long, 1928	250	375	600
Military Cannon Truck	olive drab paint, cannon shoots marbles, 10" long, 1939	125	200	250
Milk Truck	white truck, 4" long, 1940	60	90	120
Miniature Mayflower Moving Van	operating lights	60	90	125
Motor Market Truck	10" long, 1939	100	175	225
Navy Jeep	windup	50	75	100
North American Van Lines Tractor Trailer	windup, 13" long, 1940's	100	175	225
Panel Wagon Truck		30	55	75
Pet Shop Truck	plastic, 6 compartments with 6 different vinyl dogs, 11" long	125	200	250
Pickup Truck	blue and yellow with wood tires, 9" long, 1940's	50	75	150
Polar Ice Co. Ice Truck	13" long, 1940's	100	175	275
Police Patrol Mack Truck	windup, 10" long	200	300	475
Popeye Dippy Dumper Truck	Popeye figure is celluloid, tin windup	325	500	800
Pure Milk Dairy Truck	glass bottles, pressed steel, 1940	55	80	250
R.C.A. Television Service Truck	plastic Ford panel truck, 8 1/2" long, 1948-50	150	225	300
Railway Express Agency Truck	green closed van truck, 1940's	90	135	300
Range Rider	tin windup, 1930's	250	375	500
Reversing Road Roller	tin windup	60	100	125
Road Builder Tank	1950	125	200	250
Rocker Dump Truck	17 1/2" long	60	90	120
Roy Rogers and Trigger Cattle Truck	metal, 15" long, 1950's	60	100	225
Royal Oil Co. Mack Truck	dark red cab, medium green tank, windup, 8 1/4" long, 1927	200	300	500
Royal Van Co. Mack Truck	1927	250	375	600
Royal Van Co. Mack Truck	red cab, tin litho and paint, windup, 9" long, 1928	210	315	550
Sand and Gravel Truck "Builders Supply Co."	tin windup, 1920	150	225	400
Sand Truck	tin litho, 12 1/2" long, 1940's	125	200	275
Sand Truck	9" long, 1948	100	150	200
Sand-Gravel Dump Truck	tin litho, 12" long, 1950	150	225	300
Sand-Gravel Dump Truck	blue cab, yellow dump with "Gravel" on side, tin, 1930's	140	225	285

VEHICLES

VEHICLES

Top to bottom: Reversible Six-Wheel Farm Tractor-Truck, 1950; Roll-over Airplane, 1920s; Charlie McCarthy Private Car, 1935; A&P Super Markets Truck, all Marx.

Marx

NAME	DESCRIPTION	GOOD	EX	MINT
Sanitation Truck	1940's	135	200	270
Searchlight Truck	pressed steel, 9 3/4" long, 1930's	160	250	325
Side Dump Truck	1940	90	130	275
Side Dump Truck	10" long, 1930's	140	225	280
Side Dump Truck and Trailer	15" long, 1935	150	225	300
Sinclair Tanker	tin, 14" long, 1940's	150	225	300
Stake Bed Truck	rubber stamped chicken on one side of truck bunny on other	100	150	200
Stake Bed Truck	pressed steel, wooden wheels, 7" long, 1936	60	90	200
Stake Bed Truck	red cab, green stake bed, 20" long, 1947	100	150	200
Stake Bed Truck	medium blue cab, red stake bed, 10" long, 1940	75	130	185
Stake Bed Truck	red cab, yellow and red trailer, 14" long	100	150	200
Stake Bed Truck	red cab, blue stake bed, 6" long, 1930's	75	125	150
Stake Bed Truck and Trailer	red truck, silver stake bed	125	200	250
Streamline Mechanical Hauler, Van, and Tank Truck Combi	heavy guage steel, 10 3/8" long, 1936	170	275	350
Sunshine Fruit Growers Truck	red cab, yellow and white trailer with blue roof, 14" long	100	150	225
Tipper Dump Truck	windup, 9 3/4" long, 1950	75	130	175
Tow Truck	aluminum finsih, tin litho windup, 6 1/4" long	75	125	150
Tow Truck	aluminum finish, windup, 6 1/4" long	95	150	190
Tow Truck	10" long, 1935	125	200	250
Tow Truck	red cab, yellow towing unit, 6" long, 1930's	137	200	275
Toyland Dairy Truck	10" long	140	210	280
Toyland's Farm Products Mack Milk Truck	with 12 wooden milk bottles, 10 1/4" long, 1931	200	300	400
Toytown Express Truck	plastic cab	45	70	95
Tractor Trailer with Dumpster	blue and yellow hauler, tan dumpster	125	200	250
Truck	prewar, pressed steel, windup	100	175	225
Truck Train	stake hauler and five trailers, 41" long, 1933	250	400	525
Truck Train	stake hauler and four trailers, 41" long, 1938	350	550	725
Truck with Electric Lights	15" long, 1930's	110	165	300
Truck with Electric Lights	battery operated lights, 10" long, 1935	125	200	375
Truck with Searchlight	toolbox behind cab, 10" long, 1930's	150	225	400
U.S. Air Force Willys Jeep	tin body, plastic figures	70	100	140
U.S. Army Jeep w/Trailer		100	150	200
U.S. Mail Truck	metal, 14" long, 1950's	225	350	450
U.S. Trucking Co. Mack Truck	dark maroon cab, friction, 5 1/2" long, 1930	100	150	200
Van Truck	plastic, 10" long, 1950's	40	60	80
Western Auto Truck	steel, 25" long	60	90	125
Willys Jeep	steel, 12" long, 1938	125	200	250
Willys Jeep and Trailer	1940's	100	160	215
Wrecker Truck	1930's	145	225	290

Wagons and Carts

NAME	DESCRIPTION	GOOD	EX	MINT
Bluto, Brutus' Horse and Cart	celluloid figure, metal, 1938	325	500	750
Busy Delivery	open three-wheel cart, windup, 9" long, 1939	175	275	350
Farm Wagon	horse pulling wagon, 10" long, 1940's	55	80	110
Horse and Cart	with driver, 9 1/2" long, 1950's	45	75	90
Horse and Cart	windup, 7" long, 1934	100	150	225
Horse and Cart with Clown Driver	windup, 7 5/8" long, 1923	75	125	200
Pinocchio Busy Delivery	on unicycle facing 2-wheel cart, windup, 7 3/4" long, 1939	125	200	485
Popeye Horse and Cart	tin windup	200	300	600
Rooster Pulling Wagon	tin litho, 1930's	100	150	225

Marx

NAME	DESCRIPTION	GOOD	EX	MINT
Toylands Farm Products Milk Wagon	tin windup, 10 1/2" long, 1930's	65	100	225
Toylands Milk and Cream Wagon	balloon tires, tin litho, windup 10" long, 1931	125	200	400
Toytown Dairy Horsedrawn Cart	tin windup, 10 1/2" long, 1930's	100	175	400
Two Donkeys Pulling Cart	with driver, tin litho, windup, 10 1/4" long, 1940's	50	75	295
Wagon with Two-Horse Team	late 1940's, tin windup	35	55	200
Wagon with Two-Horse Team	1950, tin windup	30	50	195

Clockwise from upper left: Snowtrac Tractor, 1964; Coca Cola Lorry, 1960; Dodge funny car, 1971; boxed Matchbox including Dumper Truck, 1966; Aveling Barford Tractor Shovel, 1962, all Lesney Products.

VEHICLES

Matchbox

NO.	NAME	DESCRIPTION	GOOD	EX	MINT
43-F	0-4-0 Steam Loco	red cab, bin and tanks, black boiler, coal, and base, 3" long, 1978	4	5	7
47-A	1 Ton Trojan Van	red body, no windows, 2 1/4" long, 1958	25	35	45
73-A	10 Ton Pressure Refueller	bluish gray body, six gray plastic wheels, 2 5/8" long, 1959	20	25	35
4-H	1957 Chevy	metallic rose body, chrome interior, large 5 arch rear wheels, 2 15/16"long, 1979	2	8	12
42-G	1957 Ford T-Bird	red convertible, white interior, silver grill and trunk mounted spare, 1982	3	5	6
30-B	6-Wheel Crane Truck	silver body, orange crane, metal or plastic hook, gray wheels, 2 5/8" long, 1961	20	30	40
30-C	8 Wheel Crane Truck	green body, orange crane, red or yellow hook, 8 black wheels on 4 axles, 3" long, 1965	15	20	25
30-D	8-Wheel Crane Truck	red body, yellow plastic hook, five spoke thin wheels, 3" long, 1970	5	7	10
51-C	8-Wheel Tipper	blue tinted windows, eight black plastic wheels, 3" long, 1969	7	10	15
51-D	8-Wheel Tipper	yellow cab, silver/gray tipper, blue windows, five spoke thin wheels, 3" long, 1970	4	6	8
65-E	Airport Coach	white top and roof, metallic blue bottom, amber windows, yellow interior, comes with varying airline logo decals, 3" long, 1977	5	10	15
51-A	Albion Chieftan	yellow body, tan and light tan bags, small round decal on doors, 2 1/2" long, 1958	10	20	25
75-D	Alfa Carabo	pink body, ivory interior, black trunk, five spoke wide wheels, 3" long, 1971	3	4	5
61-B	Alvis Stalwart	white body, yellow plastic removable canopy, green windows, six plastic wheels, 2 5/8" long, 1966	20	30	50
69-E	Armored Truck	red body, white plastic roof, silver/gray base and grill, "Wells Fargo" on sides, 2 13/16" long, 1978	3	5	8
50-E	Articulated Truck	purple tinted windows, small wide wheels, five spoke wide wheels, 2 3/4" long, 1973	6	8	12
30-G	Articulated Truck	blue cab, white grill, silver/gray dumper, 5 spoke accent wheels, 3" long, 1981	3	5	6
53-A	Aston Martin DB2 Saloon	metallic light green, 2 1/2" long, 1958	20	30	40
19-C	Aston Martin Racing Car	metallic green, metal steering wheel and wire wheels, black plastic tires, 2 1/2" long, 1961	20	30	45
15-B	Atlantic Tractor Super	orange body, tow hook, spare wheel behind cab on body, 2 5/8" long, 1959	20	30	40
16-A	Atlantic Trailer	tan body, six metal wheels, tan tow bar, 3 1/8" long, 1956	15	25	35
16-B	Atlantic Trailer	orange body, eight gray plastic wheels with knobby treads, 3 1/4" long, 1957	25	50	80
32-G	Atlas Extractor	red/orange body, gray platform, turret and treads, black wheels, 3" long 1981	2	4	6
23-E	Atlas Truck	metallic blue cab, silver interior, orange dumper, red and yellow labels on doors, 3" long, 1975	6	8	10
23-G	Audi Quatro	white body, red and black print sides, clear windows, "Audi Sport" on doors, 1982	2	3	5
71-A	Austin 200 Gallon Water Truck	olive green body, four black plastic wheels, 2 3/8" long, 1959	15	25	30
36-A	Austin A50	silver grill, with or without silver rear bumper, no windows, 2 3/8" long, 1957	15	20	30
29-B	Austin A55 Cambridge Sedan	two-tone green, light green roof and rear top half of body, dark metallic green hood and lower body, 2 3/4" long, 1961	10	20	25
68-A	Austin MK II Radio Truck	olive green body, four black plastic wheels, 2 3/8" long, 1959	20	25	30

Matchbox

NO.	NAME	DESCRIPTION	GOOD	EX	MINT
17-B	Austin Taxi Cab	maroon body, 2 1/4" long, 1960	35	50	70
1-D	Aveling Barford Road Roller	green body, canopy, tow hook, 2 5/8" long, 1962	7	10	15
43-B	Aveling Barford Tractor Shovel	yellow body, yellow or red driver, four large plastic wheels, 2 5/8" long, 1962	10	15	25
16-E	Badger Exploration Truck	metallic red body, silver grill, 2 1/4" long, 1974	3	4	6
13-F	Baja Dune Buggy	metallic green, orange interior, silver motor, 2 5/8" long, 1971	3	4	6
58-A	BEA Coach	blue body, four wheels with small knobby treads, 2 1/2" long, 1958	20	25	35
30-E	Beach Buggy	pink, yellow paint splatters, clear windows, 2 1/2" long, 1970	3	4	6
47-E	Beach Hopper	dark metallic blue body, hot pink splattered over body, bright orange interior, tan driver, 2 5/8" long, 1974	3	4	5
14-C	Bedford Ambulance	white body, silver trim, two rear doors open, 2 5/8" long, 1962	40	60	80
28-A	Bedford Compressor Truck	silver front and rear grills, metal wheels, 1 3/4" long, 1956	25	35	45
42-A	Bedford Evening News Van	yellow/orange body, silver grill, 2 1/4" long, 1957	25	35	45
27-A	Bedford Low Loader	light blue cab, dark blue trailer, silver grill and side gas tanks, four metal wheels on cab, two metal wheels on trailer, 3 1/8" long, 1956	225	350	450
27-B	Bedford Low Loader	green cab, tan trailer, silver grill, four gray wheels on cab, two wheels on trailer, 3 3/4" long, 1959	35	55	75
29-A	Bedford Milk Delivery Van	tan body, white bottle load, 2 1/4" long, 1956	15	25	30
17-A	Bedford Removals Van	maroon body, peaked roof, gold grill, 2 1/8" long, 1956	65	115	150
40-A	Bedford Tipper Truck	red cab, silver grill, two front wheels, four dual rear wheels, 2 1/8" long, 1957	25	40	50
3-B	Bedford Ton Tipper	gray cab, gray wheels, dual rear wheels, 2 1/2" long, 1961	10	15	20
13-A	Bedford Wreck Truck	tan body, red metal crane and hook, 2" long, 1955	30	40	55
13-B	Bedford Wreck Truck	tan body, red metal crane and hook, crane attached to rear axle, 2 1/8" long, 1958	30	45	60
23-A	Berkeley Cavalier Trailer	decal on lower right rear of trailer, metal wheels, flat tow hook, 2 1/2" long, 1956	20	30	40
26-E	Big Banger	red body, blue windows, small front wheels, large rear wheels, 3" long, 1972	4	6	8
12-F	Big Bull	orange body, green plow blade, base and sides, chrome seat and engine, orange rollers, 2 1/2" long, 1975	4	6	8
22-F	Blaze Buster	red body, silver interior, yellow label, 5 spoke slicks, 3" long, 1975	4	6	8
61-C	Blue Shark	metallic dark blue, white driver, clear glass, four spoke wide wheels, 3" long, 1971	3	4	6
23-B	Bluebird Dauphine Trailer	green with gray plastic wheels, decal on lower right rear of trailer, door on left rear side opens, 2 1/2" long, 1960	75	135	200
56-C	BMC 1800 Pininfarina	clear windows, ivory interior, five spoke wheels, 2 3/4" long, 1970	7	10	15
45-D	BMW 3.0 CSL	orange body, yellow interior, 2 7/8" long, 1976	3	5	7
53-B	BMW M1	silver gray metallic body with plastic hood, red interior, black stripes and "52" on sides, 2 15/16" long, 1981	2	4	5
9-C	Boat and Trailer	white, hull, blue deck, clear windows, five spoke wheels on trailer, 3 1/4" long, 1966	4	6	9
9-D	Boat and Trailer	white, hull, blue deck, clear windows, five spoke wheels on trailer, 3 1/4" long, 1970	3	5	6

VEHICLES

Matchbox

NO.	NAME	DESCRIPTION	GOOD	EX	MINT
72-E	Bomag Road Roller	yellow body, base and wheel hubs, black plastic roller, 2 15/16" long, 1979	2	4	6
44-E	Boss Mustang	yellow body, amber windows, silver interior, clover leaf wide wheels, 2 7/8" long, 1972	3	4	6
25-C	BP Petrol Tanker	yellow hinged cab, white tanker body, six black plastic wheels, 3" long, 1964	8	15	20
52-B	BRM Racing Car	blue or red body, white plastic driver, yellow wheels, 2 5/8" long, 1965	10	20	25
54-C	Cadillac Ambulance	white body, blue tinted windows, white interior, red cross labels on sides, 2 3/8" long, 1970	5	7	10
27-C	Cadillac Sedan	with or without silver grill, clear windows, white roof, silver wheels, 2 3/4" long, 1960	20	30	40
38-G	Camper	red body, off white camper, unpainted base, 3" long 1980	3	5	7
16-D	Case Tractor Bulldozer	red body, yellow base, motor and blade, black plastic rollers, 2 1/2" long, 1969	5	9	12
18-C	Caterpillar Bulldozer	yellow body with driver, green rubber treads, 2 1/4" long, 1961	10	15	25
18-B	Caterpillar Bulldozer	yellow body and driver, yellow blade, No. 18 cast on back of blade, metal rollers, 2" long, 1958	30	45	65
18-D	Caterpillar Crawler Bulldozer	yellow body, no driver, green rubber treads, 2 3/8" long, 1964	30	45	60
18-A	Caterpillar D8 Bulldozer	yellow body and driver, red blade and side supports, 1 7/8" long, 1956	25	35	45
8-A	Caterpillar Tractor	driver has same color hat as body, metal rollers, rubber treads, crimped axles, 1 1/2" long, 1955	40	60	80
8-B	Caterpillar Tractor	yellow body and driver, large smoke stack, metal rollers, rubber treads, crimped axles, 1 5/8" long, 1959	20	35	45
8-C	Caterpillar Tractor	yellow body and driver, large smoke stack, metal rollers, rubber treads, rounded axles, 1 7/8" long, 1961	10	15	25
8-D	Caterpillar Tractor	yellow body, no driver, plastic rollers, rubber treads, rounded axles, 2" long, 1964	10	15	20
37-D	Cattle Truck	yellow body, gray plastic box with fold down rear door, black plastic wheels, green tinted windows, 2 1/4" long, 1966	7	10	15
37-E	Cattle Truck	gray plastic box, white plastic cattle inside, five spoke thin wheels, green tinted windows, 2 1/2" long, 1970	5	7	10
71-F	Cattle Truck	metallic brown body, yellow/orange cattle carrier, 3" long, 1976	4	6	7
25-H	Celica GT	blue body, black base, white racing stripes and "78" on roof and doors, 2 15/16" long, 1978	3	7	12
25-J	Celica GT	yellow body, blue interior, red "Yellow Fever" on hood, side racing stripes, clear windows, large rear wheels, 1982	2	4	6
3-A	Cement Mixer	blue body and rotating barrel, orange metal wheels, 1 5/8" long, 1953	25	35	45
19-G	Cement Truck	red body, yellow plastic barrel with red stripes, large wide arch wheels, 3" long, 1976	4	5	7
41-F	Chevrolet Ambulance	white body, blue windows and dome light, gray interior, 2 15/16" long, 1978	4	7	10
57-B	Chevrolet Impala	pale blue roof, metallic blue body, green tinted windows, 2 3/4" long, 1961	20	25	35
20-C	Chevrolet Impala Taxi Cab	orange/yellow or bright yellow body, ivory or red interior and driver, 3" long, 1965	10	15	20
44-G	Chevy 4x4 Van	green body and windows, white "Ridin High" with horse and fence on sides, 1982	2	3	5
34-G	Chevy Pro Stocker	white body, red interior, clear front and side windows, frosted rear window, 3" long, 1981	2	3	5
68-E	Chevy Van	orange body, unpainted base and grill, large rear wheels, 3" long, 1979	3	5	7

VEHICLES

Matchbox

NO.	NAME	DESCRIPTION	GOOD	EX	MINT
49-D	Chop Suey Motorcycle	metallic dark red body, yellow bull's head on front handle bars, 2 3/4" long, 1973	5	7	10
12-G	Citroen CX	metallic body, silver base and lights, blue plastic hatch door, 3" long, 1979	4	8	12
66-A	Citroen DS 19	light or dark yellow body, with or without silver grill, four plastic wheels, 2 1/2" long, 1959	20	30	40
51-E	Citroen SM	clear windows, frosted rear windows, five spoke wheels, 3" long, 1972	3	4	6
65-C	Claas Combine Harvester	red body, yellow plastic rotating blades and front wheels, black plastic front tires, solid rear wheels, 3" long, 1967	7	10	15
39-D	Clipper	metallic dark pink, amber windows, bright yellow interior, 3" long, 1973	3	4	5
11-H	Cobra Mustang	orange body, "The Boss" on doors, 1982	2	3	5
37-B	Coca Cola Lorry	orange/yellow body, uneven case load, open base, metal rear fenders, 2 1/4" long, 1957	35	75	100
37-C	Coca Cola Lorry	yellow body of various shades, even case load, silver wheels, black base, 2 1/4" long, 1960	35	55	75
51-F	Combine Harvester	red body, black painted base, yellow plastic grain chute, 2 3/4" long, 1978	3	5	8
69-A	Commer 30 CWT Van	silver grill, sliding left side door, four plastic wheels, yellow "NESTLE'S" decal on upper rear panel, 2 1/4" long, 1959	20	30	40
47-B	Commer Ice Cream Canteen	metallic blue body, cream or white plastic interior with man holding ice cream cone, black plastic wheels, 1963	60	95	125
21-C	Commer Milk Truck	pale green body, clear or green tinted windows, ivory or cream bottle load, 2 1/4" long, 1961	20	30	40
50-A	Commer Pickup Truck	with or without silver grill and bumpers, four wheels, 2 1/2" long, 1958	35	55	75
62-G	Corvette	metallic red body, unpainted base, gray interior, 1979	2	4	6
40-F	Corvette T Roof	white body and interior, black "09" on door, red and black racing stripes, 1982	2	3	5
26-G	Cosmic Blues	white body, blue "COSMIC BLUES" and stars on sides, 2 7/8" long, 1970	2	3	4
74-E	Cougar Village	metallic green body, yellow interior, unpainted base, 3 1/16" long, 1978	3	4	6
41-A	D-Type Jaguar	dark green body, tan driver, open air scoop, 2 13/16" long, 1957	20	30	45
41-B	D-Type Jaguar	dark green body, tan driver, silver wheels, open and closed air scoop, 2 7/16" long, 1960	70	100	145
58-C	D.A.F. Girder Truck	cream body shades, green tinted windows, six black wheels, red plastic girders, 3" long, 1968	7	10	15
58-D	D.A.F. Girder Truck	green windows, five spoke thin wheels, red plastic girders, 2 7/8" long, 1970	5	7	10
47-C	DAF Tipper Container Truck	aqua or silver cab, yellow tipper box with light gray or dark gray plastic roof, 3" long, 1968	6	8	12
47-D	DAF Tipper Container Truck	silver cab, yellow tipper box, five spoke thin wheels, 3" long, 1970	4	6	8
14-A	Daimler Ambulance	cream body, silver trim, no number cast on body, "AMBULANCE" cast on sides, 1 7/8" long, 1956	20	35	45
14-B	Daimler Ambulance	silver trim, "AMBULANCE" cast on sides, red cross on roof, 2 5/8" long, 1958	40	65	85
74-B	Daimler Bus	double deck, white plastic interior, four black plastic wheels, 3" long, 1966	10	15	20
74-C	Daimler Bus	double deck, white plastic interior, five spoke thin wheels, 3" long, 1970	7	10	15
67-E	Datsun 260Z 2+2	metallic burgundy body, black base and grill, yellow interior, 3" long, 1978	3	4	6
24-G	Datsun 280ZX	black body and base, clear windows, 5 spoke wheels, 2 7/8" long, 1979	2	3	5

VEHICLES

Matchbox

NO.	NAME	DESCRIPTION	GOOD	EX	MINT
33-E	Datsun or 126X	yellow body, amber windows, silver interior, 3" long, 1973	5	8	10
9-A	Dennis Fire Escape Engine	red body, metal wheels, no front bumper, 2 1/4" long, 1955	15	20	30
20-F	Desert Dawg Jeep 4x4	white body, red top and stripes, white "Jeep" and yellow, red and green, "Desert Dawg" decal, 1982	2	3	5
1-A	Diesel Road Roller	dark green body, flat canopy, tow hook, driver, 1 7/8" long, 1953	15	25	35
1-H	Dodge Challenger	red body, white plastic top, silver interior, wide five spoke wheels, 2 15/16" long, 1976	3	5	8
63-G	Dodge Challenger	green body, black base, bumpers and grill, clear windows, 2 7/8" long, 1980	3	5	7
52-C	Dodge Charger	clear windows, black interior, five spoke wide wheels, 2 7/8" long, 1970	4	6	8
63-C	Dodge Crane Truck	yellow body, green windows, six black plastic wheels, rotating crane cab, 3" long, 1968	7	10	15
63-D	Dodge Crane Truck	yellow body, green windows, four spoke wide wheels, yellow plastic hook, 2 3/4" long, 1970	5	7	10
70-D	Dodge Dragster	pink body, clear windows, silver interior, five spoke wide front wheels, 3" long, 1971	7	10	15
13-D	Dodge Wreck Truck	green cab and crane, yellow body, green windows, 3" long, 1965	300	500	700
13-E	Dodge Wreck Truck	yellow cab, rear body, red plastic hook, green windows, 3" long, 1970	10	15	20
43-E	Dragon Wheels Volkswagen	light green body, amber windows, silver interior, orange on black "Dragon Wheels" on sides, large rear wheels, 2 13/16" long, 1972	5	7	9
58-B	Drott Excavator	red or orange body, movable front shovel, green rubber treads, 2 5/8" long, 1962	35	50	70
2-A	Dumper	green body, red dumper, gold trim, thin driver, green painted wheels, 1 5/8" long, 1953	35	50	70
2-B	Dumper	green body, red dumper, no trim color, fat driver, 1 7/8" long, 1957	20	30	40
48-C	Dumper Truck	red body, green tinted windows, 3" long, 1966	10	20	25
48-D	Dumper Truck	bright blue cab, yellow body, green windows, 3" long, 1970	5	7	10
25-A	Dunlop Truck	dark blue body, silver grill, 2 1/8" long, 1956	10	20	25
57-E	Eccles Trailer Caravan	orange roof, green plastic interior, five spoke thin wheels, 3" long, 1970	5	7	10
20-B	ERF 686 Truck	dark blue body, silver radiator, eight plastic silver wheels, No. 20 cast on black base, 2 5/8" long, 1959	25	45	60
6-C	Euclid Quarry Truck	yellow body, 3 round axles, 2 frnt black plastic wheels, 2 solid rear dual wheels, 2 5/8" long, 1964	15	25	30
6-B	Euclid Quarry Truck	yellow body, four ribs on dumper sides, plastic wheels, 2 1/2" long, 1957	10	20	25
35-D	Fandango	white body, red interior, chrome rear engine, large 5 spoke rear wheels, 3" long, 1975	3	5	8
58-F	Faun Dump Truck	yellow cab and dumper, black base, 2 7/8" long, 1976	5	10	15
70-F	Ferrari 308 GTB	red body and base, black plastic interior, side stripe, 2 15/16" long, 1981	2	3	5
75-B	Ferrari Berinetta	metallic green body of various shades, ivory interior and tow hook, four wire or silver plastic wheels, 3" long, 1965	10	20	25
75-C	Ferrari Berinetta	ivory interior, five spoke thin wheels, 2 3/4" long, 1970	5	7	10
73-B	Ferrari F1 Racing Car	light and dark red body, plastic driver, white and yellow "73" decal on sides, 2 5/8" long, 1962	15	25	30
61-A	Ferret Scout Car	olive green, tan driver faces front or back, four black plastic wheels, 2 1/4" long, 1959	10	20	25

VEHICLES

Matchbox

NO.	NAME	DESCRIPTION	GOOD	EX	MINT
56-B	Fiat 1500	silver grill, red interior and tow hook, brown or tan luggage on roof, 2 1/2" long, 1965	10	15	20
9-G	Fiat Abarth	white body, red interior, 1982	2	3	5
18-E	Field Car	yellow body, tan plastic roof, ivory interior and tow hook, green plastic tires, 2 5/8" long, 1969	75	145	200
18-F	Field Car	yellow body, tan roof, red wheels, ivory interior and tow hook, 2 5/8" long, 1970	5	7	10
29-C	Fire Pumper Truck	red body, metal grill, white plastic hose and ladders, 3" long, 1966	3	7	10
29-D	Fire Pumper Truck	red body, metal grill, white plastic hose and ladders, 3" long, 1970	2	5	8
53-G	Flareside Pick-up	blue body, white interior, grill and pipes, clear windshield, lettered with "326", "Baja Bouncer" and "B.F. Goodrich", 1982	2	4	5
11-F	Flying Bug	metallic red, gray windows, small five spoke front wheels, large five spoke rear wheels, 2 7/8" long, 1972	5	7	10
63-B	Foamite Fire Fighting Crash Tender	red body, six black plastic wheels, white plastic hose and ladder on roof, 2 1/4" long, 1964	10	15	20
21-D	Foden Concrete Truck	orange/yellow body and rotating barrel, green tinted windows, eight plastic wheels, 3" long, 1968	3	5	7
21-E	Foden Concrete Truck	red body, orange barrel, green base and windows 5 spoke wheels, 2 7/8" long, 1970	2	3	5
26-A	Foden Ready Mix Concrete Truck	orange body and rotating barrel, silver or gold grill, four silver plastic wheels, 1 3/4" long, 1956	65	100	130
26-B	Foden Ready Mix Concrete Truck	orange body, gray plastic rotating barrel, with or without silver grill, six gray wheels, 2 1/4" long, 1961	85	130	175
7-B	Ford Anglia	blue body, green tinted windows, 2 5/8" long, 1961	10	15	20
54-D	Ford Capri	ivory interior and tow hook, clear windows, five spoke wide wheels, 3" long, 1971	3	4	6
45-B	Ford Corsair with Boat	pale yellow body, red interior and tow hook, green roof rack with green plastic boat, 2 3/8" long, 1965	10	15	20
25-E	Ford Cortina	clear windows, ivory interior and tow hook, thin five spoke wheels, 2 3/4" long, 1970	3	6	8
55-H	Ford Cortina	metallic gold/green body, unpainted base and grill, wide multispoke wheels, 3 1/16" long, 1979	3	4	5
55-I	Ford Cortina	metallic tan body, yellow interior, blue racing stripes, 1982	2	4	6
25-D	Ford Cortina G.T.	light brown body in various shade, ivory interior and tow hook, 2 7/8" long, 1968	4	7	10
31-A	Ford Customline Station Wagon	yellow body, no windows, with or without red painted tail lights, 2 5/8" long, 1957	20	30	40
9-F	Ford Escort RS2000	white body, black base and grill, tan interior, wide multispoke wheels, 3" long, 1978	3	5	7
59-B	Ford Fairlane Fire Chief's Car	red body, ivory interior, clear windows, four plastic wheels, 2 5/8" long, 1963	50	75	100
55-B	Ford Fairlane Police Car	silver grill, ivory interior, clear windows, four plastic wheels, 2 5/8" long, 1963	35	55	75
31-B	Ford Fairlane Station Wagon	green or clear windows, with or without red painted tail lights, 2 3/4" long, 1960	20	30	35
59-C	Ford Galaxie Fire Chief's Car	red body, ivory interior, driver and tow hook, clear windows, four black plastic wheels, 2 7/8" long, 1966	7	10	15
59-D	Ford Galaxie Fire Chief's Car	red body, ivory interior and tow hook, clear windows, four spoke thin wheels, 2 7/8" long, 1970	5	7	10
55-C	Ford Galaxie Police Car	white body, ivory interior, driver and tow hook, clear windows, 2 7/8" long, 1966	10	15	20

VEHICLES

Matchbox

NO.	NAME	DESCRIPTION	GOOD	EX	MINT
45-C	Ford Group 6	metallic green body, ivory interior, clear windows, five spoke wide wheels, 3" long, 1970	5	7	10
41-C	Ford GT	white or yellow body, red interior, clear windows, yellow or red plastic wheels, 2 5/8" long, 1965	20	30	40
41-D	Ford GT	white body, red interior, clear windows, five spoke wheels, 2 5/8" long, 1970	5	7	10
71-C	Ford Heavy Wreck Truck	red cab, white bumper, amber or green windows, 3" long, 1968	50	75	100
71-D	Ford Heavy Wreck Truck	red cab, white body, green windows and dome light, four spoke wide wheels, 3" long, 1970	15	25	30
8-F	Ford Mustang	wide five spoke wheels, interior and tow hook same color, 2 7/8" long, 1970	7	10	15
8-E	Ford Mustang Fastback	white body, red interior, clear windows, 2 7/8" long, 1966	7	10	15
6-E	Ford Pick-up	red body, white removable canopy, five spoke wheels, 2 3/4" long, 1970	5	7	10
6-D	Ford Pick-up	red body, white removable plastic canopy, four black plastic wheels, 2 3/4" long, 1968	10	15	20
30-A	Ford Prefect	blue body, metal wheels, silver grill, black tow hook, 2 1/4" long, 1956	40	60	80
7-C	Ford Refuse Truck	orange cab, gray plastic dumper, silver metal loader, 3" long, 1966	7	10	15
7-D	Ford Refuse Truck	gray plastic body, silver metal dumper, 3" long, 1970	6	8	12
63-A	Ford Service Ambulance	olive green body, four plastic wheels, round white circle on sides with red cross, 2 1/2" long, 1959	15	20	30
70-A	Ford Thames Estate Car	yellow upper, bluish/green lower, four plastic wheels, 2 1/8" long, 1959	10	15	20
59-A	Ford Thames Van	silver grill, four plastic knobby wheels, 2 1/8" long, 1958	50	75	100
75-A	Ford Thunderbird	cream top half, pink bottom half, green tinted windows, 2 5/8" long, 1960	25	35	45
39-C	Ford Tractor	blue body, black plastic steering wheel and tires, with or without yellow hood, 2 1/8" long, 1967	4	8	12
46-F	Ford Tractor	blue body, black base, large black plastic rear wheels, 2 3/16" long, 1987	3	5	8
66-F	Ford Transit	orange body, unpainted base, yellow interior, green windows, 2 3/4" long, 1977	2	3	4
61-D	Ford Wreck Truck	red body, black base and grill, frosted amber windows, 3" long, 1978	3	5	7
33-B	Ford Zephyr 6MKIII	blue/green body shades, clear windows, ivory interior, 2 5/8" long, 1963	15	20	30
53-D	Ford Zodiac	clear windows, ivory interior, five spoke wheels, 2 3/4" long, 1970	5	7	10
39-A	Ford Zodiac Convertible	peach/pink body shades, tan driver, metal wheels, silver grill, 2 5/8" long, 1957	35	65	90
53-C	Ford Zodiac MK IV	metallic silver blue body, clear windows, ivory interior, four black plastic wheels, 2 3/4" long, 1968	7	10	15
33-A	Ford Zodiac MKII Sedan	with or without silver grill, with or without red painted tail lights, 2 5/8" long, 1957	20	30	45
72-A	Fordson Tractor	blue body with tow hook, 2" long, 1959	10	20	25
15-F	Fork Lift Truck	red body, yellow hoist, 2 1/2" long, 1972	3	4	6
34-E	Formula 1 Racing Car	metallic pink, white driver, clear glass, wide four spoke wheels, 2 7/8" long, 1971	7	10	15
36-F	Formula 5000	orange body, silver rear engine, large clover leaf rear slicks, 3" long, 1975	5	7	10
28-H	Formula Racing Car	gold body, silver engine and pipes, white driver and "Champion", black "8" on front and sides, large clover leaf rear wheels, 1982	2	3	5
22-E	Freeman Inter-City Commuter	clear windows, ivory interior, five spoke wide wheels, 3" long, 1970	5	7	10

VEHICLES

Matchbox

NO.	NAME	DESCRIPTION	GOOD	EX	MINT
63-E	Freeway Gas Truck	red cab, purple tinted windows, small wide wheels on front, clover leaf design, 3" long, 1973	10	15	20
62-A	General Service Lorry	olive green body, six black wheels, 2 5/8" long, 1959	15	25	30
44-D	GMC Refrigerator Truck	red ribbed roof cab, turquoise box with gray plastic rear door that opens, green windows, 1967	7	10	15
44-D	GMC Refrigerator Truck	green windows, four spoke wheels, gray plastic rear door, 2 13/16" long, 1970	5	7	10
26-C	GMC Tipper Truck	red tipping cab, silver tipper body with swinging door, four wheels, 2 5/8" long, 1968	5	7	10
26-D	GMC Tipper Truck	red cab, silver/gray tipper body, four spoke wide wheels, 2 1/2" long, 1970	10	15	20
66-C	Greyhound Bus	silver body, white plastic interior, clear or dark amber windows, six black plastic wheels, 3" long, 1967	25	35	45
66-D	Greyhound Bus	silver body, white interior, amber windows, five spoke thin wheels, 3" long, 1970	5	7	10
70-B	Grit Spreader Truck	dark red cab, four black plastic wheels, 2 5/8" long, 1966	5	7	10
70-C	Grit Spreader Truck	red cab, yellow body, green windows, gray plastic rear pull, 2 5/8" long, 1970	6	8	12
4-F	Gruesome Twosome	metallic gold body, wide five spoke wheels, 2 7/8" long, 1971	5	7	10
23-F	GT 350	white body, blue stripes on hood, roof and rear deck, 2 7/8" long, 1970	3	4	5
7-E	Hairy Hustler	metallic bronze, silver interior, five spoke front wheels, clover leaf rear wheels, 2 7/8" long, 1971	5	7	10
50-G	Harley Davidson Motorcycle	silver/brown metallic frame and tank, chrome engine and pipes, brown rider, 2 11/16" long, 1980	2	3	5
66-B	Harley Davidson Motorcycle/Sidecar	metallic bronze body, three wire wheels, 2 5/8" long, 1962	30	45	65
69-B	Hatra Tractor Shovel	orange or yellow movable shovel arms, four plastic tires, 3" long, 1965	20	30	40
40-C	Hay Trailer	blue body with tow bar, yellow plastic racks, yellow plastic wheels, 3 3/4" long, 1967	3	5	8
55-G	Hellraiser	white body, unpainted base and grill, silver rear engine, 3" long, 1975	3	5	7
15-G	Hi Ho Silver	metallic pearl gray body, 2 1/2" long, 1971	7	10	15
56-D	Hi-Tailer	white body, silver engine and windshield, wide five spoke front wheels, wide clover leaf rear wheels, 3" long, 1974	4	6	8
43-A	Hillman Minx	with or without silver grill, with or without red painted tail lights, 2 5/8" long, 1958	15	20	30
38-C	Honda Motorcycle and Trailer	metallic blue/green cycle with wire wheels, black plastic tires, orange trailer, 2 7/8" long, 1967	10	15	20
38-D	Honda Motorcycle and Trailer	yellow trailer with five spoke thin wheels, 2 7/8" long, 1970	4	7	10
18-G	Hondarora Motorcycle	red frame and fenders chrome bars, fork, engine, black seat, 2 3/8" long, 1975	5	15	25
17-E	Horse Box	blue tinted windows, five spoke thin wheels, white plastic horses inside box, 2 3/4" long, 1970	4	6	8
40-E	Horse Box	orange cab, off white van with tan plastic door, small wheels, 2 13/16" long, 1977	4	5	7
17-D	Horse Box, Ergomatic Cab	red cab, green plastic box, gray side door, 1969	10	15	20
7-A	Horse Drawn Milk Float	orange body, white driver and bottle load, brown horse with white mane and hoofs, 2 1/4" long, 1954	35	55	75

VEHICLES

VEHICLES

Clockwise from upper left: Horse Trailer, Crane, and Tractor; Ferrari Formula One racer, 1962; Pipe Truck, 1966; Volkswagen Microvan, 1962; boxed Matchbox toys including Pipe Truck/Mercede-Benz truck/Articulated Truck/trailer; BMC 1800 Pininfarina, 1970; 1908 Mercedes Grand Prix racer; boxed Matchbox toys including Denver fire engine, BP exploration rig, Greyhound bus, and English bus, all Lesney Products.

Matchbox

NO.	NAME	DESCRIPTION	GOOD	EX	MINT
46-G	Hot Chocolate	metallic brown front lid and sides, black roof, 2 13/"16" long, 1972	3	4	5
67-d	Hot Rocker	metallic lime/green body, white interior and tow hook, five spoke wide wheels, 3" long, 1973	3	5	7
36-E	Hot Rod Draguar	metallic red body, clear canopy, wide five spoke wheels, 2 13/16" long, 1970	4	6	8
2-G	Hovercraft	metallic green top, tan base, silver engine, yellow windows, 3 1/8" long, 1976	4	8	12
72-D	Hovercraft	white body, black bottom and base, red props, 3" long, 1972	4	7	10
17-C	Hoveringham Tipper	red body, orange dumper, 2 7/8" long, 1963	7	10	15
42-C	Iron Fairy Crane	red body, yellow/orange crane, black plastic wheels, yellow plastic single cable hook, 3" long, 1969	7	10	15
42-D	Iron Fairy Crane	four spoke wheels, yellow plastic hook, 3" long, 1970	25	35	45
14-D	Iso Grifo	blue body, light blue interior and tow hook, clear windows, 3" long, 1968	5	7	10
14-E	Iso Grifo	five spoke wheels, clear windows, 3" long, 1969	3	4	5
65-A	Jaguar 3.4 Litre Saloon	silver grill, silver or black bumpers, four gray plastic wheels, 2 1/2" long, 1959	7	10	15
65-B	Jaguar 3.8 Litre Sedan	red body shades, green tinted windows, four plastic wheels, 2 5/8" long, 1962	5	7	10
28-C	Jaguar Mark 10	light brown body, off white interior, working hood, gray motor and wheels, 2 3/4" long, 1964	35	65	90
32-A	Jaguar XK 140 Coupe	with or without silver grill, with or without red painted tail lights, 2 3/8" long, 1957	20	30	40
32-B	Jaguar XKE	metallic red body, ivory interior, clear or tinted windows, 2 5/8" long, 1962	15	20	35
38-F	Jeep	olive green body, black base and interior, wide 5 spoke reverse accent wheels, no hubs, 2 3/8" long, 1976	5	8	12
5-H	Jeep 4x4 Golden Eagle	brown body, wide 4 spoke wheels, eagle decal on hood, 1982	2	5	8
72-B	Jeep CJ5	yellow body, red plastic interior/tow hook, four yellow wheels, black plastic tires, 2 3/8", 1966	10	15	20
72-C	Jeep CJ5	red interior and tow hook, eight spoke wheels 2 3/8" long, 1970	5	7	10
53-F	Jeep CJ6	red body, unpainted base, bumper and winch, 5 spoke rear accent wheels, 2 15/16" long, 1977	2	3	5
71-B	Jeep Gladiator Pickup Truck	red body, clear windows, green or white interior, four black plastic wheels, fine treads, 2 5/8" long, 1964	18	25	30
2-F	Jeep Hot Rod	cream seats and tow hook, large wide four spoke wheels, 2 5/16" long, 1971	7	10	15
50-B	John Deere Tractor	green body and tow hook, yellow plastic wheels, 2 1/8" long, 1964	10	20	25
51-B	John Deere Trailer	green tipping body with tow bar, two small yellow wheels, three plastic barrels, 2 5/8" long, 1964	25	35	45
11-C	Jumbo Crane	yellow body, black plastic wheels, 3" long, 1965	5	10	15
71-E	Jumbo Jet Motorcycle	dark metallic blue body, red elephant head on handle bars, wide wheels, 2 3/4" long, 1973	4	6	8
38-A	Karrier Refuse Collector	silver grill headlights and bumper, 2 3/8" long, 1957	15	25	30
50-C	Kennel Truck	metallic green body, clear or blue tinted canopy, four plastic dogs, 2 3/4" long, 1969	7	10	15
50-D	Kennel Truck	green windows, light blue tinted canopy, four plastic dogs, 2 3/4" long, 1970	5	7	10
45-E	Kenworth Caboner Aerodyne	white body with blue and brown side stripes, silver grill, tanks and pipes, 1982	2	3	5
41-G	Kenworth Conventional Aerodyne	red cab and chassis, silver tanks and pipes, black and white stripes on cab, 1982	2	3	5

VEHICLES

Matchbox

NO.	NAME	DESCRIPTION	GOOD	EX	MINT
27-F	Lamborghini Countach	yellow body, silver interior and motor, five spoke wheels, 2 7/8" long, 1973	5	7	10
20-D	Lamborghini Marzel	amber windows, ivory interior, 2 3/4" long, 1969	15	20	35
33-C	Lamborghini Miura	metal grill, silver plastic wheels, red or white interior, clear or frosted back window, 2 3/4" long, 1969	10	15	20
33-D	Lamborghini Miura	clear windows, frosted rear window, five spoke wheels, 2 3/4" long, 1970	25	40	50
36-B	Lambretta TV 175 Motor Scooter and Sidecar	metallic green, three whells, 2" long, 1961	25	35	45
12-A	Land Rover	olive green body, tan driver, metal wheels, 1 3/4" long, 1955	20	25	35
12-B	Land Rover	olive green body, no driver, tow hook, 2 1/4" long, 1959	35	55	75
57-C	Land Rover Fire Truck	red body, blue tinted windows, white plastic ladder on roof, 2 1/2" long, 1966	10	15	20
57-D	Land Rover Fire Truck	red body, blue tinted windows, white plastic removable ladder, 2 1/2" long, 1970	5	7	10
32-C	Leyland Petrol Tanker	green cab, white tank body, blue tinted windows, eight plastic wheels, 3" long, 1968	25	40	50
32-D	Leyland Petrol Tanker	green cab, white tank body, blue tinted windows, five spoke thin wheels, 3" long, 1970	10	20	25
40-B	Leyland Royal Tiger Coach	silver/gray body, green tinted windows four plastic wheels, 3" long, 1961	10	15	20
31-C	Lincoln Continental	clear windows, ivory interior, black plastic wheels, 2 7/8" long, 1964	10	15	20
31-D	Lincoln Continental	clear windows, ivory interior, five spoke wheels, 2 3/4" long, 1970	5	7	10
28-G	Lincoln Continental MK-V	red body, tan interior, 3" long, 1979	10	15	20
5-C	London Bus	red body, silver grill and headlights, 2 9/16" long, 1961	10	20	25
5-A	London Bus	red body, gold grill, metal wheels, 2" long, 1954	30	45	60
5-D	London Bus	red body, white plastic seats, black plastic wheels, 2 3/4" long, 1965	7	10	15
5-B	London Bus	red body, 2 1/4" long, 1957	30	45	60
56-A	London Trolley Bus	red body, two trolley poles on top of roof, six wheels, 2 5/8" long, 1958	45	70	90
17-F	Londoner Bus	red body, white interior, five spoke wide wheels, 3" long, 1972	10	15	20
21-A	Long Distance Coach	light green body, black base, "London to Glasgow" orange decal on sides, 2 1/4" long, 1956	10	20	25
21-B	Long Distance Coach	green body, black base, No. 21 cast on baseplate, "London to Glasgow" orange decal on sides, 2 5/8" long, 1958	20	60	45
5-E	Lotus Europa	metallic blue body, clear windows, ivory interior and tow hook, 2 7/8" long, 1969	5	7	10
19-D	Lotus Racing Car	white driver, large rear wheels, 2 3/4" long, 1966	10	15	20
19-E	Lotus Racing Car	metallic purple, white driver, five spoke wide wheels with clover leaf design, 2 3/4" long, 1970	5	10	15
60-D	Lotus Super Seven	butterscotch, clear windshield, black interior and trunk, four spoke wide wheels, 2 7/8" long, 1971	5	7	10
49-A	M3 Army Personnel Carrier	olive green body, gray rubber treads, 2 1/2" long, 1958	15	25	30
28-D	Mack Dump Truck	orange body, green windows, four large plastic wheels, 2 5/8" long, 1968	4	7	10
28-E	Mack Dump Truck	pea green body, green windows, large ballon wheels with clover leaf design, 2 5/8" long, 1970	2	4	6

VEHICLES

Matchbox

NO.	NAME	DESCRIPTION	GOOD	EX	MINT
35-A	Marshall Horse Box	red cab, brown horse box, silver grill, three rear windows in box, 2" long, 1957	20	25	35
52-A	Maserati 4 Cl. T/1948	red or yellow body, cream or white driver with or without circle on left shoulder, 2 3/8" long, 1958	10	17	25
32-E	Maserati Bora	metallic burgandy, clear windows, bright yellow interior, wide five spoke wheels, 3" long, 1972	5	7	10
4-A	Massey Harris Tractor	red body with rear fenders, tan driver, four spoke metal front wheels, 1954	30	40	55
4-B	Massey Harris Tractor	red body, no fenders, tan driver, solid metal front wheels, hollow inside rear wheels, 1 5/8" long, 1957	25	40	50
72-F	Maxi Taxi	yellow body, black "MAXI TAXI" on roof, five spoke wheels, 3" long, 1973	2	3	4
66-E	Mazda RX 500	orange body, purple windows, silver rear engine, five spoke wide wheels, 3" long, 1971	3	4	5
31-G	Mazda RX-7	white body, black base, burgundy stripe, black "RX-7", 3" long, 1979	3	4	5
31-H	Mazda RX-7	gray body with sunroof, black interior, 1982	15	25	35
10-A	Mechanical Horse and Trailer	red cab with three metal wheels, gray trailer with two metal wheels, 2 3/8" long, 1955	15	25	35
10-B	Mechanical Horse and Trailer	red cab, ribbed bed in trailer, metal front wheels on cab, 2 15/16" long, 1958	25	35	45
6-F	Mercedes 350 SL	orange body, black plastic convertible top, light yellow interior, 3" long, 1973	5	7	10
56-E	Mercedes 450 SEL	metallic blue body, unpainted base and grill, 3" long, 1979	3	4	5
3-D	Mercedes Ambulance	ivory interior, red cross label on side doors, 2 7/8" long, 1970	2	5	8
53-B	Mercedes Benz 220 SE	silver grill, clear windows, ivory interior, four wheels, 2 3/4" long, 1963	15	25	30
27-D	Mercedes Benz 230 SL	unpainted metal grill, red plastic interior and tow hook, black plastic wheels, 3" long, 1966	3	6	8
27-E	Mercedes Benz 230 SL	metal grill, blue tinted windshield, five spoke wheels, 2 7/8" long, 1970	7	10	15
46-C	Mercedes Benz 300 SE	clear windows, ivory interior, black plastic wheels, 2 7/8" long, 1968	4	7	10
46-D	Mercedes Benz 300 SE	clear windows, ivory interior, five spoke thin wheels, 2 7/8" long, 1970	5	10	15
3-C	Mercedes Benz Ambulance	varying body colors, white interior and stretcher, blue windows and dome light, metal grill, black plastic wheels, 2 7/8" long, 1968	4	7	10
1-F	Mercedes Benz Lorry	metallic gold, removable orange or yellow canopy, 3" long, 1970	4	6	8
1-E	Mercedes Benz Lorry	pale green body, removable orange plastic canopy, 3" long, 1967	5	7	10
68-B	Mercedes Coach	white plastic top half, white plastic interior, clear windows, four black plastic wheels, 2 7/8" long, 1965	30	40	55
42-F	Mercedes Container Truck	red body, black base and grill, removable ivory container with red top and back door, six wheels, 3" long, 1977	4	5	7
56-F	Mercedes Taxi	tan plastic interior, unpainted base, clear plastic windows, red "Taxi" sign on roof, 3" long, 1980	3	4	5
2-D	Mercedes Trailer	pale green body, removable orange canopy, tow hook, black plastic wheels, 3 1/2" long, 1968	5	7	10
2-E	Mercedes Trailer	metallic gold body, removable canopy, rotating tow bar, 3 1/4" long, 1970	3	4	5
49-B	Mercedes Umimog	silver grill, four black plastic tires, 2 1/2" long, 1967	10	15	20
62-C	Mercury Cougar	metallic lime green body shades, red plastic interior and tow hook, silver wheels, 3" long, 1968	7	10	15

VEHICLES

303

Matchbox

NO.	NAME	DESCRIPTION	GOOD	EX	MINT
62-D	Mercury Cougar	red interior and tow hook, five spoke thin wheels, 3" long, 1970	3	4	6
62-E	Mercury Cougar "Rat Rod"	red interior and tow hook, small five spoke front wheels, larger five spoke wide rear windows, 3" long, 1970	4	6	8
59-E	Mercury Fire Chief's Car	red body, ivory interior, two occupants, clear windows, five spoke wide wheels, 3" long, 1971	5	7	10
55-D	Mercury Police Car	white body, ivory interior with two figures, clear windows, four silver wheels with black plastic tires, 3" long, 1968	10	15	20
55-E	Mercury Police Car	white body, ivory interior, two occupants, five spoke thin wheels, 3" long, 1970	5	7	10
55-F	Mercury Police Station Wagon	white body, ivory interior, no occupants, five spoke wide wheels, 3" long, 1971	5	7	10
73-C	Mercury Station Wagon	metallic lime green body shades, ivory interior with dogs in rear, 3 1/8" long, 1968	7	10	15
73-D	Mercury Station Wagon	red body, ribbed rear roof, ivory interior with two dogs, 3" long, 1970	4	6	8
73-E	Mercury Station Wagon	red body, ribbed rear roof, ivory interior with two dogs, 3" long, 1972	3	5	7
35-C	Merryweather Fire Engine	metallic red body, blue windows, white removable ladder on roof, five spoke thin wheels, 3" long, 1969	5	7	10
48-A	Meteor Sports Boat and Trailer	metal boat with tan deck and blue hull, black metal trailer with tow bar, 2 3/8" long, 1958	30	45	60
19-A	MG Midget	white body, tan driver, red seats, spare tire on trunk, 2" long, 1956	50	60	75
19-A	MG Sports Car	silver grill and headlights, tan driver, red painted seats, 2" long, 1956	35	50	70
19-B	MG Sports Car	silver or gold grills, tan driver, 2 1/4" long, 1958	30	45	65
64-B	MG-1100	green body, ivory interior, driver, dog and tow hook, clear windows, four black plastic wheels, 2 5/8" long, 1966	5	7	10
64-C	MG-1100	ivory interior and tow hook, one occupant and dog, clear windows, 2 5/8" long, 1970	7	10	15
19-B	MGA Sports Car	white body variation, silver wheels, tan driver, silver or gold grills, 2 1/4" long, 1958	50	95	125
51-G	Midnight Magic	black body, silver stripes on hood, five spoke front wheels, clover leaf rear windows, 1972	2	3	4
14-F	Mini Haha	red body, pink driver, silver engine, large spoke rear slicks, 2 3/8" long, 1975	5	9	12
74-A	Mobile Refreshment Canteen	cream, white, or silver body, upper side door opens with interior utensils, "Refreshment" on front side, 2 5/8" long, 1959	20	40	60
1-G	Mod Rod	yellow body, tinted windows, red or black wheels, 2 7/8" long, 1971	10	15	20
25-F	Mod Tractor	metallic purple, orange/yellow seat and tow hook, 2 1/8" long, 1972	10	15	20
73-G	Model A Ford	off white body, black base, green fenders and running boards, 1979	2	3	5
3-E	Monteverdi Hai	dark orange body, blue tinted windows, ivory interior, 2 7/8" long, 1973	3	6	8
60-A	Morris J2 Pickup	blue body, open windshield and side door windows, four plastic wheels, 2 1/4" long, 1958	15	25	30
46-A	Morris Minor 1000	dark green body, metal wheels, no windows, 2" long, 1958	20	30	45
2-C	Muirhill Dumper	red cab, green dumper, black plastic wheels, 2 1/6" long, 1961	10	20	25
54-G	NASA Tracking Vehicle	white body, silver radar screen, red windows, blue "Space Shuttle Command Center", red "NASA" on roof, 1982	2	3	5

VEHICLES

Matchbox

NO.	NAME	DESCRIPTION	GOOD	EX	MINT
36-C	Opel Diplomat	metallic light gold body, white interior and tow hook, clear windows, black plastic wheels, 2 3/5" long, 1966	10	15	20
36-D	Opel Diplomat	ivory interior and tow hook, clear windows, five spoke thin wheels, 2 7/8" long, 1970	5	7	10
74-F	Orange Peel	white body, wide orange and black stripe and black "ORANGE PEEL" on each side, 3" long, 1971	3	4	5
47-F	Pannier Tank Loco	green body, black base and insert, 6 large plastic wheels, 3" long, 1979	3	5	7
8-H	Pantera	white body, blue base, red/brown interior, five spoke rear slicks, 3" long, 1975	35	45	60
54-E	Personnel Carrier	olive green body, green windows, black base and grill, tan men and benches, 3" long, 1976	4	5	7
43-G	Perterbilt Conventional	black cab and chassis, silver grill, fenders and tanks, red and white side stripes, 6 wheels, 3" long, 1982	2	3	5
19-H	Peterbilt Cement Truck	green body, orange barrel, "Big Pete" decal on hood, 1982	2	3	5
30-H	Peterbilt Quarry Truck	yellow body, gray dumper, silver tanks, "Dirty Dumper" on sides, 6 wheels, 1982	2	4	6
56-G	Peterbilt Tanker	blue cab, white tank with red "Milks's the One", silver tanks, grill, and pipes, 1982	15	25	40
48-E	Pi-Eyed Piper	metallic blue body, amber windows, small front wheels, large rear wheels, 2 7/8" long, 1972	5	7	10
46-B	Pickford Removal Van	green body, with or without silver grills, 2 5/8" long, 1960	15	30	50
10-D	Pipe Truck	red body, gray pipes, "Leyland" or "Ergomatic" on front base, eight black plastic wheels, 2 7/8", 1966	7	10	15
10-E	Pipe Truck	black pipe racks, eight five spoke thin wheels, 2 7/8" long, 1970	4	6	8
10-F	Piston Popper	metallic blue body, white interior, 2 7/8" long, 1973	4	6	8
60-F	Piston Popper	yellow body, red windows, silver engine, labels top and sides, large rear wheels, 1982	2	3	5
59-F	Planet Scout	metallic green top, green bottom and base, silver interior, grill and roof panels, large multispoke rear wheels, 2 3/4" long 1975	4	5	7
10-G	Plymouth Gran Fury Police Car	white body w/black detailing, "Police" on doors, white interior, 3" long, 1979	3	4	5
52-D	Police Launch	white deck, blue hull and men, 3" long, 1976	2	4	6
33-F	Police Motorcyclist	white frame, seat and bags, silver engine and pipes, wire wheels, 2 1/2" long, 1977	5	7	10
20-E	Police Patrol	white body, "Police" on orange side stripe, orange interior, 2 7/8" long, 1975	6	8	12
39-B	Pontiac Convertible	purple body, with or without silver grill, cream or ivory interior, silver wheels, 2 3/4" long, 1962	30	50	75
4-G	Pontiac Firebird	metallic blue body, silver interior, slick tires, 2 7/8" long, 1975	2	7	12
22-C	Pontiac Gran Prix Sports Coupe	light gray interior and tow hook, clear windows, four black plastic wheels, 3" long, 1964	6	9	12
22-D	Pontiac Gran Prix Sports Coupe	light gray interior, clear windows, five spoke thin wheels, 3" long, 1970	2	4	6
16-G	Pontiac Trans Am	white body, red interior, clear windows, blue eagle decal, 1982	2	3	4
35-F	Pontiac Trans Am T Roof	black body, red interior, yellow "Turbo" on doors, yellow eagle on hood, 1982	2	3	5
43-C	Pony Trailer	yellow body, clear windows, gray plastic rear fold-down door, four plastic wheels, 2 5/8" long, 1968	7	10	15
43-D	Pony Trailer	yellow body, clear windows, gray rear door, five spoke thin wheels, 2 5/8" long, 1970	3	5	7

VEHICLES

Matchbox

NO.	NAME	DESCRIPTION	GOOD	EX	MINT
68-C	Porsche 910	amber windows, ivory interior, five spoke wheels, 2 7/8" long, 1970	7	10	15
3-F	Porsche Turbo	metallic brown body, black base, yellow interior, wide five arch wheels, 3" long, 1978	4	7	10
15-A	Prime Mover	silver trim on grill and tank, tow hook same color as body, 2 1/8" long, 1956	25	35	45
59-G	Prosche 928	metallic brown body, black base, wide 5 spoke wheels, 3" long, 1980	3	5	6
6-A	Quarry Truck	orange cab, gray dumper with six vertical ribs, metal wheels, 2 1/8" long, 1954	20	30	40
29-E	Racing Mini	clear windows, five spoke wide wheels, 2 1/4" long, 1970	5	7	10
44-F	Railway Passenger Car	cream plastic upper and roof, red metal lower, black base, 3 1/16" long, 1978	3	5	7
14-G	Rallye Royal	metallic pearl gray body, black plastic interior, five spoke wide wheels, 2 7/8" long, 1973	3	4	5
48-G	Red Rider	red body, white "Red Rider" and flames on sides, 2 7/8" long, 1972	2	3	4
15-C	Refuse Truck	blue body, gray dumper with opening door, 2 1/2" long, 1963	10	15	20
36-G	Refuse Truck	red metallic body, silver/gray base, orange plastic container, 3" long, 1980	2	3	4
62-F	Renault 17TL	white interior, green tinted windows, green "9" in yellow and black circle, 3" long, 1974	5	7	10
21-G	Renault 5TL	yellow body and removable rear hatch, tan interior, silver base and grill, 2 11/16" long, 1978	4	9	15
1-I	Revin' Rebel	orange body, blue top, black interior, large five spoke rear wheels, 1982	2	3	5
19-F	Road Dragster	ivory interior, silver plastic motor, 2 7/8" long, 1970	3	4	6
1-B	Road Roller	pale green body, canopy, tow hook, dark tan or light tan driver, 2 1/4" long, 1953	25	45	65
1-C	Road Roller	light green or dark green body, canopy, metal rollers, tow bar, driver, 2 3/8" long, 1958	20	25	35
21-F	Road Roller	yellow body, red seat, black plastic rollers, 2 5/8" long, 1973	7	10	15
11-A	Road Tanker	green body, flat base between cab and body, gold trim on front grill, gas tanks, metal wheels, no number cast, 2" long, 1955	175	265	350
11-B	Road Tanker	red body, gas tanks, "11" on baseplate, black plastic wheels, 2 1/2" long, 1958	30	55	75
44-B	Rolls Royce Phantom V	clear windows, ivory interior, black plastic wheels, 2 7/8" long, 1964	15	20	30
44-A	Rolls Royce Silver Cloud	metallic blue body, no windows, with or without silver grill, 2 5/8" long, 1958	15	20	25
24-C	Rolls Royce Silver Shadow	metallic red body, ivory interior, clear windows, silver hub caps or solid silver wheels, 3" long, 1967	10	15	20
24-D	Rolls Royce Silver Shadow	ivory interior, clear windows, five spoke wheels, 3" long, 1970	5	7	10
69-C	Rolls Royce Silver Shadow Coupe	amber windshield, five spoke wheels, 3" long, 1969	5	7	10
39-E	Rolls Royce Silver Shadow II	metallic silver gray body, red interior, clear windshield, 3 1/16" long, 1979	3	5	7
7-G	Rompin' Rabbit	white body, red windows, yellow lettered "Rompin Rabbit" on side, 1982	2	3	5
54-B	S & S Cadillac Ambulance	white body, blue tinted windows, white interior, red cross decal on front doors, 2 7/8" long, 1965	10	15	20
65-D	Saab Sonnet	metallic blue body, amber windows, light orange interior and hood, five spoke wide wheels, 2 3/4" long, 1973	5	7	10

VEHICLES

Matchbox

NO.	NAME	DESCRIPTION	GOOD	EX	MINT
12-C	Safari Land Rover	clear windows, white plastic interior and tow hook, black plastic wheels, 2 1/3" long, 1965	7	10	15
12-D	Safari Land Rover	metallic gold, clear windows, tan luggage, five spoke thin wheels, 2 13/16" long, 1970	30	45	65
67-A	Saladin Armoured Car	olive green body, rotating gun turret, six black plastic wheels, 2 1/2" long, 1959	15	20	25
48-F	Sambron Jacklift	yellow body, black base and insert, no window, orange and yellow fork and boom combinations, 3 1/16" long, 1977	4	7	10
54-A	Saracen Personnel Carrier	olive green body, six black plastic wheels, 2 1/4" long, 1958	10	17	25
11-D	Scaffolding Truck	silver body, green tinted windows, black plastic wheels, 2 1/2" long, 1969	4	7	10
11-E	Scaffolding Truck	silver/gray body, green tinted windows, yellow pipes, 2 7/8" long, 1969	4	6	8
64-A	Scammel Breakdown Truck	olive green, double cable hook, six black plastic wheels, 2 1/2" long, 1959	15	25	30
16-C	Scammel Mountaineer Dump Truck/Snow Plow	gray cab, orange dumper body, six plastic wheels, 3" long, 1964	10	20	25
5-F	Seafire Boat	white deck, blue hull, silver engine, red pipes, 2 15/16" long, 1975	6	8	10
75-F	Seasprite Helicopter	white body, red base, black blades, 1977	3	5	7
12-E	Setra Coach	clear windows, ivory interior, five spoke thin wheels, 3" long, 1970	5	7	10
29-F	Shovel Nose Tractor	yellow body and base, red plastic shovel, silver engine, 2 7/8" long, 1976	8	15	20
24-F	Shunter	metallic green body, red base, tan instruments, no window, 3" long, 1978	3	5	7
26-F	Site Dumper	yellow body and dumper, black base, 2 5/8" long, 1976	2	3	5
60-B	Site Hut Truck	blue body, blue windows, four black plastic wheels, 2 1/2" long, 1966	7	10	15
60-C	Site Hut Truck	blue cab, blue windows, five spoke thin wheels, 2 1/2" long, 1970	5	7	10
41-E	Siva Spider	metallic red body, cream interior, clear windows, wide five spoke wheels, 3" long, 1972	5	7	10
37-G	Skip Truck	red body, yellow plastic bucket, light amber windows, silver interior, 2 11/16" long, 1976	3	5	7
64-D	Slingshot Dragster	pink body, white driver, five spoke thin front wheels, eight spoke wide rear wheels, 3" long, 1971	7	10	15
13-G	Snorkel Fire Engine	red body, yellow plastic snorkel and fireman, 3" long, 1977	3	5	7
35-B	Snowtrac Tractor	red body, silver painted grill, green windows, white rubber treads, 2 3/8" long, 1964	10	15	20
37-F	Soopa Coopa	metallic blue, amber windows, yellow interior, 2 7/8" long, 1972	3	4	5
48-B	Sports Boat and Trailer	plastic boat, red or white deck, hulls in red, white or cream, gold or silver motors, blue metal 2 wheel trailers, boat 2 3/8" long, trailer 2 5/8" long, 1961	35	65	80
4-E	Stake Truck	cab colors vary, 2 7/8" long, 1970	5	7	10
4-D	Stake Truck	yellow cab, green tinted windows, 2 7/8" long, 1967	6	8	12
20-A	Stake Truck	gold trim on front grill and side gas tanks, ribbed bed, metal wheels, 2 3/8" long, 1956	50	75	100
38-E	Stingeroo Cycle	metallic purple body, ivory horse head at rear of seat, five spoke wide rear wheels, 3" long, 1973	4	6	8
46-E	Stretcha Fetcha	white body, blue windows, pale yellow interior, 2" long, 1972	6	8	12

VEHICLES

Matchbox

VEHICLES

NO.	NAME	DESCRIPTION	GOOD	EX	MINT
28-F	Stroat Armored Truck	metallic gold body, brown plastic observer coming out of turret, five spoke wide wheels, 2 5/8" long, 1974	8	15	25
42-B	Studebaker Lark Wagonaire	blue body, sliding rear roof panel, white plastic interior and tow hook, 3" long, 1965	10	15	20
10-C	Sugar Container Truck	blue body, eight gray plastic wheels, "Tate & Lyle" decals on sides and rear, 2 5/8" long, 1961	30	55	75
37-H	Sun Burner	black body, red and yellow flames on hood and sides, 3" long, 1972	2	3	4
30-F	Swamp Rat	green deck, yellow plastic hull, tan soldier, black engine and prop, 3" long, 1976	2	4	6
27-G	Swing Wing Jet	red top and fins, white belly and retractable wings, 3" long, 1981	2	3	5
62-B	T.V. Service Van	cream body, green tinted windows with roof window, four plastic wheels, 2 1/2" long, 1963	25	40	50
53-E	Tanzara	orange body, silver interior, small front wheels, larger rear wheels, 3" long, 1972	3	4	5
24-E	Team Matchbox	white driver, silver motor, wide clover leaf wheels, 2 7/8" long, 1973	15	20	25
28-B	Thames Trader Compressor Truck	yellow body, black wheels, 2 3/4" long, 1959	20	25	35
13-C	Thames Wreck Truck	red body, bumper and parking lights, 2 1/2" long, 1961	15	25	30
74-D	Toe Joe	metallic lime green body, yellow interior, five spoke wide wheels, 2 3/4" long, 1972	3	4	6
23-C	Trailer Caravan	yellow or pink body with white roof, blue removable interior, 2 7/8" long, 1965	4	7	10
4-C	Triumph Motorcycle and Sidecar	silver/blue body, wire wheels, 2 1/8" long, 1960	25	40	60
42-E	Tyre Fryer	metallic red body, cream interior, clear windows, wide five spoke wheels, 3" long, 1972	3	4	6
5-G	US Mail Jeep	blue body, white base and bumpers, black plastic seat, white canopy, wide 5 arch rear wheels, 2 3/8" long, 1978	5	10	15
34-F	Vantastic	orange body, white base and interior, silver engine, large rear slicks, 2 7/8" long, 1975	4	7	9
22-B	Vauxhall Cresta	with or without silver grill, tow hook, plastic wheels, 2 5/8" long, 1958	25	40	50
40-D	Vauxhall Guildsman	pink body, light green windows, light cream interior and tow hook, wide five spoke wheels, 3" long, 1971	3	4	5
22-A	Vauxhall Sedan	dark red body, cream or off white roof, tow hook, 2 1/2" long, 1956	20	30	35
38-B	Vauxhall Victor Estate Car	yellow body, red or green interior, clear windows, 2 5/8" long, 1963	10	18	25
45-A	Vauxhall Victor Saloon	yellow body, with or without green tinted windows, with or without silver grill, 2 3/8" long, 1958	10	15	20
31-E	Volks Dragon	red body, purple tinted windows, 2 1/2" long, 1971	3	4	5
25-B	Volkswagen 1200 Sedan	silver-blue body, clear or tinted windows, 2 1/2" long, 1960	25	40	50
15-D	Volkswagen 1500 Saloon	off white body and interior, clear windows, "137" on doors, 2 7/8" long, 1968	10	20	30
15-E	Volkswagen 1500 Saloon	clear windows, "137" on doors, red decal on front, 2 7/8" long, 1968	7	15	20
67-B	Volkswagen 1600 TL	ivory interior, four black plastic tires, 2 3/4" long, 1967	10	15	20
67-C	Volkswagen 1600 TL	ivory interior, clear windows, five spoke wheels, 2 5/8" long, 1970	5	7	10
34-C	Volkswagen Camper Car	silver body, orange interior, black plastic wheels, raised roof, six windows, 2 5/8" long, 1967	15	20	30
34-D	Volkswagen Camper Car	silver body, orange interior, black plastic wheels, short raised sun roof, 2 5/8" long, 1968	10	20	25

Matchbox

NO.	NAME	DESCRIPTION	GOOD	EX	MINT
7-F	Volkswagen Golf	green body, black base and grill, 2 7/8" long, 1976	4	8	12
34-B	Volkswagen Microvan	light green body, dark green interior, flat roof window tinted green, 2 3/5" long, 1962	20	30	35
34-A	Volkswagen Microvan	blue body, gray wheels, "Matchbox International Express" on sides, 2 1/4" long, 1957	30	40	50
23-D	Volkswagon Camper	orange top, clear windows, 5 spoke wheels, 2 1/8" long, 1970	5	7	10
73-F	Weasel	metallic green body, large five spoke slicks, 2 7/8" long, 1974	3	4	6
24-A	Weatherhill Hydraulic Excavator	metal wheels, "Weatherhill Hydraulic" decal on rear, 2 3/8" long, 1956	20	25	35
24-B	Weatherhill Hydraulic Excavator	yellow body, small and medium front wheels, large rear wheels, 2 5/8" long, 1959	10	15	20
57-F	Wild Life Truck	yellow body, red windows, light tinted blue canopy, 2 3/4" long, 1973	3	4	6
57-A	Wolseley 1500	with or without grills, four wheels, 2 1/8" long, 1958	20	30	35
58-E	Woosh-n-Push	yellow body, red interior, large rear wheels, 2 7/8" long, 1972	3	4	5
35-E	Zoo Truck	1981	4	7	10

VEHICLES

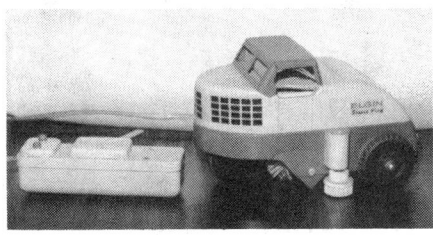

Clockwise from upper left: Mystery Taxi in box; Mystery Taxi, Wolverine; Minneapolis Moline tractor (rubber), 1950, Auburn; GMC Bank of America Armored Truck, 1949, Smith-Miller; Mack Bulldog Gasoline tanker, A.C. Williams; Elgin Street King, Ny-lint; Nash sedan, Wyandotte; Kenton double decker bus.

VEHICLES

Nylint

Cars

NAME	DESCRIPTION	GOOD	EX	MINT
Howdy Doody Pump Mobile	8 1/2" long	250	450	650

Emergency Vehicles

Ladder Truck	post war, 30" long	100	175	250

Farm and Construction Equipment

Michigan Shovel	bright yellow, bucket tips automatically when raised to boom, boom raises and lowers, 10 wheels, steerable front wheels, 31 1	150	225	275
Payloader	bright red, 3 3/4" rubber tires, 18" long, 1955	125	187	250
Road Grader	sturdy blade can be raised, lowered, or tilted, tandem-pivoted rear wheels, 3 3/4" steel wheels, 19 1/4" long, 1955	100	175	225
Speed Swing Pettibone	orange, raise or lower bucket and tip to dump, steerable wheels, 3 3/4" rubber tires, "Pettibone" decal on sides, 19" long	200	300	400
Street Sweep	windup, 8 1/4" long	175	275	350
Tournahopper Dozer	huge hopper, pull lever at rear opens wide clamshell jaws for bottom dumping, 3 3/4" rubber-tired steel wheels, 22 1/2" long,	100	150	200
Tournarocker Dozer	oversize hopper, crank action hoist, 3 3/4" rubber-tired steel wheels, 18" long, 1955	75	125	175
Tournatractor Dozer	yellow, big powerful adjustable blade on front, pivoted tow-bar on rear, 14 3/4" long, 1955	100	150	200
Traveloader	orange, synchronized feeders, buckets, and rubber conveyor belt, hand crank, steel wheels with 3 3/4" rubber tires, 30" long,	200	300	400

Trucks

Guided Missile Launcher	1957	75	125	175
Tournahauler	dark green, tractor with enclosed cab, platform trailer, slid-out ramps, 41 1/2" long with ramp extended, 1955	125	150	250
U-Haul Ford Truck and Trailer	with twin I-Beam suspension	125	187	250

Schuco

Cars

1902 Mercedes Simplex 32PS	windup, 8 1/2" long	125	187	250
1913 Mercer	windup, 7 1/2" long	87	130	175
Renault 6CV Model 1911	open two-seater, 7" long	125	187	250
Sedan	blue, tin litho, windup, 4 1/2" long, 1950's	200	300	400

Sets

Highway Patrol Official Squad Car Road Set	1958	100	150	200

Tanks

Military Miniature Tank	keywind	37	55	75

Trucks

Van	battery operated, 4" long	75	112	150

Smith-Miller

NAME	DESCRIPTION	GOOD	EX	MINT
\	**Emergency Vehicles**			
'L' Mack Aerial Ladder	all red with gold lettering and polished aluminum surface, 'S-M-F-D' decals on hood and trailer sides, 6 wheeler, 1950	375	475	795
	Trucks			
'B' Mack Associated Truck Lines	red cab, polished aluminum trailer, decals on trailer sides, 6 wheel tractor, 8 wheel trailer, 1954	500	850	1200
'B' Mack Blue Diamond Dump	all white truck with blue decals, hydraulic piston, 10 wheeler, 1954	600	950	1300
'B' Mack Lumber Truck	yellow cab and timber deck, 3 rollers, loading bar and 2 chains, 6 wheeler, load of 9 timbers, 1954	450	650	1000
'B' Mack Orange Dump Truck	construction orange all over, no decals, hydraulic piston, 10 wheeler, 1954	650	1150	1650
'B' Mack P.I.E.	red cab, polished trailer, 6 wheel tractor, 8 wheel trailer, 1954	375	600	850
'B' Mack Searchlight	dark red paint schemes, fully rotating and elevating searchlight, battery operated, 1954	500	775	1100
'B' Mack Silver Streak	yellow cab, unpainted, unpolished trailer sides, 'Silver Streak' decal on both sides, 6 wheel tractor, 8 wheel trailer, 1954	450	775	1050
'B' Mack Watson Bros.	yellow cab, polished aluminum trailer, decals on trailer sides and cab doors, 10 wheel tractor, 8 wheel trailer, 1954	650	1100	1500
'L' Mack Army Materials Truck	all Army green, flatbed with dark green canvas, 10 wheeler, load of 3 wood barrels, 2 boards, large and small crate, 1952	375	500	750
'L' Mack Army Personnel Carrier	all Army green, wood sides, Army seal on door panels, military star on roof, 10 wheeler, 1952	375	500	750
'L' Mack Bekins Van	all white, covered with 'Bekins' decals of all descriptions, 6 wheel tractor, 4 wheel trailer, 1953	1000	1650	2000
'L' Mack Blue Diamond Dump	white cab, white dump bed, blue fenders and chassis, hydraulically operated, 10 wheeler, 1952	425	750	1050
'L' Mack International Paper Co.	white tractor cab, 'International Paper Co' decals on sides, 6 wheel tractor, 4 wheel trailer, 1952	375	650	900
'L' Mack Lyon Van	silver gray cab, dark blue fenders and frame, silver-gray van box with 'Lyon' decal in blue lettering, 6 wheeler, 1950	425	800	1100
'L' Mack Material Truck	light metallic green cab, dark green fenders and frame, wood flatbed, 6 wheeler, load of 2 barrels and 6 timbers, 1950	400	600	875
'L' Mack Merchandise Van	red cab, black fenders and frame, 'Smith-Miller' decals on both sides of van box, double rear doors, 6 wheeler, 1951	425	695	1000
'L' Mack Mobil Tandem Tanker	all red cab, 'Mobilgas' and 'Mobiloil' decals on tank sides, 6 wheel tractor, 6 wheel trailer, 1952	450	725	1000
'L' Mack Orange Hydraulic Dump	orange cab, orange dump bed, hydraulically operated, 10 wheeler, may or may not have 'Blue Diamond' decals, 1952	850	1500	1950
'L' Mack Orange Materials Truck	all orange, flatbed with canvas, 10 wheeler, load of 3 barrels, 2 boards, large and small crate, 1952	400	650	900
'L' Mack P.I.E.	all red tractor, polished aluminum trailer, 'P.I.E.' decals on sides and front, 6 wheel tractor, 8 wheel trailer, 1950	395	550	850
'L' Mack Sibley Van	dark green cab, black fenders and frame, dark green van box with 'Sibley's' decal in yellow on both sides, 6 wheeler, 1950	850	1375	1850
'L' Mack Tandem Timber	red and black cab, 6 wheeler, load of 6 wood lumber rollers, 2 loading bars, 4 chains and 18 or 24 boards, 1950	400	550	725

VEHICLES

Smith-Miller

NAME	DESCRIPTION	GOOD	EX	MINT
'L' Mack Tandem Timber	two-tone green cab, 6 wheeler, load of 6 wood lumber rollers, 2 loading bars, 4 chains, and 18 timbers, 1953	400	550	725
'L' Mack Telephone Truck	all dark or two-tone green truck, 'Bell Telephone System' decals on truck sides, 6 wheeler, 1952	475	750	975
'L' Mack West Coast Fast Freight	silver with red and black or silver cab and chassis, 'West Coast-Fast Freight' decals on sides of box, 6 wheeler, 1952	475	775	1000
Chevy Arden Milk Truck	red cab, white wood body, 4 wheeler, 1945	275	465	800
Chevy Bekins Van	blue die cast cab, all white trailer, 14 wheeler, 1945	275	350	750
Chevy Coca-Cola Truck	red cab, wood body painted red, 4 wheeler, 1945	300	600	850
Chevy Flatbed Tractor-Trailer	unpainted wood trailer, unpainted polished cab, 14 wheeler, 1945	250	300	500
Chevy Heinz Grocery Truck	yellow cab, load of 4 waxed cases, 1946	225	325	475
Chevy Livestock Truck	polished, unpainted tractor cab and trailer, 1946	175	275	375
Chevy Lumber	green cab, load of 60 polished boards and 2 chains, 1946	150	195	275
Chevy Lyon Van	blue cab, silver trailer, 1946	165	325	500
Chevy Material Truck	green cab, no side rails, load of 3 barrels, 2 cases and 18 boards, 1946	135	185	225
Chevy Stake	yellow tractor cab	185	250	425
Chevy Tampa Anmature Works	royal blue cab, white wood body, 4 wheeler, 1945	300	500	850
Chevy Transcontinental Vanliner	blue tractor cab, white trailer, 'Bekins' logos and decals on trailer sides, 1946	200	350	495
Chevy Union Ice Truck	blue cab, white body, load of 8 waxed blocks of ice, 1946	300	495	800
Ford Bekins Van	red sand-cast tractor, gray sheet metal trailer, 14 wheeler, 1944	275	500	750
Ford Coca-Cola Truck	red sandcast cab, wood body painted red, 4 wheeler, 1944	400	650	900
GMC 'Be Mac' Tractor-Trailer	red cab, plain aluminum frame, 'Be Mac Transport Co.' in white letters on door panels, 14 wheeler, 1949	250	350	700
GMC 'Drive-O'	red cab, red dump body, runs forward and backward with handturned control at end of 5 1/2 ft. cable, 6 wheeler, 1949	175	300	450
GMC Arden Milk Truck	red cab, white painted wood body with red stakes, 4 wheeler, 1947	200	425	650
GMC Bank of America Truck	dark brownish green cab and box, 'Bank of America' decal on box sides, 4 wheeler, 1949	115	165	275
GMC Bekins Van-Liner	blue cab, metal trailer painted white, 14 wheeler, 1947	175	275	425
GMC Coca-Cola Truck	red cab, yellow wood body, 4 wheeler, load of 16 Coca-Cola cases, 1947	400	675	895
GMC Coca-Cola Truck	all yellow truck, red Coca-Cola decals, five spoke hubs, 4 wheeler, load of 6 cases each with 24 plastic bottles, 1954	275	450	750
GMC Dump Truck	all red truck, 6 wheeler, 1950	150	200	285
GMC Emergency Tow Truck	white cab, red body and boom, 'Emergency Towing Service' on body side panels, 4 wheeler, 1953	185	250	400
GMC Furniture Mart	blue cab, off-white body, "Furniture Mart", Complete Home Furnishings' markings on body sides, 4 wheeler, 1953	135	275	295
GMC Heinz Grocery Truck	yellow cab, wood body, 6 wheeler, 1947	250	325	450
GMC Hi-Way Freighter Tractor-Trailer	red tractor cab, hardwood bed on trailer with full length wood fences, 'Fruehauf' decal on trailer, 14 wheeler, 1948	150	210	325
GMC Kraft Foods	yellow cab, yellow steel box, large 'Kraft' decal on both sides, 4 wheeler, 1948	200	300	450
GMC Lumber Tractor-Trailer	metallic blue cab and trailer, 3 rollers and 2 chains, 14 wheeler, 1949	185	250	350
GMC Lumber Truck	green cab, 6 wheeler, 1947	165	215	300

VEHICLES

Smith-Miller

NAME	DESCRIPTION	GOOD	EX	MINT
GMC Lyon Van Tractor-Trailer	blue tractor cab, 'Lyons Van' decals on both sides, fold down rear door, 14 wheeler, 1948 (add 20% for drop-side trailer)	165	250	400
GMC Machinery Hauler	construction orange cab and lowboy trailer, 'Fruehauf' decal on gooseneck, 13 wheeler, 1949	150	225	335
GMC Machinery Hauler	all construction orange, 2 loading ramps, 10 wheeler, 1953	200	295	425
GMC Marshall Field & Company Tractor-Trailer	dark green cab and trailer, double rear doors, never had Smith-Miller decals, 10 wheeler, 1949	295	395	500
GMC Material Truck	green cab, wood body, 6 wheeler, load of 3 barrels, 3 cases and 18 boards, 1947	115	150	250
GMC Material Truck	yellow cab, natural finish hardwood bed and sides, 4 wheeler, load of 4 barrels and 2 timbers, 1949	125	175	265
GMC Mobilgas Tanker	red cab and tanker trailer, large "Mobilgas", "Mobiloil" emblems on sides and rear panel of tanker, 14 wheeler, 1949	135	225	400
GMC Oil Truck	orange cab, rear body unpainted, 6 wheeler, load of three barrels, 1947	115	185	265
GMC P.I.E.	red cab, polished aluminum box trailer, double rear doors, 'P.I.E.' decals on sides and front panels, 14 wheeler, 1949	150	265	350
GMC People's First National Bank and Trust Company	dark brownish green cab and box, 'People's First National Bank & Trust Co.' decals on box sides, 1951	165	250	385
GMC Rack Truck	red or yellow cab, natural finish wood deck, red stake sides, 6 wheeler, 1948	135	200	325
GMC Redwood Logger Tractor-Trailer	green or maroon cab, unpainted aluminum trailer with 4 hardwood stakes, load of 3 cardboard logs, 1948	365	585	700
GMC Rexall Drug Truck	orange cab and closed steel box body, 'Rexall' logo on both sides and on front panel of box, 4 wheeler, 1948	500	750	1000
GMC Scoop Dump	rack and pinion dump with a scoop, five spoke wheels, 6 wheeler, 1954	275	350	575
GMC Searchlight Truck	four wheel truck pulling four wheel trailer, color schemes vary, 'Hollywood Film Ad' on truck body side panels, 1953	300	415	695
GMC Silver Streak	unpainted polished cab and trailer, wrap around sides and shield, some had tail gate, 1950	140	200	300
GMC Sunkist Special Tractor-Trailer	cherry-maroon tractor cab, natural mahogany trailer bed, 14 wheeler, 1947	165	275	475
GMC Super Cargo Tractor-Trailer	silver-gray tractor cab, hardwood bed on trailer with red wraparound side rails, 14 wheeler, load of 10 barrels, 1948	150	225	395
GMC Timber Giant	green or maroon cab, unpainted aluminum trailer with 4 hardwood stakes, load of 3 cardboard logs, 1948	175	285	495
GMC Tow Truck	white cab, red body and boom, five spoke cast hubs, 'Emergency Towing Service' on body side panels, 4 wheeler, 1954	95	135	200
GMC Transcontinental Tractor-Trailer	red tractor cab, hardwood bed on trailer with full length wood fences, 'Fruehauf' decal on trailer, 14 wheeler, 1948	150	210	325
GMC Triton Oil Truck	blue cab, mahogany body unpainted, 6 wheeler, load of 3 Triton Oil drum (banks) and side chains, 1947	115	185	265
GMC U.S. Treasury Truck	gray cab and box, 'U.S. Treasury' insignia and markings on box sides, 4 wheeler, 1952	235	325	475

Structo

Cars

NAME	DESCRIPTION	GOOD	EX	MINT
DeLuxe Auto	solid disc wheels, rounded fenders, pressed steel, 16" long, 1921	350	550	750

Structo

NAME	DESCRIPTION	GOOD	EX	MINT
Roadster	pressed steel, 10 1/2" long, 1919	200	295	450

Emergency Vehicles

NAME	DESCRIPTION	GOOD	EX	MINT
Fire Engine	pressed steel, 18" long truck with 18" long ladders, 1927	200	300	400
Fire Pumper Truck	steel, 21" long, 1920's	225	350	450
Hook and Ladder Fire Truck	36" long, 1940's	62	93	125
Hook and Ladder Fire Truck	32" long	62	93	125

Farm and Construction Equipment

NAME	DESCRIPTION	GOOD	EX	MINT
Crawler	10" long, 1928	175	250	400
Road Grader	orange, single-blade, 18 1/2" long, 1960's	20	35	55
Steam Shovel	green steel, wood wheels	40	55	75

Trucks

NAME	DESCRIPTION	GOOD	EX	MINT
Auto Haulaway Truck	20" long	100	150	200
C.O.E. Auto Transport Tractor-Trailer	with cars, 1950's	30	40	55
C.O.E. Auto Transport Tractor-Trailer	with cars, 1960's	20	25	35
Camper Truck	11" long, 1960's	50	75	100
Cattle Farms Inc. Trailer	orange trailer, green cab	87	130	175
Cement Mixer	steel, 22" long	150	225	300
Cement Truck	20" long, 1950's	95	150	200
Communications Truck	blue, 21" long	50	85	165
Dispatch Truck	green and gray	87	130	175
Dump Truck	15" long, 1960's	15	20	30
Dump Truck	13" long, 1960's	15	20	28
Excavation Truck		175	262	350
Flat Bed Tractor Truck	27" long, 1950's	125	200	275
Flat Bed Truck	20 1/2" long, 1940's	75	125	150
Flat Bed Truck	20 1/2" long, 1950's	50	75	125
Garage Truck	1940's	60	90	150
Hi-Lift Dump Truck	12" long	50	85	135
Highway Maintenance Platform Truck	12" long	50	75	125
Hydraulic Dump Truck	red, pressed steel, 12 1/2" long, 1950's	75	100	175
Hydraulic Sanitation Truck	1960's	50	75	100
Overland Freight Lines Truck	21" long, 1950s	125	200	275
Pickup Truck	17" long, 1950's	40	55	85
Police Patrol Truck	17" long	137	205	275
Power & Light Turbine Utility Truck	17 1/2" long	75	112	150
Road Tug Service Truck	1966	87	130	175
Scoop/Dump Truck	20" long, 1950's	125	185	275
Steel Dump Truck	1940's	100	150	200
Tractor-Trailer Truck	1960's	75	125	175
Transport Truck	red trailer, blue cab	137	205	275
U.S. Mail Truck	17" long, 1930's	250	350	550
Wrecker Truck	21" long, 1930's	100	150	200

Tonka

Cars

NAME	DESCRIPTION	GOOD	EX	MINT
Dune Buggy	1970	15	20	35
Volkswagen Bug	blue with white interior, 1960s	25	35	60

Tonka

NAME	DESCRIPTION	GOOD	EX	MINT
Volkswagen Bug	black, 1968	15	20	35

Construction Vehicles

NAME	DESCRIPTION	GOOD	EX	MINT
Clark Melroe 1399 Hydrostatic Bobcat	black, 1/24 scale, 1979	15	20	30

Emergency Vehicles

NAME	DESCRIPTION	GOOD	EX	MINT
Aerial Ladder Fire Truck	1960	100	150	250
Fire Truck	32" long, 1950	250	375	500
Ford Aerial Ladder Truck	red, cast siren on right fender, 1955	150	225	300
Ford Hydraulic Aerial Ladder Truck	red, 1958	90	135	180
Ford Hydraulic Aerial Ladder Truck	white, 1959	125	185	250
Ford Rescue Squad	white, 1959	95	145	190
Ford Suburban Pumper	white, blackwalls, 1959	120	180	240
Ford Suburban Pumper	red, blackwheels, 1958	90	135	180
Ford Suburban Pumper	red, whitewalls, 1960	95	145	190
Ford Suburban Pumper	red, with thread fittings, 1956	135	225	275
Ford Suburban Pumper	red, without thread fittings, 1957	115	175	230
Ford T.F.D. Aerial Ladder Truck	red, "T.F.D." decals on tractor, 1956	160	250	325
Ford T.F.D. Tanker	white, 1958	200	300	400
Ford Tonka Tanker	red, hard plastic tank trailer with hoses, 1960	100	150	225
International Rescue Squad Metro Van	white, red "Rescue Squad" and large red cross on side of body, 1956	85	150	250
Pumper Truck	c1950	175	265	350

Farm and Construction Equipment

NAME	DESCRIPTION	GOOD	EX	MINT
Aerial Sand Loader	red	375	565	750
Ford Aerial Sand Loader	overhead travelling crane with clam bucket and loading hopper	80	120	160
Ford Crane and Clam	yellow and black, heavy steel, 26" long, 1947	165	250	350
Ford Giant Dozer and Trailer	orange, 1961	95	145	190
Ford Road Grader	orange	40	55	95
Ford Steam Shovel	orange and black, heavy steel, 16" long, 1947	145	200	275
Tractor	1963	75	140	220

Sets

NAME	DESCRIPTION	GOOD	EX	MINT
Construction Set	mobile clam, state hi-way dept. hydraulic dump, giant bulldozer, 1961	150	250	375
Fire Department	hydraulic aerial ladder in white, suburban pumper in white, fire chiefs badge, 1959	500	750	1100
Fire Department	hydraulic aerial ladder in red, suburban pumper in red, rescue squad, fire chief's badge, 1959	425	700	1000
Hi-Way Construction Set	road grader, dragline and trailer, state hi-way dump, 1959	325	475	700
Road Builder Set	road grader, state hi-way dump, shovel and carry-all trailer, Big Mike with plow, 6 highway signs, 2 road barrels, 1958	500	700	1000
State Hi-Way Department	road grader, state hi-way dump, state hi-way pickup, hi-way service, 6 highway signs, 2 road barriers, 1959	300	575	800
State Highway Department	road grader, state hi-way pickup, state hi-way dump, hi-way service truck, 6 highway signs, 2 road barrels, 1956	500	700	1000
State Turnpike	small bulldozer, state hi-way pickup, mobile dragline, state hi-way dump, 2 highway signs, 1 barrier, 1960	275	4500	7000
Tonka Fire Department	hydraulic aerial ladder, suburban pumper, rescue squad, fire chief's badge, 1957	400	650	900

Tonka

NAME	DESCRIPTION	GOOD	EX	MINT
Tonka Fire Department	hydraulic aerial ladder, suburban pumper, T.F.D. tanker, fire chief's badge, 1958	300	550	700

Trucks

NAME	DESCRIPTION	GOOD	EX	MINT
Allied Moving Van	1958	100	165	215
Boat Transport	1960	75	125	175
Camper	1973	20	35	50
Camper Pickup Truck	1962	60	95	125
Camper Truck G	c1950	50	75	100
Carnation Milk Truck		225	335	450
Dump Truck	1970	20	35	50
Federal Allied Van In Storage	c1950	275	425	550
Ford Ace Tractor and Trailer	all red, with "Ace Hardware" decals	110	175	225
Ford Air Express Truck	midnight blue, steel box, 1959	95	145	190
Ford Allied Van	orange, 1959	80	120	160
Ford Allied Van	orange, 1954	120	180	240
Ford Allied Van	orange, with duck decal, 1957	100	150	200
Ford Allied Van	orange, 1951	150	225	300
Ford American Wrecker	red	75	115	150
Ford Big Mike	orange, extra long dump bed with plow, 1958	225	335	450
Ford Big Mike	orange, extra long dump bed without plow, 1958	200	300	400
Ford Big Mike	orange, long dump bed, with plow, 1957	225	335	450
Ford Big Mike	orange, long dump bed, without plow, 1957	200	300	400
Ford Boat Transport Truck	metallic blue, 4 plastic boats and 2 outboard motors, 1959	135	225	275
Ford Boat Transport Truck	metallic blue, 4 plastic boats and 2 outboard motors, white walls, 1961	125	185	250
Ford Car Carrier	cream, 3 plastic autos on carrier, 1959	120	180	240
Ford Car Carrier	yellow, 3 plastic autos on carrier, whitewalls, 1961	60	90	120
Ford Cement Mixer	red truck, white mixing barrel and water tank, 1960	100	200	300
Ford Coast to Coast Utility Truck	red cab, yellow utility body	75	115	150
Ford Cross Country Freight Semi-Truck	white	150	200	250
Ford Dragline and Trailer	lime green and black, 1959	130	195	260
Ford Dump Truck	red cab, green dump body, 1949	60	95	130
Ford Dump Truck	red cab and frame, green body, 1958	50	75	100
Ford Dump Truck	red cab, green dump body, 1954	60	95	125
Ford Express Truck	green cab, red box, fold down end gate, 1950	115	175	230
Ford Farm Stake Truck	six separate side stake assemblies, 1958	70	125	140
Ford Fisherman Truck	blue and white, steel cap, 1960	60	90	120
Ford Flatbed Semi	red tractor, plywood trailer with four metal posts, 1953	85	135	170
Ford Gambles Pickup Truck	white	85	165	225
Ford Gambles Semi Truck	white	150	275	350
Ford Gasoline Truck	red, 1958	200	300	400
Ford Grain Hauler	red tractor, aluminum trailer, 1954	100	150	200
Ford Grain Hauler	red tractor, aluminum trailer, plywood floor in trailer, 1952	100	150	200
Ford Green Giant Transport	all white, Green Giant decals are everywhere, with refrigeration unit, 1954	110	165	275
Ford Green Giant Transport	all white, the giant is holding a pea pod and yellow ear of corn, mounted with refrigeration unit, 1953	120	180	285
Ford Green Giant Utility Truck	white, solid rubber wheels, "Green Giant Co." on truck doors, 1954	75	115	150
Ford Hardware Hank Van		150	225	300
Ford Hi-Way Service Truck	orange, sides fold down on dump bed, with plow, 1956	125	185	250
Ford Hi-Way Service Truck	orange, sides fold down on dump bed, without plow, 1956	200	300	400

Tonka

NAME	DESCRIPTION	GOOD	EX	MINT
Ford Hi-Way Service Truck'n	orange, without plow, 1958	80	120	160
Ford Hi-Way Service Truck'n	orange, with plow, 1958	110	165	220
Ford Hydraulic Dump Truck	bronze, 1958	55	85	110
Ford Hydraulic Dump Truck	bronze, 1957	65	95	130
Ford Hydraulic Land Rover	orange, 1959	200	300	400
Ford J. & R. Fox Express Truck	all blue	125	185	250
Ford Janney Semple Hill & Co. Tractor and Trailer	red tractor, red and white trailer	125	185	250
Ford Jewel Tea Semi	dark brown tractor and trailer, wood trailer floor, 1955	150	225	300
Ford Livestock Van	red, 1958	100	150	200
Ford Livestock Van	all red, drop down door back of trailer, 1954	110	165	220
Ford Livestock Van	all red, with steer head decal, 1956	100	150	200
Ford Livestock Van	all red, "Livestock" decal on trailer's front panel, 1952	100	150	200
Ford Logger Truck	red cab, aluminum trailer, load of 4 logs, 4 semi-finished timbers and 2 chains, 1959	100	150	200
Ford Logger Truck	red tractor, aluminum trailer, load of 9 logs and 2 chains, 1954	120	180	240
Ford Logger Truck	red cab, aluminum trailer, load of 9 logs and 2 chains, 1953	110	165	220
Ford Lumber Truck	red cab and frame, load of 36 finished boards and 2 chains to secure load, 1955	90	125	225
Ford Lumber Truck	red cab and frame, load of 36 finished boards and 1 chain to secure load, 1957	90	125	225
Ford Marshall Field & Co. Semi	forest green	175	225	400
Ford Marshall Field & Co. Tractor and Trailer	one color greenish-brown	150	200	375
Ford Meier and Frank Co. Tractor and Trailer	two-toned tractor; blue-green upper half and black bottom	125	200	300
Ford Minute Maid Semi		200	350	450
Ford Minute Maid Truck	all white, 1955	300	400	650
Ford Mobile Clam	orange, 1961	80	120	160
Ford Mobile Dragline	orange, 1960	85	130	170
Ford Nationwide Moving Van	white, 1958	150	225	300
Ford Our Own Hardware Tractor and Trailer	two-tone trailer	165	235	400
Ford Our Own Hardware Utility Truck	orange	75	115	150
Ford Pickup Truck	tailgate secured with chains and hooks, 1958	50	75	115
Ford Pickup Truck	snap-shut tailgate, whitewalls and solid wheel discs, 1959	35	55	100
Ford Pickup Truck	red, flare side rear fenders, 1955	60	90	145
Ford Pickup Truck	midnight blue, flare side rear fenders, 1956	50	75	135
Ford Pickup Truck with Tow Hitch	midnight blue, 1957	60	90	120
Ford Platform Stake Truck	white walls, 1959	90	175	250
Ford Power Boom Loader	1960	95	145	190
Ford Republic Van Lines Semi		150	225	300
Ford Sanitary Service Truck	rectangular body, white with black loading apparatus, 2 black refuse bins, black loading scoop, 1959	125	185	250
Ford Sanitary Service Truck	rectangular body, 1959	100	150	200
Ford Sanitary Truck	curved body, white with black loading apparatur, black refuse bin, 1960	150	175	350

Tonka

NAME	DESCRIPTION	GOOD	EX	MINT
Ford Sanitary Truck	curved body, 1960	100	150	200
Ford Service Truck	metallic blue, steel box and aluminum ladder, white walls, 1959	75	115	150
Ford Shovel and Carry-All Trailer	orange, 1957	135	225	270
Ford Shovel and Carry-All Truck	orange or lime green, 1958	135	225	275
Ford Shovel and Carry-All Truck	1954	125	185	250
Ford Sportsman Truck	steel cap, blackwalls, no boat, 1958	95	150	225
Ford Sportsman Truck	steel cap, whitewalls, with boat, 1959	75	125	195
Ford Stake Truck	red cab, frame and flatbed, green stakes, 1955	110	165	220
Ford Star-Kist Utility Truck	green cab and frame, white body, with can decals on body side panels	90	135	180
Ford Star-Kist Utility Truck	green cab and frame, white body, no decals	75	115	150
Ford Star-Kist Van	red cab, blue box, 1954	225	345	450
Ford State Hi-Way Dept. Dump Truck	orange, "975" decal on door panel, 1956	65	95	130
Ford State Hi-Way Dept. Dump Truck	black pumper, no number on decal, 1957	55	85	110
Ford State Hi-Way Dept. Hydraulic Dump Truck	orange, 1960	75	115	150
Ford State Hi-Way Dept. Pickup Truck	orange, 1956	70	125	250
Ford State Hi-Way Dump Truck	orange or lime green, 1958	60	90	120
Ford State Hi-Way Pickup Truck	orange, 1958	45	67	90
Ford Steel Carrier	orange tractor, green trailer, 1954	110	165	220
Ford Steel Carrier	orange tractor, green trailer, 1950	75	112	150
Ford Stock Rack Truck	white cab and frame, red livestock rack, 1958	85	125	170
Ford Stock Rack Truck	midnight blue cab and frame, 1957	100	150	200
Ford Tandem Air Express Truck	midnight blue, 1959	120	180	240
Ford Tandem Platform Stake Truck	bronze, 1959	110	165	220
Ford Terminix Service Truck	orange	80	120	160
Ford Thunderbird Express	white truck, decal wraps around front of trailer, 1958	110	165	220
Ford Thunderbird Express	red and white truck, decal only on side of trailer, 1960	110	165	220
Ford Thunderbird Express	white truck, single axle trailer, 1957	110	195	275
Ford Tonka Cargo King	red tractor, aluminum trailer, 1956	110	165	220
Ford Tonka Freighter	orange and red tractor, green trailer	110	165	220
Ford Tonka Gasoline Truck	red, 1957	200	300	400
Ford Tonka Tanker Standard Oil Semi		125	185	250
Ford Tonka Toy Transport	1949	110	165	220
Ford Tractor-Carry-All Trailer with Crane and Clam	green trailer, yellow tractor, 1949	200	325	450
Ford Tractor-Carry-All Trailer with Steam Shovel	blue trailer, red tractor, 1949	185	250	350
Ford United Van Lines Semi		125	185	250
Ford Utility Truck	1958	60	95	125
Ford Utility Truck	orange cab and frame, green body, 1954	75	115	150
Ford Utility Truck	with end chain, 1950	85	130	175
Ford Wheaton Semi Truck	white	100	150	200
Ford Wheaton Van Lines Semi		110	175	225
Ford Wrecker	white with black boom, 1958	50	75	100

Tonka

NAME	DESCRIPTION	GOOD	EX	MINT
Ford Wrecker	white with black boom, whitewalls, 1960	50	75	100
Ford Wrecker	red cab, frame and boom, white body, 1954	75	115	150
Ford Wrecker	white body, red boom, 1955	75	115	150
Ford Wrecker	blue truck, red boom, "Official Service Truck" on side, 1949	60	95	125
Ford Yonkers Truck and Trailer Truck	all black, yellow decal with red outline	75	115	150
Hydraulic Dump Truck	"Mighty Tonka" Series, 1976	20	35	50
International Carnation Milk Truck Metro Van	white, 1955	75	115	150
International Frederick & Nelson Metro Van	forest green	85	130	175
International Midwest Milk Truck Metro Van	white, c1950	100	150	200
International Parcel Delivery Metro Van	dark brown, 1954	65	100	135
International Parcel Delivery Metro Van	dark brown, with aluminum step, 1957	75	115	150
Jeep	blue	20	30	40
Pickup Truck	steel, 12" long, c1950	20	25	35
Pickup Truck	1958	200	300	400
State Highway Crane Truck		75	115	150
Winnebago Camper	"Mighty Tonka" Series, 1973	45	75	100
Wrecker	1973	35	65	95

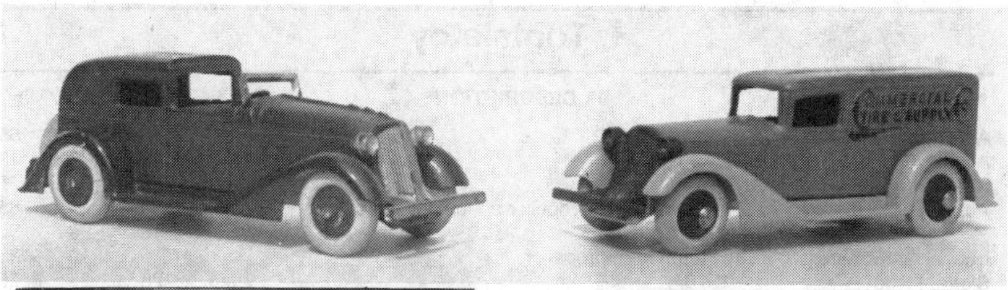

Top to bottom: Graham town car with rear spare and Graham Commercial Tire Van, 1933-35, Tootsietoys; Tootsietoy Freight Train (five piece set), 1940; 1913 Mercer, Schuco; 1956 Ford Thunderbird (friction), Japan.

Tootsietoy

VEHICLES

NAME	DESCRIPTION	GOOD	EX	MINT
Aero-Dawn	1928	15	30	55
Atlantic Clipper	2" long	5	7	12
Autogyro	1934	25	35	75
Autogyro Plane	helicopter type propellor on top, front propellor	30	40	85
B-Wing Seaplane	1926	15	20	50
Beechcraft Bonanza	orange, front propellor	6	10	25
Bleriot Plane	1910	25	35	75
Crusader		50	85	100
Curtis P-40	light green	85	135	185
Dirigible U.S.N. Los Angeles		25	35	75
Douglas D-C 2 TWA Airliner	1935	15	30	55
F-94 Starfire	green, 4 engines, 1970's	5	7	20
F9F-2 Panther Shooting Star		10	15	25
Fly-N-Gyro	1938	30	40	85
KOP-1 USN		15	20	45
Low Wing Plane	miniature	15	25	35
Navion	red, front propellor	6	10	25
Navy Jet	red, 1970's	5	7	25
Navy Jet Cutlass	red with silver wings	7	15	25
P-38 Plane	9 3/4" wingspan	40	60	85
Piper Cub	blue, front propellor	7	15	25
S-58 Sikorsky Helicopter	1970's	15	30	45
Snow Skids Airplane	rotating prop, 4' wingspan	40	60	80
Supermainliner		25	35	60
Top Wing Plane	miniature	15	25	35
Transport Plane	1941	20	30	55
Tri-Motor Plane	three propellors	50	85	125
TWA Electra		20	35	75
Twin Engine Airliner	10 windows	20	30	65
U.S. Army Plane	1936	20	25	45
UX214 Monoplane	4', 1930's	45	70	85
Waco Bomber	blue bottom half and silver upper half or silver bottom half and red upper half	50	85	125

Boats and Ships

NAME	DESCRIPTION	GOOD	EX	MINT
Battleship	silver with a little red on top, 6" long, 1939	15	20	30
Carrier	silver with a little red on top	15	20	30
Cruiser	silver with a little red on top, 6" long, 1939	15	20	50
Destroyer	4" long, 1939	7	10	25
Freighter	6" long, 1940	15	20	30
Submarine	4" long, 1939	7	10	25
Tanker	all black, 6" long, 1940	15	20	30
Tender	4" long, 1940	7	10	25
Transport	6" long, 1939	15	20	30
Yacht	4" long, 1940	7	10	35

Buses

NAME	DESCRIPTION	GOOD	EX	MINT
Fageol Bus	1927-33	30	40	85
GMC Greyhound Bus	blue and silver, 1948	25	40	55
GMC Scenicruiser Bus	blue and silver, raised passenger roof with windows, 6" long, 1957	25	45	55
Greyhound Bus	blue, 1937-41	30	60	80
Overland Bus	1929-33	35	45	125
Twin Coach Bus	red with solid black tires, 3" long, 1950	20	25	50

Cannons and Tanks

NAME	DESCRIPTION	GOOD	EX	MINT
Army Tank	1931-41	35	50	75
Army Tank		7	10	25

Tootsietoy

NAME	DESCRIPTION	GOOD	EX	MINT
Four Wheel Cannon	4" long, 1950's	7	10	40
Long Range Cannon		7	10	25
Six Wheel Army Cannon	1950's	10	15	45

Cars

NAME	DESCRIPTION	GOOD	EX	MINT
Andy Gump 348 Car	pot metal, 3" long	225	325	450
Armored Car	'U.S. Army' on sides, camouflage, solid black tires, 1938-41	20	30	50
Auburn Roadster		15	20	40
Austin-Healy	light brown open top, 6" long, 1956	20	25	45
Baggage Car		10	15	25
Bluebird Daytona Race Car		25	35	65
Boat Tail Roadster	red, open top, 6" long	25	35	55
Brougham		65	100	130
Buick Brougham		20	35	70
Buick Coupe	blue with solid white wheels, 1924	20	35	7070
Buick Coupe		20	35	45
Buick Estate Wagon	yellow and maroon with solid black wheels, 6" long, 1948	35	55	65
Buick Experimental Car	blue with solid black wheels, detailed tin bottom, 6" long, 1954	25	40	75
Buick LaSabre	red open top, solid black wheels, 6" long, 1951	25	45	65
Buick Roadmaster	blue with solid black wheels, 4-door, 1949	25	40	75
Buick Roadster		20	35	55
Buick Roadster	yellow open top, solid black wheels, 4" long, 1938	20	35	55
Buick Sedan	6" long	25	35	55
Buick Special	4" long, 1947	15	25	45
Buick Station Wagon	green with yellow top, solid black wheels, 6" long, 1954	20	35	55
Buick Tourer	red with solid white wheels, 1925	20	35	55
Buick Touring		20	35	65
Cadillac	HO series, blue car with white top, 2" long, 1960	15	20	25
Cadillac 60	reddish-orange with solid black wheels, 4-door, 1948	20	35	55
Cadillac 62	reddish-orange with white top, solid black wheels, 4-door, 6" long, 1954	20	35	110
Cadillac Brougham		25	35	110
Cadillac Coupe	blue and tan, solid black wheels	25	35	110
Cadillac Sedan		25	35	90
Cadillac Touring Car	1926	25	35	90
Chevrolet Brougham		20	35	90
Chevrolet Coupe		20	35	90
Chevrolet Roadster		20	35	90
Chevrolet Sedan		20	35	90
Chevrolet Touring		100	150	200
Chevy Bel Air	yellow with solid black wheels, 3" long, 1955	10	15	30
Chevy Coupe	green with solid black wheels	25	35	55
Chevy Fastback	blue with solid black wheels, 3" long, 1950	15	20	35
Chrysler Convertible	bluish-green with solid black wheels, 4" long, 1960	15	25	30
Chrysler Experimental Roadster	orange open top, solid black wheels	25	40	55
Chrysler New Yorker	blue with solid black wheels, 4-door, 6" long, 1953	25	35	50
Chrysler Windsor Convertible	green open top, solid black wheels, 4" long, 1941	25	35	45
Chrysler Windsor Convertible	open top, solid black wheels, 6" long, 1950	60	90	110
Classic Series 1906 Cadillac or Studebaker	green and black, spoke wheels	15	20	25
Classic Series 1907 Stanley Steamer	yellow and black, spoke wheels, 1960-65	15	20	25
Classic Series 1912 Ford Model T	black with red seats, spoke wheels	15	20	25

Tootsietoy

NAME	DESCRIPTION	GOOD	EX	MINT
Classic Series 1919 Stutz Bearcat	black and red, solid wheels	15	20	25
Classic Series 1929 Ford Model A	blue and black, solid black tread wheels, 1960-65	15	20	25
Corvair	red, 4" long, 1960's	35	55	75
Corvette Roadster	blue open top, solid black wheels, 4" long, 1954-55	20	25	40
Coupe	metal, 1921	35	45	65
Coupe	miniature	20	25	40
DeSoto Airflow	green with solid white wheels	20	35	60
Doodlebug	Same as Buick Special	60	75	100
Ferrari Racer	red with gold driver, solid black wheels, 6" long, 1956	10	15	45
Ford	red with open top, solid black wheels, 6" long, 1940	20	25	35
Ford and Trailer	powder blue car with solid white wheels, 2-wheel white trailer with 3 windows on each side	25	35	50
Ford B Hotrod	1931	7	10	15
Ford Convertible Coupe	1934	35	50	70
Ford Convertible Sedan	red with solid black wheels, 3" long, 1949	10	18	22
Ford Coupe	powder blue with tan top, solid white wheels, 1934	35	50	70
Ford Coupe	blue or red with solid white wheels, 1935	25	35	40
Ford Customline	blue with solid black wheels, 1955	12	16	20
Ford Fairlane 500 Convertible	red with solid black wheels, 3" long, 1957	10	12	15
Ford Falcon	red with solid black wheels, 3" long, 1960	7	10	15
Ford LTD	blue with solid black wheels, 4" long, 1969	15	20	25
Ford Mainliner	red with solid black wheels, 4-door, 3" long, 1952	12	18	22
Ford Model A Coupe	blue with solid white wheels	25	35	45
Ford Model A Sedan	green with solid black wheels	25	35	45
Ford Ranch Wagon	green with yellow top, 4-door, 4" long, 1954	15	25	30
Ford Ranch Wagon	red with yellow top, 4-door, 3" long, 1954	12	18	22
Ford Roadster	powder blue with open top, solid white wheels	30	40	55
Ford Sedan	1934	35	50	70
Ford Sedan	powder blue with white solid wheels, 1935	25	35	45
Ford Sedan	lime green with solid black wheels, 4-door, 3" long, 1949	12	18	22
Ford Station Wagon	powder blue with white top, solid black wheels, 6" long, 1959	15	20	25
Ford Station Wagon	blue with solid black wheels, 3" long, 1960	10	15	20
Ford Station Wagon	red with white top, solid black wheels, 4-door, 6" long, 1962	25	40	50
Ford Tourer	open top, red with silver spoke wheels	20	30	40
Ford V-8 Hotrod	red with open top, solid black wheels, open silver motor, 6" long, 1960	15	20	25
Graham Convertible Coupe	rear spare tire, 1933-35	60	115	135
Graham Convertible Coupe	side spare tire, 1933-35	60	115	135
Graham Convertible Sedan	rear spare tire, 1933-35	60	115	135
Graham Convertible Sedan	side spare tire, 1933-35	60	115	135
Graham Coupe	rear spare tire, 1933-35	60	115	135
Graham Coupe	side spare tire, 1933-35	60	115	135
Graham Roadster	rear spare tire, 1933-35	60	115	135
Graham Roadster	side spare tire, 1933-35	60	115	135
Graham Sedan	rear spare tire, 1933-35	60	115	135
Graham Sedan	side spare tire, 1933-35	60	115	135
Graham Towncar	rear spare tire, 1933-35	60	115	135
Graham Towncar	side spare tire, 1933-35	60	115	135
Insurance Patrol	miniature	20	25	35
International Station Wagon	4" long, 1940's	30	45	50

VEHICLES

Tootsietoy

NAME	DESCRIPTION	GOOD	EX	MINT
International Station Wagon	red and yellow, solid white wheels, 1939-41	20	25	40
International Station Wagon	red and yellow, 3" long	20	25	40
International Station Wagon	orange with solid black wheels, post-war	15	20	25
Jaguar Type D	green with solid black wheels, 3" long, 1957	7	10	15
Jaguar XK 120 Roadster	green open top, solid black wheels, 3" long	7	10	15
Jaguar XK 140 Coupe	blue with solid black wheels, 6" long	20	30	40
Kaiser Sedan	blue with solid black wheels, 6" long, 1947	30	40	50
Kayo Ice		250	325	400
La Salle Coupe		125	170	225
La Salle Sedan	red with solid black wheels, 3" long	15	20	25
Lancia Racer	dark green with solid black wheels, 6" long, 1956	7	10	15
Large Bluebird Racer	green with yellow solid wheels	25	35	45
LaSalle Convertible		135	195	250
LaSalle Convertible Sedan		135	195	250
LaSalle Sedan		125	170	225
Limousine	blue with silver spoke wheels	25	40	55
Lincoln Capri	red with yellow top, solid black wheels, 2-door, 6" long	20	35	45
Mercedes 190 SL Coupe	powder blue with solid black wheels, 6" long, 1956	25	40	50
Mercury	red with black wheels, 4-door, 4" long, 1952	20	30	35
Mercury Custom	blue with solid black wheels, 4-door, 4" long, 1949	20	30	35
Mercury Fire Chief Car	red with solid black wheels, 4" long, 1949	25	35	45
MG TF Roadster	red open top, solid black wheels, 6" long, 1954	20	25	35
MG TF Roadster	blue open top, solid black wheels, 3" long, 1954	16	20	25
Moon Mullins Police Car	1930's	250	300	375
Nash Metropolitan Convertible	red with solid black tires, 1954	30	35	45
Observation Car		10	15	20
Offenhauser Racer	dark blue with solid black wheels, 4" long, 1947	15	20	25
Oldsmobile 88 Convertible	yellow with solid black wheels, 4" long, 1949	15	25	35
Oldsmobile 88 Convertible	bright green with solid black wheels, 6" long, 1959	20	25	35
Oldsmobile 98	white body with blue top, skirted fenders, solid black wheels, 4" long, 1955	20	25	35
Oldsmobile 98	red body with yellow top, open fenders, solid black wheels, 4" long, 1955	20	25	35
Oldsmobile 98 Staff Car		20	25	35
Oldsmobile Brougham		25	35	45
Oldsmobile Coupe		25	35	45
Oldsmobile Roadster	orange and black, solid white wheels	25	35	45
Oldsmobile Sedan		25	35	45
Oldsmobile Touring		25	35	45
Open Touring	green with open top, solid white wheels	25	35	45
Packard	white body with blue top, solid black wheels, 4-door, 6" long, 1956	25	35	45
Plymouth	dark blue with solid black wheels, 2-door, 3" long, 1957	10	15	20
Plymouth Sedan	blue with solid black wheels, 4-door, 3" long, 1950	12	18	22
Pontiac Fire Chief	red with solid black wheels, 4" long, 1950	20	35	45
Pontiac Sedan	green with solid black wheels, 2-door, 4" long, 1950	15	25	35
Pontiac Star Chief	red with solid black wheels, 4-door, 4" long, 1959	15	25	30
Porsche Roadster	red with open top, solid black wheels, 2-door, 6" long, 1956	18	25	35
Pullman Car		10	15	20
Racer	miniature	25	40	50
Racer	orange with solid black wheels, 3" long, 1950's	10	15	20
Rambler Wagon	dark green with yellow top, black wheels with yellow insides, 1960's	16	22	30

Tootsietoy

NAME	DESCRIPTION	GOOD	EX	MINT
Rambler Wagon	blue with solid black wheels, 4" long, 1960	16	22	30
Roadster		65	100	130
Roadster	miniature	20	25	35
Sedan		65	100	130
Sedan	miniature	20	25	35
Small Racer	blue with driver, solid white wheels, 1927	60	95	125
Smitty		250	375	500
Studebaker Coupe	green with solid black wheels, 3" long, 1947	25	35	45
Studebaker Lark Convertible	lime green with solid black wheels, 3" long, 1960	7	10	15
Tank Car	miniature	20	25	35
Thunderbird Coupe	powder blue with solid black wheels, 4" long, 1955	15	30	35
Thunderbird Coupe	blue with solid black wheels, 3" long, 1955	15	20	25
Torpedo Coupe		15	20	25
Triumph TR 3 Roadster	solid black wheels, 3" long, 1956	7	10	15
Uncle Walt in a Roadster	1932	275	325	400
Uncle Willie		275	325	400
VW Bug	metallic gold with solid black tread wheels, 6" long, 1960	15	25	30
VW Bug	lime green with solid black tread wheels, 3" long, 1960	5	7	10
Yellow Cab Sedan	green with solid white wheels, 1921	10	20	25

Emergency Vehicles

NAME	DESCRIPTION	GOOD	EX	MINT
American LaFrance Pumper	all red, 3" long, 1954	15	20	25
Chevy Ambulance	army green, red cross on roof top, army star on top of hood, 4" long, 1950	15	25	35
Chevy Ambulance	yellow, red cross on top, 4" long, 1950	15	25	35
Fire Hook and Ladder	red and blue with side ladders	25	40	50
Fire Water Tower Truck	blue and orange, red water tower	40	60	80
Graham Ambulance	white with red cross on sides	65	95	125
Graham Army Ambulance		65	95	125
Hook and Ladder	with driver, 1937-41	30	45	60
Hook and Ladder	red and silver	25	40	50
Hook and Ladder		20	25	35
Hose Car	with figure driving and figure standing in back by water gun, 1937-41	30	40	55
Hose Wagon	all red except silver hose, solid white rubber wheels, 3" long, pre-war	20	25	35
Hose Wagon	all red, solid black rubber wheels, postwar	20	25	35
Insurance Patrol	all red, solid white wheels, prewar	20	25	35
Insurance Patrol	all red, solid black rubber wheels, postwar	20	25	35
Insurane Patrol	with driver	30	40	55
Mack L-Line Fire Pumper	red with ladders on sides	35	65	75
Mack L-Line Hook and Ladder	red with silver ladder	35	65	75

Farm and Construction Equipment

NAME	DESCRIPTION	GOOD	EX	MINT
Cat Bulldozer	yellow, 6" long	25	45	55
Cat Scraper	yellow with solid black wheels, silver blade, 6" long, 1956	15	25	35
Caterpillar Tractor	miniature	15	25	30
Caterpillar Tractor	1931	10	15	20
D7 Crawler with Blade	1/50 scale, die cast, 1956	25	35	50
D8 Crawler with Blade	1/87 scale, die cast	20	30	40
Farm Tractor	with driver	70	100	145
Ford Tractor	red with loader, die cast, 1/32 scale	30	40	60
Grader	1/50 scale, die cast, 1956	20	30	45
International Tractor		7	10	15
Steamroller	1931-34	100	150	200

VEHICLES

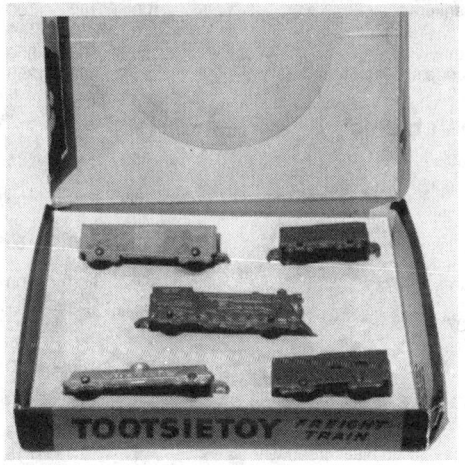

Top to bottom: The Waltons Truck, 1975, Mego; Tootsietoy Freight Train, 1940; Racing motorcycle, Schuco; Ford GT-40, Bandai.

Tootsietoy

NAME	DESCRIPTION	GOOD	EX	MINT
Sets				
Box Trailer and Road Scraper Set	with driver on road scraper	125	200	255
Contractor Set	pickup truck with three wagons	65	100	130
Four Car Transport Set	tractor-trailer, flatbed trailer carries cars	60	90	135
Freight Train	five piece set	40	60	80
Grand Prix #1687 Set	seven vehicles, 1969	60	95	125
Midget Series	yellow stake truck, red limo, green doodlebug, yellow railcar, blue racer, red fire truck, 1" long, 1936-41	5	7	10
Midget Series	green cannon, blue tank, green armored car, green tow truck, green camelback van, 1" long, 1936-41	5	7	10
Midget Series	assorted ships, 1" long, 1936-41	5	7	10
Midget Series	single engine plane, St. Louis, bomber, Atlantic Clipper, 1" long, 1936-41	5	7	10
Milk Trailer Set	tractor with three milk tankers	100	150	200
Passenger Train	five piece set	40	60	80
Playtime Set	6 cars, 2 trucks, 2 planes	400	500	850
Tractor with Scoop Shovel and Wagon	red tractor with silver scoop shovel, flatbed trailer, 1946-52	125	185	250
Space Vehicles				
Buck Rogers Attack Cruiser	cast metal, 5" long, 1930's	90	140	185
Buck Rogers Battle Cruiser	1937	50	75	100
Buck Rogers Blast Attack Ship	cast metal, 4 1/2" long, 1937	75	100	150
Buck Rogers Rocket Ships	set of four with two figures, 1937	350	800	1100
Buck Rogers Venus Duo Destroyer	cast metal, 5" long, 1937	50	75	100
Rocket Launcher		40	60	80
Trailers				
Boat Trailer	2-wheel	7	10	15
Horse Trailer	red with white top, 2-wheel, solid black tread wheels	7	10	15
House Trailer	powder blue with solid black wheels, 2-wheel, door opens	10	15	20
Restaurant Trailer	yellow with solid black tread wheels, 2-wheel, open sides	20	30	40
Small House Trailer	2-wheel, three side windows, 1935	25	40	50
U-Haul Trailer	red with solid black tread wheels, 2-wheel, U-Haul logo on sides	10	15	20
Trains				
Borden's Milk Tank Car	white embossed metal painted	10	15	25
Box Car		10	15	20
Caboose	all red	7	10	15
Coal Car		10	15	20
Cracker Jack Tootsietoy Railroad Car	embossed white metal, painted orange, black rubber tires, 3" long, 1930's	85	130	175
Fast Freight Set	five piece set, 1940	40	60	80
Log Car	silver body, red wheels, load of logs chained on flatbed car	10	15	20
Milk Tank Car	yellowish top with narrow red strip along bottom	10	15	20
Oil Tank Car	silver top with narrow red strip along bottom, 'Sinclair' on sides	10	15	20
Passenger Train Set	four-piece set, 1925	65	100	130
Pennsylvania Engine		20	30	45
Refrigerator Car	yellowish sides with narrow red strip along bottom, black roof	10	15	20

VEHICLES

Tootsietoy

NAME	DESCRIPTION	GOOD	EX	MINT
Santa Fe Engine		15	20	25
Stock Car	all red	10	15	20
Stock Car		10	15	20
Tootsietoy Flyer	three piece set, 1937	30	45	60
Wrecking Crane	green crane on silver flatbed car with red wheels	10	15	20
Zephyr Railcar	dark green, 4' long, 1935	35	50	70

Trucks

NAME	DESCRIPTION	GOOD	EX	MINT
Army Half Truck	1941	35	55	75
Army Jeep	windshield up, 6" long, 1950's	15	25	30
Army Jeep CJ3	extended back, windshield down, 4" long, 1950	7	10	15
Army Jeep CJ3	no windshield, 3" long, 1950	10	15	20
Army Supply Truck	with driver	25	40	50
Box Truck	red with solid white wheels, 3" long	10	15	20
Buick Delivery Van		25	35	45
Cadillac Delivery Van		25	35	45
Chevrolet Delivery Van		25	35	45
Chevy Cameo Pickup	green with solid black wheels, 4" long, 1956	15	25	35
Chevy El Camino	red	20	25	35
Chevy El Camino Camper and Boat	blue vehicle with red camper, black and white boat on top of camper	25	35	45
Chevy Panel Truck	light green with solid black wheels, 4" long, 1950	25	30	35
Chevy Panel Truck	green, 3" long, 1950	12	18	22
Chevy Panel Truck	green, front fenders opened, 3" long, 1950's	12	18	22
Civilian Jeep	burnt orange, open top, solid black wheels, 3" long, 1950	7	10	15
Civilian Jeep	red, open top, solid black wheels, 4" long, 1950	15	20	25
Civilian Jeep	blue with solid black tread wheels, 6" long, 1960	15	20	25
CJ3 Army Jeep	open top, no steering wheel cast on dashboard, 3" long, 1950	10	15	20
CJ5 Jeep	red with solid black tread wheels, windshield up, 6" long, 1960's	15	20	25
CJ5 Jeep	red with solid black tread wheels, windshield up, 6" long, 1950's	15	25	30
Coast to Coast Van	9" long	50	75	100
Commercial Tire Van	'Commercial Tire & Supply Co.' on sides	100	150	200
Diamond T K5 Dump Truck	yellow cab and chassis, green dump body, 6" long	25	35	45
Diamond T K5 Grain Semi	red tractor and green trailer	30	45	55
Diamond T K5 Stake Truck	orange, open sides, 6" long, 1940	25	35	45
Diamond T K5 Stake Truck	orange, closed sides, 6" long, 1940	25	35	45
Diamond T K5 Tootsietoy Semi	red tractor and light green closed trailer	25	45	55
Diamond T Metro Van	powder blue, 6" long	65	75	100
Diamond T Tow Truck	red with silver tow bar	25	35	45
Dodge D100 Panel	green and yellow, 6" long	30	40	55
Dodge Pickup	lime green, 4" long	20	30	35
Federal Bakery Van	black with solid cream wheels, 1924	55	85	110
Federal Florist Van	black with solid cream wheels, 1924	115	165	200
Federal Grocery Van	black with solid cream wheels, 1924	45	70	90
Federal Laundry Van	black with solid cream wheels, 1924	55	85	110
Federal Market Van	black with solid cream wheels, 1924	55	85	110
Federal Milk Van	black with solid cream wheels, 1924	55	85	110
Ford C600 Oil Tanker	bright yellow, 3" long	7	10	15
Ford C600 Oil Tanker	red, 4" long, 1962	15	29	25
Ford Econoline Pickup	red 1962	15	29	25
Ford F1 Pickup	orange, closed tailgate, 3" long, 1949	12	18	22
Ford F1 Pickup	orange, open tailgate, 3" long, 1949	12	18	22
Ford F6 Oil	orange, 4" long, 1949	10	15	20

VEHICLES

Tootsietoy

NAME	DESCRIPTION	GOOD	EX	MINT
Ford F6 Oil Tanker	red with Texaco, Sinclair, Shell or Standard on sides, 6" long, 1949	30	45	55
Ford F6 Pickup	red, 4" long, 1949	15	25	30
Ford F600 Army Anti-Aircraft Gun	tractor-trailer flatbed, guns on flatbed	20	25	35
Ford F600 Army Radar	tractor-trailer flatbed, yellow radar unit on flatbed, 6" long, 1955	20	25	35
Ford F600 Army Stake Truck	tractor-trailer box, army star on top of trailer box roof and 'U.S. Army' on sides, 6" long, 1955	25	40	50
Ford F600 Stake Truck	light green, 6" long, 1955	15	25	30
Ford Pickup	3" long, 1935	25	40	50
Ford Shell Oil Truck		35	50	65
Ford Styleside Pickup	orange, 3" long, 1957	10	15	20
Ford Texaco Oil Truck		35	50	65
Ford Wrecker	3" long, 1935	30	40	45
Graham Wrecker	red and black	65	95	125
Hudson Pickup	red, 4" long, 1947	25	40	50
International Bottle Truck	lime green	30	45	55
International Car Transport Truck	red tractor orange double-deck trailer with cars	35	50	65
International Gooseneck Trailer	orange tractor and flatbed trailer	30	40	50
International K1 Panel Truck	blue, 4" long	20	30	35
International K11 Oil Truck	green, comes with oil brands on sides, 6" long	30	45	55
International RC180 Grain Semi	green tractor and red trailer	15	25	30
International Sinclair Oil Truck	6" long	35	50	65
International Standard Oil Truck	6" long	35	50	65
Jeepster	bright yellow with open top, solid black wheels, 3" long, 1947	20	30	35
Jumbo Pickup	6" long, 1936-41	25	35	45
Jumbo Wrecker	6" long, 1941	30	40	55
Mack Antiaircraft Gun		25	40	50
Mack B-Line Cement Truck	red truck with yellow cement mixer, 1955	20	35	45
Mack B-Line Oil Tanker	red tractor and trailer, 'Mobil' decal on side of trailer	20	35	45
Mack B-Line Stake Trailer	red tractor, orange closed trailer, 1955	20	35	45
Mack Coal Truck	'City Fuel Company' on sides of box	60	100	135
Mack Coal Truck	orange cab with blue bed, 1925	25	40	50
Mack Coal Truck	red cab with black bed, 1928	30	40	55
Mack Dairy Tanker	1930's	75	100	150
Mack L-Line Dump Truck	yellow cab and chassis, light green dump body, 6" long, 1947	20	35	45
Mack L-Line Semi and Stake Trailer	red tractor and trailer	75	95	125
Mack L-Line Semi-Trailer	red tractor cab, silver semi-trailer, 'Gerard Motor Express' on sides	85	115	145
Mack L-Line Stake Truck	red with silver bed inside	25	35	45
Mack L-Line Tow Truck	red with silver tow bar	25	35	45
Mack Log Hauler	red cab, trailer with load of logs, 1940's	75	95	135
Mack Long Distance Hauling Truck	1930's	75	115	150
Mack Mail Truck	red cab with light brown box, 'U.S. Mail Airmail Service' on sides, 3" long, 1920's	45	55	85
Mack Milk Truck	enclosed cab, 'Tootsietoy Dairy' on side of milk tanker	60	95	125
Mack Oil Tanker	'DOMACO' on side of tanker	60	95	125
Mack Oil Truck	red cab with orange tanker, 1925	25	40	50
Mack Searchlight Truck	1931-41	25	40	50

VEHICLES

Tootsietoy

NAME	DESCRIPTION	GOOD	EX	MINT
Mack Stake Trailer-Truck	enclosed cab, open stake trailer, 'Express' on sides of trailer	55	90	115
Mack Stake Truck	orange cab with red stake bed, 1925	30	40	55
Mack Trailer-Truck		55	85	110
Mack Transport	enclosed cab with flatbed trailer	100	150	200
Mack Transport	yellow, 1941, with cars at angle	250	375	500
Mack Van Trailer-Truck	enclosed cab and box trailer	75	100	135
Mack Wrigley's Spearmint Gum Truck	4" long	90	125	200
Model T Pickup	3" long, 1914	25	40	50
Oil Tanker	green with solid white wheels	15	25	30
Oil Tanker	blue and silver, two caps on top of tanker, 3" long	18	20	25
Oil Tanker	all orange, four caps on top of tanker, 3" long, postwar	18	20	25
Oil Tanker	blue, three caps on top, 2" long, 1932	20	25	35
Oldsmobile Delivery Van		25	35	45
Sinclair Oil Truck	6" long	35	50	65
Special Delivery	1936	20	25	35
Stake Truck	miniature	25	40	50
Tootsietoy Dairy	enclosed cab with attached milk tanker plus tanker trailer	60	95	125
Tootsietoy Oil Tanker	red cab, silver tanker, "Tootsietoy Line" on side, 1950's	60	95	125
Wrecker		30	50	70
Wrigley's Box Van	with or without decal, 1940's	45	60	75

Trucks

Marcrest Dairy Truck	white truck with 3 glass bottles, 14" long	150	275	425

Williams, A.C.

Cars

Dream Car	cast iron, 4 7/8" long, 1930	75	150	250
Ford Roadster	1936	450	550	650
Lincoln Touring Cars	spoked wheels, cast iron, 8 3/4" long, 1924	400	600	800
Racer	yellow, cast iron, 8 1/2" long, 1932	300	450	600
Taxi	cast iron, 5 1/4" long, 1920	200	350	500
Touring Cars	disc wheels, cast iron, 9 1/8" long, 1922	400	600	800
Touring Cars	solid wheels, cast iron, 11 3/4" long, 1917	500	750	1200

Trucks

Austin Transport Set	with three vehicles, cast iron, 12 1/2" long, 1930	500	850	1250
Interchangeable Delivery Truck	cast iron, 7 1/4" long, 1932	175	250	350
Moving Van	cast iron, 4 3/4" long 1930	150	225	300
Pickup Truck	cast iron, 4 3/4" long, 1926	100	150	200

Wyandotte

Airplanes

Airliner Plane	metal, two engines, wooden wheels	35	55	195
American Airlines Flagship Plane	28" wingspan	70	125	300
Army Bomber	pressed steel, two engines	35	55	200
China Clipper	13" wingspan	55	85	150
High Wing Passenger Monoplane	18" wingspan	70	125	235
Military Air Transport Plane	13" wingspan	60	95	225
P-38 Plane	9 3/4" wingspan	25	40	135
Twin Engine Airliner	4 3/4" wingspan	20	30	125

Wyandotte

NAME	DESCRIPTION	GOOD	EX	MINT
Boats and Ships				
Pocket Battleship	tin litho, 7" long	25	40	100
S.S. America	7" long, 1930's	35	55	125
Sand O' Land	tin litho, sand toy, 10" long, 1940's	25	40	95
Buses				
Era Tractor-Trailer Wyandotte Truck Lines	red and green, 25" long	150	200	300
Cannons				
Cannon	shoots marbles, 14" long	35	55	125
Cars				
Cadillac Station Wagon	steel, 1941	250	400	500
Cord Coupe Model 810	pressed steel, 13" long, 1936	200	300	575
Humphrey Mobile (Joe Palooka)	tin litho, windup, 1940's	200	300	1150
Sedan	blue, rubber wheels, pressed steel, 6" long	35	55	95
Woody Station Wagon	steel	250	400	500
Emergency Vehicles				
Ambulance	steel, wood tires, 11" long, 1930's	60	95	250
Farm and Construction Equipment				
Sturdy Construction Co. Steam Shovel	litho, 20" long	75	115	275
Space Vehicles				
Flash Gordon Strat-O-Wagon	9" long	60	95	200
Trucks				
Auto Service Truck	red plastic cab, blue and white bed, 15" long, 1950's	50	75	110
Car Hauler	yellow and red metal tractor, 22" long	100	150	200
Car Hauler	red plastic tractor, 22" long	50	75	100
Circus Truck and Trailer	embossed wooden wheels, litho steel, 19" long, 1936	250	400	500
Construction Truck	red and yellow cab, blue trailer, wood wheels, 24" long	85	130	175
Dump Truck	red and white cab, blue bed, 13" long, 1950's	75	115	150
Dump Truck	pressed steel, 6" long, 1930's	75	115	150
Era Express Open Truck	blue and white bed, 22" long, 1940's	75	150	225
Flatbed with Steam Shovel	24" long	125	175	300
Highway Freight Truck	blue tractor, red trailer, 17" long	100	150	200
L Tanker	orange, pressed steel, 11" long, 1930	90	130	180
Military Amphibian	21" long	95	150	225
Motor W Fleet Side Dump Truck	yellow cab, blue dump bed, 18" long	85	130	175
Nationwide Air Rail Service Truck	red and white, 12" long	100	150	200
Scoop Dump Truck	red and yellow cab, red bed, 16" long	65	85	125
Service Car Truck	red cab, white bed, 12" long, 1950's	50	75	125
Side Dump	red and white cab, green bed, wood wheels, 20" long	100	150	200
Stake Truck	10" long, 1930's	90	140	185
Stake Truck	green, battery operated lights, 10" long	70	125	145
Stake Truck	all red, black wood wheels, pressed steel, 1930's	50	75	100
Stake Truck	red cab, turquoise stake bed, black spoke wheels, 10" long	90	140	185
Woody Convertible	steel, with retractable hard top, 1941	125	200	250
Wrecker	red cab, yellow bed sides, 22" long	100	150	200
Wrecker	red cab and bed, orange boom, 13" long, 1940's	50	100	145
Wrecker	blue cab, red bed, 12" long, 1950's	35	65	90
Wyandotte Truck Lines	cab-tractor-trailer, red and yellow, litho, 23" long	125	185	250

VEHICLES

Collecting Games

Just a few years ago, except for a handful of visionaries, no one considered board and card games as worthy of collecting, which in hindsight is itself a strong indicator of their future collectibility. Games were playthings only, and were treated with the casual disregard given to thousands of other now highly collectible products of disposable America. Their entire value was in the playing, in how much fun and social interaction they afforded. They were casually played, their parts cheerfully smudged, crumpled, and misplaced. Once their "fun quotient" was exhausted, they were just as casually tossed in closets, in attics, or out with the trash.

While the idea of games as collectibles is a recent one in the world of antiques, the field is making up for that oversight with a vengeance. This is particularly true in the three most visible groupings, Victorian games, sports games, and character/TV games.

Victorian games are loosely defined as those produced between the Civil War and World War I. During this period manufacturing technologies advanced from the hand tinted games of the pre-1860s to chromo-lithography, which ushered in the age of mass production.

The premier marque of Victorian games belongs to the McLoughlin Brothers, which progressed from hand tinting to lithography in grand style, producing as they went many of the most beautifully illlustrated — and most valuable — games ever made. In many cases, simply the McLoughlin name on the box will make a game considerably more valuable than the identical item produced by a competitor.

McLoughlin was bought out by Milton Bradley in 1920, which reissued numerous old McLoughlin games under the MB banner. As would be expected, these reissues have only a fraction of the value of their McLoughlin originals.

Milton Bradley Company's "World's Fair Game" board game, 1964-65.

Another noteworthy game maker was Parker Brothers, which from 1883 on produced many games that rival the beauty and collectibility of McLoughlin's. Bliss, Clark & Sowdon and J.H.Singer are other names to watch for, as they also held to high standards of creativity and artistry in the execution of their games.

However, if you do come across a Victorian game in your travels, the odds are most likely that it will be by Milton Bradley, by far the most prolific manufacturer of the age. While its games were generally neither as beautiful nor now as valuable as those of its competitors, they are still collected for both their artwork and subject matter.

Sports games, and there are many, have enjoyed immense gains in value along with the boom in other sports memorabilia, which essentially began in the late 1970s. This area also illustrates the issue of relative value across collecting fields.

In general, sports collectors assign a higher value to sports games than do game collectors, and asking prices for the same game can vary widely from a toy show to a sports show, or even from one table to another at the same show for that matter. But the point is repeatedly proven that collectors within a specialized field will often pay more for a particular item than generalized collectors in a related field.

Still, no hobby operates in a vacuum. Increasing demand by sports collectors will also drive up prices for game collectors, just as demand by Christmas collectors for Santa games will drive up prices for both Christmas and game collectors. Game collectors might always assign a lower value to the same game than either of these more specialized groups, but whatever its origin, demand drives value. As collectors grow more sophisticated and knowledgeable about various sources of items in their special areas, these cross-field price variances begin to fade, but they will probably never disappear altogether, particularly when one field can experience a boom time when another may not.

Sports games are collected either by sport depicted, such as baseball or horse racing, or by personality, such as Milton Bradley's Babe Ruth's Baseball Game (1926), or Gardner's Mickey

Mantle's Big League Baseball (1955). Personality-based games typically have higher values than their generic counterparts, and older generic games are normally more valuable than newer ones.

Many collectors consider World War II as the breaking point between antique and modern games. However, this turning point can be defined perhaps just as logically as either before or after the advent of television. The fact that World War II and the rise of television took place in the same decade is largely convenient coincidence. Which event wrought the most lasting changes on American life can make for a rousing evening's debate, but TV's impact on modern marketing and consumer habits is hard to overestimate.

Character games in particular lend themselves to this theory of boob tube classification. Pre-TV games can be seen as evolving from Victorian nursery rhyme games such as McLoughlin's Little Goldenlocks And The Three Bears (1890), numerous variations on Mother Goose, and Milton Bradley's Little Jack Horner (1910) to name a few.

These were supplemented by popular comic characters such as Winnie Winkle, The Katzenjammer Kids, Chester Gump, Dick Tracy, and also by literary entries such as Parker Brother's classic Wonderful Game of Oz (1921) and Kerk Guild's A. A. Milne's Winnie The Pooh Game (1931).

The 1930s also saw the tapping of a limitless well of Disney character games, beginning with numerous Mickey Mouse games and following with Ferdinand The Bull, Snow White, Alice In

Milton Bradley Company's "eye Guess" board game, 1966.

Wonderland and many more. Disney's merchandising grew more sophisticated with each new film, setting precedents that have become today's standard practice.

Post-TV games of the 1950s saw a general decline in the quality of materials and execution and an increasing reliance on character affiliation. Games such as Lowell's Gunsmoke (1950), Milton Bradley's Annie Oakley (1950), and Parker Brothers' Bing Crosby — Call Me Lucky (1954) set a new, albeit lower standard for game production and marketing.

TV-based games of the 1950s and 1960s are currently the most eagerly sought after of all modern games, and their values reflect that demand. Each issue of Toy Shop magazine turns up more ad space devoted to games, and prices continue to enjoy steady appreciation.

Finding Games

While the national awareness of all things old as collectible increases on a daily basis, it is still possible to make those blood rushing garage sale finds that lend themselves to many retellings of the event. One recent yard sale find reported to us involved two games, one valued at $200 and the other a previously unknown sports game, now valued by experts at roughly $1,000. Both games were purchased for less than $10.

Flea markets and collector shows also offer excellent prospecting opportunities, as do the major national magazines, Toy Shop and Toy Collector And Price Guide.

Interested collectors are well advised to seek out the current standard reference book on games, Bruce Whitehill's Games, American Boxed Games and Their Makers 1822-1992, published by Wallace Homestead Books.

The American Game Collectors Association, boasting a worldwide membership, offers a direct line to all games from Victorian to modern. Interested collectors can write to AGCA at 49 Brooks Avenue, Lewiston, ME 04240.

While the "super finds" are largely a thing of the past, they are not entirely vanished — you just need to look.

Trends

Today's TV, film, and cartoon character games are strong bets for future appreciation. The same holds true for sports games, particularly if they are celebrity-based, and those games that address current historical or cultural affairs. Also worth seeking out are the limited run products of smaller players in the game field, again, especially those with character affilation.

Classic Games and Mayfair Games are two smaller companies that publish limited runs of popular character based games, often in the range of 10,000 to 15,000 copies. Currently available games from these companies include Classic Games' Star Trek: The Next Generation Trivia Game and Mayfair's The Elfquest Gameboard, Thieves' World Sanctuary board game and Dragonriders of Pern, all based on long running science fiction franchises.

Word games, generic strategy games, and non-character related games have historically demonstrated little or no appreciation, and their heritage as poor investments will likely be inherited by their modern counterparts. Exceptions are games in genres such as science fiction, all of which have better than normal potential as long as sci-fi remains in vogue. Another exception is in games that utilize unique, complex, or intricate playing pieces and apparatus. A good example of this can be seen in Transogram's Ka-Bala (1965). This game would be moderately collectible simply by virtue of its fortune telling theme, but its board-dominating glow-in-the-dark eyeball centerpiece is thrillingly ghoulish, thus its current value of $150.

While the burgeoning field of game collecting offers a wealth of benefits including investment potential, history, and artistry, it also offers a less quantifiable return, play value. While most Victorian and modern character games are sought primarily for box and board art, many games over the years have been designed to fulfill the mission of fun.

Game collecting lends a unique and richly historical body of evidence to the importance of pleasure, relaxation, and social contact. If you ever have the chance to move your pewter playing piece around the gorgeous and rare Parker Brothers' The Wonderful Game of Oz, you might feel the same way. The beauty is in the box, but the magic is in the play.

The Top 25 Prewar Character (Nonsport) Games (in Excellent condition)

Bulls and Bears, 1896, McLoughlin Brothers	$20,000
Teddy's Ride From Oyster Bay, 1899, J. Crandall	6,800
Little Fireman Game, 1897, McLoughlin Brothers	6,000
National Game of The American Eagle, 1844, W. & S.B. Ives	4,000
Man in the Moon, 1901, McLoughlin Brothers	3,500
Rival Policemen, 1896, McLoughlin Brothers	3,000
Darrow Monopoly, 1934, Charles Darrow	2,500
Fire Alarm Game, 1899, Parker Brothers	2,300
The Game of Detective, 1889, R. Bliss	2,000
Game of the Newsboy, 1890, R. Bliss	2,000
Watermelon Patch Game, 1896, McLoughlin Brothers	2,000
Merry-Go-Round, 1898, Chaffee & Selchow	2,000
Hand of Fate, 1901, McLoughlin Brothers	2,000
The New Game of the American Revolution, 1844, L. Burge	1,600
Game of Shopping, 1891, R. Bliss	1,500
The Cake Walk Game, 1900s, Parker Brothers	1,500
Hi-Way Henry, 1928, All-Fair	1,500
The World's Fair Game, 1892, Parker Brothers	1,400
Amusing Game of Conundrums, 1853, John McLoughlin	1,300
Columbus, 1892, Parker Brothers	1,300
The Game of Napoleon, 1895, Parker Brothers	1,300
The Game of Playing Department Store, 1898, McLoughlin Brothers	1,300

The Wonderful Game of Oz (pewter), 1921, Parker Brothers 1,300
The Jolly Game of Goose, 1851, J.P. Beach... 1,250
Yankee Pedlar, Or What Do You Buy, 1850s, John McLoughlin............................... 1,200

The Top 10 Prewar Sports Games
(in Excellent condition)

The Championship Baseball Game, 1889, Schultz $6,800
Egerson R. Williams Baseball Game, 1889, Popular Indoor Baseball Co. 5,000
Golf, 1900, Schoenhut 5,000
Great Mails Baseball Game, 1919 4,100
American League Fan Craze Card Game, 1904, Fan Craze Co. 3,250
Parlor Base Ball, 1878 2,900
Zimmer Baseball Game, 1885, McLoughlin Brothers 2,150
The Diamond Game of Baseball, 1900, McLoughlin Brothers.................................. 2,150
Chicago Game Series Baseball, 1890s, George Doan 1,950
The Yale-Harvard Game, 1890, McLoughlin Brothers.. 1,775

The Top 25 Postwar Character Games
(in Mint Condition)

Mickey Mouse Haunted House Bagatelle, 1950s $550
Green Hornet Quick Switch Game, 1966, Milton Bradley... 475
James Bond Message From M Game, 1966, Ideal 450
Monster Lab, 1964, Ideal 450
New Avengers Shooting Game, 1976, Denys Fisher 440
Batman Batarang Toss, 1966, Pressman 400
Gilligan's Island, 1965, T. Cohn 325
Wolfman Mystery Game, 1963, Hasbro 320
Thing Ding Robot Game, 1961, Schaper.. 310
Boom Or Bust, 1951, Parker Brothers 300
Jackie Gleason's Story Stage Game, 1955, Utopia Enterprises............................... 300
Creature From The Black Lagoon, 1963, Hasbro 280
Outer Limits, 1964, Milton Bradley.. 275
Justice League of America, 1967, Hasbro 275
The Phantom Game, 1965, Transogram 255
The Untouchables, 1950s, Marx 250
Little Black Sambo, 1952, Cadaco.. 250
Groucho's You Bet Your Life, 1955, Lowell 250
Rocket Race, 1958, Stone Craft 250
Hawaiian Eye, 1960, Transogram 250
Fugitive, 1966, Ideal 240
The Time Tunnel Game, 1966, Ideal 240
The Weird-Ohs Game, 1964, Ideal... 230
Wendy, The Good Little Witch, 1966, Milton Bradley 230
Jetson's Out Of This World Game, 1963, Transogram... 225

Top 10 Postwar Sports Games
(in Mint Condition)

Red Barber's Big League Baseball Game, 1950s, G&R Anthony $900
Win A Card Trading Card Game, 1965, Milton Bradley 900
Strike Three, 1948, Tone Products Corp. 725
Willie Mays "Say Hey", 1954, Toy Development 525
Big Six Sports Games, 1950s, Gardner.. 450
Pee Wee Reese Marble Game, 1956, Pee Wee Enterprises................................... 450
Willie Mays "Say Hey" Baseball, 1958, Centennial Games 450
Bart Starr Quarterback Game, 1960s... 450
Official Baseball Card Game, 1965, Milton Bradley.. 450
Willie Mays Push Button Baseball, 1965, Eldon Champion..................................... 450

Top to bottom: Peg Base Ball Game, 1924, Parker Brothers; The Captive Princess, 1800s, McLoughlin Brothers.

Games, Prewar

NAME	TYPE	YEAR	COMPANY	GOOD	EX
21st Century Football T	B	1930s	Kerger Co.	55	90
400 Game, The	B	1890s	J.H. Singer	100	150
400, Aristrocrat of Games, The	S	1933	Morris Systems Publishing Co.	15	25
A&P Relay Boat Race Coast to Coast (cereal box)	B	1930s	A&P	30	40
ABC	C	1900s	Parker Brothers	25	40
ABC Baseball Game	B	1910s		430	715
ABC, Game of	B	1914		60	100
Abcdarian, The	B	1899	Chaffee & Selchow	40	65
Across The Channel	B	1926	Wolverine	50	85
Across The Continent	B	1892	Parker Brothers	100	175
Across The Continent	B	1922	Parker Brothers	150	250
Across The Sea Game	B	1914		60	100
Across The Yalu	B	1905	Milton Bradley	65	110
Add Too	B	1940	All Fair	10	15
Admiral Byrd's South Pole Game Little America	B	1930s	Parker Brothers	125	200
Admirals, The Naval War Game	B	1939	Merchandisers, Inc.	75	120
ADT Delivery Boy	B	1890	Milton Bradley	120	200
ADT Messenger Boy (Small Version)	B	1915	Milton Bradley	40	70
Advance And Retreat, Game of	B	1900s	Milton Bradley	95	175
Aero BAll	S	1940s	Game Makers, Inc.	30	50
Aero Chute	B	1940	American Toy Works	50	75
Aeroplane Race	B	1922	Wolverine	60	95
After Dinner	B	1937	Frederick H. Beach (Beachcraft)	10	20
Air Base Checkers	B	1942	Einson Freeman Publishing Corp.	20	30
Air Mail, The	B	1930	Archer Toy Co.	75	125
Air Mail, The Game of	B	1927	Milton Bradley	95	155
Air Ship Game, The	B	1904	McLoughlin Brothers	300	500
Air Ship Game, The	B	1912	McLoughlin Brothers	150	300
Airplane Speedway Game	B	1941	Lowe	20	30
Airship Game, The.	C	1916	Parker Brothers	30	50
Aldjemma	B	1944	Corey Games	30	45
Alee Oop	B	1937	Royal Toy Co.	30	50
Alexander's BaseBAll	B	1940s		245	400
Alice in Wonderland	B	1930s	Parker Brothers	60	100
Alice In Wonderland, Game of	B	1923	Stoll & Edwards	50	85
All American BasketbAll	B	1941	Corey Games	55	90
All American FootbAll	B	1935		35	55
All American Football Game	B	1925	Parker Brothers	100	165
All Star Baseball Game	B	1935	Whitman	100	165
Allegrando	C	1884	Theodore Presser	40	60
Allie Patriot Game	C	1917	McDowell And Mellor	30	50
Alpha Football Game	B	1940s	Replica Mfg. Co.	70	115
Amateur Golf	B	1928	Parker Brothers	145	245
Ambuscade, Constellations And Bounce	B	1877	McLoughlin Brothers	195	325
America's FootbAll	B	1939	Trojan Games	55	90
America's Yacht Race	B	1904	McLoughlin Brothers	450	750
American Boy Game	B	1920s	Milton Bradley	75	125
American Derby	B	1931	Henschel Co.	55	90
American Football Game	B	1930	Ace Leather Goods Co.	70	115
American History, The Game of	C	1890s	Parker Brothers	30	50
American League Fan Craze Card Game	C	1904	Fan Craze Co.	1950	3250
American Revolution, The New Game of The	B	1844	Lorenzo Burge	960	1600
American Sports	B	1880s		110	180
American Story and Glory, The Game of	B	1846	Langdon	0	0

Games, Prewar

NAME	TYPE	YEAR	COMPANY	GOOD	EX
Amusing Game of Conundrums	C	1853	John Mcloughlin	750	1300
Amusing Game of Innocence Abroad, The	B	1888	Parker Brothers	135	225
Amusing Game of the Corner Grocery	C	1890s		100	150
Anagrams	C	1885	Peter G. Thompson	20	35
Ancient Game of The Mandarins, The	B	1923	Parker Brothers	45	75
Andy Gump, His Game	B	1924	Milton Bradley	60	120
Anex A Gram	B	1938	The Embossing Co.	15	25
Animal & Bird Lotto	B	1926	All Fair	15	20
Apple Pie	C	1895	Parker Brothers	35	50
Arena	B	1896	R. Bliss Mfg. Co.	120	200
Astronomy	C	1905	Cincinnati Game Co.	20	35
Athletic Sports	B	1900	Parker Brothers	145	245
Atkins Real BasebAll	B	1915	Atkins & Co.	450	750
Attack, Game of	B	1889	R. Bliss Mfg. Co.	300	500
Auction Letters	C	1900	Parker Brothers	20	35
Authors	B	1861	Whipple & Smith	50	100
Authors	B	1890s	J.H. Singer	25	50
Authors Illustrated	C	1893	Clark & Sowdon	35	45
Authors, Game of Standard	C	1890s	McLoughlin Brothers	25	50
Authors, The Game of	C	1890s	Parker Brothers	15	35
Auto Game, The	B	1906	Milton Bradley	60	100
Auto Race Electro Game	B	1929	Knapp Electric & Novelty Co.	125	210
Auto Race Game	B	1925	Milton Bradley	125	200
Auto Race Jr.	B	1925	All Fair	125	175
Auto Race, Army, Navy, Game Hunt (4 game set)	B	1920s	Wilder	110	180
Auto Race, Game Of	B	1920s	Orotech	105	175
Auto Play Baseball Game	B	1911	Auto Play Co.	425	700
Automobile Race, Game of The	B	1904	McLoughlin Brothers	725	1200
Avilude	C	1873	West & Lee Co.	65	100
Aydelott's Parlor BasebAll	B	1910		195	325
Babe Ruth National Game of BasebAll	B	1929	Keiser Fry	550	910
Babe Ruth's Baseball Game	B	1926	Milton Bradley	200	500
Babe Ruth's Official Baseball Game	B	1940s	Toytown	430	715
Baby Barn Yard	B	1940s	B.L. Fry Products Co.	15	25
Bag of Fun	S	1932	Rosebud Art Co.	15	20
Bagatelle, Game of	B	1898	McLoughlin Brothers	200	350
Bagdad, The Game of The East	B	1940	Clover Games	25	40
Balance The Budget	C	1938	Elten Game Corp.	30	45
Balloonio	S	1937	Frederick H. Beach	20	40
Bally Hoo	C	1931	Sam'l Gabriel Sons & Co.	45	75
Bambino	S	1934	Bambino Products Co.	75	125
Bambino (baseball, Chicago World's Fair)	B	1933	Johnson Store Equipment Co.	295	490
Bambino Baseball Game	B	1940	Mansfield Zesiger	145	250
Bamboozle, Or The Enchanted Isle	B	1876	Milton Bradley	175	300
Bang Bird	S	1924	Doremus Schoen & Co.	20	30
Bang, Game of	B	1903	McLoughlin Brothers	135	225
Banner Lye Checkerboard	B	1930s	Geo E. Schweig & Son	15	20
Barage	B	1941	Corey Games	20	30
Barber Pole	S	1908	Parker Brothers	50	75
Barn Yard Tiddledy Winks	S	1910s	Parker Brothers	50	85
Barney Google And Spark Plug Game	B	1923	Milton Bradley	100	200
Baron Munchausen Game, The	B	1933	Parker Brothers	35	55
Base Hit	B	1944	Games, Inc.	55	90

GAMES

Games, Prewar

NAME	TYPE	YEAR	COMPANY	GOOD	EX
Base Ball, Game of	B	1886	McLoughlin Brothers	750	1300
BasebAll	B	1942	Lowe	15	25
Baseball & Checkers	B	1925	Milton Bradley	75	150
Baseball Dominoes	B	1910	Evans	250	400
Baseball Game	B	1930	All Fair	100	125
Baseball Game & G Man Target Game	B	1940	Marks Brothers	100	165
Baseball Game, New	B	1885	Clark & Martin	165	275
Baseball Wizard Game	B	1916	Morehouse Mfg.	265	450
Baseball, Game of	B	1890	J.H. Singer	325	550
BasebAll itis Card Game	C	1909	Baseballitis Card Co.	125	205
Bases Full	S	1930		45	70
Basilinda	B	1890	E.I. Horsman	105	175
Basket BAll	S	1929	Russell Mfg. Co.	150	250
BasketbAll	B	1942	Lowe	15	25
Basketball Card Game	C	1940s	Warren Built Rite	15	25
Basketball Game, Official	B	1940	Toy Creations	55	90
Batter Up, Game of	C	1914	Fenner Game Co.	100	165
Battle Checkers	B	1925	Pen Man	15	30
Battle Game, The	B	1890s	Parker Brothers	120	200
Battle of Ballots	B	1931	All Fair	55	125
Battle of Manila	B	1899	Parker Brothers	300	550
Battles, Or Fun For Boys, Game of	S	1889	McLoughlin Brothers	425	700
Bean Em	S	1931	All Fair	250	500
Bear Hunt, Game of	B	1923	Milton Bradley	45	70
Beauty And The Beast, Game of	B	1905	Milton Bradley	45	75
Bee Gee Baseball Dart Target	B	1935s	Bee Gee	70	115
Bell Boy Game, The	B	1898	Chaffee & Selchow	425	700
Belmont Park	B	1930	Marks Brothers	75	125
Bengalee	B	1940s	Advance Games	20	35
Benny Goodman Swings	B	1930s	Toy Creations	65	100
Benson Football Game, The	B	1930s	Benson	85	140
Betty Boop Co ed Bridge	C	1930s		50	75
Bible ABC's And Promises	C	1940s	Judson Press	5	10
Bible Authors	C	1895	Evangelical Pub. Co.	15	25
Bible Boys	B	1901	Zondervan Publishing House	10	15
Bible Characters	B	1890s	Decker & Decker	10	25
Bible Cities	C	1920s	Nellie T. Magee	10	15
Bible Lotto	B	1933	Goodenough And Woglom Co.	10	15
Bible Quotto	B	1932	Goodenough And Woglom Co.	6	10
Bible Rhymes	B	1933	Goodenough And Woglom Co.	10	15
Bicycle Cards	C	1898	Parker Brothers	150	200
Bicycle Game	B	1896	Donaldson Brothers	205	350
Bicycle Game, The New	B	1894	Parker Brothers	400	700
Bicycle Game, The New	B	1894	Parker Brothers	345	575
Bicycle Race	B	1910	Milton Bradley	85	140
Bicycle Race Game	B	1895	McLoughlin Brothers	450	700
Bicycle Race Game, The	B	1898	Chaffee & Selchow	430	715
Bicycle Race, A Game for the Wheelmen	B	1891	McLoughlin Brothers	700	1175
Bicycling, The Merry Game of	B	1900	Parker Brothers	100	165
Big Apple	B	1938	Rosebud Art Co.	30	50
Big Bad Wolf Game	B	1930s	Parker Brothers	75	125
Big Business	B	1936	Parker Brothers	75	125
Big Business	B	1937	Transogram	20	35
Big League Baseball Card Game	C	1940s	State College Game Lab	35	60
Big League BasketbAll	B	1920s	Baumgarten & Co.	145	245
Big Six; Christy Mathewson Indoor Baseball Game	B	1922	Piroxloid Prod. Corp.	150	250
Big Ten Football Game	B	1936	Wheaties	55	90
Bike Race Game, The	B	1930s	Master Toy Co.	35	60
Bild A Word	B	1929	Educational Card & Game Co.	20	35

Games, Prewar

NAME	TYPE	YEAR	COMPANY	GOOD	EX
Billy Bump's Visit To Boston	C	1888	Parker Brothers	20	45
Billy Whiskers	B	1923	Saalfield	45	75
Billy Whiskers	B	1924	Russell Mfg. Co.	45	75
Bilt Rite Miniature Bowling Alley	B	1930s	Atwood Momanus	70	115
Bingo	B	1925	Rosebud Art Co.	15	25
Bingo	S	1929	All Fair	15	25
Bingo Or Beano	B	1940s	Parker Brothers	10	15
Bird Center Etiquette	C	1904	Home Game Co.	30	45
Bird Lotto	B	1940s	Sam'l Gabriel Sons & Co.	20	35
Birds, Game of	C	1899	Cincinnati Game Co.	30	50
Black Beauty	B	1921	Stoll & Edwards	40	65
Black Cat Fortune Telling Game, The	C	1897	Parker Brothers	65	110
Black Falcon of The Flying G Men, The	B	1939	Ruckelshaus Game Corp.	175	300
Black Sambo, Game of	B	1939	Sam'l Gabriel Sons & Co.	90	145
Blackout	B	1939	Milton Bradley	60	90
Block	C	1905	Parker Brothers	15	20
Blockade	B	1941	Corey Games	35	60
Blondie Goes To Leisureland	B	1935	Westinghouse	30	45
Blondie Playing Game	C	1941	Whitman	25	40
Blow Football Game	B	1912		30	50
Blox O	B	1923	Lubbers & Bell	15	25
Bluff	B	1944	Games Of Fame	15	25
Bo Bang & Hong Kong	B	1890	Parker Brothers	275	450
Bo McMillan's Indoor FootbAll	B	1939	Indiana Game Co.	55	90
Bo Peep Game	B	1895	McLoughlin Brothers	195	325
Bo Peep, The Game of	B	1890	J.H. Singer	90	150
Boake Carter's Star Reporter	B	1937	Parker Brothers	105	175
Bobb, Game of	S	1898	McLoughlin Brothers	200	400
Bomb The Navy	B	1940s	J. Pressman & Co.	20	30
Bombardment, Game of	B	1898	McLoughlin Brothers	100	200
Bomber BAll	S	1940s	Game Makers, Inc.	40	60
Bombs Away	B	1944	Toy Creations	30	50
Bookie	B	1931	Bookie Games Co.	55	90
Boston Baseball Game	B	1906	Boston Game Co.	495	825
Boston Globe Bicycle Game of Circulation	B	1895	Boston Globe	55	90
Boston New York Motor Tour	B	1920s	American Toy Mfg.	90	150
Botany	C	1900s	G.H. Dunston	50	100
Bottle Imps, Game of	S	1907	Milton Bradley	300	500
Bottle Quoits	B	1897	Parker Brothers	50	85
Bottoms Up	B	1934	The Embossing Co.	20	35
Bourse, Or Stock Exchange	C	1903	Flinch Card Co.	20	30
Bow O Winks	S	1932	All Fair	60	120
Bowl 'em	B	1930s	Parker Brothers	20	35
Bowling	B	1896	Parker Brothers	80	135
Bowling Alley	S	1921	N.D. Cass	20	35
Bowling Board Game	B	1896	Parker Brothers	350	575
Box Hockey	B	1941	Milton Bradley	35	60
Boxing Game, The	B	1928	Stoll & Edwards	85	140
Boy Hunter, The	S	1925	Parker Brothers	60	100
Boy Scouts	B	1910s	McLoughlin Brothers	125	200
Boy Scouts, The Game of	C	1912	Parker Brothers	50	85
Boy Scouts, The Game of	B	1926	Parker Brothers	200	350
Boys Own Football Game	B	1900s	McLoughlin Brothers	325	575
Bradley's Circus Game	B	1882	Milton Bradley	60	100
Bradley's Telegraph Game	B	1900s	Milton Bradley	85	145
Bradley's Toy Town Post Office	B	1910s	Milton Bradley	90	150
Bringing Up Father Game	B	1920	Embee Distributing Co.	50	125
Broadway	B	1917	Parker Brothers	75	125
Brownie Auto Race	B	1920s	Jeanette Toy & Novelty	115	195

Games, Prewar

NAME	TYPE	YEAR	COMPANY	GOOD	EX
Brownie Character Ten Pins Game	S	1890s		125	200
Brownie Horseshoe Game	B	1900s	M.H. Miller Co.	30	50
Brownie Kick In Top	S	1910s	M.H. Miller Co.	35	60
Brownie Ring Toss	B	1920s	M.H. Miller Co.	30	50
Buck Rogers And His Cosmic Rocket Wars Game	B	1934		200	350
Buck Rogers In The 25th Century	C	1936	All Fair	160	200
Buck Rogers Seige of Gigantica Game	B	1934		200	400
Bucking Bronco	B	1930s	Transogram	30	50
Buffalo Bill, The Game of	B	1898	Parker Brothers	100	200
Buffalo Hunt	B	1898	Parker Brothers	175	250
Bugle Horn Or Robin Hood	C	1850s	McLoughlin Brothers	300	495
Bugle Horn Or Robin Hood, Game of	B	1895	McLoughlin Brothers	350	650
Bugville Games	B	1915	Animate Toy Co.	45	75
Bula	S	1943	Games Of Fame	30	45
Bull In The China Shop	S	1937	Milton Bradley	20	40
Bulls And Bears	B	1896	McLoughlin Brothers	13000	20000
Bulls And Bears	B	1936	Parker Brothers	75	150
Bunco	C	1904	Home Game Co.	10	20
Bunker Golf	B	1932		115	195
Bunny Rabbit, Or Cottontail & Peter, The Game of	B	1928	Parker Brothers	85	145
Buried Treasure, The Game of	B	1930s	Russell Mfg. Co.	35	60
Buster Brown At Coney Island	B	1890s	J. Ottmann Lith. Co.	225	350
Buster Brown At The Circus	C	1900s	Selchow & Righter	75	125
Buster Brown Hurdle Race	B	1890s	J. Ottmann Lith. Co.	330	550
Buster Brown, Pin The Tail On The Tiger Game	S	1900s		60	100
Busto	S	1931	All Fair	20	45
Buying And Selling Game	B	1903		20	35
Buzzing Around	S	1924	Parker Brothers	40	65
Cabby	B	1940	Selchow & Righter	40	65
Cabin Boy	B	1910	Milton Bradley	60	100
Cadet Game, The	B	1905	Milton Bradley	50	80
Cake Walk Game, The	B	1900s	Parker Brothers	750	1500
Cake, Walk, The	B	1900s	Anglo American	600	1000
Calling All Cars	B	1938	Parker Brothers	45	75
Camelot	B	1930	Parker Brothers	30	45
Camouflage, The Game of	C	1918	Parker Brothers	25	40
Canoe Race	B	1910	Milton Bradley	35	60
Capital Cities Air Derby, The	B	1929	All Fair	150	250
Captain and the (Katzenjammer) Kids	B	1940s	Milton Bradley	50	100
Captain Hop Across Junior	B	1928	All Fair	100	200
Captain Jinks	C	1900s	Parker Brothers	20	30
Captain Kidd And His Treasure	B	1896	Parker Brothers	195	350
Captain Kidd Junior	B	1926	Parker Brothers	50	75
Captive Princess	B	1880	McLoughlin Brothers	135	225
Captive Princess	B	1890s	Milton Bradley	150	250
Captive Princess	B	1899	McLoughlin Brothers	65	85
Captive Princess, Tournament And Pathfinders, Games of	B	1888	McLoughlin Brothers	100	200
Capture The Fort	B	1914	Valley Novelty Works	45	75
Car Race & Game Hunt	B	1920s	Wilder	100	165
Cargo For Victory	B	1943	All Fair	50	75
Cargoes	B	1934	Selchow & Righter	50	75
Carnival, The Show Business Game	B	1937	Milton Bradley	25	40
Cat	B	1915	Carl F. Doerr	15	25
Cat And Witch	S	1940s	Whitman	25	45

Games, Prewar

NAME	TYPE	YEAR	COMPANY	GOOD	EX
Cat, Game of	B	1900	Chaffee & Selchow	270	450
Catching Mice, Game of	B	1888	McLoughlin Brothers	200	300
Cats And Dogs	B	1929	Parker Brothers	50	150
Cavalcade	B	1930s	Selchow & Righter	40	75
Cavalcade Derby Game	S	1930s	Wyandotte	50	85
Century Ride	B	1900	Milton Bradley	60	150
Century Run Bicycle Game, The	B	1897	Parker Brothers	210	350
Champion Baseball Game, The	B	1889	Schultz	4100	6800
Champion Game of Baseball, The	B	1890s	Proctor Amusement Co.	60	100
Champion Road Race	B	1934	Champion Spark Plugs	100	165
Championship Base Ball Parlor Game	B	1914	Grebnelle Novelty Co.	150	250
Championship Fight Game	B	1940s	Frankie Goodman	20	40
Champs, The Land of Brawno	B	1940	Selchow & Righter	30	50
Characteristics	B	1845	Ives	180	300
Characters, A Game of	C	1889	Decker & Decker	25	40
Charge, The	B	1898	E.O. Clark	180	300
Charlie Chan Game	C	1939	Whitman	30	75
Charlie Chan, The Great Charlie Chan Detective Game	B	1937	Milton Bradley	90	225
Charlie McCarthy Game of Topper	B	1938	Whitman	20	45
Charlie McCarthy Put And Take Bingo Game	B	1938	Whitman	30	45
Charlie McCarthy Question And Answer Game	C	1938	Whitman	30	45
Charlie McCarthy Rummy Game	C	1938	Whitman	20	35
Charlie McCarthy's Flying Hats	B	1938	Whitman	25	40
Chasing Villa	B	1920	Smith, Kline & French	65	110
Checkered Game of Life	B	1860	Milton Bradley	600	1000
Checkered Game of Life	B	1866	Milton Bradley	240	400
Checkered Game of Life	B	1911	Milton Bradley	120	300
Checkers & Avion	B	1925	American Toy Works	30	50
Chee Chow	B	1939	Sam'l Gabriel Sons & Co.	15	25
Cheerios Hook The Fish	B	1930s	General Mills	15	25
Cherrios Bird Hunt	B	1930s	General Mills	15	25
Chessindia	B	1895	Clark & Sowdon	55	95
Chestnut Burrs	C	1896	Fireside Game Co.	15	25
Chevy Chase	B	1890	Hamilton Myers	75	125
Chicago Game Series BaseBall	B	1890s	Doan & Co.	1175	1950
Chin Chow And Sum Flu	B	1925	Novitas Sales Co.	6	10
China	B	1905	Wilkens Thompson Co.	50	85
Chinaman Party	S	1896	Selchow & Righter	75	130
Ching Gong	B	1937	Sam'l Gabriel Sons & Co.	20	40
Chiromagica, Or The Hand of Fate	B	1901	McLoughlin Brothers	225	375
Chistmas Goose	B	1890	McLoughlin Brothers	500	850
Chivalrie Lawn Game	B	1875		60	100
Chivalry	B	1925	Parker Brothers	30	50
Chivalry, The Game of	S	1888	Parker Brothers	100	200
Chocolate Splash	B	1916	Willis G. Young	51	85
Christmas Jewel, Game of The	B	1899	McLoughlin Brothers	360	600
Christmas Mail	B	1890s	J. Ottmann Lith. Co.	390	650
Chutes And Ladders	B	1943	Milton Bradley	15	35
Cinderella	C	1895	Parker Brothers	35	55
Cinderella	C	1905	Milton Bradley	30	45
Cinderella	C	1921	Milton Bradley	15	25
Cinderella	B	1923	Stoll & Edwards	40	95
Cinderella Or Hunt The Slipper	C	1887	McLoughlin Brothers	50	85
Circus Game	B	1914		75	125
Citadel	B	1940	Parker Brothers	65	110
Cities	B	1932	All Fair	25	40
City Life, Or The Boys of New York, The Game of	C	1889	McLoughlin Brothers	40	100

GAMES

Games, Prewar

NAME	TYPE	YEAR	COMPANY	GOOD	EX
City of Gold	B	1926	Zulu Toy Mfg. Co.	50	85
Classic Derby	B	1930s	Doremus Schoen & Co.	30	50
Click	S	1930s	Akro Agate Co.	50	85
Clipper Race	B	1930	Sam'l Gabriel Sons & Co.	25	60
Clown Tenpins Game	B	1912		60	100
Clown Winks	S	1930s	Sam'l Gabriel Sons & Co.	15	25
Coast To Coast	B	1940s	Master Toy Co.	15	25
Cock Robin	C	1895	Parker Brothers	25	60
Cock Robin And His Tragical Death, Game of	C	1885	McLoughlin Brothers	35	85
Cock A Doodle Doo Game	B	1914		60	100
Cocked Hat, Game of	B	1892	J.H. Singer	150	250
College Baseball Game	B	1890s	Parker Brothers	350	900
College Boat Race, Game of	B	1896	McLoughlin Brothers	350	575
Colliwogg	C	1907	Milton Bradley	125	200
Columbia's Presidents And Our Country, Game of	C	1886	McLoughlin Brothers	195	325
Columbus	B	1892	Milton Bradley	775	1300
Combination Board Games	B	1922	Wilder	40	65
Combination Tiddledy Winks	S	1910	Milton Bradley	40	65
Comic Conversation Cards	C	1890	J. Ottmann Lith. Co.	55	90
Comic Leaves of Fortune The Sibyl's Prophecy	C	1850s	Charles Magnus	345	575
Comical Animals Ten Pins	B	1910	Parker Brothers	185	310
Comical Game of 'Who', The	C	1910s	Parker Brothers	35	50
Comical Game of Whip, The	C	1920s	Russell Mfg. Co.	20	45
Comical History of America	C	1924	Parker Brothers	20	40
Comical Snap, Game of	C	1903	McLoughlin Brothers	35	55
Commanders of Our Forces, The	C	1863	E.C. Eastman	87	145
Commerce	C	1900s	J. Ottmann Lith. Co.	39	65
Competition, Or Department Store	C	1904	Flinch Card Co.	21	35
Cones & Corns	S	1924	Parker Brothers	39	65
Conette	S	1890	Milton Bradley	45	75
Coney Island Playland Park	B	1940	Vitaplay Toy Co.	54	90
Conflict	B	1942	Parker Brothers	75	125
Conquest of Nations, Or Old Games With New Faces, The	C	1853	Willis P. Hazard	54	90
Construction Game	B	1925	Wilder	75	100
Contack	S	1939	Parker Brothers	5	10
Coon Hunt Game, The	B	1903	Parker Brothers	450	750
Corn & Beans	B	1875	E.G. Selchow	51	85
Corner The Market	B	1938	Whitman	24	40
Cortella	B	1915	Atkins & Co.	21	35
Costumes And Fashions, Game of	C	1881	Milton Bradley	20	40
Cottontail And Peter, The Game of	B	1922	Parker Brothers	72	120
Country Club Golf	B	1920s	Hustler Toy Corp.	75	125
Country Store, The	B	1890s	J.H. Singer	75	125
County Fair, The	C	1891	Parker Brothers	45	75
Cousin Peter's Trip To New York, Game of	C	1898	McLoughlin Brothers	36	60
Covered Wagon	B	1927	Zulu Toy Mfg. Co.	51	85
Cowboy Game, The	B	1898	Chaffee & Selchow	210	350
Cows In Corn	B	1889	Stirn & Lyon	7	11
Crash, The New Airplane Game	B	1928	Nucraft Toys	30	70
Crazy Traveler Game	B	1892		60	100
Crazy Traveller	B	1908	Parker Brothers	36	60
Crazy Traveller	S	1920s	Parker Brothers	36	60
Crickets In The Grass	S	1920s	Madmar Quality Co.	35	50
Crime & Mystery	B	1940s	Frederick H. Beach (Beachcraft)	15	25
Criss Cross Words	B	1938	Alfred Butts	90	150
Crooked Man Game	B	1914		45	75
Cross Country	B	1941	Lowe	20	30

GAMES

Top to bottom: Round The World With Nellie Bly, 1890, McLoughlin Brothers; Phoebe Snow, 1899, McLoughlin Brothers; Old Maid & Old Bachelor or Beaux And Belles, 1800s, McLoughlin Brothers.

Games, Prewar

NAME	TYPE	YEAR	COMPANY	GOOD	EX
Cross Country Marathon	B	1920s	Milton Bradley	50	125
Cross Country Marathon Game	B	1930s	Rosebud Art Co.	50	125
Cross Country Racer	B	1940	Automatic Toy Co.	45	75
Cross Country Racer (w/windup cars)	B	1940s		75	130
Crossing The Ocean	B	1893	Parker Brothers	87	145
Crow Cards, 12 Great Games In 1	C	1910	Milton Bradley	9	15
Crow Hunt	B	1904	Parker Brothers	40	60
Crow Hunt	S	1930	Parker Brothers	50	85
Crows In The Corn	S	1930	Parker Brothers	45	75
Crusade	B	1930s	Sam'l Gabriel Sons & Co.	27	45
Cuckoo, A Society Game	B	1891	J.H. Singer	40	100
Curly Locks Game	B	1910		60	100
Cycling, Game of	B	1910	Parker Brothers	100	165
Daisy Clown Ring Game	B	1927	Schacht Rubber Mfg. Co.	9	15
Daisy Horseshoe Game	B	1927	Schacht Rubber Mfg. Co.	9	15
Danny McFayden's Stove League Baseball Game	B	1935	National Game Co.	295	490
Darrow Monopoly	B	1934	Charles Darrow	1500	2500
Day At The Circus, Game of	B	1898	McLoughlin Brothers	300	400
Deck Derby	B	1920s	Wolverine	36	60
Deck Ring Toss Game	S	1910		30	50
Decoy	B	1940	Selchow & Righter	45	75
Defenders of The Flag	C	1922	Stoll & Edwards	27	45
Defenders of The Flag Game	B	1920s		24	40
Democracy	B	1940	Toy Creations	15	25
Department Store, Game of Playing	B	1898	McLoughlin Brothers	520	1300
Derby Day	C	1900s	Parker Brothers	21	35
Derby Day	B	1930	Parker Brothers	45	75
Derby Steeple Chase	B	1888	McLoughlin Brothers	100	165
Detective, The Game of	B	1889	R. Bliss Mfg. Co.	1200	2000
Dewey At Manila	C	1899	Chaffee & Selchow	39	65
Dewey's Victory	B	1900s	Parker Brothers	120	200
Diamond Game of Baseball, The	B	1900	McLoughlin Brothers	1275	2150
Diamond Heart	B	1902	McLoughlin Brothers	111	185
DicebAll	B	1938	Ray Fair Co.	90	145
Dicex Baseball Game, The	B	1925	Chester S. Howland	195	325
Dick Tracy Detective Game	B	1933	Einson Freeman Publishing Corp.	45	75
Dick Tracy Detective Game	B	1937	Whitman	39	65
Dick Tracy Playing Card Game	C	1934	Whitman	39	65
Dick Tracy Super Detective Mystery Card Game	C	1937	Whitman	24	40
Dig	S	1940	Parker Brothers	5	15
Dim Those Lights	S	1932	All Fair	200	400
Din	C	1905	E.I. Horsman	15	25
Discretion	B	1942	Volume Sprayer Mfg. Co.	30	45
Disk	S	1900s	Madmar Quality Co.	35	55
District Messenger Boy, Game of	B	1886	McLoughlin Brothers	250	400
District Messenger Boy, Game of	B	1904	McLoughlin Brothers	80	200
Diving Fish	S	1920s	C.E. Bradley Corp.	20	30
Dixie Land, Game of	C	1897	The Fireside Game Co.	40	100
Doctor Busby Card Game	C	1910		25	40
Doctor Quack, Game of	C	1922	Russell Mfg. Co.	25	40
Doctors And The Quack	C	1890s	Parker Brothers	35	60
Dodging Donkey, The	S	1920s	Parker Brothers	45	75
Dog Race	B	1937	Transogram	20	50
Dog Show	B	1890s	J.H. Singer	85	140
Dog Sweepstakes	B	1935	Stoll & Einsen	45	75
Donald Duck Game	C	1930s	Whitman	10	20
Donald Duck Party Game	B	1938	Parker Brothers	80	200

Games, Prewar

NAME	TYPE	YEAR	COMPANY	GOOD	EX
Donald Duck Playing Game	C	1941	Whitman	25	40
Donald Duck's Own Game	B	1930s	Walt Disney	75	100
Donkey Party	S	1887	Selchow & Righter	15	25
Double Eagle Anagrams	C	1890	McLoughlin Brothers	35	55
Double Flag Game, The	C	1904	McLoughlin Brothers	45	75
Double Game Board (Baseball)	B	1925	Parker Brothers	165	275
Double Header BasebAll	B	1935	Redlich Mfg. Co.	145	250
Down And Out	S	1928	Milton Bradley	50	100
Down The Pike With Mrs. Wiggs At The St. Louis Exposition	C	1904	Milton Bradley	30	50
Dr. Busby	C	1890s	J.H. Singer	40	80
Dr. Busby	C	1900s	J. Ottmann Lith. Co.	40	60
Dr. Busby	C	1937	Milton Bradley	20	50
Dr. Fusby, Game of	C	1890s	McLoughlin Brothers	35	100
Dreamland Wonder Resort Game	B	1914	Parker Brothers	275	700
Drive 'n Putt	B	1940s	Carrom Industries	50	90
Drummer Boy Game	B	1914		60	100
Drummer Boy Game, The	B	1890s	Parker Brothers	100	150
Dubble Up	B	1940s	Sam'l Gabriel Sons & Co.	15	25
Dudes, Game of The	B	1890	R. Bliss Mfg. Co.	225	375
Durgin's New Baseball Game	B	1885	Durgin & Palmer	425	700
Eagle Bombsight	B	1940s	Toy Creations	25	60
East Is East And West Is West	B	1920s	Parker Brothers	80	200
Easy Money	B	1936	Milton Bradley	35	55
Ed Wynn The Fire Chief	B	1937	Selchow & Righter	45	75
Eddie Cantor's Tell It To The Judge	B	1930s	Parker Brothers	30	75
Edgar Bergen's Charlie McCarthy Game of Topper	B	1938	Whitman	15	35
Edgar Bergen's Charlie McCarthy Put And Take Bingo Game	B	1938		20	35
Edgar Bergen's Charlie McCarthy Rummy Game	C	1939	Whitman	20	45
Egerson R. Williams Baseball Game	C	1889	Popular Indoor Baseball Co.	2500	5000
Election	D	1896	Fireside Game Co.	20	35
Electric BasebAll	B	1935	Einson Freeman Publishing Corp.	35	60
Electric FootbAll	B	1930s	Electric Football Co.	55	90
Electric Magnetic BasebAll	B	1900		175	295
Electric Questioner	B	1920	Knapp Electric & Novelty Co.	20	35
Electric Speed Classic	B	1930	J. Pressman & Co.	390	650
Electro Gameset	B	1930	Knapp Electric & Novelty Co.	30	45
Elementaire Musical Game	B	1896	Theodore Presser	20	35
Elite Conversation Cards	C	1887	McLoughlin Brothers	20	35
Ella Cinders	B	1944	Milton Bradley	40	80
Elmer Layden's Scientific Football Game	B	1936	Cadaco Ltd.	50	100
Elsie (The Cow) Game, The	B	1941	Selchow & Righter	50	100
Enchanted Forest Game	B	1914		120	200
Endurance Run	B	1930	Milton Bradley	60	150
Errand Boy, The	B	1891	McLoughlin Brothers	120	300
Ethan Allen's All Star Baseball Game	B	1942	Cadaco Ellis	60	150
Evening Parties, Game of	B	1910s	Parker Brothers	180	300
Excursion To Coney Island	C	1880s	Milton Bradley	35	55
Excuse Me!	C	1923	Parker Brothers	15	35
Faba Baga Or Parlor Quiots	S	1883	Morton E. Converse Co.	40	65
Fairies' Cauldron Tiddledy Winks Game, The	S	1925	Parker Brothers	35	50
Fairyland Game	B	1880s	Milton Bradley	60	95
Famous Authors	C	1910	Parker Brothers	40	65
Famous Authors	C	1943	Parker Brothers	10	15

Games, Prewar

NAME	TYPE	YEAR	COMPANY	GOOD	EX
Fan Craze Card Game	C	1904	Fan Craze Co.	175	295
Fan i Tis	B	1913	C.W. Marsh	110	180
Fan Tel	B	1937	Schoenhut	20	40
Farmer Jones' Pigs	B	1890	McLoughlin Brothers	165	275
Fascination	S	1890	Selchow & Righter	35	50
Fashionable English Sorry Game, The	B	1934	Parker Brothers	20	50
Fast Mail Game	B	1910	Milton Bradley	105	175
Fast Mail Railroad Game	B	1930s	Milton Bradley	50	85
Favorite Art, Game of	C	1897	Parker Brothers	30	50
Favorite Steeple Chase	B	1895	J.H. Singer	60	100
Ferdinand The Bull Chinese Checkers Game	B	1930s		60	100
Fibber McGee	B	1936	Milton Bradley	25	40
Fibber McGee And The Wistful Vista Mystery	B	1940	Milton Bradley	30	45
Fiddlestix	S	1937	Plaza Mfg. Co.	10	15
Fig Mill	B	1916	Willis G. Young	25	40
Finance	B	1937	Parker Brothers	25	40
Finance And Fortune	B	1936	Parker Brothers	35	50
Fire Alarm Game	B	1899	Parker Brothers	1300	2300
Fire Department	B	1930s	Milton Bradley	80	125
Fire Fighters Game	B	1909	Milton Bradley	120	200
Fish Pond	B	1890	E.O. Clark	50	100
Fish Pond	B	1920s	Wilder	25	60
Fish Pond Game, Magnetic	B	1891	McLoughlin Brothers	175	300
Fish Pond, Game of	S	1910s	Wescott Brothers	30	50
Fish Pond, New And Improved	B	1890s	McLoughlin Brothers	75	125
Fish Pond, The Game of	S	1890	McLoughlin Brothers	150	250
Fishing Game	S	1899	Martin Co.	30	45
Five Hundred, Game of	C	1900s	Home Game Co.	20	35
Five Little Pigs	C	1890s	J.H. Singer	25	40
Five Wise Birds, The	S	1923	Parker Brothers	25	60
Flag Travelette	B	1895	Archarena Co.	45	75
Flags	C	1899	Cincinnati Game Co.	15	35
Flap Jacks	S	1931	All Fair	80	200
Flapper Fortunes	B	1929	The Embossing Co.	20	30
Flash	S	1940s	J. Pressman & Co.	20	50
Flight To Paris	B	1927	Milton Bradley	150	250
Flinch	C	1902	Flinch Card Co.	6	10
Fling A Ring	B	1930s	Wolverine	20	35
Flip It	B	1925	American Toy Works	40	60
Flip It	B	1940	Deluxe Game Corp.	20	30
Flip It, Auto Race & Transcontinental Tour	B	1920s	Deluxe Game Corp.	35	90
Flitters	S	1899	The Martin Co.	45	75
Flivver	B	1927	Milton Bradley	75	150
Floor Croquet Game	S	1912		60	100
Flowers, Game of	B	1899	Cincinnati Game Co.	45	75
Flying Aces	B	1940s	Selchow & Righter	50	75
Flying The Beam	B	1941	Parker Brothers	75	125
Flying The United States Airmail	B	1929	Parker Brothers	70	175
Fobaga (football)	B	1942	American Football Co.	50	80
Follow the Stars	B	1922	G.H. Allen Watts	225	375
Foolish Questions	C	1920s	Wallie Dorr Co.	30	75
Foot Race, The	B	1900s	Parker Brothers	60	100
FootAll	B	1930s	Wilder	50	80
Football Game	B	1898	Parker Brothers	295	495
Football Knapp Electro Game Set	B	1929	Knapp Electric & Novelty Co.	125	205
Football, The Game of	B	1895	George A. Childs	75	125
FootAll As You Like It	B	1940	Wayne W. Light Co.	85	145
Fore Country Club Game of Golf	B	1929	Wilder	175	295

Games, Prewar

NAME	TYPE	YEAR	COMPANY	GOOD	EX
Fort Sumter	B	1870	W. S. Reed	0	0
Fortune	B	1938	Parker Brothers	40	70
Fortune Teller, The	B	1905	Milton Bradley	20	50
Fortune Telling	C	1920s	All Fair	25	60
Fortune Telling & Baseball Game	B	1889		85	140
Fortune Telling Game	C	1930s	Stoll & Edwards	25	40
Fortune Telling Game	B	1934	Whitman	100	200
Fortune Telling Game, The	C	1890s	Parker Brothers	40	65
Fortunes, Game of	C	1902	Cincinnati Game Co.	45	75
Forty Niners Gold Mining Game	B	1930s	National Games, Inc.	25	40
Foto World	B	1935	Cadaco Ltd.	90	150
Foto Electric FootBAll	B	1930s	Cadaco Ltd.	45	75
Foto Finish Horse Race	B	1940s	J. Pressman & Co.	30	45
Four And Twenty Blackbirds	S	1890s		650	1100
Four Dare Devils, The	S	1933	Marx, Hess & Lee, Inc.	40	65
Fox & Hounds	B	1900	Parker Brothers	85	140
Fox And Geese	B	1903	McLoughlin Brothers	60	150
Fox And Geese, The New	C	1888	McLoughlin Brothers	45	75
Fox And Hounds Game	B	1912		60	100
Fox Hunt	B	1905	Milton Bradley	40	65
Fox Hunt	B	1930s	Lowe	20	35
Foxy Grandpa At The World's Fair	C	1904	J. Ottmann Lith. Co.	150	250
Foxy Grandpa Hat Party	B	1906	Selchow & Righter	55	90
Fractions	C	1902	Cincinnati Game Co.	10	15
Frank Buck's Bring 'em Back Alive Game	C	1937	All Fair	30	70
Frisko	B	1937	The Embossing Co.	20	30
Frog He Would A Wooing Go, The	B	1898	McLoughlin Brothers	40	100
Frog School Game	B	1914		45	75
Frog Who Would A Wooing Go, The	B	1920s	United Game Co.	45	75
Fun At The Circus	B	1897	McLoughlin Brothers	360	600
Fun At The Zoo; A Game	B	1902	Parker Brothers	120	200
Fun Kit	B	1939	Frederick H. Beach (Beachcraft)	15	20
Ful BAll	B	1940s	Ful Bal Co.	35	60
G Men	C	1936	Milton Bradley	20	30
G Men Clue Games	B	1935	Whitman	30	50
Games You Like To Play	B	1920s	Parker Brothers	95	175
Gamevelope	C	1944	Morris Systems Publishing Co.	20	35
Gang Busters Game	B	1938	Lynco	150	250
Gang Busters Game	B	1939	Whitman	50	80
Gavitt's Stock Exchange	C	1903	W.W. Gavitt	20	30
Gee Wiz Horse Race	S	1928	Wolverine	50	85
General Headquarters	B	1940s	All Fair	50	75
Genuine Steamer Quoits	S	1924	Milton Bradley	15	25
Geographical Cards	C	1883	Peter G. Thompson	20	50
Geographical Lotto Game	B	1921		20	30
Geography Game	B	1910s	A. Flanagan Co.	15	25
Geography Up To Date	C	1890s	Parker Brothers	30	50
George Washington's Dream	C	1900s	Parker Brothers	20	35
Ges It Game	B	1936	Knapp Electric & Novelty Co.	20	45
Get The Balls Baseball Game	B	1930		20	30
Gly Dor	B	1931	All Fair	50	125
Go Bang	B	1898	J.H. Singer	30	75
Go To The Head of The Class	B	1938	Milton Bradley	15	35
Goat, Game of	C	1916	Milton Bradley	15	25
Going To The Fire Game	B	1914		90	150
Gold Hunters, The	B	1900s	Parker Brothers	105	175
Gold Rush, The	C	1930s	Cracker Jack	10	15
Golden Egg	C	1850s	McLoughlin Brothers	285	475
Golden Egg, The	C	1845	R.H. Pease	165	275
Goldenlocks & The Three Bears	B	1890	McLoughlin Brothers	320	800

Games, Prewar

NAME	TYPE	YEAR	COMPANY	GOOD	EX
Golf	B	1900	Schoenhut	2700	5000
Golf Tokalon Series, The Game of	B	1890s	E.O. Clark	350	575
Golf, A Game of	B	1930	Milton Bradley	145	245
Golf, Game of	B	1896	McLoughlin Brothers	425	715
Golf, The Game of	B	1898	J.H. Singer	100	250
Golf, The Game of	B	1905	Clark & Sowdon	325	550
Gonfalon Scientific BasebAll	B	1930	Pioneer Game Co.	110	180
Good Old Aunt, The	B	1892	McLoughlin Brothers	150	250
Good Old Game of Corner Grocery, The	C	1900s	Parker Brothers	40	60
Good Old Game of Dr. Busby	C	1900s	Parker Brothers	40	60
Good Old Game of Dr. Busby, The	C	1920s	United Game Co.	25	50
Good Old Game of Innocence Abroad, The	B	1888	Parker Brothers	180	300
Good Things To Eat Lotto	B	1940s	Sam'l Gabriel Sons & Co.	15	25
Goose Goslin Scientific BasebAll	B	1935	Wheeler Toy Co.	250	400
Goose, The Jolly Game of	B	1851	J.P. Beach	750	1250
Goosey Gander, Or Who Finds The Golden Egg, Game of	B	1890	J.H. Singer	675	1150
Goosy Goosy Gander	B	1896	McLoughlin Brothers	300	500
Graham McNamee World Series Scoreboard Baseball Game	B	1930	Radio Sports	250	400
Grand National Sweepstakes	B	1937	Whitman	20	50
Grande Auto Race	B	1920s	Atkins & Co.	50	85
Grandma's Game of Useful Knowledge	C	1910s	Milton Bradley	20	30
Grandmama's (Improved) Arithmetical Game	C	1887	McLoughlin Brothers	30	45
Grandmama's (Improved) Game of Useful Knowledge	C	1887	McLoughlin Brothers	30	45
Grandmama's Sunday Game: Bible Questions, Old Testament	C	1887	McLoughlin Brothers	20	35
Graphic BasebAll	B	1930s	Northwestern Products	165	275
Great American Baseball Game, The	B	1906	William Dapping	145	250
Great American Flag Game, The	B	1940	Parker Brothers	35	60
Great American Game	B	1910	Neddy Pocket Game Co.	145	250
Great American Game of Baseball, The	B	1907	Pittsburgh Brewing Co.	145	250
Great American Game, Baseball, The	B	1923	Hustler Toy Corp.	105	175
Great American Game, The	B	1925	Frantz	110	180
Great American War Game	B	1899	J.H. Hunter	600	1000
Great Battlefields	C	1886	Parker Brothers	70	120
Great Composer, The	C	1901	Theodore Presser	25	40
Great Family Amusement Game, The	B	1889	Einson Freeman Publishing Corp.	20	35
Great Horse Race Game, The	B	1925	Selchow & Righter	70	115
Great Mails Baseball Game	B	1919	Walter Mails Baseball Game Co.	2475	4100
Gregg Football Game	B	1924	Albert A. Gregg	175	285
Greyhound Racing Game	B	1938	Rex Manufacturing Co.	15	25
Guess Again, The Game of	C	1890s	McLoughlin Brothers	40	65
Gump, Chester Gump Game	B	1938	Milton Bradley	65	95
Gump, Chester Hops Over the Pole	B	1930s	Milton Bradley	65	95
Gumps at the Seashore, The	B	1930s	Milton Bradley	65	95
Gym Horseshoes	B	1930	Wolverine	30	45
Gypsy Fortune Telling Game	C	1909	McLoughlin Brothers	50	75
Gypsy Fortune Telling Game, The	B	1895	McLoughlin Brothers	135	220
H.M.S. Pinafore	C	1880	McLoughlin Brothers	75	125
Halma	B	1885	Milton Bradley	30	45
Halma	B	1885	E.I. Horsman	35	60

Games, Prewar

NAME	TYPE	YEAR	COMPANY	GOOD	EX
Hand of Fate	B	1901	McLoughlin Brothers	1200	2000
Happitime Bagatelle	S	1933	Northwestern Products	30	45
Happy Family, The	B	1910	Milton Bradley	15	25
Happy Hooligan Bowling Type Game	B	1925		60	100
Happy Landing	S	1938	Transogram	30	45
Hardwood Ten Pins Wooden Game	B	1889		60	100
Hare & Hound	B	1895	Parker Brothers	245	400
Hare And Hounds	B	1890	Selchow & Righter	150	250
Harlequin, The Game of The	B	1895	McLoughlin Brothers	150	200
Harold Teen Game	B	1930s	Milton Bradley	50	85
Have U It?	C	1924	Selchow & Righter	15	25
Heads And Tails	C	1900s	Parker Brothers	20	30
Heedless Tommy	B	1893	McLoughlin Brothers	240	400
Hel Lo Telephone Game	B	1898	J.H. Singer	95	150
Helps To History	B	1885	A. Flanagan Co.	20	35
Hen That Laid The Golden Egg, The	B	1900	Parker Brothers	105	175
Hendrik Van Loon's Wide World Game	B	1935	Parker Brothers	35	65
Hening's In Door Game of Professional BasebAll	B	1889	Inventor's Co.	525	875
Hens And Chickens, Game of	C	1875	McLoughlin Brothers	105	175
Heroes of America	B	1920	Educational Card & Game Co.	20	35
Hey What?	C	1907	Parker Brothers	10	20
Hi Way Henry	B	1928	All Fair	600	1500
Hialeah Horse Racing Game	B	1940s	Milton Bradley	40	65
Hickety Pickety	B	1924	Parker Brothers	20	50
Hidden Titles	C	1908	Parker Brothers	10	20
Hide And Seek, Game of	B	1895	McLoughlin Brothers	200	500
Hippodrome Circus Game	B	1895	Milton Bradley	80	200
Hippodrome, The	B	1900s	E.O. Clark	150	200
Historical Cards	C	1884	Peter G. Thompson	20	30
History Up To Date	C	1900s	Parker Brothers	35	50
Hit that Line	B	1930s	LaRue Sales	100	165
Hockey	B	1942		15	25
Hockey, Official	B	1940	Toy Creations	50	75
Hokum	C	1927	Parker Brothers	15	25
Hold The Fort	B	1895	Parker Brothers	125	225
Hold Your Horses	B	1930s	Klauber Novelty	10	20
Hollywood Movie Bingo	C	1937	Whitman	40	70
Home Baseball Game	B	1900	McLoughlin Brothers	900	1700
Home Defenders	B	1941	Saalfield	15	25
Home Diamond	C	1913	Phillips	100	165
Home Diamond, The Great Baseball Game	B	1925	Phillips	175	295
Home Games	B	1900s	The Martin Co.	105	175
Home History Game	C	1910s	Milton Bradley	35	50
Home Run King	B	1930s	Selrite Prod.	275	450
Home Run with Bases Loaded	B	1935	T.V. Morrison	205	350
Honey Bee Game	B	1913	Milton Bradley	50	85
Hood's Spelling School	B	1897	C.I. Hood	20	35
Hood's War Game	C	1899	C.I. Hood	25	60
Hoop O Loop	B	1930	Wolverine	20	30
Hoot	C	1926	Saalfield	30	50
Hop Over Puzzle	S	1930s	J. Pressman & Co.	20	35
Hornet	B	1941	Lowe	30	45
Horse Race	B	1943	Lowe	10	20
Horse Racing	B	1935	Milton Bradley	35	60
Horses	B	1927	Modern Makers	45	75
Hounds & Hares	B	1894	J.W. Keller	35	60

Games, Prewar

NAME	TYPE	YEAR	COMPANY	GOOD	EX
House That Jack Built	C	1900s	Parker Brothers	30	55
House That Jack Built, The	C	1887	McLoughlin Brothers	35	75
Household Words, Game of	C	1916	Household Words Game Co.	50	85
How Good Are You	B	1937	Whitman	10	15
How Silas Popped The Question	C	1915	Parker Brothers	20	50
Howard H. Jones Collegiate FootAll	B	1932	Municipal Service	40	100
Huddle All American Football Game	B	1931		100	165
Hungry Willie	S	1930s	Transogram	40	70
Hunting Hare, Game of	B	1891	McLoughlin Brothers	205	350
Hunting In The Jungle	S	1920s	A. Gropper	30	45
Hunting The Rabbit	B	1895	Clark & Sowdon	70	115
Hunting, The New Game of	B	1904	McLoughlin Brothers	110	180
Hunting, The New Game of	B	1904	McLoughlin Brothers	360	600
Hurdle Race	B	1905	Milton Bradley	75	125
Hymn Quartets	B	1933	Goodenough And Woglom Co.	10	15
I Doubt It	C	1910	Parker Brothers	20	35
Ice Hockey	B	1942	Milton Bradley	25	40
Illustrated Mythology	C	1896	Cincinnati Game Co.	15	25
Improved Geographical Game, The	B	1890s	Parker Brothers	60	100
Improved Historical Cards	C	1900	McLoughlin Brothers	20	35
In And Out The Window	B	1940s	Sam'l Gabriel Sons & Co.	20	35
In Door BaseAll	B	1926	E. Bommer Foundation	100	180
India	B	1940	Parker Brothers	15	20
India Bombay	B	1910s	Cutler & Saleeby Co.	25	40
India, An Oriental Game	B	1890s	McLoughlin Brothers	100	165
India, Game of	B	1910s	Milton Bradley	15	40
Indianapolis 500 Mile Race Game	B	1938	Shaw	350	575
Indians And Cowboys	B	1940s	Sam'l Gabriel Sons & Co.	40	65
Indoor FootAll	B	1919	Underwood	145	250
Indoor Golf Dice	S	1920s	W.P. Bushell	20	35
Indoor Horse Racing	B	1924	Man O War	70	115
Industries, Game of	C	1897	A.W. Mumford Co.	15	25
Inside Baseball Game	B	1911	Popular Games	300	500
Intercollegiate FootAll	B	1923	Hustler Toy Corp.	125	245
International Automobile Race	B	1903	Parker Brothers	875	1450
International Spy, Game of	B	1943	All Fair	50	85
Ivanhoe	C	1886	Parker Brothers	30	50
Jack And Jill	B	1890s	Parker Brothers	55	95
Jack And Jill	B	1909	Milton Bradley	60	100
Jack And The Bean Stalk	B	1895	Parker Brothers	110	175
Jack And The Bean Stalk, The Game of	B	1898	McLoughlin Brothers	550	900
Jack Spratt Game	B	1914		45	75
Jack Straws, The Game of	S	1901	Parker Brothers	25	35
Jack Be Nimble	B	1940s	The Embossing Co.	20	35
Jackie Robinson Baseball Game	B	1940		425	725
Jackpot	B	1943	B.L. Fry Products Co.	15	25
Jamboree	S	1937	Selchow & Righter	50	120
Japan, The Game of	B	1903	J. Ottmann Lith. Co.	180	300
Japanese Ball Game	S	1930s	Girard	35	65
Japanese Games of Cash And Akambo	B	1881	McLoughlin Brothers	150	250
Japanese Oracle, Game of	C	1875	McLoughlin Brothers	60	100
Japanola	S	1928	Parker Brothers	35	60
Jaunty Butler	S	1932	All Fair	75	150
Jav Lin	S	1931	All Fair	105	175
Jeep Board, The	B	1944	Lowe	15	35

Clockwise from upper left: The Air Ship Game, 1904, McLoughlin Brothers; Winnie-The-Pooh, 1933, Parker Brothers; Original Game of Zoom, All-Fair; The Bicycle Race, McLoughlin Brothers; "Pitch Em," 1929, Wolverine.

GAMES

Games, Prewar

NAME	TYPE	YEAR	COMPANY	GOOD	EX
Jeffries Championship Playing Cards	B	1904		35	55
Jig Chase	B	1930s	Game Makers, Inc.	35	60
Jig Race	B	1930s	Game Makers, Inc.	35	60
Jockey	B	1920s	Carrom Industries	35	60
Joe "Ducky" Medwick's Big League Baseball Game	C	1930s	Johnson	125	200
John Gilpin, Rainbow Backgammon And Bewildered Travelers	B	1875	McLoughlin Brothers	175	300
Johnny Get Your Gun	B	1928	Parker Brothers	45	75
Johnny's Historical Game	C	1890s	Parker Brothers	30	75
Jolly Clown Spinette	S	1932	Milton Bradley	30	45
Jolly Pirates	B	1938	Russell Mfg. Co.	20	35
Jolly Robbers	S	1929	Wilder	50	75
Journey To Bethlehem, The	B	1923	Parker Brothers	95	160
Jumping Frog, Game of	C	1890	J.H. Singer	45	75
Jumping Jupiter	S	1940s	Sam'l Gabriel Sons & Co.	30	50
Jumpy Tinker	B	1920s	Toy Creations	20	30
Jungle Hunt	S	1940	Gotham Pressed Steel Corp.	30	45
Jungle Hunt	B	1940s	Rosebud Art Co.	30	50
Jungle Jump Up Game	S	1940s	Judson Press	30	45
Junior Auto Race Game	B	1925		30	50
Junior Baseball Game	B	1915	Benjamin Seller Mfg. Co.	100	165
Junior Basketball Game	B	1930s	Rosebud Art Co.	55	90
Junior Bicycle Game, The	B	1897	Parker Brothers	250	400
Junior Combination Board	B	1905	McLoughlin Brothers	105	175
Junior FootBAll	B	1944	Deluxe Game Corp.	30	45
Junior Motor Race	B	1925	Wolverine	40	70
Just Like Me, Game of	C	1899	McLoughlin Brothers	35	85
Kan Oo Win It	B	1893	McLoughlin Brothers	345	575
Kate Smith's Own Game America	B	1940s	Toy Creations	40	65
Katzenjammer Kids Hockey	S	1940s	Jaymar	40	65
Katzy Party	S	1900s	Selchow & Righter	70	120
Keeping Up With The Jone's	B	1921	Parker Brothers	50	85
Keeping Up With The Jone's, The Game of	B	1921	Phillips	75	100
Kellogg's Boxing Game	B	1936	Kellogg's	35	55
Kellogg's Football Game	B	1936	Kellogg's	25	40
Kellogg's Golf Game	B	1936	Kellogg's	25	40
Kentucky Derby Racing Game	B	1938	Whitman	20	30
Kilkenny Cats, The Amusing Game of	B	1890	Parker Brothers	60	100
Kindergarten Lotto	S	1904	Strauss	50	80
King's Quoits, New Game of	B	1893	McLoughlin Brothers	210	350
Kings	B	1931	Akro Agate Co.	55	95
Kings, The Game of	B	1845	Josiah Adams	70	115
Kitty Kat Cup BAll	B	1930s	Rosebud Art Co.	35	90
Klondike Game	B	1890s	Parker Brothers	345	575
Knockout	B	1937	Scame Games	55	90
Knockout Andy	S	1926	Parker Brothers	35	90
Knute Rockne Football Game, Official	B	1930	Radio Sports	250	400
Ko Ko The Clown	B	1940	All Fair	20	30
Komical Konversation Kards	C	1893	Parker Brothers	30	50
Kriegspiel Junior	B	1915	Parker Brothers	50	80
Kuit Kuts	S	1922	The Regensteiner Corp.	20	30
La Haza	B	1923	Supply Sales Co.	10	20
Lame Duck, The	B	1928	Parker Brothers	60	100
Land And Sea War Games	B	1941	Lowe	40	65

Games, Prewar

NAME	TYPE	YEAR	COMPANY	GOOD	EX
Lasso The Jumping Ring	B	1912		60	100
Lawson's Baseball Card Game	C	1910		85	130
Lawson's Patent Game Baseball	C	1884	Lawson's Card Co.	350	700
Le Choc	B	1919	Milton Bradley	50	85
League Parlor Base Ball	B	1889	R. Bliss Mfg. Co.	600	1000
Leap Frog Game	B	1900	McLoughlin Brothers	165	275
Leap Frog, Game of	B	1910	McLoughlin Brothers	45	75
Leaping Lena	S	1920s	Parker Brothers	60	90
Lee At Havana	B	1899	Chaffee & Selchow	55	90
Leslie's Baseball Game	B	1909	Perfection Novelty & Adv. Co.	145	250
Let's Go To College	B	1944	Einson Freeman Publishing Corp.	30	45
Let's Play Games, Golf	B	1939	American Toy Works	50	80
Let's Play Polo	B	1940	American Toy Works	45	75
Letter Carrier, The	B	1890	McLoughlin Brothers	80	135
Letters	B	1878	E.I. Horsman	30	45
Letters Improved For The Logomachist	C	1878	Noyes & Snow	20	35
Letters Or Anagrams	B	1890s	Parker Brothers	30	50
Lew Fonseca Baseball Game, The	B	1920s	Carrom Industries	525	875
Library of Games	B	1938	American Toy Works	15	25
Library of Games	C	1939	Russell Mfg. Co.	15	25
Lid's Off, The	S	1937	Atwo	35	75
Life In The Wild West	B	1894	R. Bliss Mfg. Co.	300	500
Life of The Party	B	1940s	Rosebud Art Co.	35	50
Life's Mishaps & Bobbing 'Round The Circle, The Games of	B	1891	McLoughlin Brothers	330	550
Light Horse H. Cooper Golf Game	B	1943	Trojan Games	175	295
Limited Mail & Express Game, The	B	1894	Parker Brothers	80	200
Lindy Flying Game	C	1927	Parker Brothers	20	50
Lindy Flying Game, The New	C	1927	Nucraft Toys	30	45
Lindy Flying Game, The New	C	1927	Nucraft Toys	25	45
Lindy Hop Off	B	1927	Parker Brothers	200	400
Lion & The Eagle, Or The Days of '76	C	1883	E.H. Snow	50	80
Literature Game	B	1897	L.J. Colby & Co.	15	25
Little Black Sambo, Game of	B	1934	Einson Freeman Publishing Corp.	75	150
Little Bo Beep Game	B	1914		60	100
Little Boy Blue	B	1910s	Milton Bradley	50	85
Little Colonel	B	1936	Selchow & Righter	75	100
Little Cowboy Game, The	B	1895	Parker Brothers	105	175
Little Fireman Game	B	1897	McLoughlin Brothers	4000	6000
Little Jack Horner Golf Course	B	1920s		145	250
Little Jack Horner, A Game	B	1910s	Milton Bradley	45	75
Little Nemo Game	B	1914		60	100
Little Orphan Annie Bead Game	S	1930s		20	35
Little Orphan Annie Game	B	1927	Milton Bradley	125	250
Little Orphan Annie Rummy Cards	C	1937	Whitman	30	75
Little Orphan Annie Shooting Game	S	1930s	Milton Bradley	30	50
Little Red Riding Hood Game	B	1914		90	150
Little Shoppers	B	1915	Gibson Game Co.	150	250
Little Soldier Game	B	1914		60	100
Little Soldier, The	B	1900s	United Game Co.	95	160
Logomachy Or War of Words Game	C	1903		25	40
London Bridge	B	1899	J.H. Singer	100	150
London Game, The	B	1898	Parker Brothers	165	275
Lone Ranger Game, The	B	1938	Parker Brothers	40	60
Lone Ranger Hi Yo Silver!! Target Game	S	1939	Marx	70	115

Games, Prewar

NAME	TYPE	YEAR	COMPANY	GOOD	EX
Looping The Loop	B	1940s	Advance Games	25	40
Los Angeles Rams Football Game	B	1930s	Zondine Game Co.	175	295
Lost Heir, Game of The	C	1910	Milton Bradley	20	35
Lost Heir, The Game of	C	1893	McLoughlin Brothers	45	75
Lost In The Woods	B	1895	McLoughlin Brothers	660	1100
Lotto	B	1932	Milton Bradley	6	10
Lou Gehrig's Official PlayBAll	B	1930s	Christy Walsh	525	875
Lowell Thomas' World Cruise	B	1937	Parker Brothers	50	75
Luck, The Game of	B	1892	Parker Brothers	60	100
Lucky 7 Baseball Game	B	1937	Ray Fair Co.	55	90
Mac Baseball Game	B	1930s	Mc Dowell Mfg. Co.	145	250
Macy's Pirate Treasure Hunt	B	1942	Einson Freeman Publishing Corp.	20	35
Madrap, The New Game of	B	1914		45	75
Magic Race	B	1942	Habob Co.	55	90
Magnetic Jack Straws	B	1891	E.I. Horsman	27	45
Magnetic Treasure Hunt	B	1930s	American Toy Works	15	25
Mail, Express Or Accommodation, Game of	B	1895	McLoughlin Brothers	450	750
Mail, Express Or Accommodation, Game of	B	1920s	Milton Bradley	135	225
Major League BAll	B	1921	National Game Makers	325	550
Major League Base Ball Game	B	1912	Philadelphia Game Mfg. Co.	650	1000
Major League Baseball Game, The	C	1910		85	145
Make A Million	C	1934	Rook Card Co.	10	15
Mammoth Conette	S	1898	Milton Bradley	90	150
Man Hunt	B	1937	Parker Brothers	80	200
Man In The Moon	B	1901	McLoughlin Brothers	2100	3500
Mansion of Happiness	B	1843	Ives	300	500
Mansion of Happiness	B	1864	Ives	180	300
Mansion of Happiness, The	B	1895	McLoughlin Brothers	510	850
Mar Juck	S	1923	The Regensteiner Corp.	20	30
Marathon Game, The	B	1930s	Rosebud Art Co.	75	100
Marble Muggins	S	1920s	American Toy Mfg.	60	100
Marriage, The Game of	B	1899	J.H. Singer	75	125
Match 'em	B	1926	All Fair	12	20
Mathers Parlor Baseball Game	B	1909	McClurg & Co.	30	50
Mayflower, The	C	1897	Fireside Game Co.	15	25
Meet The Missus	B	1937	Fitzpatrick Brothers	45	75
Mental Whoopee	B	1936	Simon & Schuster	10	15
Merry Steeple Chase	B	1890s	J. Ottmann Lith. Co.	45	70
Merry Go Round	B	1898	Chaffee & Selchow	1500	2000
Messenger Boy Game	B	1910	J.H. Singer	50	125
Messenger, The	B	1890	McLoughlin Brothers	75	125
Meteor Game	S	1916	A.C. Gilbert	25	45
Mexican Pete I Got It	B	1940s	Parker Brothers	20	40
Mickey Mouse BaseAll	B	1936	Post Cereal	55	90
Mickey Mouse Big Box of Games & Things To Color	B	1930s		45	75
Mickey Mouse Bridge Game	C	1935	Whitman	25	40
Mickey Mouse Circus Game	B	1930s	Marks Brothers	180	300
Mickey Mouse Coming Home Game	B	1930s	Marks Brothers	80	200
Mickey Mouse Miniature Pinball Game	S	1930s	Marks Brothers	20	35
Mickey Mouse Old Maid Game	C	1930s	Whitman	25	45
Mickey Mouse Roll'em Game	B	1930s	Marks Brothers	90	150
Mickey Mouse Shooting Game	S	1930s	Marks Brothers	120	200
Mickey Mouse Skittle Ball Game	S	1930s	Marks Brothers	60	100
Mickey Mouse Soldier Target Set	S	1930s	Marks Brothers	60	100
Midget Auto Race	B	1930s	Cracker Jack	10	15

Games, Prewar

NAME	TYPE	YEAR	COMPANY	GOOD	EX
Midget Speedway, Game of	B	1942	Whitman	55	90
Miles At Porto Rico	B	1899	Chaffee & Selchow	50	80
Miniature Golf	B	1930s	Miniature Golf Co.	35	60
Miss Muffet Game	B	1914		60	100
Mistress Mary, Quite Contrary	B	1905	Parker Brothers	60	105
Modern Game Assortment	B	1930s	J. Pressman & Co.	25	40
Moneta: 'Money Makes Money', Game of	B	1889	F.A. Wright Publishers	90	150
Monkey Shines	B	1940	All Fair	20	30
Monopolist, Mariner's Compass And Ten Up	B	1878	McLoughlin Brothers	500	1000
Monopoly	B	1935	Parker Brothers	20	50
Monopoly Jr. Edition	B	1936	Parker Brothers	15	30
Moon Mullins Automobile Race	B	1927	Milton Bradley	60	150
Mother Goose Bowling Game	B	1884	Charles M. Crandall Co.	510	850
Mother Goose, Game of	B	1914		75	125
Mother Goose, Game of	B	1921	Stoll & Edwards	30	50
Mother Hubbard	C	1875	McLoughlin Brothers	40	65
Mother Hubbard Game	B	1914		60	100
Motor Boat Race, An Exciting	B	1930	American Toy Works	110	180
Motor Cycle Game	B	1905	Milton Bradley	100	165
Motor Race	B	1922	Wolverine	90	150
Movie Inn	B	1917	Willis G. Young	45	75
Movie Millions	B	1938	Transogram	100	200
Movie Land Keeno	C	1929	Wilder	100	200
Movie Land Lotto	B	1920s	Milton Bradley	45	75
Moving Picture Game, The	B	1920s	Milton Bradley	70	120
Mr. Ree	B	1937	Selchow & Righter	60	100
Mumbly Peg	S	1920s	All Fair	25	50
Musical Lotto	C	1936	Tudor Metal Products Corp.	25	40
Mutuels	B	1938	Mutuels, Inc.	85	150
My Word, Horse Race	B	1938	American Toy Works	60	100
Mythology	C	1900	Cincinnati Game Co.	15	25
Mythology, Game of	B	1884	Peter G. Thompson	25	45
Napoleon, Game of	B	1895	Parker Brothers	775	1300
National American Baseball Game	C	1910	Parker Brothers	110	180
National Baseball Game, The	C	1913	National Baseball Playing Card Co.	700	1200
National Derby Horse Race	B	1938	Whitman	20	35
National Game of Baseball, The	B	1900s		500	875
National Game of The American Eagle, The	B	1844	Ives	2000	4000
National Game of the Star Spangled Banner	B	1844	L.I. Cohen & Co.	0	0
National Game, The	B	1900s	National Game Co.	875	1450
National League Ball Game	B	1885	Yankee Novelty Co.	350	575
Nations Or Quaker Whist, Game of	C	1898	McLoughlin Brothers	50	80
Nations, Game of	C	1908	Milton Bradley	15	25
Naughty Molly	C	1905	McLoughlin Brothers	45	75
Naval Maneuvers	B	1920	McLoughlin Brothers	240	600
Navigator	B	1938	Whitman	45	75
Navigator Boat Race	B	1890s	McLoughlin Brothers	115	195
Nebbs on the Air, A Radio Game	B	1930s	Milton Bradley	50	125
Nebbs, Game of The	B	1930s	Milton Bradley	40	75
Neck And Neck	B	1929	The Embossing Co.	25	40
Neck And Neck	B	1930	Wolverine	50	80
Nellie Bly	B	1898	J.H. Singer	100	240
New York Recorder Newspaper Supplement Baseball Game	B	1896		430	715
Newsboy, Game of The	B	1890	R. Bliss Mfg. Co.	1200	2000
NFL Strategy	B	1935	Tudor	45	70

GAMES

Games, Prewar

NAME	TYPE	YEAR	COMPANY	GOOD	EX
Nine Men Morris	B	1930s	Milton Bradley	25	45
Ninteenth Hole Golf Game	B	1930s	Einson Freeman Publishing Corp.	85	140
Nip & Tuck Hockey	B	1928	Parker Brothers	115	195
No Joke	B	1941	Volume Sprayer Mfg. Co.	12	20
Nok Out Baseball Game	B	1930	Dizzy & Daffy Dean	350	575
North Pole Game, The	B	1907	Milton Bradley	45	75
Nosey, The Game of	C	1905	McLoughlin Brothers	160	400
Object Lotto	B	1940s	Sam'l Gabriel Sons & Co.	15	25
Obstacle Race	B	1930s	Wilder	70	115
Ocean To Ocean Flight Game	B	1927	Wilder	55	95
Office Boy, The	B	1889	Parker Brothers	150	250
Official Radio Basketball Game	B	1939	Toy Creations	45	75
Official Radio Football Game	B	1940	Toy Creations	45	75
Oh, Blondie!	C	1940s		30	45
Old Curiosity Shop	C	1869	Novelty Game Co.	135	225
Old Hunter & his Game	B	1870		175	295
Old Maid	C	1890s	J.H. Singer	12	20
Old Maid	B	1898	Chaffee & Selchow	25	60
Old Maid & old Bachelor, The Merry Game of	B	1898	McLoughlin Brothers	180	300
Old Maid As Played By Mother Goose, Game of	B	1892	Clark & Sowdon	20	30
Old Maid Card Game	C	1889		20	50
Old Maid Fun Full Thrift Game	C	1940s	Russell Mfg. Co.	20	30
Old Maid Or Matrimony, Game of	B	1890	McLoughlin Brothers	150	250
Old Maid, Game of	C	1870	McLoughlin Brothers	50	80
Old Maid, With Characters From Famous Nursery Rhymes	C	1920s	All Fair	30	50
Old Mother Goose	B	1898	Chaffee & Selchow	105	175
Old Mother Hubbard, Game of	B	1890s	Milton Bradley	60	100
Old Mrs. Goose, Game of	B	1910	Milton Bradley	50	85
Old Time Shooting Gallery	S	1940	Warren Built Rite	15	25
Oldtimers	B	1940	Frederick H. Beach (Beachcraft)	15	20
Oliver Twist, The Good Old Game of	C	1888	Parker Brothers	90	153
Ollo	B	1944	Games Of Fame	15	25
Olympic Runners	B	1930	Wolverine	85	140
On The Mid Way	B	1925	Milton Bradley	45	75
One Two Button Your Shoe	B	1940s	Master Toy Co.	15	25
Open Championship Golf Game	B	1930s	Beacon Hudson Co.	45	75
Opportunity Hour	B	1940	American Toy Works	20	35
Ot O Win FootbAll	B	1920s	Ot O Win Toys & Games	55	90
Ouija	B	1920	William Fuld	20	45
Our Bird Friends	C	1901	Sarah H. Dudley	20	30
Our Defenders	B	1944	Master Toy Co.	40	65
Our Gang Tipple Topple Game	S	1930	All Fair	160	400
Our National Ball Game	B	1887	McGill & DeLang	425	650
Our National Life	C	1903	Cincinnati Game Co.	15	20
Our No. 7 Baseball Game Puzzle	B	1910	Satisfactory Co.	110	180
Our Union	B	1896	Fireside Game Co.	25	40
Outboard Motor Race, The	B	1930s	Milton Bradley	35	55
Overland Limited, The	B	1920s	Milton Bradley	45	75
Owl And The Pussy Cat, The	B	1900s	E.O. Clark	210	350
Pan Cake Tiddly Winks	B	1920s	Russell Mfg. Co.	55	90
Pana Kanal, The Great Panama Canal Game	B	1913	Chaffee & Selchow	65	110
Panama Canal Game	B	1910	Parker Brothers	135	225
Par Golf Card Game	B	1920	National Golf Services	115	195
Par, The New Golf Game	B	1926	Russell Mfg. Co.	115	195
Parcheesi	B	1880s	H.B. Chaffee	90	150
Parker Brothers Post Office Game	B	1910s	Parker Brothers	105	175

Games, Prewar

NAME	TYPE	YEAR	COMPANY	GOOD	EX
Parlor Base BAll	B	1878	1750	2900	
Parlor Baseball Game	B	1908	Mathers	200	325
Parlor Croquet	B	1940	J. Pressman & Co.	20	35
Parlor Football Game	B	1890s	McLoughlin Brothers	525	875
Parlor Golf	B	1897	Chaffee & Selchow	55	90
Pat Moran's Own Baseball Game	B	1919	Smith, Kline & French	325	550
Patch Word	C	1938	All Fair	10	15
Patent Parlor Bowling Alley	B	1899	Thomas Kochka	70	115
Paws & Claws	C	1895	Clark & Sowdon	35	60
Pe Ling	B	1923	Cookson & Sullivan	25	45
Pedestrianism	B	1879		350	575
Peeza	S	1935	Toy Creations	25	45
Peg At My Heart	B	1914	Willis G. Young	20	35
Peg Base BAll	B	1924	Parker Brothers	105	175
Peg BasebAll	B	1915	Parker Brothers	145	250
Peg'ity	B	1925	Parker Brothers	10	25
Peggy	B	1923	Parker Brothers	35	55
Pegpin, Game of	B	1929	Stoll & Edwards	25	45
Pennant Puzzle	B	1909	L.W. Hardy	250	400
Pennant Winner	B	1930s	Wolverine	175	295
Penny Post	B	1892	Parker Brothers	150	250
Pepper	C	1906	Parker Brothers	10	15
Peter Coddle And His Trip To New York	C	1890s	J.H. Singer	20	50
Peter Coddle Tells of His Trip To Chicago	C	1890	Parker Brothers	25	45
Peter Coddle's Trip To New York	C	1925	Milton Bradley	15	25
Peter Coddle's Trip To New York, The Game of	C	1888	Parker Brothers	25	45
Peter Coddle's Trip to the World's Fair	C	1939	Parker Brothers	50	75
Peter Coddle, Improved Game of	C	1900	McLoughlin Brothers	25	40
Peter Coddles	C	1890s	J. Ottmann Lith. Co.	25	45
Peter Pan	B	1927	Selchow & Righter	100	200
Peter Peter Pumpkin Eater	B	1914	Parker Brothers	60	100
Peter Rabbit Game	B	1910	Milton Bradley	55	95
Peter Rabbit Game	B	1940s	Sam'l Gabriel Sons & Co.	45	75
Philadelphia Inquirer Baseball Game, The	B	1896		145	250
Philo Vance	B	1937	Parker Brothers	70	175
Phoebe Snow, Game of	B	1899	McLoughlin Brothers	150	250
Piggies, The New Game	B	1894	Selchow & Righter	330	550
Pigskin	B	1940	Parker Brothers	20	50
Pigskin, Tom Hamilton's Football Game	B	1934	Parker Brothers	25	60
Pike's Peak Or Bust	S	1890s	Parker Brothers	40	100
Pilgrim's Progress, Going To Sunday School, Tower of Babel	B	1875	McLoughlin Brothers	120	300
Pinafore	B	1879	Fuller, Upham & Co.	45	75
Pinch Hitter	B	1930s		110	180
Pines, The	B	1896	Fireside Game Co.	15	25
Ping Pong	S	1902	Parker Brothers	50	75
Pinocchio Pitfalls Marble Game	B	1940		30	50
Pinocchio Playing Card Game	C	1939	Whitman	35	60
Pinocchio Ring The Nose Game	B	1940		20	30
Pinocchio Target Game	S	1938	American Toy Works	90	150
Pinocchio The Merry Puppet Game	S	1939	Milton Bradley	55	95
Pioneers of The Santa Fe Trail	B	1935	Einson Freeman Publishing Corp.	25	40

GAMES

Clockwise from upper left: Intercollegiate Football, 1923, Hustler Toy Corp.; Knockout Electronic Boxing Game, 1950, Northwestern Products; Carnival-The Show Business Game, Milton Bradley Co.; Game of Mail Express or Accommodation, 1895, McLoughlin Brothers; Glydor, 1931, All-Fair.

Games, Prewar

NAME	TYPE	YEAR	COMPANY	GOOD	EX
Pirate & Traveller	B	1936	Milton Bradley	15	40
Pirate Ship	B	1940	Lowe	15	25
Pitch Em, The Game of Indoor Horse Shoes	S	1929	Wolverine	25	35
Pla Golf Board Game	B	1938	Pla Golf Co.	775	1300
Play BAll	B	1920	National Game Co.	145	250
Play FootbAll	B	1934	Whitman	55	90
Pocket BasebAll	B	1940	Toy Creations	15	25
Pocket Edition Major League Baseball Game	B	1943	Anderson	85	140
Pocket FootbAll	B	1940	Toy Creations	25	40
Polar Ball BasebAll	B	1940	Bowline Game Co.	90	145
Politics, Game of	B	1859	Davis & Co.	0	0
Pool, Game of	B	1898		700	1175
Pop The Hat	S	1930s	Milton Bradley	35	50
Popeye, Play Hockey Fun with Popeye & Wimpy	B	1935	Barnum Mfg.	205	350
Popular Indoor Baseball Game	B	1896	Egerton R. Williams	500	850
Posting, A Merry Game of	B	1890s	J.H. Singer	180	300
Pro BasebAll	B	1940		70	115
Psychic BasebAll	C	1927	Psychic Baseball Corp.	60	150
Psychic Baseball Game	B	1935	Parker Brothers	175	295
Quarterback	B	1914	Littlefield Mfg. Co.	100	165
Rabbit Hunt, Game of	B	1870	McLoughlin Brothers	175	295
Race for the Cup	B	1910s	Milton Bradley	125	200
Race, The Game of the	B	1860s		425	715
Races, The Game of the	B	1844	William Crosby	850	1400
Racing Stable, Game of	B	1936	D & H Games	115	195
Radio Game	B	1926	Milton Bradley	40	100
Raggedy Ann's Magic Pebble Game	B	1941	Milton Bradley	30	75
Rainy Day Golf	B	1920	Selchow & Righter	60	100
Rambles	B	1881	American Publishing	195	275
Ranger Commandos	S	1942	Parker Brothers	35	60
Razz O Dazz O Six Man FootbAll	B	1938	Gruhn & Melton	60	100
Realistic BasebAll	B	1925	Realistic Game & Toy Corp.	205	350
Realistic Golf	B	1898	Parker Brothers	875	1450
Red Riding Hood And The Wolf, The New Game	C	1887	McLoughlin Brothers	50	85
Red Riding Hood, Game of	B	1898	Chaffee & Selchow	200	350
Red Ryder 'Whirli Crow' Target Game	S	1940s	Daisy	150	250
Red Ryder Target Game	B	1939	Whitman	50	125
Rex	C	1920s	J. Ottmann Lith. Co.	25	40
Rex And The Kilkenny Cats Game	B	1892	Parker Brothers	45	75
Ride 'em Cowboy	S	1939	Gotham Pressed Steel Corp.	25	45
Ring My Nose	S	1926	Milton Bradley	60	100
Ring Scaling	S	1900	Martin Co.	25	45
Ring A Peg	B	1885	E.I. Horsman	25	45
Rip Van Winkle	B	1890s	Clark & Sowdon	70	175
Rival Policemen	B	1896	McLoughlin Brothers	1200	3000
Road Race, Air Race (2 game set)	B	1928	Wilder	145	250
Robinson Crusoe For Little Folks, Game of	C	1900s	E.O. Clark	25	45
Robinson Crusoe, Game of	B	1909	Milton Bradley	30	75
Roll O FootbAll	B	1923	Supply Sales Co.	35	60
Roll O Golf	B	1923	Supply Sales Co.	35	60
Roll O Junior Baseball Game	B	1922	Roll O Mfg.	325	550
Roll O Motor Speedway	B	1922	Supply Sales Co.	65	110
Roly Poly Game	B	1910		30	50
Roodles	C	1912	Flinch Card Co.	20	30
Rook	C	1906	Rook Card Co.	5	10

Games, Prewar

NAME	TYPE	YEAR	COMPANY	GOOD	EX
Roosevelt At San Juan	C	1899	Chaffee & Selchow	50	85
Rose Bowl Championship Football Game	B	1940s	Lowe	40	100
Rough Riders, The Game of	B	1898	Clark & Sowdon	450	700
Roulette Baseball Game	B	1929	W. Barthonomae	115	195
Round The World Game	B	1914	Milton Bradley	60	100
Round The World With Nellie Bly	B	1890	McLoughlin Brothers	210	350
Royal Game of Kings And Queens	B	1892	McLoughlin Brothers	375	650
Rube Bressler's Baseball Game	B	1936	Bressley	130	215
Rube Walker & Harry Davis Baseball Game	B	1905		875	1450
Rummy FootbAll	B	1944	Milton Bradley	35	60
Runaway Sheep	B	1892	R. Bliss Mfg. Co.	165	275
Running The Blockade	B	1870s		0	0
Sabotage	C	1943	Games Of Fame	25	45
Sailor Boy Game	B	1910		60	100
Saratoga Horse Racing Game	B	1920	Milton Bradley	40	100
Saratoga Steeple Chase	B	1900	J.H. Singer	120	300
Scout, The	B	1900s	E.O. Clark	105	175
Scouting, Game of	B	1930s	Milton Bradley	50	125
Scrambles	B	1941	Frederick H. Beach (Beachcraft)	15	20
Shadow Game, The	B	1940s	Toy Creations	200	400
Shopping, Game of	B	1891	R. Bliss Mfg. Co.	900	1500
Shuffle Board, The New Game of	S	1920	Sam'l Gabriel Sons & Co.	50	85
Shufflebug, Game of	B	1921		20	30
Siege of Havana, The	B	1898	Parker Brothers	180	300
Simba	S	1932	All Fair	75	150
Sippa Fish	B	1936	Frederick H. Beach (Beachcraft)	20	30
Six Nations	B	1867	McLoughlin Brothers	0	0
Skating Race Game, The	B	1900	Chaffee & Selchow	250	400
Skeezix and the Air Mail	B	1930s	Milton Bradley	65	125
Skeezix Visits Nina	B	1930s	Milton Bradley	65	125
Ski Hi New York To Paris	B	1927	Cutler & Saleeby Co.	100	200
Skippy, A Card Game	C	1936	All Fair	45	75
Skippy, Game of	B	1932	Milton Bradley	75	150
Skirmish At Harper's Ferry	B	1891	McLoughlin Brothers	330	550
Skit Scat	C	1905	McLoughlin Brothers	40	70
Skot It Bagatelle	B	1930s	Northwestern Products	145	250
Sky Hawks	B	1931	All Fair	120	200
Skyscraper	B	1937	Parker Brothers	120	300
Slide Kelly! Baseball Game	B	1936	B.E. Ruth Co.	70	115
Slugger Baseball Game	B	1930	Marks Brothers	110	180
Smitty Game	B	1930s	Milton Bradley	40	75
Smitty Speed Boat Race Game	B	1930s	Milton Bradley	50	125
Smitty Target Game	S	1930s	Milton Bradley	40	75
Snake Game	B	1890s	McLoughlin Brothers	50	85
Snap	C	1883	E.I. Horsman	20	50
Snap Dragon	B	1903	H.B. Chaffee	135	225
Snap, Game of	C	1892	McLoughlin Brothers	40	65
Snap, Game of	C	1910s	Milton Bradley	20	35
Snap, The Game of	C	1905s	Parker Brothers	25	40
Snap Jacks	S	1940s	Sam'l Gabriel Sons & Co.	15	25
Sniff	B	1940s	The Embossing Co.	20	30
Snow White And The Seven Dwarfs	S	1938	American Toy Works	150	250
Snow White And The Seven Dwarfs Game	B	1938	Parker Brothers	125	200
Snow White And The Seven Dwarfs The Game of	B	1938	Milton Bradley	125	200
Snug Harbor	B	1930s	Milton Bradley	50	85
Socko The Monk, The Game of	B	1935	Einson Freeman Publishing Corp.	15	25

Games, Prewar

NAME	TYPE	YEAR	COMPANY	GOOD	EX
Soldier Boy Game	B	1914		60	100
Speculation	B	1885	Parker Brothers	40	65
Spedem Junior Auto Race Game	B	1929	All Fair	70	175
Speed Boat	B	1920s	Parker Brothers	85	140
Speed Boat Race	B	1926	Wolverine	70	115
Speed King, Game Of	B	1922	Russell Mfg. Co.	115	195
Speedway Motor Race	B	1920s	Smith, Kline & French	145	250
Spider's Web	B	1898	McLoughlin Brothers	50	125
Spin 'em Target Game	S	1930s	All Metal Product Co.	25	40
Spin It	S	1910s	Milton Bradley	20	50
Squails	B	1870s	Adams & Co.	40	65
Squails	B	1877	Milton Bradley	35	60
Stage	C	1904	C.M. Clark	40	60
Stak, International Game of	S	1937	Marks Brothers	30	60
Stanley In Africa	B	1891	R. Bliss Mfg. Co.	420	700
Star Baseball Game	C	1941	W.P. Ullrich	70	115
Star BasketbAll	B	1926	Star Paper Products	125	205
Star Reporter	B	1937	Parker Brothers	50	125
Star Ride	B	1934	Einson Freeman Publishing Corp.	15	25
Stars on Stripes Football Game	B	1941	Stars & Stripes Games Co.	55	90
Stax	S	1930s	Marks Brothers	12	20
Steeple Chase	B	1890	J.H. Singer	40	100
Steeple Chase & Checkers	B	1910	Milton Bradley	55	90
Steeple Chase, Game of	B	1900s	E. O. Clark	60	100
Steeple Chase, Game of	B	1910s	Milton Bradley	40	65
Steeple Chase, Improved Game of	B	1890s	McLoughlin Brothers	195	325
Steeple Chasing, Game of	B	1903	McLoughlin Brothers	350	575
Steps To Health Coke Game	B	1938	CDN	40	70
Sto Auto Race	B	1920s	Stough Co.	65	110
Sto Quoit	B	1920s	Stough Co.	10	15
Stook Exchange	B	1936	Parker Brothers	30	50
Stock Exchange, The Game of	B	1940s	Stox, Inc.	40	65
Stop & Go	B	1936	Einson Freeman Publishing Corp.	25	50
Stop And Go	B	1928	All Fair	75	150
Stop And Shop	B	1930	All Fair	100	150
Stop, Look, And Listen, Game of	B	1926	Milton Bradley	40	100
Strat: The Great War Game	B	1915	Strat Game Co., Inc.	25	45
Strategy, Game of	B	1891	McLoughlin Brothers	240	400
Strategy, Game of Armies	B	1938	Corey Games	40	50
Stratosphere	B	1930s	Parker Brothers	30	75
Stratosphere	B	1936	Whitman	100	200
Street Car Game, The	B	1890s	Parker Brothers	120	300
Strike Out	B	1920s	All Fair	175	295
Strike Like	B	1940s	Saxon Toy Corp.	55	90
Stunt Box	B	1941	Frederick H. Beach (Beachcraft)	12	20
Submarine Drag	B	1917	Willis G. Young	40	100
Substitute Golf	B	1906	John Wanamaker	120	200
Suffolk Downs	B	1930s	Corey Game Co.	85	140
Superman Action Game	S	1940	American Toy Works	60	100
Superman, Adventures of	B	1942	Milton Bradley	145	240
Susceptibles, The	B	1891	McLoughlin Brothers	325	550
Sweep	B	1929	Selchow & Righter	25	40
Sweeps	B	1930s	E.E. Fairchild	25	60
Sweepstakes	B	1930s	Haras Mfg. Co.	55	90
Swing A Peg	B	1890s	Milton Bradley	30	50
T.G.O. Klondyke	B	1899	J.H. Singer	150	250
Table Croquet	S	1890s	Milton Bradley	30	75
Table Golf	B	1909	McClurg & Co.	15	25
Tackle	B	1933	Tackle Game Co.	70	115
Tactics	S	1940	Northwestern Products	25	45

Games, Prewar

NAME	TYPE	YEAR	COMPANY	GOOD	EX
Tait's Table Golf	B	1914	John Tait	350	575
Tak Tiks, Basketball	B	1939	Midwest Products	15	25
Take It And Double	B	1943	Frederick H. Beach (Beachcraft)	20	35
Take It Or Leave It	B	1942	Zondine Game Co.	25	45
Take Off	C	1930s	Russell Mfg. Co.	15	20
Teddy's Bear Hunt	B	1907	Bowers & Hard	375	650
Teddy's Ride From Oyster bay to Albany	B	1899	Jesse Crandall	3000	5500
Tee Off	B	1935	Donogof	115	195
Telegrams	B	1941	Whitman	40	70
Telegraph Boy, Game of The	B	1888	McLoughlin Brothers	250	350
Telegraph Messinger Boy, Game of The	C	1886	McLoughlin Brothers	35	60
Telepathy	B	1939	Cadaco Ellis	65	110
Tell Bell, The	B	1928	Knapp Electric & Novelty Co.	50	75
Ten Pins	B	1920	Mason & Parker	35	60
Tennis & Baseball	B	1930		70	115
Terry and the Pirates	B	1930s	Whitman	50	100
Tete A Tete	B	1892	Clark & Sowdon	40	100
They're Off, Race Horse Game	B	1930s	Parker Brothers	15	20
Thorobred	B	1940s	Lowe	55	90
Thorton W. Burgess Animal Game	B	1925	Saalfield	70	115
Three Bears	B	1910s	Milton Bradley	20	50
Three Bears, The	C	1922	Stoll & Edwards	20	35
Three Blind Mice, Game of	B	1930s	Milton Bradley	25	45
Three Little Kittens	B	1910s	Milton Bradley	60	100
Three Little Pigs Game	B	1933	Einson Freeman Publishing Corp.	55	95
Three Little Pigs, The Game of The	B	1933	Kenilworth Press	100	165
Three Men In A Tub	B	1935	Milton Bradley	40	65
Three Men On A Horse	B	1936	Milton Bradley	20	50
Three Merry Men	C	1865	Amsdan & Co.	40	65
Three Point Landing	B	1942	Advance Games	35	55
Thrilling Indoor Football Game	B	1933	Cronston Co.	70	115
Through The Clouds	B	1931	Milton Bradley	75	125
Through The Locks To The Golden Gate	B	1905	Milton Bradley	70	175
Ticker	B	1929	Glow Products Co.	45	75
Tiddledy Wink Tennis	S	1890	E.I. Horsman	45	75
Tiddledy Winks, Game of	S	1910s	Parker Brothers	35	55
Tiddley Golf Game	B	1928	Milton Bradley	80	135
Tiddley Winks Game	B	1920s	Wilder	15	35
Tiger Hunt, Game of	B	1899	Chaffee & Selchow	270	450
Tiger Tom, Game of	B	1920s	Milton Bradley	30	75
Ting A Ling, The Game of	B	1920	Stoll & Edwards	25	45
Tinker Toss	S	1920s	Toy Creations	25	40
Tinkerpins	B	1916	Toy Creations	55	90
Tip The Bellboy	S	1929	All Fair	120	200
Tip Top Fish Pond	S	1930s	Milton Bradley	20	45
Tip Top Boxing	B	1922	LaVelle	200	350
Tipit	B	1929	Wolverine	15	20
Tit for Tat Indoor Hockey	B	1920s	Lemper Novelty Co.	45	75
Tit Tat Toe	B	1929	The Embossing Co.	15	25
Tit Tat Toe, Three In A Row	B	1896	Austin & Craw	45	75
To The Aid of Your Party	B	1942	Leister Game Co.	15	25
Tobagganing At Christmas, Game of	B	1899	McLoughlin Brothers	400	600
Toboggan Slide	B	1890s	Hamilton Myers	225	385
Toboggan Slide	B	1890s	J.H. Singer	200	325
Toll Gate, Game of	B	1890s	McLoughlin Brothers	280	700
Tom Barker Card Game	C	1913		1400	2300

Games, Prewar

NAME	TYPE	YEAR	COMPANY	GOOD	EX
Tom Hamilton's Pigskin	B	1935	Parker Brothers	70	115
Tom Sawyer And Huck Finn, Adventures of	B	1925	Stoll & Edwards	70	120
Tom Sawyer On The Mississippi	B	1935	Einson Freeman Publishing Corp.	40	100
Tom Sawyer, The Game of	B	1937	Milton Bradley	45	75
Toonerville Trolley Game	B	1927	Milton Bradley	175	300
Toonin Radio Game	B	1925	All Fair	150	300
Toot	C	1905	Parker Brothers	25	45
Top Hockey	B	1943	Corey Games	0	0
TopOgraphy	B	1941	Cadaco Ltd.	35	60
Topsy Turvey, Game of	B	1899	McLoughlin Brothers	150	250
Tortoise And The Hare	B	1922	Russell Mfg. Co.	45	85
Toss O	S	1924	Lubbers & Bell	15	25
Totem	C	1873	West & Lee Co.	45	75
Toto, The New Game	B	1925	Baseball Toto Sales Co.	55	90
Touchdown	S	1930s	Milton Bradley	150	250
Touchdown	B	1937	Cadaco Ltd.	65	110
Touchdown Football Game	B	1920s	Wilder	100	165
Touchdown or Parlor Football, Game of	B	1897	Union Mutual Life Ins. Premium	85	140
Touchdown, The New Game	B	1920	Hartford Mfg. Co.	70	115
Touring	C	1906	Wallie Dorr Co.	25	40
Touring	C	1926	Parker Brothers	15	30
Tourist, The, A Railroad Game	B	1900s	Milton Bradley	75	125
Tournament	B	1858	Mayhew & Baker	180	300
Town HAll	B	1939	Milton Bradley	20	30
Toy Town Bank	B	1910	Milton Bradley	90	150
Toy Town Conductors Game	B	1910	Milton Bradley	105	175
Toy Town Target With Repeating Pistol	S	1911	Milton Bradley	55	95
Toy Town Telegraph Office	B	1910s	Parker Brothers	75	150
Trackle Lite	B	1940s	Saxon Toy Corp.	55	90
Traffic Hazards	B	1939	Trojan Games	20	35
Trailer Trails	B	1937	Offset Gravure Corp.	35	60
Train For Boston	B	1900	Parker Brothers	500	1000
Traits, The Game of	C	1933	Goodenough And Woglom Co.	15	25
Transatlantic Flight, Game of The	B	1925	Milton Bradley	125	250
Transport Pilot	B	1938	Cadaco Ltd.	25	40
Trap A Tank	B	1920s	Wolverine	40	70
Traps & Bunkers	B	1926?	Milton Bradley	115	195
Traps And Bunkers, A Game of Golf	S	1930s	Milton Bradley	25	40
Travel, The Game of	B	1894	Parker Brothers	225	375
Treasure Hunt	B	1940	All Fair	20	50
Treasure Island	B	1923	Stoll & Edwards	75	125
Treasure Island	B	1934	Stoll & Einson	60	90
Treasure Island, Game of	B	1923	Gem Publishing Co.	40	65
Triangular Dominos	B	1885	Frank H. Richards	35	60
Trilby	B	1894	E.I. Horsman	270	450
Trip Around The World, A	B	1920s	Parker Brothers	25	45
Trip Round The World, Game of	B	1897	McLoughlin Brothers	300	500
Trip Through Our National Parks: Game of Yellowstone, A	C	1910s	Cincinnati Game Co.	15	35
Trip To Washington	B	1884	Milton Bradley	100	175
Triple Play	B	1930s	National Games, Inc.	12	20
Trips of Japhet Jenkens & Sam Slick	C	1871	Milton Bradley	20	35
Trolley	C	1904	Snyder Brothers	35	60
Trolley Came Off, The	C	1900s	Parker Brothers	45	75
Trolley Ride, The Game of The	B	1890s	Hamilton Myers	210	350
Trunk Box Lotto Game	B	1890s	McLoughlin Brothers	15	35

GAMES

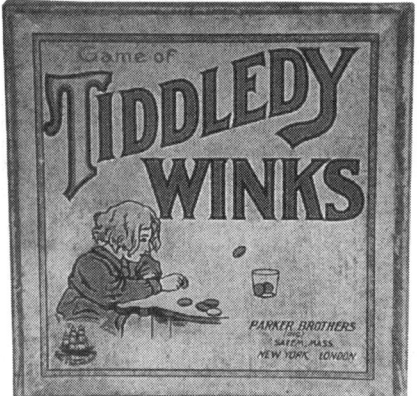

Top to bottom: Zippy Zepps Air Game, 1930, All-Fair; Game of Tiddledy Winks, 1910, Parker Brothers; Games You Like To Play for Home, Travel and Camp, 1920, Parker Brothers; The Limited Mail And Express Game, 1894, Parker Brothers.

Games, Prewar

NAME	TYPE	YEAR	COMPANY	GOOD	EX
Tumblin Five Acrobats	B	1925	Doremus Schoen & Co.	12	20
Turn Over	B	1908	Milton Bradley	75	100
Turnover	B	1898	Chaffee & Selchow	50	85
Tutoom, Journey To The Treasures of Pharoah	B	1923	All Fair	120	200
Twentieth Century Limited	B	1900s	Parker Brothers	90	150
Twenty Five, Game of	C	1925	Milton Bradley	6	10
Ty Cobb's Own Game of BasebAll	B	1910s	National Novelty Co.	350	650
U Bat It	B	1920s	Schultz Star Co.	70	115
U.S. Postman Game	B	1914		60	100
Uncle Jim's Question Bee	B	1938	Kress	20	30
Uncle Sam At War With Spain, Great Game of	B	1898	Rhode Island Game Co.	325	550
Uncle Sam's Baseball Game	B	1890	J.C. Bell	525	875
Uncle Sam's Mail	B	1893	McLoughlin Brothers	210	350
Uncle Wiggily's New Airplane Game	B	1920s	Milton Bradley	50	100
Uncle Wiggily's Woodland Games	B	1936		6	10
Union Games, The	B	1868	Milton Bradley	0	0
United States Air Mail Game, The	S	1930s	Parker Brothers	50	85
United States History, The Game of	C	1903	Parker Brothers	15	25
Va Lo Football Card Game	B	1930s		85	140
Van Loon Story of Mankind Game, The	B	1931	Kerk Guild	50	85
Vanderbelt Cup Race Game	B	1910	Bowers & Hard	650	850
Vanderbilt Cup Race	B	1906	Bowers & Hard	200	400
Varsity Football Game	B	1942	Cadaco Ellis	45	75
Varsity Football Playing Cards	B	1925	Kerger Co.	55	90
Varsity Race	B	1899	Parker Brothers	425	725
Vassar Boat Race, The	B	1899	Chaffee & Selchow	420	700
Venetian Fortune Teller, Game of	C	1898	Parker Brothers	75	125
Verborum	C	1883	Peter G. Thompson	20	30
Vest Pocket Checker Set	B	1929	The Embossing Co.	15	25
Vest Pocket Quoits	B	1944	Colorful Creations	25	45
Victo	B	1943	Spare Time	15	25
Victory	B	1920s	Klak Co., New Haven, Ct.	105	175
Vignette Author	B	1874	E.G. Selchow	35	60
Visit of Santa Claus, Game of The	B	1899	McLoughlin Brothers	300	500
Visit To The Farm	B	1893	R. Bliss Mfg. Co.	300	500
Vox Pop	B	1938	Milton Bradley	25	45
Voyage Around The World, Game of	B	1930s	Milton Bradley	105	175
Wa Hoo Pick Em Up Sticks	S	1936	Doremus Schoen & Co.	15	25
Wachter's Parlor BasebAll	B	1925	Ragetelle	145	250
Walk the Plank	B	1925	Milton Bradley	70	115
Walking The Tightrope	B	1897	McLoughlin Brothers	125	250
Walking The Tightrope	B	1920	Milton Bradley	50	100
Walt And Skeezix Gasoline Alley Game	C	1927	Milton Bradley	50	100
Walt Disney's Game Parade	B	1930s		30	50
Walt Disney's Ski Jump Target Game	B	1930s	American Toy Works	175	295
Walt Disney's Uncle Remus Game	B	1930s	Parker Brothers	50	85
Walter Johnson Baseball Game	B	1930s		195	325
Waner's Baseball Game	B	1939	Waner's Baseball Game Inc.	350	575
Wang, Game of	B	1892	Clark & Sowdon	25	45
War And Diplomacy	C	1899	Chaffee & Selchow	50	85
War of Nations	B	1915	Milton Bradley	40	65
War of Words	C	1910	McLoughlin Brothers	35	60
Ward Cuff's Football Game	B	1938	Continental Sales Co.	125	200
Washington's Birthday Party	S	1911	Russell Mfg. Co.	55	95

Games, Prewar

NAME	TYPE	YEAR	COMPANY	GOOD	EX
Watch On De Rind	B	1931	All Fair	150	250
Waterloo	B	1895	Parker Brothers	325	550
Watermelon Frolic	B	1900	E.I. Horsman	135	225
Watermelon Patch	B	1940s	Craig Hopkins	25	45
Watermelon Patch Game	S	1896	McLoughlin Brothers	900	2000
Way To The White House, The	B	1927	All Fair	75	125
We, The Magnetic Flying Game	B	1928	Parker Brothers	100	200
West Point	B	1902		90	150
What Would You Do?	C	1933	Geo E. Schweig & Son	10	15
What's My Name?	B	1920s	Jaymar	15	25
When My Ship Comes In	C	1888	Parker Brothers	35	65
Where Do You Live	C	1890s	J.H. Singer	35	65
Where's Johnny	C	1885	McLoughlin Brothers	45	75
Which Is It? Speak Quick Or Pay	C	1889	McLoughlin Brothers	45	75
Whip, The Comical Game of	C	1930	Russell Mfg. Co.	25	65
Whippet Race	B	1940s	J. Pressman & Co.	20	35
Whirlpool Game	B	1890s	McLoughlin Brothers	65	100
White Wings	B	1930s	Glevum Games Inc.	45	70
Who Is The Thief	C	1937	Whitman	25	40
Whyoo!	C	1906	Milton Bradley	30	50
Wide Awake, Game of	B	1899	McLoughlin Brothers	150	250
Wide World And A Journey Round It	B	1896	Parker Brothers	165	275
Wild West Cardboard Game	B	1914		90	150
Wild West, Game of The	B	1889	R. Bliss Mfg. Co.	500	800
Wilder's Football Game	B	1930s	Wilder	55	90
William's Popular Indoor BasebAll	C	1889	Hatch Co.	700	1175
Win, Place & Show	B	1940s	3M	25	40
Winko BasebAll	B	1940	Milton Bradley	45	70
Winnie Winkle Glider Race Game	B	1930s	Milton Bradley	65	100
Winnie The Pooh Game	B	1933	Parker Brothers	75	125
Winnie The Pooh Game, A. A. Milne's	B	1931	Kerk Guild	50	85
Witzi Wits	B	1926	All Fair	80	125
Wizard, The	B	1921	Fulton Specialty Co.	15	25
Wogglebug Game of Conumdrums, The	C	1905	Parker Brothers	100	200
Wonder Tiddley Winks	S	1899	Martin Co.	20	35
Wonderful Game of Oz (pewter pieces)	B	1921	Parker Brothers	600	1300
Wonderful Game of Oz (wooden pieces)	B	1921	Parker Brothers	165	275
Wordy	B	1938	J. Pressman & Co.	25	45
World Flyers, Game of The	B	1926	All Fair	150	300
World Series Baseball Game	B	1940s	Radio Sports	205	350
World Series Parlor BasebAll	B	1916	Cliffton E. Hooper	150	250
World's Championship BasebAll	B	1910	Champion Amusement Co.	175	295
World's Championship Golf Game	B	1930s	Beacon Hudson Co.	145	250
World's Columbian Exposition, Game of The	B	1893	R. Bliss Mfg. Co.	450	750
World's Educator Game	B	1889		45	75
World's Fair Game	B	1939		90	150
World's Fair Game, The	B	1892	Parker Brothers	800	1400
Worth While	C	1907	Doan & Co.	25	40
WPA, Work, Progress, Action	B	1935	All Fair	100	200
Wyhoo	C	1906	Milton Bradley	25	45
Wyntre Golf	B	1920s	All Fair	175	295
X Plor US	B	1922	All Fair	100	150
Yacht Race	B	1890s	Clark & Sowdon	200	350
Yacht Race	B	1930s	J. Pressman & Co.	150	250
Yachting	B	1890	J.H. Singer	70	115
Yale Harvard Football Game	B	1922	LaVelle	425	725

Games, Prewar

NAME	TYPE	YEAR	COMPANY	GOOD	EX
Yale Harvard Game	B	1890	McLoughlin Brothers	1050	1775
Yale Princeton Foot Ball Game	B	1895	McLoughlin Brothers	575	950
Yankee Doodle!	B	1940	Cadaco Ellis	30	50
Yankee Doodle, A Game of American History	B	1895	Parker Brothers	285	475
Yankee Pedlar, Or What Do You Buy	C	1850s	John Mcloughlin	725	1200
Yankee Trader	B	1941	Corey Games	35	60
Yellowstone, Game Of	C	1895	Fireside Game Co.	45	85
You're Out! Baseball Game	B	1941	Corey Games	85	140
Young Athlete, The	B	1898	Chaffee & Selchow	425	700
Young Folks Historical Game	C	1890s	McLoughlin Brothers	25	35
Young Peddlers, Game of The	C	1859	Mayhew & Baker	55	95
Young People's Geographical Game	C	1900s	Parker Brothers	15	20
Yuneek Game	B	1889	McLoughlin Brothers	450	750
Zimmer Baseball Game	B	1885	McLoughlin Brothers	1300	2150
Zip Top	B	1940	Deluxe Game Corp.	35	55
Zippy Zepps	B	1930s	All Fair	350	600
Zoo Hoo	B	1924	Lubbers & Bell	125	200
Zoom	C	1941	Whitman	35	55
Zoom, Original Game of	B	1940s	All Fair	45	85
Zulu Blowing Game	B	1927	Zulu Toy Mfg. Co.	50	100

GAMES

Top to bottom: Eliot Ness and the Untouchables Game, 1960s, Transogram; James Bond Secret Agent 007 Game, 1964, Milton Bradley; Hawaii Five-O Game, 1960s, Remco; The Man From U.N.C.L.E. Game, 1966, Ideal.

Games, Postwar

NAME	TYPE	YEAR	COMPANY	GOOD	EX	MINT
$25,000 Pyramid	B	1980s	Cardinal Industries	10	15	25
$64,000 Question Quiz Game	B	1955	Lowell	30	60	90
1-2-3 Game Hot Spot!	B	1961	Parker Brothers	5	10	15
12 0'Clock High	C	1966	Milton Bradley	15	35	55
12 O'clock High Game	B	1965	Ideal	20	35	50
2 For The Money	B	1955	Hasbro	20	30	45
25 Ghosts	B	1969	Lakeside	15	25	40
3 Stooges Fun House Game	B	1950s	Lowell	60	100	160
300 Mile Race	B	1955	Warren	30	50	80
4000 A.D. Interstellar Conflict Game	B	1972	House Of Games	10	15	25
77 Sunset Strip	B	1960	Lowell	35	50	75
77 Sunset Strip Game	B	1960	Warner	40	65	100
A-Team	B	1984	Parker Brothers	5	10	15
Abbott & Costello Who's On First?	B	1978	Selchow & Righter	5	10	15
ABC Monday Night Football	B	1972	Aurora	25	40	65
ABC Monday Night Football Roger Staubach Edition	B	1973	Aurora	35	60	90
ABC Sports Winter Olympics	B	1987	Mindscape	10	15	25
Acquire	B	1961	Avalon Hill	5	15	25
Across The Continent	B	1952	Parker Brothers	30	50	80
Action Baseball	B	1965	Pressman	35	50	75
Addams Family	C	1965	Milton Bradley	30	45	75
Addams Family	B	1973	Milton Bradley	40	65	95
Addams Family Game	B	1965	Ideal	65	100	175
Admirals	B	1960s	Parker Brothers	120	1756	225
Advance To Boardwalk	B	1985	Parker Brothers	10	15	20
Adventure In Science, An	B	1950	Jacmar Mfg. Co.	20	30	50
Agent Zero-M Spy Detector	B	1964	Mattel	30	50	80
Aggravation	B	1970	Lakeside	5	10	15
Air Assault On Crete	B	1977	Avalon Hill	10	15	25
Air Empire	B	1961	Avalon Hill	15	25	40
Air Race Around The World	B	1950s	Lido	25	45	70
Airways	S	1950s	Lindstrom Tool & Toy Co.	30	50	75
Alfred Hitchcock 'Why?'	B	1965	Milton Bradley	10	20	30
Alfred Hitchcock Presents Mystery Game	B	1958	Milton Bradley	20	35	55
Alien	B	1979	Kenner	30	45	70
All In The Family	B	1972	Milton Bradley	10	20	30
All My Children	B	1985		5	10	20
All Pro Baseball	B	1950	Ideal	45	70	110
All Pro Basketball	B	1969	Ideal	20	35	50
All Star Baseball	B	1960	Cadeco Ellis	40	65	90
All Star Basketball	B	1950s	Gardner	55	90	135
All The King's Men	B	1979	Parker Brothers	6	10	15
All Time Greats Baseball Game	B	1971	Midwest Research	15	25	35
All-American Football	B	1969	Cadaco	35	60	90
All-Pro Football	B	1967	Ideal	20	35	50
All-Star Baseball	B	1989	Cadaco	12	20	30
All-Star Baseball Fame	B	1962	Cadaco-Ellis	15	25	40
All-Star Electric Baseball & Football	B	1955	Harett-Gilmar	35	60	90
All-Star Football	B	1950	Gardner	55	90	135
Alpha Baseball Game	B	1950s	Realistic Mfg. Co.	110	180	275
American Derby, The	B	1951	Cadaco-Ellis	30	50	75
Angry Donald Duck Game	S	1970s	Mexico	40	65	100
Animal Crackers	B	1970s	Milton Bradley	4	7	11
Annette's Secret Passage	B	1958	Parker Brothers	25	40	65
Annie Oakley	B	1950	Milton Bradley	45	75	120
Annie Oakley	B	1955	Milton Bradley	30	45	70
Annie-The Movie Game	B	1981	Parker Brothers	4	6	10
Anti-(Monopoly)	B	1970s	Anti-Monopoly, Inc.	20	35	55
Anti-Monopoly	B	1973	Anti-Monopoly, Inc.	25	45	70

GAMES

Games, Postwar

NAME	TYPE	YEAR	COMPANY	GOOD	EX	MINT
APBA Baseball Master Game	B	1975	APBA	35	60	90
APBA League Football	B	1980s	APBA	12	20	30
APBA Pro League Football	B	1964	APBA	55	90	135
APBA Saddle Racing Game	B	1970s	APBA	15	25	40
Apple's Way	B	1974	Milton Bradley	15	25	40
Archie Bunker	C	1972		10	15	20
Archies, The	B	1969	Whitman	25	45	65
Arnold Palmer's Inside Golf	B	1961	D.B. Remson	50	85	130
Around The World In 80 Days Game	B	1957	Transogram	25	40	65
Art Lewis Football Game	B	1955	Morgantown Game Co.	70	115	175
Art Linkletter's House Party	B	1968		20	35	55
Art Linkletter's People Are Funny Party Game	C	1960s		10	25	35
As The World Turns	B	1966	Parker Brothers	25	40	65
ASG Baseball	B	1989	3M	12	20	30
ASG Major League Baseball	B	1973	Gerney Games	55	90	135
Assembly Line	B	1953	Selchow & Righter	35	60	95
Atom Ant Game	B	1966	Transogram	45	70	110
Aurora Pursuit! Game	B	1973	Aurora	5	10	15
Autograph Baseball Game	B	1948	Philadelphia Inquirer	110	180	275
B-17 Queen of The Skies	B	1983	Avalon Hill	6	10	15
B.T.O. (Big Time Operator)	B	1956	Bettye-B Co.	30	45	70
Bali	C	1954	I-S Ultd.	15	25	40
Ballplayer's Baseball Game	B	1955	Jon Weber	30	50	75
Bamboozle	B	1962	Milton Bradley	20	35	50
Banana Tree	B	1977	Marx	10	15	25
Barbapapa Takes A Trip	B	1977	Selchow & Righter	3	5	8
Barbie's Little Sister Skipper Game	B	1964	Mattel	35	50	75
Barbie, Queen of The Prom	B	1960		40	60	85
Baretta	B	1976	Milton Bradley	15	25	40
Barnabas Collins Game	B	1969	Milton Bradley	35	60	95
Barney Miller	B	1977	Parker Brothers	10	15	25
Barnstormer	B	1970s	Marx	20	35	55
Bart Starr Quarterback Game	B	1960s		175	295	450
Baseball	S	1960s	Tudor	25	40	60
Baseball Card All Star Game	C	1987	Captoys	6	10	15
Baseball Card Game	C	1950s	Ed-U-Cards	20	35	50
Baseball Card Game, Official	C	1965	Milton Bradley	175	295	450
Baseball Challenge	B	1980	Tri-Valley Games	15	25	35
Baseball Game, Official	B	1953	Milton Bradley	100	165	250
Baseball Game, The	B	1988	Horatio	12	20	30
Baseball Strategy	B	1973	Avalon Hill	10	15	25
Baseball, A Sports Illustrated Game	B	1975	Time Inc.	25	40	65
Baseball, Football & Checkers	B	1957	Parker Brothers	35	60	90
Bash!	S	1967	Ideal	10	15	25
Basketball Strategy	B	1974	Avalon Hill	10	15	25
Bat Masterson	B	1958	Lowell	45	75	120
Batman	B	1978	Hasbro	15	30	50
Batman And Robin Game	B	1965	Hasbro	45	75	120
Batman Batarang Toss	S	1966	Pressman	150	250	400
Batman Card Game	C	1966	Ideal	25	45	65
Batman Game	B	1966	Milton Bradley	45	75	100
Batman Pin Ball	S	1966	Marx	55	95	150
Bats in the Belfry	S	1964	Mattel	30	45	70
Batter Up	B	1946	M. Hopper	30	50	75
Batter Up Card Game	C	1949	Ed-U-Cards	25	40	60
Batter-Rou Baseball Game (Dizzy Dean)	B	1950s	Memphis Plastic	100	165	250
Battle Cry	B	1962	Milton Bradley	35	60	85
Battle Line	B	1960s	Ideal	25	50	75

Games, Postwar

NAME	TYPE	YEAR	COMPANY	GOOD	EX	MINT
Battle of The Planets	B	1970s	Milton Bradley	15	25	35
Battleship	B	1965	Milton Bradley	10	15	25
Battlestar Galactica	B	1978	Parker Brothers	10	15	25
Beany & Cecil Match It	B	1960s	Mattel	30	50	80
Beat Inflation	B	1961	Avalon Hill	15	25	40
Beat The Buzz	B	1958	Kenner	10	15	25
Beat The Clock	B	1954	Lowell	35	60	95
Beat The Clock	B	1960s	Milton Bradley	6	10	16
Beatles Flip Your Wig Game	B	1964	Milton Bradley	75	125	175
Beetle Bailey, The Old Army Game	B	1963	Milton Bradley	20	30	50
Behind The '8' Ball Game	B	1969	Selchow & Righter	15	20	35
Ben Casey MD Game	B	1961	Transogram	15	30	45
Bermuda Triangle	B	1976	Milton Bradley	10	15	25
Betsy Ross And The Flag	B	1950s	Transogram	20	30	50
Beverly Hillbillies	B	1963	Standard Toycraft, Inc.	40	65	100
Beverly Hillbillies Game, The	C	1963	Milton Bradley	25	35	50
Beverly Hills Game	C	1963		15	20	35
Bewitched	B	1965	T. Cohn Inc.	40	60	90
Bewitched Stymie Game	C	1960s		20	30	50
Bible Quiz Lotto	C	1949	Jack Levitz	20	30	50
Big 5 Poosh-M Up	S	1950s	Knickerbocker	25	40	60
Big Foot	B	1977	Milton Bradley	10	15	25
Big Game Hunt, The	S	1947	Carrom Industries	15	25	40
Big League Baseball	B	1959	Saalfield	55	90	135
Big League Baseball Game	B	1966	3M	15	25	40
Big League Manager Football	B	1965	BLM	20	35	50
Big Payoff	B	1984	Payoff Enterprises Co.	6	10	15
Big Six Sports Games	B	1950s	Gardner	175	295	450
Big Sneeze Game, The	S	1968	Ideal	10	15	20
Big Time Colorado Football	B	1983	B.J. Tall	6	10	15
Big Town	B	1954	Lowell	50	80	125
Billionaire	B	1973	Parker Brothers	5	15	20
Bing Crosby's Game, Call Me Lucky	B	1954	Parker Brothers	55	90	135
Bingo-Matic	B	1954	Transogram	5	10	20
Bionic Crisis	B	1975	Parker Brothers	10	15	20
Bionic Woman	B	1976	Parker Brothers	5	10	15
Bird Brain	B	1966	Milton Bradley	10	15	25
Bird Watcher	B	1958	Parker Brothers	75	110	150
Birdie Golf	B	1964	Barris Corp.	70	115	175
Black Beauty	B	1957	Transogram	25	50	65
Black Box	B	1978	Parker Brothers	5	10	15
Blade Runner	B	1982		40	65	100
Blast Off	B	1953	Selchow & Righter	50	100	150
Blitzkrieg	B	1965	Avalon Hill	15	25	40
Blockhead	B	1954	Russell Mfg. Co.	10	15	25
Blondie	B	1970s	Parker Brothers	6	10	15
Blondie And Dagwood's Race For The Office	B	1950	Jaymar	30	45	70
Blue Line Hockey	B	1968	3M	25	40	60
BMX Cross Challenge Action Game	B	1988	Cross Challenge Corp.	6	10	15
Bob Feller's Big League Baseball	B	1950	Saalfield	75	150	250
Bobbsey Twins	B	1957	Milton Bradley	30	45	70
Bobby Shantz Baseball Game	B	1955	Realistic Games	80	150	225
Body Language	B	1975	Milton Bradley	15	25	40
Boggle	B	1976	Parker Brothers	5	15	20
Bonanza Michigan Rummy Game	B	1964	Parker Brothers	30	45	70
Booby Trap	S	1965	Parker Brothers	10	15	25
Boom Or Bust	B	1951	Parker Brothers	100	200	300
Booth's Pro Conference Football	B	1977	Sher-Co	10	15	25
Boots and Saddles	B	1960	Chad Valley	35	65	100
Bop The Beetle	S	1963	Ideal	20	35	55
Boris Karloff's Monster Game	B	1965	Gems	75	125	200

Games, Postwar

NAME	TYPE	YEAR	COMPANY	GOOD	EX	MINT
Boston Marathon Game, Official	B	1978	Perf Products	15	25	35
Boston Red Sox Game	C	1964	Ed-U-Cards	55	90	135
Bottoms Up	B	1956	Bettye-B Co.	15	25	40
Bottoms Up	B	1970s		3	5	8
Bowl & Score	B	1974	Lowe	10	15	25
Bowl And Score	B	1962	Lowe	6	10	15
Bowl Bound!	B	1973	Sports Illustrated	15	25	40
Bowl-A-Matic	S	1963	Eldon	45	75	120
Brain Waves	B	1977	Milton Bradley	15	20	30
Branded	B	1966	Milton Bradley	30	60	90
Brass Monkey Game, The	B	1973	U.S. Game Systems	15	30	50
Break Par Golf Game	B	1950s	Warren-Built Rite	15	30	50
Break The Bank	B	1955	Bettye-B Co.	20	50	75
Breaker 1-9	B	1976	Milton Bradley	5	10	15
Brett Ball	B	1981	9th Inning	15	30	45
Bride Bingo	B	1957	Leister Game Co.	20	35	55
Broadside	B	1962	Milton Bradley	20	45	70
Bruce Jenner Decathlon Game	B	1979	Parker Brothers	4	7	11
Buck Fever	B	1984	L & D Robton	12	20	30
Buckaroo	B	1947		20	35	55
Bucket Ball	B	1972	Marx	10	15	25
Bug-A-Boo	B	1968	Whitman	10	15	20
Bugaloos	B	1971		10	20	35
Bugs Bunny Game (bagatelle)	S	1975	Ideal	15	25	40
Bugs Bunny Under The Cawit Game	B	1972	Whitman	15	25	40
Building Boom	B	1950s	Kohner	6	10	15
Built-Rite Frisky Flippers Slide Bar Game	B	1950s		5	10	15
Built-Rite Swish Basketball Game	B	1950s		7	15	25
Bullwinkle Card Game	C	1962	Ed-U-Cards	15	25	40
Bullwinkle Hide & Seek Game	B	1961	Milton Bradley	25	40	65
Bullwinkle's Super Market Game	B	1970s	Whitman	15	25	40
Buster Brown Game and Play Box	B	1950s	Buster Brown Shoes	40	75	125
Cabbage Patch Kids	B	1984	Parker Brothers	5	10	15
Call It Golf	B	1966	Strauss	15	25	40
Call My Bluff	B	1965	Milton Bradley	15	20	30
Calling All Cars	B	1950s		25	45	65
Calvin & The Colonel High Spirits	B	1962	Milton Bradley	10	20	30
Camelot	B	1955	Parker Brothers	20	30	50
Camp Granada Game, Allan Sherman's	B	1968	Milton Bradley	15	30	45
Camp Runamuck	B	1965	Ideal	30	65	85
Can You Catch It Charlie Brown?	B	1976	Ideal	10	15	20
Candid Camera Target Shot	S	1950s	Lindstrom Tool & Toy Co.	35	60	90
Candyland	B	1949	Milton Bradley	25	50	90
Caper	B	1970	Parker Brothers	20	30	50
Capital Punishment	B	1981	Hammerhead	40	70	125
Captain America	B	1977	Milton Bradley	10	15	25
Captain America Game	B	1966	Milton Bradley	40	65	105
Captain Caveman And The Teen Angels	B	1981	Milton Bradley	6	10	15
Captain Gallant Desert Fort Game	B	1956	Transogram	20	40	65
Captain Kangaroo	B	1956	Milton Bradley	40	85	125
Captain Video Game	B	1952	Milton Bradley	75	125	200
Cardino	B	1970	Milton Bradley	10	15	25
Careers	B	1957	Parker Brothers	25	40	55
Careers	B	1965	Parker Brothers	10	20	30
Careful The Toppling Tower	S	1967	Ideal	10	15	25
Carl Hubbell Mechanical Baseball	B	1950	Gotham	100	200	300
Carl Yastrzemski's Action Baseball	B	1968	Pressman	90	145	195
Carrier Strike	B	1977	Milton Bradley	15	25	45
Case of The Elusive Assassin, The	B	1967	Ideal	40	65	105

Games, Postwar

NAME	TYPE	YEAR	COMPANY	GOOD	EX	MINT
Casey Jones	B	1959	Saalfield	30	45	70
Casper The Friendly Ghost Game	B	1959	Milton Bradley	10	20	30
Casper the Friendly Ghost Game	B	1974	The Cottie Co.	5	15	25
Casper The Friendly Ghost Game	B	1974	Schaper	10	25	40
Cat & Mouse	B	1964	Parker Brothers	7	15	20
Cattlemen, The	B	1977	Selchow & Righter	10	15	25
Centipede	B	1983	Milton Bradley	6	10	15
Century of Great Fights	B	1969	Research Games	40	75	110
Challenge	C	1947	John Scarne Games, Inc.	10	20	30
Challenge Golf at Pebble Beach	B	1972	3M	15	25	35
Challenge the Yankees	B	1960s	Hasbro	125	225	325
Challenge Yahtzee	B	1974	Milton Bradley	7	15	20
Championship Baseball	B	1966	Championship Games Inc.	10	20	30
Championship Basketball	B	1966	Championship Games Inc.	10	20	30
Championship Golf	B	1966	Championship Games Inc.	10	20	30
Charlie Brown's All Star Baseball Game	B	1965	Parker Brothers	35	60	90
Charlie's Angels	B	1977	Milton Bradley	6	10	15
Charlie's Angels, (Farrah Fawcett box)	B	1977	Milton Bradley	10	20	45
Chess	B	1977	Milton Bradley	3	5	8
Chex Ches Football	B	1971	Chex Ches Games	15	25	40
Cheyenne	B	1958	Milton Bradley	25	60	95
Chicago Sports Trivia Game	B	1984	Sports Trivia Inc.	6	10	15
Chicken In Every Plot, A	B	1980s	Animal Town Game Co.	20	30	50
CHIPS	B	1981	Ideal	7	10	20
CHIPS Game	B	1977	Milton Bradley	4	7	10
Chit Chat Game	B	1963	Milton Bradley	6	10	15
Chopper Strike	B	1976	Milton Bradley	10	15	25
Chute-5	B	1973	Lowe	3	5	8
Chutes & Ladders	B	1956	Milton Bradley	10	15	25
Chutzpah	B	1967	Middle Earth	25	50	75
Chutzpah	B	1967	Cadaco	10	15	25
Cimarron Strip	B	1967	Ideal	50	85	135
Circle Racer Board Game	B	1988		6	10	15
Circus Game	B	1947		15	25	40
Civil War Game	B	1961	Parker Brothers	20	40	65
Civilization	B	1982	Avalon Hill	7	10	20
Clean Sweep	B	1960s	Schaper	20	30	50
Close Encounters	B	1977	Parker Brothers	7	15	20
Clue	B	1949	Parker Brothers	25	40	65
Clue	B	1972	Parker Brothers	4	7	11
College Basketball	B	1954	Cadaco-Ellis	20	30	55
Columbo	B	1973	Milton Bradley	6	10	15
Combat	B	1963	Ideal	35	60	100
Combat	C	1964	Milton Bradley	20	45	70
Comin' Round The Mountain	B	1954	Einson-Freeman	30	50	80
Computer Baseball	B	1966	Epoch Playtime	25	40	65
Computer Basketball	B	1969	Electric Data Corp.	45	75	115
Concentration 25th Anniversary Edition	B	1982	Milton Bradley	6	10	16
Concentration 3rd Edition	B	1960	Milton Bradley	12	20	35
Coney Island Penny Pitch	S	1950s	Novel Toy	33	55	88
Conspiracy	B	1982	Milton Bradley	4	7	11
Cootie	B	1949	Schaper	10	25	35
Countdown	B	1967	Lowe	25	45	75
Counter Point	B	1976	Hallmark	10	15	25
Cowboy Roundup	B	1952	Parker Brothers	15	25	40
Cowboys & Indians	C	1949	Ed-U-Cards	15	25	40

Games, Postwar

NAME	TYPE	YEAR	COMPANY	GOOD	EX	MINT
Crazy Clock Game	S	1964	Ideal	40	60	90
Creature Features	B	1975	Athol	15	25	50
Creature From The Black Lagoon	B	1963	Hasbro	105	175	280
Cribb Golf	B	1980s		30	50	75
Crosby Derby, The	B	1947	Fishlove Ind.	35	60	90
Cross the Board Horse Racing Game	B	1975	MPH Co.	15	25	40
Cross Up	B	1974	Milton Bradley	15	25	40
Crosswords	B	1954	National Games, Inc.	12	20	32
Crusader Rabbit Game	B	1960s		50	100	175
Curious George Game, The	B	1977	Parker Brothers	4	6	10
Curse of The Cobras Game	B	1982	Ideal	12	20	32
Cut Up Shopping Spree Game	B	1968	Milton Bradley	6	10	16
Dallas (Television Role Playing)	B	1980	SPI	4	7	11
Dallas Game	C	1980	Mego	5	8	13
Daniel Boone Trail Blazer	B	1964		25	60	85
Daniel Boone Wilderness Trail	C	1964	Transogram	15	35	50
Dark Shadows Game	B	1968	Whitman	39	65	104
Dastardly & Muttley	B	1969	Milton Bradley	25	45	65
Data Prog. Computerized Pro Football	B	1971	Data Prog. Game Co.	15	25	40
Dating Game, The	B	1967	Hasbro	15	25	40
Davy Crocket Rescue Race Game	B	1950s	Gabriel	57	95	152
Davy Crockett Adventure Game	B	1956	Gardner	45	75	120
Davy Crockett Frontierland Game	B	1955	Parker Brothers	45	75	120
Davy Crockett Radar Action Game	B	1955	Ewing Mfg. & Sales Co.	51	85	136
Dawn of The Dead	B	1978		21	35	56
Daytona 500 Race Game	B	1989	Milton Bradley	10	15	25
Dealer's Choice	B	1972	Parker Brothers	20	40	60
Decathalon	B	1972	Sports Illustrated	15	25	35
Dee Vs Meade: Battle of Gettysburg	S	1974	Gamut Of Games	60	100	160
Deluxe Wheel of Fortune	B	1986	Pressman	5	8	13
Dennis The Menace Baseball Game	B	1960		20	50	70
Denny McLain Magnetik Game, Official	B	1968	Gotham	115	195	295
Deputy Dawg Hoss Toss	S	1973		15	25	40
Deputy Dawg TV Lotto	B	1961		21	35	56
Deputy Game, The	B	1960	Milton Bradley	30	50	80
Derby Day	B	1959		21	35	56
Detectives Game, The	B	1961	Transogram	30	50	80
Dick Tracy Crime Stopper	B	1963	Ideal	57	95	152
Dick Tracy The Master Detective Game	B	1961	Selchow & Righter	30	50	80
Dick Van Dyke Board Game	B	1964	Standard Toycraft, Inc.	45	75	120
Diplomacy	B	1961	Games Research	21	35	56
Diplomacy	B	1976	Avalon Hill	15	25	40
Direct Hit	B	1950s	Northwestern Products	40	70	110
Dirty Water-The Water Polution Game	B	1970	Urban Systems	8	15	20
Disney Dodgem Bagatelle	S	1960s	Marx	30	65	90
Disney Mouseketeer	B	1964	Parker Brothers	40	65	100
Disneyland Game	B	1965	Transogram	15	35	50
Dispatcher	B	1958	Avalon Hill	20	35	55
Doctor Kildare Game	B	1967		10	15	25
Doctor Who	B	1980s	Denys Fisher	25	40	75
Dogfight	B	1962	Milton Bradley	25	60	85
Dollar A Second	B	1955	Lowell	20	50	75
Dollars & Sense	B	1946	Sidney Rogers	100	150	200
Domain	B	1983	Parker Brothers	2	3	5
Don Carter's Strike Bowling Game	B	1964	Saalfield	55	90	135
Donald Duck Big Game Box	B	1979	Whitman	10	15	20
Donald Duck Pins & Bowling Game	B	1955s	Pressman	35	60	90
Donald Duck Tiddly Winks Game	B	1950s		6	10	15
Donald Duck's Party Game	B	1950s	Parker Brothers	10	20	35
Dondi Potato Race Game	B	1950s	Hasbro	15	35	50

Top to bottom: The Twilight Zone Game, 1960s, Ideal; Weird-ohs Game, 1964, Ideal; The Six Million Dollar Man Game, 1975, Parker Brothers; The Mad Magazine Game, 1979, Parker Brothers.

Games, Postwar

NAME	TYPE	YEAR	COMPANY	GOOD	EX	MINT
Donkey Party Game	B	1950	Saalfield	15	25	40
Double Trouble	B	1987	Milton Bradley	3	5	8
Dr. Kildare's Perilous Night	B	1962	Ideal	25	40	65
Dracula Mystery Game	B	1960s	Hasbro	40	100	145
Dracula's 'I Vant To Bite Your Finger' Game	B	1981	Hasbro	10	20	30
Dragnet	B	1955	Parker Brothers	70	115	185
Dragnet	B	1955	Transogram	35	60	95
Dragnet Badge 714 Triple Fire Target Game	S	1955		15	25	40
Dragon's Lair	B	1983	Milton Bradley	7	10	20
Driver Ed	B	1973	Cadaco	6	10	20
Dukes of Hazzard	B	1981	Ideal	6	10	15
Dune	B	1984	Parker Brothers	6	10	15
Dungeon Dice	B	1977	Parker Brothers	5	8	15
Dungeons & Dragons	B	1980	Mattel	6	10	15
Duplicate Ad-Lib	B	1976	Lowe	5	10	15
Duran Duran Game	B	1985	Milton Bradley	15	25	40
Dynamite Shack Game	S	1968	Milton Bradley	10	20	35
Dynomutt	B	1977		10	15	25
E.T. The Extra Terrestrial	B	1982	Parker Brothers	6	10	15
Earl Gillespie Baseball Game	B	1961	Wei-Gill Inc.	25	40	65
Electra Woman And Dyna Girl	B	1977	Ideal	10	15	25
Electric Sports Car Race	B	1959	Tudor	35	60	90
Electronic Detective Game	B	1970s	Ideal	20	30	45
Electronic Radar Search	B	1967	Ideal	10	15	25
Ellsworth Elephant Game	B	1960	Selchow & Righter	30	45	70
Emenee Chocolate Factory	B	1966		6	10	15
Emergency	B	1971	Milton Bradley	10	15	25
Emily Post Popularity Game	B	1970	Selchow & Righter	15	20	35
Empire Auto Races	B	1950s	Empire Plastics	20	30	50
Enemy Agent	B	1976	Milton Bradley	15	20	30
Entertainment Trivia Game	B	1984	Lakeside	5	10	15
Entre's Fun & Games In Accounting	B	1988	Entrepreneurial Games	4	7	12
Escape From New York	B	1980	TSR	10	15	25
Escape From The Death Star	B	1977	Parker Brothers	20	35	55
Extra Innings	B	1975	J. Kavanaugh	15	25	35
Eye Guess	B	1960s	Milton Bradley	15	20	35
F-Troop	B	1965	Ideal	40	100	155
F.B.I.	B	1958	Transogram	35	55	90
F/11 Armchair Quarterback	B	1964	James R. Hock	15	25	40
Fact Finder Fun	B	1963	Milton Bradley	10	15	25
Facts In Five	S	1967	3M	5	15	25
Fall Guy, The	B	1981	Milton Bradley	10	15	25
Falls	B	1950s	National	85	140	215
Family Affair	B	1967	Whitman	25	40	65
Family Feud	B	1977	Milton Bradley	10	15	25
Family Ties Game, The	B	1986	Apple Street	10	20	25
Famous 500 Mile Race	B	1988		8	13	20
Fang Bang	B	1966	Milton Bradley	10	15	25
Fangface	B	1979	Parker Brothers	5	8	13
Fantastic Voyage Game	B	1968	Milton Bradley	15	25	40
Fascination	S	1962	Remco	15	35	50
Fast Golf	B	1977	Whitman	15	25	35
Fat Albert	B	1973	Milton Bradley	15	25	40
Feeley Meeley Game	S	1967	Milton Bradley	7	15	20
Felix The Cat Dandy Candy Game	B	1957		10	15	25
Felix The Cat Game	B	1960	Milton Bradley	25	40	65
Felix The Cat Game	B	1968	Milton Bradley	15	25	40
Fighter Bomber	B	1977	Cadaco	15	25	40

Games, Postwar

NAME	TYPE	YEAR	COMPANY	GOOD	EX	MINT
Finance	B	1962	Parker Brothers	10	20	35
Fire Fighters!	B	1957	Russell Mfg. Co.	15	25	40
Fireball XL-5	B	1963	Milton Bradley	40	100	145
Fireball XL-5 Magnetic Dart Game	S	1963	Magic Wand	75	125	200
First Down	B	1970	TGP Games	50	80	125
Fish Pond	B	1950s	National Games, Inc.	45	75	125
Fishbait	B	1965	Ideal	40	60	80
Flagship Airfreight The Airplane Cargo Game	B	1946	Milton Bradley	40	70	115
Flash Gordon	B	1970s	House Of Games	15	20	35
Flea Circus Magnetic Action Game	S	1968	Mattel	15	25	40
Flintstone's Cut Ups Game	C	1963	Whitman	20	30	50
Flintstone's Dino The Dinosaur Game	B	1961	Transogram	45	75	120
Flintstone's Hoppy The Hopperoo Game	B	1964	Transogram	45	75	120
Flintstones	B	1971	Milton Bradley	15	25	40
Flintstones	B	1980	Milton Bradley	15	20	35
Flintstones Animal Rummy	C	1960	Ed-U-Cards	6	10	15
Flintstones Brake Ball	S	1962	Whitman	45	75	120
Flintstones Mechanical Shooting Gallery	S	1962	Marx	75	125	200
Flintstones Mitt-Full Game	B	1962	Whitman	40	65	100
Flintstones Stone Age Game	B	1961	Transogram	20	45	65
Flip 'N Skip	B	1971	Little Kennys	5	10	15
Flip Flop Go	B	1962	Mattel	6	10	15
Flipper Flips	B	1960s	Mattel	30	50	80
Flying Nun Game, The	B	1968	Milton Bradley	20	30	50
Flying Nun Marble Maze Game, The	S	1967	Hasbro	20	30	50
Fonz Game, The	B	1976	Milton Bradley	15	25	40
Fooba-Roo Football Game	B	1955	Memphis Plastic	25	40	65
Football Fever	B	1985	Hansen	20	35	50
Fore	B	1954	Artcraft Paper Products	20	35	50
Formula One Car Race Game	B	1968	Parker Brothers	35	65	100
Fortress America	B	1986	Milton Bradley	25	40	65
Fox & Hounds, Game of	B	1948	Parker Brothers	25	40	65
Frank Cavanaugh's American Football	B	1955	F. Cavanaugh Assoc.	25	40	60
Frankenstein Game	B	1962	Hasbro	60	100	160
Fu Manchu's Hidden Hoard	B	1967	Ideal	35	55	90
Fuedel	B	1967	3M	15	25	40
Fugitive	B	1966	Ideal	90	150	240
Funky Phantom Game	B	1971	Milton Bradley	10	15	25
Funny Bones Game	C	1968	Parker Brothers	5	7	11
G.I. Joe	B	1982	International Games	15	25	40
G.I. Joe Adventure	B	1982	Hasbro	20	30	50
G.I. Joe Bagatelle Gun Action Game	S	1970s	Hasbro	7	12	15
G.I. Joe Card Game	B	1965	Whitman	10	15	25
G.I. Joe Marine Paratrooper	B	1965	Milton Bradley	30	45	70
Gambler's Golf	B	1975	Gammon Games	10	16	25
Gammonball	B	1980	Fun-Time Products	10	16	25
Gang Way For Fun	B	1964	Transogram	25	40	65
Gardner's Championship Golf	B	1950s	Gardner	55	90	135
Garfield	B	1981	Parker Brothers	4	6	10
Garrison's Gorillas	B	1967	Ideal	45	75	120
Gay Puree	B	1962		25	40	65
General Hospital	B	1974	Parker Brothers	10	15	25
General Hospital	B	1980s	Cardinal Industries	15	20	30
Gentle Ben Animal Hunt Game	B	1967	Mattel	15	25	40
George Brett's 9th Inning Baseball Game	B	1981	Brett Ball	16	25	40
George of the Jungle Game	B	1968	Parker Brothers	40	65	125

GAMES

Games, Postwar

NAME	TYPE	YEAR	COMPANY	GOOD	EX	MINT
Get Beep Beep The Road Runner Game	B	1975	Whitman	15	25	35
Get Smart Game	C	1966	Ideal	30	75	105
Get The Picture	B	1987	Worlds Of Wonder	6	10	15
Gettysburg	B	1960	Avalon Hill	20	30	50
Ghosts	B	1985	Milton Bradley	4	6	10
Giant Wheel Thrills'n Spills Horse Race	B	1958	Remco	30	50	75
Gidget	C	1966	Milton Bradley	15	35	50
Gil Hodges' Pennant Fever	B	1970	Research Games	65	100	150
Gilligan's Island	B	1965	T. Cohn Inc.	120	200	325
Gilligan, The New Adventures of	B	1974	Milton Bradley	15	35	50
Globetrotter Basketball, Official	B	1950s	Meljak	60	100	150
Go For Broke	B	1965	Selchow & Righter	10	15	25
Go for the Green	B	1973	Sports Illustrated	45	70	110
Go Go Go	C	1950s	Arrco Playing Card Co.	6	10	15
Goal Line Stand	B	1980	Game Shop Inc.	12	20	30
Godfather	B	1971	Family Games	6	10	15
Godzilla Game	B	1963	Mattel	30	50	80
Going, Going, Gone!	B	1975	Milton Bradley	9	15	25
Golden Trivia Game	B	1984	Western Publishing	6	10	15
Gomer Pyle Game	B	1960s	Transogram	20	45	65
Gong Hee Fot Choy	C	1948	Zondine Game Co.	15	20	30
Gong Show	B	1975	Milton Bradley	15	25	40
Good Ol'Charlie Brown Game	B	1971	Milton Bradley	15	20	30
Goofy's Mad Maze	B	1970s	Whitman	6	10	15
Goonies	B	1980s	Milton Bradley	10	15	20
Gotham Professoinal Basketball	B	1950s	Gotham	35	50	70
Gotham's Ice Hockey	S	1960s	Gotham	25	60	85
Grab A Loop	S	1968	Milton Bradley	7	12	20
Grand Master of Martial Arts	B	1986	Hoyle	6	10	15
Great Grape Ape Game, The	B	1975	Milton Bradley	15	25	40
Green Acres Game, The	B	1960s	Standard Toycraft, Inc.	25	65	90
Green Ghost Game	B	1965	Transogram	65	95	150
Green Hornet Playing Cards	C	1960s		45	75	125
Green Hornet Quick Switch Game	S	1966	Milton Bradley	180	300	475
Gremlins	B	1984	International Games	10	15	25
Greyhound Pursuit	B	1985	N/N Games	8	13	20
Grizzly Adams Game	B	1978	House Of Games	15	25	40
Groucho's TV Quiz Game	B	1954	Pressman	50	75	150
Groucho's You Bet Your Life	B	1955	Lowell	75	125	250
Guinness Book of World Records Game, The	B	1979	Parker Brothers	5	9	15
Gulf Strike	B	1983		15	25	40
Gunsmoke Game	B	1950s	Lowell	40	65	100
Gusher	B	1946	Carrom Industries	40	100	135
Half-Time Football	B	1979	Lakeside	5	9	15
Handicap Harness Racing	B	1978	Hall of Fame Games	15	25	35
Hands Down	S	1965	Ideal	10	15	25
Hang On Harvey	B	1969	Ideal	15	20	30
Hangman	B	1976	Milton Bradley	5	8	15
Hank Aaron Baseball Game	B	1970	Ideal	50	80	125
Hank Bauer's "Be a Manager"	B	1960s	Barco Games	75	125	175
Happiness	B	1972	Milton Bradley	7	11	20
Happy Days	B	1976	Parker Brothers	15	25	45
Hardy Boys Mystery game, The	B	1968	Milton Bradley	15	25	40
Hardy Boys Treasure	B	1960	Parker Brothers	20	30	50
Harlem Globetrotter Official Edition Basketball	B	1970s	Cadaco-Ellis	45	75	115
Harlem Globetrotters Game	B	1971	Milton Bradley	30	50	75
Harry's Glam Slam	C	1962	Harry Obst	35	60	90
Hashimoto	B	1963	Transogram	30	45	70

Games, Postwar

NAME	TYPE	YEAR	COMPANY	GOOD	EX	MINT
Haunted House Game	B	1963	Ideal	65	100	175
Haunted Mansion	B	1970s	Lakeside	30	50	80
Have Gun Will Travel Game	B	1959	Parker Brothers	50	85	135
Hawaii Five-O	B	1960s	Remco	15	25	40
Hawaiian Eye	B	1960	Transogram	95	160	250
Hector Heathcote	B	1963	Transogram	50	85	135
Hi Pop	S	1946	Advance Games	20	35	55
Hi-Ho! Cherry-O	B	1960	Whitman	6	10	15
Hide N Seek	B	1967	Ideal	10	15	25
Hip Flip	B	1968	Parker Brothers	6	10	15
Hippety Hop	B	1947	Corey Game Co.	25	40	65
Hit The Beach	B	1965	Milton Bradley	40	60	90
Hocus Pocus	B	1960s	Transogram	30	45	70
Hogan's Heroes Game	B	1966	Transogram	35	85	120
Holiday	B	1958	Replogie Globes Inc.	40	65	100
Hollywood Go	B	1954	Parker Brothers	30	45	75
Hollywood Squares	B	1974	Ideal	6	10	15
Hollywood Squares	B	1980	Milton Bradley	4	6	10
Home Court Basketball	B	1954		145	250	375
Home Game	B	1950s	Pressman	30	50	80
Home Stretch Harness Racing	B	1967	Lowe	35	60	90
Home Team Baseball Game	B	1957	Selchow & Righter	35	60	90
Honey West	B	1965	Ideal	50	85	135
Honeymooners Game, The	B	1986	TSR	7	12	20
Hoodoo	B	1950	Tryne	7	12	20
Hookey Go Fishin'	B	1974	Cadaco	10	16	25
Hopalong Cassidy Bean Bag Toss Game	S	1950s		20	40	60
Hopalong Cassidy Chinese Checkers Game	B	1950s		20	50	75
Hopalong Cassidy Game	B	1950s	Milton Bradley	40	100	145
Hoppity Hooper Pin Ball Game	S	1965	Lido	40	75	115
Hot Rod	B	1953	Harett-Gilmar	30	50	75
Hot Spot	B	1961	Parker Brothers	10	20	30
Hot Wheels Game	B	1982	Whitman	8	13	20
Hot Wheels Wipe-Out Game	B	1968	Mattel	30	50	75
Houndcats Game	B	1970s	Milton Bradley	8	15	25
House Party	B	1968	Whitman	10	25	35
Houston Astros Baseball Challenge Game	B	1980	Croque Ltd.	15	25	35
How To Succeed In Business Without Really Trying	B	1963	Milton Bradley	10	15	25
Howdy Doody Dominoes Game	S	1951	Ed-U-Cards	60	100	160
Howdy Doody Game	C	1954	Russell Mfg. Co.	20	30	50
Howdy Doody's 3 Ring Circus	B	1950	Harett-Gilmar	45	75	120
Howdy Doody's Own Game	B	1949	Parker Brothers	75	125	200
Huckleberry Hound	B	1981	Milton Bradley	10	20	35
Huckleberry Hound Bumps	B	1960	Transogram	20	50	75
Huckleberry Hound Spin-O-Game	B	1959		45	75	120
Huckleberry Hound Western Game	B	1959	Milton Bradley	25	40	65
Huggin' The Rail	S	1948	Selchow & Righter	45	65	100
Hullabaloo	B	1965	Remco	50	85	135
Humor Rumor	B	1969		10	20	30
Humpty Dumpty Game	B	1950s	Lowell	6	10	15
Hunt For Red October	B	1988	TSR	5	15	25
I Dream of Jeanie Game	B	1965	Milton Bradley	30	75	105
I Spy	B	1965	Ideal	50	95	150
I Survived New York!	C	1981	City Enterprises	4	7	12
I'm George Gobel, And Here's The Game	B	1955	Schaper	40	60	80
I-Qubes	S	1948	Capex Co. Inc.	10	15	25
Identipops	B	1969	Playvalue	75	175	250

Games, Postwar

NAME	TYPE	YEAR	COMPANY	GOOD	EX	MINT
Incredible Hulk	B	1978	Milton Bradley	6	10	15
Indiana Jones Raiders of The Lost Ark	B	1981	Kenner	20	35	55
Indianpolis 500 75th Running Race Game	B	1991	International Games	8	13	20
Input	B	1984	Milton Bradley	4	7	15
Inspector Gadget	B	1983	Milton Bradley	15	25	40
Instant Replay	B	1987	Parker Brothers	8	13	20
International Grand Prix	B	1975	Cadaco	30	50	75
Interpretation of Dreams	B	1969	Hasbro	10	15	25
Intrigue	B	1954	Milton Bradley	20	40	60
Inventors, The	B	1974	Parker Brothers	10	25	35
Ipcress File	B	1966	Milton Bradley	20	50	75
Ironside	B	1976	Ideal	55	95	150
Is the Pope Catholic?!	B	1970s		35	55	85
Jace Pearson's Tales of The Texas Rangers	B	1955	E.E. Fairchild	50	75	125
Jack And The Beanstalk	B	1946	National Games, Inc.	30	45	75
Jack And The Beanstalk Adventure Game	B	1957	Transogram	25	50	75
Jack Barry's Twenty One	B	1956	Lowe	20	30	50
Jackie Gleason's And AW-A-A-A-Y We Go!	B	1956	Transogram	75	125	200
Jackie Gleason's Story Stage Game	B	1955	Utopia Enterprises, Inc.	100	200	300
Jackpot	B	1975	Milton Bradley	7	11	20
Jacmar Big League Baseball	B	1950s		100	175	250
James Bond (Live and Let Die) Tarot Game	C	1973	US Games Systems	15	35	50
James Bond 007 Goldfinger Game	B	1966	Milton Bradley	50	85	135
James Bond 007 Thunderball Game	B	1965	Milton Bradley	50	85	135
James Bond Message From M Game	S	1966	Ideal	165	275	450
James Bond Secret Agent 007 Game	B	1964	Milton Bradley	15	35	50
James Bond You Only Live Twice	B	1984	Victory Games	5	8	13
Jan Murray's Charge Account	B	1961	Lowell	25	40	65
Jan Murray's Treasure Hunt	B	1950s		10	25	35
JDK Baseball	B	1982	JDK Baseball	10	16	25
Jeane Dixon's Game of Destiny	B	1968	Milton Bradley	7	12	20
Jeopardy	B	1964	Milton Bradley	10	15	25
Jerry Kramer's Instant Replay	B	1970	EMD Enterprises	15	25	40
Jetson's Out of this World Game	B	1963	Transogram	85	140	225
Jetsons Fun Pad Game	B	1963	Milton Bradley	40	100	145
Jetsons Race Through Space Game	B	1985	Milton Bradley	5	10	15
Jimmy the Greek Oddsmaker Football	B	1974	Aurora	15	25	35
Jockette	B	1950s	Jockette Co.	25	40	60
Joe Palooka Boxing Game	B	1950s	Lowell	35	65	100
John Drake Secret Agent	B	1966	Milton Bradley	30	45	70
Johnny Apollo Moon Landing (bagatelle)	S	1969	Marx	20	35	55
Johnny Ringo	B	1959	Transogram	75	125	200
Johnny Unitas Football Game	B	1970	Pro Mentor	25	40	65
Joker's Wild	B	1973	Milton Bradley	5	10	15
Jonathan Livingston Seagull	B	1973	Mattel	6	10	15
Jonny Quest Game	B	1964	Transogram	60	100	160
Jose Canseco's Perfect Baseball Game	B	1991	Perfect Game Co.	8	13	20
Jubilee	B	1950s	Cadaco	6	10	15
Jumbo Jet	B	1963	Jumbo	6	10	15
Junior Bingo-Matic	B	1968	Transogram	6	10	15
Junior Executive	B	1963	Whitman	7	15	20
Junior Quarterback Football	B	1950s	Warren-Built-Rite	30	50	75
Justice	B	1954	Lowell	30	45	75
Justice League of America	B	1967	Hasbro	105	175	275

Games, Postwar

NAME	TYPE	YEAR	COMPANY	GOOD	EX	MINT
Ka Bala	B	1965	Transogram	65	100	150
KaBoom!	S	1965	Ideal	10	15	25
Kar-Zoom	B	1964	Whitman	15	20	35
Kennedys, The	B	1962	Transogram	40	100	140
Kentucky Derby	B	1960	Whitman	15	25	40
Kentucky Jones	B	1964	T. Cohn Inc.	25	40	65
Ker-Plunk	S	1967	Ideal	5	10	15
Keyword	B	1954	Parker Brothers	5	10	15
Kimbo	S	1950s	Parker Brothers	10	15	25
King Arthur	S	1950s	Northwestern Products	25	40	65
King Kong Game	B	1966	Ideal	10	15	30
King Kong Game	B	1966	Milton Bradley	10	15	25
King Leonardo And His Subjects Game	B	1960	Milton Bradley	30	50	80
King Pin Deluxe Bowling Alley	S	1947	Baldwin Mfg. Co.	10	20	30
King Zor, The Dinosaur Game	B	1964	Ideal	20	35	55
Kismet	B	1971	Lakeside	3	5	8
KISS On Tour Game	B	1978	Aucoin	20	30	50
Knight Rider	B	1983	Parker Brothers	7	12	20
Knockout, Electronic Boxing Game	S	1950s	Northwestern Products	90	150	240
Know Your States	C	1955	Garrard Press	10	15	25
Kojack	B	1975	Milton Bradley	7	12	20
Kommisar	B	1960s	Selchow & Righter	10	15	25
Korg 70,000 BC	B	1974	Milton Bradley	10	15	25
Kreskin's ESP	B	1966	Milton Bradley	10	20	30
Krokay	S	1955	Transogram	30	50	75
Krull	B	1983	Parker Brothers	6	10	15
KSP Baseball	B	1983	Koch Sports Products	15	25	35
Kukla & Ollie	B	1962	Parker Brothers	30	45	70
Lancer	B	1968	Remco	70	120	190
Land of The Giants	B	1968	Ideal	60	100	160
Land of The Lost	B	1975	Milton Bradley	20	50	75
Land of The Lost Pinball	S	1975	Larami	10	20	30
Laramie	B	1960	Lowell	35	85	120
Las Vegas Baseball	B	1987	Samar Enterprises	8	13	20
Last Straw	B	1966	Schaper	5	10	15
Laugh-In's Squeeze Your Bippy Game	B	1968	Hasbro	60	100	160
Laverne & Shirley Game	B	1977	Parker Brothers	9	15	25
Leave It To Beaver Ambush Game	B	1959		20	45	65
Leave It To Beaver Money Maker	B	1959	Hasbro	20	45	65
Leave It To Beaver Rocket To The Moon	B	1959	Hasbro	20	45	65
Lee Vs. Meade: Battle of Gettysburg	B	1974	Gamut Of Games	15	25	50
Legend of Jessie James Game, The	B	1965	Milton Bradley	75	125	200
Lemans	B	1961	Avalon Hill	25	40	65
Let's Bowl a Game	B	1960	DMR	20	35	50
Let's Make A Deal Game	B	1970s	Ideal	10	15	25
Let's Play Basketball	C	1965	The D.M.R. Co.	10	15	25
Let's Play Golf "The Hawaiian Open"	B	1968	Burlu	25	40	60
Let's Play Safe Traffic Game	B	1960s	X-acto	25	50	90
Leverage	B	1982	Milton Bradley	4	6	10
LF Baseball	B	1980	Len Feder	12	20	30
Li'l Abner's Spoof Game	C	1950	Milton Bradley	40	65	100
Lie Detector Game	B	1961	Mattel	30	50	75
Lieutenant	B	1963	Transogram	85	140	225
Life, The Game of	B	1960	Milton Bradley	10	15	25
Line Drive	B	1953	Lord & Freber Inc.	60	100	150
Linebacker Football	B	1990	Linebacker Inc.	12	20	30
Linus the Lionhearted Uproarious Game	B	1965	Transogram	50	85	135
Lippy the Lion Game	B	1963	Transogram	25	45	70
Little Black Sambo	B	1952	Cadaco	80	150	250
Little Creepies Monster Game	B	1974	The Toy Factory	6	10	15
Little House On The Prairie	B	1978	Parker Brothers	15	35	50

GAMES

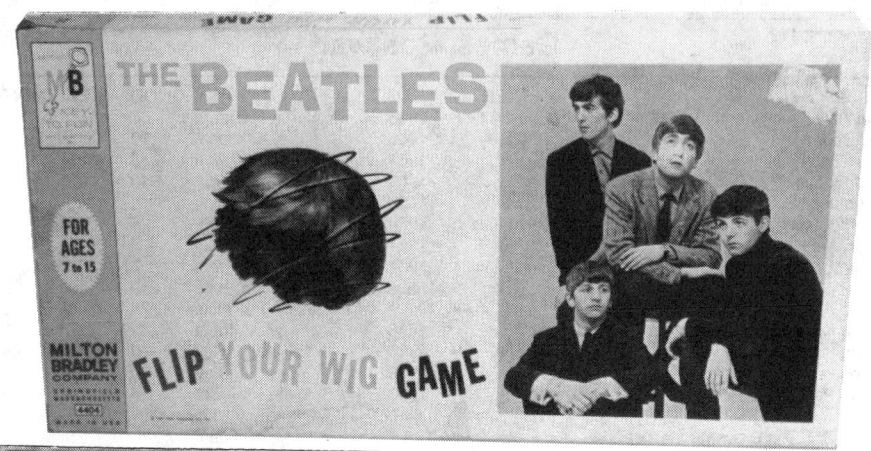

Top to bottom: The Beatles Flip Your Wig Game, 1964, Milton Bradley; M Squad, Bell; Huggin' The Rail, 1948, Selchow & Righter; The Game Dragnet, 1955, Transogram.

Games, Postwar

NAME	TYPE	YEAR	COMPANY	GOOD	EX	MINT
Little League Baseball Game	B	1950s	Standard Toycraft, Inc.	25	45	70
Little Orphan Annie	B	1981	Parker Brothers	10	20	30
Little Red Schoolhouse	B	1952		20	35	55
Lone Ranger And Tonto Spin Game, The	S	1967	Pressman	15	25	40
Long Shot	B	1962	Parker Brothers	45	75	125
Longball	B	1975	Ashburn Industries	30	50	75
Look ALI-Star Baseball Game	B	1960	Progressive Research	35	60	90
Looney Tunes Game	B	1968	Milton Bradley	50	75	100
Los Angeles Dodgers Baseball Game	B	1964	Ed-U-Cards	30	50	80
Lost In Space Game	B	1965	Milton Bradley	45	75	120
Love Boat World Cruise	B	1980		5	15	20
Loving Game, The	B	1987	R.J.E. Enterprises	4	6	10
Lucan, The Wolf Boy	B	1977	Milton Bradley	5	10	15
Lucky Break	B	1975	Gabriel	10	20	30
Lucky Strike	B	1972	International Toy Co.	5	10	15
Lucy Show Game, The	B	1962	Transogram	60	100	160
Lucy's Tea Party Game (Peanuts)	B	1971	Milton Bradley	20	35	55
Ludwig Von Drake Ball Toss Game	B	1960		6	10	15
Luftwaffe	B	1971	Avalon Hill	15	25	40
M Squad	B	1958	Bell Toys	35	75	125
M*A*S*H Game	B	1981	Milton Bradley	15	35	50
MacDonald's Game, The	B	1975	Milton Bradley	15	20	30
Mad Magazine Game, The	B	1979	Parker Brothers	15	25	40
Mad, What Me Worry?	B	1987	Milton Bradley	6	10	15
Magilla Gorilla	B	1964	Ideal	40	65	100
Magnetic Fish Pond	S	1948	Milton Bradley	15	25	65
Magnetic Flying Saucers	B	1950s	Pressman	21	35	55
Magnificent Race	B	1975	Parker Brothers	10	25	40
Main Street Baseball	B	1989	Main St. Toy Co.	20	35	55
Major League Baseball	B	1965	Cadaco	10	20	30
Major League Baseball Magnetic Dart Game	B	1958	Pressman	55	90	135
Man from U.N.C.L.E.	C	1965	Milton Bradley	15	35	50
Man From U.N.C.L.E. Illya Kuryakin Card Game	C	1966	Milton Bradley	20	30	40
Man From U.N.C.L.E. Pinball Game	S	1966		80	135	215
Man from U.N.C.L.E. Thrush Ray Gun Affair Game	B	1966	Ideal	50	85	135
Manage Your Own Team	B	1950s	Warren	55	90	135
Maniac	B	1979	Ideal	7	15	20
Marathon Game	S	1978	Sports Games Co.	10	20	35
Margie, The Game of Whoopie	B	1961	Milton Bradley	25	50	75
Marlin Perkin's Zoo Parade	B	1965	Cadaco-Ellis	35	60	95
Martin Luther King Jr.	B	1980	Cadaco	6	10	15
Marx-O-Matic All Star Basketball	S	1950s	Marx	150	250	400
Mary Hartman Mary Hartman	B	1976	Reiss Games	20	35	55
Mary Poppins Carousel Game	B	1964	Parker Brothers	20	40	65
Masquerade Party	B	1955	Bettye-B Co.	45	75	120
Mastermind	B	1970s	Invicta	5	10	15
Masterpiece, The Art Auction Game	B	1971	Parker Brothers	10	20	30
Match	C	1953	Garrard Press	10	15	25
Match Game 3rd Edition, The	B	1963	Milton Bradley	15	25	40
Matchbox Traffic Game	B	1960s		25	45	70
McDonald's Farm	B	1948	Selchow & Righter	20	45	65
McDonald's Game, The	B	1980s	Milton Bradley	8	15	20
McHale's Navy Game	B	1962	Transogram	30	50	80
Mechanical Shooting Gallery	S	1950s	Wyandotte	70	135	195
Meet The Presidents	B	1953	Selchow & Righter	15	25	40
Melvin The Moon Man	B	1960s	Remco	50	85	135
Men Into Space	B	1960	Milton Bradley	50	100	150

Games, Postwar

NAME	TYPE	YEAR	COMPANY	GOOD	EX	MINT
Merry Milkman, The	B	1955	Hasbro	60	120	200
Merv Griffin's Word For Word	B	1963	Mattel	5	10	15
Miami Vice: The Game	B	1984	Pepperlane	10	25	35
Mickey Mantle's Action Baseball	B	1960	Pressman	50	125	175
Mickey Mantle's Big League Baseball	B	1958	Gardner	125	250	325
Mickey Mouse	B	1950	Jacmar Mfg. Co.	35	55	90
Mickey Mouse	B	1976	Parker Brothers	6	10	15
Mickey Mouse Basketball	B	1950s	Gardner	55	90	135
Mickey Mouse Canasta Jr.	C	1950	Russell Mfg. Co.	20	45	75
Mickey Mouse Haunted House (bagatelle)	S	1950s		210	350	550
Mickey Mouse Jr. Royal Rummy	C	1970s	Whitman	4	7	15
Mickey Mouse Library of Games	C	1946	Russell Mfg. Co.	25	45	70
Mickey Mouse Lotto Game	B	1950s	Jaymar	10	15	25
Mickey Mouse Pop Up Game	B	1970s	Whitman	7	15	20
Mickey Mouse Slugaroo	B	1950s		20	30	50
Mid Life Crisis	B	1982	Gameworks Inc.	5	15	20
Mighty Hercules Game	B	1963	Hasbro	60	100	160
Mighty Heroes On the Scene Game	B	1960s	Transogram	20	45	85
Mighty Mouse	B	1978	Milton Bradley	15	20	30
Mighty Mouse Rescue Game	B	1960s	Harett-Gilmar	35	60	125
Mighty Mouse Target Game	S	1960s	Parks	30	60	90
Milton The Monster	B	1966	Milton Bradley	25	45	70
Mind Over Matter	B	1968	Transogram	10	15	25
Miss Popularity Game	B	1961	Milton Bradley	20	35	55
Mission Impossible	B	1967	Ideal	40	100	135
Mission Impossible	B	1975	Berwick	10	15	25
Mister Ed Game	B	1962	Parker Brothers	25	50	100
Mister Magoo Maddening Misadventures Game, The	B	1970	Transogram	45	75	120
Monday Morning Quarterback	B	1963	Zbinden	20	35	50
Monkees Game	B	1968	Transogram	30	75	105
Monkeys And Coconuts	B	1965	Schaper	10	15	20
Monopoly Library Edition	B	1960s	Parker Brothers	7	12	20
Monster Game	B	1977	Ideal	45	75	120
Monster Game, The	B	1965	Milton Bradley	15	25	40
Monster Lab	S	1964	Ideal	175	275	450
Monster Mansion	B	1981	Milton Bradley	7	12	20
Monster Squad	B	1977	Milton Bradley	35	65	95
Moon Shot	B	1960s	Cadaco	35	55	90
Moon Tag, Game of	B	1957	Parker Brothers	75	125	200
Mork And Mindy	B	1978	Milton Bradley	10	15	25
Mostly Ghostly	B	1975	Cadaco	15	20	30
Mouse Trap	S	1963	Ideal	25	45	70
Movie Moguls	B	1970	RGI	6	10	15
Mr. Bug Goes To Town	B	1955	Milton Bradley	30	75	105
Mr. Ed Board Game	B	1962	Parker Brothers	20	50	75
Mr. Machine Game	B	1961	Ideal	40	65	100
Mr. Magoo Visits The Zoo	B	1961	Lowell	25	45	70
Mr. Ree	B	1957	Selchow & Righter	25	45	70
Mr. T Game	C	1983	Milton Bradley	2	4	6
Mt. Everest	B	1955	Sam'l Gabriel Sons & Co.	15	25	40
Mug Shots	B	1975	Cadaco	7	12	20
Munsters Game	C	1966	Milton Bradley	25	45	70
Munsters Game	B	1966	Milton Bradley	65	95	150
Muppet Show	B	1977	Parker Brothers	10	15	25
Murder She Wrote	B	1985	Warren	4	6	10
Mushmouse & Punkin Puss	B	1964	Ideal	45	75	120
MVP Baseball, The Sports Card Game	B	1989	Ideal	8	13	20
My Fair Lady	B	1960s	Standard Toycraft, Inc.	15	20	30

Games, Postwar

NAME	TYPE	YEAR	COMPANY	GOOD	EX	MINT
My Favorite Martian	B	1963	Transogram	50	90	125
My First (Walt Disney Character) Game	B	1963	Gabriel	20	35	60
Mystery Checkers	B	1950s	Creative Designs	15	20	30
Mystery Date	B	1966	Milton Bradley	40	60	100
Mystery Mansion	B	1984	Milton Bradley	7	12	20
Mystic Skull The Game of Voodoo	B	1965	Ideal	30	50	80
Mystic Wheel of Knowledge	B	1950s	Novel Toy	15	25	40
Name That Tune	B	1959	Milton Bradley	20	30	50
Napoleon Solo Man From U.N.C.L.E. Game	B	1965	Ideal	25	45	70
Nascar Daytona 500	B	1990	Milton Bradley	8	13	20
National Football League Quarterback, Official	B	1965	Toy Craft	50	80	125
National Inquirer	B	1991	Tyco	10	15	20
National Lampoon's Sellout	B	1970s	Cardinal Industries	4	7	12
National Pro Football Hall of Fame Game	B	1965	Cadaco	15	25	40
National Pro Hockey	B	1985	Sports Action	15	25	40
National Velvet Game	B	1950s	Transogram	20	40	65
NBA Basketball Game, Official	B	1970s	Gerney Games	45	70	110
NBC Game of the Week	B	1969	Hasbro	25	40	65
NBC Peacock	B	1966	Selchow & Righter	12	20	30
NBC Pro Playoff	B	1969	Hasbro	25	40	65
NBC TV News	B	1960	Dadan	15	25	40
Nebula	B	1976	Nebula Inc.	4	6	10
Neck & Neck	B	1981	Yaquito	8	13	20
Negamco Basketball	B	1975	Nemadji Game Co.	10	16	25
New Avengers Shooting Game	B	1976	Denys Fisher	165	275	440
New Frontier	B	1962	Colorful Products, Inc.	30	50	80
New York World's Fair	B	1964	Milton Bradley	20	50	75
New York World's Fair Children's Game	C	1964	Ed-U-Cards	15	25	40
Newlywed Game 1st Edition	B	1967	Hasbro	10	14	17
NFL All-Pro Football Game	S	1967	Ideal	15	25	40
NFL Armchair Quarterback	B	1986	Trade Wind Inc.	8	13	20
NFL Franchise	B	1982	Rohrwood	10	16	25
NFL Game Plan	B	1980	Tudor	6	10	15
NFL Quarterback	B	1977	Tudor	14	23	35
NHL All-Pro Hockey	B	1969	Ideal	10	16	25
Nightmare On Elm Street	B	1989	Cardinal Industries	15	30	45
Nixon Ring Toss	S	1970s		20	30	50
No Respect, The Rodney Dangerfield Game	B	1985	Milton Bradley	15	25	40
Nok-Hockey	B	1947	Carrom Industries	20	35	55
Noma Party Quiz	B	1947	Noma Electric Corp.	20	35	50
Northwest Passage	B	1969		12	20	30
Nuclear War	C	1965	Douglas Malewicki	20	30	50
Number Please TV Quiz	B	1961	Parker Brothers	15	25	40
Numble	B	1968	Selchow & Righter	12	20	30
Nutty Mads (bagatelle)	S	1963	Marx	25	45	85
Nutty Mads Target Game	S	1960s	Marx	35	70	125
NY Mets Baseball Card Game, Official	C	1961	Ed-U-Cards	40	65	100
Obsession	B	1978	Mego	5	10	15
Off To See The Wizard	B	1968		10	15	25
Official NFL Football Game	S	1968	Ideal	20	30	50
Oh Magoo Game	B	1960s	Warren	15	30	45
Oh, Nuts! Game	B	1968	Ideal	10	15	25
Oh-Wah-Ree	B	1966	Avalon Hill	4	7	12
On Guard	B	1967	Parker Brothers	6	10	15
Operation	B	1965	Milton Bradley	10	15	25
Orbit	B	1959	Parker Brothers	25	55	100
Original Home Jai-Alai Game, The	B	1984	Design Origin	15	25	35

Games, Postwar

NAME	TYPE	YEAR	COMPANY	GOOD	EX	MINT
Oscar Robertson's Pro Basketball Strategy	B	1964	Research Games	70	115	175
Our Gang Bingo	B	1958		50	85	135
Outdoor Survival	B	1972	Avalon Hill	6	10	15
Outer Limits	B	1964	Milton Bradley	110	180	275
Outwit	B	1978	Parker Brothers	5	10	15
Ozark Ikes	B	1956	Stephen Stesinger	55	90	135
P.T. Boat 109 Game	B	1963	Ideal	30	50	90
Pac-Man	B	1980	Milton Bradley	5	15	20
Panic Button	B	1978	Mego	7	15	20
Panzer Blitz	B	1970	Avalon Hill	12	20	30
Panzer Leader	B	1974	Avalon Hill	10	15	25
Par '73	B	1961	Big Top Games	15	25	40
Par Golf	B	1950s	National Games, Inc.	40	100	145
Par-A-Shoot Game	S	1947	Baldwin Mfg. Co.	15	25	40
Parcheesi (Gold Seal Edition)	B	1964	Selchow & Righter	7	12	20
Pari Horse Race Card Game	B	1959	Pari Sales Co.	20	35	50
Paris Metro Basketball	B	1981	Infinity Games	10	16	25
Park & Shop	B	1952	Traffic Game Co.	40	70	110
Park and Shop Game	B	1960	Milton Bradley	25	60	85
Parker Bros. Baseball Game	B	1955	Parker Brothers	45	70	110
Parlor Baseball	B	1980s		1275	2150	3300
Partridge Family	B	1974	Milton Bradley	7	15	20
Password	B	1963	Milton Bradley	10	16	25
Password 25th Anniversary Edition	B	1986	Milton Bradley	3	5	8
Pathfinder	B	1977	Milton Bradley	5	15	20
Patty Duke Game	B	1963	Milton Bradley	25	40	75
Paul Brown's Football Game	B	1947	Trikilis	110	180	275
Paydirt	B	1979	Avalon Hill	10	18	30
Paydirt!	B	1973	Sports Illustrated	10	16	25
Peanuts The Game of Charlie Brown And His Pals	B	1959	Selchow & Righter	20	30	50
Pebbles Flintstone Game	B	1962	Transogram	20	35	55
Pee Wee Reese Marble Game	B	1956	Pee Wee Enterprises	175	295	450
Pennant Chasers Baseball Game	B	1946	Craig Hopkins	25	45	70
Pennant Drive	B	1980	Accu-Stat Game Co.	8	13	20
People Trivia Game	B	1984	Parker Brothers	7	11	20
Perquackey	B	1970	Lakeside	3	5	8
Perry Mason Case of The Missing Suspect Game	B	1959	Transogram	20	30	50
Personalysis	B	1957	Lowell	15	25	40
Peter Gunn Dectective Game	B	1960		20	40	60
Peter Pan	B	1953	Transogram	40	65	105
Peter Potamus Game	B	1964	Ideal	30	50	80
Petticoat Junction	B	1963	Standard Toycraft, Inc.	35	55	90
Phalanx	B	1964	Whitman	35	55	90
Phantom Game, The	B	1965	Transogram	95	160	255
Phantom's Complete 3 Game Set, The	B	1955	Built-Rite	65	100	175
Phil Silvers' You'll Never Get Rich Game	B	1955	Gardner	30	75	105
Philip Marlowe	B	1960	Transogram	20	50	75
Photo-Electric Baseball	B	1951	Cadaco-Ellis	55	90	135
Pigskin Vegas	B	1980	Inkari/US	0	10	15
Pinbo Sport-o-Rama	B	1950s		35	60	90
Pink Panther	B	1981	Cadaco	4	7	12
Pink Panther Game	B	1977	Warren	20	40	60
Pinky Lee and the Runaway Frankfurters	B	1950s		30	75	105
Pinocchio	B	1977	Parker Brothers	5	8	15
Pinocchio Board Game, Disney's	B	1960	Parker Brothers	10	15	25
Pinocchio, The New Adventures of	B	1961	Lowell	25	45	70
Pirate And Traveller	B	1953	Milton Bradley	10	25	35

Games, Postwar

NAME	TYPE	YEAR	COMPANY	GOOD	EX	MINT
Planet of the Apes	B	1970s	Milton Bradley	25	40	70
Play Basketball with Bob Cousy	B	1950s	National Games, Inc.	115	195	295
Playoff Football	B	1970s	Crestline Mfg. Co.	20	35	50
Pocket Size Bowling Card Game	B	1950s	Warren-Built-Rite	15	25	40
Pocket Whoozit	B	1985	Trivia, Inc.	4	7	10
Pole Position	B	1983	Parker Brothers	8	13	20
Pony Polo	S	1960s	Remco	10	20	35
Poosh-em-up Slugger Bagatelle	B	1946	Northwestern Products	30	75	105
Popeye, Adventures of	B	1957	Transogram	50	75	125
Power Play Hockey	B	1970	Romac Ind.	30	50	75
Pro Bowl Live Action Football	S	1960s	Marx	35	65	95
Pro Draft	B	1974	Parker Brothers	15	25	40
Pro Football	B	1980s	Strat-O-Matic	6	10	15
Pro Foto-Football	B	1977	Cadaco	25	40	60
Pro Franchise Football	B	1987	Rohrwood	10	16	25
Pro Golf	B	1982	Avalon Hill	7	11	17
Pro Quarterback	B	1964	Tod Lansing	20	40	60
Pro-Baseball Card Game	C	1980s	Just Games	6	10	15
Pug-i-Lo	B	1960	Pug-i-Lo Games	55	90	135
Pursue the Pennant	B	1984	Pursue the Pennant	25	40	60
Quarterback Football Game	B	1969	Transogram	25	40	65
Race-O-Rama	B	1960	Warren-Built-Rite	35	60	90
Raceway	B	1950s	B & B Toy	30	50	75
Ralph Edwards This Is Your Life	B	1950s		6	10	15
Rat Patrol Game	B	1966	Transogram	45	75	120
Rat Patrol Spin Game	S	1967	Pressman	40	65	105
Rawhide	B	1959	Lowe	50	85	135
Raymar of The Jungle	B	1952	Dexter Wayne	40	100	135
Razzle Dazzle Football Game	B	1954	Texantics Unlimited	50	80	125
React-Or	B	1979		15	25	40
Ready-Clown 3-Ring Circus Game	B	1952	Parker Brothers	25	40	65
Real Action Baseball Game	B	1966	Real-Action Games	20	35	50
Real Baseball Card Game	B	1990	National Baseball	110	180	275
Real Life Basketball	B	1974	Gamecraft	10	16	25
Realistic Football	B	1976	Match Play	15	25	35
Rebel, The	B	1961	Ideal	50	85	135
Red Barber's Big League Baseball Game	B	1950s	G & R Anthony	350	575	900
Reddy Clown 3-Ring Circus Game	B	1952	Parker Brothers	15	25	55
Reese's Pieces Game	B	1983	Ideal	5	8	15
Regatta	B	1946		40	70	110
Replay Series Baseball	B	1983	Bond Sports Ent.	6	10	15
Restless Gun	B	1950s	Milton Bradley	15	25	40
Return To Oz Game	B	1985	Western Publishing	10	15	25
Rich Uncle	B	1946	Parker Brothers	35	50	80
Rich Uncle The Stock Market Game	B	1955	Parker Brothers	30	50	80
Richie Rich	B	1982	Milton Bradley	3	5	10
Ricochet Rabbit Game	B	1965	Ideal	45	75	120
Rifleman Game	B	1959	Milton Bradley	40	100	135
Rin-Tin-Tin Game	B	1950s	Transogram	20	50	65
Ripley's Believe It Or Not	B	1979	Whitman	6	10	15
Risk	B	1959	Parker Brothers	25	50	75
Riverboat Game	B	1950s	Parker Brothers (Disney)	20	35	55
Road Runner Game	B	1968	Milton Bradley	20	40	60
Road Runner Pop Up Game	B	1982	Whitman	20	30	50
Robert Schuller's Possibility Thinkers Game	B	1977	Selchow & Righter	3	5	10
Robin Hood	B	1955	Harett-Gilmar	30	50	80
Robin Hood Game	B	1970s	Parker Brothers	7	15	20
Robin Hood, Adventures of	B	1956	Bettye-B Co.	45	75	120

Games, Postwar

NAME	TYPE	YEAR	COMPANY	GOOD	EX	MINT
Robin Roberts Sports Club Baseball Game	B	1960	Dexter Wayne	100	165	250
Robot Sam The Answer Man	B	1950	Jacmar Mfg. Co.	25	45	70
Rock Trivia	B	1984	Pressman	5	10	15
Rocket Patrol Magnetic Target Game	S	1950s	American Toy Products Co.	45	75	120
Rocket Race	B	1958	Stone Craft	100	165	250
Rocket Race To Saturn	B	1950s	Lido	15	20	35
Rocket Sock-It Rifle and Target Game	S	1960s	Kenner	20	35	75
Roger Maris' Action Baseball	B	1962	Play-Rite	50	125	175
Rol-A-Lite	B	1947	Durable Toy & Novelty	45	75	120
Roll And Score Poker	B	1977	Lowe	4	7	12
Roll-A-Par	B	1964	Lowe	15	25	40
Route 66 Game	B	1960	Transogram	75	125	200
Roy Rogers Game	B	1950s		20	35	55
Russian Campaign, The	B	1976	Avalon Hill	3	5	10
S.O.S.	B	1947	Durable Toy & Novelty	40	70	110
S.W.A.T. Game	B	1970s	Milton Bradley	10	15	25
Saddle Racing Game	B	1974	APBA	35	60	90
Sail Away	B	1962	Howard Mullen	35	60	90
Salvo	B	1961	Ideal	15	25	40
Samsonite Basketball	B	1969	Samsonite	15	25	35
Samsonite Football	B	1969	Samsonite	20	35	50
Sandlot Slugger	B	1960s		35	60	90
Saratoga: 1777	S	1974	Gamut Of Games	30	50	80
Scavenger Hunt	B	1983	Milton Bradley	3	5	10
Scooby-Doo And Scrappy Doo	B	1983	Milton Bradley	20	30	50
Scoop	B	1956	Parker Brothers	25	60	85
Scott's Baseball Card Game	C	1989	Scott's Baseball Cards	12	20	30
Scrabble	B	1953	Selchow & Righter	5	12	20
Screwball The Mad Mad Mad Game	B	1960	Transogram	30	75	105
Scrimmage	B	1973	SPI	15	25	35
Scruples	B	1986	Milton Bradley	5	10	15
Sealab 2020 Game	B	1973	Milton Bradley	6	10	15
Secret Agent Man	B	1966	Milton Bradley	25	40	65
Secret of NIMH	B	1982	Whitman	6	10	15
Seduction	B	1966	Createk	20	30	45
Sergeant Preston Game	B	1950s	Milton Bradley	15	35	50
Set Point	B	1971	XV Productions	25	40	65
Seven Seas	B	1960	Cadaco-Ellis	30	50	80
Seven Up	B	1960s	Transogram	7	12	20
Shazam, Captian Marvel's Own Game	B	1950s	Reed & Associates	30	60	95
Shotgun Slade	B	1960	Milton Bradley	20	45	65
SI: The Sporting Word Game	B	1961	Time Inc.	10	16	25
Silly Carnival	B	1969	Whitman	7	12	20
Silly Safari	B	1966	Topper	40	65	105
Simpsons Mystery of Life, The	B	1990	Cardinal Industries	5	7	10
Sinbad	B	1978	Cadaco	20	30	50
Sinking of The Titanic, The	B	1976	Ideal	25	45	70
Sir Lancelot, Adventures of	B	1975	Lisbeth Whiting	40	70	110
Six Million Dollar Man	B	1975	Parker Brothers	5	10	15
Skatterbug, Game of	B	1951	Parker Brothers	30	50	80
Skeeter	C	1950s	Arrco Playing Card Co.	6	10	15
Ski Gammon	B	1962	American Publishing Corp.	10	16	25
Skins Golf Game, Official	B	1985	O'Connor Hall	12	20	30
Skip-A-Cross	B	1953	Cadaco	10	15	25
Skipper Race Sailing Game	B	1949	Cadaco-Ellis	55	90	135
Skirmish	B	1975	Milton Bradley	20	35	55
Skully	B	1961	Ideal	3	5	10
Skunk	B	1950s	Schaper	7	12	20

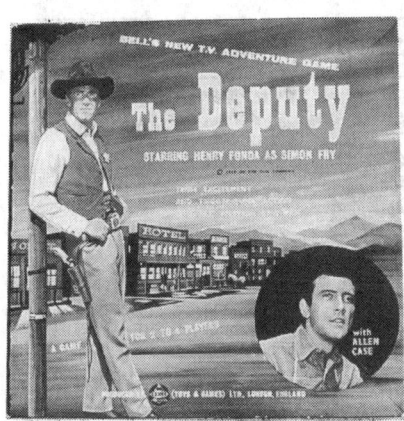

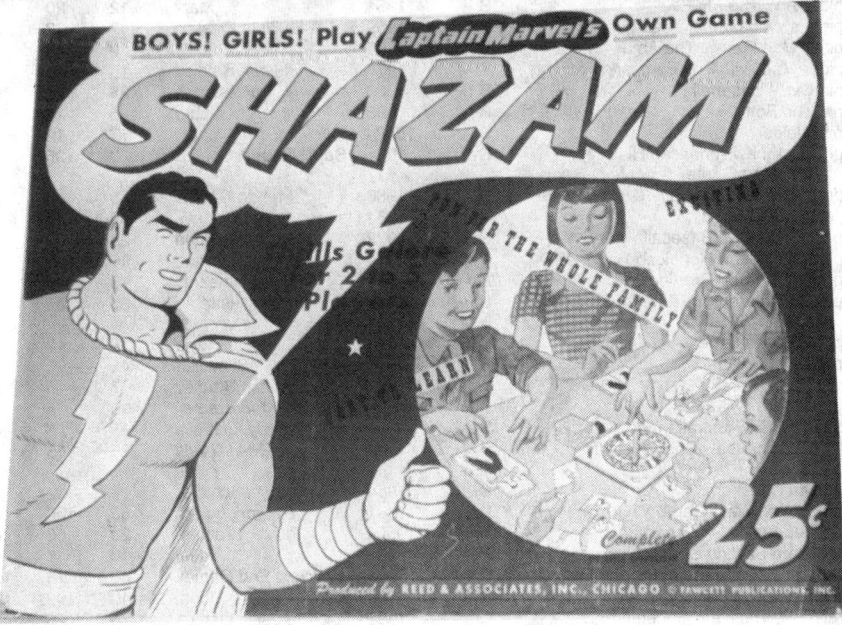

Clockwise from upper left: The Deputy, 1960, Milton Bradley; Donald Duck's Party Game For Young Folks, 1950s, Parker Brothers; The Jetsons Fun Pad Game, 1962, Milton Bradley; The Jetsons Out Of This World Game, 1963, Transogram; Captain Marvel's Shazam. 1950s, Reed & Associates.

Games, Postwar

NAME	TYPE	YEAR	COMPANY	GOOD	EX	MINT
Sky's The Limit, The	B	1955	Kohner	15	25	40
Sla-lom Ski Race Game	B	1957	Cadaco-Ellis	55	90	135
Slapshot	B	1982	Avalon Hill	6	10	15
Smog	B	1970	Urban Systems	9	15	25
Smokey: The Forest Fire Prevention Bear	B	1961	Ideal	40	65	105
Smurf Game	B	1984	Milton Bradley	4	7	12
Snagglepuss Fun At The Picnic Game	B	1961	Transogram	40	60	100
Snakes & Ladders	B	1974	Summmerville Ind. Ltd./Canada	4	7	10
Snakes In The Grass	B	1960s	Kohner	10	15	25
Snappet Catch Game with Harmon Killebrew	B	1960	Killebrew Inc.	55	90	135
Snoopy & The Red Baron	B	1970	Milton Bradley	15	35	50
Snoopy Game (Peanuts)	B	1960	Selchow & Righter	25	45	70
Snow White Boardgame	B	1970s		6	10	15
Snuffy Smith Game	B	1970s	Milton Bradley	10	30	45
Sod Buster	B	1980	Santee	10	16	25
Solarquest	B	1986	Western Publishing	6	10	15
Solid Gold Music Trivia	B	1984	Ideal	6	10	15
Solitaire (Lucille Ball)	B	1973	Milton Bradley	5	15	20
Soupy Sales Sez Go-Go-Go Game	B	1960s	Milton Bradley	55	95	150
Space 1999 Game	B	1975	Milton Bradley	10	15	25
Space Age Game	B	1953	Parker Brothers	45	75	120
Space Angel Game	B	1966	Transogram	20	35	55
Space Shuttle 101	B	1978	Media-Ungame	15	25	40
Speed Circuit	S	1971	3M	15	30	45
Speedorama	B	1950s	Jacmar Mfg. Co.	30	50	80
Speedway, Big Bopper Game	B	1961	Ideal	35	60	90
Spider And The Fly	B	1981	Marx	12	20	30
Spider's Web Game, The	B	1969	Multiple Plastics	7	12	20
Spider-Man Game, The Amazing	B	1967	Milton Bradley	20	45	85
Spider-Man With The Fantastic Four	B	1977	Milton Bradley	10	15	25
Spin Cycle Baseball	B	1965	Pressman	25	40	65
Spin The Bottle	B	1968	Hasbro	6	10	15
Spin Welder	B	1960s	Mattel	7	12	20
Spiro T. Agnew American History Challenge Game	B	1971	Sam'l Gabriel Sons & Co.	20	35	55
Sporting News Baseball	B	1986	Mundo Games	8	13	20
Sports Arena No. 1	B	1954	Rennoc Games & Toys	35	60	90
Sports Illustrated Baseball	B	1972	Sports Illustrated	25	40	65
Sports Illustrated Handicap Golf	B	1971	Sports Illustrated	15	25	35
Sports Illustrated Pro Football	B	1970	Time Inc.	15	25	40
Sports Illustrated, All Time All Star Basketball	B	1973	Sports Illustrated	20	35	50
Sports Trivia Game	B	1984	Hoyle	6	10	15
Sports Yesteryear	B	1977	Skor-Mor	15	25	35
Spot Cash	B	1959	Milton Bradley	7	15	20
Spy Vs Spy	B	1986	Milton Bradley	10	15	25
Squatter: The Australian Wool Game	B	1960s	John Sands, Australia	15	25	45
St. Louis Cardinals Baseball Card Game	B	1964	Ed-U-Cards	35	55	85
Stadium Checkers	B	1954	Schaper	7	12	20
Stagecoach West Game	B	1961	Transogram	60	100	160
Star Reporter	B	1950s	Parker Brothers	20	50	65
Star Team Battling Spaceships	B	1968	Ideal	10	15	25
Star Trek Adventure Game	B	1985	West End Games	45	75	120
Star Trek Game	B	1960s	Ideal	40	100	135
Star Trek: The Next Generation	B	1993	Classic Games	15	25	50
Star Wars Adventures of R2D2 Game	B	1977	Kenner	12	20	30
Star Wars Battle At Sarlacc's Pit	B	1983	Parker Brothers	10	15	25
Star Wars ROTJ Ewoks Save The Trees	B	1984	Parker Brothers	10	15	25

Games, Postwar

NAME	TYPE	YEAR	COMPANY	GOOD	EX	MINT
Star Wars Wicket the Ewok	B	1983	Parker Brothers	7	12	20
Star Wars X-Wing Aces Target Game	B	1978		20	30	50
State Capitals, Game of	B	1952	Parker Brothers	12	20	30
States, Game of The	B	1975	Milton Bradley	4	6	10
Statis Pro Football	B	1970s	Statis-Pro	25	40	65
Stay Alive	B	1971	Milton Bradley	4	6	10
Steps of Toyland	B	1954	Parker Brothers	20	35	55
Steve Canyon	B	1959	Lowell	35	65	100
Steve Scott Space Scout Game	B	1952	Transogram	30	75	105
Sting, The	B	1976	Ideal	25	45	70
Stock Car Race	B	1950s	Gardner	65	110	165
Stock Car Racing Game	B	1956	Whitman	30	50	75
Stock Car Racing Game (w/Petty/Yarborough)	B	1981	Ribbit Toy Co.	17	30	45
Stock Car Speedway, Game Of	B	1965	Johnstone	55	90	135
Stock Market	B	1970	Avalon Hill	7	12	20
Stock Market Game	B	1955	Sam'l Gabriel Sons & Co.	20	30	50
Stoney Burk	B	1963	Transogram	25	60	85
Straight Arrow	B	1950	Selchow & Righter	25	45	70
Straightway	B	1961	Selchow & Righter	45	70	110
Strat-O-Matic Baseball	B	1961	Strat-O-Matic	100	165	250
Strat-O-Matic College Football	B	1976	Strat-O-Matic	25	40	60
Strat-O-Matic Hockey	B	1978	Strat-O-Matic	35	55	85
Strat-O-Matic Sports "Know-How"	B	1984	Strat-O-Matic	6	10	15
Strategic Command	B	1950s	Transogram	25	45	70
Stratego	B	1961	Milton Bradley	15	25	40
Strategy Manager Baseball	B	1967	McGuffin-Ramsey	20	35	50
Strategy Poker Fine Edition	C	1967	Milton Bradley	5	7	12
Stretch Call	B	1986	Sevedeo A. Vigil	12	20	30
Strike Three	B	1948	Tone Products Corp.	275	475	725
Suffolk Downs Racing Game	B	1947	Corey Game Co.	60	100	150
Sugar Bowl	B	1950s	Transogram	20	45	65
Summit	B	1961	Milton Bradley	20	45	65
Sunken Treasure	B	1948	Parker Brothers	15	25	45
Sunken Treasure	B	1976	Milton Bradley	7	12	20
Super Coach TV Football	B	1974	Coleco	25	40	65
Super Powers	B	1984	Parker Brothers	15	25	40
Super Spy	B	1971	Milton Bradley	15	25	40
Superboy Game	B	1960s	Hasbro	45	75	135
Supercar Road Race	B	1962	Standard Toycraft, Inc.	40	65	105
Supercar To The Rescue Game	B	1962	Milton Bradley	35	60	95
Superheroes Card Game	C	1978	Milton Bradley	10	20	30
Superman & Superboy	B	1967	Milton Bradley	40	65	105
Superman Game	B	1965	Hasbro	45	75	120
Superman Game	B	1966	Merry Manufacturing Co.	35	55	90
Superman Game	C	1966	Whitman	30	50	80
Superman II	B	1981	Milton Bradley	10	20	35
Superman III	B	1982	Parker Brothers	20	30	50
Superman Spin Game	S	1967	Pressman	40	65	105
Superman, Adventures of	B	1950s	Milton Bradley	25	40	65
Superman, Calling	B	1954	Transogram	50	125	175
Superstar Baseball	B	1966	Sports Illustrated	30	50	75
Superstar Pro Wrestling Game	B	1984	Super Star Game Co.	8	13	20
Superstar TV Sports	B	1980	ARC	6	10	15
Superstition	B	1977	Milton Bradley	10	15	25
Sure Shot Hockey	B	1970	Ideal	15	25	40
Surfside 6	B	1961	Lowe	60	120	200
Surprise Package	B	1961	Ideal	12	20	30
Suspense	S	1950s	Northwestern Products	15	20	30

Games, Postwar

NAME	TYPE	YEAR	COMPANY	GOOD	EX	MINT
Swahili Game	B	1968	Milton Bradley	15	20	30
Swat Baseball	B	1948	Milton Bradley	20	35	50
Swayze	B	1954	Milton Bradley	20	35	55
Swish	B	1948	Jim Hawkers Games Mfg.	45	75	115
Swoop	B	1969	Whitman	7	12	20
Sword In The Stone Game	B	1960s	Parker Brothers	10	15	25
Syllable	C	1948	Garrard Press	10	15	25
T.V. Bingo	B	1970	Selchow & Richter	3	5	10
Tabit	B	1954	John Norton Co.	25	35	75
Tactics II	B	1984	Avalon Hill	6	10	15
Taffy's Party Game	B	1960s	Transogram	10	15	25
Tales of Wells Fargo	B	1959	Milton Bradley	40	65	105
Talking Football	B	1971	Mattel	12	20	30
Talking Monday Night Football	B	1977	Mattel	8	13	20
Tally Ho!	B	1950s	Whitman	25	40	65
Tank Battle	B	1975	Milton Bradley	10	20	35
Tantalizer	B	1958	Northern Signal Co.	25	50	85
Tarzan	B	1984	Milton Bradley	5	10	15
Tarzan To The Rescue	B	1976	Milton Bradley	10	15	25
Tee Off by Sam Snead	B	1973	Glenn Industries	115	195	295
Teed Off!	B	1966	Milton Bradley	15	25	40
Teeko	B	1948	John Scarne Games, Inc.	20	30	50
Television	B	1953	National Novelty	35	75	100
Tell It To The Judge	B	1959	Parker Brothers	50	75	100
Temple of Fu Manchu Game, The	B	1967	Pressman	20	30	50
Ten-Four, Good Buddy	B	1976	Parker Brothers	5	7	12
Tennessee Tuxedo	B	1963	Transogram	75	125	200
Tennis	B	1975	Parker Brothers	10	16	25
Tension	B	1970	Kohnes	7	12	20
Terrytoons Hide N' Seek Game	B	1960	Transogram	20	35	55
Texas Millionaire	B	1955	Texantics	45	75	120
The Egg And I	B	1947	Capex Co. Inc.	30	50	80
The Kennedy's	B	1962	Transogram	40	100	135
The My Fair Lady Game	B	1962	Standard Toycraft, Inc.	25	40	65
The Space Shuttle	B	1981	The Ungame Company	15	25	40
They're at the Post	B	1976	MAAS Marketing	25	40	65
Thing Ding Robot Game	B	1961	Schaper	115	195	310
Thinking Man's Football	B	1969	3M	15	20	30
Thinking Man's Golf	B	1966	3M	20	35	50
Third Reich	B	1974	Avalon Hill	7	12	20
This Is Your Life	B	1954	Lowell	15	35	50
Three Musketeers	B	1958	Milton Bradley	35	55	90
Thunder Road	B	1986	Milton Bradley	10	16	25
Thunderbirds Game	B	1965	Waddington (England)	40	70	115
Tic-Tac Dough	B	1957	Transogram	20	30	45
Tickle Bee	S	1956	Schaper	15	20	35
Tiddle Flip Baseball	B	1949	Modern Craft Ind.	20	35	50
Time Bomb	S	1965	Milton Bradley	35	50	75
Time Machine	B	1961	American Toy Mfg.	70	115	184
Time Tunnel Game, The	B	1966	Ideal	90	150	240
Time Tunnel Spin Game, The	S	1967	Pressman	60	100	160
Tiny-Tim Game of Beautiful Things, The	B	1970	Parker Brothers	25	65	90
Tipp Kick	S	1970s	Top Set	15	25	40
Tom & Jerry	B	1977	Milton Bradley	10	15	25
Tom & Jerry Adventure In Blunderland	B	1965	Transogram	25	45	70
Tom Seaver's Action Baseball	B	1970	Pressman	50	125	175
Tomorrowland Rocket To Moon	B	1956		50	80	130
Tootsie Roll Train Game	B	1969	Hasbro	20	30	50

Games, Postwar

NAME	TYPE	YEAR	COMPANY	GOOD	EX	MINT
Top Cat Game	B	1962	Transogram	25	45	70
Top Cop	B	1961	Cadaco-Ellis	40	65	105
Top Pro Basketball Quiz Game	B	1970	Ed-U-Cards	15	25	40
Top Pro Football Quiz Game	B	1970	Ed-U-Cards	15	25	40
Top Ten College Basketball	B	1980	Top Ten Game Co.	10	16	25
Touche Turtle Game	B	1964	Ideal	75	100	175
Tournament Labyrinth	S	1980s	Pressman	10	15	25
Town & Country Traffic Game	B	1950s	Ranger Steel Products Corp.	75	125	200
Track Meet	B	1972	Sports Illustrated	15	25	35
Trade Winds: The Caribbean Sea Pirate Treasure Hunt	B	1959	Parker Brothers	20	30	50
Traffic Game	B	1968	Matchbox	35	55	90
Traffic Jam	B	1954	Harett-Gilmar	40	60	80
Trail Drive	C	1950s	Arrco Playing Card Co.	10	15	25
Trapped	B	1956	Bettye-B Co.	50	75	100
Traps, The Game of	B	1950s	Traps Mfg.	75	125	200
Travel America	B	1950	Jacmar Mfg. Co.	15	25	40
Travel-Lite	B	1946	Saxon Toy Corp.	45	75	120
Treasure Island	B	1954	Harett-Gilmar	30	50	80
Tri Ominoes, Deluxe	B	1978	Pressman	3	5	10
Triple Play	B	1978	Milton Bradley	5	10	15
Triple Yahtzee	B	1972	Lowe	4	6	10
Tru-Action Electric Baseball Game	B	1955	Tudor	25	50	80
Tru-Action Electric Basketball	S	1965	Tudor	25	50	75
Tru-Action Electric Harness Race Game	S	1950s	Tudor	15	40	65
Trust Me	B	1981	Parker Brothers	5	8	13
Try-It Maze Puzzle Game	S	1965	Milton Bradley	7	12	20
TSG I: Pro Football	B	1971	TSG	35	55	85
TV Guide Game	B	1984		8	13	20
Twiggy, Game of	B	1967	Milton Bradley	40	60	85
Twilight Zone Game	B	1960s	Ideal	50	100	150
Twinkles Trip to the Star Factory	B	1960	Milton Bradley	45	75	120
Twister	B	1966	Milton Bradley	10	15	25
Two For The Money	B	1950s	Lowell	10	25	40
U.N. Game of Flags	B	1961		12	20	30
U.S. Air Force, Game of	B	1950s	Transogram	25	45	70
Ultimate Golf	B	1985	Ultimate Golf Inc.	20	35	55
Uncle Wiggly	B	1979	Parker Brothers	4	7	12
Undercover: The Game of Secret Agents	B	1960	Cadaco-Ellis	20	35	55
Underdog	B	1964	Milton Bradley	90	150	225
Underdog Save Sweet Polly	B	1972	Whitman	25	45	70
Undersea World of Jacques Cousteau	B	1968	Parker Brothers	30	45	60
Ungame, The	B	1975	The Ungame Company	4	6	10
Untouchables, The	S	1950s	Marx	95	160	250
Uranium Rush	B	1955	Gardner	50	100	150
USAC Auto Racing	B	1980	Avalon Hill	45	70	110
Vallco Pro Drag Racing Game	B	1975	Zyla	15	25	35
VCR Basketball Game	B	1987		6	10	15
VCR Quarterback Game, The	B	1986	Interactive VCR Games	8	13	20
Veda, The Magic Answer Man	B	1960s	Pressman	20	30	45
Verne Gagne World Champion Wrestling	B	1950	Gardner	70	115	175
Video Village	B	1960	Milton Bradley	25	50	75
Vietnam	B	1984	Victory Games	10	15	25
Vince Lombardi's Game	B	1970	Research Games	40	75	120
Virginian, The	B	1962	Transogram	50	90	150
Visit To Walt Disney World Game	B	1970	Milton Bradley	15	20	35

Games, Postwar

NAME	TYPE	YEAR	COMPANY	GOOD	EX	MINT
Voice of The Mummy	B	1960s	Milton Bradley	20	35	55
Voodoo Doll Game	B	1967	Schaper	30	45	65
Wackiest Ship In The Army	B	1964	Ideal	40	65	105
Wacky Races Game	B	1970s	Milton Bradley	15	25	40
Wagon Train	B	1960	Milton Bradley	25	40	65
Wahoo	B	1947	Zondine Game Co.	15	20	30
Wally Gator Game	B	1963	Transogram	50	85	130
Walt Disney 101 Dalmatians	B	1960	Whitman	20	35	55
Walt Disney's 20,000 Leagues Under The Sea	B	1954	Jacmar Mfg. Co.	45	75	120
Walt Disney's Jungle Book	B	1967	Parker Brothers	15	25	45
Walt Disney's Official Frontier Land	B	1950s	Parker Brothers	25	45	70
Walt Disney's Sleeping Beauty Game	B	1958	Whitman	30	50	80
Walt Disney's Swamp Fox Game	B	1960	Parker Brothers	30	50	80
Waltons	B	1974	Milton Bradley	15	30	45
Wanted Dead or Alive	B	1959	Lowell	50	75	125
War At Sea	B	1976	Avalon Hill	10	20	30
Watergate Scandal, The	B	1973	American Symbolic Corp.	15	20	30
Waterloo	B	1962	Avalon Hill	30	50	80
Waterworks	B	1972	Parker Brothers	6	10	15
Weird-Ohs Game, The	B	1964	Ideal	85	145	230
Welcome Back Kotter	B	1977	Ideal	15	25	35
Welcome Back Kotter Game	C	1976	Milton Bradley	15	25	40
Wendy, The Good Little Witch	B	1966	Milton Bradley	85	145	230
What Shall I Be	B	1966	Selchow & Righter	10	15	25
What's My Line Game	B	1950s	Lowell	25	45	65
Wheel of Fortune	B	1985	Pressman	5	8	13
Where's The Beef? (Wendy's)	B	1984	Milton Bradley	6	10	15
Which Witch?	B	1970	Milton Bradley	25	35	55
Whirl-A-Ball	B	1978	Pressman	10	15	25
Whirly Bird Play Catch	B	1960s	Innovation Industries	20	35	50
White Shadow Basketball Game, The	B	1980	Cadaco	10	20	30
Who Framed Roger Rabbit?	B	1987	Milton Bradley	20	35	55
Who What Or Where?	B	1970	Milton Bradley	5	8	13
Who, Game of	B	1951	Parker Brothers	30	50	80
Whodunit	B	1972	Selchow & Righter	10	15	25
Whosit?	B	1976	Parker Brothers	6	10	15
Wide World	B	1962	Parker Brothers	15	25	40
Wide World of Sports Golf	B	1975	Milton Bradley	45	70	110
Wil-Croft Baseball	B	1971	Wil-Croft	10	16	25
Wild Bill Hickock	B	1955	Built-Rite	45	75	125
Wild Kingdom Game	B	1977	Teaching Concepts	20	35	50
Wildlife	B	1971	Lowe	30	50	80
Willie Mays "Say Hey"	B	1954	Toy Development Co.	200	350	525
Willie Mays "Say Hey" Baseball	B	1958	Centennial Games	190	295	450
Willie Mays Push Button Baseball	B	1965	Eldon	175	295	450
Willow	B	1988	Parker Brothers	6	10	15
Win A Card Trading Card Game	B	1965	Milton Bradley	350	575	900
Winky Dink Official TV Game Kit	B	1950s		20	30	50
Winnie The Pooh	B	1979	Parker Brothers	5	8	13
Winnie The Pooh Game	B	1959	Parker Brothers	30	50	80
Wiry Dan's Electric Baseball Game	B	1953	Harett-Gilmar	25	40	65
Wiry Dan's Electric Football Game	B	1953	Harett-Gilmar	15	40	65
Wise Old Owl	C	1950s	Novel Toy	20	35	55
Witch Pitch Game	B	1970	Parker Brothers	15	25	40
Wizard of Oz Game	B	1962	Lowe	20	30	50
Wizard of Oz Game, The	B	1974	Cadaco	10	15	25
Wolfman Mystery Game	B	1963	Hasbro	120	200	320
Wonder Woman Game	B	1967	Hasbro	20	35	65
Wonderbug Game	B	1977	Ideal	6	10	15

Top to bottom: The Bugs Bunny Game, 1975, Ideal; Walt Disney's Fantasyland Game, 1956, Parker Brothers; Story Stage Starring Jackie Gleason and his TV Troupe, 1955, Utopia Enterprises.

Games, Postwar

NAME	TYPE	YEAR	COMPANY	GOOD	EX	MINT
Woody Woodpecker Game	B	1959	Milton Bradley	60	95	145
Woody Woodpecker's Crazy Mixed Up Color Factory	B	1972	Whitman	12	20	30
Woody Woodpecker's Moon Dash Game	B	1976	Whitman	12	20	30
Woody Woodpecker, Travel With	B	1950s	Cadaco-Ellis	35	65	100
World Champion Wrestling Official Slam o'Rama	B	1990	International Games	5	8	12
World of Micronauts	B	1978	Milton Bradley	10	15	25
World Wide Travel	B	1957	Parker Brothers	25	45	70
World's Fair Game, The Official New York	B	1964	Milton Bradley	25	45	85
World's Greatest Baseball Game	B	1977	J. Woodlock	30	50	80
Wow! Pillow Fight For Girls Game	S	1964	Milton Bradley	10	20	30
Wrestling Superstars	B	1985	Milton Bradley	8	13	20
WWF Wrestling Game	B	1991	Colorforms	5	8	12
Wyatt Earp Game	B	1958	Transogram	35	60	85
Yacht Race	B	1961	Parker Brothers	60	100	160
Yahtzee	B	1956	Lowe	6	10	15
Yertle, The Game of	B	1960	Revell	55	95	150
Yogi Bear Break A Plate Game	B	1960s	Transogram	50	80	130
Yogi Bear Cartoon Game	B	1950s		3	5	10
Yogi Bear Game	B	1971	Milton Bradley	20	35	65
Yogi Bear Go Fly A Kite Game	B	1961	Transogram	40	65	125
Yours For A Song	B	1962	Lowell	20	35	55
Zaxxon	B	1982	Milton Bradley	6	10	15
Zig Zag Zoom	B	1970	Ideal	10	20	30
Ziggy Game, A Day With	B	1977	Milton Bradley	15	20	30
Zingo	B	1950s	Empire Plastics	15	20	30
Zip Code Game	B	1964	Lowell	35	55	90
Zorro Game, Walt Disney's	B	1966	Parker Brothers	45	75	120
Zorro Target Game W/Dart Gun	B	1950s	Knickerbocker Plastic Co.	20	30	50
Zowie Horseshoe Game	S	1947	James L. Decker	20	35	55

Food-Related Collectibles

Nothing fascinates humanity like itself. This fascination takes endless forms, some grand and some modest, but to the true collector nothing is without value. Cast iron banks may hold no interest for a PEZ collector, just as a bank collector may have no interest in PEZ, but this is a matter of personal preference, not a reflection on the collectible validity of an entire field. Most collectors seek the toys of their youth, and the toys of choice change from generation to generation as much as they do from wallet to wallet. The toy universe is a big place and all fields from Marklin to Mattel have their place in it.

In this section we look at several fields that can be grouped together by the common denominator of food. Additionally, they are all relatively new fields, with beginnings traceable to 1935, 1952, and 1979, respectively. As recent entries to the time line of toys, these fields hold much promise in terms of the high numbers of potential collectors they can attract. Young fields are also more easily manipulated, and collectors must be on guard against being misled by manufactured market booms. For these reasons alone they warrant close watching. The next decade in particular will tell which, if any, of these fields will grow into a major constellation in the toy firmament.

Lunch Boxes

In 1935, the firm of Geuder, Paeschke and Frey produced a small oval lunch tin with a lid and wire handle, which they called the Mickey Mouse Lunch Kit. Decorated with an early long-nosed Mickey on the lid and other Disney characters on the side band, this is considered the first true American character lunch box. Lunch kits had been in manufacture since the 1920s, but this was the first kit to use an established children's character as a selling point.

It took the star power of television to launch the lunch box industry out of the domed steel domain of workmen into the colorful art boxes generations of children carried to school each day.

Aladdin's "Land of the Lost" steel lunch box and matching plastic thermos, 1965; "The Man From U.N.C.L.E." steel lunch box from King Seeley Thermos, 1966.

As World War II ended, Aladdin Industries returned to providing millions of workmen with sturdy if uninspired lunch kits designed to take the beating of the workplace. The great change came in 1950 when Aladdin released a pair of rectangular steel boxes, one red and one blue, sporting scalloped color decals of the TV western hero of the day, Hopalong Cassidy. In short order, 600,000 Hoppy boxes were being carried to school by proud young owners. The youth market had been found and it would never be ignored again.

The envious classmates of those first Hoppy boxers would not be denied. American Thermos, Aladdin's chief competitor, would not be denied either. It went one up on Aladdin by introducing the 1953 Roy Rogers box in full color lithography. Aladdin responded by issuing a new 1954 Hoppy box in full color litho, and the lunch box era officially began.

Throughout the latter 1950s, the box wars were fought in earnest between Aladdin and American Thermos, with occasional challenges by ADCO Liberty, Ohio Art, and Okay Industries.

The smaller firms produced some classic boxes, notably Mickey Mouse and Donald Duck (1954), Howdy Doody (1954) and Davy Crockett (1955) from ADCO Liberty; and Captain Astro (1966), Bond XX (l967), Snow White (1980) and Pit Stop (1968) from Ohio Art. Okay Industries weighed in briefly later on with the now highly prized Wake Up America (1973) and Underdog (1974) boxes, but from the beginning it had always been a two-horse race.

The popular boxes of each year mirrored the stars, heroes, and interests of the times. From the westerns and space explorations of the late 1950s through the 1960s, Americans enjoyed a golden age of cartoon and film heroes such as the "Flintstones" (1962), Dudley Do-Right (1962), Bullwinkle and Rocky (l962) and Mary Poppins (l965). As the decade progressed, America grew more aggressive, turning towards such violent heroes as the "Man From U.N.C.L.E." (1966), "Rat Patrol" (1967), and G.I. Joe (1967) before Vietnam changed the national consciousness.

The early 1970s brought us such innocuous role models as H.R. Pufnstuf (1970), "The Partridge Family" (1971), and Bobby Sherman (l972), and by decade's end we were greeting both the promise and the threat from beyond in Close Encounters (1978) and Star Wars (1978).

The metal box reigned supreme through the mid-1980s when parental groups began calling for a ban on metal boxes as "deadly weapons." The industry capitulated, and by 1986 both Aladdin and American Thermos were producing all their boxes in plastic. Both firms continue production today.

The switch to plastic was not nearly as abrupt as might be expected. Aladdin and Thermos had been making plastic and vinyl boxes since the late 1950s. These included many character boxes that had no counterparts in metal, which is presently their major saving grace in the collector market.

Vinyl boxes were made of lower cost materials, consisting basically of cardboard sheathed in shower curtain-grade vinyl. They were not as popular as metal boxes, and their poor construction combined with lower unit sales have resulted in a field with higher rarity factors than the metal box arena. Additionally, vinyl was more affordable to small companies, which produced numerous limited-run boxes for sale or use as premiums.

Vinyl box collecting is an emerging field with few firmly established prices compared to the relative maturity of the metal box market, so any price guide such as this will be more open to debate. As the field matures, the pricing precedents of sales and time will build into a stronger body of knowledge. Alert readers will be quick to note the substantial devaluation of both the plastic and vinyl markets from last year's levels.

For the reader's benefit, this book will list lunch kits alphabetically in three categories according to box composition: plastic, steel, and vinyl.

PEZ

Nineteen fifty-two saw the inauspicious introduction to American shores of an Austrian mint in a handy dispenser. Long popular in the homeland, the pocket candy lost something in the translation from German to English. The marketing cure for this was successful beyond all expectations.

PEZ was created in 1927 as a peppermint candy and breath mint sold in a clever package, which dispensed the candies one at a time. Highly successful in Europe, it became the fashionable adult candy of its time. But its launch in America found a disinterested public. It was quickly decided that PEZ would be reinvented for the American market as a children's candy with fruit-flavored candies replacing the staid pfefferminz of old. The dispensers were redesigned and given colorful heads in the shapes of popular cartoon characters, and American children quickly claimed the new candy as their own.

Today PEZ is available everywhere from your local K Mart to the corner convenience store, and few Americans can handle a dispenser without evoking a few childhood memories in the process. This ability to reconnect us, either with our own childhoods or with our national past, is central to collectibility in any field, and a PEZ dispenser holds a rich postwar legacy in its little plastic container.

PEZ collectors nationwide have formed clubs, published newsletters, and now hold national conventions each year. Long ago, kids threw away the dispensers because they only meted one piece of candy at a time. These once lowly candy holders have grown in popularity and respect to the point that rare dispensers are now highly prized collectibles and have been recently sold by firms such as Christies Auction House in New York.

The PEZ market has developed some noteworthy variations on standard collecting procedures. In many fields a toy still in the original package commands a premium over the same toy with no package. This is not usually the case in PEZ collecting. Pre-blister card era PEZ dispensers were packaged in boxes or cellophane bags, which did not allow for either display or handling of the toys themselves. Experts hold that condition of the dispenser itself is paramount, and that a dispenser in original packaging commands no premium over a loose one, with some exceptions. These exceptions apply to certain multiple piece dispensers or sets, such as "Make A Face," which consists of several detachable parts, and dispensers on unusual display cards such as the "Space Gun." Another notable exception is the Stand By Me dispenser, which was packaged with a pack of candy and a miniature poster from the movie of the same name.

Circus clown series of PEZ dispensers.

Since dispenser stems are easily interchangeable, PEZ authorities hold that only variations in head configuration or coloring affect value. There is no difference in value between dispensers with different colored stems but the same head.

Finally there is the matter of feet and no feet. This refers to the presence or absence of a flattened rounded base on the stem resembling flat shoes. PEZ dispensers released in America before 1987 were all of the no-feet variety, so a dispenser with feet was made after that year. However, certain older molds continued to be produced with no feet after 1987 as well, but these are common dispensers with little variance in value between feet and no-feet varieties. The major difference in value here applies to older no-feet dispensers that were discontinued and perhaps reissued after 1987 with feet.

The field continues to grow, and PEZ continues to produce a limited number of new dispensers each year, but the total universe of PEZ still numbers less than three hundred-fifty dispensers. The serious collector can find common dispensers in abundance, scarce ones with persistence, and can still hope to amass a complete collection. Additionally, many dispensers remain easily affordable.

One last note is called for here. According to our sources in the field, the hot growth we reported here last year seems to be continuing. Market volatility is a factor that must be considered when making purchases, particularly when a collectible item or category is in a hot period. Even a modest influx of collectors into a small market can cause prices to rocket, as can a run of national attention or simple speculation by dealers and investors. Only time can tell if these periods of frenetic activity will establish new market levels or dissipate as the market returns to prior levels, leaving some lucky sellers with large profits and some unlucky buyers with overvalued goods.

McDonald's Happy Meal Toys and Other Fast Food Premiums

Collecting McDonald's Happy Meal and other fast food toys is a recent but already highly developed area. There are numerous branches of a national McDonald's Collector's Club, and a convention is held in Chicago each year. Several books have been written on the subject in spite of the fact that the first national-campaign Happy Meal was not issued until 1979.

The typical Happy Meal customer is under twelve years old, an age range not renowned for gentle play habits. Thus, condition of toys is the critical factor. Only rare toys hold any value at all if found in less than perfect condition. This price guide will classify condition in only two grades: Mint In Package (MIP), and Mint, No Package (MNP). MIP toys have never been removed from their packages and are valued on average at two hundred percent of MNP toys. MNP toys may exhibit minor evidence of play, but they remain clean and intact.

Rarity is also a primary factor in determining value. The age of a toy plays a role in this, as does popularity and cross collectibility. Modern Disney movie tie-in toys are frequently worth more than older toys because of the strong Disneyana collector market. The same holds true for popular cartoon or comic character items.

Another factor affecting rarity is distribution. Some McDonald's and other fast food company campaigns were run only in certain regions of the country. The toys of these campaigns are known as regionals, and command higher than average prices due to their limited release areas.

Several years back some toys had to be recalled, resulting in the design of special one-piece toys for younger children, commonly called "Under 3" toys. "Under 3" toys are not produced for each campaign, and are not normally advertised in the in-store displays. Lower numbers of these toys are released, again resulting in premiums typically twenty percent higher than the regular toys of the same campaign.

One more factor deserves mention--the international toy. Major film and comic character toy campaigns sometimes run worldwide with little or no change from country to country. Sometimes only the package printing is changed. But occasionally foreign market toys are never released in America. These toys are highly valued by some collectors simply due to their foreign status. Other collectors also consider aspects such as popular character affiliation when calculating the value of these toys. Again, as the market matures, these values and item inventories will establish a track record.

Like PEZ, the market for fast food toys is new and is presently in a period of sustained growth.

While the universe of collectible PEZ items is much smaller than the fast food sector, PEZ has the advantage of nearly thirty more years of exposure to the American public, falling into the period of greatest nostalgia for the majority of baby boom collectors. The ten-year-old McDonald's customer of 1979 will not turn thirty years old until 1999, but it seems likely that nostalgia will accompany following generations into their middle years, as it has for baby boomers. As McDonald's is the first global restaurant, and popular film and TV tie-ins are now the rule of the day, the future of this field looks secure.

Trends
Lunch Boxes

We talked with noted collector/dealer Jon Shapiro about the lunch kit market and here is his assessment.

"The past year has not been kind to plastic and vinyl lunch boxes," he said. "It seems that all at once collectors decided to concentrate on metal boxes, and plastic and vinyl values have plummeted. The reason? Probably overexposure. Prices just shot up too quickly and collectors got tired of shelling out big bucks for these relatively unattractive kits. Also, there turned out to be a lot more different vinyl and plastic boxes out there than anyone realized. There aren't nearly as many different metal kits.

Which brings us to the good news. For the most part, metal boxes have maintained their value. Character boxes have stayed especially strong, and continue to be popular among non-lunch box collectors as well. Rare generic metal boxes have fared less well, as they lack cross-over value.

As it appears now, the hobby is moving toward metal character boxes that appeal to a larger collecting community than just kit collectors. Baby boomer boxes can only go up, and your money should be safe in "Jetsons," "Flintstones," "Star Trek," "Gunsmoke," and "Laugh-In" kits. For the time being, though, don't spend tons on less interesting kits, no matter how rare they are. Right now, rarity is out and characters are in."

Fast Food

McDonald's continues to lead the way, with even recent issues such as last year's Snow White eight-piece set commanding strong prices and high demand. Also particularly strong is the 1991 Barbie set. Across fast food companies, character toys lead all others, with certain series standing heads above their peers. The Nightmare Before Christmas franchise has caught fire in the toy collecting community and last November's Burger King's four-piece premium watch set is one of the hottest items going at toy show and flea markets across the country. Prices are now over $20 per set at most shows and dealers quickly sell out of all their stock. Overall, the fast food toy market shows no sight of fatigue and continues on a strong course.

PEZ

Nationally, PEZ dispensers have seen mixed performance at shows, but the upward trend continues. Low-end common dispensers have been selling in large quantities at shows, but many new collectors are hedging at the price points of dispensers in the $75 and up range, indicating this is still a strongly collector-based market. Most pieces are holding the values established last year, with the strongest increases being seen in rare dispensers and sets such as the "Make a Face." The top of the market appears healthy, with reports of dispensers achieving two and three times prior estimates at auction.

The Top 10 Metal Lunch Boxes
(in Mint Condition)

1. 240 Robert, 1978, Aladdin $1,000
2. Jetsons, 1968, Aladdin 850
3. Superman, 1954, Universal 800
4. Underdog, 1974, Okay Industries 800
5. Jetsons (Dome), 1963, Aladdin 675
6. Casey Jones, 1960, Universal 650
7. Action Jackson, 1973, Okay Industries 600
8. Bullwinkle and Rocky, 1962, Universal 600
9. Dudley Do-Right, 1962, Universal 600
10. Toppie Elephant, 1957, American Thermos 600

The Top 10 McDonald's Toy Sets
(in Mint in Package condition, per toy)

1. McDonaldland Express, 1984, set of four $38
2. Star Trek, 1979, set of five 30
3. Ship Shape #1, 1983, set of four 23
4. Berenstain Bear Test Market Set, 1986, set of four holiday figures 20
5. Castle Maker/Sand Castle, 1987 20
6. Ship Shape #2, 1985, set of four 18
7. Playmobile, 1982, set of five 15
8. Astrosnicks #3, 1985 14
9. Runaway Robots, 1988 14
10. Commandrons, 1985 13

The Top 10 PEZ Dispensers
(in Mint condition)

1. Make-A-Face $1,500
2. Mueslix, licensed characters series 1,200
3. Lion's Club Lion 900
4. Pineapple, Crazy Fruit series 900
5. Bride, Pez Pals series 800
6. Pear, Crazy Fruit series 800
7. Alpine, Olympics Series 600
8. Easter Bunny Die Cut 550
9. Donkey Kong, Jr. 500
10. Hippo 500

FOOD TOYS

Top to bottom: Psychedelic Eye and Psychedelic Flower variations; Superheroes including Spider-Man, Batgirl, Batman, and Captain America; Disney characters including Huey, Dopey, Snow White, and Jiminy Cricket; assorted figures including Wile E. Coyote in three variations.

FOOD RELATED

PEZ

Christmas

NAME	DESCRIPTION	NM	MINT
Angel A*	no feet, hair and halo	15	35
Rudolph	no feet, brown deer head, red nose	15	35
Santa Claus A	no feet, ivory head with painted hat	80	150
Santa Claus B	no feet, small head with flesh painted face, black eyes, red hat	85	195
Santa Claus C	no feet, large head with white beard, flesh face, red open mouth and hat	1	10
Santa Claus C	with feet, removable red hat, white beard	1	3
Snowman	no feet, black hat, white face, removable black facial features	3	15
Snowman A	no feet, black hat, white head, removable facial features	5	25

Circus

Big Top Elephant With Hair*	no feet, yellow head and red hair	125	225
Big Top Elephant, Flat Hat*	no feet, gray-green head, red flat hat	35	60
Big Top Elephant, Pointed Hat*	no feet, orange head with blue pointed hat	35	75
Clown with Chin*	no feet, long chin clown face with hat and hair	25	45
Clown with Collar	no feet, yellow collar, red hair, clown face and green hat	15	40
Giraffe	no feet, orange head with horns, black eyes	40	75
Gorilla	no feet, black head with red eyes and white teeth	15	40
Lion with Crown	no feet, black mane, green head with yellow cheeks and red crown	40	70
Little Lion	no feet, yellow head with brown mane	15	35
Monkey Sailor	no feet, cream face, brown hair, white sailor cap	20	45
Monkey with Baseball Cap*	no feet, monkey head and ball cap, white eyes	35	75
Pony-Go-Round	no feet, orange head, white harness, blue hair	25	60

Crazy Fruit

Orange	no feet, orange head with face and leaves on top	65	145

Crazy Fruits

Pear	no feet, yellow pear face, green visor	500	800
Pineapple	no feet, pineapple head with greenery and sunglasses	600	900

Die Cuts

Bozo the Clown	die cut Bozo and Butch on stem, no feet, white face, red hair and nose	75	125
Casper The Friendly Ghost		100	145
Donald Duck		100	175
Easter Bunny Die Cut	no feet, die cut	400	550
Mickey Mouse Die-cut	no feet, die-cut stem with Minnie, die cut face mask	75	120

Disney

Baloo*	with feet, blue head	10	35
Baloo*	no feet, blue head	15	30
Bambi	with feet,	5	25
Bouncer Beagle		5	25
Captain Hook	no feet, black hair, flesh face winking with right eye open	20	45
Chip (Chip & Dale)*	no feet, black top hat, tan head with white cheeks, brown nose, foreign issue	15	65
Dalmation Pup	with feet, white head with left ear cocked, foreign issue	15	35
Dewey	no feet, blue hat, white head, yellow beak, small black eyes	5	25
Donald Duck A	no feet, blue hat, one-piece head and bill, open mouth	5	15
Donald Duck B	with feet, blue hat, white head and hair with large eyes, removable beak	1	3
Dopey	no feet, flesh colored die cut face with wide ears, orange cap	100	200

FOOD TOYS

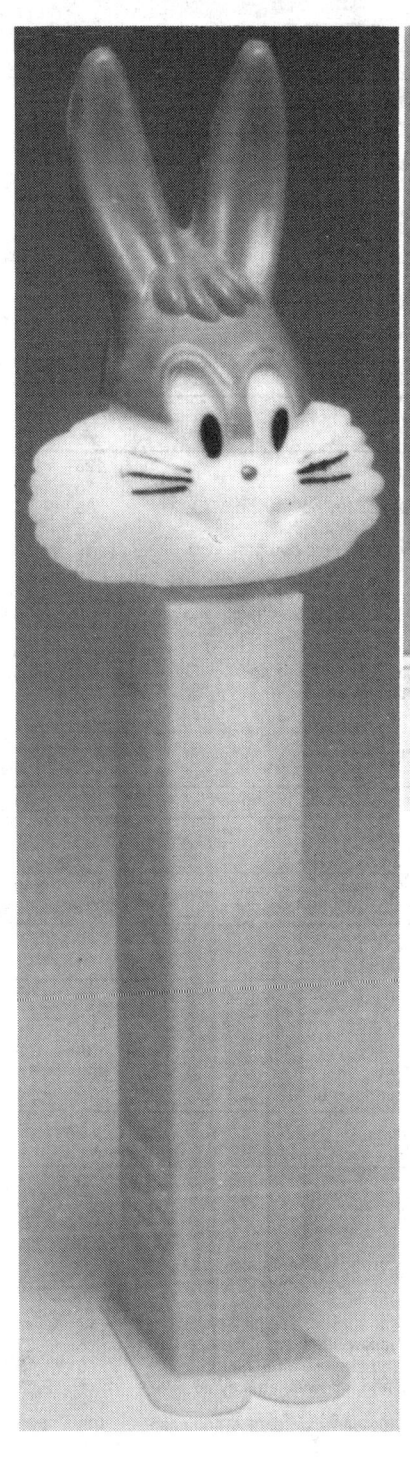

Top to bottom: PEZ Bugs Bunny with feet, Warner Brothers; PEZ Space Gun and display and Astronaut variations; PEZ store display with various Regular dispensers including With Regular (second from left) and Arithmetic (fourth from left).

Disney

NAME	DESCRIPTION	NM	MINT
Duck Child	no feet, blue or green hat, yellow beak, small eyes	10	25
Dumbo*	with feet, blue head with large ears, yellow hat	15	60
Dumbo*	no feet, gray head with large ears, red hat	35	65
Goofy A*	no feet, red hat, painted nose, removable white teeth	5	50
Goofy B*	same as version A except teeth are part of head	25	50
Goofy D	with feet, beige snout, green hat	1	3
Gyro Gearloose		5	25
Huey	no feet, red hat, white head, yellow beak, small black eyes	5	25
Huey, Dewey or Louie Duck	with feet, red, blue, or green stem and matching cap, white head and orange beak	10	20
Jiminy Cricket	no feet, green hatband and collar, flesh face, black top hat	30	65
King Louie	no feet, brown hair and 'sideburns' over light brown head	15	35
King Louie	with feet, brown hair and 'sideburns' over light brown head	10	25
Li'l Bad Wolf	no feet, black ears, white face, red tongue	15	25
Li'l Bad Wolf	with feet, black ears, white face, red tongue	12	25
Louie	no feet, green hat, white head, yellow beak, small black eyes	5	25
Mary Poppins	no feet, flesh face, reddish hair, lavender hat	350	500
Mickey Mouse A	no feet, black head and ears, pink face, mask with cut out eyes and mouth, nose pokes through mask	40	60
Mickey Mouse B	no feet, painted face, non-painted black eyes and mouth	50	100
Mickey Mouse C	no feet, flesh face, removable nose, painted eyes	5	15
Mickey Mouse D	no feet, flesh face mask embossed white and black eyes	3	15
Mickey Mouse E	with feet, flesh face, bulging black and white eyes, oval nose	1	3
Mowgli	no feet, black hair over amber-brown head	15	35
Mowgli	with feet, black hair over amber-brown head	2	25
Peter Pan	no feet, green hat, flesh face, orange hair	80	160
Pinocchio A*	no feet, red or yellow cap, pink face, black painted hair	75	125
Pinocchio B	no feet, black hair, red hat	45	85
Pluto A	no feet, yellow head, long black ears, small painted eyes	5	20
Pluto C	with feet, yellow head, long painted black ears, large white and black decal eyes	1	5
Practical Pig A	no feet, blue hat, pointed up ears, small cheeks, round nose	10	35
Practical Pig B	no feet, blue hat, large cheeks, half nose	10	35
Practical Pig C	with feet, blue hat, large cheeks, half round nose	5	15
Scrooge McDuck A	no feet, white head, yellow beak, black top hat and glasses, white sideburns	10	25
Scrooge McDuck B	with feet, white head, removable yellow beak, tall black top hat and glasses, large eyes	5	20
Snow White*	no feet, flesh face, black hair with ribbon and matching collar	75	125
Thumper	no feet, orange face	40	75
Tinkerbell	no feet, pale pink stem, white hair, flesh face with blue and white eyes	75	150
Winnie the Pooh	with feet, yellow head	10	45
Zorro	no feet, flesh face, black mask and hat	20	50

Easter

NAME	DESCRIPTION	NM	MINT
Bunny 1990	with feet, long ears, white face	1	3
Bunny Original A	no feet, narrow head and tall ears	300	425
Bunny Original B	no feet, tall ears and full face, smiling buck teeth	300	425
Bunny W/Fat Ears	no feet, wide ear version	1	15
Chick in Egg, No Hat*	no feet, yellow chick in egg shell, no hat	75	150
Chick in Egg, With Hat*	no feet, yellow chick in egg shell, red hat	5	20
Duckie with Flower*	no feet, flower, duck head with beak	25	55
Lamb	no feet, white head with a pink bow	3	10
Rooster*	no feet, head, comb, and wattle	15	40

Eerie Spectres

NAME	DESCRIPTION	NM	MINT
Air Spirit	no feet, reddish triangular fish face	25	45
Diabolic	no feet, soft orange monster head with black and red tints	25	45
Scarewolf	no feet, soft head with orange painted hair, and ears	25	55

FOOD TOYS

Eerie Spectres

NAME	DESCRIPTION	NM	MINT
Vamp	no feet, light gray head on black collar, green tinted hair and face, red teeth	35	55
Zombie	no feet, burgundy and black soft head	25	55

Full Bodied

Santa, Full Bodied	full body stem with painted Santa suit and hat	110	150
Space Trooper*	robotic figure with backpack	150	225

Halloween

Dr. Skull A	no feet, black cowl, white head	5	15
Dr. Skull B	with feet, black collar	1	5
Jack O Lantern*	no feet, orange stem, carved face	3	15
Mr. Ugly*	no feet, black hair, green head, red eyes and buck teeth	15	60
Mr. Ugly*	no feet, black hair, yellow face, red eyes and buck teeth	45	125
Octopus*	no feet, black, orange or red head	25	60
One-Eyed Monster*	no feet, gorilla head with one eye missing	25	80
Witch 1 Piece*	no feet, black stem with witch embossed on stem, orange 1 piece head	100	165
Witch 3 Piece A	with feet, red head and hair, green mask, black hat	1	5
Witch 3 Piece B*	no feet, chartreuse face, black hair, orange hat	35	80

Humans

Astronaut A*	no feet, helmet, yellow visor, small head	135	250
Astronaut B	no feet, green stem, white helmet, yellow visor, large head	60	100
Betsy Ross	no feet, dark hair and white hat, Bicentennial issue	40	80
Captain (Paul Revere)	no feet, blue hat, Bicentennial issue	45	85
Cowboy	no feet, human head, brown hat	200	400
Daniel Boone	no feet, light brown hair under dark brown hat, Bicentennial issue	85	150
Football Player	no feet, white stem, red helmet with white stripe	45	125
Indian Brave*	no feet, small human head, indian headband with one feather, Bicentennial issue	100	200
Indian Chief*	no feet, warbonnet, Bicentennial issue	50	85
Indian Squaw	no feet, black hair in braids with headband	45	85
Pilgrim	no feet, pilgrim hat, blond hair, hat band	85	125
Pilot	no feet, blue hat, grey headphones	45	95
Spaceman	no feet, clear helmet over flesh-color head	75	125
Stewardess	no feet, light blue flight cap, blond hair	45	100
Uncle Sam	no feet, stars and stripes on hat band, white hair and beard, Bicentennial issue	50	100
Wounded Soldier	Bicentennial Series, no feet, white bandage, brown hair	80	150

Kooky Zoo

Cockatoo*	no feet, yellow beak and green head, red head feathers	25	60
Cow A*	no feet, cow head, separate nose	20	45
Cow B	no feet, blue head, separate snout, horns, ears and eyes	65	125
Crocodile	no feet, green head with red eyes	45	100
Panda A*	no feet, yellow head with black eyes and ears	125	200
Panda A*	no feet, white head with black eyes and ears	5	15
Panda B*	with feet, white head with black eyes and ears	1	3
Panther	no feet, blue head with pink nose	45	65
Puzzy Cat*	no feet, cat head with hat	25	50
Raven*	no feet, black head, beak, and glasses	20	35
Yappy Dog*	no feet, black floppy ears and nose, green or orange head	35	85

Licenced Characters

Arlene	with feet, pink head, Garfield's 'girlfriend'	5	15
Asterix	no feet, blue hat with wings, yellow mustache, European	250	350

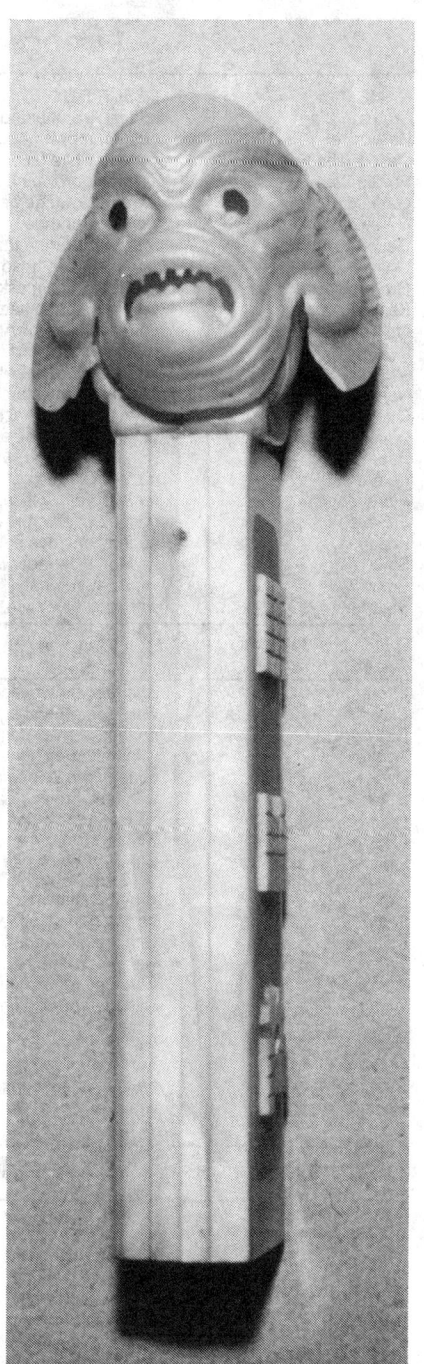

FOOD TOYS

Clockwise from upper left: PEZ Christmas including Angel with feet and halo, Snowman with feet, Santa Claus A, B, and C, and Rudolph; PEZ Creature From The Black Lagoon, Universal; PEZ Football Player and Baseball dispenser; PEZ Circus all with feet including Big Top Elephant with Hair, Flat Hat, and Pointed Hat.

Licenced Characters

NAME	DESCRIPTION	NM	MINT
Brutus	no feet, black beard and hair	85	150
Bullwinkle	no feet, brown head, yellow antlers	150	200
Casper The Friendly Ghost	no feet, white face	50	80
Fozzie Bear	with feet, brown head, bow tie, small brown hat	1	5
Garfield	with feet, orange head	1	3
Garfield With Teeth	with feet, orange head, wide painted toothy grin	5	18
Garfield With Visor	with feet, orange face, green visor	1	8
Gonzo	with feet, blue head, yellow eyelids bow tie	1	3
Green Hornet	no feet, green mask and hat	150	300
Kermit	with feet, green head	1	5
Little Orphan Annie	no feet, light brown hair, flesh face with black painted features	40	65
Miss Piggy	with feet, pink face, yellow hair	1	5
Mueslix	no feet, white beard, moustache and eyebrows, European	850	1200
Nermal	with feet, gray stem and head	5	20
Obelix	no feet, red mustache and hair, blue hat, European	250	400
Olive Oyl	no feet, black hair and flesh painted face	95	150
Papa Smurf	with feet, red hat, white beard, blue face	3	15
Peter PEZ	no feet, blue top hat that says PEZ, white face, yellow hair	65	100
Popeye A	no feet, yellow face, painted hat	45	95
Popeye B	no feet, removable white sailor cap	20	35
Popeye C	no feet, one eye painted, removable pipe and cap	35	75
Smurf	no feet, blue face, white hat	3	15
Smurf	with feet, red hat	25	60
Smurfette	with feet, blue face, yellow hair, white hat	3	15

Merry Music Makers

Camel	with feet, brown face with red fez hat	15	25
Clown	with feet, green hat, with a clown face, foreign issue	5	20
Dog	no feet, orange dog face with black ears, foreign issue	10	25
Donkey	with feet, gray head with pink nose	10	25
Duck	no feet, brown head with yellow beak	10	25
Duck	with feet, brown head with yellow beak	10	18
Frog	no feet, yellow and green head with black eyes, foreign issue	25	40
Frog	with feet, yellow and green head with black eyes, foreign issue	12	25
Indian	with feet, black hair and green headband with feather, foreign issue	5	15
Koala	with feet, brown head with a black nose, foreign issue	15	40
Lamb	no feet, pink stem, white head	10	25
Monkey	with feet, tan monkey face in brown head, foreign issue	10	25
Panda	with feet, white head, foreign issue	6	15
Parrot	with feet, red hair, yellow beak and green eyes	5	25
Penguin	with feet, penguin head with yellow beak and red hat, foreign issue	5	25
Pig	with feet, pink pig head	20	40
Rhino	with feet, green head, red horn, foreign issue	15	35
Rooster	no feet, head, comb	10	25
Rooster	with feet, head, comb	5	20
Tiger	with feet, tiger head with white snout	5	15

MGM

Barney Bear	no feet, brown head, white cheeks and snout, black nose	15	35
Barney Bear	with feet, brown head, white cheeks and snout, black nose	15	35
Droopy Dog	no feet, white face, flesh snout, black ears and red hair	5	25
Jerry (Tom & Jerry)	no feet, brown face, pink lining in ears	20	40
Jerry (Tom & Jerry)	with feet, brown face, pink lining in ears	10	25
Spike	with feet, brown face, pink snout	5	25
Tike	with feet, brown head	10	25
Tom A (Tom & Jerry)	no feet, gray cat head with painted black features	20	35
Tom B (Tom & Jerry)	with feet, gray cat head with removable facial features	5	20
Tyke	with feet, brown face	5	15

FOOD TOYS

MGM

NAME	DESCRIPTION	NM	MINT
Arithmetic*	no feet, headless dispenser with white top, side of body has openings with columns of numbers	150	400
Baseball Dispenser Set	no feet, baseball glove with ball, white homeplate marked "PEZ" and bat	300	325
Baseball Glove	no feet, brown baseball glove with white ball	150	250
Candy Shooter	black PEZ Gun with PEZ monogram on stock	100	200
Candy Shooter	red body, white grip, with German license and Doppel (Double) PEZ candy	100	150
Make-A-Face	no feet, oversized head with 18 different facial parts	750	1500
Personalized (Regular)	no feet, no head, stem with label for monogramming, and top	150	225
Psychedelic Eye*	no feet, decal design on stem, beige hand with green eye	250	450
Psychedelic Flower*	no feet, stem with decal on side, green eye in flower center	300	500
Regular*	no feet, no head, stem with top only	100	225
Space Gun 1950s*	various color bodies in red, dark blue, black, maroon, green, yellow, silver or light blue, with white triggers, butts	150	225
Space Gun 1980s*	red space gun with black handgrips, on blister pack	50	125
Whistles*	stems with police whistles on top	1	25
Zorro with Logo	no feet, Zorro mask and black hat, says Zorro on stem	75	125

Olympics

Alpine	no feet, green hat with beige plume, black mustache	350	600
Vucko Wolf with Bobsled Hat*	1984 Yugoslavia Olympics issue, with feet, gray or brown face with bobsled helmet	200	500
Vucko Wolf with Ski Hat*	1984 Yugoslavia Olympics issue, with feet, gray or brown face with ski hat	200	500
Vucko Wolf*	1984 Yugoslavia Olympics issue, with feet, gray or brown face	200	500
Winter Olympics Snowman	1976 Innsbruck, red nose and hat, white head with arms extended, black eyes, blue smile	250	400

Peanuts

Charlie Brown	with feet, crooked smile, blue cap	1	5
Charlie Brown W/Tongue	with feet, blue cap, smile with red tongue at corner	15	25
Charlie Brown, Eyes Closed	with feet, blue cap	20	50
Lucy	with feet, black hair,	1	5
Snoopy	with feet, with white head and black ears	1	3
Woodstock	with feet, yellow head	1	3

PEZ Pals

Boy with Cap*	no feet, white hair, blue cap	15	40
Boy*	no feet, brown hair	3	15
Bride*	no feet, white veil, light brown, blond or red hair	450	800
Doctor	no feet, white hair and mustache, gray reflector on white band, black stethoscope	35	55
Engineer	no feet, blue hat	30	50
Fireman	no feet, black mustache, red hat with gray #1 insignia	25	35
Girl*	no feet, blond pigtails	5	20
Girl*	with feet, pigtails	1	15
Groom	no feet, black top hat, white bow tie	100	300
Knight*	no feet, gray helmet with plume	80	145
Maharajah	no feet, green turban with red inset	25	40
Mexican	no feet, yellow sombrero, black beard and mustache combination	20	45
Nurse*	no feet, girl's hair, white nurse's cap	40	65
Pirate	no feet, red cap, patch over right eye	30	45
Policeman	no feet, blue hat with gray badge	15	40
Ringmaster	no feet, white bow tie, white hat with red hatband, and black handlebar moustache	65	130
Sailor	no feet, blue hat, white beard	75	100
Sheik*	no feet, white head drape, headband	25	55
Sheriff	no feet, brown hat with badge	50	95

FOOD TOYS

PEZ Pals

NAME	DESCRIPTION	NM	MINT
Cocoa Marsh Spaceman	no feet, clear helmet on small male head, with Cocoa Marsh embossed on side	85	150
Donkey Kong Jr.	no feet, blond monkey face, dark hair, white cap with J on it	275	500
Golden Glow	no feet, no head, gold stem and top	125	250
Hippo	no feet, green stem with "Hippo" imprinted on side, hippo on top, foreign issue	400	500
Lions Club Lion	no feet, stem imprinted Lions Club Inter'l Convention writing, yellow roaring lion head	400	900
Sparefroh (foreign issue)	no feet, green stem, red triangle hat, coin glued on stem	175	300
Stand By Me	dispenser packed with mini poster of film	125	275

Superheroes

Batgirl	Soft Head Superhero, no feet, blue mask, black hair	45	115
Batman	Soft head Superhero, no feet, blue mask	65	115
Batman	with feet	1	5
Batman with Cape	no feet, blue cape, mask and hat	65	150
Batman*	no feet, blue hat and black face mask	1	15
Captain America*	no feet, blue cowl, black mask with white letter A	30	70
Hulk A	no feet, dark green head, black hair	15	35
Hulk B	no feet, light green head, dark green hair	5	15
Hulk B	with feet, light green head, tall dark green hair	1	5
Joker	Soft Head Superhero, no feet, green painted hair	60	100
Penguin	Soft Head Superhero, no feet, yellow top hat, black painted monocle	45	100
Spiderman A	no feet, red head with black eyes	5	15
Spiderman B	with feet, bigger red head	1	5
Thor	no feet, yellow hair, gray winged helmet	75	150
Wonder Woman	Soft Head Superhero, no feet, black hair and yellow band with star	40	80
Wonder Woman	with feet, red stem	1	5
Wonder Woman*	no feet, black hair and yellow band with red star	1	5

Trucks

Truck A*	cab, stem body, single rear axle	35	100
Truck B*	cab, stem body, dual rear axle and dual arch fenders	25	65
Truck C*	cab, stem body, dual rear wheels with single arch fender, movable wheels	1	15
Truck D*	cab, stem body, dual rear wheels, single arch fender, nonmovable wheels	1	5

Universal Monsters

Creature From Black Lagoon	no feet, green head and matching stem	150	250
Frankenstein	no feet, black hair, gray head	150	250
Wolfman	no feet, black stem, gray head	150	250

Warner Brothers

Bugs Bunny*	no feet, dark gray head with white cheeks	1	20
Bugs Bunny*	with feet, gray head with white cheeks	1	5
Cool Cat	with feet, orange head, blue snout, black ears	10	30
Daffy Duck A	no feet, black head, yellow beak, removable white eyes	1	15
Daffy Duck B	with feet, black head, yellow beak	1	5
Foghorn Leghorn	no feet, brown head, yellow beak, red wattle	25	45
Foghorn Leghorn	with feet, brown head, yellow beak, red wattle	10	15
Henry Hawk	no feet, light brown head, yellow beak	15	35
Merlin Mouse	no feet, gray head with flesh cheeks, green hat	10	35
Merlin Mouse	with feet, gray head with flesh cheeks, green hat	10	35
Petunia Pig	no feet, black hair in pig tails	30	60
Roadrunner A	no feet, purple head, yellow beak	15	30
Roadrunner B	with feet, purple head, yellow beak	10	25

FOOD TOYS

Warner Brothers

NAME	DESCRIPTION	NM	MINT
Speedy Gonzales	no feet, brown head, yellow sombrero	5	25
Speedy Gonzales	with feet, greenish head, yellow sombrero	5	25
Sylvester	no feet, black head, white whiskers, red nose	5	10
Sylvester	with feet, black head, white whiskers, red nose	1	5
Tweety Bird	no feet, yellow head	1	15
Tweety Bird	with feet, yellow head	1	3
Wile E. Coyote	no feet, brown head	5	25
Wile E. Coyote	with feet, brown head	5	25

FOOD TOYS

FOOD TOYS

Clockwise from upper left: Mickey Mouse head plastic lunch kit, 1989, Aladdin; Jabberjaw plastic box, 1977, Thermos; The Incredible Hulk plastic box, 1976, Aladdin of Canada; Thunderbirds plastic lunch kit, 1991, Entertainment Group Ltd.; Yogi Bear plastic box, Aladdin of Canada.

FOOD RELATED

Lunch Boxes

Plastic

NAME, YEAR, COMPANY, DESCRIPTION	BOX NM	BOTTLE NM
*with matching bottles unless noted		
101 Dalmations, 1990, Aladdin,	6	3
18 Wheeler, 1978, Aladdin,	30	10
ALF, unknown, red plastic	18	0
Animalympics Dome, 1979, Thermos,	20	5
Astronauts, 1986, Thermos,	8	3
Atari Missile Command Dome, 1983, Aladdin,	20	5
Back to School, 1980, Aladdin,	10	4
Back to the Future, 1989, Thermos,	15	6
Bang Bang, 1982, Thermos,	15	5
Barbie with Hologram Mirror, 1990, Thermos,	10	4
Batman (dark blue), 1989, Thermos,	10	4
Batman (light blue), 1989, Thermos,	20	4
Batman Returns, 1991, Thermos,	10	3
Beach Bronto, 1984, Aladdin, no bottle	25	0
Beach Party (blue/pink), 1988, Deka, with generic plastic bottle	15	5
Bear with Heart (3-D), 1987, Servo,	12	0
Beauty & the Beast, 1991, Aladdin,	6	3
Bee Gees, 1978, Thermos,	30	10
Beetlejuice, 1980, Thermos,	10	4
Big Jim, 1976, Thermos,	40	20
Bozostuffs, 1988, Deka,	20	5
C.B. Bears, 1977, Thermos,	20	5
Care Bears, 1986, Aladdin,	5	2
Centurions, 1986, Thermos,	10	3
Chiclets, 1987, Thermos, no bottle	20	0
Chipmunks, (Alvin and the), 1983, Thermos,	15	5
ChiPs, 1977, Thermos,	30	5
Cindorolla, 1992, Aladdin,	5	2
Civil War, The, 1961, Universal, generic "Thermax" bottle	250	10
Colonial Bread Van, 1984, Moldmark Industries,	50	10
Crestman Tubular!, 1980, Taiwan,	20	3
Days of Thunder, 1988, Thermos,	10	5
Deka 4X4, 1988, Deka, with generic plastic bottle	25	5
Dick Tracy, 1989, Aladdin,	10	5
Dino Riders, 1988, Aladdin,	10	4
Dinobeasties, 1988, Thermos,	5	0
Dinorocker with Radio & Headset, 1986, Fundes,	30	0
Disney on Parade, 1970, Aladdin, plastic bottle, glass liner	30	15
Disney's Little Mermaid, 1989, Thermos, with generic plastic bottle	8	3
Duck Tales (4X4/Game), 1986, Aladdin,	10	4
Dukes of Hazzard, 1981, Aladdin,	15	5
Dukes of Hazzard Dome, 1981, Aladdin,	20	5
Dune, 1984, Aladdin,	20	8
Dunkin Munchkins, 1972, Thermos,	25	8
Ecology Dome, 1980, Thermos,	20	5
Ed Grimley, 1988, Aladdin,	20	5
Entenmann's, 1989, Thermos,	10	0
Ewoks, 1983, Thermos,	20	5
Fame, 1972, Thermos,	20	5
Fievel Goes West, 1991, Aladdin,	10	4
Fire Engine Co. 7, 1985, D.A.S., with generic plastic bottle	15	2
Fisher Price Mini Lunch Box, 1962, Fisher Price, red with barnyard scenes, matching bottle	10	2
Flash Gordon Dome, 1979, Aladdin,	40	10
Flintstone Kids, 1987, Thermos,	10	5
Flintstones, unknown, premium, Denny's Restaurants	15	0

FOOD TOYS

Plastic

NAME, YEAR, COMPANY, DESCRIPTION	BOX NM	BOTTLE NM
Food Fighters, 1988, Aladdin,	10	5
Fraggle Rock, 1987, Thermos,	12	3
Frito Lay's, 1982, Thermos, no bottle	25	0
G.I. Joe, (Space Mission), 1989, Aladdin,	20	5
G.I. Joe, Live the Adventure, 1986, Aladdin,	20	5
Garfield (food fight), 1979, Thermos,	15	5
Garfield (lunch), 1977, Thermos,	15	5
Geoffrey, 1981, Aladdin,	30	10
Get Along Gang, 1983, Aladdin,	8	2
Ghostbusters, 1986, Deka,	10	3
Go Bots, 1984, Thermos,	8	2
Golden Girls, 1984, Thermos,	10	3
Goonies, 1985, Aladdin,	15	5
Gumby, 1986, Thermos,	15	5
Hot Wheels, 1984, Thermos,	8	3
Howdy Dowdy Dome, 1977, Thermos,	25	5
Incredible Hulk Dome, 1980, Aladdin,	30	10
Inspector Gadget, 1983, Thermos,	20	8
It's Not Just the Bus -Greyhound, 1980, Aladdin,	40	10
Jabber Jaw, 1977, Thermos,	50	10
Jetsons (paper picture), 1987, Servo,	40	5
Jetsons 3-D, 1987, Servo,	25	5
Jetsons, The Movie, 1990, Aladdin,	10	5
Kermit the Frog, Lunch With, 1988, Thermos,	18	5
Kermit's Frog Scout Van, 1989, Superseal, no bottle	15	0
Kool-Aid Man, 1986, Thermos,	15	5
Lisa Frank, 1980, Thermos,	10	3
Little Orphan Annie, 1973, Thermos,	40	10
Looney Tunes Birthday Party, 1989, Thermos, blue or red	10	5
Looney Tunes Dancing, 1977, Thermos,	20	5
Looney Tunes Playing Drums, 1978, Thermos,	20	5
Looney Tunes Tasmanian Devil, 1988, Thermos, with generic plastic bottle	12	5
Los Angeles Olympics, 1984, Aladdin,	20	5
Lucy's Luncheonette, 1981, Thermos, Peanuts characters	15	5
Lunch Man with Radio, 1986, Fun Design, with built-in radio, no bottle	18	0
Lunch Time with Snoopy Dome, 1981, Thermos,	15	5
Lunch'n Tunes Safari, 1986, Fun Design, with built-in radio, no bottle	25	0
Lunch'n Tunes Singing Sandwich, 1986, Fun Design, with built-in radio, no bottle	25	0
Mad Balls, 1986, Aladdin,	10	3
Marvel Super Heroes, 1990, Thermos,	15	5
Max Headroom (Coke), 1985, Aladdin,	20	5
McDonald's Happy Meal, 1986, Fisher Price,	15	0
Menudo, 1984, Thermos,	12	5
Mickey & Minnie Mouse in Pink Car, 1988, Aladdin,	10	3
Mickey Mouse & Donald Duck, 1984, Aladdin,	10	3
Mickey Mouse & Donald Duck See-Saw, 1986, Aladdin,	10	3
Mickey Mouse at City Zoo, 1985, Aladdin,	10	3
Mickey Mouse Head, 1989, Aladdin,	20	5
Mickey on Swinging Bridge, 1987, Aladdin,	10	2
Mickey Skate Boarding, 1980, Aladdin,	18	5
Mighty Mouse, 1979, Thermos,	35	10
Miss Piggy's Safari Van, 1989, Superseal, no bottle	15	0
Monster in My Pocket, 1990, Aladdin,	8	3
Movie Monsters, 1979, Universal,	35	12
Mr. T, 1984, Aladdin,	12	3
Munchie Tunes Bear with Radio, 1986, Fun Design, with built-in radio	25	5
Munchie Tunes Punchie Pup w/Radio, 1986, Fun Design, with built-in radio	25	5
Munchie Tunes Robot with Radio, 1986, Fun Design, with built-in radio	25	5
Muppets (blue), 1982, Thermos,	12	4
Muppets Dome, 1981, Thermos, plastic red box with matching bottle	20	5

Plastic

NAME, YEAR, COMPANY, DESCRIPTION	BOX NM	BOTTLE NM
New Kids on the Block (pink/orange), 1990, Thermos,	5	2
Nosy Bears, 1988, Aladdin,	8	2
Official Lunch Football, 1974, unknown, football shaped box, red or brown	100	0
Peanuts, Wienie Roast, 1985, Thermos,	10	4
Pee Wee's Playhouse, 1987, Thermos, with generic plastic bottle	10	5
Peter Pan Peanut Butter, 1984, Taiwan,	30	10
Pickle, 1972, Fesco, plastic pickle box, no bottle	100	0
Popeye & Son, 1987, Servo, plastic red box, flat paper label, with matching bottle	10	4
Popeye & Son 3-D, 1987, Servo, plastic box, red or yellow, with matching bottle	25	5
Popeye Dome, 1979, Aladdin,	20	5
Popeye, Truant Officer, 1964, King Seeley Thermos, plastic red box, matching metal bottle (Canada)	250	35
Punky Brewster, 1984, Deka,	15	4
Q-Bert, 1983, Thermos,	10	2
Race Cars, 1987, Servo,	20	0
Raggedy Ann & Andy, 1988, Aladdin,	12	5
Rainbow Bread Van, 1984, Moldmark Industries,	50	10
Rainbow Bright, 1983, Thermos,	5	2
Robot Man and Friends, 1984, Thermos,	10	4
Rocketeer, 1990, Aladdin,	5	2
Rocky Roughneck, 1977, Thermos,	20	5
Roller Games, 1989, Thermos,	15	3
S.W.A.T. Dome, 1975, Thermos,	30	8
Scooby Doo, 1984, Aladdin,	25	8
Scooby Doo, 1973, Thermos,	35	10
Scooby-Doo, A Pup Named, 1988, Aladdin,	10	5
Sesame Street, 1985, Aladdin, Canada,	10	5
Shirt Tales, 1981, Thermos,	10	2
Sky Commanders, 1987, Thermos, with generic plastic bottle	10	5
Smurfette, 1984, Thermos,	10	5
Smurfs, 1984, Thermos,	10	5
Smurfs Dome, 1981, Thermos,	20	5
Smurfs Fishing, 1984, Thermos,	10	5
Snak Shot Camera, 1987, Hummer, plastic camera shaped box, blue or green, with generic plastic bottle	20	2
Snoopy Dome, 1978, Thermos,	20	5
Snorks, 1984, Thermos,	8	3
Snow White, 1980, Aladdin,	20	5
Spare Parts, 1982, Aladdin, with generic plastic bottle	20	5
Sport Billy, 1982, Thermos,	15	5
Sport Goofy, 1986, Aladdin,	30	5
Star Com. U.S. Space Force, 1987, Thermos,	10	5
Star Trek Next Generation, 1988, Thermos, blue box, group picture, matching bottle	10	5
Star Trek Next Generation, 1989, Thermos, red box, Picard, Data, Wesley, matching bottle	20	5
Star Wars, Droids, 1985, Thermos,	20	5
Strawberry Shortcake, 1980, Aladdin,	5	2
Superman II Dome, 1986, Aladdin,	30	5
Superman, This is a Job For, 1980, Aladdin, no bottle	20	0
Talespin, 1986, Aladdin,	5	2
Tang Trio, 1988, Thermos, red or yellow box with generic plastic bottle	10	2
Teenage Mutant Ninja Turtles, 1990, Thermos, with generic plastic bottle	6	3
Thundarr the Barbarian Dome, 1981, Aladdin, plastic dome box with matching bottle	20	5
Timeless Tales, 1989, Aladdin,	5	2
Tiny Toon Adventures, 1990, Thermos,	5	2
Tom & Jerry, 1989, Aladdin,	15	5
Transformers, 1985, Aladdin,	12	3

Plastic

NAME, YEAR, COMPANY, DESCRIPTION	BOX NM	BOTTLE NM
Transformers Dome, 1986, Aladdin, Canada, dome box, generic plastic bottle	15	5
Tweety & Sylvester, 1986, Thermos,	15	5
V (TV Series), 1984, Aladdin,	140	40
Wayne Gretzky, 1980, Aladdin,	40	10
Wayne Gretzky Dome, 1980, Aladdin,	55	10
Where's Waldo, 1990, Thermos,	5	1
Who Framed Roger Rabbit, 1987, Thermos, red or yellow, with matching bottle	15	5
Wild Fire, 1986, Aladdin,	5	2
Wizard of Oz, 50th Anniversary, 1989, Aladdin,	10	5
Woody Woodpecker, 1972, Aladdin, yellow box, red bottle	50	40
World Wrestling Federation, 1986, Thermos,	10	5
Wrinkles, 1984, Thermos,	8	3
Wuzzles, 1985, Aladdin,	8	3
Yogi's Treasure Hunt, 1987, Servo, plastic box, flat paper label, with matching bottle	25	5
Yogi's Treasure Hunt 3-D, 1987, Servo, plastic 3-D box, green or pink, with matching bottle	20	5

Steel

NAME, YEAR, COMPANY, DESCRIPTION	BOX NM	BOTTLE NM
240 Robert, 1978, Aladdin, plastic bottle	1000	250
A-Team, The, 1985, King Seeley Thermos, plastic bottle	10	3
Action Jackson, 1973, Okay Industries, matching steel bottle	600	200
Adam-12, 1973, Aladdin, matching plastic bottle	50	5
Addams Family, 1974, King Seeley Thermos, matching steel bottle	80	10
Airline, 1969, Ohio Art, no bottle	60	0
All American, 1954, Universal, steel/glass bottle	350	65
America on Parade, 1976, Aladdin, matching plastic bottle	30	5
Americana, 1958, King Seeley Thermos, steel/glass bottle	325	125
Animal Friends, 1978, Ohio Art, steel, either yellow or red background behind name	25	0
Annie Oakley & Tagg, 1955, Aladdin, matching steel bottle	175	50
Annie, The Movie, 1982, Aladdin, plastic bottle	25	5
Apple's Way, 1975, King Seeley Thermos, plastic bottle	75	5
Archies, 1969, Aladdin, matching plastic bottle	70	20
Astronaut Dome, 1960, King Seeley Thermos, steel dome box, steel/glass bottle	125	35
Astronauts, 1969, Aladdin, matching plastic bottle	65	20
Atom Ant/Secret Squirrel, 1966, King Seeley Thermos, matching steel bottle	200	50
Auto Race, 1967, King Seeley Thermos, matching steel bottle	50	25
Back in '76, 1975, Aladdin, plastic bottle	30	5
Barbie Lunch Kit, 1962, King Seeley Thermos, tall steel/glass bottle, lunch box	150	40
Basketweave, 1968, Ohio Art, no bottle	20	0
Batman, 1966, Aladdin, matching steel bottle	145	55
Battle Kit, 1965, King Seeley Thermos, matching steel bottle	75	30
Battle of the Planets, 1979, King Seeley Thermos, matching plastic bottle	40	10
Battlestar Galactica, 1978, Aladdin, matching plastic bottle	40	10
Beatles, 1966, Aladdin, blue, matching bottle	300	100
Bedknobs & Broomsticks, 1972, Aladdin, plastic bottle	50	40
Bee Gees, 1978, King Seeley Thermos, steel Barry back box, matching plastic bottle	30	10
Bee Gees, 1978, King Seeley Thermos, steel Maurice back box with matching plastic bottle	40	5
Bee Gees, 1978, King Seeley Thermos, steel Robin back box, matching plastic bottle	40	5
Bernstein Bears, 1983, American Thermos, matching plastic bottle	25	5
Beverly Hillbillies, 1963, Aladdin, matching steel bottle	130	35
Bionic Woman, with Car, 1977, Aladdin, plastic bottle	35	10
Bionic Woman, with Dog, 1978, Aladdin, matching plastic bottle	35	10

FOOD TOYS

Steel

NAME, YEAR, COMPANY, DESCRIPTION	BOX NM	BOTTLE NM
Black Hole, 1979, Aladdin, matching plastic bottle	40	10
Blondie, 1969, King Seeley Thermos, matching steel bottle	100	35
Boating, 1959, American Thermos, matching steel bottle	400	125
Bobby Sherman, 1972, King Seeley Thermos, matching steel bottle	55	25
Bonanza, 1968, Aladdin, steel black rim box, steel bottle	120	50
Bonanza, 1963, Aladdin, steel brown rim box with matching steel bottle	95	45
Bonanza, 1963, Aladdin, steel green rim box with steel bottle	80	40
Bond XX, 1967, Ohio Art, no bottle	150	0
Bond-XX Secret Agent, 1966, Ohio Art, no bottle	175	0
Boston Bruins, 1973, Okay Industries, steel/glass bottle	350	125
Bozo the Clown Dome, Aladdin, steel dome box and steel bottle	125	50
Brady Brunch, 1970, King Seeley Thermos, matching steel bottle	150	60
Brave Eagle, 1957, American Thermos, red, blue, gray or green band, matching steel bottle	200	25
Bread Box Dome, 1968, Aladdin, steel dome with Campbell's Soup bottle	250	100
Buccaneer Dome, 1957, Aladdin, steel dome box, matching bottle	200	60
Buck Rogers, 1979, Aladdin, matching plastic bottle	20	5
Bugaloos, 1971, Aladdin, matching plastic bottle	70	10
Bullwinkle & Rocky, 1962, Universal, steel blue box with matching steel bottle	600	220
Cabbage Patch Kids, 1984, King Seeley Thermos, matching plastic bottle	12	2
Cable Car Dome, 1962, Aladdin, steel dome box and steel/glass bottle	175	60
Campbell Kids, 1973, Okay, matching steel bottle	120	140
Campus Queen, 1967, King Seeley Thermos, matching steel bottle	35	15
Canadian Pacific Railroad, 1970, Ohio Art, no bottle	5	0
Captain Astro, 1966, Ohio Art, no bottle	190	0
Care Bear Cousins, 1985, Aladdin, matching plastic bottle	10	2
Care Bears, 1984, Aladdin, plastic bottle	7	3
Carnival, 1959, Universal, matching steel bottle	425	175
Cartoon Zoo Lunch Chest, 1962, Universal, steel/glass bottle	300	125
Casey Jones, 1960, Universal, steel dome box, steel/glass bottle	650	125
Chan Clan, The, 1973, King Seeley Thermos, plastic bottle	85	15
Charlie's Angels, 1978, Aladdin, matching plastic bottle	35	10
Chavo, 1979, Aladdin, matching plastic bottle	40	5
Children's, 1984, Ohio Art, no bottle	10	0
Children, Blue, 1974, Okay Industries, matching steel bottle	160	40
Children, Yellow, 1974, Okay Industries, matching steel bottle	110	40
Chitty Chitty Bang Bang, 1969, King Seeley Thermos, matching steel bottle	45	25
Chuck Wagon Dome, 1958, Aladdin, steel dome box, matching bottle	140	55
Circus Wagon, 1958, King Seeley Thermos, steel dome box, steel/glass bottle	300	110
Clash of the Titans, 1981, King Seeley Thermos, matching plastic bottle	30	8
Close Encounters of the Third Kind, 1978, King Seeley Thermos, plastic bottle	80	5
Color Me Happy, 1984, Ohio Art, no bottle	40	0
Corsage, 1958, American Thermos, matching steel bottle	50	20
Cowboy in Africa, Chuck Connors, 1968, King Seeley Thermos, matching steel bottle	190	35
Cracker Jack, 1969, Aladdin, matching plastic bottle	55	10
Curiosity Shop, 1972, King Seeley Thermos, matching steel bottle	45	20
Cyclist, The: Dirt Bike, 1979, Aladdin, plastic bottle	45	10
Daniel Boone, 1955, Aladdin, matching steel bottle	200	50
Daniel Boone, 1965, Aladdin, matching steel bottle	120	40
Dark Crystal, 1982, King Seeley Thermos, matching plastic bottle	10	3
Davy Crockett, 1955, Holtemp, matching steel bottle	100	40
Davy Crockett, 1955, Kruger, no bottle	350	0
Davy Crockett/Kit Carson, 1955, Adco Liberty,	200	
Debutante, 1958, Aladdin, matching steel bottle	30	5
Denim Diner Dome, 1975, Aladdin, steel dome, matching plastic bottle	20	5
Dick Tracy, 1967, Aladdin, matching steel bottle	120	55
Disco, 1979, Aladdin, matching plastic bottle	15	5
Disco Fever, 1980, Aladdin, matching plastic bottle	15	5
Disney Express, 1979, Aladdin, matching plastic bottle	10	5

FOOD TOYS

Steel

NAME, YEAR, COMPANY, DESCRIPTION	BOX NM	BOTTLE NM
Disney Fire Fighters Dome, 1974, Aladdin, steel dome box, matching plastic bottle	80	20
Disney School Bus Dome, 1968, Aladdin, steel dome box, steel/glass bottle	35	20
Disney World, 1972, Aladdin, matching plastic	20	10
Disney's Magic Kingdom, 1980, Aladdin, plastic bottle	12	5
Disney's Rescuers, The, 1977, Aladdin, plastic bottle	18	5
Disney's Robin Hood, 1974, Aladdin, plastic bottle	55	20
Disney, Wonderful World of, 1982, Aladdin, plastic bottle	12	5
Disneyland (Castle), 1957, Aladdin, matching steel bottle	125	35
Disneyland (Monorail), 1968, Aladdin, matching steel bottle	150	35
Doctor Doolittle, 1968, Aladdin, matching steel bottle	80	35
Donald Duck, 1980, Cheinco, no bottle	15	0
Double Decker, 1970, Aladdin, matching plastic bottle	60	25
Dr. Doolittle, Aladdin, steel/glass bottle	90	40
Dr. Seuss, 1970, Aladdin, matching plastic bottle	70	20
Drag Strip, 1975, Aladdin, matching plastic bottle	35	10
Dragon's Lair, 1983, Aladdin, matching plastic bottle	20	4
Duchess, 1960, Aladdin, steel/glass bottle	30	10
Dudley Do-Right, 1962, Universal, matching steel bottle	600	225
Dukes of Hazzard, 1983, Aladdin, matching plastic bottle	20	5
Dutch Cottage, 1958, King Seeley Thermos, steel dome box, steel/glass bottle	450	150
Dyno Mutt, 1977, King Seeley Thermos, plastic bottle	45	7
E.T., The Extra Terrestrial, 1982, Aladdin, matching plastic bottle	30	8
Early West Oregon, 1982, Ohio Art, no bottle	25	0
Early West Pony Express, 1982, Ohio Art, no bottle	25	0
Emergency!, 1973, Aladdin, plastic bottle	45	10
Emergency!, 1977, Aladdin, steel dome box, plastic bottle	80	12
Evel Knievel, 1974, Aladdin, plastic bottle	40	12
Exciting World of Metrics, The, 1976, King Seeley Thermos, plastic bottle	25	5
Fall Guy, 1981, Aladdin, matching plastic bottle	15	5
Family Affair, 1969, King Seeley Thermos, matching steel bottle	60	30
Fat Albert and the Cosby Kids, 1973, King Seeley Thermos, plastic bottle	25	6
Fess Parker, 1965, King Seeley Thermos, matching steel bottle	140	35
Fireball XL5, 1964, King Seeley Thermos, steel/glass bottle	140	45
Firehouse Dome, 1959, American Thermos, steel dome box, steel/glass bottle	300	100
Flag-O-Rama, 1954, Universal, steel/glass bottle	325	60
Flintstones, 1973, Aladdin, matching plastic bottle	45	15
Flintstones, 1962, Aladdin, orange, 1st issue, matching bottle	75	30
Flintstones, 1963, Aladdin, yellow, 2nd issue, matching bottle	95	35
Flipper, 1966, King Seeley Thermos, matching steel bottle	125	50
Floral, 1970, Ohio Art, no bottle	12	0
Flying Nun, 1968, Aladdin, matching steel bottle	40	20
Fonz, The, 1978, King Seeley Thermos, plastic bottle	25	4
Fox and the Hound, 1981, Aladdin, plastic bottle	30	5
Fraggle Rock, 1984, King Seeley Thermos, matching plastic bottle	15	5
Fritos, 1975, King Seeley Thermos, generic bottle	85	3
Frontier Days, 1957, Ohio Art, no bottle	210	0
Frost Flowers, 1962, Ohio Art, no bottle	40	0
Fruit Basket, 1975, Ohio Art, no bottle	15	0
Funtastic World of Hanna Barbera, 1978, King Seeley Thermos, Flintstones & Yogi, plastic bottle	25	5
Funtastic World of Hanna Barbera, 1977, King Seeley Thermos, Huck Hound, plastic bottle	25	5
G.I. Joe, 1982, King Seeley Thermos, plastic bottle	12	3
G.I. Joe, 1967, King Seeley Thermos, steel/glass bottle	75	40
Gene Autry, 1954, Universal, steel/glass bottle	275	120
Gentle Ben, 1968, Aladdin, plastic bottle, glass liner	75	20
Get Smart!, 1966, King Seeley Thermos, steel/glass bottle	110	40
Ghostland, 1977, Ohio Art, spinner game, no bottle	35	0

FOOD TOYS

Steel

NAME, YEAR, COMPANY, DESCRIPTION	BOX NM	BOTTLE NM
Globe-Trotter Dome, 1959, Aladdin, steel dome box, matching steel/glass bottle	120	40
Gomer Pyle, 1966, Aladdin, matching steel bottle	60	35
Goober and the Ghostchasers, 1974, King Seeley Thermos, matching plastic bottle	35	8
Great Wild West, 1959, Universal, matching steel bottle	350	100
Green Hornet, 1967, King Seeley Thermos, matching steel bottle	275	60
Gremlins, 1984, Aladdin, matching plastic bottle	20	8
Grizzly Adams Dome, 1977, Aladdin, plastic bottle	60	10
Guns of Will Sonnett, The, 1968, King Seeley Thermos, steel/glass bottle	85	40
Gunsmoke, 1972, Aladdin, steel "mule splashing" box with matching bottle	55	15
Gunsmoke, 1973, Aladdin, steel "stagecoach" box, matching bottle	100	15
Gunsmoke, 1959, Aladdin, plastic bottle	120	35
Gunsmoke, Double L Version, 1959, Aladdin, double L error version, matching bottle	375	40
Gunsmoke, Marshal Matt Dillon, 1962, Aladdin, matching steel bottle	160	55
H.R. Pufnstuf, 1970, Aladdin, matching plastic bottle	90	20
Hair Bear Bunch, 1972, King Seeley Thermos, plastic bottle	35	10
Hansel and Gretel, 1982, Ohio Art, no bottle	55	0
Happy Days, 1977, American Thermos, matching plastic bottle	20	4
Hardy Boys Mysteries, 1977, King Seeley Thermos, matching plastic bottle	30	5
Harlem Globetrotters, 1971, King Seeley Thermos, matching steel bottle in either blue or purple uniforms	45	35
Have Gun, Will Travel, 1960, Aladdin, steel	250	0
He-Man & Masters of the Universe, 1984, Aladdin, matching plastic bottle	5	5
Heathcliff, 1982, Aladdin, matching plastic bottle	20	4
Hector Heathcote, 1964, Aladdin, matching steel bottle	170	55
Hee Haw, 1971, King Seeley Thermos, matching steel bottle	65	20
Highway Signs, 1972, Ohio Art, no bottle	50	0
Hogan's Heroes Dome, 1966, Aladdin, steel dome box, steel/glass bottle	175	70
Holly Hobby, 1968, Aladdin, red rim, matching plastic bottle	10	5
Holly Hobbie, 1970, Aladdin, matching plastic bottle	5	1
Holly Hobbie, 1979, Aladdin, matching plastic bottle	5	1
Home Town Airport Dome, 1960, King Seeley Thermos, steel dome box, steel/glass bottle	600	150
Hong Kong Phooey, 1975, King Seeley Thermos, steel/glass bottle	25	15
Hopalong Cassidy, 1954, Aladdin, black rim, steel/glass bottle	175	50
Hopalong Cassidy, 1952, Aladdin, full litho, matching steel bottle	210	70
Hopalong Cassidy, 1950, Aladdin, red or blue, steel/glass bottle	95	35
Hot Wheels, 1969, King Seeley Thermos, matching steel bottle	60	20
How the West Was Won, 1979, King Seeley Thermos, matching plastic bottle	35	5
Howdy Doody, 1954, Adco Liberty,	225	0
Huckleberry Hound, 1961, Aladdin, steel/glass bottle	60	30
Incredible Hulk, The, 1978, Aladdin, plastic bottle	30	10
Indian Territory, 1982, Ohio Art, plastic bottle	20	
Indiana Jones, 1984, King Seeley Thermos, matching plastic bottle	15	5
Indiana Jones Temple of Doom, 1984, King Seeley Thermos, matching plastic bottle	15	5
It's About Time Dome, 1967, Aladdin, steel dome box, matching bottle	200	50
It's About Time Dome, 1967, Aladdin, steel dome box, steel/glass bottle	140	50
Jack and Jill, 1982, Ohio Art,	400	0
James Bond 007, 1966, Aladdin, matching steel bottle	130	35
Jet Patrol, 1957, Aladdin, matching steel bottle	250	55
Jetsons, 1968, Aladdin, steel dome, steel/glass bottle	850	160
Jetsons Dome, 1963, Aladdin, steel dome box, matching bottle	675	175
Joe Palooka, 1949, Continental Can, no bottle	30	0
Johnny Lightning, 1970, Aladdin, plastic bottle	45	5
Jonathan Livingston Seagull, 1973, Aladdin, matching plastic bottle	50	20
Julia, 1969, King Seeley Thermos, matching steel bottle	65	30
Jungle Book, 1968, Aladdin, matching steel bottle	55	30
Junior Miss, 1978, Aladdin, matching plastic bottle	30	10

FOOD TOYS

FOOD TOYS

Top to bottom: Munsters steel bottle, 1965, King Seeley Thermos; Peanuts steel bottle, 1980, King Seeley Thermos; Tom Corbett Space Cadet steel box with matching bottle, 1952, Aladdin.

Steel

NAME, YEAR, COMPANY, DESCRIPTION	BOX NM	BOTTLE NM
Kellogg's Breakfast, 1969, Aladdin, plastic bottle	75	20
King Kong, 1977, King Seeley Thermos, plastic bottle	29	6
KISS, 1977, King Seeley Thermos, plastic bottle	65	8
Knight in Armor, 1959, Universal, matching steel bottle	425	150
Knight Rider, 1984, King Seeley Thermos, matching plastic bottle	20	5
Korg, 1975, King Seeley Thermos, matching plastic bottle	30	5
Krofft Supershow, 1976, Aladdin, matching plastic bottle	15	8
Kung Fu, 1974, King Seeley Thermos, matching plastic bottle	50	8
Lance Link, Secret Chimp, 1971, King Seeley Thermos, matching steel bottle	75	30
Land of the Giants, 1968, Aladdin, plastic bottle	75	50
Land of the Lost, 1975, Aladdin, matching plastic bottle	50	8
Laugh-In (Helmet), 1969, Aladdin, helmet on back, matching plastic bottle	90	25
Laugh-In (Tricycle), 1969, Aladdin, trike on back, matching plastic bottle	100	15
Lawman, 1961, King Seeley Thermos, generic bottle	80	25
Legend of the Lone Ranger, 1980, Aladdin, plastic bottle	35	8
Lidsville, 1971, Aladdin, matching plastic bottle	90	25
Little Dutch Miss, 1959, Universal, matching steel bottle	80	35
Little Friends, 1982, Aladdin, matching plastic bottle	300	50
Little House on the Prairie, 1978, King Seeley Thermos, matching plastic bottle	50	10
Little Red Riding Hood, 1982, Ohio Art, no bottle	25	0
Lone Ranger, 1955, Adco Liberty, red rim, blue band, no bottle	450	0
Looney Tunes TV Set, 1959, King Seeley Thermos, steel/glass bottle	135	60
Lost in Space Dome, 1967, King Seeley Thermos, steel dome box, steel/glass bottle	550	15
Ludwig Von Drake, 1962, Aladdin, steel/glass bottle	175	50
Luggage Plaid, 1955, Adco Liberty, no bottle	75	0
Luggage Plaid, 1957, Ohio Art, no bottle	40	22
Magic of Lassie, 1978, King Seeley Thermos, matching plastic bottle	35	10
Major League Baseball, 1968, King Seeley Thermos, matching bottle	60	20
Man from U.N.C.L.E., 1966, King Seeley Thermos, matching steel bottle	140	35
Marvel Super Heroes, 1976, Aladdin, black rim, matching plastic bottle	35	5
Mary Poppins, 1965, Aladdin, steel/glass bottle	70	25
Masters of the Universe, 1983, Aladdin, matching plastic bottle	10	2
Mickey Mouse & Donald Duck, 1954, Adco Liberty, matching steel bottle	240	400
Mickey Mouse Club, 1977, Aladdin, red rim, sky boat, matching bottle	30	3
Mickey Mouse Club, 1976, Aladdin, white, matching steel bottle	40	20
Mickey Mouse Club, 1963, Aladdin, yellow, steel/glass bottle	45	20
Miss America, 1972, Aladdin, matching plastic bottle	30	10
Mod Floral, 1975, Okay Industries, matching steel bottle	180	150
Monroes, 1967, Aladdin, matching steel bottle	125	35
Mork & Mindy, 1979, American Thermos, matching plastic bottle	25	5
Mr. Merlin, 1982, King Seeley Thermos, matching plastic bottle	20	5
Munsters, 1965, King Seeley Thermos, matching steel bottle	160	60
Muppet Babies, 1985, King Seeley Thermos, matching plastic bottle	12	5
Muppet Movie, 1979, King Seeley Thermos, plastic bottle	30	6
Muppet Show, 1978, King Seeley Thermos, plastic bottle	16	6
Muppets, 1979, King Seeley Thermos, back shows Animal, Fozzie or Kermit, matching plastic bottle	20	5
My Lunch, 1976, Ohio Art, no bottle	30	
Nancy Drew, 1978, King Seeley Thermos, plastic bottle	20	5
NFL, 1978, King Seeley Thermos, blue rim, matching plastic bottle	20	5
NFL, 1976, King Seeley Thermos, red rim, matching plastic bottle	20	5
NFL, 1975, King Seeley Thermos, yellow rim, plastic bottle	20	10
NFL, 1972, Okay, black rim, steel/glass bottle	110	125
NFL Quarterback, 1964, Aladdin, matching steel bottle	70	35
NHL, 1970, Okay Industries, plastic bottle	525	225
Orbit, 1963, King Seeley Thermos, matching steel bottle	150	35
Oregon Trail, 1982, Ohio Art, plastic bottle	20	5
Osmonds, The, 1973, Aladdin, matching plastic bottle	50	15
Our Friends, 1982, Aladdin, matching plastic bottle	300	150

FOOD TOYS

Steel

NAME, YEAR, COMPANY, DESCRIPTION	BOX NM	BOTTLE NM
Pac-Man, 1980, Aladdin, matching plastic bottle	10	4
Para-Medic, 1978, Ohio Art, no bottle	45	0
Partridge Family, 1971, King Seeley Thermos, plastic or steel bottle	45	30
Pathfinder, 1959, Universal, matching steel bottle	375	125
Patriotic, 1974, Ohio Art, no bottle	45	0
Peanuts, 1976, King Seeley Thermos, red steel "pitching" box, plastic bottle	25	5
Peanuts, 1980, King Seeley Thermos, steel "pitching" box, yellow face, green band, matching bottle	20	5
Peanuts, 1966, King Seeley Thermos, orange rim, matching steel bottle	35	15
Peanuts, 1973, King Seeley Thermos, steel red rim "psychiatric" box, plastic bottle	35	5
Pebbles & Bamm Bamm, 1971, Aladdin, matching plastic bottle	55	20
Pelé, 1975, King Seeley Thermos, matching plastic bottle	40	8
Pennant, 1950, Ohio Art, steel basket type box, no bottle	30	0
Pete's Dragon, 1978, Aladdin, matching plastic bottle	35	10
Peter Pan, 1969, Aladdin, matching plastic bottle, Disney	80	30
Pets'n Pals, 1961, King Seeley Thermos, matching steel bottle	55	20
Pigs In Space, 1977, King Seeley Thermos, matching plastic bottle	20	5
Pink Gingham, 1976, King Seeley Thermos, matching plastic bottle	25	4
Pink Panther & Sons, 1984, King Seeley Thermos, matching plastic bottle	20	4
Pinocchio, 1971, Aladdin, plastic bottle	60	20
Pinocchio, 1938, unknown, square	150	0
Pinocchio, 1938, unknown, steel round tin with handle	150	0
Pit Stop, 1968, Ohio Art, generic bottle	175	10
Planet of the Apes, 1974, Aladdin, matching plastic bottle	65	15
Play Ball, 1969, King Seeley Thermos, game on back, steel bottle	60	25
Police Patrol, 1978, Aladdin, plastic bottle	140	10
Polly Pal, 1975, King Seeley Thermos, matching plastic bottle	10	4
Pony Express, 1982, Ohio Art,	19	
Popeye, 1980, Aladdin, steel "arm wrestling" box, plastic bottle	25	5
Popeye, 1964, King Seeley Thermos, steel "Popeye in boat" box with matching steel bottle	100	30
Popeye, 1962, Universal, steel "Popeye socks Bluto" box, matching bottle	450	250
Popples, 1986, Aladdin, plastic bottle	10	1
Porky's Lunch Wagon, 1959, King Seeley Thermos, steel dome box, steel/glass bottle	350	65
Pro Sports, 1974, Ohio Art, no bottle	50	0
Psychedelic Dome, 1969, Aladdin, steel dome box, plastic bottle	210	30
Racing Wheels, 1977, King Seeley Thermos, plastic bottle	50	5
Raggedy Ann & Andy, 1973, Aladdin, plastic bottle	25	5
Rambo, 1985, King Seeley Thermos, matching plastic bottle	8	3
Rat Patrol, 1967, Aladdin, steel/glass bottle	80	40
Red Barn Dome, 1957, King Seeley Thermos, steel dome box, closed door version, plain Holtemp bottle	50	20
Red Barn Dome, 1958, King Seeley Thermos, steel dome box, open door version, matching steel bottle	35	20
Red Barn Dome, Cutie, 1972, Thermos, matching steel/glass bottle	25	10
Rifleman, The, 1961, Aladdin, steel/glass bottle	200	50
Road Runner, 1970, King Seeley Thermos, lavender or purple rim, steel or plastic bottle	50	25
Robin Hood, 1956, Aladdin, matching bottle	100	40
Ronald McDonald, Sheriff, 1982, Aladdin, plastic bottle	30	5
Rose Petal Place, 1983, Aladdin, plastic bottle	20	3
Rough Rider, 1973, Aladdin, plastic bottle	30	5
Roy Rogers & Dale Dbl R Bar Ranch, 1954, American Thermos, blue or red band, woodgrain tall bottle	90	40
Roy Rogers & Dale Dbl R Bar Ranch, 1955, American Thermos, steel eight-scene box, red or blue band, matching bottle	60	40
Roy Rogers & Dale Dbl R Bar Ranch, 1953, King Seeley Thermos, steel/glass bottle	80	40

Top to bottom: Roy Rogers And Dale Evans Double R Bar Ranch steel box, 1954, American Thermos; Mork & Mindy steel box, 1979, American Thermos; Huckleberry Hound and his friends steel/glass bottle, 1961, Aladdin; Gremlins steel box, 1984, Aladdin; The Astronauts, 1969/The Rat Patrol, 1967/Batman and Robin, 1966, all steel boxes, all Aladdin.

FOOD TOYS

Steel

NAME, YEAR, COMPANY, DESCRIPTION	BOX NM	BOTTLE NM
Roy Rogers & Dale Evans, 1955, American Thermos, steel cowhide back box, red or blue band, matching bottle	80	40
Roy Rogers & Dale on Rail, 1957, American Thermos, red or blue band, matching bottle	90	40
Roy Rogers Chow Wagon Dome, 1958, King Seeley Thermos, steel dome box, steel/glass bottle	165	45
Saddlebag, 1977, King Seeley Thermos, generic plastic bottle	60	5
Satellite, 1958, American Thermos, matching bottle	75	25
Satellite, 1960, King Seeley Thermos, steel bottle	110	35
Scooby Doo, 1973, King Seeley Thermos, yellow or orange rim, plastic bottle	30	5
Secret Agent T, 1968, King Seeley Thermos, matching bottle	70	30
Secret of NIMH, 1982, Aladdin, plastic bottle	45	5
Secret Wars, 1984, Aladdin, plastic bottle	45	10
See America, 1972, Ohio Art, no bottle	35	0
Sesame Street, 1983, Aladdin, yellow rim, plastic bottle	10	7
Sigmund and the Sea Monsters, 1974, Aladdin, plastic bottle	35	15
Six Million Dollar Man, 1974, Aladdin, plastic bottle	35	5
Six Million Dollar Man, 1978, Aladdin, plastic bottle	35	5
Skateboarder, 1977, Aladdin, plastic bottle	35	5
Sleeping Beauty, 1960, General Steel Ware, Canada, generic steel bottle	450	55
Smokey Bear, 1975, Okay Industries, plastic bottle	225	200
Smurfs, 1983, King Seeley Thermos, steel blue box, plastic bottle	110	5
Snoopy Dome, 1968, King Seeley Thermos, steel dome, yellow, "Have Lunch With Snoopy", matching bottle	35	5
Snow White, Disney, 1975, Aladdin, orange rim, plastic bottle	35	5
Snow White, with Game, 1980, Ohio Art, no bottle	35	0
Space Explorer Ed McCauley, 1960, Aladdin, matching steel bottle	250	65
Space Ship, 1950, unknown, Decoware, dark blue square	250	0
Space Shuttle Orbiter Enterprise, 1977, King Seeley Thermos, plastic bottle	45	10
Space:1999, 1976, King Seeley Thermos, plastic bottle	45	10
Speed Buggy, 1974, King Seeley Thermos, red rim, plastic bottle	100	35
Spiderman & Hulk, 1980, Aladdin, Captain America on back, plastic bottle	30	5
Sport Goofy, 1983, Aladdin, yellow rim, plastic bottle	25	9
Sport Skwirts, 1982, Ohio Art, All four sports box	35	0
Sport Skwirts, Jimmy Blooper, 1982, Ohio Art, no bottle	25	0
Sport Skwirts-Freddie Face Off, 1982, Ohio Art, no bottle	25	0
Sport Skwirts-Willie Dribble, 1982, Ohio Art, no bottle	25	0
Sports Afield, 1957, Ohio Art, no bottle	130	0
Star Trek Dome, 1968, Aladdin, steel dome box, matching bottle	575	175
Star Trek, The Motion Picture, 1980, King Seeley Thermos, matching bottle	50	10
Star Wars, 1978, King Seeley Thermos, cast or stars on band, matching plastic bottle	40	5
Star Wars Return of the Jedi, 1983, King Seeley Thermos, plastic bottle	40	10
Star Wars, Empire Strikes Back, 1980, King Seeley Thermos, plastic bottle	40	10
Stars and Stripes Dome, 1970, King Seeley Thermos, steel dome box, matching plastic bottle	50	11
Steve Canyon, 1959, Aladdin, steel/glass bottle	10	40
Strawberry Shortcake, 1980, Aladdin, plastic bottle	10	4
Strawberry Shortcake, 1981, Aladdin, plastic bottle	10	4
Street Hawk, 1985, Aladdin, plastic bottle	125	40
Submarine, 1960, King Seeley Thermos, steel/glass bottle	75	30
Super Friends, 1976, Aladdin, matching plastic bottle	40	10
Super Powers, 1983, Aladdin, plastic bottle	30	5
Supercar, 1962, Universal, steel/glass bottle	250	125
Superman, 1978, Aladdin, red rim "Daily Planet Office" on back, matching bottle	25	10
Superman, 1967, King Seeley Thermos, red rim, "under fire" art on back, matching steel/glass bottle	75	50
Superman, 1954, Universal, blue rim	800	
Tapestry, 1963, Ohio Art, no bottle	60	0
Tarzan, 1966, Aladdin, steel/glass bottle	70	35

Steel

NAME, YEAR, COMPANY, DESCRIPTION	BOX NM	BOTTLE NM
Teenager, 1957, King Seeley Thermos, generic bottle	70	25
Teenager Dome, 1957, King Seeley Thermos, steel dome box, generic bottle	120	10
Three Little Pigs, 1982, Ohio Art, red rim, generic/plastic bottle	45	0
Thundercats, 1985, Aladdin, plastic bottle	20	3
Tom Corbett Space Cadet, 1952, Aladdin, steel blue or red paper decal box, steel/glass bottle	200	50
Tom Corbett Space Cadet, 1954, Aladdin, full litho, matching bottle	475	50
Toppie Elephant, 1957, American Thermos, yellow, matching bottle	600	250
Track King, 1975, Okay Industries, matching steel bottle	180	100
Train, 1971, Ohio Art, steel box, no bottle made	10	0
Transformers, 1986, Aladdin, steel red box, matching plastic bottle	10	5
Traveler, 1962, Ohio Art, no bottle	40	0
Trigger, 1956, King Seeley Thermos, no bottle	130	
U.S. Mail Dome, 1969, Aladdin, steel dome box, plastic bottle	40	10
U.S. Space Corps, 1961, Universal, plastic bottle	350	95
UFO, 1973, King Seeley Thermos, plastic bottle	90	10
Underdog, 1974, Okay Industries, plastic bottle	800	300
Universal's Movie Monsters, 1980, Aladdin, plastic bottle	40	5
Voyage to the Bottom of the Sea, 1967, Aladdin, steel/glass bottle	300	50
VW Bus Dome, 1960, King Seeley, steel dome box, plastic bottle	450	125
Wagon Train, 1964, King Seeley Thermos, generic bottle	100	30
Wags 'n Whiskers, 1978, King Seeley Thermos, matching plastic bottle	18	4
Wake Up America, 1973, Okay Industries, matching steel bottle	600	250
Waltons, The, 1973, Aladdin, plastic bottle	30	5
Washington Redskins, 1970, Okay Industries, steel bottle	200	100
Wee Pals Kid Power, 1974, American Thermos, matching plastic bottle	20	5
Welcome Back Kotter, 1977, Aladdin, flat or embossed face, red rim, matching plastic bottle	40	5
Western, 1963, King Seeley Thermos, tan or red rim, steel/glass bottle	80	45
Wild Bill Hickock, 1955, Aladdin, steel/glass bottle	100	45
Wild Frontier, 1977, Ohio Art, spinner game on back, no bottle	30	0
Wild, Wild West, 1969, Aladdin, plastic bottle	150	20
Winnie the Pooh, 1976, Aladdin, blue rim, plastic bottle	150	35
Yankee Doodles, 1975, King Seeley Thermos, plastic bottle	15	2
Yellow Submarine, 1968, King Seeley Thermos, steel/glass bottle	250	150
Yogi Bear & Friends, 1961, Aladdin, black rim, matching steel bottle	95	30
Zorro, 1958, Aladdin, black band, steel/glass bottle	90	40
Zorro, 1966, Aladdin, red band, steel/glass bottle	90	40

Vinyl

NAME, YEAR, COMPANY, DESCRIPTION	BOX NM	BOTTLE NM
Alice in Wonderland, 1972, Aladdin, matching plastic bottle	125	15
All American, 1976, Bayville, styrofoam bottle	90	5
All Dressed Up, 1970s, Bayville, styrofoam bottle	90	5
All Star, 1960, Aladdin,	400	40
Alvin & Chipmunks, 1963, King Seeley Thermos, matching plastic bottle	275	60
Annie 1, 1981, Aladdin, matching plastic bottle	40	15
Bach's Lunch, 1975, Volkwein Bros., Inc., red, styrofoam bottle	120	10
Ballerina, 1962, Aladdin, pink, steel/glass bottle	90	30
Ballerina, 1960s, Universal, black, Thermax bottle	500	10
Ballet, 1961, Universal, red, plastic generic bottle	500	20
Banana Splits, 1969, King Seeley Thermos, matching steel/glass bottle	200	40
Barbarino Brunch Bag, 1977, Aladdin, zippered bag, plastic bottle	40	8
Barbie & Francie, 1965, King Seeley Thermos, black, matching steel/glass bottle	90	30
Barbie & Midge, 1965, King Seeley Thermos, black, matching steel/glass bottle	95	30
Barbie & Midge Dome, 1964, King Seeley Thermos, matching glass/steel bottle	170	30
Barbie Softy, 1988, King Seeley Thermos, pink, plastic bottle is unmatched	6	2

FOOD TOYS

Vinyl

NAME, YEAR, COMPANY, DESCRIPTION	BOX NM	BOTTLE NM
Barbie, World of, 1971, King Seeley Thermos, blue box, matching steel/glass bottle	35	5
Barbie, World of, 1971, King Seeley Thermos, pink box, matching steel/glass bottle	35	5
Barnum's Animals, 1978, Adco Liberty, no bottle	50	0
Beany & Cecil, 1963, King Seeley Thermos, steel/glass bottle	300	50
Beatles Brunch Bag, 1966, Aladdin, zippered bag, matching bottle	350	100
Beatles Kaboodles Kit, The, 1965, Standard Plastic Products, no bottle	600	0
Beatles, The, 1965, Air Flite, no bottle	500	0
Betsey Clark Munchies Bag, 1977, King Seeley Thermos, zippered bag, plastic bottle	25	7
Betsy Clark, 1977, King Seeley Thermos, yellow box, matching plastic bottle	25	5
Blue Gingham Brunch Bag, 1975, Aladdin, zippered box and plastic bottle	10	5
Bobby Soxer, 1959, Aladdin,	300	0
Boston Red Sox, 1960s, Universal,	30	0
Boy on the Swing, Abeama Industries,	50	0
Buick 1910, 1974, Bayville, styrofoam bottle	90	5
Bullwinkle, 1963, King Seeley Thermos, blue steel/glass bottle	400	100
Bullwinkle, 1963, King Seeley Thermos, yellow, generic steel bottle	300	40
Calico Brunch Bag, 1980, Aladdin, zippered bag, plastic bottle	10	5
Captain Kangaroo, King Seeley Thermos, steel/glass bottle	225	40
Carousel, 1962, Aladdin, matching steel/glass bottle	300	30
Cars, 1960, Universal,	90	0
Casper the Friendly Ghost, 1966, King Seeley Thermos, blue box, orange steel bottle	300	40
Challenger, Space Shuttle, 1986, Babcock, puffy box, no bottle	160	0
Charlie's Angels Brunch Bag, 1978, Aladdin, zippered bag, plastic bottle	40	10
Coca-Cola, 1947, Aladdin, styrofoam bottle	100	10
Coco the Clown, 1970s, Gary, styrofoam bottle	40	5
Combo Brunch Bag, 1967, Aladdin, zippered bag, steel/glass bottle	400	40
Corsage, 1970, King Seeley Thermos, steel/glass bottle	75	20
Cottage, 1974, King Seeley Thermos,	50	0
Cowboy, 1960, Universal, plain "syro" plastic bottle	170	5
Dateline Lunch Kit, 1960, Hasbro, box in blue/pink, no bottle	175	0
Dawn, 1972, Aladdin, matching plastic bottle	90	15
Dawn, 1971, Aladdin, matching plastic bottle	90	15
Dawn Brunch Bag, 1971, Aladdin, zippered bag, plastic bottle	90	10
Denim Brunch Bag, 1980, Aladdin, zippered bag, plastic bottle	30	5
Deputy Dawg, 1964, Thermos, no bottle	350	0
Deputy Dawg, King Seeley Thermos, vinyl box, steel/glass bottle	200	35
Donny & Marie, 1977, Aladdin, long hair version, matching plastic bottle	50	10
Donny & Marie, 1978, Aladdin, short hair version, matching plastic bottle	50	20
Donny & Marie Brunch Bag, 1977, Aladdin, zippered bag, plastic bottle	50	10
Dr. Seuss, 1970, Aladdin, plastic bottle	275	15
Dream Boat, 1960, Feldco, styrofoam bottle	275	10
Eats 'n Treats, King Seeley Thermos, blue steel/glass bottle	200	20
Fess Parker Kaboodle Kit, 1960s, Aladdin, matching steel bottle	350	35
Fishing, 1970, Universal, styrofoam bottle	120	5
Frog Flutist, 1975, Aladdin, matching plastic bottle	30	10
Fun to See'n Keep Tiger, 1960, unknown, no bottle	150	0
G.I. Joe, 1989, King Seeley Thermos, generic plastic bottle	12	2
Gigi, 1962, Aladdin, matching steel/glass bottle	90	35
Girl & Poodle, 1960, Universal, styrofoam bottle	90	5
Glamour Gal, 1960, Aladdin, steel/glass bottle	150	35
Go-Go Brunch Bag, 1966, Aladdin, plastic bottle	100	10
Goat Butt Mountain, 1960, Universal, plain "styro" plastic bottle	100	5
Happy Powwow, 1970s, Bayville, red or blue, with styrofoam bottle	40	5
Highway Signs Snap Pack, 1988, Avon,	30	0
Holly Hobbie, 1972, Aladdin, white bag, matching plastic bottle	30	5
I Love a Parade, 1970, Universal, plain "styro" plastic bottle	50	5
Ice Cream Cone, 1975, Aladdin, matching plastic bottle	30	10

FOOD TOYS

Vinyl

NAME, YEAR, COMPANY, DESCRIPTION	BOX NM	BOTTLE NM
It's a Small World, 1968, Aladdin, matching steel/glass bottle	125	40
Jonathan Livingston Seagull, 1974, Aladdin, matching plastic bottle	100	10
Junior Deb, 1960, Aladdin, steel/glass bottle	125	35
Junior Miss Safari, 1962, Prepac, no bottle	120	0
Junior Nurse, 1963, King Seeley Thermos, steel/glass bottle	160	35
Kaboodle Kit, 1960s, Aladdin, pink or white, no bottle	150	0
Kewtie Pie, Aladdin, steel/glass bottle	60	30
Kodak Gold, 1970s, Aladdin,	35	15
Kodak II, 1970s, Aladdin,	35	15
L'il Jodie (Puffy), 1985, Babcock,	45	0
Lassie, 1960s, Universal, styrofoam bottle	90	5
Liddle Kiddles, 1969, King Seeley Thermos, matching steel/glass bottle	125	35
Linus the Lion-Hearted, 1965, Aladdin, steel/glass bottle	325	60
Little Ballerina, 1975, Bayville, styrofoam bottle	50	5
Little Kiddles, unknown,	50	0
Little Old Schoolhouse, 1974, Dart,	30	0
Love, 1972, Aladdin, matching plastic bottle	75	15
Lunch 'n Munch, 1959, American Thermos, corsage bottle	300	35
Lunch'n Munch, 1959, King Seeley Thermos, vinyl box, steel/glass bottle	400	25
Mam'zelle, 1971, Aladdin, plastic bottle	100	10
Mardi-gras, 1971, Aladdin, matching plastic bottle	40	10
Mary Ann, 1960, Aladdin, matching steel/glass bottle	70	20
Mary Ann Lunch 'N Bag, 1960, Universal, bag, no bottle	100	0
Mary Poppins, 1973, Aladdin, matching plastic bottle	90	25
Mary Poppins Brunch Bag, 1966, Aladdin, bag, steel/glass bottle	130	30
Mod Miss Brunch Bag, 1969, Aladdin, plastic bottle	40	10
Monkees, 1967, King Seeley Thermos, matching steel/glass bottle	250	35
Moon Landing, 1960, Universal, styro bottle	140	5
Mr. Peanut Snap Pack, 1979, Dart, snap close bag, no bottle	40	0
Mushrooms, 1972, Aladdin, matching plastic bottle	40	20
New Zoo Review, 1975, Aladdin, plastic bottle	160	30
Pac-Man (Puffy), 1985, Aladdin,	10	0
Peanuts, 1971, King Seeley Thermos, green "baseball" box, steel bottle	50	20
Peanuts, 1969, King Seeley Thermos, red "baseball" box, steel bottle	50	20
Peanuts, 1967, King Seeley Thermos, red "kite" box, steel/glass bottle	50	20
Peanuts, 1973, King Seeley Thermos, white "piano" box, steel bottle	35	20
Pebbles & Bamm-Bamm, 1973, Gary, matching plastic bottle	150	10
Penelope & Penny, 1970s, Gary, yellow box with styrofoam bottle	50	5
Peter Pan, 1969, Aladdin, white box, matching plastic bottle	140	25
Pink Panther, 1980, Aladdin, matching plastic bottle	50	10
Pony Tail, 1965, King Seeley Thermos, white box, fold over lid, steel/glass bottle	200	30
Pony Tail Tid-Bit-Kit, 1962, King Seeley Thermos, steel/glass satellite bottle	200	30
Pony Tail with Gray Border, 1960s, Thermos, white box, original art with gray border added, no bottle	200	0
Ponytails Poodle Kit, 1960, King Seeley Thermos, steel/glass bottle	150	20
Princess, 1963, Aladdin, steel/glass bottle	90	30
Psychedelic (yellow), 1969, Aladdin, matching steel/glass bottle	90	30
Pussycats, The, 1968, Aladdin, plastic bottle.	90	10
Ringling Bros. Circus, 1970, King Seeley Thermos, orange box with matching steel/glass bottle	275	35
Ringling Bros. Circus, 1971, King Seeley Thermos, puffy blue box, steel/glass bottle	50	40
Robo Warriors, 1970, unknown, no bottle	35	0
Roy Rogers Saddlebag, Brown, 1960, King Seeley Thermos, steel/glass bottle	150	40
Roy Rogers Saddlebag, Cream, 1960, King Seeley Thermos, steel/glass bottle	300	40
Sabrina, 1972, Aladdin, yellow box with matching plastic bottle	90	15
Sesame Street, 1979, Aladdin, orange, matching plastic bottle	35	10
Sesame Street, 1981, Aladdin, yellow, matching plastic bottle	35	5

Vinyl

NAME, YEAR, COMPANY, DESCRIPTION	BOX NM	BOTTLE NM
Shari Lewis, 1963, Aladdin, matching steel/glass bottle	200	35
Sizzlers, Hotwheels, 1971, King Seeley Thermos, matching steel/glass bottle	150	35
Skipper, 1965, King Seeley Thermos, steel/glass bottle	125	35
Sleeping Beauty, Disney, 1970, Aladdin, white box, matching plastic bottle	120	25
Smokey the Bear, 1965, King Seeley Thermos, steel/glass bottle	275	40
Snoopy Munchies Bag, 1977, King Seeley Thermos, plastic bottle	40	8
Snoopy, Softy, 1988, King Seeley Thermos, matching plastic bottle	12	2
Snow White, 1975, Aladdin, white box with matching plastic bottle	140	15
Snow White, Disney, 1967, unknown, fold over lid tapered box, no bottle	150	0
Soupy Sales, 1966, King Seeley Thermos, blue box, no bottle	500	0
Spirit of '76, unknown, red	70	0
Sports Kit, 1960, Universal,	400	20
Stewardess, 1962, Aladdin, steel/glass bottle	300	35
Strawberry Shortcake, 1980, Aladdin, matching plastic bottle	35	5
Tammy, 1964, Aladdin, matching steel/glass bottle	120	35
Tammy & Pepper, 1965, Aladdin, matching steel/glass bottle	120	35
Tinkerbell, Disney, 1969, Aladdin, plastic bottle	200	35
Twiggy, 1967, King Seeley Thermos, steel/glass bottle	125	40
Twiggy, 1967, Aladdin, vinyl box with matching steel/glass bottle	200	35
U.S. Mail Brunch Bag, 1971, Aladdin, zippered bag, plastic bottle	200	10
Winnie the Pooh, Aladdin, steel/glass bottle	400	45
Wonder Woman (blue), 1977, Aladdin, matching plastic bottle	60	10
Wonder Woman (yellow), 1978, Aladdin, matching plastic bottle	60	10
Wrangler, 1982, Aladdin, steel/glass bottle	50	35
Yosemite Sam, 1971, King Seeley Thermos, matching steel/glass bottle	250	40
Ziggy's Munch Box, 1979, Aladdin, plastic bottle	60	10

Top to bottom: Cars, 1992, Wendy's; Barney Rubble, 1993, Denny's; Definitely Dinosaurs Bead Game, 1989, Wendy's; Simpsons (four of five), 1991, Burger King.

FOOD TOYS

FOOD RELATED

Fast Food

Kids' Meals Toys

NAME	DESCRIPTION	YEAR	MNP	MIP
Arby's				
Babar at the Beach Summer Sippers	set of three squeezie bottles; orange, yellow or purple top	1991	1	5
Babar Figures	set of four; Babar with sunglasses, elephant with binoculars, Babar with monkey, and with camera	1990	1	3
Babar License Plates	set of four: Paris, Brazil, USA, North Pole	1990	1	2
Babar Puzzles	set of four; Cousin Arthur's New Camera, Babar's Gondola Ride, Babar and the Haunted Castle, Babar's Trip to Greece	1990	3	5
Babar Stampers	set of three; Babar, Flora, Arthur	1990	2	4
Babar Storybooks	set of three; Read Get Ready, Set, Go, Calendar-Read and Have Fun-Read and Grow and Grow	1991	2	2
Babar World Tour Vehicles	set of three vehicles; Babar in helicopter, Arthur on trike, Zephyr in car	1990	2	4
Little Miss Figures	set of seven; Giggles, Shy, Splendid, Late, Naughty, Star, Sunshine	1981	2	4
Looney Tunes Car-Tunes	set of six, Sylvester's Cat-illac, Daffy's Dragster, Yosemite Sam's Rackin Frackin Wagon, Taz' Slush Musher, Bugs' Buggy, Roadrunner's Racer	1990	1	3
Looney Tunes Characters	set of three, Tazmanian Devil as pilot, Daffy as student, Sylvester as fireman	1991	2	4
Looney Tunes Christmas Ornament	Bugs, Porky Pig		3	6
Looney Tunes Figures	Bugs, Daffy, Taz, Elmer, Roadrunner, and Wile E. Coyote	1988	2	4
Looney Tunes Figures	stiff legged figures; Elmer, Roadrunner, Bugs, Daffy, Coyote, Taz	1988	2	4
Looney Tunes Figures	figures on oval base, Tasmanian Devil, Tweetie, Porky, Bugs, Yosemite Sam, Sylvester, Pepe Le Pew	1987	3	5
Looney Tunes Pencil Toppers	Sylvester, Yosemite, Porky, Bugs, Taz, Daffy, Tweety	1988	2	4
Mr. Men Figures	set of 12; Bump, Clever, Daydream, Funny, Greedy, Grumpy, Happy, Lazy, Noisy, Rush, Strong, Tickle	1981	2	6
Big Boy				
Helicopters	set of plastic vehicles; Ambulance, Police, Fire Department	1991	0	3
Monster In My Pocket	various secret monster packs	1991	2	4
Burger King				
Adventure Kits	set of four activity kits with crayons; Passport, African Adventure, European Escapades, Worldwide Treasure Hunt	1991	1	3
Aladdin	set of five figures; Jafar and Iago, Genie in Lamp, Jasmine and Rajah, Abu, Aladdin and the Magic Carpet	1992	2	4
Alf	Joke & riddle disc, door knob card, sand mold, refrigerator magnet	1987	1	2
Alf Puppets	puppets with records; Sporting with Alf, Cooking with Alf, Born to Rock, Surfing with Alf	1987	2	3
Alvin and the Chipmunks	set of three toys; super ball, stickers, pencil topper	1987	1	2
Animal Boxes	set of four activity booklets; bear, hippo, lion, one more	1986	1	2
Aquaman Tub Toy	green		3	5
Archie Cars	set of four, Archie in red car, Betty in aqua car, Jughead in green car, Veronica in purple car	1991	2	4
Barnyard Commandos	set of four; Major Legger Mutton in boat, Sgt. Shoat & Sweet in plane, Sgt. Wooley Pullover in sub, Pvt. Side O'Bacon in truck	1991	2	3
Batman Toothbrush Holder			2	4

Kids' Meals Toys

NAME	DESCRIPTION	YEAR	MNP	MIP
Beauty & the Beast PVC Figures	set of four PVc figures; Belle, Beast, Chip, Cogsworth	1991	2	4
Beetlejuice	set of six figures; Uneasy Chair, Head Over Heels, Ghost to Ghost TV, Charmer, Ghost Post, Peek A Boo Doo	1990	2	4
Bicycle License Plate			2	4
Bicycle Safety Fun Booklet			2	4
BK Kids Action Figures	set of four; Boomer, I.Q., Jaws, Kid Vid	1991	2	4
Bone Age Skeleton Kit	set of four dinos; T-Rex, Dimetron, Mastadon, Similodon	1989	2	4
Bone Age Skeleton Kit Boxes	The Past is a Blast, The Greatest Mystery in History	1989	1	2
Burger King Clubhouse	full size for kids to play in		15	35
Burger King Socks	rhinestone accents		2	5
Calendar "20 Magical Years" Walt Disney World		1992	2	4
Capitol Critters	set of four; Hemmet for Prez in White House, Max at Jefferson Memorial, Muggle at Lincoln Memorial, Presidential Cat	1992	2	4
Capitol Critters Cartons	punch out masks; dog, chicken, duck, panda, rabbit, tiger, turtle	1992	1	2
Captain Planet	set of 4 flip over vehicles; Captain Planet & Hoggish Greedily, Linka, Ma-Ti & Dr. Blight ecomobile, Verminous Skumm & Kwane helicopter, Wheeler and Duke Nukem snowmobile	1991	2	4
Captain Planet Cartons	Containers; Powerbase Spaceship, Biodread Patroller Spaceship, Powerjet XT-7	1991	2	3
Christmas Cassette Tapes	set of three Christmas sing-a-long tapes; Joy to the World/ Silent Night, We Three Kings/ O Holy Night, Deck the Haals/ Night Before Christmas	1989	2	4
Christmas Crayola Bear Plush Toys	red, yellow, blue or purple	1986	2	4
Coloring Book	Keep Your World Beautiful		2	4
Crayola Coloring Books	set of six books; Boomer's Color Chase, I.Q.'s Computer Code, Kids Club Poster, Jaws' Colorful Clue, Snaps' Photo Power, Kid Vid's Video Vision	1990	2	4
Dino Meals	punch-out sheets; Stegosaurus, Woolly Mammoth, T-Rex, Triceratops	1987	3	6
Disney 20th Anniversary Figures	set of four wind up vehicles with connecting track; Minnie, Donald, Roger Rabbit, Mickey	1992	3	5
Fairy Tale Cassette Tapes	set of four fairy tale cassettes; Goldilocks, Jack in the Beanstalk, Three Little Pigs, Hansel and Gretel	1989	1	3
Freaky Fellas	set of four, blue, green, red, yellow, each with a roll of Life Savers candy	1992	2	4
Go-Go Gadget Gizmos	set of four Inspector Gadget toys; gray copter, black inflatable, orange scuba, green surfer	1991	2	4
Golden Junior Classic Books	set of four; Roundabout Train, The Circus Train, Train to Timbucktoo, My Little Book of Trains		1	2
Goof Troop Bowlers	set of four; Goofy, Max, Pete, P.J.	1992	1	2
It's Magic	set of four; Magic Trunk, Disappearing Food, Magic Frame, Remote Control	1992	1	2
Kid Transformers	set of six; Bloomer w/Super Show, Kid Vid w/SEGA Gamestar, I.Q. w/World Book Mobile, Snaps w/Camera Car, Jaws w/Burger Racer and Wheels w/Turbo Wheelchair	1990	1	3
Lickety Splits	plastic wheeled food items, Apple Pie Man, Flame Broiler Buggy, Drink Man, French Fry Man, Croissant Man, Chicken Tenders, French Toast Man	1990	2	4
Masters of the Universe Cups	Thunder Punch He-Man Saves the Day, He-Man and Roboto to the Rescue, He-Man Takes on the Evil Horde, Skeletor	1985	1	2
Matchbox Boxes (Buildings)	punch-out buildings; apartment, fire station, engine and firemen, restaurant, barn horse whirl and wheel game	1989	2	3
Matchbox Cars	set of four vehicles; blue Mountain Man 4X4,, yellow Corvette, red Ferrari, Black & white police car	1989	3	6

433

Kids' Meals Toys

NAME	DESCRIPTION	YEAR	MNP	MIP
Mealbots	paper masks with 3-D lenses, red Broil Master, blue Winter Wizard, gray Beta Burger, yellow Galactic Guardians	1986	4	8
Nerfuls	rubber characters, interchangeable, set of four, Bitsy Ball, Fetch, Officer Bob, Scratch	1985	2	4
Pilot Paks	set of four styrofoam airplanes; two seater, sunburst, lightening, one more	1988	4	8
Pinocchio	set of five toys, pail and following inflatables, Beach ball, Figaro, Jiminy Cricket, Monstro	1992	1	2
Purrtenders	set of four plush toys; Hop-purr, Flop-purr, Scamp-purr, Romp-purr	1988	22	4
Purrtenders	set of four; Free Wheeling Cheese Rider, Flip-Top Car, Radio Bank, Storybook	1988	2	3
Record Breakers	set of six cars; Aero, Indy, Dominator, Accelerator, Fastland, Shockwave	1989	3	6
Rodney & Friends, Reindeer	plush toys with holiday fun booklets; Ramona Holiday Sweets and Treats, Rodney Holiday Fun and Games Box, Rhonda Holiday decorating box	1987	4	8
Sea Creatures	terrycloth wash mitts; Stella Starfish, Dolly Dolphin, Sammy Seahorse, Ozzie Octopus	1989	2	3
Simpsons	set of five figures, Bart with backpack, Homer with skunk, Lisa with sax, Maggie with turtle, Marge with birds	1991	2	4
Simpsons Cups	set of four	1991	1	22
Simpsons Dolls	set of five soft plastic dolls; Bart, Homer, Lisa, Marge, Maggie	1991	2	4
Spacebase Racers	set of five plastic vehicles; Moon Man Rover, Skylab Cruiser, Starship Viking, Super Shuttle, Cosmic Copter	1989	2	4
Super Heroes Cups	set of five cups with figural handles; Batman, Robin, Wonder Woman, Darkseid, Superman	1984	3	4
Teenage Mutant Ninja Turtles Badges	six different, Michaelanglo, Leonardo, Raphael, Donetello, Heroes In a Half Shell, Shredder	1990	2	4
Thundercats	set of four toys; cup/bank, Snarf strawholder, light switch plate, secret message ring	1986	3	8
Top Kids	set of four spinning tops with figural heads; Wheels, Kid Vid, two more	1992	1	2
Tricky Treaters Boxes	Monster Manor, Creepy Castle, Haunted House	1989	1	3
Tricky Treaters PVC Figures	set of three; Frankie Steen, Gourdy Goblin, Zelda Zoom Broom	1989	3	6
Water Club Mates	set of four; Lingo's Jet Ski, Snaps in Boat, Wheels on raft, I.Q. on dolphin	1991	2	4

Chucky E Cheese Pizza

NAME	DESCRIPTION	YEAR	MNP	MIP
Chucky E Cheese PVC Figures		1980s	4	7

Dairy Queen

NAME	DESCRIPTION	YEAR	MNP	MIP
Suction Cup Throwers	set of four		2	4

Denny's

NAME	DESCRIPTION	YEAR	MNP	MIP
Dino-Makers	set of six, including blue dino, purple elephant, orange bird	1991	2	4
Flintstones Dino Racers	Set of six: Fred, Bam-Bam, Dino, Pebbles, Barney, Wilma	1991	2	4
Flintstones Fun Squirters	set of six; Fred w/telephone, Wilma w/camera, Dino w/flowers, Bam Bam w/soda, Barney, Pebbles	1991	1	2
Flintstones Glacier Gliders	set of six; Bam Bam, Barney, Fred, Dino, Hoppy, Pebbles	1990	1	4
Flintstones Mini Plush	in packages of 2: Fred/Wilma, Betty/Barney, Dino/Hoppy, Pebbles/Bamm Bamm, each 4" tall	1989	2	4
Flintstones Rock & Rollers	Set of six: Fred w/guitar, Barney w/sax, Bam Bam, Dino w/piano, Elephant, Pebbles	1990	2	4
Flintstones Stone-Age Cruisers	Set of six: Fred in green car, Wilma in red car, Dino in blue car, Pebbles in in purple bird, Bam Bam in orange car, Barney in yellow car w/sidecar	1991	2	4
Flintstones Vehicles	Set of eight: Fred, Barney, Pebbles, Wilma, Betty, Bamm Bamm, Dino	1990	2	4

FOOD TOYS

Top to bottom: Hulk, one of four of the Super Heroes in vehicles series, 1990, Hardee's; Noids (two of seven), 1987, Domino's Pizza; Food Squirters, 1990, Hardee's; Beauty & The Beast PVC Figures, 1991, Burger King.

Kids' Meals Toys

NAME	DESCRIPTION	YEAR	MNP	MIP
Jetson's Game Packs	Set of six: George, Elroy, Judy, Astro, Rosie, Jane	1992	1	2
Jetson's Go Back to School	Set of six school tools: mini dictionary, folder, message board, pencil & topper, pencil box, triangle & curve	1992	1	2
Jetson's Puzzle Ornaments	Set of twelve: six shapes in two colors each green/clear or purple/clear	1992	1	1
Jetson's Space Balls (Planets)	Set of six: Jupiter, Neptune, Earth, Saturn, Mars, glow-in-the-dark Moon	1992	1	2
Jetson's Space Cards	Set of five: Spacecraft, Phenomenon, Astronomers, Constellations, Planets	1992	1	2
Jetson's Space Travel Coloring Books	Set of six books: each with four crayons	1992	1	2

Dominos Pizza

NAME	DESCRIPTION	YEAR	MNP	MIP
Noids	Set of seven figures: Boxer, Clown, He Man, Holding Bomb, Holding Jack Hammer, Hunchback, Magician	1987	1	2
Quarterback Challenge Cards	pack of four cards	1991	1	1

Hardee's

NAME	DESCRIPTION	YEAR	MNP	MIP
California Raisins	third set, of four; Alotta Stile in pink boots, Anita Break w/package under her arm, Benny bowling, Buster w/skateboard	1991	1	2
California Raisins	first set, of four; dancer w/blue & white shoes, singer w/mike, sax player, fourth with sunglasses	1987	3	6
California Raisins	second set, of six; w/guitar in orange sneakers, on rollerskate w/yellow sneakers, with radio w/ yellow sneakers, surfer w/red sneakers, trumpet player w/ blue sneakers	1988	2	4
California Raisins Plush	set of four, each 6" tall; Lady in yellow shoes, Dancer in yellow hat, with mike in white shoes, in sunglasses w/orange hat	1988	3	6
Days of Thunder Racers	Set of four cars: Mello Yellow #51, Hardees #18 orange, City Chevy #46, Superflo in pink/white	1990	2	4
Disney's Animated Classics Plush Toys	Pinocchio, Bambi		3	5
Ertl Camero	Marked Hardee's Roadrunner	1990	3	6
Fender Bender 500 Racers	Set of five: Quickdraw/Babalouie's Covered Wagon, Huckleberry/Snagglepuss's truck, Wally Gater/Magilla's Toilet, Dick Dastardly/Mutley's rocket racer, Yogi/Booboo's basket	1990	2	4
Finger Crayons	Set of four: not marked Hardee's		2	3
Flintstones First 30 Years	Set of five: Fred w/TV, Barney w/grill, Pebbles w/ phone, Dino w/juke box, Bam Bam w/pinball	1991	1	3
Food Squirters	Set of four: cheeseburger, hot dog, shake, fries	1990	1	2
Gremlin Adventures	Set of five book and record sets: Gift if the Mogwai, Gismo & the Gremlins, Escape from the Gremlins, Gremlins Trapped, The Late Gremlin	1989	2	4
Halloween Hideaways	Set of four: Goblin in blue caldron, Ghost in yellow bag, cat in pumpkin, Bat in stump	1989	2	4
Home Alone 2	Set of four	1992	1	2
Kazoo Crew Sailors	Set of four: brown bear, monkey, rabbit, rhino	1991	1	2
Little Golden Books	Set of four: The Little Red Caboose, The Three Bears, Old MacDonald Had a Farm, Three Little Kittens		1	2
Pound Puppies	Set of six: white w/black spots, black, brown, gray w/ brown spots, gray w/black ears, tan w/black ears	1991	2	3
Shirt Tales Plush Dolls	Set of five, each 7" tall: Bogey, Pammy, Tyg, Digger, Rick		2	4
Smurfin' Smurfs	Set of four surfers: Papa w/red board, boy w/orange board, girl w/purple board, dog w/blue board	1990	2	3
Smurfs Box	Smurf Solution, Smurf Angle, Smurf Surprise, Stop & Smurf the Flowers	1990	1	2
Super Heroes	Marvel figures in vehicles; She Hulk, Hulk, Captain America, Spiderman	1990	3	6
Tang Mouth Figures	Set of four: Lance, Tag, Flap, Annie	1989	2	4
Waldo's Straw Buddies	Set of four: Waldo, Girl, Wizard	1990	2	3

FOOD TOYS

Kids' Meals Toys

NAME	DESCRIPTION	YEAR	MNP	MIP
Waldo's Travel Adventure	complete set of four		2	3

International House of Pancakes

Pancake Kid Refrigerator Magnets	Chocolate Chip Charlie and Bonnie Blueberry	1992	3	6
Pancake Kids	set of six		2	4
Pancake Kids	set of eight	1992	1	2
Pancake Kids Stuffed Figures	Chocolate Chip Charlie and Bonnie Blueberry		0	9

Jack-In-The-Box

Jack Pack Puzzle Books	set of three		3	6
Magnets	set of three		2	3
Scratch and Sniff	Pineapple/lilac		0	3

Kentucky Fried Chicken

Alvin and the Chipmunks	Canadian issues; Alvin and Theodore	1991	3	5
Alvin and the Chipmunks	Canadian issues; Alvin and Simon	1992	2	4

Long John Silver's

Adventure on Volcano Island	paint with water activity book	1991	2	4
Fish Cars	red and yellow cars shaped like fish, w/stickers	1989	2	3
Sea Walkers	set of four packaged with string; Parrot, Penguin, Turtle and Sylvia	1990	3	5
Sea Walkers	set of four packaged w/o string; Parrot, Penguin, Turtle and Sylvia	1990	3	5
Sea Watchers Kaleidoscopes	Set of three: orange, yellow, pink	1991	3	5
Treasure Trolls		1992	0	3
Water Blasters	Set of three: Billy Bones, Captain Flint, Ophelia Octopus	1990	3	5

Pizza Hut

Beauty & the Beast Puppets	Set of four: Belle, Beast, Chip, Cogsworth	1992	3	6
Eureeka's Castle Puppets	Set of three: Batly, Eureeka, Magellan	1991	3	6
Land Before Time Puppets	Set of six: Spike, Sharptooth, Petri, Little Foot, Cera, Ducky	1988	2	4

Rax

Optic Top	spinning optic top		3	6
Sipper	yellow		2	4

Roy Rogers

Critters	Set of eight: blue eyes-yellow, blue eyes-orange, blue eyes-purple, blue eyes-red, yellow eyes-orange, yellow eyes-yellow, pink eyes, orange, pink eyes-purple		1	2
Gator Tales	set of four		1	2
Gumby	Gumby, blue girl		4	8
Ickky Stickky Bugs	set of sixteen		2	3
Skateboard Kids (figurines)	yellow knee pads, orange knee pads, purple knee pads, red knee pads		3	5
Star Searchers	saucer w/green top and orange pilot, sled w/orange top and purple bottom, sled w/purple top and orange bottom, sled w/green top and orange bottom, robot w/orange head, green front		2	3

Sonic Drive In

Paper Bag Man	set of two (?); Marbles, Bookworm	1989	2	4

Taco Bell

Happy Talk Sprites	yellow Spark, white Twink		2	4
Hugga Bunch Plush Dolls			0	4

FOOD TOYS

Kids' Meals Toys

NAME	DESCRIPTION	YEAR	MNP	MIP
Wendy's				
Alf Tales	Set of six: Sleeping Alf, Alf Hood, Little Red Riding Alf, Alf of Arabia, Three Little Pigs, Sir Gordon of Melmac	1990	2	4
Alien Mix-Ups	Set of six: red Crimsonoid, blue blueziod, green Limetoid, orange Spotasoid, yellow Yellowboid, Purple Purpapoid	1990	2	4
All Dogs Go To Heaven	Set of six: Anne Marie, Car Face, Charlie, Flo, Itchy, King Gator	1989	2	4
Definitely Dinosaurs	Set of four: blue Apatosaurus, gray T-Rex, yellow Anatosaurus, green Triceratops	1988	3	6
Definitely Dinosaurs	Set of five: green Ankylosaurus, blue Parasaurolophus, green Ceratosaurus, yellow Stegosaurus, pink Apatosaurus	1989	2	4
Fast Food Racers	Set of five: hamburger, fries, shake, salad, kid's meal	1990	2	4
Furskins Plush dolls	Set of three, 7" tall: Boone in plaid shirt and red pants, Farrell in plaid shirt and blue jeans, Hattie in pink and white dress	1988	2	4
Glass Hangers	yellow turtle, yellow frog, yellow penguin and purple gator		3	5
Glo Friends	set of 12 (?); Book Bug, Bop Bug, Butterfly, Clutter Bug, Cricket, Doodle Bug, Globug, Granny Bug, Skunk Bug, Snail Bug, Snug Bug	1988	1	2
Good Stuff Gang	Cool Stuff, Cat, Hot Stuff, Overstuffed, Bear, Penguin	1985	4	8
Jetson's	Set of six figures in spaceships: George, Judy, Jane, Elroy, Astro, Spacely	1989	2	4
Jetson's, The Movie, Space Gliders	Set of six PVC figures on wheeled bases; Astro, Elroy, Judy, Fergie, Grunchee, George	1990	2	4
Micro Machines Super Sky Carriers	set of six kits that connect to form Super Sky Carrier	1990	0	5
Mighty Mouse	Set of six: Bat Bat, Cow, Mighty Mouse, Pearl Pureheart, Petey, Scrappy	1989	3	5
Play-Doh Fingles	Set of three finger puppet molding kits: green dough with black mold, blue dough with green mold, yellow dough with white mold	1989	3	6
Potato Head Kids	Set of six: Captain Kid, Daisy, Nurse, Policeman, Slugger, Sparky	1987	4	8
Speed Writers	Set of six car shaped pens: black, blue, fushia, green, orange red	1991	2	4
Summer Fun	float pouch, sky saucer	1991	2	4
Teddy Ruxpin	Set of five: Professor Newton Gimmick, Teddy, Wolly Wahts- It, Fob, Grubby Worm	1987	2	4
Too Kool for School	set of five		2	4
Tricky Tints	set of four		2	4
Wacky Wind-Ups	Set of five: Milk Shake, Biggie French Fry, Stuff Potato, Hamburger, Hamburger in box	1991	2	3
World Wildlife Foundation	Set of four plush toys: panda, snow leopard, koala, tiger	1988	4	8
World Wildlife Foundation	Set of four books: All About Koalas, All About Tigers, All About Snow Leopards, All About Pandas	1988	1	2
Yogi Bear & Friends	Set of six: Ranger Smith in kayak, Boo Boo on skateboard, Yogi on skates, Cindy on red scooter, Huckleberry in inner tube, Snagglepuss with surf board	1990	2	4
Whataburger				
Posable Animals	Horse, camel, giraffe, dog, cat, rabbit and monkey		2	3
White Castle				
Camp White Castle			3	6
Camp White Castle Bowls	Bowls, heavy orange plastic		3	6
Castle Creatures			0	5
Castle Friends Bubble Makers	set of four		2	3

FOOD TOYS

Top to bottom: Flintstones Fun Squirters (five of six), 1991, Denny's; Land Before Time Puppets, 1988, Pizza Hut; She Hulk, one of four of the Super Heroes in vehicles series, 1990, Hardee's; All Dogs Go To Heaven (two of six), 1989, Wendy's; E.T., 1985, McDonald's; Jetsons, The Movie, Space Gliders (three of six), 1990, Wendy's.

FOOD TOYS

Kids' Meals Toys

NAME	DESCRIPTION	YEAR	MNP	MIP
Castle Meal Friends	set of six	1991	2	4
Castle Meal Friends	set of five	1992	2	4
Castleburger Dudes Wind Up Toys	set of four		2	4
Cosby Kids	set of four		2	3
Easter Pals	Rabbit with carrot, rabbit with purse		2	4
Glow in the Dark Pull Apart Monsters	set of three		1	2
Godzilla Squirter			3	6
Holiday Huggables	Candy Canine, Kitty Lights, Holly Hog		0	5
Nestle's Quik Plush Bunny			0	7
Push 'N GO GO GO!	set of three		2	4
Silly Putty	set of three		2	4
Stunt Grip Geckos	set of four		2	4
Tiara	Ballerina's Tiara		3	6
Tootsie Roll Express	set of four		2	4
Totally U Back To School	pencil, pencil case		2	4
Willis the Dragon	Christmas giveaway		3	6
Willis the Dragon Sunglasses			2	4

McDonald's Happy Meal Toys

FOOD TOYS

NAME	DESCRIPTION	YEAR	MNP	MIP
101 Dalmations	Set of four PVC figures, Lucky, Pongo, Sergeant Tibbs, Cruella	1991	2	4
101 Dalmations	3 & Under toy, each	1991	6	3
3-D Happy Meal	set of four cartons with 3-D designs with 3-D glasses inside; Bugsville, High Jinx, Loco Motion, Space Follies	1981	3	5
Adventures of Ronald McDonald	set of seven rubber figures, Ronald, Birdie, Big Mac, Captain Crook, Mayor McCheese, Hamburglar, Grimace	1981	1	3
Airport	Birdie Bentwing Blazer, Fry Guy Flyer, Grimace Bi-Plane, Big Mac Helicopter (green), Ronald Sea Plane	1986	4	8
Alvin & The Chipmunks	set of four figures; Simon, Theodore, Brittany and Alvin	1991	2	4
American Tail	set of four books; Fievel and Tiger, Fievel's Friends, Fievel's Boat Trip, Tony and Fievel	1986	2	3
Animal Riddles	eight different rubber figures 2-2 1/2" tall, condor, snail, turtle, mouse, anteater, alligator, pelican, dragon, in various colors	1979	1	2
Astronauts	four different: Lunar Rover, Satellite dish, Command Module & Space Shuttle	1991	3	5
Astronauts	under 3 toy: Ronald McDonald in Lunar Rover	1991	6	3
Astrosnicks 1	eight different 3" rubber space creatures, Scout, Thirsty, Robo, Laser, Snickapotomus, Sport, Ice Skater, Astralia	1983	5	10
Astrosnicks 2	six different rubber space creatures, Copter, Drill, Ski, Racing, Perfido, Commander	1984	5	10
Astrosnicks 3	total of 14 figures, many same as 1984 series, but without "M" logo; Commander, Robo, Perfido, Galaxo, Laser, Copter, Scout, Snikapotamus, boy, Pyramido, Racer, Astrosnick rocket	1985	7	14
Back to the Future	4 different figures, Marty, Einstein, Verne, Doc	1992	2	4
Bambi	Set of four, Owl, Flower, Thumper, Bambi	1988	2	4
Barbie	8 different plastic dolls, Ice Capades, All American, Lights & Lace, Hawaiian Fun, Happy Birthday, Costume Ball, Wedding Day Midge, My First Barbie	1991	4	8
Barbie	Set of eight dolls, Sparkle Eyes, Roller Blade, Rappin Rockin, My First Ballerina, Snap-On, Sun Sensation, Birthday Surprises, Rose Bride	1992	2	4
Barbie	under 3 toy: Costume Ball Barbie	1991	10	5
Barbie	eight different: My First Ballerina, Birthday Party, Western Stampin, Romantic Bride, Hollywood Hair, Paint'n Dazzle, Twinkle Lights, Secret Hearts	1993	3	6

McDonald's Happy Meal Toys

NAME	DESCRIPTION	YEAR	MNP	MIP
Batman Returns Press & Go Vehicles	set of four: Catwoman, Batman, Penguin in vehicles and Batmobile	1992	2	4
Beach Ball	set of three balls tied to Olympics; Grimace in kayak, Ronald, Birdie on sailboat	1984	3	6
Beach Ball	set of three inflatables; red Ronald, blue Birdie, yellow Grimace	1986	2	4
Beach Toys	Set of eight, four inflatables; Fry Kid Super Sailer, Birdie Seaside Submarine, Grimace Bouncin' Beach ball, Ronald Fun Flyer, four sand toys; 2 buckets, shovel and rak	1989	2	4
Beachcomber	white pail with blue lid, and shovel	1986	5	10
Bedtime	set of four items with boxed tube of Crest toothpaste, Ronald toothbrush, Ronald bath mitt, Ronald glow-in-the-dark star figure, Ronald cup	1989	4	8
Berenstain Bear Story Books	set of four, Attic Treasure, Substitute Teacher, Eager Beavers, Life With Papa	1990	2	4
Berenstein Bears II	set of four figures, Papa with wheelbarrow, Mama with shopping cart, Brother with scooter, Sister with wagon	1987	3	5
Bernstein Bear Test Market Set	set of four Christmas figures, Papa with wheelbarrow, Mama with shopping cart, Brother on scooter, Sister on sled	1986	10	20
Bigfoot	set of four Ford trucks with two wheel sizes, Bigfoot, Ms. Bigfoot, Shuttle, Bronco	1987	3	5
Black History	set of two coloring books	1988	3	5
Boats'n Floats	set of four plastic container boats, Chicken McNugget lifeboat, Birdie float, Fry Kids raft, Grimace power boat	1987	5	10
Bobby's World	set of 4: Wagon-Race Car, Innertube-Subamrine, Three Wheeler-Space Ship, Skates-Roller Coaster	1994	2	4
Bobby's World	under 3 toy: Innertube	1994	3	6
Cabbage Patch	set of five Christmas dolls,, Tiny Dancer, Holiday Pageant, Holiday Dreamer, Fun On Ice, All Dressed Up	1992	2	4
Cabbage Patch	under 3 toy: Anne Louise "Ribbons & Bows"	1992	3	6
Camp McDonaldland	set of four, utensils, Birdie camper mess kit, collapsible cup, canteen	1990	2	4
Castle Maker/Sand Castle	set of four molds: Dome, square, cylindrical and rectangle	1987	10	20
Changeables	set of six different figures that change into robots, Big Mac, Shake, Egg McMuffin, Quarter Pounder, French Fries, Chicken McNuggets	1987	2	4
Changeables	set of eight different figures that turn into robots, Large Fries, Quarter Pounder, Hot Cakes, Big Mac, Chesseburger, Shake, Soft Serve Cone, Small Fries	1989	2	4
Changeables	under 3 toy: Character Changer Cube	1989	3	6
Chip 'N Dale's Rescue Rangers	set of four figures in fanciful vehicles; Monterey Jack in Propelaphone, Gadget in Rescue Racer, Chip in Whirlicupter, Dale in Rotoroadster	1989	2	4
Chip 'N Dale's Rescue Rangers	under 3 toy: Gadget Rider & Chip's Rockin Racer	1989	3	6
Circus	set of eight; 1.fun house mirror, 2.acrobat Ronald, 3.French Fry Faller, 4.Strong Gong, 5 & 6. punchout sheets, 7 & 8. fun house and puppet show background	1983	1	3
Circus Parade	set of four; Ringmaster Donald, Birdie Rider, Grimace caliope, Fry Guy and elephant	1991	3	5
Circus Wagon	set of four rubber toys; poodle, chimp, clown, horse	1979	2	3
Commandrons	set of four robots on blister cards; Solardyn, Magna, Motron, Velocitor	1985	7	13
CosMc Crayola	Set of five coloring kits, Crayolas, two with markers, chalk sticks, paint set	1988	2	4
CosMc Crayola	under 3 toy	1988	3	6
Crayola Magic	set of three stencil kits with crayons or markers; triangles with marker, rectangles with crayons, circles with crayons	1986	4	7
Crazy Vehicles	set of four; Ronald's dune buggy, Hamburglar's train, Birdie's airplane, Grimace's car	1991	3	5

FOOD TOYS

McDonald's Happy Meal Toys

NAME	DESCRIPTION	YEAR	MNP	MIP
Design-O-Saurs	set of four plastic interchangeable parts: Ronald on Tyrannosaurus, Grimace-Pterodactyl, Fry Guy-Brontosaurus, Hamburglar-Triceratops	1987	4	8
Dink the Dinosaur	set of six dino finger puppets, each packed with diorama and description; Dink, Flapper, Amber, Crusty, Scat, Shyler	1990	3	6
Dino-Motion Dinosaurs	set of six Sinclairs: Charlene, Earl, Grandma Ethyl, Fran, Baby, Robbie	1993	2	4
Dino-Motion Dinosaurs	under 3 toy: Baby Sinclair	1993	3	6
Dinosaur Days	set of six rubber dinos in different colors; Pteranodon, Triceratops, Stegosaurus, Dimetrodon, T-Rex, Ankylosaurus	1981	2	4
Discover the Rain Forest	set of four activity books with punch out figures, Sticker Safari, Wonders in the Wild, Paint It Wild, Ronald and the Jewel of the Amazon Kingdom	1991	2	4
Disney Favorites	set of four activity books; Lady and the Tramp, Dumbo, Cinderella, The Sword in the Stone	1987	1	2
Duck Tales I	set of four toys; telescope, duck code quacker, magnifying glass, wrist decoder	1988	2	4
Duck Tales II	set of four toys; Uncle Scrooge in red car, Launchpad in plane, Huey, Dewie and Louie on jet ski, Webby on blue trike	1988	4	8
Dukes of Hazzard	set of five container vehicles, regional; white Caddy, Jeep, cop car and pickup, orange General Lee Charger	1982	3	6
Dukes of Hazzard	set of six white plastic cups, national; Luke, Boss Hogg, Bo, Sheriff Roscoe, Daisy, Uncle Jesse	1982	2	4
Earth Days	set of 4: Birdfeeder, Globe terarium, binoculars, tool carrier with shovel	1994	2	4
Earth Days	under 3 toy: tool carrier with shovel	1994	3	6
Fast Macs	set of four pull-back action cars; white Big Mac police car, yellow Ronald jeep, red Hamburglar racer, pink Birdie convertible	1984	3	5
Feeling Good	set of five grooming toys; A. Grimace soap dish, B. Fry Guy sponge, C. Birdie mirror, D. Ronald or Hamburglar toothbrush, E. Captain Crook comb	1985	1	3
Field Trip	set of four: Kaleidoscope, Leaf printer, Nature viewer, Explorer bag	1993	2	4
Field Trip	under 3 toy: Nature viewer	1993	3	6
Fitness Fun	set of eight toys	1992	2	4
Flintstone Kids	set of four figures in animal vehicles, Betty, Barney, Fred, Wilma	1987	2	4
Food Fundamentals	set of four: Slugger, Otis, Milly & Ruby	1993	2	4
Food Fundamentals	under 3 toy: Dunkan	1993	3	6
Fraggle Rock	set of four; Gobo in carrot car, Red in raddish car, Mokey in eggplant car, Wembly and Boober in pickle car	1988	2	4
Fraggle Rock Doozers	set of two; Cotterpin in forklift and Bulldoozer in bulldozer	1988	5	10
Fry Benders	set of four bendable fry figures with accessories; Grand Slam, Froggy, Roadie, Freestyle	1990	3	5
Fun To Go	set of seven cartons with games and activities;	1977	2	4
Fun with Food	set of four multi-piece toys with faces; hamburger, fries, soft drink, McNuggets	1989	3	6
Funny Fry Friends	set of eight toys; Too Tall, Tracker, Rollin' Rocker, Sweet Cuddles, ZZZ's, Gadzooks, Matey, Hoops	1990	2	4
Funny Fry Friends	under 3 toy: two different, Lil'Chief & Little Darling	1990	3	6
Garfield	set of four figures; on scooter, on skateboard, in jeep, with Odie on motorscooter	1989	2	4
Good Morning	set of four grooming items; Ronald toothbrush, McDonaldland comb, Ronald play clock, white plastic cup	1991	2	4
Gravedale High	set of four mechanical Halloween figures; Cleofatra, Frankentyke, Vinnie Stoker, Sid the Invisible Kid	1991	4	8
Halloween Boo Bags	set of three glow-in-the-dark vinyl bags; Witch, Ghost and Frankie	1991	3	5
Halloween Buckets	set of three pumpkin-shaped buckets; McGoblin, McPumpkin, McBoo	1986	2	4

FOOD TOYS

FOOD TOYS

Clockwise from upper left: Tail Spin (three of four), 1990; Super Mario Brothers, 1991; Gravedale High, 1991; Tonka loader (one of five) 1992; Cruella DeVille (one of four) from 101 Dalmatians, 1991, all McDonald's Happy Meal Premiums.

McDonald's Happy Meal Toys

NAME	DESCRIPTION	YEAR	MNP	MIP
Halloween Buckets	set of three lidded pails with black plastic strap handles, with safety stickers attached; orange pumpkin, white glow-in-dark ghost, Green witch	1990	2	4
Halloween Buckets	Ghost, Witch, Pumpkin	1992	1	2
Halloween McNugget Buddies	set of six: Pumpkin, McBoo, Monster, McNuggula, Witchie, Mummie	1993	2	4
Halloween McNugget Buddies	under 3 toy" McBoo McNugget	1993	3	6
Happy Pails, Olympics	set of four with shovels; swimming, cycling, track, Olympic Games	1984	1	3
Hook	set of four figures, Peter Pan, Mermaid, Rufio, Hook	1991	2	4
Hot Wheels	set of eight cars per coast of USA	1983	3	7
Hot Wheels	split promo with Barbie, set of 8 cars; purple or orange Z28 Camero, white or yellow '55 Chevy, green or black '63 Corvette, turquoise or red '57 T-Bird	1991	2	3
Hot Wheels	under 3 toy: Wrench-Hammer (same package)	1991	4	8
Hot Wheels	eight different: Quaker State Racer #62, McDonalds Dragster, McDonalds Thunderbrd #23, Hot Wheels Dragster, McDonalds Funny Car, Hot Wheels Funny Car, Hot Wheels Camaro #1, Duracell Racer #88	1993	3	6
Hot Wheels	under 3 toy: wrench & hammer	1993	4	8
Hot Wheels Mini-Streex	set of eight: Flame-Out, Quick Flash, Turbo Flyer, Black Arrow, Hot Shock, Racer Tracer, Night Shadow, Blade Burner	1992	2	4
Hot Wheels Mini-Streex	under 3 toy: orange arrow	1992	3	6
I Like Bikes	set of four bike accessories; Ronald Basket, Grimace mirror, Birdie spinner, Fry Guy Horn	1990	4	8
Jordan Fitness	set of four; water bottle, disc, football, baseball	1992	2	4
Jordan Fitness	set of eight: jump rope, soccer ball, water bottle, disc, football, baseball, stop watch & basketball	1992	2	4
Jungle Book	set of four wind-up figures; Baloo the bear, Shere Kahn the tiger, King Louie the orangutan, Kaa the snake	1990	2	4
Jungle Book	uner 3 toy, two different: Junior & Mowgli	1990	3	6
Kissy Fur	set of eight rubber figures, some furry surfaced; Toot, Gus, Floyd, Jolene, Lennie, Beehonie, Duane, Kissy Fur	1987	5	10
Lego Building Sets	set of four Duplo kits; airplane, ship, truck, helicopter	1984	4	8
Lego Building Sets	under 3 Duplo toys; animal or building	1984	3	6
Lego Building Sets	under 3 toys by Duplo; bird or boat	1986	3	6
Lego Little Travelers Building Sets	four different sets; blue tanker boat, green airplane, red roadster, yellow helicopter	1986	5	10
Lego Motion	eight different kits; Gyro Bird, Lightning Striker, Land Laser, Sea Eagle, Wind Whirler, Sea Skimmer, Turbo Force, Swamp Stinger	1989	2	4
Linkables	set of four: Birdie on tricycle, Ronald in soap-box racer, Grimace in wagon, Hamburglar in airplane (no under 3 toy made)	1993	3	6
Lion Circus	set of four rubber figures, bear, elephant, hippo, lion	1979	2	3
Little Engineer	Set of five train engines; Birdie Bright Lite, Fry Girl's Express, Fry Guy's Flyer, Grimace Streak, Ronald Rider	1987	3	6
Little Gardener	set of four tools and seed packets; Ronald Water Can, Birdie Shovel, Grimace Rake, Fry Guy Pail	1989	2	4
Little Golden Books	set of five books; Country Mouse and City Mouse, Tom & Jerry, Pokey Little Puppy, Benji, Monster at the End of This Block	1982	2	4
Little Mermaid	set of four figures; Flounder, Ursula, Prince Eric, Ariel with Sebastian	1989	2	4
Looney tunes Quack Up Cars	Taz Tornado Tracker, Porky Ghost Catcher, Bugs Super Stretch Limo, Daffy Splittin' Sportster	1993	2	4
Looney tunes Quack Up Cars	under 3 toy" Swingin' Sedan	1993	3	6
Lost Arches	Pericope, Flashlight, phone, camera	1991	2	4
M-Squad	set of four: Spystamper, Spytracker, Spy-Nocular, Spycoder	1993	2	4

FOOD TOYS

McDonald's Happy Meal Toys

NAME	DESCRIPTION	YEAR	MNP	MIP
M-Squad	under 3 toy: Spytracker	1993	3	6
Mac Tonight	Set of six figures in vehicles; sports car, off-roader, motorcycle, scooter, jet ski, airplane	1990	2	4
Mac Tonight	under 3 toy: Maconskateboard	1990	3	6
Magic Show	set of four tricks; string pull, disappearing hamburger patch, magic tablet, magic picture	1985	3	6
Makin' Movies	set of four: Sound effects machine, camera, clapboard with chalk, Director megaphone	1994	2	4
Makin' Movies	under 3 toy: Sound effects machine	1994	3	6
Matchbox Mini-Flexies	eight different rubber cars; cosmobile, Hairy Hustler, Planet Scout, Hi-Tailer, Datsun, Beach hopper, Baja Buggy	1979	1	3
McBunny Pails	set of three different Easter pails; Pinky, Fluffy, Whiskers	1989	5	10
McDino Changeables	set of 8 dinosaurs; Happy Mealodon, Quarter Pounder Cheesosaur, Big Macosaurus Rex, McNuggetosaurus, Hotcakesodactyl, Large Fryosaur, Trishakatops, McDino cone	1991	2	4
McDino Changeables	under 3 toy: Bronto Cheeseburger, small fry, Ceratops	1991	3	6
McDonaldland Band	set of eight music toys; Grimace saxophone, Fry Guy trumpet and whistle, Ronald harmonica, whistle and pan pipes, kazoo, Hamburglar whistle	1987	3	5
McDonaldland Carnival	set of four toys; Birdie on swing, Grimace in turn-around, Hamburglar on ferris wheel, Ronald on carousel	1990	3	5
McDonaldland Connectables	set of four toys that can be connected to form a train of vehicles; Grimace in wagon, Birdie on a trike, Hamburglar in airplane, Ronald in a race car	1991	3	5
McDonaldland Express	set of four train car containers; Ronald engine, caboose, freight car, coach car	1982	19	38
McDonaldland Junction	train set of four snap together cars; yellow Birdie's Parlor car, red or blue Ronald Engine, purple Grimace caboose, green or white Hamburglar flat car	1983	3	5
McDonaldland Play-Doh	set of eight colors, orange, purple, red, yellow, blue, green, white, pink	1986	1	3
McDrive thru Crew	set of four vehicle toys; fries in potatoe roadster, shake in milk carton, McNugget in egg roadster, hamburger in ketchup bottle	1990	4	6
McNugget Buddies	set of ten rubber figures and accessories; Sparky, Volley, Corny, Drummer, Cowpoke, Sarge, Snorkel, First Class, Rocker, Boomerang	1989	3	5
McNugget Buddies	under 3 toy, boy-Slugger, girl-Daisy	1989	3	6
Michael Jordon Fitness	set of eight toys; Soccer Ball, Squeeze Bottle, Stopwatch, Basketball, Football, Baseball, Jumprope, Frisbee	1992	2	4
Mickey's Birthdayland	set of five characters in vehicles, Minnie's convertible, Donald's train, Goofy's Jalopy, Mickey's roadster, Pluto's rumbler	1989	3	6
Mickey's Birthdayland	under 3 toy, five different, Mickey, Minnie, Goofy & Donald	1989	4	7
Mighty Mini 4 X 4s	set of four big wheel vehicles; Cargo Climber, Dune Buster, L'il Classic, Pocket Pickup	1991	3	4
Mix'em Up Monsters	Set of four monsters with interchangeable parts; Corkle, Thugger, Gropple, Blibble	1990	2	4
Moveables	set of six vinyl bendies; Birdie, Captain Crook, Fry Girl, Hamburglar, Professor, Ronald	1988	3	6
Muppet Babies	four different; Kermit on red skateboard, Fozzie on hobby horse with wheels, Gonzo on tricycle with red wheels, Piggy in pink convertible	1987	3	5
Muppet Babies	four different; Piggy on trike, Gonzo in airplane, Fozzie in wagon, Kermit on soapbox car	1991	2	4
Music (Records in Sleeves)	set of four 45 RPM records in sleeves with different songs and colored labels; labels were green, yellow, purple, blue	1985	3	6
My Little Pony	split promo with Transformers, set of six; Minty, Snuzzle, Blossom, Cotton Candy, Blue Belle, Butterscotch	1985	2	4
Mystery of Lost Arches	4 different	1991	2	4

FOOD TOYS

FOOD TOYS

Clockwise from upper left: McDino Changeables (four of eight) in ready position, 1991; Space Raiders, 1979; Back to the Future (one of four), 1992; Batman Returns Press & Go Vehicles (one of four), 1992; Peanuts, 1990, all McDonald's Happy Meal Premiums.

McDonald's Happy Meal Toys

NAME	DESCRIPTION	YEAR	MNP	MIP
Mystery of the Lost Arches	set of four; Mini-cassette, Phone, Telescope, Camera	1992	2	3
Nature's Helper	four different	1990	1	2
Nature's Helpers	set of five garden tools with seeds; 1. hinged trowel, 2. rake, 3. water can	1991	2	3
Nature's Helpers	set of five items; 4. Terrarium, 5. Bird Feeder	1991	2	3
Nature's Watch	set of four: Double Shovel-Rake, Green house with seeds, Bird Buddy, water pail (no under 3 toy made)	1991	2	4
New Archies	Set of six figures in bumper cars, Moose, Reggie, Archie, Veronica, Betty, Jughead	1988	5	9
New Food Changeables	set of eight; Krypto Cup, Fry Bot, Turbo Cone, Macro Mac, Gallacta Burger, Robo Cakes, C-2 Cheeseburger, Fry Force	1989	3	4
Nickelodeon	set of four: Blimp game, Loud-Mouth Mike, Gotcha Gusher, Applause Paws	1993	2	4
Nickelodeon	under 3 toy: Blimp Squirt toy	1993	3	6
Old McDonald's Farm	set of six figures: Farmer, Wife, rooster, pig, sheep, cow	1986	5	10
Old West	set of six rubber figures; cowboy, frontiersman, lady, Indian, Indian woman, Sheriff	1981	4	6
Oliver and Company	Set of four figures; Oliver, Georgette, Francis and Dodger	1988	3	5
On the Go	set of five games; stop light bead game, Ronald slate board lift pad, Hamburglar lift pad, stop and go bead game, decal transfer	1985	3	5
Peanuts	set of four characters in vehicles; Charlie, Snoopy, Lucy, Linus	1990	2	4
Peanuts	under 3 toys, two different, Charlie Brown's Egg Basket or Snoopy's Potato Sack	1990	3	6
Pencil Puppets	six different pencil toppers in shapes of McDonaldland characters	1978	22	4
Piggsburg Pigs	set of four figures on vehicles; Rembrandt in hotrod,, Huff & Puff wolves on catapult, Piggy & Crackers on crate car, Portly & Pighead on motorcycle	1991	3	5
Playmobile	set of five toys and accesories; Farmer, Sheriff, Indian, Umbrella Girl, Horse and saddle	1982	10	15
Popoids	set of four interconnecting constructor kits; cylinder, triangle, sphere, cube	1985	6	12
Potato Head Kids	set of eight toys;	1992	3	5
Raggedy Ann & Andy	set of four toys; Andy on slide, Grouchy on carousel, Ann with swing, camel on seesaw	1990	2	4
Rescuers Down Under	set of four slide viewing movie camera toys; Jake, Wilbur, Bernard and Bianca, Cody	1990	2	4
Rescuers Down Under	under 3 toy: Bernard	1990	3	6
Runaway Robots	set of six, dark blue Skull, green Coil, red Flame, purple Bolt, blue Beak and yellow Jab	1988	7	14
Safari Adventure	six different rubber animals; alligator, monkey, gorilla, tiger, hippo, rhino	1980	1	3
Sailors	Set of four floating toys: Hamburglar Sailboat, Ronald Airboat, Grimace Sub and Fry Kids Ferry	1987	5	10
School Days	set of five school tools; 1. pencils, Ronald, Grimace, Hamburglar, 2. erasers, Ronald, Grimace, Hamburglar, Captain Crook, Birdie, 3. pencil sharpener, 4. ruler, 5. pencil case	1984	1	3
Ship Shape 1	set of four boat containers with stickers; Hamburglar, Ronald, Grimace, Captain Crook	1983	12	23
Ship Shape 2	set of four boat containers with stickers, same as issued in 1983	1985	9	18
Sky-Busters	set of six rubber airplanes; Skyhawk AAf, Phantom f$E, Mirage F1, United DC-10, MIG-21, Tornado	1982	2	3
Smart Duck	set of six rubber figures; Duck, Cat, Donkey, Chipmunk, two rabbits	1979	1	3
Snow White	Prince, Snow White, Dock, Bashful, Sleepy, Queen Witch		2	4
Snow White	under 3 toy: Dopey & Sneezy		3	6
Sonic 3 The Hedgehog	set of 4: Sonic the Hedgehog, Miles "Tails" Power, Knuckles & Dr. Ivo Robofnik	1994	2	4
Sonic 3 The Hedgehog	under 3 toy: Sonic Ball	1994	3	6

FOOD TOYS

McDonald's Happy Meal Toys

NAME	DESCRIPTION	YEAR	MNP	MIP
Space Aliens	set of eight rubber monsters; lizard man, vampire bat, gill face, tree monster, winged fish, cyclops, veined brain, insectman	1979	2	3
Space Raiders	set of eight rubber aliens; Drak, Dard, flying saucer, Rocket Kryoo-5, Horta, Zama, Rocket Ceti-3, Rocket Altair-2	1979	2	3
Sport Ball	Four different: Basketball, Baseball, Football and Tennis Ball	1988	2	4
Sportsball	set of four balls; Baseball, Football, Basketball, Soccer	1990	2	4
Star Trek	set of five toys; 1. Rings, Kirk, Spock, Starfleet insignia, Enterprise	1979	3	6
Star Trek	set of five toys; 2. Starfleet game	1979	3	6
Star Trek	set of five toys; 3. set of five video viewers, each with a different story	1979	4	8
Star Trek	set of five toys; 4. set of four glitter iron-ons of characters, Kirk, Spock, McCoy, Ilia, packaged in pairs	1979	8	15
Star Trek	set of five toys; 5. Navigation bracelet with decals	1979	15	30
Stomper Mini 4X4	set of eight big wheel cars; Tercel, AMC Eagle, Chevy S-10 Pickup, Chevy Van, Chevy Blazer, Ford Ranger, Jeep Renegade, Dodge Ram,	1986	5	10
Super Looney Tunes	Set of four figures with costumes; Super Bugs, Bat Duck, Taz Flash, Wonder Pig	1991	2	4
Super Looney Tunes	under 3 toy: Bat Duck in ear	1991	3	6
Super Mario Brothers	Set of four action figures; Mario, Luigi, Little Gooma, Koopa	1990	2	4
Super Mario Brothers	under 3 toy: Super Mario	1990	3	6
Super Summer	Sand Castle Pail with shovel, Sand Pail with rake, Fish Sand Mold, Sailboat, Beach Ball	1988	2	4
Tail Spin	Under 3 toys; Baloo's seaplane or Wildcat's jet	1990	3	6
Tail Spin	Set of four characters in airplanes; Molly, Balloo, Kit, Wildcat	1990	2	4
Tinosaurs	set of eight figures; Link the Elf, Baby Jad, Merry Bones, Dinah, Time Traveller Fern, Tiny, Grumpy Spell, Kave Kolt Kobby	1986	5	10
Tiny Toons Flip Cars	set of four cars, each with two characters depending on which side is up; Montana Max/Gobo Dodo, Babs/Plucky Duck, Hampton/Devil, Elmyra/Buster Bunny	1991	2	4
Tiny Toons Flip Cars	under 3 toy: Sweetie	1991	3	6
Tom & Jerry Band	Set of four characters with instruments; Tom at keyboard, Jerry on drums, Spike on bass, Droopy at the mike	1990	3	5
Tom & Jerry Band	under 3 toy: Droopy	1990	5	8
Tonka	set of five; Fire Truck, Loader, Cement Mixer, Dump Truck, Backhoe	1992	1	2
Tonka	under 3 toy: dump truck	1992	3	6
Totally Toy Holiday	set of nine: Atack pack, Keyforce truck, Keyforce carm Magic Nursery, Polly Pocket, Sally Secrets (Caucasion or African-American), Lil' Miss Candistripe, Tattoo Machine, Mighty May	1993	3	6
Totally Toy Holiday	under 3 toy: Magic Nursery (candy canes)	1993	5	10
Totally Toy Holiday	under 3 toy: Magic Nursery (Holly)	1993	3	6
Totally Toy Holiday	under 3 toy: Keyforce car	1993	3	6
Turbo Macs	set of four pull back action cars with characters driving and large "M" on hood; Ronald, Grimace, Birdie, Hamburglar	1990	2	4
Under Sea	set of six cartons with undersea art, alligator, dolphin, hammerhead shark, sea turtle, seal, walrus	1980	1	2
United Airlines Friendly Skies	set of 2 airplanes with United markings, either Ronald or Grimace flying	1991	3	5
What is it	set of six rubber animals; skunk, squirrel, bear, owl, baboon, snake	1979	1	3
Wild Animal Toy Books	complete set of four plus under three	1991	2	4
Wild Animal Toy Books	under 3 toy: Giant Panda	1991	3	6
Winter World	set of five flat vinyl tree ornaments; Ronald, Hamburglar, Grimace, Mayor McCheese, Birdie	1983	4	8
Yo Yogi	set of four characters	1991	3	5

FOOD TOYS

Top to bottom: Garfield (three of four), 1989; Fraggle Rock (one of four), 1988; McDino Changeables (five of eight) in closed position, 1991; Dinosaurs, 1993, all McDonald's Happy Meal Premiums.

McDonald's Happy Meal Toys

NAME	DESCRIPTION	YEAR	MNP	MIP
Young Astronauts	set of four snap together models; Apollo Command Module, Argo Land Shuttle, Space Shuttle, Cirrus Vtol	1986	5	10
Young Astronauts	set of four vehicles	1992	2	4
Zoo Face	set of four rubber noses and makeup kits; alligator, monkey, tiger, toucan	1988	2	4

Miscellaneous Restaurant Toys

Arby's

Megaphone, Minnesota Twins 25th Anniversary		1986	1	2

Arthur Treachers

Flintstone, Pebbles Cup	Yabba Dabba Dew plastic cup	1974	10	15

Big Boy

Action Figures	complete set of four; skater, pitcher, surfer, race driver	1990	2	5
Big Boy Bank, Large	produced from 1966-1976, 18" tall, full color	1960s	175	300
Big Boy Bank, Medium	produced from 1966-1976, 9" tall, brown	1960s	75	100
Big Boy Bank, Small	produced from 1966-1976, 7" tall, painted red/white	1960s	65	100
Big Boy Board Game		1960s	75	100
Big Boy Kite	kite with image of Bog Boy	1960s	10	25
Big Boy Nodder		1960s	100	150
Big Boy Playing Cards	produced in four designs	1960s	20	40
Big Boy Stuffed Dolls	set of three, Big Boy, girl friend Dolly, both 12" tall, and dog Nuggets, appx. 7" tall (Note:all are currently being reproduced)	1960s	30	50

Burger King

Doll Cloth Cartoon King Doll	16" tall	1972	20	30
Frisbee	Small yellow or orange, embossed Burger King		2	4
Lunch Box	blue plastic box embossed with BK logo		2	5
Sports Watches	in white or yellow		2	5
Super Bowl Poster		1992	2	4
Teenage Mutant Ninja Turtles Poster		1991	2	4

Chesty Boy

Chesty Boy Squeek Toy	8" tall	1950	35	75

Chucky E Cheese

Chucky E Cheese Bank	plastic	1980	15	25

Dairy Queen

Radio Flyer Replica			4	8

Goodhumor Ice Cream

Ice Cream Bar Doll		1975	10	20

Hardee's

Backpack	orange		2	4
Beach Bunnies	set of four; girl with ball, boy with skatebaord, girl with skates and boy with frisbee	1989	2	4
Frisbee	4 3/4" wide, white with yellow imprinting		1	4
Ghostbusters Headquarters Posters			1	3
Super Bowl Cloisonne Pins	set of 25 officially licensed NFL Super Bowl pins, plus one error pin.	1991	2	3

FOOD TOYS

Miscellaneous Restaurant Toys

NAME	DESCRIPTION	YEAR	MNP	MIP
Waldo and Friends Holiday Ornaments	Set A and B		5	10

International House of Pancakes

NAME	DESCRIPTION	YEAR	MNP	MIP
Pancake Kids Lunch Box		1992	2	5

Kentucky Fried Chicken

NAME	DESCRIPTION	YEAR	MNP	MIP
Colonel Sanders Figure	9" tall	1960s	35	50
Colonel Sanders Nodder	7" tall	1960s	35	60
WWF Stampers	set of four, Canadian issues		2	4

Little Caesars

NAME	DESCRIPTION	YEAR	MNP	MIP
Stuffed Pizza Pizza Man			2	5

McDonald's

NAME	DESCRIPTION	YEAR	MNP	MIP
Astrosnicks Rocket	9 1/2" rocketship coupon with Happy Meal	1984	15	30
Baseball Cards (Donruss '92)	set of 32 cards with checklist	1992	12	24
Baseball Cards (Hoops '92)	set of 62 cards	1992	12	24
Baseball Cards (Hoops '92)	set of 70 cards, including eight extra Chicago Bulls cards	1992	18	36
Baseball Cards (Topps '91)	set of 44 cards	1991	18	36
Batman Cups w/flying lids	set of six, Batman, Penguin for Mayor, Catwoman, Batmobile, Ballroom Scene,	1992	1	2
Birdie Bike Horn-Japan	Made for kid's bike		4	9
Birdie Magic Trick	green or orange		0	5
Captain Crook Bike Reflector-Canada	blue plastic	1988	1	3
Christmas Ornaments	Fry Guy and Fry Girl cloth hanging ornaments, 3 1/2" tall	1987	2	5
Colorful Puzzles-Japan	Dumbo, Mickey & Minnie, Dumbo & Train		5	10
Coloring Stand-Ups	characters and backgrounds to color, punch out and stand	1978	4	8
Combs	Capt. Crook-red, Grimace-yellow, Ronald-yellow, blue or purple, Grimace Groomer-green	1988	1	2
Construx	set of four pieces used to make a spaceship; axle, wing, body cylinder, canopy	1986	7	15
Crayola Squeeze Bottle, Kay Bee	regional set of four		2	5
Double Bell Alarm Clock	wind up alarm clock with silver bells, hammer ringer, silver feet, image of Ronald on face with head tilted over folded hands, as if asleep		20	40
E.T. Posters	set of four posters, 17"x24"; A.Boy and ET on bike	1985	5	11
E.T. Posters	B. Boy and ET finger touch, C. ET and radio	1985	3	8
E.T. Posters	D. ET waving	1985	5	10
Favorite Friends	set of seven character punch-out cards	1978	2	5
French Fry Radio	large red fry container with fries AM/FM radio	1977	12	25
Friendship Spaceship Ring		1985	3	4
Frisbee - Canada	3" around, blue, early Ronald image		1	2
Ghostbusters	set of five school tools; Slimer pencil, Stay Puft pad and eraser, containment chamber pencil case, Stay Puft pencil sharpener, Ghostbusters ruler	1987	3	7
Glow in the Dark Yo-Yo	no markings or dates	1978	2	5
Golf Ball	McDonald's logo		1	3
Good Times Great Taste Record			2	4
Grimace Bank	ceramic bank, purple, 9" tall	1985	10	20
Grimace Enamel Pin	enamel pin		6	12
Grimace Miniature Golf		1986	3	5
Grimace Sponge	Grimace, Grimace Car Wash		2	4
Grimmace Ring		1970	8	15

FOOD TOYS

Miscellaneous Restaurant Toys

NAME	DESCRIPTION	YEAR	MNP	MIP
Halloween Certificate Book w/Roger Rabbit Puffy Sticker		1988	1	3
Halloween Pumpkin Ring	orange pumpkin face ring		1	3
Hamburglar Doll	7" stuffed doll, by Remco, part of set of seven, sold on blister card	1976	12	25
Hamburglar Hockey		1979	2	4
High Flying	set of three kites: Ronald, Birdie and Hamburglar	1987	3	8
Honey, I Shrunk the Kids Cups	set of three white 20 oz. plastic cups; Giant bee, on the dog's nose, riding the ant	1988	2	4
Looney Tunes Christmas Dolls-Canada	Set of four, Sylvester in nightgown & cap, Tasmanian Devil in Santa hat, Bugs in winter scarf and Tweetie dressed as Elf		3	6
Luggage Tags	set of four; Birdie, Grimace, Ronald, Hamburglar		2	4
Mac Tonight	Fingertronic foam puppet	1988	6	15
Mac Tonight Enamel Pin	Moonface and slogan enamel pin	1988	2	4
Mac Tonight Sunglasses	adult	1988	2	5
McDonald's All-Star Race Team (MAXX) '91	complete set of cards	1991	5	15
McDonald's All-Star Race Team (MAXX) '92	complete set of 36 cards	1992	5	15
McDonald's Playing Cards	two decks to a set		2	5
McDonald's Spinner Top-Holland			2	4
Mickey's Birthdayland	set of four Under 3 vehicles, Mickey's convertible, Goofy's car, Minnie's convertible, Donald's Jeep	1989	2	4
Minnesota Twins Baseball Glove	Twins logo on side, Coca-Cola inside glove, McDonald's satin logo on back, given to the first 100 kids at the 1984 game	1984	40	75
Minute Maid Juice Bottles	mini sueeze bottles	1991	1	2
Norman Rockwell Brass Ornament	clear acrylic, "Christmas Trio" gift boxed	1978	3	7
Norman Rockwell Brass Ornament	50th Anniversary Norman Rockwell design, gift packaged with McDonald & Coca-Cola logos	1983	3	7
On The Go Lunch Box	Three colors: Green, red or blue with arches on handles, stickers and embossed McDonaldland Characters going to schools	1988	2	4
Paint with Water	paintless coloring board with self contained frame and easel	1978	5	10
Pin, Ronald in Christmas Wreath	enamel pin		6	12
Punkin' Makins	cutouts in character faces to attach to Halloween pumpkins; Ronald, Goblin, Grimace	1977	7	15
Rescuers Down Under Xmas Ornament	Miss Bianca, Bernard	1990	2	4
Rings	set of five rings with character heads; Big Mac, Captain Crook, Grimace, Hamburglar, Ronald	1977	5	10
Roger Rabbit Scarf-Japan	McDonald's logo, Japanese writing on scrarf	1988	10	20
Ronald "Shoe Wallet" - Canada	Ronald yellow shoe wallet, attach to shoes with laces	1987	1	2
Ronald Bank	Ronald sitting with legs crossed, 7 1/2" tall		5	12
Ronald Bike Seat Pad-Japan			4	9
Ronald Cookie Cutter	Green with balloons	1987	2	3
Ronald Foldable LCD Clock-Japan			4	9
Ronald Inflatable 12"	weighted base	1990	2	5
Ronald Magic Tablet			1	3
Ronald McDonald Doll	14" vinyl head with a soft body by Dakin		15	35
Ronald McDonald Doll	7" doll, by Remco		12	25
Ronald McDonald Maze	lift up mystery game	1979	4	10
Ronald Plastic Flyers	Ronald with legs and arms extended red or yellow		1	3
Ronald Popsicle Maker-Canada	green or yellow	1984	1	3
Ronald Shoe & Sock Game-Japan	Plastic with ball and string - in Japanese writing		5	10

FOOD TOYS

Miscellaneous Restaurant Toys

NAME	DESCRIPTION	YEAR	MNP	MIP
Ronald Tote Bag-Japan	writing in Japanese		5	10
Santa Claus the Movie Reindeer Xmas Ornament		1985	1	3
Sindy Doll	dressed in older McDonald's uniform	1970	4	8
Singing Wastebasket Bank	5 1/8" white plastic basket with coin slot in top		6	10
Speedie "Touch of Service" Pin	enamel pin		6	12
Spinner Baseball Game	green plastic with four characters	1983	1	3
Sticker Club	set of five different sticker sheets; reflectors, scratch and sniff, color designs, action stickers, puffy designs	1985	1	3
Stocking, Merry Christmas To my Pal	plastic stocking	1981	3	6
Sunglasses	Hamburglar, Ronald w/yellow lenses or Ronald McDonald on stem		2	4
Tic Tac Mac Game	yellow base, Grimace is X, Ronald is O	1981	2	5
Tootler Harmonica		1985	1	3
Tops	set of three finger tops in red, blue and green	1978	3	7
Trays, Serving	set of six white plastic wedge shaped trays with pictures of characters; Ronald, Big Mac, Mayor McCheese, Hamburglar, Grimace, Captain Crook		3	7
Walt Disney Video Viewer-Cinderella - Japan			5	10
Who Framed Roger Rabbit Cups	Roger Rabbit-Hollywood, Benny the Cab, Roger Being Chased	1988	1	3
Wrist Wallets	set of four watch-type bands with coin holding dial with character faces; Ronald, Captain Crook, Big Mac, Hamburglar	1977	5	10
Yo Yo	half red, half yellow	1979	2	5

Pizza Hut

NAME	DESCRIPTION	YEAR	MNP	MIP
Universal Monster Cups/ Holograms	set of three		2	5

Roy Rogers

NAME	DESCRIPTION	YEAR	MNP	MIP
Skateboard Kids Figures			2	5

Tastee Freeze

NAME	DESCRIPTION	YEAR	MNP	MIP
Roy Campanella Figure			20	35

Wendy's

NAME	DESCRIPTION	YEAR	MNP	MIP
Fun Flyers	3 1/2" wide in red, yellow or blue		1	3
Where's the Beef Stickers	set of six	1984	1	3

Science Fiction and Space Toys

Ode to Robby

Some would trace the modern age of science fiction to 1956 and Forbidden Planet. Undoubtedly, the toy world would be poorer for the lack of Robby the Robot. But through one medium or another, science fiction has enthralled millions over the last two hundred-thirty years, back to Swift's "Gulliver's Travels." If you do not care about bending the definition almost in half, you could say science fiction goes back to Galileo Galilei.

The first universally acclaimed work of science fiction was Mary Shelley's "Frankenstein" or "The Modern Prometheus," and her vision made way for Verne, Wells, Burroughs, Lovecraft, Heinlein, Asimov, Clarke, and a host of others, whose collective imaginations led us up to today and through tomorrow.

Frankenstein, while often considered more a movie monster than a science fiction or even romantic character, has nonetheless become one of the most enduring and widely embraced characters in all fiction. Like many classic examples of the genre, Frankenstein has become symbolic of both the hopes and perils of the age of science, and his spiritual offspring are legion, including our own dear Robby the Robot.

Even with its classical pedigree, science fiction is almost exclusively a product of the twentieth century, as the hard foundation of science had to exist before fiction writers could extrapolate upon it. In particular, science fiction is a phenomenon of the atomic age. World War II, more than any other event this century, opened our eyes to the wondrous and horrific potential of applied science. Forbidden Planet juxtaposed with The Day the Earth Stood Still. The hope of Close Encounters was tempered by the pessimism of Blade Runner. We had Weird Science and Neuromancer and a thousand other visions of our future simultaneously enticing us forward and warning us away.

Just as science fiction has captivated readers of all ages, so has its toys. Buck Rogers and Flash Gordon may have told us that the heavens could be treacherous places, but they also promised some really cool tools to get there. Spaceships and rockets and ray guns indeed!

"Space:1999" walking spaceman.

Buck Rogers made his first appearance in 1928, the same year that Mickey Mouse was introduced in "Steamboat Willie." In 1929, Buck went from pulp to newsprint, becoming the first science fiction comic strip. Flash followed Buck into print in 1934 and was an immediate success. Within two years Flash was on the silver screen, played by Buster Crabbe. Buck finally made it to the screen in 1939, and Buster played him, too.

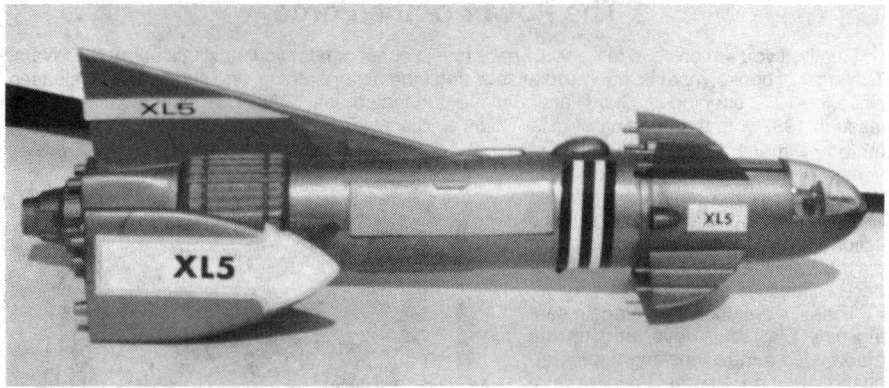

XL5 plastic space vehicle from 1960s.

During this period, Marx produced numerous toys in support of each character, including two ships that have become classics of the space toy field. Opinions vary as to which wind-up is better executed, Buck Rogers' 25th Century Rocket Ship or Flash Gordon's Rocket Fighter. Both are considered superb examples of tin character space toys.

The 1935 Buck Rogers' ship is twelve inches long and beautifully lithographed in the style of the comic strip art with vivid art deco designs. The 1939 Flash Gordon Rocket Fighter shows Flash manning the guns of a brilliantly colored red and yellow open cockpit rocketship. Both are worthy of centerpiece status in many collections.

Brzzap! You're Space Dust!

No discussion of space toys would be complete without mention of ray guns. Here again Marx is a major player, producing numerous generic space guns as well as character items such as the Flash Gordon Signal Pistol and Flash Gordon Water Pistol. Daisy Manufacturing also produced several classics, including the Buck Rogers Rocket Pistol, the 1936 XZ-44 Liquid Helium Water Pistol, and the 1936 XZ-38 Disintegrator Pistol. Other companies, such as Hubley and Wyandotte, made memorable contributions to the art as well.

Space toys run in an uninterrupted stream through most of the twentieth century. The 1930s and 1940s saw Buck and Flash. The 1950s saw fiction become reality with the growth of a new medium, television. "Captain Video" was the first space series on TV, appearing in the summer of 1949. Buzz Correy and his Space Patrol and Tom Corbett — Space Cadet would feed the appetite for adventure until 1956 when the heavens took on a visual scale and grandeur never seen before, in the panoramic wonder of Forbidden Planet.

Z-101 tin flying saucer from Japan, 1950s.

In 1966, when a low budget outer space western called "Star Trek" went on the air, few dreamed that life for millions would never be the same. Even though the original show ran only three seasons, its impact and legacy are undeniable. The phenomenon of "Star Trek" has grown far beyond "cult" status, and the extraordinary success of "Star Trek: The Next Generation" has only broadened its reach.

Star Trek toys and memorabilia of all types are eagerly traded at shows and Trek conventions nationwide, and in light of this, many Trek toys would seem undervalued.

The Power of the Force

Finally, if you are going to talk about space toys, sooner or later you must talk about Star Wars. The array of books, models, figures, playsets, and other toys, clothing, and sundry items released since the 1977 opening of Star Wars continued unabated until 1988. The license gained a new lease in 1987 with the opening of Star Tours at Disneyland and Disney World, generating still more new merchandise. In terms of diversity of toys, the universe of Star Wars is easily the most fully realized and diversely populated in all science fiction. Star Wars figures, vehicles, and playsets are the most widely traded science fiction toys on the market today.

Today, store shelves are again seeing new Star Wars toys, priming the market for a new onslaught of merchandise when George Lucas' promised new films open in 1997. Star Wars' longevity and international name recognition are excellent assurances of the continuing popularity of its toys.

Kenner's Star Wars "Darth Vader TIE Fighter," 1977.

Trends

In general, the field of science fiction toys is one with particular growth potential, given the prevalence of science fiction in today's culture and the exceptional strength of franchises such as "Star Trek" and Star Wars. Recent generations have been weaned on Luke Skywalker, Han Solo, Mr. Spock, Captain Picard, and now Benjamin Cisco. It is a foregone conclusion that as they enter the collector market, many of these new collectors will choose science fiction and space toys as the basis of their collections. Perhaps they might become the basis of yours, too.

The Top 25 Science Fiction and Space Toys (in Mint In Package condition)

1. Lost In Space Doll Set, Marusan $7,000
2. Space Patrol Monorail Set, Toys of Tomorrow, 1950s 4,000
3. Buck Rogers Roller Skates, Marx, 1935 3,500
4. Buck Rogers Figure Set (lead), Britains, 1930s 2,500
5. Lost In Space Switch 'N Go Set, Mattel, 1966 2,300
6. Buck Rogers Uniform, Sackman Bros., 1934 2,100
7. Buck Rogers Cut-Out Adventure Book, Cocomalt, 1933 2,000
8. Lost In Space Roto-Jet Gun Set, Mattel 1966 2,000
9. Buck Rogers 25th Century Scientific Laboratory, Porter Chem. Co, 1934 1,600
10. Flash Gordon Home Foundry Casting Set, 1935 1,500
11. Lost In Space Model Kit, with Chariot, Marusan 1,500
12. Lost In Space Robot, Aurora, 1966 1,500
13. Frankenstein Robot (tin/plastic, w/wired remote) 1,400
14. Chief Robotman, KO, 1965 1,200
15. Lost In Space 3-D Fun Set, Remco, 1966 1,200
16. Godzilla Combat Joe Set, 1984 1,000
17. Lost In Space Jupiter Model, Marusan 1,000
18. Buck Rogers Pocket Watch, E. Ingraham Co., 1935 1,000
19. Mechanized Robot, 1960s 925
20. Buck Rogers Repeller Ray Ring 925
21. Answer Game Machine (robotic) 900
22. Lost In Space Helmet and Gun Set, Remco, 1967 800
23. Buck Rogers 25th Century Rocket, Marx, 1939 775
24. Space Patrol Cosmic Rocket Launcher Set, 1950s 750
25. Star Trek Mission to Gama VI, Mego, 1976 750

Clockwise from upper left: Robot Control, unknown; Astronaut Space Helmet, Ideal; Battery Powered Mini-Robo Tank, Japan; Battery Powered Cragstan Mystery Action Satellite, Japan; (left) Electric Robot, 1950s, Marx, and first generation Robert the Robot, 1954-55, Ideal.

SPACE TOYS

Alien

NAME & DESCRIPTION	COMPANY	YEAR	GOOD	EX	MIB
Alien Game	Milton Bradley	1980	30	50	75
Alien Model Kit	Tsukuda	1980s	125	225	350
Alien Model Kit	MPC	1979	20	60	85

Aliens

Alien Figure (18" tall)	Kenner	1979	125	300	500
Alien Warrior (base and egg)	Halcyon		20	30	45
Aliens Commodore Computer Game	Commodore	1985	15	25	35
Colorform Aliens Set			10	15	35

Battlestar Galactica

Action Figure, Apollo	Mattel	1978	8	15	20
Action Figure, Baltar	Mattel	1978	20	35	50
Action Figure, Boray	Mattel	1978	10	15	25
Action Figure, Colonial Warrior	Mattel	1978	15	25	40
Action Figure, Commander Adama	Mattel	1978	8	15	20
Action Figure, Cylon Centurian	Mattel	1978	10	15	25
Action Figure, Cylon Centurian	Mattel	1978	25	40	60
Action Figure, Cylon Commander	Mattel	1978	30	50	75
Action Figure, Cylon Commander	Mattel	1978	25	40	65
Action Figure, Daggit	Mattel	1978	6	10	15
Action Figure, Imperious Leader	Mattel	1978	5	8	12
Action Figure, Lt. Starbuck	Mattel	1978	6	10	15
Action Figure, Lucifer	Mattel	1978	6	10	15
Action Figure, Ovion	Mattel	1978	8	15	20
Action Figures, Set of Four	Mattel	1978	25	40	60
Action Figures, Set of Six	Mattel	1978	35	60	95
Action Figures, Set of Three	Mattel	1978	25	40	65
Colonial Scarab Vehicle	Mattel	1978	15	30	45
Colonial Scarab Vehicle	Mattel	1978	15	25	35
Colonial Stellar Probe Ship	Mattel	1978	15	30	45
Colonial Stellar Probe Ship	Mattel	1978	15	25	35
Colonial Viper	Mattel	1978	15	30	45
Colonial Viper	Mattel	1978	10	15	25
Colorforms Adventure Set	Colorforms	1978	12	20	30
Costume, Cylon Warrior, boxed with mask		1978	10	15	25
Cylon Bubble Machine		1978	6	10	15
Cylon Helmet Radio		1979	10	15	25
Cylon Raider	Mattel	1978	15	30	45
Cylon Raider	Mattel	1978	15	25	35
Galactic Cruiser	Larami	1978	5	8	12
Game of Starfighter Combat	FASA	1978	10	15	25
Game, Battlestar Galactica	Parker Brothers	1978	10	15	25
Game, Space Alert	Mattel	1978	15	25	40
L.E.M. Lander	Larami	1978	5	8	12
Lasermatic Pistol	Mattel	1978	15	30	45
Lasermatic Rifle	Mattel	1978	25	40	65
Lunch Box	Aladdin	1978	20	35	50
Model Kit, Battlestar Galactica	Monogram	1979	15	25	40
Model Kit, Colonial Viper	Monogram	1979	15	25	40
Model Kit, Cylon Base Star	Monogram	1979	15	30	45
Model Kit, Cylon Raider	Monogram	1979	15	25	40
Poster Art Set	Craft Master	1978	6	10	15
Puzzles	Parker Brothers	1978	6	10	15
"The Entire Movie & New Adventures" Comics	Whitman	1979	5	7	10

Black Hole

NAME & DESCRIPTION	COMPANY	YEAR	GOOD	EX	MIB
Maxmillian & Dr. Reinhardt Puzzle	Western Pub. Co.		7	10	15
Pop-up Book			11	16	25
Poster Book			7	10	15
Press-out Book			11	16	25
Stamp Activity Book			7	10	15
Sticker Activity Book			7	10	15
Tray Puzzles			7	10	15
V. I. N. cent Puzzle	Western Pub. Co.		7	10	15
Wrist Watch			20	30	45

Buck Rogers

NAME & DESCRIPTION	COMPANY	YEAR	GOOD	EX	MIB
25th Century Scientific Laboratory (complete with three manuals)	Porter Chem. Co.	1934	600	1000	1600
Action Figure, Ardella	Mego	1979	6	10	15
Action Figure, Buck Rogers	Mego	1979	20	35	50
Action Figure, Buck Rogers	Mego	1979	12	20	30
Action Figure, Doctor Huer	Mego	1979	20	35	50
Action Figure, Doctor Huer	Mego	1979	4	7	10
Action Figure, Draco	Mego	1979	20	35	50
Action Figure, Draco	Mego	1979	4	7	10
Action Figure, Draconian Guard	Mego	1979	20	35	50
Action Figure, Draconian Guard	Mego	1979	8	13	20
Action Figure, Killer Kane	Mego	1979	20	35	50
Action Figure, Killer Kane	Mego	1979	4	7	10
Action Figure, Tiger Man	Mego	1979	20	35	50
Action Figure, Tiger Man	Mego	1979	6	10	15
Action Figure, Twiki	Mego	1979	8	15	20
Action Figure, Walking Twiki	Mego	1979	20	35	50
Action Figure, Wilma Deering	Mego	1979	10	15	25
Badge, Chief Explorer		1936	85	145	225
Badge, Solar Scouts, Member		1935	60	100	150
Badge, Spaceship Commander, Whistling		1930s	50	80	125
Banner, Spaceship Commander		1936	75	125	195
Book, Big Big, The Adventures Of	Whitman	1934	50	80	125
Book, Big Little, And The Depth Men Of Jupiter	Whitman	1935	50	80	125
Book, Big Little, And The Doom Comet	Whitman	1935	50	80	125
Book, Big Little, And The Overturned World	Whitman	1941	45	70	110
Book, Big Little, And The Planetoid Plot	Whitman	1936	50	80	125
Book, Big Little, And The Super Dwarf Of Space	Whitman	1943	45	70	110
Book, Big Little, In The 25th Century	Whitman	1933	70	115	175
Book, Big Little, In The 25th Century	Whitman	1933	50	80	125
Book, Big Little, In The City Below The Sea	Whitman	1934	75	130	200
Book, Big Little, In The City Of Floating Globes	Whitman	1935	115	195	300
Book, Big Little, In The War With The Planet Venus	Whitman	1938	45	70	110
Book, Big Little, On The Moons of Saturn	Whitman	1934	75	130	200
Book, Big Little, Vs. The Fiend Of Space	Whitman	1940	45	70	110
Book, Cut-Out Adventure Book		1933	775	1300	2000
Book, Little Golden, And the Children of Hopetown	Golden Press	1980s	4	7	10
Book, Pop-Up, Strange Adventures in the Spider Ship		1935	105	180	275
Button, Buck Rogers in the 25th Century		1935	80	130	200
Button, Century of Progress		1934	135	225	350
Button, Satelite Pioneers		1950s	20	35	50
Cadet Commission, Satelite Pioneers		1958	20	35	50
Camera, Super Foto	Norton-Honer	1955	40	65	100
Captain Action Outfit	Ideal	1967	350	575	895
Chemistry Set, Advanced	Grooper Co.	1937	275	475	725
Chemistry Set, Beginners	Grooper Co.	1937	235	390	600
Clock	Huckleberry Time	1970s	30	50	75
Colorforms Adventure Set	Colorforms	1979	10	15	25

SPACE TOYS

Buck Rogers

NAME & DESCRIPTION	COMPANY	YEAR	GOOD	EX	MIB
Communicator Set		1970s	10	15	25
Draconian Marauder	Mego	1979	20	35	50
Figure, Buck Rogers	Tootsietoy	1937	90	145	225
Figure, Wilma	Tootsietoy	1937	70	115	175
Figures, Lead, Set	Cocomalt	1934	80	130	200
Figures, Lead, Set	Britains	1930s	975	1625	2500
Films, set of six	Irwin	1936	105	180	275
Folder, Chief Explorer		1936	70	115	175
Galactic Playset	HG Toys	1980s	17	30	45
Game of the 25th Century		1934	145	245	375
Game, Battle For The 25th Century	TSR	1988	15	25	35
Game, Buck Rogers Adventures In The 25th Century	Transogram	1970s	17	30	45
Game, Buck Rogers Adventures In The 25th Century	Milton Bradley	1979	10	16	25
Game, Buck Rogers In The 25th Century	All-Fair	1936	185	310	475
Game, Martian Wars	TSR	1980s	14	25	35
Games, Interplanetary Games, Set		1934	235	390	600
Gun, Atomic Pistol U-235	Daisy	1945	90	145	225
Gun, Atomic Pistol U-238	Daisy	1946	90	145	225
Gun, Buck Rogers 25th Century Pop Gun	Daisy	1930s	135	225	350
Gun, Buck Rogers in the 25th Century Pistol Set	Daisy	1930s	165	275	425
Gun, Buck Rogers Sonic Ray Flashlight Gun	Norton-Honer	1950s	90	145	225
Gun, Combat Set (gun & holster XZ-32)	Daisy	1934	250	425	650
Gun, Combat Set (gun & holster XZ-37)	Daisy	1935	195	325	500
Gun, Combat Set XZ-40	Daisy	1935	165	275	425
Gun, Combat Set XZ-42	Daisy	1935	165	275	425
Gun, Disintegrator Pistol XZ-38	Daisy	1935	90	145	225
Gun, Liquid Helium Water Pistol XZ-44	Daisy	1936	155	260	400
Gun, Liquid Helium Water Pistol XZ-44	Daisy	1936	165	275	425
Gun, Rocket Pistol XZ-31	Daisy	1934	105	180	275
Gun, Rocket Pistol XZ-35	Daisy	1935	90	145	225
Gun, Rubber Band		1930s	30	50	75
Gun, Sonic Ray	Norton-Honer	1950s	40	65	100
Helmet and Rocket Pistol Set	Einson Freeman	1933	115	195	300
Helmet XZ-34	Daisy	1935	285	475	725
Holster XZ-33	Daisy	1934	60	100	150
Holster XZ-36	Daisy	1935	60	100	150
Holster XZ-39	Daisy	1935	60	100	150
Holster, Atomic Pistol U-238	Daisy	1946	30	50	75
Iron-on Transfers, Rocket Rangers		1940s	30	50	75
Land Rover	Mego	1979	15	25	40
Laserscope Fighter	Mego	1979	15	25	40
Lite-blaster Flashlight		1936	155	260	400
Lunch Box	Aladdin	1979	10	16	25
Manual, Solar Scouts Radio Club		1936	100	165	250
Map, Satelite Pioneers, of Solar System		1958	20	35	50
Masks, Buck and Wilma, Set	Einson Freeman	1933	115	195	300
Medallion, Century of Progess		1934	100	165	250
Membership Card, Rocket Rangers			45	75	120
Membership Card, Satelite Pioneers		1950s	30	50	75
Model Kit, Superdreadnought SD51X		1936	100	165	250
Model Kits, Interplanetary Space Fleet		1935	100	165	250
Official Utility Belt	Remco	1970s	15	25	35
Paddle Ball, Comet Socker	Lee-Tex	1935	35	60	95
Paint By Number Set	Craft Master	1980s	8	13	20
Pencil Box	American Pencil	1930s	70	115	175
Pendant Watch	Huckleberry Time	1970s	105	180	275
Pocket Knife	Adolph Kastor	1934	245	400	625
Pocket Watch	E. Ingraham Co.	1935	390	650	1000
Pocket Watch	Huckleberry Time	1970s	90	145	225
Punching Bag	Morton Salt	1942	40	65	100

SPACE TOYS

Buck Rogers

NAME & DESCRIPTION	COMPANY	YEAR	GOOD	EX	MIB
Puzzle	Milton Bradley	1950	20	35	50
Puzzle	Milton Bradley	1952	20	35	50
Puzzle	Milton Bradley	1979	6	10	15
Puzzle, Buck Rogers & his Atomic Bomber	Puzzle Craft	1945	60	100	150
Ring, Repeller Ray			365	600	925
Ring, Saturn	Post Corn Toasties	1944	125	210	325
Ring, Space Ranger, Halolight	Sylvania	1952	50	80	125
Rocket Ship	Marx	1934	245	400	625
Rocket, 25th Century Police Patrol	Marx	1935	175	295	450
Rocket, Battle Cruiser	Tootsie Toy	1937	105	180	275
Rocket, Buck Rogers 25th Century	Marx	1939	300	500	775
Rocket, Electric Caster	Marx	1930s	125	210	325
Rocket, Flash Blast Attack Ship	Tootsie Toy	1937	90	145	225
Roller Skates	Marx	1935	1350	2300	3500
School Bag			60	100	150
School Crayons Ship Box & Pencils	American Pencil	1935	90	145	225
Space Glasses	Norton-Honer	1955	40	65	100
Space Ranger Kit	Sylvania	1952	50	80	125
Spaceport Playset	Mego	1979	75	125	195
Star Fighter	Mego	1979	20	35	50
Star Fighter Command Center	Mego	1979	25	40	65
Starfinder, Satelite Pioneers		1950s	20	35	50
Starseeker	Mego	1979	25	40	60
Stationary, Spaceship Commander		1930s	50	80	125
Strato-Kite	Aero-Kite Co.	1946	17	30	45
Super Scope Telescope	Norton-Honer	1955	40	65	100
Sweater Emblem (three colors)			135	225	350
Two Way Trans-Ceiver	DA Myco	1948	80	130	200
Uniform	Sackman Bros.	1934	825	1375	2100
Venus Duo-Destroyer	Tootsietoy	1937	100	165	250
View-Master Set	View-Master	1979	4	7	10
Walkie Talkies	Remco	1950s	60	100	150
Wrist Watch	GLJ Toys	1978	10	16	25
Wrist Watch	Huckleberry Time	1970s	60	100	150
Wrist Watch	E. Ingraham	1935	390	650	1000

Captain Midnight

Air Heroes Stamp Album		1930s	30	50	75
Captain Midnight Medal		1930s	60	100	150
Cup			25	40	65
Membership Manual		1930s	30	50	75
Secret Society Decoder		1949	22	35	55

Captain Video

Figures, Interplantary Space Men		1950s	50	80	125
Game, Captain Video	Milton Bradley	1952	80	130	200
Gun, Rite-O-Lite Flashlight		1950s	25	35	55
Mysto-Coder		1950s	60	100	150
Photo, Capt. Video and Ranger		1950s	17	30	45
Riding Toy, Galaxy Space Ship		1950s	250	425	650
Ring, Flying Saucer		1950s	125	210	325
Ring, Rocket on Keychain		1950s	80	130	200
Ring, Secret Seal		1950s	80	130	200
Rocket Tank	Lido	1952	55	95	145
Space Port Playset	Superior	1950s	250	425	650
Troop Transport Ship	Lido	1950s	55	95	145

Close Encounters of the Third Kind

Alien Figure	Imperial	1977	11	16	25

SPACE TOYS

SPACE TOYS

Top to bottom: Ding A Ling Super Return Space Skyway, unknown; Cyberman from Doctor Who series, 1970s, Denys Fisher; Battery Operated Smoking Robot, Japan.

Close Encounters of the Third Kind

NAME & DESCRIPTION	COMPANY	YEAR	GOOD	EX	MIB
Postcard Book		1980	9	13	20

Defenders of the Earth

NAME & DESCRIPTION	COMPANY	YEAR	GOOD	EX	MIB
12 piece Tray Puzzle			9	13	20
Defenders Claw Copter	Galoob	1985	11	16	25
Flash Gordon Battle Action Figure (5 1/2" tall with battle action knobs)	Galoob	1985	7	10	15
Flash Swordship	Galoob	1985	11	16	25
Garax Battle Action Figure (5 1/2" tall with battle action knobs)	Galoob	1985	11	16	25
Garax Swordship	Galoob	1985	11	16	25
Gripjaw Vehicle	Galoob	1985	11	16	25
Lothar Battle Action Figure (5 1/2" tall with battle action knobs)	Galoob	1985	8	11	17
Mandrake the Magician Battle Action Figure (5 1/2" tall with battle knobs)	Galoob	1985	7	10	15
Ming the Merciless Battle Action Figure (5 1/2" tall with battle knobs)	Galoob	1985	8	11	17
Mongor (purple serpent)	Galoob	1985	16	23	35
Phantom Skull Copter	Galoob	1985	7	10	15
Sticker Fun Book			2	3	5
The Phantom Battle Action Figure (5 1/2" tall with battle action knobs)	Galoob	1985	7	10	15

Defenders of the Universe

NAME & DESCRIPTION	COMPANY	YEAR	GOOD	EX	MIB
Battling Black Lion Voltron Vehicle	LJN	1986	9	13	20
Coffin of Darkness Voltron Vehicle	LJN	1986	7	10	15
Doom Blaster Voltron Vehicle (mysterious flying machine)	LJN	1986	9	13	20
Doom Commander Figure	Matchbox	1985	5	7	10
Green Lion Voltron Vehicle	LJN	1986	9	13	20
Hagar Figure	Matchbox	1985	5	7	10
Hunk Figure	Matchbox	1985	5	7	10
Keith Figure	Matchbox	1985	5	7	10
King Zarkon Figure	Matchbox	1985	5	7	10
Lance Figure	Matchbox	1985	5	7	10
Motorized Lion Force Voltron Vehicle Set (black lion with blazing sound)	LJN	1986	9	13	20
Motorized Lion Force Voltron Vehicle Set (green & yellow lion)	LJN	1986	9	13	20
Motorized Lion Force Voltron Vehicle Set (Red lion & blue lion)	LJN	1986	9	13	20
Pidge Figure	Matchbox	1985	5	7	10
Prince Lothar Figure	Matchbox	1985	5	7	10
Princess Allura Figure	Matchbox	1985	5	7	10
Robeast Mutilor Figure	Matchbox	1985	5	7	10
Robeast Scorpious Figure	Matchbox	1985	5	7	10
Skull Tank Voltron Vehicle	LJN	1986	9	13	20
Vehicle Team Assembler (forms Voltron)	LJN	1986	9	13	20
Voltron Lion Force & Vehicle Team Assemblers Gift Set	LJN	1986	9	13	20
Voltron Motorized Giant Commander (plastic 36" tall plastic, wire remote)	LJN	1984	15	25	35
Voltron Time Keeper Robot Watch			2	3	5
Zarkon Zapper Voltron Vehicle (with galactic sound)	LJN	1986	11	16	25

Doctor Who

NAME & DESCRIPTION	COMPANY	YEAR	GOOD	EX	MIB
Ace Figure	Dapol	1986	11	16	25

Doctor Who

NAME & DESCRIPTION	COMPANY	YEAR	GOOD	EX	MIB
Anniversary Set (Dr. Who, Melanie, K-9, Tardis, base, five-side console)	Dapol	1986	275	390	600
Cyberman	Dapol	1986	11	16	25
Cyberman Robot Doll (10" tall)	Denys Fisher	1970s	250	350	550
Dalek	Dapol	1986	14	20	30
Dalek	Marx	1960s	9	13	150
Dalek Army Gift Set	Denys Fisher	1976	46	60	95
Dalek Bagatelle	Denys Fisher	1976	70	100	150
Dalek Shooting Game (8" x 20" litho board with Daleks in action)	Marx	1965	225	325	500
Davros (villain with left arm)	Dapol	1986	11	16	40
Doctor Who Card Set (set of twelve 2 x 3" color octagon cards)		1970s	14	20	30
Doctor Who Character Card Set (24 cards)	Denys Fisher	1976	18	26	40
Doctor Who Daleks Oracle Question & Answer Board Game		1965	115	165	250
Doctor Who Doll (10" tall with scarf & screwdriver)	Denys Fisher	1976	90	130	200
Doctor Who Tardis Playset	Denys Fisher	1970s	205	295	450
Doctor Who Trump Card Game		1970s	9	13	20
Doctor Who...Dodge the Daleks Board Game (robots firing at the sci-fi guy)		1965	115	165	250
Ice Warrior (villain)	Dapol	1986	9	13	20
K-9 (the Doctor's dog)	Dapol	1986	7	10	15
Mel (pink or blue jacket)	Denys Fisher	1976	9	13	20
Seventh Doctor, The (grey or brown jacket)	Denys Fisher	1976	9	13	20
Tardis (the Doctor's transporter)	Denys Fisher	1976	225	325	500

ET: The Extra Terrestrial

NAME & DESCRIPTION	COMPANY	YEAR	GOOD	EX	MIB
ET Collectible Figure Set (ET with phone, cup, flower & reading a book)	LJN	1982	7	10	15
ET Figurine (5" tall)			4	5	7
ET Reading a Book (2" tall)	LJN	1982	1	2	3
ET with Cup (2" tall)	LJN	1982	1	2	3
ET with Flower (2" tall)	LJN	1982	1	2	3
ET with Phone (2" tall)	LJN	1982	1	2	3

Flash Gordon

NAME & DESCRIPTION	COMPANY	YEAR	GOOD	EX	MIB
Action Figure, Arak	Mattel	1979	17	30	45
Action Figure, Beastman	Mattel	1979	15	25	40
Action Figure, Dale Arden	Mego	1976	30	45	70
Action Figure, Dr. Zarkov	Mego	1976	45	70	110
Action Figure, Dr. Zarkov	Mattel	1979	15	25	40
Action Figure, Flash	Mego	1976	45	70	110
Action Figure, Flash	Galoob	1986	6	10	15
Action Figure, Flash	Mattel	1979	10	16	25
Action Figure, Lizard Woman	Mattel	1979	15	25	35
Action Figure, Ming	Mego	1976	25	40	60
Action Figure, Ming	Galoob	1986	6	10	15
Action Figure, Ming	Mattel	1979	12	20	30
Action Figure, Thun, Lion Man	Mattel	1979	15	25	40
Action Figure, Vultan	Mattel	1979	15	30	45
Battle Rocket, With Space Probing Action		1976	6	10	15
Book Bag		1950s	17	30	45
Book, And the Ape Men of Mor	Dell	1942	70	115	175
Book, Big Little, And the Fiery Desert of Mongo	Whitman	1948	30	50	80
Book, Big Little, And the Monsters of Mongo	Whitman	1935	50	80	125
Book, Big Little, And the Monsters of Mongo	Whitman	1935	75	125	195
Book, Big Little, And the Perils of Mongo	Whitman	1940	35	60	90
Book, Big Little, And the Power Men of Mongo	Whitman	1943	35	55	85
Book, Big Little, And the Red Sword Invaders	Whitman	1945	30	50	80

Land of the Giants

NAME & DESCRIPTION	COMPANY	YEAR	GOOD	EX	MIB
Book, Big Little, And the Tournaments of Mongo	Whitman	1935	45	70	110
Book, Big Little, And the Tyrant of Mongo	Whitman	1941	35	60	95
Book, Big Little, And the Witch Queen of Mongo	Whitman	1936	45	70	110
Book, Big Little, In the Forest Kingdom of Mongo	Whitman	1938	40	65	100
Book, Big Little, In the Ice World of Mongo	Whitman	1942	35	60	90
Book, Big Little, In the Jungles of Mongo	Whitman	1947	35	55	85
Book, Big Little, In the Water World of Mongo	Whitman	1937	35	60	95
Book, Big Little, In the Water World of Mongo	Whitman	1937	35	60	95
Book, Big Little, On The Planet Mongo	Whitman	1934	55	95	145
Book, Flash Gordon Paint Book		1930s	60	100	150
Book, Vs. the Emperor of Mongo	Dell	1936	70	115	175
Button, Flash and Ming		1970s	4	7	10
Buttons, Flash Gordon, The Movie		1980	2	3	5
Candy Box		1970s	4	7	10
Captain Action Outfit	Ideal	1966	165	275	425
Captain Action Outfit, with videomatic ring	Ideal	1967	185	310	475
Costume, Flash Gordon Space Outfit	Esquire Novelty	1951	90	145	225
Figure, Flash Gordon		1944	115	195	300
Game, Adventure on the Moons of Mongo Game	House of Games	1977	15	25	35
Game, Flash Gordon	House Of Games	1970s	15	25	35
Gun, Flash Gordon Air Ray Gun	Budson	1950s	215	350	550
Gun, Flash Gordon Arresting Ray	Marx	1939	175	295	450
Gun, Flash Gordon Radio Repeater Clicker Pistol	Marx		215	350	550
Gun, Flash Gordon Signal Pistol	Marx	1930s	195	325	500
Gun, Flash Gordon Water Pistol	Marx		80	130	200
Gun, Space Water Gun	Nasta	1976	6	10	15
Gun, Three Color Ray Gun	Nasta	1976	8	13	20
Gun, Water Pistol	Marx	1950s	155	260	400
Hand Puppet, Flash		1950s	90	145	225
Home Foundry Casting Set, Lead		1935	575	975	1500
Kite, Flash Gordon		1950s	55	90	135
Lunch Box, Dome	Aladdin	1979	20	35	50
Medals and Insignia	Larami	1978	3	5	8
Ming's Space Shuttle	Mattel		15	25	35
Model Kit, Flash Gordon and Alien	Revell	1965	60	100	150
Pencil Box		1951	70	120	185
Playset, Flash Gordon	Mego	1976	45	70	110
Puzzle, Flash Gordon Featured Funnies Jigsaw Puzzle		1930s	55	95	145
Puzzle, Flash Gordon Tray Puzzle	Milton Bradley	1951	45	80	120
Puzzles, Set	Milton Bradley	1951	105	180	275
Rocket Fighter	Marx	1939	175	295	450
Rocket Ship		1975	10	16	25
Rocket Ship, Inflatable	Mattel	1979	20	35	50
Solar Commando Set	Premier Products	1950s	65	105	165
Space Compass		1950s	25	40	65
Sunglasses	JA RU	1981	3	5	8
Two-Way Telephone	Marx	1940s	60	100	150
View-Master Set	View-Master	1963	20	35	50
View-Master Set, In The Planet Mongo	View-Master	1976	6	10	15
Wallet		1949	70	115	175
Wrist Watch	Bradley	1979	70	115	175

Gremlins

Applause Figure (12" plush)	LJN		25	35	55
Gizmo Figure (8" tall "Don't get them Wet!")	LJN	1984	14	20	30
Set of Eight Gremlins (2" tall PVC)	LJN		20	30	45
Stripe Doll (12" tall)		1984	16	23	35
Book, Comic	Gold Key	1968	10	15	25
Book, Flight of Fear	Whitman		8	13	20

SPACE TOYS

Land of the Giants

NAME & DESCRIPTION	COMPANY	YEAR	GOOD	EX	MIB
Book, Land of the Giants	Pyramid		8	13	20
Book, Sling Shot for a David	World Dist./UK		30	50	75
Book, The Hot Spot	Pyramid		12	20	30
Book, The Mean City	World Dist./UK		30	50	75
Book, The Trap	World Dist./UK		30	50	75
Book, Unknown Danger	Pyramid		12	20	30
Books, Comic	Gold Key	1968-	8	13	20
Books, LOTG Annaul	World Dist./UK	1969	30	50	75
Colored Pencil Set	Hasbro	1969	60	100	150
Colorforms Set	Colorforms	1968	30	50	75
Coloring Book	Whitman	1968	20	35	50
Costumes	Ben Cooper	1968	60	100	150
Flashlight, Wrist	Bantam Lite	1968	30	50	75
Flying Saucer	Remco	1968	60	100	150
Game, Double Action Bagatelle	Hasbro	1969	60	100	150
Game, Land of the Giants	Ideal	1968	60	105	160
Gum Cards, Box	Topps/A & BC	1968	395	650	1000
Gum Cards, Set	Topps USA	1968	275	450	700
Gum Cards, Set	A & BC Prod./England	1968	275	450	700
Gum Cards, Wrapper	Topps/A & BC	1968	60	100	150
Lunchbox	Aladdin	1969	80	130	200
Model, Snake Kit	Aurora	1968	275	450	700
Model, Spindrift	Aurora	1968	275	450	700
Model, Spindrift	Aurora	1975	115	195	300
Model, Spindrift	Lunar Models	1989	30	50	80
Model, Spindrift	Midori/Japan		0	0	0
Model, Spindrift	Comet Miniatures		0	0	0
Model, Spindrift Interior	Lunar Models	1989	35	55	85
Motorized Flying Rocket	Remco	1968	80	130	200
Movie Viewer	Acme	1968	30	45	70
Painting Set	Hasbro	1969	40	65	100
Puzzle, Floor	Whitman	1968	35	55	85
Rub Ons	Hasbro	1969	30	50	75
Shoot & Stick Target Rifle Set	Remco	1968	90	145	225
Signal Ray Space Gun	Remco	1968	70	115	175
Space Ship Control Panel	Remco	1968	195	325	500
Space Sled	Remco	1968	195	325	500
Spindrift Toothpick Kit	Remco	1968	30	50	80
Target Set	Hasbro	1969	60	100	150
Viewmaster Set	GAF	1968	20	35	50
Walkie Talkies	Remco	1968	80	130	200

Lost in Space

NAME & DESCRIPTION	COMPANY	YEAR	GOOD	EX	MIB
Book, Comic	Innovation	1990s	1	2	3
Book, Comic Space Family Robinson, Lost in Space	Gold Key Comics	1960s	15	25	40
Book, Paperback	Pyramid	1967	13	20	30
Costume, Space Suit	Ben Cooper	1965	85	130	200
Diorama	Aurora	1966	425	650	1000
Dolls, Set	Marusan/Japanese		2900	4500	7000
Fan Cards		1960s	20	35	50
Fan Cards		1960s	15	25	35
Game	Milton Bradley	1965	65	100	150
Game, 3-D Fun Set	Remco	1966	500	775	1200
Gun, First Season Laser Pistol	Lunar Models		50	75	115
Gun, Laser Water Pistol			30	50	75
Gun, Second Season Laser Pistol	Lunar Models		45	70	105
Helmet and Gun Set	Remco	1967	325	525	800
Lunchbox and Thermos	King Seeley Thermos	1967	300	450	700
Model, Accessory Set	Lunar Models	1989	25	40	60
Model, Cast Set #3	Lunar Models	1991	25	35	55

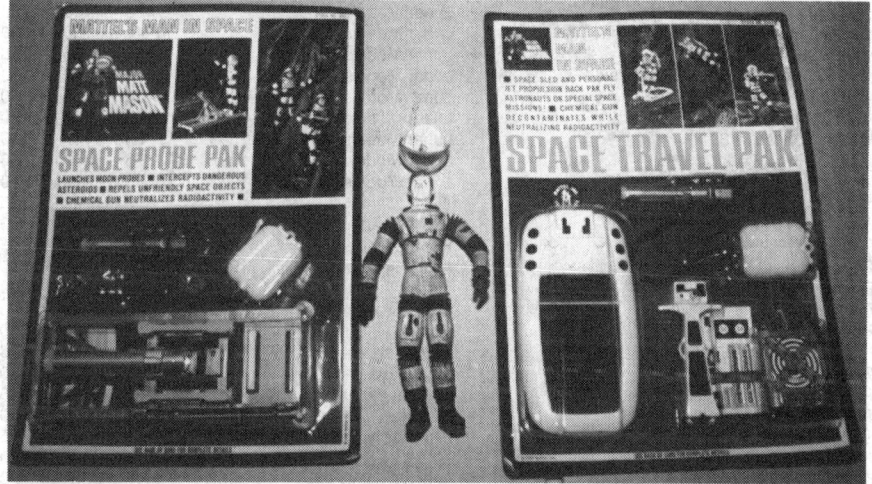

SPACE TOYS

Top to bottom: Battery Operated Mr. Zerox, Japan; Battery Operated Apollo-Z Moon Traveler, unknown; Man In Space Probe Pak and Travel Pak, Mattel; Captain Video Space Game, Milton Bradley.

Lost in Space

NAME & DESCRIPTION	COMPANY	YEAR	GOOD	EX	MIB
Model, Cave Diorama	Lunar Models	1986	50	80	120
Model, Chariot	Lunar Models	1987	35	50	80
Model, Chariot	Marusan/Japanese		625	975	1500
Model, Crash Site Diorama	Lunar Models	1993	25	40	65
Model, Derelict Ship	Lunar Models	1990	25	40	65
Model, Don West	Lunar	1993	50	80	120
Model, Figure Set #1	Lunar Models	1988	25	40	60
Model, Figure Set #2	Lunar Models	1990	20	35	50
Model, Illusion Machine Diorama	Lunar Models	1993	55	85	130
Model, Invaders' Fifth Dimension Space Ship	Lunar Models	1990	25	40	60
Model, John Robinson	Lunar Models	1993	45	70	110
Model, Jupiter	Marusan/Japanese	1966	425	650	1000
Model, Jupiter-2	Lunar Models	1989	15	25	35
Model, Jupiter-2	Lunar Models		75	120	185
Model, Jupiter-2	Marusan/Japanese	1966	425	650	1000
Model, Jupiter-2	Comet/England		8	13	20
Model, Jupiter-2	Lunar Models		35	50	80
Model, Jupiter-2 Fusion Core Light Kit	Lunar Models		40	60	90
Model, Jupiter-2 Interior	Lunar Models	1988	55	85	130
Model, Jupiter-2, Crash Site Version	Lunar Models	1989	13	20	30
Model, Penny Robinson	Lunar Models	1993	50	80	120
Model, Robot	Lunar Models	1993	65	100	150
Model, Robot YM-3	Lunar Models	1993	65	100	150
Model, Space Pod	Lunar Models	1987	35	50	80
Model, Spider Diorama	Lunar Models	1993	45	70	110
Note Pad			25	40	65
Puzzles, Frame Tray	Milton Bradley	1966	40	65	100
Robot	Remco	1965	200	310	475
Robot	AHI	1977	20	35	50
Robot	Aurora	1966	425	650	1000
Robot	Aurora	1966	625	975	1500
Robot	Remco	1966	295	450	700
Robot	Ahi/Kmart	1977	40	65	100
Robot YM-3	Masudaya	1986	85	130	200
Robot YM-3	Masudaya	1985	20	30	45
Roto-Jet Gun Set	Mattel	1966	850	1300	2000
Saucer Gun	Ahi	1977	30	50	75
Switch and Go Set	Mattel	1966	975	1500	2300
Trading Cards, Set	Topps	1966	175	260	400
Tru-View Magic Eyes Set	GAF	1967	30	50	75
Viewmaster	GAF	1967	25	40	60
Walkie Talkies	Ahi	1977	30	50	75
Weapons Set, 1/35 Scale	Lunar Models		2	3	5

Miscellaneous

NAME & DESCRIPTION	COMPANY	YEAR	GOOD	EX	MIB
Astro Base (22" tall, red & white astronaut base)	Ideal	1960	225	325	500
Astro Boy Mask (blue hair with boy smiling)		1960s	20	45	65
Astro Boy Mask/Glasses (blue glasses with Astro boy hair on top)		1960s	20	45	65
Astro-ray Space Gun (10")			20	30	45
Astronaut Halloween Costume	Collegeville	1960	18	25	40
Astronaut Halloween Costume	Ben Cooper	1962	18	25	40
Astronaut Space Commander Play Suit (green outfit & cap)	Yankeeboy	1950s	35	50	80
Atomic Disintegrator (cap gun)	Hubley	1940s	50	70	110
Cherilea Space Gun (miniature scale, die cast)	Marx		27	40	60
Dan Dare & the Aliens Ray Gun (21" color tin litho gun)		1950s	105	155	235
Martian Bobbing Head (7" tall, blue vinyl plastic)		1960s	23	35	50
Men into Space Astronaut Space Helmet (plastic helmet with visor)	Ideal	1960s	35	50	75

Miscellaneous

NAME & DESCRIPTION	COMPANY	YEAR	GOOD	EX	MIB
Moon Map Jigsaw Puzzle (10 x 14" picture of the moon's surface)	Selchow & Righter	1970	14	20	30
Rex Mars Atomix Pistol Flashlight (plastic, 1950's)	Marx		50	75	100
Rocket Gun (7" hard yellow/green plastic with spring loaded plunger)	Jak-Pak	1958	9	13	20
Rocky Jones Space Ranger (14 x 16" colorbook)	Whitman	1951	2407	39	60
Space Hopper Child's Rubbers (5 x 6" black rubber overshoes)		1950s	2300	29	45
Space Safari Planetary Playset (with vehicles, astronauts, aliens)		1969	45	65	95
Space Water Pistol	Nasta	1976	7	10	15
Sparkling Ray Gun	Nasta	1976	7	10	15
TV Space Riders Coloring Book (14 X 15")	Abbott	1952	7	10	15
V-Enemy Visitor Doll (12" tall)	LJN	1984	16	23	35

Monsters

NAME & DESCRIPTION	COMPANY	YEAR	GOOD	EX	MIB
Creature from the Black Lagoon Figure (on card)	AHI		295	425	650
Creature from the Lagoon Glow in the Dark Mini Monsters	Remco		27	40	60
Creature/Black Lagoon Official Universal Studios Figure (8" tall)	Remco	1980	90	130	200
Deadly Grell, The (bendable)	LJN	1983	5	7	10
Dracula Action Figure (with Aurora head)	AHI		65	125	200
Dracula Glow in the Dark Mini Monsters	Remco		23	30	45
Dracula Official Universal Studios Figure (8" tall)	Remco	1980	30	45	70
Dwarves of the Mountain Human/Monster Figure	LJN	1983	5	7	10
Evil Monster Figure Bugbear & Goblin (Orcs of the Broken Bone)	LJN	1983	5	7	10
Frankenstein Action Figure (with Aurora head)	AHI		65	125	200
Frankenstein Figure (poseable figure, glow-in-the-dark)	Remco	1978	23	35	50
Frankenstein Glow in the Dark Mini Monsters	Remco		20	30	45
Frankenstein Official Universal Studios Figure (8" tall)	Remco	1980	27	40	60
Freddy Krueger Maxx FX Doll			11	16	25
Freddy Krueger Stick-Up Doll (6" doll)			5	7	10
Freddy Krueger Talking Doll	Matchbox		18	26	40
Freddy Krueger YoYo	Spectra Star		2	3	5
Godzilla Combat Joe Set (vinyl set with a 12" tall combat Joe figure)		1984	450	650	1000
Godzilla Figure (13" tall, arms, legs and tail moveable)	Imperial	1985	15	23	35
Godzilla Figure (19" tall)	Mattel	1977	20	30	45
Godzilla Figure (6 1/2" tall, arms, legs and tail moveable)	Imperial	1985	5	10	15
Rocks & Bugs & Things (five different sets, each)	Ideal	1986	2	3	5

Moon McDare

NAME & DESCRIPTION	COMPANY	YEAR	GOOD	EX	MIB
Action Communication Set	Gilbert	1966	25	35	55
Moon Explorer Set	Gilbert	1966	35	50	75
Moon McDare Figure (12" tall astronaut with accessories)	Gilbert	1966	55	80	125
Space Accessory Pack	Gilbert	1966	25	35	55
Space Gun Set	Gilbert	1966	30	40	65
Space Mutt Set	Gilbert	1966	30	45	70

Munsters, The

NAME & DESCRIPTION	COMPANY	YEAR	GOOD	EX	MIB
Herman Figure	Remco	1964	105	150	230
Lilly Figure	Remco	1964	115	165	250
Lilly Munster Baby Figure	Ideal	1965	30	45	65

SPACE TOYS

Munsters, The

NAME & DESCRIPTION	COMPANY	YEAR	GOOD	EX	MIB
Set of Three Munsters (Herman, Lilly and Grandpa)	Remco		225	425	700

Other Worlds, The

Castle Zendo	Arco	1983	20	30	45
Fighting Glowgons Figure Set	Arco	1983	18	25	40
Fighting Terrans Figure Set	Arco	1983	20	30	45
Kamaro Figure	Arco	1983	8	12	18
Sharkoss Figure	Arco	1983	8	12	18

Outer Limits

Jigsaw Puzzle	Milton Bradley	1960s	50	80	125
Outer Limits Game	Milton Bradley	1964	105	180	275

Outer Space Men, The

Alpha 7 (The Man from Mars) Figure	Colorforms	1968	55	80	125
Astro-Nautilus (The Man from Neptune) Figure	Colorforms	1960s	55	80	125
Colossus Rex (The Man from Jupiter) Figure	Colorforms	1960s	70	100	150
Commander Comet (The Man from Venus) Figure	Colorforms	1960s	55	80	125
Electron (The Man from Pluto) Figure	Colorforms	1960s	55	80	125
Orbitron (The Man from Uranus) Figure	Colorforms	1960s	55	80	125
Xodiac (The Man from Saturn) Figure	Colorforms	1960s	55	80	125

Planet of the Apes

Planet of Apes "Gallen" Bank			25	35	55
Planet of the Apes Color-Vue Set	Hasbro	1970s	30	45	65
Planet of the Apes Fun-Doh Modeling Molds (molds of Zir)	Chemtoy	1974	20	30	45
Planet of the Apes Photo Puzzles (96 pc. puzzles in can)	H.G. Toys		7	10	15
Planet of the Apes Trash Can (oval can)	Chein	1967	25	35	55
Planet of the Apes View Master Reels (set of three reels with booklet)	View Master	1970s	15	25	35
Planet of the Apes Wagon (friction powered prison wagon)	AHI		20	45	65

Robots

Answer Game Machine (battery operated robot performs math tricks)			300	650	900
Attacking Martian (10")	S.H.	1960s	45	65	100
B.O. Robot (7-1/2", electric remote control)		1950s	325	450	700
Big Max & his Electronic Conveyor (9")	Remco		50	70	110
Captain Astro (6", wind-up)		1970	40	55	85
Chief Robotman (12")	KO	1965	400	850	1200
Countdown-Y (9")	Cragstan	1960s	100	145	225
Electric Robot (plastic, 15" tall)	Marx		125	200	275
Forbidden Planet Robby Figure (16" tall, talking)	Matsudaya		80	115	175
Forbidden Planet Robby Figure (5" tall, wind-up)	Matsudaya		16	23	35
Frankenstein Robot (tin and plastic battery operated)			550	900	1400
Laughing Robot	Marx		50	70	110
Launching Robot (10")	S.H.	1975	25	35	55
Lunar Robot (wind-up, companion to Thunder Robot)			225	450	650
Lunar Spaceman (12", battery operated)		1978	20	30	45
Magnor (9")	Cragstan	1975	23	35	50
Mechanical Interplanetary Explorer (8", wind-up)		1950s	180	260	400
Mechanical Moon Creature (6", wind-up)	Marx	1960	90	130	200

SPACE TOYS

Top to bottom: Steve Zodiac's Fireball XL5 Space City, Multiple; Flash Gordon Play Set, Mego; Star Trek Inter-Space Communicator, Lone Star; Space Ship X-5, China.

Robots

NAME & DESCRIPTION	COMPANY	YEAR	GOOD	EX	MIB
Mechanical Spaceman (6", wind-up)		1960s	40	60	90
Mechanical Television Spaceman (7", wind-up)	ALPS	1965	45	60	95
Mechanized Robot (13", battery operated)		1960s	425	600	925
Mekanda Robo (6 1/2")	Zncron	1981	12	26	40
Mini Robot (2 1/2" tall, blue plastic wind-up robot)			14	20	30
Moon Creature (mechanical, wind-up, 5 1/2" tall)	Marx	1960s	115	170	260
Myrobo (9", battery operated)		1970s	25	35	55
Outer Space Ape Man Robot	Ilco	1970s	14	20	30
Outer Space Robot (10", battery operated)		1979	20	30	45
Plastic Spaceman (wind-up)	Irwin	1950	45	65	100
Radio Control Robot	Bilko	1970s	25	35	55
Red, Blue & Silver Robot (plastic with antenna, metal key)		1970s	11	16	25
Rendezvous 7.8 (15")	Yanoman		170	245	375
Ro-Gun "It's A Robot" (sho-gun type robot that changes into rifle)	Arco	1984	11	16	25
Robert the Wonder Toy (14")	Ideal	1960	90	130	200
Robot (plastic and metal, 12" tall, 1967)	Marx		115	165	250
Robot 2500	Durham Industries	1979	14	20	30
Robot Commando (19")	Ideal	1960s	60	85	130
Robot Tank II (10", battery operated)		1965	150	210	325
Robot YM-3 (5" tall)	Matsudaya	1985	14	20	30
Sky Robot (8")	S.H.	1970s	20	30	45
Space Guard Pilot (8")	Asak	1975	20	30	45
Sparky Robot (7")	KO	1960	45	65	100
Star Robot (black robot)	CDO	1970s	0	0	0
Super Astronaut	S.J.M.	1981	11	16	25
TR-2 Robotank (5", battery operated)		1975	35	50	75
TV Spaceman (battery operated)		1965	90	130	200
Zerak the Blue Destroyer Zeroid (6" metal blue plastic with motor)	Ideal	1968	45	65	100

Rocky Jones, Space Ranger

Button, Space Ranger, Pinback		1954	17	30	45
Coloring Book, Rocky Jones Space Ranger	Whitman	1951	25	40	60
Pin, Wings		1954	17	30	45
Wrist Watch		1954	80	130	200

Six Million Dollar Man

Backpack Radio	Kenner		5	7	10
Bionic Adventure DSI Undercover Assignment Set	Kenner		9	13	20
Bionic Bigfoot Doll (18" tall poseable of hairy alien robot)	Kenner	1976	40	60	90
Bionic Mission Vehicle	Kenner		10	18	30
Bionic Transport & Repair Station	Kenner		18	25	40
Colonel Steve Austin with Bionic Grip (12" tall)	Kenner	1976	14	20	30
Maskatron Doll (12" tall)	Kenner		18	26	40
Oscar Goldman Doll (12" tall)	Kenner	1976	18	26	40
Porta Communicator	Kenner		11	16	25
Six Million Dollar Man Bank (vinyl, small)			8	15	20

Space Patrol

Badge, Space Patrol		1950s	60	100	150
Belt, Jet Glow Code Belt		1950s	105	180	275
Coins, Interplantary Space Patrol Credits			10	16	25
Cosmic cap		1950s	115	195	300
Cosmic Rocket Launcher Set		1950s	295	475	750
Drink Mixer, in box		1950s	60	100	150

Space Patrol

NAME & DESCRIPTION	COMPANY	YEAR	GOOD	EX	MIB
Emergency Kit, in case		1950s	295	475	750
Gun, Atomic Pistol Flashlight	Marx	1950s	80	130	200
Gun, Cosmic Smoke Gun		1950s	100	165	250
Gun, Cosmic Smoke Gun		1950s	90	145	225
Gun, Rocket Gun and Holster Set		1950s	175	295	450
Gun, Rocket Gun Set		1950s	105	180	275
Handbook		1950s	55	95	145
Helmet, Space Patrol Commander		1950s	135	225	350
Lunar Fleet Base		1950s	195	325	500
Mask, Man-From-Mars Totem Head		1950s	65	105	165
Mask, Outer Space Helmet		1950s	100	165	250
Membership Card, Space Patrol Cadet		1950s	20	35	50
Monorail Set	Toys of Tomorrow	1950s	1550	2600	4000
Pen, (rocket shaped)		1950s	105	180	275
Project-O-Scope		1950s	175	295	450
Puzzle, Frame Tray	Milton Bradley	1950s	35	60	95
Ring, Hydrogen Ray Gun Ring		1950s	70	115	175
Rocket Lite Flashlight	Ray-O-Vac	1950s	135	225	350
Rocket Port Set	Marx	1950s	115	195	300
Space Binoculars		1950s	70	115	175
Space Binoculars		1950s	105	180	275
Space Patrol Persicope		1950s	60	100	150
Space-A-Phones, Set		1950s	145	250	375
Space-O-Phones, Set		1950s	80	130	200
Wrist Watch		1950s	250	425	650

Space Ships

NAME & DESCRIPTION	COMPANY	YEAR	GOOD	EX	MIB
Eagle Lunar Module (9")		1960s	80	115	175
Friendship 7 (9 1/2", friction)			35	50	75
Inter-Planet Toy Rocketank Patrol (10")	Macrey	1950	30	45	70
Jupiter Space Station (8")	TN (Japan)	1960s	90	125	195
Moon-Rider Space Ship (tin wind-up, 1930's)	Marx		125	200	250
Mystery Space Ship (35mm astronauts and moon-men, rocket, launchers)	Marx	1960s	50	75	100
Rocket Fighter (with tail fin and sparking action, tin wind-up)	Marx	1950s	250	375	500
Rocket Fighter Spaceship (celluloid window, tin wind-up, 12" long)	Marx	1930s	125	200	250
Satelitte X-107 (9")	Cragstan	1965	90	130	200
Sky Patrol Jet (5x13x5" bump and go battery op. gunner)	TN (Japan)	1960s	295	425	650
Solar-X Space Rocket (15")	TN (Japan)		45	65	100
Space Bus (tin helicopter, battery operated with wired remote)			350	500	750
Space Pacer (7", battery operated)		1978	2300	29	45
Space Ship (bronze, hard plastic)	Marx		40	60	90
Space Survey X-09 (battery operated tin and plastic flying saucer)			175	350	525
Space Train (9" long, engine & three metallic cars)		1950s	18	26	40
Super Space Capsule (9 1/2")		1960s	70	100	150
X-3 Rocket Gyro		1950s	25	35	50

Space:1999

NAME & DESCRIPTION	COMPANY	YEAR	GOOD	EX	MIB
Action Figure, Commander Koenig	Mattel	1976	17	30	45
Action Figure, Dr. Russell	Mattel	1976	17	30	45
Action Figure, Professor Bergman	Mattel	1976	17	30	45
Action Figure, Zython	Mattel	1976	20	35	50
Adventure Playset	Amsco/Milton Bradley	1976	30	50	75
Book, Cut And Color	Saalfield	1975	6	10	15
Colorforms Adventure Set	Colorforms	1975	10	16	25

SPACE TOYS

Space:1999

NAME & DESCRIPTION	COMPANY	YEAR	GOOD	EX	MIB
Eagle Freighter	Dinky	1975	15	30	45
Eagle One Spaceship	Mattel	1976	60	100	150
Eagle Transport	Dinky	1975	17	30	45
Film Viewer TV Set, on card		1976	8	13	20
Galaxy Time Meter, on card		1976	6	10	15
Game, Space:1999	Milton Bradley	1975	10	16	25
Gun, Astro Popper, on card		1976	6	10	15
Lunch Box	King Seeley	1976	22	35	55
Model Kit, Eagle One	MPC	1976	12	20	30
Model Kit, Eagle Transporter	Airfix	1976	12	20	30
Model Kit, Moon Base Alpha	MPC	1976	17	30	45
Moonbase Alpha Playset	Mattel	1976	30	50	75
Puzzle, in box	HG Toys	1976	6	10	15
Space Expedition Dart Set, on card		1976	6	10	15
Stamping Set		1976	8	13	20
Superscope, on card		1976	6	10	15
Talking View-Master Set	View-Master	1975	6	10	15
Utility Belt Set (with disc shooting stun gun, watch, compass)	Remco	1976	12	20	30
View-Master Set	View-Master	1975	10	16	25
Walking Spaceman		1975	20	35	50

Star Trek

NAME & DESCRIPTION	COMPANY	YEAR	GOOD	EX	MIB
"Passage to Moauv" Book/Record Set			5	8	11
20th Anniversary Vulcan Ears	Ballantine		4	5	7
Action Toy Book	Random House	1976	7	10	15
Antican Alien Figure, ST:TNG (3 3/4", fully poseable)	Galoob	1988	14	20	45
Attemped Hijacking of the U.S.S. Enterprise & Officers	H.G. Toys	1974	6	8	12
Battle on the Planet Klingon Puzzle (150 pieces)	H.G. Toys	1974	5	7	10
Battle on the Planet Romulon Puzzle (150 pieces)	H.G. Toys	1974	5	7	10
Beanbag Chair, ST:TMP (group drawing)			25	35	55
Bop Bag (plastic, inflatable Spock)		1975	55	80	125
Bowl, ST:TMP (plastic)	Deka	1979	3	4	10
Bridge Punch-out Book, ST:TMP	Wanderer	1979	7	10	15
Bridge Scene Frame Tray Puzzle (8 1/2" x 11" tray)	Whitman	1978	2	3	8
Bulletin Board, ST:TMP (die cut board with four pens)	Milton Bradley	1979	6	8	12
Cartoon Puzzle	Whitman	1978	4	5	8
Clock (Enterprise orbiting planet, rectangular)		1989	23	33	50
Clock (white wall clock, 20th Anniversary logo, Star Trek Fan Club)		1986	14	20	30
Collegeville Halloween Costume (Spock, Kirk Ilia or Klingon)		1979	11	16	25
Colorforms Adventure Set (plastic stick-ons)	Colorforms	1975	14	20	30
Comb & Brush Set (6" x 3", blue, oval brush)		1977	14	20	30
Communicators (black plastic walkie talkies with flip-up grid)	McNerney	1989	35	50	75
Communicators (walkie talkies, blue plastic with flip-up grid)	Mego	1976	70	100	150
Communicators, ST:TMP (plastic wrist band walkie talkie)	Mego	1980	90	130	200
Controlled Space Flight (plastic Enterprise, battery operated)	Remco	1976	80	115	175
Data Figure, ST:TNG (3 3/4", fully poseable)	Galoob	1988	7	10	25
Digital Travel Alarm	Lincoln Enterprises		14	20	30
Enterprise Jigsaw Puzzle, ST:TMP (551 pieces)	Aviva	1979	9	13	20
Enterprise Make A Model, ST:TNG	Chatham River Press	1990	4	5	8
Enterprise Punch-out Book, ST:TMP	Wanderer	1979	9	13	20

Star Trek

NAME & DESCRIPTION	COMPANY	YEAR	GOOD	EX	MIB
Enterprise Puzzle, ST:IV (551 pieces "The Voyage Home")	Mind's Eye Press	1986	14	20	30
Enterprise Puzzle, ST:TMP (100 pieces)	Arrow		4	6	9
Enterprise Puzzle, ST:TMP (15 piece sliding puzzle)	Larami	1979	5	7	10
Enterprise Puzzle, ST:TMP (250 piece color photo)	Milton Bradley	1979	5	7	10
Enterprise Watch, ST:TMP	Bradley		20	30	45
Enterprise Watch, ST:TMP (gold plated silver men & women styles)	Rarities Mint	1989	55	80	125
Enterprise, ST:III (4" long, die cast with black plastic stand)	Ertl	1984	7	10	25
Enterprise, ST:IV (24", silver plastic, inflatable)	Sterling	1986	20	30	45
Enterprise, ST:TMP (20" long, white plastic, lights & sound)	South Bend	1979	80	115	175
Enterprise, ST:TNG (6" long, die cast with detachable saucer section)	Galoob	1988	9	13	25
Excelsior, ST:III (4" long, die cast with black plastic stand)	Ertl	1984	7	10	25
Ferengi Alien Figure, ST:TNG (3 3/4", fully poseable)	Galoob	1988	25	35	75
Ferengi Fighter, ST:TNG (orange plastic with moveable canopy & guns)	Galoob	1989	20	30	45
Ferengi Halloween Costume, ST:TNG	Ben Cooper	1988	7	10	20
Figurine Painting (plastic figurine, brush & five paints)	Milton Bradley	1979	14	20	30
Flashlight (battery operated, small phaser shape)		1976	6	8	12
Flashlight, ST:TMP (hand flashlight)	Larami	1979	6	8	12
Force Field Capture Puzzle (150 pieces)	H.G. Toys	1976	4	6	10
Giant in the Universe Pop-up Book	Random House	1977	14	20	30
Golden Trivia Game (trivia cards, game board & dice)	Western Publishing	1985	20	30	45
Helmet (plastic, electronic sound, flashing red light on top)	Remco	1976	55	80	125
Kirk & Officers Beaming Down Puzzle (150 pieces)	H.G. Toys	1974	5	7	10
Kirk & Spock or Spock Mirror (two sizes available)		1977	11	16	25
Kirk & Spock Watch, ST:TMP (LCD rectangular)	Bradley		25	35	55
Kirk Bank (12" plastic)	Play Pal	1975	25	35	55
Kirk Figure, ST:III (3 3/4" tall with communicator, poseable)	Ertl	1984	9	13	25
Kirk Figure, ST:TMP (13" tall, soft body with plastic head)	Knickerbocker	1979	16	23	35
Kirk Figure, ST:V (7" tall, posed statuette)	Galoob	1989	25	35	45
Kirk Halloween Costume (plastic mask, one piece jumpsuit)	Ben Cooper	1975	9	13	20
Kirk Needlepoint Kit (number 10 mesh canvas with white background)	Arista	1980	16	23	35
Kirk or Spock Halloween Costumes (lightweight, tie-on jumpsuit)	Ben Cooper	1967	11	16	25
Kirk Puzzle, ST:TMP (15 piece sliding puzzle)	Larami	1979	5	7	10
Kirk, Mr. Spock, Dr. McCoy Puzzle (150 pieces)	H.G. Toys	1976	4	6	10
Kite (TV Enterprise or Spock)	Hi-Flyer	1975	14	20	30
Kite, ST:III (pictures Enterprise)	Lever Bros.	1984	14	20	30
Kite, ST:TMP (picture of Spock)	Aviva	1976	11	16	25
Klaa Figure, ST:V (7" tall, posed statuette)	Galoob	1989	25	35	45
Klingon Bird of Prey, ST:III (3 1/2", die cast with blacj plastic stand)	Ertl	1984	7	10	25
Klingon Figure, ST:III (3 3/4" tall with pet, fully poseable)	Ertl	1984	7	10	25
Klingon Halloween Costume (plastic mask, one piece jumpsuit)	Ben Cooper	1975	9	13	25
Klingon Halloween Costume, ST:TNG	Ben Cooper	1988	7	10	15
LaForge Figure, ST:TNG (3 3/4", fully poseable)	Galoob	1988	3	4	10
Light Switch Cover, ST:TMP	American Tack & Hdw.	1985	6	8	12

SPACE TOYS

Top to bottom: Rocket Bank, unknown; Battery Operated Space Explorer TV Robot, 1960s, Japan; Planet Robot, Japan; Battery Operated Sky Patrol, Japan; Space Commander Walkie Talkies, Remco.

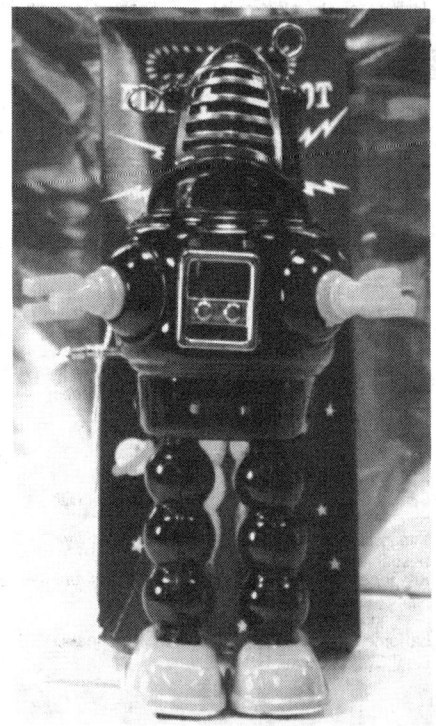

Star Trek

NAME & DESCRIPTION	COMPANY	YEAR	GOOD	EX	MIB
Magic Slates (four different designs)	Whitman	1979	7	10	15
Make-a-Game Book	Wanderer	1979	7	10	15
McCoy Figure, ST:V (7" tall, posed statuette)	Galoob	1989	25	35	45
Metal Detector (metal detector with U.S.S. Enterprise decal)	Jetco	1976	100	145	225
Mirror (2" x 3" metal, with black & white photo of crew)		1966	1	2	3
Mission to Gamma VI (18" high plastic cave creature)	Mego	1976	350	475	750
Mix 'n Mold (three separate kits: Kirk, Spock or McCoy)		1975	35	50	75
Movie Viewer (3" red & black plastic)	Chemtoy	1967	11	16	25
Paint by Numbers (large) (canvas paint, Kirk, Spock & Enterprise)	Hasbro	1972	35	50	75
Paint by Numbers (small) (canvas paint, Kirk, Spock & Enterprise)	Hasbro	1972	23	35	50
Pen & Poster Kit (line posters & felt tipped pens)	Open Door	1976	11	16	25
Pen & Poster Kit, ST:III (3-D poster "Search for Spock")	Placo	1984	9	13	20
Pennant "Paramount Pictures Adventure" (9" x 21-1/2")	Universal Studios	1988	5	7	10
Pennant, "Spock in Vulcan Robes", ST:II (12" x 30" "Spcok Lives")	Image Products	1982	6	8	12
Pennant, "The Wrath of Khan", ST:II (12" x 30" black w/ Enterprise)	Image Products	1982	6	8	12
Phaser (Astro Buzz-Ray Gun with three color flash beam)	Remco	1967	80	115	175
Phaser (black plastic, electronic sound, flashlight projects target)	Remco	1975	35	50	75
Phaser Battle Game (black plastic, 13" high battery op. target game)	Mego	1976	195	275	425
Phaser, ST:III (white & blue plastic gun with light & sound effects)	Daisy	1984	35	50	75
Phaser, ST:TNG (gray plastic light & sound hand phaser)	Galoob	1988	14	20	35
Picard Figure, ST:TNG (3 3/4", fully poseable)	Galoob	1988	3	4	15
Pinball Game, ST:TMP (12", plastic, two styles Kirk or Spock)	Azrak-Hamway		23	35	75
Pinball Game, ST:TMP (electronic pinball game)	Bally	1979	200	295	600
Pocket Flix (battery operated movie viewer & film cartridge)	Ideal	1978	18	25	40
Pop-up Book, ST:TMP	Wanderer	1980	11	16	30
Q Alien Figure, ST:TNG (3 3/4", fully poseable, black outfit)	Galoob	1988	23	35	75
Riker Figure, ST:TNG (3 3/4", fully poseable)	Galoob	1988	2	3	10
Role Playing Game, 2001 Deluxe Edition	FASA		20	30	45
Scottie Figure, ST:III (3 3/4" tall with phaser, fully poseable)	Ertl	1984	7	10	25
Selay Alien Figure, ST:TNG (3 3/4", fully poseable, green reptile)	Galoob	1988	16	23	45
Shuttlecraft Galileo, ST:TNG	Galoob	1989	16	23	40
Space Design Center, ST:TMP	Avalon	1979	70	100	150
Spock & Enterprise 20th Anniversary Digital Watch, ST:TMP	Lewco	1986	9	13	30
Spock & Enterprise Halloween Costume	Ben Cooper	1973	11	16	30
Spock Bank (12" plastic)	Play Pal	1975	25	35	55
Spock Chair, ST:TMP (inflatable)		1979	16	23	35
Spock Ears, ST:TMP	Aviva	1979	7	10	15
Spock Figure, ST:III (3 3/4" tall with phaser, fully poseable)	Ertl	1984	9	13	25
Spock Figure, ST:TMP (13" tall, soft body with plastic head)	Knickerbocker	1979	16	23	35
Spock Figure, ST:V (7" tall, posed statuette)	Galoob	1989	25	35	45
Spock Halloween Costume (plastic mask, one piece jumpsuit)	Ben Cooper	1975	9	13	25

SPACE TOYS

Star Wars

NAME & DESCRIPTION	COMPANY	YEAR	GOOD	EX	MIB
Spock in Space Suit Frame Tray Puzzle (8 1/2" x 11" tray)	Whitman	1978	2	3	8
Spock Needlepoint Kit (14" x 18" number 10 mesh canvas)	Arista	1980	16	23	35
Spock Puzzle, ST:TMP (15 piece sliding puzzle)	Larami	1979	5	7	10
Spock Puzzle, ST:TMP (551 pieces)	Aviva	1979	9	13	20
Spock Tray (17 1/2" metal lap tray with legs "Mr. Spock")	Aviva	1979	9	13	20
Spock Watch, ST:TMP	Bradley		20	30	45
Star Trek Cartoon Puzzle	Whitman	1978	4	5	8
Star Trek Color & Activity Book	Whitman	1979	2	3	8
Star Trek Coloring Book	Saalfield	1979	7	10	15
Star Trek II USS Enterprise Ship (3" metal die cast)	Corgi	1982	9	13	20
Star Trek Paint by Number Set	Hasbro	1974	11	16	25
Star Trek Puzzle	Marvel	1976	5	7	10
Star Trek Puzzle "The Alien"		1975	7	10	15
Star Trek TMP Dinnerware Set	Deka	1979	16	23	35
Star Trek TMP Poster Pen Set (14x20")	Aviva	1979	9	13	20
Star Trek Watch (Spock on dial with revolving Enterprise)	Bradley	1979	45	65	100
Star Trek Writing Tablet (8 x 10")		1967	11	16	25
Starship U.S.S. Enterprise & Its Officers Puzzle (300 pieces)	H.G. Toys	1974	9	13	20
Sybok Figure, ST:V (7" tall, posed statuette)	Galoob	1989	25	35	45
Telescreen (plastic, battery operated target game with light & sound)	Mego	1976	70	100	150
The Role Playing Game, 2004 Basic Set (three reading books)	FASA		7	10	15
The Role Playing Game, Second Deluxe Edition	FASA		14	20	30
Tracer Gun (plastic pistol with colored plastic discs)	Rayline	1966	45	65	100
Tracer Scope (rifle with discs)	Rayline	1968	55	80	125
Transporter Frame Tray Puzzle (8 1/2" x 11" tray)	Whitman	1978	2	3	5
Tricorder (blue plastic tape recorder, battery operated)	Mego	1976	70	100	150
Trillions of Trilligs Pop-up Book	Random House	1977	16	23	35
U.S.S. Enterprise Action Playset (8" dolls, stools, console)	Mego	1975	125	180	275
U.S.S. Enterprise Bridge, ST:TMP (white plastic)	Mego	1980	70	100	150
Utility Belt	Remco	1975	45	65	100
View Master (16 page story booklet with three viewer reels)	GAF	1968	7	10	25
View Master (16 page story booklet with three viewer reels)	GAF	1974	7	10	20
View Master (talking) (three viewer reels "Mr. Spock's Time Trek)	GAF	1974	11	16	25
View Master Double-Vue, ST:TMP (double plastic cassette)	GAF	1981	11	16	25
View Master Gift Pak, ST:TMP (viewer, three reels, 3-D posters)	GAF	1979	35	50	75
View Master, ST:II (view master with three reels)	View Master	1982	7	10	20
View Master, ST:TMP (story booklet with three reels)	GAF	1979	7	10	20
Waste Basket, ST:TMP (13" high, metal rainbow painting)	Chein	1979	11	16	35
Waste Paper Basket (black metal)	Chein	1977	35	50	75
Water Pistol (white plastic, shaped like U.S.S. Enterprise)	Azrak-Hamway	1976	20	30	45
Water Pistol, ST:TMP (gray plastic, early phaser)	Aviva	1979	11	16	25
Worf Figure, ST:TNG (3 3/4", fully poseable)	Galoob	1988	3	4	10
Yar Figure, ST:TNG (3 3/4", fully poseable)	Galoob	1988	7	10	18
Yo-Yo, ST:TMP (blue sparkle plastic)	Aviva	1979	7	10	20
"Assault on Death Star" Movie Cartridge			14	20	30

Star Wars

NAME & DESCRIPTION	COMPANY	YEAR	GOOD	EX	MIB
"Battle in Hyperspace" Movie Cartridge			14	20	30
"Danger at the Cantina" Movie Cartridge			14	20	30
"Destroy Death Star" Movie Cartridge			14	20	30
"May the Force be with You" Movie Cartridge			9	13	20
"Planet of the Hoojibs" Book/Record Set			4	5	7
"The Maverick Moon" Book	Random House		2	3	5
"The Mystery of the Rebellious Robot" Book	Random House		2	3	5
3-D Electronic Quartz Sceni-Clock	Bradley	1982	16	23	55
3-D Ewok Perk-Up Sticker Sets			5	7	10
ABC Fun	Random House	1985	6	8	12
Aboard the Millennium Falcon Jigsaw Puzzle (1000 pieces)			7	10	15
Admiral Ackbar Figurine Paint Set	Craftmaster		9	13	20
Attack of the Sand People Jigsaw Puzzle (140 pieces)	Kenner		5	7	10
Backpack, ROTJ			11	16	25
Bantha Jigsaw Puzzle, The (140 pieces)	Kenner		6	8	12
Battle on Hoth Paint Set			11	16	25
Ben Kenobi & Darth Vader Poster	Proctor & Gamble	1978	9	13	20
Blanket			14	20	30
Boba Fett Cake Pan			11	16	25
Boba Fett Figure (bisque)	Towle/Sigma	1983	45	65	100
Boba Fett Poster			7	10	15
Bubble Bath (four different, each)			6	8	12
Burger Chef Fun Book	Kenner	1978	6	8	12
C-3PO & R2-D2 Alarm Clock	Bradley	1980	16	23	75
C-3PO & R2-D2 Digital Watch	Bradley	1970s	55	80	125
C-3PO & R2-D2 Watch (digital, rectangular)	Bradley	1970s	30	45	65
C-3PO & R2-D2 Watch (digital, round face)	Bradley	1970s	45	65	95
C-3PO & R2-D2 Watch (digital, round, musical)	Bradley	1970s	70	100	150
C-3PO & R2-D2 Watch (vinyl band, drawing)	Bradley	1970s	45	60	95
C-3PO & R2-D2 Watch (vinyl band, photo)	Bradley	1970s	30	45	65
C-3PO & R2-D2 Watch (white border, photo)	Bradley	1970s	45	65	95
C-3PO Bank (ceramic)	Roman Ceramics	1977	25	35	55
C-3PO Bust Case			20	30	45
C-3PO Cake Pan			11	16	25
C-3PO Cookie Jar	Roman Ceramics	1977	80	115	175
C-3PO Earrings			7	10	15
C-3PO Figurine Paint Set	Craftmaster		9	13	20
C-3PO Mug (ceramic)	Sigma		16	23	35
C-3PO Pencil Tray	Sigma		23	35	50
C-3PO Pendant			9	13	20
C-3PO Picture Frame	Sigma		20	30	45
C-3PO Ring			10	14	22
C-3PO Stickpin			6	8	12
C-3PO Tape Dispenser	Sigma		23	35	50
Cantina Band Jigsaw Puzzle (500 pieces)	Kenner		6	8	12
Cast Coloring Book	Kenner		5	7	10
Charm Bracelet			7	10	15
Chewbacca & C-3PO Coloring Book	Kenner		5	7	10
Chewbacca & Leia Coloring Book	Kenner		5	7	10
Chewbacca Bandolier Strap Case			11	16	20
Chewbacca Bank	Sigma	1983	25	35	55
Chewbacca Medal	W. Berrie	1980	6	8	12
Chewbacca Mug (ceramic)	Sigma		18	25	40
Chewbacca Pendant			9	13	20
Chewbacca Poster	Burger King	1978	5	7	10
Chewbacca Punching Bag (50")	Kenner	1977	45	65	100
Chewbacca's Activity Book	Random House		2	3	8
Chewbacca, Han, Leia & Lando Coloring Book	Kenner		5	7	10
Chewbacca/Darth Vader Bookends	Sigma		25	35	55
Chewbecca Birthday Candle	Wilton		6	8	12

SPACE TOYS

Star Wars

NAME & DESCRIPTION	COMPANY	YEAR	GOOD	EX	MIB
Darth Vader & Ben Kenobi Duel Jigsaw Puzzle	Kenner		5	7	10
Darth Vader & Imperial Guard Roller Skates			16	23	35
Darth Vader & Stormtroopers Coloring Book	Kenner		5	7	10
Darth Vader & Stromtroopers Poster			7	10	15
Darth Vader Bank	Adam Joseph	1983	9	13	20
Darth Vader Bank (anodized silver plated)	Leonard Silver	1981	45	65	95
Darth Vader Bank (ceramic)	Roman Ceramics	1977	25	35	55
Darth Vader Beach Towel			6	8	15
Darth Vader Belt Buckle	Leather Shop	1977	14	20	30
Darth Vader Birthday Candle	Wilton		6	8	12
Darth Vader Cookie Jar	Roman Ceramics	1977	80	115	175
Darth Vader Digital Watch	Bradley	1970s	30	45	65
Darth Vader Duty Roster (school supplies)			5	7	10
Darth Vader Earrings			7	10	15
Darth Vader Figure (bisque)	Towle/Sigma	1983	30	45	65
Darth Vader Glow-in-the-Dark Paint Set			9	13	20
Darth Vader Mug (ceramic)	Sigma		16	23	35
Darth Vader Pendant			9	13	20
Darth Vader Picture Frame	Sigma		30	45	65
Darth Vader Pillow		1983	9	13	20
Darth Vader Poster	Burger King	1978	5	7	10
Darth Vader Poster	Nestea	1980	9	13	20
Darth Vader Poster (life size)			9	13	20
Darth Vader Premium Poster	Proctor & Gamble	1980	4	5	8
Darth Vader Punching Bag (50")	Kenner	1977	25	40	60
Darth Vader Ring	W. Berrie	1980	6	8	12
Darth Vader Speaker Phone	ATC	1983	55	80	100
Darth Vader SSP Van (black)	Kenner	1978	18	25	40
Darth Vader Stickpin			6	8	12
Darth Vader Watch (star & planet on face)	Bradley	1970s	45	65	95
Darth Vader Watch (vinyl band)	Bradley	1970s	30	45	65
Darth Vader's Activity Book	Random House		2	3	8
Darth Vader, R2-D2 & C-3PO Cookie Jar (hexagon)	Sigma		55	80	125
Death Star Poster	Proctor & Gamble	1978	9	13	20
Degobah Play-Doh Set			16	23	35
Degobah Premium Poster	Burger King	1980	5	7	10
Droid Dilemma, The (book)	Random House	1979	4	5	8
Droid Wall Clock	Bradley		20	30	45
Droids Digital Watch	Bradley	1970s	30	45	65
Droids Dinnerware Set			11	16	25
Duel Racing Set	Lionel	1978	60	85	125
Electric Toothbrush	Kenner	1980	16	23	35
Electric Toothbrush	Kenner	1978	16	23	35
Emperor's Royal Guard Bank	Adam Joseph	1983	9	13	20
Empire Strikes Back Dinnerware Set			16	23	35
Empire Strikes Back Panorama Book, The	Random House		14	20	30
Empire Strikes Back Pop-up Book, The	Random House	1980	11	16	25
Empire Strikes Back Poster Album Vol. 1			11	16	25
Empire Strikes Back Radio Program Poster			14	20	30
Empire Strikes Back Sketchbook, The	Ballantine	1980	14	20	30
Empire Strikes Back Storybook, The	Random House	1980	9	13	20
Empire Strikes Back Wall Clock	Bradley		20	30	45
Escape from the Monster Ship (book)	Random House	1985	6	8	12
Ewok Music Box Radio			18	25	40
Ewok Teaching Clock			20	30	35
Ewoks Coloring Books (each)	Kenner	1983	7	10	15
Ewoks Digital Watch	Bradley	1970s	30	45	65
Ewoks Give-A-Show Projector with Strips	Kenner	1984	18	25	40
Ewoks Play-Doh Set			11	16	30
Ewoks Watch (vinyl band)	Bradley	1970s	30	45	65
Fan Club Kit			16	23	35

SPACE TOYS

Top to bottom: Space Patrol Drink Mixer, Star-lite; Buck Rogers View-Master, GAF Corp.; Fireball XL5 Steve Zodiac Puppet, unknown; (left) Super Space Capsule and (right) Capsule Mercury/Friendship-7, both Japan.

SPACE TOYS

Star Wars

NAME & DESCRIPTION	COMPANY	YEAR	GOOD	EX	MIB
Flying R2-D2 Rocket Kit	Estes	1978	9	13	20
Fuzzy as an Ewok (book)	Random House	1985	6	8	12
Galactic Emperor Figure (bisque)	Towle/Sigma	1983	35	50	75
Gammorean Guard Figure (bisque)	Towle/Sigma	1983	25	35	55
Gammorean Guard Mug (ceramic)	Sigma		10	20	30
Gamorrean Guard Bank	Adam Joseph	1983	7	10	20
Give-A-Show Projector with Strips	Kenner	1979	23	35	50
Han & Chewbacca Jigsaw Puzzle (140 pieces)	Kenner		6	8	12
Han Solo Figure (bisque)	Towle/Sigma	1983	45	65	100
Han Solo Figurine Paint Set	Craftmaster		16	23	35
Han Solo Mug (ceramic)	Sigma		23	35	50
Hoth Premium Poster	Burger King	1980	5	7	10
How the Ewoks Saved the Trees	Random House	1985	6	8	12
Ice Planet Hoth Play-Doh Set			16	23	35
Inflatable Lightsaber	Kenner	1977	45	60	95
Interceptor (INT-4) Mini-rig			16	23	35
Introducing Yoda Press Kit		1980	20	30	45
Jabba the Hut Digital Watch	Bradley	1970s	30	45	65
Jabba the Hut Play-Doh Set			11	16	30
Jabba the Hut Puzzle	Craft Master	1983	3	4	6
Jabba the Hutt Bank	Sigma	1983	25	35	55
Jabba's Prison (Sears exclusive)			45	65	100
Jawa Punching Bag (50")	Kenner	1977	45	65	100
Jawas Capture R2-D2 Jigsaw Puzzle (140 pieces)	Kenner		4	6	9
Kneesaa Bank	Adam Joseph	1983	7	10	20
Lando Calrissian Figure (bisque)	Towle/Sigma	1983	30	45	65
Lando Calrissian Mug (ceramic)	Sigma		16	23	35
Lando Fighting Skiff Guard Coloring Book	Kenner	1983	7	10	15
Lando in Falcon Cockpit Coloring Book	Kenner	1983	7	10	15
Learn-to-Read Activity Book	Random House	1985	6	8	12
Leia & Han Solo Glow-in-the-Dark Paint Set			9	13	20
Leia Figurine Paint Set	Craftmaster		11	16	25
Leia Mug (ceramic)	Sigma		20	30	45
Lost Prince, The (book)	Random House	1985	6	8	12
Luke & Leia Leap for their Lives Jigsaw Puzzle (500 pieces)	Kenner		6	8	12
Luke & Tauntaun Figurine Paint Set	Craftmaster		14	20	30
Luke & Tauntaun Teapot Set			25	35	55
Luke Coloring Book	Kenner	1983	7	10	15
Luke Mug (ceramic)	Sigma		20	30	45
Luke Skywalker AM Headset Radio			95	135	400
Luke Skywalker Figure (bisque)	Towle/Sigma	1983	30	45	65
Luke Skywalker Glow-in-the-Dark Paint Set			9	13	20
Luke Skywalker Jigsaw Puzzle (500 pieces)	Kenner		11	16	25
Luke Skywalker Poster	Burger King	1978	5	7	10
Luke Skywalker Poster	Nestea	1980	9	13	20
Luke Skywalker Poster			18	25	40
Luke Skywalker Premium Poster	Proctor & Gamble	1980	4	5	8
Max Rebo Coloring Book	Kenner	1983	7	10	15
May the Force Be with You Ring	W. Berrie	1980	6	8	12
Movie Viewer	Kenner	1978	14	20	30
My Jedi Journal	Ballantine		7	10	15
Notebooks (school supplies, each)			5	7	10
Original Fan Club Kit			20	30	75
Original Press Kit		1977	25	35	55
Pillowcase			4	5	8
Pirates of Tarnoonga, The (book)	Random House	1985	6	8	12
Placemats (complete set)			14	20	30
Portable Clock/Radio	Bradley	1984	11	16	30
Portfolio (school supplies)			5	7	10
Princess Leia's Beauty Bag			14	20	30
R2-D2 & C-3PO Belt Buckles	Leather Shop	1977	14	20	30

Star Wars

NAME & DESCRIPTION	COMPANY	YEAR	GOOD	EX	MIB
R2-D2 & C-3PO Jigsaw Puzzle (140 pieces)			6	8	12
R2-D2 & C-3PO Premium Poster	Proctor & Gamble	1980	4	5	8
R2-D2 Bank	Adam Joseph	1983	7	10	20
R2-D2 Bank (ceramic)	Roman Ceramics	1977	25	35	55
R2-D2 Belt Buckle	Leather Shop	1977	14	20	30
R2-D2 Birthday Candle	Wilton		6	8	12
R2-D2 Coloring Book	Kenner		5	7	10
R2-D2 Cookie Jar	Roman Ceramics	1977	90	130	200
R2-D2 Earrings			7	10	15
R2-D2 Pendant			9	13	20
R2-D2 Picture Frame	Sigma		30	45	65
R2-D2 Poster	Burger King	1978	5	7	10
R2-D2 Punching Bag (50")	Kenner	1977	25	40	60
R2-D2 Ring			10	14	22
R2-D2 Stickpin			6	8	12
R2-D2 String Dispenser with Scissors	Sigma		20	30	45
Read & the Force is with You Poster (Yoda)			14	20	30
Return of the Jedi Activity Book	Happy House	1983	5	7	10
Return of the Jedi Art Portfolio			9	13	20
Return of the Jedi Belt, Illustrated	Leather Shop	1977	5	7	10
Return of the Jedi Candy Container (complete set of eighteen)	Topps	1983	45	65	100
Return of the Jedi Coloring Book		1984	2	3	8
Return of the Jedi Dinnerware Set			14	20	30
Return of the Jedi Maze Book	Happy House	1983	5	7	10
Return of the Jedi Monster Activity Book	Happy House	1983	5	7	10
Return of the Jedi Picture Puzzle Book	Happy House	1983	5	7	10
Return of the Jedi Pop-up Book	Random House	1983	11	16	25
Return of the Jedi Punch-out & Make It Book	Random House		14	20	30
Return of the Jedi Sketchbook	Ballantine	1983	11	16	25
Return of the Jedi Storybook	Random House	1983	9	13	20
Return of the Jedi Word Puzzle Book	Happy House	1983	5	7	10
Selling of Driods Jigsaw Puzzle, The (500 pieces)	Kenner		4	6	9
Sheets			11	16	25
Sleeping Bag			15	21	32
Snow Speeder Toothbrush Holder	Sigma		25	35	55
Space Battle Jigsaw Puzzle (500 pieces)	Kenner		7	10	15
Star Destroyer Poster	General Mills	1978	9	13	20
Star Tours Stamp Set			4	5	8
Star Wars Action Play-Doh Set			20	30	45
Star Wars Adventure (Movie Poster) Jigsaw Puzzle (1000 pieces)			7	10	15
Star Wars Blueprint Set			14	20	30
Star Wars Calendars (1978-1990)			7	10	15
Star Wars Dinnerware Set			20	30	45
Star Wars ESB Coloring Book		1980	2	3	8
Star Wars ESB Story Book (hardcover)		1980	4	5	8
Star Wars Intergalactic Passport & Stickers	Ballantine	1983	7	10	15
Star Wars Luke Skywalker's Activity Book	Random House		2	3	8
Star Wars Pop-up Book	Random House	1978	11	16	25
Star Wars Poster Art Coloring Set	Craftmaster	1978	9	13	20
Star Wars Question & Answer About Space Book	Random House	1979	5	7	10
Star Wars Question & Answer Book About Computers, The			5	7	10
Star Wars Radio Program Poster			23	35	50
Star Wars ROTJ "The Ewoks Join the Fight" Book	Golden		2	3	5
Star Wars Sketchbook	Ballantine	1977	14	20	30
Sticker Book (256 stickers)	Panini	1977	23	35	50
Sticker Set & Album	Burger King		7	10	15
Stickers (set of four stickers)	Trix		3	4	6
Stickers (set of four stickers)	Lucky Charms		3	4	6
Stickers (set of four stickers)	Monster Cereal		3	4	6

Star Wars

NAME & DESCRIPTION	COMPANY	YEAR	GOOD	EX	MIB
Stickers (set of four stickers)	Cocoa Puffs		3	4	6
Stormtrooper Pendant			9	13	20
Stormtroopers Stop the Landspeeder Jigsaw Puzzle (140 pieces)	Kenner		5	7	10
Sy Snootles & Rebo Band	Sigma	1983	25	35	55
The Jedi Master's Quizbook	Random House	1985	6	8	12
TIE Fighter & X-Wing Poster	General Mills	1978	9	13	20
TIE Fighter Rocket Kit	Estes	1978	11	16	25
Trapped in the Trash Compactor Jigsaw Puzzle (140 pieces)	Kenner		5	7	10
Turret Music Box with C-3PO			20	30	45
Victory Celebration Jigsaw Puzzle (500 pieces)	Kenner		6	8	12
Wicket & the Dandelion Warriors (book)	Random House	1985	6	8	12
Wicket 3 in 1 Stamp Set			4	5	8
Wicket Bank	Adam Joseph	1983	7	10	20
Wicket Figurine Paint Set	Craftmaster		9	13	20
Wicket Play Phone			23	35	40
Wicket the Ewok Dinnerware Set			11	16	25
Wicket the Ewok Roller Skates			14	20	40
Wicket the Ewok Watch (stars & planet on face)	Bradley	1970s	30	45	65
Wicket Toothbrush (battery operated)		1984	9	13	20
Wookie Doodle School Pad			5	7	10
Wookie Storybook, The	Random House	1979	7	10	15
World of Star Wars, The	Paradise Press	1981	9	13	20
X-Wing Fighter Rocket Kit	Estes	1978	11	16	25
X-Wing Fighters Prepare to Attack Jigsaw Puzzle (500 pieces)	Kenner		6	8	12
X-Wing Medal	W. Berrie	1980	6	8	12
X-Wing Ring	W. Berrie	1980	6	8	12
X-Wing with Maxi-Brutel Rocket Kit	Estes	1978	18	25	40
Yoda Backpack	Sigma		14	20	30
Yoda Bank	Sigma	1983	25	35	55
Yoda Bank (lithographed tin with combination dials)			11	16	25
Yoda Coloring Book	Kenner		5	7	10
Yoda Figurine Paint Set	Craftmaster		9	13	20
Yoda Glow-in-the-Dark Paint Set			9	13	20
Yoda Hand Puppet			14	20	40
Yoda Jedi Master Fortune Teller Ball			45	65	95
Yoda Sleeping Bag			16	23	35
Yoda Tumbler/Pencil Cup	Sigma		20	30	45
Yoda Watch	Bradley	1970s	30	45	65

Time Tunnel

Coloring Book	Saalfield	1967	15	25	40
Time Tunnel Game, The	Ideal	1966	95	155	240
Time Tunnel Spin Game, The	Pressman	1967	60	105	160

Tom Corbett

Belt, Tom Corbett Space Cadet		1950s	65	105	165
Binoculars, Tom Corbett Space Cadet			60	100	150
Book, Push-Outs	Saalfield	1952	30	50	75
Coloring Book	Saalfield	1950s	25	40	65
Flash X-1 Space Gun			55	95	145
Gun, Official Sparkling Space Gun	Marx		80	130	200
Gun, Space Cadet Atomic Rifle		1950s	100	165	250
Gun, Tom Corbett Space Cadet Offical Space Pistol	Marx	1950s	50	80	125
Gun, Tom Corbett Space Gun (9 1/2" long light blue & black sparking)		1950s	70	120	185

Tom Corbett

NAME & DESCRIPTION	COMPANY	YEAR	GOOD	EX	MIB
Lunch Box, with bottle	Aladdin	1954	250	350	550
Lunch Box, with bottle	Aladdin	1952	105	180	275
Model Craft Molding Super Set (six-figure set)	Kay Standley	1950s	100	165	250
Puzzles, Frame Tray	Saalfield	1950s	17	30	45
Ring, Portrait, Tom Corbett		1950s	20	35	50
Ring, Rocket Scout		1950s	10	16	25
Ring, Space Suit		1950s	10	16	25
Signal Siren Flashlight	Usalite	1950s	70	115	175
Space Academy Playset, #7010	Marx	1952	185	310	475
Space Academy Playset, #7012	Marx	1950s	185	310	475
Spaceship, Polaris (wind-up, 12" long, 1952)	Marx		195	325	500
Spaceship, Space Cadet 2 (tin wind-up, 12" long, 1930s)	Marx		225	375	575
View-Master Set, Tom Corbett Space Cadet	Sawyer's	1950s	30	50	75
View-Master Set, Tom Corbett, Secret From Space	Sawyer's	1950s	30	50	75
Wrist Watch	Ingraham	1950s	250	425	650

UFO

"Past & Present" Sticker Book	Whitman	1968	16	23	35

Voyage to the Bottom of the Sea

Voyage to the Bottom of the Sea Playset	Remco	1960s	45	195	300

SPACE TOYS

SPACE TOYS

Top to bottom: The Empire Strikes Back Imperial Attack Base and Imperial TIE Fighter, both 1980; Wicket W. Warrick from Ewoks series, 1985; Kez-Iban from the Star Wars Droids series, 1985; Luke Skywalker, 1985, all Kenner.

KENNER STAR WARS TOYS

Die Cast Vehicles

Series I

TOY	YEAR	MNP	MIP
Darth Vader Tie Fighter	1979	25	60
Land Speeder	1979	25	60
Tie Fighter	1979	25	60
X-Wing	1979	22	65

Series II

Imperial Cruiser	1979	50	150
Millennium Falcon	1979	50	125
Tie Bomber	1979	275	750
Y-Wing	1979	40	125

Series III

Slave I	1979	25	75
Snowspeeder	1979	25	85
Twin Pod Cloud Car	1979	25	75

Droids

3 3/4" Figures

A-Wing Pilot	1985	20	75
Boba Fett	1985	10	75
C-3PO	1985	20	40
Jann Tosh	1985	5	12
Jord Dusat	1985	5	12
Kea Moll	1985	5	12
Kez-Iban	1985	5	12
R2-D2 with pop-up lightsaber	1985	20	30
Sise Fromm	1985	5	20
Thall Joben	1985	5	12
Tig Fromm	1985	5	20
Uncle Gundy	1985	5	12

Accessories

Droids Light Saber	1985	41	150

Vehicles

A-Wing Fighter, Droids Box	1983	25	125
ATL Interceptor	1985	20	50
Imperial Side Gunner	1985	10	40

Empire Strikes Back

3 3/4" Figures

2-1B	1980	8	25
4-LOM	1981	7	35
AT-AT Commander	1980	8	25
AT-AT Driver	1981	7	20
Bespin Security Guard, black	1980	8	25
Bespin Security Guard, white	1980	6	30
Bossk, bounty hunter	1980	8	25

Empire Strikes Back

TOY	YEAR	MNP	MIP
C-3PO with removable limbs	1982	10	35
Cloud Car Pilot	1982	7	20
Dengar	1980	7	20
FX-7	1980	7	25
Han Solo in Bespin outfit	1981	10	40
Han Solo in Hoth outfit	1980	10	35
IG-88	1980	10	30
Imperial Commander	1981	7	20
Imperial Storm Trooper in Hoth battle gear	1980	10	20
Imperial Tie Fighter Pilot	1982	10	30
Lando Calrissian		5	20
Lobot	1981	5	15
Luke Skywalker in Bespin outfit	1980	10	35
Luke Skywalker in Hoth battle gear	1982	10	30
Princess Leia in Bespin gown	1980	10	30
Princess Leia in Hoth outfit	1981	10	30
R2-D2 with sensorscope		5	30
Rebel Commander	1980	8	25
Rebel Snow Soldier in Hoth battle gear	1980	10	30
Ugnaught	1981	8	25
Yoda with brown snake	1981	8	25
Yoda with orange snake	1981	8	25
Zuckuss	1982	10	35

Accessories

3 Position Laser Rifle	1980	41	135
Darth Vader Carring Case	1982	7	35
Display Arena	1980	13	40
Hoth Wompa	1982	5	30
Laser Pistol	1980	15	55
Light Saber, red or green	1980	17	50
Light Saber, yellow	1980	17	75
Mini Figure Case	1980	8	25
Tauntaun, solid belly	1980	8	30
Tauntaun, split belly	1982	10	35

Playsets

Cloud City Playset (Sears Exclusive)	1981	100	375
Dagobah	1982	12	65
Darth Vader Star Destroyer		30	100
Hoth Ice Planet	1980	26	100
Imperial Attack Base	1980	20	65
Imperial TIE Fighter	1980	35	100
Rebel Command Center	1980	66	200
Turret and Probot	1980	26	80

Vehicles

AT-AT All-Terrain Armored Transport	1980	45	200

SPACE TOYS

SPACE TOYS

Top to bottom: Star Wars Cantina Play Set (Sears exclusive), 1977, Palitoy; Luke Skywalker AM Headset Radio, Kenner; Boba Fett from The Empire Strikes Back, 1979, Kenner.

Empire Strikes Back

TOY	YEAR	MNP	MIP
Rebel Armored Snowspeeder	1980	17	90
Rebel Transport	1982	21	80
Scout Walker	1982	12	55
Slave I (Bobba Fett's Space Ship)	1980	25	95
Twin-Pod Cloud Car	1980	15	75

Ewoks

3 3/4" Figures

Dulok Scout	1985	5	12
Dulok Shaman	1985	5	12
King Gornesh	1985	5	12
Logray (Ewok Medicine Man)	1985	5	15
Urgah Lady Gorneesh	1985	5	12
Wicket W. Warrick	1985	5	15

Vehicles

Ewoks Fire Cart	1985	7	30
Ewoks Treehouse	1985	17	50
Ewoks Woodland Wagon	1985	7	30

Large Figures

12" Figures

Ben (Obi-Wan) Kenobi	1979	55	250
Boba Fett with Empire Strikes Back box	1979	80	250
Boba Fett with Star Wars box	1979	95	250
C3-PO	1979	40	130
Darth Vader	1978	50	150
Han Solo	1979	125	400
IG-88	1980	175	450
Jawa	1979	50	160
Luke Skywalker		50	200
Princess Leia Organa	1977	50	200
R2-D2	1979	40	130
Storm Trooper	1979	58	200

Mail Away Figures

Bagged Figures

AT-AT Commander	10
AT-ST Driver	10
C-3PO, removable limbs	10
Han Solo in Hoth outfit	10
Han Solo in trench coat	10
Luke Skywalker in Hoth outfit	10

Mail Away Figures

TOY	YEAR	MNP	MIP
Pruneface			10
R2-D2 with sensorscope			10

White Box

Anakin Skywalker		13	30
Nien Numb		5	10
The Emperor		5	10

Micro Series

Bespin Control Room	1982	8	35
Bespin Freeze Chamber	1982	17	75
Bespin Gantry	1982	8	35
Death Star Compactor	1982	8	35
Death Star Escape	1982	8	35
Death Star Escape (Sears white box)	1982	8	35
Death Star World	1982	20	125
Hoth Generator Attack	1982	7	35
Hoth Ion Cannon	1982	7	55
Hoth Turrent Defense	1982	7	35
Hoth Wompa Cave	1982	7	35
Hoth World	1982	17	125
Imperial Tie Fighter	1982	20	60
Millinium Falcon	1982	75	225
Snow Speeder	1982	50	175
X-Wing Fighter	1982	15	60

Mini Rigs

A3T-5	1983	3	20
CAP-2 Captivator	1982	3	20
Desert Sail Skiff	1984	3	25
Endor Forest Ranger	1984	3	20
INT-4 Interceptor	1982	3	20
ISP-6 Imperial Shuttle Pod	1983	4	20
MLC-3 Mobile Laser Cannon	1981	4	20
MTV-7 Multi-Terrain Vehicle	1981	4	25
PDT-8 Personal Deployment Transport	1981	4	25
Radar Lase Cannon	1982	4	25
Security Scout		19	75
Tatoonie Skiff		50	175
Tatoonie Skiff in Droid Box		50	175
Tri-Pod Laser Cannon	1982	4	20
Vehicle Maintence	1982	4	20
A-Wing Pilot	1985	15	45
Amanaman	1985	15	55
Anakin Skywalker	1985	26	80
B-Wing Pilot	1985	5	15
Barada	1985	12	35
Ben (Obi-Wan) Kenobi	1985	12	35
Biker Scout	1985	7	25
EV-9D9	1985	15	60

SPACE TOYS

SPACE TOYS

Clockwise from upper left: Barada from Star Wars series, 1985; Land Of The Jawas Play Set, 1977; Yoda The Jedi Master from The Empire Strikes Back series, 1981; Rebel Transport from The Empire Strikes Back series, 1982; Droid Factory from Star Wars series, 1977, all Kenner.

Mini Rigs

TOY	YEAR	MNP	MIP
Han Solo in Carbonite outfit	1985	33	100
Imperial Dignitary	1985	15	50
Imperial Gunner	1985	15	40
Jawa	1985	13	45
Lando Calrissian General Pilot	1985	12	45
Luke in stormtrooper outfit	1985	40	175
Luke Skywalker X-Wing fighter pilot	1985	10	45
Lumat	1985	10	35
Paploo	1985	10	35
R2-D2 with pop-up light saber	1985	20	40
Romba	1985	10	35
The Emperor	1985	15	40
Warok	1985	10	35
Wicket	1985	7	35
Yak Face	1985	75	250

Vehicles

Ewok Battle Wagon	1985	21	75
Imperial Sniper Vehicle	1985	40	120
One-Man Sand Skimmer	1985	25	95
Sand Skimmer		15	95
Security Scout Vehicle	1985	40	120
Tatooine Skiff	1985	60	200

Return of the Jedi

3 3/4" Figures

4-LOM		5	15
8D8	1983	5	30
Admiral Ackbar	1983	3	10
AT-ST Commander		4	12
AT-ST Driver		4	12
Biker Scout	1983	3	10
Chief Chirpa	1983	4	12
Emperor's Royal Guard		5	25
Gamorrean Guard	1983	3	8
General Madine	1983	3	10
Han Solo in trench coat	1984	5	15
Klaatu	1983	3	8
Klaatu in Skiff guard outfit	1983	10	25
Lando Calrissian in Skiff outfit	1983	5	15
Logray (Ewok Medicine Man)	1983	4	12
Luke Skywalker in Jedi Knight outfit	1983	8	25
Lumat	1983	10	25
Nein Numb	1983	5	15
Nikto	1984	3	10
Princess Leia in combat poncho	1984	10	20
Princess Leia Organa in Boushh outfit	1983	10	20
Prune Face	1984	5	10

Return of the Jedi

TOY	YEAR	MNP	MIP
Rancor Keeper	1984	5	10
Rebel Commander	1983	5	15
Rebel Commando	1983	5	15
Ree-Yees	1983	4	12
Squid Head	1983	3	10
Sy Snootles and the Reebo Band (boxed set)	1984	13	60
Teebo	1984	4	12
The Emperor	1983	7	20
Warok	1983	10	25
Weequay		3	8
Wicket W. Warrick	1984	5	15
Zuckuss		5	15

Accessories

Biker Scout Laser Pistol	1984	20	50
C-3PO Collectors Case	1983	10	45
Chewbacca Bandoiler Strap	1983	5	20
DV Head Carring Case with three figures	1983	10	35
Ewok Assault Catapult	1983	7	25
Jabba the Hutt (regular issue)		10	40
Laser Rifle Carring Case	1984	5	40
Light Saber (red or green plastic)		15	45
Rancor Monster	1983	10	45
Tri-Pod Cannon	1983	3	20
Vehicle Maintenance Energizer	1983	3	20

Playsets

Ewok Village	1983	17	65
Jabba the Hutt (loose figure)		10	
Jabba the Hutt Dungeon (Sears Exclusive)	1983	25	125
Jabba the Hutt Dungeon (Sears Exclusive)	1983	25	95

Vehicles

B-Wing Fighter	1984	12	75
Ewok Combat Glider	1984	3	25
Imperial Shuttle	1984	45	200
Speeder Bike	1983	7	20
TIE Interceptor	1984	20	75
Y-Wing Fighter	1983	20	90

Star Wars

Early Bird figures in mailing box (set of four)	1977	140	340
Early Bird Kit	1977	100	300

3 3/4" Figures

Ben (Obi-Wan) Kenobi (original 12)	1977	10	95
Boba Fett	1978	25	125
C-3PO (original 12)	1977	10	95
Chewbacca (original 12)	1977	10	95

Star Wars

TOY	YEAR	MNP	MIP
Darth Vader (original 12)	1977	10	95
Death Squad Commander (original 12)	1977	10	80
Death Star Droid	1978	15	75
Greedo	1978	13	80
Hammerhead	1978	13	75
Han Solo in vest (original 12)	1977	10	150
Jawa with cloth cape (original 12)	1977	10	95
Jawa with plastic cape (original 12)	1977	100	400
Luke Skywalker (original 12)	1977	10	95
Luke Skywalker X-Wing Pilot	1978	13	75
Power Droid	1978	17	65
Princess Leia Organa (original 12)	1977	10	125
R2-D2 (original 12)	1977	10	80
R5-D4	1978	15	65
Snaggletooth (blue body) Sears Exclusive	1978	100	
Snaggletooth (red body)	1978	10	65
Stormtrooper (original 12)	1977	10	95
Tuskan Raider, sand person (original 12)	1977	10	100
Walrus Man	1978	12	75

Accessories

Action Figure Display Stand (Mail-In Premium)	1977	50	125
Han Solo's Laser Pistol		20	75

Return of the Jedi

TOY	YEAR	MNP	MIP
Inflatible Light Saber	1977	55	75
R2-D2 (radio controlled)	1978	50	125
SW 24 Figure Vinyl Figure Case		10	35

Playsets

Cantina Adventure Set (Sears Exclusive)	1977	150	375
Creature Cantina	1977	25	100
Death Star Space Station	1977	75	150
Droid Factory	1977	18	65
Land of the Jawas	1977	30	90
Patrol Dewback	1977	20	55

Vehicles

Darth Vader TIE Fighter	1977	25	80
Imperial TIE Fighter	1977	25	80
Jawa Sand Crawler (remote controlled)	1977	125	375
Land Speeder (battery operated)	1977	20	80
Millennium Falcon Spaceship	1977	50	175
Sonic Land Speeder (J.C. Penney Exclusive)	1977	120	400
X-Wing Fighter	1977	25	85

Toy Guns

By George H. Newcomb

Almost everyone alive today who grew up in the United States remembers the toy guns of their childhood. Baby boomers who grew up watching TV westerns and space operas have been largely responsible for the increased interest in toy gun collecting. This, plus the increasing interest in related western and cowboy hero collectibles, have brought prices and demand into new realms.

Toy guns fall into a number of major categories and production periods. The first cap gun patent dates to about 1860. The preferred material of toy construction until World War II was cast iron. Most early cap guns, exploders, and figural guns were made of cast iron, though variations of plastic, tin, and stamped metal also appear.

The late 1930s was a golden age of cast iron guns. Many guns of this period have realistic revolving cylinders, fine nickel finishes, colorful plastic grips, and complicated mechanisms.

Metal toy gun production ceased in about 1940. The companies that survived in the postwar years after 1946 experimented with a variety of materials and methods of production, the most successful of which was die-casting. This involved injecting an alloy mixture into a mold, which resulted in good detail, lowered costs, and a much lighter gun. Die-cast guns dominated the 1950s and 1960s marketplace. Television brought little buckaroos daily installments of the thrilling adventures of cowboy heroes such as Roy, Gene, and Hoppy. Manufacturers such as the George Schmidt Company, Classy Products, Marx, Hubley, Nichols, Kilgore, and Wyandotte Toys all scrambled to arm these cowpokes and compete for a piece of the market. This, by some standards, was the golden age of die-cast guns.

Mattel, a relative newcomer to the business, introduced their first western-style cap gun in 1957, the "Fanner 50." This shiny, oversized gun would, by its incredible success, drive many manufacturers out of the toy gun market and some others out of business completely.

The bottom line of toy making has always been costs. The less money spent on production, the greater the profit. Mattel produced a line of low priced plastic and die-cast cap guns, rifles, and machine guns that arrived on the market at a time when TV western heroes such as Maverick, The Lawman, Sugarfoot, Bronco, and Matt Dillon were replacing the older serial cowboys.

Cap guns would never be the same. All of the major producers added plastic to their lines, and the race was on to lower production costs and remain competitive. As far as most collectors are concerned, things went downhill from there. Each year toy guns seemed to decrease in quality and detail.

While cap guns sold in the American market were largely produced by American manufacturers in both the prewar and postwar periods, other styles of toy guns flowed from overseas producers. Tin lithography became an art form in postwar Japan. The same Japanese companies that produced the robots and tin cars so widely sought by today's collectors also produced space ray guns, cowboy, "G-Men," and military-style tin litho guns for the American market.

While prewar Japanese toy guns are rare, postwar sparking ray guns, pop guns, water, and clicker guns abound. These guns were generally cheaper than their American competition and provided a profitable "low end" product for American retailers.

As the postwar Japanese economy grew and workers demanded more money, toy gun manufacturing moved from Japan to Hong Kong to Taiwan to Korea and eventually to mainland China, currently one of the major producers of toy guns and caps.

The Last Gunfighter

By the late 1960s, the majority of TV westerns had been cancelled. Cowboy heroes were forgotten, and toy guns in general were seen in a more ominous light by parents and child psychiatrists. Fads of spy guns and space guns revived the market for a while, but the golden age was over. The eventual end point of this tale can be found in the Toys R Us or Kay-Bee Toys near you today. Go there and visit the toy gun section. You will see a fair assortment of multi-colored Super-Soakers and noise making space guns, but you will not find a Roy Rogers holster set or even a Fanner 50.

Variations on the Big Bang

Walking through any toy or collectibles show or picking up any toy publication these days will bring you in contact with collectible toy guns. The prices, conditions, and varieties are confusing to say the least. To make sense of toy gun collecting, you must first understand that there are all sorts of toy guns. I refer to the hobby as "toy gun" collecting instead of "cap gun" collecting because all toy guns are not cap guns. There are cap guns, water guns, dart guns, cork guns, clicker guns, BB guns, air guns, pellet guns, guns that shoot peas, and even some that can fire four or five types of the aforementioned ammunition.

As in collecting anything, you have to decide some basic questions before you begin. What do you want to collect and why do you want to collect it? Most toy gun collectors are buying back a bit of their childhood. They played with these guns in countless childhood fantasies, and holding them again unlocks a wealth of memories. Some collectors are searching only for the toy guns they had (or wish they had) as children. These folks may actually play with their guns and (heaven forbid!) fire caps in them. They may not be as particular about condition and packaging as collectors on the other side of the spectrum, who collect for the investment. Investment collectors seek only the highest grade, unplayed with, mint-from-the-store-shelves quality pieces that will appreciate in value. They never play with their guns, and the idea of firing caps in a gun gives them shivers. Most collectors fall between these two extremes.

Some collectors amass only guns by a specific manufacturer — Hubley, Mattel, Nichols, and so on — while others search for the shootin' irons that bear the names of their cowboy heroes. Some specialize in military-style toy weapons, bullet-firing machine guns, rocket-firing bazookas, and cap-firing hand grenades. Recently, space guns have begun to increase in popularity. Toy gun collectors do not even have to be limited to the guns themselves. One collector I know only collects boxes of caps. He does not have a single gun to shoot them with but carries photos of the hundreds of different cap boxes that he has found. Another collector buys only cap gun advertising, catalogs, and store displays.

Unless you have unlimited funds and storage space, you may want to limit your collection to certain types of guns. Whatever your choice, you should be happy with what you collect.

The condition of a toy gun can be a subjective matter. More than once I have read descriptions or talked to people on the phone about a supposed "New Old Stock In Box" piece that, after a bit of investigation, was found to be nonworking, damaged, missing parts, and in a water-damaged box. Yes, it is "Old Stock" from somewhere, and, yes, it is "In Box," and it may be "New" to you, but reality has to come into play somewhere along the line. There has to be an agreement as to grading of toy guns.

It may be a Holy Grail type of quest to collect only the most perfect Mint In Box pieces. Many times guns came off the assembly line in less than perfect condition, due to bad batches of metal, lack of quality control, or poor design. Toy guns are, after all, just that: toys. They were meant to be played with, used and abused and then discarded. To expect them to exist in pristine condition for decades is often wishful thinking.

The high acid content of the paper pulp used to make toy gun boxes has damaged the finish of many guns. The Mattel Fanner guns are especially prone to this problem. Some guns have inherent problems that surface with age, such as flaking or lifting of their finish. This occurs because the gun was not properly prepared before being plated or the particular batch of plating was not properly mixed. The big Nichols Stallion .45s suffer from this problem. Sometimes a better quality of nickel just was not available and the production line had to use what was on hand. Any grading system used should be flexible and take all this into consideration.

As in collecting any toy, the Mint In Box piece will always be the most desirable. Common and lower quality guns may provide the quantity to fill collections, but the quality of any collection will always be measured by the finer and more sought-after pieces.

Variations enhance the quality and expand the depth of a collection, but they can also drive you crazy. Forced by economic considerations, most companies would continue production of a popular gun for years with only minor changes to the molds or packaging. Each of these changes, no matter how minor, is technically a variation.

A good example of variations in production can be found in the Hubley "Texan" series of cap guns. The Texan began life in the golden age of cast iron, the mid-1930s. It was a large, well-designed, sturdy pistol that fired roll caps and had a revolving cylinder and the look of the Old West that young hombres desired. The initial cast iron version was offered in a cap-firing version and a dummy version. (Most manufacturers offered dummy versions in states and cities where caps were classified as fireworks and prohibited.) The dummy version had a different hammer than the standard version; it lacked the inside of the hammer so that it never touched and could not fire caps.

For a while, Hubley boosted their sales appeal by obtaining a license to use the famous Colt Firearms rearing horse logo instead of the traditional Hubley star. This variation is a must for any complete Texan collection. After World War II, the Texan reappeared in a nearly identical die-cast version that was offered in nickel, gold, gray, or blue finishes with both standard or dummy hammers.

Then, to make matters even more complicated, Hubley changed the basic design of the gun and issued that as a "Texan," offering it in nickel, gold, and metal finishes. All of these guns had the familiar Hubley longhorn steer head plastic grips. These vary by being either all white or white with a black painted steer. There is a possibility that postwar "swirl" plastic grips of different colors may also exist.

So far, we have only talked about variations in one gun. Hubley also produced a smaller version of the Texan, called the Texan Jr. It was also produced in many variations. All of these guns were sold in different style boxes. The point is, collecting every variation of just this one gun would make a sizable collection by itself. This is why deciding what you want to collect is so important and also what makes undetailed price guides so dangerously inaccurate.

Faking It

Reproductions and misrepresentations are common in any collecting field and a hazard for both the novice and experienced collector. Reproductions of highly sought toy guns have yet to appear on the scene, mainly because the production and tooling costs for the potential market demand would be too high. But this has not stopped many small-time entrepreneurs from producing reproduction hammers, plastic grips, and laser-copy boxes. I have no problem with toy gun restoration or repair. As certain pieces become more difficult to find it is the logical approach to fill the collecting demand. But restored or repaired pieces should always be presented as such and should be priced accordingly lower than an unrestored version.

Most laser-copy boxes are easy to spot. The inside cardboard is usually whiter than an original and many times flaws, tears, and dirt from the original are visible. Remember, copy machines copy everything as it is and producing a mint copy requires a mint original.

Repro hammers are usually easy to spot, especially when placed next to an original. The original usually tends to be cleaner and has better detail. Clues such as a mint hammer and pitted pan or anvil (striking surface) are also dead giveaways.

Misrepresentations are sometimes the most difficult situations to unmask. Many dealers misrepresent pieces simply from lack of knowledge and experience. Many beginning collectors fall victim for the same reasons. Simple mistakes such as giving a gun the wrong date and misidentifying the manufacturer of a gun are common. Beware of the dealer who does not know the difference between cast iron and die-cast. Experience will show you that an "H" in a diamond or oval on the side of a gun is usually the trademark of either Leslie-Henry (the "Diamond H brand") or the Halco (J. Halpern) Company, not Hubley. Hubley guns are usually marked "Hubley" somewhere on the gun. Knowing trademarks and distinguishing styles are simply learned through experience.

A Holster for Any Occasion

One area of major confusion seems to be centered around the question of "what gun went in what holster?" Holster makers were often not the same companies that made the guns. The cowboy cap gun craze of the 1940s through the 1950s supported the growth of many small production leather holster manufacturers. Companies such as Keyston Bros., Halco, Classy Products, and Pilgrim Leather Goods produced holsters for major manufacturers and at the same time produced empty holsters. These empty sets were sold to jobbers who would buy quantities of guns — the cheaper the better — from different manufacturers to fill them. They would then sell the "married" sets to individual retailers. Remember, large chain stores such as Toys R Us and K-Mart did not exist at this time. Every local department store, hardware store, hobby shop and sporting goods store was a potential customer. Years later, it is not uncommon to find Leslie Henry Gene Autry Pistols in a Lasso 'Em Bill Holster set by Keyston Bros. What is this set? It is not really a Gene Autry set, no matter how nice it looks, and it should not be represented as such. Most character holster sets were produced by major companies that paid hefty licensing fees to use cowboy names on both guns and holsters. Though there were a number of exceptions, most of these sets were sold with matching guns and holsters. The "Gunsmoke" set would have "Gunsmoke" or "Marshall Matt Dillon" guns in it, the "Have Gun Will Travel" set would have "Paladin" guns, and so forth. Beware of the Mint In Box character set with a photo of the cowboy star on the box cover, his name on the holsters and common, no-name guns in them. Always ask questions.

Collecting toy guns can be a fascinating and rewarding hobby. The majority of collectors and dealers are friendly and good-natured. After all, we are having fun. When my Fanner 50 is in my hand, I am eight years old again.

George H. Newcomb, born in 1952, began collecting and dealing in antiques over twenty years ago as a college student. He has focused on toy guns for the last six years.

Trends

The toy gun market has grown into its own version of what can be called Collectors Frenzy Syndrome, which has been seen many times in other collecting fields. Escalating prices and reduced availability deliver a one-two punch to the market. Dealers can not find enough quality merchandise to satisfy customer demand, and this demand pushes prices beyond the means of many collectors. Market activity is reduced, and frustrated collectors either sell off their collections — flooding and devaluing the market, or else turn to new collecting interests and shelve their collections, theoretically maintaining value but further hindering the active trade that a marketplace needs to thrive. In short, a state is reached wherein dealers can not find inventory to sell and collectors can not buy.

Currently, the market in western guns has been nearly tapped out, and interest has been turning to space guns, dart guns and target sets, water pistols, and other exotic toy guns. On the bright side, riding the wave of western collectibles, character Western guns such as Roy Rogers, Lone Ranger, Hopalong Cassidy, and Gene Autry have seen the strongest growth this past year. For example, the Classy Roy Rogers Double Holster set with ten inch guns rose from $425 last year to $695 this year.

The Top 10 Toy Guns (in Mint In Box condition)

1. Man From U.N.C.L.E. THRUSH Rifle, Ideal, 1966 .... $2,500
2. Lost In Space Roto-Jet Gun Set, Mattel, 1966 2,000
3. Man From U.N.C.L.E. Attache Case, Ideal, 1965 1,500
4. Showdown Set with Three Shootin' Shell Guns, Mattel, 1958 950
5. Man From U.N.C.L.E. Attache Case, Lone Star 1966 850
6. Lost In Space Helmet and Gun Set, Remco, 1967 800
7. Man From U.N.C.L.E. Napoleon Solo Gun Set, Ideal, 1965 800
8. Roy Rogers Double Gun and Holster Set, Classy, 1950s 800
9. Roy Rogers Double Holster Set, Classy, 1950s 695
10. Shootin' Shell .45 Fanner Cap Pistol, Mattel, 1959 650

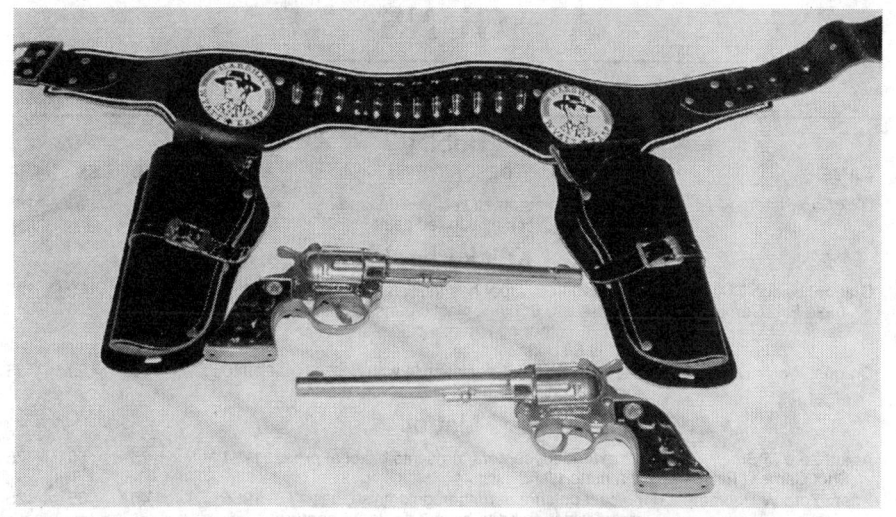

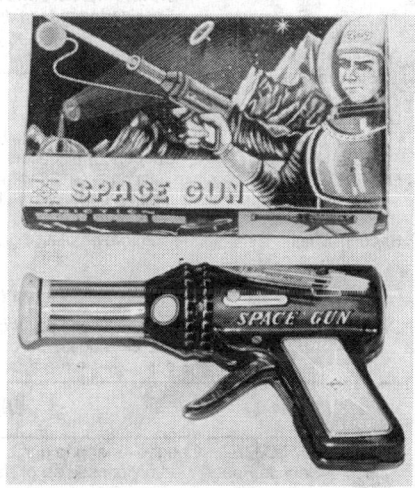

Top to bottom: Wyatt Earp Double Holster Set, 1959, Hubley; Gene Autry Cap Pistol, Jr. Model, 1940, Kenton; Baby Space Gun, 1950s, Daiya; Texan Jr. Cap Pistol, 1950s, Hubley; Stallion 45, first version in rare lift-off top box, 1950, Nichols.

GUNS

Detective

Hubley

NAME	DESCRIPTION	YEAR	GOOD	EX	MIP
Dick Cap Pistol	4 1/8" cast iron automatic style, side loading, nickel finish, Dick oval in red paint	1930	75	125	195

Knickerbocker

Dragnet Badge 714 Triple Fire Comb. Game	tin litho stand-up target has four plastic spinners, includes two 6" black plastic guns, one fires darts, the other cork gun, includes 4 plastic darts & 3 corks	1955	90	150	225
Dragnet Water Gun	6" black plastic, .38 Special-style with gold "714" shield on grip	1960s	15	20	35

Mattel

Agent Zero M Snap-Shot Camera Pistol	7 1/2" extended, camera turns into pistol at press of a button	1964	25	50	75
Agent Zero W Potshot	3" die cast potshot derringer, gold finish, has brown vinyl armband holster holds 2 Shootin' Shell cartridges, gold buckle with Agent Zero W logo	1965	40	85	125
Official Detective Shootin' Shell Snub-Nose	.38 die cast chrome with brown plastic grips, black vinyl shoulder holster, wallet, badge, ID card, Pistol Range Target & bullets	1959	95	175	250
Official Detective Shootin' Shell Snub-Nose	.38 die cast 7" chrome finish, gold cylinder, brown plastic grips, Private Detective badge & Shootin' Shell bullets	1960	60	100	165
Official Dick Tracy Shootin' Shell Snub-Nose .38	die cast chrome .38 with brown plastic grips, chrome finish with Shootin' Shell bullets & Stick-m caps	1961	75	125	185
Shootin' Shell Snub-Nose .38	.38 with brown plastic grips & silver finish	1959	65	145	200

Nichols

Detective Shell Firing Pistol	5 1/2" snub-nose pistol chambers & fires 6 3-pc. cap cartridges, cut-out badge, bullet cartridges, extra red plastic bullet heads	1950s	100	175	275

Military

Man From U.N.C.L.E.* Secret Service Pop Gun	bagged luger pop gun on header card with (*)unlicensed art of Napoleon and Illya	1960s	4	7	10

Buddy L

Spitfire Cap Firing Machine Gun	biped stand attached to muzzle, black plastic stock & grip with cap or clicker firing	1950	80	130	175

Coibel

Official James Bond 007 Thunderball Pistol	4 1/2" Walther PPK style, single shot fires plastic caps, Secret Agent ID	1985	15	25	45

Daisy

Model 12 SoftAir Gun	machine gun style, loads "SoftAir" pellets in plastic cartridge, 10 rounds, spring fired, can be cocked by barrel grip or bolt	1990	35	50	75
SA Automatic Burp Gun	10" with stock, black plastic, burp-gun style, removeable clip, loads & fires white plastic bullets	1970s	20	35	50

Military

NAME	DESCRIPTION	YEAR	GOOD	EX	MIP
Edison					
Matic 45 Cap Gun	24", plastic gun with stock, fires "Supermatic System" strip caps	1980s	10	15	20
Esquire Novelty					
7580 UZI Automatic	10 1/2", battery operated, black plastic uses 250 shot roll caps, shoulder strap	1986	20	45	60
Hubley					
Army .45 Cap Pistol	6 1/2" automatic, dark grey finish, white plastic grips, pop-up cap	1950s	45	90	165
Automatic Cap Pistol No. 290	die cast, 6 1/2", nickel finish, brown checkered grips, magazine pops up when slider is pulled back		85	150	225
Ideal					
Man From U.N.C.L.E. Attache Case	15 x 10 x 2 1/2", comes with cap firing pistol and clip, ID card, wallet, cap grenade, badge, passport and secret message sender	1965	525	975	1500
Man From U.N.C.L.E. Illya K. Special Lighter Gun	cigarette lighter gun shoots caps, has radio compartment concealed behind fake cigs, in window box	1966	125	225	350
Man From U.N.C.L.E. Illya Kuryakin Gun Set	includes clip loading, cap firing plastic pistol, badge, wallet, ID card, in window box	1966	210	390	600
Man From U.N.C.L.E. Napoleon Solo Gun Set	clip loading, cap firing plastic pistol with rifle attachments, badge, ID card, in window box	1965	280	520	800
Man From U.N.C.L.E. Pistol and Holster	7" long pistol and plastic holster, both with orange ID sticker	1965	45	85	125
Man From U.N.C.L.E. Stash Away Guns	three cap firing guns, holsters, two straps, ID card and badge, in window box	1966	210	390	600
Man From U.N.C.L.E. THRUSH Rifle	36" long	1966	875	1600	2500
Larami					
9mm Z-Matic Uzi Cap Gun	8" replica, removeable cap storage magazine, black finish, small orange plug in barrel	1984	5	10	15
Lone Star					
Girl From U.N.C.L.E. Garter Holster	"gang buster" metal gun fires plastic bullets from metal shells, checker design vinyl holster and bullet pouch, on card	1966	80	145	225
Man From U.N.C.L.E. Attache Case	small cardboard briefcase, contains diecast Mauser and parts to assemble U.N.C.L.E. Special	1966	140	260	400
Man From U.N.C.L.E. Attache Case	vinyl covered cardboard, 9mm automatic Luger, shoulder stock, sight, silencer, belt, holster, secret wrist holster and pistol, granade, wallet, passport, money	1966	300	550	850
Man From U.N.C.L.E. Attache Case	vinyl case with pistol, holster, walkie talkie, cigarette box gun, badge, passport, invisible cartridge pen, handcuffs	1966	265	475	750
Maco					
Molotov Cocktail Tank Buster Cap Bomb	6" plastic & die cast, insert caps in head & throw	1964	10	15	25
MP Holster Set	plastic pistol has removeable magazine, loads & ejects bullets, white leather belt & holster	1950s	65	120	175
Paratrooper Carbine	24" carbine, removeable magazine, fires plastic bullets, bayonet & target	1950s	75	115	165

GUNS

Military

NAME	DESCRIPTION	YEAR	GOOD	EX	MIP
USA Machine Gun	12" tripod-mounted gun fires plastic bullets, red & yellow plastic	1950s	65	130	200

Main Machine

Mustang Toy Machine Gun	25" long chrome and hard plastic paper firing gun	1950s	75	125	175

Marx

Anti Aircraft Gun	mechanical, sparks, tin litho, 16 1/2" long, 1941		35	100	195
Army Automatic Pistol	2 1/2" automatic, (ACP style), black with white grips, small leather holster with flaps (Marx Miniature)	1950s	10	15	25
Army Pistol with Revolving Cylinder	tin litho		35	60	95
Army Sparkling Pop Gun	1940-50		45	75	125
Desert Patrol Machine Pistol	plastic, 11" long		30	45	75
G-Man Gun Wind-up Machine Gun	23" tin-litho, red, black, orange & grey litho, round drum magazine, wind-up mechanism makes sparks from muzzle, uses cigarette flint, wooden stock	1948	145	200	275
Green Beret Tommy Gun	sparkles, trigger action, on card	1960s	35	55	95
Man From U.N.C.L.E. Pistol Cane	25" long, cap firing, bullet shooting aluminum cane with 8 bullets and one metal shell, on card	1966	175	325	500
Mini-M.A.G. Combat Gun	cap pistol, miniature scale, diecast		20	30	40
Siren Sparkling Airplane Pistol	heavy-guage enamel steel, 9 1/2" wingspan, 7" long		50	75	100
Sparkling Siren Machine Gun	26" long, 1949		45	75	110
Special Mission Tommy Gun			20	35	45
Tommy Gun	sparks and makes noise, 1939		65	95	150

Mattel

Burp Gun	17" plastic with die cast works, perforated roll caps fired by cranking the handle	1957	25	50	75

National

Automatic Cap Pistol	4 1/4" automatic style, grip swivels to load, black finish	1925	45	75	135
Automatic Cap Pistol	6 1/2" silver finish with simulated walnut grip	1950s	35	65	90

Nichols

Army 45 Automatic	4 1/4" all metal, side loading automatic, olive	1959	15	20	35

Parris Mfg.

M-1 Kadet Training Rifle	32" wood/metal M-1 carbine, clicker action, metal barrel, trigger guard, bolt	1960s	20	35	50

Redondo

Revolver Mauser Cap Pistol	6 1/4", Mauser style automatic pop-up magazine, silver finish, brown plastic grips	1960s	4	7	10

Stevens

Spitfire Automatic Cap Pistol	4 5/8" cast iron, side loading, silver finish, "flying airplanes" white plastic grips	1940	60	110	165

Military

Topper/Deluxe

NAME	DESCRIPTION	YEAR	GOOD	EX	MIP
Johnny Seven One Man Army-OMA	36" multi-purpose seven guns in one, removeable pistol fires caps, rifle fires white plastic bullets, bolt spring fired machine gun "tommy gun" sound, rear launcher fires grenades, forward diff. shell	1964	125	225	350

Unknown

NAME	DESCRIPTION	YEAR	GOOD	EX	MIP
MM Automatic Carbine	24" recoil red slide in muzzle & flashing light, brown & black plastic	1960s	15	30	45

Miscellaneous

Ambrit Industries

NAME	DESCRIPTION	YEAR	GOOD	EX	MIP
Spud Gun	case aluminum, pneumatic all-metal gun shoots pellets	1950s	15	35	50

Atomic Industries

NAME	DESCRIPTION	YEAR	GOOD	EX	MIP
Dynamic Automatic Repeating Bubble Gun	8" black plastic pistol projects bubbles		25	40	65

Benton Harbor Novelty

NAME	DESCRIPTION	YEAR	GOOD	EX	MIP
Dick Cap Pistol	4 3/4", automatic, side loading, black finish	1950s	20	30	45

Buddy L

NAME	DESCRIPTION	YEAR	GOOD	EX	MIP
Paper Cracker Rifle	26" steel & machined aluminum mechanism, barrel, trigger & operating lever, uses 1000 shot paper roll, brown plastic stock	1940s	95	165	225

Daisy

NAME	DESCRIPTION	YEAR	GOOD	EX	MIP
Buck Rogers in the 25th Century Pistol Set	holster is red, yellow & blue leather gun is 9 1/2" pressed steel pop gun	1930s	200	300	425
Buzz Barton Special, No. 195	blue metal finish, wood stock with ring sight	1930s	85	125	200
Jack Armstrong Shooting Propeller Plane Gun	5 1/2" gun, shoots flying disc, pressed tin	1933	35	60	95
Model No. 25 Pump Action BB Gun	plastic stock	1960s	35	65	125
Targeteer No. 18 Target Air Pistol	10" gun metal finish, push barrel to cock, Daisy BB tin with special BBs, spinner target	1949	50	85	125
Water Pistol No. 17	5 1/4" tin	1940	25	40	65
Water Pistol No. 8	5 1/4" tin	1930s	25	50	80

Edison

NAME	DESCRIPTION	YEAR	GOOD	EX	MIP
Sharkmatic Cap Gun	6" automatic, style pistol, fires "Supermatic System" strip caps	1980s	10	15	20

EMU Rififi

NAME	DESCRIPTION	YEAR	GOOD	EX	MIP
Automatic Sparking Pistol	6 1/2", plastic/metal, uses cigarette lighter flints to make sparks, available in green-red, yellow-green, red or white colors	1960s	15	30	45

Esquire Novelty

NAME	DESCRIPTION	YEAR	GOOD	EX	MIP
Hideaway Derringer	3 1/2" single shot, loads solid metal bullet, grip is removeable to store two more bullets, gold finish, white plastic grips	1950s	65	90	110

Hubley

NAME	DESCRIPTION	YEAR	GOOD	EX	MIP
Dick Cap Pistol	die cast, 4 1/4" automatic style, side loading with nickel finish	1950s	25	40	65

Miscellaneous

NAME	DESCRIPTION	YEAR	GOOD	EX	MIP
Midget Cap Pistol	5 1/2" long, diecast, all metal flintlock with silver finish	1950s	20	35	50
Pirate Cap Pistol	9 1/2" side-by-side flintlock style with die cast frame & cast double hammers & trigger, chrome finish, white plastic grips feature Pirate in red oval	1950	50	100	175
Tiger Cap Pistol	6 7/8", single action, mammoth caps, metal finish	1935	35	65	100
Trooper Cap Pistol	6 1/2" all metal, pop up cap magazine, nickel finish, black grips	1950	30	50	85
Winner Cap Pistol	4 3/8" automatic style, pop-up magazine release in front of trigger guard, nickel finish	1940	45	85	135

Kilgore

NAME	DESCRIPTION	YEAR	GOOD	EX	MIP
Clip 50 Cap Pistol	4 1/4", unusual automatic style, black Bakelite plastic frame, removeable cap magazine clip	1940	75	110	175
Mountie Automatic Cap Pistol	6", double action, automatic style with pop-up magazine, unusual nickel finish, black plastic grips		20	30	45
Presto Cap Pistol	5 1/8", pop up cap magazine nickel finish brown plastic grips	1940	65	120	185
Rex Automatic Cap Pistol	3 7/8", blue metal finish cast iron, small size automatic style, side loading, white pearlized grips	1939	45	80	125

Langson Mfg. Co.

NAME	DESCRIPTION	YEAR	GOOD	EX	MIP
Nu-Matic Paper Popper Gun	pressed steel, 7" squeeze grip trigger, mechanism pops roll of paper (reel at top of gun) to make loud noise, black finish	1940s	30	50	75

Leslie-Henry

NAME	DESCRIPTION	YEAR	GOOD	EX	MIP
Gene Autry Cap Pistol	9" break-to-front, lever release, nickel finish, black plastic horse-head grips	1950s	65	145	250

Marx

NAME	DESCRIPTION	YEAR	GOOD	EX	MIP
Automatic Repeater Paper Pop Pistol	7 3/4" long		35	65	95
Blastaway Cap Gun	50 shooter repeater		25	35	50
Burp Gun	20" battery operated, green & black plastic	1960s	20	35	50
Click Pistol	tin litho		20	35	50
Click Pistol	pressed steel, 7 3/4" long		30	45	60
Famous Firearms Deluxe Edition Collectors Album	set of 4 rifles, 5 pistols & 4 holsters miniature series, includes: Mare's Laig, Thompson machine gun, Sharps rifle, Winchester saddle rifle, Derringer, .38 snub-nose, Civil War Pistol, 6 shooter/Flink	1959	50	95	150
Marx Miniatures Famous Gun Sets	4 gun set features Tommy Gun, Civil War Revolver, "Mare's Laig" & Western Saddle Rifle	1958	25	45	65
Marxman Target Pistol	plastic, 5 1/2" long		35	50	70
Popeye Pirate Click Pistol	tin litho, 10" long, 1930's		100	150	200
Repeating Cap Pistol	aluminum		20	30	50
Sparkling Pop Gun			25	40	65
Streamline Siren Sparkling Pistol	tin litho		25	45	65

Mattel

NAME	DESCRIPTION	YEAR	GOOD	EX	MIP
Burp Gun	23" with folding stock extended, 16 1/2" from rear of breech to muzzle, wind up mechanism fires perforated roll caps, pressed steel gun body & barrel, white metal folding stock, red plastic magazine	1950s	90	150	225
Fanner 50 Cap Pistol	later version 11" fires perforated roll caps, black finish, white plastic antelope grips	1960s	30	65	95

GUNS

Miscellaneous

NAME	DESCRIPTION	YEAR	GOOD	EX	MIP
Mattel-O-Matic Air Cooled Machine Gun	16" machine gun fires perforated roll caps by cranking handle, plastic with die cast, tripod-mounted	1957	45	85	125

Meldon

NAME	DESCRIPTION	YEAR	GOOD	EX	MIP
P-38 Clicker Pistol	7 1/2" black finish, automatic style	1950s	35	50	75

Midwest

NAME	DESCRIPTION	YEAR	GOOD	EX	MIP
Long Tom Dart Gun	11" pressed steel	1950s	35	50	75

Nichols

NAME	DESCRIPTION	YEAR	GOOD	EX	MIP
Model 95 Shell Firing Rifle	35 1/2" rifle uses shell firing cartridges, holds 5 in removeable magazine & one chamber, lever action ejects cartridges, open frame box holds 6 bullets & 12 additional red bullet heads	1961	200	350	550
Pinto Cap Pistol	3 1/2", chrome finish, black plastic grips, flip out cylinder, white plastic "Pinto" holster in leather holster clip	1950s	20	35	50
Spitfire with Clip	9" mini rifle, chrome finish, plastic stock, plastic holders with 2 extra cartridges	1950s	12	25	35
Tophand 250 Cap Pistol	9 1/2" break-to-front, lever release, black finish, brown plastic grips with a roll of "Tophand 250" caps	1960	45	85	135

Palmer Plastics

NAME	DESCRIPTION	YEAR	GOOD	EX	MIP
Airplane Clicker Pistol	4 1/2" yellow & black plane, red pilot & guns	1950s	25	45	65
Ray Gun Water Pistols	5 1/2" many color variations green, orange, translucent blue, royal blue, black, yellow & red	1950s	10	15	20

Park Plastics

NAME	DESCRIPTION	YEAR	GOOD	EX	MIP
Atomee Water Pistol	4 1/4" black plastic	1960s	10	20	35

Rosvi

NAME	DESCRIPTION	YEAR	GOOD	EX	MIP
Revolver Aquila Pop Pistol	10" green finish, pop gun breaks to cock, fires cork from barrel, sparks from mechanism under barrel	1960s	15	20	35

Stevens

NAME	DESCRIPTION	YEAR	GOOD	EX	MIP
25 Jr. Cap Pistol	4 1/8" automatic, side loading, silver finish	1930	25	50	75
6 Shot Cap Pistol	6 3/4", six separate triggers revolve to deliver caps to hammer, metal finish		85	135	175
Model 25-50	4 1/2" nickel finish	1930	35	65	95
Pluck Cap Pistol	3 1/2" cast iron, single shot single action	1930	15	30	45

Unknown

NAME	DESCRIPTION	YEAR	GOOD	EX	MIP
Double Holster Set	black leather, large size, steer head conches, lots of red jewels, fringe, holsters only, no guns	1950s	65	100	150

Unknown - Hong Kong

NAME	DESCRIPTION	YEAR	GOOD	EX	MIP
Potato Gun	Spud Gun, plastic, pneumatic action fires potato pellet from muzzle	1991	3	7	10

US Plastics

NAME	DESCRIPTION	YEAR	GOOD	EX	MIP
Rocket Jet Water Pistols	5" ray gun fills through hole in top, orange or yellow plastic	1960s	6	12	17

Welco

NAME	DESCRIPTION	YEAR	GOOD	EX	MIP
Spud Gun (Tira Papas)	6" all-metal gun	1960s	10	20	35

GUNS

Police

NAME	DESCRIPTION	YEAR	GOOD	EX	MIP
Acme Novelty					
G-Boy Pistol	7" automatic (ACP) style entire left rear side of gun swings down to load	1950s	15	35	50
Hubley					
Mountie Automatic Cap Pistol	diecast, 7 1/4" automatic style, pop up lever-release magazine, blue finish	1960s	20	35	55
Kilgore					
Machine Gun Cap Pistol	5 1/8", long cast iron crank-fired gun	1938	100	175	250
Marx					
.38 Cap Pistol	with caps, miniature scale, diecast		35	75	125
Detective Snub-Nose Special	die cast, 5 3/4" top release break, unusual revolving cylinder, fires Kilgore style disc caps, chrome finish with black plastic grips	1950s	65	100	150
Dick Tracy Click Pistol			35	50	75
Dick Tracy Jr Click Pistol	aluminum, 1930's		100	150	200
Dick Tracy Siren Pistol	pressed steel, 8 1/2" long, 1935		40	75	125
Dick Tracy Sparkling Pop Pistol	tin litho		35	55	95
G-Man Automatic Silent Arm Pistol	tin		35	55	70
G-Man Automatic Sparkling Pistol	pressed steel, 4" long, 1930's		45	65	110
G-Man Machine Gun	tin miniature, windup, 1940's		25	35	65
G-Man Sparkling Sub-Machine Gun			50	75	100
G-Man Tommy Gun	sparkles when wound, 1936		55	80	110
Gang Buster Crusade Against Crime Sub-Machine Gun	litho, metal with wooden stock, 23" long, 1938		95	150	225
Marx Miniatures Detective Set	miniature cap firing brown and gray tommy gun, chrome pistol and holster on card with wood grain frame border	1950s	25	45	65
Official Detective-Type Sub-Machine Gun			35	55	75
Sheriff Signal Pistol	plastic, 5 1/2" long, 1950		20	30	50
Siren Sparkling Pistol	tin litho		20	35	45
Mattel					
Official Dick Tracy Tommy Burst Machine Gun	25" Thompson style machine gun fire perforated roll caps, single shot or in full burst when bolt is pulled back, brown plastic stock & black plastic body, lift up rear sight, Dick Tracy decal on stock	1960s	100	175	275
Pilgrim Leather					
Peter Gunn Private Eye Revolver & Holster Set	36" die cast Remington gun with six 2-pc. bullets, badge & wallet, Peter Gunn business cards, black leather shoulder holster	1959	200	325	450
T. Cohn					
Sparkling "Sure-Shot" Machine Gun	long body tin multicolored red/yellow/blue tin noise making gun, great box graphics show boy shooting sparks as pigtailed blonde girl looks	1950s	0	0	125

Space

NAME	DESCRIPTION	YEAR	GOOD	EX	MIP
### Ahi					
Lost In Space Saucer Gun	disc shooting gun, TV show tie-in	1977	30	50	75
### Arco					
Ro-Gun "It's A Robot"	sho-gun type robot that changes into a rifle, in window box, currently available	1984	8	13	20
### Arliss					
Space Dart Gun	4" solid color plastic gun, shoots standard rubber tipped darts	1950s	12	20	30
X100 Mystery Dart Gun	3 3/4" long, yellow or grey plastic gun on cardboard display card, with two yellow and blue talcum impregnated darts which create a smoke effect when striking any target	1956	25	40	60
### Asahitoy/Japan					
Space Navigator Gun	3 1/2" long, tin, looks like sawed off military .45, colorfully trimmed blue body with smiling spaceman, blasting winged rocketship and "Space Navigator" logo on grips, planets and star on body	1953	35	60	95
### Aviva					
Star Trek Water Pistol, ST:TMP	gray plastic, early pistol-grip phaser design	1979	10	16	25
### Azrak-Hamway					
Star Trek Water Pistol	white plastic, shaped like U.S.S. Enterprise	1976	15	30	45
### Beaver Toys					
Bee-Vo Bell Gun	#204, 6 1/2" long, red plastic, fires trapped marble at bell in muzzle, in box	1950s	25	40	60
### Budson					
Flash Gordon Air Ray Gun	10" long, unusual air blaster, handle on top cocks mechanism, shoots blast of air, red and silver pressed steel	1950s	175	295	450
### Chemtoy					
Pop Gun	4 1/2" long red hard plastic gun with space designs on handle	1967	25	35	50
### Chien					
Atomic Flash Gun	7 1/2" long, tin, sparking action seen through tinted elongated oval plastic muzzle, with yellow and red on turquoise body with red lettered "Atomic Flash" over trigger	1955	40	65	100
### Daisy					
Buck Rogers Holster for U-238 Atomic Pistol	leather holster only	1946	40	65	100
Buck Rogers Holster for XZ-35 Atomic Pistol	embossed leather, attached to belt by two short riveted straps	1946	40	65	100
Buck Rogers in the 25th Century XZ-35 Rocket Pisto	holster is red, yellow & blue leather, gun is 9 1/2" pressed steel Rocket Pistol with single cooling fin at barrel base	1934	155	260	400
Buck Rogers U-235 Atomic Pistol	9 1/2" long, pressed steel, makes pop noise and flash in window when trigger is pulled	1946	105	180	275

GUNS

Space

NAME	DESCRIPTION	YEAR	GOOD	EX	MIP
Buck Rogers XZ-31 Rocket Pistol	10 1/2" long, heavy blued metal, grip pumps the action, gun pops when trigger is pulled	1934	135	225	350
Buck Rogers XZ-35 Space Gun	7" long, heavy blued metal ray gun, the grip pumps the action and the gun pops when trigger is pulled, single cooling fin at barrel base, also called "Wilma Gun"	1934	105	180	275
Buck Rogers XZ-38 Disintegrator Pistol	10 1/2" long, polished copper or blued finish, four flutes on barrel, spark is produced in the window on top of the gun when the trigger is pulled	1936	115	195	300
Buck Rogers XZ-44 Liquid Helium Water Gun	7 1/2" long, red and yellow lightning bolt design stamped metal body with a leather bladder to hold water. A later version was available in copper finish	1936	145	245	375
Daisy Rocket Dart Pistol	7" long, red, blue and yellow sheet metal gun with blue body, blue grips with yellow trim, blue and yellow barrel stripes, same body as Zooka Pop Pistol but with connecting rod from gun to barrel	1954	60	100	150
Daisy Zooka Pop Pistol	7" long, colorful red, blue and yellow sheet metal gun with blue body, red grips with yellow trim and litho star reading "Its a Daisy Play Gun", yellow barrel with red stripes, and wide red muzzle, handle cock	1954	80	130	200
Star Trek Phaser, ST:III	white & blue plastic gun with light & sound effects	1984	30	50	75

Daiya/Japan

NAME	DESCRIPTION	YEAR	GOOD	EX	MIP
Baby Space Gun	6" friction siren & spark action	1950s	35	60	95
Space Gun	6" long, tin, sparking action, metallic teal finish with red grooves and muzzle, green spaceship on body above "Space Gun," small Daiya logo inside red/yellow burst on grip with "577001" at bottom of grip	1957	35	60	95
Super Sonic Space Gun	7 1/2" long, tin litho, metallic gray body with red gunsight fin, friction siren and sparking action, large oval center art with outstanding lunar scene of rockets, mountains and Earth in sky, red helmeted spaceman on grip	1957	40	65	100

Endoh/Japan

NAME	DESCRIPTION	YEAR	GOOD	EX	MIP
Super Sonic Gun	9" long, tin, sparking action with 3 red plastic spark windows and clear red plastic barrel, blue body with red lightning bolt beneath yellow "Super Sonic" on rounded gun body, small ENDOH logo printed above grips	1957	40	65	100

England

NAME	DESCRIPTION	YEAR	GOOD	EX	MIP
Space Outlaw Ray Gun	10" long, chrome plated, die cast metal, recoiling barrel action, "Cosmic", "Sonic" or "Gamma" power levels, large red clear plastic teardrop shaped window	1965	115	195	300

Futuristic Products

NAME	DESCRIPTION	YEAR	GOOD	EX	MIP
Strato Gun	9" long, gray finish die cast, cap firing, internal hammer, top of gun lifts to load	1950s	70	115	175
Strato Gun	9" long, chrome finish die cast, red cooling fins, cap firing, internal hammer, top of gun lifts to load	1950s	100	165	250
Strato Gun	12" long, die cast metal, chrome finish with red "cooling fins," shoots caps	1955	145	250	375

Galoob

NAME	DESCRIPTION	YEAR	GOOD	EX	MIP
Star Trek Phaser, ST:TNG	gray plastic light & sound hand phaser	1988	12	20	30

GUNS

Space

NAME	DESCRIPTION	YEAR	GOOD	EX	MIP
### Gilbert					
Moon McDare Space Gun Set		1966	25	40	65
### H.G. Toys					
Alien Blaster Target Game	set features large free standing cardboard Alien target and plastic dart shooting rifle, gun has large block letters "Alien" on side, based on the movie	1979	55	95	145
### H.Y. Mfg./Hong Kong					
Razer Ray Gun	plastic bronze finish body with 5 large cooling fins near red plastic barrel, friction sparking action, chrome finish muzzle tip, "Razer Ray Gun" embossed on rear of barrel	1972	10	15	25
### Haji					
Atomic Gun	9" long, red, gray and yellow tin litho gun with plastic muzzle, friction sparking action, large hollow letter "ATOMIC GUN" on body	1969	15	30	45
Over and Under Ray Gun	8 1/2" long, red, yellow, white and black tin litho gun with 2 over and under reciprocating plastic muzzles, friction sparking action	1960s	30	50	75
### Hasbro					
Jet Plane Missile Gun	jet shaped handgun shoots darts, targets supplied on box back	1968	35	60	95
### Hero Toy/Japan					
Space Gun	7" long, tin litho, friction sparking action, yellow body with blue and red trim. small Hero Toy logo by trigger	1960	35	55	85
### Hiller Mfg.					
Atom Ray Gun	5 1/2" long, sleek red body gun of aluminum and brass with bulbous water reservoir on top of gun, reads "Atom Ray Gun" between two lightning bolts on reservoir	1949	135	225	350
### Hong Kong					
Ratchet Water Pistol Ray Gun	6 1/2" unusual pull back mechanism loads pistol, ratchet forces water out when trigger is pulled	1960s	25	40	65
Satellite & Rocket Pistol	5" long, green plastic gun fires either yellow plastic darts or saucers, on card	1960s	12	20	30
Visible Sparking Ray Gun	8 1/2" long, plastic, mechanism visible, bagged with header card		15	25	40
### Hubley					
Atomic Disintegrator Ray Gun	8" in long, die cast metal with red handles, ornately embellished with dials and other equipment outcroppings, shoots caps	1949	15	260	400
### Ideal					
Ideal Flash Gun	9" long, plastic 3-color flashlight gun with red or blue body and bulbous contrasting-color blue or red rimmed flash unit, trigger switch and tail battery compartment cover, with color switch at top of flash unit	1957	80	130	200
Ratchet Sound Space Gun	7" long, red plastic with silver trim, flywheel ratchet on top of gun	1950s	30	50	75

Space

NAME	DESCRIPTION	YEAR	GOOD	EX	MIP
Star Team Ionization Nebulizer	9" water gun fires water mist, red, white, blue & black plastic, Star Team decal	1969	30	50	75

Irwin

NAME	DESCRIPTION	YEAR	GOOD	EX	MIP
Clicker Ray Gun	9" long, red plastic with deep blue cooling fins on barrel base	1960	30	45	70
Space Ship Flashlight Gun	7 1/4", blue plastic ray gun has cockpit with orange spaceman, nose unscrews for AAA batteries, pulling trigger lights nose & moves guns & spaceman	1950s	60	100	150

J. & E. Stevens

NAME	DESCRIPTION	YEAR	GOOD	EX	MIP
Atomic Jet Gun	8 1/2" long, gold chromed die cast metal, cap shooting, "Atomic Jet" and large circular "S" logo on grip	1954	90	145	225
Jet Jr. Cap Gun	6 1/2" long, fires roll caps, side loading door, silver finish, rear jet "Blast Off Fins"	1948	135	225	350
Space Police Neutron Blaster Cap Pistol	7 3/4" die cast, cap firing ray gun, lock mechanism pulls out through the top of the gun, silver finish	1949	185	310	475

Jak-Pak

NAME	DESCRIPTION	YEAR	GOOD	EX	MIP
Rocket Gun	7" hard yellow/green plastic with spring loaded plunger that shoots corks "up to fifty feet"	1958	8	13	20

Japan

NAME	DESCRIPTION	YEAR	GOOD	EX	MIP
888 Space Gun	3" long, tin, shoots caps, painted blue body and grip with stars, planets and spaceship, red barrel with "888" above grip	1955	30	50	75
Atomic Gun	5" long, gold, blue, white and red tin litho, friction sparking action, "Atomic Gun" on body sides	1960s	20	35	50
Jet Gun	6" long, tin, sparking action, red body with 3 small red tinted spark windows near muzzle, grip shows silver-suited astronaut in modern helmet and wording "JET GUN" at top of grip near trigger	1957	35	60	90
Ray Gun	6 1/2" long, tin, sparking action with 2 red tinted plastic tapered rectangle windows at muzzle, "Ray Gun" in red at top of body with rocket exhaust encircling green/blue planet against deep blue star studded background	1957	30	50	80
S-58 Space Gun	12" long, tin litho, deep metallic blue body with friction sparking action, "S-58" on muzzle, with ringed planet graphic on front sight	1957	35	55	85
Space Atomic Gun	5 1/2" long, tin, sparking action seen through red tinted plastic window, two-tone blue body with red/white atomic symbol on grip, "Space Atomic Gun" letters around oval spaceship-and-stars logo above trigger	1955	30	50	75
Space Gun	9" long, friction sparking action with 3 red tinted plastic spark windows and clear red plastic barrel, body in metallic blue with large red "SPACE GUN" letters on yellow background	1957	35	55	85
Super Space Gun	6" long, tin litho, friction sparking action, blue on blue body with white/yellow/red highlights, large red on white "SUPER SPACE" lettering on side	1960	25	40	65

Kenner

NAME	DESCRIPTION	YEAR	GOOD	EX	MIP
Star Wars ESB 3 Position Laser Rifle		1980	55	90	135
Star Wars ESB Han Solo Laser Pistol	same as original Han Solo pistol but with "ESB" decal on side	1980	20	35	50

GUNS

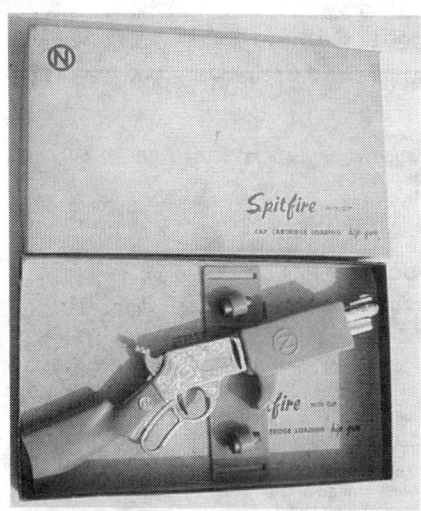

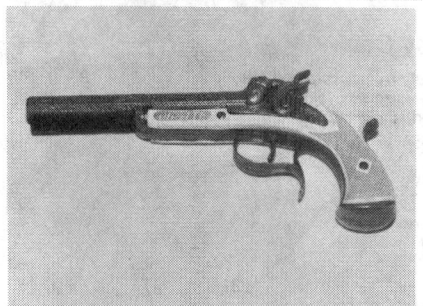

Top to bottom: Spitfire Hip Gun #100, 1950s, Nichols; Automatic (Army .45) Cap Pistol, 1950s, National; Army .45 Cap Pistol, 1950s, Hubley; Texan Cap Pistol, 1940, Hubley; 2 Guns in 1 Cap Pistol, 1950s, Hubley; Pirate Cap Pistol, 1950, Hubley; Dyna-Mite Derringer, 1955, Nichols.

GUNS

Space

NAME	DESCRIPTION	YEAR	GOOD	EX	MIP
Star Wars ESB Laser Pistol		1980	20	35	55
Star Wars Han Solo Laser Pistol	flat dark gray finish Luger-like design with realistic scope sight and barrel extensions, black and white "Star Wars" decal on side	1978	30	50	75
Star Wars ROTJ Biker Scout Laser Pistol		1983	20	35	50

Knickerbocker

NAME	DESCRIPTION	YEAR	GOOD	EX	MIP
4-Barrel Waist Space Dart Gun Belt	11" wide gun system on belt, designed to be worn on waist or chest and aimed with periscope sight, red plastic belt	1950s	30	50	75
Space Jet Water Pistol	4" long, black plastic with white "Space Jet" lettering and spaceship line art on sides, fill plug in gunsight	1957	15	30	45

KO/Japan

NAME	DESCRIPTION	YEAR	GOOD	EX	MIP
Space Jet Gun	9" long, tin, sparking action with black body, orange "Space Jet" on body with orange and red atomic symbol on grip, clear green plastic finned barrel base, clear blue plastic finned muzzle	1957	35	60	90

Larami

NAME	DESCRIPTION	YEAR	GOOD	EX	MIP
Dick Tracy Special Ray Gun	derringer with metal Dick Tracy New York Police Detective Badge	1964	30	50	75

Lido

NAME	DESCRIPTION	YEAR	GOOD	EX	MIP
Space X- Ray Gun	#46598, 8 1/2" long, plastic, friction sparking action, same body as Razer Ray Gun but with more futuristic handgrip and noisemaker at rear, sold in bag with header card	1970s	15	25	35

LJN

NAME	DESCRIPTION	YEAR	GOOD	EX	MIP
Dune Fremen Tarpel Gun	8" long, battery operated with internal light, light beam and chirping sound, plastic	1984	25	40	65
Dune Sardaukar Laser Gun	7" black plastic with flashing lights, battery operated	1984	20	35	50

Lunar Models

NAME	DESCRIPTION	YEAR	GOOD	EX	MIP
Lost In Space Laser Pistol Model Kit	#SF036, 1/1 scale, plastic, first season version, includes electronics for light effects, currently available		45	75	115
Lost In Space Laser Pistol Model Kit	#SF056, 1/1 scale, plastic, second season version, includes electronics for light effects, currently available		40	70	105

M & L Toy Co.

NAME	DESCRIPTION	YEAR	GOOD	EX	MIP
Space Rocket Gun	9" grey plastic, modern police-style pistol grip and shell chamber body with oversized barrel and muzzle sights, spring loaded, shoots rocket projectiles, in box with 2 "rockets"	1950s	55	95	145

Marx

NAME	DESCRIPTION	YEAR	GOOD	EX	MIP
Atomic Ray Gun	30" long, "Captain Space Solar Scout," blue plastic with oversized telescope sight flashlight and "electric buzzer" sound	1957	75	130	200
Cherilea Space Gun	miniature scale, die cast		25	40	60
Flash Gordon Radio Repeater Clicker Pistol	10" long, silver and red long-barrel gun with "Radio Repeater in red beneath barrel and image and Flash Gordon name in side of grips	1937	175	295	450

Space

NAME	DESCRIPTION	YEAR	GOOD	EX	MIP
Flash Gordon Signal Pistol	7" long, green bulbous teardrop sheet metal body with red flared sight and muzzle, prominent Flash Gordon decal on body, siren sounds when trigger is pulled	1930s	195	325	500
Flash Gordon Water Pistol	7 1/2" long, plastic with whistle in handle	1940s	80	130	200
Rex Mars Atomic Pistol Flashlight	plastic, battery operated	1950s	40	65	100
Rex Mars Planet Patrol 45 Caliber Machine Gun	22" long, tin and plastic, wind-up	1950s	60	100	150
Rocket Signal Pistol	same bulbous teardrop metal body as Flash Gordon Signal Pistol and Siren Sparkling Airplane Pistol but without siren hole or wings--but keeping same rear fin, red with litho of 3 horizontally stacked finned orange/yellow bombs	1930s	135	225	350
Space Patrol Atomic Flashlight Pistol	gold/bronze finish pistol with 7 large cooling fins on barrel and 3 smaller ones at back pf gun, large clear plastic diffuser on muzzle, white "Official Space Patrol on handgrip	1950s	135	225	350
Sparkling Atom Buster Pistol	aluminum		30	50	75
Sparkling Space Gun Rifle			50	80	125
Tom Corbett Official Space Cadet Gun	poorly designed composite rifle with modern military plastic stock and front grip at ends of long tin litho gun body with litho bombs and "Ray Adjuster" scale	1950s	135	225	350
Tom Corbett Space Cadet Atomic Flashlight Pistol	identical to Space Patrol Atomic Flashlight Pistol except for body colors and "Tom Corbett Space Cadet" printed upside down on handgrip	1950s	135	225	350
Tom Corbett Space Cadet Gun	10 1/2" long, sheet metal clicker based on Flash Gordon Radio repeater molds, red body, blue barrel reads "Space Cadet," handgrips show bust of Tom in front of planet with rocket ship symbol above	1952	125	210	325

Matchbox

NAME	DESCRIPTION	YEAR	GOOD	EX	MIP
Robotech Water Pistol		1985	6	10	15

Mattel

NAME	DESCRIPTION	YEAR	GOOD	EX	MIP
Battlestar Galactica Lasermatic Pistol		1978	15	30	45
Battlestar Galactica Lasermatic Rifle		1978	25	40	65
Lost In Space Roto-Jet Gun Set	TV tie-in, modular gun can be reconfigured into different variations, shoots discs	1966	775	1300	2000
Space:1999 Astro Popper Gun	on card	1976	6	10	15

Mego

NAME	DESCRIPTION	YEAR	GOOD	EX	MIP
Star Trek Phaser Battle Game	black plastic, 13" high battery operated electronic target game, LED scoring lights, sound effects & adjustable controls	1976	135	225	350

Mercury Toys

NAME	DESCRIPTION	YEAR	GOOD	EX	MIP
Planet Clicker Bubble Gun	8" long, plastic, red body with yellow accents, dip the barrel in bubble solution and pull trigger to make bubbles and produce click sound, in illustrated box	1953	40	65	100

GUNS

Space

NAME	DESCRIPTION	YEAR	GOOD	EX	MIP
Metamol/Spain					
Jack Dan Space Gun	7 1/2" long, in black, red or blue painted die cast metal cap gun with "Jack Dan" over trigger	1959	105	180	275
Mil Jo Mfg.					
Space Scout Spud Gun	7" black & white plastic	1960s	15	30	45
Nasta					
Flash Gordon Space Water Gun	water ray gun on illustrated card	1976	6	10	15
Flash Gordon Three Color Ray Gun	battery operated	1976	8	13	20
Space Water Pistol		1976	6	10	15
Sparkling Ray Gun		1976	6	10	15
Nomura/Japan					
Cosmic Gun	12" long, plastic, battery operated with a small electric motor that runs reciprocating light in clear red plastic barrel, dark blue body, red and orange lettered "COSMIC GUN" decal	1970	35	55	85
Space Control Space Gun	3" long, tin sparking gun with green body, red sights, decorated all over with stars and planets, red and yellow "Space Control" letters over trigger and spacemen firing gun and rocket flying overhead on grips	1954	30	50	75
Norton-Honer					
Buck Rogers Sonic Ray Flashlight Gun	7 1/4" black, green & yellow plastic with code signal screw	1955	70	115	175
Nu-Age Products					
Smoke Ring Gun	large, sleek gray finished breakfront pistol with red barrel and muzzle ring, used rocket shaped matches to produce smoke, trigger fired smoke rings, small engraved "Smoke Ring Gun" logo on gunsight fin	1950s	175	295	450
Ohio Art					
Astro Ray Laser Lite Beam Dart Gun	10" red & white plastic flashlight lights target with four darts	1960s	70	115	175
Palmer Plastics					
Ray Gun Water Pistols	5 1/2" many color variations green, orange, translucent blue, royal blue, black, yellow & red	1950s	8	13	20
Space Water Gun	5 1/2" long, clear red plastic body with embossed Ringed planet and star, 4 cooling fins at barrel base, hollow telescope sight, yellow plastic trigger, white plastic stopper attached by loop to red knob at gun back	1957	15	30	45
Park Plastics					
Atomee Water Pistol	4 1/4" black plastic	1960s	15	25	35
Space Water Gun	6" long, red transparent plastic, stopper at rear of gun, finned trigger guard, zeppelin shaped reservoir with single embossed lightning bolt running its length, tiny "Park Plastics" imprinted along lateral reservoir fin	1960	15	25	35
Power House Candy					
Captain Video Rite-O-Lite Flashlight Gun	3" long, red plastic gun with bulb, space map, paper, directions and order form, in mailing envelope	1950s	20	35	55

Space

NAME	DESCRIPTION	YEAR	GOOD	EX	MIP
### Quisp Cereal/Quaker					
Quisp Powered Sugar Space Gun	7" long, red, mail away premium		150	250	400
### Ranger Steel Products					
Cosmic Ray Gun	9" long, tin body with plastic barrel, boldly painted in blue, yellow and red lightning bolts	1954	50	80	125
Cosmic Ray Gun #249	8" long, plastic, blue body, yellow barrel, red tip, in box showing two space kids in bubble helmets and backpacks shooting at spaceships	1953	40	65	100
Space Pistol	large yellow and orange flint gun		40	65	100
### Rayline					
Star Trek Tracer Gun	6 1/2" plastic firing tracer gun	1966	40	65	100
Star Trek Tracer Scope	rifle with discs	1968	50	80	125
### Remco					
Jupiter 4 Color Signal Gun	9" long black, red and yellow plastic gun that lights up in 4 colors, red telescoping sight	1950s	35	55	85
Lost In Space Helmet and Gun Set	child size helmet with blue flashing light and logo decals, blue and red molded gun	1967	310	525	800
Space:1999 Utility Belt Set	with disc shooting stun gun, watch and compass	1976	12	20	30
Star Trek Phaser	Astro Buzz-Ray Gun with three color flash beam	1967	50	80	125
Star Trek Phaser	black plastic shaped like pistol, electronic sound, flashlight projects target	1975	30	50	75
### Royal Plastics					
Flash-O-Matic, The Safe Gun	7" long red and yellow plastic battery operated light beam gun	1950s	60	100	150
### S. Horikawa/Japan					
Floating Satellite Target Game	6 1/2" x 9", battery operated, includes a pistol and three rubber tipped darts, a blower supports the styrofoam ball on a column of air and the players shoot darts to knock it down	1958	175	295	450
### San/Japan					
Space Gun	3 1/4" long, tin, sparking action, aqua blue body with red and yellow highlights and "Space" in script lettering over grip, grip shows rocket shooting toward planets, circular San/Japan logo behind trigger	1955	30	50	75
### Shawnee					
Tomi Space Gun	solid red plastic with yellow barrel plug, modelled after modern .45 caliber pistol with rounded reservoir lined with 2 horizontal fins over grip; embossed logo and circular Shawnee logos on grip	1950s	50	80	125
### Shudo/Japan					
Astro Ray Gun	9" long, friction spark action, tin litho body with clear red plastic barrel, red on yellow "ASTRO RAY GUN" lettering	1968	25	40	65
Astro Ray Gun	5 7/8" long, silver finish body with red, yellow and black detailing, friction sparking action, single large spark window near muzzle, prominent "Astro Ray Gun" in center of body	1960s	15	30	45

GUNS

Space

NAME	DESCRIPTION	YEAR	GOOD	EX	MIP
Flash X-1	4" long, tin litho, friction sparking action, red body with blue and yellow inset and grips, large white "Flash X-1" on body, 4 red tinted plastic sparking windows	1967	15	25	35
Space Gun	4" long, tin litho, friction sparking action, red body with blue inset and grips, yellow block letter "SPACE GUN," large yellow and white vertical painted fins, 6 red tinted plastic sparking windows, oval Shudo logo by grip	1967	20	35	55

Spain

NAME	DESCRIPTION	YEAR	GOOD	EX	MIP
Pistola Sideral	tin litho pistol body with clear plastic barrel, planets and stars on gun with "Pistola Sideral" in read and white letters above handgrip		40	65	100

T. Cohn, Inc.

NAME	DESCRIPTION	YEAR	GOOD	EX	MIP
Space Target Game	24" tall, metal target with rubber tipped darts and dartgun to shoot down all the jet rockets and missiles	1952	40	65	100

T/Japan

NAME	DESCRIPTION	YEAR	GOOD	EX	MIP
Space Atomic Gun	4" long, tin litho, friction sparking action, silver gray finish with yellow and red trim, with "SPACE" on body in white small all caps and large yellow lower caps "atomic gun", small "T/ Made in JAPAN" logo above trigger	1960	25	40	65
Universe Gun	4" long, blue, yellow and red tin litho gun with friction sparking action, large all caps italic "Universe" on body side, sold in bag with header card	1960s	15	25	35

Tarrson Co.

NAME	DESCRIPTION	YEAR	GOOD	EX	MIP
Ray Dart Gun	9 1/2" long, blue plastic body with yellow muzzle, with 3 darts, storage compartment in red handle base	1968	15	25	35

TN/Japan

NAME	DESCRIPTION	YEAR	GOOD	EX	MIP
Space Gun	8" long, battery operated, reciprocating barrel shaft has red and blue lenses that flash when fired, also makes rat-a-tat noise, large circular "8" over handgrip, winged eagle over trigger, large block letter "SPACE GUN" on barrel	1960s	70	115	175

TNT/Hong Kong

NAME	DESCRIPTION	YEAR	GOOD	EX	MIP
Robot Raiders Space Signal Gun	6" long flashlight gun with interchangeable lenses and click sound, thinly veiled "Transformers" rip-off graphics on card	1980s	6	10	15

Tommy Toys

NAME	DESCRIPTION	YEAR	GOOD	EX	MIP
Space Gun	aluminum		55	95	145

U.S. Plastics

NAME	DESCRIPTION	YEAR	GOOD	EX	MIP
Rocket Jet Water Pistol	5" long, red, orange or yellow clear plastic body, fill plug at top of gun, large integral gunsight fin at rear, small sight fin at front	1957	12	20	30
Space Patrol Rocket Gun	black or red plastic pistol body with red trigger, grip embossed with vertically printed "Space Patrol" in irregular oval grip design showing rocket, stars and ringed planet, shoots rubber tipped darts	1954	145	250	375

GUNS

Space

NAME	DESCRIPTION	YEAR	GOOD	EX	MIP
United States					
Signal Flash Gun	6" long, plastic flashlight, black body with translucent white plastic light housing at muzzle and pearl finish plastic grip plates, modern missile type sight on top of barrel, large "SIGNAL FLASH" above trigger	1957	20	35	55
Unknown					
Atom Bubble Gun	red tubular barrel with handle attached, 2 sets of silver finish fins--at barrel base and muzzle, wire loop projects from muzzle for bubble blowing, handle embossed "Atom Trade Mark"	1940s	75	130	200
Batman Ray Gun	cap pistol with a bat symbol for the sigh	1960s	35	55	85
Bicycle Water Cannon Ray Gun	10", red plastic, swivel mount attached to bicycle handles, fired by lever	1950s	30	50	75
Buck Rogers Rubber Band Gun	cut-out paper gun, on card, advertising premium item	1930s	30	50	75
Clicker Ray Gun	5" red, blue or gray hard plastic, no boxes, sold loose	1950s	15	25	35
Clicker Ray Gun	5" green and/or rose swirl plastic	1950s	12	20	30
Clicker Whistle Ray Gun	5" plastic, two color variations: blue-green or olive-green swirl plastic, imprinted spacemen & rocket ships, back of gun is a whistle	1950s	15	25	35
Dan Dare & the Aliens Ray Gun	21" color tin litho gun	1950s	90	150	235
Lost In Space Laser Water Pistol	5" long, first season pistol style		30	50	75
Planet Patrol Saucer Gun	with spaceman motif	1950s	40	65	100
Radar Gun	5 1/2" long, mauve or silver/gray swirl plastic body with green or yellow spaceman sight and trigger, Saturn and star embossed above grip and "Radar Gun" embossed above that	1956	20	35	55
Rocket Pop Gun	wood, green and red horizontal striped body with black tri-fin pump base, cork and string stopper in nose, pump fins into body to make it pop	1955	25	40	65
Secret Squirrel Ray Gun		1960s	25	40	60
Space Atomic Gun	4" silver, orange & red litho, sparking action, tin-litho	1960s	25	40	65
Space Control Ray Gun	5 1/2" long, red plastic with yellow trigger, clicks	1956	25	40	65
Space Dart Gun	6" long, gun has one white side and one black side, both with star and lightning motif, 8 thin cooling fins on barrel	1950s	30	50	75
Space Patrol Cosmic Smoke Gun	solid color red or green plastic with "Space Patrol" on body above grip, TV show tie-in, shoots baking powder	1950s	135	225	350
Space Patrol Hydrogen Ray Gun Ring	glow-in-the-dark, cap firing ring, TV show tie-in	1950s	70	115	175
Space Patrol Rocket Gun and Holster Set	with darts	1950s	185	310	475
Space Patrol Rocket Gun Set	with darts, sold without holster, TV show tie-in	1950s	105	180	275
Space Pistol	plastic ray gun that shoots rubber tip darts	1954	20	35	50
Star Trek Phaser Flashlight	battery operated, small phaser shape	1976	8	13	20
Superior Rocket Gun	8" long, dark gray plastic all over, embossed "Superior Rocket Gun" on grip	1956	30	50	75
Tom Corbett Flash X-1 Space Gun	5" long	1950s	55	95	145

GUNS

Space

NAME	DESCRIPTION	YEAR	GOOD	EX	MIP
Tom Corbett Space Cadet Atomic Rifle	well designed silver-gray finish long rifle with futuristic styling and embossed logo on stock and above front tommy gun-styled handgrip	1950s	135	225	350

Webb Electric Co.

NAME	DESCRIPTION	YEAR	GOOD	EX	MIP
Atom Buster Mystery Gun	11" long yellow plastic gun with inner bladder, fires blast of air at tissue paper atomic mushroom target, with instructions, atomic explosion cover art on box	1950s	105	180	275

Wham-O

NAME	DESCRIPTION	YEAR	GOOD	EX	MIP
Wham-O Air Blaster	10" long plastic gun uses rubber diaphram to shoot air; styling is reminiscent of Budson Flash Gordon Air Ray Gun	1960s	70	115	175

Wyandotte

NAME	DESCRIPTION	YEAR	GOOD	EX	MIP
Pop Ray Gun	red pressed steel body with 5 widely spaced vertical round fins, unpainted trigger and muzzle with large gunsight, rod connects body to pop mechanism in muzzle	1930s	70	115	175
Ray Gun	7" stamped metal pop gun that uses a captive cork to make the pop, red body, unpainted muzzle, with connecting rod from body to barrel tip	1936	35	55	85

Yoshiya/Japan

NAME	DESCRIPTION	YEAR	GOOD	EX	MIP
Space Gun	7" long, tin with sparking action, shows a realistic white rocket blasting off over lunar terrain on side of body and atomic symbol on grip center with diamond-shaped "SY" logo and "Made in Japan" at bottom of grip	1957	40	65	100

Western

Actoy

NAME	DESCRIPTION	YEAR	GOOD	EX	MIP
Pony Cap Pistol	single shot, all-metal, nickel finish with eagle on grip	1950s	25	45	65
Wells Fargo Buntline Cap Pistol	11" long barrel, break-to-front, cream plastic stag grips	1950s	85	135	195
Wyatt Earp Buntline Special	11" barrel, die cast, friction break-to-front, white plastic grips, nickel finish	1950s	95	135	175

Buzz-Henry

NAME	DESCRIPTION	YEAR	GOOD	EX	MIP
Lone Rider Cap Pistol	8" die cast, white plastic inset rearing horse grips	1950s	35	50	90

Carnell

NAME	DESCRIPTION	YEAR	GOOD	EX	MIP
Maverick Cap Pistol	9" break-to-front, lever release, nickel finish, Maverick on sides, cream & brown swirl colored grips features notch bar with extra set of black plastic grips	1960	40	85	135
Maverick Two Gun Holster Set	9" break-to-front, lever release, nickel finish, Maverick on sides, cream & brown swirl grips features notch bar, black leather dbl. holster set with silver plates, studs & white trim, 6 loops, buckle	1960	175	250	365

Classy

NAME	DESCRIPTION	YEAR	GOOD	EX	MIP
Dale Evans Holster Set	brown & yellow leather, white fringe on holsters, stylized blue butterflies are also "DE" logo, if buckled in front, holsters are backwards; holsters only, no guns		55	100	150

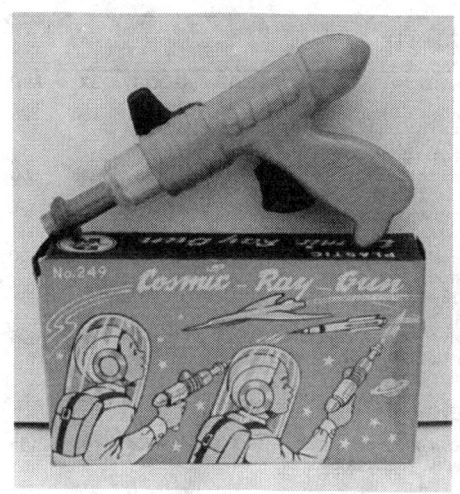
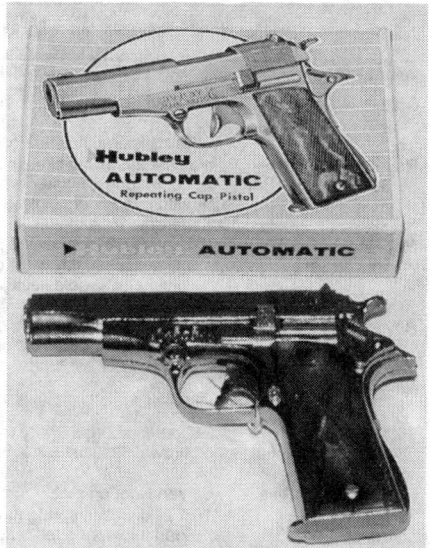

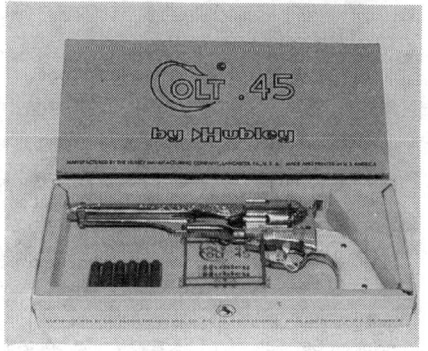
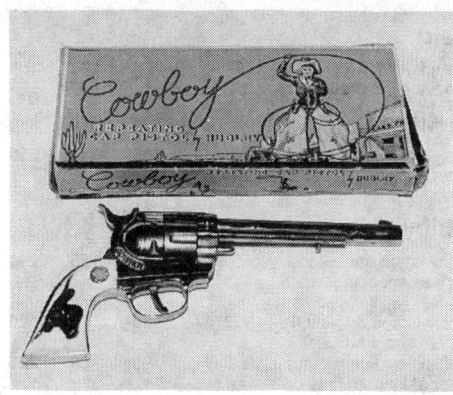
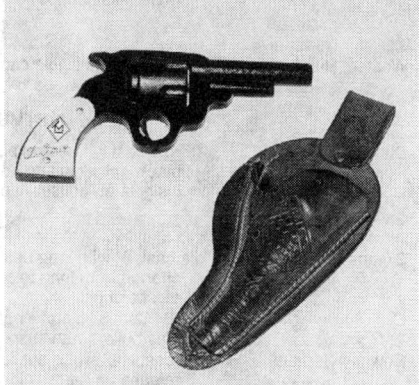

Top to bottom: Cosmic Ray Gun #249, 1950s, Ranger Steel Products; Automatic Repeating Cap Pistol #290, 1956, Hubley; Space Rocket Gun, 1950s, M&L; Colt .45 Cap Pistol, 1959, Hubley; Cowboy Cap Pistol, 1950s, Hubley; Tom Mix Wooden Gun, 1938, Ralston Purina premium item.

Western

NAME	DESCRIPTION	YEAR	GOOD	EX	MIP
Double Holster Set	imitation alligator-texture brown leather, steer-head conches on holsters, lots of studs, yellow felt backing, holsters only, no guns	1950s	100	150	225
Roy Rogers Double Gun & Holster Set	die cast 2-8 1/2" nickel finish pistols with copper figural grips, holster is brown & black leather with raised detail, plastic play bullets & leather tie-downs	1950s	275	525	800
Roy Rogers Double Holster Set	10" guns with plain nickel finish & copper grips, lever release, brown & cream leather set, silver studs, gold fleck jewels, & 4 wooden bullets		225	425	650
Roy Rogers Double Holster Set	black & white leather set, silver studs & conches, 9" classy Roy Rogers pistols with plain nickel finish & copper figural grips, friction release	1950s	240	450	695

Daisy

NAME	DESCRIPTION	YEAR	GOOD	EX	MIP
760 Rapid Fire Shotgun Air Rifle	31" pump shotgun, grey metal one piece frame, brown plastic stock & Slider grip, fires blast of air	1960s	65	95	135
Red Ryder BB Rifle	carved wooden stock	1980	25	40	60
Spittin Image Peacemaker BB Pistol	10 1/2" die cast, spring fired, single action, BBs load into spring fed magazine under barrel		25	50	75

Edison

NAME	DESCRIPTION	YEAR	GOOD	EX	MIP
Susanna 90 12 Shot Cap Pistol	9" uses ring caps, wind out cylinder, black finish, plastic wood grips	1980s	10	15	25

Esquire Novelty

NAME	DESCRIPTION	YEAR	GOOD	EX	MIP
Authentic Derringer	die cast, 2" cap firing, copper finish, twin swivel barrel	1960	15	25	45
Authentic Derringer	classic miniature Series #10, 2" cap firing, copper finish. twin swivel barrel	1960	15	25	45
Johnny Ringo, Adventure of, Gun & Holster	10 3/4" long barrel Actoy, friction break, w/blk. & gld. plastic stag grips, blk. leather two gun holster (cut out flowers over a gld. background), felt backing, loops hold 4 to 6 bullets, silver buckle	1960	200	350	485
Pony Boy Double Holster Set	brown leather double holster with bucking broncs & studs, cuffs, spurs & spur leathers, guns are Actoy "Spitfires", die cast 8 1/2" copper finish, white plastic grips	1950s	100	195	275

Haig

NAME	DESCRIPTION	YEAR	GOOD	EX	MIP
Western Buntline Pistol	13", pistol fires single caps and/or BBs, BBs are propelled down barrel sleeve by cap explosion	1963	75	120	165

Harvel-Kilgore

NAME	DESCRIPTION	YEAR	GOOD	EX	MIP
Classy The Rebel Holster & Pistol	12" die cast long barrel pistol, brown plastic grips, black leather single holster left side, Rebel insignia on holster flap,	1960s	195	350	550

Hubley

NAME	DESCRIPTION	YEAR	GOOD	EX	MIP
2 Guns in 1 Cap Pistol	die cast, 8" with long barrel, twist-off barrels to change from long to short, side loading, white plastic grips	1950s	75	125	200
Colt .38 Detective Special	4 1/2" Colt .38 pistol single shot caps & loads 6 play bullets with suspenders chest holster	1959	35	75	125
Cowboy Cap Gun	die cast, 12" swing out revolving cyclinder, release on barrel, nickel or aluminum finish, white plastic steer grips with black steer head	1950s	95	165	225
Cowboy Cap Gun	12" swing-out revolving cylinder, release on barrel, gold finish, black plastic steer grips	1950s	100	160	250

Western

NAME	DESCRIPTION	YEAR	GOOD	EX	MIP
Cowboy Cap Pistol	8" friction break-to-front nickel finish cast iron, rose swirl plastic grips w/Colt logo	1940	85	150	200
Cowboy Jr. Cap Pistol No. 225	9" die cast, revolving cylinder, side loading, release on barrel, silver finish, white plastic cow grips, lanyard ring & cord	1950s	65	125	185
Dagger Derringer	7", unusual over & under pistol has hidden red plastic dagger that slides out from between barrels, rotating barrels load & fire 2-pc. bullets	1958	55	100	145
Davy Crockett Buffalo Rifle	25" die cast & plastic, unusual flintlock, style fires single cap under pan cover, brown plastic stock, ammo storage door in stock	1950s	75	145	200
Deputy Cap Pistol	10" die cast, front breaking, release on barrel ornate scroll work, nickel finish	1950s	45	75	115
Flintlock Jr. Cap Pistol	7 1/2" single shot, double action, brown swirl plastic stock	1955	10	20	35
Flintlock Pistol	9 1/4", two shot cap shooting single action double barrel, over & under style, brown swirl plastic stock, nickel finish	1954	50	95	145
Flip Special from the Rifleman Cap Rifle	3' long rifle, resembles classic Winchester with ring lever, brown plastic stock, pop down cap magazine	1959	125	225	275
Frontier Repeating Cap Rifle	35 1/4" rifle nickel finish with brown plastic stock & forestock, blue metal barrel, red plastic choke & front sight, pop down magazine, released by catch in front of trigger, scroll work	1950s	75	165	210
Model 1860 Cal .44 Cap Pistol	13", revolving cylinder with closed chamber ends, 6 2-pc. bullets, flat aluminum finish, white plastic grips, complete with wooden display plaque	1959	125	250	350
Panther Pistol	die cast, 4" derringer style pistol snaps out from secret spring-loaded wrist holster	1957	85	130	185
Remington .36 Cap Pistol	8" long, nickel finish, black plastic grips, revolving cylinder chambers 2-pc. bullets	1950s	75	145	225
Rex Trailer Two Gun & Holster Set	9 1/2" side loading, nickel finish, stag plastic grips, brown textured tooled leather with white holsters & trim, 6 bullet loops with plastic silver bullets, plain buckle	1960	90	165	250
Rodeo Cap Pistol	7 1/2" single shot, white plastic steer grips	1950s	20	40	60
Texan .38 Cap Pistol	10" long, revolving cylinder gun chambers 6 solid brass bullets (rd. caps go into cylinder first), top release front break automatically ejects shells, plastic steer grips	1950s	100	165	225
Texan Cap Pistol	9 1/4" cast iron revolving cylinder lever release, white plastic steer grips, nickel finish, Colt rearing horse logo on grips	1940	90	175	275
Texan Cap Pistol	die cast, nickel finish, white plastic steer grips, star logo on grips	1950s	75	145	200
Texan Cap Pistol No. 285	9 1/4", cast iron, revolving cylinder, lever release, white plastic steer grips, nickel finish with star logo in grip	1940	85	185	300
Texan Dummy Cap Pistol	9 1/4", revolving cylinder, lever release, white plastic steer grips, nickel finish, star logo on grip	1950s	90	165	225
Texan Dummy Cap Pistol	9 1/4" revolving cylinder, lever release, white plastic steer grips, nickel finish, Colt rearing horse logo	1940	65	110	165
Texan Jr. Cap Pistol	10", spring button release on side of cylinder, break-to-front, nickel finish white plastic grips with black steers	1950s	65	90	125
Texan Jr. Cap Pistol	die cast, release under cylinder, nickel finish, white plastic Longhorn grips	1954	50	75	110
Texan Jr. Gold Plated Cap Pistol	9", gold finish with black longhorn steer grips, break-to-front release from cylinder	1950s	85	130	185

Western

NAME	DESCRIPTION	YEAR	GOOD	EX	MIP
Wyatt Earp Double Holster Set	black & white leather holster with silk screened "Marshal Wyatt Earp" logo, two No. 247 Hubley Wyatt Earp Buntline Specials, 10 3/4" nickel finish, purple swirl grips	1950s	175	275	385

Ideal

NAME	DESCRIPTION	YEAR	GOOD	EX	MIP
Yo Gun	7 1/4" red plastic gun releases yellow plastic ball which snaps back when trigger is pulled, functions like a yo-yo	1960s	30	45	60

Kenton

NAME	DESCRIPTION	YEAR	GOOD	EX	MIP
Gene Autry Dummy Cap Pistol	cast iron, 8 3/8", long barrel, dark grey gun, metal finish, white plastic grips	1939	70	110	185
Lawmaker Cap Pistol	8 3/8", break-to-front friction break, unusual dark grey gunmetal finish, white plastic raised grips	1941	100	175	250

Kilgore

NAME	DESCRIPTION	YEAR	GOOD	EX	MIP
Big Horn Cap Pistol	7" all metal revolving cylinder, break-to-front, disc caps, silver finish	1950s	85	125	175
Bronco Cap Pistol	8 1/2", revolving swing-out cylinder fires Kilgore disc caps, silver finish, black plastic "Bronco" grips	1950s	50	100	175
Champion Quick Draw Timer Cap Pistol	silver finish, side loading, wind up mechanism in grip records elapsed time of draw, black plastic grips	1959	65	150	225
Cheyenne Cap Pistol	9 3/4" side loading, "Sure-K" plastic stag grips, silver finish	1974	10	15	25
Fastest Gun Electronic Draw Game	die cast, wire plug into "Rangers" gun grips, gun that shoots first lights eye of plastic battery operated steer head, red & blue plastic holsters w/matching cowboy gun grips, plastic belts	1958	80	145	250
Grizzly Cap Pistol	10" revolving cylinder fires disc caps, swing out cylinder, black plastic grips with grizzly bear	1950s	95	165	230
Hawkeye Cap Pistol	4 1/4" all metal, automatic style, side loading, silver finish	1950s	20	35	55
Lone Ranger Cap Pistol	8 1/2" cast iron, large hammer, nickel finish, spring release on side for break, red-brown, Hi-Yo Silver grips	1940	85	145	225
Lone Ranger Cap Pistol	8 1/4" cast iron, small hammer, nickel finish, friction break, purple plastic "Hi-Yo Silver" grips	1938	95	175	275
Mustang Cap Pistol	9 1/2" chrome finish with "stag" plastic grips	1960s	20	35	55
Ranger Cap Pistol	8 1/2" nickel finish, brown swirl plastic grips, spring release on right side, break-to-front	1940	90	125	165
Roy Rogers Cap Pistol	10" revolving cylinder swings out to load, fires disc caps, white plastic horse-head grips with "RR" logo	1950s	125	225	350

Langson Mfg. Co.

NAME	DESCRIPTION	YEAR	GOOD	EX	MIP
Cody Colt Paper Buster Gun	7 3/4", paper popper, nickel finish, white plastic steer grips fires Cody Colt ammunition	1950s	35	65	90

Leslie-Henry

NAME	DESCRIPTION	YEAR	GOOD	EX	MIP
Gene Autry 44 Cap Pistol	11" lever release, side loading, long barrel, loads solid metal bullets, nickel finish, brown translucent plastic horse-head grips	1950s	85	175	285
Gene Autry 44 Cap Pistol	11" lever release, side loading, long barrel, nickel finish, white plastic horse-head grips	1950s	75	135	275
Gene Autry Cap Pistol	9", break-to-front lever release, nickel finish, white plastic horse-head grips	1950s	90	145	250
Gene Autry Cap Pistol	9" break-to-front lever release, copper finish, white plastic horse-head grips	1950s	65	155	250

Western

NAME	DESCRIPTION	YEAR	GOOD	EX	MIP
Gene Autry Cap Pistol	7 3/4" die cast, small size, lever release, break-to-front, nickel finish with extension scroll work, black plastic horse-head grips	1950s	65	145	250
Marshal Cap Pistol	10" revolving cylinder chambers Nichols-style bullets, white plastic grips with star ovals	1950s	35	75	110
Marshal Matt Dillon "Gunsmoke" Cap Pistol	10" pop-up cap magazine, release in front of trigger guard, scroll work, bronze steer-head grips	1950s	50	80	135
Maverick Derringer	3 1/4" with removeable cap-shooting bullets, tan vinyl holster with two bullets	1958	35	55	85
Ranger Cap Pistol	7 3/4" derringer with removeable cap, shooting bullets, tan vinyl holster with two bullets	1950s	85	110	150
Texas Cap Pistol	9" die cast, break-to-front, lever release, nickel finish, plastic horse-head grips	1950s	65	95	125
Texas Ranger Cap Pistol	8 1/4" die cast, lever release break to front, nickel finish, scroll work, vasoline colored plastic grips		65	95	125
Wagon Train Complete Western Cowboy Outfit	plastic flip ring lever rifle & wagon train pistol (late model L-H pistol) & leather holster	1960	75	145	200
Wild Bill Hickok 44 Cap Pistol Set	11" nickel finish, swing out side loading action, revolving cylinder chambers 6 metal bullets, amber plastic horse head grips, single holster black & brown leather w/silver studs, diamond conches	1950s	200	325	475
Wild Bill Hickok Cap Pistol	10" pop-up cap magazine, release in front of trigger guard, scroll work, translucent brown plastic grips with oval star inserts	1950s	90	135	190
Young Buffalo Bill Cowboy Outfit	black & white leather holster set with pistol, white grip, holster bands read Texas Ranger		100	175	225

Lone Star

NAME	DESCRIPTION	YEAR	GOOD	EX	MIP
Gun Fighter Holster Set	9" Frontier Ace, lever release, break-to-front, silver finish, brown plastic grips, holster white & red leather "Laramie" single holster with separate belt	1960s	55	85	145
Pecos Kid Cap Pistol	9" silver chrome finish, brown plastic grips, lever release	1970s	10	15	30
Pepperbox Derringer Cap Pistol	die cast, 6 1/4" rotating barrel holds four cap loads, silver finish with black plastic grips	1960	75	110	185

Long Island Die Casting Inc.

NAME	DESCRIPTION	YEAR	GOOD	EX	MIP
Texas Cap Pistol	die cast, 8 1/2" friction break, Circle "T" logo, scroll work on barrel	1950s	45	85	110

Marx

NAME	DESCRIPTION	YEAR	GOOD	EX	MIP
Bonanza Guns Outfit	25" cap firing saddle rifle, magazine pulls down to load, 9 1/2" western pistol fires 2-pc. Marx shooting bullets, wood plastic stocks & gun metal gray plastic body, tan vinyl holster	1960s	75	145	225
Buffalo	50 shooter repeater		70	100	140
Centennial Rifle	with big sound		25	35	65
Cork-Shooting Rifle			25	40	60
Double Holster Set	10" two pistols similar to 1860s Remington, fires roll caps by use of a lanyard that is pulled from the bottom of the grip, internal hammer, white plastic horse & steer grips, silver, brown vinyl holster	1960	70	120	175
Double-Barrel Pop Gun Rifle	22" long, 1935		45	75	100
Double-Barrel Pop Gun Rifle	28" long, 1935		50	85	125
Hi-Yo Silver Lone Ranger Pistol	tin gun		45	65	90

Western

NAME	DESCRIPTION	YEAR	GOOD	EX	MIP
Historic Guns Derringer	Marx Historic Guns series, Derringer with plastic presentation case, 4 1/2" long, on card	1974	15	25	30
How the West Was Won Gun Rifle	deep gray winchester model with tan stock, in box		55	85	145
Johnny Ringo Gun & Holster Set	die cast gun, white plastic head grips, vinyl quick draw holster has rawhide tie, gun is fired by lanyard which passes through grip butt & attaches to belt, when pulled lanyard trips internal hammer	1960	80	130	185
Lone Ranger 45 Flasher Flashlight Pistol			25	50	85
Lone Ranger Carbine	26" grey plastic repeater-style rifle has pull down cap magazine, western trim & Lone Ranger signature on stock	1950s	65	100	165
Lone Ranger Clicker Pistol	8", nickel finish, red jewels, inlaid white plastic grip with the Lone Ranger, Hi-Yo Silver & LR head embossed, brown leather holster	1938	60	100	150
Lone Ranger Double Target Set	9 1/2" square stand up target, tin litho, wire frame holds target upright, backed with bullseye target, 8" metal dart gun fires wooden shaft dart	1939	95	165	245
Lone Ranger Sparkling Pop Pistol	tin litho		45	65	90
Mare's Laig Rifle Pistol	13 1/2" brown plastic, black plastic body, pull down magazine		60	95	135
Official Wanted Dead or Alive Mare's Laig Rifle	19" bullet loading, cap firing saddle rifle-pistol ejects plastic bullets, with holster		100	195	300
Ranch Rifle	plastic, repeater		30	45	60
Roy Rogers Carbine	26" grey plastic repeater-style rifle has pull down cap magazine, western trim & Roy Rogers signature on stock	1950s	80	120	155
Side-By Double Barrel Pop Gun Rifle	9" long		20	35	45
Thundergun Cap Pistol	12 1/2" single action,"Thundercaps" perforated roll cap system, silver finish, brown plastic grips	1950s	100	175	250
Wanted Dead or Alive Miniature Mares Laig	Marx Miniatures series miniature cap rifle on "wood frame" card, 1959		20	35	50
Zorro Flintock Pistol			30	45	75
Zorro Rifle			60	90	120

Mattel

NAME	DESCRIPTION	YEAR	GOOD	EX	MIP
Fanner 50 "Swivelshot Trick Holster" Set	die cast bullet loading Fanner 50, leather swivel style holster, attached to any belt, gun fires in holster when swiveled, string included for last ditch draw	1958	85	165	250
Fanner 50 Cap Pistol	11" fanner non-revolving cylinder, stag plastic grips, nickel finish, black vinyl "Durahyde" holster	1960s	50	75	110
Fanner 50 Smoking Cap Pistol	10 1/2" with revolving cylinder, first version with grapefruit cylinder does not chamber bullets	1957	75	150	250
Fanner 50 Smoking Cap Pistol	10 1/2" with revolving cylinder, chambers 6 metal play bullets, die cast	1958	55	120	200
Shootin' Shell .45 Fanner Cap Pistol	11" revolving cylinder pistol shoots Mattel Shootin' Shell cartridges, shell ejector	1959	250	400	650
Shootin' Shell Buckle Gun	cap & bullet shooting copy of Remington Derringer pops out from belt buckle, 2 brass cartridges & 6 bullet head	1958	45	80	125
Shootin' Shell Fanner	9" die cast chrome finish, revolving cylinder chambers 6 Shootin' Shell bullets	1958	65	125	225
Shootin' Shell Fanner & Derringer Set	small size Shooting Shell Fanner w/chrome finish, revolving cylinder, chambers 6 Shootin' Shell bullets, brown leather holster	1958	75	145	250

Western

NAME	DESCRIPTION	YEAR	GOOD	EX	MIP
Shootin' Shell Fanner Single Holster	cowhide holster takes small size Shootin' Shell Fanner with six brass play bullets & tie-downs	1959	80	145	250
Shootin' Shell Indian Scout Rifle	29 1/2" plastic/metal Sharps rolling block rifle, chambers 2-pc. Shootin' Shell bullets, secret compartment in stock for ammo storage, plastic stock & metal barrel	1958	100	145	250
Shootin' Shell Potshot Remington Derringer	3" derringer, on card	1959	35	60	90
Showdown Set with 3 Shootin' Shell Guns	30" single shot rifle with metal barrel, die cast w/ plastic stock, Shootin' Shell Fanner sm. size, revolving cylinder, chrome finish & Imitation stag plastic grips, tan holster w/bullet loops	1958	300	600	950
Winchester Saddle Gun Rifle	33" die cast & plastic, perforated roll caps & chambers 8 play bullets loaded thru side door	1959	45	95	350

Nichols

NAME	DESCRIPTION	YEAR	GOOD	EX	MIP
Dyna-Mite Derringer	3 1/4" die cast, loads single cap cartridge, silver finish, white plastic grips	1955	15	30	45
Dyna-Mite Derringer in Clip	3 1/2" die cast, fires single cap in Nichols cartridge, nickel finish, white plastic grips with small leather holster		25	45	65
Nichols Cap Gun Store Display	24 x 14" wood board, derringer & 2 strips of Nichols bullets	1950s	300	600	950
Silver Pony Cap Pistol	7 1/2" single shot, silver metal grip & one replacement black plastic grip, silver finish	1950s	30	45	70
Spitfire Hip Gun No. 100	9" cap cartridge loading mini rifle, chrome finish, tan plastic stock	1950s	15	20	35
Stallion .22 Cap Pistol	7" revolving cylinder chambers five two piece cartridges, single action, black plastic stag grips, never came in box	1950s	30	55	85
Stallion .22 Double Action Cap Pistol	7" double action, pull trigger to fire, white plastic grips, nickel finish, cylinder revolves	1950s	45	80	125
Stallion .38 Cap Pistol	9 1/2", chambers 6 2-pc. cap cartridges, nickel finish, white plastic grips	1950s	75	115	175
Stallion .45 MK I Cap Pistol	die cast, 12" chrome finish, revolving cylinder, chambers 6 2 pc. bullets, shell ejector, white "pearlescent" plastic grips with rearing stallion, red jewels & 6 bullets & Stallion caps	1950	100	250	375
Stallion .45 MK II Cap Pistol	12" pistol, chrome finish, revolving cylinder, chambers 6 2-pc. bullets, shell ejector, extra set of white grips to replace black grips on gun & box of Stallion caps	1956	100	185	275
Stallion 32 Six Shooter	8" revolving cylinder chambers 6 2-pc. cartridges, nickel finish, black plastic grips	1955	75	110	165
Stallion 41-40 Cap Pistol	10 1/2" revolving cylinder chrome finish pistol, swing out cylinder that chambers 6 2-pc. cap cartridges, shell ejector, scroll work on frame, creme-purple swirl colored plastic grips	1950s	150	250	345

Ohio Art

NAME	DESCRIPTION	YEAR	GOOD	EX	MIP
Sheriff's Derringer Pocket Pistol	3 1/4" silver finish derringer chambers 2-pc., Nichols style cartridge, red plastic grips with an "A" logo, on card	1960s	10	20	25

Pilgrim Leather

NAME	DESCRIPTION	YEAR	GOOD	EX	MIP
Ruff Rider Western Holster Set	brown leather double holster, variety of studs & red jewels, 12 plastic silver bullets, tie-downs		100	150	200

Product Engineering Co.

NAME	DESCRIPTION	YEAR	GOOD	EX	MIP
45 Smoker	10" single cap, shoots talcum-like powder by use of bellows when trigger is pulled, aluminum finish	1950s	45	75	115

GUNS

Western

NAME	DESCRIPTION	YEAR	GOOD	EX	MIP
Frontier Smoker	9 1/2" cap pistol, die cast, pop up magazine & shoots white powder from internal bellows, all metal, black grips, silver finish, gold magazine, hammer & trigger		85	135	200

Ralston Purina

NAME	DESCRIPTION	YEAR	GOOD	EX	MIP
Tom Mix Wooden Gun	3 all-wood versions with leather holster, came in mailer, each	1930s	125	250	350

Schmidt

NAME	DESCRIPTION	YEAR	GOOD	EX	MIP
Buck 'n Bronc Marshal Cap Pistol	10" long barrel revolver style, lever release, break-to-front, plain silver finish, copper color metal grips	1950s	90	150	225

Smart Style

NAME	DESCRIPTION	YEAR	GOOD	EX	MIP
Real Texan Outfit with Nichols Stallion .22	brown & white leather double holsters have silver conches with red reflectors, silver horses at top of holster, belt with 3 bullet loops, guns are a pair of double action .22s	1950s	95	175	265

Stevens

NAME	DESCRIPTION	YEAR	GOOD	EX	MIP
49-er Cap Pistol	cast iron, 9", unusual internal hammer with revolving steel cylinder, nickel finish, white plastic figural grips	1940	100	200	350
Buffalo Bill Cap Pistol	7 3/4", silver nickel finish, side loading magazine door, white "tenite" plastic horse & cowboy grips, red jewels	1940	65	110	175
Colt Cap Pistol	6 1/2", revolver style double action	1935	15	25	50
Cowboy Cap Pistol	3 1/2" cast iron, single shot single action, sold loose	1935	20	35	50
Cowboy King Cap Pistol	9" break-to-front release, gold finish cast iron, black plastic grips, yellow jewels	1940	85	175	225

Topper/Deluxe

NAME	DESCRIPTION	YEAR	GOOD	EX	MIP
Johnny Eagle Red River Bullet Firing	over 12" double action revolving cylinder pistol, die cast hammer, trigger, blue plastic overall with wood plastic grips with gold horse, side loading, shell ejector, fires 2 piece plastic bullets	1965	65	95	150

Unknown

NAME	DESCRIPTION	YEAR	GOOD	EX	MIP
Davy Crockett Frontier Fighter Cork Gun	21" pop gun shoots cork on string & has cigarette flint mechanism at muzzle that makes sparks when fired, wood stock, leather sling	1950s	65	135	195
Gene Autry Champion Single Holster Set	leather & cardboard, red, yellow & green "jewels," four white wooden bullets, silver buckle	1940s	125	175	250
Lone Ranger Holster	9", leather & pressboard, Hi-Yo Silver & Lone Ranger printed, red jewel, belt loop		35	45	60
Tom Mix Gun & Holster Outfit (Box Only)	10 x 5 x 2" box, gun and holster unknown	1930s	55	85	125
Wyatt Earp Double Holster Set	med. size, reflectors, black leather with brown rawhide fringe, holsters only	1950s	50	70	95

Wyandotte

NAME	DESCRIPTION	YEAR	GOOD	EX	MIP
Red Ranger Jr. Cap Pistol	7 1/2" lever release, break-to-front, silver finish, white plastic horse grips	1950s	55	85	120

Young Premiums

NAME	DESCRIPTION	YEAR	GOOD	EX	MIP
Official Wyatt Earp Buntline Clicker Pistol	18 1/2" plastic		35	60	90

Model Kits

Model kits have always been popular toys for boys, and in recent years the kits have found a new following among older collectors...primarily former boys recapturing a part of their youth.

Plastic model kits were first produced shortly before World War II, but it was not until after the war that plastic kit building really began to take off. Automobiles, aircraft, and ships all became subject matter for the miniature replicas popularized by such companies as Aurora, Revell, Monogram, Frog, and Lindberg.

Each type of model kit has its own enthusiastic following, but probably the most collectible kits today are the figure and character kits produced primarily in the 1960s. These kits have seen dramatic increases in collector values over the past ten years.

The company that did the most to popularize the figure kit was Aurora, with its introduction in the early 1960s of a line of kits representing the monsters from Universal Pictures. Aurora had been producing figure kits prior to that time, but the monster craze of the period was responsible for a highly successful line of kits.

These kits have become the backbone of the hobby, as our Top 10 listing at the end of this article shows. In fact, Aurora kits occupy slots 1 through 29 in our index by price and kit name. Not until slot number 30 does Revell make an appearance, with Robbin' Hood Fink, at $400.

Starting with the Frankenstein monster in 1961, Aurora went on to produce kits of many memorable movie monsters before moving into more general monstrosities, such as its famous working guillotine kit. Such toys offended the sensibilities of some groups, who brought about political pressure that spelled the end of this line of kits. The firm also produced kits based on popular television shows, comic characters, and sports celebrities. Some of the kits originally made by Aurora were later reissued by Monogram and Revell. And resin copies of the more hard-to-get Aurora kits are still being produced and sold today by independent garage kit makers.

"Frantic Banana Punishing the Skins" from Hawk, 1965; "The Vampire" from Aurora.

Other popular monster kits were also a fad in the 1960s. These were not the traditional movie monsters, but rather an assortment of strange characters that often came in wild hot rods. Among the more popular were those based on Ed "Big Daddy" Roth's Rat Fink concept. These kits were produced by Revell. Other firms, most notably Hawk, also produced kits of this new type of monster.

Even popular celebrities of the day became the subject of model kits. Revell, for example, issued figure kits of each of the four Beatles.

Figure kits began to enjoy new popularity in the 1980s as new large-scale kits of rather limited production runs were being made in vinyl and resin. Billiken, a Japanese company, produced vinyl kits of the classic movie monsters, some of which have become highly collectible. Screamin' and Horizon are two leaders in a burgeoning "garage kit" field of of large-scale vinyl and resin kits of movie, monster, and comic book characters. This area bears watching as the limited run nature of these garage kits will no doubt translate into collectibility in the future.

Some of the kits we have characterized in the category of collectible figure/character kits do not represent actual figures. However, they are generally considered to fall into this category because they have some relationship to a popular character, personality, or historical figure.

The prices indicated are intended to provide general guidelines as to what these kits would sell for today at retail. MIB refers to a kit that is mint in box. It is in like-new condition in the original like-new box, with instructions. The box may not be in the original factory seal, but if the kit pieces were contained in bags inside the box, the bags have not been opened. Kits that remain in pristine condition in factory seals may command a slight premium. NM refers to a near-mint condition kit that is like new, complete and unassembled. The box may show some shelf wear and the interior bags may have been opened. B/U refers to a kit that has been assembled or built up. These price guidelines assume a neatly built, complete kit.

Trends

Classic 1960s character kits have flattened in the last year. One reason for this is that garage kits, with their superior detailing and more graphic subject matter, have siphoned dollars from the classic kit market. Again, the garage kit market remains in the retail stage. Only a handful of these kits are to date truly "out of print," and it is too early to tell if these kits will begin to appreciate alongside their "classic" counterparts.

The top end of the market remains firm, as do "franchise" kits such as Star Trek. The true Aurora rarities such as Godzilla's Go Cart and those kits with cult and crossover followings such as Lost In Space, continue to post steadily increasing values.

Aurora's "UFO From The Invaders" plastic model kit.

The Top 10 Classic Figural Model Kits
(in Mint In Box condition)

1. Godzilla's Go-Cart, Aurora, 1966 ....$3,000
2. Lost In Space, (Cyclops and Chariot), Aurora, 1966..1,300
3. Frankenstein, Gigantic 1/5 Scale, Aurora, 1964...1,200
4. Munsters (Living Room), Aurora, 1964 1,200
5. King Kong's Thronster, Aurora, 1966 1,000
6. Lost In Space, (small kit), Aurora, 1966 900

Aurora's "Prehistoric Scenes" series of plastic model kits, 1971-72.

7. Addams Family Haunted House, Aurora, 1964 800
8. Bride of Frankenstein, Aurora, 1965 750
9. Lost In Space, The Robot, Aurora, 1968 700
10. Godzilla, Aurora, 1964 500

Aurora's "Silver Knight" and "Blue Knight" originally issued in 1956 and re-issued in 1963.

MODELS

MODELS

Top to bottom: Allosaurus from Prehistoric Scenes, 1970s, Aurora; The Man From U.N.C.L.E., 1966, Aurora; Hitler from Born Losers series, 1965, Park's Model Kit; Barnabas Vampire Van from Dark Shadows series, MPC.

MODELS

Addar

NAME	DESCRIPTION	YEAR	B/U	NM	MIP
106	Caesar, Planet of the Apes	1974	15	40	45
101	Cornelius, Planet of the Apes	1974	12	30	35
216	Cornfield, Planet of the Apes	1975	15	40	45
102	Dr. Zaius, Planet of the Apes	1974	10	25	30
105	Dr. Zira, Planet of the Apes	1974	10	25	30
104	Gen. Aldo, Planet of the Apes		10	25	30
103	Gen. Ursus, Planet of the Apes		12	30	35
217	Jailwagon, Planet of the Apes	1975	15	40	45
270	Jaws diorama		20	50	60
107	Stallion & Soldier, Planet of the Apes	1974	25	75	100
215	Treehouse, Planet of the Apes	1975	15	40	45

Airfix

NAME	DESCRIPTION	YEAR	B/U	NM	MIP
3542	Anne Boleyn	1974	7	15	20
823	Aston Martin, James Bond	1965	60	200	225
2502	Black Prince	1973	10	25	30
212	Boy Scout	1965	7	15	20
211	Charles I	1965	10	20	25
2501	Henry VIII	1973	4	8	10
2504	Julius Caesar	1973	10	25	30
	Monkeemobile	1967	75	240	275
2508	Napolean	1978	4	8	10
3546	Queen Elizabeth I	1980	7	15	20
3544	Queen Victoria	1976	7	15	20
203	Richard I	1965	10	25	30
2507	Yeoman of the Guard	1978	4	8	10

AMT

NAME	DESCRIPTION	YEAR	B/U	NM	MIP
7701	Big Foot	1978	20	60	75
	Brute Farce		5	10	15
610	Cliff Hanger		5	10	15
	Dragula, Munsters	1965	40	200	225
497	Flintstones Rock Crusher	1974	20	50	60
495	Flintstones Sports Car	1974	20	55	65
	Girl From U.N.C.L.E. Car	1974	75	200	250
309	Graveyard Ghoul Duo (Munsters cars)	1970		100	125
2501	KISS Custom Chevy Van	1977	20	50	60
462	Laurel & Hardy '27 T Roadster	1976	20	50	60
461	Laurel & Hardy '27 T Touring Car	1976	20	50	60
	Man From U.N.C.L.E. Car	1966	75	175	200
6058	Monkee Mobile (AMT/Ertl)		20	55	65
956	Mr. Spock, large box	1973	20	125	150
	Mr. Spock, small box	1973	20	125	150
901	Munsters Koach	1964	50	150	175
904	My Mother The Car	1965	15	35	40
	Sonny & Cher Mustang		75	250	300
	Threw'd Dude		5	10	15
614	Touchdown?		5	10	15
	UFO Mystery Ship		15	60	75
950	USS Enterprise Bridge, Star Trek	1975	10	25	30
921-200	USS Enterprise w/ lights, Star Trek	1967	40	200	250
951-250	USS Enterprise, Star Trek	1966	40	125	150

Aurora

NAME	DESCRIPTION	YEAR	B/U	NM	MIP
805	Addams Family Haunted House	1964	300	750	800
409	American Astronaut	1967	15	60	75

Top to bottom: United States Sailor (Famous Fighters), 1957; Robin The Boy Wonder, 1966; Glows In The Dark The Phantom Of The Opera, 1972, all Aurora.

Aurora

NAME	DESCRIPTION	YEAR	B/U	NM	MIP
402	American Buffalo	1964	8	20	25
402	American Buffalo, re-issue	1972	8	12	15
401	Apache Warrior on Horse	1960	175	300	450
K-10	Aramis, Three Musketeers	1958	20	75	100
582	Archie's Car	1969	25	85	100
819	Aston Martin Super Spy Car		40	150	200
K-8	Athos, Three Musketeers	1958	20	75	100
832	Banana Splits Banana Buggy	1969	150	350	400
811	Batboat	1968	150	400	450
810	Batcycle	1967	125	350	400
467	Batman	1964	15	200	250
187	Batman, Comic Scenes	1974	15	40	60
486	Batmobile	1966	100	275	325
487	Batplane	1967	75	200	250
407	Black Bear and Cubs	1962	15	30	40
407	Black Bear and Cubs, re-issue	1969	15	20	25
400	Black Fury	1958	10	25	30
400	Black Fury, re-issue	1969	10	13	15
K-3	Black Knight	1956	10	30	35
473	Black Knight, re-issue	1963	10	13	15
463	Blackbeard	1965	75	200	225
K-2	Blue Knight	1956	10	35	50
472	Blue Knight, re-issue	1963	10	17	20
414	Bond, James	1966	150	325	450
482	Bride of Frankenstein	1965	300	650	750
863	Brown, Jimmy	1965	75	150	175
409	Canyon, Steve	1958	75	175	250
480	Captain Action	1966	100	275	300
476	Captain America	1966	85	250	300
192	Captain America, Comic Scenes	1974	30	100	125
464	Captain Kidd	1965	25	70	80
738	Cave Bear	1971	15	35	40
416	Chinese Girl	1957	10	20	25
415	Chinese Mandarin	1957	12	25	30
213	Chinese Mandarin & Girl set	1957		200	300
828	Chitty Chitty Bang Bang	1968	30	85	100
402	Confederate Raider	1959	150	300	350
426	Creature From The Black Lagoon	1963	65	325	400
483	Creature From The Black Lagoon, Glow Kit	1969	65	175	200
483	Creature From The Black Lagoon, Glow Kit	1972	65	100	125
653	Creature, Monsters of Movies	1975	75	175	200
730	Cro Magnon Man	1971	10	30	45
731	Cro Magnon Woman	1971	7	25	35
K-7	Crusader	1959	75	150	200
410	D'Artagnan, Three Musketeers	1966	50	150	175
861	Dempsy vs Firpo	1965	20	75	75
631	Dr. Deadly	1971	25	70	80
632	Dr. Deadly's Daughter	1971	25	65	75
460	Dr. Jekyll as Mr. Hyde	1964	45	250	300
482	Dr. Jekyll, Glow Kit	1969	45	100	150
482	Dr. Jekyll, Glow Kit	1972	45	65	80
462	Dr. Jekyll, Monster Scenes	1971	40	100	125
654	Dr. Jekyll, Monsters of Movies	1975	25	60	70
424	Dracula	1962	25	225	300
466	Dracula's Dragster	1966	125	300	350
454	Dracula, Frightning Lightning	1969	20	300	450
454	Dracula, Glow Kit	1969	20	100	150
454	Dracula, Glow Kit	1972	20	60	75
641	Dracula, Monster Scenes	1971	100	150	200
656	Dracula, Monsters of Movies	1975	100	175	225
413	Dutch Boy	1957	10	25	30
209	Dutch Boy & Girl set	1957		200	300

MODELS

Aurora

NAME	DESCRIPTION	YEAR	B/U	NM	MIP
414	Dutch Girl	1957	10	20	25
817	Flying Sub	1968	35	175	200
254	Flying Sub, re-issue	1975	35	85	100
422	Forgotten Prisoner	1966	65	350	400
453	Forgotten Prisoner, Frightning Lightning	1969	65	325	450
453	Forgotten Prisoner, Glow Kit	1969	65	175	200
453	Forgotten Prisoner, Glow Kit	1972	65	150	175
423	Frankenstein	1961	20	210	250
449	Frankenstein, Frightning Lightning	1969	20	375	400
470	Frankenstein, Gigantic 1/5 scale	1964	400	1000	1200
449	Frankenstein, Glow Kit	1969	20	65	150
449	Frankenstein, Glow Kit	1972	20	50	75
633	Frankenstein, Monster Scenes	1971	50	75	100
651	Frankenstein, Monsters of Movies	1975	100	175	200
465	Frankie's Flivver	1964	150	300	350
451	Frog, Castle Creatures	1966	75	200	250
658	Ghidrah	1975	95	260	300
643	Giant Insect, Monster Scene	1971	95	350	400
469	Godzilla	1964	85	425	500
485	Godzilla's Go-Cart	1966	650	2500	3000
466	Godzilla, Glow Kit	1969	75	200	250
466	Godzilla, Glow Kit	1972	75	150	175
K-5	Gold Knight on Horse	1957	125	250	300
475	Gold Knight on Horse	1965	125	250	275
413	Green Beret	1966	75	150	175
489	Green Hornet 'Black Beauty'	1966	125	350	450
634	Gruesome Goodies	1971	25	80	100
800	Guillotine	1964	125	350	400
637	Hanging Cage	1971	20	80	100
481	Hercules	1965	125	250	275
184	Hulk, Comic Scenes	1974	25	75	85
421	Hulk, original	1966	75	250	300
460	Hunchback of Notre Dame	1964	45	250	300
481	Hunchback of Notre Dame, Glow Kit	1969	45	100	150
481	Hunchback of Notre Dame, Glow Kit	1972	45	65	75
417	Indian Chief	1957	40	90	100
212	Indian Chief & Squaw set	1957		125	150
418	Indian Squaw	1957	15	38	45
411	Infantryman	1957	20	75	100
813	Invaders UFO	1968	35	85	100
256	Invaders UFO	1975	25	65	75
853	Iwo Jima	1966	75	175	200
408	Jesse James	1966	75	175	200
851	Kennedy, John F.	1965	50	100	150
885	King Arthur	1973	100	125	200
825	King Arthur of Camelot	1967	30	65	75
468	King Kong	1964	75	350	400
484	King Kong's Thronester	1966	350	850	1000
468	King Kong, Glow Kit	1969	75	200	250
468	King Kong, Glow Kit	1972	75	150	175
816	Land of the Giants (diorama)	1968	150	360	400
830	Land of the Giants Space Ship	1968	150	300	350
808	Lone Ranger	1967	75	150	175
188	Lone Ranger, Comic Scenes	1974	20	45	50
420	Lost In Space, large kit w/ chariot	1966	450	1100	1300
419	Lost In Space, small kit	1966	300	800	900
418	Lost In Space, The Robot	1968	250	600	700
455	Mad Barber	1972	45	125	150
457	Mad Dentist	1972	45	125	150
456	Mad Doctor	1972	45	125	150
412	Man From U.N.C.L.E., Ilya Kuryakin	1966	75	150	175
411	Man From U.N.C.L.E., Napoleon Solo	1966	75	225	250

Top to bottom: Odd Job from Goldfinger (1966)/James Bond 007 (1966)/Wonder Woman (1965), all Aurora; Huey's Hut Rod from Weird-ohs series, 1963, Hawk; Glows In The Dark The Mummy, 1983, Monogram; Sail Back Reptile from Prehistoric Scenes, 1970s, Aurora.

MODELS

Aurora

NAME	DESCRIPTION	YEAR	B/U	NM	MIP
412	Marine	1959	20	80	100
860	Mays, Willie	1965	100	250	300
421	Mexican Caballero	1957	75	100	150
422	Mexican Seniorita	1957	50	100	150
583	Mod Squad Wagon	1970	35	125	150
463	Monster Customizing Kit #1	1964	35	110	125
464	Monster Customizing Kit #2	1964	65	150	175
828	Moon Bus from 2001	1968	100	275	300
655	Mr. Hyde, Monsters of Movies	1975	25	65	75
922	Mr. Spock	1972	25	100	125
427	Mummy	1963	20	275	300
459	Mummy's Chariot	1965	200	400	450
452	Mummy, Frightning Lightning	1969	20	300	350
452	Mummy, Glow Kit	1969	20	100	150
452	Mummy, Glow Kit	1972	20	50	60
804	Munsters	1964	400	900	1200
729	Neanderthal Man	1971	15	40	50
802	Neuman, Alfred E.	1965	100	250	300
806	Nutty Nose Nipper	1965	45	175	200
415	Odd Job	1966	100	225	250
635	Pain Parlor	1971	25	100	125
636	Pendulum	1971	25	65	75
416	Penguin	1967	200	450	500
428	Phantom of the Opera	1963	20	275	300
451	Phantom of the Opera, Fright'ng Light'ng	1969	20	300	350
451	Phantom of the Opera, Glow Kit	1969	20	100	150
451	Phantom of the Opera, Glow Kit	1972	20	70	80
409	Pilot USAF	1957	75	150	175
K-9	Porthos, Three Musketeers	1958	25	75	100
814	Pushmi-Pullyu, Dr. Dolittle	1968	30	75	85
340	Rat Patrol	1967	30	75	90
K-4	Red Knight	1957	15	75	100
474	Red Knight	1963	15	40	50
488	Robin	1966	40	75	90
193	Robin, Comic Scenes	1974	20	70	85
657	Rodan	1975	125	300	350
405	Roman Gladiator with sword	1959	75	150	175
406	Roman Gladiator with trident	1964	75	150	175
216	Roman Gladiators set	1959		225	250
862	Ruth, Babe	1965	100	250	300
410	Sailor, U.S.	1957	10	25	30
419	Scotch Lad	1957	10	25	30
214	Scotch Lad & Lassie set	1957		85	100
420	Scotch Lassie	1957	10	20	25
707	Seaview, Voyage to the Bottom of Sea	1966	100	250	300
253	Seaview, Voyage to the Bottom of Sea	1975	100	175	200
K-1	Silver Knight	1956	12	45	50
471	Silver Knight	1963	12	18	20
881	Sir Galahad	1973	15	45	50
826	Sir Galahad of Camelot	1967	25	100	175
882	Sir Kay	1973	20	45	50
883	Sir Lancelot	1973	20	45	50
827	Sir Lancelot of Camelot	1967	25	100	125
884	Sir Percival	1973	20	45	50
405	Spartacus (Gladiator/sword re-issue)	1964	85	200	250
477	Spider-Man	1966	85	250	300
182	Spider-Man, Comic Scenes	1974	50	75	85
923	Star Trek, Klingon Cruiser	1972	20	65	75
921	Star Trek, USS Enterprise	1972	20	85	100
478	Superboy	1964	75	225	250
186	Superboy, Comic Scenes	1974	35	50	60
462	Superman	1963	25	275	300

MODELS

534

Aurora

NAME	DESCRIPTION	YEAR	B/U	NM	MIP
185	Superman, Comic Scenes	1974	20	45	50
735	Tarpit	1972	50	100	125
820	Tarzan	1967	25	175	200
181	Tarzan, Comic Scenes	1974	15	30	35
207	Three Knights Set	1959		150	175
398	Three Musketeers Set	1958		300	350
809	Tonto	1967	10	175	200
183	Tonto, Comic Scenes	1974	10	20	25
818	Tracy, Dick	1968	75	200	250
819	Tracy, Dick, Space Coupe	1968	50	125	150
408	U.S. Marshall	1958	50	90	100
864	Unitas, Johnny	1965	75	150	175
452	Vampire, Castle Creatures	1966	60	200	250
638	Vampirella	1971	75	125	150
632	Victim	1971	20	65	75
K-6	Viking	1959	75	200	250
831	Voyager, Fantastic Voyage	1969	150	400	450
807	Wacky Back Whacker	1965	50	200	250
852	Washington, George	1965	25	65	75
865	West, Jerry	1965	50	125	150
401	White Stallion	1964	10	25	30
401	White Stallion, re-issue	1969	10	17	20
403	White Tail Deer	1962	10	25	30
403	White Tail Deer, re-issue	1969	10	17	20
204	Whoozis, Alfalfa	1966	25	65	75
203	Whoozis, Denty	1966	25	65	75
202	Whoozis, Esmerelda	1966	25	65	75
205	Whoozis, Kitty	1966	25	65	75
206	Whoozis, Snuffy	1966	25	65	75
201	Whoozis, Susie	1966	25	65	75
483	Witch	1965	75	250	300
470	Witch, Glow Kit	1969	75	150	200
470	Witch, Glow Kit	1972	75	100	125
425	Wolfman	1962	20	250	300
458	Wolfman's Wagon	1965	175	350	400
450	Wolfman, Frightening Lightning	1969	20	350	400
450	Wolfman, Glow Kit	1969	20	100	150
450	Wolfman, Glow Kit	1972	20	65	75
652	Wolfman, Monsters of Movies	1975	150	200	225
479	Wonder Woman	1965	150	450	500
801	Zorro	1965	125	275	300

Billiken

NAME	DESCRIPTION	YEAR	B/U	NM	MIP
	Batman, type A	1989	35	85	90
	Batman, type B	1989	35	90	100
	Bride of Frankenstein		100	200	225
	Colossal Beast	1986	20	35	40
	Creature From Black Lagoon	1991	50	90	100
	Dracula		60	135	150
	Frankenstein		60	100	125
	Joker	1989	35	85	90
	Mummy	1990	60	135	150
	Phantom of the Opera		125	225	250
	Predator		25	60	65
	Saucer Man		20	35	40
	She-Creature		25	45	50
	Syngenor		100	200	225
	The Thing		150	275	300
	Ultraman		20	35	40

MODELS

Top to bottom; Nutty Mad Organ Player, Marx; Bonanza, 1965, Revell; The Hunchback of Notre Dame, 1964, Aurora; Mr. Gasser, 1963, Revell.

Billiken

NAME	DESCRIPTION	YEAR	B/U	NM	MIP
542	Beach Bunny	1964	25	65	75
532	Daddy the Suburbanite	1963	30	75	85
531	Davy the Way-Out Cyclist	1963	30	75	85
530	Digger and Dragster	1963	30	75	85
	Drag Hag	1963	30	75	85
537	Endsville Eddie	1963	20	50	60
535	Francis The Foul	1963	15	35	40
548	Frantics Banana Skins	1965	20	80	100
550	Frantics Cats	1965	20	70	75
547	Frantics Steel Pluckers	1965	20	70	75
549	Frantics Totally Fab	1965	20	80	100
533	Freddy Flameout	1963	20	65	75
543	Hodad Silly Surfer	1964	20	65	75
541	Hot Dogger Hanging' Ten	1964	20	65	75
538	Huey's Hut Rod	1963	20	45	50
	Killer McBash	1963	40	125	150
534	Leaky Boat Louie	1963	25	80	90
	Riding Tandem		25	65	75
637	Sling Rave Curvette	1964	12	25	30
636	Wade A Minute	1963	12	25	30
	Weird-Oh Customizing Kit	1964	75	250	300
545	Wild Woodie Car		20	50	55
540	Woodie On Safari	1964	25	85	100

Lindberg

NAME	DESCRIPTION	YEAR	B/U	NM	MIP
6422	Bert's Bucket	1971	30	80	90
	Big Wheeler	1964	30	80	90
6420	Fat Max	1971	30	80	90
276	Road Hog	1964	30	80	90
	Satan's Crate	1964	75	125	150
	Scuttle Bucket	1964	30	80	90
6421	Sick Cycle	1971	30	80	90

Monogram

NAME	DESCRIPTION	YEAR	B/U	NM	MIP
6028	Battlestar Galactica	1979	15	35	40
6008	Dracula	1983	20	25	30
105	Flip Out	1965	50	150	175
6007	Frankenstein	1983	20	30	35
6300	Godzilla	1978	40	65	75
6010	Mummy	1983	20	30	35
	Snoopy & Motorcycle	1971	15	25	30
6779	Snoopy & Sopwith Camel	1971	20	30	35
	Snoopy as Joe Cool	1971	25	50	100
MM106	Speed Shift	1965	70	175	200
	Super Fuzz	1965	80	200	225
6301	Superman	1978	20	30	35
6012	UFO, The Invaders	1979	15	35	40
6009	Wolfman	1983	20	30	35

MPC

NAME	DESCRIPTION	YEAR	B/U	NM	MIP
1-1961	Alien	1979	25	75	100
0303	Ape Man Glow Head	1975	10	30	40
	AT-AT, Empire Strikes Back	1980	12	30	35
	Barnabus Vampire Van		75	200	225
550	Barnabus, Dark Shadows	1968	100	300	350
1702	Batman	1984	20	30	35
612	Beverly Hillbillies Truck	1968	60	175	200
0609	Bionic Bustout, Six Million $ Man	1975	12	25	30

MODELS

MPC

NAME	DESCRIPTION	YEAR	B/U	NM	MIP
0610	Bionic Repair, Bionic Woman	1976	12	25	30
5003	Condemned to Chains Forever	1974	20	45	50
103	Curls Girl		25	65	75
	Darth Vader bust	1977	20	45	50
	Darth Vader with light saber	1977	15	35	40
5005	Dead Man's Raft	1974	20	90	100
5001	Dead Men Tell No Tales	1974	20	45	50
1983	Encounter With Yoda diorama	1981	15	30	35
5053	Escape From the Crypt	1974	20	45	50
5004	Fate of the Mutineers	1974	20	45	50
0602	Fight for Survival, Six Million $ Man	1975	12	20	25
0635	Fonzie & Dream Rod	1976	12	30	35
0634	Fonzie & Motorcycle		8	15	20
5007	Freed in the Nick of Time	1974	20	70	75
5006	Ghost of the Treasure Guard	1974	20	45	50
5051	Grave Robbers Reward	1974	20	45	50
402	Hogan's Heroes Jeep	1968	25	85	100
5002	Hoist High the Jolly Roger	1974	20	45	50
101	Hot Curl		20	45	50
	Hot Shot		20	45	50
1932	Hulk	1978	20	30	35
	Jabba's Throne Room	1983	20	35	40
1925	Millenium Falcon with light	1977	35	85	100
605	Monkeemobile	1967	70	170	200
1-0702	Muldowny, Shirley, Drag Kit		20	55	65
0304	Mummy Glow Head	1975	10	20	25
	Night Crawler Wolfman Car	1971	45	100	125
622	Paul Revere & The Raiders Coach	1970	40	100	125
5052	Play It Again Sam	1974	35	90	100
1906	Raiders of the Lost Ark Chase Scene	1982	15	35	40
	Road Runner Beep Beep T		20	65	75
1931	Spider-Man	1978	20	30	35
0902	Strange Changing Mummy	1974	15	35	40
0903	Strange Changing Time Machine	1974	20	45	50
0901	Strange Changing Vampire	1974	20	45	50
100	Stroker McGurk & Surf Rod		30	85	100
102	Stroker McGurk Tall T		30	85	100
1701	Superman	1984	15	20	25
641	Sweathog Dream Machine	1976	7	15	20
5050	Vampire's Midnight Madness	1974	20	45	50
	Werewolf, Dark Shadows	1969	75	200	225
2651	Wile E. Coyote		20	55	65
617	Yellow Submarine	1968	70	170	200

Multiple

955	Automatic Baby Feeder	1965	25	65	75
958	Back Scrubber	1965	25	65	75
956	Painless False Tooth Extractor	1965	25	65	75
957	Signal for Shipwrecked Sailors	1965	25	65	75

Parks

803	Castro, Born Losers	1956	25	65	75
802	Hitler, Born Losers	1956	25	65	75
801	Napolean, Born Losers	1965	25	65	75

Precision

402	Captain Kidd		25	65	75
501	Jesus Christ		20	45	50

MODELS

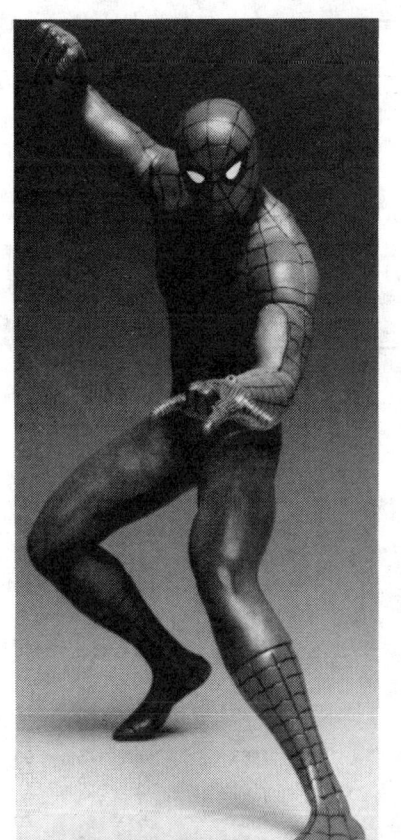

MODELS

Clockwise from upper left: Spiderman, 1980s, Horizon; Daddy the Suburbanite, 1963, Hawk; Escape From The Crypt from Walt Disney's Haunted Mansion series, 1974, MPC; Vampirella, 1971, Aurora; Glows In The Dark Dr. Jekyll As Mr. Hyde, 1972, Aurora.

Pyro

NAME	DESCRIPTION	YEAR	B/U	NM	MIP
166	Der Baron	1958	50	75	100
278	Earp, Wyatt		20	50	60
175	Gladiator Show Cycle		20	40	50
281	Indian Chief		20	50	60
282	Indian Medicine Man		20	50	60
283	Indian Warrior	1960	20	50	60
168	Lil Corporal	1970	25	65	75
276	Rawhide, Gil Favor	1958	20	50	60
277	Restless Gun Deputy	1959	20	50	60
176	Surf's Up	1970	15	35	40
286	U.S. Marsahll		20	50	60

Revell

NAME	DESCRIPTION	YEAR	B/U	NM	MIP
1307	Angel Fink	1965	40	100	125
	Beatles, George Harrison	1965	100	200	250
1352	Beatles, John Lennon	1965	100	200	250
	Beatles, Paul McCartney	1965	100	175	200
1351	Beatles, Ringo Starr	1965	100	150	185
1931	Bonanza	1965	50	125	150
1304	Brother Rat Fink	1963	20	50	55
2000	Cat in the Hat	1960	45	100	130
1397	Charlie's Angels Van	1977	10	20	25
1303	Dragnut	1963	20	50	60
1310	Fink Eliminator	1965	30	175	200
1450	Flash Gordon & Alien	1965	60	125	150
1930	Flipper	1965	75	125	150
	Horton the Elephant	1960	35	85	100
323	McHale's Navy PT-73	1965	25	65	75
1302	Mother's Worry	1963	20	60	75
1301	Mr. Gasser	1963	30	75	90
3181	Mr. Gasser BMR Racer	1964	30	75	90
1451	Phantom & Witch Doctor	1965	50	175	200
1305	Rat Fink	1963	25	60	70
	Rat Fink Lotus Racer	1964	25	65	75
	Robbin' Hood Fink	1965	200	350	400
1309	Scuz Fink	1965	200	300	350
	Super Fink	1964	150	300	350
1306	Surf Fink	1965	35	85	100
1271	Tweedy Pie & Boss Fink	1965	200	350	400

Marx Playsets and Figures

For all practical purposes, the word "playset" could have been invented by Louis Marx...at least as far as boys growing up in the 1950s and 1960s were concerned.

The words "Marx" and "playset" just went together, and they still go together today for many dedicated collectors.

A typical Marx playset included buildings, figures, and lots of realistic accessories that helped bring the miniature world to life. The Fort Apache Stockade, for example, came with a hard plastic log fort, a colorful lithographed tin cabin, and, of course, pioneers and Indians locked in deadly combat. It was no wonder millions of kids had a burning desire for these toys. The play scenarios were almost endless.

This modern version of an age-old toy was a tribute to the marketing/manufacturing talents and whimsical genius of Louis Marx, the modern-day king of toys.

Not only was he responsible for developing the playset, but he popularized the yo-yo and produced some of the most innovative tin wind-ups, guns, dolls, trains, trikes, trucks, and other types of toys that were commercially feasible. In 1955, Marx sold more than $50 million worth of toys, easily making him the largest toy manufacturer in the world.

"The Battle of Little Bighorn" from Marx, 1972.

What makes his domination even more impressive was the fact that Louis Marx rose from humble beginnings. He was born in Brooklyn in 1896 and did not learn to speak English until he started school. At age 16, Marx went to work for Ferdinand Strauss, a toy manufacturer who produced items for Abraham & Strauss Department Stores. By the age of 20, Marx was managing the company's New Jersey factory.

After being fired by Strauss, Marx started contracting with manufacturers to produce toys he designed. By the mid-1920s, Marx had three plants in this country. By 1955 there were more than 5,000 items in the Marx toy line with plants worldwide.

Mass production and mass marketing through chain stores such as Sears and Wards allowed Marx to keep his price levels low and quality high. Marx was also a master at producing new toys from the same basic components. Existing elements could be modified slightly and new lithography would produce a new building from standard stock.

Part of Marx's repackaging genius included using popular TV or movie tie-ins to breathe new life into existing products. The Rifleman Ranch, Roy Rogers Ranch, and Wyatt Earp and Wagon Train playsets were examples of toys produced by Marx where existing parts were repackaged to capture the fad of the day.

Marx enjoyed his recreation as well as his work. He had a table reserved at the "21" nightclub in Manhattan and would hand out toys from oversized pockets in his custom-made suits. He also donated truckloads of toys to churches and other charities.

Believing that toy manufacturing was a young man's business, Louis Marx sold his company to the Quaker Oats Company in 1972 for $51 million. Quaker Oats sold the Marx company four years later for $15 million after losing money every year it owned the company.

With the passing of a few short decades, what were once affordable children's toys have become highly prized collectibles. Playsets are among the price leaders in today's market for childhood treasures. And the figures that went with the playsets are also highly desired by collectors because of their quality and detail.

Most of the figures listed in this section are soft plastic in 54mm and 60mm size and were packaged with various playsets. Other Marx plastic figures are also listed here, such as their larger scale figures and the individually sold hand-painted "Warriors of the World."

A playset listed as MIB (Mint In Box) is as originally sold with all pieces untouched and unassembled in the original box. Excellent (Ex) condition means a complete well-cared for set, but the buildings are assembled and the box may be worn or damaged. Good condition means the playset shows wear and may have a few minor pieces missing.

Trends

The playset market continues on a strong course, with several sets now commanding prices as high as some of their "classic" Marx tin wind-up counterparts. Many sets saw dramatic increases this year, and while part of this is due to differences in review sources, it is clear the playset market had a good year. Expert opinions point to another strong year in 1995.

The Top 25 Marx Playsets (in Mint In Box condition)

1. Captain Gallant of the Foreign Legion, 1956, #4730 ....$2,300
2. Civil War Centennial, 1961, #5929 2,000
3. Johnny Tremain Revolutionary War, 1957, #3402 2,000
4. Ben Hur, #4710 1,800
5. Johnny Ringo Western Frontier, 1959, #4784 1,800
6. Sears Store, (Allstate) 1961, #5490 1,800
7. Gunsmoke Dodge City, 1960, #4268 1,500
8. Untouchables, 1961, #4676 1,500
9. Wagon Train, #4888 1,500
10. World War II European Theatre, #5949 1,500
11. Jungle Jim Playset, 1957, #3706 1,400
12. Zorro, Walt Disney's, 1958, #3754 1,300
13. Adventures of Robin Hood, 1956, #4722 1,250
14. Ben Hur, 1959, #4702 1,250
15. Roy Rogers Rodeo Ranch, 1958, #3986 1,250

Marx's "Captain Gallant Of The Foreign Legion," 1956.

Knight figures, including The Black Knight and Prince Valiant, in soft silver plastic, 54mm, from Marx.

16. Custer's Last Stand, 1963, #4670 1,200
17. Fire House, #4820 1,200
18. Rin Tin Tin at Fort Apache, 1956, #3658...1,200
19. Tales of Wells Fargo Train Set, 1959, #54752..1,200
20. Zorro, Walt Disney's, 1958, #3758 1,200
21. I.G.Y. Arctic Satellite Base, 1959, #4800 1,175
22. Zorro, Walt Disney's, 1958, #3753 1,150
23. Fire House, #4819 1,100
24. Captain Gallant of the Foreign Legion, 1956, #4729....................................1,000
25. Davy Crockett at the Alamo, 1955, #3544..1,000

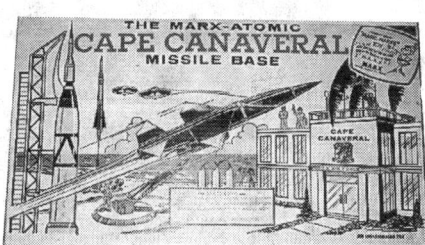

Clockwise from upper left: The Flintstones Play Set, 1962; Official Johnny Tremain Revolutionary War Set; Noah's Ark Miniature Play Set; Zorro Play Set, 1962; Wards Service Center; The Marx Atomic Cape Canaveral Missile Base, all Marx.

MARX SETS

Miniature Playsets

NAME	DESCRIPTION	NO.	GOOD	VG	MIP
20 Minutes to Berlin	1964, 174 pieces		100	300	500
8 Miniature Sports Cars		5930	5	15	25
Attack on Fort Apache	stable, cowboys, Indians		85	225	350
Battleground	1963, 170 pieces	HK-6111	20	60	100
Battleground	170 pieces	HK-6111	45	100	150
Blue and Gray	1960s, 101 individual pieces	HK-6109	90	175	275
Border Battle	Mexican American War		145	250	450
Charge of the Bengal Lancers	British/Arabs		125	325	500
Charge of the Light Brigade	Sears, 216 pieces, Lancers/Cossacks		175	325	500
Charge of the Light Brigade	2nd version, photo box art, Lancers/Arabs		110	300	475
Charge of the Light Brigade	smaller version, Lancers/Cossacks		75	225	350
Charge of the Light Brigade	2nd version, small set, Lancers/Arabs		75	225	350
Civil War Centennial	1961	5929	400	1200	2000
Covered Wagon Attack			85	200	300
Custer's Last Stand	1964, 181 pieces		125	325	500
Disney Circus Parade	Super Circus performers, Disneykins		85	225	350
Disney See and Play Castle	1st and 2nd series Disneykins		150	425	650
Disney See and Play Doll House	1st series Disneykins		100	265	400
Farm Set	farmers and animals		45	115	175
Flintstones	1962, TV Tinykins, Bedrock Village	5948	295	550	850
Fort Apache	1963, 90 pieces, Cowboys, Indians	HK-7526	85	165	250
Fort Apache	90 pieces, Cavalry, Indians		70	175	275
Fort Apache	large set, HQ bldg., Cavalry/Cowboys/Indians		115	295	450
Guerrilla Warfare	1960's, Viet Cong		275	525	800
Invasion Day	1964, 304 pieces		65	130	200
Jungle	smaller than Jungle Safari		85	165	275
Jungle Safari	260 pieces, Hunters/Natives		145	325	500
Knights and Castle	1963, 132 pieces	HK-7563	130	200	300
Knights and Castle	1964, 64 pieces	HK-7562	95	175	275
Knights and Vikings	64 pieces		95	195	300
Knights and Vikings Play Set	1964, 143 pieces		145	275	425
Munchville	vegetable characters		65	165	250
Noah's Ark	1968, 100 pieces		25	65	100
Noah's Ark	Ward's version, soft plastic figures		20	50	75
Over The Top	WW I, Germans/Doughboys		200	525	800
Revolutionary War	British/Colonials		95	250	375
Sands of Iwo Jima	1964, 296 pieces		150	295	450
Sands of Iwo Jima	1963, 205 pieces		115	210	325
Sands of Iwo Jima	1963, 88 pieces		75	145	225
See and Play Dollhouse	American Beauties/Campus Cuties		75	175	275
Sword in the Stone	British only Disney release		300	650	1000
Ten Commandments	Montgomery Ward		150	395	600
Tiger Town	ENCO-like tigers, 1960s		90	225	350

Miniature Playsets

NAME	DESCRIPTION	NO.	GOOD	VG	MIP
Trench Warfare	WW I, Germans/French		200	525	800
Troll Village			80	175	275
Western Town		48-24398	80	175	275
Western Town	1960's	48-24398	20	60	100
Western Town	over 170 pieces	48-24398	50	150	250
Wooden Horse of Troy	British only issue		125	325	500

Disneykin TV Playsets

NAME	DESCRIPTION	NO.	GOOD	VG	MIP
"The Lost Boys" Play Set	2nd Series, 1961		40	120	200
*6 original sets, 6 "New" sets, HO scale, hand painted					
101 Dalmations	1961, "The Barn Scene"		75	195	300
101 Dalmations	1961, "The Wedding Scene"		75	195	300
Alice in Wonderland Play Set	"New" series, 1961		100	225	350
Babes In Toyland	6 different scenes, each		25	65	100
Cinderella	"New" series		100	225	350
Disney 3 in 1 Set	original series		75	225	350
Donald Duck	original series, Donald, Daisy, Louie, Goofy		45	100	150
Dumbo's Circus Play Set	original series		50	100	150
Lady and the Tramp Play Set	"New" series, 1961		100	225	350
Lost Boys	"New" series		100	225	350
Ludwig Von Drake	1962, "The Nearsighted Professor"		50	100	150
Ludwig Von Drake	1962, "The Professor Misses"		50	100	150
Ludwig Von Drake	RCA premium set		65	130	200
Mickey Mouse and Friends Play Set	original series, display box		50	100	150
Panchito "Western" Play Set	original series, display box		50	100	150
Pinocchio	6 different sets, each original series, display box		65	165	250
Pinocchio 3 in 1 Set			115	295	450
Sleeping Beauty Play Set	"New" series, 1961		75	175	275
Snow White and the Seven Dwarfs	original series, display box		50	100	150
Three Little Pigs	"New" series		100	225	350

Fairykins

NAME	DESCRIPTION	NO.	GOOD	VG	MIP
Fairykin Playsets	6 different, each		30	80	125
Fairykin TV Scenes	12 different, each		8	20	30
Fairykin TV Scenes Gift Set	2 different, each with 6 scenes, each		65	165	250
Fairykins 3 in 1 Diorama Set			100	265	400
Gift Box Set of 34	in window box		40	115	175
Hardcover Booklet Boxed Set of 34			65	165	250
TV Scenes Boxed Set of 8			45	115	175

Hanna-Barbera TV-Tinykins

NAME	DESCRIPTION	NO.	GOOD	VG	MIP
*12 diff., each w/5-7 characters					
Flintstones Playsets	3 different, each		75	115	175
Huckleberry Hound Presents Playsets	2 different, each		75	115	175

MARX

Miniature Playsets

NAME	DESCRIPTION	NO.	GOOD	VG	MIP
Quick Draw McGraw Playsets	2 different, each		75	115	175
Top Cat Playsets	3 different, each		75	115	175
TV Scenes	12 different, each		12	35	50
TV-Tinykins Gift Set	set of 34 figures		115	295	450
Yogi Bear Playsets	2 different, each		75	115	175

Playsets

NAME	DESCRIPTION	NO.	GOOD	VG	MIP
Adventures of Robin Hood	1956, Richard Greene TV Series	4722	250	750	1250
Alamo Playset	1960, for 54mm figures	3534	140	250	400
Alamo Playset		3546	100	300	500
Alaska Frontier Playset	1959, 100 piecesc	3708	275	525	800
American Airline Astro Jet Port		4822	150	295	450
American Airlines International Jet Port	1962, 98 pieces	4810	150	295	450
Arctic Explorer	1960, Series 2000	3702	250	450	700
Army Combat Set	Sears, 411 pieces	6019	100	300	500
Army Combat Training Center		2654	20	55	90
Army Combat Training Center		4150	50	150	250
Atomic Cape Canaveral Missile Base	1959	4528	80	240	400
Atomic Cape Canaveral Missle Base Set		2656	90	275	450
Babyland Nursery		3379	125	225	350
Bar-M Ranch		3956	50	100	150
Battle of Iwo Jima	1964, 247 pieces	4147	80	240	400
Battle of Iwo Jima	1964, 128 pieces	6057	35	105	175
Battle of the Blue & Gray	Series 1000, small set, no house	2646	80	240	400
Battle of the Blue & Gray	Series 2000, large set	4658	250	475	750
Battle of the Blue & Gray	1963, Centennial edition	4744	200	575	900
Battle of the Blue & Gray	1959, Series 2000, 54mm	4745	175	375	600
Battle of the Little Big Horn	1972	4679MO	125	250	400
Battlefield	1958, Series 5000	4756	25	95	150
Battleground	1963, Montgomery Ward	3745	80	240	400
Battleground	U.S. and Nazi troops	4169	30	90	150
Battleground	1971, Montgomery Ward	4752	90	275	450
Battleground	1962, 200 pieces	4754	35	110	185
Battleground	1959, 180 pieces	4751	35	110	185
Battleground	1958, largest of military sets	4750	130	395	650
Battleground	1963, Sears, 160 pieces		70	210	350
Battleground	1970's	4756	40	125	250
Beachhead Landing Set	U.S. and Nazi Troops	4939	15	65	100
Ben Hur	Blister Card	2648	25	95	150
Ben Hur	1959, 132 pieces	4696	170	510	850
Ben Hur	1959, Series 2000, medium set	4702	250	750	1250
Ben Hur	Series 5000, large set	4701	350	1075	1800
Big Inch Pipeline	1964, Sears, 201 pieces, 54mm SP	4445	80	240	400
Big Inch Pipeline	1963, 200 pieces	6008	80	240	400
Big Top Circus	1952	4310	80	325	500
Boot Camp	Carry All Tin Box Set	4645	30	130	200

MARX

Playsets

NAME	DESCRIPTION	NO.	GOOD	VG	MIP
Boy Scout Playset			115	450	700
Boys Camp	1956	4103	130	395	650
Cape Canaveral	1960	4524	85	195	300
Cape Canaveral	1959, Sears Set	5963	80	325	500
Cape Canaveral Missile Center	1959	2656	50	150	250
Cape Canaveral Missile Center		4525	50	195	300
Cape Canaveral Missile Set	1958	4526	55	225	350
Cape Kennedy Carry All	1968, Tin Box Set	4625	35	45	75
Captain Gallant of the Foreign Legion	1956	4729/4730	200	600	1000
Captain Gallant of the Foreign Legion	1956	4730	450	1375	2300
Captain Space Solar Academy		7026	80	325	500
Captain Space Solar Academy	1954	7018	65	260	400
Castle and Moat Set	Sears Exclusive	4734	65	260	400
Castle Fort		4710	40	160	250
Cattle Drive	mid 1970's	3983	60	245	375
Comanche Pass	1976	3416	30	130	200
Complete Happitime Dairy Farm	Sears	5957	80	325	500
Complete U.S. Army Training Center	1954	4145	70	210	350
Construction Camp	1956, 54mm, Series 1000	4442	110	325	550
Construction Camp	1954	4439	90	275	450
Cowboy And Indian Camp	1953	3950	90	275	450
Custer's Last Stand	1956, Series 500	4779	80	325	500
Custer's Last Stand	1963, Sears, 187 pieces	4670	195	780	1200
D-Day Army Set	U.S. and Nazi troops	6027	100	300	500
D.E.W. Defense Line Arctic Satellite Base		4802	100	300	500
Daktari	1967, 110 pieces	3717	100	300	500
Daktari	1967, 140 pieces	3720	130	395	650
Daktari		3718	80	325	500
Daniel Boone Frontier Playset		1393	60	230	350
Daniel Boone Wilderness Scout Playset	1964	631	75	225	375
Daniel Boone Wilderness Scout Playset	1964	670	75	225	375
Daniel Boone Wilderness Scout Playset	1964	2640	120	360	600
Davy Crockett at the Alamo		3442	120	360	600
Davy Crockett at the Alamo	1955, Official Walt Disney, 100 pieces. first set	3530	65	260	400
Davy Crockett at the Alamo	1955, Official Walt Disney, Biggest Set	3544	160	650	1000
Desert Fox	1966, 244 pieces	4177	90	275	450
Desert Patrol	1967, U.S., Nazi Troops	4174	60	175	300
Farm Set	1958, 100 pieces, Series 2000	3948	80	250	400
Farm Set		5942	40	160	250
Farm Set		6006	50	195	300
Farm Set		6050	45	180	275
Farm Set, Deluxe	1969	3953	75	225	375

Playsets

NAME	DESCRIPTION	NO.	GOOD	VG	MIP
Farm Set, Lazy Day	1960, 100 pieces	3945	55	165	275
Fighting Knights Carry All	1966-68	4635	45	135	225
Fire House		3779	50	150	250
Fire House	with two friction vehicles	4820	195	780	1200
Fire House		4819	180	715	1100
Flintstones Set	small set	2670	80	240	400
Flintstones Set	1961, 50 pieces	4672	50	195	300
Fort Apache	1952, 60mm	3612	25	70	115
Fort Apache		3616	30	90	150
Fort Apache	1067	3681	45	135	225
Fort Apache	1976	3681	40	120	200
Fort Apache		3682	15	50	85
Fort Apache	giant set	3685	140	425	700
Fort Apache	1970's	4202	15	50	80
Fort Apache	Sears	6059	11	35	55
Fort Apache	1965, Sears, 335 pieces	6063	105	315	525
Fort Apache		6068	35	100	165
Fort Apache	1972, Sears, over 100 pieces	59093C	30	90	150
Fort Apache	1965, Sears, 147 pieces		40	120	200
Fort Apache		3681A	30	90	150
Fort Apache Carryall		4685	15	45	75
Fort Apache Rin Tin Tin		3512	95	285	475
Fort Apache Rin Tin Tin		3616	100	300	500
Fort Apache Rin Tin Tin	early, 60mm	3627	100	300	500
Fort Apache Rin Tin Tin	larger, 60mm	3628	100	300	500
Fort Apache Rin Tin Tin	54mm	3658	90	275	450
Fort Apache Rin Tin Tin	mixed scale set	3957	90	275	450
Fort Apache Stockade	1951	3610	70	210	350
Fort Apache Stockade	1960, 60mm	3660	55	165	275
Fort Apache Stockade	1953	3612	50	155	255
Fort Apache Stockade	1960, Series 2000, 60mm figures	3660	75	225	375
Fort Apache Stockade	1961, Series 5000		55	165	275
Fort Apache with Famous Americans		3636	55	165	270
Fort Dearborn	1952, with metal walls	3510	85	255	425
Fort Dearborn	larger set	3514	20	60	100
Fort Dearborn	with plastic walls	3688	80	240	400
Fort Mohawk	British, Colonials, Indians, 54mm	3751	80	325	500
Fort Pitt	1959, Series 750, 54mm	3741	65	260	400
Fort Pitt	1959, Series 1000, 54mm	3742	70	290	450
Four Level Allstate Service Station		6004	80	325	500
Four Level Parking Garage		3502	40	120	200
Four Level Parking Garage		3511	40	120	200
Freight Trucking Terminal	plstic trucks	5220	30	90	150
Freight Trucking Terminal	friction trucks	5422	30	90	150
Freight Trucking Terminal		5420	20	55	95
Galaxy Command	1976	4206	10	30	50
Gallant Men Army Playset	U.S. Troops	4632	65	260	400
Gallant Men Playset	official set from TV series	4634	70	290	450
Gunsmoke Dodge City	1960, Official, Series 2000, 80 pieces	4268	245	975	1500

Clockwise from upper left: Operation Moon Base; Roy Rogers Rodeo Ranch, 1957; 20 Minutes To Berlin Miniature Play Set, 1964; Yogi Bear at Jellystone National Park, 1962; Official Ben Hur Play Set, 1959; Battleground Play Set (Montgomery Ward exclusive), 1971; Official Davy Crockett At The Alamo, 1955, all Marx.

Playsets

NAME	DESCRIPTION	NO.	GOOD	VG	MIP
Happitime Army and Air Force Training Center	1954, 147 pieces	4159	50	150	250
Happitime Civil War Centennial	1962, Sears	5929	115	455	700
Happitime Farm Set		3480	25	95	150
Happitime Roy Rogers Rodeo Ranch	1953	3990	60	180	300
Heritage Battle of the Alamo Playset	1972, Heritage Series	59091	80	240	400
History in the Pacific	1972	4164	90	275	450
Holiday Turnpike	battery-operated with HO scale vehicles	5230	10	30	45
I.G.Y. Arctic Satellite Base	1959, Series 1000	4800	235	700	1175
Indian Warfare	Series 2000	4748	65	260	400
Irrigated Farm Set	working pump	6021	7	20	35
Johnny Apollo Moon Launch Center	1970	4630	45	135	225
Johnny Ringo Western Frontier Set	1959, Series 2000	4784	290	1170	1800
Johnny Tremain Revolutionary War	1957, Official Walt Disney, Series 1000	3402	400	1200	2000
Jungle Jim Playset	1957, Official, Series 1000	3706	280	850	1400
Jungle Playset	metal trading post, Series 500	3705	110	325	550
Jungle Playset	1960, 48 pieces, Sears, large animals	3716	25	95	150
Knights and Vikings	1972	4743	50	150	250
Knights and Vikings	1973	4733	50	150	250
Knights and Vikings		4773	30	90	150
Little Red School House	1956	3381	90	270	450
Lone Ranger Ranch	1957, Series 500	3969	85	250	425
Lone Ranger Ranch		3980	55	165	275
Lone Ranger Rodeo Set	1953	3696	30	90	150
Marx Masterbuilder Kit	the White House and 35 Presidents		10	35	55
Medieval Castle	1964, Gold Knights, Moat	4704	30	90	150
Medieval Castle	with knights and vikings	4707	35	110	180
Medieval Castle	1959, Sears, Series 2000	4708	120	350	600
Medieval Castle	1954	4709	25	95	150
Medieval Castle	with knights and vikings	4733	30	130	200
Medieval Castle	Sears, with knights and vikings	4734	75	290	450
Medieval Castle	1960, Metallic Knights	4700	75	290	450
Medieval Castle Fort	1953	4710	30	130	200
Midtown Service Station	1960	3420	50	150	250
Midtown Shopping Center		2644	30	90	150
Military Academy		4718	90	275	455
Modern Farm Set	1951, 54mm	3931	50	150	250
Modern Farm Set	1967	3932	60	180	300
Modern Service Center	1962	3471	70	210	350
Modern Service Station	1966	6044	35	105	175
Moon Base	1962	4652	90	270	450
Navarone Mountain Battleground Set	1976	3412	40	115	195
New Car Sales and Service		3466	75	290	450
One Million, B.C.	1970s		40	115	195
Operation Moonbase		4654	75	290	450
Pet Shop		4209	60	230	350
Pet Shop	1953	4210	60	230	350

Playsets

NAME	DESCRIPTION	NO.	GOOD	VG	MIP
Prehistoric	1969	3398	35	105	175
Prehistoric Dinosaur Playset	1978	4208	35	105	175
Prehistoric Times	Series 500	3389	35	105	175
Prehistoric Times		2650	30	130	200
Prehistoric Times		3388	25	75	125
Prehistoric Times	1957, Series 1000, Big Set	3390	60	230	350
Prehistoric Times		3391	20	55	95
Prince Valiant Castle	1955	4705	90	270	450
Prince Valiant Castle	1955, has figures	4706	100	300	500
Project Apollo Cape Kennedy		4523	25	75	125
Project Apollo Moon Landing		4646	50	150	250
Project Mercury Cape Canaveral	1959	4524	90	270	450
Raytheon Missile Test Center	1961	603-A	60	180	300
Real Life Western Wagon		4998	15	45	75
Red River Gang	1970s, mini set with cowboys	4104	35	105	175
Revolutionary War Set	Series 1000	3404	120	490	750
Revolutionary War Set	1959, 80 pieces, Sears	3408	100	390	600
Revolutionary War Set	1957, Series 500	3401	200	600	1000
Rex Mars Planet Patrol		7040	100	300	500
Rex Mars Space Drome	1954	7016	130	395	650
Rifleman Ranch, The	1959	3997	115	455	700
Rifleman Ranch, The	1959	3998	130	395	650
Rin Tin Tin at Fort Apache	1956, Series 5000	3686R	240	725	1200
Rin Tin Tin at Fort Apache	1956, Series 500, 60mm	3628	160	475	800
Rin Tin Tin at Fort Apache	1956, Series 1000, 54mm		180	550	900
Robin Hood Castle Set	60mm	4717	120	360	600
Robin Hood Castle Set	1958, 54mm	4718	80	325	500
Roy Rogers Double R Bar Ranch	1962	3982	100	300	500
Roy Rogers Mineral City	1958, 95 pices	4227	100	300	500
Roy Rogers Ranch		3980	20	60	100
Roy Rogers Rodeo		3689	20	60	100
Roy Rogers Rodeo Ranch	1952	3979	55	165	275
Roy Rogers Rodeo Ranch	1958	3986R	250	750	1250
Roy Rogers Rodeo Ranch	54mm	3988	65	195	325
Roy Rogers Rodeo Ranch	Series 2000	3996	130	395	650
Roy Rogers Rodeo Ranch	1952, 60mm	3985	45	135	225
Roy Rogers Western Town		4216	80	240	400
Roy Rogers Western Town	1952, large set	4258	160	475	800
Roy Rogers Western Town	official, Series 5000	4259	80	235	395
Sears Store	1961, Allstate box	5490	350	1075	1800
Service Station		5459	15	45	75
Service Station	with parking garage	3485	105	315	525

Playsets

NAME	DESCRIPTION	NO.	GOOD	VG	MIP
Service Station	with elevator	3495	30	90	150
Service Station	deluxe	3501	50	150	250
Shopping Center		3755	40	120	200
Silver City Western Town	has Custer, Boone, Carson, Buffalo Bill, Sitting Bull	4220	50	150	250
Skyscraper	working elevator	5449	155	620	950
Skyscraper	working elevator and light	5450	155	620	950
Sons of Liberty	Sears	4170	50	150	250
Space Patrol Rocket Port Set	official	7020	130	520	800
Star Station Seven	1970s		10	30	50
Strategic Air Command		6013	130	520	800
Super Circus		4220	60	180	300
Super Circus	1952, over 70 pieces	4319	80	240	400
Super Circus	1952 with Character Figures	4320	75	290	450
Tactical Air Command	1970s	4106	15	40	65
Tales of Wells Fargo		4263	80	240	400
Tales of Wells Fargo	Series 1000	4264	150	450	750
Tales of Wells Fargo		4262	150	450	750
Tales of Wells Fargo Train Set	1959, with electric train	54752	240	720	1200
Tank Battle	Sear, U.S., Nazi Troops	6056	40	120	200
Tank Battle	U.S., Nazi Troops	6060	40	120	200
Tom Corbett Space Academy	1952, Official, 45mm	7010	170	500	850
Tom Corbett Space Academy	1953	7012	80	325	500
Turnpike Service Center	1961	3460	100	300	500
U.S. Airforce Playset		4807	30	90	160
U.S. Armed Forces		4151	70	210	350
U.S. Armed Forces Training Center	1955, Series 500	4149	85	255	425
U.S. Armed Forces Training Center	Marines, soldiers, sailors, airmen, tin litho bldg, etc.	4144	40	160	250
U.S. Armed Forces Training Center	1956	4158	110	330	550
U.S. Army Mobile Set	1956, flat figures	3655	20	60	100
U.S. Army Training Center		3146	20	55	95
U.S. Army Training Center		3378	20	55	95
U.S. Army Training Center		4122	20	60	100
U.S. Army Training Center		4123	25	75	125
U.S. Army Training Center		4153	15	55	85
Untouchables	1961, 90 pieces	4676	245	975	1500
Vikings and Knights		6053	60	180	300

Playsets

NAME	DESCRIPTION	NO.	GOOD	VG	MIP
Wagon Train	Series 1000, X Team	4805	160	480	800
Wagon Train	official, Series 5000	4888	245	975	1500
Wagon Train	official, Series 2000	4788	120	360	600
Walt Disney Television Playhouse	1953	4350	100	300	500
Walt Disney Television Playhouse	1953, Peter Pan Figures	4352	105	420	650
Walt Disney Television Playhouse	1953	4352	120	360	600
Walt Disney Zorro	1972, Official, Series 1000	3758	160	475	800
Walt Disney Zorro		3753	140	550	850
Wards Service Station	1959	3488	80	240	400
Western Frontier Set			90	275	450
Western Mining Town	1950's	4266	135	405	675
Western Mining Town	1950's	4265	135	400	675
Western Ranch Set		3954	35	105	175
Western Ranch Set		3980	35	105	175
Western Stagecoach Playset	1965	1395	20	60	100
Western Town	single level	2652	60	180	300
Western Town	1952, bi-level town	4229	120	490	750
Westgate Auto Center	1968		40	120	200
White House	house with 8 figures		15	40	70
White House & Presidents	house & figures, 1/48 scale presidents	3920	15	40	70
White House & Presidents	house & figures	3921	15	40	70
Wild Animal Jungle Play Set	large animals	3716	10	30	50
World War II Battleground	1970's	4204	30	90	150
World War II European Theatre	rare big set	5949	245	975	1500
World War II European Theatre	Sears	5939	155	465	775
World War II Set	U.S., Nazi troops	5938	25	75	125
Wyatt Earp Dodge City Western Town	1957, Series 1000	4228	150	450	750
Yogi Bear Jellystone National Park	1962, 60mm	4364	115	350	575
Zorro, Walt Disney	1958, Official, Series 1000	3754	260	775	1300
Zorro, Walt Disney	1958, Official, Series 500	3753	230	695	1150
Zorro, Walt Disney	1958, Official	3758	240	725	1200

Clockwise from upper left: Animal Kingdom figures series; Warriors of the World painted figures; Fairy Tale figures including Humpty Dumpty; Famous Americans figures including Buffalo Bill and Sitting Bull, 1950s; Bavarian painted figures, 1950s, all Marx.

MARX PLASTIC FIGURES

4" Figures

Firemen

*Set of six 4" white or cream figures, valued at $7 each.
Firechief with bullhorn
In boots, overcoat, hat, with fire extinguisher
Running, putting on coat
With axe
With axe in boots, tie
With fire extinguisher
With firehose

5" Figures

Spacemen

*Set of 7 astronauts and 1 alien, with clear plastic helmets, issued in various colors, valued at $10 each.
Advancing with ray pistol
Alien
Crewman turning to fan ray pistol
In uniform, with knife
Kneeling with radio phone
Robot
Running, one foot off ground
Signalman with props
Taking careful aim with ray pistol
Walking carrying geiger counter
Walking with rifle
With flare gun
With geiger counter
With pistol
With radio antenna

54mm Astronauts

*54mm, in metallic silver or blue, valued at $5 each, unless indicated.
Climbing rock, both feet down
Climbing rock, one foot up
In separate bell space suit, $20
On side with camera separate tool box
Sitting in separate capsule
Space Walker
Space walking with wrench
Standing pointing, in Republic spacesuit, $20
Walking

6" Figures

6" Disney Figures

*From the 1960s. Fluorescent blue figures valued at $12; others lower, as indicated.
Bambi, $12
Donald Duck strutting/waddling, $5
Dopey (Snow White), $8
Goofy, $8
Mickey Mouse, left hand waving, $5
Minnie Mouse, $8
Peter Pan, $12

6" Figures

Pinocchio, catch the movie this summer, $12
Pluto, $5
Snow White, $12
Tinker Bell (Peter Pan), $15

Blame-Its

*set of 6, Blame-Its are impish looking children made of light blue soft plastic. Valued as indicated.
I Didn't Break It, $25
I Didn't Get Dirty, $40

Campus Cuties, Series 1

*Figures of college girls made of flesh-colored soft plastic. Series 1 figures valued at $13 each.
Dinner for Two
Lazy Afternoon
Lodge Party
Nightly Night

On the Beach
On the Town
Shopping Anyone
Stormy Weather

Campus Cuties, Series 2

*Figures of college girls made of flesh-colored soft plastic. Series 2 figures valued at $25 each.
Belle of the Ball
Bermuda Holiday
Day at the Races
Night at the Opera
Our Girl Friday
Saturday Afternoon
Touch of Mink
Twist Party

Cavemen Figures

*Six different, from the 1960s, valued at $5 each.
Holding rock in both hands overhead
Running with raised club and knife
Running with raised tomahawk
Standing left hand extended forward
Standing with long spear in right hand
Swinging raised club

Man from U.N.C.L.E. Figures

*set of 6 poses, with copyright on base, blue or gray soft plastic figures from 1960s. Valued at $15 each.
Agent #1, firing rifle
Agent #2, with kepi, pistol, karate chop
Illya Kuryakin
Inspector Waverly pointing
Napoleon Solo
Officer crouching with pistol

Marvel Super Heroes

*set of 6 in fluorescent colors. Valued at $20 each.
Captain America
Daredevil
Hulk
Iron Man
Spider-Man
Thor

6" Figures

Nutty Mad Soldiers
*Caricatures of military personnel. Valued at $200 each.
American Sergeant
British Colonel
Cuban Guerilla
German Officer
Japanese Lieutenant
Russian Officer

Nutty Mads, Series 1
*set of 6, soft plastic caricatures, 1960s. Series 1 and 2 valued at $10 each, unless indicated; Series 3 at $50.
Dippy the Deep Diver
Donald the Demon
Rocko the Champ
Roddy the Hot Rod
Waldo the Weight Lifter

Nutty Mads, Series 2
All Heart Hogan
Bullpen Boo Boo
Chief Lost Teepee, $15
End Zone, $15
Suburban Sidney
The Thinker

Nutty Mads, Series 3
Hippo Crit
Mudder
Now Children
Smokey Sam
U.S. Male

Secret Agents
*Brown soft plastic figures, 1960s. Valued at $10. Sold in UNCLE playset.
Attacking with flashlight and pistol
Bearded with weird pistol
Clubbing with pistol
Firing revolver
Leaving with briefcase in right hand
With pistol and walkie-talkie

60mm Astronauts

orange or white, 12 poses, each $2.
Walking carrying tool box
Walking with bag
Walking with box
Walking with hand on visor
With camera
With flag
With hammer and stake
With mine sweeper
With scoop and bag
With shovel
With square scoop

African Hunters and Natives

Characters
*54mm, Cream colored, valued at $10, except as indicated.
Daktari with stethescope and bag
Jungle Jim with rifle, pointing, $75
Kulu with knife and slung rifle, $75
Missionary with bible, $20
Paula, $20
Tamba, tan colored

Hunters
*54mm, cream colored, valued at $7 each.
Lost hunter, arms at side
Sitting, driving jeep
Standing, shooting rifle
With separate rifle
Woman with rifle, $25

Natives
*54mm, valued at $10 each in red-brown, medium brown or chocolate brown. Waxy dark brown, $4 each.
Chief in top hat, with cigar
Crouching, arms at side
Kneeling, beating separate drum
Marching with rifle on shoulder
Throwing separate spear
Throwing separate spear, with tattoos
Witch doctor in leopard head
With bow, drawing arrow
With spear, hand to cheek
With staff, hand to head

Alamo & Zorro Figures

Mexicans with Sombrero Hats
*54mm, figures from both Alamo and Zorro sets. Blue soft plastic valued at $2 each; blue flat finish at $5.
Advancing with bayoneted rifle across waist
Guard, standing with lance
Lunging with sword
Mounted with lance
Mounted with sword
Running with rifle across chest
Standing shooting rifle
Walking with rifle in right hand
With pistol
With sword overhead

Alamo Figures

Mexicans with Shako Hats
*54mm, reported in three finishes: sky blue valued at $5 each, metallic blue at $10 and cream at $15.
Advancing with rifle across waist
Climbing ladder with rifle
Clubbing with rifle
Mounted with sword and rifle
Standing shooting rifle
Walking with rifle in right hand

American Heroes

Cereal Premiums

*52mm, came in cream and white finishes. Valued at $10 each.
Daniel Boone
Davy Crockett
George Washington
J.P. Jones
N. Hale
P. Henry
Robert Lee
U.S. Grant

Generals

*60mm hard white plastic figures from the 1950s. Valued at $10 each unless indicated otherwise.
Admiral Dewey
Admiral Halsey
Admiral Radford
Colonel Roosevelt, $20
Commodore Perry, $20
General Arnold
General Bradley
General Clark
General Clay
General Doolittle
General Eisenhower
General George Patton
General Grant
General Gruenther
General Jackson
General Lee
General Lemay
General MacArthur
General Marshall
General O'Donnell
General Pershing
General Pickett, $15
General Ridgeway
General Shepherd, $200
General Sheridan, $15
General Smith
General Snyder
General Spaatz
General Stillwell
General Taylor, $15
General Vandergrift, $30
General Washington

Archies

*Characters from the Archie comic strip, valued at $40.
Archie
Betty
Jughead
Veronica

Bavarian Figures

*Hand-painted hard plastic figures valued at $25 each.
*sold as set of 12, in box, $100
Baker standing wearing white hat holding spatula
Night watchman holding lantern
Old gentleman with moustache

Bavarian Figures

Old lady hunched over with age
Shoeman standing wearing leather apron
Short man standing
Short woman
Woman standing wearing apron with stains
Young man standing, smiling, hands in pockets
Young peasant happily drunk standing, arms on pole
Young peasant walking holding pipe and umbrella

Ben Hur Figures

*54mm, unless indicated, these came in cream, brown and grey soft plastic and are valued at $5 each.
Ben Hur, purple, $45; cream $100
Chained slave
Chariot driver with spear
Chariot driver, no armor, curled whip
Chariot driver, with armor, straight whip
Citizen, thumbs down, yelling
Emperor, purple, $45; cream, $100.
Empress, purple, $45; cream $100.
Gladiator kneeling with sword and shield
Gladiator with long sword, square shield
Gladiator with long sword, round shield
Gladiator with long sword, square shield
Gladiator with net and trident
Gladiator with short sword, round shield
Man sitting
Merchant with money box
Slave master with whip
Slave woman with vase
Stopping horse, brown or black, $3
Trumpeter
Woman sitting bathing

Birds

*60mm in various colors. Valued at $3.
Baltimore Oriole
Cardinal
Gold Finch
House Sparrow
House Wren
Parakeet
Scarlet Tanager
Tufted Titmouse

Boy Scouts

*60mm blue or tan figures. Valued at $7 each.
Blowing bugle
Kneeling, making fire
On hands and knees, reading map
Playing basketball
Saluting
Scoutmaster with map and compasslight
Scoutmaster with staff and flashlight
Shooting basketball
Shooting bow, with quiver
Sitting, hands on ground
Sitting, hands on knees
Standing with paddle
With axe
With first aid book and stretcher

Boy Scouts

With rope tying knots
With signal flags

Cavemen

*45mm, came in cream or tan colors. Valued at $2 each.
Crouching with spear
Crouching, skinning rabbit
Walking with club
With club and knife
With rock and flint
With rock overhead

Champion Show Dogs

*1950s soft plastic 60mm scale in brown or white. Valued at $5 each.
Airedale
Basset Hound
Boxer
Cocker Spaniel
Doberman Pinscher
English Bulldog
French Poodle
German Sheperd
Gordon Setter
Pointer

Civil War Figures

Animals
Dead Horse, 54mm, $20
Falling Horse, 54mm, cream, $35, or grey, $50.

Characters
*54mm, cream colored figures valued at $6 each.
Abraham Lincoln
General Robert E. Lee
General U.S. Grant
Jefferson Davis, $12

Confederate Soldiers
*60mm figures from the 1950s valued at $15 each.
Advancing with bayonet and rifle
Cannoneer
Cavalryman riding with sabre
Loading rifle
Marching with rifle and pack
Officer saluting
Standing Officer
Standing shooting rifle

Confederates/Centennial Poses
*54mm figures in light grey generally valued at $5 and dark grey at $3 unless indicated otherwise.
Bayonetting downward
Calling, seated, $25
Clubbing with rifle
Crawling with rifle
Drummer boy running
Falling off horse with sword, $50
Kneeling with binoculars

Civil War Figures

Lying wounded
Mounted with felt hat and sword
Officer with sword and pistol
Running with ramrod and bucket
Shot, dropping rifle
Sitting with arm in sling
Sitting wounded
Stretcher
Stretcher bearer

Confederates/First Issue Poses
*Issued in light grey, generally valued at $1, but the flat finish issues valued at $3.
Advancing with bayoneted rifle
Calling with rifle overhead
Kneeling, shooting rifle

Marching, shouldered rifle
Mounted with kepi and sword
Running with rifle, brim hat
Shot, dropping pistol
Standing, loading rifle
Standing, shooting rifle
Standing, sword overhead
Waving with flag

Union Soldiers
Bugler
Charging with rifle and bayonet
Dispatch carrier
Kneeling, shooting pistol
Marching with rifle
Riding holding banner
Riding officer with sword
Standing, shooting rifle

Union/Centennial Poses
*Dark blue finish valued at $7; light blue at $1; flat light blue at $3.
Bayonetting downward
Calling, seated
Clubbing with rifle
Drummer boy running
Kneeling with binoculars
Lying wounded
Mounted with felt hat and sword
Mounted with kepi and sword
Mounted with whip
Officer with sword and pistol
Running with ramrod and bucket
Shot, dropping rifle
Sitting with arm in sling
Sitting wounded
Stretcher
Stretcher bearer

Union/First Issue Poses
*Light blue figures valued at $1 each; flat finish valued at $3.
Advancing, rifle across waist
Lighting cannon with fire stick
Lying, shooting rifle
Marching, shouldered rifle
Officer with pistol, arm out
Running with rifle across waist
Standing shooting rifle
Standing with flag and pistol
Standing with pistol and bugle
Standing with ramrod across waist

MARX

MARX

Civilian Figures

*54mm cream colored figures valued at $10 each.
Boy with Ping-Pong paddle
Deliveryman with box and paper
Doorman as if holding door
Doorman in overcoat with whistle
Girl with Ping-Pong paddle
Lifeguard with whistle
Man with both arms to front
Man with right hand in pocket, with paper
Newsboy with paper, paper facing downward
Pharmacist with bottle and pen
Policeman
Walking with briefcase and overcoat, $5
With briefcase, hand in pocket
Woman sitting, writing, $20
Woman with child on leash

Construction Workers

*54mm cream colored workmen valued at $2 each.
Barechested, poised with pick
Carrying 3 boards
Digging with shovel
Flagman
Gloved, kneeling with oil can
Jackhammering
Lineman straddling position for pole
Right hand up, directing traffic
Surveying, slouched
Walking with lantern
With folded map
With hammer and chisel

Disneykins, HO Scale

*Disney characters issued in the 1960s. Individual figures generally valued at $5 unless indicated otherwise.
Disneykins TV-scenes store display
Piper Pig, or 2 others, each $50
The Disneykin "Gift Box" Set
Toby Tortoise, $100
Willie the Whale, $75

101 Dalmatians

Anita Radcliff
Colonel, $25
Cruella Deville
Horace Badham
Jasper Badham
Maid
Perdita, 3 poses, each $25
Pongo, 3 poses, each $25
Puppies, 36 different, each $40
Roger
Sgt. Tibbs (cat), $25

Alice In Wonderland

Alice, $7
Mad Hatter, $7
March Hare, $7
Queen of Hearts, $50
White Rabbit, $7

Disneykins, HO Scale

Babes in Toyland

*set of 8 in 4 different poses, each valued at $5.
Soldier with bayonet
Soldier with drum
Soldier with gun
Soldier with trumpet

Bambi

Bambi, $3
Flower the Skunk, $35
Thumper, $35

Donald Duck and Friends

Daisy, $7
Dewey, $7
Donald, $7
Huey, $7
Louie, $7
Uncle Scrooge, $7

Dumbo

Dumbo, $10
Fireman clown, $30
Regular clown, $30
Ringmaster, $10
Timothy, $10

Jungle Book

Baby Elephant "Sonny," $20
Bagheera, $20
Baloo, $20
Colonel Hathi, $20
King Louie, $20
Mowgli, $20
Shere Khan, $20

Lady and the Tramp

Am (cat), $40
Boris, $40
Bull, $40
Doxie, $40
Jock, $40
Lady and the Tramp Kennel Box Set, 12-figure set, $450
Lady, the dog, $40
Pedro, $40
Peg, $40
Si (cat), $40
Toughy, $40
Tramp, the dog, $40
Trusty, $40

Mickey Mouse and Friends

Goofy, $7
Mickey Mouse, $7
Minnie, $7
Monty, $7
Pluto, $7

Panchito Western

Brer Rabbit
Joe Carioca
Panchito
Pecos Bill

Disneykins, HO Scale

Peter Pan

Captain Hook
Lost Boys, $50
Peter Pan
Smee
Tinker Bell
Wendy

Pinocchio

Blue Fairy
Cleo, $50
Figaro
Fowlfellow, $50
Geppetto
Jiminy Cricket
Lampwick, $50
Pinocchio
Stromboli, $50

Sleeping Beauty

Fauna, $30
Flora, $30
Maleficent, $75
Merryweather, $30
Prince Charming, $50
Sleeping Beauty, $40

Snow White and the Seven Dwarfs

Bashful
Doc
Dopey
Grumpy
Happy
Sleepy
Sneezy
Snow White

Doll House Figures

*60mm soft plastic figures from the 1950s, valued at $3 each.
Boy brushing teeth
Boy catching
Boy pitching
Boy with bat
Boy with sand pail
Crawling baby
Girl drying dishes
Girl sitting to play with kitten
Girl standing on toes
Girl to sit at vanity
Governess
Playful kitten
Standing brother
Standing father
Standing mother
Standing sister
Toddling baby in pajamas

English Guards

*60mm silver or cream in 5 poses, valued at $8 each.
Marching with sword on shoulder

English Guards

Standing blowing trumpet
Standing with sword held out

Eskimos, Trappers, Explorers

Eskimos

*Issued in blue and yellow; valued at $7.
Dog sled
Kayak
Kneeling with kayak paddle
On skis with poles
Sled dog to dog sled
Spearing with rope
Walking with spear and fish
With dead seal
With pelt
With sling rifle to dog sled
With snowshoes and rifle
Woman sitting to dog sled
Woman with spear and fish

Explorers

*Issued in grey and blue, valued at $8 each.
Crouching, rope only to hand
Kneeling with pan
Pilot with helmet
Pulling rope, rope separate
With flare gun
With pistol and gold bag

Trappers

*54mm, grey colored, valued at $8 each.
Bearded, gun drawn
In buckskins, walking with rifle in crook of arm
Man drawing against another
Man fighting
Wagon driver
Walking with rifle and lantern
With trap, stocking cap

Fairy Tale Figures

*60mm soft plastic figures in various colors, valued at $5 unless otherwise indicated.
Goldilocks
Goldilocks - Baby Bear
Goldilocks - Mama Bear
Goldilocks - Papa Bear
Humpty Dumpty
Jack & The Beanstalk
Jack Be Nimble
Jack of Jack and Jill
Jill of Jack and Jill
Kitten with mittens
Kitten with no mittens, crying
Little Bo Peep
Little Boy Blue
Little Jack Horner
Little Kitten, Sitting
Little Miss Muffet
Little Red Riding Hood
Little Red Riding Hood - Wolf
Mother Cat
Old Mother Hubbard
Old Mother Hubbard's Dog
Pieman

MARX

Fairy Tale Figures

Simple Simon

Famous Comic Figures

*60mm soft plastic comic characters issued in the 1950s and valued at $8-10 each, unless indicated otherwise.

Blondie Series

Alexander Bumstead
Alexander Bumstead, Revised, $25
Blondie Bumstead
Blondie Bumstead, Revised, $25
Cookie Bumstead
Cookie Bumstead, Revised, $25
Dagwood Bumstead
Daisy

Bringing Up Father Series

Jiggs
Maggie

Dick Tracy Series

B.O. Plenty
Dick Tracy, $12
Gravel Gertie
Junior
Sparkle Plenty

Lil Abner Series

Daisy Mae
Li'l Abner
Mammy Yokum
Pappy Yokum
Salomey Pig

Orphan Annie Series

Orphan Annie
Sandy, Annie's dog

Popeye Series

Olive Oyl
Popeye
Swee' Pea
Wimpy

Snuffy Smith Series

Jug Haid
Loweezie
Snuffy Smith, $12
Sut Tattersall

Farm Animals

*Cream colored 60mm figures from the 1950s, each $4.
Calf
Draft horse
Grazing colt
Lying down cow
Mule with pack
Pig
Standing cow
Standing horse

Farm People

*60mm figures, 1950s. Cream colored figures valued at $3 each; colored versions at $5.
Barefoot boy with big bucket of pig slop, 54mm
Farmer boy with pail, 60mm
Farmer standing with hoe, 60mm
Farmer working with shovel, 60mm
Farmer's wife, 60mm
Hatless, in sun, hoeing, sleeves up, 54mm
Lugging feed sack, 54mm
Man sitting to milk cow, 60mm
Moustached, retrieving chicken,
Scarecrow with arms out, 60 or 54mm
Sitting with gear shift, 60mm
Straw hat, pitchfork, 54mm
Threatening with wrench and clinched fist

Flintstones

*60mm issued in cream and blue. Values double in blue.
Baby Puss, $10
Barney Rubble, $10
Betty Rubble, $10
Construction worker with board, $3
Cop with whistle, $3
Dino, $10
Fireman with axe, $3
Fred Flintstone, $10
Gas attendant with hose, $3
Waiter with glass and menu, $6
Waitress with pad and pencil, $6
Wilma Flintstone, $6

Foreign Legion

Arabs

*60mm red-brown figures valued at $25; chocolate brown at $35; silver at $20.
Kneeling with rifle
Mounted with rifle overhead
Mounted with sword overhead
Standing shooting rifle
Standing with curved sword overhead
Standing with knife in air
Walking with rifle across waist

Characters

Camel, brown, black or tan, $40
Captain Gallant, running with rifle, blue or silver, $15
Cuffy, saluting, blue or silver, $10

Legionaires

*60mm blue or silver figures valued at $7 each.
Guard with rifle on ground
Guard with rifle, saluting
Lying shooting rifle
Marching, shouldered rifle
Mounted with sword overhead
Pointing, with binoculars
Standing shooting rifle
Walking with pistol

MARX

Fox Hunt

Animals

*54mm brown or cream colored, valued at $10 each.
Dog chasing fox
Dog sniffing
Fox outwitting chasing dog
Fox outwitting sniffing dog
Horse jumping, cropped mane and tail
Horse standing, cropped mane

Freight Station Figures

*60mm blue, 5 poses, cream or grey, valued at $8 each.
Checker
Man carrying a crate
Man sleeping on carton
Man with broom
Man with crate hook
Sitting on box, arms crossed
With broom
With freight hook, left hand on bibs
With papers, hand on hat

Gas Attendants

*60mm, in 5 pppposes, issued in blue, yellow and grey, valuod at $8 each.
Changing tire
Crouching on knee
Sitting on box
With rag, waving
With wrench and rag

Ground Crew Figures

*54mm, Cape Personnel, issued in blue, cream or tan, valued at $2 each.
Crouching, with wrench
Officer with hands clasped behind back
Running with firehose, in hood
Standing with firehose, in hood
With fire extinguisher, in hood
With gas tank, in hood
With geiger counter
With microphone looking up
With paper, waving gloves
With wrench pointing up

Howdy Doody

*60mm white figures valued at $35 each; mint green versions at $45.
Clarabell
Dilly Dally
Howdy Doody
Mr. Bluster
The Princess

Knights

*54mm silver soft plastic, valued at $2 each.
Horse, running, $3
Horse, stopping, $3
Horse, walking, $3
Mounted with lance
Pointing, with shield
Stabbing with spear
Standing with lance
Swinging ball and chain
With ax overhead
With shield and sword

Lassie

*60mm, cream colored, valued at $25 each.
Gramps
Jeff
Lassie

Lone Ranger

*60mm, cream colored, valued at $20 each.
Lone Ranger riding
Lone Ranger standing
Tonto standing

Mexican War Figures

*Metallic blue figures valued at $20 each.
Marching with rifle
Running with rifle across waist
Standing shooting
With paper

Military Figures

Air Force

*52mm in metallic blue, grey or cream colors, valued at $2 each.
Bending over with wrench
Crouching with rocket overhead
Pilot crouching with paper
Pilot walking in high altitude suit and helmet
Pilot walking with left hand at chest
Pilot walking, swinging arms
With air hose in right hand
With ammo belt on shoulders
With gas hose across waist
With signal light

British WWII

*54mm, khaki colored figures valued at $15 each.
Advancing with rifle across waist
British with pistol and ammo can
Officer with pistol
Standing shooting rifle
Throwing grenade
With walkie-talkie

French WWII

*54mm, sky blue figures valued at $8 each.

Military Figures

Advancing with rifle across waist
Kneeling shooting rifle
Officer with pistol
Running with machine gun in left hand
Standing shooting rifle
Walking, slung rifle

Frontier Cavalry

*60mm, 8 poses, in light blue, metallic blue or cream, valued as listed.
Kneeling shooting pistol, $10
Kneeling shooting rifle, $20
Mounted shooting pistol, $15
Mounted with sword, $15
Officer with sword and pistol, $7
Officer with sword and pistol, tan, $10
Officer with sword and pistol, light grey, $20
Standing shooting rifle, $10
With pistol and bugle, $12
With sword, no hat, $7
With sword, no hat, light grey, $20
With sword, no hat, tan, $12

Germans WWII

*Light grey figures valued at $1 each; dark grey versions at $2.
Advancing with rifle across waist
Goose-stepping
Kneeling shooting rifle
Lying dead, machine gun across legs
Motorcycle with separate sidecar
Officer pointing down
Running with machine gun in right hand
Running with pistol
Sitting with right arm out
Standing shooting rifle
Throwing grenade
Walking with bazooka across chest
Walking with machine gun across shoulder
With binoculars
With machine gun across waist

Horses

*54mm in various cream, tan, brown or black, valued at $3 unless indicated otherwise.
Cavalry horse
Dead horse, $20
Wagon horse

Japanese

*Khaki colored, valued at $2 each; flat finish valued at $3.
Advancing with rifle across chest
Bayonetting down
Being shot, dropping machine gun
Crouched with rifle
Firing machine gun from hip
Kneeling with radio and pistol
Running with flag
Running with long knife
Running with pistol and sword
Throwing grenade
With hands behind neck
With machine gun waving

Marching Army Band

*60mm vinyl in olive drab green, valued at $7 each.
Clarinet player

Military Figures

Cymbal player
Drum major with baton
Drummer
Trumpet player
Tuba player

Marines

*60mm green versions valued at $1, olive drab green and khaki at $2, blue and grey at $3.
Advancing with rifle across waist
Guard with rifle across chest
Guard with rifle on ground
Guard with rifle on shoulder
Kneeling with pistol and radio
Kneeling with rifle
Kneeling with rifle and ammo clip
Lying with rifle at right side
Marching with flag
Marching with rifle
Marching with rifle on shoulder
Running with machine gun on shoulder
Running with pistol, waving
Running with slung rifle and ammo can
Sitting with right arm out
Standing at attention
Standing at ease rifle on ground
Standing presenting arms
Throwing grenade with tommy gun
Walking swinging arms
With flamethrower
With pistol, in life jacket
With rifle across chest

Military Cadets

*White figures valued at $10 each.
Blowing bugle
Drummer
Guard with rifle along leg
Guard with rifle out from leg
Guard with rifle to front
Marching with flag
Marching with rifle, dress uniform
Marching with rifle, in overcoat
Marching with sword on shoulder
Saluting
Walking to class with notebook

Russians WWII

*Green figures valued at $10 each.
Being shot, dropping pistol
Butting with rifle
Officer with pistol
Running with machine gun in right hand
Standing shooting rifle
Throwing grenade

Sailors

*60mm in white and various blue finishes, valued at $5.
Heaving on rope
In life vest, with pistol
Marching
Marching with rifle on shoulder
Officer at attention
Officer with binoculars
Pulling rope
Scuba diver with knife
Shore patrol on guard with night stick

Military Figures

Signaling with flags
Slung rifle, with arms crossed
Swabbing the deck with mop and bucket
Walking with duffle bag
With arms crossed, slung rifle
With duffle bag
With duffle bag on shoulder
With pistol, hand on helmet
With signal flags
With signal light
With tommy gun and life vest

Soldiers in Training

*60mm olive drab from 1950s, valued at $5.
At parade rest with rifle
Inspecting officer
Kneeling with field glasses
Kneeling with walkie-talkie
Military policeman standing at parade rest
Sergeant reading orders
Sitting to shoot machine gun
Standing at port arms
With mine detector

U.S. GIs

*60mm in shades of green and khaki. So many produced hard to pinpoint values. Generally from $.25-3 depending on colors.
Advancing with bayonetted rifle
Advancing with rifle across chest, with grenade
Advancing with rifle across chest
Advancing with rifle across waist
Attacking with knife
Being shot, dropping pistol
Blowing bugle
Butting with rifle
Carrying ammunition box
Carrying separate wounded GI
Clubbing rifle
Crawling with rifle
Crawling with tommy gun
Crawling wounded, with hand to chest
Crouching with rifle in right hand
Crouching with tommy gun
Flagbearer
Guard with rifle across chest
Guard with rifle on ground
Kneeling shooting rifle
Kneeling with bazooka
Kneeling with binoculars
Kneeling with m.g. on tripod
Kneeling with mortar shell
Kneeling with radio and pistol
Kneeling with shovel
Kneeling with walkie-talkie
Lying shooting rifle
Lying with rifle on bipod
Lying wounded
M.P. with hands behind back
Marching with flag
Marching with rifle on shoulder
Nurse, kneeling with canteen
Officer marching, swinging arms
Officer with right hand up
Running with pistol and ammo can
Running with rifle across chest
Saluting
Sergeant holding paper up
Sergeant marching
Sergeant marching with side arms

Military Figures

Sitting drinking from cup
Sitting paddle to left
Sitting right wrist on knee
Sitting with hands on knees
Sitting with rifle with telescopic sight
Sitting with scoped rifle
Sitting, paddle to right
Soldier marching with rifle and helmet
Squatting with elbow on knee
Standing shooting rifle
Stretcher
Stretcher bearer
Throwing grenade
Throwing grenade, with tommy gun
Walking swinging arms
Walking with slung rifle
Waving, with pistol in right hand
Waving, with walkie-talkie
With artillery shell across waist
With ax, pouch on belt
With binoculars
With box
With gas mask and bayonetted rifle
With gas mask and slung rifle
With mess kit and cup
With mine sweeper
With pistol in right hand
With pistol, waving
With radio and rifle
With rifle across chest
With rifle overhead
With rings for parachute
With rings, gathering separate parachute
With tommy gun across waist
With two gas cans
With walkie-talkie
Wounded GI being carried

U.S. Medical Corps

*60mm olive drab from 1950s, valued at $7 unless indicated otherwise.
Army nurse administering plasma, $10
Being carried by GI
Chaplain with bible, $10
Digging foxhole
Kneeling peeling potatoes
Kneeling with rolled bandages
Kneeling with shovel
Nurse kneeling
Sitting and eating
Stretcher
Stretcher bearer - front, $10
Stretcher bearer - rear
To carry wounded soldier, $10
Wounded for stretcher, $10
Wounded to be carried, $10

WWI Soldiers

*60mm olive drab or light tan, from 1950s, valued at $15 each.
Charging with rifle and gas mask
Marching officer with sword
Marching with rifle and campaign hat
Marching with rifle and overcoat
Throwing a grenade

Paint Your Own Series

Louis Marx, "The Toy King," $25
Simple Simon and Pieman

Paint Your Own Series

English & European Personalities

*60mm white hard plastic, valued at $20 each.
Anthony Eden, $75
Duke of Edinburgh
Duke of Windsor
Pierre Mendes (France), $40
Prince Charles
Princess Anne
Princess Margaret
Queen Elizabeth II
Queen Mother
Winston Churchill, $30

Military Figures

*60mm white figures, valued at $12 unless indicated otherwise.
Admiral Dewey
Admiral Radford
Commodore Perry
General Arnold
General Clay
General George Pickett, $20
General Gruenther
General John Pershing
General Marshall
General O'Donnell
General Phil Sheridan
General Ridgeway
General Robert E. Lee, $20
General Taylor
General U.S. Grant
Marshal Zhukov
Napoleon I, $35

U.S. Political Figures

*36 different, 60mm hard white plastic, valued at $5 each.
Abraham Lincoln
Arthur
B. Harrison
Barry Goldwater, $40
Bobby Kennedy, $50
Buchanan
Charles Percy, $50
Cleveland
Coolidge
F.D. Roosevelt
Filmore
Garfield
George Washington
Grant
Harding
Harry Truman
Hayes
Hoover
Hubert Humphrey, $20
Ike and Mamie Eisenhower
J.Q. Adams
Jackie Kennedy, $75
Jackson
Jefferson
John Adams
Johnson
Madison
Martha and George Washington
Mary Todd and Abraham Lincoln
McKinley

Paint Your Own Series

Monroe
Pierce
Polk
Ronald Reagan, $30
Taft
Taylor
Theodore Roosevelt
Tyler
W.H. Harrison
Wilson

Pirates

*60mm cream, blue, yellow and white figures from the 1950s, valued at $12 each; hand-painted versions at $7.
Charging figure with belaying pin and dagger
Dueling figure
Hands on pistol and sword
Lunging with sword
Peg-leg, with sword and crutch
Searching figure holding oar
Standing figure, hands on belt
With club, knife in teeth
With hand on chest, parrot and sword
With pistol, hand on sword hilt
With shovel digging

Prehistoric Creatures

*60mm, issued in brown, tan, green and grey. Valued at $8 unless indicated otherwise.
Allosaurus
Ankylosaurus
Brontosaurus, $12
Cynognaurus
Dimetrodon
Hadrosaurus
Iguanodon
Kronosaurus
Megatherium, $12
Moschops
Parasaurolophus
Plateosaurus
Pteranodon
Smilodon, $12
Sphenagodon
Stegosaurus
Struthiomimus, $12
Styracosaurus
Trachodon
Triceratops
Tyrannosaurus, flat belly, $12
Tyrannosaurus, sleek
Wooly Mammoth, $12

Prince Valiant & Knights

Characters

*60mm silver figures, valued at $20 each.
Aleta
Boltar
Prince Valiant
Sir Gawain
The Black Knight

Prince Valiant & Knights

Knights

*54mm silver figures valued at $2 each.
Guard with spear
Guard with sword
Mounted with flag
Mounted with separate lance
Mounted with shield and separate lance
Pointing, with shield
Separate lance
Swinging club
Swinging sword, shield with horse head
With sword and round shield
With sword and tricorner shield

Race Track Figures

*54mm cream or blue figures valued at $7 each.
Crouching with arm out
Holding up sign
Man sitting
Sitting with Coke bottle
Waving flag
With binoculars
With gas can and rags
With stop watch in right hand
Woman sitting

Railroad Figures

*45mm cream or grey figures valued at $2 each.
Boy with newspaper
Conductor with lantern
Engineer with oil can
Hobo
Lady with dog
Little boy with toy locomotive
Man with briefcase
Man with newspaper
Porter
Trainman with signal flag

Railroad/Gas Station/ Airport Figures

*HO scale in cream plastic or off-white vinyl, valued at $1 unless indicated otherwise.
1940's paperboy
Abestos suited, extinguishing, $3
Bargain shopper leaving Woolworth's
Clod pushing luggage cart, $2
Crazed body man with chisel, hammer, $2

Executive hurrying into propwash, looking at watch
Fueler dispensing kerosene
Ground crewman parking plane, $2
Hardhat shirking with light load on shoulder, $3
Hiding something in hatbox
Kid in argyle sweater with pencil, "nerd" pose
Mechanic in glasses feeling cowling
Mother and child walking
Mother with child in baby carriage, $2
Motorcycle cop, 40's-50's uniform, Harley, $3
Pilot seated, talking into mike, $4

Railroad/Gas Station/ Airport Figures

Pilot walking in high altitude suit, $3
Pilot walking out of lounge
Policeman halting mother with child in carriage
Secretary waiting with paper sack of groceries
Seductive woman in '50's fashion fur stole
Spy in suit with two parcels awaiting metal detector
Stewardess walking away from propwash
Supervisor with pink slip consoling workman
Sweeping with push broom, improperly
Tiny, tiny girl with ice cream cone, $2
Workman in coveralls with wood crate, weeping

Religious Figures

*60mm white or cream colored, valued at $8 each.
Andrew
Bartholomew
James the Greater
James the Less
Jesus, left hand raised in a blessing, $35
Jesus, right hand raised in a blessing, $15
John
Judas
Jude
Matthew
Matthias
Paul
Peter
Philip
Simon
Thomas

Revolutionary War

*60mm white or cream valued at $8; hand-painted versions at $5.
Drummer
Fife player
In winter dress
Marching drummer
Marching fifer
Marching soldier with flag
Marching soldier with rifle
Marching with rifle on shoulder
Mounted, pointing, in overcoat
Officer with walking stick at side
Standing officer with sword
Standing sentry with rifle
Walking with rifle in right hand
With flag

Continental Soldiers

*Blue figures valued at $2; flat finish at $3.
Bayonetting
Drummer
Kneeling shooting rifle
Marching with rifle on shoulder
Mounted, arms at side
Running with rifle in right hand
Standing shooting rifle
Walking with pistol
With flag
With sword, left arm up

Revolutionary War

English

*Red figures valued at $3; flat finish at $5.
Kneeling shooting rifle
Marching with rifle on shoulder
Mounted with sword
Officer afoot with sword, right arm up
Running with rifle at waist
Standing shooting rifle

Johnny Tremain Figures

Celia, $20
James Otis, cream color, $20
Johnny Tremain, cream color, $20
Paul Revere riding, white or cream, $20
Samuel Adams, $20

Rex Mars Space Figures

*60mm figures, valued at $15 each.
Crouching with ray pistol with helmet
Helmets, $6
Kneeling with space phone with helmet
Robot
Running with helmet
Standing, shooting flare gun with helmet
Walking with geiger counter with helmet
Walking with space rifle with helmet
Wiring equipment with helmet

Robin Hood

Characters

*Came in 54mm and 60mm sizes; valued as indicated.
Friar Tuck with scroll, 54mm silver $6, other versions $15
Friar Tuck with staff, 60mm green or brown, $20
Little John with staff across chest, 60mm green or brown $20
Little John with staff, $10
Maid Marian, 54mm cream or silver $10, 60mm versions $20
Minstrel with harp, 54mm cream $10, other versions $6
Robin Hood shooting bow, 60mm, $20
Robin Hood with bow and sword, 54mm silver $6, cream $10
Robin Hood with bow and sword, 60mm $20
Sheriff of Nottingham with sword, 54mm silver $6, cream $10
Sheriff of Nottingham with sword, 60mm $20
Sheriff of Nottingham with sword and knife, 60mm, $20
Stag, tan, $810

Knights

*60mm silver figures valued at $7.
Mounted with club and shield, $15
Mounted with separate lance, $15
Mounted with shield and separate lance
Running with sword and shield
Standing, shooting crossbow
With axe overhead in both hands
With club and shield

Robin Hood

With shield, sword overhead
With spear across waist
With sword and knife
With sword and long axe

Merry Men

*54 and 60mm versions; 54mm valued at $3 each, 60mm at $5 each.
Blowing horn
Blowing horn with axe
Calling, hand on hip
Carrying rock
Jumping down with sword and knife
Lunging with knife
Lunging with sword and knife
Running with chest and bow
Shooting bow
Squatting with bow
Standing with sword
Walking with bow and arrow
With bow and arrow
With bow, drawing arrow
With staff and knife
With staff overhead

Roy Rogers

*60mm cream colored figures from the 1950s, valued at $10 each unless indicated otherwise.
Dale Evans with hat in hand
Lt. Rip Masters with pistol, $20
Pat Brady, hand on hat, $12
Rin Tin Tin and Bullet
Roy Rogers mounted waving
Roy Rogers standing with hands on holsters, $15
Roy Rogers standing with pistols
Rusty with rifle, $25

Royal Canadian Mounted Police

*60mm figures from the 1950s, valued at $20 each.
Mountie riding
Mountie standing
Mountie walking
Mountie with binoculars

Sea

*1950s figures in silver, grey, cream or metallic blue, valued at $7.
Barracuda, $15
Marlin
Porpoise
Rainbow Trout, $10
Sailfish
Sea Horse
Shark
Skin Diver with Equipment, $40
Swordfish
Tuna

Spacemen, Aliens and Cadets

Aliens

*45mm tan, grey or blue, valued at $8 each.
Lying with pistol, $20
Robot, walking
With arms crossed
With big ears, waving
With face mask and pistol
With pistol
With translator

Cadets

*45mm tan, grey or blue, valued at $4.
Arms in air
Climbing ladder
Crouching with pistol
Fighting
Kneeling with pistol
Looking through sextant
Pushing something
Running wearing cap
Sitting with book in lap
Sitting, hand to head
Sitting, no cap
Sitting, wearing cap
Standing at attention
Standing hands on belt
Throwing punch
With microphone
With paper in right hand
Woman climbing ladder
Woman wearing cap
Woman with hands on hips

Spacemen

*45mm tan, grey or blue figures valued at $3; cream colored versions at $6.
Arms around wounded
Carrying equipment
Carrying walkie-talkie near head
Floating with light
Helmet for figures, clear plastic
Holding up hoops
Kneeling with left hand to head
Looking through device
Pointing
Putting on helmet
Throwing punch
Walking with geiger counter
Walking with large pistol
Walking with small pistol
With hose
With large pistol
With small pistol left arm out
Woman walking
Wounded, hand at chest

Spanish American War Rough Riders

*Brown 60mm figures, valued at $20 each.
Marching with rifle
Mounted, sword overhead
Officer standing
Running with rifle across waist
Running, rifle across waist
Standing, shouldered rifle

Sports Figures

*60mm in cream, white, light blue and yellow, valued at $10.
Baseball batter
Baseball pitcher
Basketball player
Bowler
Boxer, jabbing with left
Boxer, jabbing with right
Figure skater
Football player
Golfer
Hockey player
Polo player
Runner
Skiing
Soccer Player
Swimmer
Tennis player

Football Players

*54mm Dallas Cowboy figures red ot gray, from 1967, valued at $5 unless indicated otherwise.
Ball carrier stiff-arming #22
Coach, Tom Landry, $12
Defensive end slow getting up #78
Defensive left cornerback, incredulous #40
Defensive left linebacker angered at referee #66
Defensive receiver #42
Defensive right end misreading sweep #73
Defensive right linebacker signaling #61
Defensive tackle in 3-point stance #63
Defensive tackle in 3-point stance #72
Down center, with ball #51
Down guard #60
Down tackle # 70
End going out of bounds #85
Fullback, slow getting up #33
Going out for pass #87
Guard pulling #65
Holding football, #44, $8
Middle linebacker #67
Offensive flanker back yelling #23
Place kicker #27, $7
Punting, jersey #22
Quarterback in shotgun calling signals #16, $7
Quarterback passing, #14, $8
Referee signaling a clipping penalty
Referee signaling a touchdown
Referee with watch and whistle
Referee with whistle and penalty flag, $8
Tackle cross-blocking #62

Super Circus

*45mm, issued in yellow, orange-brown or tan, valued at $4 each.
Acrobat lying on back
Balloon vendor
Barker with cane
Boy carrying bucket
Boy lifting tent
Clown cop
Clown walking, hands on hips
Clown with hole in hat
Clown with umbrella
Elephant trainer with hook
Fat lady
Father with son
Hula dancer
Lion tamer with whip and gun

Super Circus

Man hanging by knees with rope
Man on stilts with cane and hat
Man on trapeze
Mother with daughter
Mr. Tom Thumb with cane
Mrs. Tom Thumb with umbrella
Policeman grabbing boy lifting tent
Popcorn vendor
Ringmaster, holding top hat
Siamese twins
Snake charmer
Strongman with barbells
Sword swallower
Woman hanging by neck
Woman on trapeze
Woman rider for horse
Woman sitting, waving (for elephant howdah)
Woman with arms out
Woman with hands on hips
Woman with stick across waist

Animals

*Issued in grey or green, valued at $4 unless indicated otherwise.
Baby bear
Baby giraffe
Buffalo
Camel
Crocodile
Dog for barrel act
Dressed monkey
Elephant sitting
Elephant walking
Large bear
Large giraffe
Leopard
Lion sitting, right paw out, $8
Monkeys (6 poses), $20 for all
Performing bear, hands up
Performing dog, in clothes, walking
Performing dog, in clothes, sitting
Performing dog, in clothes, sitting with cane
Performing gorilla, hands up, $6
Performing horse, prancing, $8
Performing seal with ball on nose, $8
Tiger
Zebra

Characters

*Off-white figures valued at $10 each.
Cliffy Clown
Juggler
Mary Hartline
Nicky Clown
Ringmaster Kirchner
Scampy Clown

Train Depot Figures

*45mm grey figures, valued at $3 each.
Businessman walking with paper, briefcase
Businessman with paper and pipe
Conductor with lantern, glasses fogged, checking water
Flagman
Newsboy peddling papers
Oiler stopping
Porter, carrying bags
Small boy with toy engine, waving

Train Depot Figures

Tramp, staggering with sack and stick
Woman waving with dog on leash

TV-Tinykins

Hokey Wolf, $12
Pixie Mouse, $15
Quick Draw McGraw, $19

Flintstones

Baby Puss, $25
Fire Chief, $25

TV-Tinykins, Hanna-Barbera

*HO scale, 36 different, valued at $12 each.

Flintstones

*HO scale, 8 different, valued at $10 each.

Untouchables

Characters

Al Capone, 54mm cream color, $20
Eliot Ness, 54mm cream color, $20

Cops and Robbers

*54mm flat finish versions valued at $12.
Being shot, dropping hat and pistol
Cop shooting machine gun
Cop shooting pistol
Crouching with pistol
Handcuffed with hands behind back
In shirt with pistol and shoulder holster
In trenchcoat with machine gun
Lady 'flapper' with purse
Lying dead
Pointing pistol, wearing mask
Pulling shotgun from golf bag
Reaching right hand in coat
Running with pistol and axe
Shooting down with pistol
Surrendering, hands in air
With machine gun across waist

Vikings

*54mm, green, valued at $1 each; flat finish versions at $5.
Axe overhead in both hands
Running with knife and shield
Running with sword and ax
Swinging ax with both hands
Walking with sword on ground
With knife and shield
With spear and shield
With sword and round shield

West Point Cadets

*60mm, issued in white and grey, valued at $5 each.
Marching in full dress with sword

West Point Cadets

Marching in full dress with banner
Marching in full dress with rifle
Marching with overcoat and rifle
Standing at attention

Western Figures

Boonesborough Pioneers

*54mm caramel color, valued at $15 each.
Clubbing with rifle
Daniel Boone with rifle across chest
Running with rifle, felt hat
Shot with arrow in shoulder
Shouldered rifle with pheasant in right hand
Standing shooting pistol, knife in left hand
Standing shooting rifle
Walking with knife poised and axe
Walking, rifle butt on ground, left hand raised

Cavalry

*54mm, turquoise versions valued at $1; silver, tan and yellow at $2.
Butting down with rifle
Clubbing with pistol
Kneeling shooting rifle
Marching, rifle on shoulder
Mounted blowing bugle
Mounted with flag
Mounted with flag and pistol
Mounted with hand on belt
Mounted with sword
Mounted, arms at sides
Running with pistol
Running with rifle in right hand
Running with rifle, hand on hat
Running with sword
Shot, dropping pistol
Sitting with whip, wagon driver
Standing shooting rifle
Standing with rifle
Standing, arms to back
Walking, swinging arms
With pistol and bugle
With pistol, arrow in chest
With rifle across chest
With rifle, blowing bugle
With sword, holding scabbard

Characters

*54 and 60mm cream colored figures, valued at $10 unless indicated otherwise.
Bullet, $5
Chester, $150
Custer with pistol, waving, turquoise, $100
Dale Evans waving, $5
Davy Crockett, $15
Doc, $150
Flint McCullough, $50
Jim Hardy, $75
Johnny Ringo, $300
Kitty, $150
Lone Ranger afoot, drawing pistol, $10
Lone Ranger mounted shooting pistol, $8
Lone Ranger mounted with pistol, $15
Lt. Rip Masters, $25
Lucas McCain, $60
Major Seth Adams, $50
Mark McCain, $50

Western Figures

Matt Dillon, $200
Pat Brady with hammer, $5
Rin-Tin-Tin, $15
Roy Rogers, mounted with pistol, 54mm
Rusty, $35
Tonto with feather, 60mm
Tonto, no feather, 54mm, $20
Wyatt Earp, $25

Cowboys

*54mm and 60mm, various colors, valued at $2-3 depending on color.
Arms forward, standing on fence
Arms out, leg up mounting horse
Bad man knocked down
Bad man with hands up
Bandit fanning pistol
Bandit with arms in air
Bandit with pistol and bag
Bandit with pistol and strongbox
Being punched, wearing scarf
Boy on knees, wrestling
Clenched fists, head turned
Drawing pistol
Fighting cowboy
Fighting, no hat
Fighting, with hat
Hands in air
Hands to mouth, calling
Kneeling cowboy with carbine
Kneeling with branding iron
Kneeling with separate rifle
Lady with bonnet and basket
Mounted bandit shooting pistol
Mounted cowgirl waving hat
Mounted falling off horse
Mounted left hand waving
Mounted shooting rifle
Mounted waving hat
Mounted with pistol
Mounted with right arm at side
Mounted with rope
Mounted, right arm out
Mounted, with rope
Pointing pistol, no hat
Riding cowboy for bucking bronco
Riding cowboy shooting to rear
Riding cowboy with lasso
Right hand on belt, left arm out
Running bad man shooting to rear
Running with pistol
Running with rifle
Seated with hat, waving
Sheriff in hat with pistol
Sheriff with gun and money sack
Sheriff with pistol and bag
Sheriff with pistol drawn
Sheriff with two pistols
Sheriff with two pistols drawn
Sitting on box with guitar
Sitting on ground, being punched
Sitting wagon driver
Sitting with arm out, leg up
Sitting with right arm up
Standing cowboy with harness
Standing in lasso
Standing punching
Standing with rope
Standing with whip
Turned, fanning pistol
Turned, shooting pistol

Western Figures

Walking with arms at side
Walking with harness
Walking with rope
Walking with saddle
Walking, swinging arms
With branding iron and rope
With cradled rifle across waist
With rope
With two pistols drawn
Woman with basket
Woman with hands on hips
Woman with right arm out

Early Frontier

*60mm Figures
Man crouching with pistol, $5
Man kneeling shooting rifle, $5
Man running with musket, $5
Man smashing with rifle butt, $5
Woman loading musket, $5

Indians

*45, 54 and 60mm, flesh colored valued at $1; others at $3 unless indicated otherwise.
Advancing with tomahawk overhead
Charging with rifle and tomahawk
Chief dancing with rattles, red or light rust version $5
Chief sitting with pipe
Chief sitting with right hand extended, $5
Chief with knife
Chief with knife and tomahawk
Chief with raised club
Chief with rifle
Climbing wall with club
Climbing wall with tomahawk
Crawling with knife
Crawling with knife and tomahawk
Crawling with tomahawk
Crouched with tomahawk, arrow in side
Dancing with bow and flail
Kneeling beating drum
Kneeling shooting bow
Kneeling shot with arrow
Medicine man with drum and beater
Mounted chief waving
Mounted chief with feathered lance
Mounted with bow, drawing arrow
Mounted with rifle and knife
Mounted with spear
Rider with spear
Running with tomahawk and rifle
Scout rifle in left hand
Shooting arrow, bow on ground
Shooting bow, Mohawk haircut
Squaw crawling with baby
Squaw sitting on knees, $5
Squaw standing with bowl, $5
Squaw walking with papoose on back, $5
Stabbing with knife
Standing chief with bow, lance and shield
Standing shooting bow
Standing shooting rifle
Standing with bow
Throwing spear
Throwing spear with shield
With bow and club
With bow, drawing arrow
With knife in right hand
With spear and shield
With tomahawk

Western Figures

With tomahawk and rifle
With tomahawk and scalp, $10
With tomahawk and shield
With tomahawk overhead
With tomahawk, bow and shield

Long Coat Cavalry

*54mm sky blue, valued at $10 unless indicated otherwise.
Mounted blowing bugle
Mounted turning arms down
Mounted with flag, $6
Mounted with pistol
Mounted with sword in air

Miners and Trappers

*54mm butterscotch color valued at $5 each; tan, brown or grey versions at $7.
Standing with fur
Walking with cradled rifle
Walking with pan on stick
With lantern, carrying separate sack, $15
With pistol and bag
With rifle and lantern
With shovel and jacket

Pioneers

*54mm most colors valued at $1-3; brown, grey, cream at $5; metallic blue at $8.
Butting with rifle
Clubbing with rifle
Kneeling shooting rifle
Mounted bandit with pistol
Mounted with sword overhead
Running with axe and rifle
Running with rifle in buckskins
Running with rifle in left hand
Sitting on ground with pipe
Standing shooting rifle
Walking with rifle and turkey
Walking with rifle in right hand
With pistol and powder horn
With pistol, no base
With rifle across waist
With rifle and powder horn
With rifle, no base
With sword and bugle
Woman loading rifle

Zorro

*54mm, cream colored versions (and black Zorro) valued at $20 each; grey versions at $35.
Bernardo with lantern
Don Alejandro with hat in hand
Don Diego with walking stick
El Commandante with sword
Sergeant Garcia with hands on belt
Zorro, mounted with sword, in cape

Barbie & Friends

By Marcie Melillo

The Barbie doll was issued in 1959, and after thirty-five years is one of the most popular toys of all time. Mattel was somewhat innovative for the early 1960s, and extensively used television to advertise its merchandise, particularly the Barbie doll. A tremendous marketing campaign was launched that displayed Barbie dressed for dating, school, and a variety of careers. The seeds were planted in childrens' minds and imaginations shifted into overdrive. There was a quick response to these television commercials and the toy became a phenomenon.

Modern American culture is uniquely chronicled through the doll's beautifully detailed wardrobe. What a range of styles! From the early 1960s Jackie Kennedy suits with pill box hats to the wild mod fashions. Remember the 1970s peasant dresses and shiny polyester disco attire? The Barbie doll wore all of those styles and currently is wearing Benison separates and carrying a Caboodle's bag. Truly impressive is the line of Bob Mackie sequined gowns that are waiting for the next opening night extravaganza. This is a wardrobe that would make Imelda Marcos jealous.

Today, the Barbie doll, her friends, and accessories, whether new or old, are highly collectible. One reason for the popularity of collecting and selling Barbie dolls is that they were favorite playthings for most Baby Boomers. These "grown-ups" are now attempting to replace their discarded toys and recapture a part of their childhood. Many are also interested in the investment potential of Barbie paraphernalia.

Of course, not all Barbie dolls increase in value. Predicting which dolls will rise, fall, or remain static is work for the Las Vegas odds makers. The 1970s dolls and accessories were of poorer quality than the 1960s dolls and thus, slow to increase in value. The limited edition dolls released in the 1980s and 1990s show mixed results dependent upon the size of the production run.

Mattel's "Dramatic New Living Barbie," 1970.

The 1988 Holiday Barbie doll is a perfect example of the effect that size of release has on price appreciation. First in a series, the doll has rapidly increased in value both because of its beauty and limited issue. Due to the success of the 1988 doll, the 1989-1992 Holiday Barbie dolls were released in large quantities. While a popular series, subsequent dolls have not matched the meteoric rise of the 1988 Holiday Barbie, which retailed for $29.95 and now fetches $500 plus in the secondary market.

A couple of years ago, a problem arose for collectors and dealers alike. Mattel reissued three dolls from the International Barbie series. These dolls were packaged differently, and the boxes were easily identifiable as a reissue, but collectors perceived the originals as less valuable and prices began to drop for the entire group. Additional dolls from the series have been reissued and the valuation on the entire group remains depressed. Until the reissue, the International Barbie dolls had increased in value at amazing speed. Whether the reduction is permanent or temporary is anyone's guess.

The Teen Talk Barbie is another doll worthy of mention. Mattel issued an estimated 350,000 of these talking dolls in 1992 for a retail of $34.99. Approximately one percent have a voice chip that says, among other things, "Math is tough." This phrase was met with disdain by many, so the use of the chip was halted. In fact, Mattel offered a refund for any doll voicing the offending phrase. Well, returning that Barbie would have been a costly mistake for its current value on the secondary market is approximately $350. New dolls that become valuable quickly are one reason why this collectible is so popular.

Before using the pricing guidelines in this section, a few terms should be clarified. The values given are for dolls Mint In Box (MIB), defined as never played with and still in the original boxes with all accessories, and Mint No Package (MNP), defined as mint condition dolls and outfits missing only the original packaging.

The earliest Barbie dolls were packaged in two-piece boxes and held in place

Mattel's "The Yachtsman" outfit for Ken doll, 1962.

by a cardboard liner. The doll could be taken out of the liner with no difficulty. Thus, the popular Not Removed From Box (NRFB) term that is frequently used with the older Barbie dolls is accurate only for later issue dolls that were sealed in the box. For consistency, the MIB term is used throughout this portion of the price guide and it indicates a doll in the condition that it left the Mattel factory. You may expect to receive between fifty and seventy percent of the book value when selling to a dealer. Fifty percent is the norm with the higher percentage reserved for items of great value and scarcity.

Pricing Barbie dolls that are not MIB or MNP is usually left to the parties involved. A doll that has been played with may have a myriad of flaws, i.e. green ears, neck splits, missing fingers, hair cut or missing or re-styled, etc., thus pricing is difficult and usually reached by hard fought mutual agreement.

Remember that MIB and MNP condition dolls and outfits are rare and the price for these toys reflects that scarcity. Played with dolls and fashions are plentiful and command a small percentage of the MIB or mint price.

The following is a brief analysis of the investment side of this collectible:

Market Update 1995
Overall Annual Increase For 1960's Dolls 15%
Overall Annual Increase For 1960's Fashions 10%
Overall Annual Increase For 1970's Dolls 1/2%
Overall Annual Increase For 1980's Dolls 0%

Trends

Collectors are concentrating on 1960s dolls and purchasing Mattel's new lines of high end collector dolls. Some of these dolls have experienced a rapid increase in value. For example, the 1990 Gold Bob Mackie retailed for around $200 and now is selling for $500 plus on the secondary market. Also, vintage Barbie dolls and clothing in mint condition continue to increase in value at a respectable rate each year.

What to Avoid

Avoid regular issue Mattel dolls from 1970s to present. With few exceptions, these dolls show flat returns on investment. This is projected to continue for many years because of the poor quality of the 1970s dolls and the enormous production runs of the 1980s.

The Top 10 Barbie Dolls
(in MIB condition)

Doll	Year	#	Price
Ponytail Barbie #1, Brunette	1959	#850	$4,500
Ponytail Barbie #1, Blonde	1959	#850	3,500
American Girl, Side Part	1965	#1070	3,500
Ponytail Barbie #2, Brunette	1959	#850	3,000
Ponytail Barbie #2, Blonde	1959	#850	2,200
Color Magic Barbie, Black	1966	#1150	2,200
Pink Jubilee Barbie	1989		1,900
American Girl Barbie, Color Magic Face	1965		1,800
Color Magic Barbie, Blonde	1966	#1150	1,300
American Girl Barbie, Brunette	1966	#1070	1,100

The Top 10 Barbie Fashions
(in NRFB condition)

Fashion	#	Price
Roman Holiday	#968	2,500
Easter Parade	#971	1,500
Pan American Stewardess	#1678	1,320
Gay Parisienne	#964	1,000
Shimmering Magic	#1664	950
Here Comes The Bride	#1665	825
Gold 'N Glamour	#1647	750
Commuter Set	#916	715
Riding In The Park	#1668	715
Saturday Matinee	#1615	715

Top to bottom: Teen Talk Barbie, 1992; #1 Blonde Barbie ponytail wearing rare 1959 Easter Parade, 1959; Color Magic Barbie (left), 1966, and Fashion Queen Barbie with Color Magic Wig, 1966, all Mattel.

Barbie Vintage Fashions 1959-1966

NO.	OUTFIT	MNP	MIE
1631	Aboard Ship	140	275
934	After Five	55	140
984	American Airlines Stewardess	70	190
917	Apple Print Sheath	40	140
0874	Arabian Nights	110	220
989	Ballerina	50	165
953	Barbie Baby-Sits	80	190
1605	Barbie in Hawaii	70	140
0823	Barbie in Holland	80	190
0821	Barbie in Japan	220	440
0820	Barbie in Mexico	80	190
0822	Barbie in Switzerland	70	165
1634	Barbie Learns to Cook	110	275
1608	Barbie Skin Diver	40	90
962	Barbie-Q Outfit	70	165
1651	Beau Time	50	110
1698	Beautiful Bride	385	660
1667	Benefit Performance	385	660
1609	Black Magic	70	165
947	Bride's Dream	90	190
1628	Brunch Time	100	220
981	Busy Gal	140	275
956	Busy Morning	100	220
1616	Campus Sweetheart	220	385
0889	Candy Striper Volunteer	55	140
954	Career Girl	70	190
1687	Caribbean Cruise	80	190
0876	Cheerleader	80	190
0872	Cinderella	140	275
1672	Club Meeting	140	275
1670	Coffee's On	80	165
916	Commuter Set	385	715
1627	Country Club Dance	90	220
1603	Country Fair	40	165
1604	Crisp'n Cool	80	190
918	Cruise Stripes	55	140
1626	Dancing Doll	110	220
1666	Debutante Ball	275	495
946	Dinner At Eight	55	165
1633	Disc Date	110	220
1613	Dog n' Duds	110	250
1669	Dreamland	70	165
0875	Drum Majorette	80	190
971	Easter Parade	875	1500
983	Enchanted Evening	110	220
1695	Evening Enchantment	275	495
1660	Evening Gala	130	250
961	Evening Splendor	90	190
1676	Fabulous Fashion	110	275
943	Fancy Free	30	70
1635	Fashion Editor	300	605
1656	Fashion Luncheon	275	550
1691	Fashion Shiner	90	220
1696	Floating Gardens	220	410
921	Floral Petticoat	40	110
1697	Formal Occasion	190	410
1638	Fraternity Dance	195	400
979	Friday Night Date	60	165
1624	Fun At The Fair	140	220
1619	Fun n' Games	140	220
931	Garden Party	50	140
1606	Garden Tea Party	70	165
1658	Garden Wedding	110	220
964	Gay Parisienne	500	1000
1647	Gold n' Glamour	350	750
992	Golden Elegance	90	190
1610	Golden Evening	105	190
911	Golden Girl	70	140
1645	Golden Glory	165	330
945	Graduation	40	80
0873	Guinevere	140	275
1665	Here Comes The Bride	410	825
1639	Holiday Dance	165	275
942	Icebreaker	55	140
1653	International Fair	55	330
1632	Invitation To Tea	190	385
0819	It's Cold Outside, brown	50	110
0819	It's Cold Outside, red	80	165
1620	Junior Designer	140	275
1614	Junior Prom	190	385
1621	Knit Hit	80	165
1602	Knit Separates	55	140
957	Knitting Pretty, blue	80	190
957	Knitting Pretty, pink	140	250
978	Let's Dance	55	140
0880	Little Red Riding Hood & The Wolf	220	440
1661	London Tour	150	300
1600	Lunch Date	40	110
1649	Lunch On The Terrace	80	165
1673	Lunchtime	90	190
1646	Magnificience	275	550
944	Masquerade	60	165
1640	Matinee Fashion	300	615
1617	Midnight Blue	190	385
1641	Miss Astronaut	330	660
1625	Modern Art	190	330
940	Mood For Music	80	190
933	Movie Date	280	110
1633	Music Center Matinee	250	495
965	Nightly Negligee	55	140
1644	On The Avenue	330	660
985	Open Road	140	550
987	Orange Blossom	55	165
1650	Outdoor Art Show	190	385
1637	Outdoor Life	110	220
1601	Pajama Party	20	55
1678	Pan American Stewardess	660	1320
958	Party Dance	80	165
1692	Patio Party	140	275
915	Peachy Fleecy	70	150

Barbie Vintage Fashions 1959-1966

NO.	OUTFIT	MNP	MIB
1648	Photo Fashion	140	275
967	Picnic Set	140	250
1694	Pink Moonbeams	110	220
966	Plantation Belle	165	330
1643	Poodle Parade	220	410
1652	Pretty As A Picture	110	220
1686	Print Aplenty	140	275
949	Rain Coat	50	140
1654	Reception Line	190	385
939	Red Flare	80	165
991	Registered Nurse	80	165
963	Resort Set	50	140
1668	Riding In The Park	300	715
968	Roman Holiday	1500	2500
1611	Satin n' Rose	90	190
1615	Saturday Matinee	385	715
951	Senior Prom	55	165
986	Sheath Sensation	55	165
1664	Shimmering Magic	550	950
977	Silken Flame	50	165
988	Singing In The Shower	50	140
1629	Skater's Waltz	140	275
948	Ski Queen	80	165
1636	Sleeping Pretty	100	190
1674	Sleepytime Gal	105	220
1642	Slumber Party	105	220
982	Solo In The Spotlight	40	275
993	Sophisticated Lady	165	300
937	Sorority Meeting	55	165
1671	Sporting Casuals	70	165
0949	Stormy Weather	50	140
1622	Student Teacher	165	330
1690	Studio Tour	110	220
969	Suburban Shopper	70	190
1675	Sunday Visit	275	550
1683	Sunflower	140	250
976	Sweater Girl	55	165
973	Sweet Dreams, pink	165	330
973	Sweet Dreams, yellow	55	165
955	Swingin' Easy	55	165
941	Tennis Anyone	55	165
959	Theatre Date	70	190
1612	Theatre Date	50	190
1688	Travel Togethers	110	220
1655	Under Fashions	70	165
919	Undergarments	40	140
1685	Underprints	55	140
1623	Vacation Time	80	190
972	Wedding Day Set	140	275
1607	White Magic	80	190
975	Winter Holiday	55	165

Francie Fashions 1966

NO.	OUTFIT	MNP	MIB
1259	Checkmates	80	165
1258	Clam Diggers	80	165
1256	Concert In The Park	55	140
1257	Dance Party	55	140
1260	First Formal	80	165
1252	First Things First	55	140
1254	Fresh As A Daisy	55	140
1250	Gad-About	55	140
1251	It's A Date	55	140
1255	Polka Dots N' Raindrops	55	140
1261	Shoppin' Spree	80	165
1253	Tuckered Out	55	140

Ken Vintage Fashions 1961-1966

NO.	OUTFIT	MNP	MIB
0779	American Airlines Captain, two versions	110	220
797	Army and Air Force	60	125
1425	Best Man	165	330
1424	Business Appointment	165	330
1410	Campus Corduroys	20	40
770	Campus Hero	30	70
0782	Casuals, striped shirt	55	140
782	Casuals, yellow shirt	30	70
1416	College Student	110	220
1400	Country Clubbin'	55	110
793	Dr. Ken	60	125
785	Dreamboat	30	70
0775	Drum Major	75	150
1407	Fountain Boy	80	165
1408	Fraternity Meeting	30	80
791	Fun On Ice	50	110
1403	Going Bowling	30	70
1409	Going Huntin'	50	100
795	Graduation	20	50
1426	Here Comes The Groom	250	500
1412	Hiking Holiday	80	165
1414	Holiday	110	220
780	In Training	20	60
1420	Jazz Concert	80	165
1423	Ken A Go Go	165	330
0774	Ken Arabian Nights	75	150
1404	Ken In Hawaii	70	125
0777	Ken In Holland	75	150
0778	Ken In Mexico	75	150
0776	Ken In Switzerland	75	150
1406	Ken Skin Diver	30	55
0773	King Arthur	110	220
794	Masquerade (Ken)	60	140
1427	Mountain Hike	140	275
1415	Mr. Astronaut	220	495
1413	Off To Bed	80	165
	Pak Items Various Shirts, Slacks, Sweaters	5	10
792	Play Ball	50	110
788	Rally Day	20	70
1405	Roller Skate Date, with hat	40	80

Ken Vintage Fashions 1961-1966

NO.	OUTFIT	MNP	MIE
1405	Roller Skate Date, with slacks	40	80
1417	Rovin' Reporter	90	190
796	Sailor	60	125
786	Saturday Date	30	70
1421	Seein' The Sights	80	165
798	Ski Champion	60	125
0781	Sleeper Set, blue	55	140
781	Sleeper Set, brown	30	70
1401	Special Date	75	150
783	Sport Shorts	20	60
1422	Summer Job	140	275
784	Terry Togs	30	80
0772	The Prince	110	220
789	The Yachtsman, no hat	55	110
0789	The Yachtsman, with hat	80	165
790	Time For Tennis	30	80
1418	Time To Turn In	70	140
799	Touchdown	40	80
787	Tuxedo	50	125
1419	TV's Good Tonight	70	140
1411	Victory Dance	40	110

Ricky Fashions 1965-1966

NO.	OUTFIT	MNP	MIE
1506	Let's Explore	55	110
1501	Lights Out	40	80
1504	Little Leaguer	40	80
1502	Saturday Show	40	80
1505	Skateboard Set	55	110
1503	Sunday Suit	40	80

Skipper Vintage Fashions 1964-1966

NO.	OUTFIT	MNP	MIE
1905	Ballet Class	30	70
1923	Can You Play?	40	80
1926	Chill Chasers	40	110
1912	Cookie Time	55	110

Skipper Vintage Fashions 1964-1966

NO.	OUTFIT	MNP	MIE
1933	Country Picnic	165	275
1911	Day At The Fair	90	165
1929	Dog Show	70	140
1909	Dreamtime	30	70
1906	Dress Coat	30	70
1904	Flower Girl	30	70
1920	Fun Time	55	110
1919	Happy Birthday	80	190
1934	Junior Bridesmaid	165	275
1917	Land & Sea	50	110
1935	Learning To Ride	140	275
1932	Let's Play House	80	190
1930	Loungin' Lovelies	30	70
1903	Masquerade (Skipper)	40	80
1913	Me N' My Doll	90	165
1915	Outdoor Casuals	55	110
1914	Platter Party	40	80
1916	Rain Or Shine	30	80
1928	Rainy Day Checkers	55	110
1901	Red Sensation	30	70
1907	School Days	40	80
1921	School Girl	80	165
1918	Ship Ahoy	60	110
1902	Silk N' Fancy	30	70
1908	Skating Fun	40	80
1936	Sledding Fun	70	140
1910	Sunny Pastels	40	80
1924	Tea Party	55	110
1922	Town Togs	70	140
1900	Under-Pretties	30	55
1925	What's New At The Zoo?	55	110

Tutti Fashions 1966

NO.	OUTFIT	MNP	MIE
3601	Puddle Jumpers	30	55
3603	Sand Castles	30	55
3602	Ship Shape	30	55
3604	Skippin' Rope	30	55

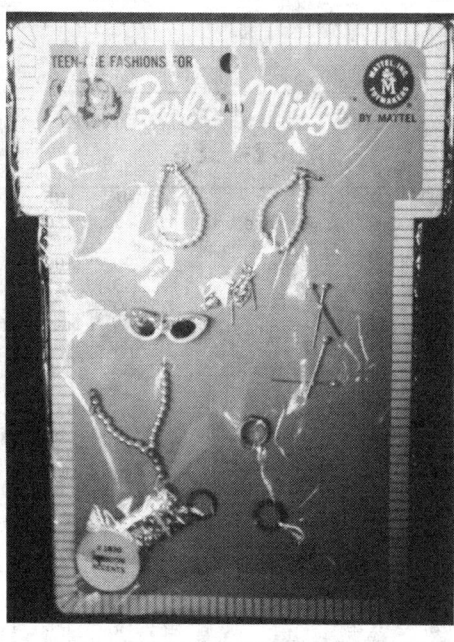

Clockwise from upper left: #3 Brunette Barbie ponytail wearing 1959 Plantation Belle, 1960; "Fashion Accents" Barbie accessory pack, 1964; 30th Anniversary Porcelain Ken, 1991; "Morning Workout" Ken accessory pack, 1964, all Mattel.

BARBIE

Barbie & Friends

NO.	TOY NAME	YEAR	MNP	MIP
9423	All American Barbie	1991	4	10
9425	All American Christie	1991	4	10
9424	All American Ken	1991	4	10
9427	All American Kira	1991	4	10
9426	All American Teresa	1991	4	10
3553	All Star Ken	1981	7	20
9099	All Stars Barbie	1989	5	12
9352	All Stars Christie	1989	5	12
9361	All Stars Ken	1989	5	12
9360	All Stars Midge	1989	5	12
9353	All Stars Teresa	1989	5	15
1010	Allan, bendable leg	1966	125	250
1000	Allan, straight leg	1964	55	135
4930	American Beauty Mardi Gras Barbie	1987	40	85
3137	American Beauty Queen	1991	5	10
3245	American Beauty Queen, black version	1991	5	10
	American Girl Barbie "Color Magic Face"	1965	550	1800
1070	American Girl Barbie, all blondes, titian	1965	375	975
1070	American Girl Barbie, brunette	1966	375	1100
1070	American Girl Side-Part Barbie, brunette, blonde, titian	1965	2225	3500
5640	Angel Face Barbie	1982	8	25
4828	Animal Lovin' Barbie, black version	1988	5	15
1350	Animal Lovin' Barbie, white version	1988	5	12
1395	Animal Lovin' Ginger Giraffe	1988	7	15
1351	Animal Lovin' Ken	1988	5	12
1352	Animal Lovin' Nikki	1988	7	15
1393	Animal Lovin' Zizi Zebra	1988	7	15
1207	Astronaut Barbie, black version	1985	30	80
2449	Astronaut Barbie, white version	1985	25	65
9434	Babysitter Courtney	1991	4	10
9433	Babysitter Skipper	1991	4	10
1599	Babysitter Skipper, black version	1991	4	10
9000	Baggy Casey, blonde	1975	30	75
9093	Ballerina Barbie	1976	15	40
9613	Ballerina Barbie on Tour, gold	1978	65	155
9528	Ballerina Cara	1976	15	45
9805	Barbie & Her Fashion Fireworks	1976	17	40
2751	Barbie & the Beat	1990	5	15
2752	Barbie & the Beat Christie	1990	5	15
2754	Barbie & the Beat Midge	1990	6	20
1144	Barbie with Growin' Pretty Hair	1971	65	175
9601	Bathtime Fun Barbie	1991	3	8
9603	Bathtime Fun Barbie, black version	1991	3	8
3237	Beach Blast Barbie	1988	3	10
3253	Beach Blast Christie	1988	4	9
3238	Beach Blast Ken	1988	4	9
3244	Beach Blast Miko	1988	4	9
3242	Beach Blast Skipper	1988	4	9
3251	Beach Blast Steven	1988	4	9
3249	Beach Blast Teresa	1988	5	10
9907	Beautiful Bride Barbie	1978	75	225
1290	Beauty Secrets Barbie, 1st issue	1979, 1980	20	45
1295	Beauty Secrets Christie	1979	20	45
1018	Beauty, Barbie's Dog	1979	12	30
9404	Benetton Barbie	1991	6	15
9407	Benetton Christie	1991	6	15
9409	Benetton Marina	1991	6	15
1293	Black Barbie, 1st issue black	1979	15	50
1142	Brad, bendable leg	1970	50	100
850	Bubblecut Barbie Brownette	1961	150	375

Barbie & Friends

NO.	TOY NAME	YEAR	MNP	MIP
850	Bubblecut Barbie, all blondes	1961	85	250
850	Bubblecut Barbie, black haired	1961	95	275
850	Bubblecut Barbie, brunette	1961	90	260
850	Bubblecut Barbie, Sable Brown	1961	225	450
850	Bubblecut Barbie, titian	1961	90	275
850	Bubblecut Barbie, white ginger	1961	95	225
3577	Buffy and Mrs. Beasley	1968	65	150
3311	Busy Hand Barbie	1972	95	200
3313	Busy Hand Francie	1972	90	175
3314	Busy Hand Ken	1971	25	75
3312	Busy Hand Steffie	1971	90	175
1195	Busy Talking Barbie	1972	120	200
1196	Busy Talking Ken	1972	100	175
1186	Busy Talking Steffie	1972	120	175
4547	Calgary Olympic Skating Barbie	1987	25	60
4439	California Dream Barbie	1987	5	15
4443	California Dream Christie	1987	6	13
4441	California Dream Ken	1987	8	12
4442	California Dream Midge	1987	3	12
4440	California Dream Skipper	1987	13	50
4403	California Dream Teresa	1987	15	35
7377	Carla	1976	65	125
1180	Casey Titian Hair	1967	150	400
1180	Casey, brunette, blonde	1967	85	300
1180	Casey, Titian	1967	150	425
3570	Chris, titian, blonde, brunette	1967	35	100
1150	Color Magic Barbie, black haired	1966	700	2200
1150	Color Magic Barbie, blonde	1966	455	1300
3022	Cool Times Barbie	1988	5	15
3217	Cool Times Christie	1988	5	15
3219	Cool Times Ken	1988	5	10
3216	Cool Times Midge	1988	7	15
3218	Cool Times Teresa	1988	9	17
7079	Cool Tops Courtney	1989	7	15
9351	Cool Tops Kevin	1989	5	13
4989	Cool Tops Skipper	1989	7	15
5441	Cool Tops Skipper, black version	1989	5	13
7123	Costume Ball Barbie	1991	6	15
7134	Costume Ball Barbie, black version	1991	6	15
7154	Costume Ball Ken	1991	6	15
7160	Costume Ball Ken, black version	1991	6	15
4859	Crystal Barbie, black version	1983	10	35
4598	Crystal Barbie, white version	1983	10	35
9036	Crystal Ken, black version	1983	15	50
4898	Crystal Ken, white version	1983	8	25
3509	Dance Club Barbie	1989	5	12
3513	Dance Club Devon	1989	5	12
3512	Dance Club Kayla	1989	5	12
3511	Dance Club Ken	1989	5	12
4836	Dance Magic Barbie	1990	7	25
7080	Dance Magic Barbie, black version	1990	7	25
7081	Dance Magic Ken	1990	6	20
7082	Dance Magic Ken, black version	1990	6	20
7945	Day-to-Night Barbie, black version	1984	10	30
7944	Day-to-Night Barbie, hispanic version	1984	17	60
7929	Day-to-Night Barbie, white version	1984	10	30
9018	Day-to-Night Ken, black version	1984	8	25
9019	Day-to-Night Ken, white version	1984	8	25
9217	Deluxe Quick Curl Barbie	1973	15	70
9219	Deluxe Quick Curl Cara	1973	15	70
	Deluxe Quick Curl P.J.	1973	25	75
9428	Deluxe Quick Curl Skipper	1973	12	65

Barbie & Friends

NO.	TOY NAME	YEAR	MNP	MIP
3850	Doctor Barbie	1987	8	25
4118	Doctor Ken	1987	5	20
1116	Dramatic New Living Barbie, all hair colors	1969	45	175
1117	Dramatic New Living Skipper	1969	50	110
1623	Dream Bride	1991	10	30
9180	Dream Date Barbie	1982	7	30
4077	Dream Date Ken	1982	5	20
5869	Dream Date P.J.	1982	8	40
4817	Dream Date Skipper	1990	5	15
2242	Dream Glow Barbie, black version	1985	12	40
1647	Dream Glow Barbie, hispanic version	1985	25	65
2248	Dream Glow Barbie, white version	1985	12	40
2421	Dream Glow Ken, black version	1985	13	20
2250	Dream Glow Ken, white version	1985	13	20
9180	Dream Time Barbie	1985	10	20
9180	Dream Time Barbie, pink	1985	10	20
7093	Fabulous Fur Barbie	1983	20	40
5313	Fashion Jeans Barbie	1981	15	45
5316	Fashion Jeans Ken	1981	12	45
2210	Fashion Photo Barbie	1978	20	80
	Fashion Photo Christie	1978	20	80
2323	Fashion Photo P.J.	1978	35	85
7193	Fashion Play Barbie	1983	10	20
4835	Fashion Play Barbie	1987	10	20
9429	Fashion Play Barbie	1990	2	5
9629	Fashion Play Barbie	1991	2	5
5953	Fashion Play Barbie, black version	1990	2	5
5953	Fashion Play Barbie, black version	1991	2	5
5954	Fashion Play Barbie, hispanic version	1990	2	5
5954	Fashion Play Barbie, hispanic version	1991	2	5
870	Fashion Queen Barbie	1963	125	500
1189	Feelin' Fun Barbie	1987	5	10
3421	Feelin' Groovy Barbie	1985	75	160
9916	Flight Time Barbie, black vorcion	1909	5	15
2066	Flight Time Barbie, hispanic version	1989	7	18
9584	Flight Time Barbie, white version	1989	5	15
9600	Flight Time Ken	1989	5	15
1143	Fluff	1971	35	85
1122	Francie Hair Happenins'	1970	85	225
1170	Francie Twist and Turn, bendable leg, all hair colors	1967	75	225
1129	Francie with Growin' Pretty Hair	1971	65	200
1170	Francie, bendable leg, black version	1967	400	800
1130	Francie, bendable leg, white version, blonde, brunette	1966	75	275
1140	Francie, straight leg, brunette, blonde	1966	125	350
	Free Moving Barbie	1974	10	35
	Free Moving Cara	1974	10	35
7280	Free Moving Ken	1974	8	25
7281	Free Moving P.J.	1974	10	40
7668	Fun to Dress Barbie, black version	1987	3	6
1373	Fun to Dress Barbie, black version	1988	3	6
4939	Fun to Dress Barbie, black version	1989	2	6
7373	Fun to Dress Barbie, hispanic version	1989	3	8
4558	Fun to Dress Barbie, white version	1987	3	6
4372	Fun to Dress Barbie, white version	1988	3	6
4808	Fun to Dress Barbie, white version	1989	3	6
1739	Funtime Barbie, black version	1986	5	12
1738	Funtime Barbie, white version	1986	5	12
7194	Funtime Ken	1974	7	15
1953	Garden Party Barbie	1989	8	20
1922	Gift Giving Barbie	1985	5	15
1205	Gift Giving Barbie	1988	5	15
7262	Gold Medal Olympic Barbie Skater	1975	15	75

Barbie & Friends

NO.	TOY NAME	YEAR	MNP	MIP
7264	Gold Medal Olympic Barbie Skier	1975	12	75
7263	Gold Medal Olympic P.J. Gymnast	1974	12	75
7261	Gold Medal Olympic Skier Ken	1974	15	75
7274	Gold Medal Olympic Skipper	1975	15	75
1974	Golden Dreams Barbie	1980	8	35
3533	Golden Dreams Barbie Glamerous Night	1980	25	85
3249	Golden Dreams Christie	1980	10	40
7834	Great Shapes Barbie, black version	1983	5	12
7025	Great Shapes Barbie, white version	1983	5	12
7025	Great Shapes Barbie, with walkman	1983	12	25
7310	Great Shapes Ken	1983	5	12
7417	Great Shapes Skipper	1983	5	12
4253	Groom Todd	1982	15	45
	Growing Up Ginger	1977	35	85
7259	Growing Up Skipper	1977	10	55
	Guardian Goddess Ice Empress	1979	60	200
	Guardian Goddess Lion Queen	1979	85	225
2757	Guardian Goddess Moonmystic	1979	55	145
	Guardian Goddess Soaring Eagle	1979	85	225
2757	Guardian Goddess Sunspell	1979	35	120
1922	Happy Birthday Barbie	1983	8	30
1922	Happy Birthday Barbie	1980	8	35
9561	Happy Birthday Barbie	1991	8	25
9561	Happy Birthday Barbie, black version	1991	8	25
4098	Happy Holidays Barbie, pink gown	1990	15	40
4543	Happy Holidays Barbie, pink gown, black version	1990	15	40
1703	Happy Holidays Barbie, red velvet gown	1988	200	500
3253	Happy Holidays Barbie, white satin gown	1989	50	125
1871	Happy Holidays Green	1991	30	75
2696	Happy Holidays Green-Black Doll	1991	30	75
7470	Hawaiian Barbie	1977	25	100
7470	Hawaiian Barbie	1975	30	125
5040	Hawaiian Fun Barbie	1991	3	6
5044	Hawaiian Fun Christie	1991	3	6
9294	Hawaiian Fun Jazzie	1991	3	6
5041	Hawaiian Fun Ken	1991	3	6
5043	Hawaiian Fun Kira	1991	3	6
5042	Hawaiian Fun Skipper	1991	3	6
5045	Hawaiian Fun Steven	1991	3	6
2960	Hawaiian Ken	1978	13	45
7495	Hawaiian Ken	1983	7	15
3698	High School Chelsie	1989	5	15
3600	High School Dude, Jazzie's boyfriend	1989	5	15
3635	High School Jazzie	1989	5	15
3636	High School Stacie	1989	5	15
1292	Hispanic Barbie	1979	15	50
2249	Home Pretty Barbie	1990	8	20
2390	Homecoming Queen Skipper, black version	1988	8	15
1952	Homecoming Queen Skipper, white version	1988	12	25
1757	Horse Lovin' Barbie	1982	10	45
3600	Horse Lovin' Ken	1982	8	30
5029	Horse Lovin' Skipper	1982	8	30
7927	Hot Stuff Skipper	1984	5	18
7365	Ice Capades Barbie	1990	5	15
7348	Ice Capades Barbie, black version	1989	5	12
7348	Ice Capades Barbie, black version	1990	5	15
7365	Ice Capades Barbie, white version	1989	5	12
7375	Ice Capades Ken	1990	5	15
4928	International Barbie Canadian	1988	12	25
3898	International Barbie Eskimo	1982	50	100
3188	International Barbie German	1986	35	75
2997	International Barbie Greek	1986	20	45

Clockwise from upper left: Bendable Leg Midge, 1965; Sweet Dreams in yellow Barbie Vintage Fashions; International Series Italian Barbie, 1980; Ballerina Barbie Vintage Fashions, all Mattel.

Barbie & Friends

NO.	TOY NAME	YEAR	MNP	MIP
3189	International Barbie Iceland	1987	27	60
3897	International Barbie India	1982	40	80
7517	International Barbie Irish	1984	45	100
1601	International Barbie Italian	1980	65	185
9481	International Barbie Japanese	1985	55	125
4929	International Barbie Korean	1988	10	25
1917	International Barbie Mexican	1989	10	25
3262	International Barbie Oriental	1981	55	125
1600	International Barbie Parisienne	1980	65	185
2995	International Barbie Peru	1986	30	60
1602	International Barbie Royal	1980	65	185
1916	International Barbie Russian	1989	12	27
3263	International Barbie Scottish	1981	50	150
4031	International Barbie Spanish	1983	40	120
4032	International Barbie Swedish	1983	35	75
7451	International Barbie Swiss	1984	35	75
9094	International Barbie, Brazillian version	1990	8	20
7376	International Barbie, Nigerian version	1990	8	20
7330	International Czechoslovakian	1991	25	50
9844	International Eskimo Re-Issue	1991	8	20
7329	International Malaysian	1991	10	25
9843	International Parisian Re-Issue	1991	8	20
9845	International Scottish Re-Issue	1991	8	20
4061	Island Fun Barbie	1987	3	10
4092	Island Fun Christie	1987	3	10
4060	Island Fun Ken	1987	3	10
4064	Island Fun Skipper	1987	3	10
4093	Island Fun Steven	1987	3	10
4117	Island Fun Teresa	1987	3	10
3633	Jazzie Workout	1989	5	12
1756	Jewel Secrets Barbie, black version	1986	6	18
1737	Jewel Secrets Barbie, white version	1986	6	18
3232	Jewel Secrets Ken, black version	1986	6	18
1719	Jewel Secrets Ken, rooted hair	1986	6	18
3133	Jewel Secrets Skipper	1986	6	18
3179	Jewel Secrets Whitney	1986	8	25
1124	Ken, bendable leg	1970	20	65
750	Ken, bendable leg, brunette, blonde	1965	125	250
750	Ken, flocked hair, brunette, blonde	1961	60	150
750	Ken, painted hair, brunette, blonde	1962	40	110
9325	Kevin	1991	4	10
2597	Kissing Barbie	1978	8	50
2955	Kissing Christie	1978	10	55
9725	Lights & Lace Barbie	1991	4	10
9728	Lights & Lace Christie	1991	4	10
9727	Lights & Lace Teresa	1991	4	10
1155	Live Action Barbie	1971	60	140
1175	Live Action Christie	1971	60	135
1159	Live Action Ken	1971	55	120
1172	Live Action Ken on Stage	1971	30	85
1156	Live Action P.J.	1971	65	145
1153	Live Action PJ on Stage	1971	60	130
1116	Living Barbie	1970	65	130
1117	Living Skipper	1970	40	80
7072	Lovin' You Barbie	1983	10	55
3989	Magic Curl Barbie, black version	1981	8	20
3856	Magic Curl Barbie, white version	1981	10	30
3137	Magic Moves Barbie, black version	1985	6	20
2126	Magic Moves Barbie, white version	1985	6	20
1067	Malibu Barbie	1975	10	25
1067	Malibu Barbie	1971	15	55
7745	Malibu Christie	1975	10	25

Barbie & Friends

NO.	TOY NAME	YEAR	MNP	MIP
7745	Malibu Christie	1977	15	40
1068	Malibu Francie	1971	15	50
1088	Malibu Ken	1976	8	20
1087	Malibu P.J.	1975	5	20
1069	Malibu Skipper	1977	8	20
1080	Midge, bendable leg, blonde, titian	1965	200	400
1080	Midge, bendable leg, brownette	1965	325	650
860	Midge, straight leg, blonde, titian	1964	55	175
860	Midge, straight leg, brunette	1964	65	200
1080	Midge, with teeth, all hair colors	1965	125	300
1060	Miss Barbie (Sleep-eye)	1964	200	750
1060	Miss Barbie (Sleep-eye), silver-haired	1964	395	975
4224	Mod Hair Ken	1972	45	100
9988	Music Lovin' Barbie	1985	15	45
2388	Music Lovin' Ken	1985	15	45
2854	Music Lovin' Skipper	1985	20	75
9942	My First Barbie	1991	3	10
1875	My First Barbie, aqua and yellow dress	1980	10	35
9944	My First Barbie, black version	1990	4	10
9943	My First Barbie, black version	1991	3	10
9943	My First Barbie, hispanic version	1990	5	12
9944	My First Barbie, hispanic version	1991	3	10
1875	My First Barbie, pink checkered dress	1982	5	25
1801	My First Barbie, pink tutu, black version	1986	5	15
1788	My First Barbie, pink tutu, white version	1986	5	15
9858	My First Barbie, white dress, black version	1984	7	30
1875	My First Barbie, white dress, white version	1984	5	25
1281	My First Barbie, white tutu, black version	1988	6	14
1282	My First Barbie, white tutu, hispanic version	1988	6	20
1280	My First Barbie, white tutu, white version	1988	5	14
9942	My First Barbie, white version	1990	4	10
9940	My First Ken	1990	4	10
9940	My First Ken	1991	3	10
1389	My First Ken, 1st issue	1080	4	10
9342	New Look Ken	1973	23	70
7807	Newport Barbie	1974	25	75
1170	No Bangs Francie	1971	500	1000
1127	Nurse Julia, 1 piece nurse outfit	1970	70	150
1127	Nurse Julia, 2 piece nurse outfit	1969	75	200
4405	Nurse Whitney	1987	20	45
4885	Party Treats Barbie	1989	8	25
9516	Peaches n' Cream Barbie, black version	1984	8	40
7926	Peaches n' Cream Barbie, white version	1984	8	45
4869	Pepsi Spirit Barbie	1989	10	35
4867	Pepsi Spirit Skipper	1989	10	33
4555	Perfume Giving Ken, black version	1989	6	20
4554	Perfume Giving Ken, white version	1989	6	20
4552	Perfume Pretty Barbie, black version	1989	8	25
4551	Perfume Pretty Barbie, white version	1989	8	25
4557	Perfume Pretty Whitney	1987	8	35
	Pink Jubilee Barbie, only 1200 made	1989		1900
3551	Pink n' Pretty Barbie	1981	13	37
3554	Pink n' Pretty Christie	1981	8	35
5336	Playtime Barbie	1983	15	20
850	Ponytail Barbie #1, blonde	1959	1700	3500
850	Ponytail Barbie #1, brunette	1959	2700	4500
850	Ponytail Barbie #2, blonde	1959	1500	2200
850	Ponytail Barbie #2, brunette	1959	2100	3000
850	Ponytail Barbie #3, blonde	1960	450	750
850	Ponytail Barbie #3, brunette	1960	525	1000
850	Ponytail Barbie #4, blonde	1960	200	425
850	Ponytail Barbie #4, brunette	1960	225	475

Barbie & Friends

NO.	TOY NAME	YEAR	MNP	MIP
850	Ponytail Barbie #5, blonde	1961	175	275
850	Ponytail Barbie #5, brunette	1961	200	300
850	Ponytail Barbie #5, titian	1961	250	400
850	Ponytail Barbie #6, blonde	1962	160	270
850	Ponytail Barbie #6, brunette	1962	195	295
850	Ponytail Barbie #6, titian	1962	195	395
850	Ponytail Barbie #7, blonde	1963	160	270
850	Ponytail Barbie #7, brunette	1963	195	295
850	Ponytail Barbie #7, titian	1963	195	395
850	Ponytail Swirl Style Barbie, blonde	1964	225	450
850	Ponytail Swirl Style Barbie, brunette	1964	275	550
850	Ponytail Swirl Style Barbie, platinum	1964	375	775
850	Ponytail Swirl Style Barbie, titian	1964	275	575
1117	Pose n' Play Skipper	1973	75	300
2598	Pretty Changes Barbie	1978	8	40
7194	Pretty Party Barbie	1983	12	30
4220	Quick Curl Barbie	1973	15	105
7291	Quick Curl Cara	1975	15	60
4222	Quick Curl Francie	1973	15	55
4221	Quick Curl Kelley	1973	15	75
8697	Quick Curl Miss America, blonde	1974	35	75
8697	Quick Curl Miss America, brunette	1973	45	175
4223	Quick Curl Skipper	1973	20	50
1090	Ricky	1965	55	135
1140	Rocker Barbie, 1st issue	1985	7	35
3055	Rocker Barbie, 2nd issue	1986	7	20
1196	Rocker Dana, 1st issue	1985	7	35
3158	Rocker Dana, 2nd issue	1986	7	20
1141	Rocker Dee-Dee, 1st issue	1985	7	20
3160	Rocker Dee-Dee, 2nd issue	1986	7	20
2428	Rocker Derek, 1st issue	1985	7	20
3173	Rocker Derek, 2nd issue	1986	7	20
2427	Rocker Diva, 1st issue	1985	7	20
3159	Rocker Diva, 2nd issue	1986	7	20
3131	Rocker Ken, 1st issue	1985	7	20
1880	Rollerskating Barbie	1980	8	40
1881	Rollerskating Ken	1980	8	43
4973	Safari Barbie	1983	8	30
1019	Scott	1979	15	50
9109	Sea Lovin' Barbie	1984	8	30
9110	Sea Lovin' Ken	1984	8	35
4931	Sensations Barbie	1987	5	12
4977	Sensations Becky	1987	5	12
4976	Sensations Belinda	1987	5	12
4967	Sensations Bopsy	1987	5	12
4960	Sensations Bopsy Bibops	1987	15	95
7799	Show and Ride Barbie	1988	8	35
7511	Ski Fun Barbie	1991	6	15
7512	Ski Fun Ken	1991	6	15
7513	Ski Fun Midge	1991	15	25
1030	Skipper, bendable leg, brunette, blonde, titian	1965	6800	185
950	Skipper, straight leg, brunette, blonde, titian	1964	45	100
950	Skipper, straight leg, re-issues, brunette, blonde, titian	1970	40	100
1120	Skooter, bendable leg, brunette, blonde, titian	1965	80	200
1040	Skooter, straight leg, brunette, blonde, titian	1965	55	125
1294	Sport n' Shave Ken	1980	8	40
1190	Standard Barbie, all hair colors	1967	100	300
1190	Standard Barbie, all hair colors	1970	125	250
1281	Starr Kelley	1979	8	35
1283	Starr Shaun	1979	8	40
1280	Starr Starr	1979	8	35
1282	Starr Tracy	1979	8	35

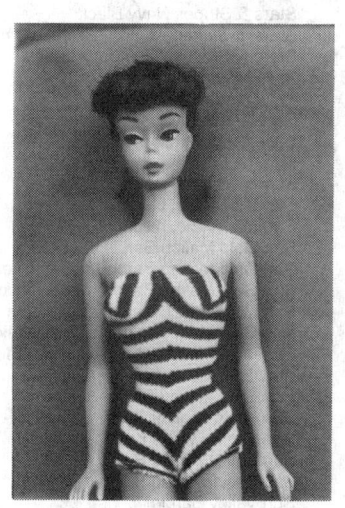

Barbie, top to bottom: PJ and Talking Barbie, 1968; Montgomery Ward Anniversary Barbie, 1972; Swirl Ponytail Barbie, 1964: Bendable Leg Midge, 1965, and American Girl Barbie, 1965.

Barbie & Friends

NO.	TOY NAME	YEAR	MNP	MIP
3360	Stars & Stripes Air Force Barbie	1991	12	30
9693	Stars & Stripes Navy	1991	10	25
9694	Stars & Stripes Navy Black	1991	10	25
3360	Stars 'n Stripes Air Force Barbie	1990	15	45
3966	Stars n' Stripes Army Barbie	1989	10	40
1283	Style Magic Barbie	1989	5	15
1288	Style Magic Christie	1989	5	15
1915	Style Magic Skipper	1989	10	30
1290	Style Magic Whitney	1989	5	15
7027	Summit Barbie	1990	8	20
7029	Summit Barbie, Asian version	1990	10	35
7028	Summit Barbie, black version	1990	8	20
7030	Summit Barbie, hispanic version	1990	10	30
7745	Sun Gold Malibu Barbie, black version	1983	5	18
4970	Sun Gold Malibu Barbie, hispanic version	1985	3	7
1067	Sun Gold Malibu Barbie, white version	1983	5	15
3849	Sun Gold Malibu Ken, black version	1983	3	18
	Sun Gold Malibu Ken, hispanic version	1985	3	7
1088	Sun Gold Malibu Ken, white version	1983	3	7
1187	Sun Gold Malibu P.J.	1983	5	15
1069	Sun Gold Malibu Skipper	1983	5	15
1067	Sun Lovin' Malibu Barbie	1978	5	15
1088	Sun Lovin' Malibu Ken	1978	5	15
1187	Sun Lovin' Malibu P.J.	1978	5	18
1069	Sun Lovin' Malibu Skipper	1978	5	15
7806	Sun Valley Barbie	1974	15	55
7809	Sun Valley Ken	1974	10	50
4970	Sunsational Hispanic Barbie	1983	20	40
1067	Sunsational Malibu Barbie	1981	6	15
4970	Sunsational Malibu Barbie, hispanic version	1981	8	20
7745	Sunsational Malibu Christie	1981	6	15
	Sunsational Malibu Ken, black version	1981	15	35
1187	Sunsational Malibu P.J.	1981	6	15
1069	Sunsational Malibu Skipper	1981	5	12
7745	Sunset Malibu Christie	1971	20	45
1068	Sunset Malibu Francie	1971	25	60
1088	Sunset Malibu Ken	1971	15	35
1187	Sunset Malibu P.J.	1977	10	25
1069	Sunset Malibu Skipper	1971	20	45
3296	Super Hair Barbie, black version	1986	8	20
3101	Super Hair Barbie, white version	1986	8	20
5839	Super Sport Ken	1982	8	20
2756	Super Teen Skipper	1978	7	15
	Supersize Barbie	1976	60	150
	Supersize Christie	1976	60	150
	Supersize Super Hair Barbie	1978	65	140
4983	Superstar Ballerina Barbie	1983	33	60
9720	Superstar Barbie	1977	33	60
1605	Superstar Barbie 30th Anniversary, black version	1989	6	13
1604	Superstar Barbie 30th Anniversary, white version	1989	8	17
9950	Superstar Christie	1977	25	60
2211	Superstar Ken	1978	17	40
1550	Superstar Ken 30th Anniversary, black version	1989	5	13
1535	Superstar Ken 30th Anniversary, white version	1989	7	17
1067	Superstar Malibu Barbie	1977	10	25
7796	Sweet 16 Barbie	1974	25	65
7635	Sweet Roses Barbie	1989	7	20
7455	Sweet Roses P.J.	1983	15	40
2064	Tahiti, Barbie's Pet Parrot	1985	3	10
1115	Talking Barbie, all hair colors	1968	80	225
1114	Talking Brad	1970	65	175
1126	Talking Christie	1968	60	150

Barbie & Friends

NO.	TOY NAME	YEAR	MNP	MIP
1128	Talking Julia	1968	55	150
1111	Talking Ken	1968	40	125
1113	Talking P.J.	1968	65	175
1125	Talking Stacey, blonde, titian	1968	70	200
1107	Talking Truely Scrumptious	1969	125	400
3634	Teen Dance Jazzie	1989	7	15
5893	Teen Fun Skipper Cheerleader	1987	5	15
5899	Teen Fun Skipper Party Teen	1987	5	15
5889	Teen Fun Skipper Workout	1987	5	15
3631	Teen Looks Jazzie Cheerleader	1989	4	10
3633	Teen Looks Jazzie Workout	1989	4	10
3634	Teen Scene Jazzie	1989	4	15
5507	Teen Scene Jazzie	1990	5	15
4855	Teen Sweetheart Skipper	1987	5	15
1950	Teen Time Courtney	1988	5	10
1951	Teen Time Skipper	1988	5	10
1760	Tennis Barbie	1986	5	22
1761	Tennis Ken	1986	5	22
1160	TNT Barbie Titian	1967	225	450
3590	Todd	1966	50	110
	Tracey Bride	1982	8	40
1022	Tropical Barbie, black version	1985	3	10
1017	Tropical Barbie, white version	1985	3	10
1023	Tropical Ken, black version	1985	3	10
4060	Tropical Ken, white version	1985	3	10
2056	Tropical Miko	1985	3	10
4064	Tropical Skipper	1985	3	10
1108	Truly Scrumptious, straight leg	1969	175	300
8128	Tutti	1976	40	90
3580	Tutti, all hair colors	1967	45	75
3550	Tutti, all hair colors	1966	50	110
1185	Twiggy	1967	125	300
5723	Twirley Curls Barbie, black version	1982	8	25
5724	Twirley Curls Barbie, hispanic version	1982	10	30
5579	Twirley Curls Barbie, white version	1982	8	25
1160	Twist and Turn Barbie, blondes and brunettes	1967	90	350
1119	Twist and Turn Christie	1969	75	195
1160	Twist and Turn Flip Barbie, all hair colors	1969	95	350
1118	Twist and Turn P.J., blonde	1970	50	175
1105	Twist and Turn Skipper, all hair colors	1969	50	125
1165	Twist and Turn Stacey, blonde, titian	1967	95	275
4774	UNICEF Barbie, asian version	1989	17	40
4770	UNICEF Barbie, black version	1989	7	20
4782	UNICEF Barbie, hispanic version	1989	7	20
1920	UNICEF Barbie, white version	1989	7	20
1675	Vacation Sensation Barbie, pink	1988	12	45
1675	Vacation Sensations Barbie, blue	1986	10	32
1182	Walk Lively Barbie	1972	60	150
1184	Walk Lively Ken	1972	30	80
3200	Walk Lively Miss America Barbie, brunette	1972	60	150
1183	Walk Lively Steffie	1972	65	145
7011	Wedding Fantasy Barbie, black version	1989	7	25
2125	Wedding Fantasy Barbie, white version	1989	7	25
9607	Wedding Party Alan	1991	7	15
9608	Wedding Party Barbie	1991	7	15
9852	Wedding Party Kelly & Todd	1991	15	35
9609	Wedding Party Ken	1991	7	15
9606	Wedding Party Midge	1991	7	15
3469	Western Barbie	1980	8	30
2930	Western Fun Barbie, black version	1989	5	15
9932	Western Fun Barbie, white version	1989	5	15
9934	Western Fun Ken	1989	5	15

BARBIE

Top to bottom: Ken in "Mr. Astronaut" outfit, 1965; Skipper in "Masquerade" outfit, 1964; Ski Queen Barbie Vintage Fashions, all Mattel.

Barbie & Friends

NO.	TOY NAME	YEAR	MNP	MIP
9933	Western Fun Nia	1989	5	15
3600	Western Ken	1980	7	20
5029	Western Skipper	1980	8	25
4103	Wet n' Wild Barbie	1989	8	18
4121	Wet n' Wild Christie	1989	3	7
4104	Wet n' Wild Ken	1989	3	7
4120	Wet n' Wild Kira	1989	3	7
4138	Wet n' Wild Skipper	1989	3	7
4137	Wet n' Wild Steven	1989	3	7
4136	Wet n' Wild Teresa	1989	3	7
	Wig Wardrobe Midge	1964	125	350
7808	Yellowstone Kelley	1974	60	195

Limited Editions, Department Store Specials, Gift Sets

NO.	TOY NAME	YEAR	MNP	MIP
3712	All American Barbie & Starstepper	1991	12	30
2909	Ames Party in Pink	1991	10	25
3406	Applause Barbie Holiday	1991	10	30
5313	Applause Style Barbie	1990	8	20
9613	Ballerina on Tour Gift Set	1976	25	85
5531	Barbie and Friends Gift Set	1982	25	55
4431	Barbie and Friends: Ken, Barbie, PJ	1982	25	75
	Barbie and Ken Camping Out	1983	25	95
892	Barbie and Ken Tennis Gift Set	1962	275	750
3303	Barbie Beautiful Blues Gift Set	1967	350	850
1596	Barbie Pink Premier Gift Set		300	650
	Barbie Twinkle Town Set	1968	250	500
1013	Barbie's Round the Clock Gift Set	1964	275	750
1011	Barbie's Sparkling Pink Gift Set	1963	250	600
1017	Barbie's Wedding Party Gift Set	1964	450	1000
1702	Beauty Secrets Barbie Pretty Reflections Gift Set	1979	40	85
5405	Bob Mackie Designer Gold	1990	250	500
2703	Bob Mackie Platinum	1991	100	200
2704	Bob Mackie Starlight Splendor	1001	200	400
3304	Casey Goes Casual Gift Set	1967	400	1000
4385	Childs World Disney Barbie	1990	15	35
9835	Childs World Disney Barbie, black version	1990	15	35
2366	Club-Doll-Jewl Jubilee	1991	10	25
4893	Cool City Blues: Barbie, Ken, Skipper	1989	20	45
4917	Dance Club Barbie Gift Set	1989	25	60
5409	Dance Magic Gift Set	1985	27	65
5409	Dance Magic Gift Set Barbie & Ken	1990	15	35
9058	Dance Sensation Barbie Gift Set	1984	15	40
2996	Deluxe Tropical Barbie	1986	10	30
2921	Fao Schwartz Night Sensation	1991	75	150
5946	Fao Schwartz Winter Fantasy	1990	100	200
7734	FAO Schwarz: Golden Greetings Barbie	1989	65	155
863	Fashion Queen Barbie & Her Friends	1964	250	650
864	Fashion Queen Barbie & Ken Trousseau Gift Set	1964	100	500
1194	Francie Rise n' Shine Gift Set	1971	175	450
1042	Francie Swingin' Separates Gift Set	1966	450	900
9519	Happy Birthday Barbie Gift Set	1984	35	85
	Hawaiian Barbie	1982	25	100
3274	Hills Evening Sparkle	1990	10	25
3549	Hills Moonlight Rose	1991	7	17
4843	Hills: Party Lace Barbie	1989	15	30
2702	JC Penney Enchanted Evening	1991	40	90
7057	JC Penney's Evening Elegance	1990	40	80
	Julia Simply Wow Gift Set	1976	145	475
4870	K-Mart: Peach Pretty Barbie	1989	10	30
2977	Kissing Barbie Gift Set	1978	25	65
1585	Living Barbie Action Accents Gift Set	1970	250	500

Barbie & Friends

NO.	TOY NAME	YEAR	MNP	MIP
7583	Loving You Barbie Gift Set	1983	45	100
1703	Malibu Barbie "The Beach Party", with case	1979	17	35
1248	Malibu Ken Surf's Up Gift Set	1971	75	250
4983	Mervyns: Ballerina Barbie	1983	30	75
1012	Midge's Ensemble Gift Set	1964	100	700
	Mix n' Match Gift Set	1962	215	895
3210	Montgomery Wards: Barbie	1972	350	600
	Montgomery Wards: Ken	1972	30	85
2483	My First Barbie Gift Set	1991	8	20
1979	My First Barbie Gift Set, pink tutu	1986	15	35
5386	My First Barbie Gift Set, pink tutu	1987	18	40
1875	My First Barbie, pink tutu, zayres hispanic	1987	8	45
1014	On Parade Gift Set, Barbie, Ken, Midge	1964	500	1000
5239	Pink & Pretty Barbie Gift Set	1981	35	85
	PJ's Swinging Silver Gift Set	1970	250	750
2598	Pretty Changes Barbie Gift Set	1978	35	85
5588	Sears Lavender Surprise	1990	8	20
9049	Sears Lavender Surprise, black version	1990	8	20
2586	Sears Southern Belle	1991	10	25
2998	Sears: 100th Celebration Barbie	1986	20	70
3596	Sears: Evening Enchantment Barbie	1989	10	20
7669	Sears: Lilac and Lovely Barbie	1988	10	25
	Sears: Perfectly Plaid Gift Set	1962	125	450
4550	Sears: Star Dream Barbie	1987	10	65
1364	Service Merchandise Blue Rhapsody	1991	75	150
3142	Shopko/Venture Blossom Beauty	1991	10	25
1021	Skipper Party Time Gift Set	1964	100	450
1172	Skipper Swing 'a' Rounder Gym Gift Set	1972	100	250
	Skooter Cut n' Button Gift Set	1967	150	500
3347	Spiegel Sterlin Wishes	1991	45	100
1648	Swan Lake Barbie	1991	50	100
	Talking Barbie Perfectly Plaid Gift Set	1971	155	495
2954	Target Cute'n Cool	1991	8	20
2587	Target Golden Evening	1991	6	15
5955	Target Party Pretty Barbie	1990	6	15
7476	Target: Gold and Lace Barbie	1989	10	25
7801	Tennis Star Barbie & Ken	1988	18	40
5949	Toy R Us Winter Fun Barbie	1990	10	25
9342	Toys R Us Beauty Pagent Skipper	1991	10	25
2721	Toys R Us Fun School	1991	6	15
2917	Toys R Us Sweet Romance	1991	8	20
7799	Toys-R-Us: Show n' Ride Barbie	1989	10	37
2996	Tropical Barbie Deluxe Gift Set	1985	20	45
3556	Tutti and Todd Sundae Treat Set	1966	120	300
	Tutti Me n' My Dog	1966	40	125
	Tutti Nighty Night Sleep Tight	1965	65	175
4097	Twirley Curls Barbie Gift Set	1982	30	85
4589	Wal-Mart 25th year: Pink Jubilee Barbie	1987	20	60
3678	Wal-Mart Ballroom Beauty	1991	8	20
7335	Wal-Mart Dream Fantasy	1990	8	20
1374	Wal-Mart: Frills and Fantasy Barbie	1988	7	30
3963	Wal-Mart: Lavender Look Barbie	1989	7	20
	Walking Jamie Strollin' in Style Gift Set	1972	250	550
9852	Wedding Party Gift Set Six Dolls	1991	45	125
5408	Western Fun Gift Set Barbie & Ken	1990	12	30
5408	Western Fun Gift Set Barbie & Ken	1989	12	30
5410	Winn Dixie Pink Sensation	1990	6	15
3284	Winn Dixie Southern Beauty	1991	10	25
7637	Winn-Dixie: Party Pink Barbie	1989	7	18
2582	Woolsworth's Special Expressions	1991	4	10
2583	Woolsworth's Special Expressions, black version	1991	4	10
5504	Woolworth's Special Expressions	1990	3	10

Barbie & Friends

NO.	TOY NAME	YEAR	MNP	MIP
5505	Woolworth's Special Expressions, black version	1990	4	10
7326	Woolworth: Special Expressions Barbie, black version	1989	5	15
4842	Woolworth: Special Expressions Barbie, white version	1989	5	10

Porcelain Barbies

NO.	TOY NAME	YEAR	MNP	MIP
1110	30th Anniversary Ken	1991	100	200
5475	Benefit Performance Barbie	1988	115	375
1708	Blue Rhapsody Barbie	1986	350	750
3415	Enchanted Evening Barbie	1987	175	350
9973	Gay Parisienne Blonde	1991	200	400
9973	Gay Parisienne Brunette	1991	100	200
9973	Gay Parisienne Redhead	1991	200	400
7613	Solo in the Spotlight	1990	100	200
5313	Sophisticated Lady	1990	100	200
2641	Wedding Party Barbie	1989	300	650

Erector Sets
I Build, Therefore I Am

Building things

The drive to build is one of our most basic, tucked away deep in our primal memory right beside food, shelter, and that procreation thing. The ability to build is highly valued by Americans, and we have long sought to develop that talent in our children--particularly with construction toys.

Most American boys can recall the number, colors, heft, and smell of their first construction set, even if they may not remember when they received it or who gave it to them. Those unfortunates who never got one no doubt made do with sets of their own design, wood scraps, string, old wheels, and nails.

Few joys of childhood rival that of laboriously constructing some mechanical wonder, some working machine, watching it work, and then gleefully tearing it all apart.

From a parental perspective, a construction set was and still is the ideal toy. It taught us youngsters basic mechanics, logic, fine motor coordination, and a host of other skills, but most importantly, it kept us out of our parents' hair for hours on end.

Now we give modern versions to our kids for the same reason--to keep them out of our hair. Okay, okay, also because they are fun and educational, but let's not say the "e" word too loudly, shall we?

Construction toys do not have the glamor of Barbie dolls or the celebrity quotient of Teenage Mutant Ninja Turtle action figures, but they still account for a mouthwatering slice of the annual toy pie. Americans buy in the neighborhood of $500 million in construction toys each year--which put another way would be enough to build a real town of 4,000 real homes, each with a $125,000 price tag on its shiny new doorknob. By any reckoning, that is a pretty big stack of blocks.

And there is no end in sight. The construction toy market has entered another growth period, spurred by new retailing techniques, new product lines such as K' Nex, ever expanding "old standards" such as Lego, and more products designed for 1990s girls.

So what of the Erector Set, the slumbering granddaddy of the field? Meccano-Erector revived the revered name in 1992 for the U.S. market, and sets have reportedly sold out each Christmas season since then.

That same public affection that led Meccano-Erector to revive the cherished Erector hallmark has also given birth to an active collector market in the original Gilbert Erector sets.

The Erector Set

Alfred Carlton Gilbert, born in 1884, was both a product of and producer for his time, the industrial coming of age in America. In his own way he helped our growing country become the greatest industrialized nation on Earth, and in the process inspired legions of bright young tinkerers to grow up to become architects of the future, engineers, scientists, and craftsmen of all types.

Young Alfred loved magic tricks, a fascination that would later pay off almost "like magic." Also an aspiring sports star, he decided on a career in health education, and worked his way through Yale medical school by doing magic shows. At Yale he also won a berth on the 1908 U.S. Olympic team and later won the Gold Medal in the Pole Vault.

While at Yale, Gilbert began giving magic lessons and became frustrated by the lack of available magic props for his students. A fellow amateur magician named John Petrie was also a mechanic, and he and Gilbert began making small magic kits and selling them.

After the Olympics, Gilbert returned to med school and the magic business, which soon grew into a mail order magic company, supplying both amateurs and professionals. Gilbert and Petrie named themselves The Mysto Magic Company.

By 1909, the newly degreed Dr. Gilbert decided to chuck his medical career and pursue his magic business. The magic business was good. In his promotional travels across America Gilbert began selling to toy store buyers, and from dealing with them he was struck by the need for quality American made toys. He went home to think on it, and the result of that thought became the Erector Set.

The origin of the first Erector Set is the stuff of debate and myth, and even Gilbert told several versions of it during his life. What is fact is that Petrie was not impressed with Gilbert's new toy idea, and this eventually led Gilbert to buy out Petrie's interest in the Mysto Magic Company.

Gilbert showed his new Mysto-Erector toy at the 1913 Toy Fair and retail orders poured in, threatening to swamp his new company. Gilbert rolled out a major national promotion for his new line, including an ingenious model building contest that provided him hundreds of new and free model configurations that he promoted in future catalogs.

The 1913 holiday sales season was a resounding success and Gilbert turned his sights on refining and improving the product line, upgrading the girders and providing factory-assembled motors for the 1914 kits. He also continued the model building contests with huge success, offering real cars as prizes, and those new model ideas were incorporated into product catalogs.

A new plant in 1915 allowed Gilbert to expand the line into better motors for his Erector sets, tin toys and small appliances to fill the off-season work void.

In 1916 Gilbert renamed his firm the The A.C. Gilbert Company, but kept the Mysto name for his magic sets. He also created the Gilbert Engineering Institute For Boys, a national club that became a highly successful promotional vehicle.

In 1917 World War I threatened to sideline the entire toy industry as "nonessential" production. Gilbert was called by the fledgling Toy Manufacturers Association, which he helped create, to lobby Congress for permission for the toy industry to continue making toys. Armed with his own toys, Gilbert triumphed and was hailed nationwide as The Man Who Saved Christmas.

Gilbert introduced his now famous chemistry sets in 1917 and proceeded to dramatically expand the Gilbert line in the years to come, introducing tool chests, a line of working tools called Miniature Machinery, more tin toys, and books.

Nineteen twenty was a year of great change, with a near total overhaul of the Gilbert lines and the introduction of a series of more advanced scientific, mechanical and radio sets. That year also saw the introduction of the "No. 10" deluxe Erector sets, which are now by far the most highly desirable of all Gilbert construction sets.

A true overhaul came in 1924, when Gilbert introduced The New Erector line. He completely redesigned the Erector system with thinner girders and debuted curved girders and numerous other advancements. The 1924 line also saw streamlining across all other divisions, with many toys dropped from production.

In 1927 Gilbert expanded his control over the construction toy market by buying Trumodel, a line of high quality toys that were redesigned and rolled into his deluxe Erector sets.

Gilbert's promotional activities took another turn in 1929 when he entered into a number of exclusive contracts with Sears, Macy's, J.C. Penney and other national retailers. Each company was allowed to market low-end Gilbert products, but not under the Erector name. They were sold under the names Trumodel, Steel-Tech, and Little Jim among others, with each retailer having a different "exclusive" line. These sets have since become collectible in their own right.

Also in 1929, Gilbert essentially monopolized the American construction toy market by acquiring his chief rival, Meccano. For 1930 the Erector and Meccano lines were partially merged in the New American Meccano line, now commonly called Erector-Meccano.

The Great Depression finally hit the Gilbert company, but not before the Erector series hit its all time pinnacle with the Erectro Hudson Locomotive model, which was included in the top of the line sets. Nineteen thirty-one also saw the most expansive set ever produced, the massive No. 10 set. In addition to containing parts to build all models from lower numbered sets (which included the White Truck model, the Steam Shovel, and the Zeppelin), it also included parts for both the Hudson Locomotive and its tender. Gilbert would never offer an Erector set this grand again.

The Depression forced major downsizing at Gilbert, but the company would survive. The Erector line was acquired by Gabriel Industries in 1967. As previously noted, the Meccano-Erector line has recently been revived with resounding success.

Trends

While it is a relatively small subset of the toy collecting field, construction set collecting, particularly of Erector and Meccano products, is enjoying moderate and steady growth. Each issue of Toy Shop contains up to a dozen ads from dealers and collectors who specialize in the field. A new series of books has recently been released, and such works often focus collector attention on underdeveloped fields, creating interest and spurring market growth. However, not all collecting fields have their day in the sun, but this one, rich as it is in history and variety, presents a strong case for higher visibility.

The Top 10 Erector Sets
(in Near Mint In Box condition)

1. #10, 1932, The Climax of Erector Glory Set $12,000
2. #10, 1931, Deluxe Set 10,000
3. #150, 1930, Master Engineer's Outfit 9,500
4. #10, 1926 Set 8,500
5. #10, 1927, Erector Deluex Set 8,000
6. #10, 1928, The Complete Deluxe Set 8,000
7. #9, Mechanical Wonders Set 7,000
8. #F, Complete Assembled Zeppelin, 1929 6,500
9. #10, 1925 Set 5,800
10. #75, Combination Airkraft & Zeppelin Set 4,500

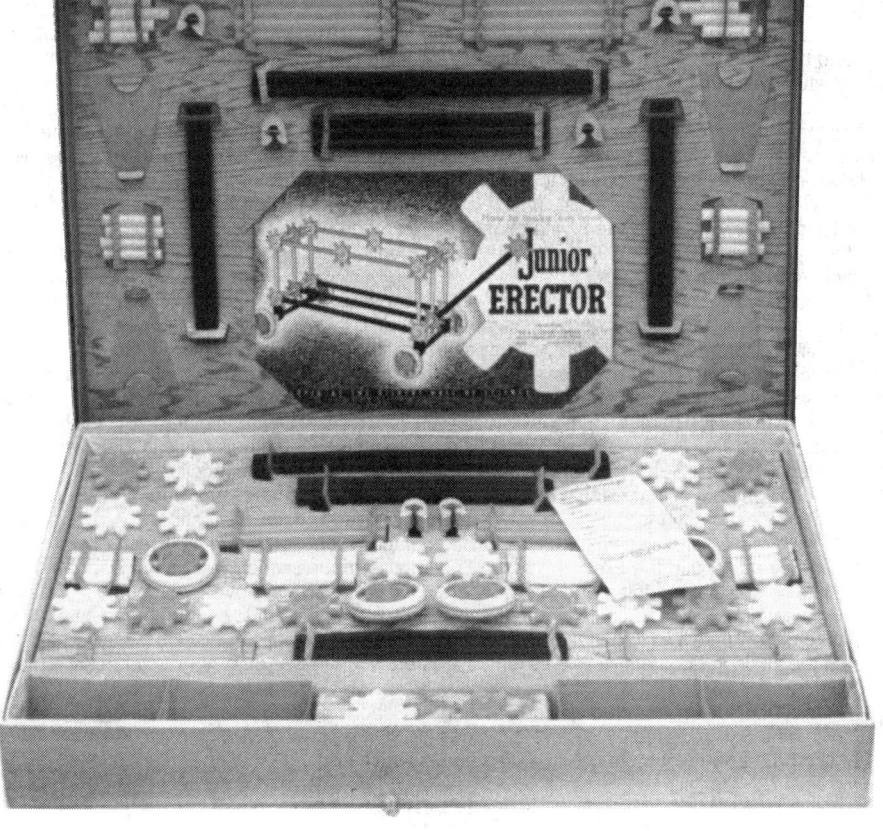

All plastic "Junior Erector" set from Gilbert Company, 1949.

ERECTOR SETS

ERECTOR SETS

NO.	NAME	DESCRIPTION	GOOD	EX	MINT
45	Complete Airplane Construction Set	built 19 different airplane models, in a cardboard box	750	1800	3000
7	The Super No. 7 Erector The Set That Builds The Steam Shovel	two layer set with removable cardboard parts tray, built 533 different models and contained 473 parts, featured the Steam Shovel in a stained varnished wooden box	75	175	295
75	Erector Combination Airkraft & Zepplin Set	built 22 different models one featured the airplane and Zepplin all parts, in a heavy cardboard box	1125	2700	4500

1913

NO.	NAME	DESCRIPTION	GOOD	EX	MINT
0		built 13 different models, in cardboard box	150	365	600
1		built 27 different models, in cardboard box	165	395	650
2		built 39 different models, large wheels, in cardboard box	190	450	750
3		built 55 different models, gears, in cardboard box	200	495	825
4		built 65 different models, parts to build electric motor, in cardboard box	300	725	1200
5		built 76 different models, in cardboard box	315	750	1250
6		built 88 different models, in cardboard box, two layer set with a removable tray	325	810	1350
7		built 92 different models, propeller blades, in cardboard box, two layer set with a removable wooden tray	450	1050	1750
8		built 100 different models, in wooden box, three layer set with two removable wood trays	665	1600	2650

1914

NO.	NAME	DESCRIPTION	GOOD	EX	MINT
0		single layer set, built 69 different models with 98 parts in a cardboard box	40	90	150
1		single layer set, built 88 different models with 140 parts and small wheels in a cardboard box	50	120	200
2		single layer set, built 120 different models with 205 parts and large wheels in a cardboard box	50	120	200
3		built 176 different models with 345 parts, gears and round plate, in a cardboard box	65	150	250
4		double layer set with removable cardboard parts tray, built 207 different models with 571 parts, assembled electric motor and contact strip in a cardboard box	95	225	375
5		double layer set with removable cardboard parts tray, built 229 different models with 679 parts with a propeller blade in a cardboard box	105	255	425
6		small two layer wooden set with removable wooden parts trays, built 264 different models with 1000 parts, large grooved wheels in wooden box	130	315	525
7		small two layer wooden set with removable wooden parts trays, built 278 different models with 1291 parts and a triangle piece in a wooden box	200	495	825
7 1/2	Motorized Erector	two layer set with removable wooden parts tray, built 543 different models and contained 870 parts, White truck model, cast iron wheels, rubber tires, in a wooden box	625	1500	2500

Erector Sets

NO.	NAME	DESCRIPTION	GOOD	EX	MINT
8		small three layer set with two removable wooden parts trays, built 304 different models with 1800 parts, gear wheel with holes and the crown gear with holes in a wooden box	325	775	1300

1915

NO.	NAME	DESCRIPTION	GOOD	EX	MINT
	Toy Builder	small set, built 61 toys with 50 parts, in a cardboard box, box lid is red, yellow and black	115	275	450
1		small set, built 88 different models with 140 parts, in a cardboard box	40	90	150
1M		small set, built 88 different models with 140 parts and included an electric motor, in a cardboard box	50	115	195
2		small set, built 120 different models with 205 parts with large wheels in a cardboard box	45	105	175
2M		small set, built 120 different models with 205 parts with large wheels and a electric motor in a cardboard box	55	135	225
3		small two layer set with removable parts tray, built 176 different models with 345 parts with gears and the round plate in a cardboard box	55	135	225
3M		small two layer set with removable parts tray, built 176 different models with 345 parts with gears, electric motor and the round plate in a cardboard box	75	180	300
4		two layer set with removable wooden parts tray, built 207 different models, 571 parts, electric motor and contact strip in a wooden box	80	185	310
5		two layer set with removable wooden parts tray, built 229 different models, 679 parts and propeller blade in a wooden box	90	210	350
6		two layer set with removable wooden parts tray, built 264 different models, 1000 parts and reverse baseplate in a wooden box	95	225	375

1916

NO.	NAME	DESCRIPTION	GOOD	EX	MINT
1		built 111 different models and contained 140 parts in a cardboard box	40	90	150
1		built 111 different models with 140 parts in a cardboard box	40	90	150
1M		built 111 different models and contained 140 parts, with an electric motor in a cardboard box	50	120	200
1M		built 111 different models with 140 parts and a electric motor in a cardboard box	50	120	200
2		built 152 different models and contained large number of parts from No. 1, with large wheels in a cardboard box	45	105	175
2		built 152 different models with large no. of parts from No. 1 with large wheels in a cardboard box	45	105	175
2M		built 152 different models and contained large number of parts from No. 1, with an electric motor in a cardboard box	65	150	250
2M		built 152 different models with large no. of parts from No. 1 with an electric motor in a cardboard box	55	135	225

Erector Sets

NO.	NAME	DESCRIPTION	GOOD	EX	MINT
3		two layer set with removable parts tray, built 197 different models and contained large number of parts from No. 2, with gears and the round plate in a cardboard box	65	150	250
3		two layer set with removable parts tray, built 197 different models with large no. of parts from No. 2 with gears and the round plate in a cardboard box	55	135	225
3M		two layer set with removable parts tray, built 197 different models and contained large number of parts from No. 2, with an electric motor in a cardboard box	75	180	300
4		two layer set with removable parts tray, built 278 different models and contained large number of parts from No. 3, with an electric motor and contact strip in a cardboard box	90	210	350
4		two layer set with removable parts tray, built 278 different models and contained large number of parts from No. 3, with an electric motor and contact strip in a wooden box	75	180	300
4		two layer set with removable parts tray, built 278 different models with large no. of parts from No. 3 with an electric motor and contract strip in a cardboard box	90	210	350
4		two layer set with removable parts tray, built 278 different models with large no. of parts from No. 3 with an electric motor and contract strip in a wooden box	70	165	275
4		two layer set with removable parts tray with a special 12 x 2" griders for building I & T beams, three sect. gear, mitre gears, pulley belt & bands in a wooden box	275	675	1100
5		two layer set with removable parts tray, built 317 different models and contained large number of parts from No. 4, and a propeller in a wooden box	90	210	350
5		two layer set with removable parts tray, built 317 different models with large no. of parts from No. 4 with a propeller in a wooden box	90	210	350
6		two layer set with removable parts tray, built 382 different models and contained large number of parts from No. 5, and a reverse plate in a wooden box	95	225	375
6		two layer set with removable parts tray, built 382 different models with large no. of parts from No. 5 with the reverse baseplate in a wooden box	100	250	400
7		two layer set with removable parts tray, built 410 different models and contained large no. of parts from No. 6, with triangle pieces in a wooden box	175	405	675
7		two layer set with removable parts tray, built 410 different models with large no. of parts from No. 6 with the triangle pieces in a wooden box	175	405	675
8		three layer set with two removable parts tray, built 454 different models and contained large no. of parts from No. 7, gear wheel & crown gear with holes in a wooden box with lock	275	675	1100

Erector Sets

NO.	NAME	DESCRIPTION	GOOD	EX	MINT
8		three layer set with two removable parts tray, built 454 different models with large no. of parts from No. 7 with gear wheel & crown whell with holes in a wooden box with lock/ key	275	675	1100

1918

4		two layer set with removable parts tray, with a special 12 x 2" girders for building I & T beams, three sect. gear, mitre gears, pulley belt & bands in a wooden box	250	625	1025

1920

1		built 111 different models in a cardboard box	40	90	150
1		built 152 different models with all parts in set No. 1, large wheels and round plate in a cardboard box	50	115	195
10		three layer set with removable wooden parts tray, 3rd layer pf parts mounted on cardboard to inside of lid, built 410 different models with all parts in set No. 8 in a wooden box	95	225	375
3		two layer set with removable parts tray, built 197 different models with all parts in set No. 2, and gears in a cardboard box	55	135	225
6		two layer set with removeable wooden parts tray, built 278 different models with all parts in set No. 3, electric motor, reverse baseplate, gearbox sideplates, etc. in wooden box	95	225	375
7		two layer set with removable wooden parts tray, built 317 different models with all parts in set No. 6, propeller blades in a wooden box	130	315	525
8		two layer set with removable wooden parts tray, built 382 different models with all parts in set No. 7 in a wooden box	175	400	675

1921

1		built 111 different models in a cardboard box	40	90	150
10		three layer set with removable wooden parts tray & 3rd layer of parts mounted on cardboard attached inside of lid, built 454 different models with parts found in No. 8, wooden box	325	775	1300
2		built 152 different models with parts found in No. 1, large wheels in a cardboard box	50	115	195
3		built 197 different models with parts found in No. 2, round plate & gearbox sideplates in a cardboard box	50	120	200
4		built 278 different models with parts found in No. 3, unassembled electric motor in a cardboard box	90	210	350
6		two layer set with removable cardboard parts tray, built 317 different models with parts found in No. 4, lg. gear wheel, assembled electric motor, in a wooden box	195	465	775
7		two layer set with removable wooden parts tray, built 382 different models with parts found in No. 6, pulley bands, rev. baseplate & propeller blades, in a wooden box	115	275	450

Erector Sets

NO.	NAME	DESCRIPTION	GOOD	EX	MINT
8		two layer set with removable wooden parts tray, built 410 different models with parts found in No. 7 in a wooden box	130	325	525

1922

NO.	NAME	DESCRIPTION	GOOD	EX	MINT
1		built 111 different models in cardboard box	40	90	150
10		three layer set with removable wooden parts tray, built 454 different models, contained a lg. number of parts found in No. 8 with 110V electric motor, in a wooden box	425	1050	1750
3		built 197 different models, contained a lg. number of parts found in No. 2, lg. wheels, rd. plate, gearbox sideplate & gears in cardboard box	50	120	200
4		two layer set with removable cdbd. parts tray, built 278 different models, contained a lg. number of parts found in No. 3, assembled electric motor in wooden box	75	175	295
7		two layer set with removable wooden parts tray, built 382 different models, contained a lg. number of parts found in No. 4, lg. wheel, pulley bands, etc. in a wooden box	115	275	450
8		two layer set with removable wooden parts tray, built 410 different models, contained a lg. number of parts found in No. 7 in a wooden box	130	325	525

1923

NO.	NAME	DESCRIPTION	GOOD	EX	MINT
1		built 111 different models, contains car trucks, in a cardboard box	40	90	150
3		built 197 different models, contains a lg. number of parts from No. 1, gears and round plate, in a cardboard box	50	120	200
4		two layer set with removable cdbd. parts tray, built 278 different models, contains a lg. number of parts from No. 3, electric motor & gearbox side plate, in a wooden box	75	175	295
5	The Erector Machinery Outfit	two layer set with a layer of parts on cdbd., built special machine shop models from cast iron miniature model machinery (lathes, saws, etc.) and many erector parts in a wooden box	325	775	1300
7	The Erector Machinery Outfit	three layer set with two removable cdbd. parts trays, built 382 different models, parts from No. 4, lg. gear wheel, reverse baseplate, propeller blade in a wooden box	195	450	750
8		two layer set with removable wooden parts trays, built 410 different models, parts from No. 7, in a wooden box	130	325	525

1924

NO.	NAME	DESCRIPTION	GOOD	EX	MINT
8		multi-layer set with removable pigeonhold parts compartment, built 559 different models and contained 1226 parts plus elect. parts & clamshell scoop in a stained & varnished box	575	1400	2350

1925

NO.	NAME	DESCRIPTION	GOOD	EX	MINT
1		built 278 different models and contained 104 parts in a cardboard box	40	90	150

Erector Sets

NO.	NAME	DESCRIPTION	GOOD	EX	MINT
10		multi-layer set with 3 removable wooden parts tray, built 677 different models and contained 2469 parts, plus lg. fly wheel, reverse baseplate, dial plates, etc. in a oak box	1450	3500	5800
3		built 381 different models and contained 169 parts in a cardboard box	45	110	185
4		two layer set with removable cardboard parts tray, built 500 different models and contained 235 parts plus an electric motor in a cardboard box	70	165	275
7		two layer set with removable cardboard parts tray, built 533 different models and contained 473 parts plus a Steam Shovel model in a stained & varnished box	75	180	300
8		built 559 different models and contained 1226 parts, removable wooden parts tray, plus elect. parts & clamshell scoop in a stained & varnished box, brass side grips, steel straps, etc	500	1200	2000

1926

NO.	NAME	DESCRIPTION	GOOD	EX	MINT
1	Dandy Beginner's Set	built 278 different models and contained 104 parts in a cardboard box	40	90	150
10		built over 700 different models, contained 2500 parts, fly wheels, tooth gears, propeller blades, etc. in varnished wooden box	2100	5100	8500
3	The Set with the Big Red Wheels	built 381 different models and contained 169 parts in a cardboard box	45	110	185
4	The Famous No. 4 Erector	two layer set with removable cardboard parts tray, built 500 different models and contained 235 parts, an electric motor in a cardboard box	70	165	275
8	The Set with Dredge & Electric Motor	multi-layer set with removable wooden parts tray, built 570 different models and contained 1226 parts, elect. parts & clamshell scoop in a stained/varnished wooden box	525	1250	2100

1927

NO.	NAME	DESCRIPTION	GOOD	EX	MINT
1	Dandy Beginner's Set	built 278 different models and contains 104 parts in a cardboard box	40	90	150
10	Erector Deluxe In All Its Glory	built 714 models, 2000 parts, fly wheel, rev. switch, dial plates, pointer, beam girders, etc. in eight drawer oak chest	2000	4800	8000
3		built 371 different models and contains 169 parts, gears and the big red wheels in a cardboard box	45	110	185
4		two layer set with removable cardboard trays, built 500 different models and contains 235 parts, electric motor in a cardboard box	70	165	275
7		two layer set with removable cardboard parts tray, built 533 different models and contains 473 parts, featured the Steam Shovel model in a stained/varnished wooden box	75	180	300
7 1/2	The Wonderful New 7 1/2	two layer set with removable wooden parts tray, built 554 different models and contains 627 parts, featured the "chassis" truck model in a stained/varnished wooden box	175	400	675

Erector Sets

NO.	NAME	DESCRIPTION	GOOD	EX	MINT
8	The 1927 Set-The Trumodel Set	The Trumodel Set, two layer set with wooden parts tray, 922 parts, Trumodel parts & boiler plates, featuring the Stiff Leg Derrick model, in varnished wood box	625	1550	2550
A	Big Girder Set	two layer set with removable cardboard parts tray, built 42 different models, contained 442 parts in a cardboard box	375	900	1500

1928

NO.	NAME	DESCRIPTION	GOOD	EX	MINT
1	Beginner's Set	built 240 different models in a cardboard box, illustrated on the lid with a picture of White truck and a coal loader	55	135	225
10	The Complete Deluxe Set	built 1000 different models, contains the giant fly wheel, truck & ferris wheel parts, machine frames, pile driver wt., wood handle screwdriver, etc. in a 9 drawer oak chest	2000	4800	8000
3		built 562 different models in a cardboard box, illustrated on the lid with a picture of a railroad bridge	50	120	200
4	With Powerful Electric Motor	two layer set with removable cardboard parts tray, built 680 different models, electric motor & gearbox in a cardboard box pictured of a railroad bridge	70	165	275
7	Builds the Steam Shovel	two layer set with removable cardboard parts tray, built 719 different models, featuring the Steam Shovel in a stained/varnished wooden box with brass side grips & suitcase locks	75	180	300
7 1/2	Builds the Chassis	two layer set with removable metal parts tray, built 749 different models, featuring the White truck model in a wooden box finished in red with black brass corners	130	325	525
77	Builds the Steam Shovel	three layer set with 2 removable metal parts tray, built 727 different models, featuring the Steam Shovel with nickel disc wheels in a red with blk corner, brass side grips & locks	95	225	375
8	The Trumodel Set with 110 Volt Motor	multi-layer set with 2 removable parts trays, built 827 different models, including the Stiff Leg Derrick, the Trumodel parts, machine frame, cams, & 110V motor, in a wooden box	425	1050	1750
A	Giant Girder Set	two layer set with removable cardboard parts trays, built 42 different models, contains 442 parts, featuring the beam griders in a cardboard box	375	900	1500
B	Ferris Wheel Set	two layer set with removable metal parts tray, built 54 different models, featured the ferris wheel & the big channel girders in a wooden box red with black corners	275	675	1100
C	Builds the Ferris Wheel	built 19 different airplane models, contains new airplane parts with a 110 volt motor & ceiling swivel in a wooden box red with black corners, brass grips & suitcase locks	800	1900	3200

1929

NO.	NAME	DESCRIPTION	GOOD	EX	MINT
1	Dandy Beginner's Set	built 492 different models, in a cardboard box	55	135	225
10	Complete, Erector in All Its Glory	multi layer set with 3 removable metal parts tray & a multiplewooden dividers, 250 pg manual, built all models no. 9 including ferris wheel, fly wheel models, etc. in a wooden box	375	900	1500

Erector Sets

NO.	NAME	DESCRIPTION	GOOD	EX	MINT
3	The Set with the Long Girders	built 611 different models with small disc wheels, in a cardboard box	70	165	275
6	The Big Red Chest with the Steam Boiler	two layer set with removable metal parts tray, built 765 different models including boilers, in a wooden box	70	165	275
7	The Set that Builds the Steam Shovel	two layer set with removable metal parts tray, built 777 different models including Steam Shovel, in a wooden box	75	180	300
7 1/2	The Set that Builds the Steam Shovel	two layer set with removable metal parts tray, built 809 different models including White truck model, in a wooden box finished with red & black corners	130	325	525
8	Trail Blazing Set that Builds the Zepplin	two layer set with removable metal parts tray, built all models in no. 7 1/2 including Zepplin model, in a wooden box finished with red & black corners	650	1550	2600
9	Mechanical Wonders Set with 110 Volt Motor	multi layer set with 2 removable parts tray & a wooden divider, built all models in no. 8 including mechanical wonder models, and 110 volt motor in a wooden box	1500	3600	6000
A	Giant Girder Set	two layer set with removable cardboard parts tray, built 42 different models, featuring the beam girders, & contained 442 parts in a cardboard box	375	900	1500
B		two layer set with removable metal parts tray, built 54 different models, featuring the ferris wheel, nickel disc wheels, etc. in a wooden box	250	625	1025
C	Complete Airplane Construction Set	built 19 different airplane models, featuring airplant parts including a 110 volt motor and ceiling swivel in a cardboard box	875	2100	3500
D	Completely Assembled All Metal Airplane	contains airplane construction with rivets & 110 volt motor & ceiling swivel	1125	2700	4500
F	Complete Assembled Zepplin	contains Zepplin constructed with rivets & swivel unit in an unfinished wooden box	1600	3900	6500

1930

NO.	NAME	DESCRIPTION	GOOD	EX	MINT
1	Apprentice Outfit	built over 100 different models, in a cardboard box	60	150	245
110	Big Motor Car Outfit	two layer set with removable parts tray in red, featuring the truck & car models and large rubber tires & radiator in a wooden box	165	395	650
125	Ship Building Outfit	two layer set with removable metal parts tray in red, built over 350 different models, contains a hull & other ship parts, in a wooden box	1125	2700	4500
150	Master Engineer's Outfit	Master Engineer's Outfit, multi layer set with two metal parts trays, built over 500 different models, contains a ferris wheel & giant revolving crane, etc. in a wooden box	2400	5700	9500
3	Junior Engineer's Outfit	built over 200 different models, contained a large pulley wheels, braced girder & segment plates, in a cardboard box	70	165	275
5	Meccano Super Power Outfit	one layer set builts over 300 different models with electric motor, in a cardboard box	115	275	450

1931

NO.	NAME	DESCRIPTION	GOOD	EX	MINT
1	Dandy Beginner's Set	built over 100 different models, in a cardboard box	55	135	225
10	Deluxe Set	the ultimate set, 3 removable metal parts trays, built over 500 different models, including Steam Shovel, White truck, Hudson locomotive, etc. in oak box	2500	6000	10000

Erector Sets

NO.	NAME	DESCRIPTION	GOOD	EX	MINT
3	The Set with the Long Girders	built over 200 different models, included the small disc wheels in a cardboard box	70	165	275
4	Famous No. 4 Erector with Powerful Electric Motor	two layer set with removable cardboard parts tray, built over 300 different models, included electric motor & disc wheels in a cardboard box	95	225	375
45	A Complete Airplane Construction Set	built 19 different airplane models, in cardboard box	625	1500	2500
6	The Wonderful No. 6 Erector	two layer set with removable metal parts tray, built over 325 different models, featuring boiler models in a cardboard box	95	225	375
7	The New Big No. 7 Erector	two layer set with removable metal parts tray, built over 350 different models, in a wooden box	115	275	450
7 1/2	The Sensational No. 7 1/2 Erector	two layer set with removable metal parts tray, built over 400 different models, in a wooden box	195	450	750
8	The Twentieth Century No. 8	two layer set with removable metal parts tray finished in red, built over 399 different models, including all models from No. 7 plus the Erector Hudson & Hudson parts in wooden box	200	600	1000
9	Mechanical Wonders Set with 110 Volt Motor	two layer set with removable metal parts tray, built over 451 different models, including Zepplin, Mechanical Wonders models, and featured models in wooden box	1750	4200	7000
A	The Mile-A-Minute	two layer set with removable metal parts tray finished in red, built the Erector Hudson locomotive and 12 other locomotive parts in wooden box	225	525	875
B	The Big Girder Set	two layer set with removable metal parts tray, built 58 differnet models and contained nickel wheels, the big channel griders, and parts to build ferris wheel, in wooden box	300	725	1200
D	Completely Assembled All-Metal Airplane	contained factory assembled airplane constructed with rivets and 110 volt motor and ceiling swivel	1000	2400	4000
L	Locomotive & Tender Model	contained factory assembled Erector Hudson locomotive & tender constructed with rivets and display truck	1250	3000	5000
T	Erector Tender Accessory Set	two layer set with removable cardboard parts tray, contains tender parts only	675	1650	2750

1932

NO.	NAME	DESCRIPTION	GOOD	EX	MINT
7	The New Big No. 7 Erector	two layer set with removable metal parts tray, built over 350 different models, in a wooden box	150	375	600
7 1/2	The Sensational No. 7 1/2 Erector	two layer set with removable metal parts tray, built over 400 different models, in a wooden box	195	475	775

Character Toys

This year's character toy section features expanded listings for Popeye, Dick Tracy, Oz, Man From U.N.C.L.E., Captain Marvel, and Superman among others. These expanded sections will give you an idea of how future editions of Toys & Prices will look, as we will continue fleshing out sections for each new edition, not only in character toys, but throughout the book. We welcome reader submissions of unlisted items for future inclusion.

Everybody In?

Character toy collecting is the broadest field in the hobby. If time and space would permit, we could easily fill this book with character toys and still only skim its colorful surface. From Gomez Addams to Mickey Mouse, from Popeye to Zorro and anywhere in between, the choices are boggling. Collectors are restrained only by their budgets, as the variety even within single categories such as Mickey Mouse or Popeye is wide enough to comprise entire collections. Tin wind-ups, bisques, dolls, books, puppets, playsets, puzzles, models, battery-ops, games--it never ends.

Beginning collectors are soon faced with either narrowing their collecting into specialties or being overwhelmed by choice. The options are so tempting that many collectors "go eclectic" and never specialize at all. Flash Gordon and Mickey Mouse not only look just fine together, they belong together. Dick Tracy, Doctor Who, and Donald Duck look great side by side on a shelf, and whatever configuration a developing toy collection takes, it will tell a fascinating story of its time and its relationship to all other toys around it. For this reason, those "eclectic" collections can be the most satisfying of all.

Twenty-four inch stuffed Santa Claus from the late 1950s.

For those who are already engaged in or are considering specialization, a few words of introduction to some of the major classes in this section are in order.

Going Up--The Universe of Walt

Walt Disney has often been called the single greatest contributor to American culture. Disneyana is one of the largest and most vigorous areas of toy collecting, and it shows no signs of slowing.

No comic creations have ever been as honored (or as frequently pirated) as Disney characters. From the rat-snouted mouse of 1928 to the virtually human version circa 1995, the various stages of Mickey Mouse have adorned more toys, trinkets, and whatnots than any other image in modern history. And he is hardly alone. The Universe of Walt has grown full, is richly populated, and the theory of cosmic expansion holds true--each new Disney film brings new planets of characters into being. And across our planet, millions of budding little collectors line up, first for their movie tickets, then for their popcorn, and then for every piece of merchandise their allowances will buy.

The implication is clear. The future belongs to little mice, cranky ducks, and wooden puppets. The market demand for the classic toys of Mickey, Donald, Pinocchio, and other early Disney characters is secure and will stay that way until the end of the world as we know it. And we can only guess what those first 1930s Disney toys will be worth when they finally turn 100 years old, or what The Lion King toys will be worth in 2094.

Going Down--Aarf!

But not all toys endlessly escalate in value. Enduring popularity is a key element in predicting future demand for pop culture collectibles. How popular was it when it was new? How long did it stay at the top? These and other factors all impact on future collectibility.

As collectibles mature and their active collecting public increases, their prices can be reasonably expected to rise. But once the collector base begins to erode, once it gets old or loses interest or gets interrupted by a war or replaced by the next big thing, then values can begin to slide. When collectors go away, sometimes an entire hobby goes with them. The best insurance against this is the continued popularity of some aspect of their identity, particularly if it is character based.

Another prime example of this longevity is Popeye, who made his first appearance in 1929 in a comic strip called the "Thimble Theatre," created by E.C. Segar. He rapidly became one of the most popular characters of his day, bringing fame to his cartoon companions as well.

That fame manifested itself in some of the most beautiful and well designed toys of the 1930s. Classic Marx toys such as the Popeye the Champ and Popeye and Olive Oyl Jiggers have risen to their respective mint values of $4,600 ($3,200 in last year's edition) and $1,800 ($1,500 last year) not only because they are superb examples of the toy makers art, but also because their characters continue to charm to this day.

Even though they are often over fifty years old, Popeye cartoons are still seen every day by children over their morning cereal, and the average four-year-old is well acquainted with the pipe smoking sailor man and his string bean girl friend. These new fans may be too young to read, but they all know Popeye's theme song by heart.

It is this durability that translates into collectibility, and the power to draw new collectors from each maturing generation. The universal and continuing appeal of Mickey and Minnie Mouse, Donald Duck, Snoopy, Popeye, Snow White, and others such as Little Orphan Annie and Dick Tracy are assurances that future generations of collectors will seek out their toys and place them proudly on their shelves, someday perhaps right next to their fifty-year-old Teenage Mutant Ninja Turtle and "rare" Nightmare Before Christmas action figures.

Trends

Of toy collecting's blue chip investments, character toys remain the "blue-est." Demand for a particular character may ebb and flow according to times and market whims, but overall this is one of the strongest and most reliable areas of toy collecting. Throughout our listings this year, roughly half the listings held their value, and half saw increases, older toys and tin toys in particular. A nominal percentage of items were reduced in value, and these can be attributed to regional or other variations according to reviewer.

The Top 50 Character Toys
(in Mint Condition)

1. Popeye the Heavy Hitter, Chein $6,500
2. Popeye the Acrobat, Marx 5,500
3. Popeye The Champ, Marx, 1936..4,600
4. Superman Hood Ornament, Lee Mfg. Co., 1940s 4,000
5. Captain Marvel Sirocco Figurine, Fawcett, 1940s 3,500
6. Captain Marvel Statuette, Fawcett, 1940s.................................3,200
7. Dick Tracy Table Lamp, 1950s..3,000
8. Man From U.N.C.L.E. THRUSH Rifle, Ideal, 1966 2,500
9. Popeye and His Punching Bag, Chein, 1930s...........................2,500
10. Popeye Express, with airplane, Marx, 1936 2,400
11. Beany and Cecil Animated Clock, 1960s 2,000
12. Superman Doll, Super Babe, Imperial Crown, 1947...............2,000
13. The Wonderful Game of OZ, Parker Brothers, 1921 2,000
14. Mickey Mouse Radio, Emerson, 1934......................................2,000
15. Mickey Mouse Tricycle Toy, Steiff, 1932 2,000

Mr. T twelve-inch doll from galoob, 1983; Brooke Shields doll from Ljn, 1982.

16. Dick Tracy Pedal Car, Murray, 1950s 2,000
17. Snow White/Seven Dwarfs Radio, Emerson, 1938 2,000
18. Superman Roll-Over Plane, Marx, 1940 2,000
19. Superman Sirocco Statue, Syracuse Ornament Co., 1942 2,000
20. Mechanical Superman Turnover Tank, Linemar, 1940 2,000
21. Superman Tank, (large) Linemar, 1958 2,000
22. Superman Turn-Over Tank, Marx, 1940 2,000
23. Popeye Handcar, Marx 2,000
24. Superman Doll, Ideal, 1940 1,800
25. Popeye and Olive Oyl Jiggers, Marx, 1936 1,800
26. Amos & Andy Wind-ups, Marx, 1930, each 1,650
27. Amos & Andy Fresh Air Taxicab, Marx, 1930s 1,600
28. Funny Fire Fighters (Popeye, Bluto) Marx, 1930s 1,600
29. Mickey Mouse Lionel Circus Train, Lionel, 1935 1,600
30. Popeye Flyer, (w/Wimpy) Marx, 1936 1,600
31. Man From U.N.C.L.E. Attache Case, Ideal, 1965 1,500
32. Barney Google and Spark Plug, Nifty, 1920s 1,500
33. Superman Wrist Watch, New Haven, 1940s 1,500
34. Charlie Chaplin Wind-Up 1,250
35. Popeye Flyer, Marx, 1936 1,250
36. Popeye Jigger, Marx 1,250
37. Andy Gump Automobile, Arcade, 1930s 1,200
38. Donald Duck Driving Pluto 1,200
39. Donald Duck Duet, Marx, 1946 1,200
40. Snow White Doll Set, Deluxe, 1940s 1,200
41. Popeye The Pilot, Marx, 1936 1,200
42. Superman Supertime Wrist Watch, National Comics, 1950s 1,200
43. Judy Garland as Dorothy Doll (Oz), Ideal, 1939 1,100
44. Mickey Mouse Riding Toy, Mengel, 1930s 1,100
45. Disney Ferris Wheel, Chein 1,000
46. Jack Pumpkinhead Doll (Oz), Oz Doll & Toy Co., 1924 1,000
47. Scarecrow Doll (Oz), Oz Doll & Toy Co., 1924 1,000
48. The Strawman By Ray Bolger Doll (Oz), Ideal 1939 1,000
49. Tin Man Doll (Oz), Oz Doll & Toy Co., 1924 1,000
50. White Rabbit Figure (Alice) Disneyland promo, 1984 1,000

Top to bottom: Bugs Bunny Ceramic Figure, 1975, Warner Brothers; Popeye Spinach Can Pop-Up, 1957, Mattel; Howdy Doody Ventriloquist's Dummy, 1970s, Goldberger; Lone Ranger Sheriff Jail Keys, 1945, Esquire Novelty; Roy Rogers and Dale Evans Western Dinner Set, 1950s, Ideal; Yogi Bear Bubble Pipe, 1963, Transogram.

101 Dalmations

TOY	COMPANY	YEAR	DESCRIPTION	GOOD	EX	MIB
101 Dalmatians Snow Dome	Marx	1961	3x5x3 1/2" tall	35	65	100
101 Dalmatians Wind-up Toy	Linemar	1959		60	115	375
Dalmatian Pups Figurine Set of Three	Enesco	1960s	4 1/2" tall	12	23	125
Lucky Figure "101 Dalmatians"	Enesco	1960s	4" tall	35	65	100
Lucky Squeeze Toy	Dell		7" tall, squeakers in the bottom	5	10	35

Addams Family

TOY	COMPANY	YEAR	DESCRIPTION	GOOD	EX	MIB
Addams Family Bank		1964	The Thing mechanical plastic battery operated bank, hand takes money	35	65	95
Addams Family Movie Key Chains				2	4	6
Lurch Figure	Remco	1964		55	105	160

Amos & Andy

TOY	COMPANY	YEAR	DESCRIPTION	GOOD	EX	MIB
Amos & Andy Card Party		1930	6"x8", score pads & tallies	14	25	75
Amos & Andy Fresh Air Taxicab Co.	Marx	1930s	5"x8" long, wind-up, tin	650	1000	1600
Amos & Andy Jigsaw Puzzle	Pepsodent	1932	8 1/2"x10", pictured Amos, Andy, Brother Crawford, Lightin! & King Fish	20	35	55
Amos Wind-up	Marx	1930	12" tall, tin, wind-up	575	1075	1650

Andy Gump

TOY	COMPANY	YEAR	DESCRIPTION	GOOD	EX	MIB
Andy Gump Automobile	Arcade		7"x6", cast iron with a large figure	350	650	1200
Andy Gump Brush & Mirror			4" diameter, red on ivory colored surface of brush	18	35	95
Chester Gump Playstone Funnies Mold Set		1940s		50	95	145
Chester Gump/Herby Nodders		1930s	ceramic 2 1/4" string nodders, each	55	100	225

Archie

TOY	COMPANY	YEAR	DESCRIPTION	GOOD	EX	MIB
Archie Picture Puzzle	Jaymar	1960s	"Swinging Malt Shop" complete	9	16	25
Veronica Figure	Mattel	1977	6 1/2" tall, huggable cloth vinyl head	9	16	45

Atom Ant

TOY	COMPANY	YEAR	DESCRIPTION	GOOD	EX	MIB
Atom Ant Kite	Roalex			11	20	30
Atom Ant Punch-out Set	Whitman	1966		16	30	75
Atom Ant Push Button Puppet	Kohner			14	25	75

Babes in Toyland

TOY	COMPANY	YEAR	DESCRIPTION	GOOD	EX	MIB
Morocco Mole Bubble Club Soap Container	Purex	1960s	7" hard plastic	11	20	45
Squiddley Diddly Bubble Club Soap Container	Purex	1960s	10 1/2" hard plastic	11	20	75

CHARACTER

Babes in Toyland

TOY	COMPANY	YEAR	DESCRIPTION	GOOD	EX	MIB
Winsome Witch Bubble Club Soap Container	Purex	1960s	10 1/2" hard plastic	11	20	45
Babes in Toyland Go Mobile Friction Car	Linemar	1961	4"x5"x6"	60	115	375
Babes In Toyland Hand Puppets	Gund		Silly Dilly Clown, Soldier, or Gorgonzo figures, each	35	65	100
Babes in Toyland Puzzle	Jaymar	1961	7 1/2"x8 /12"	7	13	35
Babes in Toyland Twist'n Bend Toy	Marx	1963	4" tall flexible toy with Private Valiant holding a baton	11	20	30
Babes in Toyland Wind-up Toy	Linemar	1950s	tin wind-up toy 66"	75	135	375
Cadet Doll "Babes in Toyland"	Gund		15 1/2" tall, fabric	11	20	75

Bambi

TOY	COMPANY	YEAR	DESCRIPTION	GOOD	EX	MIB
Bambi & Thumper Lamp			figural lamp	11	20	125
Bambi & Thumper Throw Rug		1960s	21"x39"	25	50	75
Bambi Book	Grosset & Dunlap	1942	black & white illustrations	12	23	35
Bambi Prints	New York Graphic Soc	1947	11"x14" framed, 8"x10" matted	25	50	75
Bambi Soakie				6	12	25
Flower Bank		1940s	5"x5"x7" tall, plaster bank	45	90	150
Thumper Ashtray	Goebel	1950s	4" tall	45	85	150
Thumper Ceramic Figural Bank	Leeds	1950s		45	80	125
Thumper Doll			16" tall, plush	12	23	35
Thumper Pull Toy	Fisher-Price	1942	#533, 7 1/2"x12", wood & metal, Thumper's tail rings the bell	18	35	150
Thumper Soakie				9	16	35
Thumper Story Book	Grosset & Dunlap	1942	color & black & white illustrations	12	23	35

Barbie

TOY	COMPANY	YEAR	DESCRIPTION	GOOD	EX	MIB
Barbie Wristwatch	Mattel	1973	3/4 figure illustration, articulated arms, vinyl band	23	40	125

Barnacle Bill

TOY	COMPANY	YEAR	DESCRIPTION	GOOD	EX	MIB
Barnacle Bill in a Barrel	Chein	1932	7 1/2" tall, tin, wind-up	225	400	800

Barney Google

TOY	COMPANY	YEAR	DESCRIPTION	GOOD	EX	MIB
Barney Google & Spark Plug	Nifty	1920s	7 1/2" tall, wind-up	400	750	1500
Barney Google & Spark Plug Figurines			3"x3", bisque, on white bisque pedestal	45	80	125
Barney Google Doll	Schoenhut	1922	8 1/2" tall, wood & wood composition	155	295	850
Spark Plug Doll	Schoenhut	1922	9" long & 6 1/2" tall, jointed wood construction with fabric	155	295	850
Spark Plug Pull Toy			10"x8" tall, wood	80	145	225
Spark Plug Squeaker Toy		1923	5" long, rubber with squeaker in mouth	14	25	125
Spark Plug Toy		1920s	5" tall, wood construction on wheels	45	90	225

Batman

TOY	COMPANY	YEAR	DESCRIPTION	GOOD	EX	MIB
Bat Bomb	Mattel	1966		40	70	75
Bat Cycle	Toy Biz			5	10	15
Batboat	Duncan	1987	12x8" on card	18	35	35
Batboat Pullstring Toy	Eidai (Japan)			60	115	175
Batman & Robin Society Button	Button World	1966	large size pinback button	11	20	20
Batman Action Figure	Takara	1989	12" tall doll in box	45	80	175
Batman Action Figure	Biken	1989	8" tall on card	11	20	20
Batman Baterang Bagatelle	Marx	1966	colorful bagatelle game in long, illustrated box showing Batman, Robin, and Batman knocking down a crook in a bright yellow suit	60	110	125
Batman Bendie Action Figure	Bully	1989	7" tall bendie	14	25	15
Batman Bendie Figure	Deline	1960s	on card	45	80	75
Batman Candy Cigarettes	(England)	1960s		25	45	50
Batman Cast and Paint Set		1960s	plaster casting mold and paint set	25	50	75
Batman Cereal Bank		1989	figural bank given away with Batman Cereal	4	7	10
Batman Charm Bracelet			on card	25	50	125
Batman Christmas Ornament	Presents	1989		7	13	15
Batman Colorforms		1976		12	23	35
Batman Crazy Foam		1974	metal car, plastic top, artwork on both	25	50	50
Batman Figure	Applause		12" tall with stand	18	35	25
Batman Give-A-Show Projector Cards	Kenner	1960s	set of four slide cards in package	45	80	50
Batman Kid's Belt		1960s	elastic belt with bronze logo buckle	25	45	50
Batman Kite	Hiflyer	1982		4	7	10
Batman Oil Paint By Numbers Set	Hasbro	1966	in 9 1/2x13" box	16	30	75
Batman Pencil Box	Empire Pencil Co.	1966	gun shaped pencil box with set of Batman pencils, on card	23	40	65
Batman Pinball Game	Marx	1960s	tin litho with plastic casing	25	50	125
Batman Radio Belt and Buckle		1966		25	50	125
Batman Raygun		1960s	7" long blue and black futuristic space gun with bat sights and bats on handgrip	80	145	350
Batman Road Race Set		1960s	slot car racing set	90	165	350
Batman Shooting Arcade	AHI	1970s	graphics of Joker, Catwoman and Penguin against brightly colored Gotham City background	25	50	75
Batman Slot Car	Magicar (England)	1966	5" long Batmobile being driven by Batman and Robin in illustrated display window box	115	210	325
Batman Soakie				23	45	85
Batman Standing Figure on Base	Presents		15 1/2" tall vinyl and cloth figure	9	16	25
Batman String Puppet	Madison	1977		25	50	75
Batman Stuffed Doll Plush	Commonwealth	1960s		105	195	175
Batman Super Powers Stain & Paint Set		1984		5	10	15
Batman Superfriends Lite Brite Refill Pack		1980		9	16	25
Batman Superhero Stamp Set		1970s		9	16	25

Batman

TOY	COMPANY	YEAR	DESCRIPTION	GOOD	EX	MIB
Batman Talking Alarm Clock	Janex	1975	molded plastic clock with bat logo at center of face and Batman figure running beside Batmobile with Robin driving	20	35	125
Batman Utility Belt	Remco	1979	includes handcuffs, communicator, decoder glasses, watch, Gotham City decoder map, ID card and secret message	23	45	65
Batman, Robin and Superman Costume Patterns	McCalls	1960s	complete patterns for making all three costumes in paper envelope	12	23	25
Batmobile	Duncan	1987	12x8" on card	18	35	25
Batmobile	Corgi	1966	5" long in display box	85	130	175
Batmobile	Toy Biz			9	16	25
Batmobile	Simms	1960s	plastic car on card	18	33	50
Batmobile	AHI	1972	11" long tin litho battery op. mystery action car with blinking light and jet engine noise	90	165	250
Batmobile & Batboat Set	Corgi			200	300	475
Batmobile AM Radio	Bandai			45	80	125
Batmobile, Motorized Kit	Aoshinu (Japan)	1980s		30	55	85
Batmobile, Motorized Kit	Aoshinu (Japan)	1980s	smaller snap kit	20	40	60
Batmobile, Pullback	Bandai	1980s	pullback vehicle with machine guns	35	60	95
Batmobile, Radio Controlled	Matsushiro			60	115	175
Batmobile, Radio Controlled	Apollo (Japan)			60	115	175
Batwing	Toy Biz			12	23	35
DC Comics Robin Figure	Palitoy		8" figure on card	25	45	70
Inflatable TV Chair		1982	in box	16	30	25
Joker Cycle	Toy Biz			5	10	15
Joker Figure	Presents		15" vinyl figure	11	20	25
Joker Mobile	Corgi	1975	model #99, white van with Joker decals, on card	16	30	75
Joker Van	Ertl	1989	die cast vehicle on card	4	8	10
Joker Watch	Fossil	1980s		25	50	75
Joker Wind-up	Billiken	1989		45	80	125
Joker YoYo		1980s	on card	4	7	10
Penguin Mobile	Corgi	1970s		35	55	75
Riddler Music Box		1978	ceramic figural music box	25	50	175
Robin Christmas Ornament	Presents	1989		7	13	20
Robin Shuttle	Mego	1979	sized for British made Mego figures, in box	9	16	75
Robin Soakie				23	45	125
Robin Standing Figure on Base	Presents		cloth and vinyl standing figure on base	9	16	25
Super Accelerator Batmobile	AHI	1970s	on card	11	20	50
View-Master Set	View-Master	1960s	set of three reels in package	7	13	35

Beany & Cecil

TOY	COMPANY	YEAR	DESCRIPTION	GOOD	EX	MIB
Beanie Doll			Closed eyes version	90	165	250
Beany & Cecil & Their Pals Record Player	Vanity Fair	1961		19	35	225

Beany & Cecil

TOY	COMPANY	YEAR	DESCRIPTION	GOOD	EX	MIB
Beany & Cecil Animated Clock			wood	700	1300	2000
Beany & Cecil Carrying Case			9" diameter with strap, vinyl covered cardboard	9	16	100
Beany & Cecil Play Luggage Set	Mattel	1962		18	35	150
Beany & Cecil Propeller Disks	Mattel	1961		35	65	100
Beany & Cecil Round Travel Case		1960s	8" tall red vinyl case with zipper and strap	25	50	75
Beany & Cecil Skill Ball		1960s	colorful tin with wood frame	16	30	75
Beany & Cecil Square Travel Case		1960s	4 1/2x3 1/2x3" red vinyl case with carrying strap, illustrated with characters on four sides	35	65	100
Beany Figure	Caltoy	1984	8" tall	5	10	15
Beany Talking Doll	Mattel	1960s	17" tall, stuffed cloth, vinyl head with a pull string voice box	55	105	160
Bob Clampetts' Beany Coloring Book	Whitman	1960s		18	35	35
Captain Huffenpuff Puzzle		1961	large puzzle	16	30	50
Cecil and His Disguise Kit	Mattel	1962	17" tall plush Cecil with disguise wigs, mustaches, etc.	30	55	125
Cecil in the Music Box	Mattel	1961		45	90	135
Cecil Soakie			8 1/2" tall, plastic	18	35	50
Cecil Talking Doll	Mattel	1965	17" tall, stuffed cloth, vinyl head with a pull string voice box	55	105	160
Leakin' Lena Plastic Boat	Irwin	1962		60	115	175
Leakin' Lena Pound 'n Pull Toy	Pressman	1960s	wood	50	90	150

Betty Boop

TOY	COMPANY	YEAR	DESCRIPTION	GOOD	EX	MIB
Betty Boop	NJ Croce	1988	bend-n-flex 9" figure	4	8	12
Betty Boop	M-Toy	1986	outfits for 12" dolls high fashion boutique, each	4	7	10
Betty Boop 3" Figure			3" collectible PVC figure eight different poses and outfits, each	2	3	5
Betty Boop Delivery Truck	Schylling Ass., Inc.	1990	tin litho	16	30	45
Betty Boop Doll 12"	M-Toy	1986	vinyl jointed with six different outfits, fur coat, winter woolens, Mae West-pink gown, Flapper, business suit or ballerina	9	16	25

Blondie & Dagwood

TOY	COMPANY	YEAR	DESCRIPTION	GOOD	EX	MIB
Blondie Featured Funnies Jigsaw Puzzle		1930s	9 1/2"x14" puzzle	23	40	35
Blondie Paint Book	Whitman	1947	8 1/2"x11"	18	35	50
Blondie Paint Set	American Crayon Co.	1946		16	30	50
Blondie Paper Doll Book	Whitman	1944		45	90	135
Blondie Paper Doll Book	Whitman	1955	paper dolls and clothes	25	50	80
Blondie's Peg Board Set	King Features	1934	9"x15 1/2", multi-colored pegs, hammer, cut-outs of Dagwood, Blondie, etc.	55	100	125

Blondie & Dagwood

TOY	COMPANY	YEAR	DESCRIPTION	GOOD	EX	MIB
Blondie's Presto Slate	Presto Productions	1944	10"x13" illustration of Blondie & Dagwood and other characters	20	35	55
Dagwood & Kids Figures	K.F.S.	1944	Dagwood 5" tall, included are: Alexander & Cookie	20	35	75
Dagwood Marionette		1945	14"	65	120	185
Dagwood's Solo Flight Airplane	Marx	1935	12" wingspan, plane 9" in length	325	625	950
Lucky Safety Card		1953	2"x4" cards, Dagwood warns you about playing in safe areas	9	16	25
Miniature Blondie Figure		1940s	2 1/2" tall, lead	7	13	20
Miniature Dagwood Figure			2 3/4" tall, lead	7	13	20

Bonanza

TOY	COMPANY	YEAR	DESCRIPTION	GOOD	EX	MIB
Ben Cartwright with Palomino	American Character	1966		50	90	175
Bonanza Figures on Buckboard with Three Horses and Accessories	American Character			80	145	325
Bonanza Jigsaw Puzzle	Milton Bradley	1964	jigsaw puzzle	12	23	35
Four in One Wagon	American Character			105	195	300
Hoss Cartwright with Stallion	American Character	1966		50	90	175
Little Joe Cartwright with Pinto	American Character	1966		55	100	175
Mustang	American Character		The Outlaw's horse	23	45	65
Outlaw, The	American Character	1966		40	70	110
Palomino	American Character		Ben's horse	23	45	65
Pinto	American Character		Little Joe's horse	23	45	65
Stallion	American Character		Hoss's horse	23	45	65

Bozo

TOY	COMPANY	YEAR	DESCRIPTION	GOOD	EX	MIB
Bozo the Clown			stuffed doll	9	16	50
Bozo the Clown Figure		1970s	vinyl figure 5" tall	5	10	15
Bozo the Clown Push Button Marionette	Knickerbocker	1962		35	65	125
Bozo the Clown Slide Puzzle		1960s	2 1/2"x2 1/2"	16	30	45
Bozo the Clown Soakie	Stephen Riley Co.			11	20	35
Bozo the Clown Towel		1960s	beach towel 16"x24"	16	30	45
Bozo Trick Trapeze		1960s	red base	20	35	75

Bringing Up Father

TOY	COMPANY	YEAR	DESCRIPTION	GOOD	EX	MIB
Bringing Up Father Paint Book with Jiggs & Maggie			8 1/2"x11"	15	30	75
Maggie Statue			12" tall	30	55	125

Bugs Bunny

TOY	COMPANY	YEAR	DESCRIPTION	GOOD	EX	MIB
Bugs Bunny			wearing Uncle Sam outfit	11	20	30

Bugs Bunny

TOY	COMPANY	YEAR	DESCRIPTION	GOOD	EX	MIB
Bugs Bunny	Dakin			7	13	25
Bugs Bunny "Chatter Chum"	Mattel	1982		9	16	25
Bugs Bunny & Tweety Bird Costume	Collegeville	1960s	thin plastic mask & one piece costume	9	16	25
Bugs Bunny Bank		1940s	5 3/4"x5 1/2", pot metal barrel bank with figure on base	45	80	175
Bugs Bunny Bank	Dakin	1971	on a basket of carrots	16	30	45
Bugs Bunny Bendable Figure	Applause	1980s	4" tall	4	8	12
Bugs Bunny Charm Bracelet		1950s	brass charms of Bugs Bunny, Tweety, Sniffles, Fudd, etc.	20	35	75
Bugs Bunny Clock	Litech	1972	12"x14"	23	45	125
Bugs Bunny Colorforms		1958	Bugs Bunny, Tweety Bird and Elmer	23	45	50
Bugs Bunny Figure	Warner Bros.	1975	2 3/4" tall, ceramic	12	23	35
Bugs Bunny Figure	Dakin	1971	10" tall	11	20	45
Bugs Bunny Figure	Dakin	1976	yellow globes in "Cartoon Theater" box	9	16	25
Bugs Bunny Figure Holding Carrot	Warner Bros.	1975	5 1/2" tall, ceramic	12	23	50
Bugs Bunny in Uncle Sam Outfit	Dakin	1976	distributed through Great America Theme Park, Illinois	23	45	65
Bugs Bunny Mini Snowdome	Applause	1980s		5	10	15
Bugs Bunny Musical Ge-tar	Mattel	1977		9	16	35
Bugs Bunny Nitelite	Applause	1980s		5	10	15
Bugs Bunny Soakie			soft rubber	9	16	25
Bugs Bunny Talking Alarm Clock				18	35	125
Bugs Bunny Talking Doll	Mattel	1971		25	50	75
Bugs Bunny Wristwatch	Lafayette	1978		14	25	95

Captain America

TOY	COMPANY	YEAR	DESCRIPTION	GOOD	EX	MIB
Captain America Bendie Figure	Lakeside			45	80	125
Captain America Rocket Racer	Buddy-L	1984	Secret Wars remote controlled battery operated car	14	25	125

Captain Marvel

TOY	COMPANY	YEAR	DESCRIPTION	GOOD	EX	MIB
Adventures of Captain Marvel Ink Blotter/Ruler	Republic/Fawcett	1940s	6" blotter with ruler advertises the 12 part serial, was a theatre premium	60	120	200
Boy Who Never Heard Of Captain Marvel Mini Comic	Bond Bread	1940s	Bond Bread premium mini comic	45	90	150
Captain Marvel and Billy's Big Game Mini Comic		1940s		115	225	375
Captain Marvel Bean Bags		1940s	Captain Marvel, Mary Marvel or Hoppy, each	25	45	75
Captain Marvel Beanie		1940s	cap shows image of Captain Marvel flying laterally toward word "Shazam"	45	90	150
Captain Marvel Booklet	Fawcett Publications	1940s		25	45	75
Captain Marvel Buzz Bomb	Fawcett Publications	1950s	paper airplane in envelope	10	20	35
Captain Marvel Celluloid Pinback		1940s		25	55	90

CHARACTER

Captain Marvel

TOY	COMPANY	YEAR	DESCRIPTION	GOOD	EX	MIB
Captain Marvel Club Button		1941	tin litho, showing Captain Marvel in bust 3/4 view, with "Shazam" in lightening bolts at bottom	25	50	85
Captain Marvel Club Felt Shoulder Patches	Fawcett Publications	1940s	blue or yellow background patches show Captain Marvel diving toward the Earth	25	45	75
Captain Marvel Club Membership Card	Fawcett Publications	1940s		15	30	50
Captain Marvel Code Finder		1943		125	250	400
Captain Marvel Comic Hero Punch-Outs	Lowe	1942	cardboard punch-out figures of Captain Marvel and supporting characters	115	225	375
Captain Marvel Felt Pennant	Fawcett Publications	1940s	blue and yellow pennant shows Captain Marvel flying	40	80	135
Captain Marvel Film Viewer Gun		1940s	gun shaped movie viewer with film strips from Paramount series	105	210	350
Captain Marvel Flannel Patch	Fawcett Publications	1940s		25	45	75
Captain Marvel Glow Pictures	Fawcett Publications	1940s	set of four	225	450	750
Captain Marvel Illustrated Soap	Fawcett Publications	1947	set of three soap bars in box	90	180	300
Captain Marvel Iron-Ons, Sheet	Fawcett Publications	1950s		15	30	50
Captain Marvel Jr. Die Cast Figure	Fawcett Publications	1940s	unpainted die cast, only two known, no value established	0	0	0
Captain Marvel Jr. Shoulder Patch	Fawcett Publications	1940s		45	90	150
Captain Marvel Jr. Ski Jump	Reed & Assocciates	1947	paper, in envelope	10	20	35
Captain Marvel Jr. Wristwatch		1940s	blue band, round dial with blue costumed Jr. flying up toward 1:00, no box, price higher with box	150	300	500
Captain Marvel Keychain	Fawcett Publications	1940s		35	65	110
Captain Marvel Lightning Wind-Up Race Car	Fawcett Publications	1947	4" long tin wind-up car, in green, yellow, orange or blue, each	45	90	145
Captain Marvel Magic Dime Register Bank	Fawcett Publications	1948	available in three colors	55	105	175
Captain Marvel Magic Flute		1940s	on die cut card, shows Captain Marvel on side	40	80	135
Captain Marvel Magic Lightning Box	Fawcett Publications	1940s		50	100	165
Captain Marvel Magic Membership Card	Fawcett Publications	1940s		15	30	50
Captain Marvel Magic Picture	Reed & Associates	1940s		15	30	50
Captain Marvel Magic Picture	Reed & Associates	1940s	paper item shows Billy Batson 'transforming' into Captain Marvel	15	30	50
Captain Marvel Magic Whistle	Fawcett Publications	1948	seed company premium, has color picture of Captain Marvel on both sides, on card	40	75	125
Captain Marvel Meets the Weatherman Mini Comic	Bond Bread	1940s	bread premium comic book	45	90	150
Captain Marvel Neck Tie	Fawcett Publications	1940s		30	60	100
Captain Marvel Overseas Cap		1940s		60	120	200

Captain Marvel

TOY	COMPANY	YEAR	DESCRIPTION	GOOD	EX	MIB
Captain Marvel Paint Set		1940s	paint set with five chalk figurines	150	300	500
Captain Marvel Paper Horn	Fawcett Publications	1940s		15	30	50
Captain Marvel Picture Puzzle	Reed & Associates	1940s	in envelope	35	75	125
Captain Marvel Picture Puzzle	Fawcett Publications	1941		25	45	75
Captain Marvel Pinback	Fawcett Publications	1940s		25	40	65
Captain Marvel Pinback Pattern	Fawcett Publications	1940s	pattern for original pinback	8	15	25
Captain Marvel Portrait	Whiz Comics/ Fawcett	1940s		45	90	150
Captain Marvel Portrait	Republic	1940s	different version than Whiz Comics portrait	45	90	150
Captain Marvel Power Siren	Fawcett Publications	1940s		45	90	150
Captain Marvel Secret Code Sheet	Fawcett Publications	1940s		6	12	20
Captain Marvel Sirocco Figurine	Fawcett Publications	1940s		1000	2100	3500
Captain Marvel Skull Cap		1940s		65	135	225
Captain Marvel Stationary	Fawcett Publications	1940s	color note paper and envelopes in box	60	120	200
Captain Marvel Statuette	Fawcett Publications	1940s	hand painted plastic statue showing Captain Marvel standing with arms crossed, on base with name engraved	950	1900	3200
Captain Marvel Suspenders	Fawcett Publications	1940s		40	75	125
Captain Marvel Sweater		1940s	white or off-white sweater with Red Captain marvel logo	55	105	175
Captain Marvel Tattoo Transfers	Fawcett Publications	1940s		25	45	75
Captain Marvel Tie Bar		1940s	on card	30	60	100
Captain Marvel Vinyl Brunch Bag		1940s	red rectangular vinyl bag with strap handle	30	60	100
Captain Marvel Wristwatch		1948	in box, shows Captain Marvel holding an airplane	195	395	650
Captain Marvel, Jr. Booklet	Fawcett Publications	1940s		25	45	75
Cloth Patches	Fawcett Publications	1940s	patches show Captain Marvel, Captain Marvel Jr. and Mary Marvel, each	25	45	75
Fawcett's Comic Stars Christmas Tree Ornaments	Fawcett Publications	1940s	metal star-shaped ornaments with art of Captain Marvel and Hoppy	25	40	65
Good Humor Man, The, Lobby Card	Columbia Pictures	1950	film revolved around a Captain Marvel club, card shows club members wearing club sweaters, in clubhouse	30	60	100
Mary Marvel Illustrated Soap	Fawcett Publications	1947	set of three soap bars in box	90	180	300
Mary Marvel Pinback	Fawcett Publications	1940s		12	25	40
Mary Marvel Stationary	Fawcett Publications	1940s	color note paper and envelopes in box	60	120	200
Mary Marvel Wristwatch		1940s	in box	175	350	600
Membership Secret Code Card	Fawcett Publications	1940s		17	35	55

Captain Midnight

TOY	COMPANY	YEAR	DESCRIPTION	GOOD	EX	MIB
Puzzle, "One Against Many"	Reed & Associates	1940s	jigsaw puzzle in envelope	8	15	25
Rocket Raider	Fawcett Publications	1940s	paper airplane in envelope	10	20	35
Air Heroes Stamp Album		1930s	twelve stamps	25	50	75
Captain Midnight Cup			plastic, 4" tall, "Ovaltine-The Heart of a Hearty Breakfast"	23	45	75
Captain Midnight Medal		1930s	gold medal pin with centered wings and words "Flight Commander" embossed. Capt. is embossed on top with medal dangling beneath	55	100	275
Captain Midnight Secret Society Decoder		1949	used to decode messages only known by "Society" members	20	35	125
Membership Manual		1930s	Secret Squadron official code and manual guide	25	50	175

Captain Video

TOY	COMPANY	YEAR	DESCRIPTION	GOOD	EX	MIB
Rocket Launcher	Lido	1952		50	95	250

Cartoon & Comic Characters

TOY	COMPANY	YEAR	DESCRIPTION	GOOD	EX	MIB
Alfred E. Neuman Bust Figurine		1960s	base says "What Me Worry?"	55	105	165
Andy Panda Bank	Walter Lantz Prod.	1977	7" tall hard plastic bank	14	25	40
Banana Splits Mug		1969	plastic yellow dog mug	5	10	35
Banana Splits Record	Kelloggs	1969		9	16	25
Banana Splits Stuffed Figure	General Mills	1960s	12" tall Drooper figure	23	45	75
Bloom County Opus Doll		1986	10" tall, plush Christmas Cheer doll Opus wearing a Santa Claus cap	5	10	15
Cadbury the Butler Figure, (Ritchie Rich)	DFC	1981	3 1/2" figure on illustrated card	7	13	20
Casper, Popeye, Chipmunks, Dick Tracy, Bozo, Mr. Ed	Kenner	1965	Easy Show movie projector movies, each	9	16	25
Chilly Willy Figure	Walter Lantz Prod.	1982	plush	5	10	15
Cool Cat Figure		1969	vinyl 9" tall	23	45	65
Daffy Dog in "The Morning After" Poster			10"x13"	9	16	25
Dan Dunn Pin Back Button		1930s	1 1/4"	25	50	75
Doggie Daddy Metal Trivet		1960s	says "You have to work like a dog to live like one"	12	23	35
Dudley Doright Doll	Wham-O	1972	bendable	7	13	35
Dudley Doright Jigsaw Puzzle	Whitman	1975	Dudley and Snidley	7	13	25
Favorite Funnies Printing Set		1930s	Set no. 4004, Orphan Annie, Herby & Dick Tracy, six stamps, pad, paper and instructions	55	105	165
Geoffrey Jack-In-The-Box	Toys-R-Us	1970s	1/2"x5 1/2"x5 1/2" tin litho	12	23	45
Goober Bank			vinyl figural bank	20	35	55
Hagar the Horrible Doll		1983	12" tall	9	16	25
Hair Bear Bunch Square Bear Figural Mug		1978		4	8	12

CHARACTER

Cartoon & Comic Characters

TOY	COMPANY	YEAR	DESCRIPTION	GOOD	EX	MIB
Hair Bear Bunch Wristwatch		1972	medium gold tone case, base metal back, articulated hands, red leather snap down band	60	115	125
Harold Teen Playstone Funnies Mold Set		1940s		35	60	125
Henry on Trapeze Toy	G. Borgfeldt Co.		6"x9", celluloid, wind-up, jointed Henry suspended from trapeze	155	295	850
Herman and Katnip Punch Out Kite	Saalfield	1960s	folds into a kite	14	25	40
Hollie Hobbie Wristwatch	Bradley	1982	small gold tone case, base metal back, yellow plastic band,	5	10	35
Houndcats Board Game	Milton Bradley	1970s		9	16	25
Incredible Hulk Action Figure	Palitoy		8" figure on card	11	20	30
Jets Photo Album & Cards		1950s	64 cards	23	45	65
Katzenjammer Kids Featured Funnies Jigsaw Puzzle		1930s	9 1/2"x14" puzzle	30	55	85
King Leonardo and his Loyal Subjects Board Game	Milton Bradley	1960		25	50	80
King Leonardo Doll	Holiday Fair	1960s	cloth stuffed doll dressed in his royal robe	35	60	125
Lil' Abner Mugs		1940s	four ceramic mugs: Abner, Daisy, Pappy & Mammy Yokum	90	165	250
Little Audrey Dress Designer Kit	Saalfield	1962	die cut doll and 29 clothing accessories in illustrated box	12	23	50
Little Audrey Shoulder Bag Leathercraft Kit	Jewel	1961		30	55	85
Little Lulu Bank			8" tall hard plastic with black fire hydrant	11	20	45
Little Lulu Dish		1940s	5 1/2" hand painted ceramic dish with pictures of Lulu, Tubby and her friend	50	95	145
Mary Marvel Wristwatch	Marvel Imp.	1948	small chrome case, face shows profile of Mary flying up toward one o'clock	60	115	475
Merlin Mouse Figure	Dakin			7	13	20
Mush Mouse Pull Toy	Ideal	1960s	pull toy with vinyl figure	45	80	125
Nancy Music Box	UFS Inc.	1968	ceramic music box	30	55	125
Peter Potamus Bubbles Soakie			11" tall	12	23	35
Rosie's Beau Featured Funnies Jigsaw Puzzle		1930s	9 1/2"x14" puzzle	20	40	60
Scrappy Bank			3"x3 1/2" metal bank with an embossed illustration of Scrappy and his dog	35	60	95
Smilin' Jack Featured Funnies Jigsaw Puzzle		1930s	9 1/2"x14" puzzle	23	45	65
Smokey the Bear Bank			8" tall, Smokey waves and holds shovel	12	23	35
Smokey the Bear Figure	Dakin		8 1/2" tall plastic figure with cloth parts and shovel	12	23	35
Spider-Man Bend'em Figure	Just Toys	1991	#12056, Marvel Super Heroes Series, on card	2	4	6
Spider-Man Spider Cycle, Spider Copter, Spider Van Set	Buddy-L	1984	Secret Wars vehicle set	16	30	45
Spider-Man Spider Racer Car	Buddy-L	1984	Secret Wars remote controlled battery operated car	14	25	40

Cartoon & Comic Characters

TOY	COMPANY	YEAR	DESCRIPTION	GOOD	EX	MIB
Strawberry Shortcake Wristwatch		1970s		9	16	25
Super Heroes Flashy Flickers Filmstrip Reel		1960s	filmstrip cartoons with Wonder Woman, Aquaman & Tomahawk	9	16	25
Supercar Moulding Colour Kit	Sculptorcraft	1960s	set of rubber plaster casting models of vehicle and show characters, including Mike Mercury, Beaker and Popkiss, Jimmy and Mitch, Masterspy	35	60	125
Touche Turtle Soakie				16	30	45
Winnie Winkle Playstone Funnies Mold Set		1940s		45	80	125
Wonder Woman Standing Figure on Base	Presents		14" tall cloth and vinyl figure on base	9	16	25
Yipee Pull Toy	Ideal	1960s	pull toy with vinyl figure of Yipee, Yapee & Yahoee	50	95	145

Casper the Friendly Ghost

TOY	COMPANY	YEAR	DESCRIPTION	GOOD	EX	MIB
Casper Doll	Sutton & Sons	1972	rubber squeeze doll with logo	12	23	35
Casper Doll		1960s	15" cloth	25	45	70
Casper Figure Lamp	Archlamp Mfg.	1950	17" tall	25	50	75
Casper Halloween Costume	Collegeville	1960s	mask and costume	16	30	45
Casper Hand Puppet			8" tall, cloth and plastic head	18	35	50
Casper Light Shade		1960s	features Casper & his friends	12	23	35
Casper Night Light	Duncan	1975	6 1/2" tall	18	35	50
Casper Puffy Sticker	Chex Cereal (Ralston Purina)	1970s	glows in the dark	9	16	5
Casper Soakie				12	23	50
Casper Spinning Top		1960s	blue top with figure of Casper inside	16	30	45
Casper Stand Up		1960s	18"x10"	30	55	85
Casper the Friendly Ghost Vinyl Doll			7 3/4" tall squeeze doll holds black spotted puppy	30	55	85
Casper Wind-up Toy	Linemar	1950s	tin	70	130	375
Casper, The Friendly Ghost Four Jigsaw Puzzles	Ja-Ru	1988		2	4	6
Casper, The Friendly Ghost Game	Schaper	1974		7	13	20
Casper, The Friendly Ghost Pinball Game	Ja-Ru	1988		2	4	6
I'm Casper the Friendly Ghost Talking Doll	Mattel	1961	15" tall, terrycloth, plastic head with a pull string voice box	45	80	125
Wendy the Good Witch Soakie				9	16	35

Charlie Chaplin

TOY	COMPANY	YEAR	DESCRIPTION	GOOD	EX	MIB
Charlie Chaplin Character			8 1/2" tall, tin with cast iron feet, wind-up	325	600	1250
Charlie Chaplin Doll			11 1/2" tall, wind-up	155	295	850
Charlie Chaplin Novelty Toy			4" tall, spring mechanism tips his hat when string is pulled	12	23	125
Charlie Chaplin Pencil Case			8" long	19	35	85

Charlie Chaplin

TOY	COMPANY	YEAR	DESCRIPTION	GOOD	EX	MIB
Charlie Chaplin Quartz Watch	Bradley	1985	oldies series, large black plastic case and band, sweep seconds, shows Chaplin as Little Tramp	16	30	75
Charlie Chaplin Toy			7" tall, plastic	23	45	125
Charlie Chaplin Wristwatch	Bubbles/ Cadeaux	1972	Swiss, large chrome case, black and white dial, articulated sweep cane second hand, black leather band	45	80	175
Miniature Charlie Chaplin			2 1/2" tall, lead	19	35	85

Charlie's Angels

TOY	COMPANY	YEAR	DESCRIPTION	GOOD	EX	MIB
Cheryl Ladd Figure	Mattel	1978	12" tall	16	30	45
Farrah Fawcett-Majors Figure	Remco	1977	in swim suit	16	30	45
Kate Jackson Figure	Mattel	1978	12" tall	16	30	45
Kelly	Hasbro	1977		8	15	22
Kris	Hasbro	1977		8	14	22
River Race Outfits	Palitoy	1977		16	30	45
Sabrina	Hasbro	1977		8	14	22
Sabrina, Kelly and Kris Gift Set	Hasbro	1977		16	30	45
Slalom Caper Outfits	Palitoy			14	25	40
Underwater Intrigue Outfits	Palitoy			14	25	40

Chipmunks

TOY	COMPANY	YEAR	DESCRIPTION	GOOD	EX	MIB
Alvin Plush Doll	Knickerbocker	1963	14" tall plush doll with vinyl head	16	30	45
Alvin Soakie		1960s	8" tall	9	16	25
Chipmunks Toothbrush		1984	battery operated	7	13	20
Chipmunks Wallet		1959	vinyl	5	10	15
Soakie, Alvin, Theodore or Simon		1960s	10" tall, plastic & vinyl, each	5	10	15

Cinderella

TOY	COMPANY	YEAR	DESCRIPTION	GOOD	EX	MIB
Cinderella & Prince Wind-up Toy	Irwin		5" tall, plastic	35	65	175
Cinderella Alarm Clock	Westclox		2 1/2"x4 1/2"x4' tall	35	65	150
Cinderella Bank		1950s	ceramic with Cinderella holding a magic wand	16	30	95
Cinderella Charm Bracelet		1950	golden brass link with five charms, Cinderella, Fairy Godmother, slipper, pumpkin coach and Prince	25	50	75
Cinderella Doll	Horsman		8" tall classic doll in illustrated box	12	23	75
Cinderella Doll			11" tall, blue stain ballgown with white bridal gown, glass slippers, holding Little Little Golden Book	20	35	75
Cinderella Figurine			5" tall, ceramic	9	16	50
Cinderella Figurine			5" tall, plastic	5	10	25
Cinderella Molding Set	Model Craft	1950s	set of character molds in illustrated box	45	80	125
Cinderella Musical Jewelry Box			mahogany musical box which plays "So This Is Love"	20	35	75
Cinderella Paper Dolls	Whitman	1965	included are: Cinderella, Stepmother, Anastasia, Drizella, Prince plus clothes for each doll	23	45	65

Cinderella

TOY	COMPANY	YEAR	DESCRIPTION	GOOD	EX	MIB
Cinderella Puzzle	Jaymar	1960s	9 1/2"x12 1/2"	12	23	35
Cinderella Soakie		1960s	11" tall blue bubble bath container	11	20	30
Cinderella Story Book	Whitman	1950		5	10	15
Cinderella Wind-up Toy	Irwin	1950	5" tall with Cinderalla & Prince dancing	55	100	175
Cinderella Wristwatch	US Time	1950		45	80	125
Cinderella Wristwatch	Timex	1958	small chrome case shows Cinderella in foreground with castle at 12 o'clock, pink leather band	45	80	125
Cinderella Wristwatch	Bradley	1970s	basemetal, small gold bezel, picture and "Cinderella" on face, pink leather band	16	30	75
Fairy Godmother Pitcher			7" tall figural	25	50	75
Gus Doll	Gund	1950s	13" tall, gray doll with dark red shirt & green felt hat	60	115	175
Gus/Jaq Serving Set	Westman	1960s	creamer, pitcher and sugar bowl	45	80	125
Prince Charming Hand Puppet	Gund	1959	10" tall	25	50	75

Crusader Rabbit

TOY	COMPANY	YEAR	DESCRIPTION	GOOD	EX	MIB
Crusader Rabbit Book	Wonder Book	1958		12	23	35
Crusader Rabbit in Bubble Trouble Book	Whitman	1960		9	16	25
Crusader Rabbit Trace & Color Book	Whitman	1959		18	35	50
Paint Set			13x19"	20	35	55

Daffy Duck

TOY	COMPANY	YEAR	DESCRIPTION	GOOD	EX	MIB
Daffy Duck Bendable Figure	Applause	1980s	4" tall	9	16	25
Daffy Duck Figural Bank	Applause	1980s		9	16	25
Daffy Duck Figure	Dakin	1968	8 1/2" tall	11	20	30
Daffy Duck Mug			large size plastic	2	5	10

Danger Mouse

TOY	COMPANY	YEAR	DESCRIPTION	GOOD	EX	MIB
Danger Mouse Doll	Russ	1988	15" tall	18	35	50
Danger Mouse I.D. Set	Gordy	1985		7	13	20
Danger Mouse Pendant Necklace	Gordy	1986		5	10	15

Davy Crockett

TOY	COMPANY	YEAR	DESCRIPTION	GOOD	EX	MIB
Davy Crockett Figure	Marx	1964	6" tall, vinyl	60	115	175
Davy Crockett Figure	Marx		2" tall, rubber figure, cream color with "Official Davy Crockett As Portrayed by Fess Parker" under base	25	50	75
Davy Crockett Guitar		1955	11x24x1 1/2" varnished plywood guitar shaped like a bell	45	80	125
Davy Crockett Jigsaw Puzzle	Marx		14"x19" titled "Siege on the Fort"	16	30	45
Davy Crockett Official Wallet		1955	wallet contains calendar card for 1966 and 1956, in box	18	35	50
Davy Crockett Puzzle	Whitman	1955	11 1/4"x15"	11	20	30
Davy Crockett Puzzle	Jaymar		8"x10"x2"	16	30	45
Davy Crockett Travel Bag	Neevel	1950s	6 1/2"x12"x10" heavy cardboard with brass hinges & plastic handle	35	65	100

Dennis the Menace

TOY	COMPANY	YEAR	DESCRIPTION	GOOD	EX	MIB
Dennis the Menace & Ruff Book Ends		1974	ceramic	18	35	125
Dennis the Menace Colorform Kit	Colorforms	1961		16	30	45
Dennis the Menace Giant Mischief Kit	Hasbro	1950s	snap gum, spilled ink, floating sugar and etc.	35	60	95
Dennis the Menace Inlay Jigsaw Puzzle	Whitman	1960		12	23	35
Dennis the Menace Paint Set	Pressman	1954	paints, crayons, brush and trays	35	60	95

Deputy Dawg

TOY	COMPANY	YEAR	DESCRIPTION	GOOD	EX	MIB
Deputy Dawg Figure	Dakin	1977	6 " tall, plastic body with vinyl head	25	50	75
Deputy Dawg Soakie Bath Soap Container		1966	9 1/2" tall, plastic	9	16	25
Deputy Dawg Stuffed Doll	Ideal	1960s	14" tall, cloth with plush arms & vinyl head	25	50	75

Dick Tracy

TOY	COMPANY	YEAR	DESCRIPTION	GOOD	EX	MIB
Alarm Clock				25	40	65
Auto Magic Picture Gun		1950s	6 1/2" x 9" metal picture gun & filmstrip	30	50	80
Automatic Police Station	Marx	1950s	tin litho police station and car	195	325	500
Badge, Air Detective, Member		1938	brass, wing shape	40	65	100
Badge, Detective	Quaker	1938	leather "secret" pouch	50	80	125
Badge, Detective Club Belt Badge		1938	leather pouch back on brass badge	40	65	100
Badge, Detective Club, "Crime Stoppers"	Guild	1940s		30	50	75
Badge, Dick Tracy Crime Stopper		1960s	star shape giveaway badge from WGN "9 Official Dick Tracy Crimestopper" TV Badge	40	65	100
Badge, Dick Tracy Detective Club	Quaker	1938	brass belt badge with a leather pouch back	40	65	100
Badge, Secret Service Patrol - Captain	Quaker	1938	brass badge	75	130	200
Badge, Secret Service Patrol - Girl's Division	Quaker	1938	brass badge	30	50	75
Badge, Secret Service Patrol - Inspector General	Quaker	1938		195	325	500
Badge, Secret Service Patrol - Lieutenant	Quaker	1938		60	100	150
Badge, Secret Service Patrol - Member		1938	2nd yr. Chevron	15	25	35
Badge, Secret Service Patrol - Sergeant		1938		30	50	75
Baking Set	Pillsbury	1937	cookie cutter, 6 bright colored press out sheets with pictures of Dick Tracy and his pals	75	130	200
Bank, Sparkle Plenty Savings Bank	Jayess	1940s	12" tall, base measures 4 1/2 x 9" with a medallion of Dick Tracy as "Godfather" also medallions on the either side of Gravel Gertie and B.O. Plenty	195	325	500
Belt Badge, Dick Tracy Detective Club			premium	40	65	100
Belt, Detective Club		1937	leather "secret" pouch	60	100	150

Dick Tracy

TOY	COMPANY	YEAR	DESCRIPTION	GOOD	EX	MIB
Black Light Magic Kit, Dick Tracy	Stroward	1952	ultra-violet bulb, cloth, invisible pen, brushes & fluorescent dyes	75	130	200
Bombsight, Dick Tracy Jr.	Miller Bros. Hat Co.	1940s	cardboard	40	65	100
Bonny Braids	Ideal	1951	14" tall with toothbrush, "America's New Darling, She Sobs! She Cries! She Coos!"	135	225	350
Bonny Braids Store Contest Card		1951	5 1/2" x 5 1/2" punch out a name & write it on the back winner wins a free Bonny Braids doll	30	50	75
Bonny Braids Stroll Toy	Charmore	1951	tin litho, Bonny doll in carriage	60	100	150
Book, Ace Detective	Whitman	1943		10	15	25
Book, Big Big, Dick Tracy and the Mystery of the Purple Cross	Whitman	1938	7 x 9 1/2," 320 pages, hardcover	115	195	300
Book, Big Big, The Adventures of Dick Tracy	Whitman	1934	7 x 9 1/2," 320 pages, hardcover	75	130	200
Book, Big Little, Detective Dick Tracy and the Spider Gang	Whitman	1937	240 pages, with Republic movie photos	35	55	85
Book, Big Little, Dick tracy and His G-Men	Whitman	1941	432 pages, hardcover with flip pictures	30	50	75
Book, Big Little, Dick Tracy and the Bicycle Gang	Whitman	1948	288 pages, hardcover	30	50	75
Book, Big Little, Dick Tracy and the Boris Arson Gang	Whitman	1935	432 pages, hardcover	35	55	85
Book, Big Little, Dick Tracy and the Hotel Murders	Whitman	1937	432 pages, hardcover	30	50	75
Book, Big Little, Dick Tracy and the Invisible Man	Whitman	1939	Quaker promium, 132 pages, softcover	60	100	150
Book, Big Little, Dick Tracy and the Mad Killer	Whitman	1947	288 pages, hardcover	25	40	65
Book, Big Little, Dick Tracy and the Phantom Ship	Whitman	1940	432 pages, hardcover	30	50	75
Book, Big Little, Dick Tracy and the Racketeer Gang	Whitman	1936	432 pages, hardcover	35	55	85
Book, Big Little, Dick Tracy and the Stolen Bonds	Whitman	1934	320 pages, hardcover	35	55	85
Book, Big Little, Dick Tracy and the Tiger Lilly Gang	Whitman	1949	288 pages, hardcover	25	40	65
Book, Big Little, Dick Tracy and the Wreath Kidnapping Case	Whitman	1945	432 pages, hardcover	30	45	70
Book, Big Little, Dick Tracy and Yogee Yamma	Whitman	1946	352 pages, hardcover	25	45	70
Book, Big Little, Dick Tracy Encounters Facey	Whitman	1967	260 pages, hardcover, cover price 39 cents	10	15	25
Book, Big Little, Dick Tracy From Colorado to Nova Scotia	Whitman	1933	320 pages, hardcover	35	55	85

Dick Tracy

TOY	COMPANY	YEAR	DESCRIPTION	GOOD	EX	MIB
Book, Big Little, Dick Tracy in Chains of Crime	Whitman	1936	432 pages, hardcover	30	50	80
Book, Big Little, Dick Tracy Meets a New Gang	Whitman	1939	Quaker premium, 132 pages, softcover	75	130	200
Book, Big Little, Dick Tracy on the High Seas	Whitman	1939	432 pages, hardcover	30	50	75
Book, Big Little, Dick Tracy on the Trail of Larceny Lu	Whitman	1935	432 pages, hardcover	35	55	85
Book, Big Little, Dick Tracy on Voodoo Island	Whitman	1944	352 pages, hardcover	30	45	70
Book, Big Little, Dick Tracy Out West	Whitman	1933	300 pages, hardcover	35	55	85
Book, Big Little, Dick Tracy Returns	Whitman	1939	432 pages, hardcover, Republic movie serial tie-in	30	50	80
Book, Big Little, Dick Tracy Solves the Penfield Mystery	Whitman	1934	320 pages, hardcover	35	55	85
Book, Big Little, Dick Tracy Special FBI Operative	Whitman	1943	432 pages, hardcover	25	45	70
Book, Big Little, Dick Tracy the Man with No Face	Whitman	1938	432 pages, hardcover	30	50	75
Book, Big Little, Dick Tracy the Super Detective	Whitman	1939	432 pages, hardcover	30	50	75
Book, Big Little, Dick Tracy VS. Crooks in Disguise	Whitman	1939	352 pages, hardcover with flip pictures	30	50	75
Book, Big Little, Dick Tracy's Ghost Ship	Whitman	1939	Quaker premium, 132 pages, softcover	60	100	150
Book, Big Little, The Adventures of Dick Tracy and Dick Tracy Jr.	Whitman	1933	320 pages, hardcover	100	165	250
Book, Big Little, The Adventures of Dick Tracy the Detective	Whitman	1933	first book on the Big Little Book Series, hardcover	145	245	375
Book, Bonny Braids Paper Dolls	Saalfield	1951	#1559, 11" x 14" Dick Tracy's New Daughter and Tess	20	35	50
Book, Dick Tracy Junior Detective Kit	Golden Press	1962	punchout book of Tracy tools, including badges, revolver, wrist radio	20	35	50
Book, Dick Tracy Little Golden Book	Golden Press	1962	6 3/4 x 8" features characters like Joe Jitsu and Hemlock Holmes from the TV show	15	25	40
Book, Dick Tracy Paint Book	Saalfield	1930s	10 3/4" x 13 3/4" x 3/8", contains 96 full pages	100	165	250
Book, Dick Tracy Super Detective Book	Whitman	1941		8	13	20
Book, Secret Code	Quaker	1938	prernium	40	65	100
Book, The Capture of Boris Arson Pop-Up Story Book	Pleasure Books, Inc.	1935	pop-up book	115	195	300
Booklet, Gum	Big Thrill Chewing Gum	1934	5 different premiums, 8 pages each, Cross Country Race, Valuable Information, Saved in the Nick of Time, Uncovers the Dope Ring, each	40	65	100

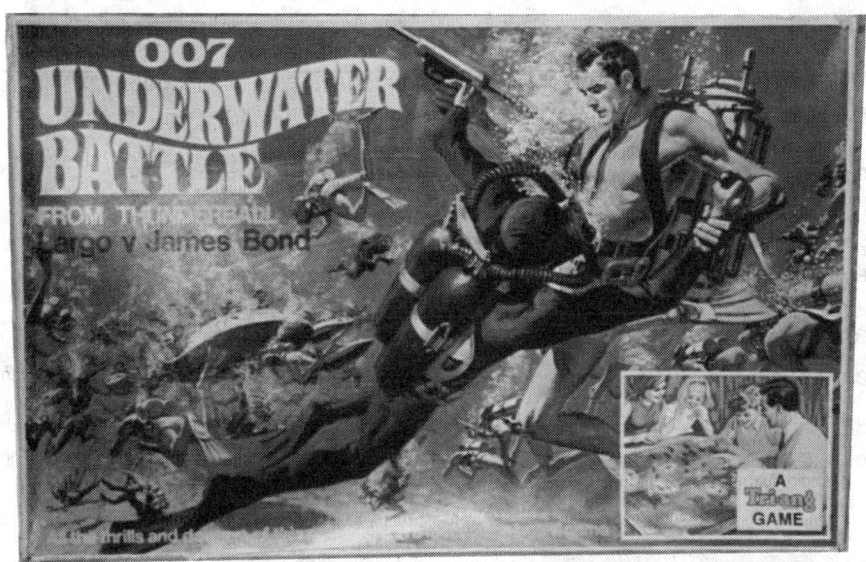

Top to bottom: James Bond Thunderball Underwater Battle, Triang; Joan Palooka Stringless Marionette, 1952, National Mask & Puppet; Alice in Wonderland Mad Hatter Marionette, 1950s, Peter Puppet; Howdy Doody Air-O-Doodle, 1950s.

CHARACTER

Dick Tracy

TOY	COMPANY	YEAR	DESCRIPTION	GOOD	EX	MIB
Booklet, Gum	Big Thrill Chewing Gum	1934	8 page premium, The Vault of Death	25	35	50
Bracelet, Dick Tracy Wing Bracelet	Quaker	1938	cereal premium offer with proof of five boxtops, girl's bracelet with airplane and raised lettering that says Dick Tracy Air Detective	40	65	100
Bracelet, Secret Service Patrol	Quaker	1938	chain bracelet with small head of Dick Tracy & Junior & f leaf clover	115	195	300
Button, Detective		1930s	celluloid pinback with portrait, newspaper premium	30	50	75
Button, Dick Tracy & Little Orphan Annie	Genung Promo			100	165	250
Button, Secret Service Patrol Member	Quaker	1938	1 1/4" blue & silver, pinback	20	35	50
Camera, Dick Tracy Candid Camera	Seymour Sales Co.	1950s	3 x 3 x 5 1/4" with 50mm lens, plastic carrying case & 127 film	75	130	200
Candy, Box	Novel Package Co.	1940s	1 x 2 1/2 x 3 3/4" box with cartoons and story on back, red, white and yellow four panel comic strips on bottom, when removed are used as playing cards	50	80	125
Cap, Air Detective	Quaker	1938		60	100	150
Car, Convertible Squad Car	Marx	1948	20", friction power with flashing lights	155	260	400
Car, Copmobile	Ideal	1963	24" long, white and blue plastic marking of Dick Tracy Copmobile on the drivers side, battery operated with a microphone with amplified speaker on top	60	100	150
Car, Dick Tracy	Marx	1950s	6 1/2" long, light blue with machine gun pointing out of the front window	115	195	300
Car, Dick Tracy Police Car	Marx		9" long, green litho tin, friction drive when wond up siren on the side of station turns rapidly	75	130	200
Car, Dick Tracy Police Dept. Squad Car No. 1	Marx	1950s	11" long, with characters painted on windows, Dick Tracy badge on door, equipped with red light and machine gun	135	225	350
Car, Dick Tracy Police Dept. Squad Car No. 1	Marx	1950s	7" long, dark green body with invisible dashboard	115	195	300
Car, Dick Tracy Police Squad Car	Marx	1950s	7" long, litho tin with electric flashing light, friction powered motor	115	195	300
Car, Dick Tracy Police Station Riot Car	Marx	1950s	friction, sparkling, 7 1/2" long, 1946	115	195	300
Car, Dick Tracy Space Coupe	Aurora	1968	assembly required, all plastic	75	130	200
Car, Dick Tracy Squad Car	Marx		yellow flashing light, tin litho, windup, 11" long, 1940's	145	245	375
Car, Dick Tracy Squad Car	Marx		battery operated, tin litho, 11 1/4" long, 1949	145	245	375
Car, Dick Tracy Squad Car	Marx	1950s	20" long, green litho tin, convertible with friction drive with a battery operated flashing light to the drivers side	155	260	400
Car, Get Away Car	Playmates	1990		15	25	40
Car, Police Car, Dick Tracy	Linemar	1949	3 1/4x6x9" tin, battery operated remote control	225	400	600

Dick Tracy

TOY	COMPANY	YEAR	DESCRIPTION	GOOD	EX	MIB
Car, Riot Car		1950s	7 1/2", friction power	115	195	300
Car, Squad Car	Marx		18", tin, friction power	125	210	325
Christmas Tree Light Bulb, Dick Tracy		1930s	early painted figure of Dick Tracy	30	50	75
Coloring Book, Baby Sparkle Plenty Coloring Book	Saalfield	1948	#1015, 11" x 14" front cover has Baby Sparkle sitting in a chair with foot stool pictured around her are a rabbit, bear and toys as she appears to be reading a book	30	50	75
Coloring Book, Bonnie Braids	Saalfield		11" x 14," #1174, Dick Tracy's New Daughter	20	35	50
Coloring Book, Dick Tracy	Saalfield	1946	9" x 11"	20	35	50
Comic Book	Popped Wheat Cereal	1947	premium	6	10	15
Comic Book, Dick Tracy in 3-D, Ocean Death Trap	Blackthorne	1986	comic book in 3-D	2	3	5
Comic Book, Motorola Presents Dick Tracy Comics	Motorola	1953	premium comic book with paper mask and vest	40	65	100
Crimestopper Club Kit	Chicago Tribune	1961	premium kit containing badge, whistle, decoder, magnifying glass, fingerprinting kit and identification card and a crimestopper textbook	15	25	40
Crimestopper Playset	Hubley	1970s	Dick Tracy cap gun, holster, handcuffs, wallet, flashlight, badge and magnifying glass	50	80	125
Crimestoppers Set	Larami	1973	handcuffs, nightstick & badge	20	35	50
Decoder Card	Post Cereal		cereal premium, red or green	12	20	30
Detective Kit		1944	Dick Tracy Junior Detective Manual, Secret Decoder, rulor, Certificate of Membership and badge	195	325	500
Dick Tracy Braces	Deluxe	1940s	Chicago Tribune premium, suspenders on colorful card	40	65	100
Dick Tracy Braces for Smart Boys and Girls	Deluxe	1950s	Police badge, whistle, suspenders with a Dick Tracy badge as a holder and magnifying glass	60	100	150
Dick Tracy Braces for Smart Boys and Girls	Deluxe	1950s	Police badge, metal handcuffs, whistle, suspenders with a Dick Tracy badge as a holder and magnifying glass	60	100	150
Dick Tracy Cartoon Kit	Colorforms	1962		30	50	75
Dick Tracy Crime Lab	Ja-Ru	1980s	click pistol, fingerprint pad, badge and magnifying glass, available in orange and bright yellow	8	13	20
Dick Tracy Crime Stoppers Laboratory	Porter Chemical Co.	1955	60 power microscope, fingerprint pack, glass slides and magnifying glass and text book	115	195	300
Dick Tracy Detective Set	Pressman	1930s	color graphics of Junior and Dick Tracy, ink roller, glass plate, and Dick Tracy fingerprint record paper	100	165	250
Dinnerware Set	Zak Designs	1980s	three pieces, plate, cup, bowl	10	15	25
Dinnerware Set, Dick Tracy	Homer Laughlin	1950s	bowl 5 1/2" wide by 2" deep, 9" dinner plate and mug	105	175	275
Doll, Bonny Braids		1950s	6" tall, plastic walking wobble doll	20	35	50

Dick Tracy

TOY	COMPANY	YEAR	DESCRIPTION	GOOD	EX	MIB
Doll, Crawling Bonny Braids	Ideal	1952	8" tall, 11" long, crawls when wound	115	195	300
Doll, Dick Tracy		1930s	13" tall, composition, grey trench coat with moveable head and mouth that operates when the back pull string, came in either grey or yellow coat, each	195	325	500
Doll, Little Honey Moon	Ideal	1965	16" space baby, bubble helmet and outfit with white pigtails, doll sitting on half a moon with stars in the background	100	165	250
Doll, Sparkle Plenty Doll	Ideal	1947	12" tall, with yarn-like hair that can be restyled	155	260	400
Dolls, Baby Sparkle Plenty Paper Dolls	Saalfield	1948	11" x 14", #1510, Baby Sparkle cover is standing by a clothes line with clothes and laundry basket behind her	30	50	75
Famous Funnies Printing Set Deluxe, Dick Tracy		1930s	14 stamps, paper and stamp pad, in illustrated box	70	115	175
Famous Funnies Printing Set, Dick Tracy		1930s	has fewer stamps than deluxe set, paper and stamp pad, in illustrated box	60	100	150
Favorite Funnies Printing Set	Stampercraft	1935	rubber stamp setf for creating comic strips, features Tracy and other cartoon characters	50	80	125
Figure Set, Dick Tracy	Marx	1950s	figures range from 2 1/2" to 3 1/2" tall, Tracy, Junior, Sparkle Plenty, B.O. Plenty and Gravel Gertie	40	65	100
Figure, B.O. Plenty	Marx	1950s	Famous Comic Figures series, waxy cream, pink, 60mm tall	4	7	10
Figure, Bendie, Dick Tracy	Lakeside			20	35	50
Figure, Big Boy	Playmates	1990		4	7	10
Figure, Breathless Mahoney	Applause	1990	14" tall	4	7	10
Figure, Chalk, Dick Tracy	Professional Art	1940s	7" unpainted detailed white chalk figure	100	165	250
Figure, Dick Tracy	Marx	1950s	Famous Comic Figures series, waxy cream, pink, 60 mm tall	6	10	15
Figure, Gravel Gertie	Marx	1950s	Famous Comic Figures series, waxy cream, pink, 60 mm tall	4	7	10
Figure, Junior	Marx	1950s	Famous Comic Figures series, waxy cream, pink, 60 mm tall	4	7	10
Figure, Sparkle Plenty	Marx	1950s	Famous Comic Figures series, waxy cream, pink, 60 mm tall	4	7	10
Figure, Steve the Tramp	Playmates	1990		4	7	10
Figure, The Blank	Playmates	1990	figure with gun and hat with featureless face attached	70	115	175
Film Print		1950s	16mm color-sound 30-40 minutes	20	35	50
Film Strip Viewer	Tru-Vue, Inc.	1940s	3-D image film strip	60	100	150
Film Strip Viewer, Dick Tracy In Movie Style	Acme	1948	film strip viewer and 2 films in colorful illustrated box	70	115	175
Film Strip Viewer, Jumbo Movie Style	Acme	1964	film strip viewer with 2 boxes of film, on card, rack toy	30	50	75
Film Viewer, Mini Color Tele-Viewer	Larami	1973	rack toy with two paper film strips to thread through viewer	20	35	50
Film Viewer, Moviescope Viewer		1940s	two films included	60	100	150
Film, Super 8 Color Film	Republic	1965	b/w cartton "Trick or Treat"	20	35	50
Fingerprint Outfit, Dick Tracy	Pressman	1933	microscope, fingerprint pad, magnifying glass and badge	155	260	400
Flashlight	Quaker	1939	3" pen light, black	30	50	75

Dick Tracy

TOY	COMPANY	YEAR	DESCRIPTION	GOOD	EX	MIB
Flashlight, 3 Color	Bantam Lite	1961	wrist light, o cardhand size, metal	40	65	100
Flashlight, Pocket	Quaker	1939	bullet shaped with shield tag, colored red, green and black	60	100	150
Game, Dick Tracy Bingo, Lock Them Up in Jail and Harmonize with Tracy		1940s	object of each is to roll BBs into different holes on the face of a glas framed game card for points	40	65	100
Game, Dick Tracy Crime Stopper	Ideal	1963	board game, play workstation contains crime indicator dial, decoder knobs, criminal buttons, clue cards and holders and clue windows	40	65	100
Game, Dick Tracy Detective Game	Einson-Freeman	1937	board game	60	100	150
Game, Dick Tracy Master Detective Game	Selchow & Righter	1961	board game	30	50	80
Game, Dick Tracy Pinball	Marx	1967	14 x 24", pinball game shows characters from TV show pilot	60	100	150
Game, Dick Tracy Playing Card Game	Whitman	1937		30	50	75
Game, Dick Tracy Playing Card Game	Whitman	1934	card game	40	65	100
Game, Dick Tracy Playing Card Game	Esquire Novelty	1939		30	50	75
Game, Dick Tracy Pop-Pop Game	Ja-Ru	1980s	Diet Smith and Flattop are targets	10	15	25
Game, Dick Tracy Super Detective Mystery Card Game	Whitman	1973		25	40	65
Game, Dick Tracy Super Detective Mystery Card Game	Whitman	1941	card game	40	65	100
Game, Dick Tracy Super Detective Playing Card Game	Whitman	1941		30	50	75
Game, Dick Tracy Target Game	Marx	1940s	character targets with dart gun	70	115	175
Game, Sunday Funnies Board Game	Ideal	1972		30	50	75
Gun, 45 Special Water Handgun	Tops Plastics	1950s	plastic	40	65	100
Gun, Automatic Target Range Gun	Marx	1967	BB gun mounted in an enclosed plastic shooting gallery	75	130	200
Gun, Camera Dart Gun	Larami	1971	rack toy, 8mm camera-shaped toy with dart shooting viewer	25	40	60
Gun, Click Pistol	Marx	1930s		60	100	150
Gun, Dick Tracy Jr. Click Pistol #78	Marx	1930s	aluminum	40	65	100
Gun, Dick Tracy Sparkling Pop Pistol	Marx	1930s	tin litho	75	130	200
Gun, Dick Tracy Sub-Machine Gun	Tops Plastics	1950s	12" long, red, green or blue, water gun " holds over 500 shots on one filling," Dick Tracy decal on magazine	78	130	200
Gun, Luger Water Gun	Larami	1971		20	35	50
Gun, Official Dick Tracy Shootin' Shell Snub-Nose .38	Mattel	1961	die cast chrome .38 with brown plastic grips, chrome finish with Shootin' Shell bullets & Stick-m caps	70	120	185

Dick Tracy

TOY	COMPANY	YEAR	DESCRIPTION	GOOD	EX	MIB
Gun, Official Dick Tracy Tommy Burst Machine Gun	Mattel	1962	25" Thompson style machine gun fire perforated roll caps, single shot or in full burst when bolt is pulled back, brown plastic stock & black plastic body, lift up rear sight, Dick Tracy decal on stock	105	180	275
Gun, Pop	Tip Top Bread	1944	7 1/2 x 4 1/2" paper pop gun premium for bread and radio show	60	100	150
Gun, Power Jet Squad Gun	Mattel	1962	29" long cap and water rifle	80	130	200
Gun, Rapid Fire Tommy Gun	Parker Johns	1940s	20" long tommy gun with Tracy on stock	100	165	250
Gun, Remington .41 Derringer "Dick Tracy Special Ray Gun"	Larami	1964	derringer with a metal Dick Tracy New York Police Detective Badge	30	50	75
Gun, Repeater Cap Gun	Larami	1972		15	25	40
Gun, Siren Pistol	Marx	1930s	pressed steel, 8 1/2" long	100	165	250
Gun, Sparkling Pop Pistol No. 96	Marx			75	130	200
Hand Puppet, Dick Tracy	Ideal	1961	10 1/2" tall, fabric & vinyl, with record	70	115	175
Hand Puppet, Hemlock Holmes	Ideal	1961	includes record	60	100	150
Hand Puppet, Joe Jitsu	Ideal	1961	10 1/2" tall, fabric & vinyl, includes record	60	100	150
Handcuffs for Junior	John Henry Prod.	1946	metal toy handcuffs on display header card	40	65	100
Hat, Dick Tracy	Miller Bros. Hat Co.	1940s	"100% Wool Dick Tracy Model Snap-Brim Fedora", blue-grey felt	60	100	150
Hingees "Dick Tracy and his Friends to Life" Punch-outs	Reed & Associates	1944	6 1/2" tall figures, Tess Trueheart, Chief Brandon, Junior, Pat Patton and Tracy	30	50	75
Junior Detective Kit	Sweets Co. of Am.	1944	certificate, secret code dial, wall chart, file cards & tape measure	70	115	175
Junior Detective Kit	Golden Press	1962	punch-out book with detective badge, three-way wrist radio, holster, fingerprint chart, disguises, handcuffs and detective revolver	20	35	50
Junior Dick Tracy Crime Detection Folio		1942	radio premium, contained detective's notebook, decoder with 3 mystery sheets, and puzzle	80	130	200
Lamp, Table		1950s	painted ceramic bust of Tracy in black coat, yellow hat and red tie, holding pistol, round base is embossed Dick Tracy, shade shows Tracy and Sparkle Plenty	1150	2000	3000
Lunch Box, Dick Tracy	Aladdin	1967	steel lunch box with steel bottle	100	165	250
Mask, Dick Tracy	Philadelphia Inquirer	1933	full color paper mask	115	195	300
Mobile Commander	Larami	1973	toy telephone with plastic connecting tube, plastic gun and badge	25	40	60
Model Kit, Dick Tracy in Action	Aurora	1968	plastic model kit	100	165	250
Nodder, Dick Tracy		1960s	6 1/2" tall, ceramic nodding head bust doll	390	650	1000

Dick Tracy

TOY	COMPANY	YEAR	DESCRIPTION	GOOD	EX	MIB
Numbered Pencil Coloring Set	Hasbro	1967	six pre-sketched pictures to color	40	65	100
Offical Holster Outfit	Classy Products	1940s	leather holster with painted Tracy profile	80	130	200
Pedal Car, Dick Tracy Squad Car No.1	Murray	1950s	deep green with yellow markings and white plastic light on hood	775	1300	2000
Pin, Bar, Secret Service Patrol Leader	Quaker	1938	litho bar pin	155	260	400
Pin, Detective Club		1942	yellow, tab back	30	50	80
Pin, Dick Tracy Gang		1940s	premium ring given to theatre goers, Davis Theatre	50	80	125
Pin, Dick Tracy's Bonny Braids	Charmore	1951	1 1/4" figure plastic pin on full color on a 3 1/2 x 5 1/2" card	30	50	75
Pin, Pep, B.O. Plenty	Kellogg's	1945	tin litho pinback button	15	25	40
Pin, Pep, Chief Brandon	Kellogg's	1945	tin litho pinback button	8	13	20
Pin, Pep, Dick Tracy	Kellogg's	1945	tin litho pinback button	20	35	50
Pin, Pep, Flattop	Kellogg's	1945	tin litho pinback button	20	35	50
Pin, Pep, Gravel Gertie	Kellogg's	1945	tin litho pinback button	15	30	45
Pin, Pep, Junior Tracy	Kellogg's	1945	tin litho pinback button	12	20	30
Pin, Pep, Pat Patten	Kellogg's	1945	tin litho pinback button	8	13	20
Pin, Pep, Tess Trueheart	Kellogg's	1945	tin litho pinback button	12	20	30
Pin, Pep, Vitamin Flintheart	Kellogg's	1945	tin litho pinback button	10	15	25
Play Set, Dick Tracy	Ideal	1973	contains eighteen cardboard figures that measure 3 1/2 to 5" tall, self contained in its own carrying case that measures 17 x 7 1/2 x 6"	60	100	150
Playset, Dick Tracy	Placo	1982	plastic dart gun, targets of different villians and a set of handcuffs	30	50	80
Playstone Funnies Kasting Kit Molds	Allied Mfg.	1930s	molds of Dick Tracy's cast, tin measures 3 x 8 1/2 x 1"	30	50	80
Playtsone Funnies Kasting Kit	Allied Mfg.	1930s	molds for casting figures of Tracy and other characters	60	100	150
Portrait, Dick Tracy	Pillsbury	1940s	part of set of 8, each 7 x 10" in mat, shows Tracy and Junior, originally sold for 10 cents plus one box top	75	130	200
Puzzle, "The Bank Holdup" Jigsaw Puzzle	Jaymar	1960s	triple-thick interlocking pieces featuring the TV cartoon	20	35	50
Puzzle, Dick Tracy Tray Puzzle		1952	11 x 14" Dick Tracy standing outside on a sidewalk with a door directly behind him and name in large letters	60	100	150
Puzzle, Dick Tracy's New Daughter Tray Puzzle	Saalfield	1951	Bonny Braids 10 1/4 x 11 1/2"	30	50	75
Puzzle, Dick Tracy's Two in One Mystery Puzzle	Jaymar	1958	one puzzle shows the crime and the other the solution	40	65	100
Puzzles, Dick Tracy Big Little Book Picture Puzzles	Whitman	1938	8 x 10 x 2" contains two puzzles of BLB scenes	75	130	200
Radio Receiver, Two Transistor	American Doll & Toy	1961	shoulder holster and secret ear plug with two transistors radio receiver	40	65	100
Record, Dick Tracy Original Radio Broadcast Album	Coca-Cola	1972	presents the cast from "The Case of the Firebug Murders" radio show	30	50	75

Dick Tracy

TOY	COMPANY	YEAR	DESCRIPTION	GOOD	EX	MIB
Record, Flattop Story Double Record Set	Mercury Records	1947	listen to the record, write the story and color the comics	80	130	200
Ring, Dick Tracy Enameled Portrait	Miller Bros. Hat Co.	1940s		80	130	200
Ring, Dick Tracy Secret Compartment Ring	Quaker	1938	removable cover picturing Tracy and good luck symbols	100	165	250
Ring, Dick Tracy Service Patrol		1966	premium item	15	25	35
Ring, Dick Tracy, Air Detective		1938		35	55	85
Ring, Dick Tracy, Monogram	Quaker	1938	premium ring shows initials only, no Tracy name or picture	155	260	400
Secret Code Writer and Pencil		1939		50	80	125
Secret Detective Methods and Magic Tricks	Quaker	1939	5 x 7 1/2 x 1/2" cereal premium book	30	50	75
Secret Detector Kit	Quaker	1938	Secret Formula Q-11 & negatives	60	100	150
Secret Service Phones, Dick Tracy	Quaker	1938	cardboard phones, walkie talkie type	60	100	150
Shoulder Holster Set	J. Hapern Co.	1950s	leather holster with Dick Tracy's profile embossed	60	100	150
Soakie Bottle, Dick Tracy	Colgate-Palmolive	1965	10" tall bottle contained bubble bath	30	50	75
Sparkle Paints, Dick Tracy	Kenner	1963	five glitter colors in unbreakable plastic container, brushes and six pictures of Dick Tracy with Jo Jetsu, Go Go Gomez, Flattop and Mumbles to paint	30	50	80
Sparkle Plenty Christmas Tree Lights	Mutual Equip.	1940s	7 multiple lights in colorful box	40	65	100
Sparkle Plenty Christmas Tree Lights	Mutual Equip.	1940s	15 multiple lights in colorful box	60	100	150
Sparkle Plenty Islander Ukette	Styron	1950	musical instrument, junior size, with instruction book	100	165	250
Sparkle Plenty Washing Machine	Kalon Radio Corp.	1940s	12" tall, litho tin, inside are the components of a real wash machine, pictured outside on tub is Gravel Gertie doing the wash as B.O. Plenty holds baby Sparkle	115	195	300
Talking Phone	Marx	1967	green with an ivory handle, battery operated plays ten different sayings such as "Make your arrest," etc.	40	65	100
Target Set, Dick Tracy	Larami	1969	available in red, green or blue, shoots rubber bands	20	35	50
Target, Dick Tracy	Marx	1940s	17" circular cardboard target (similar to a bulls-eye), with dart gun and box	85	145	225
Target, Dick Tracy	Marx	1941	10" square tin litho with "Recovery" and "Rescueing" points on the front side and bullseye target on the back side	60	100	150
Tie Clasp, Dick Tracy Jr. Detective Agency		1930s	two versions, silver or brass, each	30	50	75
Wall Clock		1990s	16 x 20" battery power quartz, face shows Disney movie Tracy takling into wrist radio	12	20	30
Wallpaper, Section		1950s	shows comic strip scenes of Tracy and 7 other characters	30	50	75

Dick Tracy

TOY	COMPANY	YEAR	DESCRIPTION	GOOD	EX	MIB
Watch, Pocket	Bradley	1959		70	115	175
Whistle, Siren Police Whistle No. 64	Marx		tin	40	65	100
Wind-Up, B.O. Plenty	Marx	1940s	8 1/2" tall holding baby Sparkle, litho tin, walks, hat tips up and down when key is wound	145	245	375
Wrist Radio, Dick Tracy Wrist Band AM Radio	Creative Creations	1976	complete with earphone and two mercury batteries in colorful box showing Tracy and Flattop	40	65	100
Wrist Radio, Dick Tracy Wrist Radio	Da-Myco Products	1947	crystal set with receiver on a leather band, 30" wires and connectors for aerial and ground, no batteries, no tubes and no electric	195	325	500
Wrist Radios, 2-Way Wrist Radio Set	American Doll & Toy	1960s	plastic with power pack, battery operated, works on citizen's bank with 1,500 foot range	50	80	125
Wrist Radios, Dick Tracy 2-Way Electronic Wrist Radios	Remco	1950s	2 1/2 x 9 1/2 x 13 1/2", plastic battery operated wrist radios, two stations with buzzer that works up to 1/4 mile range	60	100	150
Wrist Radios, Dick Tracy Detective Club	Gaylord	1945	phone receiver shaped, nonworking toy on band, in 3 versions, green/black wood with strap, red or blue plastic, shows Tracy, on card with Detective Club badge, on card	100	165	250
Wrist Radios, Dick Tracy, 2-Way	Ertl	1990	working pair of wrist radios, use 9 volt battery in each set, box shows Tracy	10	15	25
Wrist TV, Dick Tracy Wrist TV	Larami	1973	paper roll of cartoon strips are threaded through the TV viewer	20	35	50
Wrist TV, Dick Tracy Wrist TV	Ja-Ru	1980s	paper roll of cartoon strips are threaded through the TV viewer	10	15	25
Wrist Watch	New Haven	1937	oblong or square face, in box	145	245	375
Wrist Watch	New Haven	1937	round face, in box	145	245	375
Wrist Watch with Animated Gun	New Haven	1951		125	210	325
Wrist Watch, Dick Tracy Two-Way Wrist Watch	Playmates	1990	on illustrated card, no radio function, toy is a watch only	4	7	10
Wrist Watch, Digital Watch	Omni	1981	digital watch ccomes in a cardboard police car packaging	30	50	80

Disney

TOY	COMPANY	YEAR	DESCRIPTION	GOOD	EX	MIB
20,000 Leagues Under the Sea Board Game	Gardner	1950s	8x16x1 1/2"	12	23	75
2nd National Duck Bank	Chein		3 1/2" tallx7" long	85	145	225
Aristocats Thomas O'Malley Figure	Enesco	1967	8" tall ceramic figure	35	65	125
Black Hole Puzzle	Whitman	1979	jigsaw puzzle 9"x11", V.I.N.C.E.N.T. or Cygnus	4	8	12
Carousel	Linemar		7" tall with 3" figures, wind-up	50	95	145
Casey Jr. Disneyland Express Train	Marx	1950s	12" long, tin, wind-up	175	275	250
Character Molding & Coloring Set		1950s	red rubber molds of Bambi, Thumper, Dumbo, Goofy, Flower and Joe Carioca to make plaster figures	40	70	125

Disney

TOY	COMPANY	YEAR	DESCRIPTION	GOOD	EX	MIB
Chitty Chitty Bang Bang Candy Card Set		1960s	1 1/2"x2 1/2" set of fifty color photos	11	20	30
Disney "Sea Scouts" Puzzle	Williams Ellis & Co.	1930s	5x8x1"	16	30	45
Disney Fantasy Jigsaw Puzzle	Whitman	1981	large size 22x33" puzzle with many characters shown	4	7	10
Disney Ferris Wheel	Chein		17" tall, wind-up	490	625	1000
Disney Figure Golf Balls			set of twelve	12	23	35
Disney Filmstrips	Craftman's Guild	1940s	set of thirteen color filstrips	95	180	275
Disney Metal Tub		1960s	20" in diameter 11" tall pictures of Donald, Mickey and Pluto putting toys into their tub	25	50	125
Disney Rattle	Noma	1930s	4" tall with Mickey & Minnie, Donald & Pluto carrying a Christmas tree	60	115	275
Disney Shooting Gallery	Welso Toys	1950s	8"x 12"x1 1/2" tin target with molded figures of Mickey, Donald, Goofy and Pluto	60	115	225
Disney Tray	Ohio Art		8"x10", tin, pictures are: Mickey & Minnie Mouse, Goofy, Horace, Pluto, Donald Duck & Clarabelle	23	45	65
Disney Treasure Chest Set	Craftman's Guild	1940s	red plastic film viewer and filmstrips stored in a blue box designed like a chest	65	125	190
Disney World Globe	Rand McNally	1950s	6 1/2" metal base, 8" diameter with Disney characters	25	50	225
Disney's Bunnies	Fisher-Price	1936	2x3x2 1/2" tall wooden toys titled Wee Bunny, Big Bunny and Little Bunny	60	115	175
Disneyland ashtray		1950s	5" diameter, china ashtray with Tinker Bell & castle	11	20	125
Disneyland Auto Magic Picture Gun & Theater		1950s	battery operated metal gun with oval filmstrip	45	85	130
Disneyland Bagatelle	Wolverine	1970s	large size bagatelle game with Disneyland graphics	11	20	75
Disneyland China Ash Tray	Eleanor Welborn Art	1955	3 1/2x4 1/2x1 1/2" light green outer rim, with Tinkerbell in white middle with yellow hair, green suit and wings	11	20	125
Disneyland Electric Light	Econlite Corp.	1950s	picture of Disney characters leaving a bus on a drum base	45	80	225
Disneyland F.D. Fire Truck	Linemar		18" long, battery operated, moveable, Donald Duck fireman climbs the ladder	70	130	250
Disneyland Felt Banner	Disney	1960s	"The Magic Kingdom" 24 1/2" red/white/blue coat of arms	25	50	75
Disneyland Give-A-Show Projector Color Slides		1960s	112 color slides	60	115	175
Disneyland Haunted House Bank	(Japan)	1960s		35	65	175
Disneyland Metal Craft Tapping Set	Pressman	1950s		16	30	85
Disneyland Miniature License Plates	Marx	1966	2x4" plates with Mickey, Minnie and Pluto, or Snow White, Donald and Goofy, each	12	23	35
Disneyland Pen		1960s	6" long with a picture of a floating riverboat in liquid	11	20	30
Disneyland Tray Puzzle	Whitman	1956	11 1/2x14 1/2" w/ Mickey & friends in Fantasyland tea cup ride	14	25	40

Disney

TOY	COMPANY	YEAR	DESCRIPTION	GOOD	EX	MIB
Disneyland Tray Puzzle	Whitman	1956	11 1/2x14 1/2" with Mickey & friends riding the stagecoach through Frontierland	14	25	40
Disneyland View-Master Set		1960s	scenes of Fantasyland	12	23	75
Disneyland Wind-up Roller Coaster	Chein		8"x19"x10", tin	425	650	950
Disneyland Wood Pencil		1970s	11x1/2" thick	5	10	15
Duck Tales Travel Tote		1980s	travel agency premium	4	7	10
Early Settlers Log Set	Halsam	1960s	log building set based on Disneyland's Tom Sawyer's Island	25	45	125
Fantasia Bowl	Vernon Kilns	1940	12" diameter & 2 1/2" tall, pink bowl with a winged nymph from Fantasia	115	210	325
Fantasia Ceramic Unicorn Figure	Vernon Kilns	1940s	black-winged unicorn	45	80	175
Fantasia Cup & Saucer Set	Vernon Kilns	1940	6 1/4" diameter saucer & 2" tall cup	60	115	175
Fantasia Figure	Vernon Kilns		half-woman, half-zebra centaur	80	145	225
Fantasia Musical Jewelry Box	Schmid Bros.	1990	box features Mickey and plays "The Sorcerer's Apprentice"	30	55	85
Fantasyland Tray Puzzle	Whitman	1957	11 1/2x14 1/2" Mickey & Donald on an amusement ride	12	23	35
Figural Light Switch Plates	Monogram		hand painted switch plates, Goofy, Donald, Mickey, on card, each	5	10	15
Golf Club Guards			Mickey, Minnie, Donald, Pluto and Goofy, each	4	7	10
Grasshopper & the Ants Album	Disney	1949	45 RPM record-reader album	14	25	40
Happy Birthday/Pepsi Placemats	Pepsi Co.	1978	set of four mats: Goofy, Uncle Scrooge, Mickey at a party and Mickey & Goofy fishing	11	20	30
Hayley Mills Paper Doll Kit	Whitman	1963	9 1/2" tall, "Summer Magic"	25	50	75
Horace Horsecollar Hand Puppet	Gund	1950s		25	50	125
Horace Horsecollar/ Clarabelle Cow Set	Pepsi Co.	1977	glass "Pepsi Collector/Happy Birthday Mickey" Set	11	20	30
Johnny Tremain Figure & Horse	Marx	1957	plastic 9 1/2" tall horse and 5 1/2" tall Johnny	60	115	175
Jose Carioca Figure	Marx	1960s	5 1/2" tall plastic figure, wire arms and legs, in box	45	80	125
Jose Carioca Figure	Marx		2" tall, plastic	30	55	50
Jose Carioca Wind-up Toy	France	1940s	3 1/2x5x7 1/2" tall	115	210	325
King Brian Hand Puppet	Gund	1959	10" tall	25	50	75
Lap Trays	Hasko	1960s	set of four features Donald, Goofy & Pluto, Peter Pan and the Seven Dwarfs	35	65	100
Lil' Hiawatha Charm		1960s	laminated/sterling silver charm	16	30	45
Merry-Go-Round Lamp			10" tall, metal & plastic, when light heats up it makes cylinder	18	35	50
Mother Goose Hand Puppet	Gund	1950s	11" tall	25	50	75
Nautilis Expanding Periscope	Pressman	1954	Inspired by 20,000 Leagues Under the Sea, 19" long	35	65	100
Nautilis Wind-up Submarine	Sutcliffe/ England	1950s		80	145	225
Official Santa Fe & Disneyland R.R. Scale Model Train	Tyco	1966	Model Ho electric train set 16x19x1 1/2"	155	295	450
Pecos Bill Wind-up Toy	Marx	1950s	10" tall, riding his horse Widowmaker and holding a metal lasso	100	150	200

Disney

TOY	COMPANY	YEAR	DESCRIPTION	GOOD	EX	MIB
Peculiar Penguins	Disney	1934	storybook	25	50	75
Pedro Hand Puppet	Gund			16	30	45
Robin Hood Colorforms	Colorforms	1973		5	10	15
Rocketeer Doll	Applause		9" tall	7	13	20
Sand Pail & Shovel	Ohio Art	1930s	features pie-eyed Mickey selling cold drinks to Pluto, Minnie & Clarabell	50	95	225
Shaggy Dog Figures	Enesco	1959	three versions: blue hat & jacket with a white "S" green/black base; white pajamas with blue stripes, brown base; blue hat holding a red steering wheel, black base, each	25	50	75
Shaggy Dog Hand Puppet	Gund	1959	9" tall, red cloth body with white felt hands, yellow ribbons tied around his neck	16	30	75
Silly Symphony Fan			wooden handle	25	50	125
Silly Symphony Lights	Noma		8 original lights	55	100	350
Sketchagraph	Ohio Art			14	25	75
Swamp Fox Board Game	Parker Brothers	1960	18 1/2x18 1/2" game board, from the TV series with Leslie Nielson	30	50	80
Swamp Fox Coloring Book	Whitman	1961		7	13	45
Toby Tyler Circus Playbook	Whitman	1959	punch out character activity book	25	50	75
Walt Disney Movie Viewer & Cartridge	Action Films	1972	action set #9312 with the cartridge "Lonesome Ghosts"	9	16	25
Walt Disney Paint Book	Whitman	1937	11"x14"	20	35	55
Walt Disney's Character Scramble	Plane Facts Co.	1940s	10 cardboard figures 6" tall	23	40	125
Walt Disney's Clock Cleaners, Picture Book	Whitman	1938	linen-like illustrated book	40	70	110
Walt Disney's Game/ Parade/Academy Award Winners	American Toy Works		15 games for all ages	60	115	175
Walt Disney's Jimmie Dodd Coloring Book	Whitman	1956		11	20	30
Walt Disney's Realistic Noah's Ark	W.H. Greene Co.	1940s	6"x7"x18" Ark, 101 2" animals, and 4" human figures on cardboard	185	345	525
Walt Disney's Silly Symphony Bells	Noma		Christmas tree bells pictured: Three Little Pigs, Elmer the Elephant, The Tortoise & The Hare, etc.	55	100	250
Walt Disney's Snap-Eeze Set	Marx	1963	12 1/2x15x1" box with 12 flat plastic figures: Peter Pan, Pinocchio, Donald, Jiminy Cricket, Gepetto, Bambi, Dewey, Goofy, Pluto, Mickey, Joe Carioca and Bere Rabbit	60	115	275
Walt Disney's Television Car	Marx		8" long, friction toy lights up a picture on the roof when motor turns	125	200	425

Donald Duck

TOY	COMPANY	YEAR	DESCRIPTION	GOOD	EX	MIB
Carpet Sweeper		1940s	red wood and metal sweeper shows Donald sweeping while Minnie Watches	35	65	125
Daisy Duck Watch	US Time	1948	Daisy is in black & white, yellow & light blue	105	195	475
Disneyland Frame Tray Puzzle	Whitman	1955	11x15" with Donald & his nephews on a pirate ship	16	30	45
Donald & His Gang Puzzle	Ontex	1940s		25	50	75

Donald Duck

TOY	COMPANY	YEAR	DESCRIPTION	GOOD	EX	MIB
Donald Drum Major Doll	Knickerbocker	1938	17" tall, red jacket with yellow piping & a black plush hat	175	325	850
Donald Duck & Minnie Mouse Sweeper	Ohio Art	1930s	3" with wooden handle	55	100	225
Donald Duck & Nephews Puzzle	Whitman	1960s	acrobatics themed illustration	7	13	20
Donald Duck & Pluto Car	Sun Rubber Co.		6 1/2" long, hard rubber	18	35	125
Donald Duck Alarm Clock	Glen Clock/ Scotland	1950s	5 1/2x5 1/2x2", Donald pictured with blue bird on his hand	95	180	350
Donald Duck Alarm Clock	Bayard	1960s	2x4 1/2x5"	80	150	350
Donald Duck and Mickey Mouse Crayon Box	Transogram	1946	tin illustrated crayon box	25	50	75
Donald Duck Bank	Crown Toy	1938	6" tall, composition, head is moveable	55	100	375
Donald Duck Bank, Ceramic			3 1/2x4x7" tall, Donald in a cowboy outfit	35	65	125
Donald Duck Bank, Ceramic		1940s	4 1/2x4 1/2x6 1/2" tall, Donald holding a rope with a large brown fish by his side	35	65	125
Donald Duck Bank, China		1940s	5 1/2x6x7 1/2" tall, Donald seated holding a coin in one hand	70	130	275
Donald Duck Bank, Plastic			4x4x8 1/2" tall, Donald seated on a treasure chest dressed as a cowboy	9	16	50
Donald Duck Bathtub		1960s		25	50	125
Donald Duck Bubble Bath		1950s	3 1/2" diameter, 7" tall	9	16	25
Donald Duck Camera	Herbert-George Co.	1950s	3x4x3"	25	50	125
Donald Duck Choo Choo Pull Toy	Fisher-Price	1940	#450, Donald in red cap rings bell as toy is pulled	90	165	275
Donald Duck Disney Dipsy Car	Marx	1953	6" tall, wind-up with a spring necked Donald	425	650	895
Donald Duck Driving Pluto Toy			9" long, wind-up, celluloid	275	500	1200
Donald Duck Duet Dancing Toy	Marx	1946	10 1/2" tall, wind-up, Goofy dances & Donald Duck plays drum	425	775	1200
Donald Duck Dump Truck	Linemar	1950s	2x5x2" tall	80	145	375
Donald Duck Electric Lamp	Dolly Toy Co.	1970s	Donald on a tug boat	25	50	125
Donald Duck Figure, Celluloid	(Japan)	1930s	long billed Donald walking	55	100	450
Donald Duck Figurine			3 1/2" tall, celluloid, jointed arms & leg, winking	35	60	300
Donald Duck Framed Picture		1950s	8 1/2x10 1/2", glow-in-the-dark with Donald on a bike	9	16	25
Donald Duck Fun-e-Flex Figure	Fun-E-Flex	1930s	wooden Donald on red sled with rope	40	80	275
Donald Duck Funnee Movee Set	Transogram	1940	box features Donald, Mickey, and the nephews	45	90	225
Donald Duck Funnee Movie Set	Irwin	1949	hand crank movie 'camera' viewer and 4 films in box	80	145	225
Donald Duck Hair Brush	Disney		2x3 1/2x1 1/2"	18	35	75
Donald Duck Jack-in-the-Box			paper covering on box, figure made of fabric with composition head	45	80	275
Donald Duck Lamp, China		1940s	6x7x9" tall, Donald holding an axe standing next to a tree trunk	55	100	225

Donald Duck

TOY	COMPANY	YEAR	DESCRIPTION	GOOD	EX	MIB
Donald Duck Light Switch Cover	Dolly Toy Co.	1976	plastic light switch cover, Donald on a boat	4	7	10
Donald Duck Marionette	Peter Puppet	1950s	6 1/2" tall	25	50	125
Donald Duck Moving Eye Clock	Disney	1960s	3x5x9" tall, small brass/plastic pendulum moves back & forth below Donald's feet while his eyes follow it	25	50	125
Donald Duck Music Box	Anri	1971	Donald with guitar, music "My Way"	35	65	125
Donald Duck Nodder		1960s	5 1/2" tall on green base	12	23	125
Donald Duck Paint Box	Transogram	1938	8" long, paint set	16	30	125
Donald Duck Pencil Sharpener			1 1/2" tall, red celluloid	16	30	45
Donald Duck Pitcher, Ceramic		1940s	4x4 1/2x6" tall, shaped like Donald Duck's body with a small handle & spout on top	30	60	90
Donald Duck Plate, Ceramic	Disney	1960s	1 1/4' deep & 9" diameter, light green background with a dark blue hat	14	25	75
Donald Duck Pocket Watch	Ingersoll	1939		60	115	375
Donald Duck Projector	Stephens Prod.	1950s	projector in box with 4 films	65	120	185
Donald Duck Pull Toy	Fisher-Price	1941	4 1/2x11x10" tall, wood, Donald's arms swing back and forth while in a forward motion	90	165	250
Donald Duck Pull Toy	Fisher-Price	1953	10", wood, baton twirling Donald with a white sailor hat	60	115	175
Donald Duck Puppet	Pelham Puppets	1960s	10" tall, hollow composition	25	50	175
Donald Duck Push Cart			18" plastic, colorful	11	20	75
Donald Duck Push Figure	Kohner	1950s	wood jointed push toy, Donald is steering a ship dressed as a sailor	45	90	175
Donald Duck Push Toy	Gong Bell	1950s	4 1/2x8x1" thick with an 18" handle	30	55	85
Donald Duck Puzzle	Jaymar	1940s	7x10x2", Donald & his nephews having a picnic	14	25	40
Donald Duck Ramp Walker	Marx	1950s	1 1/4x3 1/2x3" tall, Donald is pulling red wagon with his nephews	100	125	250
Donald Duck Rubber Car	Sun Rubber Co.	1950s	2 1/2x3 1/2x6 1/2" long	35	65	125
Donald Duck Rubber Figure	Seiberling		3x3 1/2x5" tall	45	80	125
Donald Duck Rubber Figure	Dell	1950s	7" tall	35	65	125
Donald Duck Rubber Figure	Seiberling		6" tall, solid rubber with moveable head	40	70	125
Donald Duck Rubber Figure	Seiberling		6" tall, hollow rubber with squeaker in the base	35	65	125
Donald Duck Sand Pail	Ohio Art	1939	4 1/2" tall, Donald at beach playing tug-of-war with his two nephews	55	100	175
Donald Duck Scooter	Marx	1960s	tin wind-up, 4x4x2" tall	80	140	225
Donald Duck Skating Rink Toy	Mettoy	1950s	4" diameter, Donald is skating while other Disney characters circle the rink	35	65	100
Donald Duck Snow Shovel	Ohio Art		wood, tin, litho	60	115	175
Donald Duck Soakie			large size bottle	12	25	35
Donald Duck Soap Figure	Disney		castile soap	35	65	85

Donald Duck

TOY	COMPANY	YEAR	DESCRIPTION	GOOD	EX	MIB
Donald Duck Squeeze Toy	Dell	1960s	8" tall, rubber	7	13	35
Donald Duck Sweeper	Ohio Art		6" wide base	25	50	75
Donald Duck Talking Figure	Mattel	1976	4x5x6 1/2" tall, Donald says "I'm Donald Duck" and sneezes when string is pulled	25	50	75
Donald Duck Tea Set	Ohio Art		7 1/2" long tray, 2 1/4" diameter cups & 2 1/2" & 4 1/4" diameter sizes of saucers with lithography	40	80	150
Donald Duck Telephone Bank	N.N. Hill Brass Co.	1938	5" tall with cardboard figure of character	70	130	275
Donald Duck the Drummer	Marx	1940s	5x7x10" tall, Donald as a drummer beating on a metal drum	150	250	350
Donald Duck Tin Sand Pail	Ohio Art	1950s	3 1/2" diameter & 3 1/2" tall, Donald in life preserver fighting off seagulls	30	55	85
Donald Duck Toothbrush Holder		1930s	ceramic 4 1/2" tall, with detail of Donald holding the toothbrush holder, orange base	90	165	250
Donald Duck Toothbrush Holder		1930s	bisque, two Donald Duck figures stand side by side, with faces in opposite directions, toothbrush hole is behind figures	55	100	150
Donald Duck Toothbrush Holder, Bisque		1935	2x3x5" tall, holds two toothbrushes	90	165	250
Donald Duck Toy	Schuco		5 1/2" tall, wind-up with a bellows quacking sound	300	550	850
Donald Duck Toy Raft	Ideal	1950s	2" tall, blue plastic raft with yellow sail with Donald looking through a telescope	35	65	100
Donald Duck Trapeze Toy	Linemar		5" tall, celluloid, wind up	55	105	450
Donald Duck Umbrella Handle		1930s	3 1/4" tall	35	60	125
Donald Duck Watch	US Time	1940s		125	225	350
Donald Duck Watering Can	Ohio Art	1938	6" tall, tin, litho	18	35	95
Donald Duck Wind-up	Durham Plastic Co.	1972	6 1/2" tall hard plastic wind-up toy	14	25	50
Donald Duck Wooden Xylophone Pull Toy	Fisher-Price	1938	11x12 1/2" pulltoy has Donald play xylophone, red wheels, dark blue base, Donald in blue cap	105	195	275
Donald Duck WW I Pencil Box	Dixon		5x8 1/2x1 1/4" deep, Donald flying a plane, holding a tomahawk	35	60	125
Donald Duck, Mickey & Minnie Mouse Toothbrush Holder			4 1/2" tall, bisque, Donald Duck hugging Mickey & Minnie Mouse	55	100	250
Donald Tricycle Toy	Linemar	1950s	tin	235	440	675
Donald's Hockey Bowl	Ontex	1940s	6 1/2x11x2"	18	35	75
Donald's Olympic Try-Out Puzzle	Jaymar	1960s		9	16	25
Frontierland Donald Figure	Arco		bendable	9	16	25
Huey, Dewey and Louie Stuffed Dolls	Gund	1950s	set of three 8" tall dolls	105	195	300
Louie Stuffed Doll "Huey, Dewey & Louie"	Gund	1940s	8" tall, body is white & light green plush with yellow felt on the legs, beak & tail	35	65	100

Donald Duck

TOY	COMPANY	YEAR	DESCRIPTION	GOOD	EX	MIB
Louie Wrist Watch	US Time			130	245	375
Ludwig Von Drake Mug		1961	handle mug with raised face on mug	11	20	50
Ludwig Von Drake Wonderful World of Color Pencil Box	Hasbro	1961	box shows Ludwig and the nephews	25	50	75
Professor Ludwig Von Drake Figure Toy	Marx	1961	3" tall, from the "Snap-Eeze" Collection	9	16	25
Professor Ludwig Von Drake in Go Cart	Linemar	1960s	tin and plastic	130	245	450
Professor Ludwig Von Drake Mug		1961	3 1/2" white china	7	13	35
Professor Ludwig Von Drake Squeeze Toy	Dell	1961	8" tall, rubber	11	20	45
Professor Ludwig Von Drake Tiddly Winks	Whitman	1961	10x10x1 1/2"	9	16	45
Sled, Donald and Nephews	S. L. Allen & Co.	1935	36" long wooden slat and metal runner sled with character decals	140	260	400
Uncle Scrooge "Frame Tray Puzzle Funnies"	Whitman	1980s	11 1/2x14 1/2", showing 12 panel Uncle Scrooge and Donald Duck comic	11	20	30
Uncle Scrooge Charm		1960s	laminated/sterling silver charm	16	30	45
Uncle Scrooge Wallet		1970s	3x4", Uncle Scrooge tossing coins	9	16	25
W.D. Easter Parade #475	Fisher-Price	1936	Donald Duck on wooden wheels	105	195	350
Walt's Disney Easter Parade Push Toy Play Set	Fisher-Price	1930s	on wooden ball wheels, three rabbits, a hen & 4" tall Donald Duck	130	245	375

Dr. Doolittle

TOY	COMPANY	YEAR	DESCRIPTION	GOOD	EX	MIB
Dr. Doolittle Figure	Mattel	1967	7" tall	14	25	75
Dr. Doolittle Figure	Mattel	1967	5" tall with parrot	12	23	50
Dr. Doolittle Giraffe in the Box				11	20	75

Dr. Seuss

TOY	COMPANY	YEAR	DESCRIPTION	GOOD	EX	MIB
Dr. Seuss "Cat in the Hat" Doll	Coleco	1983	stuffed	20	35	55
Dr. Seuss "Yertle the Turtle" Doll	Coleco	1983	12"	18	35	125

Dumbo

TOY	COMPANY	YEAR	DESCRIPTION	GOOD	EX	MIB
Dumbo 50th Anniversary Christmas Ornament			2" porcelain bisque	5	10	15
Dumbo Figure	Dakin			9	16	25
Dumbo Milk Pitcher		1940s	6" tall	25	50	125
Dumbo Plush Figure			12" tall	9	16	25
Dumbo Roll Over Wind-up Toy	Marx	1941	4" tall, tin with tumbling action	150	225	450
Dumbo Squeak Toy	Dakin			11	20	45
Dumbo Squeeze Toy	Dell	1950s	3x4 1/2x5" tall	18	35	45

Elmer Fudd

TOY	COMPANY	YEAR	DESCRIPTION	GOOD	EX	MIB
Elmer Fudd Figural Mug	Applause	1980s		7	13	25
Elmer Fudd Figure	Dakin	1968	8" tall	14	25	40
Elmer Fudd Figure	Dakin	1971	in a red hunting outfit	23	40	65

Elmer Fudd

TOY	COMPANY	YEAR	DESCRIPTION	GOOD	EX	MIB
Elmer Fudd Figure		1950s	metal 5" high on a 3X5 1/2" green base with his name embossed, next to him a brown bucket, Elmer dressed in hunting outfit	55	105	160
Elmer Fudd Fun Farm Figure	Dakin	1977		11	20	30
Elmer Fudd Mini Snowdome	Applause	1980s		7	13	35
Elmer Fudd Pull Toy Car	Brice Toys	1940s	wooden, Elmer in the Fire Chief's car 9" long, pull and he rings the bell	60	105	225

Felix The Cat

TOY	COMPANY	YEAR	DESCRIPTION	GOOD	EX	MIB
Felix Cartoon Lamp Shade			6" tall	25	50	125
Felix Soakie Soap Bottle			10" tall, plastic	16	30	45
Felix Squeaker Toy			6" tall soft rubber	16	30	45
Felix the Cat Doll		1920s	13" tall, jointed arms	135	250	385
Felix the Cat Doll		1920s	8" tall, wood, fully jointed	45	80	275
Felix the Cat Figure	Schoenhut	1920s	4" tall, wood, leather ears, stands on a white wood base	35	60	175
Felix the Cat Flashlight			contains whistle	9	16	75
Felix the Cat on a Scooter	Nifty	1924	tin, wind-up	165	310	650
Felix the Cat Pencil Case		1950s		25	50	75
Felix the Cat Pull Toy	Nifty	1920s	5 1/2" tall, 8" long, tin, cat is chasing two red mice on the front of the cart, litho pictures of Felix on side	115	210	650
Felix the Cat Punch Bag		1960s	11" tall inflatable bobber	25	45	125
Felix the Cat Sip-a-Drink Cup			5" tall	14	25	40
Felix Wrist Watch		1960s		45	80	225

Ferdinand the Bull

TOY	COMPANY	YEAR	DESCRIPTION	GOOD	EX	MIB
Ferdinand Card Game	Whitman	1938	5x6 1/2x1" deep, set of black, white & red cards picturing Ferdinand, the matador, picador, trumpeter & banderilleo	25	50	125
Ferdinand Ceramic Figure	Delco	1938	4 1/2" tall, ceramic figure seated with a purple garland around his neck	35	65	125
Ferdinand Figure	Knickerbocker	1938	5x 9x8 1/2" tall, joint composition with cloth tail & flower stapled in his mouth	95	180	275
Ferdinand Hand Puppet	Crown Toy	1938	9 1/2" tall	45	80	125
Ferdinand Plush Doll	Knickerbocker	1930s	10" tall, 14" long, with a flower in mouth	80	145	225
Ferdinand Rubber Figure	Seiberling	1930s	3x5 1/2x4" tall	25	50	125
Ferdinand the Bull & the Matador	Marx	1938	5 1/2" high by 8" long, wind-up action between the matador & Ferdinand "bull fight"	225	425	750
Ferdinand the Bull Bisque Figure			3 1/2" bisque	11	20	85
Ferdinand the Bull Book	Whitman	1938	linen picture book	18	35	125
Ferdinand the Bull Figure	Disney	1940s	composition figure	70	130	275

Ferdinand the Bull

TOY	COMPANY	YEAR	DESCRIPTION	GOOD	EX	MIB
Ferdinand the Bull Plastic Figure			9" tall	16	30	125
Ferdinand the Bull Savings Bank	Crown Toy		5" tall, wood composition with silk flower with metal trap door	20	40	175
Ferdinand the Bull Toy	Knickerbocker	1940	wood composition with jointed head & legs with flower in his mouth	40	75	250
Ferdinand the Bull Wind-up Tail Spinning Toy	Marx	1938	6" long, tin, wind-up, when wounded, the wire tail of the figure spins & causes him to jump around	150	350	525

Flintstones

TOY	COMPANY	YEAR	DESCRIPTION	GOOD	EX	MIB
Baby Pebbles Doll	Ideal	1963	15" tall	45	80	125
Baby Puss Figure	Knickerbocker	1961	10" tall, vinyl	55	100	150
Bamm Bamm Bank			11" tall, hard plastic figure sitting on turtle	16	30	45
Bamm Bamm Bubble Pipe	Transogram	1963	figural pipe on illustrated card	18	35	50
Bamm Bamm Doll	Ideal	1962	15" tall	45	80	125
Bamm Bamm Figure	Dakin	1970	7" tall	12	25	35
Bamm Bamm Soakie				9	16	25
Barney Figure	Knickerbocker	1961	10" tall, vinyl	45	80	125
Barney Finger Puppet	Knickerbocker	1972		4	8	12
Barney Policeman Action Figure	Flintoys	1986		4	8	12
Barney Rubble Action Figure	Flintoys	1986		4	7	10
Barney Rubble Bank		1973	solid plastic, Barney holding a bowling ball	9	16	25
Barney Rubble Doll		1962	6" tall, soft vinyl doll, moveable arms & head	18	35	50
Barney Rubble Figural Night Light	Electricord	1979		4	8	12
Barney Rubble Figure	Dakin	1970	7 1/4" tall	18	35	50
Barney Rubble Riding Dino Toy	Marx	1960s	8" long, metal & vinyl, wind-up	175	295	475
Barney Rubble Wind-up	Marx	1960s	3 1/2" tall figure, tin	70	130	325
Barney's Car	Flintoys	1986		7	13	20
Betty Figure	Knickerbocker	1961	10" tall, vinyl	60	115	175
Betty Rubble Action Figure	Flintoys	1986		4	7	10
Dino Action Figure	Flintoys	1986		4	7	10
Dino Bank			hard vinyl, blue with Pebbles on his back	9	16	25
Dino China Bank			Dino carrying a golf bag	60	115	175
Dino Doll			moveable head and arms	9	16	25
Dino Figure	Dakin	1970	7 3/4" tall	25	50	75
Dino the Dinosaur Toy Bath Puppet Sponge			bath glove	7	13	20
Dino Wind-up	Marx	1960s	3 1/2" tall figure, tin	90	165	350
Fang Figure	Dakin	1970	7" tall	25	50	75
Flintmobile	Flintoys	1986		9	16	25
Flintmobile with Fred Action Figure	Flintoys	1986		16	30	45
Flintstones 3" Colored Figures	Empire	1976	solid figures of Fred, Barney, Wilma and Betty	7	13	35
Flintstones Ashtray		1960	ceramic with Wilma	9	16	75
Flintstones Bank		1971	19" tall with Barney & Bamm Bamm	16	30	75
Flintstones Bank		1961	8" tall	25	50	75
Flintstones Car	Remco	1964	battery operated car with Barney, Fred, Wilma and Betty	45	80	350
Flintstones Eight Figure Set	Spoontigues	1981		30	55	85

Flintstones

TOY	COMPANY	YEAR	DESCRIPTION	GOOD	EX	MIB
Flintstones Figures	Imperial	1976	eight acrylic figures, Fred, Barney, Wilma, Betty, Pebbles, Bamm Bamm, Dino and Baby Puss	14	25	40
Flintstones House	Flintoys	1986		12	25	35
Flintstones Lamp			9 1/2" tall, plastic Fred with lampshade picturing characters	60	115	175
Flintstones Party Place Set	Reed	1969		18	35	75
Flintstones Pillowcase		1960		12	25	35
Flintstones Roto Draw	(England)	1969	British	19	35	55
Flintstones Tru-Vue Stereo Film Card	Tru-Vue	1962	viewer card #T-37, with strips of Fred	16	30	45
Flintstones Tru-Vue Stereo Film Card	Tru-Vue	1962	viewer card #T-58, with strips of Pebbles and Bamm Bamm	15	25	40
Fred Figure	Knickerbocker	1961	10" tall, vinyl	45	80	125
Fred Flintstone Action Figure	Flintoys	1986		4	8	12
Fred Flintstone Bubble Blowing Bust Pipe			soft vinyl with curved stem	5	10	15
Fred Flintstone Doll		1960	13" soft vinyl doll with moveable head	45	90	135
Fred Flintstone Doll	Perfection Plastic	1972	11" tall	12	23	35
Fred Flintstone Figural Night Light		1970		9	16	25
Fred Flintstone Figure	Dakin	1970	8 1/4" tall	12	23	35
Fred Flintstone Figure	Knickerbocker	1960	15" tall	50	95	145
Fred Flintstone Gum Ball Machine			in the shape of Fred's head	7	13	35
Fred Flintstone Push Puppet	Kohner			12	23	35
Fred Flintstone Riding Dino	Marx	1962	18 " long battery operated with Fred in Howdah	250	350	575
Fred Flintstone Riding Dino	Marx	1962	8" long, tin & vinyl, wind-up	200	350	550
Fred Flintstone's Bedrock Bank	Alps	1962	9", tin & vinyl battery operated	155	295	650
Fred Flintstone's Lithograph Wind-up	Marx	1960s	3 1/2" tall figure, metal	90	165	350
Fred Loves Wilma Bank			ceramic	60	115	175
Fred Playing Xylophone	Fisher-Price	1962		60	105	165
Fred Policeman Action Figure	Flintoys	1986		4	8	12
Great Big Punch-out Book	Whitman	1961		14	25	40
Just For Kicks Target Game				85	160	245
Motorbike	Flintoys	1986		5	10	15
Pebbles Bank			9" tall vinyl with Pebbles sitting in chair	9	16	25
Pebbles Figure	Dakin	1970	8" tall with blonde hair and purple velvet shirt	11	20	30
Pebbles Flintstone Doll	Mighty Star	1982	vinyl head, arms and legs, cloth stuffed body 12" tall	15	30	45
Pebbles Flintstones Cradle	Ideal	1963	for a 14" doll	15	25	125
Pebbles Soakie				9	16	35
Police Car	Flintoys	1986		7	13	20
Wilma Figure	Knickerbocker	1961	10" tall, vinyl	60	115	175
Wilma Flintstone Action Figure	Flintoys	1986		4	7	10
Wilma Friction Car	Marx	1962	metal	90	165	350

Foghorn Leghorn

TOY	COMPANY	YEAR	DESCRIPTION	GOOD	EX	MIB
Foghorn Leghorn Figure	Dakin	1970	6 1/4" tall	25	45	65
Foghorn Leghorn Hand Puppet		1960s	9" hand puppet, fabric with vinyl head	12	25	35
Foghorn Leghorn PVC Figure	ApplauseA	1980s		3	5	8

Fontaine Fox

TOY	COMPANY	YEAR	DESCRIPTION	GOOD	EX	MIB
Cast Metal Toonerville Trolley			4" tall, red pot metal	75	135	475
Metal Toonerville Trolley			3" tall	70	130	375
Miniature Toonerville Trolley	Nifty		2" tall	130	245	375
Powerful Katrinka Figure		1923	5 1/2" tall, wind-up, pushing a wheel barrow with Jimmy	245	450	950
The Toonerville Trolley		1922	7 1/2" tall, wind-up	265	495	950

Garfield

TOY	COMPANY	YEAR	DESCRIPTION	GOOD	EX	MIB
Garfield 3-D Light Switch Plate	Prestigeline	1978		4	8	12
Garfield Chair Bank	Enesco	1981		12	23	35
Garfield Easter Figure	Enesco	1978		4	7	10
Garfield Figural Music Box/Dancing	Enesco	1981		18	35	50
Garfield Figure Bank	Enesco	1981	4 3/4"	12	23	35
Garfield Graduate Figurine	Enesco	1978		4	7	10
Garfield Large Mug, Soup Mug & Snack Dish				9	16	25

Gasoline Alley

TOY	COMPANY	YEAR	DESCRIPTION	GOOD	EX	MIB
Skeezix Comic Figure		1930s	6" chalk statue	4	7	35
Skeezix Stationery		1926	6"x8 1/2"	9	16	25
Uncle Walt & Skeezix Figure Set			bisque, Uncle Walt, Skeezix, Herby & Smitty, heights range from 3 1/2" tall to 2 1/4"	70	130	200
Uncle Walt & Skeezix Pencil Holder	F.A.S.		5" tall, bisque	25	50	150

Girl from U.N.C.L.E.

TOY	COMPANY	YEAR	DESCRIPTION	GOOD	EX	MIB
Book, 1967 Annual	World Distributors /England	1967	hardcover, 95 pages, photo cover	11	20	30
Book, 1968 Annual	World Distributors /England	1968	hardcover, 95 pages, photo cover	11	20	30
Book, 1969 Annual	World Distributors /England	1969	hardcover, 95 pages, photo cover	11	20	30
Comic Book, #1	Gold Key Comics	1966	32 color pages, "The Fatal Accidents Affair"	11	20	30
Comic Book, #2	Gold Key Comics	1966	32 color pages, "The Kid Commandos' Caper"	5	10	15
Comic Book, #3	Gold Key Comics	1967	32 color pages, "The Captain Kid Affair"	5	10	15
Comic Book, #4	Gold Key Comics	1967	32 color pages, "The One-Way Tourist Affair"	5	10	15
Comic Book, #5	Gold Key Comics	1967	32 color pages, "The Harem-Scarem Affair"	5	10	15

Top to bottom: Mickey Mouse Club Mouseketeer Dolls, 1960s, Horsman; Roy Rogers Composition Nodder, Japan; Bullwinkle Bank.

CHARACTER

Girl from U.N.C.L.E.

TOY	COMPANY	YEAR	DESCRIPTION	GOOD	EX	MIB
Costume, Halloween	Halco	1967	came with either transparent or painted mask, dress-style costume has show logo and silhouette image of Girl spy holding smoking gun, in illustrated window box	80	145	225
Digest Magazine Volume 1-#1 December	Leo Marguiles Corp.	1966	144 pages, contains "The Sheik from Araby Affair"	7	13	20
Digest Magazine Volume 1-#2, February	Leo Marguiles Corp.	1967	144 pages, contains "The Velvet Voice Affair"	5	10	15
Digest Magazine Volume 1-#3, April	Leo Marguiles Corp.	1967	144 pages, contains "The Burning Air Affair"	5	10	15
Digest Magazine Volume 1-#4, June	Leo Marguiles Corp.	1967	144 pages, contains "The Deadly Drug Affair"	5	10	15
Digest Magazine Volume 1-#5, August	Leo Marguiles Corp.	1967	144 pages, contains "The October Affair"	5	10	15
Digest Magazine Volume 1-#6, October	Leo Marguiles Corp.	1967	144 pages, contains "The Stolen Spaceman Affair"	5	10	15
Digest Magazine Volume 2-#1, December	Leo Marguiles Corp.	1967	144 pages, contains "The Sinister Satellite Affair"	5	10	15
Doll, The Girl From U.N.C.L.E.	Marx	1967	11" tall with 30 accessory pieces, in art illustrated box	315	575	900
Garter Holster	Lone Star	1966	"gang buster" metal pistol fires small plastic bullets from metal shells, checker design vinyl holster and bullet pouch, on card	80	145	225
Model Car Kit	AMT	1967	contains same kit as Man From U.N.C.L.E. car kit, but with different box graphics, photo box cover	210	390	600
Record, Music from the Television Series	M.G.M. Records	1966	photo cover shows Stephanie against a wall	11	20	30
Wrist Watch, Secret Agent Watch	Bradley	1966	watch has pink face with April Dancer image, in case	125	225	350

Goofy, Disney

TOY	COMPANY	YEAR	DESCRIPTION	GOOD	EX	MIB
Backwards Goofy Watch	Helbros	1972		265	490	750
Backwards Goofy Watch	Pedre		silver case 2nd edition	35	65	150
Fantasyland Goofy Figure	Arco		bendable	9	16	25
Goofy Laughing Doll		1970s	5x6x13" tall, fabric & vinyl doll of Goofy with recording of his laugh	25	50	75
Goofy Lil' Headbobber	Marx			16	30	75
Goofy Nite Lite	Horsman	1973	green figural nite light	16	30	45
Goofy Safety Scissors	Monogram	1973	on card	4	7	10
Goofy Snap-eeze Figure	Marx		on a white plastic base, background of 3 apples hanging from the sky with green grass along the bottom	18	35	50
Goofy Toothbrush	Pepsodent	1970s		4	7	10

Goofy, Disney

TOY	COMPANY	YEAR	DESCRIPTION	GOOD	EX	MIB
Goofy Walt Disney's Twist'n Bend Flexible Toy	Marx	1963	4" tall	11	20	30
Goofy with Bump 'n Go Action Lawn Mower	Illfelder	1980s	3 1/2x10x11", plastic figure pushing lawn mower with silver handle	35	65	100
Goofy-Rolykins Figures	Marx		1x1x1 1/2" tall, plastic, one of the "Walt Disney Rolykins with ball bearing action set	25	50	75

Green Hornet

TOY	COMPANY	YEAR	DESCRIPTION	GOOD	EX	MIB
Black Beauty Car			12", battery operated	210	390	600
Black Beauty H.O. Scale Race Car	Aurora	1966		70	130	275
Coloring Book, Kato's Revenge				14	25	40
Green Hornet Agent Wall Clock				25	50	75
Green Hornet Bendie Figure	Lakeside	1966	on card	70	130	200
Green Hornet Bendie Figure		1966		30	60	90
Green Hornet Bubble Gum Ring	Frito Lay		rubber ring, in cello pack	25	50	75
Green Hornet Charm Bracelet	Grenway Prod.	1966	gold finish chain with five charms, Hornet, Van, Kato, Pistol, Black Beauty, on 3x7 1/2" illustrated card	45	80	175
Green Hornet Colorforms Set	Colorforms			105	195	300
Green Hornet Coloring Book				9	16	35
Green Hornet Costume Set	Den Cooper	1960s	mask, cape, in box	55	100	250
Green Hornet Cutlery Set			set of fork and spoon	11	20	50
Green Hornet Flasher Ring Store Display	Chemtoy	1960s	on illustrated display card	200	375	650
Green Hornet Flasher Rings	Chemtoy	1960s	eight designs, Hornet Sting, Kato and GH in action, GH running w/hostage, Hornet logo, Black Beauty/TV logo, Kato running down thief, GH and Miss Case, each	25	45	70
Green Hornet Flasher, Large			7"	30	55	50
Green Hornet Flicker Ring			plastic	5	10	25
Green Hornet Frame Tray Puzzle		1960s		23	45	65
Green Hornet Frame Tray Puzzles	Whitman	1960s	set of four puzzles in illustrated box	45	80	125
Green Hornet Milk Mug		1966		23	45	65
Green Hornet Movie Viewer				35	60	95
Green Hornet Squeeze Candy		1960s		175	325	500
Green Hornet Sticker Packs				25	50	75
Green Hornet Strikes!, The	Whitman		book	7	13	20
Green Hornet Thingmaker Plate	Mattel	1966		25	50	125

Green Hornet

TOY	COMPANY	YEAR	DESCRIPTION	GOOD	EX	MIB
Green Hornet View-Master Reels	View-Master		set of 3 reels in illustrated envelope	35	60	95
Green Hornet Wallet		1966		25	50	75
Green Hornet Wallet Store Display Set		1966	one wallet and header card	125	225	350
Green Hornet Wrist Radios	Remco	1960s	set of two batt. op plastic wrist radios, send and receive messages by voice or code	130	145	250

Gumby

Adventures of Gumby Electric Drawing Set	Lakeside	1966		12	23	50
Gumby Adventure Costume	Lakeside	1960s	fireman, cowboy, knight and astronaut, each	23	45	75
Gumby Bendable Figure	Applause	1980s	5 1/2 " tall three different kinds	4	8	15
Gumby Hand Puppet	Lakeside	1965	9" tall with vinyl head	9	16	35
Gumby Modeling Dough	Chemtoy	1960s		23	45	75
Gumby Poseable Figure	Applause	1980s	12" tall	9	16	25
Gumby's Jeep	Lakeside	1960s	yellow tin litho, Gumby & Pokey's names are printed on seat,	95	175	265
Gumby's Pal Pokey Figure	Lakeside	1960s		23	45	65
Gumby's Pal Pokey Modeling Dough	Chemtoy	1960s		23	45	75

Happy Hooligan

Happy Hooligan Nesting Toy Set	Anri		4" tall, wooden set of four pieces, three being smaller and fit into the larger one	60	105	165
Happy Hooligan Toy	Chein	1932	6" tall, wind-up, walking figure	250	275	500

Hardy Boys

Hardy Boys Figures	Kenner	1979	12" tall Joe Hardy (Shaun Cassidy) or Frank Hardy (Parker Stevenson)	7	13	20

Heathcliff

Heathcliff "Sonja" Friction-Powered Mover	Talbot Toys	1982		4	8	12
Heathcliff Schoolhouse Game	Hourtou	1983	game with figures	7	13	20

Heckle & Jeckle

Heckle & Jeckle Figures			7" tall soft foam figures	12	23	35
Heckle & Jeckle Storybook	Wonder Book	1957		9	16	25
Little Roquefort Figure		1959	8 1/2" tall, wood	23	45	65

Honey West

Accessory Set	Gilbert	1965	telephone purse, lipstick, handcuffs and telescope lens necklace	23	45	65

Honey West

TOY	COMPANY	YEAR	DESCRIPTION	GOOD	EX	MIB
Accessory Set	Gilbert	1965	cap-firing pistol, binoculars, shoes and glasses	23	45	65
Formal Outfit	Gilbert	1965		25	50	75
Honey West Doll	Gilbert	1965	12" tall with black leotards, belt, shoes, binoculars and gun	105	195	250
Karate Outfit	Gilbert	1965		25	50	75
Pet Set with Ocelot	Gilbert	1965		35	65	100
Secret Agent Outfit	Gilbert	1965		25	50	75

How the West Was Won

TOY	COMPANY	YEAR	DESCRIPTION	GOOD	EX	MIB
Dakota Figure	Mattel	1978		7	13	20
Lone Wolf Figure	Mattel	1978		7	13	20
Zeb Macahan Doll	Mattel	1978		7	13	20

Howdy Doody

TOY	COMPANY	YEAR	DESCRIPTION	GOOD	EX	MIB
Clarabell Jumping Toy	Linemar	1950s	7" tall tin litho Clarabell, squeeze lever to make figure hop forward and squeak	265	490	750
Flub A Dub Flip A Ring Game		1950s	9" long ring toss game, object is to toss ring over Flub A Dub's nose.	12	23	50
Howdy Doody Air-O-Doodle		1950s	red and yellow plastic combination train, boat and plane toy on card with cut out character passengers	16	30	75
Howdy Doody Alarm Clock		1971	Howdy centered in clock face, pink	45	80	175
Howdy Doody Bubble Pipe	Lido	1950s	4" long silver plastic pipe with bowl shaped like Howdy's face	35	60	125
Howdy Doody Coloring Books	Whitman	1955	set of six 8"x8" coloring books in box	25	50	75
Howdy Doody Dominos		1950s		45	80	125
Howdy Doody Figure	Stahlwood Co.		5"x7" rubber squeeze figure on airplane	235	440	675
Howdy Doody Fingertronic Puppet Theater	Sutton's	1970s		4	8	12
Howdy Doody Paint Set	Milton Bradley	1950s	11"x16" set in box	45	80	125
Howdy Doody Puppet Show Set	Kagran	1950s	includes, Howdy, Clarabell, Mr. Bluster, Flub, Dillie Dally	90	165	250
Howdy Doody Ranch House Toolbox	Liberty Steel	1950s	14x6x3" illustrated steel box with handle	45	80	125
Howdy Doody Ukelele	Emenee	1950s	small plastic guitar labelled with Howdy art	45	80	125
Howdy Doody Ventriloquist's Dummy	Goldberger	1970s	30" tall Howdy dressed in blue pants and red plaid shirt	55	100	150
Howdy Doody Vinyl Doll		1950s	7" tall vinyl squeeze toy of Howdy in blue pants and red shirt	50	90	135
Howdy Doody Wrist Watch	Ingraham	1954	deep blue band with blue and white dial showing character faces, came with a box showing Howdy holding the watch	175	325	500
Howdy Doody Wristwatch		1987		23	45	50

Huckleberry Hound

TOY	COMPANY	YEAR	DESCRIPTION	GOOD	EX	MIB
Hokey Wolf Figure	Dakin	1970		30	55	85
Hokey Wolf TV-Tinkykin Figure	Marx	1961		9	16	25

CHARACTER

Huckleberry Hound

TOY	COMPANY	YEAR	DESCRIPTION	GOOD	EX	MIB
Huckleberry Hound China Figure		1960s	6" tall, glazed china	18	35	35
Huckleberry Hound Doll	Knickerbocker	1959	18" tall, stuffed plush doll with vinyl hands & face	45	80	65
Huckleberry Hound Figural Bank	Dakin	1980	5" tall figural bank of Huck sitting	19	35	35
Huckleberry Hound Figural Bank	Knickerbocker	1960	10" tall, hard plastic figure bank	9	16	35
Huckleberry Hound Figure	Dakin		8" tall figure	16	30	35
Huckleberry Hound Figure	Dakin		7" tall	35	60	35
Huckleberry Hound Go Cart	Linemar	1960s	6 1/2" tall, in go-cart, friction	90	165	275
Huckleberry Hound Tiddleywinks	Milton Bradley	1959	samll board game, tennis tiddleywinks, Huck and Mr. Jinks on cover	16	30	45
Huckleberry Hound TV Scenes Miniature Figure Set	Marx	1961		18	35	75
Huckleberry Hound TV- Tinykin Figure	Marx	1961		18	35	50
Huckleberry Hound Wind-up Toy	Linemar	1962	4" tall, tin	70	130	300
Huckleberry Hound Wristwatch	Bradley	1965	medium size chrome case, wind-up mechanism, grey leather band, face shows Huck in full view	45	90	135
Huckleberry Hound's Huckle Chuck Target Game	Transogram	1961	target game with plastic rings, beanbags & darts	25	50	75
Mr. Jinks Bubble Soap Container	Purex	1960s	10" tall, Pixie & Dixie hard plastic container	9	16	25
Mr. Jinks Stuffed Doll	Knickerbocker	1959	13" tall, stuffed plush doll with vinyl face	25	50	75
Pixie & Dixie Dolls	Knickerbocker	1960	12" tall, each	25	50	75
Pixie & Dixie Magic Slate		1959		12	23	35

Indiana Jones

TOY	COMPANY	YEAR	DESCRIPTION	GOOD	EX	MIB
3-D Indiana Jones View-Master Gift Set	View-Master		View-Master	16	30	45
Indiana Jones Sticker Sheet			two sets	12	23	35
Indiana Jones The Legend			coffee mug	5	10	15
Indy "Pepsi" Backpack				18	35	50
Last Crusade Pepsi Retailer Button				5	10	15
Temple of Doom Calendar				5	10	15
Temple of Doom Storybook			hardbound	7	13	20

James Bond

TOY	COMPANY	YEAR	DESCRIPTION	GOOD	EX	MIB
Disguise Kit	Gilbert	1965		45	80	125
Disguise Kit #2	Gilbert	1965		45	80	125
James Bond	Gilbert	1965	large figure	105	195	250
James Bond Aston Martin Car	Gilbert	1965	12", battery operated	225	425	375
James Bond Aston Slot Car	Gilbert	1965		30	60	125

James Bond

TOY	COMPANY	YEAR	DESCRIPTION	GOOD	EX	MIB
James Bond Hand Puppet	Gilbert	1965		60	115	125
James Bond Harpoon Gun (Thunderball)	Lone Star	1960s	box is illustrated with undersea fight scene graphics	45	80	125
James Bond Secret Attache Case	MPC	1965		225	400	450
James Bond View-Master Pack, Live & Let Die	View-Master	1973		11	20	30
Jaws Doll	Mego	1979		145	270	500
Moonraker Doll	Gilbert	1979		25	50	75
Odd Job	Gilbert	1965	12" doll in karate outfit, with hat	140	260	450
Scuba Outfit #2	Gilbert	1965		45	80	125
Scuba Outfit #3	Gilbert	1965		20	40	60
Scuba Outfit #4	Gilbert	1965		20	40	60
Scuba Outfit Deluxe	Gilbert	1965		45	80	125
Ski Outfit	Gilbert	1965		60	115	175
Thunderball Set	Gilbert	1965		55	100	150
Tuxedo Outfit	Gilbert	1965		60	115	175

Jetsons

TOY	COMPANY	YEAR	DESCRIPTION	GOOD	EX	MIB
Jetsons Colorforms Kit		1963		35	65	100
Jetsons Elroy Toy	Transogram	1963		45	80	125
Jetsons Jigsaw Puzzle	Whitman	1962	70 pieces	12	23	35
Judy Jetson Figure	Applause	1990	10" tall figure	5	10	15
Rosie Doll	Applause	1980s	10" tall	9	16	25
The Jetsons Birthday Surprise	Whitman	1963	book	12	23	35

Jungle Book

TOY	COMPANY	YEAR	DESCRIPTION	GOOD	EX	MIB
Baghera the Tiger Flasher	Disney	1966		5	10	15
Baloo Doll			12" tall, plush	9	16	25
Jungle Book Carrying Case	Ideal	1966	5x14x8" tall	25	50	75
Jungle Book Dinner Set			vinyl placemat, 6 1/2" bowl, 8" plate & 8 oz. cup	7	13	20
Jungle Book Fork & Spoon Set			flatware with melamine handles	4	7	10
Jungle Book Fun-L Tun-L	New York Toy Corp.	1966	108" long by 2 feet wide tunnel	35	65	100
Jungle Book Magic Slate	Watkins-Strathmore	1967	8 1/2x13 1/2"	9	16	25
Jungle Book Sand Pail and Shovel	Chein	1966	tin litho pail and shovel are illustrated with Jungle Book characters	23	45	65
Jungle Book Tea Set	Chein	1966	tin litho set of three 5" plates, 4" saucers & 1 1/2" tea cups and a 7x10 1/2" serving tray	25	50	75
Mowgli Figure	Holland Hill	1967	8" vinyl figure	19	35	55
Mowgli's Hut Mobile Toy & Figures	Multiple Toymakers	1968	2x3x3" mobile with Baloo & King Louis figures	35	65	100
Mowgli/Baloo Digital Watch			clear plastic band	5	10	15
Shere Kahn Figure	Enesco	1965	5" tall, ceramic	12	23	35

Lady & the Tramp

TOY	COMPANY	YEAR	DESCRIPTION	GOOD	EX	MIB
Lady Doll	Woolikin	1955	5x8x8 1/2", light tank with brunt orange accents on face, ears, stomach & tail, plastic eyes, nose & a white silk ribbon around neck	45	80	125
Modeling Clay	Pressman	1955	10 1/2x14 1/2x2"	25	50	75
Perri Plush Doll	Steiff		3x5x6" tall	35	65	100
Plastic Figures	Marx	1955	Lady is 1 1/2" tall & white, Tramp is 2" tall & tan	25	50	75
Plush Dolls	Schuco	1955		60	115	175
Toy Bus	Modern Toys/ Japan	1966	3 1/2x4x14" long	140	260	400
Tramp Plush Doll	Schuco	1955	4x9x8" tall, brown with a white underside & face, hard plastic eyes & nose	60	115	175
Tray Puzzle	Whitman	1954	11x15"	9	16	25

Laurel & Hardy

TOY	COMPANY	YEAR	DESCRIPTION	GOOD	EX	MIB
Laurel & Hardy Die Cut Puppets	Larry Harmon	1970s	moveable, each	11	20	60
Laurel & Hardy Squeeze Toy	Dell	1982	soft vinyl, squeeze and hat pops up on their heads	7	13	35
Oliver Hardy Doll	Dakin		5" tall wind-up dancing/shaking vinyl doll	16	30	75
Oliver Hardy Figure	Dakin	1974	7 1/2" tall	25	50	75
Stan Laurel Figural Bank		1974	8" tall hard plastic bank, brown with green pants	9	16	35
Stan Laurel Figural Bank		1972	15" tall, vinyl	16	30	35
Stan Laurel Figure	Dakin	1974	8" tall	25	50	75

Li'l Abner

TOY	COMPANY	YEAR	DESCRIPTION	GOOD	EX	MIB
Li'l Abner Dogpatch Band	Unique Art Mfg. Co.	1945	9"x9" tall, wind-up, Daisy Mae plays piano, Li'l Abner dances, Pappy Yokum plays drums & Mammy Yokum sits on top of piano smoking her pipe	225	425	650
Li'l Abner Snack Vending Machine			2" vending machine that dispenses nutritious snacks for 10 cents	115	210	325

Lippy the Lion

TOY	COMPANY	YEAR	DESCRIPTION	GOOD	EX	MIB
Lippy the Lion Game	Transogram	1963		25	45	70
Lippy the Lion Soakie	Purex	1960s	11 1/2" tall, hard plastic	12	23	35

Little Mermaid

TOY	COMPANY	YEAR	DESCRIPTION	GOOD	EX	MIB
Ariel Doll			9" tall, in gown	9	16	25
Ariel Musical Jewelry Box			5 3/4x4 1/2"	9	16	25
Ariel PVC Doll				9	16	25
Ariel Toothbrush			7 1/2x10" battery operated with holder	9	16	25
Backpack				7	13	20
Eric Doll			9 1/2" tall in full dress uniform	9	16	25
Flounder & Ariel Faucet Cover			plastic	5	10	15
Flounder Plush Figure			15" fish figure	9	16	25
Flounder Shaped Pillow			14x24"	7	13	20
Globe			4" water globe	9	16	25

Little Mermaid

TOY	COMPANY	YEAR	DESCRIPTION	GOOD	EX	MIB
Pencil Box				7	13	20
Purse			6" diameter, vinyl, canteen styles purse	5	10	15
PVC Figures	Applause		King Triton, Ariel & Sebastian on rock, Ariel sitting alone, Ariel human in dress, Ariel w/mirror, Ariel leaping from water, Eric, Ariel & Flounder, each	1	3	4
Scuttle Plush Figure			15" seagull	11	20	30
Sebastian Plush Figure			16" crab figure	9	16	25
Under the Sea Muscial Jewelry Box			4" mahogany	18	35	50

Little Orphan Annie

TOY	COMPANY	YEAR	DESCRIPTION	GOOD	EX	MIB
Annie Doll	Knickerbocker	1982	10" tall, complete with two dresses and a removable heart locket	5	10	15
Annie Doll without Locket	Knickerbocker	1982		2	5	7
Annie Miniatures Set of Seven			Annie in a red dress, Daddy Warbucks, Punjab, Grace, Rooster, Sandy and Annie in a white dress and Rooster, each	1	3	4
Annie Set of Six Miniature Figures	Knickerbocker	1982	2" tall, Annie blue dress, Punjab, Grace, Daddy Warbucks, Sandy and Miss Hannigan, each	1	3	4
Beetleware Cup			4" tall, green plastic	14	25	40
Beetleware Mug			3" tall, white	9	16	25
Daddy Warbucks	Knickerbocker	1982		9	16	25
Little Orphan Annie & Chizzler Big Little Book				23	45	65
Little Orphan Annie & Sandy Ash Tray			3" tall, ceramic	55	100	150
Little Orphan Annie & Sandy Dolls	Famous Artists Synd.	1930	9 3/4" tall	45	80	125
Little Orphan Annie & Sandy Toothbrush Holder			bisque	30	60	150
Little Orphan Annie & the Gooneyville Mystery Book	Whitman	1947		9	16	25
Little Orphan Annie & the Haunted House Book	Cupples & Leon	1928		25	45	70
Little Orphan Annie Bucking the World Book	Cupples & Leon	1929	hardcover	25	50	75
Little Orphan Annie Clothes Pins	Gold Metal Toys	1938	clothesline & pulley	23	45	65
Little Orphan Annie Colorforms Kit		1970s		9	16	25
Little Orphan Annie Costume Set			mask & slip-over paper dress	16	30	45
Little Orphan Annie Cut Out Toys	Miller Toys	1960s	cut out cook with Sandy-Grunts the Pig-Pee Wee the Elephant	25	50	75
Little Orphan Annie Doll	Well Toy Co.	1973	7" tall	11	20	30
Little Orphan Annie Famous Comics Jigsaw Puzzle	Novelty Dist.			18	35	50

Little Orphan Annie

TOY	COMPANY	YEAR	DESCRIPTION	GOOD	EX	MIB
Little Orphan Annie Figural Music Box	N.Y. News Co.	1970		18	35	95
Little Orphan Annie in the Circus Book	Cupples & Leon	1927	9"x7", 86 pages	25	50	75
Little Orphan Annie Light Up the Candles Game			3 1/2"x5"	16	30	45
Little Orphan Annie Ovaltine Cup	Harold Gray	1930		35	65	75
Little Orphan Annie Punch Outs	King, Larson, McMahon	1944	3-D toys, punch-outs of Annie, Sandy, Punjab & Daddy Warbucks	19	35	55
Little Orphan Annie Rummy Cards	Whitman	1937	5"x6", colored silhouettes of Annie on the back	19	35	55
Little Orphan Annie Shipwrecked Book	Cupples & Leon	1931	9"x7", 86 pages	25	50	75
Little Orphan Annie Stove			non-electric model, gold-brass lithographed labels of Annie & Sandy, oven doors functional	30	55	85
Little Orphan Annie Stove			electric version, 8"x9", gold metal, litho plates, functional oven doors & back burner	35	65	95
Little Orphan Annie Toothbrush Holder			3 1/2"x3", bisque	45	80	125
Little Orphan Annie Wind-up Toy	Marx	1930s	5" tall, tin, wind-up	150	275	725
Miniature Orphan Annie Figure		1940s	1 1/2" tall, lead	9	16	25
Miniature Sandy Figure		1940s	3/4" tall, lead	6	12	18
Miss Hannigan	Knickerbocker	1982		4	8	12
Molly	Knickerbocker	1982		4	8	12
Punjab	Knickerbocker	1982		5	10	15
Radio Annie's Secret Decoder Pins		1930s	used to decode messages	25	45	70
Radio Annie's Secret Society Booklet		1936	6"x9"	35	65	100
Radio Annie's Secret Society Manual		1938	6"x9", 12 pages	25	50	75
Sandy Wind-up Toy	Marx	1930s	4" tall, tin, wind-up	95	175	275

Lone Ranger

TOY	COMPANY	YEAR	DESCRIPTION	GOOD	EX	MIB
Banjo Figure	Gabriel	1979		11	20	30
Buffalo Bill Cody Figure	Gabriel	1980	3 3/4" tall	11	20	30
Butch Cavendish Figure	Gabriel	1980	3 3/4" tall	9	18	27
Dan Reid Figure	Gabriel	1979		9	18	27
Little Bear with Nama the Hawk Figure	Gabriel	1979		9	18	27
Lone Ranger & Silver Figures	Gabriel	1979		14	26	40
Lone Ranger Doll		1987	10" tall, poseable with removeable mask, costume, gun, holster, rifle, hat, shoes and bandana	11	20	30
Lone Ranger Figure	Gabriel	1980	3 3/4" tall	9	16	25
Lone Ranger Hand Puppet		1940s	cloth body puppet in blue and white polka dot shirt with bells in both hands	55	100	150
Lone Ranger Movie Film Ring	General Mills	1950s	gold ring holds silver finish viewer with adjustable focus, came with film which slid into slot at end of viewer	80	145	175

Talking Freddy Krueger doll from A Nightmare On Elm Street series, 1980s, Matchbox.

Lone Ranger

TOY	COMPANY	YEAR	DESCRIPTION	GOOD	EX	MIB
Lone Ranger Sheriff Jail Keys	Esquire Novelty	1945	5" jail keys on ring, came on 8 1/2x7" card with cut out Sheriff card	60	115	85
Lone Ranger Wooden Record Player	Dekka	1940s	12x10x6" wooden box with burned in illustrations, leather carry strap	140	260	400
Red Sleeves	Gabriel	1979		9	18	27
Silver with 8-Way Action Saddle	Gabriel	1979		11	20	30
Smoke	Gabriel	1979		11	20	30
Tonto and Scout Figures	Gabriel	1979		14	25	40
Tonto Figure	Gabriel	1980	3 3/4" tall	9	16	25

Looney Tunes

TOY	COMPANY	YEAR	DESCRIPTION	GOOD	EX	MIB
Cool Cat	Dakin	1969		23	45	40
Merlin the Magic Mouse "Goofy Gram"	Dakin	1971		18	35	50
Merlin the Magic Mouse Figure	Dakin	1970	7 3/4" tall	16	30	45
Second Banana Figure	Dakin	1970	6" tall	12	23	35

Maggie & Jiggs

TOY	COMPANY	YEAR	DESCRIPTION	GOOD	EX	MIB
Bringing Up Father Figure Set	G. Borgfeldt Co.	1934	4" tall, bisque	70	130	150
Maggie & Jiggs Figure Set		1940s	2 1/2" tall-Maggie, 1" tall Jiggs	16	30	75

Magilla Gorilla

TOY	COMPANY	YEAR	DESCRIPTION	GOOD	EX	MIB
Droop-A-Long Coyote Soakie	Purex	1960s	12", plastic	7	13	45
Magilla Gorilla Big Golden Book	Golden	1964		7	13	20
Magilla Gorilla Cannon	Ideal	1964		14	25	75
Magilla Gorilla Cereal Bowl	MB Inc.			9	16	25
Magilla Gorilla Coloring Book	Whitman	1964		11	20	30
Magilla Gorilla Doll			11" tall, moveable cloth body, hard arms & legs, hard plastic head	23	45	85
Magilla Gorilla Plate			8"	7	13	20
Magilla Gorilla Plush Doll	Ideal	1966	18 1/2" tall with vinyl head	30	55	85
Magilla Gorilla Pull Toy	Ideal	1960s	pull toy with vinyl figure	45	80	125
Magilla Gorilla Push Puppet	Kohner	1960s	brown plastic figure in pink shorts and shoes holding a stick on a yellow base with gold label	23	45	65
Punkin' Puss Soakie	Purex	1960s	11 1/2" tall, plastic	14	25	40
Ricochet Rabbit Hand Puppet	Ideal	1960s	11" tall with a vinyl head	35	65	100
Ricochet Rabbit Soakie with a Six Shooter	Purex	1960s	10 1/2" tall, plastic	23	45	50

Man from U.N.C.L.E.

TOY	COMPANY	YEAR	DESCRIPTION	GOOD	EX	MIB
Action Figure Apparel Set	Gilbert	1965	bullet proof vest, 3 targets, 3 shells, binoculars, and bazooka	60	115	175

Man from U.N.C.L.E.

TOY	COMPANY	YEAR	DESCRIPTION	GOOD	EX	MIB
Action Figure Armament Set	Gilbert	1965	for 12"figures, jacket, cap firing pistol with barrel extension, bipod stand, telescopic sight, grenade belt, binoculars, accessory pouch and beret	55	100	150
Action Figure Arsenal Set #1	Gilbert	1965	tommy gun, bazooka, 3 shells, cap firing pistol & attachments, in shallow window box	45	80	125
Action Figure Arsenal Set #2	Gilbert	1965	cap firing THRUSH rifle with telescopic sight, grenade belt and four grenades, on wrapped header card	30	55	85
Action Figure Jumpsuit Set	Gilbert	1965	for 12" figures, jumpsuit with boots, helmet with chin strap, 28" parachute & pack, cap firing tommy gun with scope, instructions	60	115	175
Action Figure Pistol Conversion Kit	Gilbert	1965	binoculars and pistol with attachments, for 12" figures, on wrapped header card	25	50	75
Action Figure Scuba Set	Gilbert	1965	used on 12" Gilbert dolls, swim trunks, air tanks, tank bracket, tubes, scuba jacket and knife	80	145	225
Action Figure, Illya Kuryakin	Gilbert	1965	12" tall plastic figure, black sweater, pants and shoes, spring loaded arm for firing cap pistol, folding badge, I.D. card and instruction sheet, in photo long box	70	130	200
Action Figure, Napoleon Solo	Gilbert	1965	11" tall plastic figure, white shirt, black pants and shoes, spring loaded arm for firing cap pistol, folding badge, I.D. card and instruction sheet	70	130	200
Arcade Cards			postcards with b/w photo fronts and bios on back, each	4	8	12
Attache Case	Ideal	1965	15 x 10 x 2 1/2" comes with a cap firing pistol & clip, I.D. card, secret wallet, cap grenade, U.N.C.L.E. badge, passport & secret message sender	525	975	1500
Attache Case	Lone Star	1966	small cardboard briefcase, contains die cast Mauser & parts to assemble U.N.C.L.E. Special	140	260	400
Attache Case, British	Lone Star	1966	cardboard covered in vinyl, 9mm automatic luger, shoulder stock, sight, silencer, belt, holster, secret wrist holster & pistol that fires cap & cork, grenade, wallet with passport, play money	300	550	850
Attache Case, British	Lone Star	1965	15 x 8 x 2" vinyl case with a pistol, holster, walkie talkie, cigarette box gun, U.N.C.L.E. badge, international passport, invisible cartridge pen and handcuffs	265	475	750
Badge, U.N.C.L.E. Badges Store Display	Lone Star	1965	illustrated card holds 12 triangular black plastic badges with gold lettering, complete with badges	55	100	150
Book, "The Affair of the Gentle Saboteur"	Whitman Publishing	1966	hardcover book	4	7	10
Book, "The Affair of the Gunrunners' Gold"	Whitman Publishing	1967	hardcover book	5	10	15

Man from U.N.C.L.E.

TOY	COMPANY	YEAR	DESCRIPTION	GOOD	EX	MIB
Book, "The Coin of El Diablo Affair"	Wonder Books	1965	softcover illustrated book, 48 pages	4	7	10
Book, 1966 Annual	World Distributors /England	1966	hardcover, 95 pages, photo cover	9	16	25
Book, 1967 Annual	World Distributors /England	1967	hardcover, 95 pages, photo cover	9	16	25
Book, 1968 Annual	World Distributors /England	1968	hardcover, 95 pages, photo cover	9	16	25
Book, 1969 Annual	World Distributors /England	1969	hardcover, 95 pages, photo cover	9	16	25
Book, Big Little Book, The Calcutta Affair	Whitman Publishing	1967	254 pages, color illustrations	4	7	10
Book, Illya, That Man From U.N.C.L.E.	Pocket Books	1966	6 x 9" paperback, 100 pages of David McCallum	11	20	30
Book, Television Picture Story Book	P.B.S. Limited	1967	hardcover, 62 pages, Gold Key reprints	11	20	30
Book, Television Picture Story Book	P.B.S. Limited	1968	hardcover, 62 pages, Gold Key reprints	11	20	30
Book, The Diving Dames Affair	Souvenir Press/ England	1967	#10 in series	4	8	12
Book, The Doomsday Affiar	Souvenir Press	1965	#2 in series	4	8	12
Book, The Power Cube Affair	Souvenir Press/ England	1968	#15 in series	4	8	12
Button	Button World	1965	3 1/2" diam. round button with portrait of Napoleon or Illya, each	9	16	25
Candy Cigarette Box	Cadet Sweets	1966	came with candy & one trading card, illustrated box	30	60	90
Candy Cigarette Counter Display Box	Cadet Sweets	1966	holds 72 of the candy cigarette boxes - nicely illustrated	350	650	1000
Car, Crime Buster Gift Set	Corgi	1966	set includes Man from U.N.C.L.E. car, James Bond Aston Martin and Batmobile with Batboat on trailer, in window box	325	625	950
Car, Die Cast	Playart	1968	2 3/4" long, die cast metal, metallic purple	105	195	300
Car, Gun Firing THRUSH-Buster	Corgi	1966	4 1/4" long die case metal blue Oldsmobile with Solo and Kuryakin leaning out of the car, makes firing noise when the roof periscope is pressed	80	150	230
Car, Gun Firing THRUSH-Buster	Corgi	1966	white version of car	210	390	600
Car, Missile Firing	Husky/Corgi	1966	blue die cast metal 3" long with small figures of Solo & Kuryakin inside, shoots small missiles from under the hood	55	100	150
Car, Thrushbuster Display Box	Corgi	1966	large display box with graphics, holds 12 cars	225	425	650
Car, U.N.C.L.E.	Corgi	1966	blue die cast metal 2 3/4" long with small figures of Solo & Kuryakin inside, shoots small missiles from under the hood	55	100	150
Code-Board		1966	chalkboard with line art illustrations	90	165	250
Coloring Book	Whitman Publishing	1967	cover shows Solo and Kuryakin in a winter setting	18	35	50

Man from U.N.C.L.E.

TOY	COMPANY	YEAR	DESCRIPTION	GOOD	EX	MIB
Coloring Book	Watkins-Strathmore Co.	1965	192 pages, contains both "Crush THRUSH Coloring Book" and "The Man from U.N.C.L.E. Coloring Book"	15	25	40
Coloring Book, Crush THRUSH	Watkins-Strathmore Co.	1965	96 pages, photo cover	9	15	25
Comic Book #1	Gold Key Comics	1965	full color issue of "The Explosive Affair"	20	35	55
Comic Book #1	Entertainment Publishing	1987	"The Number One with a Bullet Affair" b/w 24 pages	1	3	4
Comic Book #10	Gold Key Comics	1967	full color, "The Trojan Horse Affair"	7	13	20
Comic Book #10	Entertainment Publishing	1987	"The Turncoat Affair"	1	3	4
Comic Book #11	Gold Key Comics	1967	full color, "The Three-Story Giant Affair"	5	10	15
Comic Book #11	Entertainment Publishing	1987	"The Craters of the Moon Affair"	1	3	4
Comic Book #12	Gold Key Comics	1967	full color, "The Dead Man's Diary Affair"	5	10	15
Comic Book #13	Gold Key Comics	1967	full color, "The Flying Clowns Affair"	5	10	15
Comic Book #14	Gold Key Comics	1967	full color, "The Great Brain Drain Affair"	5	10	15
Comic Book #15	Gold Key Comics	1967	full color, "The Animal Agents Affair"	5	10	15
Comic Book #16	Gold Key Comics	1968	full color, "The Instant Disaster Affair"	5	10	15
Comic Book #17	Gold Key Comics	1968	full color, "The Deadly Visions Affair"	5	10	15
Comic Book #18	Gold Key Comics	1968	full color, "The Alien Affair"	5	10	15
Comic Book #19	Gold Key Comics	1968	full color, "The Knight in Shining Armor Affair"	5	10	15
Comic Book #2	Gold Key Comics	1965	full color issue of "The Fortune Cookie Affair"	11	20	30
Comic Book #2	Entertainment Publishing	1987	"The Number One with a Bullet Affair" conclusion	1	3	4
Comic Book #20	Gold Key Comics	1968	full color, "The Deep Freeze Affair"	5	10	15
Comic Book #21	Gold Key Comics	1968	full color, "The Trojan Horse Affair" reprint of #10	5	10	15
Comic Book #22	Gold Key Comics	1968	full color, "The Pixilated Puzzle Affair" reprint of #7	5	10	15
Comic Book #3	Gold Key Comics	1965	full color issue of "The Deadly Devices Affair"	7	13	20
Comic Book #3	Entertainment Publishing	1987	"The E-I-E-I-O Affair"	1	3	4
Comic Book #4	Gold Key Comics	1966	full color issue of "The Rip Van Solo Affair"	7	13	20
Comic Book #5	Gold Key Comics	1966	full color issue of "The Ten Little Uncles Affair"	7	13	20
Comic Book #5	Entertainment Publishing	1987	"The Wasp Affair"	1	3	4
Comic Book #6	Gold Key Comics	1966	full color issue of "The Three Blind Mice Affair"	7	13	20
Comic Book #6	Entertainment Publishing	1987	"The Lost City of THRUSH Affair"	1	3	4
Comic Book #7	Gold Key Comics	1966	full color issue of "The Pixilated Puzzle Affair"	7	13	20
Comic Book #7	Entertainment Publishing	1987	"The Wildwater Affair"	1	3	4

Man from U.N.C.L.E.

TOY	COMPANY	YEAR	DESCRIPTION	GOOD	EX	MIB
Comic Book #8	Gold Key Comics	1966	full color, "The Floating People Affair"	7	13	20
Comic Book #8	Entertainment Publishing	1987	"The Wilder West Affair"	1	3	4
Comic Book #9	Gold Key Comics	1966	full color, "The Spirit of St. Louis Affair"	7	13	20
Comic Book #9	Entertainment Publishing	1987	"The Canadian Lightning Affair"	1	3	4
Comic Book, Issue #1	World Distributors /England	1966	5 x 7" digest sized b/w comic book with "The Ten Little Uncles Affair"	5	10	15
Comic Book, Issue #10	World Distributors /England	1966	5 x 7" digest sized b/w comic book with "The Mad, Mad, Mad Affair"	4	8	12
Comic Book, Issue #11	World Distributors /England	1966	5 x 7" digest sized b/w comic book with "The Big Bazoom Affair"	4	8	12
Comic Book, Issue #12	World Distributors /England	1966	5 x 7" digest sized b/w comic book with "The Hot Line Affair"	4	8	12
Comic Book, Issue #13	World Distributors /England	1966	5 x 7" digest sized b/w comic book with "The Two Face Affair"	4	8	12
Comic Book, Issue #14	World Distributors /England	1966	5 x 7" digest sized b/w comic book with "The Humpty Dumpty Affair"	4	8	12
Comic Book, Issue #2	World Distributors /England	1966	5 x 7" digest sized b/w comic book with "The Three Blind Mice Affair"	4	8	12
Comic Book, Issue #3	World Distributors /England	1966	5 x 7" digest sized b/w comic book with "The Pixiliated Puzzle Affair"	4	8	12
Comic Book, Issue #4	World Distributors /England	1966	5 x 7" digest sized b/w comic book with "The Floating People Affair"	4	8	12
Comic Book, Issue #5	World Distributors /England	1966	5 x 7" digest sized b/w comic book with "The Target Blue Affair" new British Story	4	8	12
Comic Book, Issue #6	World Distributors /England	1966	5 x 7" digest sized b/w comic book with "The Hong Kong Affair"	4	8	12
Comic Book, Issue #7	World Distributors /England	1966	5 x 7" digest sized b/w comic book with "The Shufti Peanuts Affair"	4	8	12
Comic Book, Issue #8	World Distributors /England	1966	5 x 7" digest sized b/w comic book with "The Assassins Affair"	4	8	12
Comic Book, Issue #9	World Distributors /England	1966	5 x 7" digest sized b/w comic book with "The Magic Carpet Affair"	4	8	12
Costume, Illya Kuryakin	Halco	1967	painted mask, rayon costume in three colors showing Illya holding a gun, in illustrated window box	55	100	150
Costume, Napoleon Solo	Halco	1965	transparent plastic "mystery mask," costume has line art shirt, tie, shoulder holster and U.N.C.L.E. logo, in illustrated box	55	100	150
Counterspy Outfit	Marx	1966	contains trench coat with secret pockets, pistol, shoulder holster, launcher barrel, silencer, scope sight, two pair of glasses, beards, eye patch, badge case, etc., in box	160	295	450

CHARACTER

Man from U.N.C.L.E.

TOY	COMPANY	YEAR	DESCRIPTION	GOOD	EX	MIB
Counterspy Outfit (Store Display)	Marx	1966	35" x 36" wide cardboard display with one piece of each item in Counterspy Outfit	350	650	1000
Credentials and Passport Set, Napoleon Solo	Ideal	1965	silver I.D. card, badge, identification wallet, slide window passport, on header card	55	100	150
Credentials and Secret Message Sender Set, Napoleon Solo	Ideal	1965	message sender, badge, and silver I.D., on card	55	100	150
Digest Magazine Volume 1-#1, February	Leo Marguiles Corp.	1966	144 pages, contains "The Howling Teenagers Affair"	7	13	20
Digest Magazine Volume 1-#3, April	Leo Marguiles Corp.	1966	144 pages, contains "The Unspeakable Affair"	5	10	15
Digest Magazine Volume 2-#5, December	Leo Marguiles Corp.	1966	144 pages, contains "The Goliath Affair"	5	10	15
Digest Magazine Volume 3-#1, February	Leo Marguiles Corp.	1967	144 pages, contains "The Deadly Dark Affair"	5	10	15
Digest Magazine Volume 3-#4, May	Leo Marguiles Corp.	1967	144 pages, contains "The Synthetic Storm Affair"	5	10	15
Digest Magazine Volume 3-#5, June	Leo Marguiles Corp.	1967	144 pages, contains "The Ugly Man Affair"	5	10	15
Digest Magazine Volume 3-#6, July	Leo Marguiles Corp.	1967	144 pages, contains "The Electronic Frankenstein Affair"	5	10	15
Digest Magazine Volume 4-#1, August	Leo Marguiles Corp.	1967	144 pages, contains "The Genghis Khan Affair"	5	10	15
Digest Magazine Volume 4-#4, November	Leo Marguiles Corp.	1967	144 pages, contains "The Volcano Box Affair"	7	13	20
Digest Magazine Volume 4-#5, December	Leo Marguiles Corp.	1967	144 pages, contains "The Pillars of Salt Affair"	7	13	20
Digest Magazine Volume 4-#6, January	Leo Marguiles Corp.	1968	144 pages, contains "The Million Monster Affair"	20	35	55
Figure, Alexander Waverly	Marx	1966	blue plastic figure 5 3/4" tall, stamped with character's name and U.N.C.L.E. logo on the bottom of base	6	12	18
Figure, Illya Kuryakin	Marx	1966	blue or grey plastic figure, 5 3/4" tall, stamped with character's name and U.N.C.L.E. logo on the bottom of base	6	12	18
Figure, Napoleon Solo	Marx	1966	blue or grey plastic figure, 5 3/4" tall stamped with character's name and U.N.C.L.E. logo on the bottom of base	6	12	18
Figures, Generic Spies	Marx	1966	six different solid plastic, unpainted figures 5 3/4" tall, each	4	8	12
Figures, THRUSH Agents	Marx	1966	three different blue plastic figures 5 3/4" tall stamped with titles and U.N.C.L.E. logo on the bottom of each base, each	6	12	18

Man from U.N.C.L.E.

TOY	COMPANY	YEAR	DESCRIPTION	GOOD	EX	MIB
Finger Print Kit		1966	ink pad, roller, code book, magnifier, finger print records and pressure plate, in illustrated window box	175	325	500
Foto-Fantastiks Coloring Set	Eberhard Faber	1965	6 colored pencils, paint brush, and six 8x10" photos, came in 4 different versions, each	45	85	130
Game, Bagatelle	Hong Kong	1966	8 x 14" pinball game	90	165	250
Game, Card	Japan	1966	small artwork cards in illustrated box	90	165	250
Game, Illya Kuryakin Card Game	Milton Bradley	1966	18 Illya cards, 42 U.N.C.L.E. cards, chips and tray, in color artwork box	15	30	45
Game, Man From U.N.C.L.E. Target Game	Marx	1966	54 x 17" cardboard backdrop, 12 plastic figures, 3 blue plastic U.N.C.L.E. and THRUSH agents, 6 brown generic agents, 2 dart pistols, 6 darts, two grenades and instruction sheet	195	350	550
Game, Napoleon Solo, The Man from U.N.C.L.E. Board Game	Ideal	1965	4 U.N.C.L.E. agent markers, 9 THRUSH chief cards, 9 U.N.C.L.E. assignment cards and THRUSH agent tokens, 2 dice and a playing board	25	40	65
Game, Playing Cards Display Box	Ed-U-Cards	1965	Display box holds 12 packs - logo & photos	125	225	350
Game, Secret Code Wheel Pinball	Marx	1966	10 x 22 x 6" tin litho pinball game	95	180	275
Game, Shoot Out!	Milton Bradley	1965	skill and action game for two players, plastic marble game in illustrated box	95	180	275
Game, Shooting Arcade	Marx	1966	tin litho arcade with mechanical wind-up THRUSH agent targets for pellet shooting pistol, a scope and stock attachments	265	475	750
Game, Shooting Arcade	Marx	1966	smaller version with THRUSH spinner targets	175	325	500
Game, Target	Ideal	1965	cardboard "building" target game with window targets for shooting plastic THRUSH agents, includes dart gun and darts	80	145	225
Game, The Man from U.N.C.L.E. Card Game	Milton Bradley	1965	42 Solo cards, chips and a tray, in color artwork box	15	30	45
Game, The Pinball Affair	Marx	1966	12 x 24" tin litho pinball game	70	130	200
Game, The THRUSH Ray-Gun Affair Game	Ideal	1966	4 U.N.C.L.E. agent pieces, 4 Area Decoder cards, 4 3-D THRUSH hideouts, 4 THRUSH vehicles, 4 crayons, dice and a rotating "ray gun," in illustrated box	55	100	150
Gum Card Box	Topps	1965	3 3/4" x 8" x 2"	60	115	175
Gum Cards	Topps	1965	set of 55 b/w photo cards	70	130	200
Gum Cards	Cadet Sweets	1966	set of 50, 1 3/8" x 2 9/16" cards, color photos, set	20	35	55
Gun, Counter Spy Water Gun	Hong Kong	1960s	luger water gun with unlicensed Napoleon Solo illustration header card	4	7	10
Gun, Die Case Metal Gun	Lone Star	1965	die cast automatic cap pistol with plastic grips, plus cut-out badge, on card	80	150	225

CHARACTER

Man from U.N.C.L.E.

TOY	COMPANY	YEAR	DESCRIPTION	GOOD	EX	MIB
Gun, Illya K. Special Secret Lighter Gun	Ideal	1966	cigarette lighter gun shoots caps, has radio compartment concealed behind fake cigarettes, in window box	125	225	350
Gun, Illya Kuryakin Gun Set	Ideal	1966	includes clip loading, cap firing plastic pistol, U.N.C.L.E. badge, wallet, and I.D. card, in photo-illustrated window box	210	390	600
Gun, Napoleon Solo Gun Set	Ideal	1965	includes clip loading, cap firing plastic pistol with rifle attachments, U.N.C.L.E. badge, I.D. card, in photo-illustrated window box	280	520	800
Gun, Pistol Cane	Marx	1966	25" long, cap firing, bullet shooting aluminum cane with eight bullets and one metal shell, on illustrated card	175	325	500
Gun, Secret Service Gun	Ideal	1965	pistol, holster, bade and silver I.D. card, in window box	210	390	600
Gun, Secret Service Pop Gun		1960s	bagged Luger pop gun on header card with unlicensed illustration of Illya and Napoleon on header	4	7	10
Gun, Stash Away Guns	Ideal	1966	three cap firing guns, holsters, two straps, I.D. card and badge, in window box	210	390	600
Gun, THRUSH Rifle	Ideal	1966	36" long, clip loading cap firing rifle, hit switch and four targets are set up in the sight and then vanish as cap is fired	875	1600	2500
Hankerchief	England	1966	U.N.C.L.E. logo, Illya & Napoleon	55	100	150
Headquarters Transmitter	Cragstan	1965	molded gold colored plastic transmitter, amplifier and under cover case, silver I.D. card, 20 foot wire, in box	95	180	275
License Plates, Bicycle	Marx	1967	4 different metal bicycle plates: Man from U.N.C.L.E., The Girl from U.N.C.L.E., Napoleon Solo, Illya Kuryakin, each	15	30	45
Lunchbox and Thermos	King-Seeley	1966	Jack Davis art of Napoleon, Illya and Alexander Waverly, steel box and matching bottle	155	295	450
Magic Slates	Watkins-Strathmore Co.	1965	9 x 14" slate with two punch out figures of either Napoleon Solo or Illya Kuryakin, each	50	95	145
Model Car Kit	AMT	1967	1/25 scale kit with flame throwers, laser beams, rocket launchers and computers, box art shows car	140	260	400
Model Kit, Illya Kuryakin	Aurora	1966	plastic kit of Illya, interlocks with Napoleon Solo kit to create a large diorama	80	145	225
Model Kit, Napoleon Solo	Aurora	1966	plastic kit of Napoleon Solo that interlocks with Illya Kuryakin kit in a large diorama	85	160	245
Pen, Invisible Writing Cartridge Pen	Platignum/ England	1965	pen with logo, two vials of ink and two invisible ink vials - photo card	165	310	475
Playing Cards	Ed-U-Cards	1965	standard 54 card deck with action photo illustrations, on card	20	35	50
Poster, Movie, One Spy Too Many			14 x 36" insert	15	30	45
Poster, Movie, The Spy With My Face			27 x 41" one sheet	25	50	75

Man from U.N.C.L.E.

TOY	COMPANY	YEAR	DESCRIPTION	GOOD	EX	MIB
Puppet, Illya Kuryakin Action Puppet	Gilbert	1965	13" tall, soft vinyl hand puppet of Illya holding a communicator, on 10 x 16" card	125	225	350
Puppets, Finger	Dean	1966	6 vinyl finger puppets: THRUSH agent, Solo, Kuryakin, Waverly and two female agents, box back has cut outs for props and a short play, "Moon Shot," in die cut window box	175	325	500
Puzzle, "Illya Crushes THRUSH"	Milton Bradley	1966	10 x 19" puzzle with 100 pieces, part of Junior Jigsaw series	20	35	55
Puzzle, "Illya's Battle Below"	Milton Bradley	1966	10 x 19" puzzle with 100 pieces, part of Junior Jigsaw series	20	35	55
Puzzle, Jigsaw	England	1966	four different color artwork 11 x 17" puzzles, each with 340 pieces, The Getaway, Solo in Trouble, The Frogman Affair, Secret Plans, in color artwork box, each	55	100	150
Puzzle, Mystery Jigsaw Series	Milton Bradley	1965	14 x 24" puzzle, 250 pieces plus story booklet, The Loyal Groom, The Vital Observation, The Impossible Escape, The Micro-film Affair, each	25	40	65
Puzzles, Frame Tray	Jaymar	1965	three different color artwork 11 x 14" puzzles, each	9	16	25
Record	Union/Japan	1966	45 rpm with photo sleeve	55	100	150
Record, More Music from U.N.C.L.E.	RCA	1966	LP record with photo cover showing Illya, Napoleon and a woman in a wrecked car	15	25	40
Record, Original Music from...	RCA	1965	photo cover shows Napoleon holding gun on faceless buxom woman while Illya holds a golden mask	9	16	25
Record, The Man from the U.N.C.L.E.	Capitol Records	1965	45 rpm with The Man from U.N.C.L.E. theme song and "The Vagabond"	40	75	120
Record, The Man from U.N.C.L.E.	Crescendo Records	1965	by the Challengers, cover shows blonde female spy with gun	5	10	15
Record, The Man from U.N.C.L.E. and other TV Themes	Metro Records	1965	photo cover, has three songs from U.N.C.L.E. plus theme songs from Dr. Kildare, Mr. Novak, Bonanza and other shows	9	16	25
Ring, Flicker		1965	silver plastic ring with b/w photos, each	5	10	15
Ring, Flicker		1966	blue plastic with "changing portrait" of Napoleon or Illya, each	4	7	10
Secret Message Pen	American Character	1966	6 1/2" long double tipped pen for writing invisible messages, on header card	95	180	275
Secret Print Putty	Colorforms	1965	putty in a gun shaped container, print paper, display cards of Kuryakin & Solo and a book of spy & weapons illustrations, on card	15	30	45
Secret Weapon Set	Ideal	1965	clip loading cap firing pistol, holster, I.D. wallet, silver I.D. card, U.N.C.L.E. badge, two demolition grenades and holster, in window box	250	450	700
Sheet Music, Theme Song	Hastings Music Corp.	1964	6 pages and a brief description of the TV show	15	30	45

668

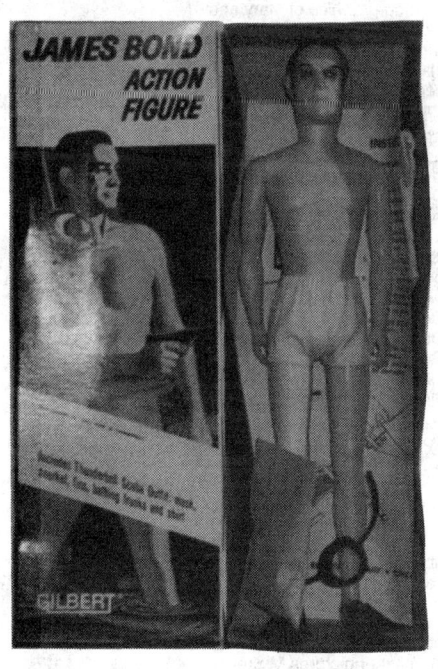

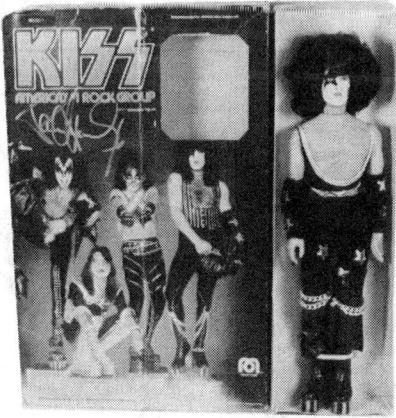

Top to bottom: Huckleberry Hound Tiddly Winks, 1960s, Chad Valley; James Bond Action Figure, 1965, Gilbert; KISS Paul Stanley Doll, 1979, Mego.

CHARACTER

Man from U.N.C.L.E.

TOY	COMPANY	YEAR	DESCRIPTION	GOOD	EX	MIB
Shirt, Official		1965	has secret pocket, glow-in-the-dark badge & ID, photo package	245	455	700
Spy Magic Tricks	Gilbert	1965	Mystery gun, Illya playing cards, many tricks & items - photo boxed	165	310	475
Sweater	Brentwood	1965	100% acrylic, U.N.C.L.E. logo on neck tag	95	180	275
Trading Cards	ABC/England	1966	25 card set, unopened pack	20	35	55
Video Cassette, Return of the Man from U.N.C.L.E.	Trans World Entertainment	1984	96 minute color video reunion of McCallum and Vaughn, in illustrated case	12	25	35
View-Master Set	Sawyers/View-Master	1966	3 reel set and story book from "The Very Important Zombie Affair"	20	40	60
Wrist Watch, Secret Agent Watch	Bradley	1966	gray watch face shows Solo holding a communicator, came with either plain "leather" or "mod" watch band, in case	140	260	400

Mary Poppins

TOY	COMPANY	YEAR	DESCRIPTION	GOOD	EX	MIB
Mary Poppins Ceramic Statue			8" tall	14	25	75
Mary Poppins Doll	Gund	1964	11 1/2" tall, bendable	25	50	150
Mary Poppins Manicure Set	Tre-Jur	1964		12	23	75
Mary Poppins Pencil Case		1964	vinyl with zipper top, shows cartoon graphics of Mary and the kids	9	16	35
Mary Poppins Tea Set	Chein	1964	creamer, tin plates, large plates, place settings, cups & a serviing tray	30	55	85

Mickey & Minnie Mouse, Disney

TOY	COMPANY	YEAR	DESCRIPTION	GOOD	EX	MIB
Adventureland Frame Tray Puzzle	Whitman	1957	11x15" with Mickey, Minnie, Donald & his nephew in a boat surrounded by jungle beasts & a alligator	16	30	45
Adventures of Mickey Mouse, The	David McKay Co.	1931	full color illustrations, softcover book	45	80	125
Characters Schmid Bisque Spirit of '76 Music Box	Schmid Bros.		3x6x7 1/2" tall, plays "Yankee Doodle", Mickey, Goofy and Donald are dressed as Revolutionary War Minutemen	60	115	175
Craftmen's Guild Mickey Mouse Viewer	Craftman's Guild	1940s	film viewer in box with set of 12 films, each individually boxed	65	120	185
Crayons From Your Favorite Funsters Mickey & Donald	Transogram	1946	crayon set	20	40	60
Hitchhiking Hobo Mickey Squeeze Toy	Dell	1950s	rubber figure with his belongings tied in a bandana around a stick which he holds over one shoulder	35	60	95
Mickey & Minnie Mouse Sled	S.L. Allen & Co.	1935	30" or 32" long, wooden slat sled with metal runners, with "Mickey Mouse" decal on steering bar	140	260	400
Mickey & Minnie Mouse Tea Set	Ohio Art	1930s	5"x8", pitcher pictures Mickey at piano & cups picture Mickey, Pluto & Minnie	45	80	125

Mickey & Minnie Mouse, Disney

TOY	COMPANY	YEAR	DESCRIPTION	GOOD	EX	MIB
Mickey & Three Pigs Top	Lackawanna Mfg. Co.		9" diameter, tin	35	65	135
Mickey and Minnie Trash Can	Chein	1970s	13" tall tin litho can shows Mickey and Minnie fixing a flat tire on one side, other side shows Mickey feeding Minnie soup	25	50	75
Mickey and Pluto Quartz LCD Wrist Watch	Bradley	1980	5 fuction LCD, base chrome bezel, black vinyl band	19	35	55
Mickey Mouse "50 Years with Mickey" Quartz Wrist Watch	Bradley	1983	small round chrome case, insription and serial number on back	45	80	125
Mickey Mouse & Donald Duck Alarm Clock	Jerger/ Germany	1960s	2 1/2x5x7" tall with a metal case on a dark brass finish, 3-D plastic figures of Mickey & Donald on either side	55	100	150
Mickey Mouse & Little Mice Sugar Bowl,Salt & Pepper Set			sugar bowl is 4" tall-Mickey with the little mice on the side of tray as salt & pepper shakers	45	80	125
Mickey Mouse & Minnie Toothbrush Holder			4 1/2" tall, bisque, toothbrush holes are located behind their heads	55	100	175
Mickey Mouse Activity Book	Whitman	1936	40 big pages of stories, coloring, gags & activities	20	35	55
Mickey Mouse Alarm Clock	House Martin	1988	5x7x2"	11	20	30
Mickey Mouse Alarm Clock	Bayard	1960s	2x4 1/2x4 1/2" tall, 1930's style Mickey with moveable head that ticks off the seconds	95	180	275
Mickey Mouse Alarm Clock	Bradley	1970s	2x4 1/2x5" tall	35	65	100
Mickey Mouse Ash Tray			3 1/2" tall, wood composition figure of Mickey	35	65	125
Mickey Mouse Baby Gift Set		1930s	boxed set of silver plated cup, fork, spoon, cup and napkin holder have inlaid Mickey, box interior shows Mickey, Minnie, and Donald sitting down to eat	90	165	250
Mickey Mouse Band Drum		1936	7" tallx14" diameter, cloth mesh & paper drum heads	50	90	225
Mickey Mouse Band Leader Bank	Knickerbocker		7 1/2" tall, plastic	9	16	25
Mickey Mouse Band Sand Pail	Ohio Art	1938	6" tall, tin, pictures of Mickey Mouse, Minnie, Horace, Pluto, & Clarabelle the cow parading down street	35	60	135
Mickey Mouse Band Top			9" diameter	35	65	95
Mickey Mouse Bank		1930s	2 1/2" diameter by 6" tall, shaped like a mailbox with Mickey holding an envelope	55	100	275
Mickey Mouse Bank	Fricke & Nacke	1978	3x5x7" tall, embossed image of Mickey in front, side panels have Minnie, Goofy, Donald and Pluto	12	23	150
Mickey Mouse Bank	Crown Toy	1938	6" tall, composition, key locked trap door on base, with figure standing next to treasure chest & head is moveable	55	100	350
Mickey Mouse Beanie Hat		1950s	blue and yellow felt hat with Mickey on front	16	30	75
Mickey Mouse Bicentennial Pocket Watch	Bradley	1976	3 1/2x4 1/2x3/4", Mickey in his Bicentennial outfit	55	100	150

Mickey & Minnie Mouse, Disney

TOY	COMPANY	YEAR	DESCRIPTION	GOOD	EX	MIB
Mickey Mouse Boxed Lantern Slides	Ensign	1930s	5x6x2", cartoons: Traffic Troubles, Gorilla Mystery, Cactus Kid, Castaway, Delivery Boy, Fishin Around, Firefighters, Moose Hunt & Mickey Steps Out	150	275	425
Mickey Mouse Bubble Buster	Kilgore Mfg.	1936	8" long, cork gun	55	100	150
Mickey Mouse Bump & Go Spaceship	Matsudaya	1980s	battery op. tin litho spaceship with clear dome, has 6 flashing lights, rotating antenna	35	65	100
Mickey Mouse Camera	Ettelson Corp.	1960s	3x3x5"	18	35	50
Mickey Mouse Camera	Child Guidance	1970s	4x7x7"	9	16	25
Mickey Mouse Camera	Helm Toy Corp.	1970s	2x5x4 1/2", Mickey in engineer's uniform riding on top of the train	11	20	30
Mickey Mouse Car	Polistil	1970s	2x4x1 1/2" tall plastic car with rubber figure, Mickey in driver's seat	25	50	75
Mickey Mouse Cardboard House	O.B. Andrews	1930s	14x 12x13" tall	115	210	325
Mickey Mouse Chatty Chums	Mattel	1979		6	12	25
Mickey Mouse China Bank		1950s	5x 5 1/2x6" shaped like Mickey's haead with slot between his ears	25	50	125
Mickey Mouse China Dinner Set	Empresa Electro	1930s	2" creamer, 3-4 1/2" plates and 3- 5" plates, 7" long dish, two oval platters, 1-5 1/2" and the other 6 1/2"	130	245	375
Mickey Mouse China Milk Pitcher	(Germany)	1930s	2" tall by 3" diameter, white with green shading around the base, Mickey on each side	60	115	175
Mickey Mouse China Music Box	(Japan)	1970s	4x6" tall, plays "Side by Side", Mickey is brushing a kitten in a washtub	12	23	55
Mickey Mouse Choo Choo Pull Toy	Fisher-Price	1938	#432, blue base, yellow wheels, Mickey in red hat rings bell as toy is pulled	115	210	325
Mickey Mouse Christmas Lights	Noma	1930s	set of 8 lamps with holiday decals of Mickey, Minnie, Donald, Clarabelle, Pluto & Mickey's nephews	115	210	325
Mickey Mouse Club Bank	Play Pal Plastics	1970s	4 1/2x6x11 1/2" tall, vinyl	16	30	45
Mickey Mouse Club Coffee Tin		1950s	illustrated lid promotes the club, tin included MM badge	30	55	85
Mickey Mouse Club Fun Box	Whitman	1957	9x11 1/2x1 1/2" deep, includes a stamp book, club scrapbook, 6 coloring books and 4 small gameboards	40	70	110
Mickey Mouse Club Magic Divider	Jacmar	1950s	arithmetic game	25	50	75
Mickey Mouse Club Magic Kit	Mars Candy	1950s	2- 8x20" punch-out sheets	25	50	75
Mickey Mouse Club Magic Subtractor	Jacmar	1950s	arithmetic game	11	20	30
Mickey Mouse Club Marionette		1950s	3x6 1/2x13 1/2" tall, composition figure of a girl with a black felt hat & mouse ears	45	80	225
Mickey Mouse Club Mouseketeer Doll	Horsman	1960s	8" tall doll in blue jumpsuit, in box	23	45	100
Mickey Mouse Club Mouseketeer Doll	Horsman	1960s	8" tall doll in red dress, in box	23	45	65

Mickey & Minnie Mouse, Disney

TOY	COMPANY	YEAR	DESCRIPTION	GOOD	EX	MIB
Mickey Mouse Club Mousketeer Ears	Kohner		7x12 1/2"	11	20	30
Mickey Mouse Club Newsreel with Sound	Mattel	1950s	4x4 1/2x9" tall, orange box, plastic projector with 2 short filmstrips, record, cardboard screen, cartoons "Touchdown Mickey" and "No Sail"	90	165	250
Mickey Mouse Club Plastic Plate	Arrowhead	1960s	9" diameter, clubhouse with Goofy, Pluto & Donald wearing mouse ears and sweaters with club emblems	16	30	45
Mickey Mouse Club Puzzle	Jaymar	1960s	Pluto's Wash and Scrub Service	12	23	35
Mickey Mouse Club Toothbrush	Pepsodent	1970s		4	7	10
Mickey Mouse Cowboy Tin Top	Chein	1950s	tin litho top features Mickey in cowboy outfit, and other characters	55	100	150
Mickey Mouse Cube Travel Alarm Clock	Bradley (Germany)	1975	large red cube case, shut off button on top, separate alarm wind, in sleeve box	30	55	85
Mickey Mouse Cup	Cavalier	1950s	3" tall, silver-plated cup with a 2 1/2" opening	25	50	75
Mickey Mouse Dart Gun Target	Marks Brothers	1930s	10" target, dart gun & suction darts	30	55	225
Mickey Mouse Disney Dipsy Car	Marx	1953	tin litho car illustrated with Disney characters, plastic Mickey with head on spring, hand out signaling a turn	425	655	875
Mickey Mouse Doll		1930s	4x7x10 1/2" tall, moveable arms & legs, swivel head, black velveteen body & red felt pants	300	575	875
Mickey Mouse Doll	Schuco	1950s	10" tall	115	210	325
Mickey Mouse Doll	Knickerbocker	1935	11" tall, stuffed, cloth, removeable shoes & jointed head	70	130	200
Mickey Mouse Drum	Ohio Art	1930s	6" diameter	30	55	150
Mickey Mouse Drum	Ohio Art	1938	6" diameter	30	55	150
Mickey Mouse Electric Casting Set	Home Foundary	1930s	9 1/2x16x2"	60	115	175
Mickey Mouse Electric Table Radio	General Electric	1960s	4 1/2x10 1/2x6" tall	25	50	125
Mickey Mouse Electric Wall Clock	Elgin	1970s	10x15x3"	11	20	30
Mickey Mouse Electric Watch	Timex	1958		115	210	325
Mickey Mouse Fabric Doll	Knickerbocker		5x7x12" tall, Mickey in checkered shorts & green jacket with a white felt flower stapled to it	175	340	525
Mickey Mouse Figure	Marx	1970	6" tall, vinyl	9	16	25
Mickey Mouse Figure	Seiberling	1930s	3 1/2" tall, latex figure	55	100	150
Mickey Mouse Figure	Seiberling	1930s	6 1/2" tall, latex figure	80	145	225
Mickey Mouse Figure		1930s	3" tall, bisque, plays saxophone	35	60	95
Mickey Mouse Figure		1930s	3" tall, bisque, holding a parade flag & sword	35	60	95
Mickey Mouse Figure	Seiberling	1930s	3 1/2" tall, black hard rubber	25	50	75
Mickey Mouse Figure			composition, part of the Lionel Circus Train Set	55	100	150
Mickey Mouse Figurine	Goebel	1930s	3 1/2" tall, in a hunting outfit reading a book	35	65	100
Mickey Mouse Figurine			4" tall, bisque, dressed in a green nightshirt	35	65	100
Mickey Mouse Fire Engine	Sun Rubber Co.		7" long, rubber, push toy	40	80	135

Mickey & Minnie Mouse, Disney

TOY	COMPANY	YEAR	DESCRIPTION	GOOD	EX	MIB
Mickey Mouse Fire Truck with Figure	Sun Rubber Co.		2 1/2x6 1/2x4", Mickey driving & mold-in image of Donald standing on the back holding onto his helmet	45	80	125
Mickey Mouse Fireman Doll	Gund	1960s	12" tall	25	50	75
Mickey Mouse Fork & Spoon Set	Wm. Rogers & Son	1947	6" fork and 5 1/2" spoon	45	80	125
Mickey Mouse Fun-E-Flex Figure	Fun-E-Flex	1930s	7" tall figure with four-finger hands, chest decal says 'Mickey Mouse', in red shorts with orange feet	245	455	700
Mickey Mouse German Tin Bank		1934	2 1/2" diameter by 3" tall, bright yellow bank shaped like a beehive, Mickey approaching door holding a honey jar in one arm & key to open the door in the other	175	340	525
Mickey Mouse Gumball Bank	Hasbro	1968		16	30	45
Mickey Mouse Has a Busy Day Storybook	Whitman	1937	16 pages	25	35	55
Mickey Mouse Imitation Digital Wristwatch	Bradley	1973	wind-up 'digital' watch, rectangular base, 2 windows show date and minutes on rotating disks, black leather band, face show Mickey to right of windows	35	65	100
Mickey Mouse in Giantland Book	David McKay Co.	1934	45 pages, hardcover	45	80	125
Mickey Mouse in the Music Box		1960s	5x5x6" tall, plays "Pop Goes the Weasel", designed like a Jack-in-the-box, Mickey pops up	30	50	80
Mickey Mouse Jack-In-the-Box		1970s	5 1/2" square tin litho box shows Mickey, Pluto, Donald and Goofy, inside is Mickey	23	45	65
Mickey Mouse Jigsaw Puzzle	Disney	1933	8"x10"	30	55	85
Mickey Mouse Lamp	Soreng-Manegold Co.	1935	10 1/2" tall	70	130	200
Mickey Mouse Lionel Circus Train	Lionel	1935	train set with 5 cars w/Mickey in #1509 tender, train is 30 inches long, 84 inches of track, circus tent, Sunoco station, truck, tickets, Mickey composition statue	550	1050	1600
Mickey Mouse Lionel Circus Train Handcar	Lionel		9" long with 6" tall figures, metal, figures of Mickey & Minnie are composition & rubber	235	440	675
Mickey Mouse Magic Slate	Watkins-Strathmore	1950s	8 1/2x14" tall	16	30	45
Mickey Mouse Map of the United States	Dixon	1930s	9 1/4x14"	35	65	100
Mickey Mouse Marbles	Monarch		marbles and Mickey bag, on card	4	7	10
Mickey Mouse Mechanical Pencil		1930s	head of Mickey on one end and decal of Mickey walking on other side of pencil	35	65	100
Mickey Mouse Mechanical Robot	Gabriel			65	120	185
Mickey Mouse Mousegetar		1960s	10x30x2 1/2" black plastic	45	80	125
Mickey Mouse Mousegetar	Mattel	1950s	8x23x2" dark red plastic front	35	65	100

Mickey & Minnie Mouse, Disney

TOY	COMPANY	YEAR	DESCRIPTION	GOOD	EX	MIB
Mickey Mouse Movie Projector	Keystone	1934	5 1/2x11 1/2x11" tall for 8mm movies	140	260	400
Mickey Mouse Movie-Fun Shows	Mastercraft Toy Co.	1940s	7 1/2" square by 4" deep, animated action movies	95	180	275
Mickey Mouse Music Box			4 1/2x5x7", plays "It's a Small World", Mickey in conductor's uniform standing on cake	18	35	50
Mickey Mouse Music Box	Anri	1971	5" tall & 3 1/2" diameter, plays "If I Were Rich Man"	45	80	125
Mickey Mouse Music Box	Schmid Bros.	1970s	3 1/2" diameter by 5 1/2" tall, plays "Mickey Mouse Club March," cereamic figure of Mickey in western clothes standing next to a cactus	30	55	85
Mickey Mouse Musical Money Box		1970s	3" diameter by 6" tall, tin box with Mickey, Pluto, Donald and Goofy	30	55	85
Mickey Mouse Nite Light	Disney	1938	4" tall, tin	90	165	250
Mickey Mouse Old Timers Fire Engine	Matsudaya	1980s	tin and plastic red fire truck with Mickey at the wheel	35	65	100
Mickey Mouse Pencil Box	Dixon	1930s	5 1/2x10 1/2x1 1/4" Mickey illustrated in a gymnasium	35	65	100
Mickey Mouse Pencil Box	Dixon	1930s	Mickey ready to hitch Horace to a carriage in which Minnie is sitting	35	65	100
Mickey Mouse Pencil Box	Dixon	1937	5 1/2x9x3/4", Mickey, Goofy & Pluto riding a rocket	35	65	100
Mickey Mouse Pencil Box	Dixon	1937	5x8 1/2x1 1/4", Mickey is a circus ringmaster & Donald riding a seal	35	65	100
Mickey Mouse Pencil Holder	Dixon	1930s	4 1/2" tall	60	115	175
Mickey Mouse Pencil Sharpener	Hasbro	1960s	pencil sharpener is the shape of Mickey's head, pencil goes into mouth	12	23	35
Mickey Mouse Pencil Sharpener			3" tall, celluloid, sharpener located on base	55	100	225
Mickey Mouse Pendant Watch	Bradley	1970s	white plastic case, bubble crystal, articulated hands, gold chain	25	50	75
Mickey Mouse Picture Gun	Stephens Prod.	1950s	6 1/2x9 1/2x3", metal picture gun, lights to show filmstrips	55	100	250
Mickey Mouse Plastic Bank	Transogram	1970s	5x7 1/2x19" tall, plastic with Mickey standing on a white chest	12	23	35
Mickey Mouse Plastic Bank	Wolverine	1960s	1 1/2x5 1/2x'1" tall	25	50	75
Mickey Mouse Pocket Watch	Ingersoll	1930s	2" diameter	195	350	700
Mickey Mouse Pocket Watch	Bradley	1970s		35	65	100
Mickey Mouse Presents a Silly Symphony Book	Whitman	1934	Big Little Book	23	45	65
Mickey Mouse Print Shop Set	Fulton Specialty Co.	1930s	6 1/2x6 1/2", ink pad, stamper, metal tweezers & a wooden tray	70	130	200
Mickey Mouse Puddle Jumper	Fisher-Price			17	30	50
Mickey Mouse Pull Toy	Fisher-Price	1936	3x4x1/2" wooden figure of Mickey running with bright color paper labels on side	90	165	250
Mickey Mouse Pull Toy	Toy Kraft		7x22x8" tall, horse cart drawn by wooden horses	300	550	850

Mickey & Minnie Mouse, Disney

TOY	COMPANY	YEAR	DESCRIPTION	GOOD	EX	MIB
Mickey Mouse Pull Toy	N.N. Hill Brass Co.	1935	14" tall, wood & metal	130	245	375
Mickey Mouse Puppet Forms	Colorforms	1960s		9	16	25
Mickey Mouse Push Toy	Fisher-Price	1930s	6" long x 4 1/2" tall, wood	95	180	275
Mickey Mouse Quartz Pocket Watch	Lorus	1988	#2202, small gold bezel, gold chain and clip fob, articulated hands	20	40	60
Mickey Mouse Radio	Philgee Int.	1970s		16	30	45
Mickey Mouse Radio	Emerson	1934	wood composition cabinet with designs of Mickey Mouse playing various musical instruments	335	625	2000
Mickey Mouse Record Player	General Electric	1970s	1 12/x14x4 1/2", Mickey in blue & white striped shirt & bow tie, the playing arm is the design of Mickey's arm	55	100	150
Mickey Mouse Registered Commemorative Ed. Wrist Watch	Bradley	1978		60	115	175
Mickey Mouse Riding Toy	Mengel	1930s	6x17x16" tall	385	715	1100
Mickey Mouse Rodeo Rider	Matsudaya	1980s	plastic wind-up cowboy Mickey rides a bucking bronco	20	40	60
Mickey Mouse Rolykins	Marx		1x1x1 1/2" tall, Walt Disney Rolykins with ball bearing action set	9	16	25
Mickey Mouse Rub'n Play Magic Transfer Set	Colorforms	1978	rub'n clothing transfers to Mickey, Donald Duck, Pluto, Peg Leg Pete, Big Bad Wolf, Pluto and other characters	14	25	40
Mickey Mouse Rubber Figure	Seiberling	1930s	6 1/2" tall, rubber	65	120	185
Mickey Mouse Safety Blocks	Halsam	1930s	set of nine blocks	80	145	225
Mickey Mouse Sand Pail	Ohio Art		4 1/2" tall, Mickey, Minnie & Pluto at the beach	45	80	125
Mickey Mouse Sand Pail	Ohio Art	1938	6" tall, Mickey, Donald & Goofy playing golf	45	80	225
Mickey Mouse Sand Pail	Ohio Art	1938	3" tall	40	70	175
Mickey Mouse Sand Shovel	Ohio Art		10" long	25	50	95
Mickey Mouse Saxophone	Harbo	1930s	3" wide at opening, 10" wide, 16" tall	115	210	325
Mickey Mouse Scissors	Disney		3" long, child's scissors with Mickey figure	20	35	55
Mickey Mouse Sewing Cards	Colorforms	1978	7 1/2"x12" cut-out card designs of Mickey, Minnie, Pluto, Clarabelle, Donald Duck and Horace	11	20	30
Mickey Mouse Silk Ornament	Hallmark	1978	4x4x5 1/2" tall, Mickey as Santa riding a stream train	12	23	35
Mickey Mouse Snowdome			figure with cake	5	10	15
Mickey Mouse Spirit of '76 Colorforms Set	Colorforms	1976	8x 12 1/2x1"	9	16	25
Mickey Mouse Squeeze Toy	Dell	1960s	8" tall	16	30	45
Mickey Mouse Squeeze Toy	Sun Rubber Co.	1950s	10" tall	20	40	60
Mickey Mouse Stamp Pad		1930s	3" long	25	50	75

Mickey & Minnie Mouse, Disney

TOY	COMPANY	YEAR	DESCRIPTION	GOOD	EX	MIB
Mickey Mouse Steamboat	Matsudaya	1988	Wind-up plastic steamboat with Mickey as Steamboat Willie, runs on floor as smokestacks go up and down, box says "60 years With You"	30	55	85
Mickey Mouse Stuffed Doll	Knickerbocker	1935	11" tall, stuffed with removeable shoes & jointed head in red shorts	115	210	325
Mickey Mouse Talking Figure	Hasbro	1970s	4x5 1/2x7 1/2" tall	18	35	50
Mickey Mouse Talking Figure	Horsman	1972	3x10x12" tall, says five different phrases	18	35	50
Mickey Mouse Tea Set	Wolverine		plastic	55	110	165
Mickey Mouse Tea Set		1930s	3" saucer, 2" pitcher, 2 1/2" sugar bowl each piece shows Mickey & Minnie in a rowboat	25	50	225
Mickey Mouse Telephone Bank	N.N. Hill Brass Co.	1938	5" tall with cardboard figure of character	70	130	275
Mickey Mouse Throw Rug	Alex. Smith Carpet	1935	26x42", Mickey, Donald & a pig are playing musical instruments	125	225	350
Mickey Mouse Tin Bank		1930s	2x3x2 1/4" tall, shaped like a treasure chest with Mickey & Minnie next to a treasure chest on the "Isle of the Thrift"	150	295	450
Mickey Mouse Tin Serving Tray		1960s	11" diameter, tin	11	20	30
Mickey Mouse Top	Fritz Bueschel	1930s	7" diameter, 7" tall, Mickey, Minnie, a nephew, Donald & Horace playing a musical instrument	125	225	350
Mickey Mouse Toy Tractor	Sun Rubber Co.		3x4x4" tall, red body & white rubber tires, Mickey sitting in seat with moveable head	30	55	85
Mickey Mouse Tractor	Sun Rubber Co.		5" long, rubber	25	50	75
Mickey Mouse Transistor Radio	Gabriel	1950s	6 1/2x7x1 1/2"	35	65	125
Mickey Mouse Tricycle Toy	Steiff	1932	8 1/2"x7", wood & metal frame, action movement	420	780	2000
Mickey Mouse Twirling Tail Toy	Marx	1950s	3x5 1/2x5 1/2" tall, with a built in key & his tail is a metal rod that spins around as the toy vibrates	115	210	325
Mickey Mouse Wall Clock	Elgin	1978	9" diameter dial, 15" long vinyl straps for watchband look, giant wrist watch with the "50 Happy Years" logo on dial	25	50	75
Mickey Mouse Washer	Ohio Art		8" tall, tin, litho with Mickey & Minnie Mouse pictured doing their wash	45	80	225
Mickey Mouse Watering Can	Ohio Art	1938	6" tall, tin, litho	65	120	225
Mickey Mouse Wind-up Musical Toy	Illco	1970s	6" tall, plays "Lullaby & Goodnight", 3-D figure of Mickey in dark red pants & yellow shirt	20	35	50
Mickey Mouse Wind-up Toy	Gabriel	1978	plastic transparent figure of Mickey with visible metal gears	9	15	35
Mickey Mouse Wind-up Trike	(Korea)	1960s	tin litho trike with plastic Mickey with flag and balloon on handle, bell on back	80	145	225
Mickey Mouse Wooden Bell Pull Toy	N.N. Hill Brass Co.		8 1/2" tall, 13" long, Mickey on roller skates	125	225	350

Mickey & Minnie Mouse, Disney

TOY	COMPANY	YEAR	DESCRIPTION	GOOD	EX	MIB
Mickey Mouse Wooden Sled	Flexible Flyer	1930s	18x30x6" tall	125	210	450
Mickey Mouse Wrist Watch	Bradley	1983	medium black octagonal case, articulated hands, no numbers on face, in plastic window box	20	35	55
Mickey Mouse Wrist Watch & Figural Stand	Ingersoll	1950s		25	50	150
Mickey Mouse Wrist Watch, Hologram (Woman's)			18K gold, electroplate with black leather band	23	45	65
Mickey Mouse Wrist Watch, Hologram Sorcerer			18K gold, electroplate with black leather band	23	45	65
Mickey Mouse Wristwatch	Bradley	1984	medium white case, articulated hands, black face, sweep seconds, white vinyl band, in plastic window box	20	35	55
Mickey Mouse Wristwatch	Timex	1960s	large round case, stainless back, articulated hands, red vinyl band	45	80	125
Mickey Mouse Wristwatch	Ingersoll	1939		115	210	325
Mickey Mouse Wristwatch	Ingersoll	1939	retangular with standard second hand between Mickey's legs	200	375	575
Mickey Mouse Xylophone Player Pull Toy	Fisher-Price	1939	#798, 11" tall, wood, pull string and his arms move up & down and he plays the five notes on the xylophone	80	145	275
Mickey Mouse Yarn Sewing Set	Marks Brothers	1930s	9x17x1 1/2"	20	35	50
Mickey's Air Mail Plane	Sun Rubber Co.	1940s	3 1/2x6" long, 5" wingspan, rubber plane	35	65	125
Mickey, Minnie & Donald Throw Rug			26 x41", Mickey & Minnie in an airplane with Donald parachuting	60	115	175
Mickey, Minnie, Pluto & Donald Sand Pail	Ohio Art		5" tall and 5" diameter, Mickey, Minnie, Pluto & Donald in boat looking across water at castle, with swivel handle	45	80	175
Mickey/Donald Crayons & Box Set	Transogram	1946	4 1/2x5 3/4x1/2", tin box, with crayons from your favorite funsters Donald Duck & Mickey Mouse	25	50	75
Mickey/Donald Jack-In-The-Box	Lakeside	1966	3 1/2x3 1/2x4 1/2", Mickey is pictured on one side of the box with a small piece of fabric which says "Pull My Tie", when it is pulled Donald pops out	25	50	75
Mickey/Minnie & Pluto Flashlight	Usalite Co.	1930s	6" long, Mickey leading Minnie through the darkness guided by flashlight & Pluto	35	65	100
Mickey/Minnie Carpet			27x41", Pegleg Pete is lassoed by Mickey; all characters in western outfits	115	210	325
Mickey/Minnie Dolls	Gund	1940s	13" tall, each	115	210	450
Mickey/Minnie Puzzle	Marks Brothers	1930s	10x12", Mickey polishing the boiler on his "Mickey Mouse R.R." train engine and Minnie waving from the cab	40	70	110
Mickey/Minnie Snow Dome	Monogram	1970s	3x4x3" tall, Mickey & Minnie with a pot of gold at the end of the rainbow	12	23	35
Mickey/Minnie Toothbrush Holder			2 1/2x4x3 1/2", Mickey & Minnie on sofa with Pluto at their feet	70	130	250

The Rocketeer bendable doll, 1990s, JusToys.

Mickey & Minnie Mouse, Disney

TOY	COMPANY	YEAR	DESCRIPTION	GOOD	EX	MIB
Mickey/Pluto Ceramic Ashtray			3x4x3" tall, Mickey & Pluto playing banjos while sitting on the edge of the ashtray	115	210	325
Minnie Mouse Alarm Clock	Bradley (Germany)	1970s	pink metal electric two-bell clock with articulated hands	23	45	65
Minnie Mouse Car	Matchbox	1979		5	10	15
Minnie Mouse Choo-Choo Train Pull Toy	Linemar	1940s	3x8 1/2x7" tall, green metal base & green wooden wheels	70	125	195
Minnie Mouse Doll	Petz	1940s	3x5x10" tall	115	210	325
Minnie Mouse Doll	Knickerbocker	1935	14" tall, stuffed, cloth, polka-dot skirt & lace pantaloons	105	195	475
Minnie Mouse Figure	Ingersoll	1958	5 1/2" tall, plastic	30	55	85
Minnie Mouse Figure			6" tall, plastic	9	16	25
Minnie Mouse Figurine			4" tall, bisque, dressed in a nightshirt	40	70	110
Minnie Mouse Fun-E-Flex Figure	Fun-E-Flex	1930s	5" Minnie with fingered hands	125	225	350
Minnie Mouse Hand Puppet		1940s	11" tall, white on red polka-dot, fabric hard cover and a pair of black & white felt hands	50	95	145
Minnie Mouse Music Box	Schmid Bros.	1970s	3 1/2" diameter, plays "Love Story"	11	20	50
Minnie Mouse Plastic Clock	Phinney-Walker	1970s	8" diameter by 1 1/2" deep, "Behind Every Great Man, There is A Woman!"	35	65	100
Minnie Mouse Rocker	Marx	1950s	tin, wind-up, rocker moves back & forth with gravity motion of her head & ears	250	450	695
Minnie Mouse Wristwatch	Timex	1958	small round chrome case, stainless back, articulated hands, yellow vinylite band	55	100	150
Minnie Mouse Wristwatch	Bradley	1978	meduim gold case, sweep seconds, red vinyl band, articulated hands	16	30	45
Minnie with Bump-n-Go Action Shopping Cart	Illfelder	1980s	4x9 1/2x11 1/2" tall, plastic, battery operated Minnie pushing cart	25	50	75
Mousegetar Jr.	Mattel	1955	hand crank play guitar with paper litho label of Mickey's face	35	60	95
Mouseketeer Cut-outs	Whitman	1957	figures & clothing sheets	23	45	65
Mouseketeer Fan Club Typewriter		1950s	lithographed tin	45	85	135
Tennis Sport Watch	Bradley	1970s	2 1/2x6x2 1/2" tall with a plastic case, white dial with Mickey playing tennis, the second hand has a tennis ball on the end of it	35	65	100
Tin Tray (Mickey & Minnie in a Rowboat)	Ohio Art	1930s	5 1/2"x7 1/4", tin, lithograph	20	40	125
Two Gun Mickey Mouse Watch			saddle tan western style band	23	45	65
Walt Disney's Mickey Mouse Play Tiles	Halsam	1964	336 tiles to create Pinocchio, Jiminy Cricket, Goofy, Donald and etc.	12	23	35
Walt Disney's Mickey Mouse Water Globes		1970s	3x4 1/2x5" tall, three dimensional plastic figures of Mickey seated with a plastic water globe between his legs	25	50	75

Mighty Mouse

TOY	COMPANY	YEAR	DESCRIPTION	GOOD	EX	MIB
Mighty Mouse & Heckle & Jeckle Sliding Puzzle	Fleetwood	1979		6	12	35

Mighty Mouse

TOY	COMPANY	YEAR	DESCRIPTION	GOOD	EX	MIB
Mighty Mouse & His TV Pals Puzzle			2 1/2" square, tile	9	16	25
Mighty Mouse Ball Game	Ja-Ru	1981		4	7	10
Mighty Mouse Charm Bracelet		1950s	brass charms of Gandy Goose, Terry Bear, Mighty Mouse and other Terrytoon characters	25	50	75
Mighty Mouse Cinema Viewer	Fleetwood	1979	has four strips	7	13	20
Mighty Mouse Dynamite Dasher	Takara	1981		5	10	15
Mighty Mouse Figure		1950s	9" vinyl, squeeze toy	16	30	45
Mighty Mouse Figure	Dakin	1977	hard & soft vinyl figure	18	35	50
Mighty Mouse Flashlight	Dyno	1979	3.5" figural light	5	10	25
Mighty Mouse Fun Farm Figure	Dakin	1978		23	45	65
Mighty Mouse Make a Face Sheet	Towne	1958	with dials to change face parts	12	23	35
Mighty Mouse Mighty Money	Fleetwood	1979		2	4	6
Mighty Mouse Money Press	Ja-Ru	1981	stamps, pads and money	2	4	6
Mighty Mouse Movie Viewer	Chemtoy	1980		5	10	15
Mighty Mouse Picture Play Lite	Janex	1983		6	11	17
Mighty Mouse PVC Figures		1988	hands on hips, taking off, hands clasped at chest, each	2	4	6
Mighty Mouse Sneakers	Randy Co.	1960s	children sizes 7 1/2 unused, graphics on box, picture on sneakers	35	65	95
Mighty Mouse Squeaker Figure		1950s	9 1/2" rubber	25	50	75
Mighty Mouse Stuffed Doll	Ideal	1950s	14" tall, stuffed, cloth	40	75	115
Mighty Mouse Vinyl Doll			9 1/2" tall	35	65	100
Mighty Mouse Wallet	Larami	1978		5	10	15
Mighty Mouse Wristwatch	Bradley	1979	medium chrome case,	30	55	85

Miscellaneous Characters

TOY	COMPANY	YEAR	DESCRIPTION	GOOD	EX	MIB
Bewitched Samantha Doll	Ideal	1967	12 1/2" tall	185	340	525
Bonzo Scooter Toy			7" in length & Bonzo 6", wind-up	140	260	400
Bruce Lee Figure	Largo	1986		6	12	18
Captain Kangaroo and Bunny Rabbit Presto Slate	Fairchild	1960s	lift-to-erase drawing slate on card illustrated with pictures of the Captain and Bunny Rabbit	14	25	25
Captain Kangaroo and Mr. Green Jeans Presto Slate	Fairchild	1960s	lift-to-erase drawing slate on card illustrated with pictures of the Captain and Mr. Green Jeans	14	25	25
Daniel Boone Figure	Remco	1964	5" tall, hard plastic body with large vinyl head, has cloth coonskin cap and longrifle	35	60	125
Debby Boone Doll	Mattel		10" tall	20	40	85
Diamond Jim		1930s	5 1/2" tall	45	80	125
Donny & Marie Country & Rock Rhythm Set		1976		7	13	20
Dr. Kildare Photo Scrapbook		1962	photo scrapbook	23	45	65

Miscellaneous Characters

TOY	COMPANY	YEAR	DESCRIPTION	GOOD	EX	MIB
Dragnet Badge 714	Knickerbocker	1955	2 1/2" bronze finish badge in yellow box with illustration of Jack Webb, box bottom has ID card	23	35	65
Flipper Tray Puzzles		1966	set of four frame tray puzzles	20	40	60
Flying Nun Doll	Kayline Co.	1970		16	30	150
Flying Nun Oil Paint By Numbers	Hasbro	1960s	boxed set of two scenes and 10 paint vials	11	20	75
Jimmy Carter Radio			peanut shaped transistor radio with Jimmy	16	30	65
Jimmy Carter Wind-up Walking Peanut			5" tall	9	16	25
Joan Palooka "Stringless Marionette"	Nat'l Mask & Puppet	1952	12 1/2" tall 'daughter of Joe Palooka' doll comes with pink blanket and birth certificate	60	115	175
Joe Penner & His Duck Goo Goo	Marx	1934	wind-up, "Wanna buy a duck?" lithographed on ths side of Joe's basket of ducks	300	450	950
Komic Kamera Film Viewer Set			5" long, Dick Tracy, Little Orphan Annie, Terry & the Pirates & The Lone Ranger	35	65	105
Little King Lucky Safety Card		1953	2"x4" cards, Little King warns you about crossing the street at night	11	20	30
Lyndon Johnson Figure	Remco	1960s		30	55	85
Mr. & Mrs. Potato Head Set	Hasbro	1960s	cars, boats, shopping trailer and etc.	30	55	85
Mr. Potato Head Frankie Frank	Hasbro	1966	companion toy to Mr. P., with accessories	25	45	70
Mr. Potato Head Frenchy Fry	Hasbro	1966	companion toy to Mr. P., with accessories	25	45	70
Mr. Potato Head Ice Pops	Hasbro	1950s	plastic head molds for freezing treats in box	16	30	45
Patton Figure	Excel Toy Corp.		poseable doll with cloth cloths and accessories	16	30	45
Pinky Lee Costume		1950s	hat, pants and shirt	50	95	145
President Bush Figure			7" tall	11	20	30
Prince Charles of Wales Figure	Goldberger	1982	13" tall, dressed in palace guard uniform	25	50	75
Red Ranger Ride 'em Cowboy	Wyandotte	1930s	tin wind-up rocker	140	260	450
Ringling Bros & Barnum & Bailey Circus Playset		1970s	vinyl fold out set with animals, trapeze personnel, clowns and assorted circus equipment	16	30	125
Robin Hood View-Master Pack		1954		7	13	20
Sir Reginald Play-N-Save Bank		1960s	7" tall plastic lion & hunter, on a 15" green plastic base, the hunter fires the coin into the lion's mouth	45	80	125
Starsky & Hutch Shoot-Out Target Set	Berwick	1970s		9	16	45
Sylvester Stallone Rambo Figure		1986	18" tall, poseable figure	9	16	25
Uncle Don's "Puzzy & Sizzy" Membership Card		1950s		12	23	35
Willie Whopper Pencil Case		1930s	green with illustrations of Willie, Pirate and his gal	40	75	120
Winky Dink Magic Crayons		1960s		16	30	45
Winky Dink Secret Message Game	Lowell	1950s		70	125	195

Moon Mullins

TOY	COMPANY	YEAR	DESCRIPTION	GOOD	EX	MIB
Moon Mullins & Kayo Railroad Handcar Toy	Marx	1930s	6" long, wind-up, both figures bendable arms & legs	300	450	850
Moon Mullins & Kayo Toothbrush Holder			4" tall, bisque	35	60	150
Moon Mullins Featured Funnies Jigsaw Puzzle		1930s	9 1/2"x14" puzzle	35	60	95
Moon Mullins Figure Set			bisque, Uncle Willie, Kayo, Moon Mullins & Emmy, heights range from 2 1/4" tall to 3 1/2"	95	180	350
Moon Mullins Playstone Funnies Mold Set		1940s		35	65	150

Mr. Magoo

TOY	COMPANY	YEAR	DESCRIPTION	GOOD	EX	MIB
Magoo Car	(Japan)		9" battery operated car	80	145	325
Magoo Soakie Soap Container		1960s	10" tall, vinyl & plastic	12	23	35
Mr. Magoo Car	Hubley	1961	7 1/2" tallx9" long, metal, battery operated	90	165	350
Mr. Magoo Doll	Ideal	1962	5" tall, vinyl head with cloth body	30	55	125
Mr. Magoo Doll	Ideal	1970	12" tall	20	35	55
Mr. Magoo Drinking Glass		1962	5 1/2" tall	12	23	35
Mr. Magoo Figure	Dakin		7" tall	25	50	75
Mr. Magoo Hand Puppet		1960s	vinyl head with cloth body	30	50	50

Munsters, The

TOY	COMPANY	YEAR	DESCRIPTION	GOOD	EX	MIB
Doll, Herman Munster	Remco	1964		105	150	230
Doll, Lilly	Remco	1964		115	165	285
Doll, Lilly Baby	Ideal	1965		60	95	150
Dolls, Set, Herman, Lilly, Grandpa	Remco	1964		225	425	750

Music

TOY	COMPANY	YEAR	DESCRIPTION	GOOD	EX	MIB
Andy Gibb Figure	Ideal	1979	6" tall	9	16	25
Andy Gibb Figure	Ideal	1979	8" tall	15	25	40
Beatles Coloring Book	Saalfield	1964	thick coloring book	23	45	125
Beatles Toy Watches		1960s	set of four nonworking play watches, tin w/plastic bands, on card	25	50	150
Boy George Doll	LJN	1980s	12" tall posable in alphabet shirt	35	60	95
Boy George Doll	LJN	1980s	15" Huggard Cute Cuddly version in polka dot shirt	25	50	75
Dolly Parton Doll	Goldberger	1970s	12" tall	25	45	70
Donny Osmond Figure	Mattel	1976	12" tall	14	25	40
Elvis Doll	World Doll	1984	18" tall doll in box	30	60	90
Elvis Doll	World Dolls	1984	21" tall	55	100	150
Elvis Presley Figure			12" tall	25	50	75
Elvis Presley Quartz Wrist Watch	Bradley	1983	white plastic case, stainless back, face shows a young Elvis, white vinyl band	16	30	45
KISS Ace Doll	Mego	1979		40	70	110
KISS Gene Simmons Doll	Mego	1979	12' tall in costume and makeup	35	65	100
KISS Paul Stanley Doll	Mego	1979	12" tall makeup and costume	35	65	100
KISS Peter Criss Doll	Mego	1979		35	65	100
Marie Osmond Figure	Mattel		12" tall	11	20	30

Music

TOY	COMPANY	YEAR	DESCRIPTION	GOOD	EX	MIB
Marie Osmond Modeling Doll			30" tall	35	60	95
Michael Jackson AM Radio	Ertl	1984		9	16	45
Michael Jackson American Music Awards Doll	LJN			18	35	50
Michael Jackson Beat It Doll	LJN			18	35	50
Michael Jackson Cordless Electronic Microphone	LJN	1984		7	13	25
Michael Jackson Grammy Awards Doll	LJN			18	35	50
Michael Jackson Thriller Doll	LJN			18	35	50
Michel Jackson Thriller Gang/Glow Bendy Set of Six				23	45	65
Paul McCartney Doll	Remco	1964		80	150	230
Ringo Starr Doll	Remco	1964		80	150	230
Toni Tennile Figure			12" tall	11	20	30

Mutt & Jeff

TOY	COMPANY	YEAR	DESCRIPTION	GOOD	EX	MIB
Mutt & Jeff Coin Bank			5" tall, cast iron, two piece construction held together by screw in the back	45	80	175
Mutt & Jeff Doll Set		1920s	8" & 6 1/2" tall, composition hands & Heads with heavy cast iron feet, moveable arms & legs, fabric clothing	165	310	475
Mutt & Jeff Statue Set	A. Steinhardt & Bros	1911	ceramic, with a coin inserted into the base of each one	70	125	195

Peanuts

TOY	COMPANY	YEAR	DESCRIPTION	GOOD	EX	MIB
Character Figures		1958	9" tall, vinyl, Charlie Brown, Snoopy and Lucy, each	18	35	50
Charlie Brown Nodder	(Japan)	1960s	5 1/2" tall, bobbing head	18	35	50
Lucy Wristwatch	Timex	1970s	small chrome case, articulated arms, sweep seconds, white vinyl band	23	45	65
Peanuts Coloring Book	Saalfield	1960s		9	16	25
Peanuts Gang Five Figural Banks	United Features	1970	set of five	45	80	125
Peanuts Gang Five Figural Music Boxes	Schmid Bros.	1970	set of five music boxes	35	65	100
Peanuts Music Box	Schmid Bros.	1980		15	25	40
Peanuts Tea Set		1961	one tray, 2 plates, 2 cups, 4 small plates all with Peanuts artwork on all of them	25	50	75
Peppermint Patty Peanuts Gang Rag Doll			14" tall	7	13	20
Snoopy as Astronaut Music Box	Schmid Bros.	1970s		20	35	55
Snoopy Belle Figure	Knickerbocker		figure	4	8	12
Snoopy Figural Bank	United Features	1968	7"	14	25	40
Snoopy Jack in the Box	Mattel	1970s	tin box with a plastic Snoopy that pops out after you crank it	16	30	45
Snoopy Nodder	(Japan)	1960s	5 1/2" tall, bobbing head	19	35	55

Peanuts

TOY	COMPANY	YEAR	DESCRIPTION	GOOD	EX	MIB
Snoopy Pop Up Figure Music Box	Mattel	1966	5", steel	16	30	45
Snoopy Snowdome			on top of doghouse	6	12	18
Snoopy Tennis Animated Wristwatch	Timex	1970s	gold bezel, tennis ball circles Snoopy on clear disk, articulated hands holding racket, denim background and band	25	50	75

Pee Wee Herman

TOY	COMPANY	YEAR	DESCRIPTION	GOOD	EX	MIB
Ball Dart Set				5	10	15
Billy Baloney Figure	Matchbox	1988	18" tall	7	13	20
Chairry Figure	Matchbox	1988	5" tall	3	5	8
Chairry Figure	Matchbox		15" tall	6	12	18
Conky Wacky Wind-up	Matchbox	1988		3	5	8
Cowboy Curtis Figure	Matchbox			5	10	15
Deluxe Colorforms				5	10	15
Globey with Randy	Matchbox	1988		4	7	10
King of Cartoons Figure	Matchbox	1988	5" tall	3	5	8
Magic Screen Figure	Matchbox	1988	5" tall poseable	3	5	8
Magic Screen Wacky Wind-up	Matchbox	1988	6" tall	3	5	8
Miss Yvonne Figure	Matchbox	1988	poseable 5" tall	3	5	8
Pee Wee Doll			15" tall, non talking	7	13	20
Pee Wee Herman	Matchbox	1988	poseable 5" tall	4	8	12
Pee Wee Herman Play Set	Matchbox	1989	deluxe play set 20"x28"x8" for use with all 5" figures, Pee Wee's bike and folds into large carrying case	12	23	35
Pee Wee Herman Ventriloquist Doll				14	25	40
Pee Wee with Scooter and Helmet	Matchbox	1988		4	7	10
Pee Wee Yo Yo				2	3	5
Pterri Doll			13" tall	14	25	40
Pterri Wacky Wind-ups	Matchbox	1988		3	5	8
Reba Figure	Matchbox	1988	poseable	3	5	8
Ricardo Figure	Matchbox	1988		3	5	8
Slumber Bag	Matchbox	1988		11	20	30
Vance the Talking Pig	Matchbox	1987		14	25	40
View Master Gift Set				5	10	15

Pepe Le Pew

TOY	COMPANY	YEAR	DESCRIPTION	GOOD	EX	MIB
Pepe Le Pew "Goofy Gram"	Dakin	1971		20	35	65
Pepe Le Pew Figure	Dakin	1971	8" tall	23	45	65

Percy Crosby

TOY	COMPANY	YEAR	DESCRIPTION	GOOD	EX	MIB
Our Beloved Skippy Jigsaw Puzzles				25	50	75
Skippy Doll	Effanbee		wood composition	55	100	150
Skippy Figurine		1930s	5" tall, bisque	45	80	125
Skippy Jigsaw Puzzle		1933	framed	15	25	40
Skippy Mug			silverplate	45	80	125
Skippy Silverware Set			spoon & fork 4 1/2" long, plate 8" diameter	55	100	150

Peter Pan

TOY	COMPANY	YEAR	DESCRIPTION	GOOD	EX	MIB
Captain Hook Figurine			8" tall, plastic	9	15	25

Peter Pan

TOY	COMPANY	YEAR	DESCRIPTION	GOOD	EX	MIB
Captain Hook Hand Puppet	Gund	1950s	9" tall	30	60	90
Peter Pan Baby Figure	Sun Rubber Co.	1950s		45	80	125
Peter Pan Cardboard Statuettes	Whitman	1952	7 1/2x12 1/2x1", set of 11 diecut cardboard figures	35	65	100
Peter Pan Charm Bracelet		1974		11	20	50
Peter Pan Doll	Duchess Doll Corp.	1953	11 1/2" tall, brown trim fabric shoes, green mesh stockings with flocked outfit, hat with a large red feather, shiny silver white metal dagger in belt, eyes, arms & head move	150	275	425
Peter Pan Doll	Ideal	1953	18" tall	95	180	275
Peter Pan Frame Tray Puzzle	Jaymar	1950s	puzzle show Peter, Wendy, John and Michael flying over Neverland	16	30	45
Peter Pan Hand Puppet	Oak Rubber Co.	1953	rubber hand puppet	45	80	95
Peter Pan Map of Neverland		1953	18x24", collectors issue, dedicated to the users of Peter Pan Beauty Bar	45	80	125
Peter Pan Nodder		1950s	6" tall	25	50	125
Peter Pan Push Puppet	Kohner	1950s	6" tall, green & flesh colored beads with plastic head & a light green thin plastic hat	30	50	80
Peter Pan Sewing Cards	Whitman	1952		14	25	40
Tinker Bell Doll	Duchess Doll Corp.	1953	8" tall, flocked green outfit with a pair of large white fabric wings with gold trim, eyes open & close, jointed arms & head moves	115	210	325
Tinker Bell Figure	A.D. Sutton & Sons	1960s	7" tall, plastic & rubber figure	25	50	75
Tinker Bell Pincushion with Figure		1960s	pincushion with 1 1/2" tall Tink figure, in clear plastic display can	25	45	65
Wendy Doll	Duchess Doll Corp.	1953	8" tall, full purple length skirt with purple bow in back of dress, eyes open & close, jointed arms & head moves	115	210	325

Peter Potamus

TOY	COMPANY	YEAR	DESCRIPTION	GOOD	EX	MIB
Breezley Soakie	Purex	1967	9" tall, plastic	25	50	75
Peter Potamus Soakie	Purex	1960s	10 1/2" tall, plastic	12	23	35

Pink Panther

TOY	COMPANY	YEAR	DESCRIPTION	GOOD	EX	MIB
Pink Panther & Inspector Hiking Puzzle	Whitman		100 pieces	4	8	12
Pink Panther & Sons Fun at the Picnic Book	Golden			5	10	15
Pink Panther & Sons Lite	Ja-Ru		mini flashlight, plastic	4	7	10
Pink Panther & Sons Target Game	Ja-Ru			5	10	15
Pink Panther & The Fancy Party Book	Golden			5	10	15

Pink Panther

TOY	COMPANY	YEAR	DESCRIPTION	GOOD	EX	MIB
Pink Panther & The Haunted House Book	Golden			5	10	15
Pink Panther at Castle Kreep Book	Whitman			4	7	10
Pink Panther at The Circus Sticker Book	Golden	1963		5	10	15
Pink Panther Coloring Book	Whitman	1976	cover shows PP roasting hot dogs	7	13	20
Pink Panther Figure	Dakin	1971	8" tall, with legs closed	16	30	45
Pink Panther Figure	Dakin	1971	8" tall, with legs open	14	25	40
Pink Panther Memo Board			write on or off memo board	5	10	15
Pink Panther Motorcycle			2 1/2" plastic	4	7	10
Pink Panther One Man Band	Illco	1980	10" tall, battery operated, plush body with vinyl head	25	50	75
Pink Panther Pool Game	Ja-Ru			4	7	10
Pink Panther Putty	Ja-Ru			4	7	10
Pink Panther Puzzle	Whitman		100 pieces "in refrigerator"	4	8	12
Pink Panther Puzzle, "Club Posh"	Whitman		100 pieces	4	8	12
Pink Panther Wind-up			3" tall, plastic, walking wind-up with trench coat & glasses	7	13	20
Xmas Music Box	Royal Orleans	1982	1982 limited edition	25	45	85
Xmas Music Box	Royal Orleans	1983	1983 limited edition	25	45	85
Xmas Music Box	Royal Orleans	1984	1984 limited edition	25	45	85

Pinocchio & Jiminy Cricket

TOY	COMPANY	YEAR	DESCRIPTION	GOOD	EX	MIB
"Walt Disney Tells the Story of Pinocchio"	Whitman	1939	4 1/4"x6 1/2" paperback, 144 pages	25	50	75
Figaro Figure	Multi-Wood Products	1940	3" tall, hand-carved wood composition	30	55	85
Figaro Roll Over Wind-up Toy	Marx	1940	5" long with ears & tail, tin	125	225	350
Figaro Tin Friction Toy	Linemar	1960s	1 1/2x3x1 1/2" tall	25	50	250
Gepetto Figure	Multi-Wood Products	1940	5 1/2" tall, wood composition	50	95	145
Gideon Figure	Multi-Wood Products		5" tall	35	65	100
Honest John Figure	Multi-Wood Products		2 1/2x3" base with a 7" tall figure	45	80	125
Jiminy Cricket Hand Puppet	Gund	1950s	11" tall	35	60	60
Jiminy Cricket Marionette	Pelham Puppets	1950s	3x6x10" tall, head is a dark green with a reddish/orange mouth and large black & white eyes with yellow accents in a gray felt hat	70	130	250
Jiminy Cricket Pushing Bass Fiddle Ramp Walker	Marx	1960s	1x3x3" tall, pushing a bass fiddle	110	150	220
Jiminy Cricket Snap-eeze Figure	Marx		3 1/2x4 3/4" tall, white plastic base with moveable arms & legs	25	50	75
Jiminy Cricket Soakie			7" tall bottle	9	16	25
Jiminy Cricket Toothbrush Set	Dupont	1950s	plastic wall hanging Jiminy holds a toothbrush	25	50	75
Jiminy Cricket Wooden Doll	Ideal	1940	wooden jointed doll	140	260	400
Jiminy Cricket Wristwatch	US Time	1948	"1948 US Time Birthday Series"	60	115	175

Pinocchio & Jiminy Cricket

TOY	COMPANY	YEAR	DESCRIPTION	GOOD	EX	MIB
Lampwick Figure	Multi-Wood Products	1940	5 1/2" tall, wood composition	40	75	115
Pin the Nose on Pinocchio Game	Parker Brothers	1939	15 1/2"x20"	45	80	125
Pinocchio Bank	Pal Plastics	1970s	7x7x10" tall, vinyl, 3-D molded head of Pinocchio	12	23	35
Pinocchio Bank	Crown Toy	1939	5" tall, wood composition with metal trap door on back	65	120	275
Pinocchio Bell-Ringing Pull Toy	Fisher-Price	1939	7" long by 9" tall, Pinocchio figure rocks back & forth on donkey and rings the bell on top of donkey's head	70	130	250
Pinocchio Big Little Book		1940		23	45	65
Pinocchio Color Box	Transogram		also known today as paint box	12	23	35
Pinocchio Cut-Out Books	Whitman	1940		40	70	110
Pinocchio Doll	Knickerbocker	1940	3 1/2x4x9 1/2" tall, jointed composition doll with moveable arms & head	185	345	525
Pinocchio Doll	Ideal	1939	12" tall with wire mesh arms & legs	105	195	300
Pinocchio Doll	Ideal	1940	19 1/2" tall, composition doll	225	425	650
Pinocchio Doll	Ideal	1940	10" tall, wood composition head & jointed arms & legs attached to body	70	130	275
Pinocchio Doll	Ideal	1940	8" tall, wood composition head, others are jointed wood	55	105	200
Pinocchio Figure	Crown Toy		9 1/2" tall, jointed arms	50	90	175
Pinocchio Figure	Crown Toy		3 1/2x4x9 1/2", joint composition figure	55	100	225
Pinocchio Figure	Multi-Wood Products	1940	5" tall, wood composition figure	45	80	125
Pinocchio Hand Puppet	Knickerbocker	1962		25	45	70
Pinocchio Hand Puppet	Crown Toy		9" tall, composition	25	50	75
Pinocchio Hand Puppet with Squeeker	Gund	1950s	10" tall	30	60	90
Pinocchio Music Box			plays "Puppet on a String"	12	23	35
Pinocchio Paint Book	Disney	1939	11"x15" heavy paper cover	30	35	55
Pinocchio Paperweight & Thermometer	Plastic Novelties	1940		25	50	75
Pinocchio Picture Book	Grosset & Dunlap	1939	9 1/2"x13", laminated cover	25	50	75
Pinocchio Plastic Bank	Play Pal Plastics	1960s	11 1/2" tall, plastic	12	23	35
Pinocchio Plastic Cup	Safetyware	1939	2 3/4" tall, plastic	35	65	100
Pinocchio Push Puppet	Kohner	1970s	5" tall	9	16	25
Pinocchio Puzzle	Jaymar	1960s	5x7", titled "Pinocchio's Expedition"	11	20	30
Pinocchio Soakie				9	16	25
Pinocchio Story Book	Whitman	1939	8 1/2"x11 1/2", 96 pages	16	30	45
Pinocchio Story Book Set	Whitman	1940	8 1/2"x11 1/2" complete set of six books 24 pages each	95	180	275
Pinocchio Tea Set	Ohio Art	1939	tin, tray, plates, saucers, serving platter, cups, bowls & smaller plates	45	85	175
Pinocchio the Acrobat Wind-up Toy	Marx	1939	2 1/2x11x17" tall, turning on a trapeze-like frame, titled "Pinocchio the Acrobat"	295	550	850
Pinocchio Tin Crayon Box	Transogram	1940s	4 1/2x5 1/2x1 1/2" deep	16	30	45
Pinocchio Vinyl Bank	Play Pal Plastics		4 1/2x5x11" tall	12	23	35

Pinocchio & Jiminy Cricket

TOY	COMPANY	YEAR	DESCRIPTION	GOOD	EX	MIB
Pinocchio Walker	Marx	1939	9" tall, animated eyes, rocking action	250	450	700
Pinocchio Water Dome	Disney	1970s	3x4 1/2x5" tall, Pinocchio is sitting and holding a plastic dome between his hands & feet, plastic small house with snowflakes creating a winter scene when shaken	25	50	75
Pinocchio Wind-Up Toy	Linemar		6" tall, wind-up, arms & legs move	80	145	350
Pinocchio/Cricket Dolls	Knickerbocker	1962	6" tall, vinyl, titled "Knixies", each	25	50	75
Pinocchio/Jiminy Push Puppet	Marx	1960s	2 1/2x5x4" tall, double puppet	25	50	75
Walt Disney's Version of Pinocchio	Random House	1939	8 1/2"x11 1/2" hardcover	20	40	60

Pluto

TOY	COMPANY	YEAR	DESCRIPTION	GOOD	EX	MIB
Miniature Pluto	Linemar		4" long, when pushed friction motor makes his tongue wag	30	55	85
Pluto	Linemar		9" friction toy, Pluto pulling red wagon	60	115	175
Pluto "Pop Up Critter" Figure	Fisher-Price	1936	wooden figure of Pluto standing on base 10 1/2" long	80	145	225
Pluto Bank	Animal Toys Plus Inc.	1970s	9" tall vinyl, Pluto standing in front of a doghouse	15	30	45
Pluto Ceramic Bank	Disney	1940s	4x4 1/2x6 1/2" tall	35	65	100
Pluto Drum Major	Marx	1950s	tin	225	355	595
Pluto Electric Alarm Clock	Allied Mfg.	1955	4x5 1/2x10" tall, eyes and hands shaped like dog bones and they glow in the dark	80	145	225
Pluto Friction Toy	Linemar		4"	35	60	95
Pluto Fun-E-Flex Figure	Fun-E-Flex	1930s	wood	35	65	100
Pluto Hand Puppet	Gund	1950s	9" tall	25	50	75
Pluto Lantern Toy	Linemar	1950s		150	275	425
Pluto Pop-A-Part Toy	Multiple Toymakers	1965	9" long, plastic	11	20	30
Pluto Purse	Gund	1940s	9x14x2"	30	55	85
Pluto Push Toy	Fisher-Price	1936	8" long, wood	70	130	200
Pluto Rolykins Figure	Marx		1x1x1 1/2" tall, with ball bearing action set	14	25	40
Pluto Rubber Figure	Seiberling	1930s	3 1/2" long	35	65	100
Pluto Rubber Figure	Seiberling	1930s	7" tall, rubber	35	65	100
Pluto Sports Car	Empire		2" long	9	16	25
Pluto the Acrobat Trapeze Toy	Linemar		10" tall, metal, celluloid, wind-up	70	125	195
Pluto Tricycle Toy	Linemar	1950s	tin	195	350	550
Pluto Watch Me Roll Over	Marx	1939	8" long, tin, Pluto turns over as his tail passes beneath him	150	275	425

Popeye

TOY	COMPANY	YEAR	DESCRIPTION	GOOD	EX	MIB
12 Giant Crayons	Dixon	1958		12	20	30
12 Giant Crayons	American Crayon	1933		15	23	35
Activity Pad	Merrigold Press	1982		6	10	15
Air Mattress	Zee Toys	1979	wet set	10	15	25
Alarm Clock	Smiths	1967	British	115	195	300
Apprentice Printer	MSS	1970s		6	10	15
Apron with Popeye's Face	Chester	1990		8	13	20
Apron w/Six Character Home of Popeye	Chester	1992		8	13	20

Top to bottom: The Flying Nun Oil Paint By Numbers Set, 1960s, Hasbro; Yogi Bear Friction Toy, 1960s; Rocky & Bullwinkle china banks, 1960s; Bugs Bunny and Tweety Chatter Chums, 1982, Mattel.

Popeye

TOY	COMPANY	YEAR	DESCRIPTION	GOOD	EX	MIB
Arcade	Fleetwood	1980		6	10	15
Ashtray (Popeye's Chicken)			aluminum	1	2	3
Baby Rattle Toy	Cribmates	1979		10	15	25
Baby Squeeze Toy (Olive)	Cribmates	1979		12	20	30
Baby Squeeze Toy (Popeye)	Cribmates	1979		12	20	30
Baby Squeeze Toy (Wimpy)	Cribmates	1979		12	20	30
Ball & Jacks Set	MSS		double	6	10	15
Ball & Paddle	BC Int'l.			6	10	15
Ball and Paddle, Biffbat-Fly Back Paddle	Unknown	1935		12	20	30
Balloon Pump (Popeye's)		1957	inflato-pump	30	50	75
Bank, 60th Year Bank	Presents	1988	metal P5988	8	13	20
Bank, Brutus Mini Bank	KFS	1979		4	7	10
Bank, Daily Dime Bank	KFS	1956		25	40	60
Bank, Daily Quarter Bank	Kalin	1950s	4 1/2" tall, metal	65	115	175
Bank, Dime Register Bank	KFS	1929	2 1/2" x 2 1/2", square, window shows total deposits in bank	50	80	125
Bank, Gum Ball Bank	Hasbro	1968		12	20	30
Bank, Gum Ball Bank	Hasbro	1981	Canada	14	23	35
Bank, Knockout Bank	Straits	1935		275	455	700
Bank, Olive Oyl Bank	Unknown	1940s	cast iron	60	100	150
Bank, Olive Oyl Mini Bank	KFS	1979		4	7	10
Bank, Popeye Bank, Cast Iron	Unknown	1940s	9" cast iron	60	100	150
Bank, Popeye Bust Bank	Play Pal	1972	shape of Popeye's head, plastic	20	35	50
Bank, Popeye Bust Bank	Renz	1970	beige	8	13	20
Bank, Popeye Bust Bank, Ceramic	Mexico	1990	ceramic	15	25	40
Bank, Popeye Bust Bank, Ceramic	Vandor	1980	ceramic	195	325	500
Bank, Popeye Collectors Bank	Chester	1980		4	7	10
Bank, Popeye in Light Blue Cap Bank			1 of 3 ceramic	6	10	15
Bank, Popeye Mini Bank	KFS	1979		4	7	10
Bank, Popeye Sitting	Leonard	1980	silver	12	20	30
Bank, Popeye Sitting on Rope	Play Pal	1970s	plastic	12	20	30
Bank, Popeye Sitting on Rope	Vandor	1980	ceramic	12	20	30
Bank, Popeye with Life Preserver	Unknown	1940s		30	50	75
Bank, Popeye with Removeable Pipe	Presents	1991	vinyl	10	15	25
Bank, Spinach Can	KFS	1975	blue spinach can with raised characters	15	25	40
Bank, Swee'Pea Figural Bank	Vandor	1980	6 1/2" tall	15	25	35
Bank, Swee'Pea Mini Bank	KFS	1979		4	7	10
Bank, with Padlock	Sanrio	1990		6	10	15
Barber Shop	Larami			6	10	15
Baseball	Ja-Ru	1983		6	10	15

Popeye

TOY	COMPANY	YEAR	DESCRIPTION	GOOD	EX	MIB
Bath Foam	Colgate-Palmolive Ltd.	1977		12	20	30
Beach Boat	H.G. Industries	1980	red or yellow, each	6	10	15
Beach Set	Peer Products	1950s		20	35	50
Bedspread	Hanna-Barbera	1978		12	20	30
Bell	Vandor	1980	Popeye on top, ceramic	10	15	25
Bendable Olive Oyl	Comic-Spain	1986	6"	6	10	15
Bendable Olive Oyl	Jesco	1988	large	6	10	15
Bendable Olive Oyl	Jesco	1988	small	4	7	10
Bendable Olive Oyl	Amscan	1980	large	6	10	15
Bendable Popeye	Jesco	1988		6	10	15
Bendable Popeye	Lakeside	1968	miniflex	10	16	25
Bendable Popeye	Lakeside	1969	superflex	12	20	30
Bendable Popeye	Jesco	1988	small	4	7	10
Bendable Popeye	Amscan	1980	small	4	7	10
Bendable Popeye	Amscan	1980	large	6	10	15
Bendable Popeye & Olive Oyl	Bronco Co.	1978	each	6	10	15
Bendable Popeye with White Pants	Comic-Spain	1986	6"	6	10	15
Bendiface	Lakeside	1967		20	35	50
Bendy Popeye	England	1950s	7" tall with yellow pants	40	65	100
Billion Bubbles	Larami	1984		6	10	15
Bingo	Nasta	1980		6	10	15
Blackboard	Bar Zim Toy Co.	1962		12	20	30
Blinky Cup	Beacon Plastics			8	13	20
Boat, Bubble Blower	Larami	1984		6	10	15
Book, 60th Anniversary Collection	Hawk Books	1990		20	35	50
Book, Adventures of Popeye	Saalfield	1934		40	65	100
Book, Big Big Thimble Theatre Book	Whitman	1935	starring Popeye	40	65	100
Book, Big Little Danger Ahoy! Book	Whitman	1969		6	10	15
Book, Big Little Deep Sea Danger	Whitman	1980		4	7	10
Book, Big Little Ghost Ship to Treasure Island	Whitman	1967		6	10	15
Book, Big Little Ghost Ship to Treasure Island	Whitman	1967	color	6	10	15
Book, Big Little in a Sock for Susan's Sake	Whitman	1940		20	35	50
Book, Big Little in Quest of Poopdeck Pappy	Whitman	1937		20	35	50
Book, Big Little Popeye & the Deep Sea Mystery	Whitman	1939		20	35	50
Book, Big Little Popeye All Picture Comic	Whitman	1942		20	35	50
Book, Big Little Popeye and the Jeep	Whitman	1937		20	35	50
Book, Big Little Popeye in Puddleburg	Saalfield	1934		20	35	50
Book, Big Little Popeye Sees the Sea	Whitman	1936	By Segar	20	35	50

Popeye

TOY	COMPANY	YEAR	DESCRIPTION	GOOD	EX	MIB
Book, Big Little Popeye the Sailor	Whitman	1942		20	35	50
Book, Big Little Popeye's Ark	Saalfield	1936		20	35	50
Book, Big Little Queen Olive Oyl	Whitman	1973		6	10	15
Book, Big Little Quest for the Rainbird	Whitman	1943		20	35	50
Book, Big Little Wimpy the Hamburger Eater	Whitman	1938		20	35	50
Book, Big Suprise Book	Wonder Books	1976		6	10	15
Book, Captain George Presents Popeye	Memory Lane	1970		8	13	20
Book, Color & Re-Color Book	Jack Built	1957	Simms/Zaboly	12	20	30
Book, Color & Recolor Popeye Goes Home to Visit	Avalon	1980		4	7	10
Book, Popeye Climbs a Mountain	Wonder Books	1983		4	7	10
Book, The Outer Space Zoo	Golden	1980		4	7	10
Book, What, no Spinach?	Golden	1918		4	7	10
Book, Wimpy Tricks Popeye	Whitman	1937		40	65	100
Book, Wimpy, What's Good to Eat?	Tuffy Books	1980		4	7	10
Bop Bag	Miner Industries	1981		10	16	25
Bowl	National Home Products	1979	plastic	4	7	10
Bowl	Deka	1971	oval, plastic	6	10	15
Bowl	Deka	1971	round plastic	6	10	15
Bowl	Vandor	1980	ceramic 1 of 3	10	16	25
Box, 60th Year	Presents	1988	metal P4248	6	10	15
Box, 60th Year	Presents	1988	metal P5987	6	10	15
Box, 60th Year	Presents	1988	metal P5987	6	10	15
Box, 60th Year Heart Shaped Candle Box	Presents	1989	metal P5979	6	10	15
Boxing Gloves	Everlast	1960s	1150	30	50	75
Brush, Clothes	KFS	1929	wooden, black or brown	20	35	50
Brutus Hi-Pop Ball	Ja-Ru	1981		6	10	15
Brutus Hookies	Tiger	1977		6	10	15
Brutus in Jeep		1950s	tiny plastic car	10	16	25
Brutus Jump-Up	Imperial			6	10	15
Bubble 'N Clean	Woolfoam	1960s		25	40	60
Bubble Blower	Transogram	1958		12	20	30
Bubble Blowing Train	Hong Kong	1970s	pink	4	7	10
Bubble Gum	Amurol Products Co.	1981	Popeye shredded	4	7	10
Bubble Pipe	Ja-Ru	1985		6	10	15
Bubble Pipe	KFS	1960s	yellow with red end	10	16	25
Bubble Set	Transogram	1936	set of two wooden pipes, tray and soap in colorful 5x7 1/2" box	20	35	50
Bubble Shooter	Ja-Ru	1985		6	10	15
Bubble Shooter	Ja-Ru	1980s	orange or yellow body	6	10	15
Bubble Tub Set	Ja-Ru	1985		6	10	15
Bubbleblaster	Carlin Playthings	1980		8	13	20

Popeye

TOY	COMPANY	YEAR	DESCRIPTION	GOOD	EX	MIB
Bubbles Blaster	Larami	1984		6	10	15
Bubbles with Dip Pow Bubbles	MSS	1986		6	10	15
Button, Bluto Getting Socked	Lisa Frank	1979	2"	2	3	5
Button, Brutus			2"	2	3	5
Button, Brutus, Gonna Eat You for Breakfast	Strand	1985	3"	2	3	5
Button, Brutus, I'm Mean	Mini Media	1983	1"	2	3	5
Button, Brutus, Movie	Factors	1980	3"	2	3	5
Button, Brutus, Ya Little Runt	Strand	1985	3"	2	3	5
Button, Cowboy Popeye & Indian Olive Oyl	Lisa Frank	1980	1"	2	3	5
Button, Famous Fried Chicken	Unknown	1980s	1"	2	3	5
Button, Famous Studio's Popeye	KFS	1950s	1"	12	20	30
Button, From Cookin' with Gags	KFS	1988	2"	2	3	5
Button, I Yam What I Yam	Lisa Frank	1979	2"	2	3	5
Button, Marine Conservation	KFS	1989	3"	2	3	5
Button, New York Evening Journal	Offset Gravure Co.	1930s	1" 558289	12	20	30
Button, No Wimps	Mini Media	1983	1"	2	3	5
Button, Olive	Factors	1980	3"	2	3	5
Button, Olive	Pep	1946		10	15	25
Button, Olive & Popeye I Love You	Unknown	1970s	1 1/2"	2	3	5
Button, Olive Kissing Popeye - I Love You	Mini Media	1983	1"	2	3	5
Button, Olive Oyl			2"	2	3	5
Button, Olive, More Than Just A Pretty Face	Mini Media	1983	1"	2	3	5
Button, Popeye			2"	2	3	5
Button, Popeye	Pep	1946		10	15	25
Button, Popeye & Olive			2"	2	3	5
Button, Popeye & Santa Cruz Boardwalk	S. Cruz	1990	2"	2	3	5
Button, Popeye for President	Lisa Frank	1980	1"	2	3	5
Button, Popeye Holds Flowers with Olive	Lisa Frank	1980	1"	2	3	5
Button, Popeye Holds Flowers with Olive	Lisa Frank	1979	2"	2	3	5
Button, Popeye Movie	Factors	1980	3"	2	3	5
Button, Popeye Pointing Finger			1"	2	3	5
Button, Popeye Pow			plastic	2	3	5
Button, Popeye the Sailor	Parisian Novelty	1929	1"	12	20	30
Button, Popeye the Sailor Man	KFS		1.5"	2	3	5
Button, Popeye White Suit & Swee'Pea	Unknown		2"	2	3	5
Button, Popeye, I'm Strong to the Finich	Strand	1985	3"	2	3	5
Button, Popeye, Shiver Me Timbers	Strand	1985	3"	2	3	5

Popeye

TOY	COMPANY	YEAR	DESCRIPTION	GOOD	EX	MIB
Button, Popeye, Shove Off	Strand	1985	3"	2	3	5
Button, Popeye/Olive Movie	Factors	1980	3"	2	3	5
Button, Popeye/Swee'Pea, No Myskery to Life	Strand	1985		2	3	5
Button, Sew-On Card	Lowe	1959		12	20	30
Button, SS Popeye	KMOX TV	1960s	1"	4	7	10
Button, Star Shaped	Lisa Frank	1979	1 1/2" Popeye inside	2	3	5
Button, Strong to the Finich	Strand	1985	3"	2	3	5
Button, Swee'Pea with Flower				2	3	5
Button, Swee'Pea with Jeep	Lisa Frank	1980	1"	2	3	5
Button, Swee'Pea with Jeep	Lisa Frank	1979	2"	2	3	5
Button, Well Blow Me Down			1" black & white	2	3	5
Button, Wimpy	Pep	1946		10	16	25
Button, Wimpy with Hamburgers	Lisa Frank	1980	1"	2	3	5
Button, Wimpy with Hamburgers	Lisa Frank	1979	2"	2	3	5
Button, Wimpy, Movie	Factors	1980	3"	2	3	5
Button, Wimpy, Must Go Home & Water the Ducks	Strand	1985	3"	2	3	5
Cabinet			mirrored	20	35	50
Cameo Doll	Cameo	1935		115	195	300
Candy & Toy Boxes	Phoenix Candy Co.	1960	3/display Popeye & his pals	15	25	40
Candy Box	Phoenix Candy Co.	1960s	green style	4	7	10
Candy Cigarettes	Primrose Confectionery - England	1959		8	13	20
Candy Rings	Alberts	1989		6	10	15
Candy Sticks	World Candies	1990	48 count	6	10	15
Candy Sticks	Hearst	1989	red box	6	10	15
Candy Treats	Alberts	1980	Bonbon	6	10	15
Candy-Candy Sticks Popeye	KFS	1989		6	10	15
Cap Gun	Ja-Ru	1981		6	10	15
Car	Vandor	1980	Popeye & Olive riding in a blue car	30	50	75
Car	Vandor	1980	Popeye & Olive riding in a pink car	30	50	75
Card Game	Ed-U-Card			8	13	20
Card Game	KFS	1934		15	25	40
Card Game	Whitman	1934		15	25	40
Cards, 60th Year - Two Decks of Cards	Presents	1988	metal box P5998	6	10	15
Cassette with Read Book Alongs A Whale of a Tale	Peter Pan	1983		6	10	15
Cereal Bowl	National Home Products	1979		6	10	15
Cereal Box	Cocoa-Puffs with Gum	1987		4	7	10

CHARACTER

695

Popeye

TOY	COMPANY	YEAR	DESCRIPTION	GOOD	EX	MIB
Chain Bubbles Maker	Larami	1984	red Popeye	6	10	15
Change Purse	Sanrio	1990		4	7	10
Charm Links with Six Character Bracelet	Peter Brams	1990	gold	10	16	25
Charm Links with Six Character Bracelet	Peter Brams	1990	silver	10	16	25
Charm, Celluloid		1930s	Popeye loop at top	12	20	30
Checker Board	Ideal	1959		12	20	30
Chimes Doll	J. Swedlin	1950s	Gundy, gray plush body-chimes	20	35	50
Chinese Jump Rope	MSS			6	10	15
Christmas Lamp Shades	General Electric	1930s	set of ten	100	165	250
Christmas Light Covers - Cheers	General Electric Textolite	1929		50	80	125
Christmas Ornament, Alice the Goon	Presents	1987	P-4237	6	10	15
Christmas Ornament, Bluto	Bully - Germany	1981		8	13	20
Christmas Ornament, Brutus	Presents	1987	P-4244	6	10	15
Christmas Ornament, Dufus	Bully - Germany	1981		8	13	20
Christmas Ornament, Olive Ice Skating	KFS	1981		6	10	15
Christmas Ornament, Olive in a Present	KFS	1981		6	10	15
Christmas Ornament, Olive Oyl	Presents	1987	P-4242	6	10	15
Christmas Ornament, Popeye	Presents	1987	P-4240	12	20	30
Christmas Ornament, Popeye in a Santa Suit	KFS	1981		6	10	15
Christmas Ornament, Season's Greetings	Presents	1989		6	10	15
Christmas Ornament, Swee'Pea	Presents	1987	P-4241	6	10	15
Christmas Ornament, Swee'Pea Next to a Candy Cane	KFS	1981		6	10	15
Christmas Ornament, Wimpy	Presents	1987	P-4243	6	10	15
Christmas Ornament, Wimpy in Wreath	KFS	1981		6	10	15
Christmas Ornaments, Set of Three	Unknown	1992	Olive Climbing, Popeye with Anchor & Swee'Pea in Tire	8	13	20
Christmas Tree Lamp Set	General Electric	1935		100	165	250
Clobber Cans	Gardner	1950s		40	65	100
Clown Stencils with Popeye Characters	Encore	1993		1	2	3
Coke-Kollect-A-Set Olive Oyl Cup	Coke	1977		4	7	10
Coke-Kollect-A-Set Popeye Cup	Coke	1977		4	7	10
Coke-Kollect-A-Set Swee'Pea Cup	Coke	1977		4	7	10
Coke-Kollect-A-Set Wimpy Cup	Coke	1977		4	7	10
Collector Pin	Chester	1990		4	7	10
Collectors Key Chain	Chester	1990		2	3	5
Color Markers	Sanrio	1990	six	4	7	10
Color TV Show	MSS	1980s		6	10	15

Popeye

TOY	COMPANY	YEAR	DESCRIPTION	GOOD	EX	MIB
Color-Me Stickers	Diamond Toymakers	1983		4	7	10
Color-Vue Pencil by Numbers	Hasbro	1979		10	16	25
Colorforms Birthday Party Set	Colorforms	1961		30	50	75
Colorforms Cartoon Kit	Colorforms	1957	big	30	50	75
Colorforms Cartoon Kit	Colorforms	1957	small	20	35	50
Colorforms Cartoon Set	KFS	1960s		20	35	50
Colorforms Movie Version	Colorforms	1980		10	15	25
Coloring Book	Lowe	1964		6	10	15
Coloring Book	Merrigold Press	1988		4	7	10
Coloring Book	Whitman - Japan	1971		6	10	15
Coloring Book with Bird	Whitman	1978		4	7	10
Coloring Book, Activity Coloring Book	Grosset & Dunlap	1978		6	10	15
Coloring Book, Comic Strip Coloring Book	Parkes Run	1980		4	7	10
Coloring Books & Crayons	Colortoons	1960	5	10	16	25
Coloring Books without Crayons	Colortoons	1960		10	16	25
Comb & Brush	KFS	1979		10	16	25
Comical Book	D. McKay Pub. Segar	1937	feature book #2	40	65	100
Comical Book	D. McKay Pub. Segar	1937	feature book #3	40	65	100
Construction Trucks	Larami	1981		6	10	15
Cook's Catch-All Wimpy	KFS	1980	ceramic	12	20	30
Cookie Jar	KFS	1980	ceramic head of Popeye	195	325	500
Cookie Jar	McCoy	1965	ceramic white suited Popeye	60	100	150
Crayon Box	American Crayon	1950s		12	20	30
Crayon Set	American Crayon	1950s		12	20	30
Crazy Foam	American Aerosol	1980		6	10	15
Cup	New Zealand	1940s	Popeye on skis ceramic	30	50	75
Cup	Deca Plastics, Inc.	1979	plastic	4	7	10
Cup	Popeye Picnic	1989	plastic	2	3	5
Decals for 1979 Anniversary Plate	Lynell	1979		4	7	10
Dice	Unknown	1990	with Popeye head	4	7	10
Dimension Weld Stickers	KFS	1979	four different	6	10	15
Dish Set	Boonton Ware	1964	three piece plastic	30	50	75
Dockside Presto Magix	APC	1980		6	10	15
Doddle Ball	Ja-Ru	1981		6	10	15
Dog Toy, Brutus	Petex	1986		6	10	15
Dog Toy, Wimpy	Petex	1986		8	13	20
Doll, Brutus	Presents	1985	large	14	23	35
Doll, Brutus	Presents	1985		15	25	35
Doll, Brutus	Presents	1985	small	8	13	20
Doll, Cloth	Quaker	1960s	approximately 12" tall	10	16	25
Doll, Jeep	Presents	1985	small	8	13	20
Doll, Jeep	Presents	1985	large	8	13	20
Doll, Olive Oyl	Uneeda	1979	removeable clothing	8	13	20
Doll, Olive Oyl	Dakin	1960s	8" tall	8	13	20

Popeye

TOY	COMPANY	YEAR	DESCRIPTION	GOOD	EX	MIB
Doll, Olive Oyl	Toy Toons	1990		4	7	10
Doll, Olive Oyl	Uneeda	1979		8	13	20
Doll, Olive Oyl	Presents	1985	large	14	23	35
Doll, Olive Oyl	Presents	1985	small	8	13	20
Doll, Olive Oyl	Dakin		hard plastic	8	13	20
Doll, Olive Oyl	Presents	1991	small molded plastic doll, # P5966	4	7	10
Doll, Olive Oyl	Rempel	1950s	small	20	35	50
Doll, Olive Oyl	Dakin	1970s	Cartoon Theatre, in box	12	20	30
Doll, Olive Oyl Christmas	Presents	1991	small	8	13	20
Doll, Olive Oyl with Swee'Pea			9" tall vinyl sqeeze doll	12	20	30
Doll, Olive Oyl, Musical	Presents	1990	small molded plastic doll, # P5948	8	13	20
Doll, Poopdeck Pappy	Presents	1985		15	25	35
Doll, Popeye	Gund	1958	20" tall	50	80	125
Doll, Popeye	Lakeside	1968	12" tall, sponge-like rubber	15	25	40
Doll, Popeye	Stack Mfg.	1936	12" tall, wood jointed with pipe	110	175	275
Doll, Popeye	Chicago Herald American	1950s		25	40	60
Doll, Popeye	Uneeda	1979	16" tall	30	50	75
Doll, Popeye	Dakin		hard plastic	12	20	30
Doll, Popeye	Toy Toons	1990		4	7	10
Doll, Popeye	Uneeda	1979		8	13	20
Doll, Popeye	Presents	1985	small	8	13	20
Doll, Popeye	Gund	1950s	20" tall	50	80	125
Doll, Popeye	Gund	1950s	laughs when cranked	50	80	125
Doll, Popeye	Presents	1985	small doll with pipe molded into hand	8	13	20
Doll, Popeye	Sears/Cameo	1957	13" in box	195	325	500
Doll, Popeye	Presents	1991	small molded plastic doll, # P5967	8	13	20
Doll, Popeye	Rempel	1950s	small	20	35	50
Doll, Popeye	Dakin	1970s	Cartoon Theatre, in box	12	20	30
Doll, Popeye		1940s	5" wooden jointed	40	65	100
Doll, Popeye	KFS	1930s	wooden jointed head, arms & legs	325	550	850
Doll, Popeye Christmas	Presents	1991	small doll	8	13	20
Doll, Popeye Collectors	Unknown		23" tall china	115	195	300
Doll, Popeye Plush	Woolikin	1950s	white plush body doll	20	35	50
Doll, Popeye with Swee'Pea			9" tall vinyl sqeeze doll	10	16	25
Doll, Popeye, Musical	Presents	1990	small molded plastic doll, # P5949	8	13	20
Doll, Popeye, Plush	Etone International	1983	8" tall	6	10	15
Doll, Popeye, Squeaker	Dakin	1974		10	15	25
Doll, Popeye, Squeaker	Chad Valley	1950s	7" tall	15	25	35
Doll, Sea Hag	Presents	1985		15	25	35
Doll, Stuffed Carnival	Play By Play	1993	2 1/2' tall	4	7	10
Doll, Stuffed Carnival	Play By Play	1993	3 1/2' tall	6	10	15
Doll, Stuffed Carnival	Play By Play	1993	5' tall	20	35	50
Doll, Stuffed Carnival	Play By Play	1993	9" tall	2	3	5
Doll, Swee'Pea	Presents	1985		10	16	25
Doll, Swee'Pea			handmade cloth	10	16	25
Doll, Swee'Pea	Presents	1991	small molded plastic doll, # P5968	8	13	20
Doll, Swee'Pea	Uneeda	1979		8	13	20
Doll, Swee'Pea	Presents	1985	large	10	16	25
Doll, Swee'Pea	Presents	1985	small	8	13	20
Doll, Swee'Pea Christmas	Presents	1991	small	8	13	20
Doll, Wimpy	Presents	1985	holding a hamburger	14	23	35
Doll, Wimpy	Presents	1985	small	8	13	20
Doll, Wimpy		1940s	5" wooden jointed	40	65	100
Doll, Wimpy, Rubber	KFS	1950s		20	35	50

Popeye

TOY	COMPANY	YEAR	DESCRIPTION	GOOD	EX	MIB
Double Action Water Gun Set	MSS			6	10	15
Drawing Desk	Carlin Playthings	1980		12	20	30
Duck Shoot	Ja-Ru	1983		6	10	15
Duck Shoot	Ja-Ru	1989		6	10	15
Ed-U-Card Games	Ed-U-Card	1960s	set of four deluxe	10	16	25
Egg Cup	Japan	1940s	Popeye sitting at table with spinach	40	65	100
Egg Cup & Mug	Magna	1989	Great Britian	30	50	75
Eras-O-Baord & Magic Screen Set	Hassenfeld Bros.	1957		30	50	75
Figure Painting Kit, Brutus	Avalon	1980		6	10	15
Figure Painting Kit, Olive Oyl	Avalon	1980		6	10	15
Figure Painting Kit, Popeye	Avalon	1980		6	10	15
Figure, Brutus	Popeye's Chicken	1991	blue plastic	1	2	3
Figure, Brutus	Presents	1990	PVC	2	3	5
Figure, Brutus	Bully	1981	pink shirt	4	7	10
Figure, Brutus in Barrel, Wooden	Japan Olympics	1962		100	165	250
Figure, Brutus with Club	Comic-Spain	1984		4	7	10
Figure, Brutus, Wooden	KFS-Hearst	1991	small	4	7	10
Figure, Jeep, Wooden	KFS-Hearst	1991	small	4	7	10
Figure, Olive Oyl	Mexico	1990	ceramic	8	13	20
Figure, Olive Oyl	Popeye's Chicken	1991	blue plastic	1	2	3
Figure, Olive Oyl	Presents	1990	PVC	2	3	5
Figure, Olive Oyl	Multiple Toymakers	1950s	2" tall	8	13	20
Figure, Olive Oyl	Bully	1981	holding flower	4	7	10
Figure, Olive Oyl	Bully	1981	with hands clasp	4	7	10
Figure, Olive Oyl	Unknown	1940s	lead	12	20	30
Figure, Olive Oyl Hanging Figure	KFS	1980	arms clamped together	6	10	15
Figure, Olive Oyl Miniature Figure			lead	10	16	25
Figure, Olive Oyl Stand-up	Presents	1990	3" tall, plastic	2	3	5
Figure, Olive Oyl with Flower	Comic-Spain	1984	PVC	4	7	10
Figure, Olive Oyl, Wooden	Japan Olympics	1962		100	165	250
Figure, Olive Oyl, Wooden	KFS-Hearst	1991	small	4	7	10
Figure, Pappy	Popeye's Chicken	1991	blue plastic	1	2	3
Figure, Popeye	Dakin		8" tall with spinach can	15	30	45
Figure, Popeye	Duncan	1970	8" tall	15	30	45
Figure, Popeye			5" tall, wood jointed held together with internal string	40	65	100
Figure, Popeye	Mexico	1990	ceramic	8	13	20
Figure, Popeye		1990	3' tall styrofoam	6	10	15
Figure, Popeye	Popeye's Chicken	1991	blue plastic	1	2	3
Figure, Popeye		1940s	celluloid with wooden feet	15	25	35
Figure, Popeye			plastic with Popeye on four wheels with telescope	15	25	35
Figure, Popeye	Japan	1950s	celluloid	20	35	50
Figure, Popeye	Unknown		red hat & pants	2	3	5

Popeye

TOY	COMPANY	YEAR	DESCRIPTION	GOOD	EX	MIB
Figure, Popeye	Presents	1990	PVC	2	3	5
Figure, Popeye	Multiple Toymakers	1950s	2" tall	8	13	20
Figure, Popeye	Comic-Spain		hand on chin	4	7	10
Figure, Popeye	Bully	1981	holding pipe	4	7	10
Figure, Popeye	Bully	1981	holding spinach	4	7	10
Figure, Popeye	Bully	1981	with hands on hip	4	7	10
Figure, Popeye	Unknown	1940s	lead	12	20	30
Figure, Popeye at Bat, Wooden	Japan Olympics	1962		195	325	500
Figure, Popeye Hanging Figure	KFS	1980	arms clamped together	6	10	15
Figure, Popeye on One Hand, Wooden	Japan Olympics	1962		100	165	250
Figure, Popeye Stand-up	Presents	1990	3" tall, plastic	2	3	5
Figure, Popeye Walking		1940s	celluloid, large feet	20	35	50
Figure, Popeye Walking	England	1960s	rubber band powered, white suit	12	20	30
Figure, Popeye with Baggage, Wooden	KFS-Hearst	1991	large	6	10	15
Figure, Popeye with Baggage, Wooden	KFS-Hearst	1991	small	4	7	10
Figure, Popeye with Flag, Wooden	KFS-Hearst	1991	large	6	10	15
Figure, Popeye with Flag, Wooden	KFS-Hearst	1991	small	4	7	10
Figure, Popeye with Spinach	Comic-Spain	1984	PVC	4	7	10
Figure, Popeye with Spinach, Wooden	KFS-Hearst	1991	small	4	7	10
Figure, Popeye, Chalk		1930s	12" tall, chalk figure with pipe, hat & ashtray base	0	0	0
Figure, Popeye, Spanish	Spain	1940s	with hanging pipe & spinach	60	100	150
Figure, Popeye, Wooden	Unknown		10" tall	12	20	30
Figure, Popeye, Wooden	Unknown		16" tall	10	16	25
Figure, Popeye, Wooden	Scirroco-KFS	1944	5"	30	50	75
Figure, Swee'Pea	Popeye's Chicken	1991	blue plastic	1	2	3
Figure, Swee'Pea	Presents	1984	PVC	2	3	5
Figure, Swee'Pea Stand-up	Presents	1990	3" tall, plastic	4	7	10
Figure, Swee'Pea with Cake	Comic-Spain	1984	PVC	4	7	10
Figure, Swee'Pea, Wooden	KFS-Hearst	1991	small	4	7	10
Figure, Wimpy	Popeye's Chicken	1991	blue plastic	1	2	3
Figure, Wimpy	Presents	1990	PVC	2	3	5
Figure, Wimpy	Buitoni	1950s	premium	39	65	100
Figure, Wimpy	Bully	1981	wearing yellow hat	4	7	10
Figure, Wimpy	Unknown	1940s	lead	12	20	30
Figure, Wimpy with Hamburger	Comic-Spain	1984		4	7	10
Figure, Wimpy, Wooden	Scirroco-KFS	1944	5"	30	50	75
Figure, Wimpy, Wooden	KFS-Hearst	1991	small	4	7	10
Figures, Chalk	Unknown	1970s	ceramic, removeable head Popeye	10	16	25
Figures, Dufus	Bully	1981	with hand on stomach	4	7	10

Bozo The Clown Push Button Marionette, knickerbocker.

Popeye

TOY	COMPANY	YEAR	DESCRIPTION	GOOD	EX	MIB
Figures, Home of Popeye, Wooden	KFS-Hearst	1991	large	6	10	15
Figures, Pewter	Ron Lee	1992	set of six different; Liberty, Men!, Ohh, Popeye!, Par Excellence, Strong To The Finish, That's My Boy, each	325	550	850
Figures, Pewter	Spoontiques	1981	set of three 1" figures; Olive w/ hands clasped, Popeye with muscles, Jeep standing, each	12	20	30
Figures, Pewter	Spoontiques	1980	two different 1" figures; Jeep lifting tail, Swee'Pea with feet showing, each	12	20	30
Figures, Pewter	Spoontiques	1980	set of 2" figures; Popeye with barbell, Popeye with parrot, Olive walking, Popeye flexing muscles, Popeye with spinach	12	20	30
Figures, Pewter	Spoontiques	1980	set of 3" figures; Brutus, Olive, Wimpy, each	12	20	30
Figures, Popeye & Brutus Jump-Up	Imperial	1979	each	6	10	15
Figurine, Brutus			1 of 5 ceramic	4	7	10
Film Card	Tru-Vue	1959	T-28	12	20	30
Film Viewer, Cartoon Story Viewer	Sawyers	1959	seven pictures	12	20	30
Film, Circus Man	Brumberger	1950s	8mm	6	10	15
Film, Circus Man	Ken Films	1960s	8mm	6	10	15
Finger Puppet Family	Denmark Plastics	1960s		30	50	75
Fishing Gane	Fleetwood	1980		6	10	15
Flashlight	KFS	1960	with Bosun's whistle-bantamlite	20	35	50
Flashlight	Larami	1983	with HD. Battery, blue	6	10	15
Flashlight	Larami	1983	with HD. Battery, red	6	10	15
Flashlight	Larami	1983	with HD. Battery, yellow	6	10	15
Flashlight	Larami	1970	yellow	6	10	15
Flashlite	Bantam-Lite	1960s	three color wrist	12	20	30
Flashlite	Bantam-Lite	1960s	super powered	12	20	30
Flea's a Crowd Record	Peter Pan	1976		6	10	15
Fleas A Crowd Record	Peter Pan	1962	78 rpm	6	10	15
Flip Book	Orbit Gum	1933	four in one	30	50	75
Fork	Unknown		blue & white Popeye	6	10	15
Fork & Spoon	Arrow Plastic Mfg.	1970s		8	13	20
Fork & Spoon	Arrow Plastic Mfg.	1972	blisterpak	10	16	25
Foto-Fun Printing Kit	Fun Bilt	1958		25	40	60
Freeze Ice Sticks	Empire Plastics	1950s		12	20	30
Freezicles	Imperial	1980		6	10	15
Funny Color Foam	Creative Aerosol	1983		6	10	15
Funny Face Maker	Jaymar	1962		15	25	40
Funny Films Viewer	Unknown	1940s		12	20	30
Funny Fire Fighters	Marx	1930s	celluloid figures, Popeye on ladder & Bluto drives fire truck--both figures wear boxing gloves from overrun of Big Fight boxing ring toy	625	1050	1600
Funtime Fiesta Coloring Book	Whitman	1979	1383-32	4	7	10
Fuzzy Face	Ja-Ru	1981		6	10	15
Game, Adventures Of Popeye	Transogram	1957		50	80	125
Game, Atari-Video Game Cartridge	Nintendo	1983		12	20	30

Popeye

TOY	COMPANY	YEAR	DESCRIPTION	GOOD	EX	MIB
Game, Ball Toss Game	KFS	1950s		30	50	75
Game, Boxing	Harmony	1981		6	10	15
Game, Comic Card Game	Milton Bradley	1978		10	16	25
Game, Jiffy Pop Fun'N Games Booklet	Spot-O-Gold Corp.	1980		4	7	10
Game, Jumbo Card Game	House of Games	1978		6	10	15
Game, Jumbo Trading Card Game	Dynamic Toy Inc.	1960s		12	20	30
Game, Magic Play Around Game Set	Amsco	1960s		30	50	75
Game, Pinball	Ja-Ru	1983		6	10	15
Game, Popeye Arcade Game	Parker Bros.	1980		8	13	20
Game, Popeye Break-A-Plate Game	Combex	1963		40	65	100
Game, Popeye Hammer Game	Holgate	1960s	Popeye stands in a wire arch with "weights" attached, when hammer strikes one of two keys by his feet he kicks the weight through hoop to other side	75	130	200
Game, Popeye Intelligence Test		1929		15	30	45
Game, Popeye Magnetic Fishing Game	Transogram	1962		20	35	50
Game, Popeye Menu Pinball Game	Durable Toy & Novelty	1935		50	80	125
Game, Popeye Mini Tennis	Nordic	1970s		8	13	20
Game, Popeye Movie Game	Milton Bradley	1980		8	13	20
Game, Popeye Nail-On Game	Colorforms	1963		25	40	60
Game, Popeye Party Game	Whitman	1937	posters, paper pipes and game box	30	50	75
Game, Popeye Pipe Toss Game	Rosebud Art Co. Inc.	1935	small version with wooden pipe	30	50	75
Game, Popeye Playing Card Game	Parkers Bros.	1983		6	10	15
Game, Popeye Playing Card Game	Whitman	1934	5x7" green box houses card game	15	25	40
Game, Popeye Playing Card Game	Whitman	1938	blue box	15	25	40
Game, Popeye Shipwreck Game	Funland	1933		40	65	100
Game, Popeye Shipwreck Game	Eison-Freeman	1933		40	65	100
Game, Popeye Skill Game			Beads, tin and glass	2455	42	65
Game, Popeye Spinach Target Game	Gardner	1960s		40	65	100
Game, Popeye the Juggler Bead Game	Bar Zim Toy Co.	1929	3 1/2" x 5", covered with glass	20	35	50
Game, Popeye The Sailor Shipwreck Game	Einson-Freeman			40	65	100
Game, Popeye's Gang Pinball Game	MSS	1970s		6	10	15
Game, Popeye's Lucky Jeep	Northwestern Products	1936		40	65	100

Popeye

TOY	COMPANY	YEAR	DESCRIPTION	GOOD	EX	MIB
Game, Popeye's Peg Board Game	Bar Zim Toy Co.	1934		60	100	150
Game, Popeye's Sliding Boards & Ladders	Warren Built-Rite	1958		12	20	30
Game, Popeye's Spinach Hunt	Whitman	1976		10	16	25
Game, Popeye's Three Game Set	Built-Rite	1956		20	35	50
Game, Popeye's Tiddly Winks	Parker Bros.	1948		20	35	50
Game, Popeye's Treasure Map Game	Whitman	1977		12	20	30
Game, Popeye's Where's Me Pipe Game				30	45	70
Game, Popeye/Olive/Wimpy Skill Games	Lido	1965		10	16	25
Game, Puzzle Game	Waddingtons House of Games	1978		10	16	25
Game, Ring the Bell with Hammer Game	Harett-Gilmer Inc.	1960s		25	40	60
Game, Ring Toss Game	Rosebud Art Co.	1933		40	65	100
Game, Ring Toss Game	Transogram	1957		30	50	75
Game, Ring Toss Stand-up Game	Transogram	1958	blisterpak	20	35	50
Game, Roly Poly & Cork Gun Game	Knickbocker	1958		50	80	125
Game, Rub 'N Win Party Game	Spot-O-Gold Corp.	1980		8	13	20
Game, Rub'N Win Party Game	Unique Ind. Inc.	1980		8	13	20
Game, Scott Fun 'N games Booklet	Spot-O-Gold Corp.	1980	set of five	10	15	25
Game, Skeet Shoot Game	Irwin	1950		50	80	125
Game, Skoozit Pick A Puzzle Game	Ideal	1966	8 1/4" round can	12	20	30
Game, Video Game & Watch	Nintendo	1983		15	25	40
Game, Water Ball Game	Nintendo	1983	one basket	6	10	15
Game, Water Ball Game	Nintendo	1983	two hooks	6	10	15
Game, Water Ball Game	Nintendo	1983	three slots	6	10	15
Giant 24 Big Picture Coloring Book	Merrigold Press	1981		4	7	10
Giant 24 Big Picture Coloring Book	Merrigold Press	1989		4	7	10
Giant Paint Book	Whitman	1937	blue	30	50	75
Giant Paint Book	Whitman	1937	red	30	50	75
Giant Sails the 7 Seas Coloring Book	Parkes Run	1980		6	10	15
Give-A-Show Projector	Kenner		112 color slides	50	80	125
Give-A-Show Slides	Chad Valley			8	13	20
Glow Whistle	Helm	1984	red, orange & green	6	10	15
Goes Fishing Book	Wonder Books	1980		4	7	10
Goes on Picnic Book	Wonder Books	1958		6	10	15
Gumball Dispenser	Superior Toys	1983	pocket pack	6	10	15

Popeye

TOY	COMPANY	YEAR	DESCRIPTION	GOOD	EX	MIB
Gumball Machine	Hasbro	1968	long neck	12	20	30
Gumball Machine	Superior Toys	1983	Popeye eating spinach	8	13	20
Gumball Machine	Superior Toys	1983	Popeye gives Olive flowers	8	13	20
Gumball Machine	Hasbro	1968	short neck	12	20	30
Gumball Machine	Hasbro	1968	short neck, clear face	12	20	30
Halloween Bucket	Renz	1979	shaped like Popeye's head, red, yellow or blue	8	13	20
Hand Puppet, Brutus	Gund	1960s		12	20	30
Hand Puppet, Olive Oyl	Gund	1960s	Olive illustrated with comic strip body	25	45	65
Hand Puppet, Popeye	Kohner	1960	4" tall	25	35	55
Hand Puppet, Popeye	Gund	1960s	Popeye's head on a cloth body with little ships on material	30	50	75
Hand Puppet, Popeye	Popeye's Chicken	1980	plastic glove like Popeye	1	2	3
Hand Puppet, Popeye	Unknown	1950s	possibly handmade Popeye	12	20	30
Hand Puppet, Popeye	Gund	1950s	fuzzy plush Popeye	15	25	40
Hand Puppet, Popeye	Unknown		three finger	12	20	30
Hand Puppet, Popeye	Mexico			12	20	30
Hand Puppet, Sea Hag	Presents	1987		8	13	20
Hand Puppet, Swee'Pea	Gund	1960s	bonnet on head with cloth body decorated with baby lambs	15	25	45
Hand Puppet, Swee'Pea	Gund	1950s		15	25	40
Hand Puppet, Swee'Pea	Presents	1987		8	13	20
Hand Puppet, Wimpy	Gund	1950s	fabric hand cover, vinyl squeaker head & voice	20	35	55
Hand Puppet, Wimpy	Gund	1960s	Wimpy's head on a cloth body with squares on material	15	25	35
Hand Puppet, Wimpy	Presents	1987		8	13	20
Hand Puppet, Wimpy	Gund	1950s	brown cap	15	25	35
Hardback Book	Random House	1980	based on movie	4	7	10
Harmonica	Larami	1973		6	10	15
Hits & Missiles Record	Americom	1965	45 rpm - 33 rpm	6	10	15
Holster Set	Halco	1960s	#2	30	50	75
Holster Set	Halco	1960s		30	50	75
Home Movies - Sockoi	Cinelab	1940s	8mm	6	10	15
Home of Popeye Corn Cob Pipe	Chester	1991		2	3	5
Hometown Pride Cup	Buena Vista Bank	1992		2	3	5
Horseshoe Magnets	Larami	1984		6	10	15
House that Popeye Built Book	Wonder Books	1960		6	10	15
House that Popeye Built Book	Wonder Books	1976		4	7	10
Hunting Knife	Larami	1973		6	10	15
I.D. Set	Gordy Int'l.	1982		6	10	15
Indian Fighter Film	Atlas Films		8mm	6	10	15
Jack Knife	Imperial	1940s	green Popeye on pearl handle	40	65	100
Jack-in-the-box	Mattel	1957		50	80	125
Jack-in-the-box	Mattel	1961		40	65	100
Jack-in-the-box	Nasta	1979		12	20	30
Jack-in-the-box	Nasta	1983		12	20	30
Jeep Lucky Spinner	KFS	1936		40	65	100
Jeep Wooden Desk Sculpture	Unknown	1989		6	10	15
Kaleidoscope	Larami	1979		6	10	15
Kazoo & Harmonica	Larami	1979		6	10	15
Kazoo & Harmonica	Larami	1981		6	10	15

Popeye

TOY	COMPANY	YEAR	DESCRIPTION	GOOD	EX	MIB
Kazoo Pipe	Northwestern Products	1934		20	35	50
Kazoo Pipe	Peerless Playthings	1960s	yellow	10	15	25
Keys & Cash	Ja-Ru	1987	Popeye & Son TV Show	4	7	10
King of the Jungle Film	Atlas Films	1960s	8mm	6	10	15
Kite	Sky-Way Prod.	1980	inflatable	6	10	15
Kite	Sky-Way Prod.	1980	regular	6	10	15
Knapsack	Fabil	1979		10	15	25
Kooky Straw	Imperial	1980		6	10	15
Lamp Mini Hurrican 60th Year	Presents	1989	P5981-1993	6	10	15
Lamp, Boat with Popeye Light Bulb	Unknown	1940s		310	525	800
Lantern	Linemar	1950s	7 1/2" tall, battery operated light in his belly	150	250	400
Liferaft	KFS	1979	large blue	10	16	25
Lite-Brite	Hasbro	1980		10	16	25
Little Pops the Ghost Book	Random House	1981		4	7	10
Little Pops the Magic Flute Book	Random House	1981		4	7	10
Little Pops the Spinach Burgers Book	Random House	1981		4	7	10
Little Pops the Treasure Hunt Book	Random House	1981		4	7	10
Lots of Bills (play money)	Ja-Ru	1987	Popeye & Son TV Show	4	7	10
Lunchbox	Aladdin Ind. Inc.	1980	plastic	30	50	75
Lunchbox & Thermos	Aladdin Ind. Inc.	1980		12	20	30
Lunchbox & Thermos	Thermos	1964		12	20	30
Lunchbox & Thermos	Thermo-Serv.	1987	Popeye & Son TV Show, red	10	15	25
Lunchbox & Thermos	Thermo-Serv.	1987	Popeye & Son TV Show, yellow	10	15	25
Lunchbox & Thermos	Countermates	1987	red no raised front, Popeye & Sons TV Show	10	15	25
Lunchbox & Thermos	Aladdin Ind. Inc.	1979	plastic	10	15	25
Mae West & Popeye Book			eight pages	6	10	15
Magic Eyes Film Card	Tru-Vue	1962	set of three	12	20	30
Magic Glow Putty	FC Famous Toys			6	10	15
Magic Slate	Lowe	1959		8	13	20
Magic Slate Paper Saver	Western Publishing	1969		6	10	15
Magic Slate Paper Saver	Whitman	1981		6	10	15
Make-A-Picture Premium	Quaker	1934		15	25	35
Marble	Unknown	1940s	1" blue/white with black & white Popeye	6	10	15
Marble	Unknown	1940s	1" blue/white with red Popeye	6	10	15
Marble Set	Imperial	1980		6	10	15
Marble Set #116	Akro Agate Co.	1935		195	325	500
Marble Shooter	Unknown	1940s	milk glass container	12	20	30
Marionette, Popeye, Wooden	Create-Japan			75	125	200
Matchbox Car, Brutus in Steamroller	Lesney Corp.	1980	Matchbox	8	13	20

Popeye

TOY	COMPANY	YEAR	DESCRIPTION	GOOD	EX	MIB
Metal Tapping Set	Carlton Dank	1950s		30	50	75
Metal Target Set	Ja-Ru	1983		6	10	15
Metal Whistle	Ja-Ru	1981		6	10	15
Micro-Movie "Popeye-Ali Baba"	Fascinations	1990		6	10	15
Mini Write On Wipe Off Board	Freelance	1980		8	13	20
Mini-Lunchbox	Sanrio	1990		6	10	15
Miniature Train Set	Larami	1980		6	10	15
Mirror	Creative Acc.	1985	Olive's head	12	20	30
Mirror	Creative Acc.	1985	Popeye holding Swee'Pea full length	25	40	60
Mirror	Creative Acc.	1985	Popeye's head	12	20	30
Mirror	Creative Acc.	1985	Popeye full length squeezing toothpaste	25	40	60
Mirror	Freelance	1978	Popeye lifting weights, large	8	13	20
Mirror	Freelance	1979	Olive with mirror, small	6	10	15
Mirror Rattle	Cribmates	1979		8	13	20
Mix or Match Storybook	Random House	1981		8	13	20
Model Kit	Tokyo Plamo	1964	Popeye #808	40	65	100
Model Kit	Carto	1970s	Popeye & Olive Oyl Moldes Para Escultura	40	65	100
Modeling Clay	American Crayon	1936		30	50	75
Mold-Metal-Chocolate-Popeye	Unknown	1940s		60	100	150
Mold-Plastic-Chocolate-Popeye	Turmic Plastics	1991		6	10	15
Mondo Popeye Book	Bobby London-St. Martin Press	1986		6	10	15
Motor Friend	Nasta	1076		12	20	30
Movie Film		1935	16mm, Fancy Skaters	6	10	15
Muscle Builder Bluto	Carlin Playthings	1980		6	10	15
Muscle Builder Popeye	Carlin Playthings	1980		6	10	15
Music Box	Vandor	1980	Wimpy on top of hamburger, ceramic	20	35	50
Music Box	Vandor	1980	Revolving Olive with Popeye Dancing, ceramic	20	35	50
Music Box	Vandor	1980	Revolving Popeye spanks Swee'Pea, ceramic	20	35	50
Music Box, Brutus Dancing	KFS	1980		8	13	20
Music Box, Brutus, 60th Year	Presents	1989	#P5984	8	13	20
Music Box, Olive Dancing	KFS	1980		8	13	20
Music Box, Olive Oyl, 60th Year	Presents	1989	#P5983	8	13	20
Music Box, Popeye Dancing	KFS	1980		8	13	20
Music Box, Popeye, 60th Year	Presents	1989	#P5992	8	13	20
Music Box, Popeye, Ceramic	KFS	1970s	Popeye waving	30	50	75
Music Box, Swee'Pea Dancing	KFS	1980		8	13	20
Music Box, Swee'Pea, 60th Year	Presents	1989	#P5986	8	13	20

CHARACTER

Top to bottom: Dick Tracy Braces, De Luxe Products; Mickey Mouse Mousegetar Jr., 1960s, Mattel; Popeye soakie bottle; Little Audrey's Dress Designer Kit, 1962, Saalfield.

Popeye

TOY	COMPANY	YEAR	DESCRIPTION	GOOD	EX	MIB
Music Box, Wimpy, 60th Year	Presents	1989	#P5985	8	13	20
Music Lovers Film	Atlas Films	1960s	8mm	6	10	15
Musical Chime Rattle	Cribmates	1979		8	13	20
Musical Hairbrush, Olive Oyl	Cribmates	1979		8	13	20
Musical Hairbrush, Popeye	Cribmates	1979		8	13	20
Musical Mug	KFS	1982	ceramic	12	20	30
Musical Popeye At The Wheel	Woolnough	1950s		150	250	400
My Popeye Coloring Kit	American Crayon	1957		25	40	60
Night Light, Popeye	Arrow Plastic Mfg.			8	13	20
Night Light, Swee'Pea	Presents	1978	bone china	10	15	25
Numbered Pencil Coloring Set	Hasbro	1960s		10	15	25
Official Popeye Pipe "It Lites, It Toots"		1958	5" stem with 2" bowl, battery operated	30	50	75
Old Time Wild West Train	Larami	1984		6	10	15
Olive & Swee'Pea Hot Water Bottle	Duarry	1970		60	100	150
Olive & Swee'Pea Thermometer	KFS	1981		8	13	20
Olive & Swee'Pea Wash Up Book	Tuffy Books	1980		6	10	15
Olive Charm			silver with dangly parts	8	13	20
Olive Hi-Pop Ball	Ja-Ru	1981		6	10	15
Olive Mug	Vandor	1980	ceramic	6	10	15
Olive on Troubled Waters Record	Peter Pan	1976	45 rpm	6	10	15
Olive Oyl			ceramic 1 of 5	4	7	10
Olive Oyl		1940s	5" wooden jointed	40	65	100
Olive Oyl Bike Bobbers	KFS	1960s		8	13	20
Olive Oyl Charm	Unknown		solid gold	60	80	150
Olive Oyl Figural Toy	KFS	1940	8" tall	55	95	145
Olive Oyl Figurine	Ben Cooper	1974	rubber	8	13	20
Olive Oyl Halloween Costume	Collegeville	1950s		20	35	50
Olive Oyl Hookies	Tiger	1977		6	10	15
Olive Oyl in a Sports Car	Lesney Corp.	1980	Matchbox	8	13	20
Olive Oyl in Airplane	Corgi			8	13	20
Olive Oyl Jump-Up	Imperial			6	10	15
Olive Oyl Marionette	Gund	1950s	11 1/2" tall	30	45	70
Olive Oyl on a Stick Squeaky Toy	Cribmates	1979		8	13	20
Olive Oyl Painting Kit	Avalon	1980		6	10	15
Olive Oyl Picture	KFS-Sears		silver foil	10	15	25
Olive Oyl Soothier Teether	Cribmates	1979		8	13	20
Olive Oyl Squeaky Foam toy	Cribmates	1979		8	13	20
Olive Oyl Squeeze Toy	Rempel	1950s	vinyl	20	35	50
Olive Oyl Statue	Cristallerie Antonio		Italy, crystal	10	15	25
Olive Oyl Swim Ring	Wet Set-Zee Toys	1979		8	13	20
Olive Oyl Tiles	Italy	1970s	3 x 5 with stand	20	35	50
Olive Oyl Toboggan	KFS	1979		6	10	15

Popeye

TOY	COMPANY	YEAR	DESCRIPTION	GOOD	EX	MIB
Olive Oyl TV "Cartoon Theater"		1976		6	10	15
Olive Oyl Walking Ring	KFS			8	13	20
Olive Oyl Wall Plaque			ceramic	4	7	10
Olive Oyl with Rolling Pin Statue	Chester	1990	10"	8	13	20
Olive Sportscar		1950s	tiny plastic car	10	15	25
Olive Squeaky Teether Toy	Cribmates	1979		8	13	20
Olive Stuffed Foam Toy	Cribmates	1979		8	13	20
Olive TV Comics Halloween Costume	Collegeville			20	35	50
Olive/Swee'Pea Snow Globe	Presents	1989	#1 Mom P-5970	4	7	10
Olive/Swee'Pea Snow Globe	Presents	1989	Mothers are Special P-5970	4	7	10
Olive/Swee'Pea Snow Globe	Presents	1989	To Mom With Love P-5970	4	7	10
Olive/Swee'Pea Snow Globe	Presents	1989	World's Best Mom P-5970	4	7	10
Olive/Swee'Pea Telephone Shoulder Rest	Comvu Corp.	1982		8	13	20
Original Radio Broadcasts Record	Golden Age	1977	33 rpm	6	10	15
Oversized Color Storybook	England	1936		25	40	60
Paddle Wagon, Jr.	Corgi	1970s		40	65	100
Paint 'N Puff Set	Art Award Co.	1979	two different versions available	10	16	25
Paint Book	Mcloughlin Bros.	1932	blue	25	40	60
Paint with Water Book	Whitman	1981		4	7	10
Painting & Crayon Book	England	1960		10	16	25
Painting Kit, Brutus	Avalon	1980		6	10	15
Paper Party Blowouts	Gala-James River Corp.	1988		4	7	10
Paper Weight, Del Monte		1937	"Popeye Eats Del Monte Spinach"	30	50	75
Pencil Case		1930s		10	15	25
Pencil Case	Eagle	1936	beige case #9027	25	40	60
Pencil Case	KFS	1950s	blue case	25	40	60
Pencil Case	Hassenfeld Bros.	1950s	red case	25	40	60
Pencil Case, 5 In 1	Sanrio	1990		8	13	20
Pencil Sharpener	KFS	1929	orange celluloid	25	40	60
Pez Dispenser, Popeye	Pez	1950s		15	25	40
Pick-Up Sticks	Lido	1957		20	35	50
Picture Disc Record	Record of America	1948	78 rpm	8	13	20
Picture Disk Record	Peter Pan	1982	33 rpm	6	10	15
Pig for a Friend Mug	Vandor	1980	ceramic	6	10	15
Pin, Back To School Days With Popeye	JC Penney	1935		12	20	30
Pin, Brutus	Lisa Frank	1979	ceramic	2	3	5
Pin, Chester Jaycees	Chester	1970s		6	10	15
Pin, Gold Employee	Popeye's Chicken	1992		2	3	5
Pin, Japanese Popeye Club		1950s	red and white pin	15	25	40
Pin, New York Evening Journal	NYEJ	1940s	#806402	12	20	30
Pin, Olive Hugs Popeye				2	3	5

Popeye

TOY	COMPANY	YEAR	DESCRIPTION	GOOD	EX	MIB
Pin, Olive in Green Blouse w/flowers	KFS	1988		2	3	5
Pin, Olive Kissing Popeye				2	3	5
Pin, Olive Oyl, Ceramic	Lisa Frank	1979		2	3	5
Pin, Olive Walking				2	3	5
Pin, Olive with Hands Folded				2	3	5
Pin, Paramount Pictures Palace Theatre Popeye Club	Paramount	1935		12	20	30
Pin, Popeye and Olive			hands folded	2	3	5
Pin, Popeye as Cowboy				2	3	5
Pin, Popeye The Sailor	Paramount	1935		12	20	30
Pin, Popeye Walking				2	3	5
Pin, Popeye with Legs Apart				2	3	5
Pin, Popeye's Head, Plastic				2	3	5
Pin, Popeye, Ceramic	Lisa Frank	1979		2	3	5
Pin, Popeye, Fashion	KFS	1980s		2	3	5
Pin, Santa Cruz Beach Boardwalk	KFS	1984		2	3	5
Pin, Star With Olive Oyl	Lisa Frank			2	3	5
Pin, Swee'Pea Holding Flower				2	3	5
Pin, Swee'Pea Pointing				2	3	5
Pin, Swee'Pea with Hands Folded				2	3	5
Pin, Swee'Pea, Plastic				2	3	5
Pin, Wimpy				2	3	5
Pipe	Harmony	1980		6	10	15
Pipe	Micro-Lite-KFS	1958		10	15	25
Pipe	KFS	1970s	plastic Kazoo, red & blue	6	10	15
Pipe	MSS	1970s	plastic, white	8	13	20
Pirate Island Presto Magix	American Pub. Corp.	1980		6	10	15
Pistol	Delcast		Super mini cap with 24 caps No. 807-BB	6	10	15
Plane & Parachute	Fleetwood	1980		6	10	15
Play Money	The Toy House	1970		8	13	20
Play Money	Newspapers	1930s	color bucks-framed	15	25	40
Pocket Pin Ball	Nintendo-Ja-Ru	1983	cups	6	10	15
Pocket Pin Ball	Nintendo-Ja-Ru	1983	holes	6	10	15
Pollution Solution Record	Peter Pan		45 rpm	6	10	15
Pool Table	Larami	1984		6	10	15
Pop Maker & Son	Ja-Ru	1987		6	10	15
Pop Pistol	Larami	1984		6	10	15
Pop-Up Popeye Book	Random House	1981		6	10	15
Popcorn	Purity Mills	1949	white hulless, large can	20	35	50
Popeye "How to Draw Cartoons" Book	Joe Musial-D. McKay Co.	1939		20	35	50
Popeye & Brutus Bookends	Vandor	1980	cera,oc	30	50	75
Popeye & Brutus Punch Me Bop Bag	Dartmore	1960s		15	25	40

Popeye

TOY	COMPANY	YEAR	DESCRIPTION	GOOD	EX	MIB
Popeye & Cast Cigar Box				10	15	25
Popeye & Friends Record	Merry Records	1981	33 rpm	6	10	15
Popeye & Olive Dish	New Zealand	1940s	ceramic	50	80	125
Popeye & Olive Oyl Figural Music Box	Schmid Bros.		8/14"	50	80	125
Popeye & Olive Oyl Flicker Ring			blue	6	10	15
Popeye & Olive Sand Set	Peer Products	1950s	bucket, shovel	25	40	60
Popeye & Olive Suspenders	KFS	1979	blue	6	10	15
Popeye & Olive Toy Watch	Unknown	1990	yellow with flicker	6	10	15
Popeye & Oscar Flicker Ring			blue	6	10	15
Popeye & Shark Swim Ring	Laurel Star - Japan	1960s		15	25	40
Popeye & Son TV Show Stamp Set	Ja-Ru	1987		4	7	10
Popeye & Son TV Show Tool Set	Ja-Ru	1987		4	7	10
Popeye & Son TV Show Whistle & Lite	Ja-Ru	1981		6	10	15
Popeye & Spinach Walker	Marx	1960s		12	20	30
Popeye & Swee'Pea Coloring Book	Whitman	1970	1056-31	6	10	15
Popeye & Swee'Pea Flicker Badge				6	10	15
Popeye & Swee'Pea Flicker Ring			blue	6	10	15
Popeye & the American Dream Book	American Life Books	1983		8	13	20
Popeye & the Time Machine Book	Quaker	1990	mini comic	2	3	5
Popeye & Wimpy Flicker Badge				6	10	15
Popeye & Wimpy Flicker Ring			blue	6	10	15
Popeye and his Jungle Pet Book	Whitman	1937		25	45	70
Popeye and the Haunted House Book	Weekly Reader Books	1980		4	7	10
Popeye and the Jeep Book	Feature Books	1982	number 3	6	10	15
Popeye and the Pet Book	Peter Haddock	1987	3 of 4	4	7	10
Popeye Annual Book	Brown Watson	1972	England	10	16	25
Popeye Annual Book	Brown Watson	1973	England	10	16	25
Popeye Annual Book	Brown Watson	1975	England	10	16	25
Popeye Annual Book	Brown Watson	1976	England	10	16	25
Popeye Annual Book	Purnell	1961	England	12	20	30
Popeye Annual Book	Purnell	1962	England	12	20	30
Popeye Annual Book	Purnell	1964	England	12	20	30
Popeye Annual Book	Stafford Pemberton	1981	England	8	13	20

Popeye

TOY	COMPANY	YEAR	DESCRIPTION	GOOD	EX	MIB
Popeye Annual Book	Stafford Pemberton	1982	England	8	13	20
Popeye Annual Book	World Distributors	1969	England	10	16	25
Popeye Annual Book "Popeye's Adventure"	Purnell	1958	England	12	20	30
Popeye Annual Book "Unicycle"	Purnell	1964	England	12	20	30
Popeye at Steering Wheel Stick Pin				4	7	10
Popeye Bike Bobbers	KFS	1960s		8	13	20
Popeye Blinky Cup	Unknown	1950s		8	13	20
Popeye Blow Me Down Airport	Marx	1935		325	550	850
Popeye Book	Wonder Books	1976		4	7	10
Popeye Book	Whitman	1937	13 x 9 1/2"	25	40	60
Popeye Bubble Liquid	M. Shimmel Sons Inc.	1970s	shaped like Popeye with a necktie similar to a sailors knot	4	7	10
Popeye Bubble Set	Transogram	1936	5" x 7 1/2", two wooden pipes, tin soap tray & a piece of soap for bubbles	35	60	90
Popeye Calls on Olive Oyl Book	Whitman	1937	8 1/2 x 9 1/2	35	55	85
Popeye Candy & Prise	Phoenix Candy Co.	1979	24 count	6	10	15
Popeye Candy Beads	Candy Novelty Works Ltd.	1980	24 count	6	10	15
Popeye Candy Pops	Candy Novelty Works Ltd.	1980		6	10	15
Popeye Carnival	Toymaster	1965	unassembled	60	100	150
Popeye Charm			silver with dangly parts	8	13	20
Popeye Charm	Unknown		solid gold	75	125	200
Popeye Chewable Vitamins	J.B. Williams Co. Inc.	1982		8	13	20
Popeye Color & Recolor Book	Jack Built	1957	color, wipe & color again	15	25	35
Popeye Coloring Book	Golden	1986		4	7	10
Popeye Coloring Book	Whitman	1983	1055-33	4	7	10
Popeye Coloring Book	Whitman	1981	1830-33	4	7	10
Popeye Coloring Book	Whitman	1971	1651-2	6	10	15
Popeye Costume & Mask	Collegeville	1950s		20	35	50
Popeye Costume & Mask	Collegeville	1960s		10	15	25
Popeye Costume & Mask	Collegeville	1980s		4	7	10
Popeye Crazy Colorfoam	American Aerosol	1980	squeeze can & white foam comes out of mouth	4	7	10
Popeye Diary	KFS	1989	Santa Cruz White with lock	6	10	15
Popeye Dog Toy	Petex	1986		6	10	15
Popeye Ear Rings	Nitrate Occurances	1988	35mm film call	4	7	10
Popeye Earring Holder		1970s	5 1/2" tall, metal	10	16	25
Popeye Express, with airplane	Marx	1936		850	1550	2400
Popeye Figural Head Pipe	Edmonton Pipe Co.	1970		15	25	40
Popeye Figural Toothbrush Holder	Vandor		5" tall	10	15	25

Popeye

TOY	COMPANY	YEAR	DESCRIPTION	GOOD	EX	MIB
Popeye Figurine	Ben Cooper	1974	rubber	8	13	20
Popeye Figurine	Combex	1960s	rubber with a can of spinach	15	25	35
Popeye Finger Rings	Post Toasties	1949		8	13	20
Popeye Flicker Badge			eating spinach	6	10	15
Popeye Flickers	Sonwell	1960s		6	10	15
Popeye Flip Book	Orbit Gum	1933		25	40	60
Popeye Floating Boat Tub Toy	Stahlwood	1960s		10	16	25
Popeye Flyer	Marx	1936	Popeye and Olive in plane, tin litho tower, wind-up	425	825	1250
Popeye Flyer	Marx	1936	Wimpy and Swee'Pea litho on tower	600	900	1600
Popeye French Record	Polygram	1981	45 rpm	6	10	15
Popeye Galley Steward	KFS	1980	ceramic	12	20	30
Popeye Getar	Mattel	1960s	14" long, shaped like Popeye's face, use crank and guitar plays "I'm Popeye the Sailor Man" or remove pipe and strum	30	50	75
Popeye Glows Putty	Larami	1984		6	10	15
Popeye Goes Fishing Book	Peter Haddock	1987	4 of 4	4	7	10
Popeye Goes Gardening Book	Peter Haddock	1987	1 of 4	4	7	10
Popeye Goes Swimming Colorforms	Colorforms	1963		20	35	50
Popeye Goes to School Television	Zaboly	1950s		12	20	30
Popeye Grow Your Own Spinach	Chestnut Hill	1976		6	10	15
Popeye Gumball Machine	Hasbro	1968	6" tall in shape of Popeye's head	10	16	25
Popeye Halloween Costume	Ben Cooper	1989		6	10	15
Popeye Halloween Costume	KFS	1986		6	10	15
Popeye Halloween Costume	Ben Cooper	1984		8	13	20
Popeye Halloween Costume	Collegeville	1950s		20	35	50
Popeye Halloween Mask	Ben Cooper	1986		4	7	10
Popeye Hat & Pipe	Empire Plastics	1950s		25	40	60
Popeye Head Pipe		1970s	ivory	40	65	100
Popeye Hi-Pop Ball	Ja-Ru	1981		6	10	15
Popeye Holds Globe Between Legs (Snow Globe)	KFS	1960s		25	40	60
Popeye Hookies	Tiger	1977		6	10	15
Popeye Hot Water Bottle	Duarry	1970		60	100	150
Popeye Huge Shoes	Eitel Plastics-Nurnberg		inflatable	10	16	25
Popeye in a Spinach Truck	Lesney Corp.	1980	matchbox	10	16	25
Popeye in Boat	Corgi			8	13	20
Popeye in Ships Wheel Watch	KFS-Japan	1990		40	65	100
Popeye in the Movies Record with Book	Peter Pan		33 rpm	6	10	15
Popeye in the Music Box	Mattel	1957	metal, crank handle with plastic Popeye pop-up in spinach can	60	100	150

Popeye

TOY	COMPANY	YEAR	DESCRIPTION	GOOD	EX	MIB
Popeye Jump-Up	Imperial			10	15	25
Popeye Lamp	Alan Jay	1959	Popeye with legs folded holding spinach	25	40	60
Popeye Lamp	Unknown	1940s	telescope with Popeye at base	50	80	125
Popeye Lamp	Idealite	1935	Popeye base & boat scene	310	525	800
Popeye Lantern	Linemar	1960s	7 1/2" tall, metal, battery operated	100	165	250
Popeye Launches His New Song Hits Record	Peter Pan	1958	45 rpm	6	10	15
Popeye Learn & Play Activity Book	Allen Canning Co.	1985		6	10	15
Popeye Magic Play Around	Amsco	1950s	Popeye characters with magnetic bases that slide across playset	40	65	95
Popeye Marionette	Gund	1950s	11 1/2" tall	30	45	70
Popeye Mechanical Pencil	Eagle	1929	10 1/2" long illustrated pencil with box	30	50	75
Popeye Meets Bigfoot Book	Quaker	1990	mini comic	2	3	5
Popeye Meets his Rival Book	Whitman	1937	8 1/2 x 11 1/2	25	40	60
Popeye Mini Winder	Durham Ind.	1980		6	10	15
Popeye Music Box		1980	plays "I'm Popeye the Sailor Man" with figure dancing the jig	20	35	50
Popeye on a Stick Squeaky Toy	Cribmates	1979		8	13	20
Popeye on Parade/Strike Me Pink Record	Cricket	1950s	45 rpm	8	13	20
Popeye on Rocket Coloring Book	Whitman - France	1980		8	13	20
Popeye on Safari Book	Quaker	1990	mini comic	2	3	5
Popeye One Man Band	Larami	1980s		6	10	15
Popeye Paddle Ball	Larami	1984	with color photo of Popeye	6	10	15
Popeye Paint and Crayon Set	Milton Bradley	1934		30	50	75
Popeye Paint and Crayon Set	Hasbro	1960s		12	20	30
Popeye Paint By Numbers	Hasbro	1960s		12	20	30
Popeye Paint By Numbers	Hasbro	1981		8	13	20
Popeye Paint Coloring Book	Whitman	1951		20	35	50
Popeye Painting Kit	Avalon	1980		6	10	15
Popeye Paints	American Crayon	1933	6", tin	25	45	65
Popeye Pencil By Numbers	Hasbro	1979		10	15	25
Popeye Picture	KFS-Sears		silver foil	10	15	25
Popeye Pipe	Unknown	1940s	red wooden	15	25	35
Popeye Pistol	Marx	1935		60	100	150
Popeye Playset	Cribmates	1979	Popeye vinyl squeak toy, Olive Oyl and Swee'Peasqueak toy, mirror, rattle and pillow	15	30	45
Popeye Popcorn Tin	Purity Mills	1949	square can	20	35	50
Popeye Presto Paints	Kenner	1961		15	25	40
Popeye Punchout Playbook	Whitman	1961		12	20	30
Popeye Puppet Show Book	Pleasure Books	1936		25	40	60
Popeye Record	Peter Pan	1960	45 rpm	6	10	15
Popeye Rubber Kick Ball				8	13	20

Popeye

TOY	COMPANY	YEAR	DESCRIPTION	GOOD	EX	MIB
Popeye Sailboat	KFS	1976		8	13	20
Popeye Service Station	Larami	1979		6	10	15
Popeye Shadow Boxer	Chein	1935		310	525	800
Popeye Shaving Mug	Schmid Bros.	1950s	no music, ceramic	12	20	30
Popeye Soakie	Colgate-Palmolive Ltd.	1960s		12	20	30
Popeye Soakie	KFS	1987	British	12	20	30
Popeye Song Folio Book	Famous Music Corp.	1936		25	40	60
Popeye Soothier Teether	Cribmates	1979		8	13	20
Popeye Speed Boat	Harmony	1981		10	16	25
Popeye Spinach Pop-up Toy	Mattel	1957	4 1/2" tall, steel spinach can with plastic figure	45	70	110
Popeye Spoon with Matching Fork				10	16	25
Popeye Sportscar		1950s	tiny plastic car	10	16	25
Popeye Sportscar	Linemar	1950s		225	375	600
Popeye Squeaky Foam toy	Cribmates	1979		8	13	20
Popeye Squeaky Teether Toy	Cribmates	1979		8	13	20
Popeye Squeeze Toy	Rempel	1950s	8" tall, vinyl	20	35	50
Popeye Stamp - Printers		1991	copper	4	7	10
Popeye Statue	Cristallerie Antionio Imperatore		Italy - crystal	10	16	25
Popeye Statue		1940s	Popeye's arm resting on knee	30	50	75
Popeye Stay in Shape Book	Tuffy Books	1980		6	10	15
Popeye Stuffed Foam Toy	Cribmates	1979		8	13	20
Popeye Supergyro	Larami	1980s		6	10	15
Popeye Surprise Present Book	Peter Haddock	1987		4	7	10
Popeye the Ladies Man Record			33 rpm	8	13	20
Popeye the Movie Book	Avon Printing	1980		6	10	15
Popeye the Movie Soundtrack Record	Paramount	1980		6	10	15
Popeye the Sailor Cup & Saucer	Japan	1930s		50	80	125
Popeye the Sailor Man & His Friends Record	Golden	1960s	33 rpm	6	10	15
Popeye the Sailor Man Record	Diplomat Records	1960	33 rpm	6	10	15
Popeye the Sailor Man Record	Peter Pan	1976	33 rpm	6	10	15
Popeye the Sailor Man Record	Golden	1959	45 rpm	6	10	15
Popeye the Sailor Thimble	Unknown	1990		4	7	10
Popeye the Weatherman Colorforms	Colorforms	1959		25	40	60
Popeye Tiles	Italy	1970s	3 x 5 with stand	20	35	50
Popeye Toboggan	KFS	1979		6	10	15
Popeye Toothbrush Holder	Vandor	1980	ceramic	10	16	25
Popeye Toothbrush Set	Nasta	1980s	Popeye toothbrush dispenser in a boat holds two toothbrushes	6	10	15

Popeye

TOY	COMPANY	YEAR	DESCRIPTION	GOOD	EX	MIB
Popeye Train Set	Larami	1973		6	10	15
Popeye Transit Company Moving Van	Linemar	1950	tin	395	650	1000
Popeye Tricky Trapeze	Kohner	1970		12	20	30
Popeye Tricky Walker	Jaymar	1960s	plastic	10	16	25
Popeye Tricycling	Linemar	1950s		310	525	800
Popeye Tugboat	Ideal	1961	inflatable	20	35	50
Popeye TV "Cartoon Theater"		1976		15	30	45
Popeye TV Cartoon Kit	Colorforms	1966		25	40	60
Popeye TV Comics Halloween Costume	Collegeville			20	35	50
Popeye vs. Bluto the Bad Book	Quaker	1990	mini comic	2	3	5
Popeye Wall Hanging	Amscan	1979	42 in. jointed	8	13	20
Popeye Water Colors	American Crayon	1933		15	25	35
Popeye Water Pistol			plastic	6	10	15
Popeye Water Sprinkler	KFS	1960s		15	25	40
Popeye Water Sprinkler	KFS	1960s	with rubbber head	15	25	40
Popeye with his Friends Book	Whitman	1937		30	50	75
Popeye with Liberty Bell - Let Freedom Ring Stick		1975		8	13	20
Popeye with Sailor Hat Belt Buckle	Pyramid Belt	1973		12	20	30
Popeye with Spinach Belt Buckle	Lee	1980		8	13	20
Popeye with Spinach Hypo Writing Tablet	KFS	1929		20	35	50
Popeye with Spinach Statue	Chester	1990	10", ceramic	6	10	15
Popeye Wrist Watch	New Haven	1938		150	260	400
Popeye's Bingo	Bar Zim Toy Co.	1929		25	40	60
Popeye's Favorite Sea Shanties Record	RCA Camden	1960		6	10	15
Popeye's Favorite Sea Songs	Peter Pan	1959	45 rpm	6	10	15
Popeye's Favorite Stories Record	RCA Camden	1960	33 rpm	6	10	15
Popeye's Offical Wallet	KFS	1959		10	15	25
Popeye's Songs About...... Record	Golden L.P.	1961	33 rpm	6	10	15
Popeye's Submarine	Larami	1973		6	10	15
Popeye, Bubble Blowing	Linemar	1950s		310	525	800
Popeye, Olive & Wimpy Decals	IGS Stores	1935		8	13	20
Popeye, Olive Oyl & Swee'Pea Lamp Shade	Unknown	1950s		40	65	100
Popeye, Olive Oyl, Swee'Pea - Jeep Lightswitch Cover				8	13	20
Popeye, Olive, Swee'Pea Pocket Knife	Smoky Mt. Knives	1992	one blade	2	3	5
Popeye/Betty Boop Film	Exclusive Films	1935	8mm film	6	10	15
Popeye/Swee'Pea Snow Globe	Presents	1989	#1 Dad P-5969	4	7	10

Popeye

TOY	COMPANY	YEAR	DESCRIPTION	GOOD	EX	MIB
Popeye/Swee'Pea Snow Globe	Presents	1989	One of a Kind P-5969	4	7	10
Popeye/Swee'Pea Snow Globe	Presents	1989	To Dad with Love P-5969	4	7	10
Popeye/Swee'Pea Snow Globe	Presents	1989	World's Best Dad P-5969	4	7	10
Popeye/Wimpy Walk-A-Way Toy	Marx	1964		25	40	60
Pops Whistle Candy	Alberts	1989		6	10	15
Popsicle Harmonica	Czech	1929		75	130	200
Presto Paints	Kenner	1961		20	35	50
Projector	Cinexin-Spain		8mm with thirteen movies	75	130	200
Pull Toy, "Boom Boom Popeye"	Fisher-Price	1937	model #491, Popeye and Swee'Pea sit on seat, Popeye strikes drum on red base as toy is pulled	135	225	350
Pull Toy, "Popeye Spinach Eater"	Fisher-Price	1939	model #488, standing Popeye drums on spinach can drum as toy is pulled	135	225	350
Pull Toy, "Popeye The Cowboy"	Fisher-Price	1937	model #705, Popeye rides a horse on a red base w/yellow wheels	135	225	350
Pull Toy, "Popeye The Sailor"	Fisher-Price	1936	model #703, Popeye sits on seat of green boat with Swee'Pea, rings beel attached to steering wheel	135	225	350
Pull Toy, Brutus & Olive Car, Popeye police Car, Wimpy Cycle, Wooden	Fisher-Price	1940s	set of three	195	325	500
Pull Toy, Choo Choo Train Pull Along	Larami			135	225	350
Pull Toy, Popeye	Fisher-Price	1935	model #700, seated Popeye on red base w/yellow wheels, rings bell as toy is pulled	135	225	350
Pull Toy, Popeye Band Wagon Xylophone, Wooden	Fisher-Price	1950s		135	225	350
Pull Toy, Popeye Sitting on Spinach Crate			heavy cardboard	135	225	350
Pull Toy, Popeye Speedboard	Unknown	1960s		235	395	600
Pull Toy, Popeye Tug & Dingy, Wooden	Fisher-Price	1950s		135	225	350
Pull Toy, Popeye Xylophone	Metal Masters		10 1/2" x 11 1/2", wood with paper litho labels & metal wheels	125	225	350
Pull Toy, Popeye, Swee'Pea & Brutus	Metal Masters	1950s	wooden	135	225	350
Punch Ball	National Latex	1970s		8	13	20
Punch Balls	KFS			6	10	15
Punch Me Punching Bag	Dartmore	1960s		10	15	25
Punch'Um Talking Rattle Toy	Sanitoy	1950s		12	20	30
Punching Bag Film	Brumberger	1950s	8mm	6	10	15
Puppet, Popeye Jumping	Kohner		pull string - top & bottom	10	15	25
Puppet, Popeye Peppy	Kohner	1970		12	20	30
Puppetforms	Colorforms	1950s		25	40	60
Push Puppet, Olive Oyl	Kohner	1960s		10	16	25
Push Puppet, Olive Oyl	Kohner Bros.		4" tall, plastic	10	16	25
Push Puppet, Popeye	Kohner	1960s	push button on the base and Popeye's arms & waist move	15	25	35
Push Puppet, Popeye	Kohner Bros.		4" tall, plastic	10	16	25

Popeye

TOY	COMPANY	YEAR	DESCRIPTION	GOOD	EX	MIB
Puxxle, Popeye 3-D	Illco	1987	11 piece	4	7	10
Puzzle		1970s	Round ball bearing-Popeye	8	13	20
Puzzle		1970s	Round ball bearing-Swee'Pea	8	13	20
Puzzle Party Book	Cinnamon House	1979		6	10	15
Puzzle, "Bath Time"	Jaymar		18x13"	4	7	10
Puzzle, "Brutus In Orbit"	Jaymar		18x13"	4	7	10
Puzzle, "Getaway"	Jaymar		18x13"	4	7	10
Puzzle, "Happy Birthday"	Jaymar		18x13"	4	7	10
Puzzle, "Knuckle Sandwich"	Jaymar		18x13"	4	7	10
Puzzle, "Swee'Pea Power"	Jaymar		18x13"	4	7	10
Puzzle, "The Picnic"	Jaymar		18x13"	4	7	10
Puzzle, "Up In Arms"	Jaymar		18x13"	4	7	10
Puzzle, Aboard Ship	Jaymar		inlaid, Brutus with club	4	7	10
Puzzle, Brutus & Pirates Coming for Treasure	Jaymar		inlaid, 10 x 13	4	7	10
Puzzle, Brutus Gets Socked at Picnic	Jaymar		inlaid, 10 x 13	4	7	10
Puzzle, Christmas Scene	Jaymar	1991	inlaid, 12 pieces	4	7	10
Puzzle, Fishing for Hamburgers at Picnic	Jaymar		inlaid, 10 x 13	4	7	10
Puzzle, Fishing Off Dock	Jaymar		inlaid, 10 x 13, Brutus with bomb	4	7	10
Puzzle, Floor	Jaymar		22 x 17	8	13	20
Puzzle, Floor	Waddington's House of Games	1978	27 x 18	8	13	20
Puzzle, Flowers & Candy For Olive	Jaymar		inlaid, 10 x 13	4	7	10
Puzzle, Jigsaw	American Pub. Corp.	1976	5 1/2" round can	10	16	25
Puzzle, Kiddie	Jaymar		flowers 13 x 10	4	7	10
Puzzle, Kiddie	Jaymar		picnic 13 x 10	4	7	10
Puzzle, Kiddie	Jaymar		Something Fishy 13 x 10	4	7	10
Puzzle, Magnetic	Larami	1973		6	10	15
Puzzle, Olive Oyl	Jaymar		extra thick inlaid	4	7	10
Puzzle, Olive Oyl 3-D	Illco	1987	11 piece	6	10	15
Puzzle, Playland	Jaymar	1991		4	7	10
Puzzle, Pocket, Birthday Cake & Ice Cream	Ja-Ru	1989		6	10	15
Puzzle, Pocket, Boating & Dancing	Ja-Ru	1989		6	10	15
Puzzle, Popeye	Jaymar	1945	22" x 13 1/2" with frame	15	25	35
Puzzle, Popeye	Jaymar		extra thick inlaid	4	7	10
Puzzle, Popeye & Car	Tower Press		wooden	12	20	30
Puzzle, Popeye & Olive Surfing	Jaymar	1991	63 pieces, jumbo	4	7	10
Puzzle, Popeye & Swee'Pea	Jaymar		extra thick inlaid	8	13	20
Puzzle, Popeye Blowing Candles	Jaymar	1991	63 pieces, Jumbo	6	10	15
Puzzle, Popeye Gang in Circle	Jaymar		inlaid, 12 pieces	4	7	10
Puzzle, Popeye Gang Swimming	Jaymar	1991	inlaid, 12 pieces	4	7	10
Puzzle, Popeye Holding Turkey	Jaymar	1991	inlaid, 12 pieces	4	7	10

Popeye

TOY	COMPANY	YEAR	DESCRIPTION	GOOD	EX	MIB
Puzzle, Popeye in 4	Saalfield	1932		35	55	85
Puzzle, Popeye Rescues Olive	Jaymar	1991	63 pieces, Jumbo	6	10	15
Puzzle, Popeye Serenading Olive	Jaymar		inlaid	4	7	10
Puzzle, Popeye's Boat	Jaymar	1991	63 pieces, Jumbo	6	10	15
Puzzle, Popeye, Wooden	Tower Press	1962		10	16	25
Puzzle, Safari with A Tiger	Jaymar		inlaid, 10 x 13	4	7	10
Puzzle, Socking a Bad Guy	Jaymar		inlaid, 10 x 13	4	7	10
Puzzle, Tile	Roalex Co.	1960s		20	35	50
Puzzle, Tile, Popeye's Riddle	Opera Mundi	1977		10	16	25
Puzzle, What a Catch	England	1959	120 pieces	25	40	60
Puzzle, Wimpy	Jaymar		extra thick inlaid	4	7	10
Puzzles, Comic	Ja-Ru	1981		6	10	15
Puzzles, Comic	Ja-Ru	1981		6	10	15
Puzzles, Pocket, Popeye & Son TV Show	Ja-Ru	1987		6	10	15
Quaker Oatmeal Box Featuring Popeye	Quaker	1990	maple	2	3	5
Quaker Oatmeal Box Featuring Popeye	Quaker	1990	strawberry	2	3	5
Race Set	Ja-Ru	1989		6	10	15
Race to Pearl Peak Book	Golden	1982		6	10	15
Rain Boots	KFS		Spinach Power	8	13	20
Recipe-Great Spinach Debate Book	Chester	1986		4	7	10
Record	Peter Pan		33 rpm four exciting stories	6	10	15
Record	Peter Pan	1977	33 rpm four Christmas stories	6	10	15
Record	Peter Pan	1977	33 rpm four fun filled stories - 1113	6	10	15
Record	Peter Pan	1977	33 rpm four fun filled stories - 1114	6	10	15
Record Player	Emerson			30	50	75
Record Player	Emerson	1960s	Dynamite Music Machine	30	50	75
Record, A Whale of a Tale	Peter Pan	1981	45 rpm	6	10	15
Ring Toss	Fleetwood	1980		6	10	15
Road Building Set	Larami	1979		6	10	15
Roll Over Tank	Linemar	1950s		150	260	400
Roller Skating Popeye	Linemar	1950s		310	525	800
Roly Poly Popeye		1940s	with beaded arms & celluloid	75	130	200
S.S. Funboat Coloring Book	Merrigold Press	1981		4	7	10
Sailboats	Larami-KFS	1981		6	10	15
Sailor & the Spinach Stalk Coloring Book	Whitman	1982	1150-1	4	7	10
Sailor's Knife	Ja-Ru	1981		6	10	15
Saucer-A Sailor's Yarn	England	1950s		30	50	75
Screen-A Show Projector	Deny's Fisher	1973		50	80	125
Secret Message Pen	Gordy Int'l.	1981		6	10	15
Set of 10 Fun Booklets	Spot-O-Gold Corp.	1980		20	35	50
Shaving Kit	Larami	1979		6	10	15
Shaving Mug	Schmid Bros.	1950s	Olive with music, ceramic	15	30	45
Six Popeye Songs	Wonderland Records		45 rpm	6	10	15

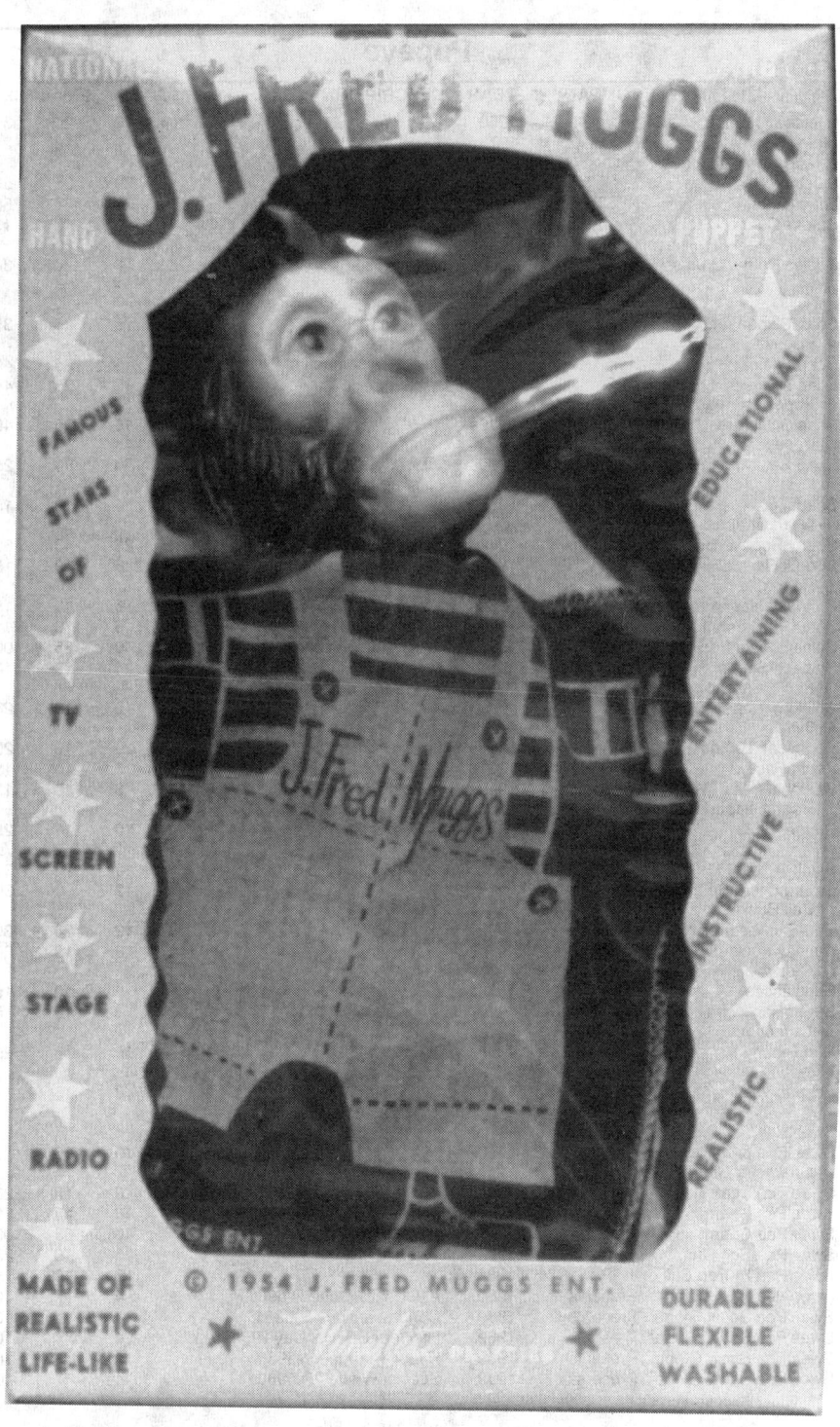

J. Fred Muggs Puppet, 1954, Vinylite Plastics.

Popeye

TOY	COMPANY	YEAR	DESCRIPTION	GOOD	EX	MIB
Sketchbook	Japan	1960		12	20	30
Slate Drawing Board with Rope Attached	KFS	1978		10	16	25
Sleeping Bag	KFS	1979		20	35	50
Sling Darts	KFS			20	35	50
Smoke Colored Storage Box	Sanrio	1990		6	10	15
Soakie Bottle, Brutus	Colgate-Palmolive Ltd.	1960s		12	20	30
Soap Dispenser, Popeye	Woolfoam	1970s		15	25	35
Soap Figure, Popeye	Kerk Guild	1940s	comes in spinach crate box	15	25	40
Soap on a Rope, Popeye's Head	KFS		white	15	25	40
Soap Sculpture, Popeye		1940s		15	25	40
Soap Set	Kerk Guild	1930s	Olive Oyl, Swee'Pea & Popeye soap figures	50	80	125
Soaps, 60th Year, Heart Shaped	Presents	1989	metal P5978	6	10	15
Soapy Popeye Boat	Kerk-Gulus	1950s		25	40	60
Song & Story Skin Diver Record	KFS	1964		6	10	15
Songs of Health Record	Golden	1960s	45 rpm	6	10	15
Songs of Safety Record	Golden	1960s	45 rpm	6	10	15
Spinach Can Pop-Up	Mattel	1957	jack in the box effect	40	65	100
Spinach Seeds Popeye's Choice F1	Erica Vale	1991	Australia	4	7	10
Spinkin - Popeye No. 3998	Kohner Bros.			10	16	25
Sportscar, Brutus		1950s	tiny plastic car	10	16	25
Squirt Face	Ja-Ru	1981		6	10	15
Stationary, 60th Year Heart Shaped Notes	Presents	1989	metal P5976	6	10	15
Statue, Bluto	Cristallerie Antonio		Italy, crystal	10	16	25
Stitch-A-Story	KFS			12	20	30
Strength thru Spinach Belt Buckle	U.S. Spinach Growers			12	20	30
Sun-Eze Pictures	Tillman Toy	1962		12	20	30
Sunday Funnies Soda Can	Flavor Valley	1970s	orange	8	13	20
Sunglasses	Larami	1980s	red, yellow or blue round	6	10	15
Super Race with Launcher	Fleetwood	1980		6	10	15
Surf Rider	Wet Set-Zee Toys	1979		10	16	25
Suspenders	Unknown	1970s	red, white & blue with plastic emblems	8	13	20
Swee'Pea			ceramic, 1 of 5	4	7	10
Swee'Pea & unknown Character Charm		1950s		10	16	25
Swee'Pea Bean Bag	Dakin	1974		10	16	25
Swee'Pea Charm	Unknown	1960s	3/4" tall blue plastic	8	13	20
Swee'Pea Charm	Unknown		solid gold	40	65	100
Swee'Pea Egg Cup	Vandor	1980	ceramic	10	16	25
Swee'Pea Hi-Pop Ball	Ja-Ru	1981		6	10	15
Swee'Pea Snow Globe	Presents	1989	A Job Well Done P-5961	4	7	10
Swee'Pea Snow Globe	Presents	1989	Congratulations P-5961	4	7	10
Swee'Pea Snow Globe	Presents	1989	For Your Special Day P-5960	4	7	10
Swee'Pea Snow Globe	Presents	1989	Happy Birthday P-5960	4	7	10
Swee'Pea Snow Globe	Presents	1989	Special Wishes P-5960	4	7	10

Popeye

TOY	COMPANY	YEAR	DESCRIPTION	GOOD	EX	MIB
Swee'Pea Snow Globe	Presents	1989	Thinking of You P-5960	4	7	10
Swee'Pea Snow Globe	Presents	1989	To Your Success P-5961	4	7	10
Swee'Pea Snow Globe	Presents	1989	You Finally Made It P-5961	4	7	10
Swee'Pea Squeak Toy	Unknown	1970s	frowning or smiling, each	12	20	30
Swee'Pea Statue	Cristallerie Antionio Imperatore		Italy - crystal	10	16	25
Swee'Pea Wall Plaque			ceramic	4	7	10
Swee'Pea's Lemonade Stand Television Film	Zaboly	1950s		10	16	25
Swim Vest	Wet Set-Zee Toys	1979		8	13	20
Swirler Flying barrel Toy	Imperial	1980		6	10	15
T.V. Magic Putty - Popeye	MSS	1970s		6	10	15
T.V. Tray	KFS	1979		8	13	20
Talking Viewmaster Reels	GAF	1962	old type	6	10	15
Tamborine	Larami	1980s		6	10	15
Tamborine	Santa Cruiz	1990		6	10	15
Target Ball	Ja-Ru	1981		6	10	15
Tatto Transfer Bubble Gum	Topps Chewing Gum-KFS	1960s		4	7	10
Telephone	Comvu Corp.	1982		15	25	35
Telescope	Larami	1973		6	10	15
Terry Toy	Cribmates	1979		8	13	20
The Hag of the 7 Seas Pop-up Book	Blue Ribbon Books	1935		30	50	75
The Outer Space Zoo Book	Golden	1980		4	7	10
The Slugger Cup	Popeye's Chicken Louisana Plastics	1989	plastic	2	3	5
Thimble Threatre Cutouts	Aldon Ind.	1950s		25	45	65
Timimight Mini Show "Popeye Ali Baba"	Fascinations	1990		6	10	15
Toy Watch with Comic Stripband	Sekonda-Japan	1987		30	50	75
Trace & Color	Fleetwood	1980		6	10	15
Training Cup	Deka	1971	with hat Popeye	8	13	20
Transistor Radio	Philgee	1960s		20	35	50
Trash Can	KFS			10	15	25
Tri-fold Wallet	Presents	1991	P-5432	6	10	15
Trick Bubble Blower	Larami	1984		6	10	15
Trinket Box	Vandor	1980	Popeye laying on top, ceramic	10	16	25
Trinket Box	Vandor	1980	Popeye's head in preserver, ceramic	10	16	25
Trumpet Bubble Blower	Larami	1984	pink or yellow pan	6	10	15
Tube-A-Loonies	Larami	1973	five small tubes, loony balloons on tubes	6	10	15
Tube-A-Loonies	Larami	1981	big green tube	6	10	15
Tube-A-Loonies	Larami	1981	big yellow tube	6	10	15
Tube-A-Loonies	Larami	1973	four small tubes on card	6	10	15
Tube-A-Loonies	Larami	1973	five small tubes on card	6	10	15
Tube-A-Loonies	Larami	1973	eight small tubes on card	6	10	15
Turn A Scope	Larami	1979		6	10	15
TV Set with Three Film Scrolls	Unknown	1957		12	20	30
Umbrella	KFS	1979	blue & white	12	20	30
View-Master	GAF	1962	three pack	8	13	20

Popeye

TOY	COMPANY	YEAR	DESCRIPTION	GOOD	EX	MIB
View-Master Reel "The Fish Story"	GAF	1959		6	10	15
View-Master Reel "The Hunting Bird"	GAF	1959		6	10	15
Wagon Works Film	Atlas Films	1960s	8mm	6	10	15
Walking Popeye	Marx		wind-up, 1930's	250	425	650
Wall Plaque			Jeep, ceramic	4	7	10
Wallet	Larami	1978		8	13	20
Wallet	Sanrio	1990		6	10	15
Watch	Bradley	1979		60	100	150
Watch	Armitron	1991		20	35	50
Watch	Bradley	1964	#308 with green case	115	195	300
What! No Spinach? Book	Golden	1981		6	10	15
Whimp Race Car		1950s	tiny plastic car #30	10	16	25
Whimp Race Car		1950s	tiny plastic car #31	10	16	25
Whistle & Lite	Ja-Ru	1981		6	10	15
Whistle Candy	Alberts	1989		6	10	15
Whistling Flashlites	Bantam-Lite	1960s	six with display	135	225	350
Whistling Wing Ding	Mego	1950s		20	35	50
White & Color Chalk Crayons	American Crayon	1953		12	20	30
White Chalk	American Crayon	1936	18 pieces	15	25	35
Wimpy			1 of 5 ceramic	4	7	10
Wimpy Finger Rings	Post Toasties	1949		8	13	20
Wimpy in Back to his First Love Book			eight pages	6	10	15
Wimpy Magnet				6	10	15
Wimpy Squeaky Foam toy	Cribmates	1979		8	13	20
Wimpy Squeeze Toy	Rempel	1950s	vinyl	20	35	50
Wimpy Statue	Cristallerie Antionio Imperatore		Italy - crystal	10	15	25
Wimpy Thermometer	KFS	1981		8	13	20
Wimpy Tricks Popeye & Roughhouse Book	Whitman	1937		35	55	85
Wimpy Tugboat	Ideal	1961	inflatable	20	35	50
Wimpy Whats Good to Eat? Book	Tuffy Books	1980		6	10	15
Wind-Up, Bluto Dippy Dumper	Marx	1935		310	525	800
Wind-Up, Bluto, Brutus Horse and Cart	Marx	1938	celluloid figure	325	500	750
Wind-Up, Brutus Mini Winder	Durham Ind.	1980		6	10	15
Wind-Up, Popeye & His Punching Bag	Chein	1930s	8" tall, tin, wind-up	875	1650	2500
Wind-Up, Popeye and Olive Oyl Jiggers	Marx		tin wind-up, on cabin roof, tin litho, 10" tall, 1936	625	1175	1800
Wind-Up, Popeye Dippy Dumper	Marx	1935	celluloid Popeye figure	310	525	800
Wind-Up, Popeye Express	Marx	1932	9" tall, wind-up with Popeye carrying a pair of parrot cages	325	550	850
Wind-Up, Popeye Handcar	Marx		Popeye and Olive Oyl rubber, metal handcar, wind-up, 1935	700	1175	2000
Wind-Up, Popeye Horse and Cart	Marx		tin windup	210	395	600
Wind-Up, Popeye in Barrel	Chein		tin wind-up	375	550	975
Wind-Up, Popeye Jigger	Marx		on cabin roof, tin wind-up, 10" tall, 1936	425	825	1250

Popeye

TOY	COMPANY	YEAR	DESCRIPTION	GOOD	EX	MIB
Wind-Up, Popeye on Tricycle	Linemar		4 1/2", tin wind-up with celluloid arms & legs, bell rings behind Popeye	75	130	200
Wind-Up, Popeye The Acrobat	Marx		tin wind-up	1900	3575	5500
Wind-Up, Popeye the Champ	Marx	1936	tin & celluloid wind-up, with Popeye & Bluto fighting until one is knocked backwards to ring a bell	1600	3000	4600
Wind-Up, Popeye the Heavy Hitter	Chein		tin wind-up, bell and mallet	2300	4300	6500
Wind-Up, Popeye the Pilot	Marx	1936	tin, #47 on side of plane, 8 1/2" wingspan	425	775	1200
Wind-Up, Popeye the Pilot	Chein	1940s	tin wind-up airplane 8" long, 8" wingspan	310	525	800
Wood Slate	Ja-Ru	1983		6	10	15
Wrist Watch, Digital	Unique Pkg.	1987	Popeye's head pop-open	12	20	30
Write on Wipe Off	Freelance	1979		6	10	15

Porky Pig

TOY	COMPANY	YEAR	DESCRIPTION	GOOD	EX	MIB
Porky & Petunia Pig Figurines	Warner Bros.	1975	4 1/2" tall	9	16	25
Porky Pig Bank		1930s	tall bisque bank of Porky, hand painted orange, blue and yellow	55	100	150
Porky Pig Doll	Mattel	1960s	17" tall, cloth doll, vinyl head	15	30	45
Porky Pig Doll	Gund	1950	14" tall	45	80	125
Porky Pig Figure	Dakin	1968	7 3/4" tall in black velvet jacket	11	20	30
Porky Pig Rotating Umbrella	Marx	1939	8" tall, tin, wind-up, umbrella with whirling action, with hat	275	525	800
Porky Pig Soakie				11	20	30
Porky Pig Umbrella		1940s	hard plastic 3" figure on end, Porky & Bugs printed in red cloth	40	75	115

Quick Draw McGraw

TOY	COMPANY	YEAR	DESCRIPTION	GOOD	EX	MIB
Auggie Doggie Soakie	Purex	1960s	10" tall, plastic	16	30	55
Augie Doggie Plush Doll	Knickerbocker	1959	10" tall, stuffed with vinyl face	16	30	45
Babalooey Bank	Knickerbocker	1960s	9" tall, vinyl bank with plastic head	16	30	35
Babalooey Plush Doll	Knickerbocker	1959	20" tall, with vinyl donkey ears & sombrero	35	65	100
Blabber Plush Toy	Knickerbocker	1959	15" tall, stuffed with vinyl face	25	50	75
Blabber Soakie	Purex	1960s	10 1/2" tall, plastic	16	30	45
Quick Draw McGraw Bank		1960	9 1/2" tall hard plastic figural bank, orange, white & blue	20	35	55
Quick Draw McGraw Playbook	Whitman	1960		15	25	40
Quick Draw McGraw Plush Toy	Knickerbocker	1959	16" tall, stuffed, vinyl face in cowboy hat	45	80	125
Quick Draw Mold & Model Cast Set		1960		25	50	75
Scooper Plush Toy	Knickerbocker	1959	20" tall, vinyl face	30	55	85

Raggedy Ann & Andy

TOY	COMPANY	YEAR	DESCRIPTION	GOOD	EX	MIB
Raggedy Andy Figure		1970s	rubber/wire figure 4" tall	5	10	15
Raggedy Ann Coloring Book		1968		5	10	15

Road Runner & Wile E. Coyote

TOY	COMPANY	YEAR	DESCRIPTION	GOOD	EX	MIB
Road Runner "Cartoon Theater" Figure	Dakin		plastic	9	16	65
Road Runner "Goofy Gram"	Dakin	1971		14	25	40
Road Runner & Coyote Lamp		1977	12 1/2" tall, with figures standing on base	20	35	125
Road Runner Bank			standing on base	5	10	15
Road Runner Figure	Dakin	1968	8 3/4" tall	12	23	35
Road Runner Hand Puppet	(Japan)	1970s	10" tall, vinyl head with plastic hand cover	4	7	10
Road Runner Stuffed Doll	Mighty Star	1971	13" tall	9	16	25
Wile E. Coyote "Cartoon Theater"	Dakin	1976		9	16	25
Wile E. Coyote "Goofy Gram"	Dakin	1971	fused bomb in right hand	14	25	40
Wile E. Coyote & the Road Runner Figurine	Royal Crown	1979	7" tall	11	20	30
Wile E. Coyote Doll	Mighty Star	1971	18" tall, stuffed	9	16	25
Wile E. Coyote Doll	Dakin	1970	on explosive box	15	30	45
Wile E. Coyote Figure	Dakin	1968	10" tall	12	23	35
Wile E. Coyote Hand Puppet	(Japan)	1970s	10" vinyl head, plastic hand cover	4	7	10
Wile E. Coyote Miniature Figure	Dakin		5 1/2" tall	9	16	25
Wile E. Coyote Nitelite	Applause	1980s		12	23	35

Rocky

TOY	COMPANY	YEAR	DESCRIPTION	GOOD	EX	MIB
Apollo Creed Doll	Phoenix Toys	1983	8" tall	3	6	9
Clubber Lang Doll	Phoenix Toys	1983	8" tall	5	10	15
Rocky Doll	Phoenix Toys	1983	8" tall	5	10	15

Rocky & Bullwinkle

TOY	COMPANY	YEAR	DESCRIPTION	GOOD	EX	MIB
Bullwinkle & Rocky Clock Bank	Larami	1969	4 1/2" tall, plastic	25	50	75
Bullwinkle & Rocky Waste Can		1961	11" tall, metal with Jay Ward cast pictured	35	65	125
Bullwinkle Bank		1960s	6" tall, glazed china	60	115	350
Bullwinkle Colorforms Cartoon Kit		1962		35	65	125
Bullwinkle Double Boomerangs	Larami	1969	two boomerangs on 11x5" card illustrated with Rocky, Bullwinkle and a chicken	12	23	35
Bullwinkle Figure "Cartoon Theater"	Dakin	1976	7 1/2" tall, plastic	25	50	75
Bullwinkle for President Bumper Sticker		1972		9	16	25
Bullwinkle Jewelry Hanger		1960s	5" tall with a suction cup on back of head	12	23	35
Bullwinkle Magic Slate		1963		15	30	45
Bullwinkle Magnetic Travel Game	Larami	1971		15	30	45
Bullwinkle Melmac Dinner Set	Boonton Molding	1960s	plate and cup illustrated with pictures of Bullwinkle and the Cheerios Kid	25	45	65
Bullwinkle Moose Figure	Dakin			25	45	65
Bullwinkle Spell & Count Board		1969		9	16	25
Bullwinkle Stamp Set	Larami	1970		11	20	30

Rocky & Bullwinkle

TOY	COMPANY	YEAR	DESCRIPTION	GOOD	EX	MIB
Bullwinkle Stickers		1984	3-2 1/2" Bullwinkle, Sherman and Peabody, Snidley Wiplash	5	10	15
Bullwinkle Talking Doll	Mattel	1970		25	50	75
Bullwinkle the Moose Paintless Paint Book	Whitman	1960		15	30	45
Bullwinkle Travel Adventure Board Game	Transogram	1960s		30	55	85
Bullwinkle's Circus Time Toy		1969	Bullwinkle on a elephant	20	35	55
Bullwinkle's Circus Time Toy		1969	Rocky on a circus horse	20	35	55
Dudley Do-Right "Cartoon Theater" Figure	Dakin	1976		15	30	45
Dudley Do-Right Flexible Figure	Wham-O	1972	5" tall	9	16	25
Mr. Peabody Bank		1960s	6" tall, glazed china	80	145	275
Mr. Peabody Flexible Figure	Wham-O	1972	4" tall	9	16	25
Natasha Figure	Wham-O	1972		7	13	20
Rocky & Bullwinkle "Presto Sparkle" Painting Set	Kenner	1962	six cartoon pictures & two comic strip panels	25	50	75
Rocky & Bullwinkle Bank		1960	5" tall, glazed china	80	145	275
Rocky & Bullwinkle Coloring Book	Watkins-Strathmore	1962		15	30	45
Rocky & Bullwinkle Movie Viewer		1960s	item #225 on card, red and white plastic viewer with three movies	23	45	65
Rocky & Bullwinkle Toothpaste & Holder		1960s	glazed china	60	114	175
Rocky & His Friends Little Golden Book	Little Golden Books	1960s	graphics of Rocky, Bullwinkle, Sherman and Peabody	9	16	25
Rocky Bank		1950s	5" tall, slot in large tail, glazed china	60	115	250
Rocky Figure "Cartoon Theater"	Dakin	1976	6 1/2" tall, plastic	18	35	50
Rocky Soakie			10 1/2" tall, plastic	12	23	35
Rocky the Flying Squirrel Coloring Book	Whitman	1960		15	30	45
Sherman Flexible Figure	Wham-O	1972	4" tall	9	16	25
Snidely Whiplash Flexible Figure	Wham-O	1972	5" tall	9	16	25

Roger Rabbit

TOY	COMPANY	YEAR	DESCRIPTION	GOOD	EX	MIB
Animates		1988	Doom, Roger, Eddie & Smart Guy, each	4	7	10
Baby Herman & Roger Rabbit Mug	Applause			4	7	10
Baby Herman Disney Ceramic Figurine	LJN	1988		7	13	20
Baby Herman Figure	LJN	1987	6" figure on card	9	16	25
Baby Herman Flexie				9	17	25
Benny the Cab	LJN			20	40	60
Benny the Cab Plush	Applause	1988	6" long	5	10	15
Book with Cassette Tape				4	7	10
Boss Weasel Animate	LJN	1988		3	5	8
Boss Weasel Flexie	LJN	1988	4" bendable	3	5	8

CHARACTER

Top to bottom: Supercar Moulding & Colour Kit, 1960s, Sculptorcraft; Radio Orphan Annie's Secret Society Manual, 1930s; Rocky & Bullwinkle Movie Viewer, 1960s.

Roger Rabbit

TOY	COMPANY	YEAR	DESCRIPTION	GOOD	EX	MIB
Deluxe Color Activity Book	Golden		#5523	4	7	10
Dip Flip Game	LJN			7	13	20
Eddie Valiant Animate	LJN	1988	6" tall poseable	3	5	8
Eddie Valiant Flexie	LJN	1988	4" bendable	3	5	8
Jessica Fashions			gold tone pendant on Jessica and Roger	11	20	30
Jessica Flexie				11	20	30
Jessica License Plate				5	10	15
Jessica Zipper Pull				5	10	15
Judge Doom Animate	LJN	1988	6" tall poseable	4	7	10
Judge Doom Flexie	LJN	1988	4" bendable	3	5	8
Paint with Water	Golden		#1702	5	10	15
Paint-A-Cel Set			Benny the Cab and Roger pictures	5	10	15
Photo Fantasy	LJN	1988	Roger, Jessica, Baby Herman and two weasels	5	10	15
Read Along Book & Tape				5	10	15
Roger Flexie				4	7	10
Roger Rabbit Animate	LJN	1988	6" poseable	4	7	10
Roger Rabbit Blow-Up Buddy			36" tall	5	10	15
Roger Rabbit Bullet Hole Wristwatch	Shiraka	1987	white case and leather band, in plastic display box	20	35	55
Roger Rabbit Figure	Applause		17" tall	5	10	15
Roger Rabbit Figure	Applause		8 1/2" tall	4	7	10
Roger Rabbit Flexie	LJN	1988	4" bendable	3	5	8
Roger Rabbit Silhouette Wrist Watch	Shiraka	1987	large gold case, black band	18	35	50
Roger Wacky Head Puppets	Applause		hand puppets	4	7	10
Roger Wind-up	Matsudaya	1988		18	35	50
Set of Four Animated Figures	LJN	1988	Roger Rabbit, Judge Doom, Eddie Valiant and Weasel	9	16	25
Smart Guy Animate	LJN	1988		3	5	8
Smart Guy Flexie	LJN	1988		3	5	8
Sticker Fun Book				4	7	10
Suction Cup Figure			cloth	5	10	15
Talking Roger in Benny the Cab			17" tall	11	20	30
Trace & Color Book	Golden		#2355	4	7	10
View-Master Gift Set				7	13	20

Rookies, The

TOY	COMPANY	YEAR	DESCRIPTION	GOOD	EX	MIB
Rookie Chris Doll	LJN	1973	8" tall	4	8	25
Rookie Mike Doll	LJN	1973	8" tall	4	8	25
Rookie Terry Doll	LJN	1973	8" tall	4	8	25
Rookie Willy Doll	LJN	1973	8" tall	4	8	25

Rootie Kazootie

TOY	COMPANY	YEAR	DESCRIPTION	GOOD	EX	MIB
Rootie Kazootie Club Button		1950s	1" tin litho	15	30	45
Rootie Kazootie Drum		1950s	8" diameter drum with Rootie on the drum head	23	45	65

Roy Rogers

TOY	COMPANY	YEAR	DESCRIPTION	GOOD	EX	MIB
Dale Evans Wrist Watch	Ingraham	1951	shows Dale inside upright horseshoe, tan background, chrome case, black leather band	60	105	165
Roy Rogers & Dale Evans Western Dinner Set	Ideal	1950s	cooking and eating utensils in 14x24" box	35	65	95
Roy Rogers Nodder	(Japan)		composition Roy stands in blue shirt, white hat and pants and red bandana and boots on green base	90	165	250

Ruff & Reddy

TOY	COMPANY	YEAR	DESCRIPTION	GOOD	EX	MIB
Ruff & Reddy Draw Cartoon Set Color	Wonder Art			45	80	125
Ruff & Reddy Go To A Party Tell-A-Tale Book	Whitman	1958		15	30	45
Ruff & Reddy Magic Rub Off Picture Set	Transogram	1958		45	80	125

Secret Squirrel

TOY	COMPANY	YEAR	DESCRIPTION	GOOD	EX	MIB
Secret Squirrel Bubble Club Soap Container	Purex	1960s		12	23	45
Secret Squirrel Frame Tray Puzzle		1967		14	25	40
Secret Squirrel Push Button Puppet	Kohner			14	25	40
Secret Squirrel Push Button Puppet	Kohner	1960s	plastic figure in white coat, blue hat holding binoculars on yellow base with gold label	18	35	50
Secret Squirrel Ray Gun				14	25	40

Sleeping Beauty

TOY	COMPANY	YEAR	DESCRIPTION	GOOD	EX	MIB
3 Fairy Godmother Hand Puppets		1958	10 1/2" tall, Flora, Merryweather & Fauna, each	35	65	100
King Huber/King Stefan Hand Puppets	Gund	1956	10" tall, molded rubber heads with fabric hand cover	25	50	75
Sleeping Beauty Alarm Clock	Phinney-Walker	1950s	2 1/2x4x4 1/2" tall, Sleeping Beauty surrounded by 3 birds and petting a rabbit	30	55	85
Sleeping Beauty Doll Crib Mattress		1960s	9x17" mattress with illustration of Sleeping Beauty and the Fairies	12	23	35
Sleeping Beauty Magic Paint Set	Whitman			25	45	65
Sleeping Beauty Musical Jack-In-The-Box	Enesco	1980s	Princess Aurora on illustrated wooden box, plays "Once Upon A Dream"	35	60	135
Sleeping Beauty Puzzle	Whitman	1958	11 12/x14 1/2", Sleeping Beauty with forest animals	15	30	45
Sleeping Beauty Puzzle	Whitman	1958	11 1/2x14 1/2". Sleeping Beauty with Prince Phillip and Three Good Fairies circling in the air around Sleeping Beauty	15	30	45
Sleeping Beauty Puzzle	Whitman	1958	11 1/2x14 1/2", Three Good Fairies circling around a baby in a crib	15	25	40

Sleeping Beauty

TOY	COMPANY	YEAR	DESCRIPTION	GOOD	EX	MIB
Sleeping Beauty Squeeze Toy	Dell	1959	4x4x5" tall, rubber, Sleeping Beauty nestling a rabbit in her arms, dressed in a yellow gown	25	45	65
Sleeping Beauty Sticker Fun	Whitman	1959	sticker push out activity book	12	25	35

Smokey the Bear

TOY	COMPANY	YEAR	DESCRIPTION	GOOD	EX	MIB
Smokey Bobbing Head Figure		1960s	6 1/4" tall	15	30	145
Smokey Figure Bank			6" tall, china	12	23	85
Smokey Soakie Soap Container		1960s	9" tall, plastic	5	10	25
Smokey the Bear Figure	Dakin	1971	figure on a tree stump	25	50	125
Smokey the Bear Wrist Watch	Hawthorne	1960s		35	65	150

Snow White & the Seven Dwarfs

TOY	COMPANY	YEAR	DESCRIPTION	GOOD	EX	MIB
Baby Rattle	Krueger	1938	Snow White at piano & the Dwarfs playing instruments	130	245	375
Bashful Doll	Ideal	1930s		55	100	150
Big Little Book	Whitman	1938		15	30	45
Dime Register Bank	Disney	1938	holds up to five dollars	50	95	145
Doc & Dopey Pull Toy	Fisher-Price	1937	12" long by 9" tall, with a chopping tree stump action when it moves back or forward	80	145	225
Doc Doll	Ideal	1930s		55	100	150
Doc Lamp	LaMode Studios	1938	8" tall, plaster	45	80	125
Dopey Bank	Crown Toy	1938	7 1/2" tall, wood composition	55	100	175
Dopey Dime Register Bank	Disney	1938	holds up to five dollars	45	80	175
Dopey Doll	Ideal	1930s		55	100	225
Dopey Doll	Chad Valley	1938	all cloth body	50	95	195
Dopey Doll	Krueger		14" tall	90	165	250
Dopey Doll	Knickerbocker	1938	3x6x11" tall composition	90	165	375
Dopey Lamp		1940s	9" tall, ceramic base with Dopey figure	35	65	175
Dopey Rolykin	Marx		2" tall, mounted on a weighted plastic half sphere so he can rock around	25	50	75
Dopey Soakie		1960s	10" tall	11	20	30
Dopey Storage Barrel		1960s	ceramic figure and barrel	15	30	75
Dopey Ventriloquist Doll	Ideal	1938	5x9 1/2x18" tall	115	210	475
Dopey Walker	Marx	1938	9" tall, tin, rocking walker	275	525	800
Grumpy Doll	Knickerbocker	1938	3x6x11" tall, composition	90	165	350
Happy Doll		1930s	5 1/2" tall, composition, holding a silver pick with a black handle	35	65	175
Happy the Dwarf Mechanical Toy	YS Toys (Taiwan)		battery op. Happy fries eggs, picture of Snow White on front	40	75	225
Radio	Emerson	1938	8x8" with characters on cabinet	325	625	2000
Seven Dwarfs Figure Set "Snow White"	Seiberling	1938	5 1/2" tall rubber figures	115	210	450
Seven Dwarfs Target Game "Snow White"	Chad Valley	1930s	6 1/2x11 1/2" target, spring locked gun that shoots wood pieces at target	115	210	325
Sled, Snow White and the Seven Dwarfs	S. L. Allen & Co.	1938	40" long wood slat and metal runner sled with character decals	140	260	400
Sneezy Doll	Krueger		14" tall	115	210	325

Snow White & the Seven Dwarfs

TOY	COMPANY	YEAR	DESCRIPTION	GOOD	EX	MIB
Snow White & the Seven Dwarfs Picture Puzzles	Whitman	1938	set of 2 puzzles in box	50	95	145
Snow White & the Seven Dwarfs Safety Blocks	Halsam	1938	7 1/2"x14 1/2"	35	65	100
Snow White & the Seven Dwarfs Sand Pail	Ohio Art	1938	8" tall, tin pictured with Snow White playing hide-n-seek with the Dwarfs	45	80	175
Snow White & the Seven Dwarfs Tea Set	Ohio Art	1937	tray 7 1/2" long, large plates 4" diameter, small saucers 2 1/2" diameter & cups are 2 1/2" across the top	65	120	250
Snow White Cut-out Dolls & Dresses	Whitman	1938	10x15x1 1/2"	80	145	225
Snow White Doll	Knickerbocker	1940	12" tall, composition	80	145	350
Snow White Doll	Horsman		8" tall classic doll in illustrated box	12	23	75
Snow White Doll	Knickerbocker	1939	3x7x3 1/2" tall, composition with moveable arms and legs	60	115	275
Snow White Doll Set	Deluxe	1940s	22" tall Snow White and 7" tall dwarfs	250	475	1200
Snow White Ironing Board	Wolverine		ironing board cover & iron	14	25	40
Snow White Lamp	LaMode Studios	1938	8 1/2" tall	35	65	150
Snow White Marionette Figure	Tony Sarg/ Alexander	1930s	12 1/2" tall	60	115	225
Snow White Mirror		1940s	9 1/2" long with plastic sculptured looking handle	23	45	65
Snow White Model Making Set	Sculptorcraft	1930s		60	115	175
Snow White Pencil Box	Venus Pencil Co.		3x8x1"	60	115	175
Snow White Puzzle	Jaymar	1960s	11x14"	18	35	50
Snow White Sewing Set	Ontex	1940s	10x14x1 1/4"	25	50	75
Snow White Sink	Wolverine	1960s	6 1/2x11x11" tall	15	25	40
Snow White Soakie				12	23	35
Snow White Stuffed Doll	Ideal	1938	3 1/2x6 1/2x16" tall, fabric face and arms, red/white dress with dwarf and forest animal design	245	450	700
Snow White Table Quoits	Chad Valley	1930s	9 1/2x21x1 1/4" deep	115	210	325
Snow White Tea Set	Wadeheath Co.	1930s	teapot 3" tall with Snow White & Dopey standing behind her skirt, 2" tall cups, saucers 4" in diameter and 2" tall creamer with a fawn all in white china	115	210	325
Snow White Tea Set	Marx	1960s	teapot 2x5x3 1/2" tall, 5 saucers, large plates and tea cups	35	65	100
Snow White Tin Refrigerator	Wolverine	1970s	15" tall single door unit in white and yellow decorated with picture of Snow White	14	25	85

Speedy Gonzales

TOY	COMPANY	YEAR	DESCRIPTION	GOOD	EX	MIB
Speedy Gonzales Figure	Dakin	1970	7 1/2" tall, vinyl	12	25	35
Speedy Gonzalez Doll	Dakin		5" tall, vinyl squeeze doll	9	16	25

Sports

TOY	COMPANY	YEAR	DESCRIPTION	GOOD	EX	MIB
Dorothy Hamil Figure	Ideal	1975	11 1/4" tall	25	50	75
Dr. J. (Julius Erving) Figure		1974		20	40	60

Sports

TOY	COMPANY	YEAR	DESCRIPTION	GOOD	EX	MIB
Evel Knievel Figure	Ideal		6" tall	15	30	45
Gretzky, Wayne	Mattel		12" tall	25	50	75
Muhammed Ali Wristwatch	Bradley	1980	chrome case, sweep seconds, brown leather band, face shows Ali in trunks and gloves, with signature beneath	25	50	75
O.J. Simpson Figure		1974		30	50	125

Steve Canyon

TOY	COMPANY	YEAR	DESCRIPTION	GOOD	EX	MIB
Steve Canyon's Costume	Halco	1959		25	50	150
Steve Canyon's Interceptor Station Punch Out	Golden	1950s		30	55	125
Steve Canyon's Membership Card & Badge			1/2"x4" Milton Caniff membership card for the Airagers, Morse code on back, 3" tin litho color badge with gold feathers with Steve's face centered	70	130	200

Sylvester the Cat & Tweety Bird

TOY	COMPANY	YEAR	DESCRIPTION	GOOD	EX	MIB
Sylvester & Tweety Bank		1972	vinyl	14	25	40
Sylvester & Tweety Figurines	Warner Bros.	1975	6" tall	9	16	25
Sylvester Figural Toy	Oak Rubber Co.	1950	6" tall	16	30	90
Sylvester Figure	Dakin	1969		11	20	30
Sylvester Figure "Cartoon Theater"	Dakin	1976		9	16	25
Sylvester Figure on Fish Crate	Dakin	1971		15	25	40
Sylvester Soakie				9	16	25
Tweety "Goofy Gram"	Dakin	1971	holding red heart	15	25	40
Tweety Doll	Dakin	1969	6", moveable head & feet	9	16	25
Tweety Figure	Dakin	1971	on bird cage	14	25	40
Tweety Figure "Cartoon Theater"	Dakin	1976		9	16	25
Tweety Soakie		1960s	8 1/2" tall, plastic	9	16	25

Tarzan

TOY	COMPANY	YEAR	DESCRIPTION	GOOD	EX	MIB
Kala Ape Figure	Dakin	1984	3" tall	4	7	10
Tarzan & Giant Ape Set				80	145	225
Tarzan & Jungle Cat Set				80	145	225
Young Tarzan Figure	Dakin	1984	4" tall and bendable	4	8	12

Tasmanian Devil

TOY	COMPANY	YEAR	DESCRIPTION	GOOD	EX	MIB
Tasmanian Devil Doll	Mighty Star	1971	13" tall, stuffed	9	16	25
Tasmanian Devil Figural Bank	Applause	1980s		5	10	15
Tasmanian Devil Figure	Superior	1989	7" tall, plastic figure on base	3	5	8

Terry & the Pirates

TOY	COMPANY	YEAR	DESCRIPTION	GOOD	EX	MIB
Terry & the Pirates Featured Funnies Jigsaw Puzzle		1930s	9 1/2"x14" puzzle	30	55	85

Terry & the Pirates

TOY	COMPANY	YEAR	DESCRIPTION	GOOD	EX	MIB
Terry & the Pirates Playstone Funnies Mold Set		1940s		25	50	75

Three Little Pigs

TOY	COMPANY	YEAR	DESCRIPTION	GOOD	EX	MIB
Big Bad Wolf Pocket Watch	Ingersoll	1934		175	340	525
Three Little Pigs "Tubby Time" Soakie Set	Drew Chemical Corp.	1960s	8" tall each: Three Little Pigs and the Big Bad Wolf	60	115	175
Three Little Pigs Puzzle	Jaymar	1940s	7x10x2"	25	50	75
Three Little Pigs Sand Pail	Ohio Art		3" tall	35	65	100
Three Little Pigs Sand Pail	Ohio Art	1930s	4 1/2" tall, tin, Three Little Pigs pictured in the woods	45	80	125
Three Little Pigs Wind-up Toy	Schuco	1930s	4 1/2" pigs playing fiddle, fife & drum	245	450	700
Three Little Pigs/Big Bad Wolf Bracelet		1930s	1/2" wide by 2 1/4" diameter, the wolf blowing down a house & a little pig running away	55	100	150
Who's Afraid of the Big Bad Wolf, Game of	Marks Brothers	1930s	9x20x1 1/2"	60	115	175
Who's Afraid of the Big Bad Wolf, Game of	Parker Brothers	1930s	13x16x1", Walt Disney's	60	115	175

Tom & Jerry

TOY	COMPANY	YEAR	DESCRIPTION	GOOD	EX	MIB
Jerry Figure	Marx	1973	4" tall	12	23	35
Tom & Jerry & Droopy Walking Figure Set		1975		15	30	45
Tom & Jerry Figural Bank	Gorham	1980	6" tall	18	35	50
Tom & Jerry Go Kart	Marx	1973	plastic, friction drive	25	50	75
Tom & Jerry on a Scooter	Marx	1971	plastic friction drive, figures on scooter	12	23	35
Tom & Jerry Quartz Wristwatch	Bradley	1985	oldies series, small white plastic case and band, sweep seconds, face shows Tom squirting Jerry with hose	9	16	25
Tom & Jerry Tray Puzzles	Whitman		set of four frame tray puzzles	15	25	40
Tom Figure	Marx	1973	6" tall	12	23	35

Top Cat

TOY	COMPANY	YEAR	DESCRIPTION	GOOD	EX	MIB
Top Cat Soakie		1960s	10" tall, vinyl	11	20	30
Top Cat TV-Tinykins Figure	Marx	1961	plastic	15	30	45
Top Cat View-Master			with three reels and booklet	12	23	35
Top Cat Viewmarx Micro-Viewer	Marx	1963	plastic with lens to view cartoon scenes	15	30	45

Underdog

TOY	COMPANY	YEAR	DESCRIPTION	GOOD	EX	MIB
Underdog Bank				12	23	35
Underdog Dot Funnies Kit	Whitman	1974		9	16	25
Underdog Figure "Cartoon Theater"	Dakin	1976	plastic	25	50	75
Underdog Jigsaw Puzzle	Whitman	1975	100 pieces	9	16	25
Barbarino Figure	Mattel	1976	with comb	15	30	45

Wizard of Oz

TOY	COMPANY	YEAR	DESCRIPTION	GOOD	EX	MIB
Epstein Figure	Mattel	1976	with bandanna	15	30	45
Horshack Doll	Mattel	1976	9" tall, with lunch box	11	20	30
Mr. Kotter Figure	Mattel	1976	with attache case	12	23	35
Sweathogs School Mechanical Bank	Fleetwood	1975	wind-up bank features Horshack and Barbarino snatching money	15	30	45
Washington Figure	Mattel	1976	with basketball	15	30	45

Winky Dink & You

TOY	COMPANY	YEAR	DESCRIPTION	GOOD	EX	MIB
Winky Dink & You Super Magic TV Kit	Standard Toy	1968		23	45	65
Winky Dink Little Golden Book	Golden	1956		9	16	25
Winky Dink Winko Magic Kit		1950s		15	30	45

Winnie the Pooh

TOY	COMPANY	YEAR	DESCRIPTION	GOOD	EX	MIB
Kanga and Roo Vinyl Squeek Toy	Holland Hill	1966	vinyl toy made in shape of stuffed doll	12	23	35
Winnie the Pooh Button		1960s	3 1/2" celluloid button	5	10	15
Winnie the Pooh Doll		1960s	12" tall	12	23	35
Winnie the Pooh Frame Tray Puzzle	Whitman	1964		7	13	20
Winnie the Pooh Jack-In-The-Box	Carnival Toys	1960s		15	30	45
Winnie the Pooh Lamp	Dolly Toy Co.	1964	7" tall	25	50	75
Winnie the Pooh Magic Slate	Western Publishing	1965	8 1/2x13 1/2"	15	30	45
Winnie the Pooh Musical Snow Globe			5 1/2"	15	30	45
Winnie the Pooh Radio	Thilgee Inter'l	1970s	5x6x1 1/2" tall	25	50	75
Winnie the Pooh/Christopher Robin Dolls	Horsman	1964	Winnie the Pooh 3 1/2" tall and Christopher 11" tall, set	60	115	175

Wizard of Oz

TOY	COMPANY	YEAR	DESCRIPTION	GOOD	EX	MIB
Alarm Clock	Westclock	1980s	white cased standing clock shows four main characters from film, same image was also used on pocket watch and other pieces	12	20	30
Bank, Cowardly Lion		1960s	ceramic, painted, squat figure with bright red nose on outsized head, with tag, in box	25	40	65
Bank, Cowardly Lion, Bust	Enesco	1989	ceramic, painted head with red ribbon over right ear	8	13	20
Bank, Dorothy		1960s	ceramic, painted, round faced Dorothy in blue dress with brown wicker basket, with tag, in box	25	40	65
Bank, Dorothy	Multi Toys	1989	12" tall, painted plastic, marked "MTC", Dorothy holding Toto and basket in her arms, dressed in blue jumper, white blouse, blue socks, blue hair ribbons & ruby slippers	10	15	25
Bank, Scarecrow		1960s	ceramic, painted, squat figure with outsized head, with tag, in box	25	40	65

Wizard of Oz

TOY	COMPANY	YEAR	DESCRIPTION	GOOD	EX	MIB
Bank, Scarecrow	Multi Toys	1989	12" tall, painted plastic, marked "MTC", shows Scarecrow standing by corn field with a crow on his shoulder, arms pointing in opposite directions, based on film scene when Dorothy asks directions	10	15	25
Bank, Scarecrow, Bust	Enesco	1989	ceramic, painted head with black hat with deep green band	8	13	20
Bank, Tin Man		1960s	ceramic, painted, squat silver finish figure with outsized head, with tag, in box	25	40	65
Bank, Tin Man	Multi Toys	1989	12" tall, painted plastic, marked "MTC", a smiling silver Tin Man sits on a tree stump holding his oil can	10	15	25
Bank, Tin Man, Bust	Enesco	1989	ceramic, painted head with silver finish	8	13	20
Bank, Wicked Witch	Multi Toys	1989	12" tall, painted plastic, marked "MTC", witch leans forward, holding broom in left hand, pointing forward with her right hand	10	15	25
Banner, The Wizard of Oz Ice Capades		1961	felt souvenir banner from Ice Capades' version of the Wizard of Oz	10	15	25
Banners, The Wizard of Oz Live!		1989	two souvenir banners showing characters and scenes from the 1989 arena tour production, identical art on either pink or blue felt, each	4	7	10
Bedding Set	Fieldcrest	1962	bedding set features primitive color characters against green background, green Tin Man, pink Scarecrow, gold Lion with green glasses and pink dressed Dorothy in white apron, blanket, pillow case, tow	15	25	35
Bells, Set of Twelve	Hamilton Collection	1988	porcelain, 5" high, 12 bells with figural handles showing a different character sitting on top of the bell, each	10	15	25
Belt Buckle	Indiana Metal Crafts	1978	rectangular buckle shows main characters leaving forest toward Emerald City with Toto in the lead, back of buckle describes scene	10	15	25
Belt Buckle, Brass	Award Design Medals, Inc.	1985	originally available only in Kansas, oval buckle shows four main characters walking toward viewer with supporting characters and scenes on rim, says, "Kansas, Land of Oz"	10	15	25
Book, Colouring, Wizard of Oz	Hutchinson (England)	1940	each picture in black & white opposite an identical color sample to be used as a guide with original drawings from the book	100	165	250
Book, Cut & Assemble the Emerald City of Oz	Dover Publishing	1980	by Dick Martin, cut out and assemble Emerald City structures	4	7	10
Book, Cut & Assemble the Wizard of Oz Toy Theater	Dover Publishing	1985	by Dick Martin, activity book with color characters & scenery	4	7	10

Wizard of Oz

TOY	COMPANY	YEAR	DESCRIPTION	GOOD	EX	MIB
Book, Dick Martin Cut & Make Masks	Dover	1982	contains eight cut-out color masks of characters	4	7	10
Book, Dorothy and Friends Visit Oz	Curtis Candy	1967	one of three premium coloring books inside boxes of candy	5	8	12
Book, Dorothy Meets the Wizard	Curtis Candy	1967	one of three premium coloring books inside boxes of candy	6	10	15
Book, Jack Pumpkinhead and the Sawhorse of Oz, Junior Edition	Rand McNally	1939	hardcover, also contains "Tik Tok and the Gnome King of Oz"	30	50	75
Book, Judy Garland in The Wizard of Oz	Ottenheimer	1977	Magic Punch Out See Thru Picture Storybook	8	13	20
Book, Little Dorothy & Toto of Oz	Rand McNally	1939	hardcover, also contains "The Cowardly Lion and the Hungry Tiger"	30	50	75
Book, Pictures from The Wonderful Wizard of Oz	George W. Ogilvie & Co.	1903	color lithograph by W. W. Denslow	115	195	300
Book, Play Mask	Watermill Press	1990	four heavy paper masks, instructions and script to help you put on your own play	3	5	7
Book, Return To Oz	Del-Rey Books	1985	paperback novelization of film by Joan Vinge with 8 pages of color stills	2	3	4
Book, Return to Oz Paper Doll Book	Western Publishing	1985	die cut figures of the Scarecrow, Tik-Tok, Jack Pumpkinhead and Dorothy with wardrobe changes	4	7	10
Book, Rinkitink in Oz, Junior Edition	Rand McNally	1939	hardcover	30	50	75
Book, Scarecrow & the Tin Man	G. W. Dillingham	1904	by Denslow	115	195	300
Book, Scarecrow & the Tin Man	Perks Publishing	1946	with two color, black & yellow pictures, art by Mary and Wallace Stover	15	25	35
Book, The Emerald City of Oz, Junior Edition	Rand McNally	1939	hardcover	30	50	75
Book, The Land of Oz, Junior Edition	Rand McNally	1939	hardcover by L. Frank Baum	30	50	75
Book, The Lost Princess of Oz, Junior Edition	Rand McNally	1939	hardcover	30	50	75
Book, The Patchwork Girl of Oz, Junior Edition	Rand McNally	1939	hardcover	30	50	75
Book, The Return To Oz Storybook	Western Publishing Co.	1985	large hardback	4	7	10
Book, The Road to Oz, Junior Edition	Rand McNally	1939	hardcover	30	50	75
Book, The Scarecrow & the Tin Wood Man of Oz, Junior Edition	Rand McNally	1939	hardcover, also contained "Princess Ozma of Oz"	30	50	75
Book, The Story of The Wizard of Oz	Whitman	1939	drawings by Henry E. Vallely softcover	15	30	45
Book, The Tin Woodsman and Dorothy	Curtis Candy	1967	one of three premium coloring books inside boxes of candy	5	8	12
Book, The Wizard of Oz	Bobbs-Merrill	1939	sepia-tone endpapers illustrated with movie stills, with a movie promo dust jacket	70	115	175
Book, The Wizard of Oz	Grossett & Dunlap	1939	illustrations by Oskar Lebeck	25	40	60

Wizard of Oz

TOY	COMPANY	YEAR	DESCRIPTION	GOOD	EX	MIB
Book, The Wizard of Oz	Saalfield Publishing Co.	1944	"Animations by Julian Wehr," spiral bound book, 24 pages, 6 pages of articulated illustrations	20	35	50
Book, The Wizard of Oz	Random House	1950	hard cover, front cover shows four main characters plus Toto standing on a hill as Dorothy adjusts a red ribbon in the Cowardly Lion's mane, back cover shows Dorothy and the Guardian of the Gates	6	10	15
Book, The Wizard of Oz	Wonder Books	1951	hard cover, forty two pages featuring a blonde Dorothy	6	10	15
Book, The Wizard of Oz	Grosset & Dunlap	1957	hard cover, illustrated by Maraja	17	30	45
Book, The Wizard of Oz	Grossett & Dunlap	1962	hard cover, illustrated by Claudine Nankivel	8	13	20
Book, The Wizard of Oz	Grossett & Dunlap	1963	hardcover unabridged version, part of Companion Library series	8	13	20
Book, The Wizard of Oz	Chicago's Children's Press	1969	contains biography of L. Frank Baum & information about gemstones	8	13	20
Book, The Wizard of Oz	Modern Productions	1970	reprint of 1969 Chicago's Children's Press edition with new cover	8	13	20
Book, The Wizard of Oz	Western Publishing	1975	Little Golden Book, #310-32, cover price 89 cents	2	3	4
Book, The Wizard of Oz	Western Publishing	1975	paperback, #5960, cover price 39 cents, part of Golden Shape Book series	3	5	8
Book, The Wizard of Oz	Grossett & Dunlap	1976	hardbound	10	15	25
Book, The Wizard of Oz	Hallmark	1977	king size pop up book with moveable 3-D pictures	12	20	30
Book, The Wizard of Oz	Doubleday Book Club	1978	hardcover edition with 16 pages of MGM movie stills	8	13	20
Book, The Wizard of Oz	Holt & Rinehart & Winston	1982	hardback, illustrated by Michael Hague	12	20	30
Book, The Wizard of Oz	Unicorn Books	1985	hardback, illustrated by Greg Hildebrandt	6	10	15
Book, The Wizard of Oz	Jellybean Press	1991	hardcover edition with illustrations by Charles Santore	6	10	15
Book, The Wizard of Oz Christmas Book	Gimbel's	1968	distributed by New York department stores for the 1968 holiday season	10	15	25
Book, The Wizard of Oz Movie Storybook	Western Publishing Co.	1989	hardback, illustrated with full color stills from the original movie	4	7	10
Book, The Wizard of Oz Paint Book	Whitman	1939	large line drawings of characters similar to their MGM screen counterparts	25	40	65
Book, The Wizard of Oz Paper Dolls	Whitman	1976		4	7	10
Book, The Wizard of Oz Picture Book	Whitman	1939	12 page abridgement of novel by Leason	25	40	65
Book, The Wizard of Oz Sticker Fun	Western Publishing	1988	stickers to be cut out as trading cards	2	4	6
Book, The Wizard of Oz Sticker Fun Book	Whitman	1976		3	5	7
Book, The Wizard of Oz Waddle Book	Blue Ribbon Books	1934	six die cut character punch-outs, a yellow brick road runway & assembly instructions at the end, rare, no price set	0	0	0

Wizard of Oz

TOY	COMPANY	YEAR	DESCRIPTION	GOOD	EX	MIB
Book, The Wizard of Oz, A Pop-Up Book	Price, Stern & Sloan	1976		5	8	12
Book, The Wizard of Oz, A Pop-Up Classic	Random House	1968	animated format with 3-D pop-up illustrations, available in two sizes	25	40	65
Book, The Wizard of Oz, A Windmill Pop-Up Book	Price, Stern & Sloan	1982	reprint of 1976 version with new cover	4	7	10
Book, The Wizard of Oz, Junior Deluxe Classic		1955		17	30	45
Book, The Wizard of Oz, Special Edition	Singer Sewing Machine Co.	1970	22 pages, photographs from the classic MGM movie	10	15	25
Book, The Wonderful Cut-Outs of Oz	Crown	1985	35 color standing figures to cut out	6	10	15
Book, The Wonderful Wizard of Oz	Whitman Publishing	1957	hard cover, abridged version	8	13	20
Book, The Wonderful Wizard of Oz	Whitman Publishing	1965	hard cover reprint of 1957 edition, with a new cover	6	10	15
Book, Walt Disney Pictures Presents Return To Oz	Random House	1985	hardback, part of Disney's Wonderful World of Reading series	4	7	10
Book, Wizard of Oz	Hutchinson (England)	1940	hard cover, illustrated with color stills from the MGM film	195	325	500
Book, Wizard of Oz Story Book	Hutchinson (England)	1940	paperback illustrated with Denslow line art, but with photo cover from film	90	145	225
Booklet, The Wizard of Oz	Hallmark	1966	paperback, oversized greeting card with three pop out illustrations	15	25	35
Booklet, Wonderful Recipes with Oz	Swift & Company	1960s	premium recipe book of recipes using Swift's Oz peanut spread	10	15	25
Books, Little Golden Book Series		1951	hardbacks, The Road to Oz, The Emerald City of Oz, & The Tin Woodman of Oz, each	6	10	15
Books, Return To Oz Little Golden Books	Western Publishing	1985	series of four books; Dorothy Returns to Oz, Escape from the Witch's Castle, Dorothy in the Ornament Room, Dorothy Saves the Emerald City, each	2	3	4
Books, The Wonderful Land of Oz Library	Rand McNally	1913	Baum's set of six Little Wizard stories, reprinted two to a volume & abridged Junior Editions of six original Oz books, in illustrated box	185	310	475
Button, "Don't Miss The Wizard of Oz at Loew's Palace"	Loew's Theater Chain	1939	yellow, red and black button illustrated with art of the Scarecrow	55	95	145
Button, "I Saw The Wizard of Oz World Arena Tour"		1989	3 1/4" blue with white and yellow print, no image	4	7	10
Button, "Read the New Baum Book The Scarecrow of Oz"	Parisian Novelty Co.	1915	celluloid button	75	130	200
Button, Aw, shucks, folks, get a glass, get a glass!	Whataburger	1989	one of six employee buttons promoting premium glasses, shows the Cowardly Lion	3	5	8
Button, Be smart..start collecting our glasses today!	Whataburger	1989	one of six employee buttons promoting premium glasses, shows the Scarecrow	3	5	8
Button, Better get a glass my pretty!	Whataburger	1989	one of six employee buttons promoting premium glasses, shows the Wicked Witch	3	5	8

Mr. Potato Head presents Frenchy Fry, 1966, Hasbro.

Wizard of Oz

TOY	COMPANY	YEAR	DESCRIPTION	GOOD	EX	MIB
Button, Get a glass before you go home!	Whataburger	1989	one of six employee buttons promoting premium glasses, shows Dorothy	3	5	8
Button, Have a Heart, start collecting our glasses today!	Whataburger	1989	one of six employee buttons promoting premiun glasses, shows the Tin Man	3	5	8
Button, MGM's The Wizard of Oz	Economy Novelty & Printing Co.	1939	promo buttons worn by theatre ushers, blue button with white reverse lettering with matching ribbons trailing	60	100	150
Button, Start at the beginning. Collect your glasses!	Whataburger	1989	one of six employee buttons promoting premium glasses, shows Glinda	3	5	8
Button, The Wizard of Oz	Economy Novelty & Printing Co.	1939	1 1/4" pin back button featuring likeness of one of film's stars, along with a number, used in theatres for contests	50	80	125
Button, Wake Up to Good Mornings	Pfizer Laboratories	1989	pin back button promoted arthritis awareness through the 1989 campaign, white background with gold photo image of a Tin Man character	4	7	10
Button, What did the Woggle Bug Say?		1904	celluloid pin back newspaper premium button, has two variations, on white or yellow backgrounds	40	65	100
Buttons, "Off to See the Wizard"	Samson Products	1967	7/8" each, set of 12 pin back buttons made for bubblegum machines, featuring Dorothy, Tin Woodsman, Scarecrow, Witch, the Cowardly Lion & the Wizard, each	4	7	10
Buttons, Land of Oz	Land of Oz Theme Park	1970s	pin-back buttons, pictures Dorothy with Toto with "Home" underneath, Scarecrow with "Brain" and the Cowardly Lion with "Courage", probably more than three were made	4	7	10
Calendar, 50th Anniversary Commemorative Edition	Cleo	1989	12 month spiral bound calendar using stills from movie in plot sequence	8	13	20
Calendar, The Wizard of Oz	Landmark	1984	12 month 1985 calendar with scenes from MGM movie, plus a laminated color print	10	15	25
Carpet Sweeper, "The New Wizard"	Bissell	1939	black body with yellow character decal of four main characters	100	165	250
Carpet Sweeper, Toy Sized	Bissell	1939	child sized carpet sweeper	100	165	250
Cast 'n Paint Set, The Wizard of Oz		1975	makes six - 6" tall molded figures	15	25	35
Chalk Board, The Wizard of Oz	Roth American	1975	wood frame, steel stand with chalk, chalk holder and eraser	16	27	40
Charm Bracelet	Vogue	1950s	featuring Dorothy, the Scarecrow, the Tin Man & the Wizard	16	27	40
Charm Bracelet, Return to Oz	General Electric	1964	featuring the Wizard, Dandy Lion, Rusty the Tin Man, Dorothy, Toto, Socrates the Scarecrow, and the Wicked Witch	16	27	40
Chess Set, 50th Anniversary Edition	Saratoga Mint	1989	32 pewter pieces, hand painted, hardwood playing board	150	260	400
Christmas Ornaments	Bradford Novelty Co.	1977	each 4 1/2" tall, in bag with header card, Dorothy, Scarecrow, Tin Man, Cowardly Lion, each	4	7	10

Wizard of Oz

TOY	COMPANY	YEAR	DESCRIPTION	GOOD	EX	MIB
Christmas Ornaments, Set of Six	Presents	1989	cloth and vinyl bust ornaments, Dorothy, Scarecrow, Tin Man, the Cowardly Lion, Glinda & the Wicked Witch, each	6	10	15
Clothes Hangers	Stemplet & Sons	1939	painted wooden hangers with wire hooks, with character decals, Dorothy, Tin Man, Winged Monkey, Toto, Glinda and probably other characters as well, each	35	60	95
Colorforms, Off to See the Wizard	Colorforms	1967	includes instruction booklet, thirty eight colored plastic character pieces & accessories with laminated display board	17	30	45
Coloring Book, Judy Garland as Dorothy in the Wizard of Oz	Aero Educational Products, Inc.	1977		6	10	15
Coloring Book, MGM Wizard of Oz Giant Story Coloring Book	Parkes Run Publishing Co.	1976	17x22", 32 pages based on Marvel/DC comic book	8	13	20
Coloring Book, Tales of the Wizard of Oz Coloring Book	Whitman	1962	coloring book with art from animated TV show	10	15	25
Coloring Book, The Wizard of Oz	Swift & Co.	1955	ten pages, advertised Oz peanut butter, cover shows four main characters with a blonde Dorothy and a white Toto	10	15	25
Coloring Book, The Wizard of Oz	Lowe	1960s	paperback, #1352, cover price 10 cents	8	13	20
Coloring Book, The Wizard of Oz	Playmore	1970s	# A400-10, cover price 79 cents	3	5	7
Coloring Book, The Wizard of Oz	Unicorn Books	1988	coloring book, puzzle and a thirty five panel fold out frieze illustrated by Greg Hildebrandt	6	10	15
Coloring Book, The Wizard of Oz		1989	souvenir sold at performances of The Wizard of Oz Live!	4	7	10
Coloring Book, The Wizard of Oz Color and Activity Book	Whitman	1976		4	7	10
Coloring Book, The Wizard of Oz Color by Number Book	Karas Publishing	1962	#A-116, cover price 29 cents, part of Twinkle Books series	8	13	20
Coloring Book, The Wizard of Oz Color/ Activity Book	Western Publishing Co.	1988	coloring book with puzzles & activities, based on MGM movie	2	4	6
Coloring Book, The Wizard of Oz Coloring Book	Saalfield	1957	front cover shows Scarecrow leading a blonde Dorothy in gleaming silver slippers up a winding Yellow Brick Road to the Emerald City	10	15	25
Coloring Book, The Wizard of Oz Giant Story Coloring Book	Colorful Fundraising, Inc.	1980	contained a T-shirt iron on transfer	4	7	10
Coloring Book, The Wonderful Wizard of Oz	Dover	1974	features Denslow illustrations and abridged text	4	7	10
Coloring Book, Wizard of Oz Giant Story Coloring Book	Stoneway Ltd.	1981	reprint of the MGM Wizard of Oz Giant Story Coloring Book with new cover	5	8	12
Comic Book, Tales of the Wizard of Oz	Dell	1962	comic book, #1306, cover price 15 cents	8	13	20
Comic Book, The Wizard of Oz, Classic Illustrated Jr.	Dell	1957	comic book, # 535, original price 15 cents	10	15	25

CHARACTER

Wizard of Oz

TOY	COMPANY	YEAR	DESCRIPTION	GOOD	EX	MIB
Comic Book, The Wizard of Oz, Dell Junior Treasury	Dell	1956	comic book, edition #5, original price 15 cents	10	15	25
Comic Book: Marvel Treasury of Oz, The Marvelous Land of Oz	Marvel & DC Comics	1975	Special Collector's Issue	8	13	20
Comic Book: MGM's Marvelous Wizard of Oz	Marvel & DC Comics	1975		8	13	20
Comic Books, The Oz/Wonderland Wars	DC Comics	1986	set of 3 comic books detailing a war between Oz characters and Alice in Wonderland characters, each	6	10	15
Cookie Jar	Clay Art	1990	white with painted framed relief figures of four main characters	20	35	50
Costume Patterns	McCall's	1985	film based patterns, licensed by MGM UA Entertainment, featuring Dorothy, Tin Man, the Cowardly Lion & the Scarecrow, sold in two different pattern numbers 2202 & 2203, each	5	8	12
Costume, Deluxe Dorothy	Collegeville	1989	adult size: includes red metallic glitter chips for shoes	25	40	65
Costume, Deluxe, Cowardly Lion	Collegeville	1989	child's size: includes costume & mask	17	30	45
Costume, Deluxe, Cowardly Lion	Collegeville	1989	adult size: includes costume & mask	25	40	65
Costume, Deluxe, Dorothy	Collegeville	1989	child's size: includes red metallic glitter chips for shoes	17	30	45
Costume, Deluxe, Scarecrow	Collegeville	1989	child's size: includes straw, costume & mask	17	30	45
Costume, Deluxe, Scarecrow	Collegeville	1989	adult size: includes straw, costume & mask	25	40	65
Costume, Deluxe, Tin Man	Collegeville	1989	child's size: includes costume & mask	17	30	45
Costume, Deluxe, Tin Man	Collegeville	1989	adult size: includes costume & mask	25	40	65
Costume, Dorothy	Ben Cooper	1975	costume includes mask	12	20	30
Costume, Dorothy	Collegeville	1989	child's size, plastic face mask & vinyl body suit	6	10	15
Costume, Rusty the Tin Man	Halco	1961	masquerade costume & mask set	17	30	45
Costume, Scarecrow, "Off to See the Wizard"	Ben Cooper	1967		15	25	35
Costume, The Cowardly Lion	Ben Cooper	1975	costume includes mask	12	20	30
Costume, The Cowardly Lion	Collegeville	1989	child's size, plastic face mask & vinyl body suit	6	10	15
Costume, The Scarecrow	Ben Cooper	1968	Flick & Trick Costume with Lite-Up Mask, included a battery operated light bulb on the mask	15	25	40
Costume, The Scarecrow	Collegeville	1989	child's size, plastic face mask & vinyl body suit	6	10	15
Costume, The Tin Man	Ben Cooper	1968		15	25	35
Costume, The Tin Man	Ben Cooper	1975	costume and mask included	12	20	30
Costume, The Tin Man	Collegeville	1989	child's size, plastic face mask & vinyl body suit	6	10	15
Costume, Wizard of Oz	Halco	1961	masquerade costume & mask set	17	30	45
Crayon Box, The Wizard of Oz	Cheinco	1975	rectangular metal crayon box, illustrated with movie scenes	5	8	12

Wizard of Oz

TOY	COMPANY	YEAR	DESCRIPTION	GOOD	EX	MIB
Crystal Sculpture, Emerald City	Crystallite Company	1990	5" x 8", Austrian crystal, licensed edition designed by Charles Castelli	145	250	375
Cup, 200th Performance Souvenir, Metal	New York Majestic Theater	1903	metal with inscription on bottom: reads Wizard of Oz 200th performance at the Majestic Theater July 11, 1903	195	325	500
Decals, Sheet, Return to Oz	Waldenbooks	1985	promotional sticker premium issued to customers with any Oz book bought, sheet contains seven characters, Dorothy, Tik Tok, Jack Pumpkinhead, Billina (the chicken), The Gump in flight, a Wheeler, and	2	3	5
Decoupage Kit, The Wizard of Oz		1975	two wooden plaques, two scenes based on 1939 film and mounting materials	12	20	30
Dinnerware	Enesco	1989	three piece melamine set, plate, cup and bowl, plate and cup both showing four main characters arm in arm in forest, bowl showing them marching toward Emerald City, set	15	25	35
Display, Emerald City of Oz	Franklin Mint	1990	12" deep x 32" wide x 25 1/2" high, designed for display of large Oz figures	60	100	150
Doll Kit, Wizard of Oz Doodle Dolls	Whiting	1979	set of three doodle dolls, die cut cardboard parts, 3 skeins of yarn, 3 styrofoam balls, 2 pieces of fabric, 4 pom poms & 1 hank of yarn, Scarecrow, Tin Man, Cowardly Lion	6	10	15
Doll, Bert Lahr as the Cowardly Lion	Franklin Mint	1987	21" tall with royal robe and a badge of courage, second in the series	75	130	200
Doll, Cowardly Lion	M-D Tissue	1971	cloth, stuffed, light brown body with white snout	6	10	15
Doll, Cowardly Lion	Mego	1974	8" tall, vinyl, original costume from the MGM movie, one of six in set	15	25	40
Doll, Cowardly Lion	Mego	1975	14" tall with vinyl head and stuffed body, in window box, part of limited edition set of four	30	50	75
Doll, Cowardly Lion	Presents	1988	vinyl with brown fuzzy body and a medal of courage pinned on his chest, on a yellow brick road base	17	30	45
Doll, Cowardly Lion	Ideal	1984	9" tall, part of Character Dolls series, in window box	17	30	45
Doll, Cowardly Lion, The Original Soft Doll from Oz	Largo Toys	1989	rag type doll, in window box	8	13	20
Doll, Dandy Lion	Artistic Toy Company	1962	14" tall, cloth & vinyl, marked "Artistic Toy" & Videocraft Ltd. 1962 on doll's head	30	50	75
Doll, Dorothy	M-D Tissue	1971	cloth, stuffed, yellow haired girl with pumpkin colored jumper	6	10	15
Doll, Dorothy	Mego	1974	8" tall, vinyl, blue checkered jumper with white blouse with blue trim, one of six in set	15	25	40
Doll, Dorothy	Mego	1975	14" tall, vinyl head, stuffed body, in window box, part of limited edition set of four	30	50	75

Wizard of Oz

TOY	COMPANY	YEAR	DESCRIPTION	GOOD	EX	MIB
Doll, Dorothy	Presents	1988	vinyl with blue checkered jumper, white blouse with blue trim, ruby red slippers, blue checkered hair ribbons, on a yellow brick road base	17	30	45
Doll, Dorothy	Madame Alexander	1991	8" tall, dressed in blue checker jumper with white blouse, basket with Toto and red shoes, part of Storyland Dolls series	20	35	50
Doll, Dorothy	Madame Alexander	1990	large doll, dressed in a solid blue jumper with blouse underneath, basket with Toto and red shoes	25	40	65
Doll, Dorothy and Toto	Ideal	1984	9" tall, part of Character Dolls series, in window box	17	30	45
Doll, Dorothy, The Original Soft Doll from Oz	Largo Toys	1989	Judy Garland based stuffed doll, in window box	8	13	20
Doll, Frank Morgan as the Wizard of Oz	Franklin Mint	1991	19" tall, black suit and tie with top hat and red vest, seventh and last in series	75	130	200
Doll, Glinda	Mego	1974	8" tall, vinyl, pink dress with gold crown and brooch, one of six in set	15	25	40
Doll, Glinda	Presents	1989	vinyl, pink dress with pink crown and wand on a yellow brick road base	17	30	45
Doll, Glinda (Billie Burke)	Franklin Mint	1989	19" tall, pink dress with stars, crown and wand	100	165	250
Doll, Jack Haley as the Tin Man	Franklin Mint	1988	22" tall, third in series	75	130	200
Doll, Jack Pumpkinhead	Oz Doll & Toy Mfg. Co.	1924	one of four stuffed dolls, 13" tall, made of "art leather"	390	650	1000
Doll, Judy Garland as "Dorothy"	Ideal	1939	13" in blue checked jumper, open and close brown eyes	250	425	650
Doll, Judy Garland as "Dorothy"	Ideal	1939	15 1/2" in blue checked jumper, open and close brown eyes	325	550	850
Doll, Judy Garland as "Dorothy"	Ideal	1939	18" tall, in blue checked jumper, open and close brown eyes	425	715	1100
Doll, Judy Garland as Dorothy	Effanbee	1984	part of Hollywood Legend series	0	0	0
Doll, Judy Garland as Dorothy	Sears	1939	15 1/2" tall, in a red checked jumper with black pin curls, also came in a blue checked jumper	375	620	950
Doll, Judy Garland as Dorothy	Effanbee	1984	14 1/2" tall, vinyl with a blue dress and hair ribbons with ruby red slippers and basket with Toto, part of Legend Series	35	60	95
Doll, Judy Garland as Dorothy	Franklin Mint	1986	17" tall, Dorothy holding Toto in basket, blue checkered dress with blouse underneath, ruby red slippers and socks, first in series of seven dolls	75	130	200
Doll, Lollipop Guild Boy	Presents	1989	vinyl, in plaid shirt, green shorts and striped socks, on a yellow brick road base	17	30	45
Doll, Lullaby League Dancing Girl	Mego	1976	4" tall, in window box	60	100	150
Doll, Lullabye League Girl	Presents	1989	vinyl, pink ballerina dress and slippers with hat on a yellow brick road base	17	30	45

Wizard of Oz

TOY	COMPANY	YEAR	DESCRIPTION	GOOD	EX	MIB
Doll, Margaret Hamilton as The Wicked Witch of the West	Franklin Mint	1990	21" tall, black dress and hat with broom in hand, sixth in series	100	165	250
Doll, Mayor of Munchkinland	Mego	1976	4" tall, in window box	60	100	150
Doll, Mayor of Munchkinland	Presents	1989	vinyl, black suit and shoes on a yellow brick road base	17	30	45
Doll, Munchkin Flower Girl	Mego	1976	4" tall, in window box	60	100	150
Doll, Munchkin General	Mego	1976	4" tall, in window box	60	100	150
Doll, Patchwork Girl	Oz Doll & Toy Mfg. Co.	1924	one of four stuffed dolls, 13" tall, made of "art leather"	70	115	175
Doll, Ray Bolger as the Scarecrow	Franklin Mint	1989	22" tall, fourth in series	75	130	200
Doll, Rusty, the Tin Man	Artistic Toy Company	1962	14" tall, cloth & vinyl marked with "Artistic Toy" & Videocraft Ltd. 1962 on doll's head	30	50	75
Doll, Scarecrow	Oz Doll & Toy Mfg. Co.	1924	one of four stuffed dolls, 13" tall, made of "art leather"	390	650	1000
Doll, Scarecrow	M-D Tissue	1971	cloth, stuffed, frowning figure with x eyes, blue pants and red and white plaid jacket	6	10	15
Doll, Scarecrow	Mego	1974	8" tall, vinyl, first version has no hair, original MGM movie costume, one of six in set	20	35	50
Doll, Scarecrow	Mego	1975	14" tall with vinyl head and stuffed body, in window box, part of limited edition set of four	30	50	75
Doll, Scarecrow	Presents	1988	vinyl with brown pants, green shirt and black hat & shoes, on a yellow brick road base	17	30	45
Doll, Scarecrow	Ideal	1984	9" tall, part of Character Dolls series, in window box	17	30	45
Doll, Scarecrow Talkin' Patter Pillow	Mattel	1968	cloth stuffed pull-string doll that says 10 sentences, smiling figure with dark blue pants and sleeves, white gloves, black boots, pink sack head and yellow hat	30	50	75
Doll, Scarecrow, The Original Soft Doll from Oz	Largo Toys	1989	rag type doll, in window box	8	13	20
Doll, Socrates, the Scarecrow	Artistic Toy Company	1962	14" tall, cloth & vinyl, marked "Artistic Toy" & Videocraft Ltd. 1962 on doll's head	30	50	75
Doll, Squeaky, Cowardly Lion	Burnstein	1939	7" tall, hollow rubber squeak toy	70	115	175
Doll, Squeaky, Cowardly Lion	Burnstein	1939	7" tall, hollow rubber squeak toy	70	115	175
Doll, Squeaky, Dorothy	Burnstein	1939	7" tall, hollow rubber squeak toy	70	115	175
Doll, Squeaky, Glinda	Burnstein	1939	7" tall, hollow rubber squeak toy	75	130	200
Doll, Squeaky, Scarecrow	Burnstein	1939	7" tall, hollow rubber squeak toy	70	115	175
Doll, Squeaky, Tin Man	Burnstein	1939	7" tall, hollow rubber squeak toy, embossed with actor's name and movie on back	70	115	175
Doll, Squeaky, Wicked Witch	Burnstein	1939	7" tall, hollow rubber squeak toy	90	145	225
Doll, Squeaky, Wizard	Burnstein	1939	7" tall, hollow rubber squeak toy	75	130	200
Doll, The Strawman by Ray Bolger	Ideal	1939	17" tall, with tan or pink pants and a black or navy jacket	295	495	750
Doll, The Strawman by Ray Bolger	Ideal	1939	21" tall with tan or pink pants and black or navy jacket, very rare	390	650	1000

Wizard of Oz

TOY	COMPANY	YEAR	DESCRIPTION	GOOD	EX	MIB
Doll, The Wicked Witch	Mego	1974	8" tall, vinyl, black dress, white belt and hat with green face and hands, one of six in set	40	65	100
Doll, The Wizard of Oz Doll Collection, Cowardly Lion	Dave Grossman	1982	17" tall sculpted porcelain, hand painted with glass eyes, mounted on an interlocking yellow brick road base with certificate, one of 250 made	390	650	1000
Doll, The Wizard of Oz Doll Collection, Dorothy	Dave Grossman	1982	15" tall sculpted porcelain, hand painted with glass eyes, hand woven basket, mounted on an interlocking yellow brick road base, with certificate, one of 250 made	390	650	1000
Doll, The Wizard of Oz Doll Collection, The Scarecrow	Dave Grossman	1982	17" tall sculpted porcelain, hand painted with glass eyes, mounted on an interlocking yellow brick road base, with certificate, one of 250 made	390	650	1000
Doll, The Wizard of Oz Doll Collection, The Tin Man	Dave Grossman	1982	18" tall sculpted porcelain, silver suit is screwed together, hand painted with glass eyes, mounted on an interlocking yellow brick road base, with certificate, one of 250 made	390	650	1000
Doll, Tin Man	Oz Doll & Toy Mfg. Co.	1924	one of four stuffed dolls, 13" tall, made of "art leather"	390	650	1000
Doll, Tin Man	M-D Tissue	1971	cloth, stuffed, grey body with blue eyes and red heart	6	10	15
Doll, Tin Man	Mego	1974	8" tall, vinyl, original costume from the MGM movie, one of six in set	15	25	40
Doll, Tin Man	Mego	1975	14" tall with vinyl head and stuffed body, in window box, part of limited edition set of four	30	50	75
Doll, Tin Man	Presents	1988	vinyl, silver body with a heart clock on chain on his chest, on a yellow brick road base	17	30	45
Doll, Tin Man	Ideal	1984	9" tall, part of Character Dolls series, in window box	15	25	40
Doll, Tin Man, The Original Soft Doll from Oz	Largo Toys	1989	rag type doll, in window box	8	13	20
Doll, Toto	Presents	1988	5 1/2" tall, stuffed	8	13	20
Doll, Wicked Witch	Presents	1989	vinyl, black dress & hat with green face & hands holding broom on a yellow brick road base	20	35	50
Dolls, Set of Four	Effanbee	1985	each 11 1/2" tall, Dorothy, the Scarecrow, Tin Man & the Cowardly Lion, each	12	20	30
Dolls, Set of Six, Wonderful Wizard of Oz Collection	World Dolls	1987	each 8" tall, jointed & accompanied by certificate, Dorothy, the Scarecrow, the Tin Man, the Cowardly Lion, Good Witch & the Wicked Witch, each	12	20	30
Dolls, The Wizard of Oz Live!	Applause, Inc.		cloth souvenir dolls of the musical tour, Dorothy, Wicked Witch, Scarecrow, Tin Man, Cowardly Lion and Toto, each available in three different sizes, each	8	13	20
Dolls, Wonderful Wizard of Oz Paper Dolls	The Toy Factory	1975	set of five dolls, Dorothy, Tin Man, Scarecrow, Cowardly Lion & Toto, clothes and accessories, in box	8	13	20

Wizard of Oz

TOY	COMPANY	YEAR	DESCRIPTION	GOOD	EX	MIB
Doorknob Hangers, Set of Four	Applause, Inc.	1989	white plastic hangers with film scenes, I Don't Think We're in Kansas Anymore (Dorothy & Scarecrow), Don't Be Afraid to Knock (Lion), Yellow Brick Road Ends Here (poppy field), Caution: Entering the L	3	5	7
Dress Up Costumes	Multi Toys	1988	set designed to convert 11 1/2" dolls into Oz characters, 2 per set, Lion and Wicked Witch, Dorothy and Tin Man, Scarecrow and Glinda, in window box, each	5	8	12
Embroidery Kit, "Land of Oz"	Family Circle Magazine	1977	3 x 5 foot baby quilt design with matching bibs, available by mail from magazine	8	13	20
Erasers, Figural, Set of Six	Applause, Inc.	1989	colored soft erasers, Scarecrow, Cowardly Lion, Dorothy, Wicked Witch, Tin Man, Glinda, set	10	15	25
Figure, Bendable, Cowardly Lion	Just Toys	1989	on card	3	5	8
Figure, Bendable, Dorothy	Just Toys	1989	on card	3	5	8
Figure, Bendable, Scarecrow	Just Toys	1989	on card	3	5	8
Figure, Bendable, Tin Man	Just Toys	1989	on card	3	5	8
Figure, Composition, Scarecrow	Artisans Studio	1939	4" tall, wood composition, hand painted	60	100	150
Figure, Composition, Tin Man	Artisans Studio	1939	4" tall, wood composition, hand painted	60	100	150
Figure, Inflatable, The Original Wizard of Oz Scarecrow	Dalen Products	1984	6 foot tall inflatable figure, sold in bag with illustrated header card	10	15	25
Figure, Poseable, Cowardly Lion	Multi Toys	1989	4" tall also available in Poseable Figure Collection	2	3	5
Figure, Poseable, Dorothy	Multi Toys	1989	4" tall also available in Poseable Figure Collection	2	3	5
Figure, Poseable, Glinda	Multi Toys	1989	4" tall also available in Poseable Figure Collection	2	3	5
Figure, Poseable, Scarecrow	Multi Toys	1989	4" tall also available in Poseable Figure Collection	2	3	5
Figure, Poseable, Tin Man	Multi Toys	1989	4" tall also available in Poseable Figure Collection	2	3	5
Figure, Poseable, Wicked Witch	Multi Toys	1989	4" tall also available in Poseable Figure Collection	2	3	5
Figure, Return to Oz, Jack Pumpkinhead	Heart & Heart	1985	3-4" tall, plastic jointed figure, based on film, in window box	30	50	75
Figure, Return to Oz, The Scarecrow	Heart & Heart	1985	3-4" tall, plastic jointed figure, based on film, in window box	25	40	65
Figure, Return to Oz, The Tin Man		1985	3-4" tall, plastic jointed figure, based on film, in window box	25	40	65
Figure, Return to Oz, Tik Tok	Heart & Heart	1985	3-4" tall, plastic jointed figure, based on the film, in window box	30	50	75
Figure, Rubb'r Nik, Cowardly Lion	Multiple Toymakers	1967	6" tall, bendy on card	15	25	35
Figure, Rubb'r Nik, Dorothy	Multiple Toymakers	1967	6" tall, bendy on card	15	25	35
Figure, Rubb'r Nik, Scarecrow	Multiple Toymakers	1967	6" tall, bendy on card	15	25	35
Figure, Rubb'r Nik, Tin Man	Multiple Toymakers	1967	6" tall, bendy on card	15	25	35

Wizard of Oz

TOY	COMPANY	YEAR	DESCRIPTION	GOOD	EX	MIB
Figure, Scarecrow		1968	15" tall, ceramic figure was available painted or unpainted	15	25	35
Figure, Tin Man		1968	15" tall, ceramic figure was available painted or unpainted	15	25	35
Figures, Miniature, Set of Four		1988	based on 1987 animated film "The Wonderful Wizard of Oz," figures are Dorothy, Scarecrow, Cowardly Lion & the Tin Man, set	10	15	25
Figures, Poseable Collection	Multi Toys	1989	4" tall poseable figures included are Dorothy, Tin Man, Scarecrow, Cowardly Lion, Wicked Witch and Glinda sold in a set of six	12	20	30
Figurine, "Judy Garland as Dorothy"	The Franklin Mint	1991	8 1/2" tall hand decorated bisque porcelain, shows Dorothy standing in front of corn field fence on section of yellow brick road with Toto at her left foot	60	100	150
Figurine, Christmas Caroller, Dorothy	Kurt S. Adler	1989	dressed in a blue & white jumper with red shoes and scarf with basket on left arm	17	30	45
Figurine, Christmas Caroller, Scarecrow	Kurt S. Adler	1989	blue scarecrow outfit with black boots holding a red caroller hymn book	17	30	45
Figurine, Christmas Caroller, The Cowardly Lion	Kurt S. Adler	1989	red checkered scarf around his neck	17	30	45
Figurine, Christmas Caroller, The Tin Man	Kurt S. Adler	1989	red scarf around his neck with Christmas tree in his right hand	17	30	45
Figurine, Cold Cast Marble, Mayor of Munchkinland	Michael Roche	1992	independently produced addition to Dave Grossman set of 5 3/4- 8 1/4" figurines	30	50	75
Figurine, Cold Cast Marble, Winged Monkey	Michael Roche	1992	independently produced addition to Dave Grossman set of 5 3/4-8 1/4" figurines	30	50	75
Figurine, Cowardly Lion	Mann Porcelain	1974	porcelain figure on round base, with hands clasped, with tag	17	30	45
Figurine, Dorothy	Mann Porcelain	1974	porcelain figure on round base, walking in pink dress carrying yellow basket, with Toto beside her right foot, with tag	17	30	45
Figurine, Musical, "Billie Burke as Glinda"	Mann Porcelain	1981	plays "Over the Rainbow"	20	35	50
Figurine, Musical, "Dorothy in the Haunted Forest"	Mann Porcelain	1983	shows Dorothy in blue dress and ruby slippers standing before the Haunted Forest road sign	17	30	45
Figurine, Scarecrow	Mann Porcelain	1974	porcelain figure on round base, standing with arms at sides, with tag	17	30	45
Figurine, Tin Man	Mann Porcelain	1974	porcelain figure on round base, standing with axe in hands, with tag	17	30	45
Figurines, Cold Cast Marble, Set	Dave Grossman	1989	5 3/4" to 8 1/4" 50th anniversary figurines featuring Dorothy, Scarecrow, Tin Man and the Cowardly Lion, each	30	50	75
Figurines, Cold Cast Marble, Set	Dave Grossman	1990	additions to the 1989 set: 5 3/4" to 8 1/4" tall, The Wizard, Glinda and the Wicked Witch, each	30	50	80

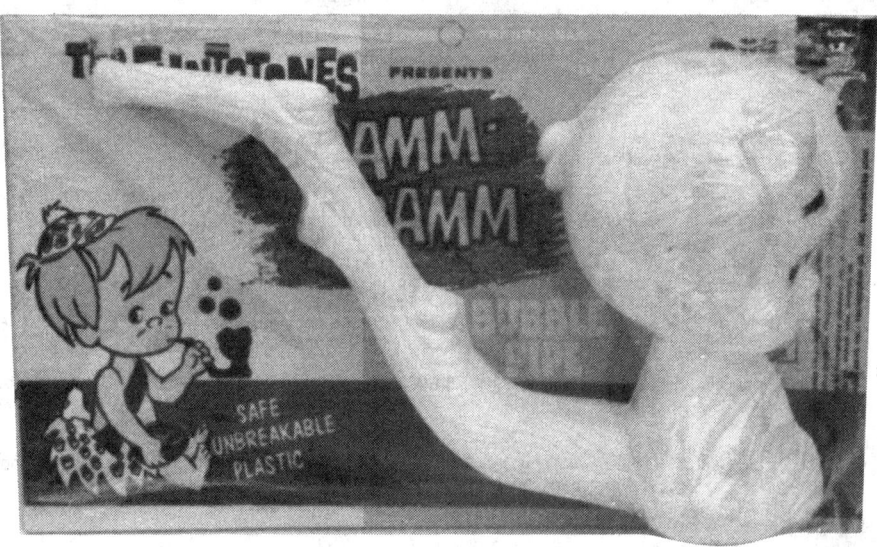

Top to bottom: The Official Mr. Magoo Car, 1961, Hubley; Flintstones Bamm-Bamm Bubble Pipe, 1963, Transogram.

Wizard of Oz

TOY	COMPANY	YEAR	DESCRIPTION	GOOD	EX	MIB
Figurines, Golden Anniversary Commemoratives	Metropolitan Guild	1989	set of four, 4" tall, layered in 24K gold, engraved with artist's signature, serial number & certificate of authenticity, Dorothy, Scarecrow, Tin Man, Cowardly Lion	25	40	65
Figurines, Pewter, Set of Four	Pewter Fancy	1970s	unpainted pewter figures on wooden base, styled after Denslow drawings, Dorothy with Toto in her arms, Cowardly Lion, Scarecrow, Tin Man, each	10	15	25
Figurines, PVC, Munchkins	Presents	1988	set of six, 1 3/4" to 2 3/4" tall, the Mayor, Lollipop Guild Boy, Sleepyhead Girl, Lady, Soldier and Ballerina, each	3	5	8
Figurines, Set of Six	Presents	1988	set of six, 3 3/4" tall, Dorothy, Scarecrow, Tin Man, Cowardly Lion, Wicked Witch, Glinda, each	2	3	5
Flasher Rings, Off to See the Wizard	Vari-Vue	1967	gumball machine prizes, early styles were silver painted resin, later gold-painted or dark or light blue plastic, each	6	10	15
Fun Shades	Multi Toys	1989	six different kid-sized sunglasses, each with a character image in bridge; Wicked Witch, Dorothy, Scarecrow, Cowardly Lion, Tin Man and Glinda, each	3	5	7
Game, Card, Wizard of Oz	Castell Brothers (England)	1940	44 hand tinted stills of MGM movie, correctly assembled cards show synopsis of film	105	175	275
Game, Off to See the Wizard	Schaper	1967	played similar to Twister, players walk along the large vinyl mat	15	25	35
Game, Off to See the Wizard	Milton Bradley	1968	contains playing board, dice, die cut markers and a "Witch" disk	17	30	45
Game, Return To Oz	Golden Press	1985	playing board, four markers, spinner and 28 ornament disks, in illustrated box on yellow background	8	13	20
Game, The Game of The Wizard of Oz	Whitman	1939	board, four wooden markers of Dorothy, Scarecrow, the Tin Woodsman & the Cowardly Lion, and dice	90	145	225
Game, The Scarecrow of Oz Answers Questions by Radio	Reilly & Lee	1924	cardboard folder with magnetized metal arrow pointer on front, actually was a promo piece for book sellers	55	95	145
Game, The Wizard of Oz	Cadaco	1974	box cover shows four main MGM characters walking arm in arm through poppy field	10	15	25
Game, The Wizard of Oz Dart Game	Dart Board Equipment Co.	1939	board illustrated with a meandering yellow brick road and circular targets of Oz characters, with three darts, in box, rare	195	325	500
Game, The Wogglebug Game of Conundrums	Parker Brothers	1905	51 each question and answer cards, with instruction in illustrated box	295	490	750
Game, The Wonderful Game of Oz	Parker Brothers	1921	original version with elaborate game board, color instructions, four pewter playing pieces of Dorothy, Scarecrow, Tin Man, and Cowardly Lion	775	1300	2000

Wizard of Oz

TOY	COMPANY	YEAR	DESCRIPTION	GOOD	EX	MIB
Game, The Wonderful Wizard of Oz	E.E. Fairchild	1957	playing board, spinner, 32 playing cards and four wooden markers	25	40	60
Give-A-Show Projector Slides	Kenner	1968	set contains 35 color slides, five different shows	15	25	40
Jacket, The Wizard of Oz	MGM/UA	1989	black satin premium jacket, ornately embroidered, given away during promotion of the 50th anniversary edition of The Wizard of Oz videotape	50	80	125
Jewelry Box, 100th Performance Souvenir	New York Majestic Theater	1903	metal with inscription on bottom: reads Majestic Theater 100th Performance, Wizard of Oz J.B. 327 Wednesday, April 15, 1903	295	490	750
Jewelry Box, Musical	Multi Toys	1989	rectangular musical box, windup key on back, illustrated with clouds, rainbows, and art of Dorothy standing before yellow brick road, when opened Dorothy revolves in front of mirror while "Over The Ra	8	13	20
Jewelry Boxes, Musical, Set of Seven	Presents	1988	wooden, decorated with images from the film, with certificate of authenticity, each	25	40	65
Jewelry Boxes, Set of Four	Hamilton Collection	1990	round porcelain boxes, available musical or non-musical, using images from commemorative plate collection, also by Hamilton, each	12	20	30
Lunch Kit	Aladdin Industries	1989	plastic lunch box & thermos, both illustrated with color art, bottle shows ruby slippers on yellow brick road	6	10	15
Magic Picture Kit, Off to See the Wizard	Jiffy Pop Popcorn	1968	a set contains a background picture from one of the following four scenes: "Dorothy Melts the Witch", "Follow the Yellow Brick Road", "In the Emerald City & "The Wizard is Exposed" and a sheet of rub-	4	7	10
Magic Slate	Whitman	1976	magic slate with primitively drawn characters on border	4	7	10
Magic Slate	Western Publishing	1989	magic slate illustrated with line art based on MGM film characters	3	5	7
Magic Slate, Return to Oz	Western Publishing	1985	magic slate illustrated with line art of film characters and scenes	4	7	10
Magic Slate, Wizard Slate from Oz	Lowe	1961	slate on illustrated header card, with art based on animated TV show	10	15	25
Magic Story Cloth	Raco	1978	38" x 44" plastic color, wipe and re-color activity sheet with 8 crayons and sponge, sheet shows various characters and Emerald City at end of yellow brick road	6	10	15
Magnet Set, Set of Six	Grynnen Barrett	1987	primitively styled painted Wicked Witch, Dorothy, Scarecrow, Toto, Cowardly Lion & Tin Man, in box	6	10	15
Magnets, Figural	Vanderbilt Products, Inc.	1989	Dorothy, Scarecrow, Tin Man, Cowardly Lion, Glinda, Wicked Witch, each	2	3	5
Mask, Cowardly Lion	Newark Mask Company	1939	pressed linen character mask, hand painted	50	80	125

Wizard of Oz

TOY	COMPANY	YEAR	DESCRIPTION	GOOD	EX	MIB
Mask, Cowardly Lion	Clay Art	1986	ceramic mask, painted, in box	25	40	60
Mask, Dorothy	Clay Art	1986	ceramic mask, painted, in box	25	40	65
Mask, Rubber, The Cowardly Lion	Don Post Studios	1983	rubber head mask	35	55	85
Mask, Rubber, The Scarecrow	Don Post Studios	1983	rubber head mask	35	55	85
Mask, Rubber, The Tin Man	Don Post Studios	1983	rubber head mask	35	55	85
Mask, Scarecrow	Newark Mask Company	1939	pressed linen character mask, hand painted	50	80	125
Mask, Scarecrow	Clay Art	1986	ceramic mask, painted, in box	25	40	60
Mask, Tin Man	Newark Mask Company	1939	pressed linen character mask, hand painted	50	80	125
Mask, Tin Man	Clay Art	1986	ceramic mask, painted, in box	25	40	60
Mask, Wicked Witch	Ben Cooper	1975		6	10	15
Masks, Set of Five	Einson-Freeman Company	1939	paper character masks: Dorothy, Tin Man, Cowardly Lion, Scarecrow & the Wizard of Oz, each	20	35	50
Message Boxes, Set of Four	Applause, Inc.	1989	square plastic box with die cut character on top, "If I Only Had A Brain", "If I Only Had A Heart", "If I Only Had Courage","Somewhere Over The Rainbow", each	3	5	8
Munchkinland Playset	Mego	1976	designed for 4" Munchkin dolls, playset contains: Dorothy's bed, Munchkin Mayor doll, and a tornado transporter that "carries" Dorothy from her Kansas bedroom to Oz	115	195	300
Music Box, "The Yellow Brick Road"	Enesco	1989	large revolving music box with all four figures arm in arm, plays "We're Off To See the Wizard"	2350	33	50
Music Box, Dorothy	Schmid	1983	plays "Over tho Rainbow"	20	35	50
Music Box, The Cowardly Lion	Schmid	1983	plays a non-Oz related tune	17	30	45
Music Box, The Scarecrow	Schmid	1983	plays a non-Oz related tune	17	30	45
Music Box, The Tin Man	Schmid	1983	plays a non-Oz related tune	17	30	45
Music Boxes, 50th Anniversary, Set of Four	Enesco	1989	Dorothy plays "Over the Rainbow," Scarecrow "If I Only Had a Brain," Tin Man "We're Off to See the Wizard" & the Cowardly Lion "If I Were King of the Forest," each	15	25	40
Musical Figurines, Set of Six	Presents	1989	PVC figures mounted on musical bases, Dorothy, Glinda, Wicked Witch, Scarecrow, Tin Man, Cowardly Lion, each	6	10	15
Musical Jack-in-the-Box, Cowardly Lion	Enesco	1988	one of six in set, limited to 7500 each, large size black and gold enamel box	60	100	150
Musical Jack-in-the-Box, Cowardly Lion	Enesco	1989	smaller version of ornate black and gold enamel limited edition box, with wind-up character action	50	80	125
Musical Jack-in-the-Box, Dorothy	Enesco	1988	one of six in set, limited to 7500 each, large size black and gold enamel box	65	115	175
Musical Jack-in-the-Box, Dorothy	Enesco	1989	smaller version of ornate black and gold enamel limited edition box, with wind-up character action	55	95	145

Wizard of Oz

TOY	COMPANY	YEAR	DESCRIPTION	GOOD	EX	MIB
Musical Jack-in-the-Box, Glinda	Enesco	1988	one of six in set, limited to 7500 each, large size black and gold enamel box	60	100	150
Musical Jack-in-the-Box, Glinda	Enesco	1989	smaller version of ornate black and gold enamel limited edition box, with wind-up character action	50	80	125
Musical Jack-in-the-Box, Scarecrow	Enesco	1989	smaller version of ornate black and gold enamel limited edition box, with wind-up character action	50	80	125
Musical Jack-in-the-Box, Scarecrow	Enesco	1988	one of six in set, limited to 7500 each, large size black and gold enamel box	60	100	150
Musical Jack-in-the-Box, The Tin Man	Enesco	1988	one of six in set, limited to 7500 each, large size black and gold enamel box	60	100	150
Musical Jack-in-the-Box, The Tin Man	Enesco	1989	smaller version of ornate black and gold enamel limited edition box, with wind-up character action	50	80	125
Musical Jack-in-the-Box, Wicked Witch	Enesco	1988	one of six in set, limited to 7500 each, large size black and gold enamel box	60	100	150
Musical Jack-in-the-Box, Wicked Witch	Enesco	1989	smaller version of ornate black and gold enamel limited edition box, with wind-up character action	50	80	125
Musical Sculpture, "Follow the Yellow Brick Road"	Franklin Mint	1991	6 1/2" high, hand painted porcelain, covered with glass dome, on wood base, shows Glinda, Dorothy, Toto and Munchkin Mayor beneath sunflower at the spiral beginning of yellow brick road	60	100	150
Musical Sculpture, "If I Were King of the Forest"	Franklin Mint	1991	6 1/2" high, hand painted porcelain, covered with glass dome, on wood base, depicts the Cowardly Lion's flower pot crowning scene from film	60	100	150
Musical Sculpture, "We're Off to See the Wizard"	Franklin Mint	1989	6 1/2" high, hand painted porcelain, covered with glass dome, on wood base, shows the four characters approaching the Emerald City	60	100	150
Needlecraft Kit, "Follow the Yellow Brick Road"	Bucilla	1977	24" x 28" picture with Dorothy, Scarecrow, Tin Man, the Cowardly Lion, Toto and the Emerald City in the background	8	13	20
Nightlight, Scarecrow	Hamilton Gifts	1989	7" tall, unpainted white bone china	10	15	25
Oil Paint by Number Set	Hasbro	1973	contains one paper canvas ready for paint with six paint vials and brush	6	10	15
Ornaments, "Treasury of Christmas Ornaments"	Enesco	1989	set of four, hand painted, Dorothy, the Scarecrow, the Tin Man & the Cowardly Lion, each	0	13	20
Ornaments, Set of Four	Adler, Inc.	1987	"Christmas Caroller" ornaments in shape of childlike Dorothy in red Santa's hat, Tin Man with tree, Scarecrow with song book, and Cowardly Lion with bell, set price	45	75	115

Wizard of Oz

TOY	COMPANY	YEAR	DESCRIPTION	GOOD	EX	MIB
Ornaments, Set of Twelve	Danbury Mint	1989	set of 12 ornaments with color scenes from the 1939 movie framed in filigreed 24k gold plate, each	8	13	20
Ornaments, Wooden	Adler, Inc.	1984	set includes the Scarecrow, Tin Man, Dorothy & the Cowardly Lion, all are jointed "Jumping Jack" pull-string figures, came with a small book version of "The Wizard of Oz", set price	15	25	40
Oz Jigglers, "Off to See the Wizard," Cowardly Lion	Diener Industries	1968	gelatin like figure, non toxic, on rounded card	17	30	45
Oz Jigglers, "Off to See the Wizard," Dorothy & Toto	Diener Industries	1968	gelatin like figure, non toxic, on rounded card	17	30	45
Oz Jigglers, "Off to See the Wizard," Scarecrow	Diener Industries	1968	gelatin like figure, non toxic, on rounded card	17	30	45
Oz Jigglers, "Off to See the Wizard," Tin Man	Diener Industries	1968	gelatin like figure, non toxic, on rounded card	17	30	45
Oz Jigglers, "Off to See the Wizard," Wicked Witch	Diener Industries	1968	gelatin like figure, non toxic, on rounded card	20	35	50
Oz Jigglers, "Off to See the Wizard," Wizard	Diener Industries	1968	gelatin like figure, non toxic, on rounded card	17	30	45
Oz Toy Book, Cut-Outs for the Kiddies		1915	sixteen pages, string bound, over four dozen character cut-outs, rare, no price set	0	0	0
Oz-kins, Set of Ten	Aurora	1967	plastic figures on card, Wizard, Soldier, Wicked Witch, Scarecrow, Tin Man, Cowardly Lion, Dorothy, Toto, Sawhorse & Glinda, with paintbrush & paints	35	60	95
Oz-kins, Set of Ten	Aurora	1967	plastic figures, Burry Biscuit premium, the Wizard, Soldier, Wicked Witch of the East, Scarecrow, Tin Man, Cowardly Lion, Dorothy, Toto, Sawhorse & Glinda, set	35	60	95
Paint by Number Set	Art Award	1989	three different versions, each with 2 pictures to color, paints, brush and glitter, each	3	5	8
Paint By Number Set, Off to See the Wizard	Craftmaster	1968	each set included six different paints, a paint brush, and one of the following pictures either the Tin Man, Cowardly Lion or the Scarecrow	15	25	40
Paint Set, The Wizard of Oz Fast Dry Acrylic Paint by Nunmber	Craft House	1979	two 10x14" panels, 15 colors, brush and instructions, pictures show Dorothy and Scarecrow oiling Tin Man, Wicked Witch and Winged Monkey viewing four main characters in crystal ball	10	15	25
Paint with Crayons Set	Art Award	1989	set includes 4 pictures to color, line art based on MGM film characters, in illustrated box	3	5	8
Paint-by-Number 'N Frame Set	Hasbro	1969	16" x 18", two plastic frames, 18 watercolors, brush and eight pictures to paint	15	25	35

Wizard of Oz

TOY	COMPANY	YEAR	DESCRIPTION	GOOD	EX	MIB
Paint-by-Number Set		1950s	acrylic paint by number featuring Dorothy, Tin Man, Scarecrow, Cowardly Lion, Tip and the Sawhorse with the Emerald City on a hill in the background	12	20	30
Patch, Emerald City Police		1980s	shield shaped shoulder patch, green background with gold trim, shows Emerald City towering over rolling fields	4	7	10
Pillow, Wizard of Oz	Modern Pillow Company	1970s	oversized pillow color art of four main characters and Munchkins on yellow brick road	15	25	40
Pin, Cloisonne, Commemorative	MGM/UA's	1989	four different 1 1/8" publicity premium pins, handed out by MGM/UA promoting the 50th anniversary video release, Scarecrow, Dorothy, Tin Man, Cowardly Lion, each	10	15	25
Pin, Costume Jewelry, Cowardly Lion	Wendy Gell Jewelry	1989	gold plated costume jewelry pin, on black rack card	17	30	45
Pin, Costume Jewelry, Scarecrow	Wendy Gell Jewelry	1989	gold plated costume jewelry pin, on black rack card	17	30	45
Pin, Costume Jewelry, Tin Man	Wendy Gell Jewelry	1989	silver plated costume jewelry pin, on black rack card	17	30	45
Pin, Dorothy	Wendy Gell Jewelry	1989	gold plated costume jewelry pin, on black rack card	17	30	45
Pin, Emerald City Police		1980s	green shield with gold rim, shows Emerald City towering over rolling fields	4	7	10
Pin, It's Oz Time at Macy's	Macy's	1989	3" button used for promotional advertisment by Macy's employees for the 50th anniversary celebration, green-rimmed pin shows four main characters approaching the Emerald City	12	20	30
Pin, Return to Oz	Disney Studios	1984	1 1/4" blue and orange pin back button used to promote the movie, says "Oz," which was the pre-release working title of "Return to Oz"	4	7	10
Pin, Ruby Slippers		1989	sparkling ruby slippers pin, sold at performances of "The Wizard of Oz Live!"	10	16	25
Placemats, Set of Two	Dow	1989	11 x 17" each, plastic product premium mats, identical central art on each, with four different color movie still inserts on each mat, set	8	13	20
Plate Collection, First Series, Set of Eight	The Hamilton Collection	1988	"There's No Place Like Home," shows Dorothy preparing to wish her way home as Glinda waves wand overhead and other characters look on	17	30	45
Plate Collection, First Series, Set of Eight	The Hamilton Collection	1988	"The Tin Man Speaks," shows Dorothy oiling Tin Man's lips as Scarecrow watches	15	25	35
Plate Collection, First Series, Set of Eight	The Hamilton Collection	1988	"We're Off to See the Wizard," shows the four characters in the woods beginning their walk to the Emerald City	15	25	35
Plate Collection, First Series, Set of Eight	The Hamilton Collection	1988	"The Great and Powerful Oz," shows the Wizard exposed as a man	15	25	35

Wizard of Oz

TOY	COMPANY	YEAR	DESCRIPTION	GOOD	EX	MIB
Plate Collection, First Series, Set of Eight	The Hamilton Collection	1988	"The Witch Casts A Spell," shows the Wicked Witch observing four characters in her crystal ball as winged monkey looks on	15	25	35
Plate Collection, First Series, Set of Eight	The Hamilton Collection	1988	"If I Were King of the Forest," shows the Cowardly Lion's "coronation" scene inside the Emerald City	15	25	35
Plate Collection, First Series, Set of Eight	The Hamilton Collection	1988	"A Glimpse of the Munchkins," shows Glinda holding Dorothy's hand as she meets the Munchkins	15	25	35
Plate, "Fifty Years of Oz"	Hamilton Collection	1990	9 1/4" diam., with 8mm wide 23K gold border, montage of characters, official commemorative plate of the Golden Anniversary	30	50	75
Plate, "Follow the Yellow Brick Road"	Knowles China Co.	1979	one of eight in first series ever devoted to any film,	10	16	25
Plate, "If I Only Had a Brain"	Knowles China Co.	1977	one of eight in first series ever devoted to any film, shows the Scarecrow in yellow brick crossroads, pointing to his head	15	25	40
Plate, "If I Only Had a Heart"	Knowles China Co.	1978	one of eight in first series ever devoted to any film,	17	30	45
Plate, "If I Were King of the Forest"	Knowles China Co.	1978	one of eight in first series ever devoted to any film, shows the Cowardly Lion in the forest, standing on the yellow brick road	17	30	45
Plate, "The Grand Finale" (We're Off to See the Wizard)	Knowles China Co.	1979	one of eight in first series ever devoted to any film, shows the four characters arm in arm leaving the forest, entering the poppy fields with the Emerald City in background	15	25	35
Plate, "Wicked Witch of the West"	Knowles China Co.	1979	one of eight in first series ever devoted to any film, shows the Wicked Witch at her window, watching the winged monkeys fly off to ambush the four characters in the woods	10	15	25
Plate, "Wonderful Wizard of Oz"	Knowles China Co.	1979	one of eight in first series ever devoted to any film, shows the Wizard in his balloon preparing to leave the Emerald City	12	20	30
Plate,"Over the Rainbow"	Knowles China Co.	1977	one of eight in first series ever devoted to any film, shows Dorothy holding Toto and basket, walking along the yellow brick road	20	35	50
Plates, The Portraits From Oz, 2nd Series	The Hamilton Collection	1989	2nd series of eight porcelain plates, 8 1/2" diam. with 23K gold "yellow brick road" border, portrait closeups of Toto, Glinda, Dorothy, Wicked Witch, Wizard, Scarecrow, Tin Man, Lion, each	15	25	40
Pocket Watch	Westclock	1980s	silver finish case shows four main characters on dial	17	30	45

Wizard of Oz

TOY	COMPANY	YEAR	DESCRIPTION	GOOD	EX	MIB
Puppet Theatre	Proctor & Gamble	1965	mail away cardboard Emerald City theater and scripts for your own plays, designed for P&G premium puppets	30	50	75
Puppet, Hand, Cowardly Lion	Presents	1989	vinyl and cloth with a mane and badge of courage on his chest	8	13	20
Puppet, Hand, Dorothy	Presents	1989	vinyl and cloth with blue and white checkered jumper and matching blouse with ribbons in her hair	8	13	20
Puppet, Hand, Glinda	Presents	1989	vinyl and cloth with pink dress and crown	8	13	20
Puppet, Hand, Gump	Welch's Jelly	1985	one of 3 in set, plush hand puppet with golden-tan fur, brown antlers and blue eyes, part of "Return To Oz" promotion	10	15	25
Puppet, Hand, Off to See the Wizard Talking Puppet	Mattel	1968	talking puppet, four vinyl heads on finger tips, Toto and Cowardly Lion on thumb pad, pull-string activates one of ten phrases	25	40	65
Puppet, Hand, Scarecrow	Welch's Jelly	1985	one of three in set, blue body, flesh colored face with red sewn-on nose and eyebrows, pale blue crown with fabric emeralds, part of "Return to Oz" promotion	10	16	25
Puppet, Hand, Scarecrow	Presents	1989	vinyl and cloth with black hat, green body, burlap scarf and straw "hair"	8	13	20
Puppet, Hand, Tik-Tok	Welch's Jelly	1985	one of three in set, copper plush body with ocher helmet band and solid green eyes, part of "Return to Oz" promotion	10	16	25
Puppet, Hand, Tin Man	Presents	1989	vinyl and cloth with silver body and head, and a red heart on his chest	8	13	20
Puppet, Hand, Wicked Witch	Presents	1989	vinyl and cloth with black hat and dress, green face	8	13	20
Puppets, Hand	Multi Toys	1989	set of six featuring Dorothy, Scarecrow, Tin Man, Cowardly Lion, Glinda and the Wicked Witch, on blister cards, each	6	10	15
Puppets, Hand, Set of Eight	Proctor & Gamble	1965	plastic hand puppets, shrink-wrapped to various detergents and other P&G products, you could also mail away for a cardboard Emerald City theater and scripts for your own plays, each	8	13	20
Puppets, Stick, The Wizard of Oz Live!		1989	set of four: Dorothy with Toto in her basket, Cowardly Lion, Scarecrow and Tin Man, each	6	10	15
Puzzle, Canister, Who will be Next?	Effanbee	1984	small canister-held puzzle of Dorothy in barnyard announcing next doll in Effanbee's Legend Series	8	13	20
Puzzle, Children's Jig Saw Puzzle	Hallmark	1983	pictured storybook characters walking toward viewer on yellow brick road	3	5	8
Puzzle, Frame Tray	Whitman	1976	pictured are: the Cowardly Lion crying with Toto at his feet, Dorothy standing with Scarecrow and the Tin Man on the yellow brick road	3	5	7

Wizard of Oz

TOY	COMPANY	YEAR	DESCRIPTION	GOOD	EX	MIB
Puzzle, Frame Tray	Western Publishing Company	1989	100 piece puzzle picturing Glinda and Dorothy in Munchkinland	4	7	10
Puzzle, Frame Tray	Western Publishing Company	1989	100 piece puzzle of Melany Taylor Kent serigraph for 50th anniversary, picturing Glinda in a bubble with the four main characters in Munchkinland surrounded by Munchkins	4	7	10
Puzzle, Frame Tray	Western Publishing Company	1989	100 piece puzzle picturing the Scarecrow, the Cowardly Lion, the Tin Man and Dorothy dancing through the poppy field	4	7	10
Puzzle, Frame Tray	Whitman	1976	puzzle shows four main characters leaving the forest and seeing the Emerald City	4	7	10
Puzzle, Interlocking Storyland Puzzle	Haret-Gilmar	1960s	10" x 14" puzzle in a canister, of highly stylized four main characters	6	10	15
Puzzle, Little Oz Books with Jig Saw Puzzles	Reilly & Lee	1932	set #1, two soft cover editions of the Scarecrow & the Tin Man, Ozma & the Little Wizard, plus two puzzles, in box	175	295	450
Puzzle, Return To Oz	Crisco Oil	1985	mail away premium, boxed 200 piece puzzle of promo art	6	10	15
Puzzle, The Wizard of Oz Classic Movie	Milton Bradley	1990	1000 interlocking jig saw puzzle pieces featuring the 1989 Norman James Company poster	6	10	15
Puzzles, Canister	American Puzzle Company	1976	200 piece canisters featuring "The Yellow Brick Road," "Dorothy and the Scarecrow" and others, each	4	7	10
Puzzles, Frame Tray	Doug Smith Productions, Inc.	1977	17" x 22" each, Dorothy with the Scarecrow and Tin Man, the Cowardly Lion wearing badge of courage, Wicked Witch and Dorothy, all four main characters, each	10	16	25
Puzzles, Frame Tray, Return To Oz	Golden Press	1985	one shows Dorothy in mirrored room with Tik Tok, one shows Dorothy and Jack Pumpkinhead entering mirrored room, each	2	3	4
Puzzles, Frame Tray, Set of Four	Jaymar	1960s	Dorothy wiping a tear from the Cowardly Lion; the Haunted Forest with all four characters; the Wizard's exit scene; Scarecrow, Lion and Tin Man in Winkie soldier costumes, each	10	16	25
Puzzles, Little Oz Books with Jig Saw Puzzles	Reilly & Lee	1932	set #2, two softcover editions of Tik-Tok & Jack Pumpkinhead & the Sawhorse, plus 25 piece puzzles, one for each book, in box	175	295	450
Puzzles, Set of Four	Jaymar	1960s	100 pieces each, Dorothy wiping a tear from the Cowardly Lion; the Haunted Forest with all four characters; the Wizard's exit scene; Scarecrow, Lion and Tin Man in Winkie soldier costumes, each	10	16	25

Wizard of Oz

TOY	COMPANY	YEAR	DESCRIPTION	GOOD	EX	MIB
Puzzles, Set, Tell-a-Tale Picture Puzzles	Whitman	1967	set of three puzzles: Peter Pan, Alice in Wonderland and The Wizard of Oz, in box	10	16	25
Puzzles, The Wizard of Oz Picture Puzzles	Whitman	1939	two sets of two puzzles each, in box, based on art by Leason, each	60	100	150
Ribbon, "The Wizard Is Coming"	Whitney Mfg.	1939	advertising ribbon worn by theatre ushers to promote movie premier, white tag with red and blue Wizard	60	100	150
Robot, Tin Man Robot	Remco	1969	21 1/2" tall, battery operated, bump & go action, lifts legs and swings arms as he walks	70	115	175
Ruby Slipper Earrings		1989	two sets: one pierced, one clipped, sold during 1989 national Wizard of Oz Live! tour, each set	12	20	30
Ruby Slippers	Multi Toys	1989	transparent red glitter embedded slippers with red bows, in window box illustrated with Turner authorized 50th Anniversary artwork	4	7	10
Sand Pails, Oz Peanut Butter	Swift & Company	1950s	red & yellow and red, yellow & white tin pails with wire handles, in 2 1/2 & 5 lb. sizes, designed for use as toys once peanut butter was gone, each	20	35	50
Scarecrow in the Box	Mattel	1967	musical hand crank operation, based on animated TV show	15	25	40
Sculpture, Dorothy with Scarecrow	Goebel Miniatures	1984	1" tall, bronze figurines, hand painted sculpted by Olszewski	35	55	85
Sculpture, Dorothy with The Cowardly Lion	Goebel Miniatures	1984	1" tall, bronze figurines, hand painted sculpted by Olszewski	35	55	85
Sculpture, Dorothy with The Tin Man	Goebel Miniatures	1984	1" tall, bronze figurines, hand painted sculpted by Olszewski	35	55	85
Sculpture, Dorothy with The Wicked Witch	Goebel Miniatures	1984	1" tall, bronze figurines, hand painted sculpted by Olszewski	35	60	95
Sculpture, Dorothy with two Munchkins	Goebel Miniatures	1984	1" tall, bronze figurines, hand painted sculpted by Olszewski	35	55	85
Showboat Accessory Packets	Remco	1962	packets of props and accessories for different Showboat plays, including Heidi, Pinocchio, Cinderella and The Wizard of Oz, each	10	16	25
Showboat Play Set	Remco	1962	pink plastic showboat with oversized central stage area, four different plays, scenery, players & scripts	35	55	85
Silver Proof Collection			Set of 12 proof "coins" in .999 silver, each 39mm diameter, with display case, certificate of authenticity & a 36 page booklet, set price	195	325	500
Snack 'N Sip Pals	Multi Toys	1989	12 red & white striped straws with detachable character figures: Dorothy, Tin Man, Glinda, Wicked Witch, Munchkin Lady, Scarecrow, Lullaby League Dancer, Lollipop Guild Boy, Drum Major, Winged Monkey,	3	5	7
Soapy Characters from the Land of Oz	Kerk Guild	1939	set of five, each 4" tall, hand painted, Wicked Witch, Wizard, Cowardly Lion, Tin Man, Scarecrow	100	165	250

Top to bottom: Green Hornet Movie Viewer, 1966, Greenway Productions; Mr. Potato Head Ice Pop Molds, Hasbro; Dragnet Badge 714, 1955, Sherry T.V. Inc.; Charlie's Angels Underwater Intrigue Outfit, 1970s, Palitoy.

Wizard of Oz

TOY	COMPANY	YEAR	DESCRIPTION	GOOD	EX	MIB
Stampers, Figural	Multi Toys	1989	12 stampers with figural handles; Dorothy, Scarecrow, Tin Man, Cowardly Lion, Glinda, Wicked Witch, Munchkin Mayor, Munchkin Lady, Lullaby League Dancer, Lollipop Guild Boy, Drum Major, Winged Monkey,	3	5	8
Stamps, Rubber, Set of Eighteen		1989	set of 18 different rubber stamps based on 1939 movie, originally sold individually	10	16	25
Stamps, Rubber, Set of Eleven		1989	eleven different rubber stamps based on characters & images from the 1939 movie, sold as a set in plastic case, set	8	13	20
Stand-Up Rub-Ons Magic Picture Transfers	Hasbro	1968	contains three full color transfer sheets, 3 stand up character outline sheets of 10 characters	15	25	40
Stationery, Wizard of Oz Children's Writing Paper	Whitman	1939	ten ruled sheets & envelopes, each sheet with a color character illustation, in illustrated box	70	115	175
Statuette, Tin Man	Tuscany Studios	1972	16" tall, chalk, silver finish	15	25	35
Statuettes, Motorized, Set of Four	Adler, Inc.	1987	set of four, each 24" tall, of Dorothy, Cowardly Lion, Scarecrow, and Tin Man, each	25	40	65
Stitch a Story Set	Hasbro	1973	two framed pictures, thread and embroidery needle	8	13	20
Storybook Playsets, Set of Six	Multi Toys	1990	all six storybooks packaged together, complete with each activity storybook with fold down yellow brick road, story, Wizard of Oz figures and accessories	25	40	60
Storybook Playsets, The Wizard of Oz	Multi Toys	1990	six individually sold sets, Dorothy, Scarecrow, Tin Man, Lion, Glinda and Wicked Witch, each with activity storybook with fold down yellow brick road, story, figures, Munchkin figure and accessories,	4	7	10
Suncatchers	Vanderbilt Products	1989	"stained glass" designed plastic oval suncatchers, Dorothy and Munchkin Mayor, Glinda, Scarecrow, Tin Man, Cowardly Lion, on card, each	4	7	10
Sunglasses, Return to Oz, Scarecrow		1985	promotional premium issued at Disney parks, yellow plastic with oval decal over bridge	5	8	12
Sunglasses, Return to Oz, Tik Tok		1985	promotional premium issued at Disney parks, blue plastic with oval decal over bridge	5	8	12
Tea Set, The Land of Oz	Ohio Art	1970s	thirty piece set, red and yellow plastic vessels with illustrated plates	25	40	65
Tin, Oz Fudge Mix	Homix Products, Inc.	1950s	"A Wiz of a Fudge Frosting," in can	15	25	35
Tin, Oz, The Wonderful Peanut Spread	Swift & Company	1940	round tin can, 12 oz. size, white with color art of characters	55	95	145

Wizard of Oz

TOY	COMPANY	YEAR	DESCRIPTION	GOOD	EX	MIB
Tin, Rectangular	Multi Toys	1989	tin, appx 8x10x2" illustrated on lid with circular art of Emerald City in center, surrounded by circular portraits of Dorothy, Cowardly Lion, Tin Man, Wicked Witch, Glinda, and Scarecrow	6	10	15
Tins, Playing Card, Set of Five	Presents	1988	five rectangular tins, each sold housing 2 decks of Oz playing cards, each	5	8	12
Towel, Beach	Metro-Goldwyn-Mayer (MGM)	1976	vertical, shows four main characters on their journey down the yellow brick road	10	16	25
Towel, Beach	Renaissance	1990	30" x 60," vertical design showing four main characters walking toward viewer through poppy fields with Emerald City in background	6	10	15
Towel, Beach	Renaissance	1991	30" x 60," horizontal design showing four main characters walking facing left through poppy fields with Emerald City far left top background	6	10	15
Towel, Calendar, 1977	Franco	1976	fabric towel imprinted with months surrounded by color art of characters and scenes from film	10	16	25
Trash Can, The Wizard of Oz	Chein	1975	oval metal can decorated with art based on the movie	17	30	45
View-Master Reel Set	Sawyer Company	1957	21 stereo pictures, three reel Wonderful Wizard of Oz packet	15	25	35
Vinyl Stick-On Playset	Multi Toys	1989	ten reusable vinyl stickers with Emerald City background, on header card	3	5	8
Vinyl Stick-Ons	Multi Toys	1989	seven reusable vinyl stickers, on header card	2	3	5
Wall Clock	Novus Industries	1989	10" round, battery operated, shows film scene of four main characters in forest	10	15	25
Wall Clock	Novus Industries	1989	10" round, battery operated, picturing Emerald City	10	15	25
Wall Decorations, "Off to See the Wizard"	Homestead Mail Order Co.	1967	set of 20 large paper cutouts based on the ABC TV show	25	40	65
Wall Decorations, Off to See the Wizard	Shepard Press	1967	set of 20 punch-out decorations, including four main characters plus Wicked Witch and the Wizard, emeralds, rainbow, yellow brick road and Emerald City	17	30	45
Wand, Glinda's Magic Wand	Multi Toys	1989	battery operated wand with red glitter star on end, lights up, on illustrated card	4	7	10
Water Guns, Wizard of Oz	Durham	1976	three different, on header cards, heads of Scarecrow, Tin Man, Cowardly Lion, mounted on stems, water squirts out of each character's nose, each	10	16	25
Wind-Up Walkers, Set of Six	Multi Toys	1989	50th anniversary editions, on illustrated cards, Dorothy, Scarecrow, Glinda, Tin Man, Cowardly Lion, Wicked Witch, each	3	5	8

Wizard of Oz

TOY	COMPANY	YEAR	DESCRIPTION	GOOD	EX	MIB
Wind-up Walking Toy, Cowardly Lion	Durham Industries	1975	on illustrated card	10	16	25
Wind-up Walking Toy, Scarecrow	Durham Industries	1975	on illustrated card	10	16	25
Wind-up Walking Toy, Tin Woodman	Durham Industries	1975	on illustrated card	10	16	25
Wind-Up, Cowardly Lion, Off to See the Wizard Dancing Toy	Louis Marx & Company	1967	plastic mechanical toy that turns in circles	15	25	35
Wind-Up, Scarecrow, Off to See the Wizard Dancing Toy	Louis Marx & Company	1967	plastic mechanical toy that turns in circles, sold in window box	15	25	35
Wind-Up, Tin Man, Off to See the Wizard Dancing Toy	Louis Marx & Company	1967	plastic mechanical toy that turns in circles	15	25	35
Wind-Up, Wizard of Oz Spin Around	Durham Industries	1975	ruby slipper with head of either Tin Woodman, Cowardly Lion or Scarecrow, shoe spins when wound up, each	12	20	30
Wind-Ups, Off to See the Wizard Dancing Toys, Set	Louis Marx & Company	1967	mechanical dancing Tin Man, the Cowardly Lion and the Scarecrow, Montgomery Wards exclusive, when wound they turned in circles	50	80	125
Witch's Castle Playset	Mego	1975	green vinyl playset with Wicked Witch doll, crystal ball and a cauldron, Sears exclusive	150	260	400
Wizard of Oz & His Emerald City Playset	Mego	1974	designed for Mego 8" Oz dolls, playset contains a folding yellow brick road, cardboard apple tree, a spinning crystal & the Wizard's throne	135	225	350
Wizard of Oz Action Forms	Lorob Industries	1964	set included record, punch out scenery and character figures	10	16	25
Wizard of Oz Portrait Sculpture Collection	The Franklin Mint	1988	set of 20 porcelain hand painted figures, sold with wood and etched glass display case, each	12	20	30
Wrist Watch, Toy		1940s	tin toy watch which featured the Scarecrow and the Tin Man on either side of the non-working dial	15	25	35
Wristwatch, LCD	EKO Corp.	1989	kid-sized LCD wristwatch, simple digital watch on red face in round yellow case with colored plastic band showing yellow brick road and Emerald City	6	10	15
Wristwatch, Oz Time	Macy's	1989	promo 50th anniversary premium, round face illustrated with art of Emerald City, black plastic band, sold for $10 with any purchase of $25 or more	25	35	55
Wristwatch, Quartz	EKO Corp.	1989	quartz wristwatch with illustrated face showing Emerald City, black plastic band	10	16	25
Wristwatch, The Wizard of Oz Live!		1989	souvenir of the 1989 arena tour production, round yellow plastic case, blue face with LCD readout, blue vinyl band with yellow brick road design	4	7	10

Woody Woodpecker

TOY	COMPANY	YEAR	DESCRIPTION	GOOD	EX	MIB
Woody Woodpecker & Andy Panda Paper Dolls	Saalfield	1968	punch out and cut	15	30	45
Woody Woodpecker Card Set		1950s	two decks in a carrying case	30	55	85
Woody Woodpecker Electric Lamp		1971	20" tall, plastic	11	20	30
Woody Woodpecker Talking Hand Puppet	Mattel	1963	pull-string voice box	30	60	90
Woody Woodpecker's Funorama Punch Out Book		1972		9	16	25
Woody's Cafe Animated Alarm Clock	Columbia Time Prod.	1959		60	115	175

Yogi Bear

TOY	COMPANY	YEAR	DESCRIPTION	GOOD	EX	MIB
Boo Boo Plush Doll	Knickerbocker	1960s	9 1/2" tall, stuffed	35	60	50
Cindy Bear Plush Doll	Knickerbocker	1959	16" tall stuffed Cindy, with vinyl face	35	65	60
Snagglepuss Figure	Dakin	1970		30	55	85
Snagglepuss Soakie	Purex	1960s	9" tall, vinyl/plastic	12	23	35
Snagglepuss Sticker Fun Book	Whitman	1963		11	20	30
Yogi and Boo Boo Coatrack	Wolverine	1979	48" tall red wood coatrack with 8 white wooden hangers, 20" tall Yogi and Boo Boo cut out in front, back of piece is marked for growth chart	35	65	95
Yogi Bear & Boo Boo Hot Water Bottle		1966		25	50	75
Yogi Bear and Pixie & Dixie Game Car	Whitman		7 1/2" pile on game in car	12	23	35
Yogi Bear Bubble Pipe	Transogram	1963	figural plastic pipe on illustrated card	15	30	45
Yogi Bear Cartoonist Stamp Set	Lido	1961		25	50	75
Yogi Bear Doll		1962	6" tall soft vinyl with moveable arms & head	20	40	60
Yogi Bear Doll	Knickerbocker	1959	10" tall	35	65	100
Yogi Bear Figural Bank	Dakin	1980	7" tall bank	5	10	15
Yogi Bear Figural Bank	Knickerbocker	1960s	22" tall figural bank	15	30	45
Yogi Bear Figure	Knickerbocker	1960s	9" tall, plastic	23	45	65
Yogi Bear Figure	Dakin	1970	7 3/4" tall	15	30	45
Yogi Bear Flashlight	Laurie	1976		4	7	10
Yogi Bear Friction Toy		1960s	friction toy of Yogi upright in yellow tie and green hat, in illustrated red box	35	65	95
Yogi Bear Hand Puppet	Knickerbocker			15	30	45
Yogi Bear Magic Slate		1963		11	20	30
Yogi Bear Plush Doll	Knickerbocker	1960s	19" tall, stuffed	25	50	75
Yogi Bear Plush Doll	Knickerbocker	1959	16" tall stuffed Yogi, with vinyl face	35	65	100
Yogi Bear Push Puppet				12	23	35
Yogi Bear Safety Scissors	Monogram	1973	on card	4	7	10
Yogi Bear Stuff & Lace Doll	Knickerbocker	1959	items to make a 13x5" stuffed Yogi, in 16x9" box with pictures of Yogi and Huck Hound	20	40	60
Yogi Bear TV-Tinykins Figure	Marx	1961		15	30	45
Yogi Bear Wrist Watch		1963		35	65	100
Yogi Bear Yo Yo with Sleeper Action		1976		9	16	25

Yogi Bear

TOY	COMPANY	YEAR	DESCRIPTION	GOOD	EX	MIB
Yogi Bear, Boo Boo & Ranger Smith Figurines		1960	12" tall, each	15	30	45
Yogi Score-A-Matic Ball Toss Game	Transogram	1960		40	75	115
Yogi Vinyl Squeeze Doll	Sanitoy	1979	12" tall Yogi	9	16	25
Yogi Wristwatch	Bradley	1967	medium base metal case, shows Yogi with hobo sack on stick, black vinyl band	35	65	100

Yosemite Sam

TOY	COMPANY	YEAR	DESCRIPTION	GOOD	EX	MIB
Yosemite Sam "Fun Farm" Figure	Dakin	1978		11	20	30
Yosemite Sam Figure	Dakin	1968	7" tall	12	23	35
Yosemite Sam Figure on Treasure Chest	Dakin	1971		14	25	40
Yosemite Sam Mini Snowdome	Applause	1980s		9	16	25
Yosemite Sam Musical Snowdome	Applause	1980s		15	30	45
Yosemite Sam Nodder		1960s	6 1/4" tall with bobbing head & spring mounted head	35	65	150

Zorro

TOY	COMPANY	YEAR	DESCRIPTION	GOOD	EX	MIB
Amigo Figure	Gabriel	1982		4	7	10
Captain Ramon Figure	Gabriel	1982		3	5	8
Picaro Figure	Gabriel	1982		4	7	10
Sergeant Gonzales Figure	Gabriel	1982		3	5	8
Tempest	Gabriel	1982		4	7	10
Zorro "Bean Bag-Dart" Game		1950s	14x16" target	14	25	85
Zorro & Horse	Lido	1950s	4 1/2" tall black plastic figure of Zorro with gun and sword, black horse has white harness and saddle, included paper mask, on card	35	65	100
Zorro Activity Box	Whitman	1965	9x12x1 1/2"	30	60	90
Zorro Board Game	Parker Brothers	1966	16 12/x16 1/2" board	18	35	50
Zorro Board Game	Whitman	1965	5 1/2x8x1 1/2"	14	25	40
Zorro Oil Painting By Numbers	Hasbro		10x13 1/2" cardboard canvas	35	60	95
Zorro Pencil Holder & Pencil Sharpener		1950s	6" tall	15	30	45
Zorro Pinwheel		1950s	red and black pinwheel with Zorro logos on petals and black mask in front	15	30	45
Zorro Puzzle "Zorro/Sgt. Garcia & Don Diego"	Jaymar	1960		15	25	40
Zorro Puzzle "Zorro/The Dual"	Jaymar	1960	10x14"	15	25	40
Zorro Secret Sight Scarf Mask	Westminster	1960	black fabric mask with 2 black & silver hard plastic eye pieces, cardboards with 2 pictures of Zorro	25	50	75
Zorro Target & Dart-Shooting Rifle Set	T. Cohn	1960	21" long plastic rifle, black plastic darts and target	60	115	200
Zorro View-Master	Sawyer's Inc.	1958	4 1/2x4 1/2" envelope with 3 reel story & booklet	15	30	45
Zorro Wristwatch	US Time	1957	chrome case, black leather band	60	115	150

YOU GET MORE WITH TOY SHOP!

MORE bargains on collectible toys
you'll find at least 17,500 toy bargains in every issue

MORE big issues
We'll send you over 300 pages of toy deals each month

MORE readability
it's easy to find what you want in our indexed classified ads

MORE dealers and collectors
network with dealers and collectors who share your interests

MORE complete toy show calendar
plan your schedule with our national & international listings

THE FACTS... THE FIGURES... THE FUN...

You get it all in TOY COLLECTOR AND PRICE GUIDE. In-depth articles give you the history of your favorite toys, along with the present-day values by condition.

Plus, you'll enjoy:
- up-to-date toy show reports
- extensive toy show calendar
- classified ad section with hundreds of hot toys for sale and wanted from coast-to-coast

KEEP UP WITH THE FASTEST GROWING HOBBY!

Subscribe today and we'll deliver TOY COLLECTOR AND PRICE GUIDE to your door every other month!

Satisfaction Guaranteed OR YOUR MONEY BACK ON ALL UNMAILED ISSUES

❑ **Send me 26 issues (1year) of TOY SHOP for $23.95.**

Name _____
Address _____
City _____
State _____ Zip _____

❑ Check or money order enclosed
(payable to Toy Shop)
❑ VISA   ❑ MasterCard
Credit card # _____
Expires: Mo. _____ Yr. _____
Signature: _____

MasterCard and VISA customers call toll-free
800-258-0929, Dept. TS7TP
Weekdays: 6:30 am - 8:00 pm, Saturdays: 8:00 am - 2:00 pm, CST

Return this form (or a photocopy of it) with payment to:

700 E. State St., Iola, WI 54990-0001

❑ **Send me 6 issues (1year) of TOY COLLECTOR & PRICE GUIDE for $12.95.**

Name _____
Address _____
City _____
State _____ Zip _____

❑ Check or money order enclosed
(payable to Toy Collector)
❑ VISA   ❑ MasterCard
Credit card # _____
Expires: Mo. _____ Yr. _____
Signature: _____

MasterCard and VISA customers call toll-free
800-258-0929, Dept. TP7TP
Weekdays: 6:30 am - 8:00 pm, Saturdays: 8:00 am - 2:00 pm, CST

Return this form (or a photocopy of it) with payment to:

700 E. State St., Iola, WI 54990-0001

THE TOY SHOP 1996 ANNUAL DIRECTORY
by the Toy Shop staff

Reserve your copy of the 1996 Annual Directory and lock into the current price of $9.95. You'll use this handy reference over and over again to locate dealers, toy clubs, toy hunting and collecting tips, and much more helpful information about this growing hobby.

RESERVE YOUR COPY TODAY!

8 1/2" x 11" ● 80+ pp ● photos

THE OFFICIAL 30th ANNIVERSARY SALUTE TO G.I. JOE
by Vincent Santelmo

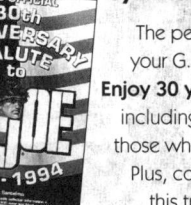

The perfect addition to your G.I. Joe collection. Enjoy 30 years of history, including interviews with those who made G.I. Joe. Plus, color photos make this tribute complete.

8 1/2" x 11" ● 208 pp ● 450 photos

1995 TOYS & PRICES
by Roger Case & Tom Hammel, Editors

Carry it with you wherever you go to buy toys. You'll find hundreds of toys listed with values in up to three conditions. From model kits...to action figures...to toy trains, no price guide is more complete than the 1995 Toys & Prices.

6" x 9" ● 778 pp ● 500+ photos

THE COMPLETE ENCYCLOPEDIA TO G.I. JOE
by Vincent Santelmo

Learn every detail about every G. I. Joe item produced from 1964-1993. Plus, find full illustrations of the original Action Series heroes, the 1970s Adventure Teams, the Force figures of the 1980s and the ultra-modern Sonic Warrior of the 1990s.

8 1/2" x 11" ● 448 pp ● 700 photos

RETAILERS: Place your bulk order by calling 1-715-445-4612, ext. 420.

ORDER TODAY!

___ copy(ies) of 1995 TOYS & PRICES.................$15.95 each $ _____
___ copy(ies) of 30th ANNIV. SALUTE TO G.I. JOE..............$34.95 each $ _____
___ copy(ies) of COMPLETE ENCYCLOPEDIA OF G.I. JOE...$24.95 each $ _____
___ copy(ies) of TOY SHOP 1996 ANNUAL DIRECTORY..........$9.95 each $ _____
$2.50 shipping for first book. $1.50 for each additional. $ _____
WI residents add 5.5% sales tax to total $ _____
TOTAL $ _____

Name _____
Address _____
City _____
State _____ Zip _____

( ) Check or money order enclosed (payable to Krause Publications)
Charge my: ( ) VISA ( ) MasterCard
Card #: _____
Expires: Mo. _____ Yr. _____
Signature _____

VISA & MASTERCARD CUSTOMERS CALL TOLL-FREE
800-258-0929 Dept. F2H1
Weekdays 6:30 am - 8 pm ● Saturdays 8 am - 2 pm, CST

Return this order form or a photocopy with payment to:
Krause Publications
Book Sales Dept. F2H1
700 E. State Street
Iola, WI 54990-0001